Albert C.J. Luo

Discontinuous Dynamical Systems

Albert C.J. Luo

Discontinuous Dynamical Systems

With 228 figures, 13 of them in color

高等教育出版社
HIGHER EDUCATION PRESS

Springer

Author

Albert C.J. Luo
Department of Mechanical and Industrial Engineering
Southern Illinois University Edwardsville
Edwardsville, IL 62026–1805, USA
Email: aluo@siue.edu

ISBN 978-7-04-031957-6

Higher Education Press, Beijing

ISBN 978-3-642-22460-7 e-ISBN 978-3-642-22461-4

Springer Heidelberg Dordrecht London New York

Library of Congress Control Number: 2011931165

Printed on acid-free paper

Springer is part of Springer Science+Business Media (www.springer.com)

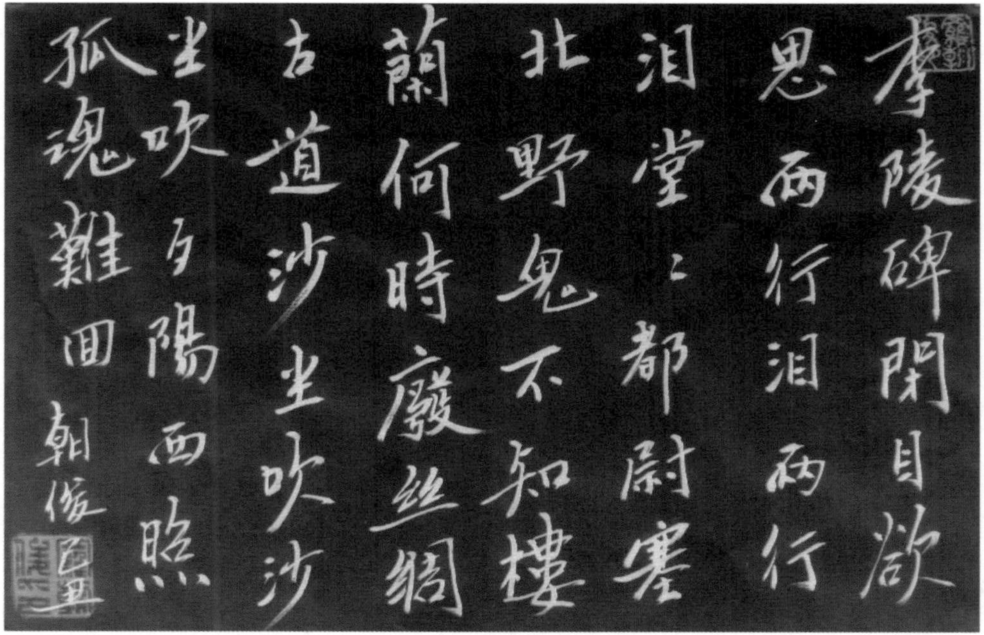

Dedicated to genuine heroes who are not recognized.

Preface

This book is about a fundamental theory for switchability, singularity and attractivity in discontinuous dynamical systems. The literature about initial development on the singularity and dynamics of discontinuous vector fields was published in 2006. The further development and refinement of singularity and dynamics of discontinuous dynamical systems were carried out since a function (G-function) was introduced. In 2009, the book on Discontinuous Dynamical Systems on Time-varying Domains was published with accumulated practical applications. In the previous development, the singularity and switchability of a flow in discontinuous dynamical systems was based on the single boundary. On the intersection of two or more boundaries, the singularity and switchability in discontinuous dynamical systems were not tackled. In this book, the theory of flow barriers on the boundary in discontinuous dynamical systems is presented systematically. With transport laws, the theory for bouncing flows on the boundary is discussed. The edge dynamics and system interaction are presented.

This book mainly discusses the switchability of a flow to a specific boundary or edge. The book consists of two parts: The first part is about the singularity, flow barriers and dynamics of discontinuous dynamical systems, and the materials are scattered in Chapters 2-5. In the second part, the edge dynamics and system interactions are addressed, which are placed in Chapters 6-9. Chapter 1 gives the purposes and objectives of this book on discontinuous dynamical systems. For the first part, the basic passability and singularity of a flow to the separation boundary of two different dynamical subsystems are presented in Chapter 2. Based on the basic theory of passability and singularity, Chapter 3 gives a generalized theory for the passability and switchability of a flow to a specific boundary. In Chapter 4, a theory of flow barriers in discontinuous dynamical systems is systematically presented for the first time, which will provide a theoretic foundation of control theory. Transport laws and multi-valued vector fields in discontinuous dynamical systems are discussed in Chapter 5. In the aforementioned chapters, the singularity and dynamics of discontinuous dynamical systems are based on the single boundary. For intersection edges of two or more boundaries, the switchability and attractivity of a flow to specific edges are very important and different from the singularity and switchability of a flow at the boundary. Thus, in the second part, the edge dynamics and singularity in discontinuous dynamical systems are presented. In Chapter 6, the switchability and attractivity of domain flows to specific edges are presented. In Chapter 7, the switchability and attractivity of $(n-1)$ -dimensional boundary flows to $(n-2)$ -dimensional edges in n -dimensional, discontinuous dynamical systems are discussed in order to help us to understand edge and vertex dynamics. To extend the ideas of singularity and switchability of the domain and boundary flows to specific edges, the switchability and attractivity of edge flows to the lower-dimensional edges are discussed in Chapter 8. Based on the theory presented in Chapters 6-8 for discontinuous dynamical systems, an interaction theory for two dynamical systems is presented in Chapter 9. The interaction of two dynamical systems is treated as a separation boundary, and such a boundary is time-varying. The system synchronization is discussed as an application of the interaction of two dynamical systems. The author believes that this book cannot involve the theoretic development for discontinuous dynamical systems completely, and typos may be scattered here and there. The author sincerely hopes readers can point out and criticize for improvement.

Finally, I would like to thank my students (Brandon M. Rapp, Brandon C. Gegg, Patrick Zwiegart, Jr., Tingting Mao, and Fuhong Min) for applying the new concepts to practical systems and completing numerical computations. Herein, I also thank my wife (Sherry X. Huang) and my children (Yanyi Luo, Robin Ruo-Bing Luo, and Robert Zong-Yuan Luo) for their tolerance, patience, understanding and support.

Albert C.J. Luo
Edwardsville, Illinois
January, 2011

Contents

Chapter 1
Introduction

Discontinuous dynamical systems exist everywhere in the real world. One used to adopt continuous models for approximate descriptions of discontinuous dynamical systems. However, such continuous modeling cannot provide adequate predictions of discontinuous dynamical systems, and also makes the problems solving be more complicated and inaccurate. In the real world, the discontinuous modeling of dynamical systems is absolute, but the continuous modeling is relative. In other words, the continuous description of dynamical systems is an approximation of the discontinuous problems. To better describe the real world, one should realize that discontinuous models can provide adequate and real predications of engineering systems. For any discontinuous dynamical system, there are many continuous subsystems in different domains or different time intervals, and the dynamical properties of any continuous subsystems are different from that of the adjacent continuous subsystems. Thus, we have two types of discontinuous dynamical systems. (i) In two different adjacent time intervals, dynamical systems are different. When a dynamical system reaches the switching time, this system will be switched to another different dynamical system. With such switching, the discontinuous dynamical system is called the switching system. (ii) In phase space, there are many different domains. On any two adjacent domains, distinct dynamical systems are defined. Thus, once a flow of a sub-system arrives to its boundary, the switchability and/or transport laws on the boundary should be addressed because of such a difference between two adjacent subsystems. A general theory for discontinuous dynamical systems will be developed. In this chapter, a brief survey on the recent development of discontinuous dynamical systems will be presented, and the purposes and objectives of this book will be described. The book layout will be presented, and the summarization of all chapters of the main body of this book will be given.

1.1. A brief history

Consider a smooth dynamical system on space $\mathscr{R}^{n+m}$

$$\dot{\mathbf{x}} = \mathbf{f}(\mathbf{x},\mathbf{u},t) \tag{1.1}$$

where the vector function $\mathbf{f} \in \mathscr{R}^n$, and state and input variable vectors are $\mathbf{x} \in \mathscr{R}^n$ and $\mathbf{u} \in \mathscr{R}^m$, respectively. In continuous dynamical systems, the sufficient condition for the existence of a solution for every initial state $\mathbf{x}(t_0)$ and input vector $\mathbf{u}(t)$ is that the vector function $\mathbf{f}(\mathbf{x},\mathbf{u},t)$ is continuous in a given domain $\Omega \subset \mathscr{R}^n$. However, such a condition cannot guarantee the uniqueness of solution. Therefore, the following Lipschitz condition should be used for guaranteeing the existence and uniqueness of the solution for the system in Eq. (1.1).

$$\| \mathbf{f}(\mathbf{x},\mathbf{u},t) - \mathbf{f}(\mathbf{y},\mathbf{u},t) \| \le K \| \mathbf{x} - \mathbf{y} \| \tag{1.2}$$

for $\mathbf{x},\mathbf{y} \in \Omega \subset \mathscr{R}^n$ and $t \in I \subset \mathscr{R}$, where K is a constant and $\|\cdot\|$ represents a vector norm. Most of the existing theories in dynamical systems are based on the Lipschitz condition (e.g., Poincaré, 1892; Birkhoff, 1927). Indeed, the existing theories are extensively adopted in science and engineering. When the dynamic behaviors of a dynamical system are constrained by engineering requirements and limitations, the dynamic system cannot satisfy the Lipschitz conditions. Thus, the traditional theory of continuous dynamical systems cannot be used. Even if one has used such a theory of continuous dynamical systems to solve discontinuous dynamical systems, the problem becomes more complicated and difficult to solve. The Lipschitz condition is very strong for practical dynamical problems, and many dynamical systems cannot satisfy such a condition. Therefore, the established dynamical system theories based on the Lipschitz condition are not adequate for such discontinuous dynamical systems. To overcome this difficulty, a theory for discontinuous dynamical systems should be developed.

The early investigation of discontinuous dynamical systems in mechanical engineering can be found (e.g., den Hartog, 1930,1931). The discontinuity in such a dynamical system is caused by friction forces. Levinson (1949) used to piecewise linear model to investigate the periodically excited van der Pol equation, and found infinitely many periodic solutions which could not be perturbed away. Levitan (1961) investigated the existence of periodic motions in a friction oscillator with a periodically driven base. Filippov (1964) investigated the motion in the Coulomb friction oscillator and presented differential equations with discontinuous right-hand sides. To determine the sliding motion along the discontinuous boundary, the differential inclusion was introduced via the set-valued analysis, and the existence and uniqueness of the solution for such a discontinuous differential equation were discussed. The detailed discussion of such discontinuous differential equations can be referred to Filippov (1988). Since the discontinuity exists widely in engineering and control systems, Aizerman and Pyatniskii (1974a,b) extended the Filippov's concepts and presented a generalized theory for discontinuous dynamical systems. From such a generalized theory, Utkin (1976) developed methods for controlling dynamic systems through the discontinuity (i.e., sliding mode control). Utkin (1978) presented sliding modes and the corresponding variable structure systems, and the theory of automatic control systems described with variable structures and sliding motions was also developed in Utkin(1981). DeCarlo *et al.* (1988) gave a review on the development of the sliding mode control. Leine *et al.* (2000) used the Filippov theory to investigate bifurcations in nonlinear discontinuous systems. The more discussion about the traditional analysis of bifurcation in non-smooth dynamical systems can be referred to Zhusubaliyev and Mosekilde (2003). However, the Filippov's theory mainly focused on the existence and uniqueness of the solutions for non-smooth dynamical systems. The local singularity caused by the separation boundary was not discussed. Such a differential equation theory with discontinuity is still difficult to be used for determining the complexity of discontinuous dynamical systems. Luo (2005a) developed a general theory to handle the local singularity of discontinuous dynamical systems. To determine the sink and source flows in discontinuous dynamical systems, the imaginary, sink and source flows were introduced in Luo (2005b). The detailed discussions can be referred to Luo (2006, 2008a). Luo and Gegg (2006a) used the local singularity theory of Luo (2005a) to develop the force criteria for motion switchability on the velocity boundary in a harmonically driven linear oscillator with dry-friction (also see, Luo and Gegg, 2006b). Lu (2007) mathematically proved the existence of such a periodic motion in such a friction oscillator. Luo and Gegg (2006c,d) investigated the dynamics of a friction-induced oscillator contacting on time-varying belts with friction. Luo and Thapa (2008) proposed a new model to describe a brake system consisting of two oscillators, and the two oscillators are connected through a contacting surface with friction. Based on this model, the nonlinear dynamical behaviors of the brake system under a periodical excitation were investigated. Luo (2008b) introduced *G*-function to measure the switchability of a flow to the discontinuous boundary. In this book, the edge dynamics and flow barriers in discontinuous dynamical systems will be tackled.

1.2. Book layout

The book will mainly discuss the switchability of a flow to boundary and edge. The book includes two parts: The first part is about the singularity, flow barriers and dynamics of discontinuous dynamical systems at specific boundary, and the materials will be scattered in Chapters 2-5. In the second part, the edge dynamics and system interaction will be discussed, which are placed in Chapters 6-9.

In Chapter 2, the basic passability of a flow to the boundary of two distinct dynamical systems will be presented in discontinuous dynamic systems. The accessible and inaccessible sub-domains will be introduced, and on the accessible domains, the corresponding dynamic systems can be defined. The passability and tangency (grazing) of a flow to the boundary between two adjacent accessible domains will be discussed, and the necessary and sufficient conditions for such passability and tangency of the flow to the boundary will be presented. The L-functions of flows will be presented, and the corresponding conditions for the flow passability to the boundary will be discussed.

In Chapter 3, a general theory for the passability of a flow to a specific boundary in discontinuous dynamical systems will be presented. The concepts of real and imaginary flows will be introduced. The G-functions for discontinuous dynamical systems will be presented to describe the general theory of the passability of a flow to the boundary. Based on the G-function, the passability of a flow from a domain to an adjacent one will be discussed. With the concepts of real and imaginary flows, the full and half sink and source flows to the boundary are discussed in detail. A flow to the boundary in a discontinuous dynamical system is either passable or non-passable. Thus, the switching bifurcations between the passable and non-passable flows will be presented. To help one understand the concept of flow passability, a discontinuous dynamical system with a parabolic boundary will be analyzed.

In Chapter 4, the theory of flow barriers in discontinuous dynamical systems will be systematically presented for the first time. The barriers vector fields in discontinuous dynamical systems will be introduced, and the passability of a flow to the boundary with flow barriers will be discussed. The coming and leaving flow barriers in passable flows to the boundary will be presented first, and flow barriers for sink and source flows will be also discussed. Once the sink flow is formed, the boundary flow barrier in the sink flow needs to be considered, and such a flow barrier is independent of vector fields in the corresponding domains. Thus, the necessary and sufficient conditions for formations and vanishing of a sink flow will be developed for a discontinuous dynamical system possessing flow barriers on the boundary. A periodically forced friction oscillator with flow barriers will be discussed.

In Chapter 5, transport laws and multi-valued vector fields in discontinuous dynamical systems will be discussed. The grazing and inflexional singular sets on the boundary will be introduced, and the real and imaginary singular sets will be also discussed. With permanent flow barriers, the forbidden boundary and the boundary channel will be presented. Further, the domain and boundary classification will be presented and sink and source domains will be addressed. Similarly, the sink and source boundary will be also presented. For the C^0-discontinuity, flow barriers, isolated domains and boundary channels, transport laws are needed to continue the flow in discontinuous dynamical systems. Multi-valued vector fields in a single domain will be introduced. With the simplest transport law (i.e., the switching rule), the bouncing flow on the boundary will be presented, and the extendable flows will be discussed as well. A controlled piecewise linear system will be discussed as an application. The vector fields on both sides of boundary in such a controlled piecewise linear system, and the bouncing flows will be illustrated.

In Chapter 6, the switchability and attractivity of domain flows to specific edges will be presented. Dynamical systems on domains, boundaries, edges, and vertexes will be introduced. The coming, leaving and tangency of a domain flow to a specific edge will be discussed through the corresponding boundaries. The switchability and passability of a flow from an accessible domain to another accessible domain will be presented with a switching rule. The convex and concave edges will be introduced for discontinuous dynamical systems. The mirror domains will also be introduced through the extension of the finite boundaries at the convex edge. The transversally grazing passability of a flow to the concave edges will be presented. The equi-measuring surface will be introduced, and the attractivity of a domain flow to the boundary will be discussed. Further, an equi-measuring edge in domain will be introduced, and the attractivity of a domain flow to a specific edge will be discussed.

In Chapter 7, the switchability and attractivity of $(n-1)$-dimensional boundary flows to $(n-2)$-dimensional edges will be discussed in order to understand edge and vertex dynamics in n-dimensional discontinuous dynamical systems. The basic properties of boundary flows to edges will be discussed first. The coming, leaving and tangency of a boundary flow to an $(n-2)$-dimensional edge will be presented. The switchability and passability of a boundary flow from an accessible boundary to another accessible boundary will be presented with a switching rule. In addition, the switchability and passability of boundary and domain flows will be discussed with a switching rule. The equi-measuring edge for boundary flows will be introduced, and the attractivity of a boundary flow to the $(n-2)$-dimensional edge will be presented. Finally, the bouncing characteristics of a boundary flow to the $(n-2)$-dimensional edge will be discussed.

In Chapter 8, the switchability and attractivity of edge flows to the lower-dimensional edges will be discussed, which are generalized from the theory of switchability and attractivity for domain and boundary flows. The basic properties of edge flows to a specific edge will be discussed. The coming, leaving and tangency of an edge flow to a specific edge will be discussed by use of the boundaries. The switchability and passability of an edge flow from an accessible edge to another accessible edge (or boundary or domain) will be discussed with a switching rule. Similarly, the equi-measuring edge for edge flows will be introduced, and the attractivity of an edge flow to the lower-dimensional edge will be presented. Finally, the bouncing characteristics of an edge flow to a specific lower dimensional edge will be discussed as well. The switchability of a flow in a 2-DOF frictional oscillator will be discussed as a sample problem to illustrate edge dynamics.

In Chapter 9, the theory presented in Chapters 6-8 will be applied to dynamical system interactions. The concept of interaction between two dynamical systems will be introduced. The interaction of two dynamical systems will be treated as a boundary in discontinuous dynamical systems, and such a boundary is time-varying. In other words, the boundary and domains for one of two dynamical systems are constrained by the other. The corresponding conditions for such an interaction will be presented via the theory for the switchabilty and attractivity of edge flows to the specific edges. The synchronization of two totally different dynamical systems will be presented as an application.

References

Aizerman, M.A., Pyatnitskii, E.S., 1974, Foundation of a theory of discontinuous systems. 1, *Automatic and Remote Control*, **35**, 1066-1079.

Aizerman, M.A., Pyatnitskii, E.S., 1974, Foundation of a theory of discontinuous systems. 2, *Automatic and Remote Control*, **35**, 1241-1262.

Birkhoff, C.D., 1927, On the periodic motions of dynamical systems, *Acta Mathematica*, **50**, 359-379.

den Hartog, J.P., 1930, Forced vibration with combined viscous and Coulomb damping, *Phil, Magazine*, **VII** (9), 801-817.

den Hartog, J.P., 1931, Forced vibrations with Coulomb and viscous damping, *Transactions of the American Society of Mechanical Engineers*, **53**, 107-115.

DeCarlo, R.A., Zak, S.H. and Matthews, G.P., 1988, Variable structure control of nonlinear multivariable systems: A tutorial, *Proceedings of the IEEE*, **76**, 212-232.

Filippov, A.F., 1964, Differential equations with discontinuous right-hand side, *American Mathematical Society Translations, Series 2,* **42**, 199-231.

Filippov, A.F., 1988, *Differential Equations with Discontinuous Righthand Sides*, Dordrecht: Kluwer Academic Publishers.

Leine, R.I., van Campen, D.H. and van de Vrande, B.L., 2000, Bifurcations in nonlinear discontinuous systems, *Nonlinear Dynamics*, **23**, 105-164.

Levinson, N., 1949, A second order differential equation with singular solutions, *Annals of Mathematics*, **50**, 127-153.

Levitan, E.S., 1960, Forced oscillation of a spring-mass system having combined coulomb and viscous damping, *Journal of the Acoustical Society of America*, **32**, 1265-1269.

Lu, C., 2007, Existence of slip and stick periodic motions in a non-smooth dynamical system, *Chaos, Solitons and Fractals,* **35**, 949-959.

Luo, A.C.J., 2005a, A theory for non-smooth dynamical systems on connectable domains, *Communication in Nonlinear Science and Numerical Simulation*, **10**, 1-55.

Luo, A.C.J., 2005b, Imaginary, sink and source flows in the vicinity of the separatrix of non-smooth dynamic system, *Journal of Sound and Vibration,* **285**, 443-456.

Luo, A.C.J., 2006, *Singularity and Dynamics on Discontinuous Vector Fields*, Amsterdam: Elsevier.

Luo, A.C.J., 2008a, *Global Transversality, Resonance and Chaotic Dynamics*, Singapore: World Scientific.

Luo, A.C.J., 2008b, A theory for flow swtichability in discontinuous dynamical systems, *Nonlinear Analysis: Hybrid Systems*, **2**, 1030-1061.

Luo, A.C.J. and Gegg, B.C., 2006a, On the mechanism of stick and non-stick periodic motion in a forced oscillator including dry-friction, *ASME Journal of Vibration and Acoustics*, **128**, 97-105.

Luo, A.C.J. and Gegg, B.C., 2006b, Stick and non-stick periodic motions in a periodically forced, linear oscillator with dry friction, *Journal of Sound and Vibration,* **291**, 132-168.

Luo, A.C.J. and Gegg, B.C., 2006c, Periodic motions in a periodically forced oscillator moving on an oscillating belt with dry friction, *ASME Journal of Computational and Nonlinear Dynamics*, **1**, 212-220.

Luo, A.C.J. and Gegg, B.C., 2006d, Dynamics of a periodically excited oscillator with dry friction on a sinusoidally time-varying, traveling surface, *International Journal of Bifurcation and Chaos,* **16**, 3539-3566.

Luo, A.C.J. and Thapa, S., 2008, Periodic motions in a simplified brake dynamical system with a periodic excitation, *Communication in Nonlinear Science and Numerical Simulation,* **14**, 2389-2412.

Poincaré, H., 1892, *Les Methods Nouvelles de la Mecanique Celeste*, Vol. 1, Paris: Gauthier-Villars.

Utkin, V. I., 1976, Variable structure systems with sliding modes, *IEEE Transactions on Automatic Control*, **AC-22**, 212-222.

Utkin, V.I., 1978, *Sliding Modes and Their Application in Variable Structure Systems*, Moscow: Mir.

Utkin, V.I., 1981, *Sliding Regimes in Optimization and Control Problem*, Moscow: Nauka.

Zhusubaliyev, Z. and Mosekilde, E., 2003, *Bifurcations and Chaos in Piecewise-smooth Dynamical Systems*, Singapore: World Scientific.

Chapter 2
Introduction to Flow Passability

In this Chapter, the passability of a flow to the boundary of two different dynamical systems will be presented. The accessible and inaccessible sub-domains will be introduced first for a theory of discontinuous dynamic systems. On the accessible domains, the corresponding dynamic systems will be introduced. The flow orientation and singular sets of boundary will be discussed. The passability and tangency (grazing) of a flow to the separation boundary between two adjacent accessible domains will be presented, and the necessary and sufficient conditions for such passability and tangency of the flow to the boundary will be presented. An L-function of flows will be introduced, and the switching bifurcation conditions for the flow passability to the boundary will be discussed. Finally, a friction-induced oscillator will be presented as an example.

2.1. Domain accessibility

For any discontinuous dynamical system, there are many vector fields defined on difference domains in phase space, and such different between two vector fields on two adjacent domains cause flows at the corresponding boundary to be non-smooth or discontinuous. To investigate the dynamics of discontinuous dynamical systems, consider a discontinuous dynamical system on a universal domain $\mho \subset \mathscr{R}^n$, and the passability of a flow from one domain to its adjacent domains will be discussed first. Thus, sub-domains Ω_α ($\alpha \in I$, $I = \{1, 2, \cdots, N\}$) of the universal domain $\mho$ will be introduced and the vector fields on the sub-domains may be defined differently. If there is a vector field on a sub-domain, this sub-domain is said an accessible domain. Otherwise, such a domain is said an inaccessible domain. Thus, the domain accessibility can provide a design possibility for discontinuous dynamical systems. The corresponding definitions of the domain accessibility are given as follows.

Definition 2.1. A sub-domain in the universal domain $\mho$ in a discontinuous dynamical system is termed an *accessible* sub-domain if at least a specific, continuous vector field can be defined on such a sub-domain.

Definition 2.2. A sub-domain in a universal domain $\mho$ in discontinuous dynamical systems is termed an *inaccessible* sub-domain if no any vector fields can be defined on such a sub-domain.

Since the accessible and inaccessible sub-domains exist in discontinuous dynamical systems, the universal domain $\mho$ is classified into the connectable and separable domains. The connectable domain is defined as follows.

Definition 2.3. A domain $\mho$ in phase space is termed a *connectable domain* if all the accessible sub-domains of the universal domain can be connected without any inaccessible sub-domain.

Similarly, a definition of the separable domain is given.

Definition 2.4. A domain $\mho$ is termed a *separable domain* if the accessible sub-domains in the universal domain are separated by inaccessible domains.

Since any discontinuous dynamical system possesses different vector fields defined on each accessible sub-domain, the corresponding dynamical behaviors in those accessible sub-domains Ω_α are distinguishing. The different behaviors in distinct sub-domains cause flow complexity in the domain $\mho$ of discontinuous dynamical systems. The boundary between two adjacent, accessible sub-domains is a bridge of dynamical behaviors in two domains for flow continuity. Any connectable domain is bounded by the universal boundary $S \subseteq \mathscr{R}^r$ ($r = n-1$), and each sub-domain is bounded by the sub-domain boundary surface $S_{\alpha\beta} \subset \mathscr{R}^r$ (α, $\beta \in I$) with or without the partial universal boundary. For instance, consider an n-dimensional connectable domain in phase space, as shown in Fig. 2.1(a) through an n_1-dimensional, sub-vector $\mathbf{x}_{n_1}$ and an n_2-dimensional, sub-vector $\mathbf{x}_{n_2}$ ($n_1 + n_2 = n$). The shaded area Ω_α is a specific sub-domain, and the other sub-domains are white. The dark, solid curve represents the original boundary of the domain $\mho$. For the separable domain, there is at least an inaccessible sub-domain to separate the accessible sub-domains. The union of inaccessible sub-domains is also called the "inaccessible sea". The inaccessible sea is the complement of the accessible sub-domains to the universal (original) domain $\mho$. That is determined by $\Omega_0 = \mho \setminus \bigcup_{\alpha \in I} \Omega_\alpha$. The accessible sub-domains in the domain $\mho$ are also called the "islands". For illustration of such a definition, a separable domain is shown in Fig. 2.1(b). The thick curve is the boundary of the universal domain, and the gray area is the inaccessible sea. The white regions are the accessible domains (or islands). The hatched region represents a specific accessible sub-domain (island).

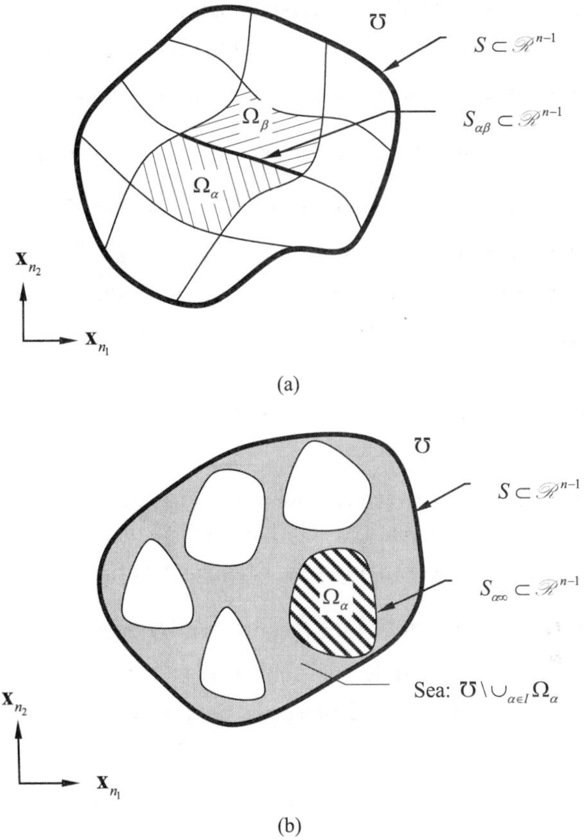

Fig. 2.1 Phase space: **(a)** connectable and **(b)** separable domains ($n_1 + n_2 = n$).

From one accessible island to another, the transport laws are needed for motion continuity, which will be discussed later in this book. The flow passability of flow from the accessible to inaccessible domains will be also discussed later.

2.2. Discontinuous dynamical systems

Consider a dynamic system consisting of N sub-dynamic systems in a universal domain $\mho \subset \mathscr{R}^n$. The universal domain is divided into N accessible sub-domains Ω_α ($\alpha \in I$) and the union of inaccessible domain Ω_0. The union of all the accessible sub-domains $\cup_{\alpha \in I} \Omega_\alpha$ and $\mho = \cup_{\alpha \in I} \Omega_\alpha \cup \Omega_0$ is the universal domain, as shown in Fig. 2.1 by an n_1-dimensional, sub-vector $\mathbf{x}_{n_1}$ and an n_2-dimensional, sub-vector $\mathbf{x}_{n_2}$ ($n_1 + n_2 = n$). For the connectable domain in Fig. 2.1(a), $\Omega_0 = \varnothing$. In Fig. 2.1(b), the union of the inaccessible sub-domains is the sea, and $\Omega_0 = \mho \setminus \cup_{\alpha \in I} \Omega_\alpha$ is the complement of the union of the accessible sub-domains. On the αth open sub-domain Ω_α, there is a C^{r_α}-continuous system ($r_\alpha \geq 1$) in form of

$$\dot{\mathbf{x}}^{(\alpha)} \equiv \mathbf{F}^{(\alpha)}(\mathbf{x}^{(\alpha)}, t, \mathbf{p}_\alpha) \in \mathscr{R}^n,\ \mathbf{x}^{(\alpha)} = (x_1^{(\alpha)}, x_2^{(\alpha)}, \cdots, x_n^{(\alpha)})^{\mathrm{T}} \in \Omega_\alpha. \tag{2.1}$$

The time is t and $\dot{\mathbf{x}} = d\mathbf{x}/dt$. In an accessible sub-domain Ω_α, the vector field $\mathbf{F}^{(\alpha)}(\mathbf{x}, t, \mathbf{p}_\alpha)$ with parameter vector $\mathbf{p}_\alpha = (p_\alpha^{(1)}, p_\alpha^{(2)}, \cdots, p_\alpha^{(l)})^{\mathrm{T}} \in \mathscr{R}^l$ is C^{r_α} continuous ($r_\alpha \geq 1$) in $\mathbf{x} \in \Omega_\alpha$ and for all time t; and the continuous flow in Eq.(2.1) $\mathbf{x}^{(\alpha)}(t) = \mathbf{\Phi}^{(\alpha)}(\mathbf{x}^{(\alpha)}(t_0), t, \mathbf{p}_\alpha)$ with $\mathbf{x}^{(\alpha)}(t_0) = \mathbf{\Phi}^{(\alpha)}(\mathbf{x}^{(\alpha)}(t_0), t_0, \mathbf{p}_\alpha)$ is C^{r+1} continuous for time t.

For discontinuous dynamical systems, the following assumptions will be adopted herein.

H2.1. The flow switching between two adjacent sub-systems is time-continuous.

H2.2. For an unbounded, accessible sub-domain Ω_α, there is a bounded domain $D_\alpha \subset \Omega_\alpha$ and the corresponding vector field and its flow are bounded, i.e.,

$$\| \mathbf{F}^{(\alpha)} \| \leq K_1(\text{const}) \text{ and } \| \mathbf{\Phi}^{(\alpha)} \| \leq K_2(\text{const}) \text{ on } D_\alpha \text{ for } t \in [0, \infty). \tag{2.2}$$

H2.3. For a bounded, accessible sub-domain Ω_α, there is a bounded domain $D_\alpha \subset \Omega_\alpha$ and the corresponding vector field is bounded, but the flow may be unbounded, i.e.,

$$\| \mathbf{F}^{(\alpha)} \| \leq K_1(\text{const}) \text{ and } \| \mathbf{\Phi}^{(\alpha)} \| \leq \infty \text{ on } D_\alpha \text{ for } t \in [0, \infty). \tag{2.3}$$

2.3. Flow passability

Since dynamical systems on different accessible sub-domains are distinguishing, the relation between flows in the two sub-domains should be developed herein for flow continuity. For a sub-domain Ω_α, there are k_α-adjacent sub-domains with k_α-pieces of boundaries ($k_\alpha \leq N-1$). Consider a boundary of any two adjacent sub-domains, formed by the intersection of the two closed sub-domains (i.e., $\partial\Omega_{ij} = \bar{\Omega}_i \cap \bar{\Omega}_j$) ($i, j \in I, j \neq i$), as shown in Fig.2.2.

Definition 2.5. The boundary in n-dimensional phase space is defined as

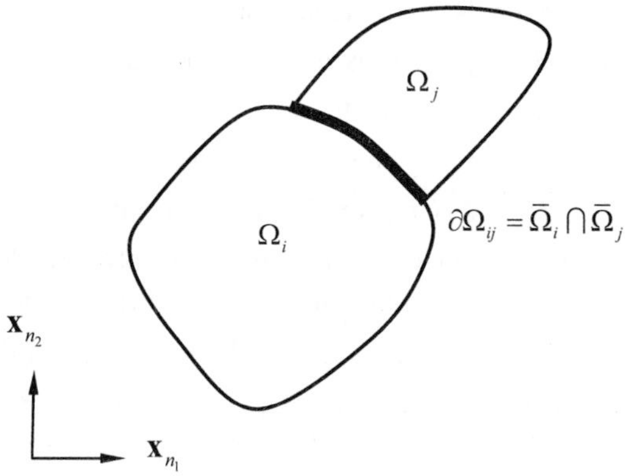

Fig. 2.2 Sub-domains Ω_α and Ω_β, the corresponding boundary $\partial\Omega_{\alpha\beta}$.

$$\begin{aligned} S_{ij} &\equiv \partial\Omega_{ij} = \bar{\Omega}_i \cap \bar{\Omega}_j \\ &= \left\{ \mathbf{x} \mid \varphi_{ij}(\mathbf{x}, t, \boldsymbol{\lambda}) = 0,\ \varphi_{ij} \text{ is } C^r\text{-continuous } (r \geq 1) \right\} \subset \mathscr{R}^{n-1}. \end{aligned} \tag{2.4}$$

Definition 2.6. The two sub-domains Ω_i and Ω_j are *disjoint* if the boundary $\partial\Omega_{ij}$ is an empty set (i.e., $\partial\Omega_{ij} = \varnothing$).

From the definition, $\partial\Omega_{ij} = \partial\Omega_{ji}$. The flow on the boundary $\partial\Omega_{ij}$ can be determined by

$$\dot{\mathbf{x}}^{(0)} = \mathbf{F}^{(0)}(\mathbf{x}^{(0)}, t) \text{ with } \varphi_{ij}(\mathbf{x}^{(0)}, t, \boldsymbol{\lambda}) = 0 \tag{2.5}$$

where $\mathbf{x}^{(0)} = (x_1^{(0)}, x_2^{(0)}, \cdots, x_n^{(0)})^{\mathrm{T}}$. With specific initial conditions, one always obtains different flows on $\varphi_{ij}(\mathbf{x}^{(0)}, t, \boldsymbol{\lambda}) = \varphi_{ij}(\mathbf{x}_0^{(0)}, t_0, \boldsymbol{\lambda}) = 0$.

Definition 2.7. If the intersection of three or more sub-domains,

$$\Gamma_{\alpha_1\alpha_2\cdots\alpha_k} \equiv \cap_{\alpha=\alpha_1}^{\alpha_k} \bar{\Omega}_\alpha \subset \mathscr{R}^r,\ (r = 0, 1, \cdots, n-2) \tag{2.6}$$

where $\alpha_k \in I$ and $k \geq 3$ is non-empty, the sub-domain intersection is termed the *singular set*.

For $r = 0$, the singular sets are singular points, which are also termed the *corner points or vertex*. In other words, any corner point is the intersection of n-linearly-independent, $(n-1)$-dimensional boundary surfaces in an n-dimensional state space. For $r = 1$, the singular sets will be curves, which are termed the 1-

dimensional singular edges to the $(n-1)$-dimensional boundary. Similarly, any 1-diemsnional singular edge is the intersection of $(n-1)$-linearly-independent, $(n-1)$-dimensional boundary surfaces in an n-dimensional state space. For $r \in \{2,3,\cdots,n-2\}$, the singular sets are the r-dimensional singular surfaces to the $(n-1)$-dimensional discontinuous boundary. In Fig. 2.3, the singular set for three closed domains $\{\bar{\Omega}_i, \bar{\Omega}_j, \bar{\Omega}_k\}$ ($i,j,k \in I$) is sketched. The circular symbols represent intersection sets. The largest solid circular symbol stands for the singular set Γ_{ijk}. The corresponding discontinuous boundaries relative to the singular set are labeled by $\partial\Omega_{ij}$, $\partial\Omega_{jk}$ and $\partial\Omega_{ik}$. The singular set possesses the hyperbolic or parabolic behavior depending on the properties of the separation boundary, which can be referred to Luo (2005, 2006). The flow on the singular sets can be similarly defined as in Eq.(2.5), by a dynamical system with the corresponding boundary constraints. The detailed discussion will be given later.

Definition 2.8. For a discontinuous dynamical system in Eq.(2.1), there is a point $\mathbf{x}(t_m) \equiv \mathbf{x}_m \in \partial\Omega_{ij}$ at time t_m between two adjacent domains Ω_α ($\alpha = i, j$). For an arbitrarily small $\varepsilon > 0$, there are two time intervals $[t_{m-\varepsilon}, t_m)$ and $(t_m, t_{m+\varepsilon}]$. Suppose $\mathbf{x}^{(i)}(t_{m-}) = \mathbf{x}_m = \mathbf{x}^{(j)}(t_{m+})$. A resultant flow of two flows $\mathbf{x}^{(\alpha)}(t)$ ($\alpha = i, j$) is called a *semi-passable flow* from domain Ω_i to Ω_j at point $(\mathbf{x}_m, t_m)$ to boundary $\partial\Omega_{ij}$ if the two flows $\mathbf{x}^{(\alpha)}(t)$ ($\alpha = i, j$) in the neighborhood of $\partial\Omega_{ij}$ possess the following properties

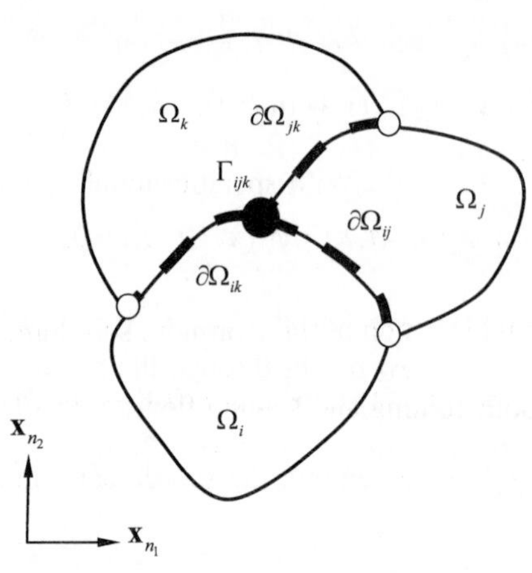

Fig. 2.3 A singular set for the intersection of three domains $\{\bar{\Omega}_i, \bar{\Omega}_j, \bar{\Omega}_k\}$ ($i,j,k \in I$). The circular circles represent intersection sets. The largest solid circular symbol stands for the singular set Γ_{ijk}. The corresponding discontinuous boundaries are marked by $\partial\Omega_{ij}$, $\partial\Omega_{jk}$ and $\partial\Omega_{ik}$.

$$\begin{aligned}&\text{either}\quad\left.\begin{aligned}&\mathbf{n}_{\partial\Omega_{ij}}^{\mathrm{T}}\cdot[\mathbf{x}^{(i)}(t_{m-})-\mathbf{x}^{(i)}(t_{m-\varepsilon})]>0\\&\mathbf{n}_{\partial\Omega_{ij}}^{\mathrm{T}}\cdot[\mathbf{x}^{(j)}(t_{m+\varepsilon})-\mathbf{x}^{(j)}(t_{m+})]>0\end{aligned}\right\}\text{ for }\mathbf{n}_{\partial\Omega_{ij}}\to\Omega_j,\\&\text{or}\quad\left.\begin{aligned}&\mathbf{n}_{\partial\Omega_{ij}}^{\mathrm{T}}\cdot[\mathbf{x}^{(i)}(t_{m-})-\mathbf{x}^{(i)}(t_{m-\varepsilon})]<0\\&\mathbf{n}_{\partial\Omega_{ij}}^{\mathrm{T}}\cdot[\mathbf{x}^{(j)}(t_{m+\varepsilon})-\mathbf{x}^{(j)}(t_{m+})]<0\end{aligned}\right\}\text{ for }\mathbf{n}_{\partial\Omega_{ij}}\to\Omega_i\end{aligned}\tag{2.7}$$

where the normal vector of the boundary $\partial\Omega_{ij}$ is

$$\mathbf{n}_{\partial\Omega_{ij}}=\nabla\varphi_{ij}\,|_{\mathbf{x}=\mathbf{x}_m}=\left(\frac{\partial\varphi_{ij}}{\partial x_1},\frac{\partial\varphi_{ij}}{\partial x_2},\cdots,\frac{\partial\varphi_{ij}}{\partial x_n}\right)^{\mathrm{T}}\Bigg|_{\mathbf{x}=\mathbf{x}_m}.\tag{2.8}$$

The notations $t_{m\pm\varepsilon}=t_m\pm\varepsilon$ and $t_{m\pm}=t_m\pm 0$ are used. $\mathbf{n}_{\partial\Omega_{ij}}\to\Omega_j$ represents that the normal vector of boundary at $(\mathbf{x}_m,t_m)$ points to domain Ω_j. In addition, a boundary $\partial\Omega_{ij}$ to semi-passable flows $\mathbf{x}^{(\alpha)}(t)$ $(\alpha=i,j)$ from domain Ω_i to domain Ω_j is called the *semi-passable* boundary (expressed by $\overrightarrow{\partial\Omega_{ij}}$). For a geometrical explanation of the semi-passable flow to the boundary, consider a flow $\mathbf{x}^{(i)}(t)$ of discontinuous dynamical system in Eq.(2.1) passing through boundary $\partial\Omega_{ij}$ from domain Ω_i to domain Ω_j. At time t_m, the flow $\mathbf{x}^{(i)}(t)$ arrives to the boundary $\partial\Omega_{ij}$, and there is a small neighborhood $(t_{m-\varepsilon},t_{m+\varepsilon})$ of time t_m, which is arbitrarily selected. Before the flow $\mathbf{x}^{(i)}(t)$ reaches to the boundary $\partial\Omega_{ij}$, a point $\mathbf{x}^{(i)}(t_{m-\varepsilon})$ lies in domain Ω_i. As $\varepsilon\to 0$, the time increment $\Delta t\equiv\varepsilon\to 0$. A point $\mathbf{x}_m$ on the boundary is the limit of $\mathbf{x}^{(i)}(t_{m-\varepsilon})$ as $\varepsilon\to 0$, and the point $\mathbf{x}_m$ must satisfy the boundary constraint of $\varphi_{ij}(\mathbf{x},t)=0$. After the flow $\mathbf{x}^{(i)}(t)$ passes through the boundary at point $\mathbf{x}_m$, the flow $\mathbf{x}^{(i)}(t)$ will switch to the flow $\mathbf{x}^{(j)}(t)$ on the side of domain Ω_j. $\mathbf{x}^{(j)}(t_{m+\varepsilon})$ is a point in the neighborhood of boundary, and a point $\mathbf{x}_m$ on the boundary is also the limit of $\mathbf{x}^{(j)}(t_{m+\varepsilon})$ as $\varepsilon\to 0$. The coming and leaving flow vectors are $\mathbf{x}^{(i)}(t_m)-\mathbf{x}^{(i)}(t_{m-\varepsilon})$ and $\mathbf{x}^{(j)}(t_{m+\varepsilon})-\mathbf{x}^{(j)}(t_m)$, respectively. Whether the flow passes through the boundary or not is dependent on the properties of both coming and leaving flows in the neighborhood of boundary. The processes of a flow passing through the boundary of $\partial\Omega_{ij}$ from domain Ω_i to Ω_j are shown in Fig. 2.4 for $\mathbf{n}_{\partial\Omega_{ij}}\to\Omega_j$ and $\mathbf{n}_{\partial\Omega_{ij}}\to\Omega_i$, respectively. Two vectors $\mathbf{n}_{\partial\Omega_{ij}}$ and $\mathbf{t}_{\partial\Omega_{ij}}$ are the normal and tangential vectors of the boundary $\partial\Omega_{ij}$, determined by $\varphi_{ij}(\mathbf{x},t)=0$. When a coming flow $\mathbf{x}^{(i)}(t)$ in domain Ω_i arrives to the semi-passable boundary $\overrightarrow{\partial\Omega_{ij}}$, the flow of $\mathbf{x}^{(i)}(t)$ can also be tangential to or bouncing on (or switching back from) the semi-passable boundary $\partial\Omega_{ij}$ .However, once a leaving flow $\mathbf{x}^{(j)}(t)$ in domain Ω_j leaves the semi-passable boundary $\overrightarrow{\partial\Omega}_{ij}$, the leaving flow cannot pass through the boundary $\overrightarrow{\partial\Omega}_{ij}$, but the leaving flow $\mathbf{x}^{(j)}(t)$ can tangentially leave the semi-passable boundary. Thus, tangential (or grazing) flows to the boundary are very important, which will be discussed later in this chapter. In the following discussion, no any control and transport laws will be inserted on the boundary. The direction of $\mathbf{t}_{\partial\Omega_{ij}} \times \mathbf{n}_{\partial\Omega_{ij}}$ is the positive direction by the right-hand rule.

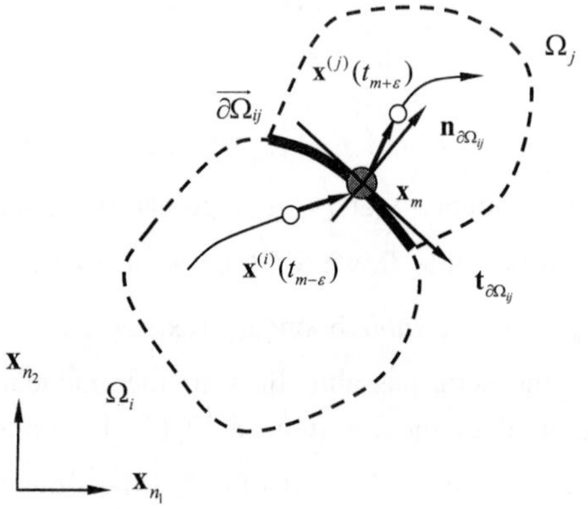

(a)

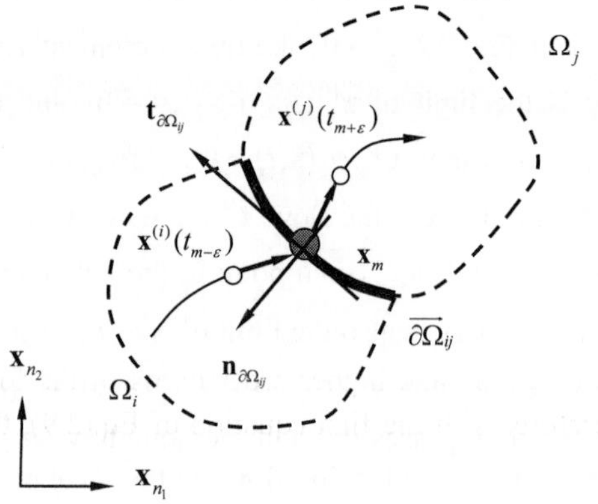

(b)

Fig. 2.4 A flow passing through the semi-passable boundary $\overrightarrow{\partial\Omega}_{ij}$ from domain Ω_i to Ω_j: **(a)** $\mathbf{n}_{\partial\Omega_{ij}} \to \Omega_j$ and **(b)** $\mathbf{n}_{\partial\Omega_{ij}} \to \Omega_i$. $\mathbf{x}^{(i)}(t_{m-\varepsilon})$, $\mathbf{x}^{(j)}(t_{m+\varepsilon})$ and $\mathbf{x}_m$ are three points in Ω_i and Ω_j and on the boundary $\partial\Omega_{ij}$, respectively. Two vectors $\mathbf{n}_{\partial\Omega_{ij}}$ and $\mathbf{t}_{\partial\Omega_{ij}}$ are the normal and tangential vectors of $\partial\Omega_{ij}$.

Theorem 2.1. *For a discontinuous dynamical system in Eq.(2.1), there is a point* $\mathbf{x}(t_m) \equiv \mathbf{x}_m \in \partial\Omega_{ij}$ *at time* t_m *between two adjacent domains* Ω_α ($\alpha = i, j$). *For an arbitrarily small* $\varepsilon > 0$, *there are two time intervals* $[t_{m-\varepsilon}, t_m)$ *and* $(t_m, t_{m+\varepsilon}]$. *Suppose* $\mathbf{x}^{(i)}(t_{m-}) = \mathbf{x}_m = \mathbf{x}^{(j)}(t_{m+})$. *Two flows* $\mathbf{x}^{(i)}(t)$ *and* $\mathbf{x}^{(j)}(t)$ *are* $C^{r_i}_{[t_{m-\varepsilon}, t_m)}$ *and* $C^{r_j}_{(t_m, t_{m+\varepsilon}]}$*-continuous* ($r_\alpha \geq 2, \alpha = i, j$) *for time t, respectively.* $\| d^{r_\alpha}\mathbf{x}^{(\alpha)}/dt^{r_\alpha} \| < \infty$ ($\alpha = i, j$). *The resultant flow of* $\mathbf{x}^{(i)}(t)$ *and* $\mathbf{x}^{(j)}(t)$ *at point* $(\mathbf{x}_m, t_m)$ *to the boundary* $\partial\Omega_{ij}$ *is semi-passable from domain* Ω_i *to* Ω_j *if and only if*

$$
\begin{aligned}
&\textit{either} \quad \left.\begin{aligned} \mathbf{n}^{\mathrm{T}}_{\partial\Omega_{ij}} \cdot \dot{\mathbf{x}}^{(i)}(t_{m-}) > 0 \\ \mathbf{n}^{\mathrm{T}}_{\partial\Omega_{ij}} \cdot \dot{\mathbf{x}}^{(j)}(t_{m+}) > 0 \end{aligned}\right\} \textit{for } \mathbf{n}_{\partial\Omega_{ij}} \to \Omega_j, \\
&\textit{or} \quad \left.\begin{aligned} \mathbf{n}^{\mathrm{T}}_{\partial\Omega_{ij}} \cdot \dot{\mathbf{x}}^{(i)}(t_{m-}) < 0 \\ \mathbf{n}^{\mathrm{T}}_{\partial\Omega_{ij}} \cdot \dot{\mathbf{x}}^{(j)}(t_{m+}) < 0 \end{aligned}\right\} \textit{for } \mathbf{n}_{\partial\Omega_{ij}} \to \Omega_i.
\end{aligned}
\tag{2.9}
$$

Proof. For a point $\mathbf{x}_m \in \partial\Omega_{ij}$ with $\mathbf{n}_{\partial\Omega_{ij}} \to \Omega_j$, suppose $\mathbf{x}^{(i)}(t_{m-}) = \mathbf{x}_m$ and $\mathbf{x}_m = \mathbf{x}^{(j)}(t_{m+})$. The two flows $\mathbf{x}^{(i)}(t)$ and $\mathbf{x}^{(j)}(t)$ are $C^{r}_{[t_{m-\varepsilon}, t_m)}$ and $C^{r}_{(t_m, t_{m+\varepsilon}]}$-continuous ($r \geq 2$) for time t, respectively. $\| \ddot{\mathbf{x}}^{(\alpha)}(t) \| < \infty$ ($\alpha \in \{i, j\}$) for $0 < \varepsilon << 1$. Consider $a \in [t_{m-\varepsilon}, t_{m-})$ and $b \in (t_{m-}, t_{m+\varepsilon}]$. Application of the Taylor series expansion of $\mathbf{x}^{(\alpha)}(t_{m\pm\varepsilon})$ with $t_{m\pm\varepsilon} = t_m \pm \varepsilon$ ($\alpha \in \{i, j\}$) to $\mathbf{x}^{(\alpha)}(a)$ and $\mathbf{x}^{(\alpha)}(b)$ gives

$$
\begin{aligned}
\mathbf{x}^{(i)}(t_{m-\varepsilon}) &\equiv \mathbf{x}^{(i)}(t_{m-} - \varepsilon) = \mathbf{x}^{(i)}(a) + \dot{\mathbf{x}}^{(i)}(a)(t_{m-} - \varepsilon - a) + o(t_{m-} - \varepsilon - a); \\
\mathbf{x}^{(j)}(t_{m+\varepsilon}) &\equiv \mathbf{x}^{(j)}(t_{m+} + \varepsilon) = \mathbf{x}^{(j)}(b) + \dot{\mathbf{x}}^{(j)}(t_{m+} + \varepsilon - b) + o(t_{m+} + \varepsilon - b).
\end{aligned}
$$

Let $a \to t_{m-}$ and $b \to t_{m+}$, the limits of the foregoing equations lead to

$$
\begin{aligned}
\mathbf{x}^{(i)}(t_{m-\varepsilon}) &\equiv \mathbf{x}^{(i)}(t_{m-} - \varepsilon) = \mathbf{x}^{(i)}(t_{m-}) - \dot{\mathbf{x}}^{(i)}(t_{m-})\varepsilon + o(\varepsilon); \\
\mathbf{x}^{(j)}(t_{m+\varepsilon}) &\equiv \mathbf{x}^{(j)}(t_{m+} + \varepsilon) = \mathbf{x}^{(j)}(t_{m+}) + \dot{\mathbf{x}}^{(j)}(t_{m+})\varepsilon + o(\varepsilon).
\end{aligned}
$$

Because of $0 < \varepsilon << 1$, the ε^2 and higher order terms of the foregoing equations can be ignored. Therefore, with the first equation of Eq.(2.9), the following relations exist,

$$
\begin{aligned}
\mathbf{n}^{\mathrm{T}}_{\partial\Omega_{ij}} \cdot [\mathbf{x}^{(i)}(t_{m-}) - \mathbf{x}^{(i)}(t_{m-\varepsilon})] &= \mathbf{n}^{\mathrm{T}}_{\partial\Omega_{ij}} \cdot \dot{\mathbf{x}}^{(i)}(t_{m-})\varepsilon > 0, \\
\mathbf{n}^{\mathrm{T}}_{\partial\Omega_{ij}} \cdot [\mathbf{x}^{(j)}(t_{m+\varepsilon}) - \mathbf{x}^{(j)}(t_{m+})] &= \mathbf{n}^{\mathrm{T}}_{\partial\Omega_{ij}} \cdot \dot{\mathbf{x}}^{(j)}(t_{m+})\varepsilon > 0.
\end{aligned}
$$

From Definition 2.8, the flow at point $(\mathbf{x}_m, t_m)$ to boundary $\partial\Omega_{ij}$ with $\mathbf{n}_{\partial\Omega_{ij}} \to \Omega_j$ is semi-passable from domain Ω_i to Ω_j under the condition in the first inequality equations of Eq.(2.9). In a similar manner, the flow at point $(\mathbf{x}_m, t_m)$ to boundary

$\partial\Omega_{ij}$ with $\mathbf{n}_{\partial\Omega_{ij}} \to \Omega_i$ is semi-passable under conditions in the second inequality equation in Eq.(2.9), and vice versa. ■

Theorem 2.2. *For a discontinuous dynamical system in Eq.*(2.1), *there is a point* $\mathbf{x}(t_m) \equiv \mathbf{x}_m \in \partial\Omega_{ij}$ *at time* t_m *between two adjacent domains* Ω_α $(\alpha = i, j)$. *For an arbitrarily small* $\varepsilon > 0$, *there are two time intervals* $[t_{m-\varepsilon}, t_m)$ *and* $(t_m, t_{m+\varepsilon}]$. $\mathbf{x}^{(i)}(t_{m-}) = \mathbf{x}_m = \mathbf{x}^{(j)}(t_{m+})$. *Two vector fields of* $\mathbf{F}^{(i)}(\mathbf{x}, t, \mathbf{p}_i)$ *and* $\mathbf{F}^{(j)}(\mathbf{x}, t, \mathbf{p}_j)$ *are* $C^{r_i}_{[t_{m-\varepsilon}, t_m)}$ *and* $C^{r_j}_{(t_m, t_{m+\varepsilon}]}$ *-continuous* ($r_\alpha \geq 1$, $\alpha = i, j$) *for time* t , *respectively.* $\| d^{r_\alpha+1}\mathbf{x}^{(\alpha)} / dt^{r_\alpha+1} \| < \infty$ ($\alpha = i, j$). *The resultant flow of two flows* $\mathbf{x}^{(i)}(t)$ *and* $\mathbf{x}^{(j)}(t)$ *at point* $(\mathbf{x}_m, t_m)$ *to boundary* $\partial\Omega_{ij}$ *is semi-passable from domain* Ω_i *to* Ω_j *if and only if*

$$
\begin{aligned}
&\text{either} && \left.\begin{aligned} \mathbf{n}^{\mathrm{T}}_{\partial\Omega_{ij}} \cdot \mathbf{F}^{(i)}(t_{m-}) > 0 \\ \mathbf{n}^{\mathrm{T}}_{\partial\Omega_{ij}} \cdot \mathbf{F}^{(j)}(t_{m+}) > 0 \end{aligned}\right\} \textit{for } \mathbf{n}_{\partial\Omega_{ij}} \to \Omega_j, \\
&\text{or} && \left.\begin{aligned} \mathbf{n}^{\mathrm{T}}_{\partial\Omega_{ij}} \cdot \mathbf{F}^{(i)}(t_{m-}) < 0 \\ \mathbf{n}^{\mathrm{T}}_{\partial\Omega_{ij}} \cdot \mathbf{F}^{(j)}(t_{m+}) < 0 \end{aligned}\right\} \textit{for } \mathbf{n}_{\partial\Omega_{ij}} \to \Omega_i
\end{aligned}
\tag{2.10}
$$

where $\mathbf{F}^{(i)}(t_{m-}) \equiv \mathbf{F}^{(i)}(\mathbf{x}_m, t_{m-}, \mathbf{p}_i)$ and $\mathbf{F}^{(j)}(t_{m+}) \equiv \mathbf{F}^{(j)}(\mathbf{x}_m, t_{m+}, \mathbf{p}_j)$.

Proof. For a point $\mathbf{x}_m \in \partial\Omega_{ij}$ with $\mathbf{n}_{\partial\Omega_{ij}} \to \Omega_j$, $\mathbf{x}^{(i)}(t_{m-}) = \mathbf{x}_m = \mathbf{x}^{(j)}(t_{m+})$. With Eq.(2.1), the first inequality equation of Eq.(2.10) gives

$$
\begin{aligned}
\mathbf{n}^{\mathrm{T}}_{\partial\Omega_{ij}} \cdot \dot{\mathbf{x}}^{(i)}(t_{m-}) = \mathbf{n}^{\mathrm{T}}_{\partial\Omega_{ij}} \cdot \mathbf{F}^{(i)}(t_{m-}) > 0, \\
\mathbf{n}^{\mathrm{T}}_{\partial\Omega_{ij}} \cdot \dot{\mathbf{x}}^{(j)}(t_{m+}) = \mathbf{n}^{\mathrm{T}}_{\partial\Omega_{ij}} \cdot \mathbf{F}^{(j)}(t_{m+}) > 0.
\end{aligned}
$$

From Theorem 2.1 and Definition 2.8, the resultant flow at point $(\mathbf{x}_m, t_m)$ to boundary $\partial\Omega_{ij}$ with $\mathbf{n}_{\partial\Omega_{ij}} \to \Omega_j$ is semi-passable. In a similar fashion, the resultant flow of two flows $\mathbf{x}^{(i)}(t)$ and $\mathbf{x}^{(j)}(t)$ at point $(\mathbf{x}_m, t_m)$ to boundary $\partial\Omega_{ij}$ with $\mathbf{n}_{\partial\Omega_{ij}} \to \Omega_i$ is semi-passable under conditions in the second inequality equations of Eq.(2.10). ■

Definition 2.9. For a discontinuous dynamical system in Eq.(2.1), there is a point $\mathbf{x}(t_m) \equiv \mathbf{x}_m \in \partial\Omega_{ij}$ at time t_m between two adjacent domains Ω_α $(\alpha = i, j)$. For an arbitrarily small $\varepsilon > 0$, there is a time interval $[t_{m-\varepsilon}, t_m)$. $\mathbf{x}^{(\alpha)}(t_{m-}) = \mathbf{x}_m$. Two flows $\mathbf{x}^{(i)}(t)$ and $\mathbf{x}^{(j)}(t)$ are called *non-passable flows of the first kind* at point $(\mathbf{x}_m, t_m)$ to boundary $\partial\Omega_{ij}$ (or termed *sink flows* at point $(\mathbf{x}_m, t_m)$ to boundary $\partial\Omega_{ij}$) if flows $\mathbf{x}^{(i)}(t)$ and $\mathbf{x}^{(j)}(t)$ in vicinity of $\partial\Omega_{ij}$ possess the following properties

$$
\begin{aligned}
&\text{either} \quad \left.\begin{aligned} \mathbf{n}^{\mathrm{T}}_{\partial\Omega_{ij}} \cdot [\mathbf{x}^{(i)}(t_{m-}) - \mathbf{x}^{(i)}(t_{m-\varepsilon})] > 0 \\ \mathbf{n}^{\mathrm{T}}_{\partial\Omega_{ij}} \cdot [\mathbf{x}^{(j)}(t_{m-}) - \mathbf{x}^{(j)}(t_{m-\varepsilon})] < 0 \end{aligned}\right\} \text{for } \mathbf{n}_{\partial\Omega_{ij}} \to \Omega_j, \\
&\text{or} \quad \left.\begin{aligned} \mathbf{n}^{\mathrm{T}}_{\partial\Omega_{ij}} \cdot [\mathbf{x}^{(i)}(t_{m-}) - \mathbf{x}^{(i)}(t_{m-\varepsilon})] < 0 \\ \mathbf{n}^{\mathrm{T}}_{\partial\Omega_{ij}} \cdot [\mathbf{x}^{(j)}(t_{m-}) - \mathbf{x}^{(j)}(t_{m-\varepsilon})] > 0 \end{aligned}\right\} \text{for } \mathbf{n}_{\partial\Omega_{ij}} \to \Omega_i.
\end{aligned}
\tag{2.11}
$$

Definition 2.10. For a discontinuous dynamical system in Eq.(2.1), there is a point $\mathbf{x}(t_m) \equiv \mathbf{x}_m \in \partial\Omega_{ij}$ at time t_m between two adjacent domains Ω_α ($\alpha = i, j$). For an arbitrarily small $\varepsilon > 0$, there is a time interval $(t_m, t_{m+\varepsilon}]$. $\mathbf{x}^{(\alpha)}(t_{m+}) = \mathbf{x}_m$. Two flows $\mathbf{x}^{(i)}(t)$ and $\mathbf{x}^{(j)}(t)$ are called *non-passable flows of the second kind* at point $(\mathbf{x}_m, t_m)$ to boundary $\partial\Omega_{ij}$ (or termed *source flows* at point $(\mathbf{x}_m, t_m)$ to boundary $\partial\Omega_{ij}$) if the flows $\mathbf{x}^{(i)}(t)$ and $\mathbf{x}^{(j)}(t)$ in neighborhood of $\partial\Omega_{ij}$ possess the following properties

$$
\begin{aligned}
&\text{either} \quad \left.\begin{aligned} \mathbf{n}^{\mathrm{T}}_{\partial\Omega_{ij}} \cdot [\mathbf{x}^{(i)}(t_{m+\varepsilon}) - \mathbf{x}^{(i)}(t_{m+})] < 0 \\ \mathbf{n}^{\mathrm{T}}_{\partial\Omega_{ij}} \cdot [\mathbf{x}^{(j)}(t_{m+\varepsilon}) - \mathbf{x}^{(j)}(t_{m+})] > 0 \end{aligned}\right\} \text{for } \mathbf{n}_{\partial\Omega_{ij}} \to \Omega_j, \\
&\text{or} \quad \left.\begin{aligned} \mathbf{n}^{\mathrm{T}}_{\partial\Omega_{ij}} \cdot [\mathbf{x}^{(i)}(t_{m+\varepsilon}) - \mathbf{x}^{(i)}(t_{m+})] > 0 \\ \mathbf{n}^{\mathrm{T}}_{\partial\Omega_{ij}} \cdot [\mathbf{x}^{(j)}(t_{m+\varepsilon}) - \mathbf{x}^{(j)}(t_{m+})] < 0 \end{aligned}\right\} \text{for } \mathbf{n}_{\partial\Omega_{ij}} \to \Omega_i.
\end{aligned}
\tag{2.12}
$$

The boundary $\partial\Omega_{ij}$ for two *sink* flows $\mathbf{x}^{(i)}(t)$ and $\mathbf{x}^{(j)}(t)$ at point $(\mathbf{x}_m, t_m)$ is called a *non-passable boundary of the first kind*, donated by $\widetilde{\partial\Omega_{ij}}$ (or termed *a sink boundary* between Ω_i and Ω_j). The boundary $\partial\Omega_{ij}$ for two *source* flows $\mathbf{x}^{(i)}(t)$ and $\mathbf{x}^{(j)}(t)$ at point $(\mathbf{x}_m, t_m)$ is called a *non-passable boundary of the second kind*, denoted by $\widehat{\partial\Omega_{ij}}$ (or termed *a source boundary* between Ω_i and Ω_j). The sink and source flows to the boundary $\partial\Omega_{ij}$ between Ω_i and Ω_j are illustrated in Fig. 2.5(a) and (b). The flows in neighborhood of boundary $\partial\Omega_{ij}$ are depicted. When a flow $\mathbf{x}^{(\alpha)}(t)$ ($\alpha = i, j$) in domain Ω_α arrives to the non-passable boundary of the first kind $\widetilde{\partial\Omega_{ij}}$, the flow can be either tangential to or sliding on the non-passable boundary $\widetilde{\partial\Omega_{ij}}$. For the non-passable boundary of the second kind $\widehat{\partial\Omega_{ij}}$, a flow $\mathbf{x}^{(\alpha)}(t)$ ($\alpha = i, j$) in the domain Ω_α can be either tangential to or bouncing on the non-passable boundary $\widehat{\partial\Omega_{ij}}$. In this chapter, only the flows tangential to the non-passable boundary will be discussed.

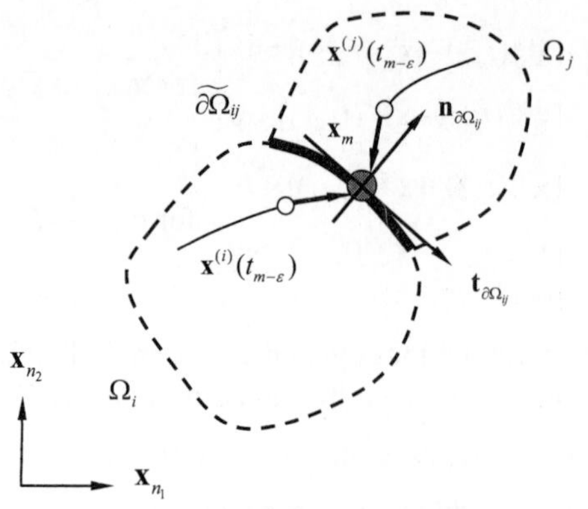

(a)

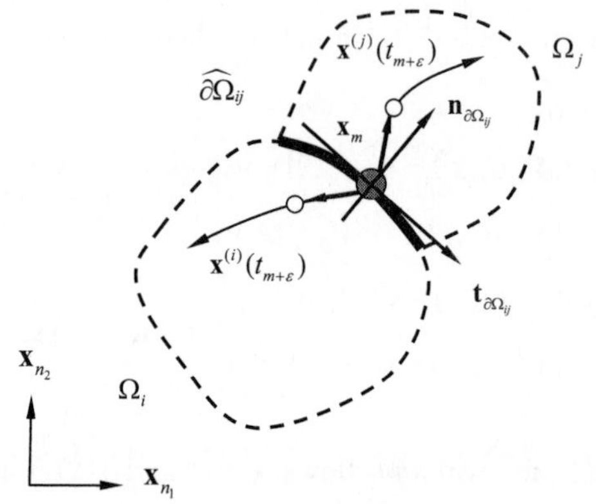

(b)

Fig. 2.5 Non-passable flow to the boundary $\partial\Omega_{ij}$ with $\mathbf{n}_{\partial\Omega_{ij}} \to \Omega_j$: **(a)** sink flow to $\widetilde{\partial\Omega}_{ij}$ (or the non-passable flow of the first kind), **(b)** source flow to $\widehat{\partial\Omega}_{ij}$ (or the non-passable flow of the second kind). $\mathbf{x}_m \equiv (\mathbf{x}_{n_1}(t_m), \mathbf{x}_{n_2}(t_m))^{\mathrm{T}}$, $\mathbf{x}^{(\alpha)}(t_{m\pm\varepsilon}) \equiv (\mathbf{x}^{(\alpha)}_{n_1}(t_{m\pm\varepsilon}), \mathbf{x}^{(\alpha)}_{n_2}(t_{m\pm\varepsilon}))^{\mathrm{T}}$ and $\alpha = \{i, j\}$ where $t_{m\pm\varepsilon} = t_m \pm \varepsilon$ for an arbitrary small $\varepsilon > 0$.

Theorem 2.3. *For a discontinuous dynamical system in Eq.*(2.1)*, there is a point* $\mathbf{x}(t_m) \equiv \mathbf{x}_m \in \partial\Omega_{ij}$ *at time* t_m *between two adjacent domains* Ω_α ($\alpha = i, j$). *For an arbitrarily small* $\varepsilon > 0$, *there is a time interval* $[t_{m-\varepsilon}, t_m)$. $\mathbf{x}^{(\alpha)}(t_{m-}) = \mathbf{x}_m$. *The flow* $\mathbf{x}^{(\alpha)}(t)$ *is* $C^{r_\alpha}_{[t_{m-\varepsilon}, t_m)}$*-continuous for time* t *and* $\| d^{r_\alpha}\mathbf{x}^{(\alpha)} / dt^{r_\alpha} \| < \infty$ ($r_\alpha \geq 2$). *Two flows* $\mathbf{x}^{(i)}(t)$ *and* $\mathbf{x}^{(j)}(t)$ *at point* $(\mathbf{x}_m, t_m)$ *to boundary* $\partial\Omega_{ij}$ *are non-passable flows of the first kind (or sink flows) if and only if*

$$\left.\begin{array}{ll} \text{either} & \begin{array}{l} \mathbf{n}^{\mathrm{T}}_{\partial\Omega_{ij}} \cdot \dot{\mathbf{x}}^{(i)}(t_{m-}) > 0 \\ \mathbf{n}^{\mathrm{T}}_{\partial\Omega_{ij}} \cdot \dot{\mathbf{x}}^{(j)}(t_{m-}) < 0 \end{array}\Bigg\} \text{for } \mathbf{n}_{\partial\Omega_{ij}} \to \Omega_j, \\ \text{or} & \begin{array}{l} \mathbf{n}^{\mathrm{T}}_{\partial\Omega_{ij}} \cdot \dot{\mathbf{x}}^{(i)}(t_{m-}) < 0 \\ \mathbf{n}^{\mathrm{T}}_{\partial\Omega_{ij}} \cdot \dot{\mathbf{x}}^{(j)}(t_{m-}) > 0 \end{array}\Bigg\} \text{for } \mathbf{n}_{\partial\Omega_{ij}} \to \Omega_i. \end{array}\right. \tag{2.13}$$

Proof. Following the proof procedure of Theorem 2.1, Theorem 2.3 can be proved. ■

Theorem 2.4. *For a discontinuous dynamical system in Eq.(2.1), there is a point* $\mathbf{x}(t_m) \equiv \mathbf{x}_m \in \partial\Omega_{ij}$ *at* t_m *between two adjacent domains* Ω_α ($\alpha = i, j$). *For an arbitrarily small* $\varepsilon > 0$, *there is a time interval* $[t_{m-\varepsilon}, t_m)$. $\mathbf{x}^{(\alpha)}(t_{m-}) = \mathbf{x}_m$. *The vector field* $\mathbf{F}^{(\alpha)}(\mathbf{x}, t, \mathbf{p}_\alpha)$ *is* $C^{r_\alpha}_{[t_{m-\varepsilon}, t_m)}$*-continuous and* $\| d^{r_\alpha+1}\mathbf{x}^{(\alpha)} / dt^{r_\alpha+1} \| < \infty$ ($r_\alpha \geq 1$). *Two flows* $\mathbf{x}^{(i)}(t)$ *and* $\mathbf{x}^{(j)}(t)$ *at point* $(\mathbf{x}_m, t_m)$ *to boundary* $\partial\Omega_{ij}$ *are non-passable flows of the first kind (or sink flows) if and only if*

$$\left.\begin{array}{ll} \text{either} & \begin{array}{l} \mathbf{n}^{\mathrm{T}}_{\partial\Omega_{ij}} \cdot \mathbf{F}^{(i)}(t_{m-}) > 0 \\ \mathbf{n}^{\mathrm{T}}_{\partial\Omega_{ij}} \cdot \mathbf{F}^{(j)}(t_{m-}) < 0 \end{array}\Bigg\} \text{for } \mathbf{n}_{\partial\Omega_{ij}} \to \Omega_j, \\ \text{or} & \begin{array}{l} \mathbf{n}^{\mathrm{T}}_{\partial\Omega_{ij}} \cdot \mathbf{F}^{(i)}(t_{m-}) < 0 \\ \mathbf{n}^{\mathrm{T}}_{\partial\Omega_{ij}} \cdot \mathbf{F}^{(j)}(t_{m-}) > 0 \end{array}\Bigg\} \text{for } \mathbf{n}_{\partial\Omega_{ij}} \to \Omega_i \end{array}\right. \tag{2.14}$$

where $\mathbf{F}^{(\alpha)}(t_{m-}) \triangleq \mathbf{F}^{(\alpha)}(\mathbf{x}, t_{m-}, \mathbf{p}_\alpha)$ ($\alpha \in \{i, j\}$).

Proof. Following the proof procedure of Theorem 2.2, Theorem 2.4 can be easily proved. ■

Theorem 2.5. *For a discontinuous dynamical system in Eq.(2.1), there is a point* $\mathbf{x}(t_m) \equiv \mathbf{x}_m \in \partial\Omega_{ij}$ *at* t_m *between two adjacent domains* Ω_α ($\alpha = i, j$). *For an arbitrarily small* $\varepsilon > 0$, *there is a time interval* $(t_m, t_{m+\varepsilon}]$. $\mathbf{x}^{(\alpha)}(t_{m+}) = \mathbf{x}_m$. $\mathbf{x}^{(\alpha)}(t)$ *is* $C^{r_\alpha}_{(t_m, t_{m+\varepsilon}]}$ *-continuous for time* t *with* $\| d^{r_\alpha}\mathbf{x}^{(\alpha)} / dt^{r_\alpha} \| < \infty$ ($r_\alpha \geq 2$). *Two flows* $\mathbf{x}^{(i)}(t)$ *and* $\mathbf{x}^{(j)}(t)$ *at point* $(\mathbf{x}_m, t_m)$ *to boundary* $\partial\Omega_{ij}$ *are non-passable flows of the second kind (or source flows) if and only if*

$$\text{either} \quad \left.\begin{array}{l} \mathbf{n}^{\mathrm{T}}_{\partial\Omega_{ij}} \cdot \dot{\mathbf{x}}^{(i)}(t_{m+}) < 0 \\ \mathbf{n}^{\mathrm{T}}_{\partial\Omega_{ij}} \cdot \dot{\mathbf{x}}^{(j)}(t_{m+}) > 0 \end{array}\right\} \text{for } \mathbf{n}_{\partial\Omega_{ij}} \to \Omega_j,$$

$$\text{or}\quad \left.\begin{array}{l}\mathbf{n}^{\mathrm{T}}_{\partial\Omega_{ij}}\cdot\dot{\mathbf{x}}^{(i)}(t_{m+})>0\\ \mathbf{n}^{\mathrm{T}}_{\partial\Omega_{ij}}\cdot\dot{\mathbf{x}}^{(j)}(t_{m+})<0\end{array}\right\}\text{for } \mathbf{n}_{\partial\Omega_{ij}}\to\Omega_i. \tag{2.15}$$

Proof. Following the procedure of the proof of Theorem 2.1, Theorem 2.5 can be proved. ■

Theorem 2.6. *For a discontinuous dynamical system in Eq.(2.1), there is a point* $\mathbf{x}(t_m)\equiv\mathbf{x}_m\in\partial\Omega_{ij}$ *at time* t_m *between two adjacent domains* Ω_α $(\alpha=i,j)$. *For an arbitrarily small* $\varepsilon>0$, *there is a time interval* $(t_m,t_{m+\varepsilon}]$. $\mathbf{x}^{(\alpha)}(t_{m+})=\mathbf{x}_m$. *The vector filed* $\mathbf{F}^{(\alpha)}(\mathbf{x},t,\mathbf{p}_\alpha)$ *are* $C^{r_\alpha}_{[t_{m-\varepsilon},t_m)}$*-continuous and* $\| d^{r_\alpha+1}\mathbf{x}^{(\alpha)}/dt^{r_\alpha+1}\|<\infty$ $(r_\alpha\geq 1)$. *Two flows* $\mathbf{x}^{(i)}(t)$ *and* $\mathbf{x}^{(j)}(t)$ *at point* $(\mathbf{x}_m,t_m)$ *to boundary* $\partial\Omega_{ij}$ *are non-passable flows of the second kind (or source flows) if and only if*

$$\begin{array}{l}\text{either}\quad \left.\begin{array}{l}\mathbf{n}^{\mathrm{T}}_{\partial\Omega_{ij}}\cdot\mathbf{F}^{(i)}(t_{m+})<0\\ \mathbf{n}^{\mathrm{T}}_{\partial\Omega_{ij}}\cdot\mathbf{F}^{(j)}(t_{m+})>0\end{array}\right\}\text{for } \mathbf{n}_{\partial\Omega_{ij}}\to\Omega_j,\\ \text{or}\quad \left.\begin{array}{l}\mathbf{n}^{\mathrm{T}}_{\partial\Omega_{ij}}\cdot\mathbf{F}^{(i)}(t_{m-})>0\\ \mathbf{n}^{\mathrm{T}}_{\partial\Omega_{ij}}\cdot\mathbf{F}^{(j)}(t_{m+})<0\end{array}\right\}\text{for } \mathbf{n}_{\partial\Omega_{ij}}\to\Omega_i\end{array} \tag{2.16}$$

where $\mathbf{F}^{(\alpha)}(t_{m+})\triangleq\mathbf{F}^{(\alpha)}(\mathbf{x},t_{m+},\mathbf{p}_\alpha)$ $(\alpha=i,j)$.

Proof. Following the proof procedure of Theorem 2.2, Theorem 2.6 can be easily proved. ■

2.4. Grazing flows

In this section, the flow local singularity and tangential flow will be discussed. The corresponding necessary and sufficient conditions will be presented.

Definition 2.11. For a discontinuous dynamical system in Eq.(2.1), there is a point $\mathbf{x}(t_m)\equiv\mathbf{x}_m\in\partial\Omega_{ij}$ at time t_m between two adjacent domains Ω_α $(\alpha=i,j)$. For an arbitrarily small $\varepsilon>0$, there are two time intervals (i.e., $[t_{m-\varepsilon},t_m)$ and $(t_m,t_{m+\varepsilon}]$). Suppose $\mathbf{x}^{(\alpha)}(t_{m\pm})=\mathbf{x}_m$ ($\alpha\in\{i,j\}$). A flow $\mathbf{x}^{(\alpha)}(t)$ is $C^{r_\alpha}_{[t_{m-\varepsilon},t_m)}$ and/or $C^{r_\alpha}_{(t_m,t_{m+\varepsilon}]}$-continuous ($r_\alpha\geq 2$). A point $(\mathbf{x}_m,t_m)$ on boundary $\partial\Omega_{ij}$ is critical to flow $\mathbf{x}^{(\alpha)}(t)$ if

$$\mathbf{n}_{\partial\Omega_{ij}}^{\mathrm{T}} \cdot \dot{\mathbf{x}}^{(\alpha)}(t_{m-}) = 0 \quad \text{and/or} \quad \mathbf{n}_{\partial\Omega_{ij}}^{\mathrm{T}} \cdot \dot{\mathbf{x}}^{(\alpha)}(t_{m+}) = 0. \tag{2.17}$$

Theorem 2.7. *For a discontinuous dynamical system in Eq.(2.1), there is a point* $\mathbf{x}(t_m) \equiv \mathbf{x}_m \in \partial\Omega_{ij}$ *at* t_m *between two adjacent domains* Ω_α ($\alpha = i, j$). *For an arbitrarily small* $\varepsilon > 0$, *there are two time intervals (i.e.,* $[t_{m-\varepsilon}, t_m)$ *and* $(t_m, t_{m+\varepsilon}]$). *Suppose* $\mathbf{x}^{(\alpha)}(t_{m\pm}) = \mathbf{x}_m$ ($\alpha \in \{i, j\}$). *A flow* $\mathbf{x}^{(\alpha)}(t)$ *is* $C^{r_\alpha}_{[t_{m-\varepsilon}, t_m)}$ *and/or* $C^{r_\alpha}_{(t_m, t_{m+\varepsilon}]}$ *-continuous (* $r_\alpha \geq 2$ *). The vector field* $\mathbf{F}^{(\alpha)}(\mathbf{x}, t, \mathbf{p}_\alpha)$ *is* $C^{r_\alpha - 1}_{[t_{m-\varepsilon}, t_m)}$ *and* $C^{r_\alpha - 1}_{(t_m, t_{m+\varepsilon}]}$ *-continuous for time t, respectively.* $\| d^{r_\alpha + 1}\mathbf{x}^{(\alpha)} / dt^{r_\alpha + 1} \| < \infty$. *A point* $(\mathbf{x}_m, t_m)$ *on the boundary* $\partial\Omega_{ij}$ *is critical to flow* $\mathbf{x}^{(\alpha)}(t)$ *if and only if*

$$\mathbf{n}_{\partial\Omega_{ij}}^{\mathrm{T}} \cdot \mathbf{F}^{(\alpha)}(t_{m-}) = 0 \ \textit{and/or}\ \mathbf{n}_{\partial\Omega_{ij}}^{\mathrm{T}} \cdot \mathbf{F}^{(\alpha)}(t_{m+}) = 0 \tag{2.18}$$

where $\mathbf{F}^{(\alpha)}(t_{m\pm}) = \mathbf{F}^{(\alpha)}(\mathbf{x}, t_{m\pm}, \mathbf{p}_\alpha)$.

Proof. Using Eq.(2.1) and Definition 2.11, Theorem 2.7 can be proved. ■

The tangential vector of the coming and leaving flows $\mathbf{x}^{(\alpha)}(t_{m\pm})$ to the boundary $\partial\Omega_{ij}$ in domain Ω_α ($\alpha \in \{i, j\}$) is normal to the normal vector of the boundary, so the coming flow is tangential to the boundary.

Definition 2.12. For a discontinuous dynamical system in Eq.(2.1), there is a point $\mathbf{x}(t_m) \equiv \mathbf{x}_m \in \partial\Omega_{ij}$ at time t_m between two adjacent domains Ω_α ($\alpha = i, j$). For an arbitrarily small $\varepsilon > 0$, there are two time intervals (i.e., $[t_{m-\varepsilon}, t_m)$ and $(t_m, t_{m+\varepsilon}]$). Suppose $\mathbf{x}^{(\alpha)}(t_{m\pm}) = \mathbf{x}_m$ ($\alpha \in \{i, j\}$). A flow $\mathbf{x}^{(\alpha)}(t)$ is $C^{r_\alpha}_{[t_{m-\varepsilon}, t_m)}$ and $C^{r_\alpha}_{(t_m, t_{m+\varepsilon}]}$ -continuous ($r_\alpha \geq 1$) for time *t*. The flow $\mathbf{x}^{(\alpha)}(t)$ in Ω_α is tangential to boundary $\partial\Omega_{ij}$ at point $(\mathbf{x}_m, t_m)$ if the following conditions hold.

$$\mathbf{n}_{\partial\Omega_{ij}}^{\mathrm{T}} \cdot \dot{\mathbf{x}}^{(\alpha)}(t_{m\pm}) = 0. \tag{2.19}$$

$$\text{either} \quad \left.\begin{aligned} \mathbf{n}_{\partial\Omega_{ij}}^{\mathrm{T}} \cdot [\mathbf{x}^{(\alpha)}(t_{m-}) - \mathbf{x}^{(\alpha)}(t_{m-\varepsilon})] > 0 \\ \mathbf{n}_{\partial\Omega_{ij}}^{\mathrm{T}} \cdot [\mathbf{x}^{(\alpha)}(t_{m+\varepsilon}) - \mathbf{x}^{(\alpha)}(t_{m+})] < 0 \end{aligned}\right\} \text{for } \mathbf{n}_{\partial\Omega_{ij}} \to \Omega_\beta, \tag{2.20}$$

$$\text{or} \quad \left.\begin{aligned} \mathbf{n}_{\partial\Omega_{ij}}^{\mathrm{T}} \cdot [\mathbf{x}^{(\alpha)}(t_{m-}) - \mathbf{x}^{(\alpha)}(t_{m-\varepsilon})] < 0 \\ \mathbf{n}_{\partial\Omega_{ij}}^{\mathrm{T}} \cdot [\mathbf{x}^{(\alpha)}(t_{m+\varepsilon}) - \mathbf{x}^{(\alpha)}(t_{m+})] > 0 \end{aligned}\right\} \text{for } \mathbf{n}_{\partial\Omega_{ij}} \to \Omega_\alpha \tag{2.21}$$

where $\alpha, \beta \in \{i, j\}$ but $\beta \neq \alpha$.

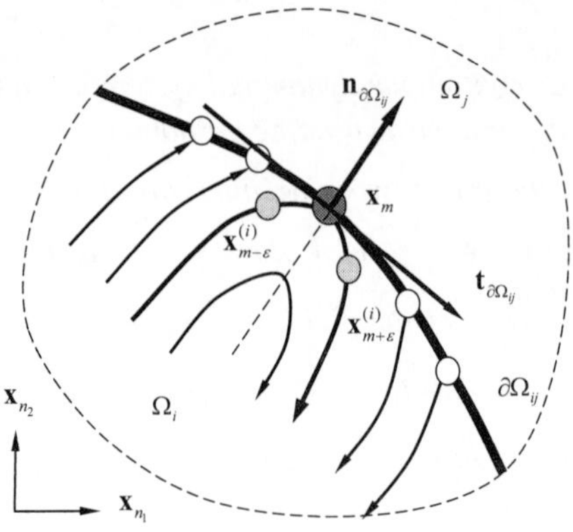

Fig. 2.6 A flow in domain Ω_i tangential to the boundary $\partial\Omega_{ij}$ with $\mathbf{n}_{\partial\Omega_{ij}} \to \Omega_j$. The gray-filled symbols represent two points ($\mathbf{x}_{m-\varepsilon}^{(i)}$ and $\mathbf{x}_{m+\varepsilon}^{(i)}$) on the flow before and after the tangency. The tangential point $\mathbf{x}_m$ on the boundary $\partial\Omega_{ij}$ is depicted by a large circular symbol.

The normal vector $\mathbf{n}_{\partial\Omega_{ij}}$ is normal to the tangential plane. Without any switching laws, Equation (2.19) gives

$$\dot{\mathbf{x}}^{(\alpha)}(t_{m-}) = \dot{\mathbf{x}}^{(\alpha)}(t_{m+}) \text{ but } \dot{\mathbf{x}}^{(\alpha)}(t_{m\pm}) \neq \dot{\mathbf{x}}^{(0)}(t_m). \tag{2.22}$$

The above equation implies that the flow $\mathbf{x}^{(\alpha)}$ on the boundary $\partial\Omega_{ij}$ is at least C^1-continuous. To demonstrate the above definition, consider a flow in domain Ω_i tangential to the boundary $\partial\Omega_{ij}$ with $\mathbf{n}_{\partial\Omega_{ij}} \to \Omega_j$, as shown in Fig. 2.6. The gray-filled symbols represent two points ($\mathbf{x}_{m\pm\varepsilon}^{(i)} = \mathbf{x}^{(i)}(t_m \pm \varepsilon)$) on the flow before and after the tangency. The tangential point $\mathbf{x}_m$ on the boundary $\partial\Omega_{ij}$ is depicted by a large circular symbol. This tangential flow is also termed a *grazing flow*.

Theorem 2.8. *For a discontinuous dynamical system in Eq.*(2.1), *there is a point* $\mathbf{x}(t_m) \equiv \mathbf{x}_m \in \partial\Omega_{ij}$ *at time* t_m *between two adjacent domains* Ω_α ($\alpha = i, j$). *For an arbitrarily small* $\varepsilon > 0$, *there are two time intervals (i.e.,* $[t_{m-\varepsilon}, t_m)$ *and* $(t_m, t_{m+\varepsilon}]$). *Suppose* $\mathbf{x}^{(\alpha)}(t_{m\pm}) = \mathbf{x}_m$ ($\alpha \in \{i, j\}$). *A flow* $\mathbf{x}^{(\alpha)}(t)$ *is* $C^{r_\alpha}_{[t_{m-\varepsilon}, t_m)}$ *and* $C^{r_\alpha}_{(t_m, t_{m+\varepsilon}]}$*-continuous* ($r_\alpha \geq 2$) *for time t.* $\| d^{r_\alpha}\mathbf{x}^{(\alpha)} / dt^{r_\alpha} \| < \infty$. *The flow* $\mathbf{x}^{(\alpha)}(t)$ *in* Ω_α *is tangential to boundary* $\partial\Omega_{ij}$ *at point* $(\mathbf{x}_m, t_m)$ *if and only if*

$$\mathbf{n}^{\mathrm{T}}_{\partial\Omega_{ij}} \cdot \dot{\mathbf{x}}^{(\alpha)}(t_{m\pm}) = 0; \tag{2.23}$$

$$\begin{array}{ll} \textit{either} & \left.\begin{array}{l} \mathbf{n}_{\partial\Omega_{ij}}^{\mathrm{T}}\cdot\dot{\mathbf{x}}^{(\alpha)}(t_{m-\varepsilon})>0 \\ \mathbf{n}_{\partial\Omega_{ij}}^{\mathrm{T}}\cdot\dot{\mathbf{x}}^{(\alpha)}(t_{m+\varepsilon})<0 \end{array}\right\} \textit{for } \mathbf{n}_{\partial\Omega_{ij}}\to\Omega_{\beta}, \\ \textit{or} & \left.\begin{array}{l} \mathbf{n}_{\partial\Omega_{ij}}^{\mathrm{T}}\cdot\dot{\mathbf{x}}^{(\alpha)}(t_{m-\varepsilon})<0 \\ \mathbf{n}_{\partial\Omega_{ij}}^{\mathrm{T}}\cdot\dot{\mathbf{x}}^{(\alpha)}(t_{m+\varepsilon})>0 \end{array}\right\} \textit{for } \mathbf{n}_{\partial\Omega_{ij}}\to\Omega_{\alpha} \end{array} \tag{2.24}$$

where $\alpha,\beta\in\{i,j\}$ *but* $\beta\neq\alpha$.

Proof. Equation (2.23) is identical to the first condition in Eq.(2.18). Consider

$$\begin{aligned} \mathbf{x}^{(\alpha)}(t_{m\pm}) &\equiv \mathbf{x}^{(\alpha)}(t_{m\pm}\pm\varepsilon\mp\varepsilon)=\mathbf{x}^{(\alpha)}(t_{m\pm}\pm\varepsilon)\mp\varepsilon\dot{\mathbf{x}}^{(\alpha)}(t_{m\pm}\pm\varepsilon)+o(\varepsilon) \\ &= \mathbf{x}^{(\alpha)}(t_{m\pm\varepsilon})\mp\varepsilon\dot{\mathbf{x}}^{(\alpha)}(t_{m\pm\varepsilon})+o(\varepsilon) \end{aligned}$$

For $0<\varepsilon<<1$, the higher order terms in the above equation can be ignored. Therefore

$$\begin{aligned} \mathbf{n}_{\partial\Omega_{ij}}^{\mathrm{T}}\cdot[\mathbf{x}^{(\alpha)}(t_{m-})-\mathbf{x}^{(\alpha)}(t_{m-\varepsilon})] &= \varepsilon\mathbf{n}_{\partial\Omega_{ij}}^{T}\cdot\dot{\mathbf{x}}^{(\alpha)}(t_{m-\varepsilon}), \\ \mathbf{n}_{\partial\Omega_{ij}}^{\mathrm{T}}\cdot[\mathbf{x}^{(\alpha)}(t_{m+\varepsilon})-\mathbf{x}^{(\alpha)}(t_{m+})] &= \varepsilon\mathbf{n}_{\partial\Omega_{ij}}^{T}\cdot\dot{\mathbf{x}}^{(\alpha)}(t_{m+\varepsilon}). \end{aligned}$$

From Eq.(2.24), the first case is

$$\mathbf{n}_{\partial\Omega_{ij}}^{\mathrm{T}}\cdot\dot{\mathbf{x}}^{(\alpha)}(t_{m-\varepsilon})>0 \text{ and } \mathbf{n}_{\partial\Omega_{ij}}^{\mathrm{T}}\cdot\dot{\mathbf{x}}^{(\alpha)}(t_{m+\varepsilon})<0$$

from which Eq.(2.20) holds for $\partial\Omega_{ij}$ with $\mathbf{n}_{\partial\Omega_{ij}}\to\Omega_{\beta}$ ($\beta\neq\alpha$). However, the second case is

$$\mathbf{n}_{\partial\Omega_{ij}}^{\mathrm{T}}\cdot\dot{\mathbf{x}}^{(\alpha)}(t_{m-\varepsilon})<0 \text{ and } \mathbf{n}_{\partial\Omega_{ij}}^{\mathrm{T}}\cdot\dot{\mathbf{x}}^{(\alpha)}(t_{m+\varepsilon})>0$$

from which Eq.(2.21) holds for $\partial\Omega_{ij}$ with $\mathbf{n}_{\partial\Omega_{ij}}\to\Omega_{\alpha}$. Therefore, from Definition 2.12, the flow $\mathbf{x}^{(\alpha)}(t)$ for time $t\in[t_{m-\varepsilon},t_{m+\varepsilon}]$ in Ω_{α} is tangential to the boundary $\partial\Omega_{ij}$. ■

Notice that the aforementioned theorem can be used for surface boundary.

Theorem 2.9. *For a discontinuous dynamical system in Eq.*(2.1), *there is a point* $\mathbf{x}(t_m)\equiv\mathbf{x}_m\in\partial\Omega_{ij}$ *at time* t_m *between two adjacent domains* Ω_{α} ($\alpha=i,j$). *For an arbitrarily small* $\varepsilon>0$, *there are two time intervals (i.e.,* $[t_{m-\varepsilon},t_m)$ *and* $(t_m,t_{m+\varepsilon}]$). *Suppose* $\mathbf{x}^{(\alpha)}(t_{m\pm})=\mathbf{x}_m$ ($\alpha\in\{i,j\}$). *A vector field* $\mathbf{F}^{(\alpha)}(\mathbf{x},t,\mathbf{p}_{\alpha})$ *is* $C^{r_{\alpha}}_{[t_{m-\varepsilon},t_m)}$ *and* $C^{r_{\alpha}}_{(t_m,t_{m+\varepsilon}]}$*-continuous* ($r_{\alpha}\geq 1$) *for time t.* $\|d^{r_{\alpha}+1}\mathbf{x}^{(\alpha)}/dt^{r_{\alpha}+1}\|<\infty$. *The flow* $\mathbf{x}^{(\alpha)}(t)$ *in* Ω_{α} *is tangential to boundary* $\partial\Omega_{ij}$ *at point* $(\mathbf{x}_m,t_m)$ *if and only if*

$$\mathbf{n}_{\partial\Omega_{ij}}^{\mathrm{T}} \cdot \mathbf{F}^{(\alpha)}(t_{m\pm}) = 0; \tag{2.25}$$

$$\begin{aligned} &\text{either} \left.\begin{aligned} \mathbf{n}_{\partial\Omega_{ij}}^{\mathrm{T}} \cdot \mathbf{F}^{(\alpha)}(t_{m-\varepsilon}) > 0 \\ \mathbf{n}_{\partial\Omega_{ij}}^{\mathrm{T}} \cdot \mathbf{F}^{(\alpha)}(t_{m+\varepsilon}) < 0 \end{aligned}\right\} \text{for } \mathbf{n}_{\partial\Omega_{ij}} \to \Omega_{\beta}, \\ &\text{or} \left.\begin{aligned} \mathbf{n}_{\partial\Omega_{ij}}^{\mathrm{T}} \cdot \mathbf{F}^{(\alpha)}(t_{m-\varepsilon}) < 0 \\ \mathbf{n}_{\partial\Omega_{ij}}^{\mathrm{T}} \cdot \mathbf{F}^{(\alpha)}(t_{m+\varepsilon}) > 0 \end{aligned}\right\} \text{for } \mathbf{n}_{\partial\Omega_{ij}} \to \Omega_{\alpha} \end{aligned} \tag{2.26}$$

where $\alpha, \beta \in \{i, j\}$ *but* $\beta \neq \alpha$.

Proof. Using Eq.(2.1) and Theorem 2.8, Theorem 2.9 can be proved. ■

For simplicity, consider $(n-1)$-dimensional *planes* in state space as the separation boundary in discontinuous dynamical systems, and the corresponding tangency to the $(n-1)$-dimensional boundary planes will be discussed as follows. Because the normal vector $\mathbf{n}_{\partial\Omega_{ij}}$ for the $(n-1)$-dimensional plane boundaries does not change with location, the corresponding conditions for a flow to tangential to such plane boundaries can help one understand the concept of a flow tangential to the general separation boundary in discontinuous dynamical systems. The $(n-1)$-dimensional surfaces as general separation boundaries will be discussed in next chapter.

Theorem 2.10. *For a discontinuous dynamical system in Eq.*(2.1), *there is a point* $\mathbf{x}(t_m) \equiv \mathbf{x}_m \in \partial\Omega_{ij}$ *at time* t_m *on the* $(n-1)$*-dimensional plane boundary* $\partial\Omega_{ij}$ *between two adjacent domains* Ω_α ($\alpha = i, j$). *For an arbitrarily small* $\varepsilon > 0$, *there are two time intervals (i.e.,* $[t_{m-\varepsilon}, t_m)$ *and* $(t_m, t_{m+\varepsilon}]$). *Suppose* $\mathbf{x}^{(\alpha)}(t_{m\pm}) = \mathbf{x}_m$ ($\alpha \in \{i, j\}$). *A flow* $\mathbf{x}^{(\alpha)}(t)$ *is* $C^{r_\alpha}_{[t_{m-\varepsilon}, t_m)}$ *and* $C^{r_\alpha}_{(t_m, t_{m+\varepsilon}]}$*-continuous* ($r_\alpha \geq 3$) *for time t and* $\| d^{r_\alpha}\mathbf{x}^{(\alpha)} / dt^{r_\alpha} \| < \infty$. *The flow* $\mathbf{x}^{(\alpha)}(t)$ *in* Ω_α *is tangential to the* $(n-1)$*-plane boundary* $\partial\Omega_{ij}$ *at point* $(\mathbf{x}_m, t_m)$ *if and only if*

$$\mathbf{n}_{\partial\Omega_{ij}}^{\mathrm{T}} \cdot \dot{\mathbf{x}}^{(\alpha)}(t_{m\pm}) = 0; \tag{2.27}$$

$$\begin{aligned} &\text{either } \mathbf{n}_{\partial\Omega_{ij}}^{\mathrm{T}} \cdot \ddot{\mathbf{x}}^{(\alpha)}(t_{m\pm}) < 0 \text{ for } \mathbf{n}_{\partial\Omega_{ij}} \to \Omega_{\beta}, \\ &\text{or } \quad \mathbf{n}_{\partial\Omega_{ij}}^{\mathrm{T}} \cdot \ddot{\mathbf{x}}^{(\alpha)}(t_{m\pm}) > 0 \text{ for } \mathbf{n}_{\partial\Omega_{ij}} \to \Omega_{\alpha} \end{aligned} \tag{2.28}$$

where $\alpha, \beta \in \{i, j\}$ *but* $\beta \neq \alpha$.

Proof. Equation (2.27) is identical to Eq.(2.19), thus the first condition in Eq.(2.19) is satisfied. From Definition 2.12, consider the boundary $\partial\Omega_{ij}$ with

$\mathbf{n}_{\partial\Omega_{ij}} \to \Omega_\beta$ $\beta \neq \alpha$) first. Suppose $\mathbf{x}^{(\alpha)}(t_{m\pm}) = \mathbf{x}_m$ ($\alpha \in \{i, j\}$) and a flow $\mathbf{x}^{(\alpha)}(t)$ is $C^{r_\alpha}_{[t_{m-\varepsilon}, t_m)}$ and $C^{r_\alpha}_{(t_m, t_{m+\varepsilon}]}$-continuous ($r_\alpha \geq 3$) for time t. For $a \in [t_{m-\varepsilon}, t_m)$ and $a \in (t_m, t_{m+\varepsilon}]$, the Taylor series expansion of $\mathbf{x}^{(\alpha)}(t_{m\pm\varepsilon})$ to $\mathbf{x}^{(\alpha)}(a)$ up to the third-order term gives

$$\begin{aligned}\mathbf{x}^{(\alpha)}(t_{m\pm\varepsilon}) \equiv \mathbf{x}^{(\alpha)}(t_{m\pm} - \varepsilon) = \mathbf{x}^{(\alpha)}(a) + \dot{\mathbf{x}}^{(\alpha)}(a)(t_{m\pm} \pm \varepsilon - a)\\ + \ddot{\mathbf{x}}^{(\alpha)}(a)(t_{m\pm} \pm \varepsilon - a)^2 + o((t_{m\pm} \pm \varepsilon - a)^2).\end{aligned}$$

As $a \to t_{m\pm}$, the limit of the foregoing equation leads to

$$\mathbf{x}^{(\alpha)}(t_{m\pm\varepsilon}) \equiv \mathbf{x}^{(\alpha)}(t_m \pm \varepsilon) = \mathbf{x}^{(\alpha)}(t_{m\pm}) \pm \dot{\mathbf{x}}^{(\alpha)}(t_{m\pm})\varepsilon + \ddot{\mathbf{x}}^{(\alpha)}(t_{m\pm})\varepsilon^2 + o(\varepsilon^2).$$

The ignorance of the ε^3 and higher-order terms, deformation of the above equation and left multiplication of $\mathbf{n}_{\partial\Omega_{ij}}$ gives

$$\mathbf{n}^{\mathrm{T}}_{\partial\Omega_{ij}} \cdot [\mathbf{x}^{(\alpha)}(t_{m-}) - \mathbf{x}^{(\alpha)}(t_{m-\varepsilon})] = \mathbf{n}^{\mathrm{T}}_{\partial\Omega_{ij}} \cdot \dot{\mathbf{x}}^{(\alpha)}(t_{m-})\varepsilon - \mathbf{n}^{\mathrm{T}}_{\partial\Omega_{ij}} \cdot \ddot{\mathbf{x}}^{(\alpha)}(t_{m-})\varepsilon^2,$$

$$\mathbf{n}^{\mathrm{T}}_{\partial\Omega_{ij}} \cdot [\mathbf{x}^{(\alpha)}(t_{m+\varepsilon}) - \mathbf{x}^{(\alpha)}(t_{m+})] = \mathbf{n}^{\mathrm{T}}_{\partial\Omega_{ij}} \cdot \dot{\mathbf{x}}^{(\alpha)}(t_{m+})\varepsilon + \mathbf{n}^{\mathrm{T}}_{\partial\Omega_{ij}} \cdot \ddot{\mathbf{x}}^{(\alpha)}(t_{m+})\varepsilon^2.$$

With Eq.(2.27), one obtains

$$\begin{aligned}\mathbf{n}^{\mathrm{T}}_{\partial\Omega_{ij}} \cdot [\mathbf{x}^{(\alpha)}(t_{m-}) - \mathbf{x}^{(\alpha)}(t_{m-\varepsilon})] = -\mathbf{n}^{\mathrm{T}}_{\partial\Omega_{ij}} \cdot \ddot{\mathbf{x}}^{(\alpha)}(t_{m-})\varepsilon^2,\\ \mathbf{n}^{\mathrm{T}}_{\partial\Omega_{ij}} \cdot [\mathbf{x}^{(\alpha)}(t_{m+\varepsilon}) - \mathbf{x}^{(\alpha)}(t_{m+})] = \mathbf{n}^{\mathrm{T}}_{\partial\Omega_{ij}} \cdot \ddot{\mathbf{x}}^{(\alpha)}(t_{m+})\varepsilon^2.\end{aligned}$$

For the plane boundary $\partial\Omega_{ij}$ with $\mathbf{n}_{\partial\Omega_{ij}} \to \Omega_\beta$, using the first inequality equation of Eq.(2.28), the foregoing two equations lead to

$$\begin{aligned}\mathbf{n}^{\mathrm{T}}_{\partial\Omega_{ij}} \cdot [\mathbf{x}^{(\alpha)}(t_{m-}) - \mathbf{x}^{(\alpha)}(t_{m-\varepsilon})] = -\mathbf{n}^{\mathrm{T}}_{\partial\Omega_{ij}} \cdot \ddot{\mathbf{x}}^{(\alpha)}(t_{m-})\varepsilon^2 > 0,\\ \mathbf{n}^{\mathrm{T}}_{\partial\Omega_{ij}} \cdot [\mathbf{x}^{(\alpha)}(t_{m+\varepsilon}) - \mathbf{x}^{(\alpha)}(t_{m+})] = \mathbf{n}^{\mathrm{T}}_{\partial\Omega_{ij}} \cdot \ddot{\mathbf{x}}^{(\alpha)}(t_{m+})\varepsilon^2 < 0.\end{aligned}$$

From Definition 2.12, the first inequality equation of Eq.(2.28) is obtained. Similarly, using the second inequality of Eq.(2.28), one obtains

$$\begin{aligned}\mathbf{n}^{\mathrm{T}}_{\partial\Omega_{ij}} \cdot [\mathbf{x}^{(\alpha)}(t_{m-}) - \mathbf{x}^{(\alpha)}(t_{m-\varepsilon})] = -\mathbf{n}^{\mathrm{T}}_{\partial\Omega_{ij}} \cdot \ddot{\mathbf{x}}^{(\alpha)}(t_{m-})\varepsilon^2 < 0,\\ \mathbf{n}^{\mathrm{T}}_{\partial\Omega_{ij}} \cdot [\mathbf{x}^{(\alpha)}(t_{m+\varepsilon}) - \mathbf{x}^{(\alpha)}(t_{m+})] = \mathbf{n}^{\mathrm{T}}_{\partial\Omega_{ij}} \cdot \ddot{\mathbf{x}}^{(\alpha)}(t_{m+})\varepsilon^2 > 0.\end{aligned}$$

for the boundary $\partial\Omega_{ij}$ with $\mathbf{n}_{\partial\Omega_{ij}} \to \Omega_\alpha$. Therefore, under Eq.(2.28), the flow $\mathbf{x}^{(\alpha)}(t)$ in domain Ω_α is tangential to the plane boundary $\partial\Omega_{ij}$, vice versa. ■

Theorem 2.11. *For a discontinuous dynamical system in Eq.*(2.1)*, there is a point* $\mathbf{x}(t_m) \equiv \mathbf{x}_m \in \partial\Omega_{ij}$ *at time* t_m *on the* $(n-1)$ *-dimensional plane boundary* $\partial\Omega_{ij}$

between two adjacent domains Ω_α ($\alpha = i, j$). *For an arbitrarily small* $\varepsilon > 0$, *there are two time intervals (i.e.,* $[t_{m-\varepsilon}, t_m)$ *and* $(t_m, t_{m+\varepsilon}]$). *Suppose* $\mathbf{x}^{(\alpha)}(t_{m\pm}) = \mathbf{x}_m$ ($\alpha \in \{i, j\}$). *The vector field* $\mathbf{F}^{(\alpha)}(\mathbf{x}, t, \boldsymbol{\mu}_\alpha)$ *is* $C^{r_\alpha}_{[t_{m-\varepsilon}, t_m)}$ *and* $C^{r_\alpha}_{(t_m, t_{m+\varepsilon}]}$*-continuous* ($r_\alpha \geq 2$) *for time t and* $\| d^{r_\alpha+1}\mathbf{x}^{(\alpha)} / dt^{r_\alpha+1} \| < \infty$. *The flow* $\mathbf{x}^{(\alpha)}(t)$ *in* Ω_α *is tangential to the plane boundary* $\partial\Omega_{ij}$ *at point* $(\mathbf{x}_m, t_m)$ *if and only if*

$$\mathbf{n}^{\mathrm{T}}_{\partial\Omega_{ij}} \cdot \mathbf{F}^{(\alpha)}(t_{m\pm}) = 0; \tag{2.29}$$

$$\begin{aligned} &\textit{either } \mathbf{n}^{\mathrm{T}}_{\partial\Omega_{ij}} \cdot D\mathbf{F}^{(\alpha)}(t_{m\pm}) < 0 \textit{ for } \mathbf{n}_{\partial\Omega_{ij}} \to \Omega_\beta, \\ &\textit{or } \quad \mathbf{n}^{\mathrm{T}}_{\partial\Omega_{ij}} \cdot D\mathbf{F}^{(\alpha)}(t_{m\pm}) > 0 \textit{ for } \mathbf{n}_{\partial\Omega_{ij}} \to \Omega_\alpha \end{aligned} \tag{2.30}$$

where $\alpha, \beta \in \{i, j\}$ *but* $\alpha \neq \beta$, *and the total differentiation* ($p, q \in \{1, 2, \cdots, n\}$)

$$D\mathbf{F}^{(\alpha)}(t_{m\pm}) = \left\{ \left[\frac{\partial F^{(\alpha)}_p(\mathbf{x}, t, \mathbf{p}_\alpha)}{\partial x_q} \right]_{n\times n} \mathbf{F}^{(\alpha)}(t_{m\pm}) + \frac{\partial \mathbf{F}^{(\alpha)}(\mathbf{x}, t, \mathbf{p}_\alpha)}{\partial t} \right\} \Bigg|_{(\mathbf{x}_m, t_{m\pm})} . \tag{2.31}$$

Proof. Using Eqs.(2.1) and (2.29), thus the first condition in Eq.(2.19) is satisfied. The derivative of Eq.(2.1) with respect to time t gives

$$\ddot{\mathbf{x}} \equiv D\mathbf{F}^{(\alpha)}(\mathbf{x}, t, \mathbf{p}_\alpha) = \left[\frac{\partial F^{(\alpha)}_p(\mathbf{x}, t, \mathbf{p}_\alpha)}{\partial x_q} \right]_{n\times n} \dot{\mathbf{x}} + \frac{\partial}{\partial t}\mathbf{F}^{(\alpha)}(\mathbf{x}, t, \mathbf{p}_\alpha).$$

For $t = t_{m\pm}$ and $\mathbf{x} = \mathbf{x}_m$, the left multiplication of $\mathbf{n}_{\partial\Omega_{ij}}$ to the above equation gives

$$\begin{aligned} &\mathbf{n}^{\mathrm{T}}_{\partial\Omega_{ij}} \cdot \ddot{\mathbf{x}}(t_{m\pm}) \\ &= \mathbf{n}^{\mathrm{T}}_{\partial\Omega_{ij}} \cdot \left\{ \left[\frac{\partial F^{(\alpha)}_p(\mathbf{x}, t, \mathbf{p}_\alpha)}{\partial x_q} \right]_{n\times n} \mathbf{F}^{(\alpha)}(t_{m\pm}) + \frac{\partial \mathbf{F}^{(\alpha)}(\mathbf{x}, t, \mathbf{p}_\alpha)}{\partial t} \right\} \Bigg|_{(\mathbf{x}_m, t_{m\pm})} \end{aligned}$$

where $\mathbf{F}^{(\alpha)}(\mathbf{x}_m, t_{m\pm}, \mathbf{p}_\alpha) \triangleq \mathbf{F}^{(\alpha)}(t_{m\pm})$. Using Eq.(2.30), the above equation leads to Eq.(2.28). From Theorem 2.10, the flow $\mathbf{x}^{(\alpha)}(t)$ in Ω_α is tangential to the plane boundary $\partial\Omega_{ij}$ at point $(\mathbf{x}_m, t_m)$, vice versa. ■

Definition 2.13. For a discontinuous dynamical system in Eq.(2.1), there is a point $\mathbf{x}(t_m) \equiv \mathbf{x}_m \in \partial\Omega_{ij}$ at time t_m on the $(n-1)$-dimensional *plane* boundary $\partial\Omega_{ij}$ between two adjacent domains Ω_α ($\alpha = i, j$). For an arbitrarily small $\varepsilon > 0$, there are two time intervals (i.e., $[t_{m-\varepsilon}, t_m)$ and $(t_m, t_{m+\varepsilon}]$). Suppose $\mathbf{x}^{(\alpha)}(t_{m\pm}) = \mathbf{x}_m$ ($\alpha \in \{i, j\}$). A flow $\mathbf{x}^{(\alpha)}(t)$ is $C^{r_\alpha}_{[t_{m-\varepsilon}, t_m)}$ and $C^{r_\alpha}_{(t_m, t_{m+\varepsilon}]}$-continuous ($r_\alpha \geq 2l_\alpha$) for time *t*. The flow $\mathbf{x}^{(\alpha)}(t)$ in Ω_α is tangential to the plane boundary $\partial\Omega_{ij}$ at point

$(\mathbf{x}_m, t_m)$ with the $(2l_\alpha - 1)$th -order if

$$\mathbf{n}_{\partial\Omega_{ij}}^{\mathrm{T}} \cdot \left. \frac{d^{k_\alpha}\mathbf{x}^{(\alpha)}(t)}{dt^{k_\alpha}} \right|_{t=t_{m\pm}} = 0 \quad \text{for } k_\alpha = 1, 2, \cdots, 2l_\alpha - 1; \tag{2.32}$$

$$\mathbf{n}_{\partial\Omega_{ij}}^{\mathrm{T}} \cdot \left. \frac{d^{2l_\alpha}\mathbf{x}^{(\alpha)}(t)}{dt^{2l_\alpha}} \right|_{t=t_{m\pm}} \neq 0; \tag{2.33}$$

$$\text{either} \left. \begin{array}{l} \mathbf{n}_{\partial\Omega_{ij}}^{\mathrm{T}} \cdot [\mathbf{x}^{(\alpha)}(t_{m-}) - \mathbf{x}^{(\alpha)}(t_{m-\varepsilon})] > 0 \\ \mathbf{n}_{\partial\Omega_{ij}}^{\mathrm{T}} \cdot [\mathbf{x}^{(\alpha)}(t_{m+\varepsilon}) - \mathbf{x}^{(\alpha)}(t_{m+})] < 0 \end{array} \right\} \text{for } \mathbf{n}_{\partial\Omega_{ij}} \to \Omega_\beta, \tag{2.34}$$

$$\text{or} \left. \begin{array}{l} \mathbf{n}_{\partial\Omega_{ij}}^{\mathrm{T}} \cdot [\mathbf{x}^{(\alpha)}(t_{m-}) - \mathbf{x}^{(\alpha)}(t_{m-\varepsilon})] < 0 \\ \mathbf{n}_{\partial\Omega_{ij}}^{\mathrm{T}} \cdot [\mathbf{x}^{(\alpha)}(t_{m+\varepsilon}) - \mathbf{x}^{(\alpha)}(t_{m+})] > 0 \end{array} \right\} \text{for } \mathbf{n}_{\partial\Omega_{ij}} \to \Omega_\alpha \tag{2.35}$$

where $\alpha, \beta \in \{i, j\}$ but $\beta \neq \alpha$.

Theorem 2.12. *For a discontinuous dynamical system in Eq.*(2.1), *there is a point* $\mathbf{x}(t_m) \equiv \mathbf{x}_m \in \partial\Omega_{ij}$ *at time* t_m *on the* $(n-1)$ *-dimensional plane boundary* $\partial\Omega_{ij}$ *between two adjacent domains* Ω_α ($\alpha = i, j$). *For an arbitrarily small* $\varepsilon > 0$, *there are two time intervals (i.e.,* $[t_{m-\varepsilon}, t_m)$ *and* $(t_m, t_{m+\varepsilon}]$ *). Suppose* $\mathbf{x}^{(\alpha)}(t_{m\pm}) = \mathbf{x}_m$ ($\alpha \in \{i, j\}$). *A flow* $\mathbf{x}^{(\alpha)}(t)$ *is* $C^{r_\alpha}_{[t_{m-\varepsilon}, t_m)}$ *and* $C^{r_\alpha}_{(t_m, t_{m+\varepsilon}]}$*-continuous* ($r_\alpha \geq 2l_\alpha + 1$) *for time t.* $\| d^{r_\alpha}\mathbf{x}^{(\alpha)} / dt^{r_\alpha} \| < \infty$. *The flow* $\mathbf{x}^{(\alpha)}(t)$ *in* Ω_α *is tangential to the plane boundary* $\partial\Omega_{ij}$ *at point* $(\mathbf{x}_m, t_m)$ *with the* $(2l_\alpha - 1)$th *-order if and only if*

$$\mathbf{n}_{\partial\Omega_{ij}}^{\mathrm{T}} \cdot \left. \frac{d^{k_\alpha}\mathbf{x}^{(\alpha)}(t)}{dt^{k_\alpha}} \right|_{t=t_{m\pm}} = 0 \quad \text{for } k_\alpha = 1, 2, \cdots, 2l_\alpha - 1; \tag{2.36}$$

$$\mathbf{n}_{\partial\Omega_{ij}}^{\mathrm{T}} \cdot \left. \frac{d^{2l_\alpha}\mathbf{x}^{(\alpha)}(t)}{dt^{2l_\alpha}} \right|_{t=t_{m\pm}} \neq 0; \tag{2.37}$$

$$\begin{array}{ll} \textit{either} & \mathbf{n}_{\partial\Omega_{ij}}^{\mathrm{T}} \cdot \left. \dfrac{d^{2l_\alpha}\mathbf{x}^{(\alpha)}(t)}{dt^{2l_\alpha}} \right|_{t=t_{m\pm}} < 0 \ \textit{for } \mathbf{n}_{\partial\Omega_{ij}} \to \Omega_\beta, \\ \textit{or} & \mathbf{n}_{\partial\Omega_{ij}}^{\mathrm{T}} \cdot \left. \dfrac{d^{2l_\alpha}\mathbf{x}^{(\alpha)}(t)}{dt^{2l_\alpha}} \right|_{t=t_{m\pm}} > 0 \ \textit{for } \mathbf{n}_{\partial\Omega_{ij}} \to \Omega_\alpha \end{array} \tag{2.38}$$

where $\beta \in \{i, j\}$ *but* $\beta \neq \alpha$.

Proof. For Eqs.(2.36) and (2.37), the first two conditions in Definition 2.13 are satisfied. Consider the boundary $\partial\Omega_{ij}$ with $\mathbf{n}_{\partial\Omega_{ij}} \to \Omega_\beta$ ($\beta \neq \alpha$) first. Choose

$a \in [t_{m-\varepsilon}, t_m)$ or $a \in (t_m, t_{m-\varepsilon}]$, and application of the Taylor series expansion of $\mathbf{x}^{(\alpha)}(t_{m\pm\varepsilon})$ to $\mathbf{x}^{(\alpha)}(a)$ and up to the $(2l_\alpha)$th -order term gives

$$\mathbf{x}^{(\alpha)}(t_{m\pm\varepsilon}) \equiv \mathbf{x}^{(\alpha)}(t_{m\pm} \pm \varepsilon) = \mathbf{x}^{(\alpha)}(a) + \sum_{k_\alpha=1}^{2l_\alpha-1} \frac{d^{k_\alpha}\mathbf{x}^{(\alpha)}(t)}{dt^{k_\alpha}}\bigg|_{t=a} (t_{m\pm} \pm \varepsilon - a)^{k_\alpha}$$
$$+ \frac{d^{2l_\alpha}\mathbf{x}^{(\alpha)}(t)}{dt^{2l_\alpha}}\bigg|_{t=a} (t_{m\pm} \pm \varepsilon - a)^{2l_\alpha} + o((t_{m\pm} \pm \varepsilon - a)^{2l_\alpha}).$$

As $a \to t_{m\pm}$, the foregoing equation becomes

$$\mathbf{x}^{(\alpha)}(t_{m\pm\varepsilon}) \equiv \mathbf{x}^{(\alpha)}(t_{m\pm} \pm \varepsilon) = \mathbf{x}^{(\alpha)}(t_{m\pm}) + \sum_{k_\alpha=1}^{2l_\alpha-1} \frac{d^{k_\alpha}\mathbf{x}^{(\alpha)}(t)}{dt^{k_\alpha}}\bigg|_{t=t_{m\pm}} (\pm\varepsilon)^{k_\alpha}$$
$$+ \frac{d^{2l_\alpha}\mathbf{x}^{(\alpha)}(t)}{dt^{2l_\alpha}}\bigg|_{t=t_{m\pm}} \varepsilon^{2l_\alpha} + o(\pm\varepsilon^{2l_\alpha}).$$

With Eqs.(2.36) and (2.37), the deformation of the above equation and left multiplication of $\mathbf{n}_{\partial\Omega_{ij}}$ produces

$$\mathbf{n}_{\partial\Omega_{ij}}^{\mathrm{T}} \cdot [\mathbf{x}^{(\alpha)}(t_{m-}) - \mathbf{x}^{(\alpha)}(t_{m-\varepsilon})] = -\mathbf{n}_{\partial\Omega_{ij}}^{\mathrm{T}} \cdot \frac{d^{2l_\alpha}\mathbf{x}^{(\alpha)}(t)}{dt^{2l_\alpha}}\bigg|_{t=t_{m-}} \varepsilon^{2l_\alpha},$$
$$\mathbf{n}_{\partial\Omega_{ij}}^{\mathrm{T}} \cdot [\mathbf{x}^{(\alpha)}(t_{m+\varepsilon}) - \mathbf{x}^{(\alpha)}(t_{m+})] = \mathbf{n}_{\partial\Omega_{ij}}^{\mathrm{T}} \cdot \frac{d^{2l_\alpha}\mathbf{x}^{(\alpha)}(t)}{dt^{2l_\alpha}}\bigg|_{t=t_{m+}} \varepsilon^{2l_\alpha}.$$

Under Eq.(2.38), the condition in Eq.(2.34) is satisfied, and vice versa. Therefore, the flow $\mathbf{x}^{(\alpha)}(t)$ in domain Ω_α is tangential to $\partial\Omega_{ij}$ with the $(2l_\alpha - 1)$th -order for $\mathbf{n}_{\partial\Omega_{ij}} \to \Omega_\beta$. Similarly, under the condition in Eq.(2.38), the flow $\mathbf{x}^{(\alpha)}(t)$ in domain Ω_α is tangential to boundary $\partial\Omega_{ij}$ at point $(\mathbf{x}_m, t_m)$ with the $(2l_\alpha - 1)$th -order for $\mathbf{n}_{\partial\Omega_{ij}} \to \Omega_\alpha$. This theorem is proved. ■

Theorem 2.13. *For a discontinuous dynamical system in Eq.(2.1), there is a point* $\mathbf{x}(t_m) \equiv \mathbf{x}_m \in \partial\Omega_{ij}$ *at time* t_m *on the* $(n-1)$ *-dimensional plane boundary* $\partial\Omega_{ij}$ *between two adjacent domains* Ω_α ($\alpha = i, j$). *For an arbitrarily small* $\varepsilon > 0$, *there are two time intervals (i.e.,* $[t_{m-\varepsilon}, t_m)$ *and* $(t_m, t_{m+\varepsilon}]$*). Suppose* $\mathbf{x}^{(\alpha)}(t_{m\pm}) = \mathbf{x}_m$ ($\alpha \in \{i, j\}$). *The vector field* $\mathbf{F}^{(\alpha)}(\mathbf{x}, t, \mathbf{p}_\alpha)$ *is* $C^{r_\alpha}_{[t_{m-\varepsilon}, t_m)}$ *and* $C^{r_\alpha}_{(t_m, t_{m+\varepsilon}]}$ *-continuous* ($r_\alpha \geq 2l_\alpha$) *for time t.* $\| d^{r_\alpha+1}\mathbf{x}^{(\alpha)} / dt^{r_\alpha+1} \| < \infty$. *The flow* $\mathbf{x}^{(\alpha)}(t)$ *in* Ω_α *is tangential to the plane boundary* $\partial\Omega_{ij}$ *at point* $(\mathbf{x}_m, t_m)$ *with the* $(2l_\alpha - 1)$th *-order if and only if*

$$\mathbf{n}_{\partial\Omega_{ij}}^{\mathrm{T}} \cdot D^{k_\alpha-1}\mathbf{F}^{(\alpha)}(t_{m\pm}) = 0 \quad for\ k_\alpha = 1, 2, \cdots, 2l_\alpha - 1; \tag{2.39}$$

$$\mathbf{n}_{\partial\Omega_{ij}}^{\mathrm{T}} \cdot D^{2l_\alpha-1}\mathbf{F}^{(\alpha)}(t_{m\pm}) \neq 0; \tag{2.40}$$

$$\begin{aligned} &\textit{either}\ \mathbf{n}_{\partial\Omega_{ij}}^{\mathrm{T}} \cdot D^{2l_\alpha-1}\mathbf{F}^{(\alpha)}(t_{m\pm}) < 0 \ \textit{for}\ \partial\Omega_{ij} \to \Omega_\beta, \\ &\textit{or}\quad\ \ \mathbf{n}_{\partial\Omega_{ij}}^{\mathrm{T}} \cdot D^{2l_\alpha-1}\mathbf{F}^{(\alpha)}(t_{m\pm}) > 0 \ \textit{for}\ \partial\Omega_{ij} \to \Omega_\alpha \end{aligned} \tag{2.41}$$

where the total differentiation

$$D^{k_\alpha-1}\mathbf{F}^{(\alpha)}(t_m) = D^{k-2}\left\{\left[\frac{\partial F_p^{(\alpha)}(\mathbf{x},t,\mathbf{p}_\alpha)}{\partial x_q}\right]_{n\times n}\dot{\mathbf{x}} + \frac{\partial \mathbf{F}^{(\alpha)}(\mathbf{x},t,\mathbf{p}_\alpha)}{\partial t}\right\}\Bigg|_{(\mathbf{x}_m,t_m)} \tag{2.42}$$

with $p,q \in \{1,2,\cdots,n\}$, $k_\alpha \in \{2,3,\cdots,2l_\alpha\}$ *and* $\beta \in \{i,j\}$ *but* $\alpha \neq \beta$.

Proof. The k_α-order derivative of Eq.(2.1) with respect to time gives

$$\begin{aligned} \left.\frac{d^{k_\alpha}\mathbf{x}^{(\alpha)}(t)}{dt^{k_\alpha}}\right|_{(\mathbf{x}_m,t_m)} &= \left.\frac{d^{k_\alpha-1}\dot{\mathbf{x}}^{(\alpha)}(t)}{dt^{k_\alpha-1}}\right|_{(\mathbf{x}_m,t_m)} = \left.\frac{d^{k_\alpha-1}\mathbf{F}^{(\alpha)}(\mathbf{x},t,\mathbf{p}_\alpha)}{dt^{k_\alpha-1}}\right|_{(\mathbf{x}_m,t_m)} \\ &\equiv D^{k_\alpha-1}\mathbf{F}^{(\alpha)}(t_m) = D^{k-2}\left\{\left[\frac{\partial F_p^{(\alpha)}(\mathbf{x},t,\mathbf{p}_\alpha)}{\partial x_q}\right]_{n\times n}\dot{\mathbf{x}} + \frac{\partial \mathbf{F}^{(\alpha)}(\mathbf{x},t,\mathbf{p}_\alpha)}{\partial t}\right\}\Bigg|_{(\mathbf{x}_m,t_m)}. \end{aligned}$$

Using the foregoing equation to the conditions in Eqs.(2.39)-(2.42), the flow $\mathbf{x}^{(\alpha)}(t)$ in Ω_α is tangential to the plane boundary $\partial\Omega_{ij}$ at point $(\mathbf{x}_m, t_m)$ with the $(2l_\alpha - 1)$th-order from Theorem 2.12. Therefore, this theorem is proved. ■

2.5. Switching bifurcations of passable flows

In this section, the switching bifurcation between the passable and non-passable flows to the boundary will be discussed. In addition, the switching bifurcation between the sink and source flows on the boundary will also be discussed. The switching bifurcations are defined first, and then the sufficient and necessary conditions for such switching bifurcations will be developed. The L-functions of flows will be introduced to develop criteria for the switching bifurcations from sufficient and necessary conditions.

Definition 2.14. For a discontinuous dynamical system in Eq.(2.1), there is a point $\mathbf{x}(t_m) = \mathbf{x}_m \in [\mathbf{x}_{m_1}, \mathbf{x}_{m_2}] \subset \overrightarrow{\partial\Omega}_{ij}$ for time t_m between two adjacent domains Ω_α ($\alpha = i, j$). For an arbitrarily small $\varepsilon > 0$, there are two time intervals (i.e., $[t_{m-\varepsilon}, t_m)$ and $(t_m, t_{m+\varepsilon}]$), and $\mathbf{x}^{(i)}(t_{m-}) = \mathbf{x}_m = \mathbf{x}^{(j)}(t_{m\pm})$. The flows $\mathbf{x}^{(i)}(t)$ and

$\mathbf{x}^{(j)}(t)$ are $C^{r_i}_{[t_{m-\varepsilon},t_m)}$ and $C^{r_j}_{[t_{m-\varepsilon},t_{m+\varepsilon}]}$-continuous ($r_\alpha \geq 1, \alpha = i,j$) for time t, respectively. The tangential bifurcation of the flow $\mathbf{x}^{(j)}(t)$ at point $(\mathbf{x}_m, t_m)$ on the boundary $\overrightarrow{\partial\Omega_{ij}}$ is termed the *switching bifurcation* of a flow from the semi-passable flow to the non-passable flow of the *first kind* (or called the *sliding bifurcation from $\overrightarrow{\partial\Omega_{ij}}$ to $\widetilde{\partial\Omega_{ij}}$*) if

$$\mathbf{n}^{\mathrm{T}}_{\partial\Omega_{ij}} \cdot \dot{\mathbf{x}}^{(j)}(t_{m\pm}) = 0 \text{ and } \mathbf{n}^{\mathrm{T}}_{\partial\Omega_{ij}} \cdot \dot{\mathbf{x}}^{(i)}(t_{m-}) \neq 0; \tag{2.43}$$

$$\text{either } \left.\begin{array}{l} \mathbf{n}^{\mathrm{T}}_{\partial\Omega_{ij}} \cdot [\mathbf{x}^{(i)}(t_{m-}) - \mathbf{x}^{(i)}(t_{m-\varepsilon})] > 0 \\ \mathbf{n}^{\mathrm{T}}_{\partial\Omega_{ij}} \cdot [\mathbf{x}^{(j)}(t_{m-}) - \mathbf{x}^{(j)}(t_{m-\varepsilon})] < 0 \\ \mathbf{n}^{\mathrm{T}}_{\partial\Omega_{ij}} \cdot [\mathbf{x}^{(j)}(t_{m+\varepsilon}) - \mathbf{x}^{(j)}(t_{m+})] > 0 \end{array}\right\} \text{ for } \mathbf{n}_{\Omega_{ij}} \to \Omega_j, \tag{2.44}$$

$$\text{or } \left.\begin{array}{l} \mathbf{n}^{\mathrm{T}}_{\partial\Omega_{ij}} \cdot [\mathbf{x}^{(i)}(t_{m-}) - \mathbf{x}^{(i)}(t_{m-\varepsilon})] < 0 \\ \mathbf{n}^{\mathrm{T}}_{\partial\Omega_{ij}} \cdot [\mathbf{x}^{(j)}(t_{m-}) - \mathbf{x}^{(j)}(t_{m-\varepsilon})] > 0 \\ \mathbf{n}^{\mathrm{T}}_{\partial\Omega_{ij}} \cdot [\mathbf{x}^{(j)}(t_{m+\varepsilon}) - \mathbf{x}^{(j)}(t_{m+})] < 0 \end{array}\right\} \text{ for } \mathbf{n}_{\Omega_{ij}} \to \Omega_i. \tag{2.45}$$

Definition 2.15. For a discontinuous dynamical system in Eq.(2.1), there is a point $\mathbf{x}(t_m) = \mathbf{x}_m \in [\mathbf{x}_{m_1}, \mathbf{x}_{m_2}] \subset \overrightarrow{\partial\Omega_{ij}}$ for time t_m between two adjacent domains Ω_α ($\alpha = i,j$). For an arbitrarily small $\varepsilon > 0$, there are two time intervals (i.e., $[t_{m-\varepsilon}, t_m)$ and $(t_m, t_{m+\varepsilon}]$), and $\mathbf{x}^{(i)}(t_{m\pm}) = \mathbf{x}_m = \mathbf{x}^{(j)}(t_{m+})$. The flows $\mathbf{x}^{(i)}(t)$ and $\mathbf{x}^{(j)}(t)$ are $C^{r_i}_{[t_{m-\varepsilon},t_{m+\varepsilon}]}$ and $C^{r_j}_{[t_{m+\varepsilon},t_m)}$-continuous ($r_\alpha \geq 1, \alpha = i,j$) for time t, respectively. The tangential bifurcation of the flow $\mathbf{x}^{(i)}(t)$ at point $(\mathbf{x}_m, t_m)$ on the boundary $\overrightarrow{\partial\Omega_{ij}}$ is termed the *switching bifurcation* of a flow from the passable flow to the non-passable flow of the *second kind* (or called the *source bifurcation from $\overrightarrow{\partial\Omega_{ij}}$ to $\widehat{\partial\Omega_{ij}}$*) if

$$\mathbf{n}^{\mathrm{T}}_{\partial\Omega_{ij}} \cdot \dot{\mathbf{x}}^{(i)}(t_{m\pm}) = 0 \text{ and } \mathbf{n}^{\mathrm{T}}_{\partial\Omega_{ij}} \cdot \dot{\mathbf{x}}^{(j)}(t_{m\pm}) \neq 0; \tag{2.46}$$

$$\text{either } \left.\begin{array}{l} \mathbf{n}^{\mathrm{T}}_{\partial\Omega_{ij}} \cdot [\mathbf{x}^{(i)}(t_{m-}) - \mathbf{x}^{(i)}(t_{m-\varepsilon})] > 0 \\ \mathbf{n}^{\mathrm{T}}_{\partial\Omega_{ij}} \cdot [\mathbf{x}^{(i)}(t_{m+\varepsilon}) - \mathbf{x}^{(i)}(t_{m+})] < 0 \\ \mathbf{n}^{\mathrm{T}}_{\partial\Omega_{ij}} \cdot [\mathbf{x}^{(j)}(t_{m+\varepsilon}) - \mathbf{x}^{(j)}(t_{m+})] > 0 \end{array}\right\} \text{ for } \mathbf{n}_{\Omega_{ij}} \to \Omega_j, \tag{2.47}$$

$$\text{or } \left.\begin{array}{l} \mathbf{n}^{\mathrm{T}}_{\partial\Omega_{ij}} \cdot [\mathbf{x}^{(i)}(t_{m-}) - \mathbf{x}^{(i)}(t_{m-\varepsilon})] < 0 \\ \mathbf{n}^{\mathrm{T}}_{\partial\Omega_{ij}} \cdot [\mathbf{x}^{(i)}(t_{m+\varepsilon}) - \mathbf{x}^{(i)}(t_{m+})] > 0 \\ \mathbf{n}^{\mathrm{T}}_{\partial\Omega_{ij}} \cdot [\mathbf{x}^{(j)}(t_{m+\varepsilon}) - \mathbf{x}^{(j)}(t_{m+})] < 0 \end{array}\right\} \text{ for } \mathbf{n}_{\Omega_{ij}} \to \Omega_i. \tag{2.48}$$

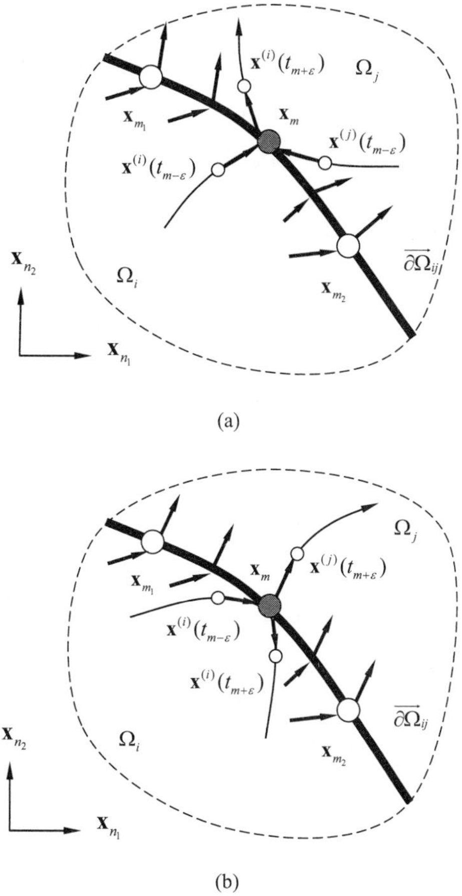

Fig. 2.7 (a) The sliding bifurcation and **(b)** the source bifurcation on the semi-passable boundary $\overrightarrow{\partial\Omega}_{ij}$. Four points $\mathbf{x}^{(\alpha)}(t_{m\pm\varepsilon})$ ($\alpha \in \{i,j\}$) and $\mathbf{x}_m$ lie in the corresponding domains Ω_α and on the boundary $\partial\Omega_{ij}$, respectively.

From the two definitions, the switching bifurcations of a flow from the semi-passable boundary to the non-passable boundaries of the first and second kinds are presented in Fig. 2.7. The source (or *sink*) bifurcation of a flow to the boundary requires the tangential bifurcation of the coming (or *leaving*) flow to the boundary. Similarly, the switching bifurcation of a passable flow from $\overrightarrow{\partial\Omega}_{ij}$ to $\overleftarrow{\partial\Omega}_{ij}$ is defined as follows.

Definition 2.16. For a discontinuous dynamical system in Eq.(2.1), there is a point $\mathbf{x}(t_m) = \mathbf{x}_m \in [\mathbf{x}_{m_1}, \mathbf{x}_{m_2}] \subset \overrightarrow{\partial\Omega}_{ij}$ for time t_m between two adjacent domains Ω_α (α =

i,j).For an arbitrarily small $\varepsilon>0$, there are two time intervals (i.e., $[t_{m-\varepsilon},t_m)$ and $(t_m,t_{m+\varepsilon}]$) and $\mathbf{x}^{(i)}(t_{m\mp})=\mathbf{x}_m=\mathbf{x}^{(j)}(t_{m\pm})$. The flows $\mathbf{x}^{(\alpha)}(t)$ ($\alpha=i,j$) are $C^{r_\alpha}_{[t_{m-\varepsilon},t_{m+\varepsilon}]}$-continuous ($r_\alpha\ge 1$) for time *t*. The tangential bifurcation of the flow $\mathbf{x}^{(i)}(t)$ and $\mathbf{x}^{(j)}(t)$ at point $(\mathbf{x}_m,t_m)$ on the boundary $\overrightarrow{\partial\Omega}_{ij}$ is termed the *switching bifurcation* of a flow from $\overrightarrow{\partial\Omega}_{ij}$ to $\overleftarrow{\partial\Omega}_{ij}$ if

$$\mathbf{n}^{\mathrm{T}}_{\partial\Omega_{ij}}\cdot\dot{\mathbf{x}}^{(\alpha)}(t_{m\pm})=0 \text{ for } \alpha=i,j; \tag{2.49}$$

$$\text{either} \quad \left.\begin{array}{l}\mathbf{n}^{\mathrm{T}}_{\partial\Omega_{ij}}\cdot[\mathbf{x}^{(i)}(t_{m-})-\mathbf{x}^{(i)}(t_{m-\varepsilon})]>0\\ \mathbf{n}^{\mathrm{T}}_{\partial\Omega_{ij}}\cdot[\mathbf{x}^{(i)}(t_{m+\varepsilon})-\mathbf{x}^{(i)}(t_{m+})]<0\\ \mathbf{n}^{\mathrm{T}}_{\partial\Omega_{ij}}\cdot[\mathbf{x}^{(j)}(t_{m-})-\mathbf{x}^{(j)}(t_{m-\varepsilon})]<0\\ \mathbf{n}^{\mathrm{T}}_{\partial\Omega_{ij}}\cdot[\mathbf{x}^{(j)}(t_{m+\varepsilon})-\mathbf{x}^{(j)}(t_{m+})]>0\end{array}\right\} \text{for } \mathbf{n}_{\Omega_{ij}}\to\Omega_j, \tag{2.50}$$

$$\text{or} \quad \left.\begin{array}{l}\mathbf{n}^{\mathrm{T}}_{\partial\Omega_{ij}}\cdot[\mathbf{x}^{(i)}(t_{m-})-\mathbf{x}^{(i)}(t_{m-\varepsilon})]<0\\ \mathbf{n}^{\mathrm{T}}_{\partial\Omega_{ij}}\cdot[\mathbf{x}^{(i)}(t_{m+\varepsilon})-\mathbf{x}^{(i)}(t_{m+})]>0\\ \mathbf{n}^{\mathrm{T}}_{\partial\Omega_{ij}}\cdot[\mathbf{x}^{(j)}(t_{m-})-\mathbf{x}^{(j)}(t_{m-\varepsilon})]>0\\ \mathbf{n}^{\mathrm{T}}_{\partial\Omega_{ij}}\cdot[\mathbf{x}^{(j)}(t_{m+\varepsilon})-\mathbf{x}^{(j)}(t_{m+})]<0\end{array}\right\} \text{for } \mathbf{n}_{\Omega_{ij}}\to\Omega_i. \tag{2.51}$$

The above definitions give the three possible switching bifurcations of the semi-passable flow to the boundary $\overrightarrow{\partial\Omega}_{ij}$. The corresponding theorems can be stated for necessary and sufficient conditions. The proofs can be completed as in Theorems 2.8-2.10.

Theorem 2.14. *For a discontinuous dynamical system in Eq.*(2.1)*, there is a point* $\mathbf{x}(t_m)=\mathbf{x}_m\in[\mathbf{x}_{m_1},\mathbf{x}_{m_2}]\subset\overrightarrow{\partial\Omega}_{ij}$ *for time* t_m *between two adjacent domains* Ω_α ($\alpha=i,j$)*. For an arbitrarily small* $\varepsilon>0$*, there are two time intervals (i.e.,* $[t_{m-\varepsilon},t_m)$ *and* $(t_m,t_{m+\varepsilon}]$*), and* $\mathbf{x}^{(i)}(t_{m-})=\mathbf{x}_m=\mathbf{x}^{(j)}(t_{m\pm})$*. The flows* $\mathbf{x}^{(i)}(t)$ *and* $\mathbf{x}^{(j)}(t)$ *are* $C^{r_i}_{[t_{m-\varepsilon},t_m)}$ *and* $C^{r_j}_{[t_{m-\varepsilon},t_{m+\varepsilon}]}$*-continuous* ($r_\alpha\ge 1$, $\alpha=i,j$) *for time t, respectively. The sliding bifurcation of the flow* $\mathbf{x}^{(i)}(t)\cup\mathbf{x}^{(j)}(t)$ *at point* $(\mathbf{x}_m,t_m)$ *from* $\overrightarrow{\partial\Omega}_{ij}$ *to* $\widetilde{\partial\Omega}_{ij}$ *occurs if and only if*

$$\mathbf{n}^{\mathrm{T}}_{\partial\Omega_{ij}}\cdot\mathbf{F}^{(j)}(t_{m\pm})=0 \quad \textit{and} \quad \mathbf{n}^{\mathrm{T}}_{\partial\Omega_{ij}}\cdot\mathbf{F}^{(i)}(t_{m-})\neq 0; \tag{2.52}$$

$$\begin{array}{ll}\textit{either} & \mathbf{n}^{\mathrm{T}}_{\partial\Omega_{ij}}\cdot\mathbf{F}^{(i)}(t_{m-})>0 \textit{ for } \mathbf{n}_{\partial\Omega_{ij}}\to\Omega_j,\\ \textit{or} & \mathbf{n}^{\mathrm{T}}_{\partial\Omega_{ij}}\cdot\mathbf{F}^{(i)}(t_{m-})<0 \textit{ for } \mathbf{n}_{\partial\Omega_{ij}}\to\Omega_i;\end{array} \tag{2.53}$$

$$\begin{aligned} &\text{either} \quad \left.\begin{array}{l} \mathbf{n}^{\mathrm{T}}_{\partial\Omega_{ij}} \cdot \mathbf{F}^{(j)}(t_{m-\varepsilon}) < 0 \\ \mathbf{n}^{\mathrm{T}}_{\partial\Omega_{ij}} \cdot \mathbf{F}^{(j)}(t_{m+\varepsilon}) > 0 \end{array}\right\} \text{for } \mathbf{n}_{\partial\Omega_{ij}} \to \Omega_j, \\ &\text{or} \quad \left.\begin{array}{l} \mathbf{n}^{\mathrm{T}}_{\partial\Omega_{ij}} \cdot \mathbf{F}^{(j)}(t_{m-\varepsilon}) > 0 \\ \mathbf{n}^{\mathrm{T}}_{\partial\Omega_{ij}} \cdot \mathbf{F}^{(j)}(t_{m+\varepsilon}) < 0 \end{array}\right\} \text{for } \mathbf{n}_{\partial\Omega_{ij}} \to \Omega_i. \end{aligned} \tag{2.54}$$

Proof. Following the proof procedures in Theorems 2.8 and 2.9, the above theorem can be easily proved. ■

Theorem 2.15. *For a discontinuous dynamical system in Eq.*(2.1), *there is a point* $\mathbf{x}(t_m) = \mathbf{x}_m \in [\mathbf{x}_{m_1}, \mathbf{x}_{m_2}] \subset \overrightarrow{\partial\Omega}_{ij}$ *at time* t_m *on the* $(n-1)$ *-dimensional plane boundary* $\partial\Omega_{ij}$ *between two adjacent domains* Ω_α $(\alpha = i, j)$. *For an arbitrarily small* $\varepsilon > 0$, *there are two time intervals (i.e.,* $[t_{m-\varepsilon}, t_m)$ *and* $(t_m, t_{m+\varepsilon}]$*), and* $\mathbf{x}^{(i)}(t_{m-}) = \mathbf{x}_m = \mathbf{x}^{(j)}(t_{m\pm})$. *The flows* $\mathbf{x}^{(i)}(t)$ *and* $\mathbf{x}^{(j)}(t)$ *are* $C^{r_i}_{[t_{m-\varepsilon}, t_m)}$ *and* $C^{r_j}_{[t_{m-\varepsilon}, t_{m+\varepsilon}]}$ *-continuous* ($r_\alpha \geq 2$, $\alpha = i, j$) *for time t, respectively. The sliding bifurcation of the flow* $\mathbf{x}^{(i)}(t) \cup \mathbf{x}^{(j)}(t)$ *at point* $(\mathbf{x}_m, t_m)$ *from* $\overrightarrow{\partial\Omega}_{ij}$ *to* $\widetilde{\partial\Omega}_{ij}$ *occurs if and only if*

$$\mathbf{n}^{\mathrm{T}}_{\partial\Omega_{ij}} \cdot \mathbf{F}^{(j)}(t_{m\pm}) = 0 \text{ and } \mathbf{n}^{\mathrm{T}}_{\partial\Omega_{ij}} \cdot \mathbf{F}^{(i)}(t_{m-}) \neq 0; \tag{2.55}$$

$$\begin{aligned} &\text{either} \quad \left.\begin{array}{l} \mathbf{n}^{\mathrm{T}}_{\partial\Omega_{ij}} \cdot \mathbf{F}^{(i)}(t_{m-}) > 0 \\ \mathbf{n}^{\mathrm{T}}_{\partial\Omega_{ij}} \cdot D\mathbf{F}^{(j)}(t_{m\pm}) > 0 \end{array}\right\} \text{for } \mathbf{n}_{\partial\Omega_{ij}} \to \Omega_j, \\ &\text{or} \quad \left.\begin{array}{l} \mathbf{n}^{\mathrm{T}}_{\partial\Omega_{ij}} \cdot \mathbf{F}^{(i)}(t_{m-}) < 0 \\ \mathbf{n}^{\mathrm{T}}_{\partial\Omega_{ij}} \cdot D\mathbf{F}^{(j)}(t_{m\pm}) < 0 \end{array}\right\} \text{for } \mathbf{n}_{\partial\Omega_{ij}} \to \Omega_i. \end{aligned} \tag{2.56}$$

Proof. Following the proof procedures of Theorems 2.10 and 2.11, the above theorem can be easily proved. ■

Theorem 2.16. *For a discontinuous dynamical system in Eq.*(2.1), *there is a point* $\mathbf{x}(t_m) = \mathbf{x}_m \in [\mathbf{x}_{m_1}, \mathbf{x}_{m_2}] \subset \overrightarrow{\partial\Omega}_{ij}$ *for time* t_m *between two adjacent domains* Ω_α $(\alpha = i, j)$. *For an arbitrarily small* $\varepsilon > 0$, *there are two time intervals (i.e.,* $[t_{m-\varepsilon}, t_m)$ *and* $(t_m, t_{m+\varepsilon}]$*), and* $\mathbf{x}^{(i)}(t_{m\pm}) = \mathbf{x}_m = \mathbf{x}^{(j)}(t_{m+})$. *The flows* $\mathbf{x}^{(i)}(t)$ *and* $\mathbf{x}^{(j)}(t)$ *are* $C^{r_i}_{[t_{m-\varepsilon}, t_{m+\varepsilon}]}$ *and* $C^{r_j}_{[t_{m+\varepsilon}, t_m)}$*-continuous* ($r_\alpha \geq 1$, $\alpha = i, j$) *for time t, respectively. The source bifurcation of the flow* $\mathbf{x}^{(i)}(t) \cup \mathbf{x}^{(j)}(t)$ *at point* $(\mathbf{x}_m, t_m)$ *from* $\overrightarrow{\partial\Omega}_{ij}$ *to* $\widehat{\partial\Omega}_{ij}$ *occurs if and only if*

$$\mathbf{n}_{\partial\Omega_{ij}}^{\mathrm{T}} \cdot \mathbf{F}^{(i)}(t_{m\pm}) = 0 \ \ and \ \ \mathbf{n}_{\partial\Omega_{ij}}^{\mathrm{T}} \cdot \mathbf{F}^{(j)}(t_{m+}) \neq 0; \tag{2.57}$$

$$\begin{aligned} &either \ \ \mathbf{n}_{\partial\Omega_{ij}}^{\mathrm{T}} \cdot \mathbf{F}^{(j)}(t_{m+}) > 0 \ \ for \ \mathbf{n}_{\partial\Omega_{ij}} \to \Omega_j, \\ &or \qquad \mathbf{n}_{\partial\Omega_{ij}}^{\mathrm{T}} \cdot \mathbf{F}^{(j)}(t_{m+}) < 0 \ \ for \ \mathbf{n}_{\partial\Omega_{ij}} \to \Omega_i; \end{aligned} \tag{2.58}$$

$$\begin{aligned} &either \ \ \left.\begin{aligned} \mathbf{n}_{\partial\Omega_{ij}}^{\mathrm{T}} \cdot \mathbf{F}^{(i)}(t_{m-\varepsilon}) > 0 \\ \mathbf{n}_{\partial\Omega_{ij}}^{\mathrm{T}} \cdot \mathbf{F}^{(i)}(t_{m+\varepsilon}) < 0 \end{aligned}\right\} for \ \mathbf{n}_{\partial\Omega_{ij}} \to \Omega_j, \\ &or \qquad \left.\begin{aligned} \mathbf{n}_{\partial\Omega_{ij}}^{\mathrm{T}} \cdot \mathbf{F}^{(i)}(t_{m-\varepsilon}) < 0 \\ \mathbf{n}_{\partial\Omega_{ij}}^{\mathrm{T}} \cdot \mathbf{F}^{(i)}(t_{m+\varepsilon}) > 0 \end{aligned}\right\} for \ \mathbf{n}_{\partial\Omega_{ij}} \to \Omega_i. \end{aligned} \tag{2.59}$$

Proof. Following the proof procedures of Theorems 2.8 and 2.9, the above theorem can be easily proved. ■

Theorem 2.17. *For a discontinuous dynamical system in Eq.*(2.1), *there is a point* $\mathbf{x}(t_m) = \mathbf{x}_m \in [\mathbf{x}_{m_1}, \mathbf{x}_{m_2}] \subset \overrightarrow{\partial\Omega}_{ij}$ *at time* t_m *on the* $(n-1)$*-dimensional plane boundary* $\partial\Omega_{ij}$ *between two adjacent domains* Ω_α $(\alpha = i, j)$. *For an arbitrarily small* $\varepsilon > 0$, *there are two time intervals (i.e.,* $[t_{m-\varepsilon}, t_m)$ *and* $(t_m, t_{m+\varepsilon}]$*), and* $\mathbf{x}^{(i)}(t_{m\pm}) = \mathbf{x}_m = \mathbf{x}^{(j)}(t_{m+})$. *The flows* $\mathbf{x}^{(i)}(t)$ *and* $\mathbf{x}^{(j)}(t)$ *are* $C^{r_i}_{[t_{m-\varepsilon}, t_{m+\varepsilon}]}$ *and* $C^{r_j}_{[t_{m+\varepsilon}, t_m)}$ *continuous* $(r_\alpha \geq 2, \alpha = i, j)$ *for time t, respectively. The source bifurcation of the flow* $\mathbf{x}^{(i)}(t) \cup \mathbf{x}^{(j)}(t)$ *at point* $(\mathbf{x}_m, t_m)$ *from* $\overrightarrow{\partial\Omega}_{ij}$ *to* $\widehat{\partial\Omega}_{ij}$ *occurs if and only if*

$$\mathbf{n}_{\partial\Omega_{ij}}^{\mathrm{T}} \cdot \mathbf{F}^{(i)}(t_{m\pm}) = 0 \ \ \text{and} \ \ \mathbf{n}_{\partial\Omega_{ij}}^{\mathrm{T}} \cdot \mathbf{F}^{(j)}(t_{m+}) \neq 0; \tag{2.60}$$

$$\begin{aligned} &either \ \ \left.\begin{aligned} \mathbf{n}_{\partial\Omega_{ij}}^{\mathrm{T}} \cdot \mathbf{F}^{(j)}(t_{m+}) > 0 \\ \mathbf{n}_{\partial\Omega_{ij}}^{T} \cdot D\mathbf{F}^{(i)}(t_{m\pm}) < 0 \end{aligned}\right\} for \ \mathbf{n}_{\partial\Omega_{ij}} \to \Omega_j, \\ &or \qquad \left.\begin{aligned} \mathbf{n}_{\partial\Omega_{ij}}^{\mathrm{T}} \cdot \mathbf{F}^{(j)}(t_{m+}) < 0 \\ \mathbf{n}_{\partial\Omega_{ij}}^{T} \cdot D\mathbf{F}^{(i)}(t_{m\pm}) > 0 \end{aligned}\right\} for \ \mathbf{n}_{\partial\Omega_{ij}} \to \Omega_i. \end{aligned} \tag{2.61}$$

Proof. Following the proof procedures of Theorems 2.10 and 2.11, the above theorem can be easily proved. ■

Theorem 2.18. *For a discontinuous dynamical system in Eq.*(2.1), *there is a point* $\mathbf{x}(t_m) = \mathbf{x}_m \in [\mathbf{x}_{m_1}, \mathbf{x}_{m_2}] \subset \overrightarrow{\partial\Omega}_{ij}$ *for time* t_m *between two adjacent domains* Ω_α $(\alpha = i, j)$. *For an arbitrarily small* $\varepsilon > 0$, *there are two time intervals (i.e.,* $[t_{m-\varepsilon}, t_m)$ *and* $(t_m, t_{m+\varepsilon}]$*), and* $\mathbf{x}^{(i)}(t_{m\mp}) = \mathbf{x}_m = \mathbf{x}^{(j)}(t_{m\pm})$. *The flows* $\mathbf{x}^{(\alpha)}(t)$ $(\alpha = i, j)$ *are*

$C^{r_\alpha}_{[t_{m-\varepsilon},t_{m+\varepsilon}]}$ *-continuous* ($r_\alpha \geq 1$) *for time t. The switching bifurcation of the flow* $\mathbf{x}^{(i)}(t) \cup \mathbf{x}^{(j)}(t)$ *at point* $(\mathbf{x}_m, t_m)$ *from* $\overrightarrow{\partial\Omega}_{ij}$ *to* $\overleftarrow{\partial\Omega}_{ij}$ *occurs if and only if*

$$\mathbf{n}^{\mathrm{T}}_{\partial\Omega_{ij}} \cdot \mathbf{F}^{(i)}(t_{m\mp}) = 0 \;\; and \;\; \mathbf{n}^{\mathrm{T}}_{\partial\Omega_{ij}} \cdot \mathbf{F}^{(j)}(t_{m\pm}) = 0; \tag{2.62}$$

$$\begin{aligned} either \quad & \left.\begin{aligned} \mathbf{n}^{\mathrm{T}}_{\partial\Omega_{ij}} \cdot \mathbf{F}^{(\alpha)}(t_{m-\varepsilon}) > 0 \\ \mathbf{n}^{\mathrm{T}}_{\partial\Omega_{ij}} \cdot \mathbf{F}^{(\alpha)}(t_{m+\varepsilon}) < 0 \end{aligned}\right\} for\ \mathbf{n}_{\partial\Omega_{ij}} \to \Omega_\beta, \\ or \quad & \left.\begin{aligned} \mathbf{n}^{\mathrm{T}}_{\partial\Omega_{ij}} \cdot \mathbf{F}^{(\alpha)}(t_{m-\varepsilon}) < 0 \\ \mathbf{n}^{\mathrm{T}}_{\partial\Omega_{ij}} \cdot \mathbf{F}^{(\alpha)}(t_{m+\varepsilon}) > 0 \end{aligned}\right\} for\ \mathbf{n}_{\partial\Omega_{ij}} \to \Omega_\alpha \end{aligned} \tag{2.63}$$

with $\alpha, \beta = i, j$ *but* $\beta \neq \alpha$.

Proof. Following the proof procedures of Theorems 2.8 and 2.9, the above theorem can be easily proved. ■

Theorem 2.19. *For a discontinuous dynamical system in Eq.*(2.1), *there is a point* $\mathbf{x}(t_m) = \mathbf{x}_m \in [\mathbf{x}_{m_1}, \mathbf{x}_{m_2}] \subset \overrightarrow{\partial\Omega}_{ij}$ *at time* t_m *on the* $(n-1)$*-dimensional plane boundary* $\partial\Omega_{ij}$ *between two adjacent domains* Ω_α ($\alpha = i, j$). *For an arbitrarily small* $\varepsilon > 0$, *there are two time intervals (i.e.,* $[t_{m-\varepsilon}, t_m)$ *and* $(t_m, t_{m+\varepsilon}]$*), and* $\mathbf{x}^{(i)}(t_{m\mp}) = \mathbf{x}_m = \mathbf{x}^{(j)}(t_{m\pm})$. *The flows* $\mathbf{x}^{(\alpha)}(t)$ ($\alpha = i, j$) *are* $C^{r_\alpha}_{[t_{m-\varepsilon},t_{m+\varepsilon}]}$ *-continuous* ($r_\alpha \geq 1$) *for time t. The switching bifurcation of the flow* $\mathbf{x}^{(i)}(t) \cup \mathbf{x}^{(j)}(t)$ *at point* $(\mathbf{x}_m, t_m)$ *from* $\overrightarrow{\partial\Omega}_{ij}$ *to* $\overleftarrow{\partial\Omega}_{ij}$ *occurs if and only if*

$$\mathbf{n}^{\mathrm{T}}_{\partial\Omega_{ij}} \cdot \mathbf{F}^{(i)}(t_{m\mp}) = 0 \;\; and \;\; \mathbf{n}^{\mathrm{T}}_{\partial\Omega_{ij}} \cdot \mathbf{F}^{(j)}(t_{m\pm}) = 0; \tag{2.64}$$

$$\begin{aligned} either \quad & \left.\begin{aligned} \mathbf{n}^{\mathrm{T}}_{\partial\Omega_{ij}} \cdot D\mathbf{F}^{(i)}(t_{m\mp}) < 0 \\ \mathbf{n}^{\mathrm{T}}_{\partial\Omega_{ij}} \cdot D\mathbf{F}^{(j)}(t_{m\pm}) > 0 \end{aligned}\right\} for\ \mathbf{n}_{\partial\Omega_{ij}} \to \Omega_j, \\ or \quad & \left.\begin{aligned} \mathbf{n}^{\mathrm{T}}_{\partial\Omega_{ij}} \cdot D\mathbf{F}^{(i)}(t_{m\mp}) > 0 \\ \mathbf{n}^{\mathrm{T}}_{\partial\Omega_{ij}} \cdot D\mathbf{F}^{(j)}(t_{m\pm}) < 0 \end{aligned}\right\} for\ \mathbf{n}_{\partial\Omega_{ij}} \to \Omega_i. \end{aligned} \tag{2.65}$$

Proof. Following the proof procedures of Theorems 2.10 and 2.11, the above theorem can be easily proved. ■

Definition 2.17 For a discontinuous dynamical system in Eq.(2.1), there is a point $\mathbf{x}(t_m) = \mathbf{x}_m \in [\mathbf{x}_{m_1}, \mathbf{x}_{m_2}] \subset \partial\Omega_{ij}$ for time t_m and $\mathbf{x}^{(\alpha)}(t_{m\pm}) = \mathbf{x}_m$ ($\alpha \in \{i, j\}$). The $L_{\alpha\beta}$ *-functions* of flows to the boundary $\partial\Omega_{ij}$ is defined as

$$L_{\alpha\beta}(\mathbf{x}_m, t_m, \mathbf{p}_\alpha, \mathbf{p}_\beta) = [\mathbf{n}^{\mathrm{T}}_{\partial\Omega_{\alpha\beta}} \cdot \mathbf{F}^{(\alpha)}(t_{m\mp})] \times [\mathbf{n}^{\mathrm{T}}_{\partial\Omega_{\alpha\beta}} \cdot \mathbf{F}^{(\beta)}(t_{m\pm})] \tag{2.66}$$

where $\beta \in \{i, j\}$ but $\beta \neq \alpha$.

From the foregoing definition, the passable flows and non-passable flows (including sink and source flows) at the boundary $\partial\Omega_{\alpha\beta}$, respectively, require the L-function satisfies

$$\begin{aligned} &L_{\alpha\beta}(\mathbf{x}_m, t_m, \mathbf{p}_\alpha, \mathbf{p}_\beta) > 0 \ \text{ on } \overrightarrow{\partial\Omega}_{\alpha\beta}; \\ &L_{\alpha\beta}(\mathbf{x}_m, t_m, \mathbf{p}_\alpha, \mathbf{p}_\beta) < 0 \ \text{ on } \overline{\partial\Omega}_{\alpha\beta} = \widetilde{\partial\Omega}_{\alpha\beta} \cup \widehat{\partial\Omega}_{\alpha\beta}. \end{aligned} \tag{2.67}$$

The switching bifurcation of a flow at point $(\mathbf{x}_m, t_m)$ on the boundary $\partial\Omega_{\alpha\beta}$ requires

$$L_{\alpha\beta}(\mathbf{x}_m, t_m, \mathbf{p}_\alpha, \mathbf{p}_\beta) = 0. \tag{2.68}$$

If the $L_{\alpha\beta}$-function of a flow is defined on one side of the neighborhood of the boundary $\partial\Omega_{\alpha\beta}$, one obtains

$$L_{\alpha\alpha}(\mathbf{x}_{m\pm\varepsilon}, t_{m\pm\varepsilon}, \mathbf{p}_\alpha) = [\mathbf{n}^{\mathrm{T}}_{\partial\Omega_{\alpha\beta}} \cdot \mathbf{F}^{(\alpha)}(t_{m-\varepsilon})] \times [\mathbf{n}^{\mathrm{T}}_{\partial\Omega_{\alpha\beta}} \cdot \mathbf{F}^{(\alpha)}(t_{m+\varepsilon})]. \tag{2.69}$$

If $L_{\alpha\alpha}(\mathbf{x}_{m\pm\varepsilon}, t_{m\pm\varepsilon}, \mathbf{p}_\alpha) < 0$ and $\mathbf{n}^{\mathrm{T}}_{\partial\Omega_{\alpha\beta}} \cdot \mathbf{F}^{(\alpha)}(t_{m-}) = 0$, the flow $\mathbf{x}^{(\alpha)}(t)$ at $(\mathbf{x}_m, t_m)$ is tangential to the boundary $\partial\Omega_{\alpha\beta}$.

Consider the $L_{\alpha\beta}$-function varying with the parameter vector $\mathbf{p}_{ij} \in \{\boldsymbol{\mu}_\alpha\}_{\alpha\in\{i,j\}}$ for the switching flow from $\overrightarrow{\partial\Omega}_{\alpha\beta}$ to $\widetilde{\partial\Omega}_{\alpha\beta}$. The $L_{\alpha\beta}$-functions of flows at different locations of the boundary are distinct. The $L_{\alpha\beta}$-functions of flows between two points $\mathbf{x}_{m_1}$ and $\mathbf{x}_{m_2}$ on the boundary $\partial\Omega_{\alpha\beta}$ are sketched in Fig. 2.8 for a parameter vector $\mathbf{p}_{\alpha\beta}$ between $\mathbf{p}^{(1)}_{\alpha\beta}$ and $\mathbf{p}^{(2)}_{\alpha\beta}$. For a specific value $\mathbf{p}^{(cr)}_{\alpha\beta}$ between $\mathbf{p}^{(1)}_{\alpha\beta}$ and $\mathbf{p}^{(2)}_{\alpha\beta}$, there is a point $\mathbf{x}_m$ on the boundary for the bifurcation of a flow switching from $\overrightarrow{\partial\Omega}_{ij}$ to $\widetilde{\partial\Omega}_{ij}$. Two points $\mathbf{x}_{k_1}$ and $\mathbf{x}_{k_2}$ are the onset and vanishing points of the sink flow for system parameter vector $\mathbf{p}_{\alpha\beta}$ on the boundary $\partial\Omega_{\alpha\beta}$. The dashed and solid curves represent $L_{\alpha\beta} < 0$ and $L_{\alpha\beta} \geq 0$, respectively. For parameter vector $\mathbf{p}_{\alpha\beta}$ varying from $\mathbf{p}^{(1)}_{\alpha\beta} \to \mathbf{p}^{(cr)}_{\alpha\beta}$, the $L_{\alpha\beta}$-function for a flow $\mathbf{x} \in (\mathbf{x}_{m_1}, \mathbf{x}_{m_2})$ on the boundary is positive (i.e., $L_{\alpha\beta} > 0$). Thus, the boundary $\partial\Omega_{\alpha\beta}$ is semi-passable. For $\mathbf{p}_{\alpha\beta}$ varying from $\mathbf{p}^{(cr)}_{\alpha\beta} \to \mathbf{p}^{(2)}_{\alpha\beta}$, there are two ranges of $L_{\alpha\beta} > 0$ for $\mathbf{x} \in [\mathbf{x}_{m_1}, \mathbf{x}_{k_1}) \cup (\mathbf{x}_{k_2}, \mathbf{x}_{m_2}]$ and a range of $L_{\alpha\beta} < 0$ for $\mathbf{x} \in (\mathbf{x}_{k_1}, \mathbf{x}_{k_2})$. From Eq.(2.67), the flow at the portion of $\mathbf{x} \in (\mathbf{x}_{k_1}, \mathbf{x}_{k_2})$ on the boundary $\partial\Omega_{\alpha\beta}$ is non-passable. The flow at the portion of boundary with $L_{\alpha\beta} > 0$ is semi-passable. For $\mathbf{p}_{\alpha\beta}$ varying from $\mathbf{p}_{\alpha\beta}^{(1)} \to \mathbf{p}_{\alpha\beta}^{(2)}$, the point $(\mathbf{x}_m, \mathbf{p}_{\alpha\beta}^{(cr)})$ on the boundary $\partial\Omega_{\alpha\beta}$ is the onset point of the non-passable flow. However, for $\mathbf{p}_{\alpha\beta}$ varying from $\mathbf{p}_{\alpha\beta}^{(2)} \to \mathbf{p}_{\alpha\beta}^{(1)}$, such a point is the vanishing point of the non-passable flow. For three critical points $\{\mathbf{x}_m, \mathbf{x}_{k_1}, \mathbf{x}_{k_2}\}$, The $L_{\alpha\beta}$-function of flows is zero (i.e., $L_{\alpha\beta} = 0$). For $L_{\alpha\beta}$ in Fig. 2.8(a), the corresponding vector fields varying with the system parameter on the boundary $\partial\Omega_{\alpha\beta}$ are illustrated in Fig. 2.8(b). $\mathbf{F}^{(\alpha)}(t_{m-})$ and $\mathbf{F}^{(\beta)}(t_{m\pm})$ are the limits of the vector fields in domains Ω_α and Ω_β to the boundary $\partial\Omega_{\alpha\beta}$, respectively. This non-passable flow on the boundary $\partial\Omega_{\alpha\beta}$ with $L_{\alpha\beta} < 0$ is a sink flow. The critical points $\{\mathbf{x}_{k_1}, \mathbf{x}_{k_2}\}$ have the same properties as point $\mathbf{x}_m$ for $\mathbf{p}_{\alpha\beta}^{(cr)}$. Namely, $L_{\alpha\beta}(\mathbf{x}_m) = 0$, $L_{\alpha\alpha}(\mathbf{x}_{m\pm\varepsilon}) < 0$ or $L_{\beta\beta}(\mathbf{x}_{m\pm\varepsilon}) > 0$.

If the two critical points have the different properties, the sliding flow between two different critical points will be discussed later. The $L_{\alpha\beta}$-functions of flows are $L_{\alpha\beta}(\mathbf{x}_{k_1}) = 0$ and $L_{\alpha\alpha}(\mathbf{x}_{k_1\pm\varepsilon}) < 0$ for point $\mathbf{x}_{k_1}$ but $L_{\alpha\beta}(\mathbf{x}_{k_2}) = 0$ and $L_{\beta\beta}(\mathbf{x}_{k_2\pm\varepsilon}) < 0$ for point $\mathbf{x}_{k_2}$. From the $L_{\alpha\beta}$-function of flows, Theorems 2.14, 2.16 and 2.18 can be restated.

Theorem 2.20. *For a discontinuous dynamical system in Eq.*(2.1)*, there is a point* $\mathbf{x}(t_m) = \mathbf{x}_m \in [\mathbf{x}_{m_1}, \mathbf{x}_{m_2}] \subset \overrightarrow{\partial\Omega}_{ij}$ *at time* t_m *between two adjacent domains* Ω_α ($\alpha = i, j$)*. For an arbitrarily small* $\varepsilon > 0$*, there are two time intervals (i.e.,* $[t_{m-\varepsilon}, t_m)$ *and* $(t_m, t_{m+\varepsilon}]$*), and* $\mathbf{x}^{(i)}(t_{m-}) = \mathbf{x}_m = \mathbf{x}^{(j)}(t_{m\pm})$. *The flows* $\mathbf{x}^{(i)}(t)$ *and* $\mathbf{x}^{(j)}(t)$ *are* $C^{r_i}_{[t_{m+\varepsilon}, t_m)}$ *and* $C^{r_j}_{[t_{m-\varepsilon}, t_{m+\varepsilon}]}$*-continuous (*$r_\alpha \geq 2$, $\alpha = i, j$*) for time t, respectively. The sliding bifurcation of the flow* $\mathbf{x}^{(i)}(t) \cup \mathbf{x}^{(j)}(t)$ *at point* $(\mathbf{x}_m, t_m)$ *from* $\overrightarrow{\partial\Omega}_{ij}$ *to* $\widetilde{\partial\Omega}_{ij}$ *occurs if and only if*

$$L_{ij}(\mathbf{x}_m, t_m, \mathbf{p}_i, \mathbf{p}_j) = 0, \tag{2.70}$$

$$\mathbf{n}^{\mathrm{T}}_{\partial\Omega_{ij}} \cdot \mathbf{F}^{(i)}(t_{m-}) \neq 0 \;\; and \;\; L_{jj}(\mathbf{x}_{m\pm\varepsilon}, t_{m\pm\varepsilon}, \mathbf{p}_j) < 0. \tag{2.71}$$

Proof. Applying the L_{ij}-functions of flows in Definition 2.17 to Theorem 2.14, the foregoing theorem can be easily proved. ∎

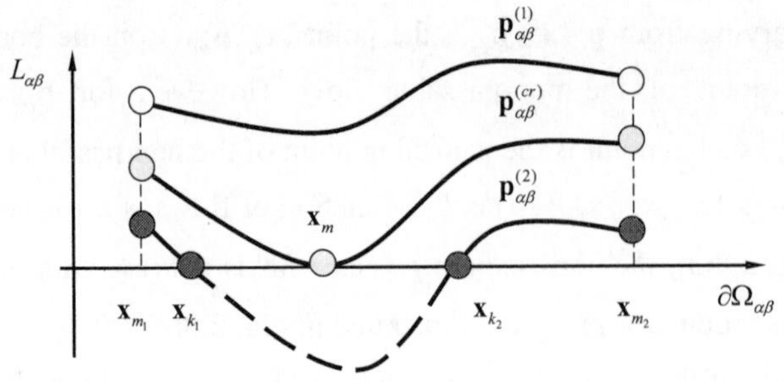

(a)

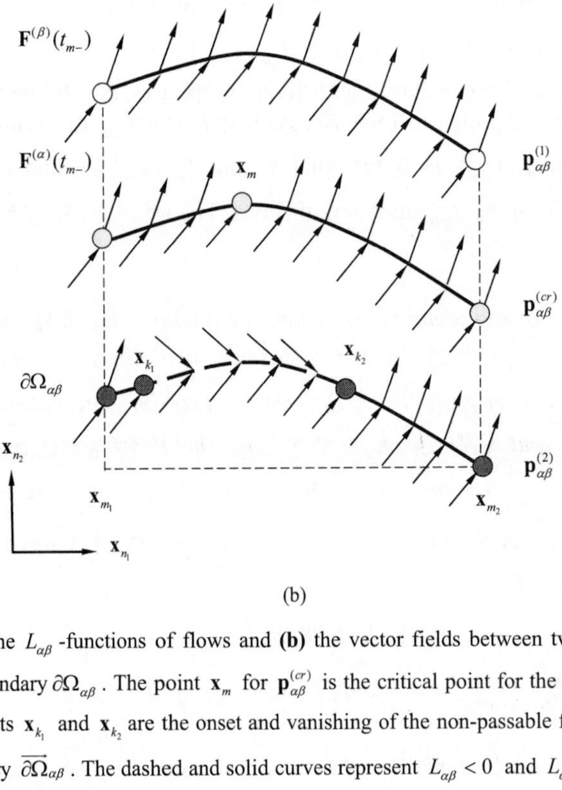

(b)

Fig. 2.8 (a) The $L_{\alpha\beta}$-functions of flows and **(b)** the vector fields between two points $\mathbf{x}_{m_1}$ and $\mathbf{x}_{m_2}$ on the boundary $\partial\Omega_{\alpha\beta}$. The point $\mathbf{x}_m$ for $\mathbf{p}_{\alpha\beta}^{(cr)}$ is the critical point for the switching bifurcation. Two points $\mathbf{x}_{k_1}$ and $\mathbf{x}_{k_2}$ are the onset and vanishing of the non-passable flow for parameter on the boundary $\overrightarrow{\partial\Omega}_{\alpha\beta}$. The dashed and solid curves represent $L_{\alpha\beta} < 0$ and $L_{\alpha\beta} \geq 0$, respectively.

Theorem 2.21. *For a discontinuous dynamical system in Eq.(2.1), there is a point* $\mathbf{x}(t_m) = \mathbf{x}_m \in [\mathbf{x}_{m_1}, \mathbf{x}_{m_2}] \subset \overrightarrow{\partial\Omega}_{ij}$ *at time* t_m *between two adjacent domains* Ω_α ($\alpha = i, j$)*.For an arbitrarily small* $\varepsilon > 0$, *there are two time intervals (i.e.,* $[t_{m-\varepsilon}, t_m)$ *and* $(t_m, t_{m+\varepsilon}]$*), and* $\mathbf{x}^{(i)}(t_{m\pm}) = \mathbf{x}_m = \mathbf{x}^{(j)}(t_{m+})$. *The flows* $\mathbf{x}^{(i)}(t)$ *and* $\mathbf{x}^{(j)}(t)$ *are*

$C^{r_i}_{[t_{m-\varepsilon},t_{m+\varepsilon}]}$ and $C^{r_j}_{[t_{m+\varepsilon},t_m)}$-continuous ($r_\alpha \geq 2, \alpha = i,j$) for time t, respectively. The source bifurcation of the flow $\mathbf{x}^{(i)}(t) \cup \mathbf{x}^{(j)}(t)$ at point $(\mathbf{x}_m, t_m)$ from $\overrightarrow{\partial\Omega}_{ij}$ to $\widehat{\partial\Omega}_{ij}$ occurs if and only if

$$L_{ij}(\mathbf{x}_m, t_m, \mathbf{p}_i, \mathbf{p}_j) = 0, \tag{2.72}$$

$$\mathbf{n}^{\mathrm{T}}_{\partial\Omega_{ij}} \cdot \mathbf{F}^{(j)}(t_{m+}) \neq 0 \text{ and } L_{ii}(\mathbf{x}_{m\pm\varepsilon}, t_{m\pm\varepsilon}, \mathbf{p}_i) < 0. \tag{2.73}$$

Proof. Applying the L-function of flows in Definition 2.17 to Theorem 2.16, this theorem can be easily proved. ■

Theorem 2.22. *For a discontinuous dynamical system in Eq.(2.1), there is a point $\mathbf{x}(t_m) = \mathbf{x}_m \in [\mathbf{x}_{m_1}, \mathbf{x}_{m_2}] \subset \overrightarrow{\partial\Omega}_{ij}$ at time t_m between two adjacent domains Ω_α ($\alpha = i, j$). For an arbitrarily small $\varepsilon > 0$, there are two time intervals (i.e., $[t_{m-\varepsilon}, t_m)$ and $(t_m, t_{m+\varepsilon}]$), and $\mathbf{x}^{(i)}(t_{m\mp}) = \mathbf{x}_m = \mathbf{x}^{(j)}(t_{m\pm})$. The flows $\mathbf{x}^{(\alpha)}(t)$ ($\alpha = i, j$) are $C^{r_\alpha}_{[t_{m-\varepsilon},t_{m+\varepsilon}]}$-continuous ($r_\alpha \geq 2$) for time t. The switching bifurcation of the flow $\mathbf{x}^{(i)}(t) \cup \mathbf{x}^{(j)}(t)$ at point $(\mathbf{x}_m, t_m)$ from $\overrightarrow{\partial\Omega}_{ij}$ to $\overleftarrow{\partial\Omega}_{ij}$ occurs if and only if*

$$L_{ij}(\mathbf{x}_m, t_m, \mathbf{p}_i, \mathbf{p}_j) = 0, \tag{2.74}$$

$$\mathbf{n}^{\mathrm{T}}_{\partial\Omega_{ij}} \cdot \mathbf{F}^{(\alpha)}(t_{m\pm}) = 0 \text{ and } L_{\alpha\alpha}(\mathbf{x}_{m\pm\varepsilon}, t_{m\pm\varepsilon}, \mathbf{p}_\alpha) < 0 \ (\alpha = i, j). \tag{2.75}$$

Proof. Applying the L-function of flows in Definition 2.17 to Theorem 2.18, the theorem can be easily proved. ■

Remark. For the $(n-1)$-dimensional *plane* boundary $\partial\Omega_{ij}$, the second conditions in Eqs.(2.71), (2.73) and (2.75) in Theorems 2.20-2.22 can be replaced by Eqs (2.56), (2.61) and (2.65) in Theorems 2.15, 2.17 and 2.19, respectively.

For the passable flow at $\mathbf{x}(t_m) \equiv \mathbf{x}_m \in [\mathbf{x}_{m_1}, \mathbf{x}_{m_2}] \subset \overrightarrow{\partial\Omega}_{ij}$ on the boundary $\overrightarrow{\partial\Omega}_{ij}$, consider the time interval $[t_{m_1}, t_{m_2}]$ for $[\mathbf{x}_{m_1}, \mathbf{x}_{m_2}]$ on the boundary, and the L-functions of flows (i.e., L_{ij}) for $t_m \in [t_{m_1}, t_{m_2}]$ and $\mathbf{x}_m \in [\mathbf{x}_{m_1}, \mathbf{x}_{m_2}]$ is also positive, i.e., $L_{ij}(\mathbf{x}_m, t_m, \mathbf{p}_i, \mathbf{p}_j) > 0$. To determine the switching bifurcation, the local minimum of $L_{ij}(\mathbf{x}_m, t_m, \mathbf{p}_i, \mathbf{p}_j)$ is introduced. Because $\mathbf{x}_m$ is a vector function of time t_m, the two total derivatives of $L_{ij}(\mathbf{x}_m, t_m, \mathbf{p}_i, \mathbf{p}_j)$ are introduced, i.e.,

$$DL_{ij}(\mathbf{x}_m,t_m,\mathbf{p}_i,\mathbf{p}_j)=\nabla L_{ij}(\mathbf{x}_m,t_m,\mathbf{p}_i,\mathbf{p}_j)\bullet\mathbf{F}_{ij}^{(0)}(\mathbf{x}_m,t_m)+\frac{\partial L_{ij}(\mathbf{x}_m,t_m,\mathbf{p}_i,\mathbf{p}_j)}{\partial t_m},\tag{2.76}$$

$$D^k L_{ij}(\mathbf{x}_m,t_m,\mathbf{p}_i,\mathbf{p}_j)=D^{k-1}(DL_{ij}(\mathbf{x}_m,t_m,\mathbf{p}_i,\mathbf{p}_j))\text{ for }k=1,2\cdots.\tag{2.77}$$

Thus, the local minimum of $L_{ij}(\mathbf{x}_m,t_m,\mathbf{p}_i,\mathbf{p}_j)$ is determined by

$$D^k L_{ij}(\mathbf{x}_m,t_m,\mathbf{p}_i,\mathbf{p}_j)=0\ \ (k=1,2,\cdots,2l-1),\tag{2.78}$$

$$D^{2l} L_{ij}(\mathbf{x}_m,t_m,\mathbf{p}_i,\mathbf{p}_j)>0.\tag{2.79}$$

Definition 2.18. For a discontinuous dynamical system in Eq.(2.1), there is a point $\mathbf{x}(t_m)=\mathbf{x}_m\in[\mathbf{x}_{m_1},\mathbf{x}_{m_2}]\subset\overrightarrow{\partial\Omega}_{ij}$ at time t_m between two adjacent domains Ω_α ($\alpha=i,j$). For an arbitrarily small $\varepsilon>0$, there are two time intervals (i.e., $[t_{m-\varepsilon},t_m)$ and $(t_m,t_{m+\varepsilon}]$). $\mathbf{x}^{(i)}(t_{m\mp})=\mathbf{x}_m=\mathbf{x}^{(j)}(t_{m\pm})$. The flows $\mathbf{x}^{(\alpha)}(t)$ are $C^{r_\alpha}_{[t_{m-\varepsilon},t_{m+\varepsilon}]}$ continuous ($r_\alpha\geq 2l$, $\alpha=i,j$) for time *t*. The *local minimum set* of $L_{ij}(\mathbf{x}_m,t_m,\mathbf{p}_i,\mathbf{p}_j)$ is defined by

$${}_{\min}L_{ij}(t_m)=\left\{L_{ij}(\mathbf{x}_m,t_m,\mathbf{p}_i,\mathbf{p}_j)\left|\begin{array}{l}\forall t_m\in[t_{m_1},t_{m_2}],\exists\mathbf{x}_m\in[\mathbf{x}_{m_1},\mathbf{x}_{m_2}],\\ \text{so that }D^kL_{ij}(\mathbf{x}_m,t_m,\mathbf{p}_i,\mathbf{p}_j)=0\\ \text{for }k=1,2,\cdots,2l-1,\text{and}\\ D^{2l}L_{ij}(\mathbf{x}_m,t_m,\mathbf{p}_i,\mathbf{p}_j)>0.\end{array}\right.\right\}\tag{2.80}$$

From the local minimum set of $L_{ij}(\mathbf{x}_m,t_m,\mathbf{p}_i,\mathbf{p}_j)$, the corresponding global minimum can be determined as follows.

Definition 2.19. For a discontinuous dynamical system in Eq.(2.1), there is a point $\mathbf{x}(t_m)=\mathbf{x}_m\in[\mathbf{x}_{m_1},\mathbf{x}_{m_2}]\subset\overrightarrow{\partial\Omega}_{ij}$ at time t_m between two adjacent domains Ω_α ($\alpha=i,j$). For an arbitrarily small $\varepsilon>0$, there are two time intervals (i.e., $[t_{m-\varepsilon},t_m)$ and $(t_m,t_{m+\varepsilon}]$), and $\mathbf{x}^{(i)}(t_{m\mp})=\mathbf{x}_m=\mathbf{x}^{(j)}(t_{m\pm})$. The flows $\mathbf{x}^{(\alpha)}(t)$ ($\alpha=i,j$) are $C^{r_\alpha}_{[t_{m-\varepsilon},t_{m+\varepsilon}]}$-continuous ($r_\alpha\geq 2l$, $\alpha=i,j$) for time *t*, respectively. The *global minimum set* of $L_{ij}(\mathbf{x}_m,t_m,\boldsymbol{\mu}_i,\boldsymbol{\mu}_j)$ is defined by

$${}_{G\min}L_{ij}(t_m)=\min_{t_m\in[t_{m_1},t_{m_2}]}\left\{\begin{array}{l}{}_{\min}L_{ij}(t_m),\ L_{ij}(\mathbf{x}_{m_1},t_{m_1},\mathbf{p}_i,\mathbf{p}_j),\\ L_{ij}(\mathbf{x}_{m_2},t_{m_2},\mathbf{p}_i,\mathbf{p}_j)\end{array}\right\}.\tag{2.81}$$

From the foregoing definition, Theorems 2.20-2.22 can be expressed through the global minimum of $L_{ij}(\mathbf{x}_m, t_m, \mathbf{p}_i, \mathbf{p}_j)$. So the following corollaries can be achieved, which give the conditions for onsets of switching bifurcations.

Corollary 2.1. *For a discontinuous dynamical system in Eq.*(2.1)*, there is a point* $\mathbf{x}(t_m) = \mathbf{x}_m \in [\mathbf{x}_{m_1}, \mathbf{x}_{m_2}] \subset \overrightarrow{\partial\Omega}_{ij}$ *for time* t_m *between two adjacent domains* Ω_α ($\alpha = i, j$)*. For an arbitrarily small* $\varepsilon > 0$*, there are two time intervals (i.e.,* $[t_{m-\varepsilon}, t_m)$ *and* $(t_m, t_{m+\varepsilon}]$*), and* $\mathbf{x}^{(i)}(t_{m-}) = \mathbf{x}_m = \mathbf{x}^{(j)}(t_{m\pm})$. *The flows* $\mathbf{x}^{(i)}(t)$ *and* $\mathbf{x}^{(j)}(t)$ *are* $C^r_{[t_{m+\varepsilon}, t_m)}$ *and* $C^r_{[t_{m-\varepsilon}, t_{m+\varepsilon}]}$*-continuous* ($r_\alpha \geq 2l$, $\alpha = i, j$) *for time t, respectively. The necessary and sufficient conditions for the sliding bifurcation onset of the flow* $\mathbf{x}^{(i)}(t) \cup \mathbf{x}^{(j)}(t)$ *at point* $(\mathbf{x}_m, t_m)$ *on the boundary* $\overrightarrow{\partial\Omega}_{ij}$ *are*

$$_{G\min} L_{ij}(t_m) = 0, \tag{2.82}$$

$$\mathbf{n}^{\mathrm{T}}_{\partial\Omega_{ij}} \cdot \mathbf{F}^{(i)}(t_{m-}) \neq 0 \ \text{ and } \ L_{jj}(\mathbf{x}_{m\pm\varepsilon}, t_{m\pm\varepsilon}, \mathbf{p}_j) < 0. \tag{2.83}$$

Proof. $L_{ij}(\mathbf{x}_m, t_m, \mathbf{p}_i, \mathbf{p}_j)$ replaced by its global minimum in Theorem 2.20 gives this corollary. This corollary is proved. ■

Corollary 2.2. *For a discontinuous dynamical system in Eq.*(2.1)*, there is a point* $\mathbf{x}(t_m) = \mathbf{x}_m \in [\mathbf{x}_{m_1}, \mathbf{x}_{m_2}] \subset \overrightarrow{\partial\Omega}_{ij}$ *for time* t_m *between two adjacent domains* Ω_α ($\alpha = i, j$)*. For an arbitrarily small* $\varepsilon > 0$*, there are two time intervals (i.e.,* $[t_{m-\varepsilon}, t_m)$ *and* $(t_m, t_{m+\varepsilon}]$*), and* $\mathbf{x}^{(i)}(t_{m\pm}) = \mathbf{x}_m = \mathbf{x}^{(j)}(t_{m+})$. *The flows* $\mathbf{x}^{(i)}(t)$ *and* $\mathbf{x}^{(j)}(t)$ *are* $C^{r_i}_{[t_{m-\varepsilon}, t_{m+\varepsilon}]}$ *and* $C^{r_j}_{[t_{m+\varepsilon}, t_m)}$*-continuous* ($r_\alpha \geq 2l$, $\alpha = i, j$) *for time t, respectively. The necessary and sufficient conditions for the source bifurcation onset of the flow* $\mathbf{x}^{(i)}(t) \cup \mathbf{x}^{(j)}(t)$ *at point* $(\mathbf{x}_m, t_m)$ *on the boundary* $\overrightarrow{\partial\Omega}_{ij}$ *are*

$$_{G\min} L_{ij}(t_m) = 0, \tag{2.84}$$

$$\mathbf{n}^{T}_{\partial\Omega_{ij}} \cdot \mathbf{F}^{(j)}(t_{m+}) \neq 0 \ \text{ and } \ L_{ii}(\mathbf{x}_{m\pm\varepsilon}, t_{m\pm\varepsilon}, \mathbf{p}_i) < 0. \tag{2.85}$$

Proof. $L_{ij}(\mathbf{x}_m, t_m, \mathbf{p}_i, \mathbf{p}_j)$ replaced by its global minimum in Theorem 2.21 gives this corollary. ■

Corollary 2.3. *For a discontinuous dynamical system in Eq.*(2.1)*, there is a point* $\mathbf{x}(t_m) = \mathbf{x}_m \in [\mathbf{x}_{m_1}, \mathbf{x}_{m_2}] \subset \overrightarrow{\partial\Omega}_{ij}$ *for time* t_m *between two adjacent domains* Ω_α ($\alpha = i, j$)*. For an arbitrarily small* $\varepsilon > 0$*, there are two time intervals (i.e.,* $[t_{m-\varepsilon}, t_m)$ *and* $(t_m, t_{m+\varepsilon}]$*), and* $\mathbf{x}^{(i)}(t_{m\mp}) = \mathbf{x}_m = \mathbf{x}^{(j)}(t_{m\pm})$. *The flows* $\mathbf{x}^{(\alpha)}(t)$ *are* $C^{r_\alpha}_{[t_{m-\varepsilon}, t_{m+\varepsilon}]}$ -

continuous ($r_\alpha \geq 2l$) *for time t. The necessary and sufficient conditions for the switching bifurcation onset of the flow* $\mathbf{x}^{(i)}(t) \cup \mathbf{x}^{(j)}(t)$ *at point* $(\mathbf{x}_m, t_m)$ *on the boundary* $\overrightarrow{\partial\Omega}_{ij}$ *are*

$$_{G\min}L_{ij}(t_m) = 0, \tag{2.86}$$

$$\mathbf{n}_{\partial\Omega_{ij}}^{\mathrm{T}} \cdot \mathbf{F}^{(t)}(t_{m\pm}) = 0 \text{ and } L_{\alpha\alpha}(\mathbf{x}_{m\pm\varepsilon}, t_{m\pm\varepsilon}, \mathbf{p}_\alpha) < 0 \text{ for } \alpha = i, j. \tag{2.87}$$

Proof. $L_{ij}(\mathbf{x}_m, t_m, \mathbf{p}_i, \mathbf{p}_j)$ replaced by its global minimum in Theorem 2.22 gives this corollary. ■

2.6. Switching bifurcations of non-passable flows

The onset and vanishing of the sliding and source flows on the boundary were discussed. The fragmentations of the sliding and source flows on the boundary are of great interest in this section. This kind of bifurcation is still a switching bifurcation. The definitions for such fragmentation bifurcations of flows on the non-passable boundary are similar to the switching bifurcations of flows from the semi-passable boundary to non-passable boundary. From the logic points of view, the necessary and sufficient conditions for the fragmentation bifurcation from a non-passable flow to a passable flow on the boundary are quite similar to the sliding and source bifurcations from a passable flow to a non-passable flow. For a clear description of the fragmentation bifurcation phenomena, the corresponding definitions and theorems are also given herein.

Definition 2.20. For a discontinuous dynamical system in Eq.(2.1), there is a point $\mathbf{x}(t_m) = \mathbf{x}_m \in [\mathbf{x}_{m_1}, \mathbf{x}_{m_2}] \subset \widetilde{\partial\Omega}_{ij}$ at time t_m between two adjacent domains Ω_α ($\alpha = i, j$). For an arbitrarily small $\varepsilon > 0$, there are two time intervals (i.e., $[t_{m-\varepsilon}, t_m)$ and $(t_m, t_{m+\varepsilon}]$), and $\mathbf{x}^{(i)}(t_{m-}) = \mathbf{x}_m = \mathbf{x}^{(j)}(t_{m\mp})$. The flows $\mathbf{x}^{(i)}(t)$ and $\mathbf{x}^{(j)}(t)$ are $C^{r_i}_{[t_{m+\varepsilon}, t_m)}$ and $C^{r_j}_{[t_{m-\varepsilon}, t_{m+\varepsilon}]}$-continuous ($r_\alpha \geq 1, \alpha = i, j$) for time t, respectively. The tangential bifurcation of the flow $\mathbf{x}^{(j)}(t)$ *at point* $(\mathbf{x}_m, t_m)$ on the boundary $\widetilde{\partial\Omega}_{ij}$ is termed the *switching bifurcation* of a non-passable flow of the *first kind* from $\widetilde{\partial\Omega}_{ij}$ *to* $\overrightarrow{\partial\Omega}_{ij}$ (or simply called the *sliding fragmentation bifurcation*) if

$$\mathbf{n}_{\partial\Omega_{ij}}^{\mathrm{T}} \cdot \dot{\mathbf{x}}^{(j)}(t_{m\mp}) = 0 \text{ and } \mathbf{n}_{\partial\Omega_{ij}}^{\mathrm{T}} \cdot \dot{\mathbf{x}}^{(i)}(t_{m-}) \neq 0, \tag{2.88}$$

$$\text{either}\quad \left.\begin{aligned} &\mathbf{n}_{\partial\Omega_{ij}}^{\mathrm{T}}\cdot[\mathbf{x}^{(i)}(t_{m-})-\mathbf{x}^{(i)}(t_{m-\varepsilon})]>0\\ &\mathbf{n}_{\partial\Omega_{ij}}^{\mathrm{T}}\cdot[\mathbf{x}^{(j)}(t_{m-})-\mathbf{x}^{(j)}(t_{m-\varepsilon})]<0\\ &\mathbf{n}_{\partial\Omega_{ij}}^{\mathrm{T}}\cdot[\mathbf{x}^{(j)}(t_{m+\varepsilon})-\mathbf{x}^{(j)}(t_{m+})]>0 \end{aligned}\right\}\ \text{for } \mathbf{n}_{\partial\Omega_{ij}}\to\Omega_j, \tag{2.89}$$

$$\text{or}\quad \left.\begin{aligned} &\mathbf{n}_{\partial\Omega_{ij}}^{\mathrm{T}}\cdot[\mathbf{x}^{(i)}(t_{m-})-\mathbf{x}^{(i)}(t_{m-\varepsilon})]<0\\ &\mathbf{n}_{\partial\Omega_{ij}}^{\mathrm{T}}\cdot[\mathbf{x}^{(j)}(t_{m-})-\mathbf{x}^{(j)}(t_{m-\varepsilon})]>0\\ &\mathbf{n}_{\partial\Omega_{ij}}^{\mathrm{T}}\cdot[\mathbf{x}^{(j)}(t_{m+\varepsilon})-\mathbf{x}^{(j)}(t_{m+})]<0 \end{aligned}\right\}\ \text{for } \mathbf{n}_{\partial\Omega_{ij}}\to\Omega_i. \tag{2.90}$$

Definition 2.21. For a discontinuous dynamical system in Eq.(2.1), there is a point $\mathbf{x}(t_m)=\mathbf{x}_m\in[\mathbf{x}_{m_1},\mathbf{x}_{m_2}]\subset\widehat{\partial\Omega}_{ij}$ at time t_m between two adjacent domains Ω_α ($\alpha=i,j$). For an arbitrarily small $\varepsilon>0$, there are two time intervals (i.e., $[t_{m-\varepsilon},t_m)$ and $(t_m,t_{m+\varepsilon}]$), and $\mathbf{x}^{(i)}(t_{m\pm})=\mathbf{x}_m=\mathbf{x}^{(j)}(t_{m+})$. The flows $\mathbf{x}^{(i)}(t)$ and $\mathbf{x}^{(j)}(t)$ are $C^{r_i}_{[t_{m-\varepsilon},t_{m+\varepsilon}]}$ and $C^{r_j}_{[t_{m+\varepsilon},t_m)}$-continuous ($r_\alpha\geq 1, \alpha=i,j$) for time t, respectively. The tangential bifurcation of flows $\mathbf{x}^{(i)}(t)$ and $\mathbf{x}^{(j)}(t)$ *at point* $(\mathbf{x}_m,t_m)$ on the boundary $\widehat{\partial\Omega}_{ij}$ is termed *the switching bifurcation* of a non-passable flow of the *second kind* from $\widehat{\partial\Omega}_{ij}$ to $\overrightarrow{\partial\Omega}_{ij}$ (or simply called *the source fragmentation bifurcation*) if

$$\mathbf{n}_{\partial\Omega_{ij}}^{\mathrm{T}}\cdot\dot{\mathbf{x}}^{(i)}(t_{m\pm})=0 \text{ and } \mathbf{n}_{\partial\Omega_{ij}}^{\mathrm{T}}\cdot\dot{\mathbf{x}}^{(j)}(t_{m+})\neq 0, \tag{2.91}$$

$$\text{either}\quad \left.\begin{aligned} &\mathbf{n}_{\partial\Omega_{ij}}^{\mathrm{T}}\cdot[\mathbf{x}^{(i)}(t_{m-})-\mathbf{x}^{(i)}(t_{m-\varepsilon})]>0\\ &\mathbf{n}_{\partial\Omega_{ij}}^{\mathrm{T}}\cdot[\mathbf{x}^{(i)}(t_{m+\varepsilon})-\mathbf{x}^{(i)}(t_{m+})]<0\\ &\mathbf{n}_{\partial\Omega_{ij}}^{\mathrm{T}}\cdot[\mathbf{x}^{(j)}(t_{m+\varepsilon})-\mathbf{x}^{(j)}(t_{m+})]>0 \end{aligned}\right\}\ \text{for } \mathbf{n}_{\partial\Omega_{ij}}\to\Omega_j, \tag{2.92}$$

$$\text{or}\quad \left.\begin{aligned} &\mathbf{n}_{\partial\Omega_{ij}}^{\mathrm{T}}\cdot[\mathbf{x}^{(i)}(t_{m-})-\mathbf{x}^{(i)}(t_{m-\varepsilon})]<0\\ &\mathbf{n}_{\partial\Omega_{ij}}^{\mathrm{T}}\cdot[\mathbf{x}^{(i)}(t_{m+\varepsilon})-\mathbf{x}^{(i)}(t_{m+})]>0\\ &\mathbf{n}_{\partial\Omega_{ij}}^{\mathrm{T}}\cdot[\mathbf{x}^{(j)}(t_{m+\varepsilon})-\mathbf{x}^{(j)}(t_{m+})]<0 \end{aligned}\right\}\ \text{for } \mathbf{n}_{\partial\Omega_{ij}}\to\Omega_i. \tag{2.93}$$

For the fragmentation bifurcation of the non-passable flow on the boundary, the vector fields near the sink and source boundaries are sketched in Figs.2.9 and 2.10, respectively. The switching from the sink or source flow to the semi-passable flow has two possibilities. Therefore, the conditions in Definitions 2.20 and 2.21 have been changed accordingly. Before the fragmentation bifurcation of the non-passable flow occurs on the boundary, the flow $\mathbf{x}^{(\alpha)}(t)$ ($\alpha\in\{i,j\}$) exists for $t\in[t_{m-\varepsilon},t_{m-})$ or $t\in(t_{m+},t_{m+\varepsilon}]$ on the sink or source boundary. Only the sliding flow exists on such a boundary. After the fragmentation bifurcation occurs, the sliding flow on the boundary will split into at least two portions of the sliding and semi-passable motions. This phenomenon is called *the fragmentation of the sliding flow on the boundary*, which can help one easily understand the sliding dynamics on the boundary. In addition, for the non-passable boundary, if flows on both sides of the non-passable boundary possess the local singularity at the boundary, the non-passable flow of the first kind switches into the non-passable flow of the second kind, and vice versa. The local singularity of such switchability is similar to the switching between the two semi-passable flows on the boundary, and the corresponding definition of the switching bifurcation is given as follows.

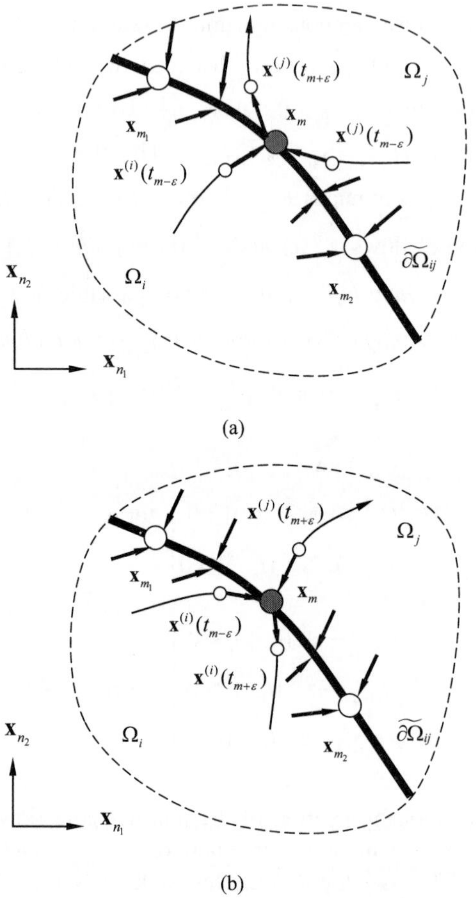

Fig. 2.9 The sliding fragmentation bifurcation to the sink boundary $\widetilde{\partial\Omega_{ij}}$ in domain: **(a)** Ω_j and **(b)** Ω_i. Four points $\mathbf{x}^{(\alpha)}(t_{m\pm\varepsilon})$, $\mathbf{x}^{(\beta)}(t_{m-\varepsilon})$ and $\mathbf{x}_m$ lie in the corresponding domains and on the boundary $\partial\Omega_{ij}$, respectively. $\alpha,\beta \in \{i,j\}$ but $\alpha \neq \beta$ and $n_1 + n_2 = n$.

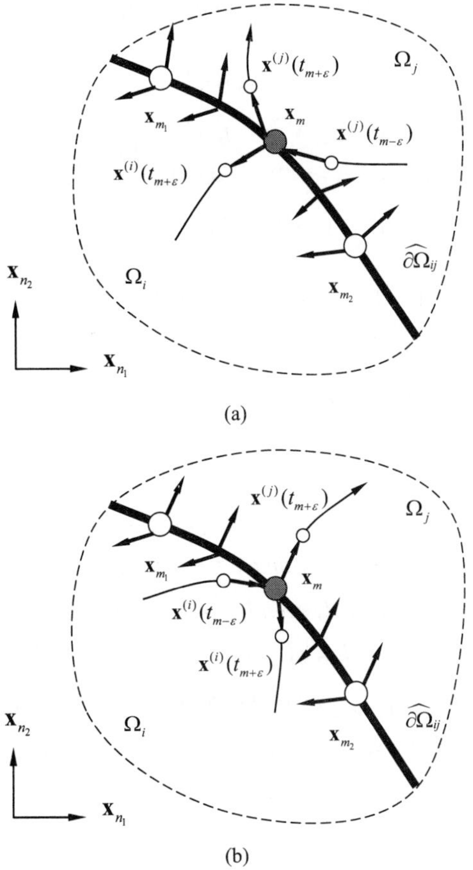

Fig. 2.10 The source fragmentation bifurcation to the source boundary $\widehat{\partial\Omega}_{ij}$ in domain: **(a)** Ω_j and **(b)** Ω_i. Four points $\mathbf{x}^{(\alpha)}(t_{m\pm\varepsilon})$, $\mathbf{x}^{(\beta)}(t_{m+\varepsilon})$ and $\mathbf{x}_m$ lie in the corresponding domains and on the boundary $\partial\Omega_{ij}$, respectively. $\alpha,\beta \in \{i,j\}$ but $\alpha \neq \beta$ and $n_1 + n_2 = n$.

Definition 2.22. For a discontinuous dynamical system in Eq.(2.1), there is a point $\mathbf{x}(t_m) = \mathbf{x}_m \in [\mathbf{x}_{m_1}, \mathbf{x}_{m_2}] \subset \widetilde{\partial\Omega}_{ij}$ (or $\widehat{\partial\Omega}_{ij}$) at time t_m between two adjacent domains Ω_α ($\alpha = i, j$). For an arbitrarily small $\varepsilon > 0$, there are two time intervals (i.e., $[t_{m-\varepsilon}, t_m)$ and $(t_m, t_{m+\varepsilon}]$), and $\mathbf{x}^{(\alpha)}(t_{m\pm}) = \mathbf{x}_m$. The flows $\mathbf{x}^{(\alpha)}(t)$ ($\alpha = i, j$) are $C^{r_\alpha}_{[t_{m-\varepsilon}, t_{m+\varepsilon}]}$-continuous ($r_\alpha \geq 1$) for time *t*. The tangential bifurcation of the flow $\mathbf{x}^{(i)}(t)$ and $\mathbf{x}^{(j)}(t)$ at point $(\mathbf{x}_m, t_m)$ on the boundary $\widetilde{\partial\Omega}_{ij}$ (or $\widehat{\partial\Omega}_{ij}$) is termed the *switching bifurcation of a non-passable flow* from $\widetilde{\partial\Omega}_{ij}$ to $\widehat{\partial\Omega}_{ij}$ (or $\widehat{\partial\Omega}_{ij}$ to $\widetilde{\partial\Omega}_{ij}$)

if

$$\mathbf{n}_{\partial\Omega_{ij}}^{\mathrm{T}} \cdot \dot{\mathbf{x}}^{(\alpha)}(t_{m\pm}) = 0 \text{ for } \alpha = i, j; \tag{2.94}$$

$$\text{either} \left.\begin{array}{l} \mathbf{n}_{\partial\Omega_{ij}}^{\mathrm{T}} \cdot [\mathbf{x}^{(i)}(t_{m-}) - \mathbf{x}^{(i)}(t_{m-\varepsilon})] > 0 \\ \mathbf{n}_{\partial\Omega_{ij}}^{\mathrm{T}} \cdot [\mathbf{x}^{(i)}(t_{m+\varepsilon}) - \mathbf{x}^{(i)}(t_{m+})] < 0 \\ \mathbf{n}_{\partial\Omega_{ij}}^{\mathrm{T}} \cdot [\mathbf{x}^{(j)}(t_{m-}) - \mathbf{x}^{(j)}(t_{m-\varepsilon})] < 0 \\ \mathbf{n}_{\partial\Omega_{ij}}^{\mathrm{T}} \cdot [\mathbf{x}^{(j)}(t_{m+\varepsilon}) - \mathbf{x}^{(j)}(t_{m+})] > 0 \end{array}\right\} \text{for } \mathbf{n}_{\partial\Omega_{ij}} \to \Omega_j, \tag{2.95}$$

$$\text{or} \left.\begin{array}{l} \mathbf{n}_{\partial\Omega_{ij}}^{\mathrm{T}} \cdot [\mathbf{x}^{(i)}(t_{m-}) - \mathbf{x}^{(i)}(t_{m-\varepsilon})] < 0 \\ \mathbf{n}_{\partial\Omega_{ij}}^{\mathrm{T}} \cdot [\mathbf{x}^{(i)}(t_{m+\varepsilon}) - \mathbf{x}^{(i)}(t_{m+})] > 0 \\ \mathbf{n}_{\partial\Omega_{ij}}^{\mathrm{T}} \cdot [\mathbf{x}^{(j)}(t_{m-}) - \mathbf{x}^{(j)}(t_{m-\varepsilon})] > 0 \\ \mathbf{n}_{\partial\Omega_{ij}}^{\mathrm{T}} \cdot [\mathbf{x}^{(j)}(t_{m+\varepsilon}) - \mathbf{x}^{(j)}(t_{m+})] < 0 \end{array}\right\} \text{for } \mathbf{n}_{\partial\Omega_{ij}} \to \Omega_i. \tag{2.96}$$

Theorem 2.23. *For a discontinuous dynamical system in Eq.*(2.1), *there is a point* $\mathbf{x}(t_m) = \mathbf{x}_m \in [\mathbf{x}_{m_1}, \mathbf{x}_{m_2}] \subset \widetilde{\partial\Omega}_{ij}$ *at time* t_m *between two adjacent domains* Ω_α ($\alpha = i, j$). *For an arbitrarily small* $\varepsilon > 0$, *there are two time intervals (i.e.,* $[t_{m-\varepsilon}, t_m)$ *and* $(t_m, t_{m+\varepsilon}]$), *and* $\mathbf{x}^{(i)}(t_{m-}) = \mathbf{x}_m = \mathbf{x}^{(j)}(t_{m\pm})$. *The flows* $\mathbf{x}^{(i)}(t)$ *and* $\mathbf{x}^{(j)}(t)$ *are* $C^{r_i}_{[t_{m-\varepsilon}, t_m)}$ *and* $C^{r_j}_{[t_{m-\varepsilon}, t_{m+\varepsilon}]}$*-continuous* ($r_\alpha \geq 1, \alpha = i, j$) *for time t, respectively. The sliding fragmentation bifurcation of the flow* $\mathbf{x}^{(i)}(t) \cup \mathbf{x}^{(j)}(t)$ *at point* $(\mathbf{x}_m, t_m)$ *from* $\widetilde{\partial\Omega}_{ij}$ *to* $\overrightarrow{\partial\Omega}_{ij}$ *occurs if and only if*

$$\mathbf{n}_{\partial\Omega_{ij}}^{\mathrm{T}} \cdot \mathbf{F}^{(j)}(t_{m\pm}) = 0 \;\; \textit{and} \;\; \mathbf{n}_{\partial\Omega_{ij}}^{\mathrm{T}} \cdot \mathbf{F}^{(i)}(t_{m-}) \neq 0; \tag{2.97}$$

$$\begin{array}{ll} \textit{either} & \mathbf{n}_{\partial\Omega_{ij}}^{\mathrm{T}} \cdot \mathbf{F}^{(i)}(t_{m-}) > 0 \textit{ for } \mathbf{n}_{\partial\Omega_{ij}} \to \Omega_j, \\ \textit{or} & \mathbf{n}_{\partial\Omega_{ij}}^{T} \cdot \mathbf{F}^{(i)}(t_{m-}) < 0 \textit{ for } \mathbf{n}_{\partial\Omega_{ij}} \to \Omega_i; \end{array} \tag{2.98}$$

$$\begin{array}{l} \textit{either} \left.\begin{array}{l} \mathbf{n}_{\partial\Omega_{ij}}^{\mathrm{T}} \cdot \mathbf{F}^{(j)}(t_{m-\varepsilon}) < 0 \\ \mathbf{n}_{\partial\Omega_{ij}}^{\mathrm{T}} \cdot \mathbf{F}^{(j)}(t_{m+\varepsilon}) > 0 \end{array}\right\} \textit{for } \mathbf{n}_{\partial\Omega_{ij}} \to \Omega_j, \\ \textit{or} \left.\begin{array}{l} \mathbf{n}_{\partial\Omega_{ij}}^{\mathrm{T}} \cdot \mathbf{F}^{(j)}(t_{m-\varepsilon}) > 0 \\ \mathbf{n}_{\partial\Omega_{ij}}^{\mathrm{T}} \cdot \mathbf{F}^{(j)}(t_{m+\varepsilon}) < 0 \end{array}\right\} \textit{for } \mathbf{n}_{\partial\Omega_{ij}} \to \Omega_i. \end{array} \tag{2.99}$$

Proof. Following the proof procedures of Theorems 2.8 and 2.9, the above theorem can be easily proved.

Theorem 2.24. *For a discontinuous dynamical system in Eq.*(2.1), *there is a point*

$\mathbf{x}(t_m) = \mathbf{x}_m \in [\mathbf{x}_{m_1}, \mathbf{x}_{m_2}] \subset \widetilde{\partial\Omega}_{ij}$ *at time* t_m *on the* $(n-1)$ *-dimensional plane boundary* $\partial\Omega_{ij}$ *between two adjacent domains* Ω_α $(\alpha = i, j)$. *For an arbitrarily small* $\varepsilon > 0$, *there are two time intervals (i.e.,* $[t_{m-\varepsilon}, t_m)$ *and* $(t_m, t_{m+\varepsilon}]$*), and* $\mathbf{x}^{(i)}(t_{m-}) = \mathbf{x}_m = \mathbf{x}^{(j)}(t_{m\pm})$. *The flows* $\mathbf{x}^{(i)}(t)$ *and* $\mathbf{x}^{(j)}(t)$ *are* $C^{r_i}_{[t_{m-\varepsilon}, t_m)}$ *and* $C^{r_j}_{[t_{m-\varepsilon}, t_{m+\varepsilon}]}$ *continuous* $(r_\alpha \geq 2, \alpha = i, j)$ *for time t, respectively. The sliding fragmentation bifurcation of the flow* $\mathbf{x}^{(i)}(t) \cup \mathbf{x}^{(j)}(t)$ *at point* $(\mathbf{x}_m, t_m)$ *from* $\widetilde{\partial\Omega}_{ij}$ *to* $\overrightarrow{\partial\Omega}_{ij}$ *occurs if and only if*

$$\mathbf{n}^{\mathrm{T}}_{\partial\Omega_{ij}} \cdot \mathbf{F}^{(j)}(t_{m\pm}) = 0 \text{ and } \mathbf{n}^{\mathrm{T}}_{\partial\Omega_{ij}} \cdot \mathbf{F}^{(i)}(t_{m-}) \neq 0; \tag{2.100}$$

$$\left.\begin{array}{ll} \textit{either} & \left.\begin{array}{l} \mathbf{n}^{\mathrm{T}}_{\partial\Omega_{ij}} \cdot \mathbf{F}^{(i)}(t_{m-}) > 0 \\ \mathbf{n}^{\mathrm{T}}_{\partial\Omega_{ij}} \cdot D\mathbf{F}^{(j)}(t_{m\pm}) < 0 \end{array}\right\} \textit{for } \mathbf{n}_{\partial\Omega_{ij}} \to \Omega_j, \\ \textit{or} & \left.\begin{array}{l} \mathbf{n}^{\mathrm{T}}_{\partial\Omega_{ij}} \cdot \mathbf{F}^{(i)}(t_{m-}) < 0 \\ \mathbf{n}^{\mathrm{T}}_{\partial\Omega_{ij}} \cdot D\mathbf{F}^{(j)}(t_{m\pm}) < 0 \end{array}\right\} \textit{for } \mathbf{n}_{\partial\Omega_{ij}} \to \Omega_i. \end{array}\right. \tag{2.101}$$

Proof. Following the proof procedures of Theorems 2.10 and 2.11, the above theorem can be easily proved. ■

Theorem 2.25. *For a discontinuous dynamical system in Eq.(2.1), there is a point* $\mathbf{x}(t_m) = \mathbf{x}_m \in [\mathbf{x}_{m_1}, \mathbf{x}_{m_2}] \subset \widehat{\partial\Omega}_{ij}$ *for time* t_m *between two adjacent domains* Ω_α $(\alpha = i, j)$. *For an arbitrarily small* $\varepsilon > 0$, *there are two time intervals (i.e.,* $[t_{m-\varepsilon}, t_m)$ *and* $(t_m, t_{m+\varepsilon}]$*), and* $\mathbf{x}^{(i)}(t_{m\pm}) = \mathbf{x}_m = \mathbf{x}^{(j)}(t_{m+})$. *The flows* $\mathbf{x}^{(i)}(t)$ *and* $\mathbf{x}^{(j)}(t)$ *are* $C^{r_i}_{[t_{m-\varepsilon}, t_{m+\varepsilon}]}$ *and* $C^{r_j}_{(t_m, t_{m+\varepsilon}]}$*-continuous* $(r_\alpha \geq 1, \alpha = i, j)$ *for time t, respectively. The source fragmentation bifurcation of the flow* $\mathbf{x}^{(i)}(t) \cup \mathbf{x}^{(j)}(t)$ *at point* $(\mathbf{x}_m, t_m)$ *on the boundary* $\widehat{\partial\Omega}_{ij}$ *occurs if and only if*

$$\mathbf{n}^{\mathrm{T}}_{\partial\Omega_{ij}} \cdot \mathbf{F}^{(i)}(t_{m\pm}) = 0 \textit{ and } \mathbf{n}^{\mathrm{T}}_{\partial\Omega_{ij}} \cdot \mathbf{F}^{(j)}(t_{m+}) \neq 0; \tag{2.102}$$

$$\begin{array}{ll} \textit{either} & \mathbf{n}^{\mathrm{T}}_{\partial\Omega_{ij}} \cdot \mathbf{F}^{(j)}(t_{m+}) > 0 \textit{ for } \mathbf{n}_{\partial\Omega_{ij}} \to \Omega_j, \\ \textit{or} & \mathbf{n}^{T}_{\partial\Omega_{ij}} \cdot \mathbf{F}^{(j)}(t_{m+}) < 0 \textit{ for } \mathbf{n}_{\partial\Omega_{ij}} \to \Omega_i; \end{array} \tag{2.103}$$

$$\begin{array}{ll} \textit{either} & \left.\begin{array}{l} \mathbf{n}^{\mathrm{T}}_{\partial\Omega_{ij}} \cdot \mathbf{F}^{(i)}(t_{m-\varepsilon}) > 0 \\ \mathbf{n}^{\mathrm{T}}_{\partial\Omega_{ij}} \cdot \mathbf{F}^{(i)}(t_{m+\varepsilon}) < 0 \end{array}\right\} \textit{for } \mathbf{n}_{\partial\Omega_{ij}} \to \Omega_j, \\ \textit{or} & \left.\begin{array}{l} \mathbf{n}^{\mathrm{T}}_{\partial\Omega_{ij}} \cdot \mathbf{F}^{(i)}(t_{m-\varepsilon}) < 0 \\ \mathbf{n}^{\mathrm{T}}_{\partial\Omega_{ij}} \cdot \mathbf{F}^{(i)}(t_{m+\varepsilon}) > 0 \end{array}\right\} \textit{for } \mathbf{n}_{\partial\Omega_{ij}} \to \Omega_i. \end{array} \tag{2.104}$$

Proof. Following the proof procedures of Theorems 2.8 and 2.9, the above theorem can be easily proved. ■

Theorem 2.26. *For a discontinuous dynamical system in Eq.*(2.1), *there is a point* $\mathbf{x}(t_m)=\mathbf{x}_m\in[\mathbf{x}_{m_1},\mathbf{x}_{m_2}]\subset\widehat{\partial\Omega}_{ij}$ *at time* t_m *on the* $(n-1)$ *-dimensional plane boundary* $\partial\Omega_{ij}$ *between two adjacent domains* Ω_α $(\alpha=i,j)$. *For an arbitrarily small* $\varepsilon>0$, *there are two time intervals (i.e.,* $[t_{m-\varepsilon},t_m)$ *and* ($t_m,t_{m+\varepsilon}]$), *and* $\mathbf{x}^{(i)}(t_{m\pm})=\mathbf{x}_m=\mathbf{x}^{(j)}(t_{m+})$. *The flows* $\mathbf{x}^{(i)}(t)$ *and* $\mathbf{x}^{(j)}(t)$ *are* $C^{r_i}_{[t_{m-\varepsilon},t_{m+\varepsilon}]}$ *and* $C^{r_j}_{(t_m,t_{m+\varepsilon}]}$ *continuous* ($r_\alpha\geq 1$, $\alpha=i,j$) *for time t, respectively. The source fragmentation bifurcation of the flow* $\mathbf{x}^{(i)}(t)\cup\mathbf{x}^{(j)}(t)$ *at point* $(\mathbf{x}_m,t_m)$ *from* $\widehat{\partial\Omega}_{ij}$ *to* $\overrightarrow{\partial\Omega}_{ij}$ *occurs if and only if*

$$\mathbf{n}^{\mathrm{T}}_{\partial\Omega_{ij}}\cdot\mathbf{F}^{(i)}(t_{m\pm})=0 \;\; and \;\; \mathbf{n}^{\mathrm{T}}_{\partial\Omega_{ij}}\cdot\mathbf{F}^{(j)}(t_{m+})\neq 0; \tag{2.105}$$

$$\begin{aligned} &either && \left.\begin{aligned}\mathbf{n}^{\mathrm{T}}_{\partial\Omega_{ij}}\cdot\mathbf{F}^{(j)}(t_{m+})>0\\ \mathbf{n}^{T}_{\partial\Omega_{ij}}\cdot D\mathbf{F}^{(i)}(t_{m\pm})<0\end{aligned}\right\} for\ \mathbf{n}_{\partial\Omega_{ij}}\to\Omega_j,\\ &or && \left.\begin{aligned}\mathbf{n}^{T}_{\partial\Omega_{ij}}\cdot\mathbf{F}^{(j)}(t_{m+})<0\\ \mathbf{n}^{T}_{\partial\Omega_{ij}}\cdot D\mathbf{F}^{(i)}(t_{m\pm})<0\end{aligned}\right\} for\ \mathbf{n}_{\partial\Omega_{ij}}\to\Omega_i.\end{aligned} \tag{2.106}$$

Proof. Following the proof procedures in Theorems 2.10 and 2.11, the above theorem can be easily proved. ■

Theorem 2.27. *For a discontinuous dynamical system in Eq.*(2.1), *there is a point* $\mathbf{x}(t_m)=\mathbf{x}_m\in[\mathbf{x}_{m_1},\mathbf{x}_{m_2}]\subset\widehat{\partial\Omega}_{ij}$ *(or* $\widetilde{\partial\Omega}_{ij}$*) at time* t_m *between two adjacent domains* Ω_α $(\alpha=i,j)$. *For an arbitrarily small* $\varepsilon>0$, *there are two time intervals (i.e.,* $[t_{m-\varepsilon},t_m)$ *and* $(t_m,t_{m+\varepsilon}]$), *and* $\mathbf{x}^{(i)}(t_{m\pm})=\mathbf{x}_m=\mathbf{x}^{(j)}(t_{m\pm})$. *The flows* $\mathbf{x}^{(\alpha)}(t)$ $(\alpha=i,j)$ *are* $C^{r_\alpha}_{[t_{m-\varepsilon},t_{m+\varepsilon}]}$ *-continuous* ($r_\alpha\geq 1$) *for time t. The switching bifurcation of the flow at point* $(\mathbf{x}_m,t_m)$ *from* $\widetilde{\partial\Omega}_{ij}$ *to* $\widehat{\partial\Omega}_{ij}$ *(or* $\widehat{\partial\Omega}_{ij}$ *to* $\widetilde{\partial\Omega}_{ij}$*) occurs if and only if*

$$\mathbf{n}^{\mathrm{T}}_{\partial\Omega_{ij}}\cdot\mathbf{F}^{(\alpha)}(t_{m\pm})=0; \tag{2.107}$$

$$\begin{aligned} &either && \left.\begin{aligned}\mathbf{n}^{\mathrm{T}}_{\partial\Omega_{ij}}\cdot\mathbf{F}^{(\alpha)}(t_{m-\varepsilon})>0,\\ \mathbf{n}^{\mathrm{T}}_{\partial\Omega_{ij}}\cdot\mathbf{F}^{(\alpha)}(t_{m+\varepsilon})<0\end{aligned}\right\} for\ \mathbf{n}_{\partial\Omega_{ij}}\to\Omega_\beta,\\ &or && \left.\begin{aligned}\mathbf{n}^{\mathrm{T}}_{\partial\Omega_{ij}}\cdot\mathbf{F}^{(\alpha)}(t_{m-\varepsilon})<0,\\ \mathbf{n}^{\mathrm{T}}_{\partial\Omega_{ij}}\cdot\mathbf{F}^{(\alpha)}(t_{m+\varepsilon})>0\end{aligned}\right\} for\ \mathbf{n}_{\partial\Omega_{ij}}\to\Omega_\alpha\end{aligned} \tag{2.108}$$

with $\alpha, \beta = i, j$ but $\alpha \neq \beta$.

Proof. Following the proof procedures of Theorems 2.8 and 2.9, the above theorem can be easily proved. ■

Theorem 2.28. *For a discontinuous dynamical system in Eq.(2.1), there is a point* $\mathbf{x}(t_m) = \mathbf{x}_m \in [\mathbf{x}_{m_1}, \mathbf{x}_{m_2}] \subset \widetilde{\partial\Omega}_{ij}$ *(or* $\widehat{\partial\Omega}_{ij}$*) at time* t_m *on the* $(n-1)$ *dimensional plane boundary* $\partial\Omega_{ij}$ *between two adjacent domains* Ω_α $(\alpha = i, j)$. *For an arbitrarily small* $\varepsilon > 0$, *there are two time intervals (i.e.,* $[t_{m-\varepsilon}, t_m)$ *and* $(t_m, t_{m+\varepsilon}]$*), and* $\mathbf{x}^{(\alpha)}(t_{m\pm}) = \mathbf{x}_m$. *The flows* $\mathbf{x}^{(\alpha)}(t)$ $(\alpha = i, j)$ *are* $C^{r_\alpha}_{[t_{m-\varepsilon}, t_{m+\varepsilon}]}$*-continuous* $(r_\alpha \geq 2)$ *for time t. The switching bifurcation of the flow at point* $(\mathbf{x}_m, t_m)$ *from* $\widetilde{\partial\Omega}_{ij}$ *to* $\widehat{\partial\Omega}_{ij}$ *(or* $\widehat{\partial\Omega}_{ij}$ *to* $\widetilde{\partial\Omega}_{ij}$*) exists if and only if*

$$\mathbf{n}^{\mathrm{T}}_{\partial\Omega_{ij}} \cdot \mathbf{F}^{(\alpha)}(t_{m\pm}) = 0 \;\; \text{for } \alpha = i, j; \tag{2.109}$$

$$\begin{aligned} &\text{either} \quad \left.\begin{aligned} \mathbf{n}^{\mathrm{T}}_{\partial\Omega_{ij}} \cdot D\mathbf{F}^{(i)}(t_{m\pm}) < 0 \\ \mathbf{n}^{\mathrm{T}}_{\partial\Omega_{ij}} \cdot D\mathbf{F}^{(j)}(t_{m\pm}) > 0 \end{aligned}\right\} \text{for } \mathbf{n}_{\partial\Omega_{ij}} \to \Omega_j, \\ &\text{or} \quad \left.\begin{aligned} \mathbf{n}^{\mathrm{T}}_{\partial\Omega_{ij}} \cdot D\mathbf{F}^{(i)}(t_{m\pm}) > 0 \\ \mathbf{n}^{\mathrm{T}}_{\partial\Omega_{ij}} \cdot D\mathbf{F}^{(j)}(t_{m\pm}) < 0 \end{aligned}\right\} \text{for } \mathbf{n}_{\partial\Omega_{ij}} \to \Omega_i. \end{aligned} \tag{2.110}$$

Proof. Following the proof procedures of Theorems 2.10 and 2.11, the above theorem can be easily proved. ■

In a similar manner, the $L_{\alpha\beta}$-functions of flows varying with $\mathbf{p}_{ij} \in \{\boldsymbol{\mu}_\alpha\}_{\alpha\in\{i,j\}}$ is used to discuss the switching of the non-passable flow from $\overline{\partial\Omega}_{\alpha\beta}$ to $\overrightarrow{\partial\Omega}_{\alpha\beta}$. The $L_{\alpha\beta}$-functions of a non-passable flow to the boundary is $L_{\alpha\beta} < 0$ with varying with boundary location. The $L_{\alpha\beta}$-function for flows between two points $\mathbf{x}_{m_1}$ and $\mathbf{x}_{m_2}$ on the sink boundary $\widetilde{\partial\Omega}_{\alpha\beta}$ are sketched in Fig. 2.11 for $\mathbf{p}_{\alpha\beta}$ between $\mathbf{p}^{(1)}_{\alpha\beta}$ and $\mathbf{p}^{(2)}_{\alpha\beta}$. The $L_{\alpha\beta}$-function is sketched in Fig. 2.11(a), and the corresponding vector fields varying with system parameters on the boundary $\partial\Omega_{\alpha\beta}$ are illustrated in Fig. 2.11(b). $\mathbf{F}^{(\alpha)}(t_{m-})$ and $\mathbf{F}^{(\beta)}(t_{m\pm})$ are limits of the vector fields to the boundary $\partial\Omega_{\alpha\beta}$ in domains Ω_α and Ω_β, respectively. The boundary relative to the non-passable flows with $L_{\alpha\beta} < 0$ is a sink boundary. There is a specific value $\mathbf{p}^{(cr)}_{\alpha\beta}$ between $\mathbf{p}^{(1)}_{\alpha\beta}$ and $\mathbf{p}^{(2)}_{\alpha\beta}$. For this specific value, a point $\mathbf{x}_m$ on the sink boundary can be found for the sliding fragmentation bifurcation on the boundary. Two points $\mathbf{x}_{k_1}$ and $\mathbf{x}_{k_2}$ are onset and vanishing points of the passable flow on the boundary $\partial\Omega_{\alpha\beta}$ for $\mathbf{p}_{\alpha\beta}$. The dashed and solid curves represent $L_{\alpha\beta} > 0$ and $L_{\alpha\beta} \leq 0$, respectively. For $\mathbf{p}_{\alpha\beta}$ varying from $\mathbf{p}_{\alpha\beta}^{(1)} \to \mathbf{p}_{\alpha\beta}^{(cr)}$, the $L_{\alpha\beta}$-functions of flows for $\mathbf{x} \in (\mathbf{x}_{m_1}, \mathbf{x}_{m_2})$ on the boundary is negative (i.e., $L_{\alpha\beta} < 0$). Therefore, the boundary $\partial\Omega_{\alpha\beta}$ is non-passable. For $\mathbf{p}_{\alpha\beta}$ varying from $\mathbf{p}_{\alpha\beta}^{(cr)} \to \mathbf{p}_{\alpha\beta}^{(2)}$, $L_{\alpha\beta} < 0$ are for $\mathbf{x} \in [\mathbf{x}_{m_1}, \mathbf{x}_{k_1}) \cup (\mathbf{x}_{k_2}, \mathbf{x}_{m_2}]$, and $L_{\alpha\beta} > 0$ are for $\mathbf{x} \in (\mathbf{x}_{k_1}, \mathbf{x}_{k_2})$. From Eq.(2.67), the flow for the portion of $\mathbf{x} \in (\mathbf{x}_{k_1}, \mathbf{x}_{k_2})$ the boundary with $L_{\alpha\beta} > 0$ is semi-passable. For $\mathbf{p}_{\alpha\beta}$ varying from $\mathbf{p}_{\alpha\beta}^{(1)} \to \mathbf{p}_{\alpha\beta}^{(2)}$, the point $(\mathbf{x}_m, \mathbf{p}_{\alpha\beta}^{(cr)})$ on the boundary $\partial\Omega_{\alpha\beta}$ is the onset point of the semi-passable flow on the boundary. The sliding flow on the boundary will be fragmentized. However, for $\mathbf{p}_{\alpha\beta}$ varying from $\mathbf{p}_{\alpha\beta}^{(2)} \to \mathbf{p}_{\alpha\beta}^{(1)}$, the sliding fragmentation disappears at such a point. At three critical points ($\mathbf{x}_m, \mathbf{x}_{k_1}, \mathbf{x}_{k_2}$), $L_{\alpha\beta} = 0$. The flow at the critical points $\{\mathbf{x}_{k_1}, \mathbf{x}_{k_2}\}$ have the same properties as at the critical point $\mathbf{x}_m$. If the two critical points have the different properties, the sliding flow between the two different critical points will be discussed later.

From the $L_{\alpha\beta}$-function of flows, the criteria for the sliding fragmentation bifurcation can be given as similar to Theorems 2.23, 2.25 and 2.27. Thus, the corresponding bifurcation conditions based on such the $L_{\alpha\beta}$-function are stated herein.

Theorem 2.29. *For a discontinuous dynamical system in Eq.(2.1), there is a point* $\mathbf{x}(t_m) = \mathbf{x}_m \in [\mathbf{x}_{m_1}, \mathbf{x}_{m_2}] \subset \widetilde{\partial\Omega}_{ij}$ *at time* t_m *between two adjacent domains* Ω_α ($\alpha = i, j$). *For an arbitrarily small* $\varepsilon > 0$, *there are two time intervals (i.e.,* $[t_{m-\varepsilon}, t_m)$ *and* $(t_m, t_{m+\varepsilon}]$), *and* $\mathbf{x}^{(i)}(t_{m-}) = \mathbf{x}_m = \mathbf{x}^{(j)}(t_{m\pm})$. *The flows* $\mathbf{x}^{(i)}(t)$ *and* $\mathbf{x}^{(j)}(t)$ *are* $C^{r_i}_{[t_{m-\varepsilon}, t_m)}$ *and* $C^{r_j}_{[t_{m-\varepsilon}, t_{m+\varepsilon}]}$*-continuous* ($r_\alpha \geq 2, \alpha = i, j$) *for time t, respectively. The sliding fragmentation bifurcation of the flow* $\mathbf{x}^{(i)}(t) \cup \mathbf{x}^{(j)}(t)$ *at point* $(\mathbf{x}_m, t_m)$ *on the boundary* $\widetilde{\partial\Omega}_{ij}$ *exists if and only if*

$$L_{ij}(\mathbf{x}_m, t_m, \mathbf{p}_i, \mathbf{p}_j) = 0, \tag{2.111}$$

$$\mathbf{n}_{\partial\Omega_{ij}}^{\mathrm{T}} \cdot \mathbf{F}^{(i)}(t_{m-}) \neq 0 \text{ and } L_{jj}(\mathbf{x}_{m\pm\varepsilon}, t_{m\pm\varepsilon}, \mathbf{p}_j) < 0. \tag{2.112}$$

Proof. Applying the L-function of flows in Definition 2.17 to Theorem 2.23, the foregoing theorem can be easily proved. ■

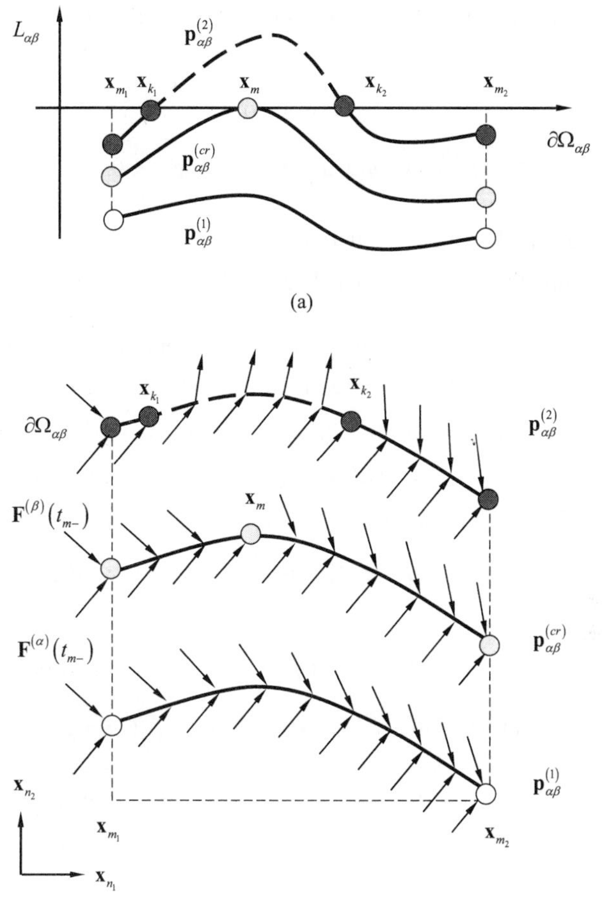

Fig. 2.11 (a) The L -function of flows ($L_{\alpha\beta}$) and **(b)** the vector fields between two points $\mathbf{x}_{m_1}$ and $\mathbf{x}_{m_2}$ on the boundary $\widetilde{\partial\Omega}_{\alpha\beta}$. The point $\mathbf{x}_m$ for $\mathbf{p}_{\alpha\beta}^{(cr)}$ is the critical point for the switching bifurcation. Two points $\mathbf{x}_{k_1}$ and $\mathbf{x}_{k_2}$ are the starting and vanishing of the passable flow on the boundary $\widetilde{\partial\Omega}_{\alpha\beta}$. The dashed and solid curves represent $L_{\alpha\beta} > 0$ and $L_{\alpha\beta} \le 0$, respectively. ($n_1 + n_2 = n$)

Theorem 2.30. *For a discontinuous dynamical system in Eq.(2.1), there is a point* $\mathbf{x}(t_m) = \mathbf{x}_m \in [\mathbf{x}_{m_1}, \mathbf{x}_{m_2}] \subset \widehat{\partial\Omega}_{ij}$ *at time* t_m *between two adjacent domains* Ω_α ($\alpha = i, j$). *For an arbitrarily small* $\varepsilon > 0$, *there are two time intervals (i.e.,* $[t_{m-\varepsilon}, t_m)$ *and* $(t_m, t_{m+\varepsilon}]$ *), and* $\mathbf{x}^{(i)}(t_{m\pm}) = \mathbf{x}_m = \mathbf{x}^{(j)}(t_{m+})$. *The flows* $\mathbf{x}^{(i)}(t)$ *and* $\mathbf{x}^{(j)}(t)$ *are*

$C^{r_i}_{[t_{m-\varepsilon},t_{m+\varepsilon}]}$ *and* $C^{r_j}_{(t_m,t_{m+\varepsilon}]}$*-continuous* ($r_\alpha \geq 2$, $\alpha = i, j$) *for time t, respectively. The source fragmentation bifurcation of the flow* $\mathbf{x}^{(i)}(t) \cup \mathbf{x}^{(j)}(t)$ *at point* $(\mathbf{x}_m, t_m)$ *on the boundary* $\widehat{\partial\Omega}_{ij}$ *occurs if and only if*

$$L_{ij}(\mathbf{x}_m, t_m, \mathbf{p}_i, \mathbf{p}_j) = 0, \tag{2.113}$$

$$\mathbf{n}^{\mathrm{T}}_{\partial\Omega_{ij}} \cdot \mathbf{F}^{(j)}(t_{m+}) \neq 0 \text{ and } L_{jj}(\mathbf{x}_{m\pm\varepsilon}, t_{m\pm\varepsilon}, \mathbf{p}_j) < 0. \tag{2.114}$$

Proof. Applying the L_{ij} and L_{jj}-functions of flows in Definition 2.17 to Theorem 2.25, the foregoing theorem can be easily proved. ■

Theorem 2.31. *For a discontinuous dynamical system in Eq.(2.1), there is a point* $\mathbf{x}(t_m) = \mathbf{x}_m \in [\mathbf{x}_{m_1}, \mathbf{x}_{m_2}] \subset \widetilde{\partial\Omega}_{ij}$ (*or* $\widehat{\partial\Omega}_{ij}$) *for time* t_m *between two adjacent domains* Ω_α ($\alpha = i, j$). *For an arbitrarily small* $\varepsilon > 0$, *there are two time intervals (i.e.,* $[t_{m-\varepsilon}, t_m)$ *and* $(t_m, t_{m+\varepsilon}]$), *and* $\mathbf{x}^{(i)}(t_{m\pm}) = \mathbf{x}_m = \mathbf{x}^{(j)}(t_{m\pm})$. *The flows* $\mathbf{x}^{(\alpha)}(t)$ ($\alpha = i, j$) *are* $C^{r_\alpha}_{[t_{m-\varepsilon},t_{m+\varepsilon}]}$*-continuous* ($r_\alpha \geq 2$) *for time t. The switching bifurcation of the flow at point* $(\mathbf{x}_m, t_m)$ *from* $\widetilde{\partial\Omega}_{ij}$ *to* $\widehat{\partial\Omega}_{ij}$ (*or* $\widehat{\partial\Omega}_{ij}$ *to* $\widetilde{\partial\Omega}_{ij}$) *occurs if and only if*

$$L_{ij}(\mathbf{x}_m, t_m, \mathbf{p}_i, \mathbf{p}_j) = 0, \tag{2.115}$$

$$\mathbf{n}^{\mathrm{T}}_{\partial\Omega_{ij}} \cdot \mathbf{F}^{(\alpha)}(t_{m\pm}) = 0 \text{ and } L_{\alpha\alpha}(\mathbf{x}_{m\pm\varepsilon}, t_{m\pm\varepsilon}, \mathbf{p}_\alpha) < 0 \text{ for } \alpha = i, j. \tag{2.116}$$

Proof. Applying the L_{ij} and $L_{\alpha\alpha}$-functions of flows in Definition 2.17 to Theorem 2.27, the foregoing theorem can be easily proved. ■

For the non-passable flow at $\mathbf{x}(t_m) = \mathbf{x}_m \in [\mathbf{x}_{m_1}, \mathbf{x}_{m_2}] \subset \widetilde{\partial\Omega}_{ij}$ (or $\widehat{\partial\Omega}_{ij}$), consider a time interval $[t_{m_1}, t_{m_2}]$ for $[\mathbf{x}_{m_1}, \mathbf{x}_{m_2}]$ on the boundary, for $t_m \in [t_{m_1}, t_{m_2}]$ and $\mathbf{x}_m \in [\mathbf{x}_{m_1}, \mathbf{x}_{m_2}]$, $L_{ij}(\mathbf{x}_m, t_m, \mathbf{p}_i, \mathbf{p}_j) < 0$. To determine the switching bifurcation, the local maximum of $L_{ij}(\mathbf{x}_m, t_m, \mathbf{p}_i, \mathbf{p}_j)$ can be determined. With Eqs.(2.78) and (2.79), the local maximum of $L_{ij}(\mathbf{x}_m, t_m, \mathbf{p}_i, \mathbf{p}_j)$ is determined by

$$D^k L_{ij}(\mathbf{x}_m, t_m, \mathbf{p}_i, \mathbf{p}_j) = 0 \quad (k = 0, 1, 2, \cdots, 2l-1), \tag{2.117}$$

$$D^{2l} L_{ij}(\mathbf{x}_m, t_m, \mathbf{p}_i, \mathbf{p}_j) < 0. \tag{2.118}$$

Definition 2.23. For a discontinuous dynamical system in Eq.(2.1), there is a point $\mathbf{x}(t_m) = \mathbf{x}_m \in [\mathbf{x}_{m_1}, \mathbf{x}_{m_2}] \subset \widetilde{\partial\Omega}_{ij}$ (or $\widehat{\partial\Omega}_{ij}$) at time t_m between two adjacent domains Ω_α ($\alpha = i, j$). For an arbitrarily small $\varepsilon > 0$, there are two time intervals (i.e.,

$[t_{m-\varepsilon}, t_m)$ and $(t_m, t_{m+\varepsilon}]$), and $\mathbf{x}^{(i)}(t_{m\pm}) = \mathbf{x}_m = \mathbf{x}^{(j)}(t_{m\mp})$. The flows $\mathbf{x}^{(i)}(t)$ and $\mathbf{x}^{(j)}(t)$ are $C^r_{[t_{m-\varepsilon}, t_{m+\varepsilon}]}$-continuous ($r \geq 2l$) for time *t*. The *local maximum set* of $L_{ij}(\mathbf{x}_m, t_m, \mathbf{p}_i, \mathbf{p}_j)$ is defined by

$$
{}_{\max}L_{ij}(t_m) = \left\{ L_{ij}(\mathbf{x}_m, t_m, \mathbf{p}_i, \mathbf{p}_j) \left| \begin{array}{l} \forall t_m \in [t_{m_1}, t_{m_2}], \exists \mathbf{x}_m \in [\mathbf{x}_{m_1}, \mathbf{x}_{m_2}], \\ \text{so that } D^k L_{ij}(\mathbf{x}_m, t_m, \mathbf{p}_i, \mathbf{p}_j) = 0 \\ \text{for } k = \{1, 2, \cdots, 2l-1\}, \text{and} \\ D^{2l} L_{ij}(\mathbf{x}_m, t_m, \mathbf{p}_i, \mathbf{p}_j) < 0. \end{array} \right. \right\} \tag{2.119}
$$

From the local maximum set of $L_{ij}(\mathbf{x}_m, t_m, \mathbf{p}_i, \mathbf{p}_j)$, the corresponding global maximum can be determined as follows.

Definition 2.24. For a discontinuous dynamical system in Eq.(2.1), there is a point $\mathbf{x}(t_m) = \mathbf{x}_m \in [\mathbf{x}_{m_1}, \mathbf{x}_{m_2}] \subset \widetilde{\partial\Omega}_{ij}$ (or $\widehat{\partial\Omega}_{ij}$) at time t_m between two adjacent domains Ω_α ($\alpha = i, j$). For an arbitrarily small $\varepsilon > 0$, there are two time intervals (i.e., $[t_{m-\varepsilon}, t_m)$ and $(t_m, t_{m+\varepsilon}]$), and $\mathbf{x}^{(i)}(t_{m\pm}) = \mathbf{x}_m = \mathbf{x}^{(j)}(t_{m\mp})$. The *global maximum set* of $L_{ij}(\mathbf{x}_m, t_m, \mathbf{p}_i, \mathbf{p}_j)$ is defined by

$$
{}_{G\max}L_{ij}(t_m) = \max_{t_m \in [t_{m_1}, t_{m_2}]} \left\{ \begin{array}{l} {}_{\max}L_{ij}(t_m), L_{ij}(\mathbf{x}_{m_1}, t_{m_1}, \mathbf{p}_i, \mathbf{p}_j), \\ L_{ij}(\mathbf{x}_{m_2}, t_{m_2}, \mathbf{p}_i, \mathbf{p}_j) \end{array} \right\}. \tag{2.120}
$$

From the foregoing definition, Theorems 2.23, 2.25 and 2.27 can be expressed through the global minimum of $L_{ij}(\mathbf{x}_m, t_m, \mathbf{p}_i, \mathbf{p}_j)$. So the following corollaries can be achieved, which give the condition of sliding fragmentation bifurcation.

Corollary 2.4. *For a discontinuous dynamical system in Eq.(2.1), there is a point* $\mathbf{x}(t_m) = \mathbf{x}_m \in [\mathbf{x}_{m_1}, \mathbf{x}_{m_2}] \subset \widetilde{\partial\Omega}_{ij}$ *at time* t_m *between two adjacent domains* Ω_α ($\alpha = i, j$). *For an arbitrarily small* $\varepsilon > 0$, *there are two time intervals (i.e.,* $[t_{m-\varepsilon}, t_m)$ *and* $(t_m, t_{m+\varepsilon}]$), *and* $\mathbf{x}^{(i)}(t_{m-}) = \mathbf{x}_m = \mathbf{x}^{(j)}(t_{m\pm})$. *The flows* $\mathbf{x}^{(i)}(t)$ *and* $\mathbf{x}^{(j)}(t)$ *are* $C^{r_i}_{[t_{m-\varepsilon}, t_m)}$ *and* $C^{r_j}_{[t_{m-\varepsilon}, t_{m+\varepsilon}]}$*-continuous* ($r_\alpha \geq 2l$, $\alpha = i, j$) *for time t, respectively. The sliding fragmentation bifurcation of the flow* $\mathbf{x}^{(i)}(t) \cup \mathbf{x}^{(j)}(t)$ *at point* $(\mathbf{x}_m, t_m)$ *on the boundary* $\widetilde{\partial\Omega}_{ij}$ *occurs if and only if*

$$
{}_{G\max}L_{ij}(t_m) = 0, \tag{2.121}
$$

$$
\mathbf{n}^{\mathrm{T}}_{\partial\Omega_{ij}} \cdot \mathbf{F}^{(i)}(t_{m-}) \neq 0 \;\; and \;\; L_{jj}(\mathbf{x}_{m\pm\varepsilon}, t_{m\pm\varepsilon}, \mathbf{p}_j) < 0. \tag{2.122}
$$

Proof. In Theorem 2.29, $L_{ij}(\mathbf{x}_m, t_m, \mathbf{p}_i, \mathbf{p}_j)$ replaced by its global maximum value ${}_{G\max}L_{ij}(t_m)$ gives the above corollary. ■

Corollary 2.5. *For a discontinuous dynamical system in Eq.*(2.1), *there is a point* $\mathbf{x}(t_m) = \mathbf{x}_m \in [\mathbf{x}_{m_1}, \mathbf{x}_{m_2}] \subset \widehat{\partial\Omega}_{ij}$ *at time* t_m *between two adjacent domains* Ω_α ($\alpha = i, j$). *For an arbitrarily small* $\varepsilon > 0$, *there are two time intervals (i.e.,* $[t_{m-\varepsilon}, t_m)$ *and* $(t_m, t_{m+\varepsilon}]$), *and* $\mathbf{x}^{(i)}(t_{m\pm}) = \mathbf{x}_m = \mathbf{x}^{(j)}(t_{m+})$. *The flows* $\mathbf{x}^{(i)}(t)$ *and* $\mathbf{x}^{(j)}(t)$ *are* $C^{r_i}_{[t_{m-\varepsilon}, t_{m+\varepsilon}]}$ *and* $C^{r_j}_{(t_m, t_{m+\varepsilon}]}$*-continuous* ($r_\alpha \geq 2l$, $\alpha = i, j$) *for time t, respectively. The source fragmentation bifurcation of the flow* $\mathbf{x}^{(i)}(t) \cup \mathbf{x}^{(j)}(t)$ *at point* $(\mathbf{x}_m, t_m)$ *on the boundary* $\widehat{\partial\Omega}_{ij}$ *occurs if and only if*

$$ {}_{G\max}L_{ij}(t_m) = 0, \tag{2.123}$$

$$\mathbf{n}^{\mathrm{T}}_{\partial\Omega_{ij}} \cdot \mathbf{F}^{(i)}(t_{m+}) \neq 0 \text{ and } L_{jj}(\mathbf{x}_{m\pm\varepsilon}, t_{m\pm\varepsilon}, \mathbf{p}_j) < 0. \tag{2.124}$$

Proof. In Theorem 2.30, $L_{ij}(\mathbf{x}_m, t_m, \mathbf{p}_i, \mathbf{p}_j)$ replaced by the global maximum ${}_{G\max}L_{ij}(t_m)$ gives the above corollary. ■

Corollary 2.6. *For a discontinuous dynamical system in Eq.*(2.1), *there is a point* $\mathbf{x}(t_m) = \mathbf{x}_m \in [\mathbf{x}_{m_1}, \mathbf{x}_{m_2}] \subset \widetilde{\partial\Omega}_{ij}$ *(or* $\widehat{\partial\Omega}_{ij}$*) at time* t_m *between two adjacent domains* Ω_α ($\alpha = i, j$). *For an arbitrarily small* $\varepsilon > 0$, *there are two time intervals (i.e.,* $[t_{m-\varepsilon}, t_m)$ *and* $(t_m, t_{m+\varepsilon}]$), *and* $\mathbf{x}^{(i)}(t_{m\pm}) = \mathbf{x}_m = \mathbf{x}^{(j)}(t_{m\pm})$. *The flows* $\mathbf{x}^{(\alpha)}(t)$ ($\alpha = i, j$) *are* $C^{r_\alpha}_{[t_{m-\varepsilon}, t_{m+\varepsilon}]}$ *-continuous* ($r_\alpha \geq 2l$) *for time t. The switching bifurcation of the flow at point* $(\mathbf{x}_m, t_m)$ *from* $\widetilde{\partial\Omega}_{ij}$ *to* $\widehat{\partial\Omega}_{ij}$ *(or* $\widehat{\partial\Omega}_{ij}$ *to* $\widetilde{\partial\Omega}_{ij}$*) occurs if and only if*

$$ {}_{G\max}L_{ij}(t_m) = 0, \tag{2.125}$$

$$\mathbf{n}^{\mathrm{T}}_{\partial\Omega_{ij}} \cdot \mathbf{F}^{(\alpha)}(t_{m\pm}) = 0 \text{ and } L_{\alpha\alpha}(\mathbf{x}_{m\pm\varepsilon}, t_{m\pm\varepsilon}, \mathbf{p}_\alpha) < 0 \text{ for } \alpha = i, j. \tag{2.126}$$

Proof. In Theorem 2.31, replacing $L_{ij}(\mathbf{x}_m, t_m, \boldsymbol{\mu}_i, \boldsymbol{\mu}_j)$ through its global maximum ${}_{G\max}L_{ij}(t_m)$ gives the above corollary.

2.7. An application: A frictional oscillator

To understand the above concepts, as in Luo and Gegg (2006a,b,c; 2007), consider a periodically forced oscillator consisting of a mass (m), a spring of stiffness (k)

and a damper of viscous damping coefficient (r), as shown in Fig. 2.12(a). The grazing and sliding flows for this problem were discussed. This oscillator also rests on the horizontal belt surface traveling with a constant speed (V). The absolute coordinate system (x,t) is for the mass. Consider a periodical force $Q_0 \cos\Omega t$ exerted on the mass where Q_0 and Ω are excitation amplitude and frequency, respectively. Since the mass contacts the moving belt with friction, the mass can move along or rest on the belt surface. Once the non-stick and stick motions exist, a kinetic friction force shown in Fig. 2.12(b) is described as

$$\bar{F}_f(\dot{x})\begin{cases} = \mu_k F_N, & \dot{x}\in[V,\infty) \\ \in[-\mu_k F_N, \mu_k F_N], & \dot{x}=V \\ = -\mu_k F_N, & \dot{x}\in(-\infty,V] \end{cases} \tag{2.127}$$

where $\dot{x} \triangleq dx/dt$, μ_k and F_N are friction coefficient and a normal force to the contact surface, respectively. For the model in Fig. 2.12, the normal force is $F_N = mg$ where g is the gravitational acceleration.

For the mass moving with the same speed of the belt surface, the non-friction forces per unit mass in the x-direction is defined as

$$F_s = A_0 \cos\Omega t - 2dV - cx, \quad \text{for } \dot{x}=V \tag{2.128}$$

where $A_0 = Q_0/m$, $d = r/2m$ and $c = k/m$. This force cannot overcome the friction force during the stick motion (i.e., $|F_s| \le F_f$ and $F_f = \mu_k F_N/m$). Thus, the mass does not have any relative motion to the belt. No acceleration exists, i.e.,

$$\ddot{x} = 0, \qquad \text{for } \dot{x}=V. \tag{2.129}$$

If $|F_s| > F_f$, the non-friction force will overcome the static friction force on the mass and the non-stick motion will appear. For the non-stick motion, the total force acting on the mass is

$$F = A_0 \cos\Omega t - F_f \,\mathrm{sgn}(\dot{x}-V) - 2d\dot{x} - cx, \quad \text{for } \dot{x}\ne V; \tag{2.130}$$

$\mathrm{sgn}(\cdot)$ is the sign function. Therefore, the equation of the non-stick motion for this oscillator with friction is

$$\ddot{x} + 2d\dot{x} + cx = A_0 \cos\Omega t - F_f \,\mathrm{sgn}(\dot{x}-V), \quad \text{for } \dot{x}\ne V. \tag{2.131}$$

Since the friction force is dependent on the direction of the relative velocity, the phase plane is partitioned into two regions in which the motion is described through two continuous dynamical systems, as shown in Fig. 2.13. The two regions are expressed by Ω_α ($\alpha \in \{1,2\}$). The following vectors are introduced as

$$\mathbf{x} \triangleq (x,\dot{x})^{\mathrm{T}} \equiv (x,y)^{\mathrm{T}} \text{ and } \mathbf{F} \triangleq (y,F)^{\mathrm{T}}. \tag{2.132}$$

The mathematical description of the regions and boundary is

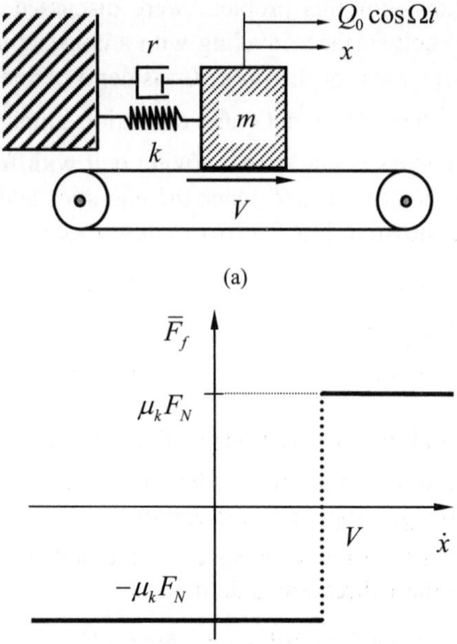

Fig. 2.12 (a) Schematic of mechanical model and **(b)** friction force.

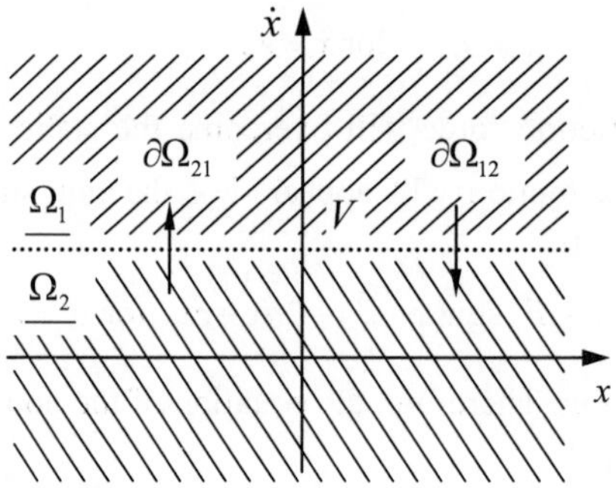

Fig. 2.13 Domain partitions in phase plane.

$$
\begin{aligned}
&\Omega_1 = \{(x,y) \mid y \in (V,\infty)\},\ \ \Omega_2 = \{(x,y) \mid y \in (-\infty,V)\},\\
&\partial\Omega_{12} = \{(x,y) \mid \varphi_{12}(x,y) \equiv y - V = 0\},\\
&\partial\Omega_{21} = \{(x,y) \mid \varphi_{21}(x,y) \equiv y - V = 0\}.
\end{aligned}
\tag{2.133}
$$

The boundary $\partial\Omega_{\alpha\beta}$ gives a motion from Ω_α to Ω_β ($\alpha,\beta \in \{1,2\}$ and $\alpha \neq \beta$). The equations of motion in Eqs.(2.129) and (2.130) can be described as

$$\dot{\mathbf{x}} = \mathbf{F}^{(\alpha)}(\mathbf{x},t) \text{ in } \Omega_\alpha \; (\alpha \in \{1,2\}), \tag{2.134}$$

where

$$\mathbf{F}^{(\alpha)}(\mathbf{x},t) = (y, F_\alpha(\mathbf{x},\Omega t))^{\mathrm{T}}, \tag{2.135}$$

$$F_\alpha(\mathbf{x},\Omega t) = A_0 \cos\Omega t - b_\alpha - 2d_\alpha y - c_\alpha x. \tag{2.136}$$

Note that $b_1 = \mu g$, $b_2 = -\mu g$, $d_\alpha = d$ and $c_\alpha = c$ for the model in Fig. 2.12.

2.7.1. Grazing phenomena

Since the boundary is a straight *line*, from Theorem 2.11, the grazing motion to the boundary is guaranteed by

$$\begin{aligned} &\mathbf{n}^{\mathrm{T}}_{\partial\Omega_{\alpha\beta}} \cdot \mathbf{F}^{(\alpha)}(\mathbf{x}_m, t_{m\pm}) = 0,\; \alpha \in \{1,2\} \\ &(-1)^\alpha \mathbf{n}^{\mathrm{T}}_{\partial\Omega_{\alpha\beta}} \cdot D\mathbf{F}^{(\alpha)}(\mathbf{x}_m, t_{m\pm}) < 0 \end{aligned} \tag{2.137}$$

where

$$D\mathbf{F}^{(\alpha)}(\mathbf{x},t) = (F_\alpha(\mathbf{x},t), \nabla F_\alpha(\mathbf{x},t)\cdot \mathbf{F}^{(\alpha)}(\mathbf{x},t) + \frac{\partial F_\alpha(\mathbf{x},t)}{\partial t})^{\mathrm{T}}. \tag{2.138}$$

where $\nabla = (\partial/\partial x, \partial/\partial y)^{\mathrm{T}}$ is the Hamilton operator. The time t_m represents the time for the motion on the velocity boundary. $t_{m\pm} = t_m \pm 0$ reflects the responses in the regions rather than on the boundary. Using the third and fourth equations of Eq.(2.133), the normal vector of the boundary are

$$\mathbf{n}_{\partial\Omega_{12}} = \mathbf{n}_{\partial\Omega_{21}} = (0,1)^{\mathrm{T}}. \tag{2.139}$$

Therefore,

$$\begin{aligned} \mathbf{n}^{\mathrm{T}}_{\partial\Omega_{\alpha\beta}} \cdot \mathbf{F}^{(\alpha)}(\mathbf{x},t) &= F_\alpha(\mathbf{x},\Omega t), \\ \mathbf{n}^{\mathrm{T}}_{\partial\Omega_{\alpha\beta}} \cdot D\mathbf{F}^{(\alpha)}(\mathbf{x},t) &= \nabla F_\alpha(\mathbf{x},\Omega t)\cdot \mathbf{F}^{(\alpha)}(\mathbf{x},t) + \frac{\partial F_\alpha(\mathbf{x},\Omega t)}{\partial t} = DF_\alpha. \end{aligned} \tag{2.140}$$

From Eqs.(2.139) and (2.140), the necessary and sufficient conditions for grazing motions are from Theorem 2.10.

$$F_\alpha(\mathbf{x}_m, \Omega t_{m\pm}) = 0,\; F_\alpha(\mathbf{x}_m, \Omega t_{m-\varepsilon}) \times F_\alpha(\mathbf{x}_m, \Omega t_{m+\varepsilon}) < 0. \tag{2.141}$$

or precisely,

$$\begin{gathered} F_\alpha(\mathbf{x}_m, \Omega t_{m\pm}) = 0,\ \alpha \in \{1,2\} \\ (-1)^\alpha F_\alpha(\mathbf{x}_m, \Omega t_{m-\varepsilon}) > 0 \text{ and } (-1)^\alpha F_\alpha(\mathbf{x}_m, \Omega t_{m+\varepsilon}) < 0. \end{gathered} \tag{2.142}$$

However, from Theorem 2.11, the necessary and sufficient conditions for grazing is given by

$$\begin{gathered} F_\alpha(\mathbf{x}_m, \Omega t_{m\pm}) = 0 \ \text{ for } \alpha \in \{1,2\} \\ (-1)^\alpha \left[\nabla F_\alpha(\mathbf{x}_m, \Omega t_{m\pm}) \cdot \mathbf{F}^{(\alpha)}(t_{m\pm}) + \frac{\partial F_\alpha(\mathbf{x}_m, \Omega t_{m\pm})}{\partial t} \right] < 0. \end{gathered} \tag{2.143}$$

A sketch of grazing motions in domain Ω_α ($\alpha = 1,2$) is illustrated in Fig. 2.14 (a) and (b). The grazing conditions are also presented, and the vector fields in Ω_1 and Ω_2 are expressed by the dashed and solid arrow-lines, respectively. The condition in Eq.(2.141) for grazing motions in Ω_α is presented by the vector fields of $\mathbf{F}^{(\alpha)}(t)$. In addition to $F_\alpha(\mathbf{x}_m, t_{m\pm}) = 0$, the condition requires $F_1(\mathbf{x}_m, \Omega t_{m-\varepsilon}) < 0$ and $F_1(\mathbf{x}_m, \Omega t_{m+\varepsilon}) > 0$ in domain Ω_1; and $F_2(\mathbf{x}_m, \Omega t_{m-\varepsilon}) > 0$ and $F_2(\mathbf{x}_m, \Omega t_{m+\varepsilon}) < 0$ in domain Ω_2. The detailed discussion can be refereed to Luo and Gegg (2006a, b).

Direct integration of Eq.(2.129) with initial condition (t_i, x_i, V) gives the sliding motion, i.e.,

$$x = V \times (t - t_i) + x_i. \tag{2.144}$$

Substitution of Eq.(2.144) into Eq.(2.136) gives the forces for a small δ neighborhood of the stick motion ($\delta \to 0$) in the two domains Ω_α ($\alpha \in \{1,2\}$), i.e.,

$$F_\alpha(\mathbf{x}_m, \Omega t_{m-}) = -2d_\alpha V - c_\alpha [V \times (t_m - t_i) + x_i] + A_0 \cos \Omega t_m - b_\alpha. \tag{2.145}$$

For the non-stick motion, select the initial condition on the velocity boundary (i.e., $\dot{x}_i = V$), and the basic solutions given in Luo and Gegg (2006a,b) will be used for construction of mappings.

In phase plane, the trajectories in Ω_α starting and ending at the velocity boundary (i.e., from $\partial\Omega_{\beta\alpha}$ to $\partial\Omega_{\alpha\beta}$) are illustrated in Fig. 2.15. The starting and ending points for mappings P_α in Ω_α are (x_i, V, t_i) and (x_{i+1}, V, t_{i+1}), respectively. The stick mapping is P_0. Define the switching planes as

$$\begin{aligned} \Xi^0 &= \{(x_i, \Omega t_i) \mid \dot{x}_i(t_i) = V\}, \\ \Xi^1 &= \{(x_i, \Omega t_i) \mid \dot{x}_i(t_i) = V^+\}, \\ \Xi^2 &= \{(x_i, \Omega t_i) \mid \dot{x}_i(t_i) = V^-\}; \end{aligned} \tag{2.146}$$

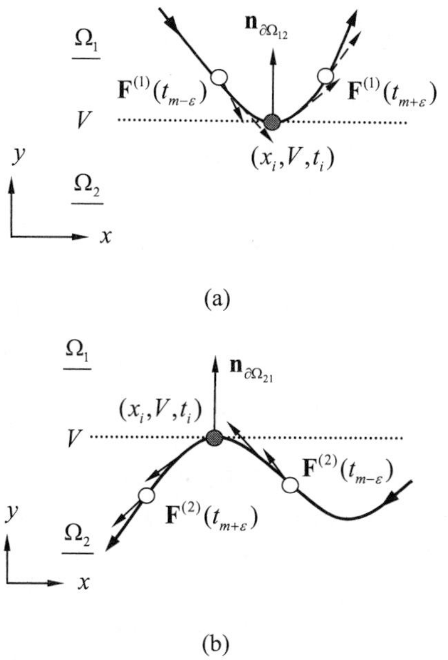

Fig. 2.14 Vector fields for grazing motion in Ω_α ($\alpha = 1,2$) and $V > 0$.

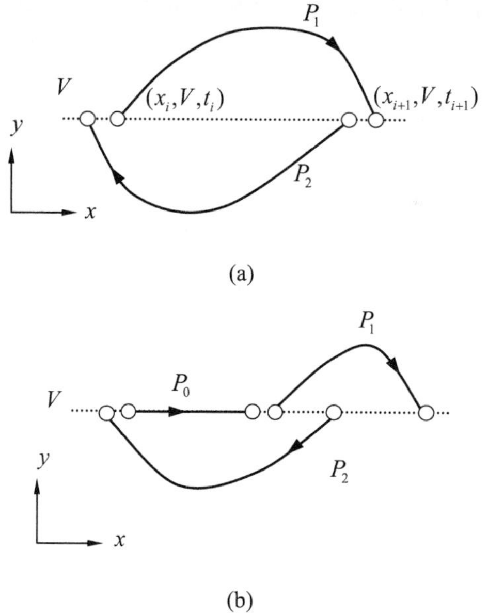

Fig. 2.15 Regular mappings (P_1 and P_2) and stick mapping P_0 .

where $V^{-}=\lim_{\delta\to 0}(V-\delta)$ and $V^{+}=\lim_{\delta\to 0}(V+\delta)$ for arbitrarily small $\delta>0$. Thus,

$$P_1:\Xi^1\to\Xi^1,\ P_2:\Xi^2\to\Xi^2,\ P_0:\Xi^0\to\Xi^0. \tag{2.147}$$

From the foregoing two equations,

$$\begin{aligned}&P_0:(x_i,V,t_i)\to(x_{i+1},V,t_{i+1}),\\&P_1:(x_i,V^{+},t_i)\to(x_{i+1},V^{+},t_{i+1}),\\&P_2:(x_i,V^{-},t_i)\to(x_{i+1},V^{-},t_{i+1}).\end{aligned} \tag{2.148}$$

The governing equations for P_0 with $\alpha\in\{1,2\}$ are

$$\begin{aligned}-x_{i+1}+V\times(t_{i+1}-t_i)+x_i&=0,\\2d_\alpha V+c_\alpha[V\times(t_{i+1}-t_i)+x_i]-A_0\cos\Omega t_{i+1}+b_\alpha&=0.\end{aligned} \tag{2.149}$$

The mapping P_0 described the starting and ending of the stick motion, the disappearance of stick motion requires $F_\alpha(\mathbf{x}_{i+1},\Omega t_{i+1})=0$. This section will not use the stick mapping to discuss the grazing flow, which is presented herein as a generic mapping. For sliding motion, the stick mapping will be used in next section. From this problem, the two domains Ω_α ($\alpha=1$ or 2) are unbounded. The flows of the dynamical systems on the corresponding domains should be bounded from Hypothesis (H2.1)-(H2.3) in discontinuous dynamical systems. Therefore, for the non-stick motion of the oscillator, there are three possible stable motions in two domains Ω_α ($\alpha\in\{1,2\}$), the governing equations of mapping P_α ($\alpha\in\{1,2\}$) are obtained from displacement and velocity responses for the three cases of motions in Luo and Gegg (2006a,b). Thus, the governing equations of mapping P_α ($\alpha\in\{0,1,2\}$) can be expressed by

$$\begin{aligned}&f_1^{(\alpha)}(x_i,\Omega t_i,x_{i+1},\Omega t_{i+1})=0;\\&f_2^{(\alpha)}(x_i,\Omega t_i,x_{i+1},\Omega t_{i+1})=0.\end{aligned} \tag{2.150}$$

If the grazing for the two non-stick mappings occurs at the final state (x_{i+1},V,t_{i+1}), from Eq.(2.143), the grazing conditions based on mappings are obtained, i.e.,

$$\begin{aligned}&F_\alpha(x_{i+1},V,\Omega t_{i+1})=0,\ \alpha\in\{1,2\},\\&(-1)^\alpha\left[\nabla F_\alpha(\mathbf{x}_{i+1},\Omega t_{i+1})\cdot\mathbf{F}^{(\alpha)}(t_{i+1})+\frac{\partial F_\alpha(\mathbf{x}_{i+1},\Omega t_{i+1})}{\partial t}\right]<0.\end{aligned} \tag{2.151}$$

With Eq.(2.138), the grazing condition becomes

$$\left.\begin{aligned}&A_0\cos\Omega t_{i+1}-b_\alpha-2d_\alpha V-c_\alpha x_{i+1}=0,\\&(-1)^\alpha\left[-c_\alpha V-A_0\Omega\sin\Omega t_{i+1}\right]<0.\end{aligned}\right\}\text{for }\alpha\in\{1,2\} \tag{2.152}$$

To ensure the initial switching sets to be passable, the passable motion requires that the initial switching sets of mapping P_α ($\alpha \in \{1,2\}$) should satisfy the following conditions (also see, Luo and Gegg, 2006a,b)

$$\begin{aligned} &F_1(x_i, V^-, \Omega t_i) < 0 \text{ and } F_2(x_i, V^-, \Omega t_i) < 0 \text{ for } \Omega_1 \to \Omega_2; \\ &F_1(x_i, V^+, \Omega t_i) > 0 \text{ and } F_2(x_i, V^-, \Omega t_i) > 0 \text{ for } \Omega_2 \to \Omega_1. \end{aligned} \tag{2.153}$$

To make motions of mappings P_α ($\alpha = 1,2$) exist, the initial switching force product $F_1 \times F_2$ at the boundary should be non-negative. The comprehensive discussion of the foregoing condition can be referred to Luo and Gegg (2006c). The condition of Eq.(2.153) guarantees the motion relative to the initial switching sets of mapping P_α ($\alpha \in \{1,2\}$) is passable on the discontinuous boundary (i.e., $y_i = V$). The force product for the initial switching sets is also illustrated to ensure the non-stick mapping exists. The force conditions for the final switching sets of mapping P_α ($\alpha \in \{1,2\}$) are presented in Eq.(2.141). The equivalent grazing conditions based on Eq.(2.143) give the inequality condition in Eq.(2.152).

To illustrate the analytical prediction of grazing motions, the motion of the oscillator will be demonstrated through time-history responses and trajectories in phase plane. The grazing strongly depends on the force responses in this discontinuous dynamical system. The force responses will be presented to illustrate the force criteria for grazing motions in such a friction-induced oscillator. The starting and grazing points of mapping P_α ($\alpha \in \{1,2\}$) are represented by large, hollow and dark-solid circular symbols, respectively. The switching points from domain Ω_α to Ω_β ($\alpha, \beta \in \{1,2\}, \alpha \neq \beta$) are depicted by smaller circular symbols. In Fig. 2.16, phase trajectories, force distributions along displacement, velocity time-history and force distributions on velocity are presented for the grazing motion of mapping P_1. The parameters ($\Omega = 8$, $V = 1$, $b_1 = -b_2 = 3$) plus the initial conditions $(x_i, y_i) = (-1,1)$ and ($\Omega t_i \approx 1.3617$, 1.6958, 1.4830) corresponding to (A_0 = 15, 18, 21) are adopted. In phase plane, the three grazing trajectories are tangential to the boundary (i.e., $y = V$), as shown in Fig. 2.16(a). In Fig. 2.16(b), the thick and thin solid curves represent the forces $F_1(t)$ and $F_2(t)$, respectively. From the force distribution along displacement, the force $F_1(t)$ has a sign change from negative to positive. This indicates that the grazing conditions in Eq.(2.141) are satisfied. The forces $F_1(t)$ and $F_2(t)$ at the switching points from domain Ω_1 to Ω_2 have a jump with the same sign, which satisfies Eq.(2.142). In the velocity time-history plot, the velocity curves are tangential to the velocity separation boundary (see Fig. 2.16(c)). Finally, the forces distributions along velocity are presented in Fig. 2.16(d). The force $F_1(t)$ at the grazing points is zero. The force jump from domain Ω_1 to Ω_2 is observed as well.

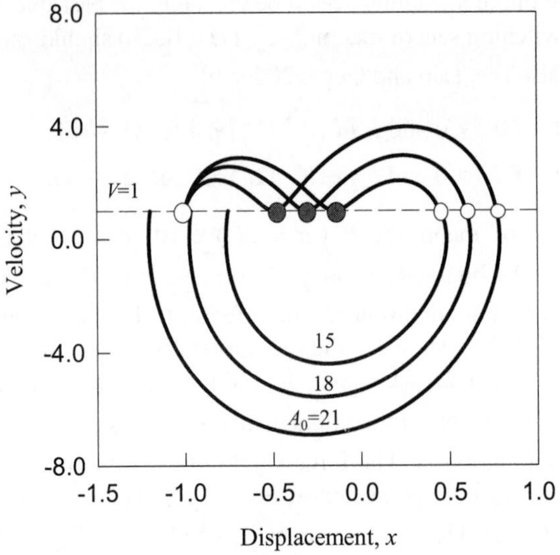

(a)

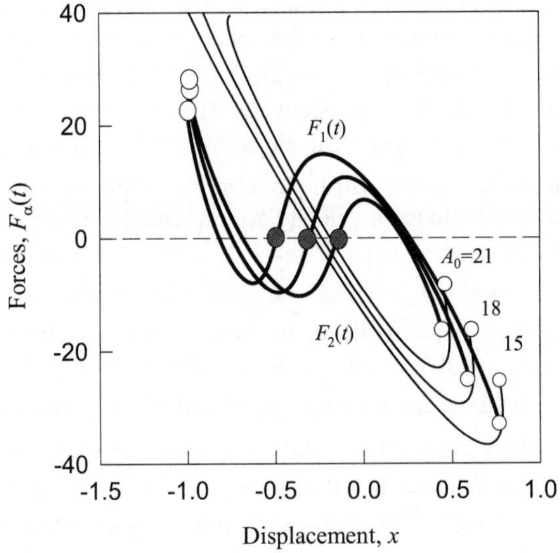

(b)

Fig. 2.16 Grazing motion of mapping P_1 for $A_0 = 15, 18, 21$: **(a)** phase trajectory, **(b)** forces distribution along displacement, **(c)** velocity-time history and **(d)** force distribution on velocity. ($\Omega = 8,\ V = 1,\ d_1 = 1,\ d_2 = 0,\ b_1 = -b_2 = 3,\ c_1 = c_2 = 30$). The initial conditions are $(x_i, y_i) =$ (−1.,1.) and $\Omega t_i \approx 1.3617, 1.6958, 1.4830$, accordingly.

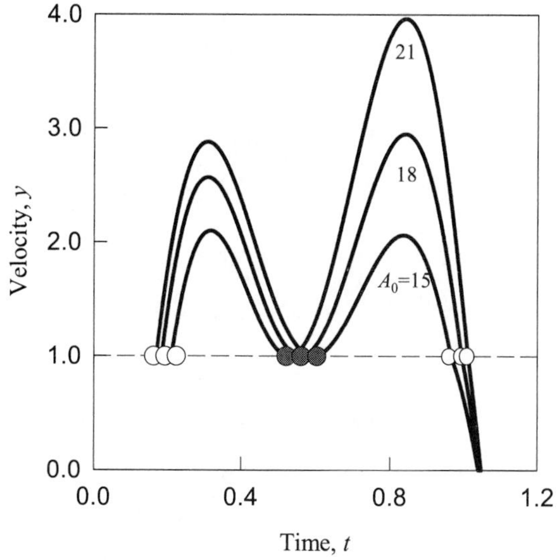

(c)

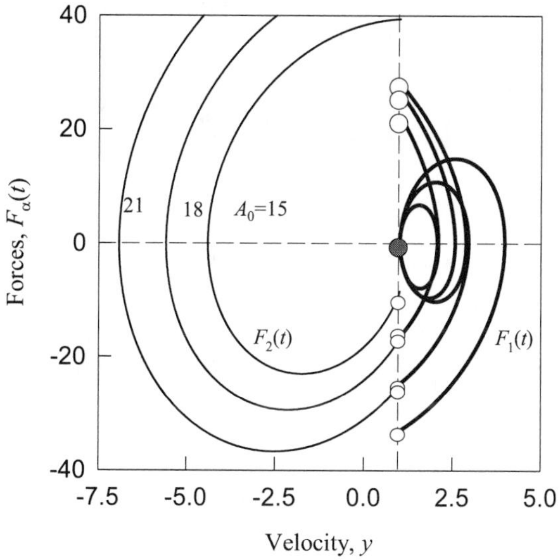

(d)

Fig. 2.16 Continued.

2.7.2. Sliding motion

For sliding motion, the equations of motion in Eqs.(2.129) and (2.131) can be described as

$$\dot{\mathbf{x}} = \mathbf{F}^{(\lambda)}(\mathbf{x},t), \lambda \in \{0,\alpha\} \tag{2.154}$$

where

$$\left.\begin{aligned}
&\mathbf{F}^{(\alpha)}(\mathbf{x},t) = (y, F_\alpha(\mathbf{x},\Omega t))^{\mathrm{T}} \ \text{ in } \Omega_\alpha \ (\alpha \in \{1,2\});\\
&\mathbf{F}^{(0)}_{\alpha\beta}(\mathbf{x},t) = (V,0)^{\mathrm{T}} \ \text{ on } \widetilde{\partial\Omega}_{\alpha\beta} \ (\alpha,\beta \in \{1,2\}, \alpha \neq \beta),\\
&\mathbf{F}^{(0)}_{\alpha\beta}(\mathbf{x},t) \in [\mathbf{F}^{(\alpha)}(\mathbf{x},t), \mathbf{F}^{(\beta)}(\mathbf{x},t)] \text{ on } \overrightarrow{\partial\Omega}_{\alpha\beta}.
\end{aligned}\right\} \tag{2.155}$$

and $F_\alpha(\mathbf{x},\Omega t)$ is given in Eq.(2.136). For $\widetilde{\partial\Omega}_{\alpha\beta}$, the critical initial and final states of a sliding motion are $(\Omega t_c, x_c, V)$ and $(\Omega t_f, x_f, V)$. Consider a sliding motion starting at $(\Omega t_i, x_i, V)$ and ending at $(\Omega t_{i+1}, x_{i+1}, V) \triangleq (\Omega t_f, x_f, V)$. From Theorem 2.4, the sliding motion on the boundary is guaranteed for $t_m \in [t_i, t_{i+1}) \subseteq [t_c, t_f)$ by

$$\begin{aligned}
&\text{either} \quad \left.\begin{aligned}
&\mathbf{n}^{\mathrm{T}}_{\partial\Omega_{\alpha\beta}} \cdot \mathbf{F}^{(\alpha)}(\mathbf{x}_m, t_{m-}) < 0\\
&\mathbf{n}^{\mathrm{T}}_{\partial\Omega_{\alpha\beta}} \cdot \mathbf{F}^{(\beta)}(\mathbf{x}_m, t_{m-}) > 0
\end{aligned}\right\} \text{ for } \mathbf{n}_{\partial\Omega_{\alpha\beta}} \to \Omega_\alpha,\\
&\text{or} \quad \left.\begin{aligned}
&\mathbf{n}^{\mathrm{T}}_{\partial\Omega_{\alpha\beta}} \cdot \mathbf{F}^{(\alpha)}(\mathbf{x}_m, t_{m-}) > 0\\
&\mathbf{n}^{\mathrm{T}}_{\partial\Omega_{\alpha\beta}} \cdot \mathbf{F}^{(\beta)}(\mathbf{x}_m, t_{m-}) < 0
\end{aligned}\right\} \text{ for } \mathbf{n}_{\partial\Omega_{\alpha\beta}} \to \Omega_\beta.
\end{aligned} \tag{2.156}$$

In other words, one obtains

$$[\mathbf{n}^{\mathrm{T}}_{\partial\Omega_{\alpha\beta}} \cdot \mathbf{F}^{(\alpha)}(\mathbf{x}_m, t_{m-})] \times [\mathbf{n}^{\mathrm{T}}_{\partial\Omega_{\alpha\beta}} \bullet \mathbf{F}^{(\beta)}(\mathbf{x}_m, t_{m-})] < 0. \tag{2.157}$$

For a boundary $\overrightarrow{\partial\Omega}_{\alpha\beta}$ with non-zero measure, the starting and ending states of the passable motion are $(\Omega t_s, x_s, V)$ and $(\Omega t_e, x_e, V)$ accordingly. From Theorem 2.2, the non-sliding motion (or called passable motion to the boundary in Luo (2005, 2006)) is guaranteed for $t_m \in (t_s, t_e)$ by

$$\begin{aligned}
&\text{either} \quad \left.\begin{aligned}
&\mathbf{n}^{\mathrm{T}}_{\partial\Omega_{\alpha\beta}} \cdot \mathbf{F}^{(\alpha)}(\mathbf{x}_m, t_{m-}) < 0\\
&\mathbf{n}^{\mathrm{T}}_{\partial\Omega_{\alpha\beta}} \cdot \mathbf{F}^{(\beta)}(\mathbf{x}_m, t_{m+}) < 0
\end{aligned}\right\} \text{ for } \mathbf{n}_{\partial\Omega_{\alpha\beta}} \to \Omega_\alpha,\\
&\text{or} \quad \left.\begin{aligned}
&\mathbf{n}^{\mathrm{T}}_{\partial\Omega_{\alpha\beta}} \cdot \mathbf{F}^{(\alpha)}(\mathbf{x}_m, t_{m-}) > 0\\
&\mathbf{n}^{\mathrm{T}}_{\partial\Omega_{\alpha\beta}} \cdot \mathbf{F}^{(\beta)}(\mathbf{x}_m, t_{m+}) > 0
\end{aligned}\right\} \text{ for } \mathbf{n}_{\partial\Omega_{\alpha\beta}} \to \Omega_\beta.
\end{aligned} \tag{2.158}$$

In other words, one gets

$$[\mathbf{n}^{\mathrm{T}}_{\partial\Omega_{\alpha\beta}} \cdot \mathbf{F}^{(\alpha)}(\mathbf{x}_m, t_{m-})] \times [\mathbf{n}^{\mathrm{T}}_{\partial\Omega_{\alpha\beta}} \cdot \mathbf{F}^{(\beta)}(\mathbf{x}_m, t_{m+})] > 0. \tag{2.159}$$

Let $t_e = t_c$, and $t_f = t_s$ for a flow switching from $\overrightarrow{\partial\Omega}_{\alpha\beta}$ to $\widetilde{\partial\Omega}_{\alpha\beta}$. Suppose the starting state is a switching state of the sliding motion. From Theorem 2.15, the switching condition of the sliding motion from $\overrightarrow{\partial\Omega}_{\alpha\beta}$ to $\widetilde{\partial\Omega}_{\alpha\beta}$ at $t_m = t_c$ is

$$\left.\begin{aligned}
&\text{either } \mathbf{n}^{\mathrm{T}}_{\partial\Omega_{\alpha\beta}} \cdot \mathbf{F}^{(\alpha)}(\mathbf{x}_m, t_{m-}) < 0, \text{ for } \mathbf{n}_{\partial\Omega_{\alpha\beta}} \to \Omega_\alpha,\\
&\text{or } \quad \mathbf{n}^{\mathrm{T}}_{\partial\Omega_{\alpha\beta}} \cdot \mathbf{F}^{(\alpha)}(\mathbf{x}_m, t_{m-}) > 0, \text{ for } \mathbf{n}_{\partial\Omega_{\alpha\beta}} \to \Omega_\beta;\\
&\mathbf{n}^{\mathrm{T}}_{\partial\Omega_{\alpha\beta}} \cdot \mathbf{F}^{(\beta)}(\mathbf{x}_m, t_{m\pm}) = 0;\\
&\mathbf{n}^{\mathrm{T}}_{\partial\Omega_{\alpha\beta}} \cdot D\mathbf{F}^{(\beta)}(\mathbf{x}_m, t_{m\pm}) > 0 \ \text{ for } \mathbf{n}_{\partial\Omega_{\alpha\beta}} \to \Omega_\beta,\\
&\mathbf{n}^{\mathrm{T}}_{\partial\Omega_{\alpha\beta}} \cdot D\mathbf{F}^{(\beta)}(\mathbf{x}_m, t_{m\pm}) < 0 \ \text{ for } \mathbf{n}_{\partial\Omega_{\alpha\beta}} \to \Omega_\alpha.
\end{aligned}\right\} \tag{2.160}$$

From Theorem 2.24, the sliding motion vanishing from $\widetilde{\partial\Omega}_{\alpha\beta}$ to $\overrightarrow{\partial\Omega}_{\alpha\beta}$ and going into the domain Ω_β at $t_m = t_f$ requires

$$\left.\begin{aligned}
&\text{either } \mathbf{n}^{\mathrm{T}}_{\partial\Omega_{\alpha\beta}} \cdot \mathbf{F}^{(\alpha)}(\mathbf{x}_m, t_{m-}) < 0, \text{ for } \mathbf{n}_{\partial\Omega_{\alpha\beta}} \to \Omega_\alpha,\\
&\text{or } \quad \mathbf{n}^{\mathrm{T}}_{\partial\Omega_{\alpha\beta}} \cdot \mathbf{F}^{(\alpha)}(\mathbf{x}_m, t_{m-}) > 0, \text{ for } \mathbf{n}_{\partial\Omega_{\alpha\beta}} \to \Omega_\beta;\\
&\mathbf{n}^{\mathrm{T}}_{\partial\Omega_{\alpha\beta}} \cdot \mathbf{F}^{(\beta)}(\mathbf{x}_m, t_{m\mp}) = 0;\\
&\mathbf{n}^{\mathrm{T}}_{\partial\Omega_{\alpha\beta}} \cdot D\mathbf{F}^{(\beta)}(\mathbf{x}_m, t_{m\mp}) > 0 \ \text{ for } \mathbf{n}_{\partial\Omega_{\alpha\beta}} \to \Omega_\beta,\\
&\mathbf{n}^{\mathrm{T}}_{\partial\Omega_{\alpha\beta}} \cdot D\mathbf{F}^{(\beta)}(\mathbf{x}_m, t_{m\mp}) < 0 \ \text{ for } \mathbf{n}_{\partial\Omega_{\alpha\beta}} \to \Omega_\alpha.
\end{aligned}\right\} \tag{2.161}$$

From Eqs.(2.160) and (2.161), the switching conditions for the sliding motion to the passable motion and vice versa are summarized as

$$\left.\begin{aligned}
&[\mathbf{n}^{\mathrm{T}}_{\partial\Omega_{\alpha\beta}} \cdot \mathbf{F}^{(\alpha)}(\mathbf{x}_m, t_{m-})] \times [\mathbf{n}^{\mathrm{T}}_{\partial\Omega_{\alpha\beta}} \cdot \mathbf{F}^{(\beta)}(\mathbf{x}_m, t_{m\pm})] = 0.\\
&\mathbf{n}^{\mathrm{T}}_{\partial\Omega_{\alpha\beta}} \cdot D\mathbf{F}^{(\beta)}(\mathbf{x}_m, t_{m\pm}) > 0 \ \text{ for } \mathbf{n}_{\partial\Omega_{\alpha\beta}} \to \Omega_\beta,\\
&\mathbf{n}^{\mathrm{T}}_{\partial\Omega_{\alpha\beta}} \cdot D\mathbf{F}^{(\beta)}(\mathbf{x}_m, t_{m\pm}) < 0 \ \text{ for } \mathbf{n}_{\partial\Omega_{\alpha\beta}} \to \Omega_\alpha.
\end{aligned}\right\} \tag{2.162}$$

The normal vector of the boundary $\partial\Omega_{12}$ and $\partial\Omega_{21}$ is given in Eq.(2.139). Therefore, the dot product of the vector field and the normal vector of the boundary is

$$\mathbf{n}^{\mathrm{T}}_{\partial\Omega_{\alpha\beta}} \cdot \mathbf{F}^{(\alpha)}(\mathbf{x}, t) = \mathbf{n}^{\mathrm{T}}_{\partial\Omega_{\beta\alpha}} \cdot \mathbf{F}^{(\alpha)}(\mathbf{x}, t) = F_\alpha(\mathbf{x}, \Omega t). \tag{2.163}$$

which is the force for the oscillator in Eq.(2.154). For a better understanding of the force characteristic of the sliding motion, the conditions for sliding and non-sliding motions in Eqs.(2.156) and (2.158) can be rewritten by

$$F_1(\mathbf{x}_m, \Omega t_{m-}) < 0 \text{ and } F_2(\mathbf{x}_m, \Omega t_{m-}) > 0 \ \text{ for } \widetilde{\partial\Omega}_{12}; \tag{2.164}$$

and

$$\begin{aligned}&F_1(\mathbf{x}_m,\Omega t_{m-})<0\ \text{ and }\ F_2(\mathbf{x}_m,\Omega t_{m+})<0\ \text{ for } \overrightarrow{\partial\Omega_{12}};\\&F_2(\mathbf{x}_m,\Omega t_{m-})>0\ \text{ and }\ F_1(\mathbf{x}_m,\Omega t_{m+})>0\ \text{ for } \overrightarrow{\partial\Omega_{21}}.\end{aligned}\tag{2.165}$$

Equations (2.160) and (2.161), respectively, give the switching conditions for the sliding motion in Eq.(2.156), i.e.,

$$\begin{aligned}&F_1(\mathbf{x}_m,\Omega t_{m-})<0, F_2(\mathbf{x}_m,\Omega t_{m\pm})=0 \text{ and } DF_2(\mathbf{x}_m,t_{m\pm})<0\\&\quad\text{for } \overrightarrow{\partial\Omega_{12}}\to\widetilde{\partial\Omega_{12}},\\&F_2(\mathbf{x}_m,\Omega t_{m-})>0,\ F_1(\mathbf{x}_m,\Omega t_{m\pm})=0 \text{ and } DF_1(\mathbf{x}_m,t_{m\pm})>0\\&\quad\text{for } \overrightarrow{\partial\Omega_{21}}\to\widetilde{\partial\Omega_{21}}\end{aligned}\tag{2.166}$$

at $t_m = t_c$, and

$$\begin{aligned}&F_1(\mathbf{x}_m,\Omega t_{m-})<0,\ \ F_2(\mathbf{x}_m,\Omega t_{m\mp})=0 \text{ and } DF_2(\mathbf{x}_m,t_{m\mp})<0\\&\quad\text{for } \widetilde{\partial\Omega_{12}}\to\Omega_2,\\&F_2(\mathbf{x}_m,\Omega t_{m-})>0,\ \ F_1(\mathbf{x}_m,\Omega t_{m\mp})=0 \text{ and } DF_1(\mathbf{x}_m,t_{m\mp})>0\\&\quad\text{for } \widetilde{\partial\Omega_{12}}\to\Omega_1\end{aligned}\tag{2.167}$$

at $t_m = t_f$.

A sketch of the vector field and classification of the sliding motion is illustrated in Fig. 2.17. In Fig. 2.17(a), the switching condition for sliding motions is presented through the vector fields of $\mathbf{F}^{(1)}(\mathbf{x},t)$ and $\mathbf{F}^{(2)}(\mathbf{x},t)$. The ending condition for sliding (or stick) motion along the velocity boundary is illustrated by $F_2(t_{m-})=0$. However, the starting point of the sliding motion may not be switching points from the passable motion boundary to the sliding motion boundary. When the flow arrives to the boundary and Eq.(2.164) holds, the sliding motion along the boundary will be formed. From the switching conditions in Eqs.(2.166) and (2.167), there are four possible sliding motions (I)-(IV), as shown in Fig. 2.17(b). The switching condition in Eq.(2.166) is the critical condition for formation of the sliding motion. The above switching conditions for onset, forming and vanishing of the sliding motions along the boundary at a certain velocity are strongly dependent on the total force acting on the oscillator. Because the total force can be contributed from the linear or nonlinear continuous forces from spring and/or damper, such switching conditions can be applied to dynamical systems possessing nonlinear, continuous spring and viscous damping forces with a nonlinear friction with a C^0-discontinuity.

The three generic mappings in Eq.(2.147) are governed by algebraic equations in Eq.(2.148). For non-sliding mapping, two equations are given. Two equations with four unknowns in Eq.(2.150) cannot give unique solutions for the sliding motion. Once an initial state is given, the final state of the sliding motion is uniquely determined. Consider two switching states $(\Omega t_c, x_c, V)$ and $(\Omega t_f, x_f, V)$ as the critical, initial and final conditions of the sliding motion, respectively. For a time interval $[t_i, t_{i+1}] \subseteq [t_c, t_f]$ of any sliding motion, the sliding motion requires $(\Omega t_{i+1}, x_{i+1}, V) \triangleq (\Omega t_f, x_f, V)$. From Eqs.(2.163) and (2.166), the initial condition of the sliding motion for $t_i \in (t_c, t_f)$ and on the boundary $\widetilde{\partial\Omega_{\alpha\beta}}$ satisfies

$$L_{12}(t_i) = F_1(x_i, V, \Omega t_i) \times F_2(x_i, V, \Omega t_i) < 0. \tag{2.168}$$

The foregoing equation is the force product from the L-function in Eq.(2.156). The switching condition in Eqs.(2.162) or (2.168) for the sliding motion at the critical time $t_i = t_c$ also gives the *initial force product condition*, i.e.,

$$\begin{aligned} &L_{12}(t_c) = F_1(x_c, V, \Omega t_c) \times F_2(x_c, V, \Omega t_c) = 0, \\ &\left.\begin{aligned} &DF_2(\mathbf{x}_c, t_{c\pm}) < 0 \quad \text{for } \overrightarrow{\partial\Omega_{12}} \to \widetilde{\partial\Omega_{12}}, \\ &DF_1(\mathbf{x}_c, t_{c\pm}) > 0 \quad \text{for } \overrightarrow{\partial\Omega_{21}} \to \widetilde{\partial\Omega_{21}}. \end{aligned}\right\} \end{aligned} \tag{2.169}$$

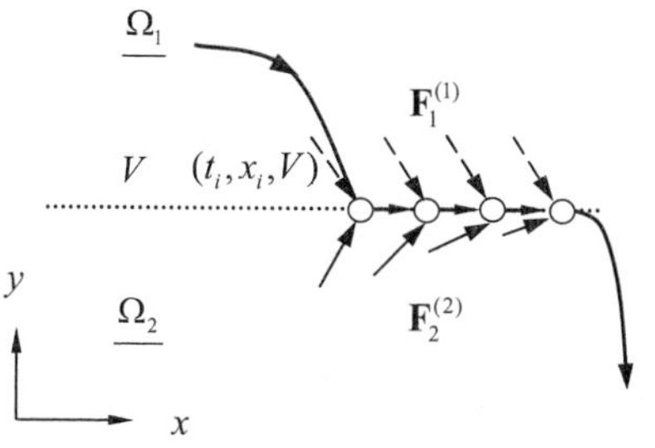

(a)

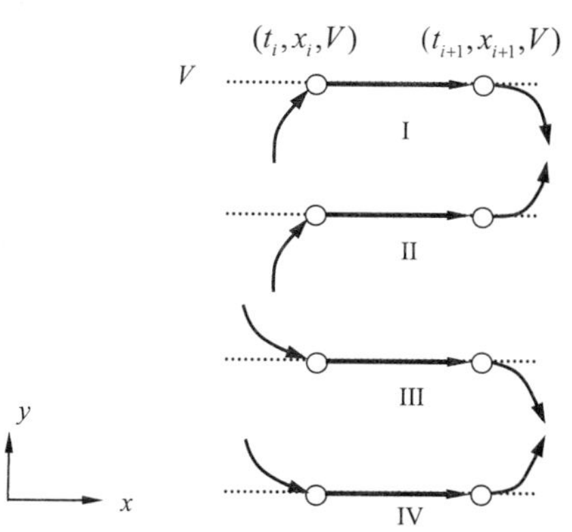

(b)

Fig. 2.17 (a) The vector field and **(b)** classification of sliding motions for belt speed $V > 0$.

If the initial condition satisfies Eq.(2.169), the sliding motion is called the critical sliding motion. The condition in Eqs.(2.161) or (2.167) for the sliding motion at the time $t_{i+1} = t_f$ gives the *final force product condition*, i.e.,

$$\begin{aligned} &L_{12}\left(t_{i+1}\right) = F_1\left(x_{i+1}, V, \Omega t_{i+1}\right) \times F_2\left(x_{i+1}, V, \Omega t_{i+1}\right) = 0.\\ &DF_2(\mathbf{x}_{i+1}, t_{(i+1)\mp}) < 0 \quad \text{for } \widetilde{\partial\Omega_{12}} \to \Omega_2 \text{ or}\\ &DF_1(\mathbf{x}_{i+1}, t_{(i+1)\mp}) > 0 \quad \text{for } \widetilde{\partial\Omega_{12}} \to \Omega_1 \end{aligned} \tag{2.170}$$

From Eq.(2.170), it is observed that the force product on the boundary is also very significant for the friction oscillator. Once the force product of the sliding motion for $t_m \in (t_i, t_{i+1})$ changes its sign, the sliding motion for the friction induced oscillator will vanish. So the characteristic of the force product for the sliding motion should be further discussed. To explain the force mechanism of the sliding motion, the force product for sliding motion starting at $(\Omega t_i, x_i, V)$ and ending at $(\Omega t_{i+1}, x_{i+1}, V)$ is sketched in Fig. 2.18 for given parameters with the belt speed $V > 0$. x_i and x_{i+1} represent the switching displacements of the stating and ending points of the sliding motion, respectively.

From the L-function, the local peak force product relative to the domains Ω_1 and Ω_2, is defined as

$$_{\max}L_{12}(t_m) = \left\{ L_{12}(\mathbf{x}_m, t_m) \left| \begin{array}{l} \forall t_m \in (t_i, t_{i+1}), \exists \mathbf{x}_m \in (\mathbf{x}_i, \mathbf{x}_{i+1}) \\ DL_{12}(\mathbf{x}, t)\,|_{(\mathbf{x}_m, t_m)} = 0 \text{ and} \\ D^2 L_{12}(\mathbf{x}, t)\,|_{(\mathbf{x}_m, t_m)} < 0, \end{array} \right. \right\} \tag{2.171}$$

and the global maximum force product is defined as

$$_{\mathrm{Gmax}}L_{12}(t_k) = \max_{t_k \in (t_i, t_{i+1})} \left\{ L_{12}(\mathbf{x}_i, t_i),\ _{\max}L_{12}(\mathbf{x}_m, t_m) \right\}. \tag{2.172}$$

Using the chain rule and $dx/dt = V$ for the sliding motion, we have

$$_{\max}L_{12}(x_m) = \left\{ L_{12}(x_m) \left| \begin{array}{l} \forall x_m \in (x_i, x_{i+1}), \exists t_m = \frac{x_m - x_i}{V} \in [t_i, t_{i+1}], \\ \frac{d}{dx} L_{12}(x)\,|_{x=x_m} = 0 \text{ and } \frac{d^2}{dx^2} L_{12}(x)\,|_{x=x_m} < 0 \end{array} \right. \right\} \tag{2.173}$$

$$_{\mathrm{Gmax}}L_{12}(x_m) = \max_{x_m \in (x_i, x_{i+1})} \left\{ L_{12}(x_i),\ _{\max}L_{12}(x_m) \right\}. \tag{2.174}$$

Once one of the global maximum force products is greater than zero, the sliding motion will disappear. Further, the sliding motion will be fragmentized. The corresponding critical condition is

$$_{\max}L_{12}(t_k) = 0 \text{ or } _{\mathrm{Gmax}}L_{12}(x_k) = 0, \text{ for } t_m \in (t_i, t_{i+1}) \tag{2.175}$$

The foregoing condition is termed the *global maximum force product condition* of

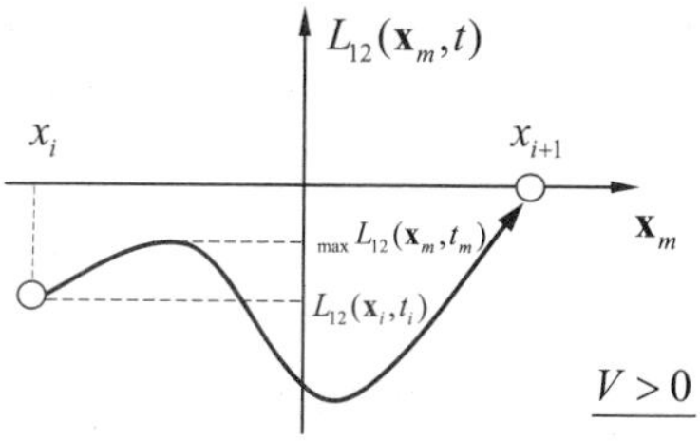

Fig. 2.18 A sketch of a force product for sliding motion under given parameters for belt speed $V>0$. ${}_{\max}L_{12}(\mathbf{x}_m,t_m)$ and $L_{12}(\mathbf{x}_i,t_i)$ represents the local *maximum* and *initial* force products, respectively. x_i and x_{i+1} represent the switching displacements of the stating and vanishing points of the sliding motion, respectively.

the sliding fragmentation. For simplicity, it is also called the *sliding fragmentation condition*. After fragmentation, consider the starting to ending points of the two sliding motions to be $(\Omega t_i, x_i, V)$ to $(\Omega t_{i+1}, x_{i+1}, V)$ and $(\Omega t_{i+2}, x_{i+2}, V)$ to $(\Omega t_{i+3}, x_{i+3}, V)$, respectively. For $t_i < t_{i+1} \le t_{i+2} < t_{i+3}$, From Eq.(2.159), the non-sliding motion between the two fragmentized motions require

$$L_{12}(t_m) > 0 \text{ for } t_m \in (t_{i+1}, t_{i+2}). \tag{2.176}$$

The inverse process of the sliding motion fragmentation is the merging of the two adjacent sliding motions. If the two sliding motions merge together, the condition of the global maximum force product in Eq.(2.175) will be satisfied at the gluing point of the two sliding motions. Once the force products of the two sliding motions are monitored, the peak force product condition similar to Eq.(2.176) can be observed. Before the merging of the two sliding motions, consider the fragmentized sliding motions with $(\Omega t_{i+1}, x_{i+1}, V) \ne (\Omega t_{i+2}, x_{i+2}, V)$. Extending definitions of Eqs.(2.170) and (2.171) to the starting and ending points of the sliding motions. The merging condition for the two adjacent sliding motions is

$$\begin{gathered}{}_{\max}L_{12}(t_k) = 0 \text{ or } {}_{\text{Gmax}}L_{12}(x_k) = 0 \text{ for } k \in \{i+1, i+2\},\\ (\Omega t_{i+1}, x_{i+1}, V) = (\Omega t_{i+2}, x_{i+2}, V).\end{gathered} \tag{2.177}$$

To explain the above condition of the fragmentation and merging of the sliding motions, the corresponding force product characteristics and phase plane of the sliding motion are sketched in Fig. 2.19. The sliding, critical sliding and sliding fragmentation motions are depicted for given parameters with belt speed $V>0$ and excitation amplitude $A_0^{(1)} < A_0^{(2)} < A_0^{(3)}$. The sliding motion vanishes from the boundary and goes into domain Ω_1. The displacements x_i and x_{i+1} represent the starting and vanishing switching points of the sliding motion, respectively. The point $x_k \in (x_i, x_{i+1})$ represents the critical point for the fragmentation or merging of the sliding motion. After the fragmentation (or before the merging), the critical point x_k is split into two new points x_{i+1} and x_{i+2}. However, after fragmentation (or merging), the index of x_{i+1} (or x_{i+3}) will be shifted as x_{i+3} (or x_{i+1}). Suppose the peak force product increases with increasing excitation amplitude. The motions from the sliding to the sliding fragmentation are given for $A_0^{(1)} \xrightarrow{\text{increase}} A_0^{(3)}$. The dashed and thin-solid curves represent non-sliding motion. It is clear that the non-sliding portion of the sliding fragmentation needs $L_{12}(t_m) > 0$ as in Eq.(2.176) for $t_m \in U_n = (t_j^n, t_{j+1}^n) \subset (t_i, t_{i+1})$ where $L_{12}(t_k) = 0$ for $t_k = \{t_j^n, t_{j+1}^n\}$. However, for the sliding portion, the force products keep the relation $L_{12}(t) < 0$ for $t \in (t_i, t_{i+1}) \setminus \cup_n U_n$. Otherwise, suppose the peak force products at the potential merging points of the two sliding motions decrease with decreasing excitation. The merging of the two sliding motions can be observed for $A_0^{(3)} \xrightarrow{\text{decreases}} A_0^{(1)}$. Whatever the merging or fragmentation of the sliding motion occurs, the old sliding motion will be destroyed. Thus, the conditions for the fragmentation and merging of the sliding motion are called the *vanishing* conditions for the sliding motion.

Consider the sliding motion with the non-zero measure of $\widetilde{\partial\Omega_{\alpha\beta}}$ with starting and ending points $(\Omega t_i, x_i, V)$ and $(\Omega t_{i+1}, x_{i+1}, V)$. Define $\delta = \sqrt{\delta_{x_i}^2 + \delta_{t_i}^2}$ with $\delta_{x_i} = x_{i+1} - x_i$ and $\delta_{t_i} = \Omega t_{i+1} - \Omega t_i \geq 0$. Suppose the starting and ending points satisfy the initial and final force product conditions, as $\delta \to 0$, the force product condition is termed the *onset* condition for the sliding motion.

$$
\begin{gathered}
L_{12}(t_m) = 0 \text{ for } m \in \{i, i+1\}, \\
(\Omega t_{i+1}, x_{i+1}, V) = \lim_{\delta \to 0}(\Omega t_i + \delta_{t_i}, x_i + \delta_{x_i}, V).
\end{gathered} \tag{2.178}
$$

The foregoing condition is also called the *onset force product condition*.

The onset condition has four possible cases from Eqs.(2.166) and (2.167), as sketched in Fig. 2.20. The four cases are: $F_2(t_i) = F_2(t_{i+1}) = 0$ (case I), $F_2(t_i) = F_1(t_{i+1}) = 0$ (case II), $F_1(t_i) = F_2(t_{i+1}) = 0$ (case III) and $F_1(t_i) = F_1(t_{i+1}) = 0$ (case IV). For $\delta \neq 0$, the four sliding motions exist. The onset conditions of the sliding motion for cases I and IV is the same as for the grazing motion.

The detailed discussion can be referred to Luo and Gegg [2006c]. The two onsets of the sliding motions can be also called the grazing onsets. The other onset conditions based on the cases II and III are the *inflexed* onsets for the sliding motions. In this subsection, the force product criteria for onset, forming and vanishing of the sliding motions along the velocity boundary have been developed. From the expressions of such criteria, the achieved criteria can be applied for a nonlinear, continuous spring and viscous damper oscillator including a nonlinear friction with a C^0-discontinuity. The numerical illustrations of sliding motions are given in Fig. 2.21.

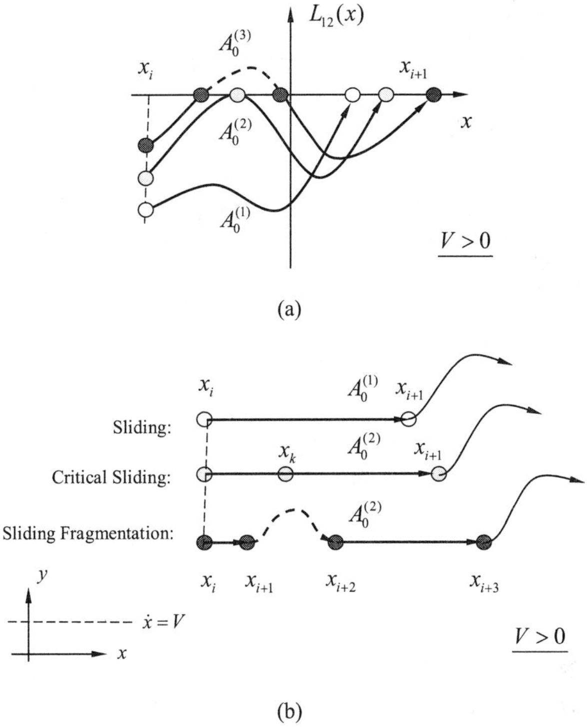

Fig. 2.19 A sketch of **(a)** force product and **(b)** phase plane for sliding, critical sliding and sliding fragmentation motions under given parameters for belt speed $V > 0$ and $A_0^{(1)} < A_0^{(2)} < A_0^{(3)}$. The ending point of the sliding motion vanishes on the boundary and goes to the domain Ω_1.

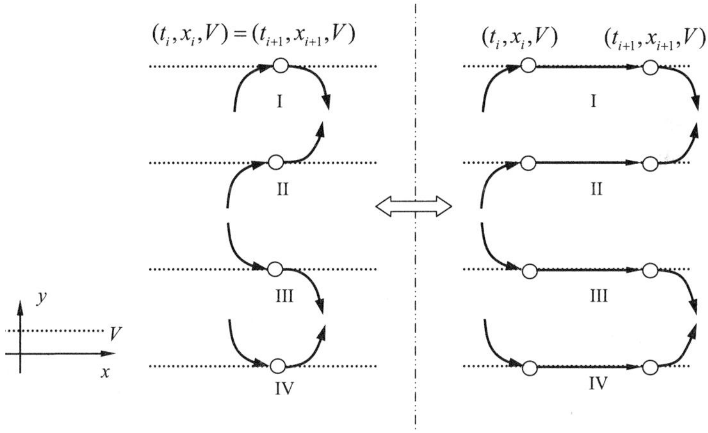

Fig. 2.20 The four onsets of the sliding motion and the corresponding sliding motion in phase plane for belt speed $V > 0$.

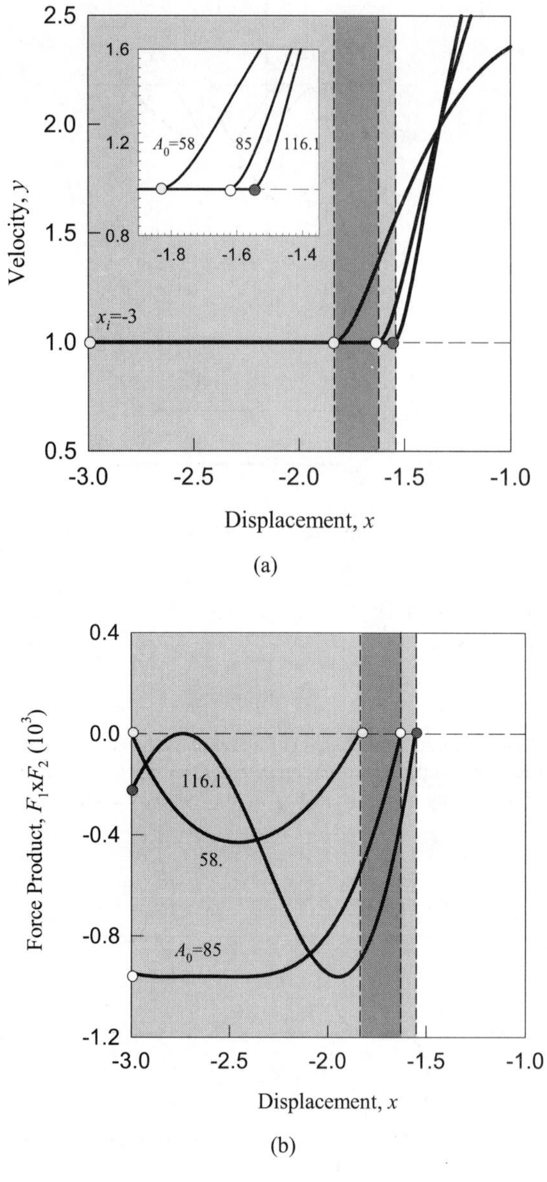

Fig. 2.21 Sliding motion vanishing on the boundary and going into the domain Ω_1 for ($A_0 = 58.0, 85.0, 116.1$): **(a)** phase plane, **(b)** force product versus displacement, **(c)** velocity time history **(d)** force product time history. ($V = 1,\ \Omega = 1,\ d_1 = 1,\ d_2 = 0,\ b_1 = -b_2 = 30,\ c_1 = c_2 = 30,$ $x_i = -3,\ \ \Omega t_i = \pi$).

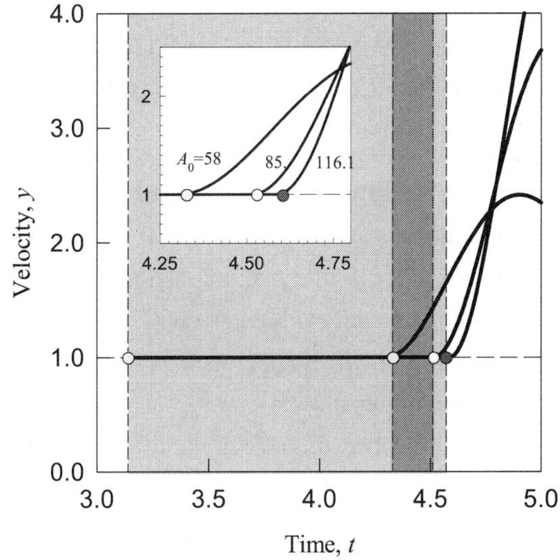

(c)

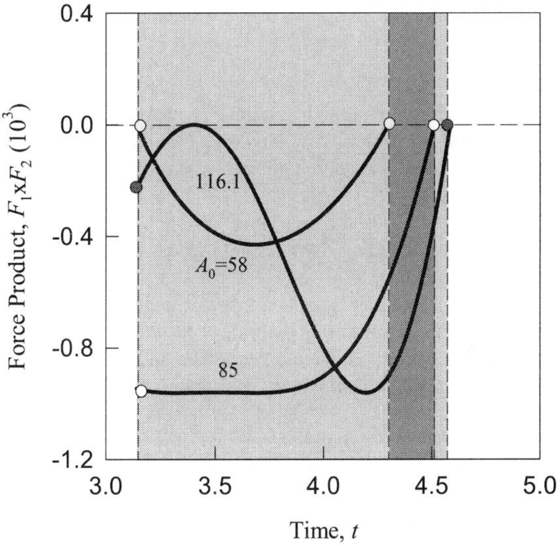

(d)

Fig. 2.21 Continued.

Consider the parameters (i.e., $V=1$, $\Omega=1$, $d_1=1$, $d_2=0$, $b_1=-b_2=30$, $c_1=c_2=30$) with initial condition ($x_i=3$, $y_i=1$ and $\Omega t_i=\pi$) for demonstration. For specific excitation amplitudes $A_0=\{58.0,\ 85.0,\ 116.1\}$, the trajectories for the corresponding sliding motions are plotted in Fig. 2.21(a). The force product versus displacement for the sliding motion are presented in Fig. 2.21(b), and it is observed that the sliding motions satisfy $F_1\times F_2\leq 0$. For $A_0=58.0$, the initial force product is zero, it implies that $A_{0\min}\approx 58.0$. If $A_0<A_{0\min}$, such a sliding motion cannot be observed. For $A_0\approx 116.1$, the maximum force product is tangential to zero. This value will be the maximum value for this sliding motion $A_{0\max}\approx 116.1$. If $A_0>A_{0\max}$, the sliding motion will be fragmentized into two parts. Thus, under the above parameters, the sliding motion starting at $(x_i,y_i,\Omega t_i)=(-3,1,\pi)$ will exist for $A_{0\min}\leq A_0\leq A_{0\max}$. Further, the velocity and force product time-histories are also presented in Fig. 2.21(c) and (d). It is interesting that the force product time history is similar to the force product versus the displacement. The equivalency of definitions in Eqs.(2.171) and (2.173) is verified numerically for non-zero constant belt speed.

References

Luo, A.C.J., 2005, A theory for non-smooth dynamic systems on the connectable domains, *Communications in Nonlinear Science and Numerical Simulation*, **10**,1-55.

Luo, A.C.J., 2006, *Singularity and Dynamics on Discontinuous Vector Fields*, Amsterdam: Elsevier.

Luo, A.C.J. and Gegg, B.C., 2006a, On the mechanism of stick and non-stick, periodic motions in a forced linear oscillator including dry friction, *ASME Journal of Vibration and Acoustics,* **128**, 97-105.

Luo, A.C.J. and Gegg, B.C., 2006b, Stick and non-stick, periodic motions of a periodically forced, linear oscillator with dry friction, *Journal of Sound and Vibration*, **291**, 132-168.

Luo, A.C.J. and Gegg, B.C., 2006c, Grazing phenomena in a periodically forced, friction-induced, linear oscillator, *Communications in Nonlinear Science and Numerical Simulation*, **11**, 777-802.

Luo, A.C.J. and Gegg, B.C., 2007, An analytical prediction of sliding motions along discontinuous boundary in non-smooth dynamical systems, *Nonlinear Dynamics*, **49**, 401-424.

Chapter 3
Singularity and Flow Passability

In this chapter, a general theory for the passability of a flow to a specific boundary in discontinuous dynamical systems will be presented. The concepts of real and imaginary flows will be introduced. The *G*-functions for discontinuous dynamical systems will be developed to describe the general theory of the passability of a flow to the boundary. Based on the *G*-function, the passability of a flow from a domain to an adjacent one will be discussed. With the concepts of real and imaginary flows, the full and half sink and source flows to the boundary will be discussed in detail. A flow to the boundary in a discontinuous dynamical system can be passable or non-passable. Thus, all the switching bifurcations between the passable and non-passable flows will be presented. To understand the concept of flow passability, a discontinuous dynamical system with a parabolic boundary is presented as an example.

3.1. Real and imaginary flows

As in Luo (2005a,b; 2006), consider a dynamic system consisting of *N*-sub-dynamic systems in a universal domain $\mho \subset \mathscr{R}^n$. The universal domain is divided into *N* accessible sub-domains Ω_i ($i \in I$, and $I = \{1, 2, \cdots, N\}$), and the union of all the accessible sub-domains $\cup_{i\in I}\Omega_i$ and a universal domain $\mho = \cup_{i\in I}\Omega_i \cup \Omega_0$, as shown in Fig. 2.1. Ω_0 is the union of the inaccessible domains. For an accessible domain Ω_i, the vector field $\mathbf{F}^{(i)}(\mathbf{x}_i^{(i)}, t, \mathbf{p}_i)$ is defined on such a domain. The dynamical system in Eq. (2.1) just satisfies the aforementioned condition. The corresponding flow given by such a dynamical system is called a *real* flow in the accessible domain Ω_i, which is defined as follows.

Definition 3.1. The C^{r_i+1}-continuous flow $\mathbf{x}_i^{(i)}(t) = \mathbf{\Phi}^{(i)}(\mathbf{x}_i^{(i)}(t_0), t, \mathbf{p}_i)$ is a *real flow* in the *i*th open sub-domain Ω_i, which is determined by a C^{r_i}-continuous system ($r_i \geq 1$) on Ω_i in a form of

$$\dot{\mathbf{x}}_i^{(i)} = \mathbf{F}^{(i)}(\mathbf{x}_i^{(i)}, t, \mathbf{p}_i) \in \mathscr{R}^n, \quad \mathbf{x}_i^{(i)} = (x_{i1}^{(i)}, x_{i2}^{(i)}, \cdots, x_{in}^{(i)})^{\mathrm{T}} \in \Omega_i, \tag{3.1}$$

with the initial condition

$$\mathbf{x}_i^{(i)}(t_0) = \mathbf{\Phi}^{(i)}(\mathbf{x}_i^{(i)}(t_0), t_0, \mathbf{p}_i). \tag{3.2}$$

Notice that time is t and $\dot{\mathbf{x}}_i^{(i)} = d\mathbf{x}_i^{(i)}/dt$. In domain Ω_i, the vector field $\mathbf{F}^{(i)}(\mathbf{x}_i^{(i)}, t, \mathbf{p}_i) \equiv \mathbf{F}_i^{(i)}(t)$ with parameter vectors $\mathbf{p}_i = (p_i^{(1)}, p_i^{(2)}, \cdots, p_i^{(l)})^{\mathrm{T}} \in \mathscr{R}^l$ is C^{r_i}-continuous ($r_i \geq 1$) in $\mathbf{x}$ and for time t. $\mathbf{\Phi}^{(i)}(\mathbf{x}_i^{(i)}(t), t_0, \mathbf{p}_i) \equiv \mathbf{\Phi}_i^{(i)}(t)$. $\mathbf{x}_i^{(t)}(t)$ denotes the flow in the ith sub-domain Ω_i, governed by a dynamical system defined on the ith sub-domain Ω_i. The real flow $\mathbf{x}_i^{(i)}(t)$ in Ω_i is governed by a dynamical system on its own domain. However, another flow $\mathbf{x}_i^{(j)}$ in Ω_i is governed by a dynamical system defined on the jth sub-domain Ω_j, which is of great interest herein. If the imaginary flow $\mathbf{\Phi}_i^{(j)}(t)$ of a real flow $\mathbf{\Phi}_j^{(j)}(t)$ cannot exist on a domain Ω_i, the domain can be called the *unextendable* domains of the real flow $\mathbf{\Phi}_j^{(j)}(t)$. If the imaginary flow $\mathbf{\Phi}_i^{(j)}(t)$ of a real flow $\mathbf{\Phi}_j^{(j)}(t)$ can exist on a domain Ω_i, the domain can be called the *extendable* domains of the real flow $\mathbf{\Phi}_j^{(j)}(t)$. The definitions are given as follows.

Definition 3.2. Consider the vector field $\mathbf{F}^{(j)}(\mathbf{x}_j^{(j)}, t, \mathbf{p}_j)$ of a real flow $\mathbf{\Phi}_j^{(j)}(t)$ in the domain Ω_j. The real flow $\mathbf{\Phi}_j^{(j)}(t)$ is not extendable to the domain Ω_i if the vector field $\mathbf{F}^{(j)}(\mathbf{x}_j^{(j)}, t, \mathbf{p}_j)$ on the domain Ω_j cannot be defined on the domain Ω_i (i.e., $\mathbf{F}^{(j)}(\mathbf{x}_i^{(j)}, t, \mathbf{p}_j)$). The domain Ω_i is called the *unextendable* domain of the flow $\mathbf{\Phi}_j^{(j)}(t)$.

Definition 3.3. Consider the vector field $\mathbf{F}^{(j)}(\mathbf{x}_j^{(j)}, t, \mathbf{p}_j)$ of a real flow $\mathbf{\Phi}_j^{(j)}(t)$ in the domain Ω_j. The real flow $\mathbf{\Phi}_j^{(j)}(t)$ is extendable to the domain Ω_i if the vector field $\mathbf{F}^{(j)}(\mathbf{x}_j^{(j)}, t, \mathbf{p}_j)$ on the domain Ω_j can be defined on the domain Ω_i (i.e., $\mathbf{F}^{(j)}(\mathbf{x}_i^{(j)}, t, \mathbf{p}_j)$). The domain Ω_i is called the *extendable* domain of the flow $\mathbf{\Phi}_j^{(j)}(t)$.

Similar to Luo (2005a,b), the hypothesis for the theory of discontinuous dynamical systems in Eq. (3.1) for the real flow become as follows:

H3.1: The switching between two adjacent sub-systems possesses time-continuity.

H3.2: For an unbounded, accessible sub-domain Ω_i, there is an open domain $D_i \subset \Omega_i$. The vector field and flow on D_i are bounded, i.e.,

$$\| \mathbf{F}_i^{(i)} \| \leq K_1(\text{const}) \text{ and } \| \mathbf{\Phi}_i^{(i)} \| \leq K_2(\text{const}) \text{ on } D_i \text{ for } t \in [0,\infty). \tag{3.3}$$

H3.3: For a bounded, accessible domain Ω_i, there is an open domain $D_i \subseteq \Omega_i$. The vector field on D_i is bounded but the corresponding flow may be unbounded, i.e.,

$$\| \mathbf{F}_i^{(i)} \| \leq K_1(\text{const}) \text{ and } \| \mathbf{\Phi}_i^{(i)} \| < \infty \text{ on } D_i \text{ for } t \in [0,\infty). \tag{3.4}$$

H3.4: The real flows $\mathbf{\Phi}_i^{(i)}(t)$ and the vector fields $\mathbf{F}_i^{(i)}$ in domain Ω_i are extendable to the adjacent domain Ω_j in the neighborhood of the boundary $\partial\Omega_{ij}$.

As stated before, a flow $\mathbf{x}_i^{(j)}$ in Ω_i is governed by a dynamical system defined on the jth sub-domain Ω_j. This kind of flows is called the *imaginary flow* because the flow is not determined by the dynamical system on its own domain. To further understand dynamical behaviors of discontinuous dynamical systems, it is necessary to introduce imaginary flows. Consider the jth imaginary flow in the ith domain Ω_i is a flow in Ω_i governed by the dynamical system defined on the jth sub-domain Ω_j. The two sub-domains can be either adjacent or separable. Thus, the mathematical definition of imaginary flows is as follows.

Definition 3.4. The C^{r_j+1} ($r_j \geq 1$)-continuous flow $\mathbf{x}_i^{(j)}(t)$ is termed the jth *imaginary flow* in the ith open sub-domain Ω_i if the flow $\mathbf{x}_i^{(j)}(t)$ is determined by an application of a C^{r_j}-continuous system on the jth open sub-domain Ω_j to the ith open sub-domain Ω_i, i.e.,

$$\dot{\mathbf{x}}_i^{(j)} = \mathbf{F}^{(j)}(\mathbf{x}_i^{(j)}, t, \mathbf{p}_j) \in \mathscr{R}^n,\ \mathbf{x}_i^{(j)} = (x_{i1}^{(j)}, x_{i2}^{(j)}, \cdots, x_{in}^{(j)})^{\mathrm{T}} \in \Omega_i, \tag{3.5}$$

with the initial conditions

$$\mathbf{x}_i^{(j)}(t_0) = \mathbf{\Phi}^{(j)}(\mathbf{x}_i^{(j)}(t_0), t_0, \mathbf{p}_j). \tag{3.6}$$

The imaginary flows $\mathbf{\Phi}_i^{(j)}(t)$ and the imaginary vector fields $\mathbf{F}_i^{(j)}$ do not necessarily satisfy the hypothesis in (H3.1)-(H3.3) in domain Ω_i indeed. However, the imaginary flows and vector fields should satisfy the above hypothesis *in vicinity of the boundary* if they are the extensions of a real flow and vector field in domain Ω_j.

Real and imaginary flows in two adjacent sub-domains are sketched in Figs.

3.1 and 3.2 for a semi-passable flow, sink and source flows to the boundary. A point $\mathbf{x}_m \in \partial\Omega_{ij}$ is at time t_m, and $\mathbf{x}_\alpha^{(\alpha)}(t_m) = \mathbf{x}_m = \mathbf{x}_\beta^{(\alpha)}(t_m)$ for $\alpha, \beta = i, j$ and $\alpha \neq \beta$. The real and imaginary flows $\mathbf{x}_\alpha^{(\alpha)}(t)$ and $\mathbf{x}_\alpha^{(\beta)}(t)$ are represented by solid and dashed curves, respectively. $\mathbf{x}_\alpha^{(\alpha)}(t_{m\pm\varepsilon})$ and $\mathbf{x}_\alpha^{(\beta)}(t_{m\pm\varepsilon})$ in vicinity of $\mathbf{x}_m \in \partial\Omega_{ij}$ are the *real* and *imaginary* flows at $t_{m\pm\varepsilon} = t_m \pm \varepsilon$ for $\varepsilon > 0$. As $\varepsilon \to 0$, $\mathbf{x}_\alpha^{(\alpha)}(t_{m\pm\varepsilon})$ and $\mathbf{x}_\alpha^{(\beta)}(t_{m\pm\varepsilon})$ approach the point $\mathbf{x}_m$ (i.e., $\mathbf{x}_\alpha^{(\alpha)}(t_{m\pm\varepsilon}) \to \mathbf{x}_m$ and $\mathbf{x}_\alpha^{(\beta)}(t_{m\pm\varepsilon}) \to \mathbf{x}_m$).

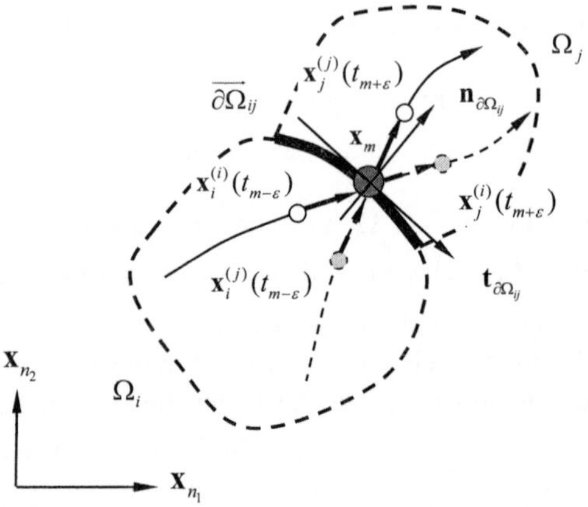

(a)

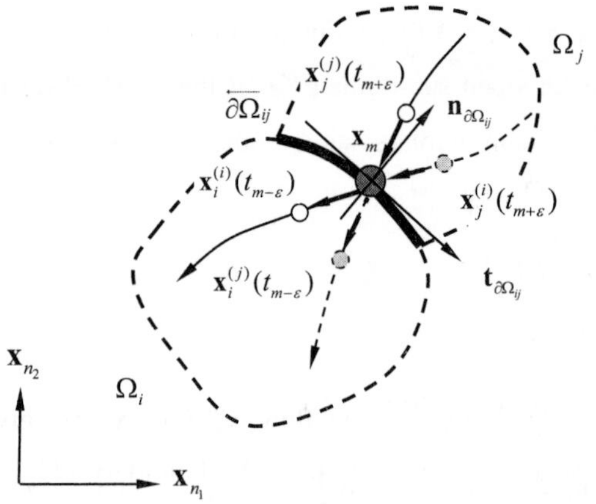

(b)

Fig. 3.1 Real and imaginary flows in vicinity of a boundary $\partial\Omega_{ij}$: **(a)** from Ω_i to Ω_j and **(b)** from Ω_j to Ω_i. The boundary point $\mathbf{x}_m$ is at time t_m. Solid and dashed curves represent real and imagery flows, respectively.

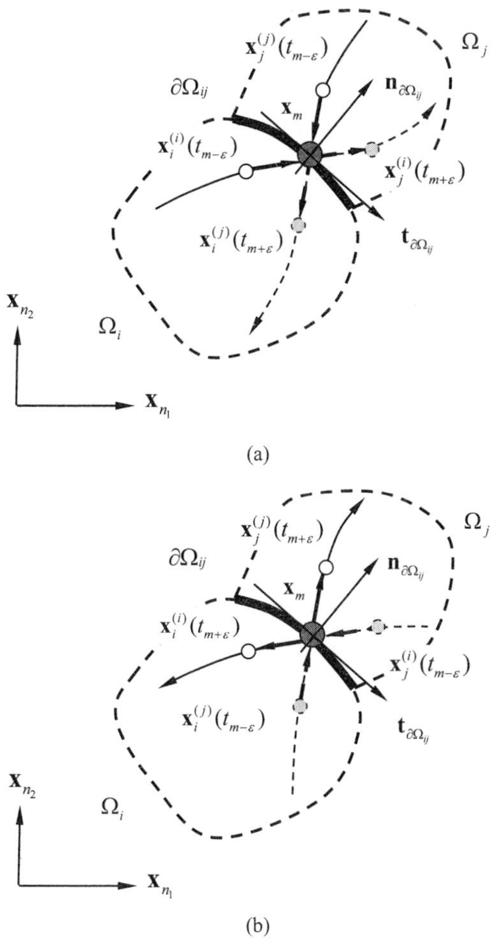

Fig. 3.2 Real and imaginary flows in vicinity of a boundary $\partial\Omega_{ij}$: **(a)** sink flow and **(b)** source flow. The boundary point $\mathbf{x}_m$ is at time t_m . Solid and dashed curves represent the real and imagery flows, respectively.

From the foregoing definition, the flow $\mathbf{x}_{\alpha}^{(\alpha)}(t) \cup \mathbf{x}_{\beta}^{(\alpha)}(t)$ gives a continuous flow in two sub-domains Ω_i and Ω_j plus the boundary $\partial\Omega_{ij}$. Once imaginary vector fields and flows are introduced, the theory for continuous dynamical systems can be employed to investigate dynamical behaviors. With varying time or parameters, the real and imaginary flows to a specific boundary can be switched. In this chapter, the singularity and generalized flow passability will be presented. Without the imaginary flows and singularity, only three types of fundamental flow passability were presented in Chapter 2. The passability of imaginary flows to the boundary can be similarly described as in Chapter 2 for real flows. To real and imaginary flows, the boundary in Eq. (2.4) should be same. The corresponding dynamical systems on the boundary $\partial\Omega_{ij}$ with $\varphi_{ij}(\mathbf{x},t,\boldsymbol{\lambda})=0$ can be given as in Eq. (2.5), i.e.,

$$\dot{\mathbf{x}}^{(0)}=\mathbf{F}^{(0)}(\mathbf{x}^{(0)},t,\boldsymbol{\lambda}) \text{ with } \varphi_{ij}(\mathbf{x}^{(0)},t,\boldsymbol{\lambda})=0 \tag{3.7}$$

where $\mathbf{x}^{(0)}=(x_1^{(0)},x_2^{(0)},\cdots,x_n^{(0)})^{\mathrm{T}}$. The flow $\mathbf{x}^{(0)}(t)=\mathbf{\Phi}^{(0)}(\mathbf{x}^{(0)}(t_0),t,\boldsymbol{\lambda})$ is C^{r_0+1}-continuous for time t with an initial condition $\mathbf{x}^{(0)}(t_0)=\mathbf{\Phi}^{(0)}(\mathbf{x}^{(0)}(t_0),t_0,\boldsymbol{\lambda})$.

3.2. *G*-functions and vector field decomposition

Before the general theory for flow passability to a specific boundary in discontinuous dynamical systems is discussed, a concept of *G*-function will be introduced to measure behaviors of discontinuous dynamical systems in the normal direction of the boundary. The real flow is used to define the *G*-functions, and such G-functions are also applicable to the imaginary flows. For simplicity, as in Luo (2008a,b), consider two infinitesimal time intervals $[t-\varepsilon,t)$ and $(t,t+\varepsilon]$. There are two flows in domain Ω_α $(\alpha=i,j)$ and on the boundary $\partial\Omega_{ij}$ determined by Eqs. (2.1) and (2.5), respectively. As in Luo (2008a,b), the vector difference between two flows for three time instants are given by $\mathbf{x}_{t-\varepsilon}^{(\alpha)}-\mathbf{x}_{t-\varepsilon}^{(0)}$, $\mathbf{x}_t^{(\alpha)}-\mathbf{x}_t^{(0)}$ and $\mathbf{x}_{t+\varepsilon}^{(\alpha)}-\mathbf{x}_{t+\varepsilon}^{(0)}$. The normal vectors of boundary relative to the corresponding flow $\mathbf{x}^{(0)}(t)$ are expressed by ${}^{t-\varepsilon}\mathbf{n}_{\partial\Omega_{ij}}$, ${}^{t}\mathbf{n}_{\partial\Omega_{ij}}$ and ${}^{t+\varepsilon}\mathbf{n}_{\partial\Omega_{ij}}$ and the corresponding tangential vectors of the flow $\mathbf{x}^{(0)}(t)$ on the boundary are expressed ${}^{t-\varepsilon}\mathbf{t}_{\partial\Omega_{ij}}$ ${}^{t}\mathbf{t}_{\partial\Omega_{ij}}$ and ${}^{t+\varepsilon}\mathbf{t}_{\partial\Omega_{ij}}$, respectively. From the normal vectors of the boundary $\partial\Omega_{ij}$, the dot product functions of the normal vector and the position vector difference between the two flows in domain and on the boundary are defined by

$$\begin{aligned} d_{t-\varepsilon}^{(\alpha)} &= {}^{t-\varepsilon}\mathbf{n}_{\partial\Omega_{ij}}^{\mathrm{T}}\cdot(\mathbf{x}_{t-\varepsilon}^{(\alpha)}-\mathbf{x}_{t-\varepsilon}^{(0)}),\\ d_{t}^{(\alpha)} &= {}^{t}\mathbf{n}_{\partial\Omega_{ij}}^{\mathrm{T}}\cdot(\mathbf{x}_{t}^{(\alpha)}-\mathbf{x}_{t}^{(0)}),\\ d_{t+\varepsilon}^{(\alpha)} &= {}^{t+\varepsilon}\mathbf{n}_{\partial\Omega_{ij}}^{\mathrm{T}}\cdot(\mathbf{x}_{t+\varepsilon}^{(\alpha)}-\mathbf{x}_{t+\varepsilon}^{(0)}) \end{aligned} \tag{3.8}$$

where the normal vector of the boundary surface $\partial\Omega_{ij}$ at point $\mathbf{x}^{(0)}(t)$ is

$$\begin{aligned} {}^{t}\mathbf{n}_{\partial\Omega_{ij}} &\equiv \mathbf{n}_{\partial\Omega_{ij}}(\mathbf{x}^{(0)},t,\boldsymbol{\lambda})=\nabla\varphi_{ij}(\mathbf{x}^{(0)},t,\boldsymbol{\lambda})\\ &=\left(\frac{\partial\varphi_{ij}}{\partial x_1^{(0)}},\frac{\partial\varphi_{ij}}{\partial x_2^{(0)}},\cdots,\frac{\partial\varphi_{ij}}{\partial x_n^{(0)}}\right)^{\mathrm{T}}. \end{aligned} \tag{3.9}$$

For time t, the normal component is the distance of the two points of two flows in the normal direction of the boundary surface.

Definition 3.5. Consider a dynamic system in Eq. (2.1) in domain Ω_α ($\alpha \in \{i,j\}$) which has a flow $\mathbf{x}^{(\alpha)} = \mathbf{\Phi}(t_0, \mathbf{x}_0^{(\alpha)}, \mathbf{p}_\alpha, t)$ with an initial condition $(t_0, \mathbf{x}_0^{(\alpha)})$, and on the boundary $\partial\Omega_{ij}$, there is an enough smooth flow $\mathbf{x}^{(0)} = \mathbf{\Phi}(t_0, \mathbf{x}_0^{(0)}, \boldsymbol{\lambda}, t)$ with an initial condition $(t_0, \mathbf{x}_0^{(0)})$. For an arbitrarily small $\varepsilon > 0$, there are two time intervals $[t-\varepsilon, t)$ or $(t, t+\varepsilon]$ for flow $\mathbf{x}^{(\alpha)}$ ($\alpha \in \{i,j\}$). The G-functions ($G_{\partial\Omega_{ij}}^{(\alpha)}$) of the domain flow $\mathbf{x}^{(\alpha)}$ to the boundary flow $\mathbf{x}^{(0)}$ on the boundary in the normal direction of the boundary $\partial\Omega_{ij}$ are defined as

$$\begin{aligned} & G_{\partial\Omega_{ij}}^{(\alpha)}(\mathbf{x}_t^{(0)}, t_-, \mathbf{x}_{t_-}^{(\alpha)}, \mathbf{p}_\alpha, \boldsymbol{\lambda}) \\ &= \lim_{\varepsilon\to 0} \frac{1}{\varepsilon}[{}^{t}\mathbf{n}_{\partial\Omega_{ij}}^{\mathrm{T}} \cdot (\mathbf{x}_{t_-}^{(\alpha)} - \mathbf{x}_t^{(0)}) - {}^{t-\varepsilon}\mathbf{n}_{\partial\Omega_{ij}}^{\mathrm{T}} \cdot (\mathbf{x}_{t-\varepsilon}^{(\alpha)} - \mathbf{x}_{t-\varepsilon}^{(0)})], \\ & G_{\partial\Omega_{ij}}^{(\alpha)}(\mathbf{x}_t^{(0)}, t_+, \mathbf{x}_{t_+}^{(\alpha)}, \mathbf{p}_\alpha, \boldsymbol{\lambda}) \\ &= \lim_{\varepsilon\to 0} \frac{1}{\varepsilon}[{}^{t+\varepsilon}\mathbf{n}_{\partial\Omega_{ij}}^{\mathrm{T}} \cdot (\mathbf{x}_{t+\varepsilon}^{(\alpha)} - \mathbf{x}_{t+\varepsilon}^{(0)}) - {}^{t}\mathbf{n}_{\partial\Omega_{ij}}^{\mathrm{T}} \cdot (\mathbf{x}_{t_+}^{(\alpha)} - \mathbf{x}_t^{(0)})]. \end{aligned} \tag{3.10}$$

From Eq. (3.10), since $\mathbf{x}_{t\pm}^{(\alpha)}$ and $\mathbf{x}_t^{(0)}$ are the solutions of Eqs. (2.1) and (2.5), their derivatives exist. Further, by use of the Taylor series expansion, equation (3.10) gives

$$\begin{aligned} & G_{\partial\Omega_{ij}}^{(\alpha)}(\mathbf{x}_t^{(0)}, t_{t\pm}, \mathbf{x}_{t\pm}^{(\alpha)}, \mathbf{p}_\alpha, \boldsymbol{\lambda}) \\ &= D_0\, {}^{t}\mathbf{n}_{\partial\Omega_{ij}}^{\mathrm{T}} \cdot (\mathbf{x}_{t\pm}^{(\alpha)} - \mathbf{x}_t^{(0)}) + {}^{t}\mathbf{n}_{\partial\Omega_{ij}}^{\mathrm{T}} \cdot (\dot{\mathbf{x}}_{t\pm}^{(\alpha)} - \dot{\mathbf{x}}_t^{(0)}) \end{aligned} \tag{3.11}$$

where the total derivative operators are defined as

$$D_0(\cdot) \equiv \frac{\partial(\cdot)}{\partial \mathbf{x}^{(0)}}\dot{\mathbf{x}}^{(0)} + \frac{\partial(\cdot)}{\partial t} \text{ and } D_\alpha(\cdot) \equiv \frac{\partial(\cdot)}{\partial \mathbf{x}^{(\alpha)}}\dot{\mathbf{x}}^{(\alpha)} + \frac{\partial(\cdot)}{\partial t}. \tag{3.12}$$

Using Eqs. (2.1) and (2.5), the *G-function* in Eq. (3.11) becomes

$$\begin{aligned} & G_{\partial\Omega_{ij}}^{(\alpha)}(\mathbf{x}_t^{(0)}, t_\pm, \mathbf{x}_{t_\pm}^{(\alpha)}, \mathbf{p}_\alpha, \boldsymbol{\lambda}) \\ &= D_0\, {}^{t}\mathbf{n}_{\partial\Omega_{ij}}^{\mathrm{T}} \cdot (\mathbf{x}_{t_\pm}^{(\alpha)} - \mathbf{x}_t^{(0)}) + {}^{t}\mathbf{n}_{\partial\Omega_{ij}}^{\mathrm{T}} \cdot [\mathbf{F}^{(\alpha)}(\mathbf{x}_{t_\pm}^{(\alpha)}, t_\pm, \mathbf{p}_\alpha) - \mathbf{F}^{(0)}(\mathbf{x}_t^{(0)}, t, \boldsymbol{\lambda})]. \end{aligned} \tag{3.13}$$

Consider the flow contacting with the boundary at time t_m (i.e., $\mathbf{x}_m^{(\alpha)} = \mathbf{x}_m^{(0)}$). Because a flow $\mathbf{x}^{(\alpha)}(t)$ approaches the separation boundary with the zero-order $\mathbf{x}^{(\alpha)}(t_{m\pm}) = \mathbf{x}_m = \mathbf{x}^{(0)}(t_m)$, the G-function is defined as

$$
\begin{aligned}
&G_{\partial\Omega_{ij}}^{(\alpha)}(\mathbf{x}_m,t_{m\pm},\mathbf{p}_\alpha,\boldsymbol{\lambda})\\
&\equiv \mathbf{n}_{\partial\Omega_{ij}}^{\mathrm{T}}(\mathbf{x}^{(0)},t,\boldsymbol{\lambda})\cdot[\dot{\mathbf{x}}^{(\alpha)}(t)-\dot{\mathbf{x}}^{(0)}(t)]\big|_{(\mathbf{x}_m^{(0)},\mathbf{x}_{m\pm}^{(\alpha)},t_{m\pm})}\\
&=[\mathbf{n}_{\partial\Omega_{ij}}^{\mathrm{T}}(\mathbf{x}^{(0)},t,\boldsymbol{\lambda})\cdot\dot{\mathbf{x}}^{(\alpha)}(t)+\frac{\partial\varphi_{ij}(\mathbf{x}^{(0)},t,\boldsymbol{\lambda})}{\partial t}]\big|_{(\mathbf{x}_m^{(0)},\mathbf{x}_{m\pm}^{(\alpha)},t_{m\pm})}\\
&=[\nabla\varphi_{ij}(\mathbf{x}^{(0)},t,\boldsymbol{\lambda})\cdot\dot{\mathbf{x}}^{(\alpha)}(t)+\frac{\partial\varphi_{ij}(\mathbf{x}^{(0)},t,\boldsymbol{\lambda})}{\partial t}]\big|_{(\mathbf{x}_m^{(0)},\mathbf{x}_{m\pm}^{(\alpha)},t_{m\pm})}.
\end{aligned}
\tag{3.14}
$$

With Eqs. (2.1) and (2.5), Equation (3.13) can be rewritten as

$$
\begin{aligned}
&G_{\partial\Omega_{ij}}^{(\alpha)}(\mathbf{x}_m,t_{m\pm},\mathbf{p}_\alpha,\boldsymbol{\lambda})\\
&=\mathbf{n}_{\partial\Omega_{ij}}^{\mathrm{T}}(\mathbf{x}^{(0)},t,\boldsymbol{\lambda})\cdot[\mathbf{F}(\mathbf{x}^{(\alpha)},t,\mathbf{p}_\alpha)-\mathbf{F}^{(0)}(\mathbf{x}^{(0)},t,\boldsymbol{\lambda})]\big|_{(\mathbf{x}_m^{(0)},\mathbf{x}_{m\pm}^{(\alpha)},t_m\pm)}\\
&=[\mathbf{n}_{\partial\Omega_{ij}}^{\mathrm{T}}(\mathbf{x}^{(0)},t,\boldsymbol{\lambda})\cdot\mathbf{F}(\mathbf{x}^{(\alpha)},t,\mathbf{p}_\alpha)+\frac{\partial\varphi_{ij}(\mathbf{x}^{(0)},t,\boldsymbol{\lambda})}{\partial t}]\big|_{(\mathbf{x}_m^{(0)},\mathbf{x}_{m\pm}^{(\alpha)},t_m\pm)}\\
&=[\nabla\varphi_{ij}(\mathbf{x}^{(0)},t,\boldsymbol{\lambda})\cdot\mathbf{F}(\mathbf{x}^{(\alpha)},t,\mathbf{p}_\alpha)+\frac{\partial\varphi_{ij}(\mathbf{x}^{(0)},t,\boldsymbol{\lambda})}{\partial t}]\big|_{(\mathbf{x}_m^{(0)},\mathbf{x}_{m\pm}^{(\alpha)},t_m\pm)}.
\end{aligned}
\tag{3.15}
$$

$G_{\partial\Omega_{ij}}^{(\alpha)}(\mathbf{x}_m,t_{m\pm},\mathbf{p}_\alpha,\boldsymbol{\lambda})$ is a time rate of the inner product of displacement difference and the normal direction $\mathbf{n}_{\partial\Omega_{ij}}(\mathbf{x}_m,t_m,\boldsymbol{\lambda})$, and $t_{m\pm}\equiv t_m\pm 0$ is to represent the quantity in the domain rather than on the boundary. If a flow in a discontinuous system crosses over the boundary $\partial\Omega_{ij}$, one obtains $G_{\partial\Omega_{ij}}^{(i)}\neq G_{\partial\Omega_{ij}}^{(j)}$. However, without the boundary, the dynamical system is continuous. Thus, $G_{\partial\Omega_{ij}}^{(i)}=G_{\partial\Omega_{ij}}^{(j)}$. Because the corresponding imaginary flow is the extension of a real flow to the boundary, the real flow and corresponding imaginary flow are continuous. Therefore, the G-functions to both the real and imaginary flows on the boundary $\partial\Omega_{ij}$ are same.

Definition 3.6. Consider a dynamic system in Eq. (2.1) in domain Ω_α ($\alpha\in\{i,j\}$) which has the flow $\mathbf{x}_t^{(\alpha)}=\boldsymbol{\Phi}(t_0,\mathbf{x}_0^{(\alpha)},\mathbf{p}_\alpha,t)$ with an initial condition $(t_0,\mathbf{x}_0^{(\alpha)})$, and on the boundary $\partial\Omega_{ij}$, there is an enough smooth flow $\mathbf{x}_t^{(0)}=\boldsymbol{\Phi}(t_0,\mathbf{x}_0^{(0)},\boldsymbol{\lambda},t)$ with an initial condition $(t_0,\mathbf{x}_0^{(0)})$. For an arbitrarily small $\varepsilon>0$, there are two time intervals $[t-\varepsilon,t)$ and $(t,t+\varepsilon]$ for a domain flow $\mathbf{x}_t^{(\alpha)}$ ($\alpha\in\{i,j\}$). The vector fields $\mathbf{F}^{(\alpha)}(\mathbf{x}^{(\alpha)},t,\mathbf{p}_\alpha)$ and $\mathbf{F}^{(0)}(\mathbf{x}^{(0)},t,\boldsymbol{\lambda})$ are $C_{[t-\varepsilon,t+\varepsilon]}^{r_\alpha}$-continuous ($r_\alpha\geq k$) for time t with $\|d^{r_\alpha+1}\mathbf{x}_t^{(\alpha)}/dt^{r_\alpha+1}\|<\infty$ and $\|d^{r_\alpha+1}\mathbf{x}_t^{(0)}/dt^{r_\alpha+1}\|<\infty$. The kth *-order G-functions* of the domain flow $\mathbf{x}_t^{(\alpha)}$ to the boundary flow $\mathbf{x}_t^{(0)}$ in the normal direction of $\partial\Omega_{ij}$ are defined as

$$
\begin{aligned}
&G_{\partial\Omega_{ij}}^{(k,\alpha)}(\mathbf{x}_t^{(0)},t_-,\mathbf{x}_{t_-}^{(\alpha)},\mathbf{p}_\alpha,\boldsymbol{\lambda})\\
&=\lim_{\varepsilon\to 0}\frac{(-1)^{k+2}}{\varepsilon^{k+1}}[{}^{t}\mathbf{n}_{\partial\Omega_{ij}}^{\mathrm{T}}\cdot(\mathbf{x}_{t_-}^{(\alpha)}-\mathbf{x}_t^{(0)})-{}^{t-\varepsilon}\mathbf{n}_{\partial\Omega_{ij}}^{\mathrm{T}}\cdot(\mathbf{x}_{t-\varepsilon}^{(\alpha)}-\mathbf{x}_{t-\varepsilon}^{(0)})\\
&\quad+\sum\nolimits_{s=0}^{k-1}G_{\partial\Omega_{ij}}^{(s,\alpha)}(\mathbf{x}_t^{(0)},t,\mathbf{x}_{t_-}^{(\alpha)},\mathbf{p}_\alpha,\boldsymbol{\lambda})(-\varepsilon)^{s+1}]\\
&=\lim_{\varepsilon\to 0}\frac{1}{\varepsilon^{k+1}}[{}^{t+\varepsilon}\mathbf{n}_{\partial\Omega_{ij}}^{\mathrm{T}}\cdot(\mathbf{x}_{t+\varepsilon}^{(\alpha)}-\mathbf{x}_{t+\varepsilon}^{(0)})-{}^{t}\mathbf{n}_{\partial\Omega_{ij}}^{\mathrm{T}}\cdot(\mathbf{x}_{t_+}^{(\alpha)}-\mathbf{x}_t^{(0)})\\
&\quad-\sum\nolimits_{s=0}^{k-1}G_{\partial\Omega_{ij}}^{(s,\alpha)}(\mathbf{x}_t^{(0)},t,\mathbf{x}_{t_+}^{(\alpha)},\mathbf{p}_\alpha,\boldsymbol{\lambda})\varepsilon^{s+1}].
\end{aligned}
\tag{3.16}
$$

Again, the Taylor series expansion applying to Eq. (3.16) yields

$$
\begin{aligned}
&G_{\partial\Omega_{ij}}^{(k,\alpha)}(\mathbf{x}_t^{(0)},t_\pm,\mathbf{x}_{t_\pm}^{(\alpha)},\mathbf{p}_\alpha,\boldsymbol{\lambda})\\
&=\sum\nolimits_{s=0}^{k+1}C_{k+1}^{s}D_0^{k+1-s}\,{}^{t}\mathbf{n}_{\partial\Omega_{ij}}^{\mathrm{T}}\cdot\left(\frac{d^s\mathbf{x}^{(\alpha)}}{dt^s}-\frac{d^s\mathbf{x}^{(0)}}{dt^s}\right)\Bigg|_{(\mathbf{x}_t^{(0)},\mathbf{x}_{t_\pm}^{(\alpha)},t_\pm)}.
\end{aligned}
\tag{3.17}
$$

Using Eqs. (2.1) and (2.5), the *k*th-order *G*-function of the flow $\mathbf{x}_t^{(\alpha)}$ to the boundary $\partial\Omega_{ij}$ is computed by

$$
\begin{aligned}
&G_{\partial\Omega_{ij}}^{(k,\alpha)}(\mathbf{x}_t^{(0)},t_\pm,\mathbf{x}_{t_\pm}^{(\alpha)},\mathbf{p}_\alpha,\boldsymbol{\lambda})\\
&=\sum\nolimits_{s=1}^{k+1}C_{k+1}^{s}D_0^{k+1-s}\,{}^{t}\mathbf{n}_{\partial\Omega_{ij}}^{\mathrm{T}}\cdot[D_\alpha^{s-1}\mathbf{F}^{(\alpha)}(\mathbf{x}^{(\alpha)},t,\mathbf{p}_\alpha)\\
&\quad-D_0^{s-1}\mathbf{F}^{(0)}(\mathbf{x}^{(0)},t,\boldsymbol{\lambda})]\big|_{(\mathbf{x}_t^{(0)},\mathbf{x}_{t_\pm}^{(\alpha)},t_\pm)}+D_0^{k+1}\,{}^{t}\mathbf{n}_{\partial\Omega_{ij}}^{\mathrm{T}}\cdot(\mathbf{x}_{t_\pm}^{(\alpha)}-\mathbf{x}_t^{(0)}),
\end{aligned}
\tag{3.18}
$$

where

$$
C_{k+1}^{s}=\frac{(k+1)k(k-1)\cdots(k+2-s)}{s!}
\tag{3.19}
$$

with $C_{k+1}^{0}=1$ and $s!=1\times 2\times\cdots\times s$.

The *G*-function $G_{\partial\Omega_{ij}}^{(k,\alpha)}$ is the time rate of $G_{\partial\Omega_{ij}}^{(k-1,\alpha)}$. If a flow contacting with $\partial\Omega_{ij}$ at time t_m (i.e., $\mathbf{x}_{m\pm}^{(\alpha)}=\mathbf{x}_m^{(0)}$) and ${}^{t}\mathbf{n}_{\partial\Omega_{ij}}^{\mathrm{T}}\equiv\mathbf{n}_{\partial\Omega_{ij}}^{\mathrm{T}}$, the *k*th-order *G*-function is

$$
\begin{aligned}
&G_{\partial\Omega_{ij}}^{(k,\alpha)}(\mathbf{x}_m,t_{m\pm},\mathbf{p}_\alpha,\boldsymbol{\lambda})\\
&=\sum\nolimits_{r=1}^{k+1}C_{k+1}^{r}D_0^{k+1-r}\mathbf{n}_{\partial\Omega_{ij}}^{\mathrm{T}}\cdot\left[\frac{d^r\mathbf{x}^{(\alpha)}}{dt^r}-\frac{d^r\mathbf{x}^{(0)}}{dt^r}\right]\Bigg|_{(\mathbf{x}_m^{(0)},\mathbf{x}_{m\pm}^{(\alpha)},t_{m\pm})}\\
&=\sum\nolimits_{r=1}^{k+1}C_{k+1}^{r}D_0^{k+1-r}\mathbf{n}_{\partial\Omega_{ij}}^{\mathrm{T}}\cdot[D_\alpha^{r-1}\mathbf{F}(\mathbf{x}^{(\alpha)},t,\mathbf{p}_\alpha)-D_0^{r-1}\mathbf{F}^{(0)}(\mathbf{x}^{(0)},t,\boldsymbol{\lambda})]\,\big|_{(\mathbf{x}_m^{(0)},\mathbf{x}_{m\pm}^{(\alpha)},t_{m\pm})}.
\end{aligned}
\tag{3.20}
$$

For $k=0$, one obtains

$$
G_{\partial\Omega_{ij}}^{(k,\alpha)}(\mathbf{x}_m,t_{m\pm},\mathbf{p}_\alpha,\boldsymbol{\lambda})=G_{\partial\Omega_{ij}}^{(\alpha)}(\mathbf{x}_m,t_{m\pm},\mathbf{p}_\alpha,\boldsymbol{\lambda}).
\tag{3.21}
$$

From now on, $\mathbf{n}_{\partial\Omega_{ij}}(\mathbf{x}^{(0)}) \equiv \mathbf{n}_{\partial\Omega_{ij}}(\mathbf{x}^{(0)}, t, \boldsymbol{\lambda})$.

The *G*-function is developed to measure the time-change rates of vector fields to the boundary. To measure changes on the boundary surface and determine the switching complexity, the tangential components on the tangential planes of points on the boundary surface should be discussed. Any vector field can be decomposed on the normal and tangential directions of a point on the boundary. Consider a point $\mathbf{x}_m^{(0)}$ of the flow $\mathbf{x}^{(0)}(t)$ on the boundary $\partial\Omega_{ij}$ with a normal vector $\mathbf{n}_{\partial\Omega_{ij}}(t_m)$ to measure the flow $\mathbf{x}^{(\alpha)}(t)$ at time t_m. For $t_{m+\varepsilon} = t_m + \varepsilon$, the flow $\mathbf{x}^{(\alpha)}(t)$ moves from $\mathbf{x}^{(\alpha)}(t_m)$ to $\mathbf{x}^{(\alpha)}(t_{m+\varepsilon})$, and the boundary flow $\mathbf{x}^{(0)}(t)$ moves from $\mathbf{x}^{(0)}(t_m)$ to $\mathbf{x}^{(0)}(t_{m+\varepsilon})$. Because the boundary surface is not flat, measuring characteristics of $\mathbf{x}^{(\alpha)}(t_{m+\varepsilon})$ on the non-flat boundary surface is based on a tangential plane at point $\mathbf{x}_{m+\varepsilon}^{(0)}$ with a normal vector $\mathbf{n}_{\partial\Omega_{ij}}(t_{m+\varepsilon})$, as shown in Fig. 3.3. To generalize this idea, the vector fields projected on the boundary $\partial\Omega_{ij}$ is defined as follows.

Definition 3.7. Consider a dynamical system in Eq. (2.1) in domain Ω_α ($\alpha \in \{i, j\}$) which has the flow $\mathbf{x}^{(\alpha)} = \boldsymbol{\Phi}(t_0, \mathbf{x}_0^{(\alpha)}, \mathbf{p}_\alpha, t)$ with an initial condition $(t_0, \mathbf{x}_0^{(\alpha)})$. On the boundary $\partial\Omega_{ij}$, a flow $\mathbf{x}^{(0)} = \boldsymbol{\Phi}(t_0, \mathbf{x}_0^{(0)}, \boldsymbol{\lambda}, t)$ with an initial condition $(t_0, \mathbf{x}_0^{(0)})$ is smooth. For an arbitrarily small $\varepsilon > 0$, there is a time interval $[t-\varepsilon, t)$ or $(t, t+\varepsilon]$. The *projected vector fields* on the tangential plane of the boundary $\partial\Omega_{ij}$ at time t are defined as

$$
\begin{aligned}
{}^{\mathrm{t}}\mathbf{F}^{(\alpha)}(\mathbf{x}^{(0)}, t_+) &= \lim_{\varepsilon\to 0}\frac{1}{\varepsilon}[\mathbf{H}^{(\alpha)}(\mathbf{x}^{(0)}, t+\varepsilon) - \mathbf{H}^{(\alpha)}(\mathbf{x}^{(0)}, t)], \\
{}^{\mathrm{t}}\mathbf{F}^{(\alpha)}(\mathbf{x}^{(0)}, t_-) &= \lim_{\varepsilon\to 0}\frac{1}{\varepsilon}[\mathbf{H}^{(\alpha)}(\mathbf{x}^{(0)}, t) - \mathbf{H}^{(\alpha)}(\mathbf{x}^{(0)}, t-\varepsilon)]
\end{aligned}
\tag{3.22}
$$

and the *projection* of the position vector ($\mathbf{x}^{(\alpha)}(t) - \mathbf{x}^{(0)}(t)$) on the tangential plane at point $\mathbf{x}^{(0)}(t)$ is

$$
\mathbf{H}^{(\alpha)}(\mathbf{x}^{(0)}, t) \equiv \mathbf{x}^{(\alpha)}(t) - \mathbf{x}^{(0)}(t) - \{\mathbf{N}_{\partial\Omega_{ij}}^{\mathrm{T}}(t)\cdot[\mathbf{x}^{(\alpha)}(t) - \mathbf{x}^{(0)}(t)]\}\mathbf{N}_{\partial\Omega_{ij}}(t) \tag{3.23}
$$

with the unit normal vector $\mathbf{N}_{\partial\Omega_{ij}}(t)$ at point $\mathbf{x}^{(0)}(t)$ given by

$$
\mathbf{N}_{\partial\Omega_{ij}}(t) = \frac{\mathbf{n}_{\partial\Omega_{ij}}(t)}{\|\mathbf{n}_{\partial\Omega_{ij}}(t)\|}. \tag{3.24}
$$

From the definition, one obtains ${}^{\mathrm{t}}\dot{\mathbf{x}}^{(\alpha)}(t) = {}^{\mathrm{t}}\mathbf{F}^{(\alpha)}(\mathbf{x}^{(0)}, t)$, i.e.,

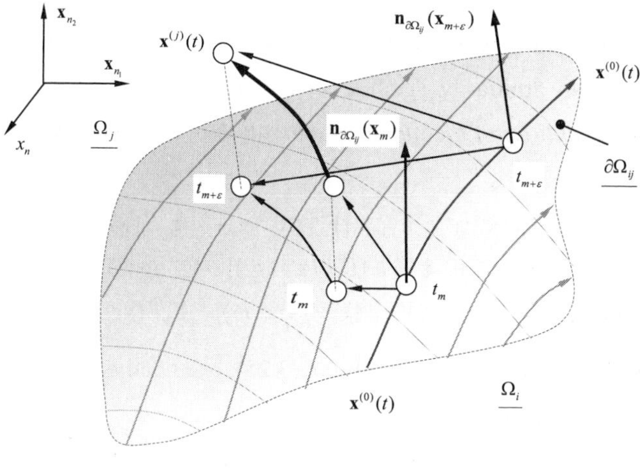

Fig. 3.3 An infinitesimal element of a real flow $\mathbf{x}^{(j)}(t)$ for time interval $(t_m, t_{m+\varepsilon}]$. The tangential component of the vector field is described through the tangential planes of a reference flow $\mathbf{x}^{(0)}(t)$ on the boundary $\partial\Omega_{ij}$ at point t_m and $t_{m+\varepsilon}$. The normal vector $\mathbf{n}_{\partial\Omega_{ij}}$ of the boundary $\partial\Omega_{ij}$ is depicted. ($n_1 + n_2 + 1 = n$)

$$\begin{aligned}
{}^{\mathrm{t}}\dot{\mathbf{x}}^{(\alpha)}(t) &= \dot{\mathbf{x}}^{(\alpha)}(t) - \{\mathbf{N}_{\partial\Omega_{ij}}^{\mathrm{T}}(t)\cdot[\mathbf{x}^{(\alpha)}(t) - \mathbf{x}^{(0)}(t)]\}D\mathbf{N}_{\partial\Omega_{ij}}^{\mathrm{T}}(t) \\
&\quad - \{D\mathbf{N}_{\partial\Omega_{ij}}^{\mathrm{T}}(t)\cdot[\mathbf{x}^{(\alpha)}(t) - \mathbf{x}^{(0)}(t)]\}\mathbf{N}_{\partial\Omega_{ij}}(t) \\
&\quad - \{\mathbf{N}_{\partial\Omega_{ij}}^{\mathrm{T}}(t)\cdot[\dot{\mathbf{x}}^{(\alpha)}(t) - \dot{\mathbf{x}}^{(0)}(t)]\}\mathbf{N}_{\partial\Omega_{ij}}(t).
\end{aligned} \tag{3.25}$$

If $\mathbf{x}^{(\alpha)}(t)$ and $\mathbf{x}^{(0)}(t)$ at time $t = t_m$ are of contact (i.e., $\mathbf{x}^{(\alpha)}(t_m) = \mathbf{x}^{(0)}(t_m)$), Equation (3.25) becomes

$${}^{\mathrm{t}}\dot{\mathbf{x}}^{(\alpha)}(t_{m\pm}) = \dot{\mathbf{x}}^{(\alpha)}(t_{m\pm}) - \{\mathbf{N}_{\partial\Omega_{ij}}^{\mathrm{T}}(t_m)\cdot[\dot{\mathbf{x}}^{(\alpha)}(t_{m\pm}) - \dot{\mathbf{x}}^{(0)}(t_m)]\}\mathbf{N}_{\partial\Omega_{ij}}^{\mathrm{T}}(t_m). \tag{3.26}$$

Using Eq. (3.1), the forgoing equation gives

$${}^{\mathrm{t}}\dot{\mathbf{x}}^{(\alpha)}(t_{m\pm}) = {}^{\mathrm{t}}\mathbf{F}^{(\alpha)}(t_{m\pm}) \equiv \mathbf{F}^{(\alpha)}(t_{m\pm}) - \{\mathbf{N}_{\partial\Omega_{ij}}^{\mathrm{T}}(t_m)\cdot\mathbf{F}^{(\alpha)}(t_{m\pm})\}\mathbf{N}_{\partial\Omega_{ij}}^{\mathrm{T}}(t_m), \tag{3.27}$$

which is for the tangential component of the vector fields $\mathbf{F}^{(\alpha)}(t_{m\pm})$ at points contacting with the boundary surface. If the vector field $\mathbf{F}^{(\alpha)}(t_{m\pm})$ is not on the boundary surface, the tangential component should be determined by Eq. (3.25). To describe the derivation of a flow $\mathbf{x}^{(\alpha)} = \boldsymbol{\Phi}(t_0, \mathbf{x}_0^{(\alpha)}, \mathbf{p}_\alpha, t)$ from the reference flow $\mathbf{x}^{(0)} = \boldsymbol{\Phi}(t_0, \mathbf{x}_0^{(0)}, \boldsymbol{\lambda}, t)$, a new quantity to measure the direction derivation of $\dot{\mathbf{x}}^{(\alpha)}$ from $\dot{\mathbf{x}}^{(0)}$ is defined as follows.

Definition 3.8. Consider a dynamic system in Eq. (2.1) in domain Ω_α ($\alpha \in \{i,j\}$) which has the flow $\mathbf{x}^{(\alpha)} = \mathbf{\Phi}(t_0, \mathbf{x}_0^{(\alpha)}, \mathbf{p}_\alpha, t)$ with an initial condition $(t_0, \mathbf{x}_0^{(\alpha)})$. On the boundary $\partial\Omega_{ij}$, a flow $\mathbf{x}^{(0)} = \mathbf{\Phi}(t_0, \mathbf{x}_0^{(0)}, \boldsymbol{\lambda}, t)$ with an initial condition $(t_0, \mathbf{x}_0^{(0)})$ is smooth. For an arbitrarily small $\varepsilon > 0$, there is a time interval $[t-\varepsilon, t)$ or $(t, t+\varepsilon]$. The *quantities of the direction derivation* of $\dot{\mathbf{x}}^{(\alpha)}$ from $\dot{\mathbf{x}}^{(0)}$ at time t are defined as

$$\begin{aligned}
{}^{\mathrm{t}}H^{(\alpha,0)}(\mathbf{x}^{(0)}, t_\pm) &= (\dot{\mathbf{x}}^{(0)})^{\mathrm{T}} \cdot {}^{\mathrm{t}}\dot{\mathbf{x}}^{(\alpha)} = [\mathbf{F}^{(0)}(\mathbf{x}^{(0)}, t_\pm]^{\mathrm{T}} \cdot {}^{\mathrm{t}}\mathbf{F}^{(\alpha)}(\mathbf{x}^{(0)}, t_\pm),\\
H^{(\alpha,0)}(\mathbf{x}^{(0)}, t_\pm) &= (\dot{\mathbf{x}}^{(0)})^{\mathrm{T}} \cdot \dot{\mathbf{x}}^{(\alpha)} = [\mathbf{F}^{(0)}(\mathbf{x}^{(0)}, t_\pm)]^{\mathrm{T}} \cdot \mathbf{F}^{(\alpha)}(\mathbf{x}^{(\alpha)}, t_\pm),\\
H^{(\alpha,\beta)}(\mathbf{x}^{(0)}, t_\pm) &= (\dot{\mathbf{x}}^{(\alpha)})^{\mathrm{T}} \cdot \dot{\mathbf{x}}^{(\beta)} = [\mathbf{F}^{(\alpha)}(\mathbf{x}^{(\alpha)}, t_\pm)]^{\mathrm{T}} \cdot \mathbf{F}^{(\beta)}(\mathbf{x}^{(\beta)}, t_\pm).
\end{aligned} \tag{3.28}$$

From the foregoing definition, with Eq. (3.25), one obtains

$$\begin{aligned}
{}^{\mathrm{t}}H^{(\alpha,0)}(\mathbf{x}_t^{(0)}, t) &= [\dot{\mathbf{x}}_t^{(0)}]^{\mathrm{T}} \cdot {}^{\mathrm{t}}\mathbf{F}^{(\alpha)}(\mathbf{x}_t^{(0)}, t)\\
&= [\dot{\mathbf{x}}_t^{(0)}]^{\mathrm{T}} \cdot \dot{\mathbf{x}}^{(\alpha)}(t) - \{\mathbf{N}_{\partial\Omega_{ij}}^{\mathrm{T}}(t) \cdot [\mathbf{x}^{(\alpha)}(t) - \mathbf{x}^{(0)}(t)]\}\\
&\quad \times \{(\dot{\mathbf{x}}_t^{(0)})^{\mathrm{T}} \cdot D\mathbf{N}_{\partial\Omega_{ij}}(t)\}\\
&= H^{(\alpha)}(\mathbf{x}_t^{(0)}, t) - \{\mathbf{N}_{\partial\Omega_{ij}}^{\mathrm{T}}(t) \cdot [\mathbf{x}^{(\alpha)}(t) - \mathbf{x}^{(0)}(t)]\}\\
&\quad \times \{(\dot{\mathbf{x}}_t^{(0)})^{\mathrm{T}} \cdot D\mathbf{N}_{\partial\Omega_{ij}}(t)\}.
\end{aligned} \tag{3.29}$$

If $\mathbf{x}^{(\alpha)}(t)$ and $\mathbf{x}^{(0)}(t)$ at time $t = t_m$ are of contact, the foregoing equation gives

$$\begin{aligned}
{}^{\mathrm{t}}H^{(\alpha,0)}(\mathbf{x}_m^{(0)}, t_{m\pm}) &= [\dot{\mathbf{x}}_m^{(0)}]^{\mathrm{T}} \cdot {}^{\mathrm{t}}\dot{\mathbf{x}}^{(\alpha)}(t_{m\pm}) = [\dot{\mathbf{x}}_m^{(0)}]^{\mathrm{T}} \cdot {}^{\mathrm{t}}\mathbf{F}^{(\alpha)}(\mathbf{x}_{m\pm}^{(\alpha)}, t_{m\pm})\\
&= [\dot{\mathbf{x}}_m^{(0)}]^{\mathrm{T}} \cdot \dot{\mathbf{x}}^{(\alpha)}(t_{m\pm}) = [\dot{\mathbf{x}}_m^{(0)}]^{\mathrm{T}} \cdot \mathbf{F}^{(\alpha)}(\mathbf{x}_{m\pm}^{(\alpha)}, t_{m\pm})\\
&= H^{(\alpha,0)}(\mathbf{x}_m^{(0)}, t_{m\pm}).
\end{aligned} \tag{3.30}$$

It is observed that the quantity for the direction derivation is the same for the contact point on the boundary surface. From Eq. (3.28), the quantities become

$$\begin{aligned}
{}^{\mathrm{t}}H^{(\alpha,0)}(\mathbf{x}^{(0)}, t_\pm) &= (\dot{\mathbf{x}}^{(0)})^{\mathrm{T}} \cdot {}^{\mathrm{t}}\dot{\mathbf{x}}^{(\alpha)} = (\|\dot{\mathbf{x}}^{(0)}\|)(\|{}^{\mathrm{t}}\dot{\mathbf{x}}^{(\alpha)}\|)\cos\theta_{(\alpha,0)}^{\mathrm{t}}\\
&= [\mathbf{F}^{(0)}(\mathbf{x}^{(0)}, t_\pm)]^{\mathrm{T}} \cdot {}^{\mathrm{t}}\mathbf{F}^{(\alpha)}(\mathbf{x}^{(0)}, t_\pm)\\
&= \|\mathbf{F}^{(0)}\| \cdot \|{}^{\mathrm{t}}\mathbf{F}^{(\alpha)}\| \cos\theta_{(\alpha,0)}^{\mathrm{t}},\\
H^{(\alpha,0)}(\mathbf{x}^{(0)}, t_\pm) &= (\dot{\mathbf{x}}^{(0)})^{\mathrm{T}} \cdot \dot{\mathbf{x}}^{(\alpha)} = (\|\dot{\mathbf{x}}^{(0)}\|)(\|\dot{\mathbf{x}}^{(\alpha)}\|)\cos\theta_{(\alpha,0)}\\
&= [\mathbf{F}^{(0)}(\mathbf{x}^{(0)}, t_\pm)]^{\mathrm{T}} \cdot \mathbf{F}^{(\alpha)}(\mathbf{x}^{(\alpha)}, t_\pm)\\
&= \|\mathbf{F}^{(0)}\| \cdot \|\mathbf{F}^{(\alpha)}\| \cos\theta_{(\alpha,0)},\\
H^{(\alpha,\beta)}(\mathbf{x}^{(0)}, t_\pm) &= (\dot{\mathbf{x}}^{(\alpha)})^{\mathrm{T}} \cdot \dot{\mathbf{x}}^{(\beta)} = (\|\dot{\mathbf{x}}^{(\alpha)}\|)(\|\dot{\mathbf{x}}^{(\beta)}\|)\cos\theta_{(\alpha,\beta)}\\
&= [\mathbf{F}^{(\alpha)}(\mathbf{x}^{(\alpha)}, t_\pm)]^{\mathrm{T}} \cdot \mathbf{F}^{(\beta)}(\mathbf{x}^{(\beta)}, t_\pm)\\
&= \|\mathbf{F}^{(\alpha)}\| \cdot \|\mathbf{F}^{(\beta)}\| \cos\theta_{(\alpha,\beta)}.
\end{aligned} \tag{3.31}$$

If ${}^{\mathrm{t}}H^{(\alpha,0)}(\mathbf{x}^{(0)},t_{\pm})>0$ (i.e., $\cos\theta^{\mathrm{t}}_{(\alpha,0)}>0$), the vector fields $\mathbf{F}^{(0)}(\mathbf{x}^{(0)},t_{\pm})$ and ${}^{\mathrm{t}}\mathbf{F}^{(\alpha)}(\mathbf{x}^{(0)},t_{\pm})$ have the same direction but not identical. If ${}^{\mathrm{t}}H^{(\alpha,0)}(\mathbf{x}^{(0)},t_{\pm})<0$ (i.e., $\cos\theta^{\mathrm{t}}_{(\alpha,0)}>0$), the vector fields $\mathbf{F}^{(0)}(\mathbf{x}^{(0)},t_{\pm})$ and ${}^{\mathrm{t}}\mathbf{F}^{(\alpha)}(\mathbf{x}^{(0)},t_{\pm})$ are in the opposite directions. If ${}^{\mathrm{t}}H^{(\alpha,0)}(\mathbf{x}^{(0)},t_{\pm})=0$ (i.e., $\cos\theta^{\mathrm{t}}_{(\alpha,0)}=0$), the vector fields $\mathbf{F}^{(0)}(\mathbf{x}^{(0)},t_{\pm})$ and ${}^{\mathrm{t}}\mathbf{F}^{(\alpha)}(\mathbf{x}^{(0)},t_{\pm})$ are perpendicular each other. The functions $H^{(\alpha,0)}(\mathbf{x}^{(0)},t_{\pm})$ and $H^{(\alpha,\beta)}(\mathbf{x}^{(0)},t_{\pm})$ for the corresponding vector fields have the same properties as ${}^{\mathrm{t}}H^{(\alpha,0)}(\mathbf{x}^{(0)},t_{\pm})$. To better understand the property of the flow switching at the boundary, the projection of a flow $\mathbf{x}^{(\alpha)}(t)$ in domain Ω_α ($\alpha\in\{i,j\}$) on the boundary $\partial\Omega_{ij}$ is defined as follows.

Definition 3.9. For a dynamical system in Eq. (2.1), a flow $\mathbf{x}^{(\alpha)}=\mathbf{\Phi}(t_0,\mathbf{x}_0^{(\alpha)},\mathbf{p},t)$ with an initial condition $(t_0,\mathbf{x}_0^{(\alpha)})$ in domain Ω_α ($\alpha\in\{i,j\}$) exist. For any time t, if there is a point $\mathbf{x}^{(0)}(t)$ on the boundary $\partial\Omega_{ij}$ to make $\mathbf{x}^{(0)}(t)$ and $\mathbf{x}^{(\alpha)}(t)$ on the same normal vector $\mathbf{n}_{\partial\Omega_{ij}}(\mathbf{x}^{(0)}(t))$, then the trajectory of all projected points $\mathbf{x}^{(0)}(t)$ of $\mathbf{x}^{(\alpha)}(t)$ on the boundary $\partial\Omega_{ij}$ is called the *projection* of the trajectory $\mathbf{x}^{(\alpha)}(t)$ on the boundary $\partial\Omega_{ij}$, which is expressed by $\mathbf{x}^{(\alpha,0)}(t)$.

Such a definition of the projection of the trajectory $\mathbf{x}^{(\alpha)}(t)$ ($\alpha\in\{i,j\}$) on the boundary $\partial\Omega_{ij}$ is sketched in Fig. 3.4. The flow $\mathbf{x}^{(j)}(t)$ is on the same normal vector of $\mathbf{x}^{(0)}(t)$ for time t. The two points $\mathbf{x}^{(j)}(t_m)$ and $\mathbf{x}^{(j)}(t_{m+\varepsilon})$ are projected on the boundary surface $\partial\Omega_{ij}$ with two points $\mathbf{x}_1^{(0)}(t_m)$ and $\mathbf{x}_2^{(0)}(t_{m+\varepsilon})$ on the two different flows $\mathbf{x}_1^{(0)}(t)$ and $\mathbf{x}_2^{(0)}(t)$. The normal vector $\mathbf{n}_{\partial\Omega_{ij}}$ on the boundary is depicted. It is observed all the projected points of all $\mathbf{x}_1^{(0)}(t_m)$ with varying t_m will form a trace on the boundary $\partial\Omega_{ij}$. Thus one can use the projected trajectories of the flows $\mathbf{x}^{(\alpha)}(t)$ on the two domains to be compared with the flow $\mathbf{x}^{(0)}(t)$ on the boundary $\partial\Omega_{ij}$, which help one understand the dynamics of two vectors switching in discontinuous dynamics. In Fig. 3.5(a), the trajectories of two real flows $\mathbf{x}^{(\alpha)}(t)$ ($\alpha=i,j$), projected on the boundary $\partial\Omega_{ij}$, are presented. The trajectory projection of the imaginary flows $\mathbf{x}_\alpha^{(\beta)}(t)$ is also presented in Fig. 3.5(b). The derivations of vector fields to the vector fields on the boundary can be clearly illustrated. After the flow projected on the boundary $\partial\Omega_{ij}$, the flows *before* and *after* switching on the boundary can be compared with the flow on the boundary, which can be used to investigate the characteristics of the flow in the vicinity of the switching point.

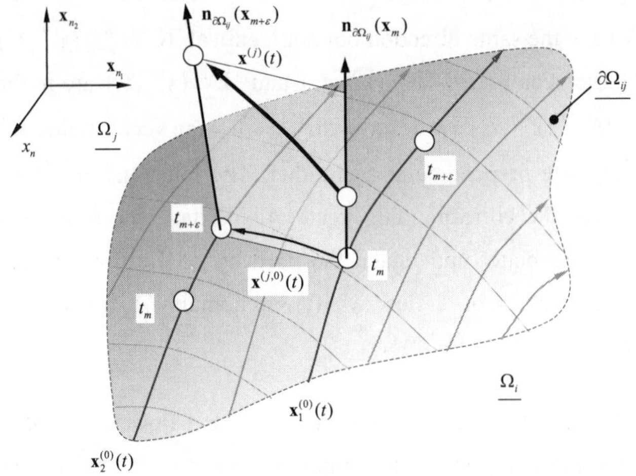

Fig. 3.4 Projection of the trajectory of a real flow $\mathbf{x}^{(j)}(t)$ on the boundary for a time interval $(t_m, t_{m+\varepsilon}]$. The flow $\mathbf{x}^{(j)}(t)$ is on the same normal vector of $\mathbf{x}^{(0)}(t)$ for time t. The two points $\mathbf{x}^{(j)}(t_m)$ and $\mathbf{x}^{(j)}(t_{m+\varepsilon})$ are projected on the boundary surface $\partial\Omega_{ij}$ with two points $\mathbf{x}_1^{(0)}(t_m)$ and $\mathbf{x}_2^{(0)}(t_{m+\varepsilon})$ on the two different flows $\mathbf{x}_1^{(0)}(t)$ and $\mathbf{x}_2^{(0)}(t)$. The normal vector $\mathbf{n}_{\partial\Omega_{ij}}$ on the boundary is depicted. ($n_1 + n_2 + 1 = n$)

3.3. Passable flows

Compared to the continuous dynamical systems, discontinuous dynamical systems possess many passable flows to the boundary $\partial\Omega_{ij}$ because $G^{(i)}_{\partial\Omega_{ij}} \neq G^{(j)}_{\partial\Omega_{ij}}$. A passable flow to a specific boundary is discussed first, as sketched in Fig. 3.6. $\mathbf{x}^{(i)}(t)$ and $\mathbf{x}^{(j)}(t)$ represent the *real* flows in domains Ω_i and Ω_j, respectively. They are depicted by the thin solid curves. $\mathbf{x}_i^{(j)}(t)$ and $\mathbf{x}_j^{(i)}(t)$ are the *imaginary* flows in domains Ω_i and Ω_j, respectively controlled by the vector fields on Ω_j and Ω_i. Such imaginary flows are depicted by dashed curves. The hollow circles are switching points, and shaded circles are starting points. The detail discussion of the real and imaginary flows can be found from Luo (2005b, 2006b). The flow on the boundary is described by $\mathbf{x}^{(0)}(t)$. The normal and tangential vectors $\mathbf{n}_{\partial\Omega_{ij}}$ and $\mathbf{t}_{\partial\Omega_{ij}}$ on the boundary are depicted. The passable flow to a specific boundary is defined as follows.

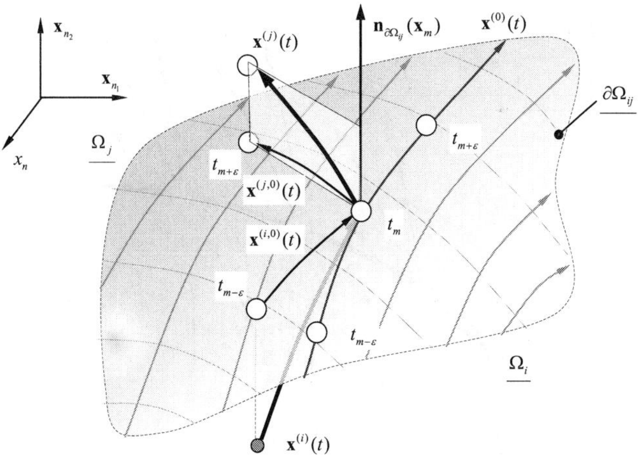

(a)

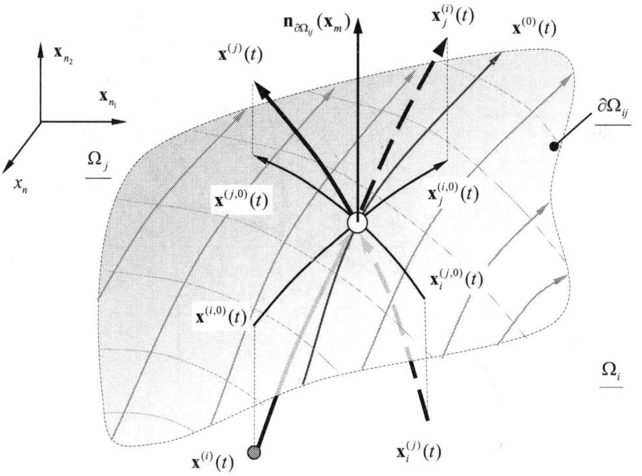

(b)

Fig. 3.5 Projection of the real and imaginary flows on the boundary surface: **(a)** real flows only, **(b)** real and imaginary flows. *Real* flows $\mathbf{x}^{(i)}(t)$ and $\mathbf{x}^{(j)}(t)$ in domains Ω_i and Ω_j are depicted by thin solid curves, respectively. *Imaginary* flows $\mathbf{x}_i^{(j)}(t)$ and $\mathbf{x}_j^{(i)}(t)$ in domains Ω_i and Ω_j, which are defined by the vector fields in Ω_j and Ω_i are depicted by dashed curves respectively. The flow on the boundary is described by $\mathbf{x}^{(0)}(t)$. The normal vector $\mathbf{n}_{\partial\Omega_{ij}}$ of the boundary is depicted. Hollow circles are for switching points on the boundary, and filled circles are for starting points. ($n_1 + n_2 + 1 = n$)

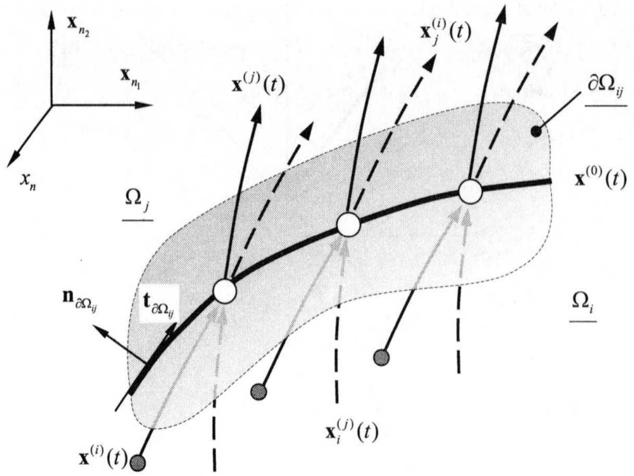

(a)

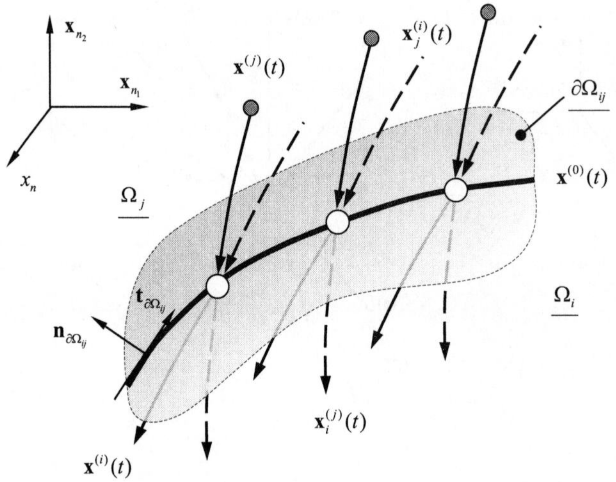

(b)

Fig. 3.6. Passable flows: **(a)** from Ω_i to Ω_j with the $(2k_i : 2k_j)$-order and **(b)** from Ω_j to Ω_i with the $(2k_j : 2k_i)$-order. *Real* flows $\mathbf{x}^{(i)}(t)$ and $\mathbf{x}^{(j)}(t)$ in domains Ω_i and Ω_j are depicted by thin solid curves, respectively. *Imaginary* flows $\mathbf{x}_i^{(j)}(t)$ and $\mathbf{x}_j^{(i)}(t)$ in domains Ω_i and Ω_j, which are defined by the vector fields in Ω_j and Ω_i are depicted by dashed curves, respectively. The flow on the boundary is described by $\mathbf{x}^{(0)}(t)$. The normal and tangential vectors $\mathbf{n}_{\partial\Omega_{ij}}$ and $\mathbf{t}_{\partial\Omega_{ij}}$ of the boundary are depicted. Hollow circles are for switching points on the boundary and filled circles are for starting points. ($n_1 + n_2 + 1 = n$).

Definition 3.10. For a discontinuous dynamical system in Eq. (2.1), there is a point $\mathbf{x}^{(0)}(t_m) \equiv \mathbf{x}_m \in \partial\Omega_{ij}$ at time t_m between two adjacent domains Ω_α ($\alpha = i, j$). For an arbitrarily small $\varepsilon > 0$, there are two time intervals $[t_{m-\varepsilon}, t_m)$ and $(t_m, t_{m+\varepsilon}]$. Suppose $\mathbf{x}^{(i)}(t_{m-}) = \mathbf{x}_m = \mathbf{x}^{(j)}(t_{m+})$. The flow $\mathbf{x}^{(i)}(t)$ and $\mathbf{x}^{(j)}(t)$ to the boundary $\partial\Omega_{ij}$ is *semi-passable* from domain Ω_i to Ω_j if

$$\begin{aligned}
&\text{either} \quad \left.\begin{aligned} \mathbf{n}^{\mathrm{T}}_{\partial\Omega_{ij}}(\mathbf{x}^{(0)}_{m-\varepsilon})\cdot[\mathbf{x}^{(0)}_{m-\varepsilon} - \mathbf{x}^{(i)}_{m-\varepsilon}] > 0 \\ \mathbf{n}^{\mathrm{T}}_{\partial\Omega_{ij}}(\mathbf{x}^{(0)}_{m+\varepsilon})\cdot[\mathbf{x}^{(j)}_{m+\varepsilon} - \mathbf{x}^{(0)}_{m+\varepsilon}] > 0 \end{aligned}\right\} \text{for } \mathbf{n}_{\partial\Omega_{ij}} \to \Omega_j, \\
&\text{or} \quad \left.\begin{aligned} \mathbf{n}^{\mathrm{T}}_{\partial\Omega_{ij}}(\mathbf{x}^{(0)}_{m-\varepsilon})\cdot[\mathbf{x}^{(0)}_{m-\varepsilon} - \mathbf{x}^{(i)}_{m-\varepsilon}] < 0 \\ \mathbf{n}^{\mathrm{T}}_{\partial\Omega_{ij}}(\mathbf{x}^{(0)}_{m+\varepsilon})\cdot[\mathbf{x}^{(j)}_{m+\varepsilon} - \mathbf{x}^{(0)}_{m+\varepsilon}] < 0 \end{aligned}\right\} \text{for } \mathbf{n}_{\partial\Omega_{ij}} \to \Omega_i.
\end{aligned} \tag{3.32}$$

Since flow properties in domains Ω_i and Ω_j are different at point $(t_m, \mathbf{x}_m)$, $G^{(i)}_{\partial\Omega_{ij}} \neq G^{(j)}_{\partial\Omega_{ij}}$ to $\partial\Omega_{ij}$. As in Chapter 2, the necessary and sufficient conditions for such a passable flow on $\partial\Omega_{ij}$ from domain Ω_i to Ω_j are given as follows.

Theorem 3.1. *For a discontinuous dynamical system in Eq.* (2.1), *there is a point* $\mathbf{x}^{(0)}(t_m) \equiv \mathbf{x}_m \in \partial\Omega_{ij}$ *at time* t_m *between two adjacent domains* Ω_α ($\alpha = i, j$). *For an arbitrarily small* $\varepsilon > 0$, *there are two time intervals* $[t_{m-\varepsilon}, t_m)$ *and* $(t_m, t_{m+\varepsilon}]$. *Suppose* $\mathbf{x}^{(i)}(t_{m-}) = \mathbf{x}_m = \mathbf{x}^{(j)}(t_{m+})$. *Two flows* $\mathbf{x}^{(i)}(t)$ *and* $\mathbf{x}^{(j)}(t)$ *are* $C^{r_i}_{[t_{m-\varepsilon}, t_m)}$ *and* $C^{r_j}_{(t_m, t_{m+\varepsilon}]}$*-continuous for time t, respectively, and* $\| d^{r_\alpha+1}\mathbf{x}^{(\alpha)}/dt^{r_\alpha+1} \| < \infty$ ($r_\alpha \geq 1$, $\alpha = i, j$). *The flow* $\mathbf{x}^{(i)}(t)$ *and* $\mathbf{x}^{(j)}(t)$ *to the boundary* $\partial\Omega_{ij}$ *is semi-passable from domain* Ω_i *to* Ω_j *if and only if*

$$\begin{aligned}
&\textit{either} \quad \left.\begin{aligned} G^{(i)}_{\partial\Omega_{ij}}(\mathbf{x}_m, t_{m-}, \mathbf{p}_i, \boldsymbol{\lambda}) > 0 \\ G^{(j)}_{\partial\Omega_{ij}}(\mathbf{x}_m, t_{m+}, \mathbf{p}_j, \boldsymbol{\lambda}) > 0 \end{aligned}\right\} \textit{for } \mathbf{n}_{\partial\Omega_{ij}} \to \Omega_j, \\
&\textit{or} \quad \left.\begin{aligned} G^{(i)}_{\partial\Omega_{ij}}(\mathbf{x}_m, t_{m-}, \mathbf{p}_i, \boldsymbol{\lambda}) < 0 \\ G^{(j)}_{\partial\Omega_{ij}}(\mathbf{x}_m, t_{m+}, \mathbf{p}_j, \boldsymbol{\lambda}) < 0 \end{aligned}\right\} \textit{for } \mathbf{n}_{\partial\Omega_{ij}} \to \Omega_i.
\end{aligned} \tag{3.33}$$

Proof. For a point $\mathbf{x}_m \in \partial\Omega_{ij}$, suppose $\mathbf{x}^{(i)}(t_{m-}) = \mathbf{x}_m = \mathbf{x}^{(j)}(t_{m+})$. Two flows $\mathbf{x}^{(i)}(t)$ and $\mathbf{x}^{(j)}(t)$ are $C^{r_i}_{[t_{m-\varepsilon}, t_m)}$ and $C^{r_j}_{(t_m, t_{m+\varepsilon}]}$-continuous ($r_\alpha \geq 1$, $\alpha = i, j$) for time t. $\| \ddot{\mathbf{x}}^{(\alpha)}(t) \| < \infty$ for $0 < \varepsilon << 1$. For $a \in [t_{m-\varepsilon}, t_{m-})$ or $a \in (t_{m+}, t_{m+\varepsilon}]$, the Taylor series expansions of $\mathbf{x}^{(\alpha)}(t_{m\pm\varepsilon})$ with $t_{m\pm\varepsilon} = t_m \pm \varepsilon$ ($\alpha \in \{i, j\}$) to $\mathbf{x}^{(\alpha)}(a)$ give

$$\mathbf{x}_{m\pm\varepsilon}^{(\alpha)} \equiv \mathbf{x}^{(\alpha)}(t_{m\pm} \pm \varepsilon) = \mathbf{x}^{(\alpha)}(a) + \dot{\mathbf{x}}^{(\alpha)}(a)(t_{m\pm} \pm \varepsilon - a) + o(t_{m\pm} \pm \varepsilon - a),$$

As $a \to t_{m\pm}$, taking the limit of the foregoing equation leads to

$$\mathbf{x}_{m\pm\varepsilon}^{(\alpha)} \equiv \mathbf{x}^{(\alpha)}(t_{m\pm} \pm \varepsilon) = \mathbf{x}^{(\alpha)}(t_{m\pm}) + \dot{\mathbf{x}}^{(\alpha)}(t_{m\pm})(\pm\varepsilon) + o(\pm\varepsilon),$$

With Eq. (2.1), one obtains

$$\mathbf{x}_{m\pm\varepsilon}^{(\alpha)} = \mathbf{x}^{(\alpha)}(t_{m\pm}) + \mathbf{F}^{(\alpha)}(t_{m\pm})(\pm\varepsilon) + o(\pm\varepsilon),$$

In a similar fashion, the flow on the boundary is expressed by

$$\begin{aligned}\mathbf{x}_{m\pm\varepsilon}^{(0)} &\equiv \mathbf{x}^{(0)}(t_{m\pm} \pm \varepsilon) = \mathbf{x}^{(0)}(t_{m\pm}) + \dot{\mathbf{x}}^{(0)}(t_{m\pm})(\pm\varepsilon) + o(\pm\varepsilon),\\ &= \mathbf{x}^{(0)}(t_{m\pm}) + \mathbf{F}^{(0)}(t_{m\pm})(\pm\varepsilon) + o(\pm\varepsilon),\end{aligned}$$

$$\mathbf{n}_{\partial\Omega_{ij}}(\mathbf{x}_m^{(0)}) = \mathbf{n}_{\partial\Omega_{ij}}(\mathbf{x}_m^{(0)}) + D_0\mathbf{n}_{\partial\Omega_{ij}}(\mathbf{x}_m^{(0)})(\pm\varepsilon) + o(\pm\varepsilon).$$

The ignorance of the ε^2-terms and high order terms, the deformation of the above equation and left multiplication of $\mathbf{n}_{\partial\Omega_{ij}}$ gives

$$\begin{aligned}\mathbf{n}_{\partial\Omega_{ij}}^{\mathrm{T}}(\mathbf{x}_{m-\varepsilon}^{(0)})\cdot[\mathbf{x}_{m-\varepsilon}^{(i)} - \mathbf{x}_{m-\varepsilon}^{(0)}] &= \mathbf{n}_{\partial\Omega_{ij}}^{\mathrm{T}}(\mathbf{x}_m^{(0)})\cdot[\mathbf{x}_{m-}^{(i)} - \mathbf{x}_m^{(0)}] - \varepsilon G_{\partial\Omega_{ij}}^{(i)}(\mathbf{x}_m, t_m, \mathbf{p}_i, \boldsymbol{\lambda}),\\ \mathbf{n}_{\partial\Omega_{ij}}^{\mathrm{T}}(\mathbf{x}_{m+\varepsilon}^{(0)})\cdot[\mathbf{x}_{m+\varepsilon}^{(j)} - \mathbf{x}_{m+\varepsilon}^{(0)}] &= \mathbf{n}_{\partial\Omega_{ij}}^{\mathrm{T}}(\mathbf{x}_m^{(0)})\cdot[\mathbf{x}_{m+}^{(j)} - \mathbf{x}_m^{(0)}] + \varepsilon G_{\partial\Omega_{ij}}^{(j)}(\mathbf{x}_m, t_m, \mathbf{p}_j, \boldsymbol{\lambda}).\end{aligned}$$

Because of $\mathbf{x}_{m\pm}^{(\alpha)} = \mathbf{x}_m^{(0)} = \mathbf{x}_m$, the foregoing equation becomes

$$\begin{aligned}\mathbf{n}_{\partial\Omega_{ij}}^{\mathrm{T}}(\mathbf{x}_{m-\varepsilon}^{(0)})\cdot[\mathbf{x}_{m-\varepsilon}^{(0)} - \mathbf{x}_{m-\varepsilon}^{(i)}] &= \varepsilon G_{\partial\Omega_{ij}}^{(i)}(\mathbf{x}_m, t_{m-}, \mathbf{p}_i, \boldsymbol{\lambda}),\\ \mathbf{n}_{\partial\Omega_{ij}}^{\mathrm{T}}(\mathbf{x}_{m+\varepsilon}^{(0)})\cdot[\mathbf{x}_{m+\varepsilon}^{(j)} - \mathbf{x}_{m+\varepsilon}^{(0)}] &= \varepsilon G_{\partial\Omega_{ij}}^{(j)}(\mathbf{x}_m, t_{m+}, \mathbf{p}_j, \boldsymbol{\lambda}).\end{aligned}$$

With Eq. (3.33), the foregoing equation gives Eq. (3.32). Using Eq. (3.32), Equation (3.33) can be obtained. This theorem is proved. ■

If the boundary $\partial\Omega_{ij}$ is independent of time, using Eq. (3.14), the above theorem is identical to Theorems 2.1 and 2.2 (also see, Luo, 2005a, 2006) owing to the zero-order contact between the flow and boundary. In Chapter 2, only the fundamental passable flow to the boundary was discussed. In Luo (2005a, 2006), the semi-passable flow with the higher order singularity to the boundary was discussed. However, the theory is only for either the *plane* boundary surface or the higher-order contact of the flow and the boundary surface. For the general case, the *G*-function in Section 3.2 should be used to describe the $(2k_i : 2k_j)$-semi-passable flow and the $(2k_i : 2k_j - 1)$-semi-passable flow to the boundary. Without any switching law or transport law on the boundary, the two semi-passable flow can be described by the $(2k_i : m_j)$-semi-passable flow ($k_i, m_j \in \mathbb{N}$) as follows.

Definition 3.11. For a discontinuous dynamical system in Eq. (2.1), there is a point $\mathbf{x}^{(0)}(t_m) \equiv \mathbf{x}_m \in \partial\Omega_{ij}$ at time t_m between two adjacent domains Ω_α ($\alpha = i, j$). For an arbitrarily small $\varepsilon > 0$, there are two time intervals $[t_{m-\varepsilon}, t_m)$ and $(t_m, t_{m+\varepsilon}]$. Suppose $\mathbf{x}^{(i)}(t_{m-}) = \mathbf{x}_m = \mathbf{x}^{(j)}(t_{m+})$. A flow $\mathbf{x}^{(i)}(t)$ is $C^{r_i}_{[t_{m-\varepsilon}, t_m)}$ continuous for time t with $\| d^{r_i+1}\mathbf{x}^{(i)} / dt^{r_i+1} \| < \infty$ ($r_i \geq 2k_i + 1$), and a flow $\mathbf{x}^{(j)}(t)$ is $C^{r_j}_{(t_m, t_{m+\varepsilon}]}$-continuous with $\| d^{r_j+1}\mathbf{x}^{(j)} / dt^{r_j+1} \| < \infty$ ($r_j \geq m_j + 1$). The flow $\mathbf{x}^{(i)}(t)$ of the $(2k_i)$th -order and $\mathbf{x}^{(j)}(t)$ of the (m_j)th -order to the boundary $\partial\Omega_{ij}$ is $(2k_i : m_j)$ *-semi-passable* from domain Ω_i to Ω_j if

$$
\begin{aligned}
&G^{(s,i)}_{\partial\Omega_{ij}}(\mathbf{x}_m, t_{m-}, \mathbf{p}_i, \boldsymbol{\lambda}) = 0 \ \text{ for } s = 0, 1, \cdots, 2k_i - 1,\\
&G^{(2k_i,i)}_{\partial\Omega_{ij}}(\mathbf{x}_m, t_{m-}, \mathbf{p}_i, \boldsymbol{\lambda}) \neq 0,
\end{aligned}
\tag{3.34}
$$

$$
\begin{aligned}
&G^{(s,j)}_{\partial\Omega_{ij}}(\mathbf{x}_m, t_{m+}, \mathbf{p}_j, \boldsymbol{\lambda}) = 0 \ \text{ for } s = 0, 1, \cdots, m_j - 1,\\
&G^{(m_j,j)}_{\partial\Omega_{ij}}(\mathbf{x}_m, t_{m+}, \mathbf{p}_j, \boldsymbol{\lambda}) \neq 0,
\end{aligned}
\tag{3.35}
$$

$$
\begin{aligned}
&\text{either} \left.\begin{aligned} \mathbf{n}^{\mathrm{T}}_{\partial\Omega_{ij}}(\mathbf{x}^{(0)}_{m-\varepsilon}) \cdot [\mathbf{x}^{(0)}_{m-\varepsilon} - \mathbf{x}^{(i)}_{m-\varepsilon}] > 0 \\ \mathbf{n}^{\mathrm{T}}_{\partial\Omega_{ij}}(\mathbf{x}^{(0)}_{m+\varepsilon}) \cdot [\mathbf{x}^{(j)}_{m+\varepsilon} - \mathbf{x}^{(0)}_{m+\varepsilon}] > 0 \end{aligned}\right\} \text{ for } \mathbf{n}_{\partial\Omega_{ij}} \to \Omega_j,\\
&\text{or} \left.\begin{aligned} \mathbf{n}^{\mathrm{T}}_{\partial\Omega_{ij}}(\mathbf{x}^{(0)}_{m-\varepsilon}) \cdot [\mathbf{x}^{(0)}_{m-\varepsilon} - \mathbf{x}^{(i)}_{m-\varepsilon}] < 0 \\ \mathbf{n}^{\mathrm{T}}_{\partial\Omega_{ij}}(\mathbf{x}^{(0)}_{m+\varepsilon}) \cdot [\mathbf{x}^{(j)}_{m+\varepsilon} - \mathbf{x}^{(0)}_{m+\varepsilon}] < 0 \end{aligned}\right\} \text{ for } \mathbf{n}_{\partial\Omega_{ij}} \to \Omega_i.
\end{aligned}
\tag{3.36}
$$

If $m_j = 2k_j$, the $(2k_i : 2k_j)$ -passable flow can be sketched as in Fig. 3.6. However, for $m_j = 2k_j - 1$, the $(2k_i : 2k_j - 1)$ -passable flow from domain Ω_i to Ω_j is sketched in Fig. 3.7(a). The tangential flow of the $(2k_j - 1)$th -order exists in domain Ω_j. The dotted curves represent the tangential curves to the boundary for time $t \in [t_{m-\varepsilon}, t_m)$. The starting point of the flow is $(t_{m-\varepsilon}, \mathbf{x}^{(i)}_{m-\varepsilon})$ in domain Ω_i. If the flow arrives to the point $(t_m, \mathbf{x}_m)$ of the boundary $\partial\Omega_{ij}$, the flow will follow the tangential flow in domain Ω_j. The $(2k_j : 2k_i - 1)$ -passable flow from domain Ω_j to Ω_i is presented in Fig. 3.7(b) with the same behavior as in Fig. 3.7(a). So, a new semi-passable flow is formed as the post-transversal, tangential flow discussed in Luo (2005a, 2006). From the definition of the $(2k_i : m_j)$ -passable flow, the corresponding necessary and sufficient conditions can be given by the following theorem.

Theorem 3.2. *For a discontinuous dynamical system in Eq.* (2.1), *there is a point*

$\mathbf{x}^{(0)}(t_m) \equiv \mathbf{x}_m \in \partial\Omega_{ij}$ *at time* t_m *between two adjacent domains* Ω_α $(\alpha = i, j)$. *For an arbitrarily small* $\varepsilon > 0$, *there are two time intervals* $[t_{m-\varepsilon}, t_m)$ *and* $(t_m, t_{m+\varepsilon}]$. *Suppose* $\mathbf{x}^{(i)}(t_{m-}) = \mathbf{x}_m = \mathbf{x}^{(j)}(t_{m+})$. *A flow* $\mathbf{x}^{(i)}(t)$ *is* $C^{r}_{[t_{m-\varepsilon}, t_m)}$ *continuous for time t with* $\| d^{r_i+1}\mathbf{x}^{(i)} / dt^{r_i+1} \| < \infty$ $(r_i \geq 2k_i + 1)$, *and a flow* $\mathbf{x}^{(j)}(t)$ *is* $C^{r_j}_{(t_m, t_{m+\varepsilon}]}$*-continuous with* $\| d^{r_j+1}\mathbf{x}^{(j)} / dt^{r_j+1} \| < \infty$ $(r_j \geq m_j + 1)$. *The flow* $\mathbf{x}^{(i)}(t)$ *of the* $(2k_i)$th *-order and* $\mathbf{x}^{(j)}(t)$ *of the* (m_j)th *-order to the boundary* $\partial\Omega_{ij}$ *is* $(2k_i : m_j)$*-semi-passable from domain* Ω_i *to* Ω_j *if and only if*

$$G^{(s,i)}_{\partial\Omega_{ij}}(\mathbf{x}_m, t_{m-}, \mathbf{p}_i, \boldsymbol{\lambda}) = 0 \;\; for \;\; s = 0, 1, \cdots, 2k_i - 1; \tag{3.37}$$

$$G^{(s,j)}_{\partial\Omega_{ij}}(\mathbf{x}_m, t_{m+}, \mathbf{p}_j, \boldsymbol{\lambda}) = 0 \;\; for \;\; s = 0, 1, \cdots, m_j - 1; \tag{3.38}$$

$$\begin{aligned} &either \quad \left.\begin{aligned} G^{(2k_i,i)}_{\partial\Omega_{ij}}(\mathbf{x}_m, t_{m-}, \mathbf{p}_i, \boldsymbol{\lambda}) > 0 \\ G^{(m_j,j)}_{\partial\Omega_{ij}}(\mathbf{x}_m, t_{m+}, \mathbf{p}_j, \boldsymbol{\lambda}) > 0 \end{aligned}\right\} for \;\; \mathbf{n}_{\partial\Omega_{ij}} \to \Omega_j, \\ &or \quad \left.\begin{aligned} G^{(2k_i,i)}_{\partial\Omega_{ij}}(\mathbf{x}_m, t_{m-}, \mathbf{p}_i, \boldsymbol{\lambda}) < 0 \\ G^{(m_j,j)}_{\partial\Omega_{ij}}(\mathbf{x}_m, t_{m+}, \mathbf{p}_j, \boldsymbol{\lambda}) < 0 \end{aligned}\right\} for \;\; \mathbf{n}_{\partial\Omega_{ij}} \to \Omega_i. \end{aligned} \tag{3.39}$$

Proof. For a point $\mathbf{x}_m \in \partial\Omega_{ij}$, suppose $\mathbf{x}^{(i)}(t_{m-}) = \mathbf{x}_m = \mathbf{x}^{(j)}(t_{m+})$. The flow $\mathbf{x}^{(i)}(t)$ is $C^{r_i}_{[t_{m-\varepsilon}, t_m)}$-continuous ($r_i \geq 2k_i + 1$) for time *t* and $\| d^{r_i+1}\mathbf{x}^{(i)} / dt^{r_i+1} \| < \infty$. The flow $\mathbf{x}^{(j)}(t)$ is $C^{r_j}_{(t_m, t_{m+\varepsilon}]}$-continuous for time *t*, and $\| d^{r_j+1}\mathbf{x}^{(j)} / dt^{r_j+1} \| < \infty$ ($r_j \geq m_j + 1$). Equations (3.37) and (3.38) are identical to the first equations of Eq. (3.34) and (3.35). Equation (3.39) implies the second equations of Eq. (3.34) and (3.35). For $a \in [t_{m-\varepsilon}, t_{m-})$ or $b \in (t_{m+}, t_{m+\varepsilon}]$, the Taylor series of $\mathbf{x}^{(i)}(t_{m-\varepsilon})$ and $\mathbf{x}^{(j)}(t_{m+\varepsilon})$ at $\mathbf{x}^{(i)}(a)$ and $\mathbf{x}^{(j)}(b)$ up to the ε^{2k_i+1} and ε^{m_j+1}-terms give

$$\begin{aligned} \mathbf{x}^{(i)}_{m-\varepsilon} &\equiv \mathbf{x}^{(i)}(t_{m-} - \varepsilon) \\ &= \mathbf{x}^{(i)}(a) + \sum\nolimits_{s=1}^{2k_i} \left.\frac{d^s\mathbf{x}^{(i)}}{dt^s}\right|_{t=a} (t_{m-} - \varepsilon - a)^s + \left.\frac{d^{2k_i+1}\mathbf{x}^{(i)}}{dt^{2k_i+1}}\right|_{t=a} (t_{m-} - \varepsilon - a)^{2k_i+1} \\ &\quad + o((t_{m-} - \varepsilon - a)^{2k_i+1}), \end{aligned}$$

$$\begin{aligned} \mathbf{x}^{(j)}_{m+\varepsilon} &\equiv \mathbf{x}^{(j)}(t_{m+} + \varepsilon) \\ &= \mathbf{x}^{(j)}(b) + \sum\nolimits_{s=1}^{m_j} \left.\frac{d^s\mathbf{x}^{(j)}}{dt^s}\right|_{t=b} (t_{m+} + \varepsilon - b)^s + \left.\frac{d^{m_j+1}\mathbf{x}^{(i)}}{dt^{m_j+1}}\right|_{t=b} (t_{m+} + \varepsilon - b)^{m_j+1} \\ &\quad + o((t_{m+} + \varepsilon - b)^{m_j+1}), \end{aligned}$$

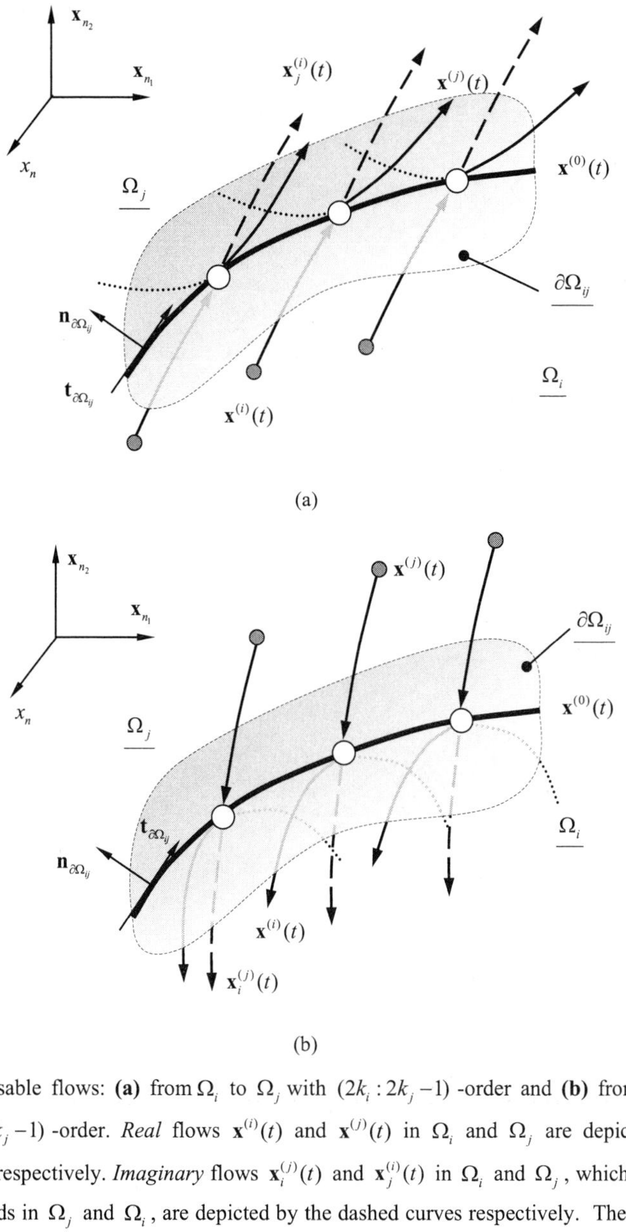

(b)

Fig. 3.7. Passable flows: **(a)** from Ω_i to Ω_j with $(2k_i : 2k_j - 1)$ -order and **(b)** from Ω_j to Ω_i with $(2k_j : 2k_j - 1)$ -order. *Real* flows $\mathbf{x}^{(i)}(t)$ and $\mathbf{x}^{(j)}(t)$ in Ω_i and Ω_j are depicted by thin solid curves, respectively. *Imaginary* flows $\mathbf{x}_i^{(j)}(t)$ and $\mathbf{x}_j^{(i)}(t)$ in Ω_i and Ω_j, which are defined by vector fields in Ω_j and Ω_i, are depicted by the dashed curves respectively. The flow on the boundary is described by $\mathbf{x}^{(0)}(t)$. The normal and tangential vectors $\mathbf{n}_{\partial\Omega_{ij}}$ and $\mathbf{t}_{\partial\Omega_{ij}}$ of the boundary are depicted. Dotted curves represent tangential flows before time t_{m+}. Hollow circles are for switching points on the boundary, and filled circles are for starting points. ($n_1 + n_2 + 1 = n$)

As $a \to t_{m-}$ and $b \to t_{m+}$, taking the limit of the foregoing equations leads to

$$\begin{aligned}\mathbf{x}_{m-\varepsilon}^{(i)} &\equiv \mathbf{x}^{(i)}(t_{m-}-\varepsilon)\\ &= \mathbf{x}^{(i)}(t_{m-}) + \sum_{s=1}^{2k_i} \frac{d^s\mathbf{x}^{(i)}}{dt^s}\Big|_{t=t_{m-}} (-\varepsilon)^s + \frac{d^{2k_i+1}\mathbf{x}^{(i)}}{dt^{2k_i+1}}\Big|_{t=t_{m-}} (-\varepsilon)^{2k_i+1} + o((-\varepsilon)^{2k_i+1})\\ &= \mathbf{x}^{(i)}(t_{m-}) + \sum_{s=1}^{2k_i} D_i^s\mathbf{F}^{(i)}(t_{m-})(-\varepsilon)^s + D_i^{2k_i+1}\mathbf{F}^{(i)}(t_{m-})(-\varepsilon)^{2k_i+1} + o((-\varepsilon)^{2k_i+1}),\end{aligned}$$

$$\begin{aligned}\mathbf{x}_{m+\varepsilon}^{(j)} &\equiv \mathbf{x}^{(j)}(t_{m+}+\varepsilon)\\ &= \mathbf{x}^{(j)}(t_{m+}) + \sum_{s=1}^{m_j} \frac{d^s\mathbf{x}^{(j)}}{dt^s}\Big|_{t=t_{m+}} \varepsilon^s + \frac{d^{m_j+1}\mathbf{x}^{(i)}}{dt^{m_j+1}}\Big|_{t=t_{m+}} \varepsilon^{m_j+1} + o(\varepsilon^{m_j+1})\\ &= \mathbf{x}^{(j)}(t_{m+}) + \sum_{s=1}^{m_j} D_j^s\mathbf{F}^{(j)}(t_{m+})\varepsilon^s + D_j^{m_j+1}\mathbf{F}^{(j)}(t_{m+})\varepsilon^{m_j+1} + o(\varepsilon^{m_j+1}),\end{aligned}$$

In a similar fashion, one obtains

$$\begin{aligned}\mathbf{x}_{m-\varepsilon}^{(0)} &\equiv \mathbf{x}^{(0)}(t_m-\varepsilon)\\ &= \mathbf{x}^{(0)}(t_m) + \sum_{s=1}^{2k_i} \frac{d^s\mathbf{x}^{(0)}}{dt^s}\Big|_{t=t_m} (-\varepsilon)^s + \frac{d^{2k_i+1}\mathbf{x}^{(0)}}{dt^{2k_{i+1}}}\Big|_{t=t_m} (-\varepsilon)^{2k_i+1} + o((-\varepsilon)^{2k_i+1})\\ &= \mathbf{x}^{(0)}(t_m) + \sum_{s=1}^{2k_i} D_0^s\mathbf{F}^{(0)}(t_m)(-\varepsilon)^s + D_0^{2k_i+1}\mathbf{F}^{(0)}(t_m)(-\varepsilon)^{2k_i+1} + o((-\varepsilon)^{2k_i+1}),\end{aligned}$$

$$\begin{aligned}\mathbf{x}_{m+\varepsilon}^{(0)} &\equiv \mathbf{x}^{(0)}(t_m+\varepsilon)\\ &= \mathbf{x}^{(0)}(t_m) + \sum_{s=1}^{m_j} \frac{d^s\mathbf{x}^{(0)}}{dt^s}\Big|_{t=t_m} \varepsilon^s + \frac{d^{m_j+1}\mathbf{x}^{(0)}}{dt^{m_{j+1}}}\Big|_{t=t_m} \varepsilon^{m_j+1} + o(\varepsilon^{m_j+1})\\ &= \mathbf{x}^{(0)}(t_m) + \sum_{s=1}^{m_j} D_0^s\mathbf{F}^{(0)}(t_m)\varepsilon^s + D_0^{m_j+1}\mathbf{F}^{(0)}(t_m)\varepsilon^{m_j+1} + o(\varepsilon^{m_j+1}),\end{aligned}$$

$$\begin{aligned}\mathbf{n}_{\partial\Omega_{ij}}(\mathbf{x}_{m-\varepsilon}^{(0)}) &= \mathbf{n}_{\partial\Omega_{ij}}(\mathbf{x}_m^{(0)}) + \sum_{s=1}^{2k_i} D_0^s\mathbf{n}_{\partial\Omega_{ij}}(\mathbf{x}_m^{(0)})(-\varepsilon)^s\\ &\quad + D_0^{2k_i+1}\mathbf{n}_{\partial\Omega_{ij}}(\mathbf{x}_m^{(0)})(-\varepsilon)^{2k_i+1} + o((-\varepsilon)^{2k_i+1}),\\ \mathbf{n}_{\partial\Omega_{ij}}(\mathbf{x}_{m+\varepsilon}^{(0)}) &= \mathbf{n}_{\partial\Omega_{ij}}(\mathbf{x}_m^{(0)}) + \sum_{s=1}^{m_j} D_0^s\mathbf{n}_{\partial\Omega_{ij}}(\mathbf{x}_m^{(0)})\varepsilon^s\\ &\quad + D_0^{m_j+1}\mathbf{n}_{\partial\Omega_{ij}}(\mathbf{x}_m^{(0)})\varepsilon^{m_j+1} + o(\varepsilon^{m_j+1}).\end{aligned}$$

The ignorance of the ε^{2k_i+2} and ε^{m_j+2}-terms and high order terms, the deformation of the above equation and left multiplication of $\mathbf{n}_{\partial\Omega_{ij}}$ gives

$$\begin{aligned}&\mathbf{n}_{\partial\Omega_{ij}}^{\mathrm{T}}(\mathbf{x}_{m-\varepsilon}^{(0)})\cdot[\mathbf{x}_{m-\varepsilon}^{(i)} - \mathbf{x}_{m-\varepsilon}^{(0)}]\\ &= \mathbf{n}_{\partial\Omega_{ij}}^{\mathrm{T}}(\mathbf{x}_m^{(0)})\cdot[\mathbf{x}_{m-}^{(i)} - \mathbf{x}_m^{(0)}] + \sum_{s=0}^{2k_i}(-\varepsilon)^s G_{\partial\Omega_{ij}}^{(s-1,i)}(\mathbf{x}_m, t_{m-}, \mathbf{p}_i, \boldsymbol{\lambda})\\ &\quad + (-\varepsilon)^{2k_i+1} G_{\partial\Omega_{ij}}^{(2k_i,i)}(\mathbf{x}_m, t_{m-}, \mathbf{p}_i, \boldsymbol{\lambda}),\end{aligned}$$

$$\begin{aligned}&\mathbf{n}^{\mathrm{T}}_{\partial\Omega_{ij}}(\mathbf{x}^{(0)}_{m+\varepsilon})\cdot[\mathbf{x}^{(j)}_{m+\varepsilon}-\mathbf{x}^{(0)}_{m+\varepsilon}]\\&=\mathbf{n}^{\mathrm{T}}_{\partial\Omega_{ij}}(\mathbf{x}^{(0)}_{m})\cdot[\mathbf{x}^{(j)}_{m+}-\mathbf{x}^{(0)}_{m}]+\sum\nolimits_{s=1}^{m_j}\varepsilon^{s}G^{(s-1,j)}_{\partial\Omega_{ij}}(\mathbf{x}_m,t_{m+},\mathbf{p}_j,\boldsymbol{\lambda})\\&\quad+\varepsilon^{m_j+1}G^{(m_j,j)}_{\partial\Omega_{ij}}(\mathbf{x}_m,t_{m+},\mathbf{p}_j,\boldsymbol{\lambda}).\end{aligned}$$

Because of $\mathbf{x}^{(\alpha)}_{m\pm}=\mathbf{x}^{(0)}_{m}=\mathbf{x}_m$, with Eqs. (3.35) and (3.36), one obtains

$$\begin{aligned}&\mathbf{n}^{\mathrm{T}}_{\partial\Omega_{ij}}(\mathbf{x}^{(0)}_{m-\varepsilon})\cdot[\mathbf{x}^{(0)}_{m-\varepsilon}-\mathbf{x}^{(i)}_{m-\varepsilon}]=(-1)^{2k_i+2}\varepsilon^{k_i}G^{(2k_i,i)}_{\partial\Omega_{ij}}(\mathbf{x}_m,t_{m-},\mathbf{p}_i,\boldsymbol{\lambda}),\\&\mathbf{n}^{\mathrm{T}}_{\partial\Omega_{ij}}(\mathbf{x}^{(0)}_{m+\varepsilon})\cdot[\mathbf{x}^{(j)}_{m+\varepsilon}-\mathbf{x}^{(0)}_{m+\varepsilon}]=\varepsilon^{m_j+1}G^{(m_j,j)}_{\partial\Omega_{ij}}(\mathbf{x}_m,t_{m+},\mathbf{p}_j,\boldsymbol{\lambda}).\end{aligned}$$

With Eq. (3.39), the foregoing equation gives Eq. (3.36). On the other hand, using Eq. (3.36), the foregoing equation gives Eq. (3.39). The proof is completed. ■

3.4. Non-passable flows

In this section, non-passable flows to a specific boundary will be discussed as in Luo (2008b,c). The initial discussion on such an issue can be found in Luo (2005a, 2006). The $(2k_i:2k_j)$-non-passable flows are sketched in Fig. 3.8 for a better understanding of non-passable flows. The non-passable flows are called the *full non-passable flows* because the flows on both sides of the boundary will approach or leave the boundary. If a flow only on one side of the boundary approaches or leaves the boundary, but the flow on the other side does not exist or is not defined, this flow to the boundary is called the *half non-passable flow*. The full non-passable flow of the first kind (sink flows) and the full non-passable flow of the second kind (source flows) are sketched in Fig. 3.8(a) and (b), respectively. The half-sink and half-source flow to the boundary will be discussed later in this chapter. For continuous systems, sink or source flows to equilibrium are just points. However, for discontinuous dynamical systems, sink and source flows will be regions on the boundary, which makes the dynamical behaviors of the dynamical systems much richer. With a little modification in Chapter 2, the sink flow is defined as follows.

Definition 3.12. For a discontinuous dynamic system in Eq. (2.1), there is a point $\mathbf{x}^{(0)}(t_m)\equiv\mathbf{x}_m\in\partial\Omega_{ij}$ at time t_m between two adjacent domains Ω_α $(\alpha=i,j)$. For an arbitrarily small $\varepsilon>0$, there is a time interval $[t_{m-\varepsilon},t_m)$. Suppose $\mathbf{x}^{(\alpha)}(t_{m-})=\mathbf{x}_m$. The flow $\mathbf{x}^{(i)}(t)$ and $\mathbf{x}^{(j)}(t)$ to the boundary $\partial\Omega_{ij}$ is *non-passable* of the first kind (or called a sink flow) if

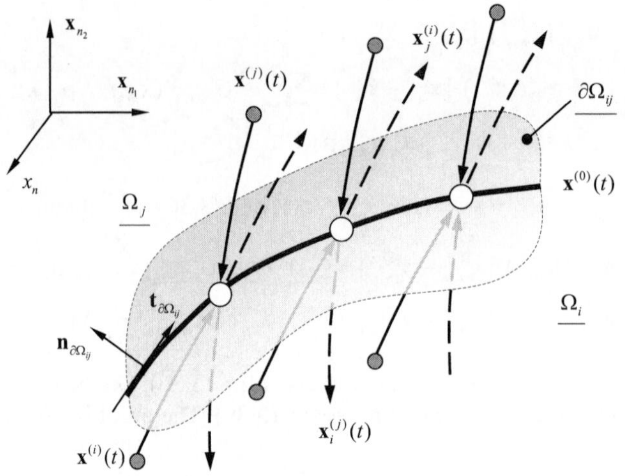

(a)

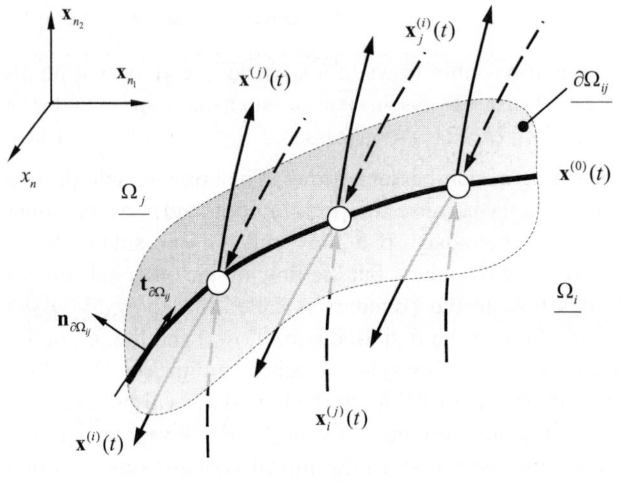

(b)

Fig. 3.8 The $(2k_i : 2k_j)$-non-passable flows: **(a)** the first kind (sink flows) and **(b)** the second kind (source flows). $\mathbf{x}^{(i)}(t)$ and $\mathbf{x}^{(j)}(t)$ represent *real* flows in domains Ω_i and Ω_j, respectively, which are depicted by thin solid curves. $\mathbf{x}_i^{(j)}(t)$ and $\mathbf{x}_j^{(i)}(t)$ represent *imaginary* flows in domains Ω_i and Ω_j, respectively controlled by the vector fields in Ω_j and Ω_i, which are depicted by dashed curves. The flow on the boundary is described by $\mathbf{x}^{(0)}(t)$. The normal and tangential vectors $\mathbf{n}_{\partial\Omega_{ij}}$ and $\mathbf{t}_{\partial\Omega_{ij}}$ of the boundary are depicted. Hollow circles are for sink and source points on the boundary, and filled circles are for starting points. ($n_1 + n_2 + 1 = n$)

$$
\begin{aligned}
&\text{either} \quad \left.\begin{array}{l}
\mathbf{n}_{\partial\Omega_{ij}}^{\mathrm{T}}(\mathbf{x}_{m-\varepsilon}^{(0)})\cdot[\mathbf{x}_{m-\varepsilon}^{(0)}-\mathbf{x}_{m-\varepsilon}^{(i)}]>0\\
\mathbf{n}_{\partial\Omega_{ij}}^{\mathrm{T}}(\mathbf{x}_{m-\varepsilon}^{(0)})\cdot[\mathbf{x}_{m-\varepsilon}^{(0)}-\mathbf{x}_{m-\varepsilon}^{(j)}]<0
\end{array}\right\}\text{for } \mathbf{n}_{\partial\Omega_{ij}}\to\Omega_j,\\
&\text{or} \quad \left.\begin{array}{l}
\mathbf{n}_{\partial\Omega_{ij}}^{\mathrm{T}}(\mathbf{x}_{m-\varepsilon}^{(0)})\cdot[\mathbf{x}_{m-\varepsilon}^{(0)}-\mathbf{x}_{m-\varepsilon}^{(i)}]<0\\
\mathbf{n}_{\partial\Omega_{ij}}^{\mathrm{T}}(\mathbf{x}_{m-\varepsilon}^{(0)})\cdot[\mathbf{x}_{m-\varepsilon}^{(0)}-\mathbf{x}_{m-\varepsilon}^{(j)}]>0
\end{array}\right\}\text{for } \mathbf{n}_{\partial\Omega_{ij}}\to\Omega_i.
\end{aligned}
\tag{3.40}
$$

From the foregoing definition, the sufficient and necessary conditions for the sink flow in Eq. (2.1) can be developed through the following theorem.

Theorem 3.3. *For a discontinuous dynamical system in Eq.* (2.1), *there is a point* $\mathbf{x}^{(0)}(t_m)\equiv\mathbf{x}_m\in\partial\Omega_{ij}$ *at time* t_m *between two adjacent domains* Ω_α $(\alpha=i,j)$. *For an arbitrarily small* $\varepsilon>0$, *there is a time interval* $[t_{m-\varepsilon},t_m)$. *Suppose* $\mathbf{x}^{(\alpha)}(t_{m-})=\mathbf{x}_m$. *A flow* $\mathbf{x}^{(\alpha)}(t)$ *is* $C^{r_\alpha}_{[t_{m-\varepsilon},t_m)}$*-continuous* ($r_\alpha\ge 1$, $\alpha=i,j$) *for time* t *with* $\|d^{r_\alpha+1}\mathbf{x}^{(\alpha)}/dt^{r_\alpha+1}\|<\infty$. *The flow* $\mathbf{x}^{(i)}(t)$ *and* $\mathbf{x}^{(j)}(t)$ *to the boundary* $\partial\Omega_{ij}$ *is non-passable of the first kind* (*or a sink flow*) *if and only if*

$$
\begin{aligned}
&\textit{either} \quad \left.\begin{array}{l}
G_{\partial\Omega_{ij}}^{(i)}(\mathbf{x}_m,t_{m-},\mathbf{p}_i,\boldsymbol{\lambda})>0\\
G_{\partial\Omega_{ij}}^{(j)}(\mathbf{x}_m,t_{m-},\mathbf{p}_j,\boldsymbol{\lambda})<0
\end{array}\right\}\textit{for } \mathbf{n}_{\partial\Omega_{ij}}\to\Omega_j,\\
&\textit{or} \quad \left.\begin{array}{l}
G_{\partial\Omega_{ij}}^{(i)}(\mathbf{x}_m,t_{m-},\mathbf{p}_i,\boldsymbol{\lambda})<0\\
G_{\partial\Omega_{ij}}^{(j)}(\mathbf{x}_m,t_{m-},\mathbf{p}_j,\boldsymbol{\lambda})>0
\end{array}\right\}\textit{for } \mathbf{n}_{\partial\Omega_{ij}}\to\Omega_i.
\end{aligned}
\tag{3.41}
$$

Proof. The proof is the same as in Theorem 3.1. ■

If the boundary $\partial\Omega_{ij}$ is independent of time, using Eq. (3.14), the above theorem is identical to Theorem 2.3 (also see, Luo, 2005a, 2006) owing to the zero-order contact between the flow and boundary. However, in Luo (2005a, 2006), a theory for the non-passable flow with the $(2k_i:2k_j)$ higher-order singularity ($k_\alpha\in\mathbb{N}$, $\alpha=i,j$) is only valid for the *plane* boundary and the $(2k_\alpha)$th*-contact* between the boundary $\partial\Omega_{ij}$ and the flow $\mathbf{x}^{(\alpha)}$ in the domain Ω_α ($\alpha=i,j$), which will be discussed later. As in Luo (2008b,c), with the higher-order singularity of a flow to the boundary, a generalized theory for the $(2k_i:2k_j)$-non-passable flow will be discussed herein.

Definition 3.13. For a discontinuous dynamic system in Eq. (2.1), there is a point $\mathbf{x}^{(0)}(t_m) \equiv \mathbf{x}_m \in \partial\Omega_{ij}$ at time t_m between two adjacent domains Ω_α $(\alpha = i, j)$. For an arbitrarily small $\varepsilon > 0$, there is a time interval $[t_{m-\varepsilon}, t_m)$. Suppose $\mathbf{x}^{(\alpha)}(t_{m-}) = \mathbf{x}_m$. A flow $\mathbf{x}^{(\alpha)}(t)$ is $C^{r_\alpha}_{[t_{m-\varepsilon}, t_m)}$-continuous ($r_\alpha \geq 2k_\alpha + 1$, $\alpha = i, j$) for time t with $\| d^{r_\alpha+1}\mathbf{x}^{(\alpha)} / dt^{r_\alpha+1} \| < \infty$. The flow $\mathbf{x}^{(i)}(t)$ of the $(2k_i)$th-order and $\mathbf{x}^{(j)}(t)$ of the $(2k_j)$th-order to the boundary $\partial\Omega_{ij}$ is $(2k_i : 2k_j)$*-non-passable* of the first kind (or called a $(2k_i : 2k_j)$-sink flow) if

$$\left.\begin{aligned} &G^{(s_\alpha,\alpha)}_{\partial\Omega_{ij}}(\mathbf{x}_m, t_{m-}, \mathbf{p}_\alpha, \boldsymbol{\lambda}) = 0 \ \text{ for } s_\alpha = 0, 1, \cdots, 2k_\alpha - 1, \\ &G^{(2k_\alpha,\alpha)}_{\partial\Omega_{ij}}(\mathbf{x}_m, t_{m-}, \mathbf{p}_\alpha, \boldsymbol{\lambda}) \neq 0 \ \ (\alpha = i, j); \end{aligned}\right\} \tag{3.42}$$

$$\begin{aligned} &\text{either} \quad \left.\begin{aligned} &\mathbf{n}^{\mathrm{T}}_{\partial\Omega_{ij}}(\mathbf{x}^{(0)}_{m-\varepsilon}) \cdot [\mathbf{x}^{(0)}_{m-\varepsilon} - \mathbf{x}^{(i)}_{m-\varepsilon}] > 0 \\ &\mathbf{n}^{\mathrm{T}}_{\partial\Omega_{ij}}(\mathbf{x}^{(0)}_{m-\varepsilon}) \cdot [\mathbf{x}^{(0)}_{m-\varepsilon} - \mathbf{x}^{(j)}_{m-\varepsilon}] < 0 \end{aligned}\right\} \text{for } \mathbf{n}_{\partial\Omega_{ij}} \to \Omega_j, \\ &\text{or} \quad \left.\begin{aligned} &\mathbf{n}^{\mathrm{T}}_{\partial\Omega_{ij}}(\mathbf{x}^{(0)}_{m-\varepsilon}) \cdot [\mathbf{x}^{(0)}_{m-\varepsilon} - \mathbf{x}^{(i)}_{m-\varepsilon}] < 0 \\ &\mathbf{n}^{\mathrm{T}}_{\partial\Omega_{ij}}(\mathbf{x}^{(0)}_{m-\varepsilon}) \cdot [\mathbf{x}^{(0)}_{m-\varepsilon} - \mathbf{x}^{(j)}_{m-\varepsilon}] > 0 \end{aligned}\right\} \text{for } \mathbf{n}_{\partial\Omega_{ij}} \to \Omega_i. \end{aligned} \tag{3.43}$$

Theorem 3.4. *For a discontinuous dynamical system in Eq.* (2.1), *there is a point* $\mathbf{x}^{(0)}(t_m) \equiv \mathbf{x}_m \in \partial\Omega_{ij}$ *at time* t_m *between two adjacent domains* Ω_α $(\alpha = i, j)$. *For an arbitrarily small* $\varepsilon > 0$, *there is a time interval* $[t_{m-\varepsilon}, t_m)$. *Suppose* $\mathbf{x}^{(\alpha)}(t_{m-}) = \mathbf{x}_m$. *A flow* $\mathbf{x}^{(\alpha)}(t)$ *is* $C^{r_\alpha}_{[t_{m-\varepsilon}, t_m)}$*-continuous* ($r_\alpha \geq 2k_\alpha + 1$, $\alpha = i, j$) *for time* t *with* $\| d^{r_\alpha+1}\mathbf{x}^{(\alpha)} / dt^{r_\alpha+1} \| < \infty$. *The flow* $\mathbf{x}^{(i)}(t)$ *of the* $(2k_i)$th*-order and* $\mathbf{x}^{(j)}(t)$ *of the* $(2k_j)$th*-order to the boundary* $\partial\Omega_{ij}$ *is* $(2k_i : 2k_j)$*-non-passable of the first kind (or a* $(2k_i : 2k_j)$*-sink flow) if and only if*

$$G^{(s_\alpha,\alpha)}_{\partial\Omega_{ij}}(\mathbf{x}_m, t_{m-}, \mathbf{p}_\alpha, \boldsymbol{\lambda}) = 0 \ \text{ for } s_\alpha = 0, 1, \cdots, 2k_\alpha - 1 \ \textit{and} \ \alpha = i, j; \tag{3.44}$$

$$\begin{aligned} &\textit{either} \quad \left.\begin{aligned} &G^{(2k_i,i)}_{\partial\Omega_{ij}}(\mathbf{x}_m, t_{m-}, \mathbf{p}_i, \boldsymbol{\lambda}) > 0 \\ &G^{(2k_j,j)}_{\partial\Omega_{ij}}(\mathbf{x}_m, t_{m-}, \mathbf{p}_j, \boldsymbol{\lambda}) < 0 \end{aligned}\right\} \textit{for } \mathbf{n}_{\partial\Omega_{ij}} \to \Omega_j, \\ &\textit{or} \quad \left.\begin{aligned} &G^{(2k_i,i)}_{\partial\Omega_{ij}}(\mathbf{x}_m, t_{m-}, \mathbf{p}_i, \boldsymbol{\lambda}) < 0 \\ &G^{(2k_j,j)}_{\partial\Omega_{ij}}(\mathbf{x}_m, t_{m-}, \mathbf{p}_j, \boldsymbol{\lambda}) > 0 \end{aligned}\right\} \textit{for } \mathbf{n}_{\partial\Omega_{ij}} \to \Omega_i. \end{aligned} \tag{3.45}$$

Proof. The proof is similar to the proof of Theorem 3.2. ■

Definition 3.14. For a discontinuous dynamic system in Eq. (2.1), there is a point $\mathbf{x}^{(0)}(t_m) \equiv \mathbf{x}_m \in \partial\Omega_{ij}$ at time t_m between two adjacent domains Ω_α $(\alpha = i, j)$. For an arbitrarily small $\varepsilon > 0$, there is a time interval $(t_m, t_{m+\varepsilon}]$. Suppose $\mathbf{x}^{(\alpha)}(t_{m+}) = \mathbf{x}_m$. The flow $\mathbf{x}^{(i)}(t)$ and $\mathbf{x}^{(j)}(t)$ to the boundary $\partial\Omega_{ij}$ is *non-passable* of the second kind (or called a source flow) if

$$
\begin{aligned}
&\text{either} && \left.\begin{aligned} \mathbf{n}^{\mathrm{T}}_{\partial\Omega_{ij}}(\mathbf{x}^{(0)}_{m+\varepsilon})\cdot[\mathbf{x}^{(i)}_{m+\varepsilon} - \mathbf{x}^{(0)}_{m+\varepsilon}] < 0 \\ \mathbf{n}^{\mathrm{T}}_{\partial\Omega_{ij}}(\mathbf{x}^{(0)}_{m+\varepsilon})\cdot[\mathbf{x}^{(j)}_{m+\varepsilon} - \mathbf{x}^{(0)}_{m+\varepsilon}] > 0 \end{aligned}\right\} \text{for } \mathbf{n}_{\partial\Omega_{ij}} \to \Omega_j, \\
&\text{or} && \left.\begin{aligned} \mathbf{n}^{\mathrm{T}}_{\partial\Omega_{ij}}(\mathbf{x}^{(0)}_{m+\varepsilon})\cdot[\mathbf{x}^{(i)}_{m+\varepsilon} - \mathbf{x}^{(0)}_{m+\varepsilon}] > 0 \\ \mathbf{n}^{\mathrm{T}}_{\partial\Omega_{ij}}(\mathbf{x}^{(0)}_{m+\varepsilon})\cdot[\mathbf{x}^{(j)}_{m+\varepsilon} - \mathbf{x}^{(0)}_{m+\varepsilon}] < 0 \end{aligned}\right\} \text{for } \mathbf{n}_{\partial\Omega_{ij}} \to \Omega_i.
\end{aligned}
\tag{3.46}
$$

Theorem 3.5. *For a discontinuous dynamical system in Eq.* (2.1), *there is a point* $\mathbf{x}^{(0)}(t_m) \equiv \mathbf{x}_m \in \partial\Omega_{ij}$ *at time* t_m *between two adjacent domains* Ω_α $(\alpha = i, j)$. *For an arbitrarily small* $\varepsilon > 0$, *there is a time interval* $(t_m, t_{m+\varepsilon}]$. *Suppose* $\mathbf{x}^{(\alpha)}(t_{m+}) = \mathbf{x}_m$ $(\alpha = i, j)$. *A flow* $\mathbf{x}^{(\alpha)}(t)$ *is* $C^{r_\alpha}_{(t_m, t_{m+\varepsilon}]}$*-continuous* ($r_\alpha \geq 1$) *for time t with* $\| d^{r_\alpha+1}\mathbf{x}^{(\alpha)} / dt^{r_\alpha+1} \| < \infty$. *The flow* $\mathbf{x}^{(i)}(t)$ *and* $\mathbf{x}^{(j)}(t)$ *to the boundary* $\partial\Omega_{ij}$ *is non-passable of the second kind (or a source flow) if and only if*

$$
\begin{aligned}
&\textit{either} && \left.\begin{aligned} G^{(i)}_{\partial\Omega_{ij}}(\mathbf{x}_m, t_{m+}, \mathbf{p}_i, \boldsymbol{\lambda}) < 0 \\ G^{(j)}_{\partial\Omega_{ij}}(\mathbf{x}_m, t_{m+}, \mathbf{p}_j, \boldsymbol{\lambda}) > 0 \end{aligned}\right\} \textit{for } \mathbf{n}_{\partial\Omega_{ij}} \to \Omega_j, \\
&\textit{or} && \left.\begin{aligned} G^{(i)}_{\partial\Omega_{ij}}(\mathbf{x}_m, t_{m+}, \mathbf{p}_i, \boldsymbol{\lambda}) > 0 \\ G^{(j)}_{\partial\Omega_{ij}}(\mathbf{x}_m, t_{m+}, \mathbf{p}_j, \boldsymbol{\lambda}) < 0 \end{aligned}\right\} \textit{for } \mathbf{n}_{\partial\Omega_{ij}} \to \Omega_i.
\end{aligned}
\tag{3.47}
$$

Proof. The proof is the same as in Theorem 3.1. ■

Definition 3.15. For a discontinuous dynamic system in Eq. (2.1), there is a point $\mathbf{x}^{(0)}(t_m) \equiv \mathbf{x}_m \in \partial\Omega_{ij}$ at time t_m between two adjacent domains Ω_α $(\alpha = i, j)$. For an arbitrarily small $\varepsilon > 0$, there is a time interval $(t_m, t_{m+\varepsilon}]$. Suppose $\mathbf{x}^{(\alpha)}(t_{m+}) = \mathbf{x}_m$. A flow $\mathbf{x}^{(\alpha)}(t)$ is $C^{r_\alpha}_{(t_m, t_{m+\varepsilon}]}$-continuous ($r_\alpha \geq m_\alpha + 1, \alpha = i, j$) for time t with $\| d^{r_\alpha+1}\mathbf{x}^{(\alpha)} / dt^{r_\alpha+1} \| < \infty$. The flow $\mathbf{x}^{(i)}(t)$ of the (m_i)th-order and $\mathbf{x}^{(j)}(t)$ of the (m_j)th-order to the boundary $\partial\Omega_{ij}$ is $(m_i : m_j)$ *-non-passable* of the second kind (or called an $(m_i : m_j)$-source flow) if

$$\left.\begin{aligned}&G_{\partial\Omega_{ij}}^{(s_i,i)}(\mathbf{x}_m,t_{m+},\mathbf{p}_i,\boldsymbol{\lambda})=0\ \text{ for } s_i=0,1,\cdots,m_i-1,\\&G_{\partial\Omega_{ij}}^{(2k_i,i)}(\mathbf{x}_m,t_{m+},\mathbf{p}_i,\boldsymbol{\lambda})\neq 0,\end{aligned}\right\}\tag{3.48}$$

$$\left.\begin{aligned}&G_{\partial\Omega_{ij}}^{(s_j,j)}(\mathbf{x}_m,t_{m+},\mathbf{p}_j,\boldsymbol{\lambda})=0\ \text{ for } s_j=0,1,\cdots,m_j-1,\\&G_{\partial\Omega_{ij}}^{(2k_j,j)}(\mathbf{x}_m,t_{m+},\mathbf{p}_j,\boldsymbol{\lambda})\neq 0,\end{aligned}\right\}\tag{3.49}$$

$$\begin{aligned}&\text{either}\ \left.\begin{aligned}&\mathbf{n}_{\partial\Omega_{ij}}^{\mathrm{T}}(\mathbf{x}_{m+\varepsilon}^{(0)})\cdot[\mathbf{x}_{m+\varepsilon}^{(i)}-\mathbf{x}_{m+\varepsilon}^{(0)}]<0\\&\mathbf{n}_{\partial\Omega_{ij}}^{\mathrm{T}}(\mathbf{x}_{m+\varepsilon}^{(0)})\cdot[\mathbf{x}_{m+\varepsilon}^{(j)}-\mathbf{x}_{m+\varepsilon}^{(0)}]>0\end{aligned}\right\}\text{for } \mathbf{n}_{\partial\Omega_{ij}}\to\Omega_j,\\&\text{or}\ \left.\begin{aligned}&\mathbf{n}_{\partial\Omega_{ij}}^{\mathrm{T}}(\mathbf{x}_{m+\varepsilon}^{(0)})\cdot[\mathbf{x}_{m+\varepsilon}^{(i)}-\mathbf{x}_{m+\varepsilon}^{(0)}]>0\\&\mathbf{n}_{\partial\Omega_{ij}}^{\mathrm{T}}(\mathbf{x}_{m+\varepsilon}^{(0)})\cdot[\mathbf{x}_{m+\varepsilon}^{(j)}-\mathbf{x}_{m+\varepsilon}^{(0)}]<0\end{aligned}\right\}\text{for } \mathbf{n}_{\partial\Omega_{ij}}\to\Omega_i.\end{aligned}\tag{3.50}$$

Note that for $m_\alpha=2k_\alpha$ ($\alpha=i,j$), the $(2k_i:2k_j)$ -source flow is obtained, which corresponds to the $(2k_i:2k_j)$ -sink flow. If $m_\alpha=2k_\alpha-1$ and $m_\beta=2k_\beta$ ($\alpha,\beta\in\{i,j\}$ and $\beta\neq\alpha$) or $m_\beta=2k_\beta-1$ ($\beta\in\{i,j\}$), because the source flow is from the boundary, three $(2k_i:2k_j-1)$, $(2k_i-1:2k_j)$, and $(2k_i-1:2k_j-1)$ source flows exist. However, the corresponding sink flows cannot be formed. Such source flows are relative to tangential flows, which will be discussed later. The question is which domain the flow will go into. If a source flow is exactly on the boundary, the source flow will keep on the boundary. However, if the flow just has a little bit perturbation on one of two domains (e.g., in domain Ω_α , $\alpha\in\{i,j\}$), the source flow will get into the domain of Ω_α with the corresponding flow of the (m_α)th -order. In fact, such a perturbed source flow is independent of the order of flow singularity to the boundary in another domain. One can say that the behavior of source flow is sensitive to the small perturbation in the vicinity of the boundary. Such a property is similar to the saddle or source points in continuous dynamical systems. However, the sink flow to the boundary is stabilized to the small perturbation to the boundary, which means the sink flow will be on the boundary whatever the perturbation of the sink flow is on the boundary or one of two domains. For a better explanation of the forces flow, the four source flows are sketched in Figs. 3.9 and 3.10. Solid and dashed curves are for real and imaginary source flows. Dotted curves represent coming flows relative to the corresponding source flows to the boundary. In Fig. 3.9, $(2k_i:2k_j)$ and $(2k_i-1:2k_j-1)$ source flows are presented. In fact, the source flow does not have any coming flow except for the source flow existing on the boundary. For a $(2k_i:2k_j)$ -source flow, the incoming flow is the imaginary flow, and for a $(2k_i-1:2k_j-1)$ -source flow, the imagined, coming flow relative to the source flow is in the same domain. As in the $(2k_i : 2k_j)$-sink flow, the $(2k_i : 2k_j)$-source flow does not have any grazing properties to the boundary. However, the $(2k_i - 1 : 2k_j - 1)$-source flow possesses the grazing characteristics. Because the grazing source flows are not important for the flow passability to the boundary, the properties of grazing source flows will not be discussed in this section. In Fig. 3.10, the $(2k_i : 2k_j - 1)$ and $(2k_i - 1 : 2k_j)$-source flows are presented.

From the foregoing definition, the sufficient and necessary conditions for the $(m_i : m_j)$-source flow in Eq. (2.1) can be developed as follows.

Theorem 3.6. *For a discontinuous dynamical system in Eq.* (2.1)*, there is a point* $\mathbf{x}^{(0)}(t_m) \equiv \mathbf{x}_m \in \partial\Omega_{ij}$ *at time* t_m *between two adjacent domains* Ω_α *(*$\alpha = i, j$*). For an arbitrarily small* $\varepsilon > 0$*, there is a time interval* $(t_m, t_{m+\varepsilon}]$*. Suppose* $\mathbf{x}^{(\alpha)}(t_{m+}) = \mathbf{x}_m$*. A flow* $\mathbf{x}^{(\alpha)}(t)$ *is* $C^{r_\alpha}_{(t_m, t_{m+\varepsilon}]}$*-continuous for time t with* $\| d^{r_\alpha+1}\mathbf{x}^{(\alpha)} / dt^{r_\alpha+1} \| < \infty$ *(*$r_\alpha \geq m_\alpha + 1$*,* $\alpha = i, j$*). The flow* $\mathbf{x}^{(i)}(t)$ *of the* (m_i)th*-order and* $\mathbf{x}^{(j)}(t)$ *of the* (m_j)th*-order to the boundary* $\partial\Omega_{ij}$ *is* $(m_i : m_j)$*-non-passable of the second kind (or* $(m_i : m_j)$*-source flow) if and only if*

$$G^{(s_i, i)}_{\partial\Omega_{ij}}(\mathbf{x}_m, t_{m+}, \mathbf{p}_i, \boldsymbol{\lambda}) = 0 \;\; for \;\; s_i = 0, 1, \cdots, m_i - 1; \tag{3.51}$$

$$G^{(s_j, j)}_{\partial\Omega_{ij}}(\mathbf{x}_m, t_{m+}, \mathbf{p}_j, \boldsymbol{\lambda}) = 0 \;\; for \;\; s_j = 0, 1, \cdots, m_j - 1; \tag{3.52}$$

$$\begin{aligned} &either \quad \left. \begin{aligned} G^{(2k_i, i)}_{\partial\Omega_{ij}}(\mathbf{x}_m, t_{m+}, \mathbf{p}_i, \boldsymbol{\lambda}) < 0 \\ G^{(2k_j, j)}_{\partial\Omega_{ij}}(\mathbf{x}_m, t_{m+}, \mathbf{p}_j, \boldsymbol{\lambda}) > 0 \end{aligned} \right\} for \;\; \mathbf{n}_{\partial\Omega_{ij}} \to \Omega_j, \\ &or \quad \left. \begin{aligned} G^{(2k_i, i)}_{\partial\Omega_{ij}}(\mathbf{x}_m, t_{m+}, \mathbf{p}_i, \boldsymbol{\lambda}) > 0 \\ G^{(2k_j, j)}_{\partial\Omega_{ij}}(\mathbf{x}_m, t_{m+}, \mathbf{p}_j, \boldsymbol{\lambda}) < 0 \end{aligned} \right\} for \;\; \mathbf{n}_{\partial\Omega_{ij}} \to \Omega_i. \end{aligned} \tag{3.53}$$

Proof. The proof is similar to the proof of Theorem 3.2. ■

Next, half -non-passable flows to the boundary will be discussed. The half-non-passable flow of the first kind is termed *a half-sink flow*. A half-sink flow to the boundary is sketched in Fig. 3.11. Such a half-sink flow in Ω_i is shown in Fig. 3.11(a). Only $\mathbf{x}^{(i)}(t)$ for time $t \in [t_{m-\varepsilon}, t_m)$ is a real flow, and imaginary flows $\mathbf{x}^{(i)}_j(t)$ for time $t \in [t_{m-\varepsilon}, t_{m+\varepsilon}]$ and $\mathbf{x}^{(j)}_i(t)$ for time $t \in (t_m, t_{m+\varepsilon}]$ are represented by dashed curves. To the same boundary $\partial\Omega_{ij}$, a half sink flow in Ω_j is sketched in Fig. 3.11(b). The coming flow $\mathbf{x}^{(j)}(t)$ for time $t \in [t_{m-\varepsilon}, \varepsilon)$ is only a real flow. The strict mathematical description is given as follows.

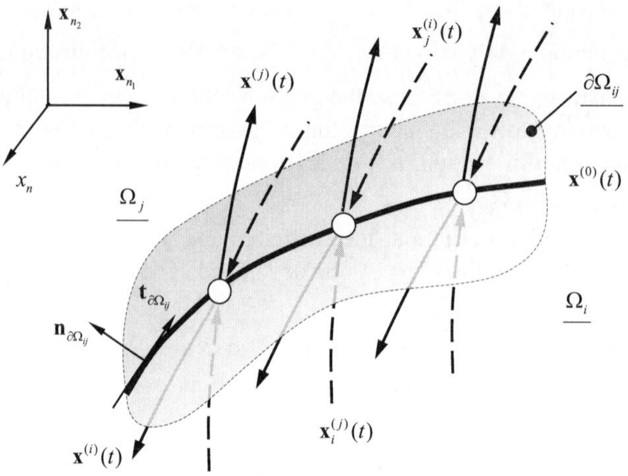

(a)

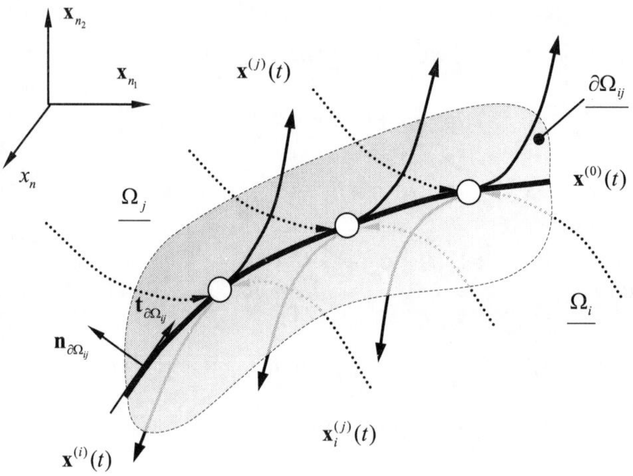

(b)

Fig. 3.9 Source flows: **(a)** the $(2k_i : 2k_j)$-source flows and **(b)** the $(2k_i - 1 : 2k_j - 1)$-source flows. $\mathbf{x}^{(i)}(t)$ and $\mathbf{x}^{(j)}(t)$ represent *real* flows in domains Ω_i and Ω_j, respectively, which are depicted by thin solid curves. $\mathbf{x}_i^{(j)}(t)$ and $\mathbf{x}_j^{(i)}(t)$ represent *imaginary* flows in domains Ω_i and Ω_j, respectively controlled by the vector fields in Ω_j and Ω_i, which are depicted by the dashed curves. The flow on the boundary is described by $\mathbf{x}^{(0)}(t)$. The normal and tangential vectors $\mathbf{n}_{\partial\Omega_{ij}}$ and $\mathbf{t}_{\partial\Omega_{ij}}$ of the boundary are depicted. Hollow circles are for sink and source points on the boundary. ($n_1 + n_2 + 1 = n$)

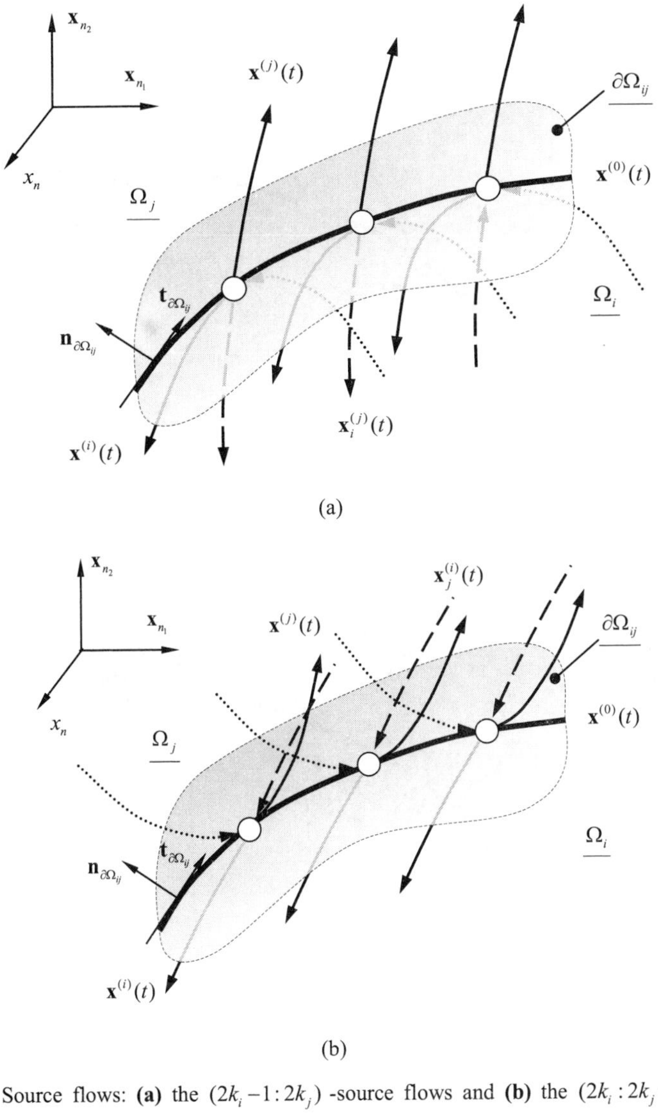

Fig. 3.10 Source flows: **(a)** the $(2k_i - 1 : 2k_j)$ -source flows and **(b)** the $(2k_i : 2k_j - 1)$ -source flows. $\mathbf{x}^{(i)}(t)$ and $\mathbf{x}^{(j)}(t)$ represent *real* flows in domains Ω_i and Ω_j, respectively, which are depicted by thin solid curves. $\mathbf{x}_i^{(j)}(t)$ and $\mathbf{x}_j^{(i)}(t)$ represent *imaginary* flows in domains Ω_i and Ω_j, respectively controlled by the vector fields in Ω_j and Ω_i, which are depicted by the dashed curves. The flow on the boundary is described by $\mathbf{x}^{(0)}(t)$. The normal and tangential vectors $\mathbf{n}_{\partial\Omega_{ij}}$ and $\mathbf{t}_{\partial\Omega_{ij}}$ of the boundary are depicted. Hollow circles are for sink and source points on the boundary. ($n_1 + n_2 + 1 = n$)

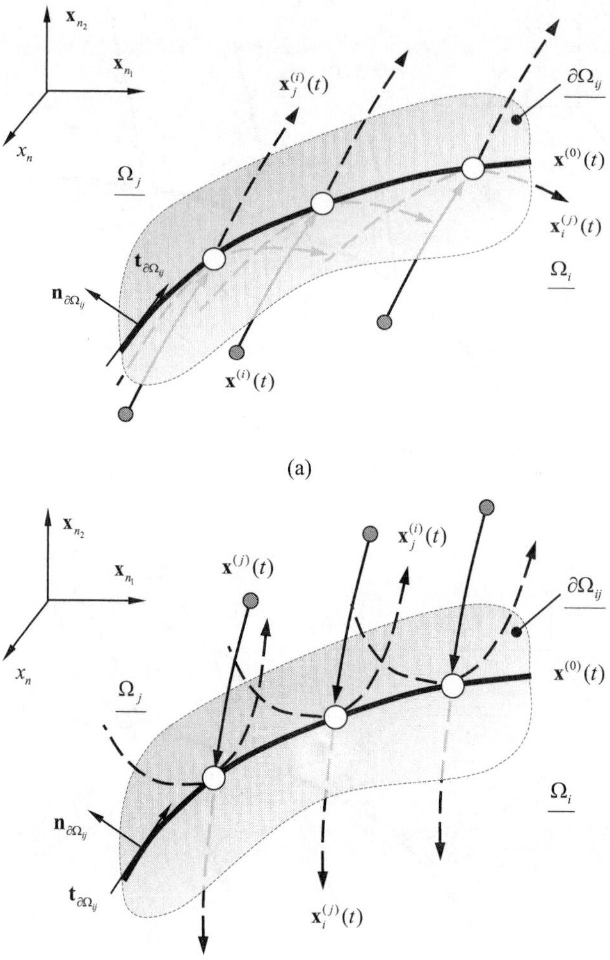

Fig. 3.11 The half sink flows: **(a)** $(2k_i : 2k_j - 1)$ -order in Ω_i and **(b)** $(2k_j : 2k_i - 1)$ -order in Ω_j. $\mathbf{x}^{(i)}(t)$ and $\mathbf{x}^{(j)}(t)$ represent *real* flows in domains Ω_i and Ω_j, respectively, which are depicted by thin solid curves. $\mathbf{x}_i^{(j)}(t)$ and $\mathbf{x}_j^{(i)}(t)$ represent *imaginary* flows in domains Ω_i and Ω_j, respectively controlled by the vector fields in Ω_j and Ω_i, which are depicted by the dashed curves. The flow on the boundary is described by $\mathbf{x}^{(0)}(t)$. The normal and tangential vectors $\mathbf{n}_{\partial\Omega_{ij}}$ and $\mathbf{t}_{\partial\Omega_{ij}}$ on the boundary are depicted. Hollow circles are for sink points on the boundary and filled circles are for starting points. ($n_1 + n_2 + 1 = n$)

Definition 3.16. For a discontinuous dynamical system in Eq. (2.1), there is a point $\mathbf{x}^{(0)}(t_m) \equiv \mathbf{x}_m \in \partial\Omega_{ij}$ at time t_m between two adjacent domains Ω_α ($\alpha = i, j$). Suppose $\mathbf{x}^{(i)}(t_{m-}) = \mathbf{x}_m = \mathbf{x}_i^{(j)}(t_{m\pm})$. For an arbitrarily small $\varepsilon > 0$, there are two time intervals $[t_{m-\varepsilon}, t_m)$ and $[t_{m-\varepsilon}, t_{m+\varepsilon}]$. A flow $\mathbf{x}^{(i)}(t)$ is $C^{r_i}_{[t_{m-\varepsilon}, t_m)}$-continuous ($r_i \geq 2k_i + 1$) with $\| d^{r_i+1}\mathbf{x}^{(i)} / dt^{r_i+1} \| < \infty$ for time t, and an imaginary flow $\mathbf{x}_i^{(j)}(t)$ is $C^{r_j}_{[t_{m-\varepsilon}, t_{m+\varepsilon}]}$-continuous ($r_j \geq 2k_j$) and $\| d^{r_j+1}\mathbf{x}_i^{(j)} / dt^{r_j+1} \| < \infty$. The flow $\mathbf{x}^{(i)}(t)$ of the $(2k_i)$ th-order and $\mathbf{x}_i^{(j)}(t)$ of the $(2k_j - 1)$th -order to the boundary $\partial\Omega_{ij}$ is $(2k_i : 2k_j - 1)$ -half-*non-passable* of the first kind in domain Ω_i (or called a $(2k_i : 2k_j - 1)$ -*half sink flow*) if

$$
\begin{aligned}
&G^{(s_i,i)}_{\partial\Omega_{ij}}(\mathbf{x}_m, t_{m-}, \mathbf{p}_i, \boldsymbol{\lambda}) = 0 \;\; \text{for } s_i = 0, 1, \cdots, 2k_i - 1,\\
&G^{(2k_i,i)}_{\partial\Omega_{ij}}(\mathbf{x}_m, t_{m-}, \mathbf{p}_i, \boldsymbol{\lambda}) \neq 0;
\end{aligned}
\tag{3.54}
$$

$$
\begin{aligned}
&G^{(s_j,j)}_{\partial\Omega_{ij}}(\mathbf{x}_m, t_{m\pm}, \mathbf{p}_j, \boldsymbol{\lambda}) = 0 \text{ for } s_j = 0, 1, \cdots, 2k_j - 2,\\
&G^{(2k_j-1,j)}_{\partial\Omega_{ij}}(\mathbf{x}_m, t_{m\pm}, \mathbf{p}_j, \boldsymbol{\lambda}) \neq 0;
\end{aligned}
\tag{3.55}
$$

$$
\begin{aligned}
&\text{either} \quad \mathbf{n}^{\mathrm{T}}_{\partial\Omega_{ij}}(\mathbf{x}^{(0)}_{m-\varepsilon}) \cdot [\mathbf{x}^{(0)}_{m-\varepsilon} - \mathbf{x}^{(i)}_{m-\varepsilon}] > 0 \;\; \text{for } \mathbf{n}_{\partial\Omega_{ij}} \to \Omega_j,\\
&\text{or} \quad \mathbf{n}^{\mathrm{T}}_{\partial\Omega_{ij}}(\mathbf{x}^{(0)}_{m-\varepsilon}) \cdot [\mathbf{x}^{(0)}_{m-\varepsilon} - \mathbf{x}^{(i)}_{m-\varepsilon}] < 0 \;\; \text{for } \mathbf{n}_{\partial\Omega_{ij}} \to \Omega_i;
\end{aligned}
\tag{3.56}
$$

$$
\begin{aligned}
&\text{either} \quad \left.\begin{aligned} &\mathbf{n}^{\mathrm{T}}_{\partial\Omega_{ij}}(\mathbf{x}^{(0)}_{m-\varepsilon}) \cdot [\mathbf{x}^{(0)}_{m-\varepsilon} - \mathbf{x}^{(j)}_{i(m-\varepsilon)}] > 0\\ &\mathbf{n}^{\mathrm{T}}_{\partial\Omega_{ij}}(\mathbf{x}^{(0)}_{m+\varepsilon}) \cdot [\mathbf{x}^{(j)}_{i(m+\varepsilon)} - \mathbf{x}^{(0)}_{m+\varepsilon}] < 0 \end{aligned}\right\} \text{for } \mathbf{n}_{\partial\Omega_{ij}} \to \Omega_j,\\
&\text{or} \quad \left.\begin{aligned} &\mathbf{n}^{\mathrm{T}}_{\partial\Omega_{ij}}(\mathbf{x}^{(0)}_{m-\varepsilon}) \cdot [\mathbf{x}^{(0)}_{m-\varepsilon} - \mathbf{x}^{(j)}_{i(m-\varepsilon)}] < 0\\ &\mathbf{n}^{\mathrm{T}}_{\partial\Omega_{ij}}(\mathbf{x}^{(0)}_{m+\varepsilon}) \cdot [\mathbf{x}^{(j)}_{i(m+\varepsilon)} - \mathbf{x}^{(0)}_{m+\varepsilon}] > 0 \end{aligned}\right\} \text{for } \mathbf{n}_{\partial\Omega_{ij}} \to \Omega_i.
\end{aligned}
\tag{3.57}
$$

Theorem 3.7. *For a discontinuous dynamical system in Eq.* (2.1), *there is a point* $\mathbf{x}^{(0)}(t_m) \equiv \mathbf{x}_m \in \partial\Omega_{ij}$ *at time* t_m *between two adjacent domains* Ω_α ($\alpha = i, j$). *Suppose* $\mathbf{x}^{(i)}(t_{m-}) = \mathbf{x}_m = \mathbf{x}_i^{(j)}(t_{m\pm})$. *For an arbitrarily small* $\varepsilon > 0$, *there are two time intervals* $[t_{m-\varepsilon}, t_m)$ *and* $[t_{m-\varepsilon}, t_{m+\varepsilon}]$. *A flow* $\mathbf{x}^{(i)}(t)$ *is* $C^{r_i}_{[t_{m-\varepsilon}, t_m)}$ *-continuous* ($r_i \geq 2k_i + 1$) *with* $\| d^{r_i+1}\mathbf{x}^{(i)} / dt^{r_i+1} \| < \infty$ *for time t, and an imaginary flow* $\mathbf{x}_i^{(j)}(t)$ *is* $C^{r_j}_{[t_{m-\varepsilon}, t_{m+\varepsilon}]}$*-continuous* ($r_j \geq 2k_j$) *with* $\| d^{r_j+1}\mathbf{x}_i^{(j)} / dt^{r_j+1} \| < \infty$. *The flow* $\mathbf{x}^{(i)}(t)$ *of the* $(2k_i)$th *-order and* $\mathbf{x}_i^{(j)}(t)$ *of the* $(2k_j - 1)$th *-order to the boundary* $\partial\Omega_{ij}$ *is* $(2k_i : 2k_j - 1)$ *-half-non-passable of the first kind in domain* Ω_i *(or a* $(2k_i : 2k_j - 1)$ *-half sink flow) if and only if*

$$G_{\partial\Omega_{ij}}^{(s_i,i)}(\mathbf{x}_m,t_{m-},\mathbf{p}_i,\boldsymbol{\lambda})=0 \ \mathit{for}\ s_i=0,1,\cdots,2k_i-1; \tag{3.58}$$

$$G_{\partial\Omega_{ij}}^{(s_j,j)}(\mathbf{x}_m,t_{m\pm},\mathbf{p}_j,\boldsymbol{\lambda})=0 \ \mathit{for}\ s_j=0,1,\cdots,2k_j-2; \tag{3.59}$$

$$\left.\begin{aligned} \mathit{either}\quad & \left.\begin{aligned} & G_{\partial\Omega_{ij}}^{(2k_i,i)}(\mathbf{x}_m,t_{m-},\mathbf{p}_i,\boldsymbol{\lambda})>0 \\ & G_{\partial\Omega_{ij}}^{(2k_j-1,j)}(\mathbf{x}_m,t_{m\pm},\mathbf{p}_j,\boldsymbol{\lambda})<0 \end{aligned}\right\} \mathit{for}\ \mathbf{n}_{\partial\Omega_{ij}}\to\Omega_j, \\ \mathit{or}\quad & \left.\begin{aligned} & G_{\partial\Omega_{ij}}^{(2k_i,i)}(\mathbf{x}_m,t_{m-},\mathbf{p}_i,\boldsymbol{\lambda})<0 \\ & G_{\partial\Omega_{ij}}^{(2k_j-1,j)}(\mathbf{x}_m,t_{m\pm},\mathbf{p}_j,\boldsymbol{\lambda})>0 \end{aligned}\right\} \mathit{for}\ \mathbf{n}_{\partial\Omega_{ij}}\to\Omega_i. \end{aligned}\right. \tag{3.60}$$

Proof. The proof is similar to the proof of Theorem 3.2. ■

Before the half-non-passable flow of the second kind is discussed, the intuitive illustration of the half-non-passable flow is sketched in Figs. 3.12 and 3.13 for a better understanding of this new concept. The half-non-passable flow of the second kind is termed *a half-source flow*. The half-source flows in Ω_i are presented in Fig. 3.12(a). $\mathbf{x}^{(i)}(t)$ for time $t\in(t_m,t_{m+\varepsilon}]$ is only a real flow. The imaginary flows $\mathbf{x}_j^{(i)}(t)$ for time $t\in[t_{m-\varepsilon},t_{m+\varepsilon}]$ and $\mathbf{x}_i^{(j)}(t)$ for time $t\in[t_{m-\varepsilon},t_m)$ are represented by dashed curves. To the same boundary $\partial\Omega_{ij}$, a half-source flow in Ω_j is sketched in Fig. 3.12(b). The leaving flow $\mathbf{x}^{(j)}(t)$ for $t\in(t_m,t_{m+\varepsilon}]$ is a real flow. Similarly, the $(2k_i-1:2k_j-1)$-half-source flow in domain Ω_i and Ω_j will be presented in Fig. 3.13 (a) and (b), respectively.

Definition 3.17. For a discontinuous dynamical system in Eq. (2.1), there is a point $\mathbf{x}^{(0)}(t_m)\equiv\mathbf{x}_m\in\partial\Omega_{ij}$ at time t_m between two adjacent domains Ω_α $(\alpha=i,j)$. Suppose $\mathbf{x}^{(\alpha)}(t_{m+})=\mathbf{x}_m=\mathbf{x}_\alpha^{(\beta)}(t_{m\pm})$. For an arbitrarily small $\varepsilon>0$, there are two time intervals $[t_{m-\varepsilon},t_m)$ and $[t_{m-\varepsilon},t_{m+\varepsilon}]$. A flow $\mathbf{x}^{(\alpha)}(t)$ is $C_{[t_{m-\varepsilon},t_m)}^{r_\beta}$-continuous $(r_\alpha\geq m_\alpha+1)$ with $\| d^{r_\alpha+1}\mathbf{x}^{(\alpha)}/dt^{r_\alpha+1}\|<\infty$ for time t, and an imaginary flow $\mathbf{x}_\alpha^{(\beta)}(t)$ is $C_{[t_{m-\varepsilon},t_{m+\varepsilon}]}^{r_\beta}$-continuous with $\| d^{r_\beta+1}\mathbf{x}_\alpha^{(\beta)}/dt^{r_\beta+1}\|<\infty$ ($r_\beta\geq 2k_\beta$, $\beta=i,j$ and $\beta\neq\alpha$). The flow $\mathbf{x}^{(\alpha)}(t)$ of the (m_α)th-order and $\mathbf{x}_\alpha^{(\beta)}(t)$ of the $(2k_\beta-1)$th-order to the boundary $\partial\Omega_{ij}$ is $(m_\alpha:2k_\beta-1)$-half-*non-passable* of the second kind in domain Ω_α (or called an $(m_\alpha:2k_\beta-1)$-*half source flow*) if

$$\left.\begin{aligned} & G_{\partial\Omega_{ij}}^{(s_\alpha,\alpha)}(\mathbf{x}_m,t_{m+},\mathbf{p}_\alpha,\boldsymbol{\lambda})=0 \ \text{for}\ s_\alpha=0,1,\cdots,m_\alpha-1, \\ & G_{\partial\Omega_{ij}}^{(2k_\alpha,\alpha)}(\mathbf{x}_m,t_{m+},\mathbf{p}_\alpha,\boldsymbol{\lambda})\neq 0, \end{aligned}\right. \tag{3.61}$$

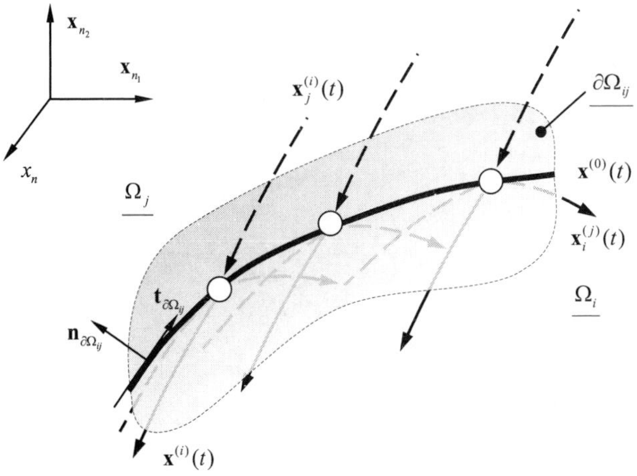

(a)

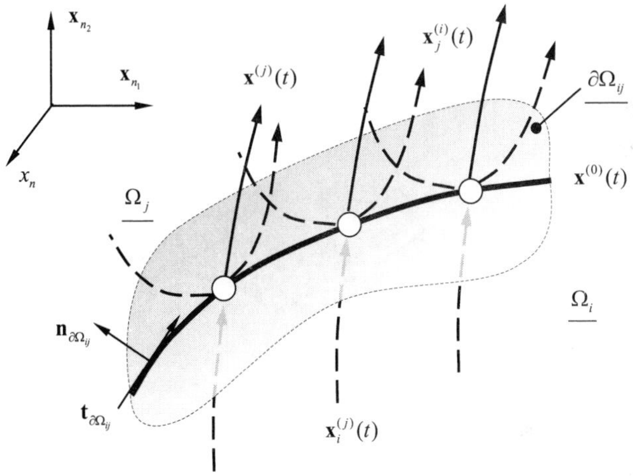

(b)

Fig. 3.12 Half-source flows: **(a)** $(2k_i : 2k_j - 1)$ -order in Ω_i and **(b)** $(2k_i - 1 : 2k_j)$ -order in Ω_j. $\mathbf{x}^{(i)}(t)$ and $\mathbf{x}^{(j)}(t)$ represent *real* flows in domains Ω_i and Ω_j, respectively, which are depicted by thin solid curves. $\mathbf{x}_i^{(j)}(t)$ and $\mathbf{x}_j^{(i)}(t)$ represent *imaginary* flows in domains Ω_i and Ω_j, respectively controlled by the vector fields in Ω_j and Ω_i, which are depicted by dashed curves. The flow on the boundary is described by $\mathbf{x}^{(0)}(t)$. The normal and tangential vectors $\mathbf{n}_{\partial\Omega_{ij}}$ and $\mathbf{t}_{\partial\Omega_{ij}}$ of the boundary are depicted. Hollow circles are for source points on the boundary. ($n_1 + n_2 + 1 = n$)

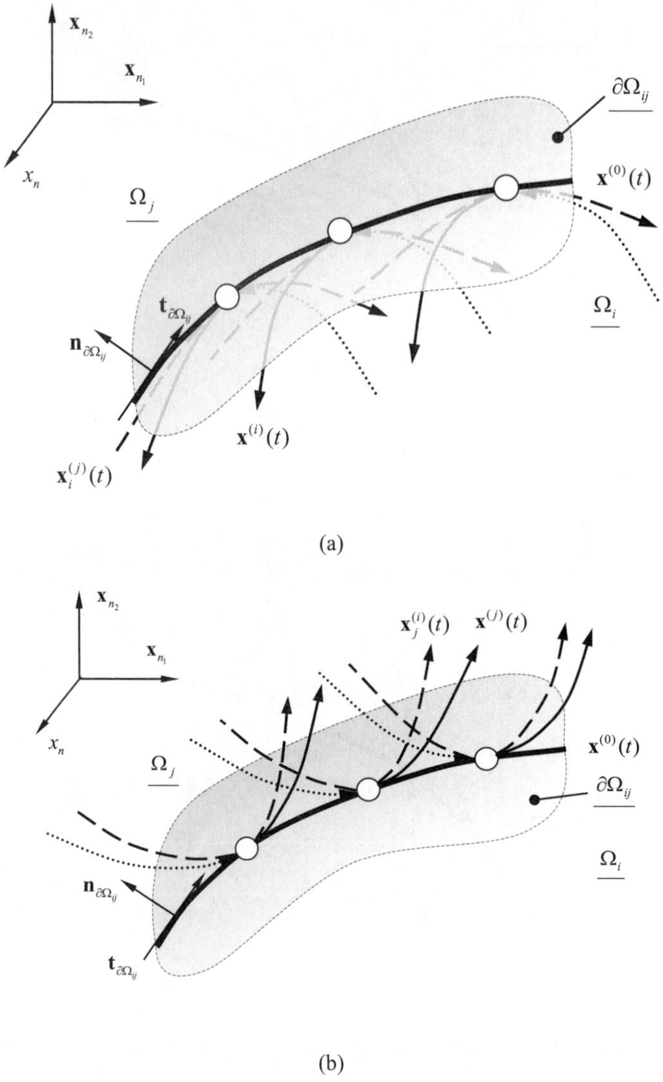

Fig. 3.13 $(2k_i-1:2k_j-1)$-half-source flows in: **(a)** domain Ω_i and **(b)** domain Ω_j. $\mathbf{x}^{(i)}(t)$ and $\mathbf{x}^{(j)}(t)$ represent *real* flows in domains Ω_i and Ω_j, respectively, which are depicted by thin solid curves. $\mathbf{x}_i^{(j)}(t)$ and $\mathbf{x}_j^{(i)}(t)$ represent *imaginary* flows in domains Ω_i and Ω_j, respectively controlled by the vector fields in Ω_j and Ω_i, which are depicted by dashed curves. The flow on the boundary is described by $\mathbf{x}^{(0)}(t)$. The normal and tangential vectors $\mathbf{n}_{\partial\Omega_{ij}}$ and $\mathbf{t}_{\partial\Omega_{ij}}$ of the boundary are depicted. Hollow circles are for source points on the boundary. ($n_1+n_2+1=n$)

$$
\begin{aligned}
&G_{\partial\Omega_{ij}}^{(s_\beta,\beta)}(\mathbf{x}_m,t_{m\pm},\mathbf{p}_\beta,\boldsymbol{\lambda})=0 \text{ for } s_\beta=0,1,\cdots,2k_\beta-2,\\
&G_{\partial\Omega_{ij}}^{(2k_\beta-1,\beta)}(\mathbf{x}_m,t_{m\pm},\mathbf{p}_\beta,\boldsymbol{\lambda})\neq 0;
\end{aligned}
\tag{3.62}
$$

$$
\begin{aligned}
&\textit{either}\quad \mathbf{n}_{\partial\Omega_{ij}}^{\mathrm{T}}(\mathbf{x}_{m+\varepsilon}^{(0)})\cdot[\mathbf{x}_{m+\varepsilon}^{(\alpha)}-\mathbf{x}_{m+\varepsilon}^{(0)}]<0 \ \textit{for}\ \mathbf{n}_{\partial\Omega_{ij}}\to\Omega_\beta,\\
&\textit{or}\quad \mathbf{n}_{\partial\Omega_{ij}}^{\mathrm{T}}(\mathbf{x}_{m+\varepsilon}^{(0)})\cdot[\mathbf{x}_{m+\varepsilon}^{(\alpha)}-\mathbf{x}_{m+\varepsilon}^{(0)}]>0 \ \textit{for}\ \mathbf{n}_{\partial\Omega_{ij}}\to\Omega_\alpha;
\end{aligned}
\tag{3.63}
$$

$$
\begin{aligned}
&\textit{either}\quad \left.\begin{aligned}&\mathbf{n}_{\partial\Omega_{ij}}^{\mathrm{T}}(\mathbf{x}_{m-\varepsilon}^{(0)})\cdot[\mathbf{x}_{m-\varepsilon}^{(0)}-\mathbf{x}_{\alpha(m-\varepsilon)}^{(\beta)}]>0\\ &\mathbf{n}_{\partial\Omega_{ij}}^{\mathrm{T}}(\mathbf{x}_{m+\varepsilon}^{(0)})\cdot[\mathbf{x}_{\alpha(m+\varepsilon)}^{(\beta)}-\mathbf{x}_{m+\varepsilon}^{(0)}]<0\end{aligned}\right\}\textit{for}\ \mathbf{n}_{\partial\Omega_{ij}}\to\Omega_\beta,\\
&\textit{or}\quad \left.\begin{aligned}&\mathbf{n}_{\partial\Omega_{ij}}^{\mathrm{T}}(\mathbf{x}_{m-\varepsilon}^{(0)})\cdot[\mathbf{x}_{m-\varepsilon}^{(0)}-\mathbf{x}_{\alpha(m-\varepsilon)}^{(\beta)}]<0\\ &\mathbf{n}_{\partial\Omega_{ij}}^{\mathrm{T}}(\mathbf{x}_{m+\varepsilon}^{(0)})\cdot[\mathbf{x}_{\alpha(m+\varepsilon)}^{(\beta)}-\mathbf{x}_{m+\varepsilon}^{(0)}]>0\end{aligned}\right\}\textit{for}\ \mathbf{n}_{\partial\Omega_{ij}}\to\Omega_\alpha.
\end{aligned}
\tag{3.64}
$$

From the above definition, the necessary and sufficient conditions for such a $(m_\alpha:2k_\beta-1)$ half-non-passable flow of the second kind (or $(m_\alpha:2k_\beta-1)$ half-source flow) are stated in the following theorem.

Theorem 3.8. *For a discontinuous dynamical system in Eq.* (2.1), *there is a point* $\mathbf{x}^{(0)}(t_m)\equiv\mathbf{x}_m\in\partial\Omega_{ij}$ *at time* t_m *between two adjacent domains* Ω_α $(\alpha=i,j)$. *For an arbitrarily small* $\varepsilon>0$, *there are two time intervals* $[t_{m-\varepsilon},t_m)$ *and* $[t_{m-\varepsilon},t_{m+\varepsilon}]$. *Suppose* $\mathbf{x}^{(\alpha)}(t_{m+})=\mathbf{x}_m=\mathbf{x}_\alpha^{(\beta)}(t_{m\pm})$. *A flow* $\mathbf{x}^{(\alpha)}(t)$ *is* $C_{[t_{m-\varepsilon},t_m)}^{r_\beta}$*-continuous* ($r_\alpha\geq m_\alpha+1$) *with* $\|d^{r_\alpha+1}\mathbf{x}^{(\alpha)}/dt^{r_\alpha+1}\|<\infty$ *for time t, and an imaginary flow* $\mathbf{x}_\alpha^{(\beta)}(t)$ *is* $C_{[t_{m-\varepsilon},t_{m+\varepsilon}]}^{r_\beta}$*-continuous with* $\|d^{r_\beta+1}\mathbf{x}_\alpha^{(\beta)}/dt^{r_\beta+1}\|<\infty$ ($r_\beta\geq 2k_\beta$). *The flow* $\mathbf{x}^{(\alpha)}(t)$ *of the* (m_α)th *-order and* $\mathbf{x}_\alpha^{(\beta)}(t)$ *of the* $(2k_\beta-1)$th *-order to the boundary* $\partial\Omega_{ij}$ *is* $(m_\alpha:2k_\beta-1)$ *-half-non-passable of the second kind in domain* Ω_α *(or an* $(m_\alpha:2k_\beta-1)$ *-half source flow) if and only if*

$$
G_{\partial\Omega_{ij}}^{(s_\alpha,\alpha)}(\mathbf{x}_m,t_{m+},\mathbf{p}_\alpha,\boldsymbol{\lambda})=0 \ \text{ for } s_\alpha=0,1,\cdots,m_\alpha-1;
\tag{3.65}
$$

$$
G_{\partial\Omega_{ij}}^{(s_\beta,\beta)}(\mathbf{x}_m,t_{m\pm},\mathbf{p}_\beta,\boldsymbol{\lambda})=0 \text{ for } s_\beta=0,1,\cdots,2k_\beta-2;
\tag{3.66}
$$

$$
\begin{aligned}
&\textit{either}\quad \left.\begin{aligned}&G_{\partial\Omega_{ij}}^{(2k_\alpha,\alpha)}(\mathbf{x}_m,t_{m+},\mathbf{p}_\alpha,\boldsymbol{\lambda})<0\\ &G_{\partial\Omega_{ij}}^{(2k_\beta-1,\beta)}(\mathbf{x}_m,t_{m\pm},\mathbf{p}_\beta,\boldsymbol{\lambda})<0\end{aligned}\right\}\textit{for}\ \mathbf{n}_{\partial\Omega_{ij}}\to\Omega_\beta,\\
&\textit{or}\quad \left.\begin{aligned}&G_{\partial\Omega_{ij}}^{(2k_\alpha,\alpha)}(\mathbf{x}_m,t_{m+},\mathbf{p}_\alpha,\boldsymbol{\lambda})>0\\ &G_{\partial\Omega_{ij}}^{(2k_\alpha,\alpha)}(\mathbf{x}_m,t_{m+},\mathbf{p}_\alpha,\boldsymbol{\lambda})>0\end{aligned}\right\}\textit{for}\ \mathbf{n}_{\partial\Omega_{ij}}\to\Omega_\alpha.
\end{aligned}
\tag{3.67}
$$

Proof. The proof is similar to the proof of Theorem 3.2. ■

3.5. Grazing flows

The tangency of a flow to the boundary in a discontinuous dynamical system is generalized. In Chapter 2, the tangential flow to the boundary is valid for the plane surface because the normal vector $\mathbf{n}_{\partial\Omega_{ij}}$ for an $(n-1)$-dimensional *plane* boundary does not change with contact location. However, the normal vector for an $(n-1)$-dimensional surface boundary changes with contact location. Thus, the theory of the tangential flow to the boundary in Chapter 2 should be generalized. Herein, a generalized tangential flow to the boundary will be presented, which also includes the tangential of the imaginary flows to the boundary.

Definition 3.18. For a discontinuous dynamical system in Eq. (2.1), there is a point $\mathbf{x}^{(0)}(t_m) \equiv \mathbf{x}_m \in \partial\Omega_{ij}$ at time t_m between two adjacent domains Ω_α ($\alpha = i, j$). Suppose $\mathbf{x}^{(\alpha)}(t_{m\pm}) = \mathbf{x}_m$ ($\alpha \in \{i, j\}$). For an arbitrarily small $\varepsilon > 0$, there is a time interval $[t_{m-\varepsilon}, t_{m+\varepsilon}]$. A flow $\mathbf{x}^{(\alpha)}(t)$ is $C^{r_\alpha}_{[t_{m-\varepsilon}, t_{m+\varepsilon}]}$-continuous ($r_\alpha \geq 2$) for time t. The flow $\mathbf{x}^{(\alpha)}(t)$ in domain Ω_α is *tangential* to the boundary $\partial\Omega_{ij}$ if

$$G^{(0,\alpha)}_{\partial\Omega_{ij}}(\mathbf{x}_m, t_m, \mathbf{p}_\alpha, \boldsymbol{\lambda}) = 0 \text{ and } G^{(1,\alpha)}_{\partial\Omega_{ij}}(\mathbf{x}_m, t_m, \mathbf{p}_\alpha, \boldsymbol{\lambda}) \neq 0; \tag{3.68}$$

$$\begin{aligned} &\text{either} \quad \left.\begin{aligned} \mathbf{n}^{\mathrm{T}}_{\partial\Omega_{ij}}(\mathbf{x}^{(0)}_{m-\varepsilon}) \cdot [\mathbf{x}^{(0)}_{m-\varepsilon} - \mathbf{x}^{(\alpha)}_{m-\varepsilon}] > 0 \\ \mathbf{n}^{\mathrm{T}}_{\partial\Omega_{ij}}(\mathbf{x}^{(0)}_{m+\varepsilon}) \cdot [\mathbf{x}^{(\alpha)}_{m+\varepsilon} - \mathbf{x}^{(0)}_{m+\varepsilon}] < 0 \end{aligned}\right\} \text{ for } \mathbf{n}_{\partial\Omega_{ij}} \to \Omega_\beta, \\ &\text{or} \quad \left.\begin{aligned} \mathbf{n}^{\mathrm{T}}_{\partial\Omega_{ij}}(\mathbf{x}^{(0)}_{m-\varepsilon}) \cdot [\mathbf{x}^{(0)}_{m-\varepsilon} - \mathbf{x}^{(\alpha)}_{m-\varepsilon}] < 0 \\ \mathbf{n}^{\mathrm{T}}_{\partial\Omega_{ij}}(\mathbf{x}^{(0)}_{m+\varepsilon}) \cdot [\mathbf{x}^{(\alpha)}_{m+\varepsilon} - \mathbf{x}^{(0)}_{m+\varepsilon}] > 0 \end{aligned}\right\} \text{ for } \mathbf{n}_{\partial\Omega_{ij}} \to \Omega_\alpha. \end{aligned} \tag{3.69}$$

Theorem 3.9. *For a discontinuous dynamical system in Eq. (2.1), there is a point* $\mathbf{x}^{(0)}(t_m) \equiv \mathbf{x}_m \in \partial\Omega_{ij}$ *at time* t_m *between two adjacent domains* Ω_α ($\alpha = i, j$). *Suppose* $\mathbf{x}^{(\alpha)}(t_{m\pm}) = \mathbf{x}_m$ ($\alpha \in \{i, j\}$). *For an arbitrarily small* $\varepsilon > 0$, *there is a time interval* $[t_{m-\varepsilon}, t_{m+\varepsilon}]$. *A flow* $\mathbf{x}^{(\alpha)}(t)$ *is* $C^{r}_{[t_{m-\varepsilon}, t_{m+\varepsilon}]}$*-continuous* ($r \geq 2$) *for time* t *with* $\| d^{r_\alpha+1}\mathbf{x}^{(\alpha)} / dt^{r_\alpha+1} \| < \infty$. *The flow* $\mathbf{x}^{(\alpha)}(t)$ *in domain* Ω_α *is tangential to the boundary* $\partial\Omega_{ij}$ *if and only if*

$$G^{(0,\alpha)}_{\partial\Omega_{ij}}(\mathbf{x}_m, t_m, \mathbf{p}_\alpha, \boldsymbol{\lambda}) = 0 \quad \text{for } \alpha \in \{i, j\}; \tag{3.70}$$

$$\begin{array}{ll} either & G_{\partial\Omega_{ij}}^{(1,\alpha)}(\mathbf{x}_m,t_m,\mathbf{p}_\alpha,\boldsymbol{\lambda})<0 \ for\ \mathbf{n}_{\partial\Omega_{ij}}\to\Omega_\beta, \\ or & G_{\partial\Omega_{ij}}^{(1,\alpha)}(\mathbf{x}_m,t_m,\mathbf{p}_\alpha,\boldsymbol{\lambda})>0 \ for\ \mathbf{n}_{\partial\Omega_{ij}}\to\Omega_\alpha. \end{array} \tag{3.71}$$

Proof. Equation (3.70) is identical to Eq. (3.68), thus the condition in Eq. (3.68) is satisfied, and vice versa. Suppose $\mathbf{x}^{(\alpha)}(t_{m\pm})=\mathbf{x}_m$ ($\alpha\in\{i,j\}$) and $\mathbf{x}^{(\alpha)}(t)$ are $C^{r_\alpha}_{[t_{m-\varepsilon},t_{m+\varepsilon}]}$-continuous ($r_\alpha\geq 2$) for time t and $\|d^{r_\alpha}\mathbf{x}^{(\alpha)}/dt^{r_\alpha}\|<\infty$ ($\alpha\in\{i,j\}$). For $a\in[t_{m-\varepsilon},t_m)$ or $a\in(t_m,t_{m+\varepsilon}]$, the Taylor series expansion of $\mathbf{x}^{(\alpha)}(t_{m\pm\varepsilon})$ to $\mathbf{x}^{(\alpha)}(a)$ up to the third-order term gives

$$\begin{aligned} \mathbf{x}_{m\pm\varepsilon}^{(\alpha)} &\equiv \mathbf{x}^{(\alpha)}\left(t_{m\pm}\pm\varepsilon\right) \\ &= \mathbf{x}^{(\alpha)}\big|_{t=a}+\dot{\mathbf{x}}^{(\alpha)}\big|_{t=a}(t_{m\pm}\pm\varepsilon-a)+\ddot{\mathbf{x}}^{(\alpha)}\big|_{t=a}(t_{m\pm}\pm\varepsilon-a)^2 \\ &\quad +o((t_{m\pm}\pm\varepsilon-a)^2). \end{aligned}$$

As $a\to t_{m\pm}$, taking the limit of the foregoing equation leads to

$$\mathbf{x}_{m\pm\varepsilon}^{(\alpha)}\equiv\mathbf{x}^{(\alpha)}(t_m\pm\varepsilon)=\mathbf{x}_{m\pm}^{(\alpha)}\pm\dot{\mathbf{x}}_{m\pm}^{(\alpha)}\varepsilon+\ddot{\mathbf{x}}_{m\pm}^{(\alpha)}\varepsilon^2+o(\varepsilon^2).$$

In a similar fashion, we have

$$\mathbf{x}_{m\pm\varepsilon}^{(0)}\equiv\mathbf{x}^{(0)}(t_m\pm\varepsilon)=\mathbf{x}_{m}^{(0)}\pm\dot{\mathbf{x}}_{m}^{(0)}\varepsilon+\ddot{\mathbf{x}}_{m}^{(0)}\varepsilon^2+o(\varepsilon^2),$$

$$\mathbf{n}_{\partial\Omega_{ij}}(\mathbf{x}_{m\pm\varepsilon}^{(0)})\equiv\mathbf{n}_{\partial\Omega_{ij}}\big|_{\mathbf{x}_m^{(0)}}\pm D_0\mathbf{n}_{\partial\Omega_{ij}}\big|_{\mathbf{x}_m^{(0)}}\varepsilon+D_0^2\mathbf{n}_{\partial\Omega_{ij}}\big|_{\mathbf{x}_m^{(0)}}\varepsilon^2+o(\varepsilon^2).$$

The ignorance of the ε^3 and high order terms, the deformation of the above equation and left multiplication of $\mathbf{n}_{\partial\Omega_{ij}}$ gives

$$\begin{aligned} \mathbf{n}_{\partial\Omega_{ij}}^{\mathrm{T}}(\mathbf{x}_{m\pm\varepsilon}^{(0)})\bullet[\mathbf{x}_{m\pm\varepsilon}^{(\alpha)}-\mathbf{x}_{m\pm\varepsilon}^{(0)}] &= \mathbf{n}_{\partial\Omega_{ij}}^{\mathrm{T}}(\mathbf{x}_{m}^{(0)})\cdot[\mathbf{x}_{m\pm}^{(\alpha)}-\mathbf{x}_{m}^{(0)}] \\ &\quad \pm\varepsilon G_{\partial\Omega_{ij}}^{(0,\alpha)}(\mathbf{x}_m^{(0)},\mathbf{x}_{m\pm}^{(\alpha)},t_m,\mathbf{p}_\alpha,\boldsymbol{\lambda}) \\ &\quad +\varepsilon^2 G_{\partial\Omega_{ij}}^{(1,\alpha)}(\mathbf{x}_m^{(0)},\mathbf{x}_{m\pm}^{(\alpha)},t_m,\mathbf{p}_\alpha,\boldsymbol{\lambda}). \end{aligned}$$

Due to $\mathbf{x}_{m\pm}^{(\alpha)}=\mathbf{x}_m^{(0)}=\mathbf{x}_m$ and $G_{\partial\Omega_{ij}}^{(0,\alpha)}(\mathbf{x}_m^{(0)},\mathbf{x}_{m\pm}^{(\alpha)},t_m,\mathbf{p}_\alpha,\boldsymbol{\lambda})\equiv G_{\partial\Omega_{ij}}^{(0,\alpha)}(\mathbf{x}_m,t_m,\mathbf{p}_\alpha,\boldsymbol{\lambda})=0$, the foregoing equation becomes

$$\begin{aligned} \mathbf{n}_{\partial\Omega_{ij}}^{\mathrm{T}}(\mathbf{x}_{m-\varepsilon}^{(0)})\cdot[\mathbf{x}_{m-\varepsilon}^{(0)}-\mathbf{x}_{m-\varepsilon}^{(\alpha)}] &= -\varepsilon^2 G_{\partial\Omega_{ij}}^{(1,\alpha)}(\mathbf{x}_m,t_m,\mathbf{p}_\alpha,\boldsymbol{\lambda}), \\ \mathbf{n}_{\partial\Omega_{ij}}^{\mathrm{T}}(\mathbf{x}_{m+\varepsilon}^{(0)})\cdot[\mathbf{x}_{m+\varepsilon}^{(\alpha)}-\mathbf{x}_{m+\varepsilon}^{(0)}] &= \varepsilon^2 G_{\partial\Omega_{ij}}^{(1,\alpha)}(\mathbf{x}_m,t_m,\mathbf{p}_\alpha,\boldsymbol{\lambda}). \end{aligned}$$

Using Eq. (3.71), Equation (3.69) is obtained. On the other hand, using Eq. (3.69), Equation (3.71) is achieved. Therefore, this theorem is proved. ■

From Eqs. (3.17) and (3.18), one obtains

$$
\begin{aligned}
&G_{\partial\Omega_{ij}}^{(1,\alpha)}(\mathbf{x}_m^{(0)},t_{m\pm},\mathbf{x}_{m\pm}^{(\alpha)},\mathbf{p}_\alpha,\boldsymbol{\lambda})\\
&=D_0^2\mathbf{n}_{\partial\Omega_{ij}}^{\mathrm{T}}\cdot(\mathbf{x}^{(\alpha)}-\mathbf{x}^{(0)})+2D_0\mathbf{n}_{\partial\Omega_{ij}}^{\mathrm{T}}\cdot(\dot{\mathbf{x}}^{(\alpha)}-\dot{\mathbf{x}}^{(0)})\\
&\quad+\mathbf{n}_{\partial\Omega_{ij}}^{\mathrm{T}}\cdot(\ddot{\mathbf{x}}^{(\alpha)}-\ddot{\mathbf{x}}^{(0)})\big|_{(\mathbf{x}_m^{(0)},t_{m\pm},\mathbf{x}_{m\pm}^{(\alpha)})}\\
&=D_0^2\mathbf{n}_{\partial\Omega_{ij}}^{\mathrm{T}}\cdot(\mathbf{x}^{(\alpha)}-\mathbf{x}^{(0)})+2D_0\mathbf{n}_{\partial\Omega_{ij}}^{\mathrm{T}}\cdot(\mathbf{F}^{(\alpha)}-\mathbf{F}^{(0)})\\
&\quad+\mathbf{n}_{\partial\Omega_{ij}}^{\mathrm{T}}\cdot(D\mathbf{F}^{(\alpha)}-D_0\mathbf{F}^{(0)})\big|_{(\mathbf{x}_m^{(0)},t_{m\pm},\mathbf{x}_{m\pm}^{(\alpha)})}.
\end{aligned}
\tag{3.72}
$$

For the first-order contact at point $(\mathbf{x}_m^{(0)},t_{m\pm},\mathbf{x}_{m\pm}^{(\alpha)})$, the followings equation holds in Kreyszig (1968) and Luo (2008a,c)

$$
\begin{aligned}
&\mathbf{x}_m^{(0)}=\mathbf{x}_{m\pm}^{(\alpha)}\ \text{and}\ \dot{\mathbf{x}}_m^{(0)}=\dot{\mathbf{x}}_{m\pm}^{(\alpha)},\\
&\mathbf{F}^{(\alpha)}=\mathbf{F}^{(0)}\big|_{(\mathbf{x}_m^{(0)},t_{m\pm},\mathbf{x}_{m\pm}^{(\alpha)})}.
\end{aligned}
\tag{3.73}
$$

Thus, one achieves

$$
\begin{aligned}
G_{\partial\Omega_{ij}}^{(1,\alpha)}(\mathbf{x}_m^{(0)},t_{m\pm},\mathbf{x}_{m\pm}^{(\alpha)},\mathbf{p}_\alpha,\boldsymbol{\lambda})&=\mathbf{n}_{\partial\Omega_{ij}}^{\mathrm{T}}\cdot(\ddot{\mathbf{x}}^{(\alpha)}-\ddot{\mathbf{x}}^{(0)})\big|_{(\mathbf{x}_m^{(0)},t_{m\pm},\mathbf{x}_{m\pm}^{(\alpha)})}\\
&=\mathbf{n}_{\partial\Omega_{ij}}^{\mathrm{T}}\cdot(D_\alpha\mathbf{F}^{(\alpha)}-D_0\mathbf{F}^{(0)})\big|_{(\mathbf{x}_m^{(0)},t_{m\pm},\mathbf{x}_{m\pm}^{(\alpha)})}.
\end{aligned}
\tag{3.74}
$$

For the n-dimensional *plane* boundary, because $\mathbf{n}_{\partial\Omega_{ij}}$ is constant, the corresponding derivatives gives

$$
D_0\mathbf{n}_{\partial\Omega_{ij}}=D_0^2\mathbf{n}_{\partial\Omega_{ij}}=0. \tag{3.75}
$$

Owing to $\mathbf{n}_{\partial\Omega_{ij}}^{\mathrm{T}}\cdot\dot{\mathbf{x}}^{(0)}=0$, the corresponding derivative becomes

$$
(D_0\mathbf{n}_{\partial\Omega_{ij}}^{\mathrm{T}}\cdot\dot{\mathbf{x}}^{(0)}+\mathbf{n}_{\partial\Omega_{ij}}^{\mathrm{T}}\cdot\ddot{\mathbf{x}}^{(0)})=0. \tag{3.76}
$$

Thus, one obtains

$$
\mathbf{n}_{\partial\Omega_{ij}}^{\mathrm{T}}\cdot\ddot{\mathbf{x}}^{(0)}=\mathbf{n}_{\partial\Omega_{ij}}^{\mathrm{T}}\cdot D_0\mathbf{F}^{(0)}=0. \tag{3.77}
$$

Finally, Equation (3.74) becomes

$$
\begin{aligned}
G_{\partial\Omega_{ij}}^{(1,\alpha)}(\mathbf{x}_m^{(0)},t_{m\pm},\mathbf{x}_{m\pm}^{(\alpha)},\mathbf{p}_\alpha,\boldsymbol{\lambda})&=\mathbf{n}_{\partial\Omega_{ij}}^{\mathrm{T}}\cdot\ddot{\mathbf{x}}^{(\alpha)}\big|_{(\mathbf{x}_m^{(0)},t_{m\pm},\mathbf{x}_{m\pm}^{(\alpha)})}\\
&=\mathbf{n}_{\partial\Omega_{ij}}^{\mathrm{T}}\cdot D_\alpha\mathbf{F}^{(\alpha)}\big|_{(\mathbf{x}_m^{(0)},t_{m\pm},\mathbf{x}_{m\pm}^{(\alpha)})},
\end{aligned}
\tag{3.78}
$$

and the zero-order G-function is

$$
\begin{aligned}
G_{\partial\Omega_{ij}}^{(0,\alpha)}(\mathbf{x}_m^{(0)},t_{m\pm},\mathbf{x}_{m\pm}^{(\alpha)},\mathbf{p}_\alpha,\boldsymbol{\lambda})&=\mathbf{n}_{\partial\Omega_{ij}}^{\mathrm{T}}\cdot\dot{\mathbf{x}}^{(\alpha)}\big|_{(\mathbf{x}_m^{(0)},t_{m\pm},\mathbf{x}_{m\pm}^{(\alpha)})}\\
&=\mathbf{n}_{\partial\Omega_{ij}}^{\mathrm{T}}\cdot\mathbf{F}^{(\alpha)}\big|_{(\mathbf{x}_m^{(0)},t_{m\pm},\mathbf{x}_{m\pm}^{(\alpha)})}.
\end{aligned}
\tag{3.79}
$$

Thus, Definition 3.18 gives a generalized definition of a grazing flow to the boundary, and compared to Theorems 2.9 and 1.10, Theorem 3.9 gives the generalized grazing conditions. The theory presented in Chapter 2 is a special case presented in this Chapter.

Definition 3.19. For a discontinuous dynamical system in Eq. (2.1), there is a point $\mathbf{x}^{(0)}(t_m) \equiv \mathbf{x}_m \in \partial\Omega_{ij}$ at time t_m between two adjacent domains Ω_α ($\alpha = i, j$). Suppose $\mathbf{x}^{(\alpha)}(t_{m\pm}) = \mathbf{x}_m$ ($\alpha \in \{i, j\}$). For an arbitrarily small $\varepsilon > 0$, there is a time interval $[t_{m-\varepsilon}, t_{m+\varepsilon}]$. A flow $\mathbf{x}^{(\alpha)}(t)$ is $C^{r_\alpha}_{[t_{m-\varepsilon}, t_{m+\varepsilon}]}$-continuous ($r_\alpha \geq k_\alpha + 1$) with $\| d^{r_\alpha+1}\mathbf{x}^{(\alpha)} / dt^{r_\alpha+1} \| < \infty$ for time t. A flow $\mathbf{x}^{(\alpha)}(t)$ in Ω_α is *tangential* to the boundary $\partial\Omega_{ij}$ of the $(2k_\alpha - 1)$th-order if

$$\begin{aligned} &G^{(s_\alpha,\alpha)}_{\partial\Omega_{ij}}(\mathbf{x}_m, t_m, \mathbf{p}_\alpha, \boldsymbol{\lambda}) = 0 \text{ for } s_\alpha = 0, 1, \cdots, 2k_\alpha - 2;\\ &G^{(2k_\alpha-1,\alpha)}_{\partial\Omega_{ij}}(\mathbf{x}_m, t_m, \mathbf{p}_\alpha, \boldsymbol{\lambda}) \neq 0; \end{aligned} \tag{3.80}$$

$$\begin{aligned} &\text{either} \quad \left. \begin{aligned} &\mathbf{n}^{\mathrm{T}}_{\partial\Omega_{ij}}(\mathbf{x}^{(0)}_{m-\varepsilon}) \cdot [\mathbf{x}^{(0)}_{m-\varepsilon} - \mathbf{x}^{(\alpha)}_{m-\varepsilon}] > 0\\ &\mathbf{n}^{\mathrm{T}}_{\partial\Omega_{ij}}(\mathbf{x}^{(0)}_{m+\varepsilon}) \cdot [\mathbf{x}^{(\alpha)}_{m+\varepsilon} - \mathbf{x}^{(0)}_{m+\varepsilon}] < 0 \end{aligned} \right\} \text{for } \mathbf{n}_{\partial\Omega_{ij}} \to \Omega_\beta,\\ &\text{or} \quad \left. \begin{aligned} &\mathbf{n}^{\mathrm{T}}_{\partial\Omega_{ij}}(\mathbf{x}^{(0)}_{m-\varepsilon}) \cdot [\mathbf{x}^{(0)}_{m-\varepsilon} - \mathbf{x}^{(\alpha)}_{m-\varepsilon}] < 0\\ &\mathbf{n}^{\mathrm{T}}_{\partial\Omega_{ij}}(\mathbf{x}^{(0)}_{m+\varepsilon}) \cdot [\mathbf{x}^{(\alpha)}_{m+\varepsilon} - \mathbf{x}^{(0)}_{m+\varepsilon}] > 0 \end{aligned} \right\} \text{for } \mathbf{n}_{\partial\Omega_{ij}} \to \Omega_\alpha. \end{aligned} \tag{3.81}$$

Theorem 3.10. *For a discontinuous dynamical system in Eq.* (2.1), *there is a point* $\mathbf{x}^{(0)}(t_m) \equiv \mathbf{x}_m \in \partial\Omega_{ij}$ *at time* t_m *between two adjacent domains* Ω_α ($\alpha = i, j$). *Suppose* $\mathbf{x}^{(\alpha)}(t_{m\pm}) = \mathbf{x}_m$ ($\alpha \in \{i, j\}$). *For an arbitrarily small* $\varepsilon > 0$, *there is a time interval* $[t_{m-\varepsilon}, t_{m+\varepsilon}]$. *A flow* $\mathbf{x}^{(\alpha)}(t)$ *is* $C^{r_\alpha}_{[t_{m-\varepsilon}, t_{m+\varepsilon}]}$*-continuous* ($r_\alpha \geq k_\alpha + 1$) *for time* t *with* $\| d^{r_\alpha+1}\mathbf{x}^{(\alpha)} / dt^{r_\alpha+1} \| < \infty$. *A flow* $\mathbf{x}^{(\alpha)}(t)$ *in* Ω_α *is tangential to the boundary* $\partial\Omega_{ij}$ *of the* $(2k_\alpha - 1)$th*-order if and only if*

$$G^{(s_\alpha,\alpha)}_{\partial\Omega_{ij}}(\mathbf{x}_m, t_m, \mathbf{p}_\alpha, \boldsymbol{\lambda}) = 0 \; for \; s_\alpha = 0, 1, \cdots, 2k_\alpha - 2; \tag{3.82}$$

$$\begin{aligned} &either \quad G^{(2k_\alpha-1,\alpha)}_{\partial\Omega_{ij}}(\mathbf{x}_m, t_m, \mathbf{p}_\alpha, \boldsymbol{\lambda}) < 0 \text{ for } \mathbf{n}_{\partial\Omega_{ij}} \to \Omega_\beta,\\ &or \quad G^{(2k_\alpha-1,\alpha)}_{\partial\Omega_{ij}}(\mathbf{x}_m, t_m, \mathbf{p}_\alpha, \boldsymbol{\lambda}) > 0 \text{ for } \mathbf{n}_{\partial\Omega_{ij}} \to \Omega_\alpha. \end{aligned} \tag{3.83}$$

Proof. Equation (3.82) is identical to Eq. (3.80), thus the condition in Eq. (3.82) is satisfied, and vice versa. Suppose $\mathbf{x}^{(\alpha)}(t_{m\pm}) = \mathbf{x}_m$ ($\alpha \in \{i, j\}$) and $\mathbf{x}^{(\alpha)}(t)$ are $C^{r_\alpha}_{[t_{m-\varepsilon}, t_{m+\varepsilon}]}$-continuous ($r_\alpha \geq 2k_\alpha + 1$) for time t and $\| d^{r_\alpha}\mathbf{x}^{(\alpha)} / dt^{r_\alpha} \| < \infty$ ($\alpha \in \{i, j\}$). For $a \in [t_{m-\varepsilon}, t_m)$ or $a \in (t_m, t_{m+\varepsilon}]$, the Taylor series expansion of

$\mathbf{x}^{(\alpha)}(t_{m\pm\varepsilon})$ to $\mathbf{x}^{(\alpha)}(a)$ up to the $(2k_\alpha+1)$th-order term gives

$$\begin{aligned}\mathbf{x}^{(\alpha)}_{m\pm\varepsilon} &\equiv \mathbf{x}^{(\alpha)}(t_{m\pm}\pm\varepsilon)\\ &= \mathbf{x}^{(\alpha)}(a)+\sum_{s_\alpha=1}^{2k_\alpha-1}\frac{d^{s_\alpha}\mathbf{x}^{(\alpha)}}{dt^{s}}\bigg|_{t=a}(t_{m\pm}\pm\varepsilon-a)^{s_\alpha}+\frac{d^{2k_\alpha}\mathbf{x}^{(\alpha)}}{dt^{2k_\alpha}}\bigg|_{t=a}\\ &\quad\times(t_{m\pm}\pm\varepsilon-a)^{2k_\alpha}+o((t_{m\pm}\pm\varepsilon-a)^{2k_\alpha}).\end{aligned}$$

As $a\to t_{m\pm}$, taking the limit of the foregoing equation leads to

$$\begin{aligned}\mathbf{x}^{(\alpha)}_{m\pm\varepsilon} &\equiv \mathbf{x}^{(\alpha)}(t_m\pm\varepsilon)\\ &= \mathbf{x}^{(\alpha)}_{m\pm}+\sum_{s_\alpha=1}^{2k_\alpha-1}\frac{d^{s_\alpha}\mathbf{x}^{(\alpha)}}{dt^{s_\alpha}}\bigg|_{\mathbf{x}^{(\alpha)}_{m\pm}}(\pm\varepsilon)^{s_\alpha}+\frac{d^{2k_\alpha}\mathbf{x}^{(\alpha)}}{dt^{2k_\alpha}}\bigg|_{\mathbf{x}^{(\alpha)}_{m\pm}}(\pm\varepsilon)^{2k_\alpha}+o(\varepsilon^{2k_\alpha}).\end{aligned}$$

In a similar fashion, one obtains

$$\begin{aligned}\mathbf{x}^{(0)}_{m\pm\varepsilon} &\equiv \mathbf{x}^{(0)}(t_m\pm\varepsilon)\\ &= \mathbf{x}^{(0)}_{m}+\sum_{s_\alpha=1}^{2k_\alpha-1}\frac{d^{s}\mathbf{x}^{(0)}}{dt^{s_\alpha}}\bigg|_{\mathbf{x}^{(0)}_{m}}(\pm\varepsilon)^{s_\alpha}+\frac{d^{2k_\alpha}\mathbf{x}^{(0)}}{dt^{2k_\alpha}}\bigg|_{\mathbf{x}^{(0)}_{m}}\varepsilon^{2k_\alpha}+o(\varepsilon^{2k_\alpha}),\end{aligned}$$

$$\begin{aligned}\mathbf{n}_{\partial\Omega_{ij}}(\mathbf{x}^{(0)}_{m\pm\varepsilon}) &\equiv \mathbf{n}_{\partial\Omega_{ij}}(\mathbf{x}^{(0)}(t_{m\pm\varepsilon}))\\ &= \mathbf{n}_{\partial\Omega_{ij}}(\mathbf{x}^{(0)}_{m})+\sum_{s_\alpha=1}^{2k_\alpha-1}D^{s_\alpha}_{\mathbf{x}^{(0)}}\mathbf{n}_{\partial\Omega_{ij}}\big|_{\mathbf{x}^{(0)}_{m}}(\pm\varepsilon)^{s_\alpha}+D^{2k_\alpha}_{\mathbf{x}^{(0)}}\mathbf{n}_{\partial\Omega_{ij}}\big|_{\mathbf{x}^{(0)}_{m}}\varepsilon^{2k_\alpha}\\ &\quad+o(\varepsilon^{2k_\alpha}).\end{aligned}$$

The ignorance of the $\varepsilon^{2k_\alpha-1}$ and high order terms, the deformation of the above equation and left multiplication of $\mathbf{n}_{\partial\Omega_{ij}}$ gives

$$\begin{aligned}\mathbf{n}^{\mathrm{T}}_{\partial\Omega_{ij}}(\mathbf{x}^{(0)}_{m\pm\varepsilon})\cdot[\mathbf{x}^{(\alpha)}_{m\pm\varepsilon}-\mathbf{x}^{(0)}_{m\pm\varepsilon}] &= \mathbf{n}^{\mathrm{T}}_{\partial\Omega_{ij}}(\mathbf{x}^{(0)}_{m})\cdot[\mathbf{x}^{(\alpha)}_{m\pm}-\mathbf{x}^{(0)}_{m}]\\ &\quad+\sum_{s_\alpha=1}^{2k_\alpha-1}(\pm\varepsilon)^{s_\alpha}G^{(s_\alpha-1,\alpha)}_{\partial\Omega_{ij}}(\mathbf{x}^{(0)}_{m},\mathbf{x}^{(\alpha)}_{m\pm},t_m,\mathbf{p}_\alpha,\boldsymbol{\lambda})\\ &\quad+\varepsilon^{2k_\alpha}G^{(2k_\alpha-1,\alpha)}_{\partial\Omega_{ij}}(\mathbf{x}^{(0)}_{m},\mathbf{x}^{(\alpha)}_{m\pm},t_m,\mathbf{p}_\alpha,\boldsymbol{\lambda}).\end{aligned}$$

Due to $\mathbf{x}^{(\alpha)}_{m\pm}=\mathbf{x}^{(0)}_{m}=\mathbf{x}_m$ and $G^{(s_\alpha,\alpha)}_{\partial\Omega_{ij}}(\mathbf{x}^{(0)}_{m},\mathbf{x}^{(\alpha)}_{m},t_m,\mathbf{p}_\alpha,\boldsymbol{\lambda})\equiv G^{(s_\alpha,\alpha)}_{\partial\Omega_{ij}}(\mathbf{x}_m,t_m,\mathbf{p}_\alpha,\boldsymbol{\lambda})=0$ for $s_\alpha=0,1,\cdots,2k_\alpha-2$, the foregoing equation becomes

$$\begin{aligned}\mathbf{n}^{\mathrm{T}}_{\partial\Omega_{ij}}(\mathbf{x}^{(0)}_{m-\varepsilon})\cdot[\mathbf{x}^{(0)}_{m-\varepsilon}-\mathbf{x}^{(\alpha)}_{m-\varepsilon}] &= -\varepsilon^{2k_\alpha}G^{(2k_\alpha-1,\alpha)}_{\partial\Omega_{ij}}(\mathbf{x}_m,t_m,\mathbf{p}_\alpha,\boldsymbol{\lambda}),\\ \mathbf{n}^{\mathrm{T}}_{\partial\Omega_{ij}}(\mathbf{x}^{(0)}_{m+\varepsilon})\cdot[\mathbf{x}^{(\alpha)}_{m+\varepsilon}-\mathbf{x}^{(0)}_{m+\varepsilon}] &= \varepsilon^{2k_\alpha}G^{(2k_\alpha-1,\alpha)}_{\partial\Omega_{ij}}(\mathbf{x}_m,t_m,\mathbf{p}_\alpha,\boldsymbol{\lambda}).\end{aligned}$$

Using Eq. (3.81), Equation (3.83) is obtained. However, using Eq. (3.83), Equation (3.81) is obtained. Therefore, this theorem is proved. ■

From the similar discussion before Definition 3.19, for the kth -order contact at point $(\mathbf{x}_m^{(0)}, t_{m\pm}, \mathbf{x}_{m\pm}^{(\alpha)})$, the followings equation holds in Kreyszig (1968) and Luo (2008a,b)

$$\begin{gathered}\mathbf{x}_m^{(0)} = \mathbf{x}_{m\pm}^{(\alpha)} \text{ and } \left.\frac{d^{s_\alpha}\mathbf{x}^{(0)}}{dt^{s_\alpha}}\right|_{(\mathbf{x}_m^{(0)}, t_{m\pm}, \mathbf{x}_{m\pm}^{(\alpha)})} = \left.\frac{d^{s_\alpha}\mathbf{x}^{(\alpha)}}{dt^{s_\alpha}}\right|_{(\mathbf{x}_m^{(0)}, t_{m\pm}, \mathbf{x}_{m\pm}^{(\alpha)})}, \\ D_\alpha^{s_\alpha-1}\mathbf{F}^{(\alpha)} = D_0^{s_\alpha-1}\mathbf{F}^{(0)}\,|_{(\mathbf{x}_m^{(0)}, t_{m\pm}, \mathbf{x}_{m\pm}^{(\alpha)})} \quad \text{for } s_\alpha = 1, 2, \cdots, k_\alpha.\end{gathered} \tag{3.84}$$

The k_αth -order G-function becomes

$$\begin{aligned}G_{\partial\Omega_{ij}}^{(k_\alpha,\alpha)}(\mathbf{x}_m^{(0)}, t_{m\pm}, \mathbf{x}_{m\pm}^{(\alpha)}, \mathbf{p}_\alpha, \boldsymbol{\lambda}) &= \mathbf{n}_{\partial\Omega_{ij}}^{\mathrm{T}} \cdot \left.\left(\frac{d^{k_\alpha}\mathbf{x}^{(\alpha)}}{dt^{k_\alpha}} - \frac{d^{k_\alpha}\mathbf{x}^{(0)}}{dt^{k_\alpha}}\right)\right|_{(\mathbf{x}_m^{(0)}, t_{m\pm}, \mathbf{x}_{m\pm}^{(\alpha)})} \\ &= \mathbf{n}_{\partial\Omega_{ij}}^{\mathrm{T}} \cdot (D_\alpha^{k_\alpha-1}\mathbf{F}^{(\alpha)} - D_0^{k_\alpha-1}\mathbf{F}^{(0)})\,|_{(\mathbf{x}_m^{(0)}, t_{m\pm}, \mathbf{x}_{m\pm}^{(\alpha)})}.\end{aligned} \tag{3.85}$$

For the n-dimensional *plane* boundary, $\mathbf{n}_{\partial\Omega_{ij}}$ is constant. So one has

$$D_0^{s_\alpha}\mathbf{n}_{\partial\Omega_{ij}} = 0 \;\text{ for } s_\alpha = 1, 2, \cdots, k_\alpha. \tag{3.86}$$

Because of $\mathbf{n}_{\partial\Omega_{ij}}^{\mathrm{T}} \cdot \dot{\mathbf{x}}^{(0)} = 0$, with Eq. (3.86), the corresponding derivatives gives

$$\mathbf{n}_{\partial\Omega_{ij}}^{\mathrm{T}} \cdot \frac{d^{s_\alpha}\mathbf{x}^{(0)}}{dt^{s_\alpha}} = \mathbf{n}_{\partial\Omega_{ij}}^{\mathrm{T}} \cdot D_0^{s_\alpha-1}\mathbf{F}^{(0)} = 0 \quad \text{for } s_\alpha = 1, 2, \cdots, k_\alpha, \tag{3.87}$$

and one achieves

$$\begin{aligned}G_{\partial\Omega_{ij}}^{(k_\alpha,\alpha)}(\mathbf{x}_m^{(0)}, t_{m\pm}, \mathbf{x}_{m\pm}^{(\alpha)}, \mathbf{p}_\alpha, \boldsymbol{\lambda}) &= \mathbf{n}_{\partial\Omega_{ij}}^{\mathrm{T}} \cdot \left.\frac{d^{k_\alpha}\mathbf{x}^{(\alpha)}}{dt^{k_\alpha}}\right|_{(\mathbf{x}_m^{(0)}, t_{m\pm}, \mathbf{x}_{m\pm}^{(\alpha)})} \\ &= \mathbf{n}_{\partial\Omega_{ij}}^{\mathrm{T}} \cdot D_\alpha^{k_\alpha-1}\mathbf{F}^{(\alpha)}\,|_{(\mathbf{x}_m^{(0)}, t_{m\pm}, \mathbf{x}_{m\pm}^{(\alpha)})}.\end{aligned} \tag{3.88}$$

Thus, Definition 3.19 gives a generalized definition of a grazing flow to the boundary with higher-order singularity, and compared to Theorems 2.11 and 2.12, Theorem 3.10 gives the generalized grazing conditions. The theory presented in Chapter 2 is a special case presented in this Chapter. In addition, the theory presented in Luo (2005a, 2006) is valid for the *plane* boundary surface and the higher-order contact with the boundary surface.

The flow grazing bifurcation to the boundary can be determined by the G-function $G_{\partial\Omega_{ij}}^{(2k_\alpha-1,\alpha)}(\mathbf{x}_m, t_m, \mathbf{p}_\alpha, \boldsymbol{\lambda})$. In other words, the conditions for a flow tangential to the boundary are $G_{\partial\Omega_{ij}}^{(s_\alpha,\alpha)}(\mathbf{x}_m, t_m, \mathbf{p}_\alpha, \boldsymbol{\lambda}) = 0$ $(s_\alpha = 0, 1, \cdots, 2k_\alpha - 2)$ and $G_{\partial\Omega_{ij}}^{(2k_\alpha-1,\alpha)}(\mathbf{x}_m, t_m, \mathbf{p}_\alpha, \boldsymbol{\lambda}) < 0$ (or $G_{\partial\Omega_{ij}}^{(2k_\alpha-1,\alpha)}(\mathbf{x}_m, t_m, \mathbf{p}_\alpha, \boldsymbol{\lambda}) > 0$) for the boundary $\partial\Omega_{ij}$ with $\mathbf{n}_{\partial\Omega_{ij}} \to \Omega_j$ (or $\mathbf{n}_{\partial\Omega_{ij}} \to \Omega_i$). To develop a uniform theory of the tangential flow with the passable and non-passable flow, the imaginary flow tangency will be introduced. To distinguish a real tangential flow from an imaginary tangential flow, the tangency of a real flow to the boundary can be restated as follows.

Definition 3.20. For a discontinuous dynamical system in Eq. (2.1), there is a point $\mathbf{x}^{(0)}(t_m) \equiv \mathbf{x}_m \in \partial\Omega_{ij}$ at time t_m between two adjacent domains Ω_α ($\alpha = i, j$). Suppose $\mathbf{x}^{(i)}(t_{m+}) = \mathbf{x}_m = \mathbf{x}_i^{(j)}(t_{m\pm})$. For an arbitrarily small $\varepsilon > 0$, there are two time intervals $[t_{m-\varepsilon}, t_m)$ and $[t_{m-\varepsilon}, t_{m+\varepsilon}]$. A flow $\mathbf{x}^{(i)}(t)$ is $C^{r_i}_{[t_{m-\varepsilon}, t_{m+\varepsilon}]}$-continuous ($r_i \geq 2k_i + 1$) for time t and $\| d^{r_i+1}\mathbf{x}^{(i)} / dt^{r_i+1} \| < \infty$, and a flow $\mathbf{x}^{(j)}(t)$ is $C^{r_j}_{[t_{m-\varepsilon}, t_m)}$ or $C^{r_j}_{(t_m, t_{m+\varepsilon}]}$-continuous with $\| d^{r_j+1}\mathbf{x}^{(j)} / dt^{r_j+1} \| < \infty$ ($r_j \geq 2k_j$). The flow $\mathbf{x}^{(i)}(t)$ of the $(2k_i - 1)$th-order with $\mathbf{x}^{(j)}(t)$ of the $(2k_j)$th-order to the boundary $\partial\Omega_{ij}$ is a $(2k_i - 1 : 2k_j)$-*tangential flow* in domain Ω_i if

$$\begin{aligned} &G^{(s_i, i)}_{\partial\Omega_{ij}}(\mathbf{x}_m, t_{m\pm}, \mathbf{p}_i, \boldsymbol{\lambda}) = 0 \;\; \text{for } s_i = 0, 1, \cdots, 2k_i - 2, \\ &G^{(2k_i, i)}_{\partial\Omega_{ij}}(\mathbf{x}_m, t_{m\pm}, \mathbf{p}_i, \boldsymbol{\lambda}) \neq 0; \end{aligned} \tag{3.89}$$

$$\begin{aligned} &G^{(s_j, j)}_{\partial\Omega_{ij}}(\mathbf{x}_m, t_{m\pm}, \mathbf{p}_j, \boldsymbol{\lambda}) = 0 \;\; \text{for } s_j = 0, 1, \cdots, 2k_j - 1, \\ &G^{(2k_j - 1, j)}_{\partial\Omega_{ij}}(\mathbf{x}_m, t_{m\pm}, \mathbf{p}_j, \boldsymbol{\lambda}) \neq 0; \end{aligned} \tag{3.90}$$

$$\left. \begin{array}{ll} \text{either} & \left. \begin{array}{l} \mathbf{n}^{\mathrm{T}}_{\partial\Omega_{ij}}(\mathbf{x}^{(0)}_{m-\varepsilon}) \cdot [\mathbf{x}^{(0)}_{m-\varepsilon} - \mathbf{x}^{(i)}_{m-\varepsilon}] > 0 \\ \mathbf{n}^{\mathrm{T}}_{\partial\Omega_{ij}}(\mathbf{x}^{(0)}_{m+\varepsilon}) \cdot [\mathbf{x}^{(i)}_{m+\varepsilon} - \mathbf{x}^{(0)}_{m+\varepsilon}] < 0 \end{array} \right\} \text{for } \mathbf{n}_{\partial\Omega_{ij}} \to \Omega_j, \\ \text{or} & \left. \begin{array}{l} \mathbf{n}^{\mathrm{T}}_{\partial\Omega_{ij}}(\mathbf{x}^{(0)}_{m-\varepsilon}) \cdot [\mathbf{x}^{(0)}_{m-\varepsilon} - \mathbf{x}^{(i)}_{m-\varepsilon}] < 0 \\ \mathbf{n}^{\mathrm{T}}_{\partial\Omega_{ij}}(\mathbf{x}^{(0)}_{m+\varepsilon}) \cdot [\mathbf{x}^{(i)}_{m+\varepsilon} - \mathbf{x}^{(0)}_{m+\varepsilon}] > 0 \end{array} \right\} \text{for } \mathbf{n}_{\partial\Omega_{ij}} \to \Omega_i; \end{array} \right\} \tag{3.91}$$

$$\left. \begin{array}{ll} \text{either} & \left. \begin{array}{l} \mathbf{n}^{\mathrm{T}}_{\partial\Omega_{ij}}(\mathbf{x}^{(0)}_{m-\varepsilon}) \cdot [\mathbf{x}^{(0)}_{m-\varepsilon} - \mathbf{x}^{(j)}_{m-\varepsilon}] < 0 \\ \mathbf{n}^{\mathrm{T}}_{\partial\Omega_{ij}}(\mathbf{x}^{(0)}_{m+\varepsilon}) \cdot [\mathbf{x}^{(j)}_{m+\varepsilon} - \mathbf{x}^{(0)}_{m+\varepsilon}] > 0 \end{array} \right\} \text{for } \mathbf{n}_{\partial\Omega_{ij}} \to \Omega_j, \\ \text{or} & \left. \begin{array}{l} \mathbf{n}^{\mathrm{T}}_{\partial\Omega_{ij}}(\mathbf{x}^{(0)}_{m-\varepsilon}) \cdot [\mathbf{x}^{(0)}_{m-\varepsilon} - \mathbf{x}^{(j)}_{m-\varepsilon}] > 0 \\ \mathbf{n}^{\mathrm{T}}_{\partial\Omega_{ij}}(\mathbf{x}^{(0)}_{m+\varepsilon}) \cdot [\mathbf{x}^{(j)}_{m+\varepsilon} - \mathbf{x}^{(0)}_{m+\varepsilon}] < 0 \end{array} \right\} \text{for } \mathbf{n}_{\partial\Omega_{ij}} \to \Omega_i. \end{array} \right\} \tag{3.92}$$

To explain the foregoing definition, such a $(2k_i - 1 : 2k_j)$-tangential flow to the boundary $\partial\Omega_{ij}$ in domain Ω_i are sketched in Fig. 3.14(a) with source in domain Ω_j and (b) with sink in domain Ω_j. The $(2k_j - 1 : 2k_i)$-tangential flow in domain Ω_j are sketched in Fig. 3.15(a) with source in domain Ω_i and (b) with sink in domain Ω_i. The sink and source flows are represented by the dotted curves. The tangential flows are presented by solid curves. The dashed curves denote the imaginary flows. If the starting point is on the flow $\mathbf{x}^{(j)}(t)$ (or $\mathbf{x}^{(j)}(t)$) in Fig. 3.14(b) (or Fig. 3.15(a)), the passable flow from domain Ω_j to Ω_i (or Ω_j to Ω_i) is formed. Such passable flows possess the post-higher-order singularity. From the above definition, the necessary and sufficient conditions for the tangential flow are given in the following theorem.

Theorem 3.11. *For a discontinuous dynamical system in Eq.* (2.1), *there is a point* $\mathbf{x}^{(0)}(t_m) \equiv \mathbf{x}_m \in \partial\Omega_{ij}$ *at time* t_m *between two adjacent domains* Ω_α ($\alpha = i, j$). *Suppose* $\mathbf{x}^{(i)}(t_{m+}) = \mathbf{x}_m = \mathbf{x}_i^{(j)}(t_{m\pm})$. *For an arbitrarily small* $\varepsilon > 0$, *there are two time intervals* $[t_{m-\varepsilon}, t_m)$ *and* $[t_{m-\varepsilon}, t_{m+\varepsilon}]$. *A flow* $\mathbf{x}^{(i)}(t)$ *is* $C^{r_i}_{[t_{m-\varepsilon}, t_{m+\varepsilon}]}$*-continuous* ($r_i \ge 2k_i + 1$) *for time t and* $\| d^{r_i+1}\mathbf{x}^{(i)} / dt^{r_i+1} \| < \infty$, *and a flow* $\mathbf{x}^{(j)}(t)$ *is* $C^{r_j}_{[t_{m-\varepsilon}, t_m)}$ *or* $C^{r_j}_{(t_m, t_{m+\varepsilon}]}$*-continuous with* $\| d^{r_j+1}\mathbf{x}^{(j)} / dt^{r_j+1} \| < \infty$ ($r_j \ge 2k_j$). *The flow* $\mathbf{x}^{(i)}(t)$ *of the* $(2k_i - 1)$th *-order and* $\mathbf{x}^{(j)}(t)$ *of the* $(2k_j)$th *-order to the boundary* $\partial\Omega_{ij}$ *is* $(2k_i - 1 : 2k_j)$ *-tangential flow in domain* Ω_i *if and only if*

$$G^{(s_i,i)}_{\partial\Omega_{ij}}(\mathbf{x}_m, t_{m\pm}, \mathbf{p}_i, \boldsymbol{\lambda}) = 0 \text{ for } s_i = 0, 1, \cdots, 2k_i - 2; \tag{3.93}$$

$$G^{(s_j,j)}_{\partial\Omega_{ij}}(\mathbf{x}_m, t_{m\pm}, \mathbf{p}_j, \boldsymbol{\lambda}) = 0 \text{ for } s_j = 0, 1, \cdots, 2k_j - 1; \tag{3.94}$$

$$\left.\begin{aligned} &\text{either} \left.\begin{aligned} &G^{(2k_i-1,i)}_{\partial\Omega_{ij}}(\mathbf{x}_m, t_{m\pm}, \mathbf{p}_i, \boldsymbol{\lambda}) < 0 \\ &\left.\begin{aligned} &G^{(2k_j,j)}_{\partial\Omega_{ij}}(\mathbf{x}_m, t_{m-}, \mathbf{p}_j, \boldsymbol{\lambda}) < 0 \text{ or} \\ &G^{(2k_j,j)}_{\partial\Omega_{ij}}(\mathbf{x}_m, t_{m+}, \mathbf{p}_j, \boldsymbol{\lambda}) > 0 \end{aligned}\right] \end{aligned}\right\} \text{for } \mathbf{n}_{\partial\Omega_{ij}} \to \Omega_j, \\ &\text{or} \left.\begin{aligned} &G^{(2k_i-1,i)}_{\partial\Omega_{ij}}(\mathbf{x}_m, t_{m\pm}, \mathbf{p}_i, \boldsymbol{\lambda}) > 0 \\ &\left.\begin{aligned} &G^{(2k_j,j)}_{\partial\Omega_{ij}}(\mathbf{x}_m, t_{m-}, \mathbf{p}_j, \boldsymbol{\lambda}) > 0 \text{ or} \\ &G^{(2k_j,j)}_{\partial\Omega_{ij}}(\mathbf{x}_m, t_{m+}, \mathbf{p}_j, \boldsymbol{\lambda}) < 0 \end{aligned}\right] \end{aligned}\right\} \text{for } \mathbf{n}_{\partial\Omega_{ij}} \to \Omega_i. \end{aligned}\right. \tag{3.95}$$

Proof. The proof is similar to the proof of Theorem 3.2. ■

Definition 3.21. For a discontinuous dynamical system in Eq. (2.1), there is a point $\mathbf{x}^{(0)}(t_m) \equiv \mathbf{x}_m \in \partial\Omega_{ij}$ at time t_m between two adjacent domains Ω_α ($\alpha = i, j$). Suppose $\mathbf{x}^{(\alpha)}(t_{m\pm}) = \mathbf{x}_m = \mathbf{x}_\alpha^{(\beta)}(t_{m\pm})$ ($\alpha, \beta \in \{i, j\}$ and $\beta \neq \alpha$). For an arbitrarily small $\varepsilon > 0$, there is a time interval $[t_{m-\varepsilon}, t_{m+\varepsilon}]$. A flow $\mathbf{x}^{(\alpha)}(t)$ is $C^{r_\alpha}_{[t_{m-\varepsilon}, t_{m+\varepsilon}]}$-continuous ($r_\alpha \ge 2k_\alpha$) and $\| d^{r_\alpha+1}\mathbf{x}^{(\alpha)} / dt^{r_\alpha+1} \| < \infty$ for time t, and an imaginary flow $\mathbf{x}_\alpha^{(\beta)}(t)$ is $C^{r_\beta}_{[t_{m-\varepsilon}, t_{m+\varepsilon}]}$-continuous with $\| d^{r_\beta+1}\mathbf{x}_\alpha^{(\beta)} / dt^{r_\beta+1} \| < \infty$ ($r_\beta \ge 2k_\beta$). The flow $\mathbf{x}^{(\alpha)}(t)$ of the $(2k_\alpha - 1)$th -order and $\mathbf{x}_\alpha^{(\beta)}(t)$ of the $(2k_\beta - 1)$th -order to the boundary $\partial\Omega_{ij}$ is a $(2k_\alpha - 1 : 2k_\beta - 1)$ *-tangential flow* in domain Ω_α if

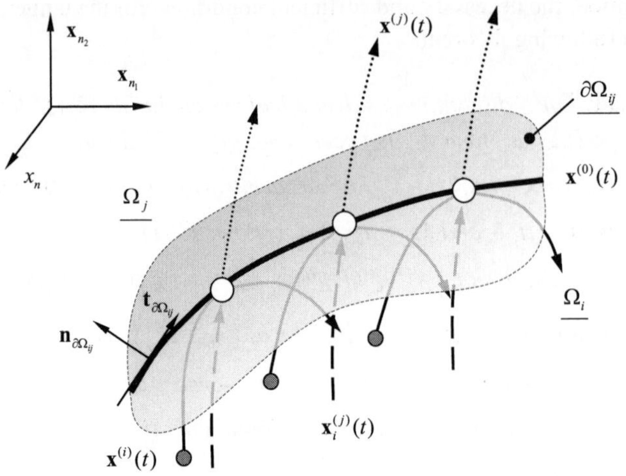

(a)

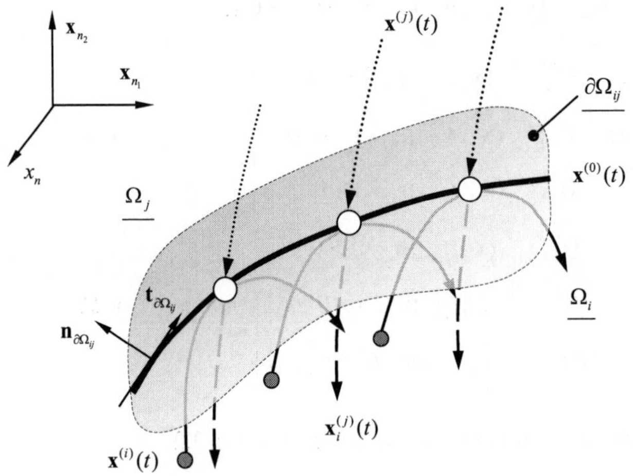

(b)

Fig. 3.14 The $(2k_i-1:2k_j)$ -tangential flows in Ω_i : **(a)** with source in Ω_j and **(b)** with sink in Ω_j . $\mathbf{x}^{(i)}(t)$ and $\mathbf{x}^{(j)}(t)$ represent *real* flows in domains Ω_i and Ω_j , depicted by thin solid and dotted curves, respectively. $\mathbf{x}_i^{(j)}(t)$ represent *imaginary* flows in domain Ω_i , controlled by the vector fields in Ω_j , which are depicted by dashed curves. The flow on the boundary is described by $\mathbf{x}^{(0)}(t)$. The normal and tangential vectors $\mathbf{n}_{\partial\Omega_{ij}}$ and $\mathbf{t}_{\partial\Omega_{ij}}$ on the boundary are depicted. Hollow circles are for grazing points on the boundary and filled circles are for starting points. ($n_1+n_2+1=n$)

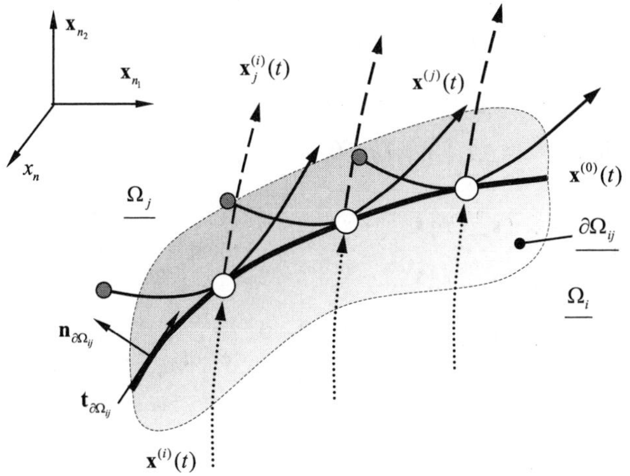

(a)

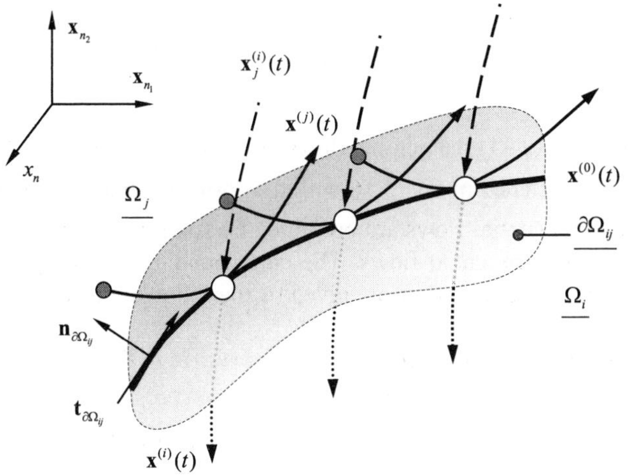

(b)

Fig. 3.15 The $(2k_j-1:2k_i)$ -tangential flows in Ω_j : **(a)** with sink in Ω_i and **(b)** with source in Ω_i. $\mathbf{x}^{(i)}(t)$ and $\mathbf{x}^{(j)}(t)$ represent the *real* flows in domains Ω_i and Ω_j, depicted by dotted and thin solid curves, respectively. And $\mathbf{x}_j^{(i)}(t)$ represents *imaginary* flows in domain Ω_j, controlled by the vector fields in Ω_i, which are depicted by dashed curves. The flow on the boundary is described by $\mathbf{x}^{(0)}(t)$. The normal and tangential vectors $\mathbf{n}_{\partial\Omega_{ij}}$ and $\mathbf{t}_{\partial\Omega_{ij}}$ on the boundary are depicted. Hollow circles are for grazing points on the boundary and filled circles are for starting points. ($n_1+n_2+1=n$)

$$\begin{aligned} & G_{\partial\Omega_{\alpha\beta}}^{(s_\alpha,\alpha)}(\mathbf{x}_m,t_{m\pm},\mathbf{p}_\alpha,\boldsymbol{\lambda})=0 \text{ for } s_\alpha=0,1,\cdots,2k_\alpha-2,\\ & G_{\partial\Omega_{\alpha\beta}}^{(2k_\alpha-1,\alpha)}(\mathbf{x}_m,t_{m\pm},\mathbf{p}_\alpha,\boldsymbol{\lambda})\neq 0; \end{aligned} \tag{3.96}$$

$$\begin{aligned} & G_{\partial\Omega_{\alpha\beta}}^{(s_\beta,\beta)}(\mathbf{x}_m,t_{m\pm},\mathbf{p}_\beta,\boldsymbol{\lambda})=0 \text{ for } s_\beta=0,1,\cdots,2k_\beta-2,\\ & G_{\partial\Omega_{\alpha\beta}}^{(2k_\beta-1,\beta)}(\mathbf{x}_m,t_{m-},\mathbf{p}_\beta,\boldsymbol{\lambda})\neq 0; \end{aligned} \tag{3.97}$$

$$\left.\begin{aligned} &\text{either} \left.\begin{aligned} &\mathbf{n}_{\partial\Omega_{ij}}^{\mathrm{T}}(\mathbf{x}_{m-\varepsilon}^{(0)})\cdot[\mathbf{x}_{m-\varepsilon}^{(0)}-\mathbf{x}_{m-\varepsilon}^{(\alpha)}]>0\\ &\mathbf{n}_{\partial\Omega_{ij}}^{\mathrm{T}}(\mathbf{x}_{m+\varepsilon}^{(0)})\cdot[\mathbf{x}_{m+\varepsilon}^{(\alpha)}-\mathbf{x}_{m+\varepsilon}^{(0)}]<0 \end{aligned}\right\}\text{for } \mathbf{n}_{\partial\Omega_{\alpha\beta}}\to\Omega_\beta,\\ &\text{or} \left.\begin{aligned} &\mathbf{n}_{\partial\Omega_{ij}}^{\mathrm{T}}(\mathbf{x}_{m-\varepsilon}^{(0)})\cdot[\mathbf{x}_{m-\varepsilon}^{(0)}-\mathbf{x}_{m-\varepsilon}^{(\alpha)}]<0\\ &\mathbf{n}_{\partial\Omega_{ij}}^{\mathrm{T}}(\mathbf{x}_{m+\varepsilon}^{(0)})\cdot[\mathbf{x}_{m+\varepsilon}^{(\alpha)}-\mathbf{x}_{m+\varepsilon}^{(0)}]>0 \end{aligned}\right\}\text{for } \mathbf{n}_{\partial\Omega_{\alpha\beta}}\to\Omega_\alpha; \end{aligned}\right\} \tag{3.98}$$

$$\left.\begin{aligned} &\text{either} \left.\begin{aligned} &\mathbf{n}_{\partial\Omega_{ij}}^{\mathrm{T}}(\mathbf{x}_{m-\varepsilon}^{(0)})\cdot[\mathbf{x}_{m-\varepsilon}^{(0)}-\mathbf{x}_{\alpha(m-\varepsilon)}^{(\beta)}]>0\\ &\mathbf{n}_{\partial\Omega_{ij}}^{\mathrm{T}}(\mathbf{x}_{m+\varepsilon}^{(0)})\cdot[\mathbf{x}_{\alpha(m+\varepsilon)}^{(\beta)}-\mathbf{x}_{m+\varepsilon}^{(0)}]<0 \end{aligned}\right\}\text{for } \mathbf{n}_{\partial\Omega_{\alpha\beta}}\to\Omega_\beta,\\ &\text{or} \left.\begin{aligned} &\mathbf{n}_{\partial\Omega_{ij}}^{\mathrm{T}}(\mathbf{x}_{m-\varepsilon}^{(0)})\cdot[\mathbf{x}_{m-\varepsilon}^{(0)}-\mathbf{x}_{\alpha(m-\varepsilon)}^{(\beta)}]<0\\ &\mathbf{n}_{\partial\Omega_{ij}}^{\mathrm{T}}(\mathbf{x}_{m+\varepsilon}^{(0)})\cdot[\mathbf{x}_{\alpha(m+\varepsilon)}^{(\beta)}-\mathbf{x}_{m+\varepsilon}^{(0)}]>0 \end{aligned}\right\}\text{for } \mathbf{n}_{\partial\Omega_{\alpha\beta}}\to\Omega_\alpha. \end{aligned}\right\} \tag{3.99}$$

The $(2k_\alpha-1:2k_\beta-1)$-tangential flows in domain Ω_α and Ω_β ($\alpha,\beta\in\{i,j\}$ and $\alpha\neq\beta$) are sketched in Fig. 3.16 with the corresponding imaginary tangential flows. The real tangential flows are presented by solid curves. Dashed curves denote the imaginary tangential flows. The corresponding necessary and sufficient conditions for the tangential flow are given by the following theorem.

Theorem 3.12. *For a discontinuous dynamical system in Eq.* (2.1), *there is a point* $\mathbf{x}^{(0)}(t_m)\equiv\mathbf{x}_m\in\partial\Omega_{ij}$ *at time* t_m *between two adjacent domains* Ω_α ($\alpha=i,j$). *Suppose* $\mathbf{x}^{(\alpha)}(t_{m\pm})=\mathbf{x}_m=\mathbf{x}_\alpha^{(\beta)}(t_{m\pm})$ ($\alpha,\beta\in\{i,j\}$ and $\alpha\neq\beta$). *For an arbitrarily small* $\varepsilon>0$, *there is a time interval* $[t_{m-\varepsilon},t_{m+\varepsilon}]$. *A flow* $\mathbf{x}^{(\alpha)}(t)$ *is* $C_{[t_{m-\varepsilon},t_{m+\varepsilon}]}^{r_\alpha}$*-continuous* ($r_\alpha\geq 2k_\alpha$) *with* $\|d^{r_\alpha+1}\mathbf{x}^{(\alpha)}/dt^{r_\alpha+1}\|<\infty$ *for time t, and an imaginary flow* $\mathbf{x}_\alpha^{(\beta)}(t)$ *is* $C_{[t_{m-\varepsilon},t_{m+\varepsilon}]}^{r_\beta}$*-continuous* with $\|d^{r_\beta+1}\mathbf{x}_\alpha^{(\beta)}/dt^{r_\beta+1}\|<\infty$ ($r_\beta\geq 2k_\beta$). *The flow* $\mathbf{x}^{(\alpha)}(t)$ *of the* $(2k_\alpha-1)$th*-order and* $\mathbf{x}_\alpha^{(\beta)}(t)$ *of the* $(2k_\beta-1)$th*-order to the boundary* $\partial\Omega_{ij}$ *is a* $(2k_\alpha-1:2k_\beta-1)$*-tangential flow in domain* Ω_α *if and only if*

$$G_{\partial\Omega_{\alpha\beta}}^{(s_\alpha,\alpha)}(\mathbf{x}_m,t_{m\pm},\mathbf{p}_\alpha,\boldsymbol{\lambda})=0 \ \textit{for}\ s_\alpha=0,1,\cdots,2k_\alpha-2; \tag{3.100}$$

$$G_{\partial\Omega_{\alpha\beta}}^{(s_\beta,\beta)}(\mathbf{x}_m,t_{m\pm},\mathbf{p}_\beta,\boldsymbol{\lambda})=0 \ \textit{for}\ s_\beta=0,1,\cdots,2k_\beta-2; \tag{3.101}$$

$$
\left.\begin{aligned}
&\textit{either}\quad \left.\begin{aligned} G^{(2k_\alpha-1,\alpha)}_{\partial\Omega_{\alpha\beta}}(\mathbf{x}_m,t_{m\pm},\mathbf{p}_\alpha,\boldsymbol{\lambda})<0\\ G^{(2k_\beta-1,\beta)}_{\partial\Omega_{\alpha\beta}}(\mathbf{x}_m,t_{m\pm},\mathbf{p}_\beta,\boldsymbol{\lambda})<0\end{aligned}\right\} \textit{for}\ \mathbf{n}_{\partial\Omega_{\alpha\beta}}\to\Omega_\beta,\\
&\textit{or}\quad \left.\begin{aligned} G^{(2k_\alpha-1,\alpha)}_{\partial\Omega_{\alpha\beta}}(\mathbf{x}_m,t_{m\pm},\mathbf{p}_\alpha,\boldsymbol{\lambda})>0\\ G^{(2k_\beta-1,\beta)}_{\partial\Omega_{\alpha\beta}}(\mathbf{x}_m,t_{m\pm},\mathbf{p}_\beta,\boldsymbol{\lambda})>0\end{aligned}\right\} \textit{for}\ \mathbf{n}_{\partial\Omega_{\alpha\beta}}\to\Omega_\alpha.
\end{aligned}\right. \tag{3.102}
$$

Proof. The proof is similar to the proof of Theorem 3.2. ■

Definition 3.22. For a discontinuous dynamical system in Eq. (2.1), there is a point $\mathbf{x}^{(0)}(t_m)\equiv\mathbf{x}_m\in\partial\Omega_{ij}$ at time t_m between two adjacent domains Ω_α ($\alpha=i,j$). Suppose $\mathbf{x}^{(\alpha)}(t_{m\pm})=\mathbf{x}_m=\mathbf{x}^{(\beta)}(t_{m\pm})$ ($\alpha,\beta\in\{i,j\}$ and $\alpha\neq\beta$). For an arbitrarily small $\varepsilon>0$, there is a time interval $[t_{m-\varepsilon},t_{m+\varepsilon}]$. A flow $\mathbf{x}^{(\alpha)}(t)$ is $C^{r_\alpha}_{[t_{m-\varepsilon},t_{m+\varepsilon}]}$-continuous ($r_\alpha\geq 2k_\alpha$) for time t with $\|d^{r_\alpha+1}\mathbf{x}^{(\alpha)}/dt^{r_\alpha+1}\|<\infty$, and the flow $\mathbf{x}^{(\beta)}(t)$ is $C^{r_\beta}_{[t_{m-\varepsilon},t_{m+\varepsilon}]}$-continuous ($r_\beta\geq 2k_\beta$) with $\|d^{r_\beta+1}\mathbf{x}^{(\beta)}/dt^{r_\beta+1}\|<\infty$. The flow $\mathbf{x}^{(\alpha)}(t)$ of the $(2k_\alpha-1)$th-order and $\mathbf{x}^{(\beta)}(t)$ of the $(2k_\beta-1)$th-order to the boundary $\partial\Omega_{\alpha\beta}$ is a $(2k_\alpha-1:2k_\beta-1)$*-double tangential flow if*

$$
\left.\begin{aligned}
&G^{(s_\alpha,\alpha)}_{\partial\Omega_{\alpha\beta}}(\mathbf{x}_m,t_{m\pm},\mathbf{p}_\alpha,\boldsymbol{\lambda})=0\ \text{for}\ s_\alpha=0,1,\cdots,2k_\alpha-2,\\
&G^{(2k_\alpha-1,\alpha)}_{\partial\Omega_{\alpha\beta}}(\mathbf{x}_m,t_{m\pm},\mathbf{p}_\alpha,\boldsymbol{\lambda})\neq 0;
\end{aligned}\right\} \tag{3.103}
$$

$$
\left.\begin{aligned}
&G^{(s_\beta,\beta)}_{\partial\Omega_{\alpha\beta}}(\mathbf{x}_m,t_{m\pm},\mathbf{p}_\beta,\boldsymbol{\lambda})=0\ \text{for}\ s_\beta=0,1,\cdots,2k_\beta-2,\\
&G^{(2k_\beta-1,\beta)}_{\partial\Omega_{\alpha\beta}}(\mathbf{x}_m,t_{m-},\mathbf{p}_\beta,\boldsymbol{\lambda})\neq 0;
\end{aligned}\right\} \tag{3.104}
$$

$$
\left.\begin{aligned}
&\text{either}\quad \left.\begin{aligned} \mathbf{n}^{\mathrm{T}}_{\partial\Omega_{ij}}(\mathbf{x}^{(0)}_{m-\varepsilon})\cdot[\mathbf{x}^{(0)}_{m-\varepsilon}-\mathbf{x}^{(\alpha)}_{m-\varepsilon}]>0\\ \mathbf{n}^{\mathrm{T}}_{\partial\Omega_{ij}}(\mathbf{x}^{(0)}_{m+\varepsilon})\cdot[\mathbf{x}^{(\alpha)}_{m+\varepsilon}-\mathbf{x}^{(0)}_{m+\varepsilon}]<0\end{aligned}\right\} \text{for}\ \mathbf{n}_{\partial\Omega_{\alpha\beta}}\to\Omega_\beta,\\
&\text{or}\quad \left.\begin{aligned} \mathbf{n}^{\mathrm{T}}_{\partial\Omega_{ij}}(\mathbf{x}^{(0)}_{m-\varepsilon})\cdot[\mathbf{x}^{(0)}_{m-\varepsilon}-\mathbf{x}^{(\alpha)}_{m-\varepsilon}]<0\\ \mathbf{n}^{\mathrm{T}}_{\partial\Omega_{ij}}(\mathbf{x}^{(0)}_{m+\varepsilon})\cdot[\mathbf{x}^{(\alpha)}_{m+\varepsilon}-\mathbf{x}^{(0)}_{m+\varepsilon}]>0\end{aligned}\right\} \text{for}\ \mathbf{n}_{\partial\Omega_{\alpha\beta}}\to\Omega_\alpha;
\end{aligned}\right\} \tag{3.105}
$$

$$
\left.\begin{aligned}
&\text{either}\quad \left.\begin{aligned} \mathbf{n}^{\mathrm{T}}_{\partial\Omega_{ij}}(\mathbf{x}^{(0)}_{m-\varepsilon})\cdot[\mathbf{x}^{(0)}_{m-\varepsilon}-\mathbf{x}^{(\beta)}_{m-\varepsilon}]<0\\ \mathbf{n}^{\mathrm{T}}_{\partial\Omega_{ij}}(\mathbf{x}^{(0)}_{m+\varepsilon})\cdot[\mathbf{x}^{(\beta)}_{m+\varepsilon}-\mathbf{x}^{(0)}_{m+\varepsilon}]>0\end{aligned}\right\} \text{for}\ \mathbf{n}_{\partial\Omega_{\alpha\beta}}\to\Omega_\beta,\\
&\text{or}\quad \left.\begin{aligned} \mathbf{n}^{\mathrm{T}}_{\partial\Omega_{ij}}(\mathbf{x}^{(0)}_{m-\varepsilon})\cdot[\mathbf{x}^{(0)}_{m-\varepsilon}-\mathbf{x}^{(\beta)}_{m-\varepsilon}]>0\\ \mathbf{n}^{\mathrm{T}}_{\partial\Omega_{ij}}(\mathbf{x}^{(0)}_{m+\varepsilon})\cdot[\mathbf{x}^{(\beta)}_{m+\varepsilon}-\mathbf{x}^{(0)}_{m+\varepsilon}]<0\end{aligned}\right\} \text{for}\ \mathbf{n}_{\partial\Omega_{\alpha\beta}}\to\Omega_\alpha.
\end{aligned}\right\} \tag{3.106}
$$

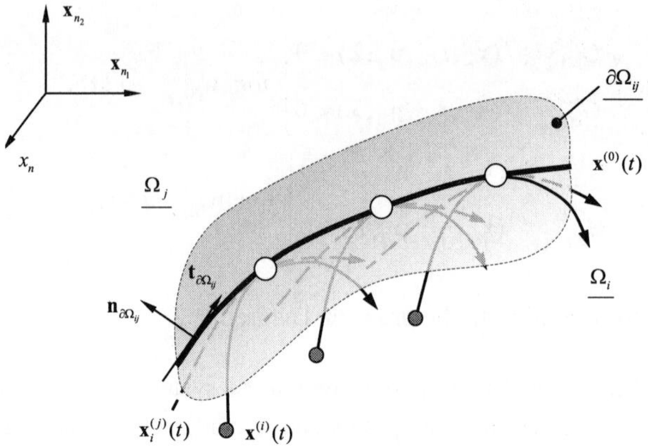

(a)

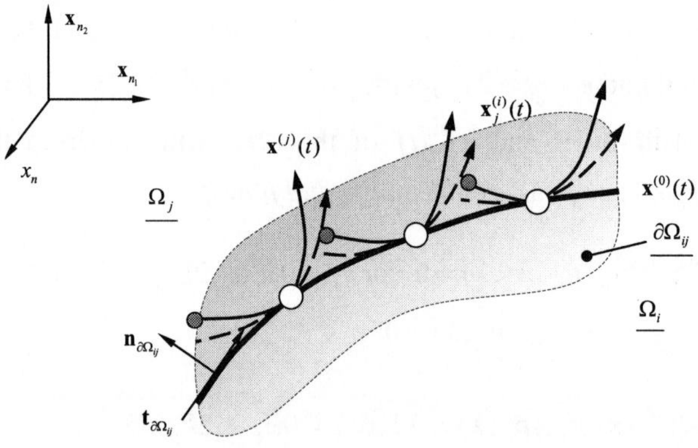

(b)

Fig. 3.16 $(2k_i-1:2k_j-1)$ real and imaginary tangential flows in: **(a)** Ω_i and **(b)** Ω_j. $\mathbf{x}^{(i)}(t)$ and $\mathbf{x}^{(j)}(t)$ represent the *real* flows in domains Ω_i and Ω_j, respectively, which are depicted by the thin solid curves. $\mathbf{x}_i^{(j)}(t)$ and $\mathbf{x}_j^{(i)}(t)$ represent the *imaginary* flows in domains Ω_i and Ω_j, respectively controlled by the vector fields in Ω_j and Ω_i, which are depicted by the dashed curves. The flow on the boundary is described by $\mathbf{x}^{(0)}(t)$. The normal and tangential vectors $\mathbf{n}_{\partial\Omega_{ij}}$ and $\mathbf{t}_{\partial\Omega_{ij}}$ of the boundary are depicted. Hollow circles are for grazing points on the boundary, and filled circles are for starting points. ($n_1+n_2+1=n$)

The $(2k_\alpha - 1 : 2k_\beta - 1)$-double tangential flows are sketched in Fig. 3.17(a) by the solid curves. The double tangential flow is formed by the two real tangential flows in both domains. The corresponding necessary and sufficient conditions for the tangential flows are given through the following theorem.

Theorem 3.13. *For a discontinuous dynamical system in Eq.* (2.1)*, there is a point* $\mathbf{x}^{(0)}(t_m) \equiv \mathbf{x}_m \in \partial\Omega_{ij}$ *at time* t_m *between two adjacent domains* Ω_α $(\alpha = i, j)$. *Suppose* $\mathbf{x}^{(\alpha)}(t_{m\pm}) = \mathbf{x}_m = \mathbf{x}^{(\beta)}(t_{m\pm})$ $(\alpha, \beta \in \{i, j\}$ *and* $\alpha \neq \beta)$. *For an arbitrarily small* $\varepsilon > 0$, *there is a time interval* $[t_{m-\varepsilon}, t_{m+\varepsilon}]$. *A flow* $\mathbf{x}^{(\alpha)}(t)$ *is* $C^{r_\alpha}_{[t_{m-\varepsilon}, t_{m+\varepsilon}]}$-*continuous* $(r_\alpha \geq 2k_\alpha)$ *and* $\| d^{r_\alpha + 1}\mathbf{x}^{(\alpha)} / dt^{r_\alpha + 1} \| < \infty$ *for time t, and a flow* $\mathbf{x}^{(\beta)}(t)$ *is* $C^{r_\beta}_{[t_{m-\varepsilon}, t_{m+\varepsilon}]}$-*continuous* $(r_\beta \geq 2k_\beta)$ *and* $\| d^{r_\beta + 1}\mathbf{x}^{(\beta)} / dt^{r_\beta + 1} \| < \infty$. *The flow* $\mathbf{x}^{(\alpha)}(t)$ *of the* $(2k_\alpha - 1)$th-*order and* $\mathbf{x}^{(\beta)}(t)$ *of the* $(2k_\beta - 1)$th-*order to the boundary* $\partial\Omega_{\alpha\beta}$ *is a* $(2k_\alpha - 1 : 2k_\beta - 1)$-*double tangential flow if and only if*

$$G^{(s_\alpha, \alpha)}_{\partial\Omega_{\alpha\beta}}(\mathbf{x}_m, t_{m\pm}, \mathbf{p}_\alpha, \boldsymbol{\lambda}) = 0 \;\; \textit{for}\; s_\alpha = 0, 1, \cdots, 2k_\alpha - 2; \tag{3.107}$$

$$G^{(s_\beta, \beta)}_{\partial\Omega_{\alpha\beta}}(\mathbf{x}_m, t_{m\pm}, \mathbf{p}_\beta, \boldsymbol{\lambda}) = 0 \;\; \textit{for}\; s_\beta = 0, 1, \cdots, 2k_\beta - 2; \tag{3.108}$$

$$\begin{aligned} &\textit{either} \left.\begin{array}{l} G^{(2k_\alpha - 1, \alpha)}_{\partial\Omega_{\alpha\beta}}(\mathbf{x}_m, t_{m\pm}, \mathbf{p}_\alpha, \boldsymbol{\lambda}) < 0 \\ G^{(2k_\beta - 1, \beta)}_{\partial\Omega_{\alpha\beta}}(\mathbf{x}_m, t_{m\pm}, \mathbf{p}_\beta, \boldsymbol{\lambda}) > 0 \end{array}\right\} \textit{for}\; \mathbf{n}_{\partial\Omega_{\alpha\beta}} \to \Omega_\beta, \\ &\textit{or} \left.\begin{array}{l} G^{(2k_\alpha - 1, \alpha)}_{\partial\Omega_{\alpha\beta}}(\mathbf{x}_m, t_{m\pm}, \mathbf{p}_\alpha, \boldsymbol{\lambda}) > 0 \\ G^{(2k_\beta - 1, \beta)}_{\partial\Omega_{\alpha\beta}}(\mathbf{x}_m, t_{m\pm}, \mathbf{p}_\beta, \boldsymbol{\lambda}) < 0 \end{array}\right\} \textit{for}\; \mathbf{n}_{\partial\Omega_{\alpha\beta}} \to \Omega_\alpha. \end{aligned} \tag{3.109}$$

Proof. The proof is similar to the proof of Theorem 3.2. ■

Definition 3.23. For a discontinuous dynamical system in Eq. (2.1), there is a point $\mathbf{x}^{(0)}(t_m) \equiv \mathbf{x}_m \in \partial\Omega_{ij}$ at time t_m between two adjacent domains Ω_α $(\alpha = i, j)$. Suppose $\mathbf{x}^{(\alpha)}_\beta(t_{m\pm}) = \mathbf{x}_m = \mathbf{x}^{(\beta)}_\alpha(t_{m\pm})$ $(\alpha, \beta \in \{i, j\}$ and $\alpha \neq \beta)$. For an arbitrarily small $\varepsilon > 0$, there is a time interval $[t_{m-\varepsilon}, t_{m+\varepsilon}]$. An imaginary flow $\mathbf{x}^{(\alpha)}_\beta(t)$ is $C^{r_\alpha}_{[t_{m-\varepsilon}, t_{m+\varepsilon}]}$-continuous $(r_\alpha \geq 2k_\alpha)$ and $\| d^{r_\alpha + 1}\mathbf{x}^{(\alpha)}_\beta / dt^{r_\alpha + 1} \| < \infty$ for time *t*, and an imaginary flow $\mathbf{x}^{(\beta)}_\alpha(t)$ is $C^{r_\beta}_{[t_{m-\varepsilon}, t_{m+\varepsilon}]}$-continuous with $\| d^{r_\beta + 1}\mathbf{x}^{(\beta)}_\alpha / dt^{r_\beta + 1} \| < \infty$ $(r_\beta \geq 2k_\beta)$. The imaginary flow $\mathbf{x}^{(\alpha)}_\beta(t)$ of the $(2k_\alpha - 1)$th-order and the imaginary $\mathbf{x}^{(\beta)}_\alpha(t)$ of the $(2k_\beta - 1)$th-order to the boundary $\partial\Omega_{ij}$ is a $(2k_\alpha - 1 : 2k_\beta - 1)$-*double inaccessible tangential flow* if

$$\left.\begin{array}{l}G_{\partial\Omega_{\alpha\beta}}^{(s_\alpha,\alpha)}(\mathbf{x}_m,t_{m\pm},\mathbf{p}_\alpha,\boldsymbol{\lambda})=0 \text{ for } s_\alpha=0,1,\cdots,2k_\alpha-2,\\ G_{\partial\Omega_{\alpha\beta}}^{(2k_\alpha-1,\alpha)}(\mathbf{x}_m,t_{m\pm},\mathbf{p}_\alpha,\boldsymbol{\lambda})\neq 0;\end{array}\right\} \tag{3.110}$$

$$\left.\begin{array}{l}G_{\partial\Omega_{\alpha\beta}}^{(s_\beta,\beta)}(\mathbf{x}_m,t_{m\pm},\mathbf{p}_\beta,\boldsymbol{\lambda})=0 \text{ for } s_\beta=0,1,\cdots,2k_\beta-2,\\ G_{\partial\Omega_{\alpha\beta}}^{(2k_\beta-1,\beta)}(\mathbf{x}_m,t_{m-},\mathbf{p}_\beta,\boldsymbol{\lambda})\neq 0;\end{array}\right\} \tag{3.111}$$

$$\left.\begin{array}{ll}\text{either} & \left.\begin{array}{l}\mathbf{n}_{\partial\Omega_{ij}}^{\mathrm{T}}(\mathbf{x}_{m-\varepsilon}^{(0)})\cdot[\mathbf{x}_{m-\varepsilon}^{(0)}-\mathbf{x}_{\beta(m-\varepsilon)}^{(\alpha)}]<0\\ \mathbf{n}_{\partial\Omega_{ij}}^{\mathrm{T}}(\mathbf{x}_{m+\varepsilon}^{(0)})\cdot[\mathbf{x}_{\beta(m+\varepsilon)}^{(\alpha)}-\mathbf{x}_{m+\varepsilon}^{(0)}]>0\end{array}\right\}\text{for } \mathbf{n}_{\partial\Omega_{\alpha\beta}}\to\Omega_\beta,\\ \text{or} & \left.\begin{array}{l}\mathbf{n}_{\partial\Omega_{ij}}^{\mathrm{T}}(\mathbf{x}_{m-\varepsilon}^{(0)})\cdot[\mathbf{x}_{m-\varepsilon}^{(0)}-\mathbf{x}_{\beta(m-\varepsilon)}^{(\alpha)}]>0\\ \mathbf{n}_{\partial\Omega_{ij}}^{\mathrm{T}}(\mathbf{x}_{m+\varepsilon}^{(0)})\cdot[\mathbf{x}_{\beta(m+\varepsilon)}^{(\alpha)}-\mathbf{x}_{m+\varepsilon}^{(0)}]<0\end{array}\right\}\text{for } \mathbf{n}_{\partial\Omega_{\alpha\beta}}\to\Omega_\alpha;\end{array}\right\} \tag{3.112}$$

$$\left.\begin{array}{ll}\text{either} & \left.\begin{array}{l}\mathbf{n}_{\partial\Omega_{ij}}^{\mathrm{T}}(\mathbf{x}_{m-\varepsilon}^{(0)})\cdot[\mathbf{x}_{m-\varepsilon}^{(0)}-\mathbf{x}_{\alpha(m-\varepsilon)}^{(\beta)}]>0\\ \mathbf{n}_{\partial\Omega_{ij}}^{\mathrm{T}}(\mathbf{x}_{m+\varepsilon}^{(0)})\cdot[\mathbf{x}_{\alpha(m+\varepsilon)}^{(\beta)}-\mathbf{x}_{m+\varepsilon}^{(0)}]<0\end{array}\right\}\text{for } \mathbf{n}_{\partial\Omega_{\alpha\beta}}\to\Omega_\beta,\\ \text{or} & \left.\begin{array}{l}\mathbf{n}_{\partial\Omega_{ij}}^{\mathrm{T}}(\mathbf{x}_{m-\varepsilon}^{(0)})\cdot[\mathbf{x}_{m-\varepsilon}^{(0)}-\mathbf{x}_{\alpha(m-\varepsilon)}^{(\beta)}]<0\\ \mathbf{n}_{\partial\Omega_{ij}}^{\mathrm{T}}(\mathbf{x}_{m+\varepsilon}^{(0)})\cdot[\mathbf{x}_{\alpha(m+\varepsilon)}^{(\beta)}-\mathbf{x}_{m+\varepsilon}^{(0)}]>0\end{array}\right\}\text{for } \mathbf{n}_{\partial\Omega_{\alpha\beta}}\to\Omega_\alpha.\end{array}\right\} \tag{3.113}$$

The $(2k_\alpha-1:2k_\beta-1)$-double inaccessible tangential flows are sketched in Fig. 3.17(b) by the dashed curves. Such a double inaccessible flow is formed by two imaginary tangential flows to the boundary. No any flows in the two domains can access the boundary. The corresponding necessary and sufficient conditions for the tangential flows are given as follows.

Theorem 3.14. *For a discontinuous dynamical system in Eq.* (2.1), *there is a point* $\mathbf{x}^{(0)}(t_m)\equiv\mathbf{x}_m\in\partial\Omega_{ij}$ *at time* t_m *between two adjacent domains* Ω_α ($\alpha=i,j$). *Suppose* $\mathbf{x}_\beta^{(\alpha)}(t_{m\pm})=\mathbf{x}_m=\mathbf{x}_\alpha^{(\beta)}(t_{m\pm})$ ($\alpha,\beta\in\{i,j\}$ *and* $\alpha\neq\beta$). *For an arbitrarily small* $\varepsilon>0$, *there is a time interval* $[t_{m-\varepsilon},t_{m+\varepsilon}]$. *An imaginary flow* $\mathbf{x}_\beta^{(\alpha)}(t)$ *is* $C_{[t_{m-\varepsilon},t_{m+\varepsilon}]}^{r_\alpha}$*-continuous* ($r_\alpha\geq 2k_\alpha$) *with* $\|d^{r_\alpha+1}\mathbf{x}_\beta^{(\alpha)}/dt^{r_\alpha+1}\|<\infty$ *for time t, and an imaginary flow* $\mathbf{x}_\alpha^{(\beta)}(t)$ *is* $C_{[t_{m-\varepsilon},t_{m+\varepsilon}]}^{r_\beta}$ *-continuous with* $\|d^{r_\beta+1}\mathbf{x}_\alpha^{(\beta)}/dt^{r_\beta+1}\|<\infty$ ($r_\beta\geq 2k_\beta$). *The imaginary flow* $\mathbf{x}_\beta^{(\alpha)}(t)$ *of the* $(2k_\alpha-1)$th*-order and the imaginary* $\mathbf{x}_\alpha^{(\beta)}(t)$ *of the* $(2k_\beta-1)$th *-order to the boundary* $\partial\Omega_{ij}$ *is a* $(2k_\alpha-1:2k_\beta-1)$*-double inaccessible tangential flow if and only if*

$$G_{\partial\Omega_{\alpha\beta}}^{(s_\alpha,\alpha)}(\mathbf{x}_m,t_{m\pm},\mathbf{p}_\alpha,\boldsymbol{\lambda})=0 \;\; \textit{for } s_\alpha=0,1,\cdots,2k_\alpha-2; \tag{3.114}$$

$$G_{\partial\Omega_{\alpha\beta}}^{(s_\beta,\beta)}(\mathbf{x}_m, t_{m\pm}, \mathbf{p}_\beta, \boldsymbol{\lambda}) = 0 \ for \ s_\beta = 0,1,\cdots,2k_\beta - 2; \tag{3.115}$$

$$\begin{aligned}
&\text{either} \quad \left.\begin{aligned} G_{\partial\Omega_{\alpha\beta}}^{(2k_\alpha-1,\alpha)}(\mathbf{x}_m, t_{m\pm}, \mathbf{p}_\alpha, \boldsymbol{\lambda}) > 0 \\ G_{\partial\Omega_{\alpha\beta}}^{(2k_\beta-1,\beta)}(\mathbf{x}_m, t_{m\pm}, \mathbf{p}_\beta, \boldsymbol{\lambda}) < 0 \end{aligned}\right\} \text{for } \mathbf{n}_{\partial\Omega_{\alpha\beta}} \to \Omega_\beta, \\
&\text{or} \quad \left.\begin{aligned} G_{\partial\Omega_{\alpha\beta}}^{(2k_\alpha-1,\alpha)}(\mathbf{x}_m, t_{m\pm}, \mathbf{p}_\alpha, \boldsymbol{\lambda}) < 0 \\ G_{\partial\Omega_{\alpha\beta}}^{(2k_\beta-1,\beta)}(\mathbf{x}_m, t_{m\pm}, \mathbf{p}_\beta, \boldsymbol{\lambda}) > 0 \end{aligned}\right\} \text{for } \mathbf{n}_{\partial\Omega_{\alpha\beta}} \to \Omega_\alpha.
\end{aligned} \tag{3.116}$$

Proof. The proof is similar to the proof of Theorem 3.2. ■

3.6. Flow switching bifurcations

In this section, the flow switching bifurcations from the passable to non-passable flow and the sliding fragmentation bifurcation from the non-passable to passable flow will be discussed. This section will extend the idea in Chapter 2 (also see, Luo, 2006b, 2008b,c). Before discussion of switching bifurcations, the product of the *G*-functions of the $(m_i : m_j)$-order on the boundary $\partial\Omega_{ij}$ is introduced.

Definition 3.24. For a discontinuous dynamical system in Eq. (2.1), there is a point $\mathbf{x}^{(0)}(t_m) \equiv \mathbf{x}_m \in \partial\Omega_{ij}$ at time t_m between two adjacent domains Ω_α $(\alpha = i, j)$. Suppose $\mathbf{x}^{(i)}(t_{m\pm}) = \mathbf{x}_m = \mathbf{x}^{(j)}(t_{m\mp})$. For an arbitrarily small $\varepsilon > 0$, there is a time interval $[t_{m-\varepsilon}, t_{m+\varepsilon}]$. A flow $\mathbf{x}^{(i)}(t)$ is $C^{r_i}_{[t_{m-\varepsilon}, t_{m+\varepsilon}]}$-continuous for time t with $\| d^{r_i+1}\mathbf{x}^{(i)} / dt^{r_i+1} \| < \infty$ ($r_i \ge m_i + 1$), and a flow $\mathbf{x}^{(j)}(t)$ is $C^{r_j}_{[t_{m-\varepsilon}, t_{m+\varepsilon}]}$-continuous ($r_j \ge m_j + 1$) with $\| d^{r_j+1}\mathbf{x}^{(j)} / dt^{r_j+1} \| < \infty$. The $(m_i : m_j)$*-product of G-functions* on the boundary $\partial\Omega_{ij}$ is defined as

$$\begin{aligned}
L_{ij}^{(m_i:m_j)}(t_m) &\equiv L_{ij}^{(m_i:m_j)}(\mathbf{x}_m, t_m, \mathbf{p}_i, \mathbf{p}_j, \boldsymbol{\lambda}) \\
&= G_{\partial\Omega_{ij}}^{(m_i,i)}(\mathbf{x}_m, t_{m-}, \mathbf{p}_i, \boldsymbol{\lambda}) \times G_{\partial\Omega_{ij}}^{(m_j,j)}(\mathbf{x}_m, t_{m+}, \mathbf{p}_j, \boldsymbol{\lambda}),
\end{aligned} \tag{3.117}$$

and for $m_i = m_j = 0$, we have $L_{ij}^{(0:0)} = L_{ij}$

$$\begin{aligned}
L_{ij}(t_m) &\equiv L_{ij}(\mathbf{x}_m, t_m, \mathbf{p}_i, \mathbf{p}_j, \boldsymbol{\lambda}) \\
&= G_{\partial\Omega_{ij}}^{(i)}(\mathbf{x}_m, t_{m-}, \mathbf{p}_i, \boldsymbol{\lambda}) \times G_{\partial\Omega_{ij}}^{(j)}(\mathbf{x}_m, t_{m+}, \mathbf{p}_j, \boldsymbol{\lambda}).
\end{aligned} \tag{3.118}$$

From the foregoing definition, the products of *G*-functions for the full passable, sink and source on the boundary $\partial\Omega_{\alpha\beta}$ are

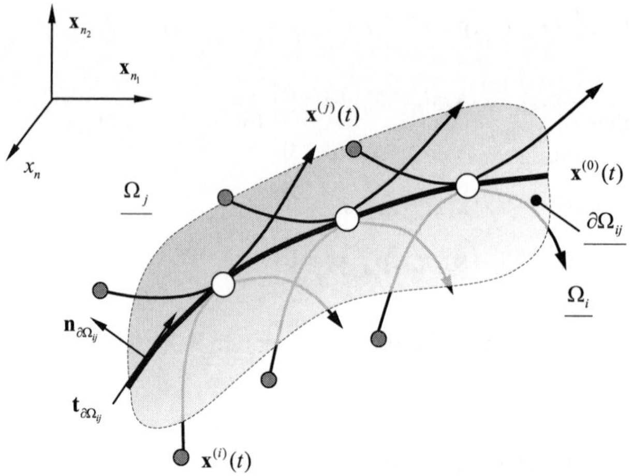

(a)

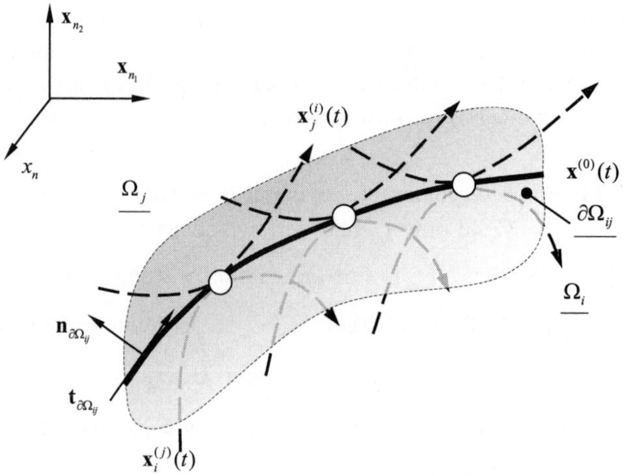

(b)

Fig. 3.17 **(a)** $(2k_i-1:2k_j-1)$ -double tangential flow in both Ω_i and Ω_j and **(b)** $(2k_i-1:2k_j-1)$ -double inaccessible tangential flow in both Ω_i and Ω_j. $\mathbf{x}^{(i)}(t)$ and $\mathbf{x}^{(j)}(t)$ represent the *real* flows in domains Ω_i and Ω_j, respectively, which are depicted by thin solid curves. $\mathbf{x}_i^{(j)}(t)$ and $\mathbf{x}_j^{(i)}(t)$ represent the *imaginary* flows in domains Ω_i and Ω_j, respectively controlled by the vector fields in Ω_j and Ω_i, which are depicted by dashed curves. The flow on the boundary is described by $\mathbf{x}^{(0)}(t)$. The normal and tangential vectors $\mathbf{n}_{\partial\Omega_{ij}}$ and $\mathbf{t}_{\partial\Omega_{ij}}$ of the boundary are depicted. Hollow circles are for grazing points on the boundary and fill circles are for starting points. ($n_1+n_2+1=n$)

$$L_{\alpha\beta}^{(2k_\alpha:2k_\beta)}(t_m) > 0 \ \text{ on } \overrightarrow{\partial\Omega}_{\alpha\beta};$$
$$L_{\alpha\beta}^{(2k_\alpha:2k_\beta)}(t_m) < 0 \ \text{ on } \overline{\partial\Omega}_{\alpha\beta} = \widetilde{\partial\Omega}_{\alpha\beta} \cup \widehat{\partial\Omega}_{\alpha\beta}. \tag{3.119}$$

The switching bifurcation of a flow at $(t_m, \mathbf{x}_m)$ on the boundary $\partial\Omega_{\alpha\beta}$ requires

$$L_{\alpha\beta}^{(2k_\alpha:2k_\beta)}(t_m) = 0. \tag{3.120}$$

For a passable flow at $\mathbf{x}(t_m) \equiv \mathbf{x}_m \in [\mathbf{x}_{m_1}, \mathbf{x}_{m_2}] \subset \overrightarrow{\partial\Omega}_{ij}$, consider a time interval $[t_{m_1}, t_{m_2}]$ for $[\mathbf{x}_{m_1}, \mathbf{x}_{m_2}]$ on the boundary and the product of G-functions for $t_m \in [t_{m_1}, t_{m_2}]$ and $\mathbf{x}_m \in [\mathbf{x}_{m_1}, \mathbf{x}_{m_2}]$ is positive, i.e., $L_{ij}^{(2k_i:2k_j)}(t_m) > 0$. To determine the switching bifurcation, the global minimum of such a product of G-functions should be determined. Because $\mathbf{x}_m$ is a function of t_m, the two total derivatives of $L_{ij}^{(2k_i:2k_j)}(t_m)$ are introduced by

$$\begin{aligned} DL_{ij}^{(2k_i:2k_j)} &= \nabla L_{ij}^{(2k_i:2k_j)}(\mathbf{x}_m, t_m, \mathbf{p}_i, \mathbf{p}_j, \boldsymbol{\lambda}) \cdot \mathbf{F}_{ij}^{(0)}(\mathbf{x}_m, t_m) \\ &\quad + \partial_{t_m} L_{ij}^{(2k_i:2k_j)}(\mathbf{x}_m, t_m, \mathbf{p}_i, \mathbf{p}_j, \boldsymbol{\lambda}), \\ D^r L_{ij}^{(2k_i:2k_j)} &= D^{r-1}\{DL_{ij}^{(2k_i:2k_j)}(\mathbf{x}_m, t_m, \mathbf{p}_i, \mathbf{p}_j, \boldsymbol{\lambda})\} \end{aligned} \tag{3.121}$$

for $r = 1, 2, \cdots$. Thus, the local minimum of $L_{ij}^{(2k_i:2k_j)}(t_m)$ is determined by

$$D^r L_{ij}^{(2k_i:2k_j)}(t_m) = 0, \ (r = 1, 2, \cdots, 2l-1) \tag{3.122}$$

$$D^{2l} L_{ij}^{(2k_i:2k_j)}(t_m) > 0. \tag{3.123}$$

Definition 3.25. For a discontinuous dynamical system in Eq. (2.1), there is a point $\mathbf{x}^{(0)}(t_m) \equiv \mathbf{x}_m \in \partial\Omega_{ij}$ at time t_m between two adjacent domains Ω_α ($\alpha = i, j$). Suppose $\mathbf{x}^{(i)}(t_{m\pm}) = \mathbf{x}_m = \mathbf{x}^{(j)}(t_{m\mp})$. For an arbitrarily small $\varepsilon > 0$, there are two time intervals (i.e., $[t_{m-\varepsilon}, t_m)$ and $(t_m, t_{m+\varepsilon}]$). A flow $\mathbf{x}^{(i)}(t)$ is $C^{r_i}_{[t_{m-\varepsilon}, t_m)}$-continuous ($r_i \geq 2k_i + 1$) with $\| d^{r_i+1}\mathbf{x}^{(i)} / dt^{r_i+1} \| < \infty$ for time t, and a flow $\mathbf{x}^{(j)}(t)$ is $C^{r_j}_{(t_m, t_{m+\varepsilon}]}$-continuous ($r_j \geq 2k_j + 1$) with $\| d^{r_j+1}\mathbf{x}^{(j)} / dt^{r_j+1} \| < \infty$. The *local minimum value set* of the $(2k_i : 2k_j)$-product of G-functions (i.e., $L_{ij}^{(2k_i:2k_j)}(t_m)$) is defined by

$${}_{\min}L_{ij}^{(2k_i:2k_j)}(t_m) = \left\{ L_{ij}^{(2k_i:2k_j)}(t_m) \,\middle|\, \begin{array}{l} \text{for } t_m \in [t_{m_1}, t_{m_2}] \text{ and } \mathbf{x}_m \in [\mathbf{x}_{m_1}, \mathbf{x}_{m_2}], \\ D^r L_{ij}^{(2k_i:2k_j)} = 0 \text{ for } r = \{1, 2, \cdots, 2l-1\}, \\ \text{and } D^{2l} L_{ij}^{(2k_i:2k_j)} > 0. \end{array} \right\} \tag{3.124}$$

From the local minimum set of $L_{ij}^{(2k_i:2k_j)}(t_m)$, the global minimum values of $L_{ij}^{(2k_i:2k_j)}(t_m)$ is defined as follows.

Definition 3.26. For a discontinuous dynamical system in Eq. (2.1), there is a point $\mathbf{x}^{(0)}(t_m) \equiv \mathbf{x}_m \in \partial\Omega_{ij}$ at time t_m between two adjacent domains Ω_α ($\alpha = i, j$). Suppose $\mathbf{x}^{(i)}(t_{m\pm}) = \mathbf{x}_m = \mathbf{x}^{(j)}(t_{m\mp})$. For an arbitrarily small $\varepsilon > 0$, there are two time intervals (i.e., $[t_{m-\varepsilon}, t_m)$ and $(t_m, t_{m+\varepsilon}]$). A flow $\mathbf{x}^{(i)}(t)$ is $C^{r_i}_{[t_{m-\varepsilon}, t_m)}$ -continuous ($r_i \geq 2k_i + 1$) with $\| d^{r_i+1}\mathbf{x}^{(i)} / dt^{r_i+1} \| < \infty$ for time *t*, and a flow $\mathbf{x}^{(j)}(t)$ is $C^{r_j}_{(t_m, t_{m+\varepsilon}]}$-continuous ($r_j \geq 2k_j + 1$) with $\| d^{r_j+1}\mathbf{x}^{(j)} / dt^{r_j+1} \| < \infty$. The *global minimum value* of the $(2k_i : 2k_j)$-product of *G*-functions (i.e., $L_{ij}^{(2k_i:2k_j)}(t_m)$) is defined by

$$_{G\min}L_{ij}^{(2k_i:2k_j)}(t_m) = \min_{t_m \in [t_{m_1}, t_{m_2}]} \{_{\min}L_{ij}^{(2k_i:2k_j)}(t_m), L_{ij}^{(2k_i:2k_j)}(t_{m_1}), L_{ij}^{(2k_i:2k_j)}(t_{m_2})\} \quad (3.125)$$

To consider the switching bifurcation varying with the system parameter $\mathbf{q} \in \{\mathbf{p}_i, \mathbf{p}_j, \boldsymbol{\lambda}\}$, $D^r L_{ij}^{(2k_i:2k_j)}$ in Eq. (3.121) is replaced by $d^r L_{ij}^{(2k_i:2k_j)} / d\mathbf{q}^r$. Similarly, the maximum set of the $(2k_i : 2k_j)$-product of *G*-functions (i.e., $L_{ij}^{(2k_i:2k_j)}(t_m)$) can be developed as follows.

Definition 3.27. For a discontinuous dynamical system in Eq. (2.1), there is a point $\mathbf{x}^{(0)}(t_m) \equiv \mathbf{x}_m \in \partial\Omega_{ij}$ at time t_m between two adjacent domains Ω_α ($\alpha = i, j$). Suppose $\mathbf{x}^{(\alpha)}(t_{m\pm}) = \mathbf{x}_m$ ($\alpha \in \{i, j\}$). For an arbitrarily small $\varepsilon > 0$, there are two time intervals ($[t_{m-\varepsilon}, t_m)$ or $(t_m, t_{m+\varepsilon}]$). The flow $\mathbf{x}^{(i)}(t)$ is $C^{r_i}_{[t_{m-\varepsilon}, t_m)}$ or $C^{r_i}_{[t_{m-\varepsilon}, t_{m+\varepsilon}]}$-continuous ($r_i \geq 2k_i + 1$) for time t and $\| d^{r_i+1}\mathbf{x}^{(i)} / dt^{r_i+1} \| < \infty$. The flow $\mathbf{x}^{(j)}(t)$ is $C^{r_j}_{[t_{m-\varepsilon}, t_{m+\varepsilon}]}$ or $C^{r_j}_{(t_m, t_{m+\varepsilon}]}$-continuous for time t and $\| d^{r_j+1}\mathbf{x}^{(j)} / dt^{r_j+1} \| < \infty$ ($r_j \geq 2k_j + 1$). *The local maximum set* of the $(2k_i : 2k_j)$ product of *G*-functions (i.e., $L_{ij}^{(2k_i:2k_j)}(t_m)$) is defined by

$$_{\max}L_{ij}^{(2k_i:2k_j)}(t_m) = \left\{ L_{ij}^{(2k_i:2k_j)}(t_m) \left| \begin{array}{l} \text{for } t_m \in [t_{m_1}, t_{m_2}] \text{ and } \mathbf{x}_m \in [\mathbf{x}_{m_1}, \mathbf{x}_{m_2}], \\ D^r L_{ij}^{(2k_i:2k_j)} = 0 \text{ for } r = \{1, 2, \cdots, 2l\}, \\ \text{and } D^{2l+1} L_{ij}^{(2k_i:2k_j)} < 0. \end{array} \right. \right\} \quad (3.126)$$

From the local maximum set of $L_{ij}^{(2k_i:2k_j)}(t_m)$, the global maximum value of $L_{ij}^{(2k_i:2k_j)}(t_m)$ is defined as follows.

Definition 3.28. For a discontinuous dynamical system in Eq. (2.1), there is a point $\mathbf{x}^{(0)}(t_m) \equiv \mathbf{x}_m \in \partial\Omega_{ij}$ at time t_m between two adjacent domains Ω_α ($\alpha = i, j$). Suppose $\mathbf{x}^{(\alpha)}(t_{m\pm}) = \mathbf{x}_m$ ($\alpha \in \{i, j\}$). For an arbitrarily small $\varepsilon > 0$, there is a time interval ($[t_{m-\varepsilon}, t_m)$ or $(t_m, t_{m+\varepsilon}]$). The flow $\mathbf{x}^{(i)}(t)$ is $C^{r_i}_{[t_{m-\varepsilon}, t_m)}$ or $C^{r_i}_{(t_m, t_{m+\varepsilon}]}$-continuous ($r_i \geq 2k_i + 1$) for time t and $\| d^{r_i+1}\mathbf{x}^{(i)} / dt^{r_i+1} \| < \infty$. The flow $\mathbf{x}^{(j)}(t)$ is $C^{r_j}_{[t_{m-\varepsilon}, t_m)}$ or $C^{r_j}_{(t_m, t_{m+\varepsilon}]}$-continuous ($r_j \geq 2k_j + 1$) and $\| d^{r_j+1}\mathbf{x}^{(j)} / dt^{r_j+1} \| < \infty$ for time t. The *global maximum* of the $(2k_i : 2k_j)$ product of *G*-functions (i.e., $L_{ij}^{(2k_i:2k_j)}(t_m)$) is defined by

$$
{}_{G\max}L_{ij}^{(2k_i:2k_j)}(t_m) = \max_{t_m \in [t_{m_1}, t_{m_2}]} \{ {}_{\max}L_{ij}^{(2k_i:2k_j)}(t_m), L_{ij}^{(2k_i:2k_j)}(t_{m_1}), L_{ij}^{(2k_i:2k_j)}(t_{m_2}) \}. \tag{3.127}
$$

Definition 3.29. For a discontinuous dynamical system in Eq. (2.1), there is a point $\mathbf{x}^{(0)}(t_m) \equiv \mathbf{x}_m \in \partial\Omega_{ij}$ at time t_m between two adjacent domains Ω_α ($\alpha = i, j$). Suppose $\mathbf{x}^{(i)}(t_{m-}) = \mathbf{x}_m = \mathbf{x}^{(j)}(t_{m\pm})$. For an arbitrarily small $\varepsilon > 0$, there are two time intervals $[t_{m-\varepsilon}, t_m)$ and $(t_m, t_{m+\varepsilon}]$. Both flows $\mathbf{x}^{(i)}(t)$ and $\mathbf{x}^{(j)}(t)$ are $C^{r_i}_{[t_{m-\varepsilon}, t_m)}$ and $C^{r_j}_{[t_{m-\varepsilon}, t_{m+\varepsilon}]}$-continuous ($r_\alpha \geq 2$ and $\alpha = i, j$) for time *t*, respectively, and $\| d^{r_\alpha+1}\mathbf{x}^{(\alpha)} / dt^{r_\alpha+1} \| < \infty$. The tangential bifurcation of the flow $\mathbf{x}^{(j)}(t)$ at point $(\mathbf{x}_m, t_m)$ on the boundary $\overrightarrow{\partial\Omega}_{ij}$ is termed the *switching bifurcation of the first kind of the non-passable flow* (or called the *sliding bifurcation*) if

$$
\begin{aligned}
&G^{(j)}_{\partial\Omega_{ij}}(\mathbf{x}_m, t_{m\pm}, \mathbf{p}_j, \boldsymbol{\lambda}) = 0,\\
&G^{(i)}_{\partial\Omega_{ij}}(\mathbf{x}_m, t_{m-}, \mathbf{p}_i, \boldsymbol{\lambda}) \neq 0,\\
&G^{(1,j)}_{\partial\Omega_{ij}}(\mathbf{x}_m, t_{m\pm}, \mathbf{p}_j, \boldsymbol{\lambda}) \neq 0;
\end{aligned} \tag{3.128}
$$

$$
\begin{aligned}
&\text{either} && \left.\begin{aligned}
&\mathbf{n}^{\mathrm{T}}_{\partial\Omega_{ij}}(\mathbf{x}^{(0)}_{m-\varepsilon}) \cdot [\mathbf{x}^{(0)}_{m-\varepsilon} - \mathbf{x}^{(i)}_{m-\varepsilon}] > 0\\
&\mathbf{n}^{\mathrm{T}}_{\partial\Omega_{ij}}(\mathbf{x}^{(0)}_{m-\varepsilon}) \cdot [\mathbf{x}^{(0)}_{m-\varepsilon} - \mathbf{x}^{(j)}_{m-\varepsilon}] < 0\\
&\mathbf{n}^{\mathrm{T}}_{\partial\Omega_{ij}}(\mathbf{x}^{(0)}_{m+\varepsilon}) \cdot [\mathbf{x}^{(j)}_{m+\varepsilon} - \mathbf{x}^{(0)}_{m+\varepsilon}] > 0
\end{aligned}\right\} \text{for } \mathbf{n}_{\partial\Omega_{ij}} \to \Omega_j,\\
&\text{or} && \left.\begin{aligned}
&\mathbf{n}^{\mathrm{T}}_{\partial\Omega_{ij}}(\mathbf{x}^{(0)}_{m-\varepsilon}) \cdot [\mathbf{x}^{(0)}_{m-\varepsilon} - \mathbf{x}^{(i)}_{m-\varepsilon}] < 0\\
&\mathbf{n}^{\mathrm{T}}_{\partial\Omega_{ij}}(\mathbf{x}^{(0)}_{m-\varepsilon}) \cdot [\mathbf{x}^{(0)}_{m-\varepsilon} - \mathbf{x}^{(j)}_{m-\varepsilon}] > 0\\
&\mathbf{n}^{\mathrm{T}}_{\partial\Omega_{ij}}(\mathbf{x}^{(0)}_{m+\varepsilon}) \cdot [\mathbf{x}^{(j)}_{m+\varepsilon} - \mathbf{x}^{(0)}_{m+\varepsilon}] < 0
\end{aligned}\right\} \text{for } \mathbf{n}_{\partial\Omega_{ij}} \to \Omega_i.
\end{aligned} \tag{3.129}
$$

Theorem 3.15. For a discontinuous dynamical system in Eq. (2.1), there is a point $\mathbf{x}^{(0)}(t_m) \equiv \mathbf{x}_m \in \partial\Omega_{ij}$ at time t_m between two adjacent domains Ω_α ($\alpha = i, j$).

Suppose $\mathbf{x}^{(i)}(t_{m-})=\mathbf{x}_m=\mathbf{x}^{(j)}(t_{m\pm})$. *For an arbitrarily small* $\varepsilon>0$, *there are two time intervals* $[t_{m-\varepsilon},t_m)$ *and* $(t_m,t_{m+\varepsilon}]$. *The flows* $\mathbf{x}^{(i)}(t)$ *and* $\mathbf{x}^{(j)}(t)$ *are* $C^{r_i}_{[t_{m-\varepsilon},t_m)}$ *and* $C^{r_j}_{[t_{m-\varepsilon},t_{m+\varepsilon}]}$ *-continuous for time* t *and* $\|d^{r_\alpha+1}\mathbf{x}^{(\alpha)}/dt^{r_\alpha+1}\|<\infty$ ($r_\alpha\geq 3$ and $\alpha=i,j$). *The sliding bifurcation of the passable flow of* $\mathbf{x}^{(i)}(t)$ *and* $\mathbf{x}^{(j)}(t)$ *at point* $(\mathbf{x}_m,t_m)$ *switching to the non-passable flow of the first kind on the boundary* $\overrightarrow{\partial\Omega}_{ij}$ *occurs if and only if*

$$\left.\begin{array}{l} G^{(j)}_{\partial\Omega_{ij}}(\mathbf{x}_m,t_{m\pm},\mathbf{p}_j,\boldsymbol{\lambda})=0,\\ L_{ij}(\mathbf{x}_m,t_m,\mathbf{p}_i,\mathbf{p}_j,\boldsymbol{\lambda})=0,\\ {}_{G\min}L_{ij}(t_m)=0; \end{array}\right\} \tag{3.130}$$

$$\left.\begin{array}{l} G^{(i)}_{\partial\Omega_{ij}}(\mathbf{x}_m,t_{m-},\mathbf{p}_i,\boldsymbol{\lambda})>0 \ \textit{for}\ \mathbf{n}_{\partial\Omega_{ij}}\to\Omega_j,\\ G^{(i)}_{\partial\Omega_{ij}}(\mathbf{x}_m,t_{m-},\mathbf{p}_i,\boldsymbol{\lambda})<0 \ \textit{for}\ \mathbf{n}_{\partial\Omega_{ij}}\to\Omega_i; \end{array}\right\} \tag{3.131a}$$

$$\left.\begin{array}{l} G^{(1,j)}_{\partial\Omega_{ij}}(\mathbf{x}_m,t_{m\pm},\mathbf{p}_j,\boldsymbol{\lambda})>0 \ \textit{for}\ \mathbf{n}_{\partial\Omega_{ij}}\to\Omega_j,\\ G^{(1,j)}_{\partial\Omega_{ij}}(\mathbf{x}_m,t_{m\pm},\mathbf{p}_j,\boldsymbol{\lambda})<0 \ \textit{for}\ \mathbf{n}_{\partial\Omega_{ij}}\to\Omega_i. \end{array}\right\} \tag{3.131b}$$

Proof. The proof is the same as in the proofs of Theorem 3.1 and Theorem 3.2. This theorem can be proved. ■

Definition 3.30. For a discontinuous dynamical system in Eq. (2.1), there is a point $\mathbf{x}^{(0)}(t_m)\equiv\mathbf{x}_m\in\partial\Omega_{ij}$ at time t_m between two adjacent domains $\Omega_\alpha\,(\alpha=i,j)$. Suppose $\mathbf{x}^{(i)}(t_{m-})=\mathbf{x}_m=\mathbf{x}^{(j)}(t_{m\pm})$. For an arbitrarily small $\varepsilon>0$, there are two time intervals (i.e., $[t_{m-\varepsilon},t_m)$ and $(t_m,t_{m+\varepsilon}]$). A flow $\mathbf{x}^{(i)}(t)$ is $C^{r_i}_{[t_{m-\varepsilon},t_m)}$-continuous ($r_i\geq 2k_i+1$) and $\|d^{r_i+1}\mathbf{x}^{(i)}/dt^{r_i+1}\|<\infty$ for time *t*, and a flow $\mathbf{x}^{(j)}(t)$ is $C^{r_j}_{[t_{m-\varepsilon},t_{m+\varepsilon}]}$-continuous ($r_j\geq 2k_j+1$) and $\|d^{r_j+1}\mathbf{x}^{(j)}/dt^{r_j+1}\|<\infty$. The bifurcation of the $(2k_i:2k_j)$-passable flow of $\mathbf{x}^{(i)}(t)$ and $\mathbf{x}^{(j)}(t)$ at point $(\mathbf{x}_m,t_m)$ on the boundary $\overrightarrow{\partial\Omega}_{ij}$ is termed the *switching bifurcation of the first kind of the* $(2k_i:2k_j)$*-non-passable flow* (or called the $(2k_i:2k_j)$*-sliding bifurcation*) if

$$\begin{array}{l} G^{(s_j,j)}_{\partial\Omega_{ij}}(\mathbf{x}_m,t_{m\pm},\mathbf{p}_j,\boldsymbol{\lambda})=0 \text{ for } s_j=0,1,\cdots,2k_j,\\ G^{(s_i,i)}_{\partial\Omega_{ij}}(\mathbf{x}_m,t_{m-},\mathbf{p}_i,\boldsymbol{\lambda})=0 \text{ for } s_i=0,1,\cdots,2k_i-1,\\ G^{(2k_i,i)}_{\partial\Omega_{ij}}(\mathbf{x}_m,t_{m-},\mathbf{p}_i,\boldsymbol{\lambda})\neq 0,\\ G^{(2k_j+1,j)}_{\partial\Omega_{ij}}(\mathbf{x}_m,t_{m\pm},\mathbf{p}_j,\boldsymbol{\lambda})\neq 0; \end{array} \tag{3.132}$$

$$\left.\begin{array}{l}\mathbf{n}^{\mathrm{T}}_{\partial\Omega_{ij}}(\mathbf{x}^{(0)}_{m-\varepsilon})\cdot[\mathbf{x}^{(0)}_{m-\varepsilon}-\mathbf{x}^{(j)}_{m-\varepsilon}]<0\\ \text{either }\ \mathbf{n}^{\mathrm{T}}_{\partial\Omega_{ij}}(\mathbf{x}^{(0)}_{m+\varepsilon})\cdot[\mathbf{x}^{(j)}_{m+\varepsilon}-\mathbf{x}^{(0)}_{m+\varepsilon}]>0\\ \mathbf{n}^{\mathrm{T}}_{\partial\Omega_{ij}}(\mathbf{x}^{(0)}_{m-\varepsilon})\cdot[\mathbf{x}^{(0)}_{m-\varepsilon}-\mathbf{x}^{(i)}_{m-\varepsilon}]>0\end{array}\right\}\text{for }\mathbf{n}_{\partial\Omega_{ij}}\to\Omega_j,$$

$$\left.\begin{array}{l}\mathbf{n}^{\mathrm{T}}_{\partial\Omega_{ij}}(\mathbf{x}^{(0)}_{m-\varepsilon})\cdot[\mathbf{x}^{(0)}_{m-\varepsilon}-\mathbf{x}^{(j)}_{m-\varepsilon}]>0\\ \text{or }\ \mathbf{n}^{\mathrm{T}}_{\partial\Omega_{ij}}(\mathbf{x}^{(0)}_{m+\varepsilon})\cdot[\mathbf{x}^{(j)}_{m+\varepsilon}-\mathbf{x}^{(0)}_{m+\varepsilon}]<0\\ \mathbf{n}^{\mathrm{T}}_{\partial\Omega_{ij}}(\mathbf{x}^{(0)}_{m-\varepsilon})\cdot[\mathbf{x}^{(0)}_{m-\varepsilon}-\mathbf{x}^{(i)}_{m-\varepsilon}]<0\end{array}\right\}\text{for }\mathbf{n}_{\partial\Omega_{ij}}\to\Omega_i. \tag{3.133}$$

Theorem 3.16. *For a discontinuous dynamical system in Eq.* (2.1)*, there is a point* $\mathbf{x}^{(0)}(t_m)\equiv\mathbf{x}_m\in\partial\Omega_{ij}$ *at time* t_m *between two adjacent domains* Ω_α ($\alpha=i,j$)*. Suppose* $\mathbf{x}^{(i)}(t_{m-})=\mathbf{x}_m=\mathbf{x}^{(j)}(t_{m\pm})$*. For an arbitrarily small* $\varepsilon>0$*, there are two time intervals* $[t_{m-\varepsilon},t_m)$ *and* $(t_m,t_{m+\varepsilon}]$*. A flow* $\mathbf{x}^{(i)}(t)$ *is* $C^{r_i}_{[t_{m-\varepsilon},t_m)}$*-continuous* ($r_i\ge 2k_i+1$) *and* $\|d^{r_i+1}\mathbf{x}^{(i)}/dt^{r_i+1}\|<\infty$ *for time t, and a flow* $\mathbf{x}^{(j)}(t)$ *is* $C^{r_j}_{[t_{m-\varepsilon},t_{m+\varepsilon}]}$*-continuous* ($r_j\ge 2k_j+1$) *and* $\|d^{r_j+1}\mathbf{x}^{(j)}/dt^{r_j+1}\|<\infty$*. The sliding bifurcation of the* $(2k_i:2k_j)$*-passable flow of* $\mathbf{x}^{(i)}(t)$ *and* $\mathbf{x}^{(j)}(t)$ *at point* $(\mathbf{x}_m,t_m)$ *switching to the* $(2k_i:2k_j)$ *-non-passable flow of the first kind on the boundary* $\overrightarrow{\partial\Omega}_{ij}$ *(a* $(2k_i:2k_j)$ *-sliding bifurcation) occurs if and only if*

$$\left.\begin{array}{l}G^{(s_j,j)}_{\partial\Omega_{ij}}(\mathbf{x}_m,t_{m\pm},\mathbf{p}_j,\boldsymbol{\lambda})=0\text{ for }s_j=0,1,\cdots,2k_j-1,\\ G^{(s_i,i)}_{\partial\Omega_{ij}}(\mathbf{x}_m,t_{m-},\mathbf{p}_i,\boldsymbol{\lambda})=0\text{ for }s_i=0,1,\cdots,2k_i-1;\end{array}\right\} \tag{3.134}$$

$$\left.\begin{array}{l}G^{(2k_j,j)}_{\partial\Omega_{ij}}(\mathbf{x}_m,t_{m\pm},\mathbf{p}_j,\boldsymbol{\lambda})=0,\\ L^{(2k_i:2k_j)}_{ij}(\mathbf{x}_m,t_m,\mathbf{p}_i,\mathbf{p}_j,\boldsymbol{\lambda})=0,\\ {}_{G\min}L^{(2k_i:2k_j)}_{ij}(t_m)=0;\end{array}\right\} \tag{3.135}$$

$$\left.\begin{array}{l}\left.\begin{array}{l}G^{(2k_i,i)}_{\partial\Omega_{ij}}(\mathbf{x}_m,t_{m-},\mathbf{p}_i,\boldsymbol{\lambda})>0\text{ for }\mathbf{n}_{\partial\Omega_{ij}}\to\Omega_j,\\ G^{(2k_i,i)}_{\partial\Omega_{ij}}(\mathbf{x}_m,t_{m-},\mathbf{p}_i,\boldsymbol{\lambda})<0\text{ for }\mathbf{n}_{\partial\Omega_{ij}}\to\Omega_i;\end{array}\right\}\\ \left.\begin{array}{l}G^{(2k_j+1,j)}_{\partial\Omega_{ij}}(\mathbf{x}_m,t_{m\pm},\mathbf{p}_j,\boldsymbol{\lambda})>0\text{ for }\mathbf{n}_{\partial\Omega_{ij}}\to\Omega_j,\\ G^{(2k_j+1,j)}_{\partial\Omega_{ij}}(\mathbf{x}_m,t_{m\pm},\mathbf{p}_j,\boldsymbol{\lambda})<0\text{ for }\mathbf{n}_{\partial\Omega_{ij}}\to\Omega_i.\end{array}\right\}\end{array}\right. \tag{3.136}$$

Proof. The proof is the same as in the proofs of Theorem 3.1 and Theorem 3.2. This theorem can be proved. ■

Definition 3.31. For a discontinuous dynamical system in Eq. (2.1), there is a point $\mathbf{x}^{(0)}(t_m) \equiv \mathbf{x}_m \in \partial\Omega_{ij}$ at time t_m between two adjacent domains Ω_α ($\alpha = i, j$). Suppose $\mathbf{x}^{(i)}(t_{m\pm}) = \mathbf{x}_m = \mathbf{x}^{(j)}(t_{m+})$. For an arbitrarily small $\varepsilon > 0$, there are two time intervals $[t_{m-\varepsilon}, t_m)$ and $(t_m, t_{m+\varepsilon}]$. Both flows $\mathbf{x}^{(i)}(t)$ and $\mathbf{x}^{(j)}(t)$ are $C^{r_i}_{[t_{m-\varepsilon}, t_{m+\varepsilon}]}$ and $C^{r_j}_{[t_{m-\varepsilon}, t_m)}$-continuous ($r_\alpha \geq 2$ and $\alpha = i, j$) for time t, respectively, and $\| d^{r_\alpha+1}\mathbf{x}^{(\alpha)} / dt^{r_\alpha+1} \| < \infty$. The tangential bifurcation of the flow $\mathbf{x}^{(i)}(t)$ at point $(\mathbf{x}_m, t_m)$ on the boundary $\overrightarrow{\partial\Omega}_{ij}$ is termed a *switching bifurcation of the non-passable flow of the second kind* (or called a *source bifurcation*) if

$$\left.\begin{aligned} G^{(j)}_{\partial\Omega_{ij}}(\mathbf{x}_m, t_{m+}, \mathbf{p}_j, \boldsymbol{\lambda}) &\neq 0,\\ G^{(i)}_{\partial\Omega_{ij}}(\mathbf{x}_m, t_{m\pm}, \mathbf{p}_i, \boldsymbol{\lambda}) &= 0,\\ G^{(1,i)}_{\partial\Omega_{ij}}(\mathbf{x}_m, t_{m\pm}, \mathbf{p}_i, \boldsymbol{\lambda}) &\neq 0; \end{aligned}\right\} \tag{3.137}$$

$$\begin{aligned} &\text{either} \left.\begin{aligned} &\mathbf{n}^{\mathrm{T}}_{\partial\Omega_{ij}}(\mathbf{x}^{(0)}_{m-\varepsilon}) \cdot [\mathbf{x}^{(0)}_{m-\varepsilon} - \mathbf{x}^{(i)}_{m-\varepsilon}] > 0\\ &\mathbf{n}^{\mathrm{T}}_{\partial\Omega_{ij}}(\mathbf{x}^{(0)}_{m+\varepsilon}) \cdot [\mathbf{x}^{(i)}_{m+\varepsilon} - \mathbf{x}^{(0)}_{m+\varepsilon}] < 0\\ &\mathbf{n}^{\mathrm{T}}_{\partial\Omega_{ij}}(\mathbf{x}^{(0)}_{m+\varepsilon}) \cdot [\mathbf{x}^{(j)}_{m+\varepsilon} - \mathbf{x}^{(0)}_{m+\varepsilon}] > 0 \end{aligned}\right\} \text{for } \mathbf{n}_{\partial\Omega_{ij}} \to \Omega_j,\\ &\text{or} \quad \left.\begin{aligned} &\mathbf{n}^{\mathrm{T}}_{\partial\Omega_{ij}}(\mathbf{x}^{(0)}_{m-\varepsilon}) \cdot [\mathbf{x}^{(0)}_{m-\varepsilon} - \mathbf{x}^{(i)}_{m-\varepsilon}] < 0\\ &\mathbf{n}^{\mathrm{T}}_{\partial\Omega_{ij}}(\mathbf{x}^{(0)}_{m+\varepsilon}) \cdot [\mathbf{x}^{(i)}_{m+\varepsilon} - \mathbf{x}^{(0)}_{m+\varepsilon}] > 0\\ &\mathbf{n}^{\mathrm{T}}_{\partial\Omega_{ij}}(\mathbf{x}^{(0)}_{m+\varepsilon}) \cdot [\mathbf{x}^{(j)}_{m+\varepsilon} - \mathbf{x}^{(0)}_{m+\varepsilon}] < 0 \end{aligned}\right\} \text{for } \mathbf{n}_{\partial\Omega_{ij}} \to \Omega_i. \end{aligned} \tag{3.138}$$

Theorem 3.17. *For a discontinuous dynamical system in Eq.* (2.1), *there is a point* $\mathbf{x}^{(0)}(t_m) \equiv \mathbf{x}_m \in \partial\Omega_{ij}$ *at time* t_m *between two adjacent domains* Ω_α ($\alpha = i, j$). *Suppose* $\mathbf{x}^{(i)}(t_{m-}) = \mathbf{x}_m = \mathbf{x}^{(j)}(t_{m+})$. *For an arbitrarily small* $\varepsilon > 0$, *there are two time intervals* $[t_{m-\varepsilon}, t_m)$ *and* $(t_m, t_{m+\varepsilon}]$. *Both flows* $\mathbf{x}^{(i)}(t)$ *and* $\mathbf{x}^{(j)}(t)$ *are* $C^{r_i}_{[t_{m-\varepsilon}, t_{m+\varepsilon}]}$ *and* $C^{r_j}_{[t_{m-\varepsilon}, t_m)}$*-continuous for time* t *with* $\| d^{r_\alpha+1}\mathbf{x}^{(\alpha)} / dt^{r_\alpha+1} \| < \infty$ ($r_\alpha \geq 2$, $\alpha = i, j$). *The source bifurcation of the passable flow of* $\mathbf{x}^{(i)}(t)$ *and* $\mathbf{x}^{(j)}(t)$ *at point* $(\mathbf{x}_m, t_m)$ *switching to the non-passable flow of the second kind on the boundary* $\overrightarrow{\partial\Omega}_{ij}$ *occurs if and only if*

$$\begin{gathered} G^{(i)}_{\partial\Omega_{ij}}(\mathbf{x}_m, t_{m\pm}, \mathbf{p}_i, \boldsymbol{\lambda}) = 0, \ \ or\\ L_{ij}(\mathbf{x}_m, t_m, \mathbf{p}_i, \mathbf{p}_j, \boldsymbol{\lambda}) = 0, \ \ or\\ {}_{G\min}L_{ij}(t_m) = 0; \end{gathered} \tag{3.139}$$

$$\left.\begin{array}{l}G_{\partial\Omega_{ij}}^{(j)}(\mathbf{x}_m,t_{m+},\mathbf{p}_j,\boldsymbol{\lambda})>0\\ G_{\partial\Omega_{ij}}^{(1,i)}(\mathbf{x}_m,t_{m\pm},\mathbf{p}_i,\boldsymbol{\lambda})<0\end{array}\right\} \text{for } \mathbf{n}_{\partial\Omega_{ij}}\to\Omega_j,$$
$$\left.\begin{array}{l}G_{\partial\Omega_{ij}}^{(j)}(\mathbf{x}_m,t_{m+},\mathbf{p}_j,\boldsymbol{\lambda})<0\\ G_{\partial\Omega_{ij}}^{(1,i)}(\mathbf{x}_m,t_{m\pm},\mathbf{p}_i,\boldsymbol{\lambda})>0\end{array}\right\} \text{for } \mathbf{n}_{\partial\Omega_{ij}}\to\Omega_i. \tag{3.140}$$

Proof. The proof is the same as in the proofs of Theorem 3.1 and Theorem 3.2. This theorem can be proved. ■

Definition 3.32. For a discontinuous dynamical system in Eq. (2.1), there is a point $\mathbf{x}^{(0)}(t_m)\equiv\mathbf{x}_m\in\partial\Omega_{ij}$ at time t_m between two adjacent domains Ω_α ($\alpha=i,j$). Suppose $\mathbf{x}^{(i)}(t_{m\pm})=\mathbf{x}_m=\mathbf{x}^{(j)}(t_{m+})$. For an arbitrarily small $\varepsilon>0$, there are two time intervals $[t_{m-\varepsilon},t_m)$ and $(t_m,t_{m+\varepsilon}]$. A flow $\mathbf{x}^{(i)}(t)$ is $C^{r_i}_{[t_{m-\varepsilon},t_{m+\varepsilon}]}$-continuous ($r_i\geq 2k_i+2$) with $\|d^{r_i+1}\mathbf{x}^{(i)}/dt^{r_i+1}\|<\infty$ for time *t*, and a flow $\mathbf{x}^{(j)}(t)$ is $C^{r_j}_{(t_m,t_{m+\varepsilon}]}$-continuous for time *t* and $\|d^{r_j+1}\mathbf{x}^{(j)}/dt^{r_j+1}\|<\infty$ ($r_j\geq 2k_j+1$). The tangential bifurcation of the $(2k_i:2k_j)$-passable flow of $\mathbf{x}^{(i)}(t)$ and $\mathbf{x}^{(j)}(t)$ at point $(\mathbf{x}_m,t_m)$ on the boundary $\overrightarrow{\partial\Omega}_{ij}$ is termed a *switching bifurcation of the* $(2k_i:2k_j)$ *-non-passable flow of the second kind* (or called a $(2k_i:2k_j)$ *-source bifurcation*) if

$$\begin{array}{l}G_{\partial\Omega_{ij}}^{(r,i)}(\mathbf{x}_m,t_{m\pm},\mathbf{p}_i,\boldsymbol{\lambda})=0 \text{ for } r=0,1,\cdots,2k_i;\\ G_{\partial\Omega_{ij}}^{(r,j)}(\mathbf{x}_m,t_{m+},\mathbf{p}_j,\boldsymbol{\lambda})=0 \text{ for } r=0,1,\cdots,2k_j-1;\\ G_{\partial\Omega_{ij}}^{(2k_j,j)}(\mathbf{x}_m,t_{m+},\mathbf{p}_j,\boldsymbol{\lambda})\neq 0;\\ G_{\partial\Omega_{ij}}^{(2k_i+1,i)}(\mathbf{x}_m,t_{m\pm},\mathbf{p}_i,\boldsymbol{\lambda})\neq 0;\end{array} \tag{3.141}$$

$$\text{either }\left.\begin{array}{l}\mathbf{n}_{\partial\Omega_{ij}}^{\mathrm{T}}(\mathbf{x}_{m-\varepsilon}^{(0)})\cdot[\mathbf{x}_{m-\varepsilon}^{(0)}-\mathbf{x}_{m-\varepsilon}^{(i)}]>0\\ \mathbf{n}_{\partial\Omega_{ij}}^{\mathrm{T}}(\mathbf{x}_{m+\varepsilon}^{(0)})\cdot[\mathbf{x}_{m+\varepsilon}^{(i)}-\mathbf{x}_{m+\varepsilon}^{(0)}]<0\\ \mathbf{n}_{\partial\Omega_{ij}}^{\mathrm{T}}(\mathbf{x}_{m+\varepsilon}^{(0)})\cdot[\mathbf{x}_{m+\varepsilon}^{(j)}-\mathbf{x}_{m+\varepsilon}^{(0)}]>0\end{array}\right\} \text{for } \mathbf{n}_{\partial\Omega_{ij}}\to\Omega_j,$$
$$\text{or }\left.\begin{array}{l}\mathbf{n}_{\partial\Omega_{ij}}^{\mathrm{T}}(\mathbf{x}_{m-\varepsilon}^{(0)})\cdot[\mathbf{x}_{m-\varepsilon}^{(0)}-\mathbf{x}_{m-\varepsilon}^{(i)}]<0\\ \mathbf{n}_{\partial\Omega_{ij}}^{\mathrm{T}}(\mathbf{x}_{m+\varepsilon}^{(0)})\cdot[\mathbf{x}_{m+\varepsilon}^{(i)}-\mathbf{x}_{m+\varepsilon}^{(0)}]>0\\ \mathbf{n}_{\partial\Omega_{ij}}^{\mathrm{T}}(\mathbf{x}_{m+\varepsilon}^{(0)})\cdot[\mathbf{x}_{m+\varepsilon}^{(j)}-\mathbf{x}_{m+\varepsilon}^{(0)}]<0\end{array}\right\} \text{for } \mathbf{n}_{\partial\Omega_{ij}}\to\Omega_i. \tag{3.142}$$

Theorem 3.18. *For a discontinuous dynamical system in Eq.* (2.1), *there is a point*

$\mathbf{x}^{(0)}(t_m) \equiv \mathbf{x}_m \in \partial\Omega_{ij}$ at time t_m between two adjacent domains Ω_α ($\alpha = i, j$). Suppose $\mathbf{x}^{(i)}(t_{m\mp}) = \mathbf{x}_m = \mathbf{x}^{(j)}(t_{m+})$. For an arbitrarily small $\varepsilon > 0$, there are two time intervals $[t_{m-\varepsilon}, t_m)$ and $(t_m, t_{m+\varepsilon}]$. A flow $\mathbf{x}^{(i)}(t)$ is $C^{r_i}_{[t_{m-\varepsilon}, t_{m+\varepsilon}]}$-continuous ($r_i \geq 2k_i + 2$) with $\| d^{r_i+1}\mathbf{x}^{(i)} / dt^{r_i+1} \| < \infty$ for time t, and a flow $\mathbf{x}^{(j)}(t)$ is $C^{r_j}_{(t_m, t_{m+\varepsilon}]}$-continuous ($r_j \geq 2k_j + 1$) with $\| d^{r_j+1}\mathbf{x}^{(j)} / dt^{r_j+1} \| < \infty$. The source bifurcation of the $(2k_i : 2k_j)$-passable flow of $\mathbf{x}^{(i)}(t)$ and $\mathbf{x}^{(j)}(t)$ at point $(\mathbf{x}_m, t_m)$ switching to the $(2k_i : 2k_j)$-non-passable flow of the second kind on the boundary $\overrightarrow{\partial\Omega}_{ij}$ (or the $(2k_i : 2k_j)$-source bifurcation) occurs if and only if

$$\left.\begin{aligned} &G^{(r,j)}_{\partial\Omega_{ij}}(\mathbf{x}_m, t_{m+}, \mathbf{p}_j, \boldsymbol{\lambda}) = 0 \ \textit{for}\ r = 0, 1, \cdots, 2k_j - 1, \\ &G^{(r,i)}_{\partial\Omega_{ij}}(\mathbf{x}_m, t_{m\pm}, \mathbf{p}_i, \boldsymbol{\lambda}) = 0 \ \textit{for}\ r = 0, 1, \cdots, 2k_i - 1; \end{aligned}\right. \tag{3.143}$$

$$\left.\begin{aligned} &G^{(2k_i,i)}_{\partial\Omega_{ij}}(\mathbf{x}_m, t_{m\pm}, \mathbf{p}_i, \boldsymbol{\lambda}) = 0, \\ &L^{(2k_i:2k_j)}_{ij}(\mathbf{x}_m, t_m, \mathbf{p}_i, \mathbf{p}_j, \boldsymbol{\lambda}) = 0, \\ &{}_{G\min} L^{(2k_i:2k_j)}_{ij}(t_m) = 0; \end{aligned}\right\} \tag{3.144}$$

$$\begin{aligned} &\left.\begin{aligned} &G^{(2k_j,j)}_{\partial\Omega_{ij}}(\mathbf{x}_m, t_{m+}, \mathbf{p}_j, \boldsymbol{\lambda}) > 0 \\ &G^{(2k_i+1,i)}_{\partial\Omega_{ij}}(\mathbf{x}_m, t_{m\pm}, \mathbf{p}_i, \boldsymbol{\lambda}) < 0 \end{aligned}\right\} \textit{for}\ \mathbf{n}_{\partial\Omega_{ij}} \to \Omega_j, \\ &\left.\begin{aligned} &G^{(2k_j,j)}_{\partial\Omega_{ij}}(\mathbf{x}_m, t_{m+}, \mathbf{p}_j, \boldsymbol{\lambda}) < 0 \\ &G^{(2k_i+1,i)}_{\partial\Omega_{ij}}(\mathbf{x}_m, t_{m\pm}, \mathbf{p}_i, \boldsymbol{\lambda}) > 0 \end{aligned}\right\} \textit{for}\ \mathbf{n}_{\partial\Omega_{ij}} \to \Omega_i. \end{aligned} \tag{3.145}$$

Proof. The proof is the same as in the proofs of Theorem 3.1 and Theorem 3.2. This theorem can be proved. ∎

Definition 3.33. For a discontinuous dynamical system in Eq. (2.1), there is a point $\mathbf{x}^{(0)}(t_m) \equiv \mathbf{x}_m \in \partial\Omega_{ij}$ at time t_m between two adjacent domains Ω_α ($\alpha = i, j$). Suppose $\mathbf{x}^{(i)}(t_{m-}) = \mathbf{x}_m = \mathbf{x}^{(j)}(t_{m+})$. For an arbitrarily small $\varepsilon > 0$, there are two time intervals $[t_{m-\varepsilon}, t_m)$ and $(t_m, t_{m+\varepsilon}]$. Both flows $\mathbf{x}^{(i)}(t)$ and $\mathbf{x}^{(j)}(t)$ are $C^{r_i}_{[t_{m-\varepsilon}, t_m)}$ and $C^{r_j}_{[t_{m-\varepsilon}, t_{m+\varepsilon}]}$-continuous for time *t*, and $\| d^{r_\alpha+1}\mathbf{x}^{(\alpha)} / dt^{r_\alpha+1} \| < \infty$ ($r_\alpha \geq 2, \alpha = i, j$). The tangential bifurcations of two flows $\mathbf{x}^{(i)}(t)$ and $\mathbf{x}^{(j)}(t)$ at point $(\mathbf{x}_m, t_m)$ on the boundary $\overrightarrow{\partial\Omega}_{ij}$ are termed a *switching bifurcation of the flow from* $\overrightarrow{\partial\Omega}_{ij}$ *to* $\overleftarrow{\partial\Omega}_{ij}$ if

$$
\begin{aligned}
&G^{(i)}_{\partial\Omega_{ij}}(\mathbf{x}_m, t_{m-}, \mathbf{p}_i, \boldsymbol{\lambda}) = 0 \text{ and } G^{(j)}_{\partial\Omega_{ij}}(\mathbf{x}_m, t_{m+}, \mathbf{p}_j, \boldsymbol{\lambda}) = 0,\\
&G^{(1,i)}_{\partial\Omega_{ij}}(\mathbf{x}_m, t_{m-}, \mathbf{p}_i, \boldsymbol{\lambda}) \neq 0 \text{ and } G^{(1,j)}_{\partial\Omega_{ij}}(\mathbf{x}_m, t_{m+}, \mathbf{p}_j, \boldsymbol{\lambda}) \neq 0;
\end{aligned}
\tag{3.146}
$$

$$
\begin{aligned}
&\text{either } \left.\begin{aligned}
&\mathbf{n}^{\mathrm{T}}_{\partial\Omega_{ij}}(\mathbf{x}^{(0)}_{m-\varepsilon})\cdot[\mathbf{x}^{(0)}_{m-\varepsilon} - \mathbf{x}^{(i)}_{m-\varepsilon}] > 0\\
&\mathbf{n}^{\mathrm{T}}_{\partial\Omega_{ij}}(\mathbf{x}^{(0)}_{m+\varepsilon})\cdot[\mathbf{x}^{(i)}_{m+\varepsilon} - \mathbf{x}^{(0)}_{m+\varepsilon}] < 0\\
&\mathbf{n}^{\mathrm{T}}_{\partial\Omega_{ij}}(\mathbf{x}^{(0)}_{m-\varepsilon})\cdot[\mathbf{x}^{(0)}_{m-\varepsilon} - \mathbf{x}^{(j)}_{m-\varepsilon}] < 0\\
&\mathbf{n}^{\mathrm{T}}_{\partial\Omega_{ij}}(\mathbf{x}^{(0)}_{m+\varepsilon})\cdot[\mathbf{x}^{(j)}_{m+\varepsilon} - \mathbf{x}^{(0)}_{m+\varepsilon}] > 0
\end{aligned}\right\} \text{for } \mathbf{n}_{\partial\Omega_{ij}} \to \Omega_j,\\
&\text{or } \left.\begin{aligned}
&\mathbf{n}^{\mathrm{T}}_{\partial\Omega_{ij}}(\mathbf{x}^{(0)}_{m-\varepsilon})\cdot[\mathbf{x}^{(0)}_{m-\varepsilon} - \mathbf{x}^{(i)}_{m-\varepsilon}] < 0\\
&\mathbf{n}^{\mathrm{T}}_{\partial\Omega_{ij}}(\mathbf{x}^{(0)}_{m+\varepsilon})\cdot[\mathbf{x}^{(i)}_{m+\varepsilon} - \mathbf{x}^{(0)}_{m+\varepsilon}] > 0\\
&\mathbf{n}^{\mathrm{T}}_{\partial\Omega_{ij}}(\mathbf{x}^{(0)}_{m-\varepsilon})\cdot[\mathbf{x}^{(0)}_{m-\varepsilon} - \mathbf{x}^{(j)}_{m-\varepsilon}] > 0\\
&\mathbf{n}^{\mathrm{T}}_{\partial\Omega_{ij}}(\mathbf{x}^{(0)}_{m+\varepsilon})\cdot[\mathbf{x}^{(j)}_{m+\varepsilon} - \mathbf{x}^{(0)}_{m+\varepsilon}] < 0
\end{aligned}\right\} \text{for } \mathbf{n}_{\partial\Omega_{ij}} \to \Omega_i.
\end{aligned}
\tag{3.147}
$$

Theorem 3.19. *For a discontinuous dynamical system in Eq.* (2.1), *there is a point* $\mathbf{x}^{(0)}(t_m) \equiv \mathbf{x}_m \in \partial\Omega_{ij}$ *at time* t_m *between two adjacent domains* Ω_α ($\alpha = i, j$). *Suppose* $\mathbf{x}^{(i)}(t_{m\pm}) = \mathbf{x}_m = \mathbf{x}^{(j)}(t_{m\pm})$. *For an arbitrarily small* $\varepsilon > 0$, *there is a time interval* $[t_{m-\varepsilon}, t_{m+\varepsilon}]$. *The flow* $\mathbf{x}^{(\alpha)}(t)$ *is* $C^{r_\alpha}_{[t_{m-\varepsilon}, t_{m+\varepsilon}]}$*-continuous for time* t *with* $\| d^{r_\alpha+1}\mathbf{x}^{(\alpha)} / dt^{r_\alpha+1} \| < \infty$ ($r_\alpha \geq 3$, $\alpha = i, j$). *The tangential bifurcations of the flow* $\mathbf{x}^{(i)}(t)$ *and* $\mathbf{x}^{(j)}(t)$ *at point* $(\mathbf{x}_m, t_m)$ *on the boundary* $\overrightarrow{\partial\Omega_{ij}}$ (or *the switching bifurcation of the flow from* $\overrightarrow{\partial\Omega_{ij}}$ to $\overleftarrow{\partial\Omega_{ij}}$) *occur if and only if*

$$
\begin{aligned}
&G^{(i)}_{\partial\Omega_{ij}}(\mathbf{x}_m, t_{m-}, \mathbf{p}_j, \boldsymbol{\lambda}) = 0 \textit{ and } G^{(j)}_{\partial\Omega_{ij}}(\mathbf{x}_m, t_{m+}, \mathbf{p}_i, \boldsymbol{\lambda}) = 0,\\
&L_{ij}(\mathbf{x}_{m_2}, t_{m_2}, \mathbf{p}_i, \mathbf{p}_j, \boldsymbol{\lambda}) = 0,\\
&{}_{G\min}L_{ij}(t_m) = 0;
\end{aligned}
\tag{3.148}
$$

$$
\begin{aligned}
&\left.\begin{aligned}
&G^{(1,i)}_{\partial\Omega_{ij}}(\mathbf{x}_m, t_{m-}, \mathbf{p}_i, \boldsymbol{\lambda}) < 0\\
&G^{(1,j)}_{\partial\Omega_{ij}}(\mathbf{x}_m, t_m, \mathbf{p}_j, \boldsymbol{\lambda}) > 0
\end{aligned}\right\} \text{for } \mathbf{n}_{\Omega_{ij}} \to \Omega_j,\\
&\left.\begin{aligned}
&G^{(1,i)}_{\partial\Omega_{ij}}(\mathbf{x}_m, t_{m-}, \mathbf{p}_i, \boldsymbol{\lambda}) > 0\\
&G^{(1,j)}_{\partial\Omega_{ij}}(\mathbf{x}_m, t_{m-}, \mathbf{p}_j, \boldsymbol{\lambda}) < 0
\end{aligned}\right\} \text{for } \mathbf{n}_{\Omega_{ij}} \to \Omega_i.
\end{aligned}
\tag{3.149}
$$

Proof. The proof is the same as in the proofs of Theorem 3.1 and Theorem 3.2. This theorem can be proved. ■

Definition 3.34. For a discontinuous dynamical system in Eq. (2.1), there is a point $\mathbf{x}^{(0)}(t_m) \equiv \mathbf{x}_m \in \partial\Omega_{ij}$ at time t_m between two adjacent domains Ω_α ($\alpha = i, j$). Suppose $\mathbf{x}^{(i)}(t_{m\pm}) = \mathbf{x}_m = \mathbf{x}^{(j)}(t_{m\pm})$. For an arbitrarily small $\varepsilon > 0$, there is a time interval $[t_{m-\varepsilon}, t_{m+\varepsilon}]$. A flow $\mathbf{x}^{(\alpha)}(t)$ is $C^{r_\alpha}_{[t_{m-\varepsilon}, t_{m+\varepsilon}]}$-continuous for time t with $\| d^{r_\alpha+1}\mathbf{x}^{(\alpha)} / dt^{r_\alpha+1} \| < \infty$ ($r_\alpha \geq 2k_\alpha + 1$, $\alpha = i, j$). The tangential bifurcation of the $(2k_i : 2k_j)$-passable flow of $\mathbf{x}^{(i)}(t)$ and $\mathbf{x}^{(j)}(t)$ at point $(\mathbf{x}_m, t_m)$ on the boundary $\overrightarrow{\partial\Omega}_{ij}$ is termed a *switching bifurcation of the* $(2k_j : 2k_i)$*-passable flow from* $\overrightarrow{\partial\Omega}_{ij}$ *to* $\overleftarrow{\partial\Omega}_{ij}$ if

$$
\begin{aligned}
&G^{(s,i)}_{\partial\Omega_{ij}}(\mathbf{x}_m, t_{m-}, \mathbf{p}_i, \boldsymbol{\lambda}) = 0 \text{ for } s = 0, 1, \cdots, 2k_i,\\
&G^{(s,j)}_{\partial\Omega_{ij}}(\mathbf{x}_m, t_{m+}, \mathbf{p}_j, \boldsymbol{\lambda}) = 0 \text{ for } s = 0, 1, \cdots, 2k_j,\\
&G^{(2k_i+1,i)}_{\partial\Omega_{ij}}(\mathbf{x}_m, t_{m-}, \mathbf{p}_i, \boldsymbol{\lambda}) \neq 0,\\
&G^{(2k_j+1,j)}_{\partial\Omega_{ij}}(\mathbf{x}_m, t_{m+}, \mathbf{p}_j, \boldsymbol{\lambda}) \neq 0;
\end{aligned}
\tag{3.150}
$$

$$
\begin{aligned}
&\text{either } \left.\begin{aligned}
&\mathbf{n}^{\mathrm{T}}_{\partial\Omega_{ij}}(\mathbf{x}^{(0)}_{m-\varepsilon}) \cdot [\mathbf{x}^{(0)}_{m-\varepsilon} - \mathbf{x}^{(i)}_{m-\varepsilon}] > 0\\
&\mathbf{n}^{\mathrm{T}}_{\partial\Omega_{ij}}(\mathbf{x}^{(0)}_{m+\varepsilon}) \cdot [\mathbf{x}^{(i)}_{m+\varepsilon} - \mathbf{x}^{(0)}_{m+\varepsilon}] < 0\\
&\mathbf{n}^{\mathrm{T}}_{\partial\Omega_{ij}}(\mathbf{x}^{(0)}_{m-\varepsilon}) \cdot [\mathbf{x}^{(0)}_{m-\varepsilon} - \mathbf{x}^{(j)}_{m-\varepsilon}] > 0\\
&\mathbf{n}^{\mathrm{T}}_{\partial\Omega_{ij}}(\mathbf{x}^{(0)}_{m+\varepsilon}) \cdot [\mathbf{x}^{(j)}_{m+\varepsilon} - \mathbf{x}^{(0)}_{m+\varepsilon}] > 0
\end{aligned}\right\} \text{for } \mathbf{n}_{\partial\Omega_{ij}} \to \Omega_j,\\
&\text{or } \left.\begin{aligned}
&\mathbf{n}^{\mathrm{T}}_{\partial\Omega_{ij}}(\mathbf{x}^{(0)}_{m-\varepsilon}) \cdot [\mathbf{x}^{(0)}_{m-\varepsilon} - \mathbf{x}^{(i)}_{m-\varepsilon}] < 0\\
&\mathbf{n}^{\mathrm{T}}_{\partial\Omega_{ij}}(\mathbf{x}^{(0)}_{m+\varepsilon}) \cdot [\mathbf{x}^{(i)}_{m+\varepsilon} - \mathbf{x}^{(0)}_{m+\varepsilon}] > 0\\
&\mathbf{n}^{\mathrm{T}}_{\partial\Omega_{ij}}(\mathbf{x}^{(0)}_{m-\varepsilon}) \cdot [\mathbf{x}^{(0)}_{m-\varepsilon} - \mathbf{x}^{(j)}_{m-\varepsilon}] < 0\\
&\mathbf{n}^{\mathrm{T}}_{\partial\Omega_{ij}}(\mathbf{x}^{(0)}_{m+\varepsilon}) \cdot [\mathbf{x}^{(j)}_{m+\varepsilon} - \mathbf{x}^{(0)}_{m+\varepsilon}] < 0
\end{aligned}\right\} \text{for } \mathbf{n}_{\partial\Omega_{ij}} \to \Omega_i.
\end{aligned}
\tag{3.151}
$$

Theorem 3.20. *For a discontinuous dynamical system in Eq.* (2.1), *there is a point* $\mathbf{x}^{(0)}(t_m) \equiv \mathbf{x}_m \in \partial\Omega_{ij}$ *at time* t_m *between two adjacent domains* Ω_α ($\alpha = i, j$). *Suppose* $\mathbf{x}^{(i)}(t_{m-}) = \mathbf{x}_m = \mathbf{x}^{(j)}(t_{m+})$. *For an arbitrarily small* $\varepsilon > 0$, *there are two time intervals* $[t_{m-\varepsilon}, t_m)$ *and* $(t_m, t_{m+\varepsilon}]$. *A flow* $\mathbf{x}^{(i)}(t)$ *is* $C^{r_i}_{[t_{m-\varepsilon}, t_m)}$*-continuous* ($r_i \geq 2k_i + 1$) *with* $\| d^{r_i+1}\mathbf{x}^{(i)} / dt^{r_i+1} \| < \infty$ *for time t, and a flow* $\mathbf{x}^{(j)}(t)$ *is* $C^{r_j}_{(t_m, t_{m+\varepsilon}]}$*-continuous* ($r_j \geq 2k_j + 2$) *with* $\| d^{r_j+1}\mathbf{x}^{(j)} / dt^{r_j+1} \| < \infty$. *The bifurcation of the* $(2k_i : 2k_j)$*-passable flow of* $\mathbf{x}^{(i)}(t)$ *and* $\mathbf{x}^{(j)}(t)$ *at point* $(\mathbf{x}_m, t_m)$ *switching to the*

$(2k_i:2k_j)$*-non-passable flow of the second kind on the boundary* $\overrightarrow{\partial\Omega}_{ij}$ *(or the switching bifurcation of the* $(2k_j:2k_i)$*-passable flow from* $\overrightarrow{\partial\Omega}_{ij}$ to $\overleftarrow{\partial\Omega}_{ij}$*) occurs if and only if*

$$\left.\begin{aligned} &G^{(s,j)}_{\partial\Omega_{ij}}(\mathbf{x}_m,t_{m+},\mathbf{p}_j,\boldsymbol{\lambda})=0 \ \textit{for}\ s=0,1,\cdots,2k_j-1,\\ &G^{(s,i)}_{\partial\Omega_{ij}}(\mathbf{x}_m,t_{m-},\mathbf{p}_i,\boldsymbol{\lambda})=0 \ \textit{for}\ s=0,1,\cdots,2k_i-1; \end{aligned}\right\} \tag{3.152}$$

$$\left.\begin{aligned} &G^{(2k_i,i)}_{\partial\Omega_{ij}}(\mathbf{x}_m,t_{m\pm},\mathbf{p}_i,\boldsymbol{\lambda})=0 \ \textit{and}\ G^{(2k_j,j)}_{\partial\Omega_{ij}}(\mathbf{x}_m,t_{m\pm},\mathbf{p}_j,\boldsymbol{\lambda})=0,\ \textit{or}\\ &L^{(2k_i:2k_j)}_{ij}(\mathbf{x}_m,t_m,\mathbf{p}_i,\mathbf{p}_j,\boldsymbol{\lambda})=0,\\ &{}_{G\min}L^{(2k_i:2k_j)}_{ij}(t_m)=0; \end{aligned}\right\} \tag{3.153}$$

$$\left.\begin{aligned} &\left.\begin{aligned} &G^{(2k_i+1,i)}_{\partial\Omega_{ij}}(\mathbf{x}_m,t_{m\pm},\mathbf{p}_i,\boldsymbol{\lambda})<0\\ &G^{(2k_j+1,j)}_{\partial\Omega_{ij}}(\mathbf{x}_m,t_{m\pm},\mathbf{p}_j,\boldsymbol{\lambda})>0 \end{aligned}\right\} \textit{for}\ \mathbf{n}_{\partial\Omega_{ij}}\to\Omega_j,\\ &\left.\begin{aligned} &G^{(2k_i+1,i)}_{\partial\Omega_{ij}}(\mathbf{x}_m,t_{m\mp},\mathbf{p}_i,\boldsymbol{\lambda})>0\\ &G^{(2k_j+1,j)}_{\partial\Omega_{ij}}(\mathbf{x}_m,t_{m\mp},\mathbf{p}_j,\boldsymbol{\lambda})<0 \end{aligned}\right\} \textit{for}\ \mathbf{n}_{\partial\Omega_{ij}}\to\Omega_i. \end{aligned}\right. \tag{3.154}$$

Proof. The proof is the same as in the proofs of Theorem 3.1 and Theorem 3.2. This theorem can be proved. ■

Following Definitions 3.27-3.34, the sliding and source fragmentation bifurcations can be similarly defined.

Definition 3.35. For a discontinuous dynamical system in Eq. (2.1), there is a point $\mathbf{x}^{(0)}(t_m)\equiv\mathbf{x}_m\in\partial\Omega_{ij}$ at time t_m between two adjacent domains Ω_α $(\alpha=i,j)$.

(i) The tangential bifurcation of the flow $\mathbf{x}^{(j)}(t)$ at point $(\mathbf{x}_m,t_m)$ on the boundary $\widetilde{\partial\Omega}_{ij}$ is termed a *fragmentation bifurcation of the non-passable flow of the first kind* (or called a *sliding fragmentation bifurcation*) if Eqs. (3.128) and (3.129) hold.

(ii) The tangential bifurcation of the flow $\mathbf{x}^{(i)}(t)$ with the $(2k_i)$th -order and $\mathbf{x}^{(j)}(t)$ of the $(2k_j)$th -order at point $(\mathbf{x}_m,t_m)$ on the boundary $\widetilde{\partial\Omega}_{ij}$ is termed a *fragmentation bifurcation of the* $(2k_i:2k_j)$*-non-passable flow of the first kind* (or called *a* $(2k_i:2k_j)$*-sliding fragmentation bifurcation*) if Eqs. (3.132) and (3.133) hold.

The necessary and sufficient conditions for the sliding fragmentation bifurcation of the non-passable flow of the first kind are given by Eqs. (3.130) and (3.131) with ${}_{G\max}L_{ij}(t_m)$ replacing ${}_{G\min}L_{ij}(t_m)$. Similarly, the necessary and sufficient conditions for the sliding fragmentation bifurcation of the $(2k_i:2k_j)$-non-passable flow of the first kind are presented by Eqs. (3.134)-(3.136) with ${}_{G\max}L_{ij}^{(2k_i:2k_j)}(t_m)$ replacing ${}_{G\min}L_{ij}^{(2k_i:2k_j)}(t_m)$.

Definition 3.36. For a discontinuous dynamical system in Eq. (2.1), there is a point $\mathbf{x}^{(0)}(t_m) \equiv \mathbf{x}_m \in \partial\Omega_{ij}$ at time t_m between two adjacent domains Ω_α $(\alpha = i, j)$.

(i) The tangential bifurcation of the flow $\mathbf{x}^{(j)}(t)$ at point $(\mathbf{x}_m, t_m)$ on the boundary $\widehat{\partial\Omega}_{ij}$ is termed a *fragmentation bifurcation of the non-passable flow of the second kind* (or called a *source fragmentation bifurcation*) if Eqs. (3.137) and (3.138) hold.

(ii) The tangential bifurcation of the flow $\mathbf{x}^{(i)}(t)$ with the $(2k_i)$th-order and $\mathbf{x}^{(j)}(t)$ of the $(2k_j)$th-order at point $(\mathbf{x}_m, t_m)$ on the boundary $\widehat{\partial\Omega}_{ij}$ is termed the *fragmentation bifurcation of the* $(2k_i:2k_j)$*-non-passable flow of the second kind* (or called a $(2k_i:2k_j)$*-source fragmentation bifurcation*) if Eqs. (3.141) and (3.142) hold.

The necessary and sufficient conditions for the source fragmentation bifurcation of the non-passable flow of the second kind are given by Eqs. (3.139) and (3.140) with ${}_{G\max}L_{ij}(t_m)$ replacing ${}_{G\min}L_{ij}(t_m)$. Similarly, the necessary and sufficient conditions for the sliding fragmentation bifurcation of the $(2k_i:2k_j)$-non-passable flow of the second kind are presented by Eqs. (3.143)-(3.145) with ${}_{G\max}L_{ij}^{(2k_i:2k_j)}(t_m)$ replacing ${}_{G\min}L_{ij}^{(2k_i:2k_j)}(t_m)$.

Definition 3.37. For a discontinuous dynamical system in Eq. (2.1), there is a point $\mathbf{x}^{(0)}(t_m) \equiv \mathbf{x}_m \in \partial\Omega_{ij}$ at time t_m between two adjacent domains Ω_α $(\alpha = i, j)$.

(i) The tangential bifurcation of the flow $\mathbf{x}^{(i)}(t)$ and $\mathbf{x}^{(j)}(t)$ at point $(\mathbf{x}_m, t_m)$ on the boundary $\widetilde{\partial\Omega}_{ij}$ (or $\widehat{\partial\Omega}_{ij}$) is termed a *switching bifurcation of the non-passable flow from* $\widetilde{\partial\Omega}_{ij}$ *to* $\widehat{\partial\Omega}_{ij}$ (or *from* $\widehat{\partial\Omega}_{ij}$ to $\widetilde{\partial\Omega}_{ij}$) if Eqs. (3.146) and (3.147) hold.

(ii) The tangential bifurcation of the flow $\mathbf{x}^{(i)}(t)$ with the $(2k_i)$th -order and $\mathbf{x}^{(j)}(t)$ with the $(2k_j)$th -order at point $(\mathbf{x}_m, t_m)$ on the boundary $\widetilde{\partial\Omega}_{ij}$ (or $\widehat{\partial\Omega}_{ij}$) is termed a *switching bifurcation of the* $(2k_i : 2k_j)$ *-non-passable flow from* $\widetilde{\partial\Omega}_{ij}$ to $\widehat{\partial\Omega}_{ij}$ (or from $\widehat{\partial\Omega}_{ij}$ to $\widetilde{\partial\Omega}_{ij}$) if Eqs. (3.150) and (3.151) hold.

The necessary and sufficient conditions for the switching bifurcation of a non - passable flow from $\widetilde{\partial\Omega}_{ij}$ to $\widehat{\partial\Omega}_{ij}$ (or from $\widehat{\partial\Omega}_{ij}$ to $\widetilde{\partial\Omega}_{ij}$) are from Eqs. (3.148) and (3.149) with ${}_{G\max}L_{ij}(t_m)$ replacing ${}_{G\min}L_{ij}(t_m)$, However, the conditions for the switching bifurcation of the $(2k_i : 2k_j)$ -non-passable flow from $\widetilde{\partial\Omega}_{ij}$ to $\widehat{\partial\Omega}_{ij}$ (or from $\widehat{\partial\Omega}_{ij}$ to $\widetilde{\partial\Omega}_{ij}$) the second kind are presented by Eqs. (3.152)-(3.154) with ${}_{G\max}L_{ij}^{(2k_i:2k_j)}(t_m)$ replacing ${}_{G\min}L_{ij}^{(2k_i:2k_j)}(t_m)$. The above conditions for the switching bifurcations of the $(2k_\alpha : 2k_\beta)$ -flows are summarized in Table 3.1. The following notations are used for simplicity.

$$L_{\alpha\beta}^{(m_\alpha : m_\beta)} \equiv L_{\alpha\beta}^{(m_\alpha : m_\beta)}(\mathbf{x}_m, t_m, \mathbf{p}_\alpha, \mathbf{p}_\beta, \boldsymbol{\lambda}), \tag{3.155}$$

$$\begin{aligned} G_{\pm}^{(m_\alpha,\alpha)} &\equiv G_{\partial\Omega_{\alpha\beta}}^{(m_\alpha,\alpha)}(\mathbf{x}_m, t_{m\pm}, \mathbf{p}_\alpha, \boldsymbol{\lambda}), \\ G_{-}^{(m_\alpha,\alpha)} &\equiv G_{\partial\Omega_{\alpha\beta}}^{(m_\alpha,\alpha)}(\mathbf{x}_m, t_{m-}, \mathbf{p}_\alpha, \boldsymbol{\lambda}), \\ G_{+}^{(m_\alpha,\alpha)} &\equiv G_{\partial\Omega_{\alpha\beta}}^{(m_\alpha,\alpha)}(\mathbf{x}_m, t_{m+}, \mathbf{p}_\alpha, \boldsymbol{\lambda}). \end{aligned} \tag{3.156}$$

Because the concept of the imaginary flow is introduced, the switching bifurcations of the $(2k_\alpha : 2k_\beta - 1)$, $(2k_\alpha - 1 : 2k_\beta)$ and $(2k_\alpha - 1 : 2k_\beta - 1)$ -flows can follow the discussion on the switching bifurcation of the $(2k_\alpha : 2k_\beta)$ -flows. The corresponding necessary and sufficient conditions for such flows can be obtained. For simplicity, the conditions for $(2k_\alpha : 2k_\beta - 1)$, $(2k_\alpha - 1 : 2k_\beta)$ and $(2k_\alpha - 1 : 2k_\beta - 1)$ flows are summarized in Tables 3.2-3.6, respectively. The switching bifurcations between a passable flow and half-non-passable flow, and between a passable flow and single tangential flow are presented. In addition, the switching bifurcations between a half-non-passable flow to a single tangential flow, and between a double tangential flow and a double inaccessible flow are given. All the conditions can help us understand the complexity in discontinuous dynamical systems.

Table 3.1 Sufficient and necessary conditions for $(2k_\alpha : 2k_\beta)$-switching bifurcations

$(2k_\alpha : 2k_\beta)$ passable flows	$\underset{G_-^{(2k_\beta)}=0,\text{ or }L_{\alpha\beta}^{(2k_\alpha:2k_\beta)}=0}{\overset{G_+^{(2k_\beta)}=0,\text{ or }L_{\alpha\beta}^{(2k_\alpha:2k_\beta)}=0}{\rightleftarrows}}$	$(2k_\alpha : 2k_\beta)$ full sink flows
$G_-^{(2k_\alpha)} > 0\,, G_+^{(2k_\beta)} > 0$ $L_{\alpha\beta}^{(2k_\alpha:2k_\beta)} > 0$	$\left.\begin{array}{l} G_\pm^{(2k_\beta+1)} > 0 \text{ for } \mathbf{n}_{\partial\Omega_{\alpha\beta}} \to \Omega_\beta; \\ G_\pm^{(2k_\beta+1)} < 0 \text{ for } \mathbf{n}_{\partial\Omega_{\alpha\beta}} \to \Omega_\alpha. \end{array}\right\}$	$G_-^{(2k_\alpha)} > 0,\ G_-^{(2k_\beta)} < 0$ $L_{\alpha\beta}^{(2k_\alpha:2k_\beta)} < 0$
$(2k_\alpha : 2k_\beta)$ passable flows	$\underset{G_+^{(2k_\alpha)}=0,\text{ or }L_{\alpha\beta}^{(2k_\alpha:2k_\beta)}=0}{\overset{G_-^{(2k_\alpha)}=0,\text{ or }L_{\alpha\beta}^{(2k_\alpha:2k_\beta)}=0}{\rightleftarrows}}$	$(2k_\alpha : 2k_\beta)$ full source flows
$G_-^{(2k_\alpha)} > 0\,, G_+^{(2k_\beta)} > 0$ $L_{\alpha\beta}^{(2k_\alpha:2k_\beta)} > 0$	$\left.\begin{array}{l} G_\pm^{(2k_\alpha+1)} < 0 \text{ for } \mathbf{n}_{\partial\Omega_{\alpha\beta}} \to \Omega_\beta; \\ G_\pm^{(2k_\alpha+1)} > 0 \text{ for } \mathbf{n}_{\partial\Omega_{\alpha\beta}} \to \Omega_\alpha \end{array}\right\}$	$G_+^{(2k_\alpha)} > 0,\ G_+^{(2k_\beta)} < 0$ $L_{\alpha\beta}^{(2k_\alpha:2k_\beta)} < 0$
$(2k_\alpha : 2k_\beta)$ passable flows	$\underset{G_+^{(2k_\alpha)}=0, G_-^{(2k_\beta)}=0;\text{ or }L_{\alpha\beta}^{(2k_\alpha:2k_\beta)}=0}{\overset{G_-^{(2k_\alpha)}=0, G_+^{(2k_\beta)}=0;\text{ or }L_{\alpha\beta}^{(2k_\alpha:2k_\beta)}=0,}{\rightleftarrows}}$	$(2k_\alpha : 2k_\beta)$ passable flows
$G_-^{(2k_\alpha)} > 0\,, G_+^{(2k_\beta)} > 0$ $L_{\alpha\beta}^{(2k_\alpha:2k_\beta)} > 0$	$\left.\begin{array}{l} G_\pm^{(2k_\alpha+1)} < 0, G_\pm^{(2k_\beta+1)} > 0 \text{ for } \mathbf{n}_{\partial\Omega_{\alpha\beta}} \to \Omega_\beta; \\ G_\pm^{(2k_\alpha+1)} > 0, G_\pm^{(2k_\beta+1)} < 0 \text{ for } \mathbf{n}_{\partial\Omega_{\alpha\beta}} \to \Omega_\alpha \end{array}\right\}$	$G_+^{(2k_\alpha)} > 0\,, G_-^{(2k_\beta)} > 0$ $L_{\alpha\beta}^{(2k_\alpha:2k_\beta)} > 0$
$(2k_\alpha : 2k_\beta)$ full sink flows	$\underset{G_+^{(2k_\alpha)}=0, G_+^{(2k_\beta)}=0;\text{ or }L_{\alpha\beta}^{(2k_\alpha:2k_\beta)}=0}{\overset{G_-^{(2k_\alpha)}=0, G_-^{(2k_\beta)}=0;\text{ or }L_{\alpha\beta}^{(2k_\alpha:2k_\beta)}=0,}{\rightleftarrows}}$	$(2k_\alpha : 2k_\beta)$ full source flows
$G_-^{(2k_\alpha)} > 0,\ G_-^{(2k_\beta)} < 0$ $L_{\alpha\beta}^{(2k_\alpha:2k_\beta)} < 0$	$\left.\begin{array}{l} G_\pm^{(2k_\alpha+1)} < 0, G_\pm^{(2k_\beta+1)} > 0 \text{ for } \mathbf{n}_{\partial\Omega_{\alpha\beta}} \to \Omega_\beta; \\ G_\pm^{(2k_\alpha+1)} > 0, G_\pm^{(2k_\beta+1)} < 0 \text{ for } \mathbf{n}_{\partial\Omega_{\alpha\beta}} \to \Omega_\alpha \end{array}\right\}$	$G_+^{(2k_\alpha)} > 0,\ G_+^{(2k_\beta)} < 0$ $L_{\alpha\beta}^{(2k_\alpha:2k_\beta)} < 0$

Table 3.2 Sufficient and necessary conditions for $(2k_\alpha : 2k_\beta - 1)$-switching bifurcations

$(2k_\alpha : 2k_\beta - 1)$ passable flows	$\xrightleftharpoons[G_{\pm}^{(2k_\beta-1)}=0,\ G_{\pm}^{(2k_\beta)}>0]{G_{+}^{(2k_\beta-1)}=0,\ G_{\pm}^{(2k_\beta)}<0}$	$(2k_\alpha : 2k_\beta - 1)$ half sink flows
$G_{-}^{(2k_\alpha)} > 0,\ G_{+}^{(2k_\beta-1)} > 0$ $L_{\alpha\beta}^{(2k_\alpha:2k_\beta-1)} > 0$	$L_{\alpha\beta}^{(2k_\alpha:2k_\beta-1)} = 0$	$G_{-}^{(2k_\alpha)} > 0,\ G_{\pm}^{(2k_\beta-1)} < 0$ $L_{\alpha\beta}^{(2k_\alpha:2k_\beta-1)} < 0$
$(2k_\alpha : 2k_\beta - 1)$ passable flows	$\xrightleftharpoons[G_{+}^{(2k_\beta)}=0,G_{\pm}^{(2k_\beta-1)}=0,G_{\pm}^{(2k_\beta)}>0]{G_{-}^{(2k_\alpha)}=0,\ G_{+}^{(2k_\beta-1)}=0,\ G_{\pm}^{(2k_\beta)}<0}$	$(2k_\alpha : 2k_\beta - 1)$ half source flows
$G_{-}^{(2k_\alpha)} > 0,\ G_{+}^{(2k_\beta-1)} > 0$ $L_{\alpha\beta}^{(2k_\alpha:2k_\beta-1)} > 0$	$L_{\alpha\beta}^{(2k_\alpha:2k_\beta-1)} = 0, G_{\pm}^{(2k_\alpha+1)} < 0$	$G_{+}^{(2k_\alpha)} < 0,\ G_{\pm}^{(2k_\beta-1)} < 0$ $L_{\alpha\beta}^{(2k_\alpha:2k_\beta-1)} < 0$
$(2k_\alpha : 2k_\beta - 1)$ passable flows	$\xrightleftharpoons[G_{+}^{(2k_\alpha)}=0]{G_{-}^{(2k_\alpha)}=0}$	$(2k_\alpha : 2k_\beta - 1)$ tangential flows
$G_{-}^{(2k_\alpha)} > 0,\ G_{+}^{(2k_\beta-1)} > 0$ $L_{\alpha\beta}^{(2k_\alpha:2k_\beta-1)} > 0$	$L_{\alpha\beta}^{(2k_\alpha:2k_\beta-1)} = 0, G_{\pm}^{(2k_\alpha+1)} < 0$	$G_{+}^{(2k_\alpha)} < 0,\ G_{\pm}^{(2k_\beta-1)} > 0$ $L_{\alpha\beta}^{(2k_\alpha:2k_\beta-1)} > 0$
$(2k_\alpha : 2k_\beta - 1)$ half sink flows	$\xrightleftharpoons[G_{+}^{(2k_\alpha)}=0]{G_{-}^{(2k_\alpha)}=0}$	$(2k_\alpha : 2k_\beta - 1)$ half source flows
$G_{-}^{(2k_\alpha)} > 0,\ G_{\pm}^{(2k_\beta-1)} < 0$ $L_{\alpha\beta}^{(2k_\alpha:2k_\beta-1)} < 0$	$L_{\alpha\beta}^{(2k_\alpha:2k_\beta)} = 0, G_{\pm}^{(2k_\alpha+1)} < 0$	$G_{+}^{(2k_\alpha)} < 0,\ G_{\pm}^{(2k_\beta-1)} < 0$ $L_{\alpha\beta}^{(2k_\alpha:2k_\beta-1)} < 0$

Table 3.3 Sufficient and necessary conditions for $(2k_\alpha : 2k_\beta - 1)$-switching bifurcations

$(2k_\alpha : 2k_\beta - 1)$ passable flows $G_-^{(2k_\alpha)} < 0, G_+^{(2k_\beta-1)} < 0$ $L_{\alpha\beta}^{(2k_\alpha:2k_\beta-1)} > 0$	$\underset{G_\pm^{(2k_\beta-1)}=0, G_\pm^{(2k_\beta)}<0}{\overset{G_+^{(2k_\beta-1)}=0, G_\pm^{(2k_\beta)}>0}{\rightleftarrows}}$ $L_{\alpha\beta}^{(2k_\alpha:2k_\beta-1)} = 0$	$(2k_\alpha : 2k_\beta - 1)$ half sink flows $G_-^{(2k_\alpha)} < 0, G_\pm^{(2k_\beta-1)} > 0$ $L_{\alpha\beta}^{(2k_\alpha:2k_\beta-1)} < 0$
$(2k_\alpha : 2k_\beta - 1)$ passable flows $G_-^{(2k_\alpha)} < 0, G_+^{(2k_\beta-1)} < 0$ $L_{\alpha\beta}^{(2k_\alpha:2k_\beta-1)} > 0$	$\underset{G_+^{(2k_\alpha)}=0, G_\pm^{(2k_\beta-1)}=0, G_\pm^{(2k_\beta)}<0}{\overset{G_-^{(2k_\alpha)}=0,\ G_+^{(2k_\beta-1)}=0, G_\pm^{(2k_\beta)}>0}{\rightleftarrows}}$ $L_{\alpha\beta}^{(2k_\alpha:2k_\beta-1)} = 0, G_\pm^{(2k_\alpha+1)} > 0$	$(2k_\alpha : 2k_\beta - 1)$ half source flows $G_+^{(2k_\alpha)} > 0, G_\pm^{(2k_\beta-1)} > 0$ $L_{\alpha\beta}^{(2k_\alpha:2k_\beta-1)} > 0$
$(2k_\alpha : 2k_\beta - 1)$ passable flows $G_-^{(2k_\alpha)} < 0, G_+^{(2k_\beta-1)} < 0$ $L_{\alpha\beta}^{(2k_\alpha:2k_\beta-1)} > 0$	$\underset{G_+^{(2k_\alpha)}=0,}{\overset{G_-^{(2k_\alpha)}=0,}{\rightleftarrows}}$ $L_{\alpha\beta}^{(2k_\alpha:2k_\beta-1)} = 0, G_\pm^{(2k_\alpha+1)} > 0$	$(2k_\alpha : 2k_\beta - 1)$ tangential flows $G_+^{(2k_\alpha)} > 0, G_\pm^{(2k_\beta-1)} < 0$ $L_{\alpha\beta}^{(2k_\alpha:2k_\beta-1)} < 0$
$(2k_\alpha : 2k_\beta - 1)$ half sink flows $G_-^{(2k_\alpha)} < 0, G_\pm^{(2k_\beta-1)} > 0$ $L_{\alpha\beta}^{(2k_\alpha:2k_\beta-1)} < 0$	$\underset{G_+^{(2k_\alpha)}=0, G_+^{(2k_\beta)}=0,}{\overset{G_-^{(2k_\alpha)}=0, G_-^{(2k_\beta)}=0}{\rightleftarrows}}$ $L_{\alpha\beta}^{(2k_\alpha:2k_\beta)} = 0, G_\pm^{(2k_\alpha+1)} > 0$	$(2k_\alpha : 2k_\beta - 1)$ half source flows $G_+^{(2k_\alpha)} > 0, G_\pm^{(2k_\beta-1)} > 0$ $L_{\alpha\beta}^{(2k_\alpha:2k_\beta-1)} > 0$

Table 3.4 Sufficient and necessary conditions for $(2k_\alpha - 1 : 2k_\beta)$-switching bifurcations

$(2k_\alpha - 1 : 2k_\beta)$ tangential flows $G_{\pm}^{(2k_\alpha - 1)} < 0\,, G_{+}^{(2k_\beta)} > 0$ $L_{\alpha\beta}^{(2k_\alpha - 1:2k_\beta)} < 0$	$\underset{G_{\pm}^{(2k_\beta - 1)} = 0, G_{-}^{(2k_\beta + 1)} > 0}{\overset{G_{+}^{(2k_\beta)} = 0, G_{\pm}^{(2k_\beta + 1)} < 0}{\rightleftarrows}}$ $L_{\alpha\beta}^{(2k_\alpha - 1:2k_\beta)} = 0$	$(2k_\alpha - 1 : 2k_\beta)$ tangential flows $G_{\pm}^{(2k_\alpha - 1)} < 0,\ G_{-}^{(2k_\beta)} < 0$ $L_{\alpha\beta}^{(2k_\alpha - 1:2k_\beta)} > 0$
$(2k_\alpha - 1 : 2k_\beta)$ tangential flows $G_{\pm}^{(2k_\alpha - 1)} < 0\,, G_{+}^{(2k_\beta)} > 0$ $L_{\alpha\beta}^{(2k_\alpha - 1:2k_\beta)} < 0$	$\underset{G_{\pm}^{(2k_\alpha - 1)} = 0, G_{-}^{(2k_\beta)} = 0; G_{-}^{(2k_\alpha)} < 0, G_{\pm}^{(2k_\beta + 1)} > 0}{\overset{G_{\pm}^{(2k_\alpha - 1)} = 0,\ G_{+}^{(2k_\beta)} = 0; G_{+}^{(2k_\alpha)} > 0, G_{\pm}^{(2k_\beta + 1)} < 0}{\rightleftarrows}}$ $L_{\alpha\beta}^{(2k_\alpha - 1:2k_\beta)} = 0,$	$(2k_\alpha - 1 : 2k_\beta)$ half sink flows $G_{\pm}^{(2k_\alpha - 1)} > 0,\ G_{-}^{(2k_\beta)} < 0$ $L_{\alpha\beta}^{(2k_\alpha - 1:2k_\beta)} < 0$
$(2k_\alpha - 1 : 2k_\beta)$ tangential flows $G_{\pm}^{(2k_\alpha - 1)} < 0\,, G_{+}^{(2k_\beta)} > 0$ $L_{\alpha\beta}^{(2k_\alpha - 1:2k_\beta)} < 0$	$\underset{G_{\pm}^{(2k_\alpha - 1)} = 0, G_{-}^{(2k_\alpha)} < 0}{\overset{G_{\pm}^{(2k_\alpha - 1)} = 0,\ G_{+}^{(2k_\alpha)} > 0}{\rightleftarrows}}$ $L_{\alpha\beta}^{(2k_\alpha - 1:2k_\beta)} = 0$	$(2k_\alpha - 1 : 2k_\beta)$ half source flows $G_{\pm}^{(2k_\alpha - 1)} > 0,\ G_{+}^{(2k_\beta)} > 0$ $L_{\alpha\beta}^{(2k_\alpha - 1:2k_\beta)} > 0$
$(2k_\alpha - 1 : 2k_\beta)$ half sink flows $G_{\pm}^{(2k_\alpha - 1)} > 0,\ G_{-}^{(2k_\beta)} < 0$ $L_{\alpha\beta}^{(2k_\alpha - 1:2k_\beta)} < 0$	$\underset{G_{+}^{(2k_\beta)} = 0; G_{\pm}^{(2k_\beta + 1)} < 0}{\overset{G_{-}^{(2k_\beta)} = 0;\ G_{\pm}^{(2k_\beta + 1)} > 0}{\rightleftarrows}}$ $L_{\alpha\beta}^{(2k_\alpha : 2k_\beta)} = 0,$	$(2k_\alpha - 1 : 2k_\beta)$ half source flows $G_{\pm}^{(2k_\alpha - 1)} > 0,\ G_{+}^{(2k_\beta)} > 0$ $L_{\alpha\beta}^{(2k_\alpha - 1:2k_\beta)} > 0$

Table 3.5 Sufficient and necessary conditions for $(2k_\alpha - 1 : 2k_\beta)$ -switching bifurcations

$(2k_\alpha - 1 : 2k_\beta)$ tangential flows $G_{\pm}^{(2k_\alpha - 1)} > 0, G_{-}^{(2k_\beta)} > 0$ $L_{\alpha\beta}^{(2k_\alpha - 1:2k_\beta)} > 0$	$\underset{G_{\pm}^{(2k_\beta)}=0;G_{-}^{(2k_\beta+1)}>0}{\overset{G_{+}^{(2k_\beta)}=0;G_{\pm}^{(2k_\beta+1)}<0}{\rightleftarrows}}$ $L_{\alpha\beta}^{(2k_\alpha - 1:2k_\beta)} = 0$	$(2k_\alpha - 1 : 2k_\beta)$ tangential flows $G_{\pm}^{(2k_\alpha - 1)} > 0,\ G_{+}^{(2k_\beta)} < 0$ $L_{\alpha\beta}^{(2k_\alpha - 1:2k_\beta)} < 0$
$(2k_\alpha - 1 : 2k_\beta)$ tangential flows $G_{\pm}^{(2k_\alpha - 1)} > 0, G_{+}^{(2k_\beta)} > 0$ $L_{\alpha\beta}^{(2k_\alpha - 1:2k_\beta)} > 0$	$\underset{G_{\pm}^{(2k_\alpha-1)}=0, G_{-}^{(2k_\beta)}=0;G_{-}^{(2k_\alpha)}>0, G_{\pm}^{(2k_\beta+1)}>0}{\overset{G_{\pm}^{(2k_\alpha-1)}=0,\ G_{+}^{(2k_\beta)}=0;G_{+}^{(2k_\alpha)}<0, G_{\pm}^{(2k_\beta+1)}<0}{\rightleftarrows}}$ $L_{\alpha\beta}^{(2k_\alpha :2k_\beta - 1)} = 0$	$(2k_\alpha - 1 : 2k_\beta)$ half source flows $G_{\pm}^{(2k_\alpha - 1)} < 0,\ G_{+}^{(2k_\beta)} < 0$ $L_{\alpha\beta}^{(2k_\alpha - 1:2k_\beta)} > 0$
$(2k_\alpha - 1 : 2k_\beta)$ tangential flows $G_{\pm}^{(2k_\alpha - 1)} > 0, G_{-}^{(2k_\beta)} > 0$ $L_{\alpha\beta}^{(2k_\alpha - 1:2k_\beta)} > 0$	$\underset{G_{\pm}^{(2k_\alpha-1)}=0, G_{-}^{(2k_\alpha)}>0}{\overset{G_{\pm}^{(2k_\alpha-1)}=0,\ G_{+}^{(2k_\alpha)}<0}{\rightleftarrows}}$ $L_{\alpha\beta}^{(2k_\alpha :2k_\beta - 1)} = 0$	$(2k_\alpha - 1 : 2k_\beta)$ half sink flows $G_{\pm}^{(2k_\alpha - 1)} < 0,\ G_{-}^{(2k_\beta)} > 0$ $L_{\alpha\beta}^{(2k_\alpha - 1:2k_\beta)} < 0$
$(2k_\alpha - 1 : 2k_\beta)$ half source flows $G_{\pm}^{(2k_\alpha - 1)} < 0,\ G_{+}^{(2k_\beta)} < 0$ $L_{\alpha\beta}^{(2k_\alpha - 1:2k_\beta)} > 0$	$\underset{G_{+}^{(2k_\beta)}=0;G_{\pm}^{(2k_\beta+1)}<0}{\overset{G_{-}^{(2k_\beta)}=0;G_{\pm}^{(2k_\beta+1)}>0}{\rightleftarrows}}$ $L_{\alpha\beta}^{(2k_\alpha :2k_\beta)} = 0,$	$(2k_\alpha - 1 : 2k_\beta)$ half sink flows $G_{\pm}^{(2k_\alpha - 1)} < 0,\ G_{-}^{(2k_\beta)} > 0$ $L_{\alpha\beta}^{(2k_\alpha - 1:2k_\beta)} < 0$

Table 3.6 Sufficient and necessary conditions for $(2k_i-1:2k_j-1)$ -switching bifurcations

$(2k_i-1:2k_j-1)$ double tangential flows $G_{\pm}^{(2k_i-1)}<0\,,G_{\pm}^{(2k_j-1)}>0$ $L_{ij}^{(2k_i-1:2k_j-1)}<0$	$\xrightleftharpoons[G_{\pm}^{(2k_i-1)}=0,G_{\pm}^{(2k_i)}<0;G_{\pm}^{(2k_j-1)}=0,G_{\pm}^{(2k_j)}>0]{G_{\pm}^{(2k_i-1)}=0,G_{\pm}^{(2k_i)}>0;G_{\pm}^{(2k_j-1)}=0,G_{\pm}^{(2k_j)}<0}$ $L_{ij}^{(2k_i-1:2k_i-1)}=0$	$(2k_i-1:2k_j-1)$ double inaccessible flows $G_{\pm}^{(2k_i-1)}>0,\ G_{\pm}^{(2k_j-1)}<0$ $L_{ij}^{(2k_i-1:2k_j-1)}<0$
$(2k_i-1:2k_j-1)$ double tangential flows $G_{\pm}^{(2k_i-1)}<0\,,G_{\pm}^{(2k_j-1)}>0$ $L_{ij}^{(2k_i-1:2k_j-1)}<0$	$\xrightleftharpoons[G_{\pm}^{(2k_j-1)}=0,G_{\pm}^{(2k_j)}>0]{G_{\pm}^{(2k_j-1)}=0,G_{\pm}^{(2k_j)}<0}$ $L_{ij}^{(2k_i-1:2k_i-1)}=0$	$(2k_i-1:2k_j-1)$ single tangential flows in Ω_i $G_{\pm}^{(2k_i-1)}<0,\ G_{\pm}^{(2k_j-1)}<0$ $L_{ij}^{(2k_i-1:2k_j-1)}>0$
$(2k_i-1:2k_j-1)$ double tangential flows $G_{\pm}^{(2k_i-1)}<0\,,G_{\pm}^{(2k_j-1)}>0$ $L_{ij}^{(2k_i-1:2k_j-1)}<0$	$\xrightleftharpoons[G_{\pm}^{(2k_i-1)}=0,G_{\pm}^{(2k_i)}<0]{G_{\pm}^{(2k_i-1)}=0,G_{\pm}^{(2k_i)}>0}$ $L_{ij}^{(2k_i-1:2k_i-1)}=0$	$(2k_i-1:2k_j-1)$ single tangential flows in Ω_j $G_{\pm}^{(2k_i-1)}>0,\ G_{\pm}^{(2k_j-1)}>0$ $L_{ij}^{(2k_i-1:2k_j-1)}>0$
$(2k_i-1:2k_j-1)$ single tangential flows in Ω_i $G_{\pm}^{(2k_i-1)}<0,\ G_{\pm}^{(2k_j-1)}<0$ $L_{ij}^{(2k_i-1:2k_j-1)}>0$	$\xrightleftharpoons[G_{\pm}^{(2k_i-1)}=0,G_{\pm}^{(2k_i)}<0;G_{\pm}^{(2k_j-1)}=0,G_{\pm}^{(2k_j)}>0]{G_{\pm}^{(2k_i-1)}=0,G_{\pm}^{(2k_i)}>0;G_{\pm}^{(2k_j-1)}=0,G_{\pm}^{(2k_j)}<0}$ $L_{ij}^{(2k_i-1:2k_i-1)}=0$	$(2k_i-1:2k_j-1)$ single tangential flows in Ω_j $G_{\pm}^{(2k_i-1)}>0,\ G_{\pm}^{(2k_j-1)}>0$ $L_{ij}^{(2k_i-1:2k_j-1)}>0$

3.7. First integral quantity increments

Consider the vector field in Eq. (2.1) for domain Ω_i as

$$\dot{\mathbf{x}}^{(i)} = \mathbf{F}^{(i)}(\mathbf{x}^{(i)}, t, \mathbf{p}_i) \equiv \mathbf{f}^{(i)}(\mathbf{x}^{(i)}, \boldsymbol{\mu}_i) + \mathbf{g}(\mathbf{x}^{(i)}, t, \boldsymbol{\pi}_i), \quad \mathbf{x}^{(i)} \in \Omega_i. \tag{3.157}$$

The time-independent and integrable dynamical system in domain Ω_i is

$$\dot{\overline{\mathbf{x}}}^{(i)} = \mathbf{f}^{(i)}(\overline{\mathbf{x}}^{(i)}, \boldsymbol{\mu}_i), \quad \overline{\mathbf{x}}^{(i)} \in \Omega_i. \tag{3.158}$$

As in Luo (2008), the first integral quantity is given by

$$F^{(i)}(\overline{\mathbf{x}}^{(i)}, \boldsymbol{\mu}_i) = E^{(i)}. \tag{3.159}$$

Once the first integrable quantity is determined for sub-domains, the flow in the corresponding sub-domains can be measured. The intuitive illustration is given in Fig. 3.18. $\mathbf{x}^{(i)}(t)$ and $\mathbf{x}^{(j)}(t)$, as depicted by the dark solid curves, represent flows in domains Ω_i and Ω_j, respectively. $\overline{\mathbf{x}}^{(i)}(t)$ and $\overline{\mathbf{x}}^{(j)}(t)$, as depicted by the light solid curves, represent integrable flows in domains Ω_i and Ω_j. The hollow circles are switching points and the shaded circles are starting and ending points. Because the dynamical system in each domain is continuous, the increment of the first integral quantity in each domain can be computed as in Luo (2008).

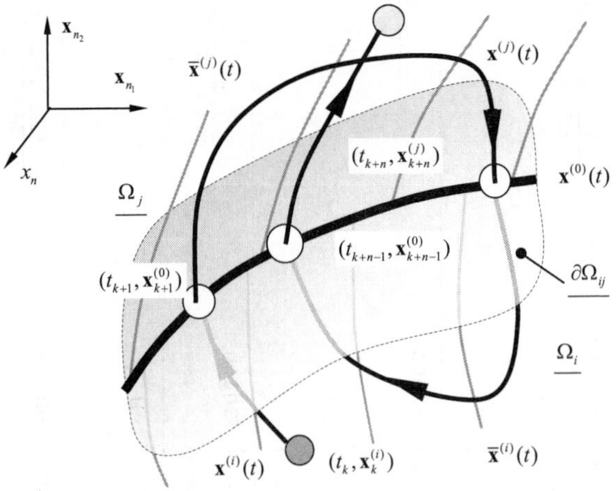

Fig. 3.18. The flow measured by different integral flows in the different domains. $\mathbf{x}^{(i)}(t)$ and $\mathbf{x}^{(j)}(t)$, as depicted by the dark solid curves, represent the flows in domains Ω_i and Ω_j, respectively. $\overline{\mathbf{x}}^{(i)}(t)$ and $\overline{\mathbf{x}}^{(j)}(t)$, as depicted by the light solid curves, represent the integrable flows in domains Ω_i and Ω_j. The hollow circles are the switching points and the shaded circles are the starting and ending points.

Definition 3.38. For time $t \in [t_k, t_{k+1}]$, the increment of the first integral quantity in domain Ω_α is defined as

$$L^{(\alpha)}(t_k, t_{k+1}) \equiv \int_{t_k}^{t_{k+1}} \nabla F(\mathbf{x}^{(\alpha)}, \boldsymbol{\mu}_\alpha) \cdot \mathbf{g}(\mathbf{x}^{(\alpha)}, t, \boldsymbol{\pi}_\alpha) dt. \tag{3.160}$$

If the investigated flow $\mathbf{x}^{(\alpha)}(t)$ for $t \in [t_k, \infty)$ is in domain Ω_α, the first integral quantity is computed for time interval $[t_k, t]$ by

$$L^{(\alpha)}(t_k, t) \equiv \int_{t_k}^{t} \nabla F(\mathbf{x}^{(\alpha)}, \boldsymbol{\mu}_\alpha) \cdot \mathbf{g}(\mathbf{x}^{(\alpha)}, t, \boldsymbol{\pi}_\alpha) dt. \tag{3.161}$$

However, for a discontinuous system, the flow will exist in finite domains Ω_α ($\alpha \in \{\alpha_1, \alpha_2, \cdots, \alpha_n\}$). The flow starts from domain Ω_{α_1} at time t_k and ending at time t_{k+n}. For time t_{k+i}, there are two points $(t_{k+i}, \mathbf{x}_{k+i}^{(\alpha_i)})$ and $(t_{k+i}, \mathbf{x}_{k+i+1}^{(\alpha_{i+1})})$. At the two points, we have

$$F^{(\alpha_i)}(\mathbf{x}_{k+i}^{(\alpha_i)}, \boldsymbol{\mu}_{\alpha_i}) = E_{k+i}^{(\alpha_i)} \text{ and } F^{(\alpha_{i+1})}(\mathbf{x}_{k+i}^{(\alpha_{i+1})}, \boldsymbol{\mu}_{\alpha_{i+1}}) = E_{k+i}^{(\alpha_{i+1})}. \tag{3.162}$$

The switching increment of the first integral quantity is determined at time t_{k+i} by

$$\Delta E_{k+i}^{\alpha_i \uparrow \alpha_{i+1}} = E_{k+i}^{(\alpha_{i+1})} - E_{k+i}^{(\alpha_i)}. \tag{3.163}$$

Such an increment is caused by the discontinuity. If the discontinuous dynamical system in two domains Ω_{α_i} and $\Omega_{\alpha_{i+1}}$ has the same integrable vector field with $\mathbf{x}_{k+i}^{(\alpha_i)} = \mathbf{x}_{k+i+1}^{(\alpha_{i+1})}$, then $\Delta E_{k+i}^{\alpha_i \uparrow \alpha_{i+1}} = 0$. The total integral quantity increment is computed by

$$\begin{aligned} L^{(\alpha_1 \cdots \alpha_n)}(t_k, t_{k+n}) &= \sum_{i=0}^{n-1} [L^{(\alpha_i)}(t_{k+i}, t_{k+i+1}) + \Delta E_{k+i}^{\alpha_i \uparrow \alpha_{i+1}}] + L^{(\alpha_n)}(t_{k+n-1}, t_{k+n}) \\ &= \sum_{i=0}^{n-1} [L^{(\alpha_i)}(t_{k+i}, t_{k+i+1}) + (E_{k+i}^{(\alpha_{i+1})} - E_{k+i}^{(\alpha_i)})] \\ &\quad + L^{(\alpha_n)}(t_{k+n-1}, t_{k+n}) \end{aligned} \tag{3.164}$$

where for $i = 1, 2, \cdots, n$

$$L^{(\alpha_i)}(t_{k+i-1}, t_{k+i}) \equiv \int_{t_{k+i-1}}^{t_{k+i}} \nabla F(\mathbf{x}^{(\alpha_i)}, \boldsymbol{\mu}_{\alpha_i}) \cdot \mathbf{g}(\mathbf{x}^{(\alpha_i)}, t, \boldsymbol{\pi}_{\alpha_i}) dt. \tag{3.165}$$

If a flow for $t \in [t_k, t_{k+n}]$ is periodic, then the total first integral quantity increment should be zero, i.e.,

$$L^{(\alpha_1 \cdots \alpha_n)}(t_k, t_{k+n}) = 0. \tag{3.166}$$

For chaotic motion, $L^{(\alpha_1 \cdots \alpha_n)}(t_k, t_{k+n}) \neq 0$ can be used to develop an iterative relations. Further, the chaos in such discontinuous dynamical system can be investigated. For a 2-D perturbed Hamiltonian system, equation (3.165) becomes

$$
\begin{aligned}
L^{(\alpha_i)}(t_{k+i},t_{k+i+1}) &\equiv \int_{t_{k+i}}^{t_{k+i+1}} \nabla F(\mathbf{x}^{(\alpha_i)},\boldsymbol{\mu}_{\alpha_i})\cdot \mathbf{g}(\mathbf{x}^{(\alpha_i)},t,\boldsymbol{\pi}_{\alpha_i})dt \\
&= \int_{t_{k+i}}^{t_{k+i+1}} \Big[f_2(\mathbf{x}^{(\alpha_i)},\boldsymbol{\mu}_{\alpha_i})\cdot g_1(\mathbf{x}^{(\alpha_i)},t,\boldsymbol{\pi}_{\alpha_i}) \\
&\quad - f_1(\mathbf{x}^{(\alpha_i)},\boldsymbol{\mu}_{\alpha_i})\cdot g_2(\mathbf{x}^{(\alpha_i)},t,\boldsymbol{\pi}_{\alpha_i}) \Big] dt.
\end{aligned}
\tag{3.167}
$$

Once Eq. (3.167) is substituted into Eq. (3.164), the $L^{(\alpha_1\cdots\alpha_n)}(t_k,t_{k+n})$-function can be obtained. In addition, for a special Poincare surface, such an L-function can be developed rather than the Melnikov functions in discontinuous dynamical systems.

3.8. An example

To further understand the complicated flows in discontinuous systems with an arbitrary boundary, an arbitrarily curved boundary will be considered. As in Luo and Rapp (2008), a mass-spring-damper oscillator in Fig. 3.19 is used as a mechanical model to discuss the switching dynamics of discontinuous systems with a parabolic boundary. The mass m is connected with a switchable spring of stiffness k_α ($\alpha = 1,2$) and a switchable damper of coefficient r_α ($\alpha = 1,2$) in the α-region. The mass of the oscillator is excited by a periodical force, i.e.,

$$
P_\alpha = Q_0 \cos(\Omega t + \phi) + U_\alpha,\ \ \alpha = 1,2 \tag{3.168}
$$

where Q_0 and Ω are excitation amplitude and frequency and U_α is constant force. The coordinate system is defined by (x,t) both t and x are time and the displacement of mass, respectively. The control law for this discontinuous dynamical system to switch is given by

$$
ax^2 + b\dot{x} = c \tag{3.169}
$$

where a,b and c are constants and $\dot{x} = dx/dt$.

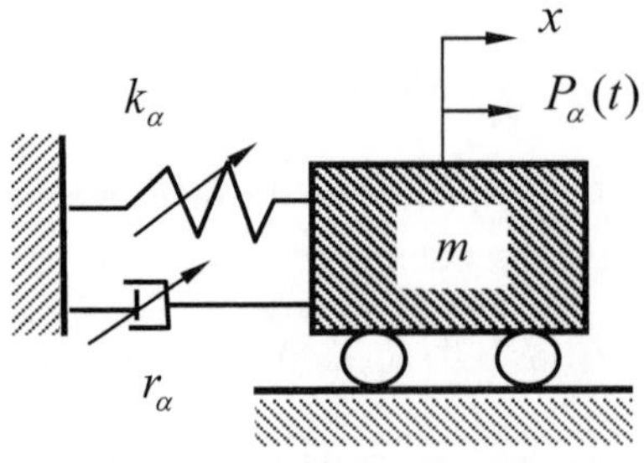

Fig. 3.19 A discontinuous dynamical system with a parabolic boundary.

The equation of motion for the aforementioned mass-spring-damper with discontinuity is:

$$\ddot{x}+2d_\alpha \dot{x}+c_\alpha x = A_0 \cos(\Omega t+\phi)+b_\alpha \tag{3.170}$$

where

$$c_\alpha = \frac{k_\alpha}{m},\ d_\alpha = \frac{r_\alpha}{2m},\ b_\alpha = \frac{U_\alpha}{m},\ A_0 = \frac{Q_0}{m}. \tag{3.171}$$

3.8.1. Conditions for sliding and grazing

To develop the analytical conditions for a flow switching and sliding on the separation boundary, the vector form of equation of motion should be introduced. In phase plane, the state vector and vector fields are defined as

$$\mathbf{x} \triangleq (x,\dot{x})^{\mathrm{T}} \equiv (x,y)^{\mathrm{T}} \text{ and } \mathbf{F} \triangleq (y,F)^{\mathrm{T}}. \tag{3.172}$$

From Eq. (3.169), this control logic creates a discontinuity in the system. In order to analyze such a system, two domains are defined as

$$\begin{aligned} \Omega_1 &= \{(x,y)\,|\, ax^2+by>c\},\\ \Omega_2 &= \{(x,y)\,|\, ax^2+by>c\}. \end{aligned} \tag{3.173}$$

The separation boundary is defined by the closure of the two domains,

$$\begin{aligned} \partial\Omega_{12} = \partial\Omega_{21} &= \bar{\Omega}_1 \cap \bar{\Omega}_2 \\ &= \{(x,y)\,|\,\varphi_{12}(x,y) \equiv ax^2+by-c=0\}. \end{aligned} \tag{3.174}$$

The domains and boundary are sketched in Fig. 3.20. The boundary is depicted by a dashed line, governed by Eq. (3.168). The two domains are shaded. The arrows crossing the boundary indicate the possible flow directions. If a flow of the motion in phase space is in domain Ω_α ($\alpha=1,2$), the vector fields in such a domain is continuous. However, if a flow of motion from domain Ω_α ($\alpha\in\{1,2\}$) switches into domain Ω_β ($\beta=\{1,2\}, \beta\neq\alpha$) through the boundary $\partial\Omega_{\alpha\beta}$, the vector field in domain Ω_α ($\alpha\in\{1,2\}$) will be changed into the one in domain Ω_β ($\beta=\{1,2\}$ accordingly. Because of the discontinuity, the flow cannot pass over the boundary under a certain condition. The flow motion along the boundary is called the sliding flow. If $a<0$ and $c<0$ (or $a>0$ and $c>0$), then there are two equilibrium points at $(\pm\sqrt{c/a},0)$. It is very easy to prove that an equilibrium is a stable node labeled by filled circle an hollow circle, as shown in Fig. 3.20(a) and (c). If $a<0$ and $c>0$ (or $a>0$ and $c<0$), no any equilibrium exists in Fig. 3.20(b) and (d). The properties of equilibrium point of the boundary is determined by

$$\dot{x}=y,\ \dot{y}=-\frac{2a}{b}xy. \tag{3.175}$$

With the boundary equation, the equilibrium points are determined. The corresponding flow on the boundary is depicted in Fig. 3.20.

The equations of motion for this discontinuous system can be described as

$$\dot{\mathbf{x}}=\mathbf{F}^{(\lambda)}(\mathbf{x},t)\ \text{ for } \lambda\in\{0,\alpha\} \tag{3.176}$$

$$\left.\begin{array}{l}\mathbf{F}^{(\alpha)}(\mathbf{x},t)=(y,F_\alpha(\mathbf{x},t))^{\mathrm{T}}\ \text{ in } \Omega_\alpha\ (\alpha\in\{1,2\}),\\ \mathbf{F}^{(0)}(\mathbf{x},t)=(y,-\frac{2a}{b}xy)^{\mathrm{T}}\ \text{ for sliding on } \partial\Omega_{\alpha\beta}\ (\alpha,\beta\in\{1,2\}),\\ \mathbf{F}^{(0)}(\mathbf{x},t)=[\mathbf{F}^{(\alpha)}(\mathbf{x},t),\mathbf{F}^{(\beta)}(\mathbf{x},t)]\ \text{ for non-sliding on } \partial\Omega_{\alpha\beta},\end{array}\right\} \tag{3.177}$$

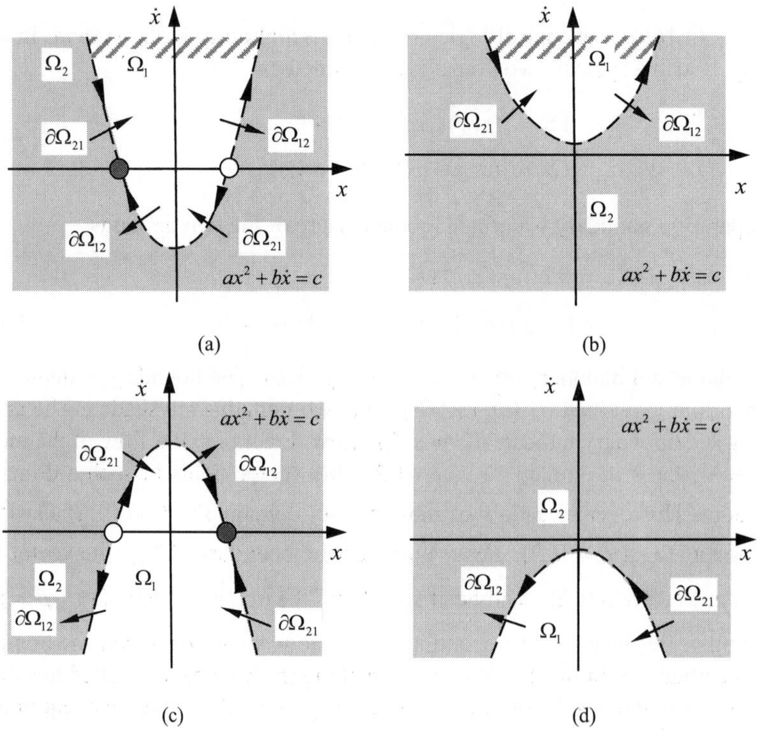

Fig. 3.20 Subdomains and parabolic boundary with static equilibriums: **(a)** $a<0,b>0,c<0$; **(b)** $a<0,b>0,c>0$; **(c)** $a>0,b>0,c>0$ and **(d)** $a>0,b>0,c<0$.

$$F_\alpha(\mathbf{x},t) = -2d_\alpha y - c_\alpha x + A_0\cos(\Omega t+\phi) + b_\alpha. \tag{3.178}$$

From Theorem 3.3, for sliding motion on $\partial\Omega_{\alpha\beta}$ with the corresponding normal vector $\mathbf{n}_{\partial\Omega_{\alpha\beta}}$ pointing to domain Ω_α (i.e., $\mathbf{n}_{\partial\Omega_{\alpha\beta}} \to \Omega_\alpha$), the necessary and sufficient conditions of the sliding motion on the switching boundary are given by

$$\begin{aligned} G^{(0,\alpha)}(\mathbf{x}_m,t_{m-}) &= \mathbf{n}^{\mathrm{T}}_{\partial\Omega_{\alpha\beta}}\cdot\mathbf{F}^{(\alpha)}(\mathbf{x}_m,t_{m-}) < 0,\\ G^{(0,\beta)}(\mathbf{x}_m,t_{m-}) &= \mathbf{n}^{\mathrm{T}}_{\partial\Omega_{\alpha\beta}}\cdot\mathbf{F}^{(\beta)}(\mathbf{x}_m,t_{m-}) > 0 \end{aligned} \tag{3.179}$$

where $\alpha,\beta\in\{1,2\}$ and $\alpha\neq\beta$ with

$$\mathbf{n}_{\partial\Omega_{\alpha\beta}} = \nabla\varphi_{\alpha\beta} = (\partial_x\varphi_{\alpha\beta}, \partial_y\varphi_{\alpha\beta})^{\mathrm{T}}_{(x_m,y_m)}. \tag{3.180}$$

where $\partial_x = \partial/\partial x$ and $\partial_y = \partial/\partial y$.

The conditions of the motion switchable to the boundary $\partial\Omega_{\alpha\beta}$ with $\mathbf{n}_{\partial\Omega_{\alpha\beta}} \to \Omega_\alpha$ are from Theorem 3.1, i.e.,

$$\begin{aligned} &\left.\begin{aligned} G^{(0,\alpha)}(\mathbf{x}_m,t_{m-}) &= \mathbf{n}^{\mathrm{T}}_{\partial\Omega_{\alpha\beta}}\cdot\mathbf{F}^{(\alpha)}(\mathbf{x}_m,t_{m-}) < 0\\ G^{(0,\beta)}(\mathbf{x}_m,t_{m+}) &= \mathbf{n}^{\mathrm{T}}_{\partial\Omega_{\alpha\beta}}\cdot\mathbf{F}^{(\beta)}(\mathbf{x}_m,t_{m+}) < 0 \end{aligned}\right\}\text{for } \Omega_\alpha \to \Omega_\beta;\\ &\left.\begin{aligned} G^{(0,\alpha)}(\mathbf{x}_m,t_{m-}) &= \mathbf{n}^{\mathrm{T}}_{\partial\Omega_{\alpha\beta}}\cdot\mathbf{F}^{(\beta)}(\mathbf{x}_m,t_{m-}) > 0\\ G^{(0,\beta)}(\mathbf{x}_m,t_{m+}) &= \mathbf{n}^{\mathrm{T}}_{\partial\Omega_{\alpha\beta}}\cdot\mathbf{F}^{(\alpha)}(\mathbf{x}_m,t_{m+}) > 0 \end{aligned}\right\}\text{for } \Omega_\beta \to \Omega_\alpha. \end{aligned} \tag{3.181}$$

Note that t_m represents switching time for the motion to the switching boundary and $t_{m\pm} = t_m \pm 0$ reflects the responses in domain rather than boundary. The grazing motion to the separation boundary $\partial\Omega_{\alpha\beta}$ is from Theorem 3.9, i.e.,

$$\begin{aligned} &G^{(0,\alpha)}(\mathbf{x}_m,t_{m\pm}) = \mathbf{n}^{\mathrm{T}}_{\partial\Omega_{\alpha\beta}}\cdot\mathbf{F}^{(\alpha)}(\mathbf{x}_m,t_{m\pm}) = 0,\\ &(-1)^\alpha G^{(1,\alpha)}(\mathbf{x}_m,t_{m\pm}) < 0 \ \text{ for } \alpha\in\{1,2\}. \end{aligned} \tag{3.182}$$

Using Eq. (3.174), Equation (3.180) gives

$$\mathbf{n}_{\partial\Omega_{12}} = \mathbf{n}_{\partial\Omega_{21}} = (2ax, b)^{\mathrm{T}}. \tag{3.183}$$

From Eq. (3.183), the normal vector always points to the domain Ω_1 and

$$\begin{aligned} G^{(0,\alpha)}(\mathbf{x}_m,t_{m\pm}) &= \mathbf{n}^{\mathrm{T}}_{\partial\Omega_{\alpha\beta}}\cdot\mathbf{F}^{(\alpha)}(\mathbf{x}_m,t_{m\pm}) = 2ax_m y_m + bF_\alpha(\mathbf{x}_m,t_{m\pm}),\\ G^{(1,\alpha)}(\mathbf{x}_m,t_{m\pm}) &= 2D\mathbf{n}^{\mathrm{T}}_{\partial\Omega_{\alpha\beta}}\cdot[\mathbf{F}^{(\alpha)}(\mathbf{x}_m,t_m) - \mathbf{F}^{(0)}(\mathbf{x}_m,t_m)]\\ &\quad + \mathbf{n}^{\mathrm{T}}_{\partial\Omega_{\alpha\beta}}\cdot[D\mathbf{F}^{(\alpha)}(\mathbf{x}_m,t_m) - D\mathbf{F}^{(0)}(\mathbf{x}_m,t_m)]\\ &= 2ax_m F_\alpha(\mathbf{x}_m,t_m) + b[\nabla F_\alpha(\mathbf{x},t)\cdot\mathbf{F}^{(\alpha)}(\mathbf{x},t)\\ &\quad + \partial_t F_\alpha(\mathbf{x},t)]_{(\mathbf{x}_m,t_m)} + 2ay_m^2. \end{aligned} \tag{3.184}$$

where $\partial_t = \partial / \partial t$

$$D\mathbf{n}_{\partial\Omega_{\alpha\beta}} = \left(\frac{2a}{b}y, 0\right)^{\mathrm{T}},$$
$$D\mathbf{F}^{(0)}(\mathbf{x},t) = \left(-\frac{2a}{b}xy, -\frac{2a}{b}y^2 + \left(\frac{2a}{b}\right)^2 x^2 y\right)^{\mathrm{T}}, \tag{3.185}$$
$$D\mathbf{F}^{(\alpha)}(\mathbf{x},t) = (F_\alpha(\mathbf{x},t), \nabla F_\alpha(\mathbf{x},t)\cdot \mathbf{F}^{(\alpha)}(\mathbf{x},t) + \partial_t F_\alpha(\mathbf{x},t))^{\mathrm{T}}.$$

The zero-order G-function is the product of vector field and the normal vector of the boundary. The first order G-function is the time change rate of the zero-order G-function.

From Eqs. (3.179) and (3.185), the conditions for sliding motion on the switching boundary are:

$$G^{(0,1)}(\mathbf{x}_m, t_{m-}) < 0 \text{ and } G^{(0,2)}(\mathbf{x}_m, t_{m-}) > 0. \tag{3.186}$$

From Eqs. (3.181) and (3.185), the switchability conditions for motion on the switching boundary are:

$$\begin{aligned} &G^{(0,1)}(\mathbf{x}_m, t_{m-}) < 0 \text{ and } G^{(0,2)}(\mathbf{x}_m, t_{m+}) < 0, \text{ for } \Omega_1 \to \Omega_2; \\ &G^{(0,1)}(\mathbf{x}_m, t_{m+}) > 0 \text{ and } G^{(0,2)}(\mathbf{x}_m, t_{m-}) > 0, \text{ for } \Omega_2 \to \Omega_1. \end{aligned} \tag{3.187}$$

From Theorem 3.2, the vanishing conditions of the sliding motion on the separation boundary are:

$$\left.\begin{aligned} &(-1)^\alpha G^{(0,\alpha)}(\mathbf{x}_m, t_{m-}) > 0 \\ &G^{(0,\beta)}(\mathbf{x}_m, t_{m\mp}) = 0 \\ &(-1)^\beta G^{(1,\beta)}(\mathbf{x}_m, t_{m\mp}) < 0 \end{aligned}\right\} \text{for } \Omega_\alpha \to \partial\Omega_{\alpha\beta} \tag{3.188}$$

with $\alpha, \beta \in \{1,2\}$ and $\alpha \neq \beta$. From Theorem 3.4, the onset condition of the sliding motion on the switching boundary is given by

$$\left.\begin{aligned} &(-1)^\alpha G^{(0,\alpha)}(\mathbf{x}_m, t_{m-}) > 0 \\ &G^{(0,\beta)}(\mathbf{x}_m, t_{m\pm}) = 0 \\ &(-1)^\beta G^{(1,\beta)}(\mathbf{x}_m, t_{m\pm}) < 0 \end{aligned}\right\} \text{for } \Omega_\alpha \to \partial\Omega_{\alpha\beta} \tag{3.189}$$

with $\alpha, \beta \in \{1,2\}$ and $\alpha \neq \beta$.

3.8.2. Periodic motions

In this section, switching planes will be introduced and basic mappings will be defined for the mapping structures to determine periodic motion. For a sliding motion on the separation boundary, it is assumed the sliding motion disappears at time t_{i+1}. For $t_m \in [t_i, t_{i+1}]$, the solution of the sliding motion is on the boundary, expressed by displacement x_m and velocity y_m. These solutions can be found analytically from Eq. (3.169). The corresponding $G^{(0,\alpha)}$-function for sliding motion is

$$G^{(0,\alpha)}(\mathbf{x}_m, t_{m-}) = 2ax_m y_m + b[-2d_\alpha y_m - c_\alpha x_m + A_0 \cos(\Omega t_{m-} + \phi) + b_\alpha]. \quad (3.190)$$

For the non-sliding motion, once the initial conditions are chosen on separation boundary, and the solution of Eq. (3.170) in all the domains Ω_α can be obtained analytically (see, Luo and Rapp, 2010). The basic solutions will be used for mapping construction. To construct the basic mappings, the switching planes in phase space will be introduced. In phase plane, trajectories in Ω_α, starting and ending at the switching boundary (i.e., from $\partial\Omega_{\beta\alpha}$ to $\partial\Omega_{\alpha\beta}$), are illustrated in Fig. 3.21. The starting and ending points for mapping P_α in Ω_α are $(\mathbf{x}_i, t_i)$ and $(\mathbf{x}_{i+1}, t_{i+1})$, respectively. The stick mapping is P_0. Define the switching planes as

$$\begin{aligned}
\Xi^0 &= \{(x_i, y_i, \Omega t_i) \mid \varphi_{\alpha\beta}(x_i, y_i) = c\},\\
\Xi^1 &= \{(x_i, y_i, \Omega t_i) \mid \varphi_{\alpha\beta}(x_i, y_i) = c^+\},\\
\Xi^2 &= \{(x_i, y_i, \Omega t_i) \mid \varphi_{\alpha\beta}(x_i, y_i) = c^-\}
\end{aligned} \quad (3.191)$$

where $c^- = \lim_{\delta\to 0}(c-\delta)$ and $c^+ = \lim_{\delta\to 0}(c+\delta)$ for arbitrary small $\delta > 0$. Therefore, three mappings are defined as

$$P_0 : \Xi^0 \to \Xi^0,\ P_1 : \Xi^1 \to \Xi^1,\ P_2 : \Xi^2 \to \Xi^2. \quad (3.192)$$

From the foregoing two equations, one obtains

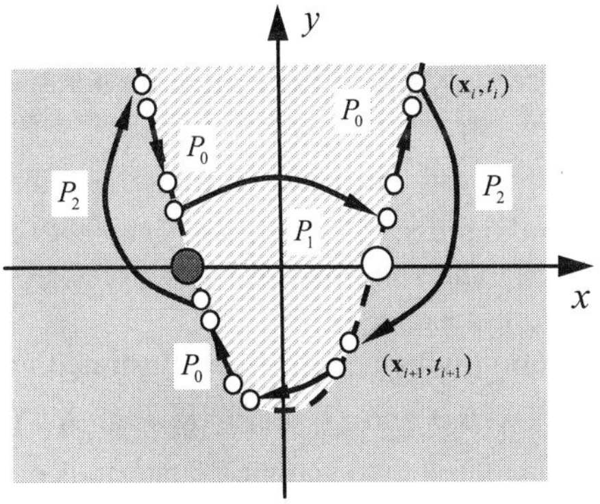

Fig. 3.21 Basic mappings for $a < 0, b > 0, c < 0$.

$$\begin{aligned} P_0 &: (x_i, y_i, t_i) \to (x_{i+1}, y_{i+1}, t_{i+1}), \\ P_1 &: (x_i^+, y_i^+, t_i) \to (x_{i+1}^+, y_{i+1}^+, t_{i+1}), \\ P_2 &: (x_i^-, y_i^-, t_i) \to (x_{i+1}^-, y_{i+1}^-, t_{i+1}). \end{aligned} \tag{3.193}$$

With Eq. (3.169), the governing equations for P_0 ($\alpha \in \{1,2\}$) are obtained from the solution of the sliding motion and

$$\begin{aligned} &G^{(0,\alpha)}(\mathbf{x}_{i+1}, t_{i+1}) = 0, \\ &G^{(0,1)}(\mathbf{x}, t) \times G^{(0,2)}(\mathbf{x}, t) \le 0 \ \text{ for } t \in [t_i, t_{i+1}) \text{ and } \mathbf{x} \in [\mathbf{x}_i, \mathbf{x}_{i+1}). \end{aligned} \tag{3.194}$$

From this problem, the two domains Ω_α ($\alpha \in \{1,2\}$) are unbounded. From the assumptions (H3.1)-(H3.3), only three possible, bounded motions exist in each domain Ω_α ($\alpha \in \{1,2\}$), from which the governing equations of mapping P_α ($\alpha \in \{1,2\}$) are obtained. With Eq. (3.169), the governing equations of each mapping P_λ ($\lambda \in \{0,1,2\}$) can be expressed by

$$\begin{aligned} f_1^{(\lambda)}(x_i, \Omega t_i, x_{i+1}, \Omega t_{i+1}) = 0, \\ f_2^{(\lambda)}(x_i, \Omega t_i, x_{i+1}, \Omega t_{i+1}) = 0. \end{aligned} \tag{3.195}$$

Consider a generalized mapping structure for periodic motion with stick as

$$\begin{aligned} P &= P_{\underbrace{(2^{k_{m2}}1^{k_{m1}}0^{k_{m0}})\cdots(2^{k_{12}}1^{k_{11}}0^{k_{10}})}_{m\text{-terms}}} \\ &= \underbrace{(P_2^{(k_{m2})} \circ P_1^{(k_{m1})} \circ P_0^{(k_{m0})}) \circ \cdots \circ (P_2^{(k_{12})} \circ P_1^{(k_{11})} \circ P_0^{(k_{10})})}_{m\text{-terms}} \end{aligned} \tag{3.196}$$

where $k_{l\alpha} \in \{0,1\}$ for $l \in \{1, 2, \cdots, m\}$ and $\lambda \in \{0,1,2\}$. $P_\lambda^{(k)} = P_\lambda \circ P_\lambda^{(k-1)}$ and $P_\lambda^{(0)} = 1$. Note that the clockwise and counter-clockwise rotations of the order of the mapping P_λ in the complete mapping P in Eq. (3.196) will not change the periodic motion. However, only the initial conditions for such a periodic motion are different. Consider the parameters ($m = 1$, $r_1 = 2$, $r_2 = 5$, $U_1 = U_2 = 1$, $\phi = 0$, $a = -5$, $b = 1$ and $c = -15$) for numerical illustrations. The trajectory in phase plane, force distribution along displacement, displacement time-history and force time-history are presented. The solid and dashed curves represent the forces for real and imaginary flows, respectively.

A simple periodic motion with $P = P_1 \circ P_2$ is illustrated in Figs. 3.22 and 3.23 for $k_1 = 70$, $k_2 = 150$, $\Omega = 10.0$ and $Q_0 = 119$. In Fig. 3.22(a), the trajectory of periodic flow in phase plane is shown and the boundary is depicted by the dashed curve. The starting points $x_0 \approx 2.8185$ and $y_0 \approx 24.7197$ with $\Omega t_0 \approx 0.8281$ on the boundary, labeled by a green circular symbol. Another switching point is labeled by a hollow circular symbol. The periodic motion of $P = P_1 \circ P_2$ is very clearly presented. The two equilibrium points are two large circular symbols. *G*-function for the inner product of the vector fields and the normal vector of the boundary are presented in Fig. 3.22(b). At the starting point, because of $G^{(0,1)} < 0$ and $G^{(0,2)} < 0$, the periodic motion will go into domain Ω_2. Once the flow in domain Ω_2 arrives to the boundary, $G^{(0,1)} > 0$ and $G^{(0,2)} > 0$. From the analytical condition, the flow will get into domain Ω_1. The flow in domain Ω_1 approaches the starting points on the boundary, which is formed a periodic motion. The displacement time-history is presented in Fig. 3.23(a). The corresponding mapping is labeled. The real and imaginary *G*-functions of the periodic flow are presented in Fig. 3.23(b) by solid and dashed curves, respectively. The motion switchability conditions at the boundary are clearly observed through the *G*-function.

Consider a periodic motion with sliding motion on the boundary for a mapping structure of $P = P_{102}$, as illustrated in Figs. 3.24 and 3.25 for $k_1 = 60$, $k_2 = 175$, $\Omega = 6.84$ and $Q_0 = 119$. In Fig. 3.24(a), the trajectory of a periodic flow in phase plane is shown and the boundary is also depicted by a dashed curve. The starting point is $x_0 \approx 2.3997$ and $y_0 \approx 13.6261$ with $\Omega t_0 \approx 6.2743$ on the boundary with the green circular symbol. The other switching points are labeled by the hollow circular symbol. Such a periodic motion with sliding motion is presented. The two equilibrium points are two large circular symbols. The periodic motion starts from the boundary and then enters domain Ω_2. Continually, the flow arrives to the boundary and slides on the boundary to the stable equilibrium point. From the equilibrium point, then the flow gets into domain Ω_1 and arrives at the starting point. The *G*-function of a flow to the boundary are presented in Fig. 3.24(b). At the starting point, because of $G^{(0,1)} < 0$ and $G^{(0,2)} < 0$, the periodic motion will go into domain Ω_2. The switchable conditions of such a flow on the boundary are the same as in Fig. 3.22. However, the sliding motion in the periodic motion requires $G^{(0,1)} < 0$ and $G^{(0,2)} > 0$, which is observed. When the motion slides on the boundary and arrives to the equilibrium, the corresponding conditions (i.e., $G^{(0,1)} \le 0$ and $G^{(0,2)} > 0$) are observed. The condition of $G^{(0,1)} = 0$ and $G^{(0,2)} > 0$. indicates the sliding motion disappears. The displacement time-history is presented in Fig. 3.25(a). The corresponding mapping is labeled. With time varying, one of $G^{(0,1)} < 0$ and $G^{(0,2)} > 0$ changes sign, the flow will move into a new domain. The real and imaginary *G*-functions of the periodic flow are presented in Fig. 3.25(b) by the solid and dashed curves, respectively. The switchability conditions of the flow at the boundary are also clearly observed Through *G*-functions. If readers are interested in other examples, the global transversality of a flow to the separatrix in the Duffing oscillator was discussed in Luo (2008b,d).

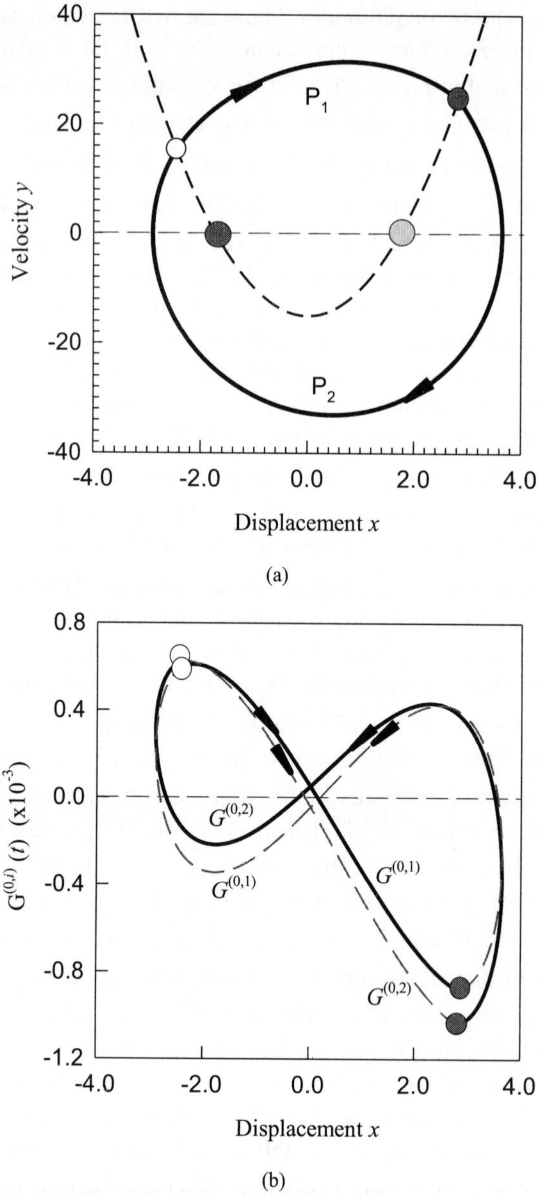

Fig. 3.22 Periodic motion of P_{12} : **(a)** phase plane, **(b)** normal vector field versus displacement. Initial condition is $\Omega t_0 \approx 0.8281$, $x_0 \approx 2.8185$ and $y_0 \approx 24.7197$. The gray dashed curves represent the G-functions for the imaginary flows ($m=1$, $r_1=2$, $k_1=60$, $r_2=5$, $k_2=175$, $U_1=1$, $U_2=1$, $Q_0=119$, $\Omega=6.84$, $\phi=0$, $a=-5$, $b=1$, $c=-15$).

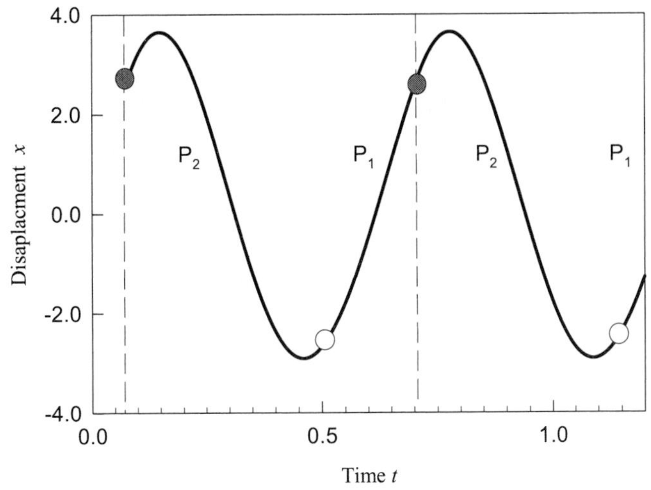

(a)

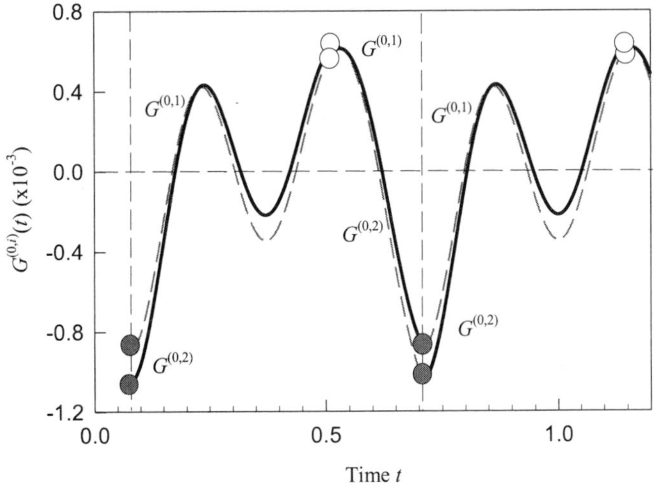

(b)

Fig. 3.23 Periodic motion of P_{12} : **(a)** displacement time-history and **(b)** time history of *G*-functions. Initial condition is $\Omega t_0 \approx 0.8281$, $x_0 \approx 2.8185$ and $y_0 \approx 24.7197$. The gray dashed curves represent the *G*-functions for the imaginary flows ($m=1$, $r_1=2$, $k_1=60$, $r_2=5$, $k_2=175$, $U_1=1$, $U_2=1$, $Q_0=119$, $\Omega=6.84$, $\phi=0$, $a=-5$, $b=1$, $c=-15$).

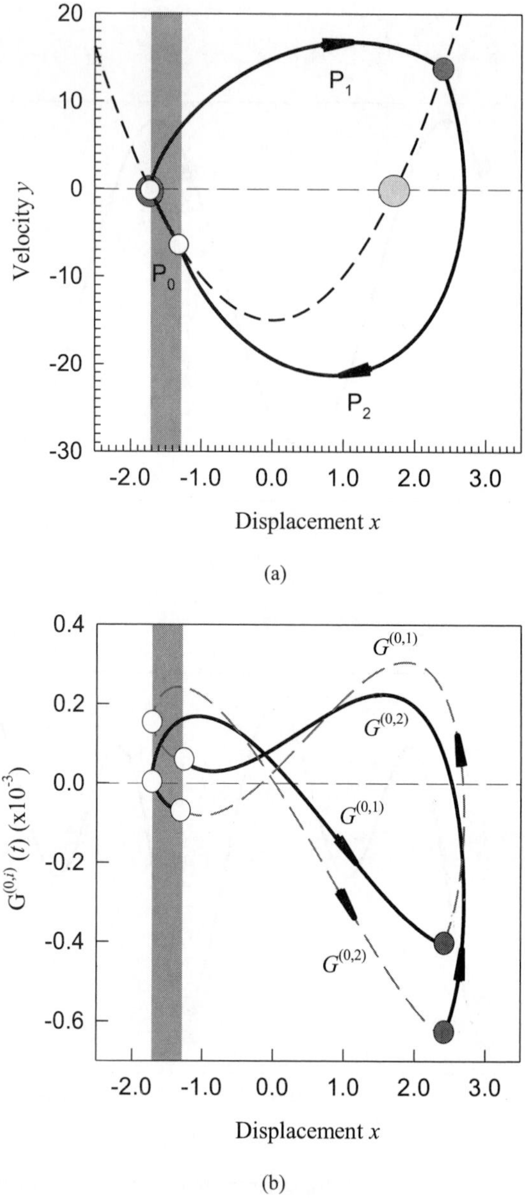

Fig. 3.24 Periodic motion of P_{102} : **(a)** phase plane and **(b)** normal vector field versus displacement. Initial condition is $\Omega t_0 \approx 6.2743$, $x_0 \approx 2.3997$ and $y_0 \approx 13.6261$. The gray dashed curves represent the G-functions for the imaginary flows ($m=1$, $r_1=2$, $k_1=60$, $r_2=5$, $k_2=175$, $U_1=1$, $U_2=1$, $Q_0=119$, $\Omega=6.84$, $\phi=0$, $a=-5$, $b=1$, $c=-15$).

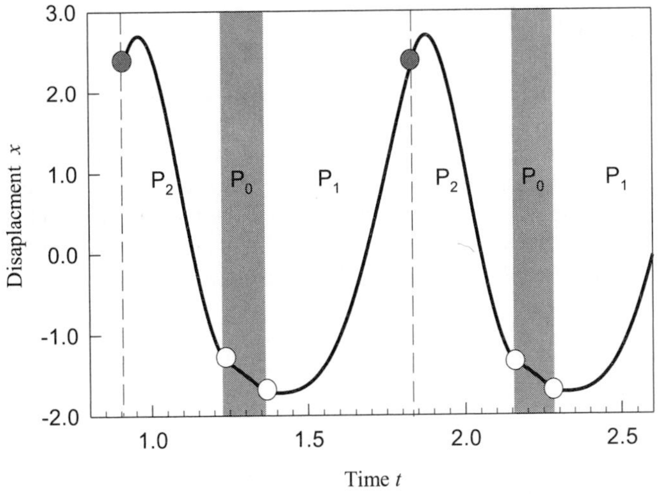

(a)

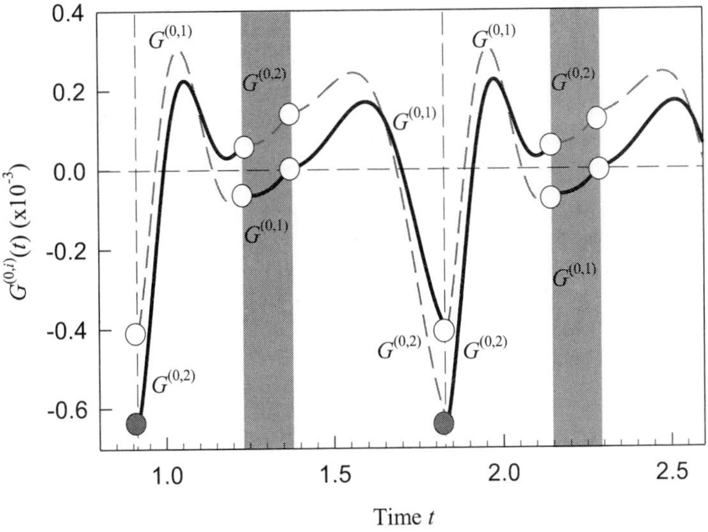

(b)

Fig. 3.25 Periodic motion of P_{102} : **(a)** displacement time history, and **(b)** time-history of G-functions. Initial condition is $\Omega t_0 \approx 6.2743$, $x_0 \approx 2.3997$ and $y_0 \approx 13.6261$. The gray dashed curves represent the G-functions for the imaginary flows ($m=1$, $r_1=2$, $k_1=60$, $r_2=5$, $k_2=175$, $U_1=1$, $U_2=1$, $Q_0=119$, $\Omega=6.84$, $\phi=0$, $a=-5$, $b=1$, $c=-15$).

3.9. Concluding remarks

In this chapter, a general theory of flow passability to a specific boundary in discontinuous dynamical systems was systematically presented. The G-functions for discontinuous dynamical systems were introduced as an important quantity. Using the G-function, the passability of a flow from a domain to an adjacent one was discussed. With help of real and imaginary flows, the *full* and *half*, sink and source flows to the boundary were discussed, and the switching bifurcations between the passable and non-passable flows were addressed as well.

References

Kreyszig, E., 1968, *Introduction to Differential Geometry and Riemannian Geometry*, Toronto: University of Toronto Press.

Luo, A.C.J., 2005a, A theory for non-smooth dynamic systems on the connectable domains, *Communications in Nonlinear Science and Numerical Simulation*, **10**, 1-55.

Luo, A.C.J., 2005b, Imaginary, sink and source flows in the vicinity of the separatrix of non-smooth dynamical systems, *Journal of Sound and Vibration*, **285**, 443-456.

Luo, A.C.J., 2006, *Singularity and Dynamics on Discontinuous Vector Fields*, Amsterdam: Elsevier.

Luo, A.C.J., 2008a, On the differential geometry of flows in nonlinear dynamic systems, ASME *Journal of Computational and Nonlinear Dynamics*, 021104-1~10.

Luo, A.C.J., 2008b, *Global Transversality, Resonance and Chaotic Dynamics*, Singapore: World Scientific.

Luo, A.C.J., 2008c, A theory for flow switchability in discontinuous dynamical systems, *Nonlinear Analysis: Hybrid Systems*, **2**(4), 1030-1061.

Luo, A.C.J., 2008d, Global tangency and transversality of periodic flows and chaos in a periodically forced, damped Duffing oscillator, *International Journal of Bifurcations and Chaos*, **18**, 1-49.

Luo, A.C.J. and Rapp, B.M., 2010, On motions and switchability in a periodically forced, discontinuous system with a parabolic boundary, *Nonlinear Analysis: Real World Applications*, **11**, 2624-2633.

Chapter 4
Flow Barriers and Switchability

In this chapter, the theory of flow barriers in discontinuous dynamical systems will be systematically presented for the first time, which help one re-think the existing theories of stability and control in dynamical systems. The concept of flow barriers in discontinuous dynamical systems will be introduced, and the passability of a flow to the boundary with flow barriers will be presented. Because the flow barriers exist on the boundary, the switchability of a flow to such a separation boundary will be changed accordingly. The coming and leaving flow barriers in passable flows will be discussed first, and the necessary and sufficient conditions for a flow to pass through the boundary with flow barrier will be developed. Flow barriers for sink and source flows will be also discussed. Once the sink flow is formed, the boundary flow barriers in the sink flow needs to be considered, and such a flow barrier is independent of vector fields in the corresponding domains. Furthermore, when the boundary flow in the sink flow disappears, the vector fields should satisfy the appropriate conditions. Thus, the necessary and sufficient conditions for formations and vanishing of a sink flow will be developed for a discontinuous dynamical system possessing flow barriers on the boundary. A periodically forced friction model will be presented as an example for a better understanding of flow barrier existence in physical problems. The flow barrier theory presented in this chapter will provide a theoretic base for one to further develop control theory and stability.

4.1. Flow barriers for passable flows

In Chapters 2 and 3, the passability of a flow to the boundary is dependent on the vector fields on both sides of the boundary. If a flow goes through the boundary $\partial\Omega_{ij}$ from domain Ω_i to domain Ω_j, there are three vectors fields to form three dynamical systems, i.e.,

$$\begin{aligned}&\dot{\mathbf{x}}^{(\alpha)} = \mathbf{F}^{(\alpha)}(\mathbf{x}^{(\alpha)}, t, \mathbf{p}_\alpha) \text{ in } \Omega_\alpha \ (\alpha = i, j),\\&\dot{\mathbf{x}}^{(0)} = \mathbf{F}^{(0)}(\mathbf{x}^{(0)}, t, \boldsymbol{\lambda}) \quad \text{with } \varphi_{ij}(\mathbf{x}^{(0)}, t, \boldsymbol{\lambda}) = 0 \text{ on } \partial\Omega_{ij}.\end{aligned} \tag{4.1}$$

For simplicity in discussion, the following sign function is introduced as

$$\hbar_\alpha = \begin{cases} +1 & \text{for } \mathbf{n}_{\partial\Omega_{ij}} \to \Omega_\beta, \\ -1 & \text{for } \mathbf{n}_{\partial\Omega_{ij}} \to \Omega_\alpha. \end{cases} \tag{4.2}$$

Without any flow barriers, the necessary and sufficient conditions for a flow to pass through the boundary are obtained from Theorem 3.1, i.e.,

$$\hbar_\alpha G^{(\alpha)}_{\partial\Omega_{ij}}(\mathbf{x}_m, t_{m-}, \mathbf{p}_\alpha, \boldsymbol{\lambda}) > 0 \text{ and } \hbar_\alpha G^{(\beta)}_{\partial\Omega_{ij}}(\mathbf{x}_m, t_{m+}, \mathbf{p}_\beta, \boldsymbol{\lambda}) > 0. \tag{4.3}$$

If the flow barrier exists on the boundary, a coming flow on the boundary may not pass through the boundary under such conditions. To investigate the flow property to the boundary with flow barriers, the G-functions for the flow barrier should be introduced as in Chapter 2. The corresponding definitions are given as follows.

Definition 4.1. For a discontinuous dynamical system in Eq. (4.1), there is a point $\mathbf{x}^{(0)}(t_m) \equiv \mathbf{x}_m \in \partial\Omega_{ij}$ at time t_m between two adjacent domains Ω_α ($\alpha = i, j$). There is a vector field of $\mathbf{F}^{(\rho \succ \gamma)}(\mathbf{x}^{(\lambda)}, t, \boldsymbol{\pi}_\lambda, q^{(\lambda)})$ for $q^{(\lambda)} \in [q_1^{(\lambda)}, q_2^{(\lambda)}]$ ($\rho, \gamma \in \{0, i, j\}$, $\lambda \in \{i, j\}$ and $\rho \neq \gamma$ if $\rho \neq 0$) on the boundary $\partial\Omega_{ij}$. For the point $\mathbf{x}^{(\rho)}(t_m) = \mathbf{x}_m$, the *G-function of the vector field* is defined as

$$\begin{aligned}&G^{(\rho \succ \gamma)}_{\partial\Omega_{ij}}(\mathbf{x}_m, t_{m\pm}, \boldsymbol{\pi}_\lambda, \boldsymbol{\lambda}, q^{(\lambda)})\\&\equiv \mathbf{n}^{\mathrm{T}}_{\partial\Omega_{ij}}(\mathbf{x}^{(0)}, t, \boldsymbol{\lambda}) \cdot \left[\mathbf{F}^{(\rho \succ \gamma)}(\mathbf{x}^{(\lambda)}, t, \boldsymbol{\pi}_\lambda, q^{(\lambda)}) - \mathbf{F}^{(0)}(\mathbf{x}^{(0)}, t, \boldsymbol{\lambda}) \right]\Big|_{(\mathbf{x}_m^{(\lambda)}, \mathbf{x}_m^{(0)}, t_{m\pm})}.\end{aligned} \tag{4.4}$$

The higher-order G-function of the vector field $\mathbf{F}^{(\rho \succ \gamma)}(\mathbf{x}^{(\lambda)}, t, \boldsymbol{\pi}_\lambda, q^{(\lambda)})$ is defined for $k_\lambda = 0, 1, 2, \cdots$ as

$$\begin{aligned}&G^{(k_\lambda, \rho \succ \gamma)}_{\partial\Omega_{ij}}(\mathbf{x}_m, t_{m\pm}, \boldsymbol{\pi}_\lambda, \boldsymbol{\lambda}, q^{(\lambda)})\\&= \sum_{r=1}^{k_\lambda + 1} C^r_{k_\lambda + 1} D_0^{k_\lambda + 1 - r} \mathbf{n}^{\mathrm{T}}_{\partial\Omega_{ij}}(\mathbf{x}^{(0)}, t, \boldsymbol{\lambda}) \cdot [D_\lambda^{r-1} \mathbf{F}^{(\rho \succ \gamma)}(\mathbf{x}^{(\lambda)}, t, \boldsymbol{\pi}_\lambda, q^{(\lambda)})\\&\quad - D_0^{r-1} \mathbf{F}^{(0)}(\mathbf{x}^{(0)}, t, \boldsymbol{\lambda})]\Big|_{(\mathbf{x}_m^{(\lambda)}, \mathbf{x}_m^{(0)}, t_{m\pm})}.\end{aligned} \tag{4.5}$$

For simplicity, the following notations are adopted.

$$\begin{aligned}&G^{(k_\alpha, \alpha)}_{\partial\Omega_{ij}}(\mathbf{x}_m, t_{m\pm}) \equiv G^{(k_\alpha, \alpha)}_{\partial\Omega_{ij}}(\mathbf{x}_m, t_{m\pm}, \mathbf{p}_\alpha, \boldsymbol{\lambda}),\\&G^{(k_\lambda, \rho \succ \gamma)}_{\partial\Omega_{ij}}(\mathbf{x}_m, q^{(\lambda)}) \equiv G^{(k_\lambda, \rho \succ \gamma)}_{\partial\Omega_{ij}}(\mathbf{x}_m, t_{m\pm}, \boldsymbol{\pi}_\lambda, \boldsymbol{\lambda}, q^{(\lambda)}).\end{aligned} \tag{4.6}$$

4.1.1. Coming flow barriers

In this section, the coming flow barriers in the semi-passable flow to the boundary will be discussed. The basic concepts of the flow barriers on the boundary will be introduced through a coming flow to the boundary.

Definition 4.2. For a discontinuous dynamical system in Eq. (4.1), there is a point $\mathbf{x}^{(0)}(t_m) \equiv \mathbf{x}_m \in \partial\Omega_{ij}$ at time t_m between two adjacent domains Ω_α ($\alpha = i, j$). Suppose there is a vector field of $\mathbf{F}^{(\alpha \succ \beta)}(\mathbf{x}^{(\alpha)}, t, \boldsymbol{\pi}_\alpha, q^{(\alpha)})$ for $q^{(\alpha)} \in [q_1^{(\alpha)}, q_2^{(\alpha)}]$ on the boundary $\partial\Omega_{ij}$ ($\alpha, \beta \in \{i, j\}$ and $\beta \neq \alpha$) with

$$\begin{aligned} \hbar_\alpha G_{\partial\Omega_{ij}}^{(\alpha\succ\beta)}(\mathbf{x}_m, q^{(\alpha)}) &\in \left[\hbar_\alpha G_{\partial\Omega_{ij}}^{(\alpha\succ\beta)}(\mathbf{x}_m, q_1^{(\alpha)}), \hbar_\alpha G_{\partial\Omega_{ij}}^{(\alpha\succ\beta)}(\mathbf{x}_m, q_2^{(\alpha)})\right] \\ &\subset [0, +\infty). \end{aligned} \tag{4.7}$$

The coming and leaving flows in the semi-passable flow satisfy

$$\hbar_\alpha G_{\partial\Omega_{ij}}^{(\alpha)}(\mathbf{x}_m, t_{m-}) > 0 \text{ and } \hbar_\alpha G_{\partial\Omega_{ij}}^{(\beta)}(\mathbf{x}_m, t_{m+}) > 0. \tag{4.8}$$

The vector field of $\mathbf{F}^{(\alpha \succ \beta)}(\mathbf{x}^{(\alpha)}, t, \boldsymbol{\pi}_\alpha, q^{(\alpha)})$ is called the coming *flow barrier* in the semi-passable flow on the α-side if the following conditions are satisfied. The critical values of $\mathbf{F}^{(\alpha \succ \beta)}(\mathbf{x}^{(\alpha)}, t, \boldsymbol{\pi}_\alpha, q_\sigma^{(\alpha)})$ ($\sigma = 1, 2$) are called the *lower and upper limits* of the coming flow barrier on the α - side.

(i) The coming flow of $\mathbf{x}^{(\alpha)}$ cannot be switched to the leaving flow of $\mathbf{x}^{(\beta)}$ at $q^{(\alpha)} \in (q_1^{(\alpha)}, q_2^{(\alpha)})$ if

$$\begin{aligned} &\mathbf{x}^{(\alpha)}(t_{m-}) = \mathbf{x}^{(\alpha\succ\beta)}(t_{m\pm}, q^{(\alpha)}) = \mathbf{x}_m, \\ &\hbar_\alpha G_{\partial\Omega_{ij}}^{(\alpha)}(\mathbf{x}_m, t_{m-}) \in \left(\hbar_\alpha G_{\partial\Omega_{ij}}^{(\alpha\succ\beta)}(\mathbf{x}_m, q_1^{(\alpha)}), \hbar_\alpha G_{\partial\Omega_{ij}}^{(\alpha\succ\beta)}(\mathbf{x}_m, q_2^{(\alpha)})\right). \end{aligned} \tag{4.9}$$

(ii) The coming flow of $\mathbf{x}^{(\alpha)}$ cannot be switched to the leaving flow of $\mathbf{x}^{(\beta)}$ at the critical points of the flow barrier (i.e., $q^{(\alpha)} = q_\sigma^{(\alpha)}$, $\sigma \in \{1, 2\}$) if

$$\begin{aligned} &\mathbf{x}^{(\alpha)}(t_{m-}) = \mathbf{x}^{(\alpha\succ\beta)}(t_{m\pm}, q_\sigma^{(\alpha)}) = \mathbf{x}_m, \\ &G_{\partial\Omega_{ij}}^{(s_\alpha, \alpha)}(\mathbf{x}_m, t_{m-}) = G_{\partial\Omega_{ij}}^{(s_\alpha, \alpha\succ\beta)}(\mathbf{x}_m, q_\sigma^{(\alpha)}) \neq 0 \text{ for } s_\alpha = 0, 1, 2, \cdots, l_\alpha - 1; \\ &(-1)^\sigma \hbar_\alpha \mathbf{n}_{\partial\Omega_{ij}}^{\mathrm{T}}(\mathbf{x}^{(0)}(t_{m+\varepsilon})) \cdot \left[\mathbf{x}^{(\alpha)}(t_{m+\varepsilon}) - \mathbf{x}^{(\alpha\succ\beta)}(t_{m+\varepsilon}, q_\sigma^{(\alpha)})\right] < 0. \end{aligned} \tag{4.10}$$

(iii) The coming flow of $\mathbf{x}^{(\alpha)}$ is switched to the leaving flow of $\mathbf{x}^{(\beta)}$ at the critical points of the flow barrier (i.e., $q^{(\alpha)} = q_\sigma^{(\alpha)}$, $\sigma \in \{1, 2\}$) if for

$$\mathbf{x}^{(\alpha)}(t_{m-}) = \mathbf{x}^{(\alpha\succ\beta)}(t_{m\pm}, q_\sigma^{(\alpha)}) = \mathbf{x}_m,$$

$$\begin{aligned}&G_{\partial\Omega_{ij}}^{(s_\alpha,\alpha)}(\mathbf{x}_m,t_{m-})=G_{\partial\Omega_{ij}}^{(s_\alpha,\alpha\succ\beta)}(\mathbf{x}_m,q_\sigma^{(\alpha)})\neq 0 \text{ for } s_\alpha=0,1,2,\cdots,l_\alpha-1;\\&(-1)^\sigma\hbar_\alpha\mathbf{n}_{\partial\Omega_{ij}}^{\mathrm{T}}(\mathbf{x}^{(0)}(t_{m+\varepsilon}))\cdot\left[\mathbf{x}^{(\alpha)}(t_{m+\varepsilon})-\mathbf{x}^{(\alpha\succ\beta)}(t_{m+\varepsilon},q_\sigma^{(\alpha)})\right]>0.\end{aligned}\tag{4.11}$$

Definition 4.3. For a discontinuous dynamical system in Eq. (4.1), there is a point $\mathbf{x}^{(0)}(t_m)\equiv\mathbf{x}_m\in\partial\Omega_{ij}$ at time t_m between two adjacent domains Ω_α ($\alpha=i,j$). Suppose there is a vector field of $\mathbf{F}^{(\alpha\succ\beta)}(\mathbf{x}^{(\alpha)},t,\boldsymbol{\pi}_\alpha,q^{(\alpha)})$ for $q^{(\alpha)}\in[q_1^{(\alpha)},q_2^{(\alpha)}]$ on the boundary $\partial\Omega_{ij}$ with the G-functions

$$\begin{aligned}&G_{\partial\Omega_{ij}}^{(s_\alpha,\alpha\succ\beta)}(\mathbf{x}_m,q^{(\alpha)})=0 \text{ for } s_\alpha=0,1,\cdots,2k_\alpha-1;\\&G_{\partial\Omega_{ij}}^{(2k_\alpha,\alpha\succ\beta)}(\mathbf{x}_m,q^{(\alpha)})\in\left[\hbar_\alpha G_{\partial\Omega_{ij}}^{(2k_\alpha,\alpha\succ\beta)}(\mathbf{x}_m,q_1^{(\alpha)}),\hbar_\alpha G_{\partial\Omega_{ij}}^{(2k_\alpha,\alpha\succ\beta)}(\mathbf{x}_m,q_2^{(\alpha)})\right]\\&\qquad\subset[0,\infty)\end{aligned}\tag{4.12}$$

($\alpha,\beta\in\{i,j\}$ and $\alpha\neq\beta$). The coming and leaving flows of the $(2k_\alpha:m_\beta)$-semi-passable flow satisfy

$$\begin{aligned}&G_{\partial\Omega_{ij}}^{(s_\alpha,\alpha)}(\mathbf{x}_m,t_{m-})=0 \text{ for } s_\alpha=0,1,\cdots,2k_\alpha-1,\\&G_{\partial\Omega_{ij}}^{(s_\beta,\beta)}(\mathbf{x}_m,t_{m+})=0 \text{ for } s_\beta=0,1,\cdots,m_\beta-1,\\&\hbar_\alpha G_{\partial\Omega_{ij}}^{(2k_\alpha,\alpha)}(\mathbf{x}_m,t_{m-})>0 \text{ and } \hbar_\alpha G_{\partial\Omega_{ij}}^{(m_\beta,\beta)}(\mathbf{x}_m,t_{m+})>0.\end{aligned}\tag{4.13}$$

The vector field of $\mathbf{F}^{(\alpha\succ\beta)}(\mathbf{x}^{(\alpha)},t,\boldsymbol{\pi}_\alpha,q^{(\alpha)})$ is called the coming *flow barrier* in the $(2k_\alpha:m_\beta)$-semi-passable flow on the α-side if the following conditions are satisfied. The critical values of $\mathbf{F}^{(\alpha\succ\beta)}(\mathbf{x}^{(\alpha)},t,\boldsymbol{\pi}_\alpha,q^{(\alpha)})$ ($\sigma=1,2$) are called the *lower and upper limits* of the coming flow barriers in the $(2k_\alpha:m_\beta)$-semi-passable flow on the α - side.

(i) The coming flow of $\mathbf{x}^{(\alpha)}$ cannot be switched to the leaving flow of $\mathbf{x}^{(\beta)}$ if

$$\begin{aligned}&\mathbf{x}^{(\alpha)}(t_{m-})=\mathbf{x}^{(\alpha\succ\beta)}(t_{m\pm},q^{(\alpha)})=\mathbf{x}_m,\\&G_{\partial\Omega_{ij}}^{(2k_\alpha,\alpha)}(\mathbf{x}_m,t_{m-})\in\left(\hbar_\alpha G_{\partial\Omega_{ij}}^{(2k_\alpha,\alpha\succ\beta)}(\mathbf{x}_m,q_1^{(\alpha)}),\ \hbar_\alpha G_{\partial\Omega_{ij}}^{(2k_\alpha,\alpha\succ\beta)}(\mathbf{x}_m,q_2^{(\alpha)})\right).\end{aligned}\tag{4.14}$$

(ii) The coming flow of $\mathbf{x}^{(\alpha)}$ cannot be switched to the leaving flow of $\mathbf{x}^{(\beta)}$ at the critical points of the flow barrier (i.e., $q^{(\alpha)}=q_\sigma^{(\alpha)}$, $\sigma\in\{1,2\}$) if

$$\begin{aligned}&\mathbf{x}^{(\alpha)}(t_{m-})=\mathbf{x}^{(\alpha\succ\beta)}(t_{m\pm},q_\sigma^{(\alpha)})=\mathbf{x}_m,\\&G_{\partial\Omega_{ij}}^{(s_\alpha,\alpha)}(\mathbf{x}_m,t_{m-})=G_{\partial\Omega_{ij}}^{(s_\alpha,\alpha\succ\beta)}(\mathbf{x}_m,q_\sigma^{(\alpha)})\neq 0 \text{ for } s_\alpha=2k_\alpha,2k_\alpha+1,\cdots,l_\alpha-1;\\&(-1)^\sigma\hbar_\alpha\mathbf{n}_{\partial\Omega_{ij}}^{\mathrm{T}}(\mathbf{x}^{(0)}(t_{m+\varepsilon}))\cdot\left[\mathbf{x}^{(\alpha)}(t_{m+\varepsilon})-\mathbf{x}^{(\alpha\succ\beta)}(t_{m+\varepsilon},q_\sigma^{(\alpha)})\right]<0.\end{aligned}\tag{4.15}$$

(iii) The coming flow of $\mathbf{x}^{(\alpha)}$ is switched to the leaving flow of $\mathbf{x}^{(\beta)}$ at the critical points of the flow barrier (i.e., $q^{(\alpha)} = q_\sigma^{(\alpha)}$, $\sigma \in \{1,2\}$) if

$$
\begin{gathered}
\mathbf{x}^{(\alpha)}(t_{m-}) = \mathbf{x}^{(\alpha \succ \beta)}(t_{m\pm}, q_\sigma^{(\alpha)}) = \mathbf{x}_m, \\
G_{\partial\Omega_{ij}}^{(s_\alpha,\alpha)}(\mathbf{x}_m, t_{m-}) = G_{\partial\Omega_{ij}}^{(s_\alpha,\alpha\succ\beta)}(\mathbf{x}_m, q_\sigma^{(\alpha)}) \neq 0 \text{ for } s_\alpha = 2k_\alpha, 2k_\alpha+1, \cdots, l_\alpha - 1; \\
(-1)^\sigma \hbar_\alpha \mathbf{n}_{\partial\Omega_{ij}}^{\mathrm{T}}(\mathbf{x}^{(0)}(t_{m+\varepsilon})) \cdot \left[\mathbf{x}^{(\alpha)}(t_{m+\varepsilon}) - \mathbf{x}^{(\alpha\succ\beta)}(t_{m+\varepsilon}, q_\sigma^{(\alpha)}) \right] > 0.
\end{gathered}
\tag{4.16}
$$

To explain the above concept, the G-functions for the coming flow barriers on the α-side of the boundary are presented in Fig. 4.1. The G-function of the leaving flow of $\mathbf{x}^{(\beta)}$, relative to the coming flow barrier on the α-side, is denoted by dashed curves. The thick line on the boundary $\partial\Omega_{ij}$ represents the G-function of the coming flow barrier. For $\mathbf{n}_{\partial\Omega_{ij}} \to \Omega_\beta$, one has $\hbar_\alpha = +1$. If $G_{\partial\Omega_{ij}}^{(2k_\alpha,\alpha)} > 0$ and $G_{\partial\Omega_{ij}}^{(m_\beta,\beta)} > 0$, without any flow barriers, the coming flow can be switched to the leaving flow from domain Ω_α to Ω_β. If there is a coming flow barrier with *lower* and *upper* limits ($G_{\partial\Omega_{ij}}^{(2k_\alpha,\alpha\succ\beta)}(q_1^{(\alpha)}) > 0$, $G_{\partial\Omega_{ij}}^{(2k_\alpha,\alpha\succ\beta)}(q_2^{(\alpha)}) > 0$) on the boundary $\partial\Omega_{ij}$. For $G_{\partial\Omega_{ij}}^{(2k_\alpha,\alpha)} \in [G_{\partial\Omega_{ij}}^{(2k_\alpha,\alpha\succ\beta)}(q_1^{(\alpha)}), G_{\partial\Omega_{ij}}^{(2k_\alpha,\alpha\succ\beta)}(q_2^{(\alpha)})]$, the coming flow in the $(2k_\alpha : m_\beta)$-semi-passable flow cannot be switched to the leaving flow in domain Ω_β. When a coming flow of $G_{\partial\Omega_{ij}}^{(2k_\alpha,\alpha)} > 0$ arrives to the boundary $\partial\Omega_{ij}$ with a flow barrier, such a coming flow can be switched from domain Ω_α to Ω_β only if $G_{\partial\Omega_{ij}}^{(2k_\alpha,\alpha)} \notin [G_{\partial\Omega_{ij}}^{(2k_\alpha,\alpha\succ\beta)}(q_1^{(\alpha)}), G_{\partial\Omega_{ij}}^{(2k_\alpha,\alpha\succ\beta)}(q_2^{(\alpha)})]$. For this case, the flow barrier is sketched in Fig. 4.1(a). In fact, the G-function for the lower limit of the coming flow barrier can be less than zero (i.e., $G_{\partial\Omega_{ij}}^{(2k_\alpha,\alpha\succ\beta)}(q_1^{(\alpha)}) < 0$). However, for $G_{\partial\Omega_{ij}}^{(2k_\alpha,\alpha)} < 0$, the coming flow of $\mathbf{x}^{(\alpha)}$ becomes a source flow, which will never pass through the boundary. Thus, the lower limit of the coming flow barrier of $G_{\partial\Omega_{ij}}^{(2k_\alpha,\alpha\succ\beta)}(q_1^{(\alpha)}) < 0$ is not important for $\mathbf{n}_{\partial\Omega_{ij}} \to \Omega_\beta$. Therefore, the lower limit of a coming flow barrier with $G_{\partial\Omega_{ij}}^{(2k_\alpha,\alpha\succ\beta)}(q_1^{(\alpha)}) = 0$ can be considered as the *lowest* limit. However, the G-function of the upper coming flow barrier can approach infinity (i.e., $G_{\partial\Omega_{ij}}^{(2k_\alpha,\alpha\succ\beta)}(q_2^{(\alpha)}) \to +\infty$). For $\mathbf{n}_{\partial\Omega_{ij}} \to \Omega_\alpha$, a coming flow barrier on the α-side can be similarly discussed, as in Fig. 4.1(b). The G-functions of the lower and upper limits of a coming flow barrier can be zero and negative infinity, respectively. The lower limit of $G_{\partial\Omega_{ij}}^{(2k_\alpha,\alpha\succ\beta)}(q_1^{(\alpha)}) > 0$ is not significant for $\mathbf{n}_{\partial\Omega_{ij}} \to \Omega_\alpha$. Of course, to block the flow from both sides of the boundary, one can simply define the coming flow barrier with negative and positive limits of the G-functions. In addition, the G-function of the coming flow barrier varies with the location of boundary.

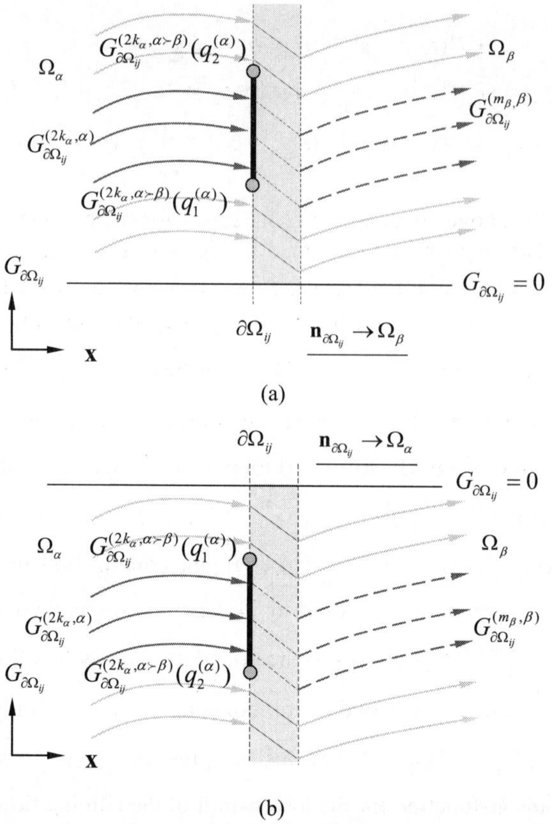

Fig. 4.1 *G*-functions for coming flow barriers in the $(2k_\alpha : m_\beta)$-semi-passable flow on the α-side of the boundary: **(a)** $\mathbf{n}_{\partial\Omega_{ij}} \to \Omega_\beta$ and **(b)** $\mathbf{n}_{\partial\Omega_{ij}} \to \Omega_\alpha$. The dashed curves are the *G*-function of the leaving flow of $\mathbf{x}^{(\beta)}$ relative to the coming flow barrier. The thick line is the *G*-function of the coming flow barriers. $G^{(2k_\alpha,\alpha\succ\beta)}_{\partial\Omega_{ij}}(q_1^{(\alpha)})$ and $G^{(2k_\alpha,\alpha\succ\beta)}_{\partial\Omega_{ij}}(q_2^{(\alpha)})$ are for *lower* and *upper* limits of the coming flow barrier ($k_\alpha, m_\beta \in \{0,1,2,\cdots\}$).

On some subsets of the boundary, no flow barriers exist. The coming flow barrier on such a boundary can be a *partial coming flow barrier.* If the coming flow barrier exists on the entire boundary, the flow barrier is a *full flow barrier* on the boundary. The strict definitions are given as follows.

Definition 4.4. For a discontinuous dynamical system in Eq. (4.1), there is a point $\mathbf{x}^{(0)}(t_m) \equiv \mathbf{x}_m \in \partial\Omega_{ij}$ at time t_m between two adjacent domains Ω_α ($\alpha = i, j$). Suppose there is a coming flow barrier $\mathbf{F}^{(\alpha\succ\beta)}(\mathbf{x}^{(\alpha)}, t, \boldsymbol{\pi}_\alpha, q^{(\alpha)})$ on the α-side in the $(2k_\alpha : m_\beta)$-semi-passable flow for $q^{(\alpha)} \in [q_1^{(\alpha)}, q_2^{(\alpha)}]$ ($\alpha, \beta \in \{i, j\}$ and $\alpha \neq \beta$).

(i) The coming flow barrier in the $(2k_\alpha : m_\beta)$-semi-passable flow is *partial* on the α-side if $\mathbf{x}_m \in S \subset \partial\Omega_{ij}$.

(ii) The coming flow barrier in the $(2k_\alpha : m_\beta)$-semi-passable flow is *full* on the α-side if $\mathbf{x}_m \in S = \partial\Omega_{ij}$.

For $k_\alpha = m_\beta = 0$, the above definitions are suitable for the fundamental flow barriers. The partial and full coming flow barriers on the α-side of the boundary $\partial\Omega_{ij}$ are sketched in Fig. 4.2. The partial coming flow barriers only exist on subsets of the boundary (i.e., $S \subset \partial\Omega_{ij}$). On the other subsets ($\partial\Omega_{ij} \setminus S$), the coming flow barriers do not exist. Once a coming flow in domain Ω_α arrives to such subsets, the coming flow in the semi-passable flow can be switched to a leaving flow on the β-side. If a coming flow barrier exists on $S = \partial\Omega_{ij}$, such a flow barrier is a *full flow barrier*. Any coming flow barriers possess the lower and upper limits. As discussed before, the lowest limit of the coming flow barrier is $G_{\partial\Omega_{ij}}^{(2k_\alpha,\alpha\succ\beta)}(q_1^{(\alpha)}) = 0$. If $G_{\partial\Omega_{ij}}^{(2k_\alpha,\alpha\succ\beta)}(q_2^{(\alpha)})$ is finite, the coming flow barrier with the lowest limit is a *flow barrier with the upper limit*. If the G-function of the *upper limit of the coming flow barrier* is infinity but $G_{\partial\Omega_{ij}}^{(2k_\alpha,\alpha\succ\beta)}(q_1^{(\alpha)}) \neq 0$, the coming flow barrier is a *flow barrier with a lower limit*. If such a flow barrier exists on $S \subseteq \partial\Omega_{ij}$, the partial and full coming flow barriers with an upper or lower limit are sketched in Figs. 4.3 and 4.4, respectively. If $G_{\partial\Omega_{ij}}^{(2k_\alpha,\alpha\succ\beta)}(q_1^{(\alpha)}) = 0$ and $G_{\partial\Omega_{ij}}^{(2k_\alpha,\alpha\succ\beta)}(q_2^{(\alpha)}) \to +\infty$ for $\mathbf{x}_m \in \partial\Omega_{ij}$, the coming flow of $\mathbf{x}^{(\alpha)}$ at a boundary point cannot pass over the boundary. The coming flow barrier at this point is *an absolute flow barrier*. Of course, if the entire boundary possesses such absolute flow barriers, the coming flow barrier is a *flow barrier wall*. If there are many partial coming flow barriers on the boundary, a coming flow barrier fence can be formed. If the G-function of the flow barrier on $S \subset \partial\Omega_{ij}$ cannot be defined for $q^{(\alpha)} \in (q_1^{(\alpha)}, q_2^{(\alpha)})$, the window of flow barrier can be formed where no flow barriers on such a portion exist. The definitions are given as follows.

Definition 4.5. For a discontinuous dynamical system in Eq. (4.1), there is a point $\mathbf{x}^{(0)}(t_m) \equiv \mathbf{x}_m \in \partial\Omega_{ij}$ at time t_m between two adjacent domains Ω_α ($\alpha = i, j$). Suppose a coming flow barrier $\mathbf{F}^{(\alpha\succ\beta)}(\mathbf{x}^{(\alpha)}, t, \boldsymbol{\pi}_\alpha, q^{(\alpha)})$ exists on the α-side in the $(2k_\alpha : m_\beta)$ semi-passable flow for $q^{(\alpha)} \in [q_1^{(\alpha)}, q_2^{(\alpha)}]$ ($\alpha, \beta \in \{i, j\}$ and $\alpha \neq \beta$).

(i) The coming flow barrier in the $(2k_\alpha : m_\beta)$-semi-passable flow is with an *upper* limit if for $\mathbf{x}_m \in S \subseteq \partial\Omega_{ij}$

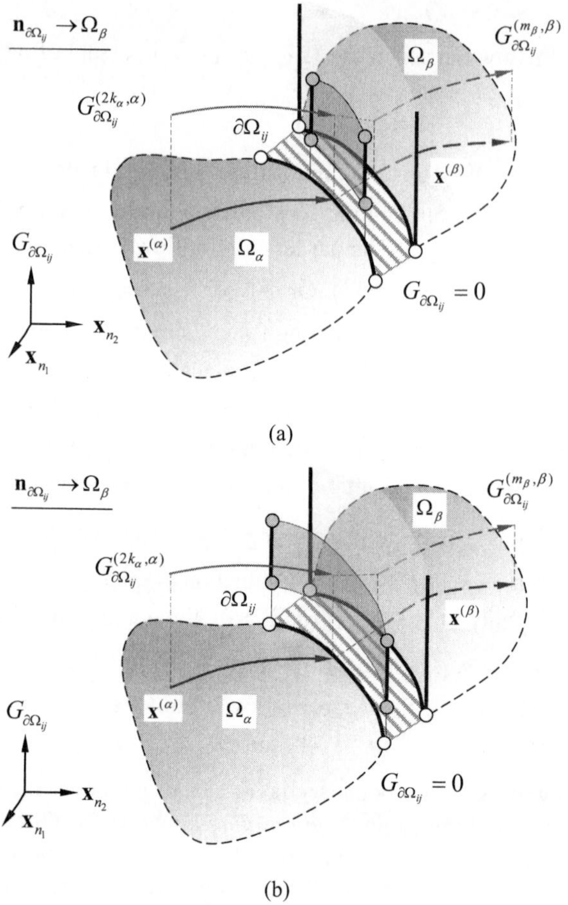

Fig. 4.2 A coming flow barrier on the α-side in the $(2k_\alpha : m_\beta)$-semi-passable flow: **(a)** partial flow barrier and **(b)** full flow barrier. The dark gray surface is for the flow barrier on $\partial\Omega_{ij}$. The upper solid and dashed curves with arrows are G-functions of flows on α and β-domains. The lower curves are semi-passable flows. The vertical surface behind the dark gray surface represents "no flow barrier". ($k_\alpha, m_\beta \in \{0,1,2,\cdots\}$). (color plot in the book end)

$$\hbar_\alpha G^{(2k_\alpha,\alpha\succ\beta)}_{\partial\Omega_{ij}}(\mathbf{x}_m, q_1^{(\alpha)}) = 0_- \text{ and } \hbar_\alpha G^{(2k_\alpha,\alpha\succ\beta)}_{\partial\Omega_{ij}}(\mathbf{x}_m, q_2^{(\alpha)}) \neq \infty. \tag{4.17}$$

(ii) The coming flow barrier in the $(2k_\alpha : m_\beta)$-semi-passable flow is with a *lower* limit for $\mathbf{x}_m \in S \subseteq \partial\Omega_{ij}$ if

$$\hbar_\alpha G^{(2k_\alpha,\alpha\succ\beta)}_{\partial\Omega_{ij}}(\mathbf{x}_m, q_1^{(\alpha)}) \neq 0_- \text{ and } \hbar_\alpha G^{(2k_\alpha,\alpha\succ\beta)}_{\partial\Omega_{ij}}(\mathbf{x}_m, q_2^{(\alpha)}) \to +\infty. \tag{4.18}$$

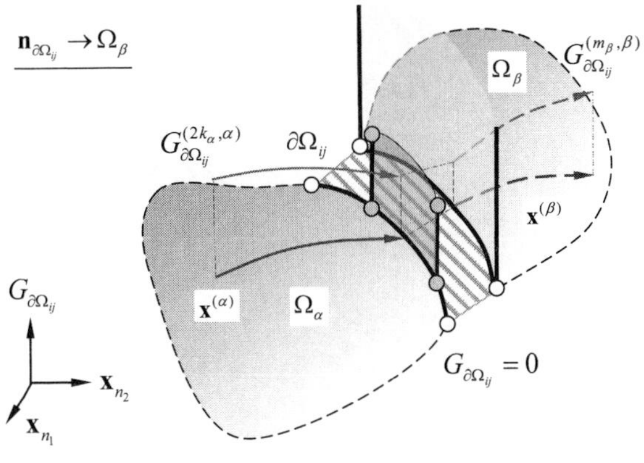

(a)

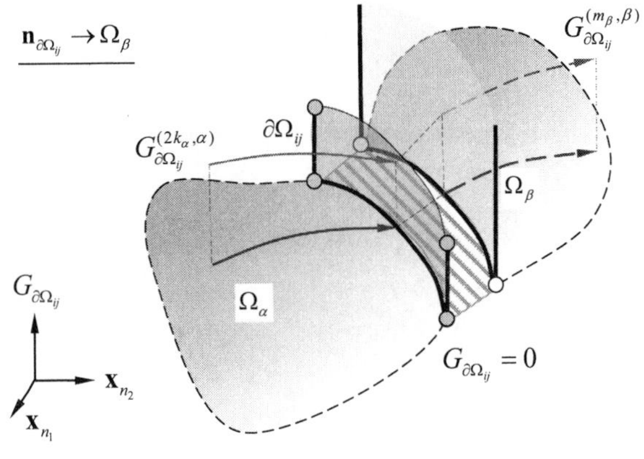

(b)

Fig. 4.3 The coming flow barrier on the α -side with upper limits in the $(2k_\alpha : m_\beta)$ -semi-passable flow: **(a)** partial flow barrier and **(b)** full flow barrier. The dark gray surface is the flow barrier surface on $\partial\Omega_{ij}$. The upper solid and dashed curves with arrows are *G*-functions of flows on α and β -domains. The lower curves are semi-passable flows. The vertical surface behind the dark gray surface represents "no flow barrier". ($k_\alpha, m_\beta \in \{0,1,2,\cdots\}$).

(iii) The coming flow barrier in the $(2k_\alpha : m_\beta)$-semi-passable flow is *absolute* for $\mathbf{x}_m \in S \subseteq \partial\Omega_{ij}$ if

$$\hbar_\alpha G_{\partial\Omega_{ij}}^{(2k_\alpha,\alpha\succ\beta)}(\mathbf{x}_m, q_1^{(\alpha)}) = 0_- \text{ and } \hbar_\alpha G_{\partial\Omega_{ij}}^{(2k_\alpha,\alpha\succ\beta)}(\mathbf{x}_m, q_2^{(\alpha)}) \to \infty. \tag{4.19}$$

(iv) The coming flow barrier in the $(2k_\alpha : m_\beta)$-semi-passable flow is a *flow barrier wall* on the α -side if the absolute flow barrier exists on $\mathbf{x}_m \in S = \partial\Omega_{ij}$.

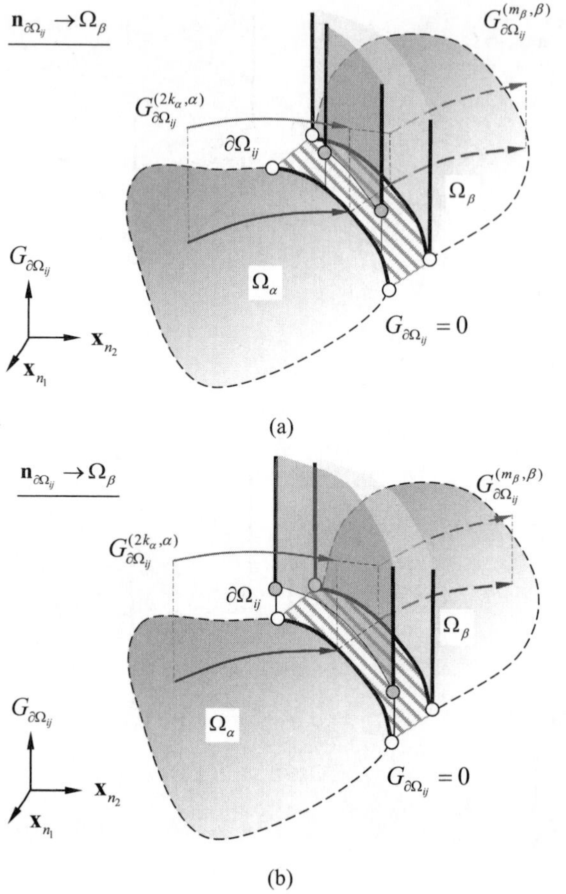

Fig. 4.4 The infinity coming flow barrier on the α -side in the $(2k_\alpha : m_\beta)$ -semi-passable flow: **(a)** partial flow barrier and **(b)** full flow barrier. The dark gray surface is the flow barrier surface on $\partial\Omega_{ij}$. The upper solid and dashed curves with arrows are *G*-functions of flows on α and β -domains. The lower curves are semi-passable flows. The vertical surface behind the dark gray surface represents "no flow barrier". ($k_\alpha, m_\beta \in \{0,1,2,\cdots\}$).

(v) The *coming* flow barrier in the $(2k_\alpha : m_\beta)$ -semi-passable flow is a *flow barrier fence* on the α -side if the flow barriers exist on $S_{k_1} \subset \partial\Omega_{ij}$ and no flow barriers on $S_{k_2} \subset \partial\Omega_{ij}$ ($k_1, k_2 \in \{1,2,\cdots\}$) for $S_{k_1} \cap S_{k_2} = \varnothing$.

Definition 4.6. For a discontinuous dynamical system in Eq. (4.1), there is a point $\mathbf{x}^{(0)}(t_m) \equiv \mathbf{x}_m \in \partial\Omega_{ij}$ at time t_m between two adjacent domains Ω_α ($\alpha = i, j$). The coming and leaving flows in the $(2k_\alpha : m_\beta)$ -semi-passable flow satisfy the conditions in Eq. (4.13). Many flow barriers $\mathbf{F}^{(\alpha\succ\beta)}(\mathbf{x}^{(\alpha)},t,\boldsymbol{\pi}_\alpha,q^{(\alpha)})$ exist on the α-side for $\mathbf{x}_m\in S\subseteq\partial\Omega_{ij}$ and $q^{(\alpha)}\in[q_{2n-1}^{(\alpha)},q_{2n}^{(\alpha)}]$ ($n=1,2,\cdots$) with $\sigma=2n-1,2n$

$$\begin{aligned}&\hbar_\alpha G_{\partial\Omega_{ij}}^{(s_\alpha,\alpha\succ\beta)}(\mathbf{x}_m,q_\sigma^{(\alpha)})=0\text{ for }s_\alpha=0,1,2,\cdots,2k_\alpha-1;\\&\hbar_\alpha G_{\partial\Omega_{ij}}^{(2k_\alpha,\alpha\succ\beta)}(\mathbf{x}_m,q^{(\alpha)})\in\left[\hbar_\alpha G_{\partial\Omega_{ij}}^{(2k_\alpha,\alpha\succ\beta)}(\mathbf{x}_m,q_{2n-1}^{(\alpha)}),\hbar_\alpha G_{\partial\Omega_{ij}}^{(2k_\alpha,\alpha\succ\beta)}(\mathbf{x}_m,q_{2n}^{(\alpha)})\right]\\&\qquad\subset[0,\infty).\end{aligned}\tag{4.20}$$

For $q^{(\alpha)}\in(q_{2n}^{(\alpha)},q_{2n+1}^{(\alpha)})$ ($n=1,2,\cdots$), no flow barriers exist on $S\subseteq\partial\Omega_{ij}$. Thus, the coming flow of $\mathbf{x}^{(\alpha)}$ can be switched to the leaving flow of $\mathbf{x}^{(\beta)}$ for $\sigma=2n,2n+1$

$$\begin{aligned}&\hbar_\alpha G_{\partial\Omega_{ij}}^{(s_\alpha,\alpha\succ\beta)}(\mathbf{x}_m,q_\sigma^{(\alpha)})=0\text{ for }s_\alpha=0,1,2,\cdots,2k_\alpha-1;\\&\hbar_\alpha G_{\partial\Omega_{ij}}^{(2k_\alpha,\alpha)}(\mathbf{x}_m,t_{m-})\in\left(\hbar_\alpha G_{\partial\Omega_{ij}}^{(2k_\alpha,\alpha\succ\beta)}(\mathbf{x}_m,q_{2n}^{(\alpha)}),\hbar_\alpha G_{\partial\Omega_{ij}}^{(2k_\alpha,\alpha\succ\beta)}(\mathbf{x}_m,q_{2n+1}^{(\alpha)})\right)\\&\qquad\subset[0,\infty).\end{aligned}\tag{4.21}$$

The G-function intervals for all $\mathbf{x}_m\in S\subseteq\partial\Omega_{ij}$ with $q^{(\alpha)}\in(q_{2n-2}^{(\alpha)},q_{2n-1}^{(\alpha)})$ are called the windows of the coming flow barrier on the α-side in the $(2k_\alpha:m_\beta)$-semi-passable flow.

Definition 4.7. For a discontinuous dynamical system in Eq. (4.1), there is a point $\mathbf{x}^{(0)}(t_m)\equiv\mathbf{x}_m\in\partial\Omega_{ij}$ at time t_m between two adjacent domains Ω_α ($\alpha=i,j$). Suppose there is a coming flow barrier of $\mathbf{F}^{(\alpha\succ\beta)}(\mathbf{x}^{(\alpha)},t,\boldsymbol{\pi}_\alpha,q^{(\alpha)})$ on the α-side in the $(2k_\alpha:m_\beta)$-semi-passable flow for $q^{(\alpha)}\in[q_1^{(\alpha)},q_2^{(\alpha)}]$ ($\alpha,\beta\in\{i,j\}$ and $\alpha\neq\beta$).

(i) The coming flow barrier in the $(2k_\alpha:m_\beta)$-semi-passable flow is *permanent* on the α-side if the flow barrier is *independent* of time $t\in[0,\infty)$.

(ii) The coming flow barrier in the $(2k_\alpha:m_\beta)$-semi-passable flow is *instantaneous* on the α-side if the flow barrier is continuously *dependent* on time $t\in[0,\infty)$.

(iii) The coming flow barrier in the $(2k_\alpha:m_\beta)$-semi-passable flow is *intermittent* on the α-side if the flow barrier exists for time $t\in[t_k,t_{k+1}]$ with $k\in\mathbb{Z}$.

(iv) The coming flow barrier in the $(2k_\alpha:m_\beta)$-semi-passable flow is *intermittent and static* on the α-side if the flow barrier is *independent* of time $t\in[t_k,t_{k+1}]$ with $k\in\mathbb{Z}$.

From the previous definition, the window of the flow barrier is sketched in Fig. 4.5. On the window area, the flow can be switched. Similarly, the permanent and instantaneous windows of the flow barrier on the boundary can be discussed. Further, the concept of the door for the flow barrier wall is also shown in Fig. 4.6.

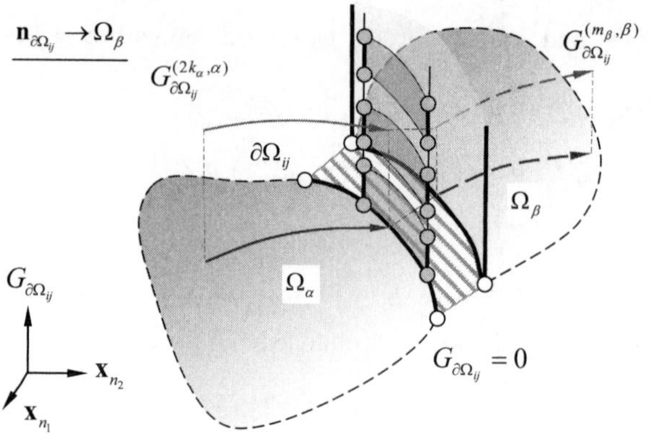

(a)

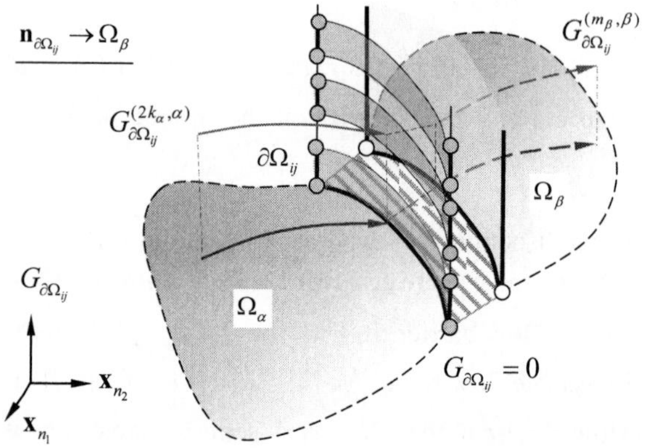

(b)

Fig. 4.5 The coming barrier windows on the α -side of $\partial\Omega_{ij}$ in the $(2k_\alpha : m_\beta)$ -semi-passable flow: **(a)** partial flow barrier and **(b)** full flow barrier. The dark gray surface is the flow barrier surface. The upper solid and dashed curves with arrows are *G*-functions of flows on α - and β - domains. The lower curves are semi-passable flows. The vertical surface behind the dark gray surface represents "no flow barrier". ($k_\alpha, m_\beta \in \{0,1,2,\cdots\}$)

Definition 4.8. For a discontinuous dynamical system in Eq. (4.1), there is a point $\mathbf{x}^{(0)}(t_m) \equiv \mathbf{x}_m \in \partial\Omega_{ij}$ at time t_m between two adjacent domains Ω_α ($\alpha = i, j$). Suppose a coming flow barrier of $\mathbf{F}^{(\alpha \succ \beta)}(\mathbf{x}^{(\alpha)}, t, \boldsymbol{\pi}_\alpha, q^{(\alpha)})$ exists on the α -side for $\mathbf{x}_m \in S \subseteq \partial\Omega_{ij}$ and $q^{(\alpha)} \in [q^{(\alpha)}_{2n-1}, q^{(\alpha)}_{2n}]$ ($n = 1, 2, \cdots$) and there is a window of the coming flow barrier for $S \subset \partial\Omega_{ij}$ and $q^{(\alpha)} \in (q^{(\alpha)}_{2n-2}, q^{(\alpha)}_{2n-1}]$ in the $(2k_\alpha : m_\beta)$ -semi-passable flow.

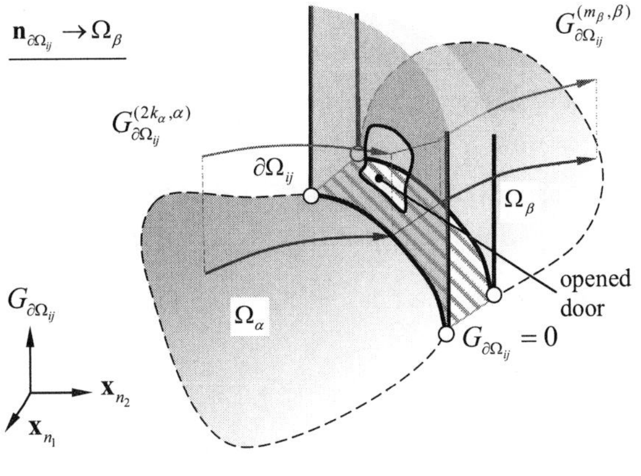

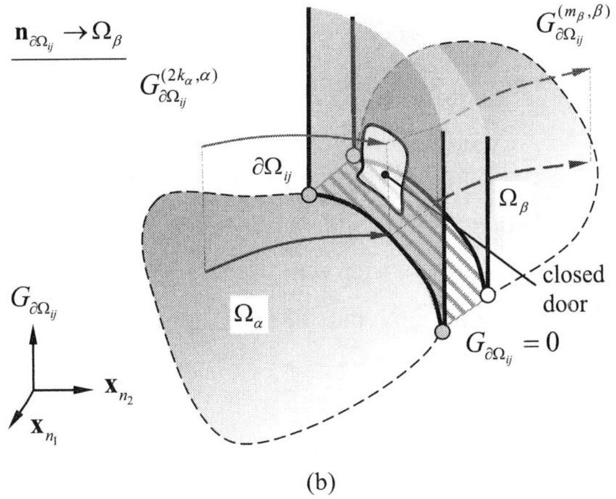

Fig. 4.6 The door of the flow barrier wall on the α -side: **(a)** opened door **(b)** closed door. The dark surface is the flow barrier surface. The upper solid and dashed curves with arrows are G-functions of flows on α and β -domains. The lower curves are semi-passable flows. The vertical surface behind the dark gray surface represents "no flow barrier". ($k_\alpha, m_\beta \in \{0,1,2,\cdots\}$).

(i) The flow barrier window in the $(2k_\alpha : m_\beta)$-semi-passable flow is *permanent* on the α -side if the window is *independent* of time $t \in [0,+\infty)$.

(ii) The flow barrier window in the $(2k_\alpha : m_\beta)$-semi-passable flow is *instantaneous* on the α -side if the window is continuously *dependent* on time $t \in [0,+\infty)$.

(iii) The flow barrier window in the $(2k_\alpha : m_\beta)$-semi-passable flow is *intermittent* on the α -side if the window exists for time $t \in [t_k, t_{k+1}]$ with $k \in \mathbb{Z}$.

(iv) The flow barrier window in the $(2k_\alpha : m_\beta)$-semi-passable flow is *intermittent and static* on the α-side if the window is *independent* of time $t \in [t_k, t_{k+1}]$ with $k \in \mathbb{Z}$.

Definition 4.9. For a discontinuous dynamical system in Eq. (4.1), there is a point $\mathbf{x}^{(0)}(t_m) \equiv \mathbf{x}_m \in \partial\Omega_{ij}$ at time t_m between two adjacent domains Ω_α ($\alpha = i, j$). Suppose a coming flow barrier wall of $\mathbf{F}^{(\alpha \succ \beta)}(\mathbf{x}^{(\alpha)}, t, \boldsymbol{\pi}_\alpha, q^{(\alpha)})$ on $S = \partial\Omega_{ij}$ exists in the $(2k_\alpha : m_\beta)$-semi-passable flow for $q^{(\alpha)} \in [0, \infty)$ and there is an *intermittent*, static window of the coming flow barrier on $S \subset \partial\Omega_{ij}$ for $q^{(\alpha)} \in [q_1^{(\alpha)}, q_2^{(\alpha)}]$ and $t \in [t_k, t_{k+1}]$ with $k \in \mathbb{Z}$.

(i) The window of the flow barrier in the $(2k_\alpha : m_\beta)$-semi-passable flow is called a *door of the flow barrier wall* on the α-side if the window and flow barriers exist alternatively.
(ii) The door of the coming flow barrier wall in the $(2k_\alpha : m_\beta)$-semi-passable flow is *open* for time $t \in [t_k, t_{k+1}]$ with $k \in \mathbb{Z}$ if the window exists.
(iii) The door of the coming flow barrier wall in the $(2k_\alpha : m_\beta)$-semi-passable flow is *closed* for time $t \in [t_{k+1}, t_{k+2}]$ with $k \in \mathbb{Z}$ if the flow barrier exists.
(iv) The door of the coming flow barrier wall in the $(2k_\alpha : m_\beta)$-semi-passable flow is *permanently open* if the window exists for time $t \in [t_k, \infty)$.
(v) The door of the coming flow barrier wall in the $(2k_\alpha : m_\beta)$-semi-passable flow is *permanently closed* if the flow barrier exists for time $t \in [t_{k+1}, \infty)$.

For a passable flow on the boundary $\partial\Omega_{ij}$, if a flow barrier exists on such a boundary with a subset $S \subseteq \partial\Omega_{ij}$, the flow cannot pass through the boundary under the following conditions

$$\hbar_\alpha G_{\partial\Omega_{ij}}^{(2k_\alpha,\alpha)}(\mathbf{x}_m, t_{m-}) \in \left(\hbar_\alpha G_{\partial\Omega_{ij}}^{(2k_\alpha,\alpha \succ \beta)}(\mathbf{x}_m, q_1^{(\alpha)}), \hbar_\alpha G_{\partial\Omega_{ij}}^{(2k_\alpha,\alpha \succ \beta)}(\mathbf{x}_m, q_2^{(\alpha)})\right) \tag{4.22}$$

for $\mathbf{x}_m \in S \subseteq \partial\Omega_{ij}$. For this case, the dynamical system on the α-side of the boundary will be constrained by the boundary $\partial\Omega_{ij}$, i.e.,

$$\begin{aligned} &\dot{\mathbf{x}}^{(\alpha)} = \mathbf{F}^{(\alpha)}(\mathbf{x}^{(\alpha)}, t, \boldsymbol{\mu}_\alpha) \text{ in } \Omega_\alpha \ (\alpha = i, j), \\ &\text{with } \varphi_{ij}(\mathbf{x}^{(\alpha)}, t, \boldsymbol{\lambda}) = 0 \text{ on } \partial\Omega_{ij}. \end{aligned} \tag{4.23}$$

The vector field of the dynamical system on the α-side of the boundary is given by the vector field on the domain of Ω_α. Because the coming flow barrier exists on the α-side of the boundary $\partial\Omega_{ij}$, the coming flow will be along the α-side of the boundary until the condition in Eq. (4.22) cannot be satisfied. If the tangential components of vector field in Eq. (4.13) on the boundary are zero, the coming flow will stay on the specific point of the boundary until the flow barrier can be passed over. The coming flow at this point is called the "*standing flow*" on the α-side of the boundary. For this case, the transport laws may exist, which transport to a different vector field in the same domain or another accessible domain or on the other boundary.

Theorem 4.1. *For a discontinuous dynamical system in Eq.* (4.1), *there is a point* $\mathbf{x}^{(0)}(t_m) \equiv \mathbf{x}_m \in \partial\Omega_{ij}$ *at time* t_m *between two adjacent domains* Ω_α ($\alpha = i, j$). *Suppose a coming flow barrier of* $\mathbf{F}^{(\alpha\succ\beta)}(\mathbf{x}^{(\alpha)}, t, \boldsymbol{\pi}_\alpha, q^{(\alpha)})$ *for* $q^{(\alpha)} \in [q_1^{(\alpha)}, q_2^{(\alpha)}]$ *in a passable flow exists on the boundary* $\partial\Omega_{ij}$ *for* $\mathbf{x}_m \in S \subseteq \partial\Omega_{ij}$

$$\begin{aligned}\hbar_\alpha G_{\partial\Omega_{ij}}^{(\alpha\succ\beta)}(\mathbf{x}_m, q^{(\alpha)}) &\in \left[\hbar_\alpha G_{\partial\Omega_{ij}}^{(\alpha\succ\beta)}(\mathbf{x}_m, q_1^{(\alpha)}), \hbar_\alpha G_{\partial\Omega_{ij}}^{(\alpha\succ\beta)}(\mathbf{x}_m, q_2^{(\alpha)})\right] \\ &\subset [0, \infty).\end{aligned} \tag{4.24}$$

The coming and leaving flows in the semi-passable flow at the boundary satisfy

$$\hbar_\alpha G_{\partial\Omega_{ij}}^{(\alpha)}(\mathbf{x}_m, t_{m-}) > 0 \;\; \textit{and} \;\; \hbar_\alpha G_{\partial\Omega_{ij}}^{(\beta)}(\mathbf{x}_m, t_{m+}) > 0. \tag{4.25}$$

(i) *The coming flow in the semi-passable flow cannot pass through the flow barrier on the* α*-side of the boundary* $\partial\Omega_{ij}$ *at* $q^{(\alpha)} \in (q_1^{(\alpha)}, q_2^{(\alpha)})$ *if and only if*

$$\hbar_\alpha G_{\partial\Omega_{ij}}^{(\alpha)}(\mathbf{x}_m, t_{m-}) \in \left(\hbar_\alpha G_{\partial\Omega_{ij}}^{(\alpha\succ\beta)}(\mathbf{x}_m, q_1^{(\alpha)}), \hbar_\alpha G_{\partial\Omega_{ij}}^{(\alpha\succ\beta)}(\mathbf{x}_m, q_2^{(\alpha)})\right). \tag{4.26}$$

(ii) *The coming flow in the semi-passable flow cannot pass over the flow barrier on the* α*-side of the boundary at* $q^{(\alpha)} = q_\sigma^{(\alpha)}$ ($\sigma \in \{1,2\}$) *if and only if*

$$\begin{aligned}&\hbar_\alpha G_{\partial\Omega_{ij}}^{(s_\alpha,\alpha)}(\mathbf{x}_m, t_{m-}) = \hbar_\alpha G_{\partial\Omega_{ij}}^{(s_\alpha,\alpha\succ\beta)}(\mathbf{x}_m, q_\sigma^{(\alpha)}) \neq 0 \\ &\textit{for } s_\alpha = 0, 1, \cdots, l_\alpha - 1; \\ &(-1)^\sigma \hbar_\alpha \left[G_{\partial\Omega_{ij}}^{(l_\alpha,\alpha)}(\mathbf{x}_m, t_{m-}) - G_{\partial\Omega_{ij}}^{(l_\alpha,\alpha\succ\beta)}(\mathbf{x}_m, q_\sigma^{(\alpha)})\right] < 0.\end{aligned} \tag{4.27}$$

(iii) *The coming flow in the semi-passable flow passes over the flow barrier on the* α*-side of the boundary at* $q^{(\alpha)} = q_\sigma^{(\alpha)}$ ($\sigma \in \{1,2\}$) *if and only if*

$$\begin{aligned}&\hbar_\alpha G_{\partial\Omega_{ij}}^{(s_\alpha,\alpha)}(\mathbf{x}_m, t_{m-}) = \hbar_\alpha G_{\partial\Omega_{ij}}^{(s_\alpha,\alpha\succ\beta)}(\mathbf{x}_m, q_\sigma^{(\alpha)}) \in (0, \infty) \\ &\textit{for } s_\alpha = 0, 1, \cdots, l_\alpha - 1; \\ &(-1)^\sigma \hbar_\alpha \left[G_{\partial\Omega_{ij}}^{(l_\alpha,\alpha)}(\mathbf{x}_m, t_{m-}) - G_{\partial\Omega_{ij}}^{(l_\alpha,\alpha\succ\beta)}(\mathbf{x}_m, q_\sigma^{(\alpha)})\right] > 0.\end{aligned} \tag{4.28}$$

Proof. (i) From Definition 4.2, the condition in Eq. (4.26) is obtained, vice versa.

(ii) An auxiliary flow determined by the flow barrier is introduced as a fictitious flow $\mathbf{x}^{(\alpha\succ\beta)}(t)$. Since $\mathbf{x}^{(\alpha\succ\beta)}(t_{m\pm}) = \mathbf{x}^{(0)}(t_{m\pm})$ and $\mathbf{x}^{(\alpha)}(t_{m\pm}) = \mathbf{x}^{(0)}(t_{m\pm})$, the G-function definition gives

$$\mathbf{n}_{\partial\Omega_{ij}}^{\mathrm{T}}(\mathbf{x}^{(0)}(t_{m+\varepsilon}))\cdot[\mathbf{x}^{(\alpha\succ\beta)}(t_{m+\varepsilon},q_{\sigma}^{(\alpha)})-\mathbf{x}^{(0)}(t_{m+\varepsilon},q_{\sigma}^{(\alpha)})]$$
$$=\sum_{s_\alpha=0}^{l_\alpha-1}G_{\partial\Omega_{ij}}^{(s_\alpha,\alpha\succ\beta)}(\mathbf{x}_m,q_{\sigma}^{(\alpha)})\varepsilon^{s_\alpha+1}+G_{\partial\Omega_{ij}}^{(l_\alpha,\alpha\succ\beta)}(\mathbf{x}_m,q_{\sigma}^{(\alpha)})\varepsilon^{l_\alpha+1}+o(\varepsilon^{l_\alpha+1}),$$

$$\mathbf{n}_{\partial\Omega_{ij}}^{\mathrm{T}}(\mathbf{x}^{(0)}(t_{m+\varepsilon}))\cdot[\mathbf{x}^{(\alpha)}(t_{m+\varepsilon})-\mathbf{x}^{(0)}(t_{m+\varepsilon})]$$
$$=\sum_{s_\alpha=0}^{l_\alpha-1}G_{\partial\Omega_{ij}}^{(s_\alpha,\alpha)}(\mathbf{x}_m,t_{m+})\varepsilon^{s_\alpha+1}+G_{\partial\Omega_{ij}}^{(l_\alpha,\alpha)}(\mathbf{x}_m,t_{m+})\varepsilon^{l_\alpha+1}+o(\varepsilon^{l_\alpha+1}).$$

Because of

$$\mathbf{x}^{(\alpha\succ\beta)}(t_{m\pm})=\mathbf{x}^{(\alpha)}(t_{m\pm}),$$
$$G_{\partial\Omega_{ij}}^{(\alpha\succ\beta)}(\mathbf{x}_m,t_{m\pm},q_{\sigma}^{(\alpha)})=G_{\partial\Omega_{ij}}^{(\alpha)}(\mathbf{x}_m,t_{m\pm})\neq 0;$$

one achieves

$$\mathbf{n}_{\partial\Omega_{ij}}^{\mathrm{T}}(\mathbf{x}^{(0)}(t_{m+\varepsilon}))\cdot[\mathbf{x}^{(\alpha)}(t_{m+\varepsilon})-\mathbf{x}^{(\alpha\succ\beta)}(t_{m+\varepsilon},q_{\sigma}^{(\alpha)})]$$
$$=[G_{\partial\Omega_{ij}}^{(l_\alpha,\alpha)}(\mathbf{x}_m,t_{m+})-G_{\partial\Omega_{ij}}^{(l_\alpha,\alpha\succ\beta)}(\mathbf{x}_m,t_{m+},q_{\sigma}^{(\alpha)})]\varepsilon^{l_\alpha+1}.$$

The definition of a coming flow not passing through the flow barrier gives

$$\mathbf{n}_{\partial\Omega_{ij}}^{\mathrm{T}}(\mathbf{x}^{(0)}(t_{m+\varepsilon}))\cdot[\mathbf{x}^{(\alpha)}(t_{m+\varepsilon})-\mathbf{x}^{(\alpha\succ\beta)}(t_{m+\varepsilon},q_{\sigma}^{(\alpha)})]<0 \text{ for } \mathbf{n}_{\partial\Omega_{ij}}\to\Omega_\beta,$$
$$\mathbf{n}_{\partial\Omega_{ij}}^{\mathrm{T}}(\mathbf{x}^{(0)}(t_{m+\varepsilon}))\cdot[\mathbf{x}^{(\alpha)}(t_{m+\varepsilon})-\mathbf{x}^{(\alpha\succ\beta)}(t_{m+\varepsilon},q_{\sigma}^{(\alpha)})]>0 \text{ for } \mathbf{n}_{\partial\Omega_{ij}}\to\Omega_\alpha$$

for the upper limit of the coming flow barrier (i.e., $\sigma=2$), and

$$\mathbf{n}_{\partial\Omega_{ij}}^{\mathrm{T}}(\mathbf{x}^{(0)}(t_{m+\varepsilon}))\cdot[\mathbf{x}^{(\alpha)}(t_{m+\varepsilon})-\mathbf{x}^{(\alpha\succ\beta)}(t_{m+\varepsilon},q_{\sigma}^{(\alpha)})]>0 \text{ for } \mathbf{n}_{\partial\Omega_{ij}}\to\Omega_\beta,$$
$$\mathbf{n}_{\partial\Omega_{ij}}^{\mathrm{T}}(\mathbf{x}^{(0)}(t_{m+\varepsilon}))\cdot[\mathbf{x}^{(\alpha)}(t_{m+\varepsilon})-\mathbf{x}^{(\alpha\succ\beta)}(t_{m+\varepsilon},q_{\sigma}^{(\alpha)})]<0 \text{ for } \mathbf{n}_{\partial\Omega_{ij}}\to\Omega_\alpha$$

for the lower limit of the coming flow barrier (i.e., $\sigma=1$). With Eq. (4.2), if a flow cannot pass over the semi-passable flow barrier, one obtains the conditions in Eq. (4.27), vice versa.

(iii) The definition of a coming flow passing through the flow barrier gives

$$\mathbf{n}_{\partial\Omega_{ij}}^{\mathrm{T}}(\mathbf{x}^{(0)}(t_{m+\varepsilon}))\cdot[\mathbf{x}^{(\alpha)}(t_{m+\varepsilon})-\mathbf{x}^{(\alpha\succ\beta)}(t_{m+\varepsilon},q_{\sigma}^{(\alpha)})]>0 \text{ for } \mathbf{n}_{\partial\Omega_{ij}}\to\Omega_\beta,$$
$$\mathbf{n}_{\partial\Omega_{ij}}^{\mathrm{T}}(\mathbf{x}^{(0)}(t_{m+\varepsilon}))\cdot[\mathbf{x}^{(\alpha)}(t_{m+\varepsilon})-\mathbf{x}^{(\alpha\succ\beta)}(t_{m+\varepsilon},q_{\sigma}^{(\alpha)})]<0 \text{ for } \mathbf{n}_{\partial\Omega_{ij}}\to\Omega_\alpha$$

for the upper limit of the coming flow barrier (i.e., $\sigma=2$), and

$$\mathbf{n}_{\partial\Omega_{ij}}^{\mathrm{T}}(\mathbf{x}^{(0)}(t_{m+\varepsilon}))\cdot[\mathbf{x}^{(\alpha)}(t_{m+\varepsilon})-\mathbf{x}^{(\alpha\succ\beta)}(t_{m+\varepsilon},q_{\sigma}^{(\alpha)})]<0 \text{ for } \mathbf{n}_{\partial\Omega_{ij}}\to\Omega_\beta,$$
$$\mathbf{n}_{\partial\Omega_{ij}}^{\mathrm{T}}(\mathbf{x}^{(0)}(t_{m+\varepsilon}))\cdot[\mathbf{x}^{(\alpha)}(t_{m+\varepsilon})-\mathbf{x}^{(\alpha\succ\beta)}(t_{m+\varepsilon},q_{\sigma}^{(\alpha)})]>0 \text{ for } \mathbf{n}_{\partial\Omega_{ij}}\to\Omega_\alpha$$

for the lower limit of the coming flow barrier (i.e., $\sigma=1$). If the flow passes over the semi-passable flow barrier, one obtains the conditions in Eq. (4.28), and vice versa.

■

Theorem 4.2. *For a discontinuous dynamical system in Eq.* (4.1), *there is a point* $\mathbf{x}^{(0)}(t_m) \equiv \mathbf{x}_m \in \partial\Omega_{ij}$ *at time* t_m *between two adjacent domains* Ω_α ($\alpha = i, j$). *Suppose a coming flow barrier of* $\mathbf{F}^{(\alpha\succ\beta)}(\mathbf{x}^{(\alpha)}, t, \boldsymbol{\pi}_\alpha, q^{(\alpha)})$ *for* $q^{(\alpha)} \in [q_1^{(\alpha)}, q_2^{(\alpha)}]$ *in the* $(2k_\alpha : m_\beta)$*-passable flow exists on the boundary* $\partial\Omega_{ij}$ *for* $\mathbf{x}_m \in S \subseteq \partial\Omega_{ij}$

$$\begin{aligned} &G_{\partial\Omega_{ij}}^{(s_\alpha,\alpha\succ\beta)}(\mathbf{x}_m, q^{(\alpha)}) = 0 \text{ for } s_\alpha = 0,1,\cdots,2k_\alpha - 1;\\ &\hbar_\alpha G_{\partial\Omega_{ij}}^{(2k_\alpha,\alpha\succ\beta)}(\mathbf{x}_m, q^{(\alpha)}) \in \left[\hbar_\alpha G_{\partial\Omega_{ij}}^{(2k_\alpha,\alpha\succ\beta)}(\mathbf{x}_m, q_1^{(\alpha)}), \hbar_\alpha G_{\partial\Omega_{ij}}^{(2k_\alpha,\alpha\succ\beta)}(\mathbf{x}_m, q_2^{(\alpha)})\right]\\ &\qquad \subset [0,\infty). \end{aligned} \tag{4.29}$$

The coming and leaving flows in the $(2k_\alpha : m_\beta)$*-semi-passable flow satisfy*

$$\begin{aligned} &G_{\partial\Omega_{ij}}^{(s_\alpha,\alpha)}(\mathbf{x}_m, t_{m-}) = 0 \ \textit{for}\ s_\alpha = 0,1,\cdots,2k_\alpha - 1;\\ &G_{\partial\Omega_{ij}}^{(s_\beta,\beta)}(\mathbf{x}_m, t_{m+}) = 0 \ \textit{for}\ s_\beta = 0,1,\cdots,m_\beta - 1;\\ &\hbar_\alpha G_{\partial\Omega_{ij}}^{(2k_\alpha,\alpha)}(\mathbf{x}_m, t_{m-}) > 0 \ \textit{and}\ \hbar_\alpha G_{\partial\Omega_{ij}}^{(m_\beta,\beta)}(\mathbf{x}_m, t_{m+}) > 0. \end{aligned} \tag{4.30}$$

(i) *The coming flow in the* $(2k_\alpha : m_\beta)$*-semi-passable flow cannot pass through the flow barrier on the* α*-side at* $q^{(\alpha)} \in (q_1^{(\alpha)}, q_2^{(\alpha)})$ *if and only if*

$$\hbar_\alpha G_{\partial\Omega_{ij}}^{(2k_\alpha,\alpha)}(\mathbf{x}_m, t_{m-}) \in \left(\hbar_\alpha G_{\partial\Omega_{ij}}^{(2k_\alpha,\alpha\succ\beta)}(\mathbf{x}_m, q_1^{(\alpha)}), \hbar_\alpha G_{\partial\Omega_{ij}}^{(2k_\alpha,\alpha\succ\beta)}(\mathbf{x}_m, q_2^{(\alpha)})\right) \tag{4.31}$$

(ii) *The coming flow in the* $(2k_\alpha : m_\beta)$*-semi-passable flow cannot pass over the flow barrier on the* α*-side at* $q^{(\alpha)} = q_\sigma^{(\alpha)}$ ($\sigma \in \{1,2\}$) *if and only if*

$$\begin{aligned} &\hbar_\alpha G_{\partial\Omega_{ij}}^{(s_\alpha,\alpha)}(\mathbf{x}_m, t_{m-}) = \hbar_\alpha G_{\partial\Omega_{ij}}^{(s_\alpha,\alpha\succ\beta)}(\mathbf{x}_m, q_\sigma^{(\alpha)}) \in (0,\infty)\\ &\qquad \textit{for}\ s_\alpha = 2k_\alpha, 2k_\alpha + 1, \cdots, l_\alpha - 1;\\ &(-1)^\sigma \hbar_\alpha \left[G_{\partial\Omega_{ij}}^{(l_\alpha,\alpha)}(\mathbf{x}_m, t_{m-}) - G_{\partial\Omega_{ij}}^{(l_\alpha,\alpha\succ\beta)}(\mathbf{x}_m, q_\sigma^{(\alpha)})\right] < 0. \end{aligned} \tag{4.32}$$

(iii) *The coming flow in the* $(2k_\alpha : m_\beta)$*-semi-passable flow passes over the flow barrier on the* α*-side at* $q^{(\alpha)} = q_\sigma^{(\alpha)}$ ($\sigma \in \{1,2\}$) *if and only if*

$$\begin{aligned} &\hbar_\alpha G_{\partial\Omega_{ij}}^{(s_\alpha,\alpha)}(\mathbf{x}_m, t_{m-}) = \hbar_\alpha G_{\partial\Omega_{ij}}^{(s_\alpha,\alpha\succ\beta)}(\mathbf{x}_m, q_\sigma^{(\alpha)}) \in (0,\infty)\\ &\qquad \textit{for}\ s_\alpha = 2k_\alpha, 2k_\alpha + 1, \cdots, l_\alpha - 1;\\ &(-1)^\sigma \hbar_\alpha \left[G_{\partial\Omega_{ij}}^{(l_\alpha,\alpha)}(\mathbf{x}_m, t_{m-}) - G_{\partial\Omega_{ij}}^{(l_\alpha,\alpha\succ\beta)}(\mathbf{x}_m, q_\sigma^{(\alpha)})\right] > 0. \end{aligned} \tag{4.33}$$

Proof. (i) From Definition 4.3, the condition in Eq. (4.31) is obtained vice versa.

(ii) To prove this theorem, an auxiliary flow determined by the flow barrier is introduced as a fictitious flow $\mathbf{x}^{(\alpha\succ\beta)}(t)$. Because $\mathbf{x}^{(\alpha\succ\beta)}(t_{m\pm}) = \mathbf{x}^{(0)}(t_{m\pm})$ and $\mathbf{x}^{(\alpha)}(t_{m\pm}) = \mathbf{x}^{(0)}(t_{m\pm})$, the *G*-function definition gives

$$\begin{aligned}&\mathbf{n}^{\mathrm{T}}_{\partial\Omega_{ij}}(\mathbf{x}^{(0)}(t_{m+\varepsilon}))\cdot[\mathbf{x}^{(\alpha\succ\beta)}(t_{m+\varepsilon},q_{\sigma}^{(\alpha)})-\mathbf{x}^{(0)}(t_{m+\varepsilon},q_{\sigma}^{(\alpha)})]\\&=\sum\nolimits_{s_\alpha=0}^{l_\alpha-1}G_{\partial\Omega_{ij}}^{(s_\alpha,\alpha\succ\beta)}(\mathbf{x}_m,q_{\sigma}^{(\alpha)})\varepsilon^{s_\alpha+1}+\sum\nolimits_{s_\alpha=2k_\alpha}^{l_\alpha-1}G_{\partial\Omega_{ij}}^{(s_\alpha,\alpha\succ\beta)}(\mathbf{x}_m,q_{\sigma}^{(\alpha)})\varepsilon^{s_\alpha+1}\\&\quad+G_{\partial\Omega_{ij}}^{(l_\alpha,\alpha\succ\beta)}(\mathbf{x}_m,q_{\sigma}^{(\alpha)})\varepsilon^{l_\alpha+1}+o(\varepsilon^{l_\alpha+1}),\end{aligned}$$

$$\begin{aligned}&\mathbf{n}^{\mathrm{T}}_{\partial\Omega_{ij}}(\mathbf{x}^{(0)}(t_{m+\varepsilon}))\cdot[\mathbf{x}^{(\alpha)}(t_{m+\varepsilon})-\mathbf{x}^{(0)}(t_{m+\varepsilon})]\\&=\sum\nolimits_{s_\alpha=0}^{2k_\alpha-1}G_{\partial\Omega_{ij}}^{(s_\alpha,\alpha)}(\mathbf{x}_m,t_{m+})\varepsilon^{s_\alpha+1}+\sum\nolimits_{s_\alpha=2k_\alpha}^{l_\alpha-1}G_{\partial\Omega_{ij}}^{(s_\alpha,\alpha)}(\mathbf{x}_m,t_{m+})\varepsilon^{s_\alpha+1}\\&\quad+G_{\partial\Omega_{ij}}^{(l_\alpha,\alpha)}(\mathbf{x}_m,t_{m+})\varepsilon^{l_\alpha+1}+o(\varepsilon^{l_\alpha+1}).\end{aligned}$$

Because of

$$\begin{gathered}\mathbf{x}^{(\alpha\succ\beta)}(t_{m\pm})=\mathbf{x}^{(\alpha)}(t_{m\pm}),\\G_{\partial\Omega_{ij}}^{(s_\alpha,\alpha\succ\beta)}(\mathbf{x}_m,q_{\sigma}^{(\alpha)})=G_{\partial\Omega_{ij}}^{(s_\alpha,\alpha)}(\mathbf{x}_m,t_{m\pm})=0,\ s_\alpha=0,1,\cdots,2k_\alpha-1,\\G_{\partial\Omega_{ij}}^{(s_\alpha,\alpha)}(\mathbf{x}_m,t_{m\pm})=G_{\partial\Omega_{ij}}^{(s_\alpha,\alpha\succ\beta)}(\mathbf{x}_m,q_{\sigma}^{(\alpha)})\neq0\text{ for }s_\alpha=2k_\alpha,2k_\alpha+1,\cdots,l_\alpha-1.\end{gathered}$$

One obtains

$$\begin{aligned}&\mathbf{n}^{\mathrm{T}}_{\partial\Omega_{ij}}(\mathbf{x}^{(0)}(t_{m+}))\cdot[\mathbf{x}^{(\alpha)}(t_{m+\varepsilon})-\mathbf{x}^{(\alpha\succ\beta)}(t_{m+\varepsilon})]\\&=[G_{\partial\Omega_{ij}}^{(l_\alpha,\alpha)}(\mathbf{x}_m,t_{m+})-G_{\partial\Omega_{ij}}^{(l_\alpha,\alpha\succ\beta)}(\mathbf{x}_m,q_{\sigma}^{(\alpha)})]\varepsilon^{l_\alpha+1}.\end{aligned}$$

From definitions, the coming flow of $\mathbf{x}^{(\alpha)}(t)$ not passing over the coming flow barrier gives

$$\begin{aligned}&\mathbf{n}^{\mathrm{T}}_{\partial\Omega_{ij}}(\mathbf{x}^{(0)}(t_{m+\varepsilon}))\cdot[\mathbf{x}^{(\alpha)}(t_{m+\varepsilon})-\mathbf{x}^{(\alpha\succ\beta)}(t_{m+\varepsilon})]<0\text{ for }\mathbf{n}_{\partial\Omega_{ij}}\to\Omega_\beta,\\&\mathbf{n}^{\mathrm{T}}_{\partial\Omega_{ij}}(\mathbf{x}^{(0)}(t_{m+\varepsilon}))\cdot[\mathbf{x}^{(\alpha)}(t_{m+\varepsilon})-\mathbf{x}^{(\alpha\succ\beta)}(t_{m+\varepsilon})]>0\text{ for }\mathbf{n}_{\partial\Omega_{ij}}\to\Omega_\alpha\end{aligned}$$

for the upper limit of the flow barrier (i.e., $\sigma=2$), and

$$\begin{aligned}&\mathbf{n}^{\mathrm{T}}_{\partial\Omega_{ij}}(\mathbf{x}^{(0)}(t_{m+\varepsilon}))\cdot[\mathbf{x}^{(\alpha)}(t_{m+\varepsilon})-\mathbf{x}^{(\alpha\succ\beta)}(t_{m+\varepsilon})]>0\text{ for }\mathbf{n}_{\partial\Omega_{ij}}\to\Omega_\beta,\\&\mathbf{n}^{\mathrm{T}}_{\partial\Omega_{ij}}(\mathbf{x}^{(0)}(t_{m+\varepsilon}))\cdot[\mathbf{x}^{(\alpha)}(t_{m+\varepsilon})-\mathbf{x}^{(\alpha\succ\beta)}(t_{m+\varepsilon})]<0\text{ for }\mathbf{n}_{\partial\Omega_{ij}}\to\Omega_\alpha\end{aligned}$$

for the lower limit of the flow barrier (i.e., $\sigma=1$). With Eq. (4.2), if a coming flow cannot pass over the flow barrier on the α-side at the critical point of the flow barrier in the $(2k_\alpha:m_\beta)$-semi-passable flow, one obtains the conditions in Eq. (4.32), vice versa.

(iii) In a similar fashion as in (ii), the definitions for the coming flow of $\mathbf{x}^{(\alpha)}(t)$ passing over the coming flow barrier lead to

$$\begin{aligned}&\mathbf{n}^{\mathrm{T}}_{\partial\Omega_{ij}}(\mathbf{x}^{(0)}(t_{m+\varepsilon}))\cdot[\mathbf{x}^{(\alpha)}(t_{m+\varepsilon})-\mathbf{x}^{(\alpha\succ\beta)}(t_{m+\varepsilon})]>0\text{ for }\mathbf{n}_{\partial\Omega_{ij}}\to\Omega_\beta,\\&\mathbf{n}^{\mathrm{T}}_{\partial\Omega_{ij}}(\mathbf{x}^{(0)}(t_{m+\varepsilon}))\cdot[\mathbf{x}^{(\alpha)}(t_{m+\varepsilon})-\mathbf{x}^{(\alpha\succ\beta)}(t_{m+\varepsilon})]<0\text{ for }\mathbf{n}_{\partial\Omega_{ij}}\to\Omega_\alpha\end{aligned}$$

for the upper limit of the flow barrier (i.e., $\sigma = 2$), and

$$\mathbf{n}_{\partial\Omega_{ij}}^{\mathrm{T}}(\mathbf{x}^{(0)}(t_{m+\varepsilon}))\cdot[\mathbf{x}^{(\alpha)}(t_{m+\varepsilon})-\mathbf{x}^{(\alpha\succ\beta)}(t_{m+\varepsilon})]<0 \text{ for } \mathbf{n}_{\partial\Omega_{ij}}\to\Omega_{\beta},$$
$$\mathbf{n}_{\partial\Omega_{ij}}^{\mathrm{T}}(\mathbf{x}^{(0)}(t_{m+\varepsilon}))\cdot[\mathbf{x}^{(\alpha)}(t_{m+\varepsilon})-\mathbf{x}^{(\alpha\succ\beta)}(t_{m+\varepsilon})]>0 \text{ for } \mathbf{n}_{\partial\Omega_{ij}}\to\Omega_{\alpha}$$

for the lower limit of the flow barrier (i.e., $\sigma = 1$). With Eq. (4.2), if a coming flow passes over the flow barrier at the critical point of the flow barrier in the $(2k_{\alpha}:m_{\beta})$-semi-passable flow, one obtains the conditions in Eq. (4.33), and vice versa. ■

4.1.2. Leaving flow barriers

For a leaving flow of the semi-passable passable flow to the boundary, as in the coming flow, there is a leaving flow barrier on the boundary.

Definition 4.10. For a discontinuous dynamical system in Eq. (4.1), there is a point $\mathbf{x}^{(0)}(t_m)\equiv\mathbf{x}_m\in\partial\Omega_{ij}$ at time t_m between two adjacent domains Ω_{α} ($\alpha = i, j$). There is a vector field $\mathbf{F}^{(\alpha\succ\beta)}(\mathbf{x}^{(\beta)},t,\boldsymbol{\pi}_{\beta},q^{(\beta)})$ for $q^{(\beta)}\in[q_1^{(\beta)},q_2^{(\beta)}]$ on $\partial\Omega_{ij}$ with

$$\begin{aligned}\hbar_{\alpha}G_{\partial\Omega_{ij}}^{(\alpha\succ\beta)}(\mathbf{x}_m,q^{(\beta)})&\in\left[\hbar_{\alpha}G_{\partial\Omega_{ij}}^{(\alpha\succ\beta)}(\mathbf{x}_m,q_1^{(\beta)}),\hbar_{\alpha}G_{\partial\Omega_{ij}}^{(\alpha\succ\beta)}(\mathbf{x}_m,q_2^{(\beta)})\right]\\&\subset[0,+\infty)\end{aligned} \tag{4.34}$$

($\alpha,\beta\in\{i,j\}$ and $\alpha\neq\beta$). The coming and leaving flows in the semi-passable flow satisfy

$$\hbar_{\alpha}G_{\partial\Omega_{ij}}^{(\alpha)}(\mathbf{x}_m,t_{m-})>0 \text{ and } \hbar_{\alpha}G_{\partial\Omega_{ij}}^{(\beta)}(\mathbf{x}_m,t_{m+})>0. \tag{4.35}$$

The vector field of $\mathbf{F}^{(\alpha\succ\beta)}(\mathbf{x}^{(\beta)},t,\boldsymbol{\pi}_{\beta},q^{(\beta)})$ is called the *leaving flow barrier* on the β-side in the semi-passable flow if the following conditions are satisfied. The critical values of $\mathbf{F}^{(\alpha\succ\beta)}(\mathbf{x}^{(\beta)},t,\boldsymbol{\pi}_{\beta},q_{\sigma}^{(\beta)})$ ($\sigma = 1,2$) are called the *lower and upper limits* of the leaving flow barrier on the β-side.

(i) The leaving flow of $\mathbf{x}^{(\beta)}$ cannot enter the domain Ω_{β} at $q^{(\beta)}\in(q_1^{(\beta)},q_2^{(\beta)})$ if

$$\begin{aligned}&\mathbf{x}^{(\beta)}(t_{m+})=\mathbf{x}^{(\alpha\succ\beta)}(t_{m\pm},q^{(\beta)})=\mathbf{x}_m,\\&\hbar_{\alpha}G_{\partial\Omega_{ij}}^{(\beta)}(\mathbf{x}_m,t_{m+})\in\left(\hbar_{\alpha}G_{\partial\Omega_{ij}}^{(\alpha\succ\beta)}(\mathbf{x}_m,q_1^{(\beta)}),\hbar_{\alpha}G_{\partial\Omega_{ij}}^{(\alpha\succ\beta)}(\mathbf{x}_m,q_2^{(\beta)})\right).\end{aligned} \tag{4.36}$$

(ii) The leaving flow of $\mathbf{x}^{(\beta)}$ cannot enter the domain Ω_{β} at the critical points of the flow barrier (i.e., $q^{(\beta)}=q_{\sigma}^{(\beta)}$, $\sigma\in\{1,2\}$) if

$$\begin{aligned}&\mathbf{x}^{(\beta)}(t_{m+})=\mathbf{x}^{(\alpha\succ\beta)}(t_{m\pm},q_{\sigma}^{(\beta)})=\mathbf{x}_m,\\&G_{\partial\Omega_{ij}}^{(\beta)}(\mathbf{x}_m,t_{m+})=G_{\partial\Omega_{ij}}^{(\alpha\succ\beta)}(\mathbf{x}_m,q_{\sigma}^{(\beta)})\neq 0,\\&(-1)^{\sigma}\hbar_{\alpha}\mathbf{n}_{\partial\Omega_{ij}}^{\mathrm{T}}(\mathbf{x}^{(0)}(t_{m+\varepsilon}))\cdot\left[\mathbf{x}^{(\beta)}(t_{m+\varepsilon})-\mathbf{x}^{(\alpha\succ\beta)}(t_{m+\varepsilon},q_{\sigma}^{(\beta)})\right]<0.\end{aligned}\tag{4.37}$$

(iii) The leaving flow of $\mathbf{x}^{(\beta)}$ enters the domain Ω_{β} at the critical points of the flow barrier (i.e., $q^{(\beta)}=q_{\sigma}^{(\beta)}$, $\sigma\in\{1,2\}$) if

$$\begin{aligned}&\mathbf{x}^{(\beta)}(t_{m+})=\mathbf{x}^{(\alpha\succ\beta)}(t_{m\pm},q_{\sigma}^{(\beta)})=\mathbf{x}_m,\\&G_{\partial\Omega_{ij}}^{(\beta)}(\mathbf{x}_m,t_{m+})=G_{\partial\Omega_{ij}}^{(\alpha\succ\beta)}(\mathbf{x}_m,q_{\sigma}^{(\beta)})\neq 0,\\&(-1)^{\sigma}\hbar_{\alpha}\mathbf{n}_{\partial\Omega_{ij}}^{\mathrm{T}}(\mathbf{x}^{(0)}(t_{m+\varepsilon}))\cdot\left[\mathbf{x}^{(\beta)}(t_{m+\varepsilon})-\mathbf{x}^{(\alpha\succ\beta)}(t_{m+\varepsilon},q_{\sigma}^{(\beta)})\right]>0.\end{aligned}\tag{4.38}$$

Definition 4.11. For a discontinuous dynamical system in Eq. (4.1), there is a point $\mathbf{x}^{(0)}(t_m)\equiv\mathbf{x}_m\in\partial\Omega_{ij}$ at time t_m between two adjacent domains Ω_{α} ($\alpha=i,j$). There is a vector field of $\mathbf{F}^{(\alpha\succ\beta)}(\mathbf{x}^{(\beta)},t,\boldsymbol{\pi}_{\beta},q^{(\beta)})$ for $q^{(\beta)}\in[q_1^{(\beta)},q_2^{(\beta)}]$ on the boundary $\partial\Omega_{ij}$ with the G-functions

$$\begin{aligned}&G_{\partial\Omega_{ij}}^{(s_{\beta},\alpha\succ\beta)}(\mathbf{x}_m,q^{(\beta)})=0\ \text{ for } s_{\beta}=0,1,\cdots,m_{\beta}-1;\\&\hbar_{\alpha}G_{\partial\Omega_{ij}}^{(m_{\beta},\alpha\succ\beta)}(\mathbf{x}_m,q^{(\beta)})\\&\in\left[\hbar_{\alpha}G_{\partial\Omega_{ij}}^{(m_{\beta},\alpha\succ\beta)}(\mathbf{x}_m,q_1^{(\beta)}),\hbar_{\alpha}G_{\partial\Omega_{ij}}^{(m_{\beta},\alpha\succ\beta)}(\mathbf{x}_m,q_2^{(\beta)})\right]\subset[0,\infty)\end{aligned}\tag{4.39}$$

($\alpha,\beta\in\{i,j\}$ and $\alpha\neq\beta$). The coming and leaving flows of the $(2k_{\alpha}:m_{\beta})$-semi-passable flow satisfy

$$\begin{aligned}&G_{\partial\Omega_{ij}}^{(s_{\alpha},\alpha)}(\mathbf{x}_m,t_{m-})=0\text{ for } s_{\alpha}=0,1,\cdots,2k_{\alpha}-1;\\&G_{\partial\Omega_{ij}}^{(s_{\beta},\beta)}(\mathbf{x}_m,t_{m+})=0\text{ for } s_{\beta}=0,1,\cdots,m_{\beta}-1;\\&\hbar_{\alpha}G_{\partial\Omega_{ij}}^{(2k_{\alpha},\alpha)}(\mathbf{x}_m,t_{m-})>0\text{ and }\hbar_{\alpha}G_{\partial\Omega_{ij}}^{(m_{\beta},\beta)}(\mathbf{x}_m,t_{m+})>0.\end{aligned}\tag{4.40}$$

The vector field of $\mathbf{F}^{(\alpha\succ\beta)}(\mathbf{x}^{(\beta)},t,\boldsymbol{\pi}_{\beta},q^{(\beta)})$ is called the *leaving flow barrier* on the β-side in the $(2k_{\alpha}:m_{\beta})$-semi-passable flow if the following conditions are satisfied. The critical values of $\mathbf{F}^{(\alpha\succ\beta)}(\mathbf{x}^{(\beta)},t,\boldsymbol{\pi}_{\beta},q_{\sigma}^{(\beta)})$ ($\sigma=1,2$) are called *the lower and upper limits* of the leaving flow barrier on the β-side.

(i) The leaving flow of $\mathbf{x}^{(\beta)}$ cannot enter the domain Ω_{β} at $q^{(\beta)}\in(q_1^{(\beta)},q_2^{(\beta)})$ if

$$\begin{aligned}&\mathbf{x}^{(\beta)}(t_{m+})=\mathbf{x}^{(\alpha\succ\beta)}(t_{m\pm},q^{(\beta)})=\mathbf{x}_m,\\&G_{\partial\Omega_{ij}}^{(m_{\beta},\beta)}(\mathbf{x}_m,t_{m-})\in\left(\hbar_{\alpha}G_{\partial\Omega_{ij}}^{(m_{\beta},\alpha\succ\beta)}(\mathbf{x}_m,q_1^{(\beta)}),\ \hbar_{\alpha}G_{\partial\Omega_{ij}}^{(m_{\beta},\alpha\succ\beta)}(\mathbf{x}_m,q_2^{(\beta)})\right).\end{aligned}\tag{4.41}$$

(ii) The leaving flow of $\mathbf{x}^{(\beta)}$ cannot enter the domain Ω_β at the critical points of the flow barrier (i.e., $q^{(\beta)} = q_\sigma^{(\beta)}$, $\sigma \in \{1,2\}$) if

$$\begin{gathered}\mathbf{x}^{(\beta)}(t_{m+}) = \mathbf{x}^{(\alpha\succ\beta)}(t_{m\pm}, q_\sigma^{(\beta)}) = \mathbf{x}_m, \\ G_{\partial\Omega_{ij}}^{(s_\beta,\beta)}(\mathbf{x}_m, t_{m+}) = G_{\partial\Omega_{ij}}^{(s_\beta,\alpha\succ\beta)}(\mathbf{x}_m, q_\sigma^{(\beta)}) \neq 0 \text{ for } s_\beta = m_\beta, m_\beta+1,\cdots,l_\beta; \\ (-1)^\sigma \hbar_\alpha \mathbf{n}_{\partial\Omega_{ij}}^{\mathrm{T}}(\mathbf{x}^{(0)}(t_{m+\varepsilon}))\cdot\left[\mathbf{x}^{(\beta)}(t_{m+\varepsilon}) - \mathbf{x}^{(\alpha\succ\beta)}(t_{m+\varepsilon}, q_\sigma^{(\beta)})\right] < 0.\end{gathered} \tag{4.42}$$

(iii) The leaving flow of $\mathbf{x}^{(\beta)}$ enters the domain Ω_β at the critical points of the flow barrier (i.e., $q^{(\beta)} = q_\sigma^{(\beta)}$, $\sigma \in \{1,2\}$) if

$$\begin{gathered}\mathbf{x}^{(\beta)}(t_{m+}) = \mathbf{x}^{(\alpha\succ\beta)}(t_{m\pm}, q_\sigma^{(\beta)}) = \mathbf{x}_m, \\ G_{\partial\Omega_{ij}}^{(s_\beta,\beta)}(\mathbf{x}_m, t_{m+}) = G_{\partial\Omega_{ij}}^{(s_\beta,\alpha\succ\beta)}(\mathbf{x}_m, q_\sigma^{(\beta)}) \neq 0 \text{ for } s_\beta = m_\beta, m_\beta+1,\cdots,l_\beta; \\ (-1)^\sigma \hbar_\alpha \mathbf{n}_{\partial\Omega_{ij}}^{\mathrm{T}}(\mathbf{x}^{(0)}(t_{m+\varepsilon}))\cdot\left[\mathbf{x}^{(\beta)}(t_{m+\varepsilon}) - \mathbf{x}^{(\alpha\succ\beta)}(t_{m+\varepsilon}, q_\sigma^{(\beta)})\right] > 0.\end{gathered} \tag{4.43}$$

To explain flow barriers on the β-side of the boundary, the corresponding *G*-functions are presented in Fig. 4.7. The *G*-function of the β-flow relative to the leaving flow barrier is denoted by the dashed curve. The thick line on the boundary represents the *G*-function of the flow barrier. For $\mathbf{n}_{\partial\Omega_{ij}} \to \Omega_\beta$, one has $\hbar_\alpha = +1$. Suppose there is a leaving flow barrier on the boundary $\partial\Omega_{ij}$ with *lower* and *upper* limits of $G_{\partial\Omega_{ij}}^{(m_\beta,\alpha\succ\beta)}(q_1^{(\beta)}) > 0$ and $G_{\partial\Omega_{ij}}^{(m_\beta,\alpha\succ\beta)}(q_2^{(\beta)}) > 0$. The leaving flow cannot leave the boundary $\partial\Omega_{ij}$ for $G_{\partial\Omega_{ij}}^{(m_\beta,\beta)} \in [G_{\partial\Omega_{ij}}^{(m_\beta,\alpha\succ\beta)}(q_1^{(\beta)}), G_{\partial\Omega_{ij}}^{(m_\beta,\alpha\succ\beta)}(q_2^{(\beta)})]$. Because no flow barriers exist on the α-side of the boundary, the α-flow with $G_{\partial\Omega_{ij}}^{(\alpha)} > 0$ can arrive to the β-side of the boundary $\partial\Omega_{ij}$. The leaving flow can leave the boundary only if $G_{\partial\Omega_{ij}}^{(m_\beta,\beta)} \notin [G_{\partial\Omega_{ij}}^{(m_\beta,\alpha\succ\beta)}(q_1^{(\beta)}), G_{\partial\Omega_{ij}}^{(m_\beta,\alpha\succ\beta)}(q_2^{(\beta)})]$, as sketched in Fig. 4.7(a). Similarly, for $\mathbf{n}_{\partial\Omega_{ij}} \to \Omega_\alpha$, the corresponding *G*-functions of the leaving flow are presented in Fig. 4.7(b). Similarly, the partial and full flow barriers on the β-side can be defined as in Section 4.1.1.

Definition 4.12. For a discontinuous dynamical system in Eq. (4.1), there is a point $\mathbf{x}^{(0)}(t_m) \equiv \mathbf{x}_m \in \partial\Omega_{ij}$ at time t_m between two adjacent domains Ω_α ($\alpha = i, j$). Suppose there is a leaving flow barrier of $\mathbf{F}^{(\alpha\succ\beta)}(\mathbf{x}^{(\beta)}, t, \boldsymbol{\pi}_\beta, q^{(\beta)})$ on the β-side in the $(2k_\alpha : m_\beta)$-semi-passable flow for $q^{(\beta)} \in [q_1^{(\beta)}, q_2^{(\beta)}]$ on the boundary $\partial\Omega_{ij}$ ($\alpha, \beta \in \{i, j\}$ and $\alpha \neq \beta$).

(i) The leaving flow barrier in the $(2k_\alpha : m_\beta)$-semi-passable flow is *partial* on the β-side if $\mathbf{x}_m \in S \subset \partial\Omega_{ij}$.

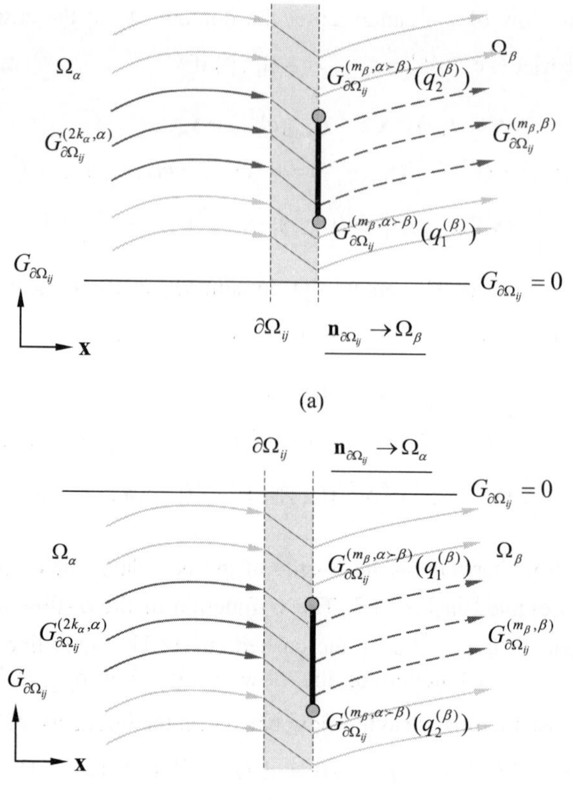

Fig. 4.7 *G*-functions for leaving flow barriers in the $(2k_\alpha;m_\beta)$-semi-passable flow on the β-side of the boundary: **(a)** $\mathbf{n}_{\partial\Omega_{ij}}\to\Omega_\beta$ and **(b)** $\mathbf{n}_{\partial\Omega_{ij}}\to\Omega_\alpha$. The dashed curves are the *G*-function of the β-flow relative to the leaving flow barrier. The thick line is the *G*-function of the flow barriers. $G^{(m_\beta,\alpha\succ\beta)}_{\partial\Omega_{ij}}(q_1^{(\beta)})$ and $G^{(m_\beta,\alpha\succ\beta)}_{\partial\Omega_{ij}}(q_2^{(\beta)})$ are for *lower* and *upper* barrier limits ($k_\alpha,m_\beta\in\{0,1,2,\cdots\}$).

(ii) The leaving flow barrier in the $(2k_\alpha:m_\beta)$-semi-passable flow is *full* on the β-side if $\mathbf{x}_m\in S=\partial\Omega_{ij}$.

The partial and full leaving flow barriers on the β-side of the boundary $\partial\Omega_{ij}$ are sketched in Fig. 4.8. The partial flow barriers only exist on subsets of the boundary (i.e., $S\subset\partial\Omega_{ij}$). On the other subsets ($\partial\Omega_{ij}\setminus S$), the leaving flow barriers do not exist, and the leaving flow from such subsets of the boundary can get into the domain Ω_β, as shown in Fig. 4.8(a). If the leaving flow barrier exists on $S=\partial\Omega_{ij}$, such a flow barrier is called the full leaving flow barrier, which is pre-

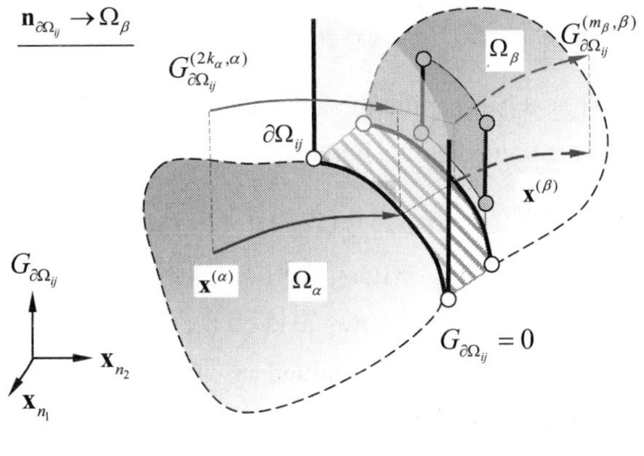

(a)

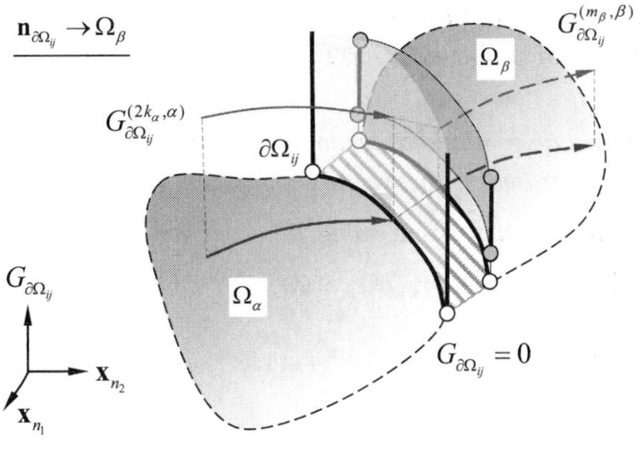

(b)

Fig.4.8 The leaving flow barrier in the $(2k_\alpha : m_\beta)$-semi-passable flow on $\partial\Omega_{ij}$: **(a)** partial flow barrier and **(b)** full flow barrier. The blue surface is for the flow barrier on $\partial\Omega_{ij}$. The red solid and dashed curves with arrows are G-functions of flows on α and β -domains. The blue curves are semi-passable flows. The light green surface represents "no flow barrier". ($k_\alpha, m_\beta \in \{0,1,2,\cdots\}$).

sented in Fig. 4.8(b). The flow barriers possess the lower and upper limits of the flow barrier on the boundary. The other discussions about the leaving flow barriers on the β -side can be similarly discussed as for the coming flow on the α -side, such as the infinity flow barrier, flow barrier fences, flow barrier widows and flow barrier doors of the leaving flow on the β -side for the semi-passable flow.

For a semi-passable flow on the boundary $\partial\Omega_{ij}$, if the leaving flow barrier exists on the β -side of such a boundary with a subset $S \subseteq \partial\Omega_{ij}$, the flow still cannot pass through the boundary under the following condition

$$\hbar_\alpha G^{(m_\beta,\beta)}_{\partial\Omega_{ij}}(\mathbf{x}_m,t_{m+}) \in \left(\hbar_\alpha G^{(m_\beta,\alpha\succ\beta)}_{\partial\Omega_{ij}}(\mathbf{x}_m,q_1^{(\beta)}), \hbar_\alpha G^{(m_\beta,\alpha\succ\beta)}_{\partial\Omega_{ij}}(\mathbf{x}_m,q_2^{(\beta)})\right) \tag{4.44}$$

for $\mathbf{x}_m \in S \subseteq \partial\Omega_{ij}$. For this case, the dynamical system along the boundary in the β-domain will be constrained by the boundary, i.e.,

$$\begin{aligned} &\dot{\mathbf{x}}^{(\beta)} = \mathbf{F}^{(\beta)}(\mathbf{x}^{(\beta)},t,\boldsymbol{\mu}_\beta) \text{ in } \Omega_\beta \ (\beta \in \{i,j\}), \\ &\text{with } \varphi_{ij}(\mathbf{x}^{(\beta)},t,\boldsymbol{\lambda}) = 0 \text{ on } \partial\Omega_{ij}. \end{aligned} \tag{4.45}$$

The vector field is given by the vector field on the domain of Ω_β. Because the flow barrier exists on the β-side of the boundary $\partial\Omega_{ij}$, the flow will be along the boundary on the β-side until the condition in Eq. (4.44) cannot be satisfied. Of course, if the tangential component of vector field in Eq. (4.45) on the boundary is zero, the system will stay on the specific point of the boundary until the normal vector field overcomes the flow barrier. This flow can be called the "standing flow" on the β-side of the boundary $\partial\Omega_{ij}$.

Theorem 4.3. *For a discontinuous dynamical system in Eq.* (4.1), *there is a point* $\mathbf{x}^{(0)}(t_m) \equiv \mathbf{x}_m \in \partial\Omega_{ij}$ *at time* t_m *between two adjacent domains* Ω_α ($\alpha = i,j$). *For* $\mathbf{x}_m \in S \subseteq \partial\Omega_{ij}$, *there is a semi-passable flow barrier* $\mathbf{F}^{(\alpha\succ\beta)}(\mathbf{x}^{(\beta)},t,\boldsymbol{\pi}_\beta,q^{(\beta)})$ *for* $q^{(\beta)} \in [q_1^{(\beta)},q_2^{(\beta)}]$ *on the boundary* $\partial\Omega_{ij}$ *with*

$$\begin{aligned} \hbar_\alpha G^{(\alpha\succ\beta)}_{\partial\Omega_{ij}}(\mathbf{x}_m,q^{(\beta)}) &\in \left[\hbar_\alpha G^{(\alpha\succ\beta)}_{\partial\Omega_{ij}}(\mathbf{x}_m,q_1^{(\beta)}), \hbar_\alpha G^{(\alpha\succ\beta)}_{\partial\Omega_{ij}}(\mathbf{x}_m,q_2^{(\beta)})\right] \\ &\subset [0,\infty) \end{aligned} \tag{4.46}$$

($\alpha,\beta \in \{i,j\}$ *and* $\alpha \neq \beta$). *The semi-passable flow satisfies*

$$\hbar_\alpha G^{(\alpha)}_{\partial\Omega_{ij}}(\mathbf{x}_m,t_{m-}) > 0 \text{ and } \hbar_\alpha G^{(\beta)}_{\partial\Omega_{ij}}(\mathbf{x}_m,t_{m+}) > 0. \tag{4.47}$$

(i) *The leaving flow in the semi-passable flow cannot pass through the flow barrier on the* β*-side at* $q^{(\beta)} \in (q_1^{(\beta)},q_2^{(\beta)})$ *if and only if*

$$\hbar_\alpha G^{(\beta)}_{\partial\Omega_{ij}}(\mathbf{x}_m,t_{m+}) \in \left(\hbar_\alpha G^{(\alpha\succ\beta)}_{\partial\Omega_{ij}}(\mathbf{x}_m,q_1^{(\beta)}), \hbar_\alpha G^{(\alpha\succ\beta)}_{\partial\Omega_{ij}}(\mathbf{x}_m,q_2^{(\beta)})\right). \tag{4.48}$$

(ii) *The leaving flow in the semi-passable flow cannot enter the domain* Ω_β *on the* β*-side at* $q^{(\beta)} = q_\sigma^{(\beta)}$ ($\sigma \in \{1,2\}$) *if and only if*

$$\begin{aligned} &\hbar_\alpha G^{(s_\beta,\beta)}_{\partial\Omega_{ij}}(\mathbf{x}_m,t_{m+}) = \hbar_\alpha G^{(s_\beta,\alpha\succ\beta)}_{\partial\Omega_{ij}}(\mathbf{x}_m,q_\sigma^{(\beta)}) \in (0,\infty) \\ &\quad \textit{for } s_\beta = 0,1,\cdots,l_\beta - 1; \\ &(-1)^\sigma \hbar_\alpha \left[G^{(l_\beta,\beta)}_{\partial\Omega_{ij}}(\mathbf{x}_m,t_{m+}) - G^{(l_\beta,\alpha\succ\beta)}_{\partial\Omega_{ij}}(\mathbf{x}_m,q_\sigma^{(\beta)})\right] < 0. \end{aligned} \tag{4.49}$$

(iii) *The leaving flow in the semi-passable flow enters the domain* Ω_β *on the* β*-side at* $q^{(\beta)} = q_\sigma^{(\beta)}$ ($\sigma \in \{1,2\}$) *if and only if*

$$\begin{aligned}
&\hbar_\alpha G_{\partial\Omega_{ij}}^{(s_\beta,\beta)}(\mathbf{x}_m,t_{m+}) = \hbar_\alpha G_{\partial\Omega_{ij}}^{(s_\beta,\alpha\succ\beta)}(\mathbf{x}_m,q_\sigma^{(\beta)}) \in (0,\infty)\\
&\quad for\ s_\beta = 0,1,\cdots,l_\beta - 1;\\
&(-1)^\sigma \hbar_\alpha \left[G_{\partial\Omega_{ij}}^{(l_\beta,\beta)}(\mathbf{x}_m,t_{m+}) - G_{\partial\Omega_{ij}}^{(l_\beta,\alpha\succ\beta)}(\mathbf{x}_m,q_\sigma^{(\beta)}) \right] > 0.
\end{aligned} \tag{4.50}$$

Proof. The proof of this theorem is similar to the proof of Theorem 4.1. ■

Theorem 4.4. *For a discontinuous dynamical system in Eq.* (4.1)*, there is a point* $\mathbf{x}^{(0)}(t_m) \equiv \mathbf{x}_m \in \partial\Omega_{ij}$ *at time* t_m *between two adjacent domains* Ω_α ($\alpha = i,j$)*. Suppose a* $(2k_\alpha : m_\beta)$*-semi-passable flow barrier* $\mathbf{F}^{(\alpha\succ\beta)}(\mathbf{x}^{(\beta)},t,\boldsymbol{\pi}_\beta,q^{(\beta)})$ *exists on the boundary* $\partial\Omega_{ij}$ *for* $q^{(\beta)} \in [q_1^{(\beta)},q_2^{(\beta)}]$ *and* $\mathbf{x}_m \in S \subseteq \partial\Omega_{ij}$ *with*

$$\begin{aligned}
&G_{\partial\Omega_{ij}}^{(s_\beta,\alpha\succ\beta)}(\mathbf{x}_m,q^{(\beta)}) = 0\ \ for\ s_\beta = 0,1,\cdots,m_\beta - 1;\\
&\hbar_\alpha G_{\partial\Omega_{ij}}^{(m_\beta,\alpha\succ\beta)}(\mathbf{x}_m,q^{(\beta)}) \in \left[\hbar_\alpha G_{\partial\Omega_{ij}}^{(m_\beta,\alpha\succ\beta)}(\mathbf{x}_m,q_1^{(\beta)}), \hbar_\alpha G_{\partial\Omega_{ij}}^{(m_\beta,\alpha\succ\beta)}(\mathbf{x}_m,q_2^{(\beta)}) \right]\\
&\qquad \subset [0,\infty)
\end{aligned} \tag{4.51}$$

($\alpha,\beta \in \{i,j\}$ *and* $\alpha \neq \beta$)*. The coming and leaving flows in the* $(2k_\alpha : m_\beta)$*-semi-passable flow satisfy*

$$\begin{aligned}
&G_{\partial\Omega_{ij}}^{(s_\alpha,\alpha)}(\mathbf{x}_m,t_{m-},\mathbf{p}_\alpha,\boldsymbol{\lambda}) = 0\ \ for\ s_\alpha = 0,1,\cdots,2k_\alpha - 1;\\
&G_{\partial\Omega_{ij}}^{(s_\beta,\beta)}(\mathbf{x}_m,t_{m+},\mathbf{p}_\beta,\boldsymbol{\lambda}) = 0\ \ for\ s_\beta = 0,1,\cdots,m_\beta - 1;\\
&\hbar_\alpha G_{\partial\Omega_{ij}}^{(2k_\alpha,\alpha)}(\mathbf{x}_m,t_{m-}) > 0\ and\ \hbar_\alpha G_{\partial\Omega_{ij}}^{(m_\beta,\beta)}(\mathbf{x}_m,t_{m+}) > 0.
\end{aligned} \tag{4.52}$$

(i) *The leaving flow in the* $(2k_\alpha : m_\beta)$*-semi-passable flow cannot pass through the flow barrier on the* β*-side at* $q^{(\beta)} \in (q_1^{(\beta)},q_2^{(\beta)})$ *if and only if*

$$\hbar_\alpha G_{\partial\Omega_{ij}}^{(m_\beta,\beta)}(\mathbf{x}_m,t_{m+}) \in \left(\hbar_\alpha G_{\partial\Omega_{ij}}^{(m_\beta,\alpha\succ\beta)}(\mathbf{x}_m,q_1^{(\beta)}), \hbar_\alpha G_{\partial\Omega_{ij}}^{(m_\beta,\alpha\succ\beta)}(\mathbf{x}_m,q_2^{(\beta)}) \right). \tag{4.53}$$

(ii) *The leaving flow in the* $(2k_\alpha : m_\beta)$*-passable flow cannot enter the domain* Ω_β *on the* β*-side at* $q^{(\beta)} = q_\sigma^{(\beta)}$ ($\sigma \in \{1,2\}$) *if and only if*

$$\begin{aligned}
&\hbar_\alpha G_{\partial\Omega_{ij}}^{(s_\beta,\beta)}(\mathbf{x}_m,t_{m+}) = \hbar_\alpha G_{\partial\Omega_{ij}}^{(s_\beta,\alpha\succ\beta)}(\mathbf{x}_m,q_\sigma^{(\beta)}) \in (0,\infty)\\
&\quad for\ s_\beta = m_\beta, m_\beta + 1,\cdots,l_\beta - 1;\\
&(-1)^\sigma \hbar_\alpha \left[G_{\partial\Omega_{ij}}^{(l_\beta,\beta)}(\mathbf{x}_m,t_{m+}) - G_{\partial\Omega_{ij}}^{(l_\beta,\alpha\succ\beta)}(\mathbf{x}_m,q_\sigma^{(\beta)}) \right] < 0.
\end{aligned} \tag{4.54}$$

(iii) *The leaving flow in the* $(2k_\alpha : m_\beta)$*-passable flow enters the domain* Ω_β *on the* β*-side at* $q^{(\beta)} = q_\sigma^{(\beta)}$ ($\sigma \in \{1,2\}$) *if and only if*

$$\begin{aligned} &\hbar_\alpha G_{\partial\Omega_{ij}}^{(s_\beta,\beta)}(\mathbf{x}_m, t_{m+}) = \hbar_\alpha G_{\partial\Omega_{ij}}^{(s_\beta,\alpha\succ\beta)}(\mathbf{x}_m, q_\sigma^{(\beta)}) \in (0,\infty) \\ &\quad for\ s_\beta = m_\beta, m_\beta+1,\cdots,l_\beta-1; \\ &(-1)^\sigma \hbar_\alpha \Big[G_{\partial\Omega_{ij}}^{(l_\beta,\beta)}(\mathbf{x}_m, t_{m+}) - G_{\partial\Omega_{ij}}^{(l_\beta,\alpha\succ\beta)}(\mathbf{x}_m, q_\sigma^{(\beta)}) \Big] > 0. \end{aligned} \tag{4.55}$$

Proof. The proof of this theorem is the same as the proof of Theorem 4.2 ■

4.1.3. Passable flows with both flow barriers

In the previous two sections, the coming and leaving flow barriers of the semi-passable flow were considered separately. In fact, the coming and leaving flow barriers for the semi-passable flow can exist together. The switchability of the semi-passable flow with two flow barriers becomes more complex, which will be discussed as follows.

Definition 4.13. For a discontinuous dynamical system in Eq. (4.1), there is a point $\mathbf{x}^{(0)}(t_m) \equiv \mathbf{x}_m \in \partial\Omega_{ij}$ at time t_m between two adjacent domains Ω_α ($\alpha = i, j$). Suppose a flow barrier of $\mathbf{F}^{(\alpha\succ\beta)}(\mathbf{x}^{(\alpha)}, t, \boldsymbol{\pi}_\alpha, q^{(\alpha)})$ at $q^{(\alpha)} \in [q_1^{(\alpha)}, q_2^{(\alpha)}]$ exists for a coming flow of $\mathbf{x}^{(\alpha)}$ on the α-side of boundary $\partial\Omega_{ij}$ with

$$\begin{aligned} G_{\partial\Omega_{ij}}^{(\alpha\succ\beta)}(\mathbf{x}_m, q^{(\alpha)}) &\in \Big[\hbar_\alpha G_{\partial\Omega_{ij}}^{(\alpha\succ\beta)}(\mathbf{x}_m, q_1^{(\alpha)}), \hbar_\alpha G_{\partial\Omega_{ij}}^{(\alpha\succ\beta)}(\mathbf{x}_m, q_2^{(\alpha)}) \Big] \\ &\subset [0, +\infty), \end{aligned} \tag{4.56}$$

and also there is a flow barrier $\mathbf{F}^{(\alpha\succ\beta)}(\mathbf{x}^{(\beta)}, t, \boldsymbol{\pi}_\beta, q^{(\beta)})$ at $q^{(\beta)} \in [q_1^{(\beta)}, q_2^{(\beta)}]$ for the leaving flow of $\mathbf{x}^{(\beta)}$ on the β-side of boundary $\partial\Omega_{ij}$ with

$$\begin{aligned} G_{\partial\Omega_{ij}}^{(\alpha\succ\beta)}(\mathbf{x}_m, q^{(\beta)}) &\in \Big[\hbar_\alpha G_{\partial\Omega_{ij}}^{(\alpha\succ\beta)}(\mathbf{x}_m, q_1^{(\beta)}), \hbar_\alpha G_{\partial\Omega_{ij}}^{(\alpha\succ\beta)}(\mathbf{x}_m, q_2^{(\beta)}) \Big] \\ &\subset [0, +\infty) \end{aligned} \tag{4.57}$$

($\alpha, \beta \in \{i, j\}$ and $\alpha \neq \beta$). The coming and leaving flows in the semi-passable flow satisfy

$$\hbar_\alpha G_{\partial\Omega_{ij}}^{(\alpha)}(\mathbf{x}_m, t_{m-}) > 0 \text{ and } \hbar_\alpha G_{\partial\Omega_{ij}}^{(\beta)}(\mathbf{x}_m, t_{m+}) > 0. \tag{4.58}$$

(i) The coming flow of $\mathbf{x}^{(\alpha)}$ cannot be switched to the leaving flow of $\mathbf{x}^{(\beta)}$ to form a semi-passable flow at the boundary $\partial\Omega_{ij}$ if

$$\text{either } \hbar_\alpha G^{(\alpha)}_{\partial\Omega_{ij}}(\mathbf{x}_m, t_{m-}) \in \left(\hbar_\alpha G^{(\alpha\succ\beta)}_{\partial\Omega_{ij}}(\mathbf{x}_m, q_1^{(\alpha)}), \hbar_\alpha G^{(\alpha\succ\beta)}_{\partial\Omega_{ij}}(\mathbf{x}_m, q_2^{(\alpha)})\right), \tag{4.59}$$

$$\text{or } \hbar_\alpha G^{(\beta)}_{\partial\Omega_{ij}}(\mathbf{x}_m, t_{m+}) \in \left(\hbar_\alpha G^{(\alpha\succ\beta)}_{\partial\Omega_{ij}}(\mathbf{x}_m, q_1^{(\beta)}), \hbar_\alpha G^{(\alpha\succ\beta)}_{\partial\Omega_{ij}}(\mathbf{x}_m, q_2^{(\beta)})\right). \tag{4.60}$$

(ii) The coming flow of $\mathbf{x}^{(\alpha)}$ cannot be switched to the leaving flow of $\mathbf{x}^{(\beta)}$ to form a semi-passable flow at the boundary $\partial\Omega_{ij}$ if $\sigma_\alpha, \sigma_\beta \in \{1,2\}$

$$\left.\begin{array}{l} \mathbf{x}^{(\alpha)}(t_{m+}) = \mathbf{x}^{(\alpha\succ\beta)}(t_{m\pm}, q^{(\alpha)}_{\sigma_\alpha}) = \mathbf{x}_m, \\ \text{either } G^{(s_\alpha,\alpha)}_{\partial\Omega_{ij}}(\mathbf{x}_m, t_{m+}) = G^{(s_\alpha,\alpha\succ\beta)}_{\partial\Omega_{ij}}(\mathbf{x}_m, q^{(\alpha)}_{\sigma_\alpha}) \neq 0 \text{ for } s_\alpha = 0,1,2,\cdots,l_\alpha - 1, \\ (-1)^{\sigma_\alpha}\hbar_\alpha \mathbf{n}^{\mathrm{T}}_{\partial\Omega_{ij}}(\mathbf{x}^{(0)}(t_{m+\varepsilon}))\cdot\left[\mathbf{x}^{(\alpha)}(t_{m+\varepsilon}) - \mathbf{x}^{(\alpha\succ\beta)}(t_{m+\varepsilon}, q^{(\alpha)}_{\sigma_\alpha})\right] < 0; \end{array}\right\} \tag{4.61}$$

$$\left.\begin{array}{l} \mathbf{x}^{(\beta)}(t_{m+}) = \mathbf{x}^{(\alpha\succ\beta)}(t_{m\pm}, q^{(\beta)}_{\sigma_\beta}) = \mathbf{x}_m, \\ \text{or } G^{(s_\beta,\beta)}_{\partial\Omega_{ij}}(\mathbf{x}_m, t_{m+}) = G^{(s_\beta,\alpha\succ\beta)}_{\partial\Omega_{ij}}(\mathbf{x}_m, q^{(\beta)}_{\sigma_\beta}) \neq 0 \text{ for } s_\beta = 0,1,2,\cdots,l_\beta - 1, \\ (-1)^{\sigma_\beta}\hbar_\alpha \mathbf{n}^{\mathrm{T}}_{\partial\Omega_{ij}}(\mathbf{x}^{(0)}(t_{m+\varepsilon}))\cdot\left[\mathbf{x}^{(\beta)}(t_{m+\varepsilon}) - \mathbf{x}^{(\alpha\succ\beta)}(t_{m+\varepsilon}, q^{(\beta)}_{\sigma_\beta})\right] < 0. \end{array}\right\} \tag{4.62}$$

(iii) The coming flow of $\mathbf{x}^{(\alpha)}$ is switched to the leaving flow of $\mathbf{x}^{(\beta)}$ to form a semi-passable flow at the boundary $\partial\Omega_{ij}$ if for $\sigma_\alpha, \sigma_\beta \in \{1,2\}$

$$\left.\begin{array}{l} \mathbf{x}^{(\alpha)}(t_{m+}) = \mathbf{x}^{(\alpha\succ\beta)}(t_{m\pm}, q^{(\alpha)}_{\sigma_\alpha}) = \mathbf{x}_m, \\ \text{both } G^{(s_\alpha,\alpha)}_{\partial\Omega_{ij}}(\mathbf{x}_m, t_{m+}) = G^{(s_\alpha,\alpha\succ\beta)}_{\partial\Omega_{ij}}(\mathbf{x}_m, q^{(\alpha)}_{\sigma_\alpha}) \neq 0 \text{ for } s_\alpha = 0,1,2,\cdots,l_\alpha - 1, \\ (-1)^{\sigma_\alpha}\hbar_\alpha \mathbf{n}^{\mathrm{T}}_{\partial\Omega_{ij}}(\mathbf{x}^{(0)}(t_{m+\varepsilon}))\cdot\left[\mathbf{x}^{(\alpha)}(t_{m+\varepsilon}) - \mathbf{x}^{(\alpha\succ\beta)}(t_{m+\varepsilon}, q^{(\alpha)}_{\sigma_\alpha})\right] > 0; \end{array}\right\} \tag{4.63}$$

$$\left.\begin{array}{l} \mathbf{x}^{(\beta)}(t_{m+}) = \mathbf{x}^{(\alpha\succ\beta)}(t_{m\pm}, q^{(\beta)}_{\sigma_\beta}) = \mathbf{x}_m, \\ \text{and } G^{(s_\beta,\beta)}_{\partial\Omega_{ij}}(\mathbf{x}_m, t_{m+}) = G^{(s_\beta,\alpha\succ\beta)}_{\partial\Omega_{ij}}(\mathbf{x}_m, q^{(\beta)}_{\sigma_\beta}) \neq 0 \text{ for } s_\beta = 0,1,2,\cdots,l_\beta - 1, \\ (-1)^{\sigma_\beta}\hbar_\alpha \mathbf{n}^{\mathrm{T}}_{\partial\Omega_{ij}}(\mathbf{x}^{(0)}(t_{m+\varepsilon}))\cdot\left[\mathbf{x}^{(\beta)}(t_{m+\varepsilon}) - \mathbf{x}^{(\alpha\succ\beta)}(t_{m+\varepsilon}, q^{(\beta)}_{\sigma_\beta})\right] > 0. \end{array}\right\} \tag{4.64}$$

Definition 4.14. For a discontinuous dynamical system in Eq. (4.1), there is a point $\mathbf{x}^{(0)}(t_m) \equiv \mathbf{x}_m \in \partial\Omega_{ij}$ at time t_m between two adjacent domains Ω_α $(\alpha = i, j)$. Suppose a flow barrier $\mathbf{F}^{(\alpha\succ\beta)}(\mathbf{x}^{(\alpha)}, t, \boldsymbol{\pi}_\alpha, q^{(\alpha)})$ at $q^{(\alpha)} \in [q_1^{(\alpha)}, q_2^{(\alpha)}]$ exists for a coming flow of $\mathbf{x}^{(\alpha)}$ on the α-side of boundary $\partial\Omega_{ij}$ with

$$\begin{array}{l} G^{(s_\alpha,\alpha\succ\beta)}_{\partial\Omega_{ij}}(\mathbf{x}_m, q^{(\alpha)}) = 0 \text{ for } s_\alpha = 0,1,\cdots,2k_\alpha - 1; \\ G^{(2k_\alpha,\alpha\succ\beta)}_{\partial\Omega_{ij}}(\mathbf{x}_m, q^{(\alpha)}) \\ \in \left[\hbar_\alpha G^{(2k_\alpha,\alpha\succ\beta)}_{\partial\Omega_{ij}}(\mathbf{x}_m, q_1^{(\alpha)}), \hbar_\alpha G^{(2k_\alpha,\alpha\succ\beta)}_{\partial\Omega_{ij}}(\mathbf{x}_m, q_2^{(\alpha)})\right] \subset [0, +\infty), \end{array} \tag{4.65}$$

and also there is a flow barrier $\mathbf{F}^{(\alpha\succ\beta)}(\mathbf{x}^{(\beta)},t,\boldsymbol{\pi}_\beta,q^{(\beta)})$ at $q^{(\beta)}\in[q_1^{(\beta)},q_2^{(\beta)}]$ for the leaving flow of $\mathbf{x}^{(\beta)}$ on the β-side of boundary $\partial\Omega_{ij}$ with

$$\begin{aligned}&G_{\partial\Omega_{ij}}^{(s_\beta,\alpha\succ\beta)}(\mathbf{x}_m,q^{(\beta)})=0 \text{ for } s_\beta=0,1,\cdots,m_\beta-1;\\&G_{\partial\Omega_{ij}}^{(m_\beta,\alpha\succ\beta)}(\mathbf{x}_m,q^{(\beta)})\in\left[\hbar_\alpha G_{\partial\Omega_{ij}}^{(m_\beta,\alpha\succ\beta)}(\mathbf{x}_m,q_1^{(\beta)}),\hbar_\alpha G_{\partial\Omega_{ij}}^{(m_\beta,\alpha\succ\beta)}(\mathbf{x}_m,q_2^{(\beta)})\right]\\&\qquad\subset[0,+\infty)\end{aligned}\tag{4.66}$$

($\alpha,\beta\in\{i,j\}$ and $\alpha\neq\beta$). The coming and leaving flows in the $(2k_\alpha:m_\beta)$ semi-passable flow satisfies

$$\begin{aligned}&G_{\partial\Omega_{ij}}^{(s_\alpha,\alpha)}(\mathbf{x}_m,t_{m-})=0 \text{ for } s_\alpha=0,1,\cdots,2k_\alpha-1,\\&G_{\partial\Omega_{ij}}^{(s_\beta,\beta)}(\mathbf{x}_m,t_{m+})=0 \text{ for } s_\beta=0,1,\cdots,m_\beta-1;\\&\hbar_\alpha G_{\partial\Omega_{ij}}^{(2k_\alpha,\alpha)}(\mathbf{x}_m,t_{m-})>0 \text{ and } \hbar_\alpha G_{\partial\Omega_{ij}}^{(m_\beta,\beta)}(\mathbf{x}_m,t_{m+})>0.\end{aligned}\tag{4.67}$$

(i) The coming flow of $\mathbf{x}^{(\alpha)}$ cannot be switched to the leaving flow of $\mathbf{x}^{(\beta)}$ to form the $(2k_\alpha:m_\beta)$-semi-passable flow at the boundary if

$$\text{either}\quad\left.\begin{aligned}&\hbar_\alpha G_{\partial\Omega_{ij}}^{(2k_\alpha,\alpha)}(\mathbf{x}_m,t_{m-})\\&\in\left(\hbar_\alpha G_{\partial\Omega_{ij}}^{(2k_\alpha,\alpha\succ\beta)}(\mathbf{x}_m,q_1^{(\alpha)}),\ \hbar_\alpha G_{\partial\Omega_{ij}}^{(2k_\alpha,\alpha\succ\beta)}(\mathbf{x}_m,q_2^{(\alpha)})\right);\end{aligned}\right\}\tag{4.68}$$

$$\text{or}\quad\left.\begin{aligned}&\hbar_\alpha G_{\partial\Omega_{ij}}^{(m_\beta,\beta)}(\mathbf{x}_m,t_{m+})\\&\in\left(\hbar_\alpha G_{\partial\Omega_{ij}}^{(m_\beta,\alpha\succ\beta)}(\mathbf{x}_m,q_1^{(\beta)}),\ \hbar_\alpha G_{\partial\Omega_{ij}}^{(m_\beta,\alpha\succ\beta)}(\mathbf{x}_m,q_2^{(\beta)})\right).\end{aligned}\right\}\tag{4.69}$$

(ii) The coming flow of $\mathbf{x}^{(\alpha)}$ cannot be switched to the leaving flow of $\mathbf{x}^{(\beta)}$ to form a $(2k_\alpha:m_\beta)$ semi-passable flow at the boundary if for $\sigma_\alpha,\sigma_\beta\in\{1,2\}$

$$\text{either}\quad\left.\begin{aligned}&\mathbf{x}^{(\alpha)}(t_{m+})=\mathbf{x}^{(\alpha\succ\beta)}(t_{m\pm},q_{\sigma_\alpha}^{(\alpha)})=\mathbf{x}_m,\\&G_{\partial\Omega_{ij}}^{(s_\alpha,\alpha)}(\mathbf{x}_m,t_{m+})=G_{\partial\Omega_{ij}}^{(s_\alpha,\alpha\succ\beta)}(\mathbf{x}_m,q_{\sigma_\alpha}^{(\alpha)})\neq0\\&\text{for } s_\alpha=2k_\alpha,2k_\alpha+1,\cdots,l_\alpha-1,\\&(-1)^{\sigma_\alpha}\hbar_\alpha\mathbf{n}_{\partial\Omega_{ij}}^{\mathrm{T}}(\mathbf{x}^{(0)}(t_{m+\varepsilon}))\cdot\left[\mathbf{x}^{(\alpha)}(t_{m+\varepsilon})-\mathbf{x}^{(\alpha\succ\beta)}(t_{m+\varepsilon},q_{\sigma_\alpha}^{(\alpha)})\right]<0;\end{aligned}\right\}\tag{4.70}$$

$$\text{or}\quad\left.\begin{aligned}&\mathbf{x}^{(\beta)}(t_{m+})=\mathbf{x}^{(\alpha\succ\beta)}(t_{m\pm},q_{\sigma_\beta}^{(\beta)})=\mathbf{x}_m,\\&G_{\partial\Omega_{ij}}^{(s_\beta,\beta)}(\mathbf{x}_m,t_{m+})=G_{\partial\Omega_{ij}}^{(s_\beta,\alpha\succ\beta)}(\mathbf{x}_m,q_{\sigma_\beta}^{(\beta)})\neq0\\&\text{for } s_\beta=m_\beta,m_\beta+1,\cdots,l_\beta-1,\\&(-1)^{\sigma_\beta}\hbar_\alpha\mathbf{n}_{\partial\Omega_{ij}}^{\mathrm{T}}(\mathbf{x}^{(0)}(t_{m+\varepsilon}))\cdot\left[\mathbf{x}^{(\beta)}(t_{m+\varepsilon})-\mathbf{x}^{(\alpha\succ\beta)}(t_{m+\varepsilon},q_{\sigma_\beta}^{(\beta)})\right]<0.\end{aligned}\right\}\tag{4.71}$$

(iii) The coming flow of $\mathbf{x}^{(\alpha)}$ is switched to the leaving flow of $\mathbf{x}^{(\beta)}$ to form a $(2k_\alpha : m_\beta)$ semi-passable flow at the boundary if for $\sigma_\alpha, \sigma_\beta \in \{1,2\}$

$$\text{both} \left.\begin{array}{l} \mathbf{x}^{(\alpha)}(t_{m+}) = \mathbf{x}^{(\alpha\succ\beta)}(t_{m\pm}, q_{\sigma_\alpha}^{(\alpha)}) = \mathbf{x}_m, \\ G_{\partial\Omega_{ij}}^{(s_\alpha,\alpha)}(\mathbf{x}_m, t_{m+}) = G_{\partial\Omega_{ij}}^{(s_\alpha,\alpha\succ\beta)}(\mathbf{x}_m, q_{\sigma_\alpha}^{(\alpha)}) \neq 0 \\ \text{for } s_\alpha = 2k_\alpha, 2k_\alpha+1, \cdots, l_\alpha - 1, \\ (-1)^{\sigma_\alpha} \hbar_\alpha \mathbf{n}_{\partial\Omega_{ij}}^{\mathrm{T}}(\mathbf{x}^{(0)}(t_{m+\varepsilon})) \cdot \left[\mathbf{x}^{(\alpha)}(t_{m+\varepsilon}) - \mathbf{x}^{(\alpha\succ\beta)}(t_{m+\varepsilon}, q_{\sigma_\alpha}^{(\alpha)})\right] > 0; \end{array}\right\} \quad (4.72)$$

$$\text{and} \left.\begin{array}{l} \mathbf{x}^{(\beta)}(t_{m+}) = \mathbf{x}^{(\alpha\succ\beta)}(t_{m\pm}, q_{\sigma_\beta}^{(\beta)}) = \mathbf{x}_m, \\ G_{\partial\Omega_{ij}}^{(s_\beta,\beta)}(\mathbf{x}_m, t_{m+}) = G_{\partial\Omega_{ij}}^{(s_\beta,\alpha\succ\beta)}(\mathbf{x}_m, q_{\sigma_\beta}^{(\beta)}) \neq 0 \\ \text{for } s_\beta = m_\beta, m_\beta+1, \cdots, l_\beta - 1, \\ (-1)^{\sigma_\beta} \hbar_\alpha \mathbf{n}_{\partial\Omega_{ij}}^{\mathrm{T}}(\mathbf{x}^{(0)}(t_{m+\varepsilon})) \cdot \left[\mathbf{x}^{(\beta)}(t_{m+\varepsilon}) - \mathbf{x}^{(\alpha\succ\beta)}(t_{m+\varepsilon}, q_{\sigma_\beta}^{(\beta)})\right] > 0. \end{array}\right\} \quad (4.73)$$

To explain the above definition of the coming and leaving flow barriers in the semi-passable flow, the flow barriers on both sides of the boundary $\partial\Omega_{ij}$ are presented through the G-functions in Fig. 4.9. The red curves represent the G-function of the flows pertaining to the flow barriers in each domain Ω_α. The thick lines denote the G-functions of the flow barriers on both sides of the boundary. The hatched area is zoomed for the boundary flow of $\mathbf{x}^{(0)}$. To more intuitively illustrate the semi-passable flow with the flow barriers for the coming and leaving flows on the boundary. The $(2k_\alpha : m_\beta)$-semi-passable flow with the partial and full flow barriers on both sides of the boundary $\partial\Omega_{ij}$ are sketched in Fig. 4.10. For $\mathbf{x}_m \in S \subseteq \partial\Omega_{ij}$, the two different colored surfaces represent the flow barriers at the α and β-side boundaries of the boundary $\partial\Omega_{ij}$. The two flow barriers can be same. The arriving and leaving flows to the boundary are depicted by the solid and dashed curves.

Theorem 4.5. *For a discontinuous dynamical system in Eq.* (4.1)*, there is a point* $\mathbf{x}^{(0)}(t_m) \equiv \mathbf{x}_m \in \partial\Omega_{ij}$ *at time* t_m *between two adjacent domains* Ω_α ($\alpha = i, j$). *Suppose a flow barrier* $\mathbf{F}^{(\alpha\succ\beta)}(\mathbf{x}^{(\alpha)}, t, \boldsymbol{\pi}_\alpha, q^{(\alpha)})$ *at* $q^{(\alpha)} \in [q_1^{(\alpha)}, q_2^{(\alpha)}]$ *exists for a coming flow of* $\mathbf{x}^{(\alpha)}$ *on the* α*-side of boundary* $\partial\Omega_{ij}$ *with*

$$G_{\partial\Omega_{ij}}^{(\alpha\succ\beta)}(\mathbf{x}_m, q^{(\alpha)}) \in \left[\hbar_\alpha G_{\partial\Omega_{ij}}^{(\alpha\succ\beta)}(\mathbf{x}_m, q_1^{(\alpha)}), \hbar_\alpha G_{\partial\Omega_{ij}}^{(\alpha\succ\beta)}(\mathbf{x}_m, q_2^{(\alpha)})\right] \subset [0, +\infty), \quad (4.74)$$

and also there is a flow barrier $\mathbf{F}^{(\alpha\succ\beta)}(\mathbf{x}^{(\beta)}, t, \boldsymbol{\pi}_\beta, q^{(\beta)})$ *at* $q^{(\beta)} \in [q_1^{(\beta)}, q_2^{(\beta)}]$ *for the leaving flow of* $\mathbf{x}^{(\beta)}$ *on the* β*-side of boundary* $\partial\Omega_{ij}$ *with*

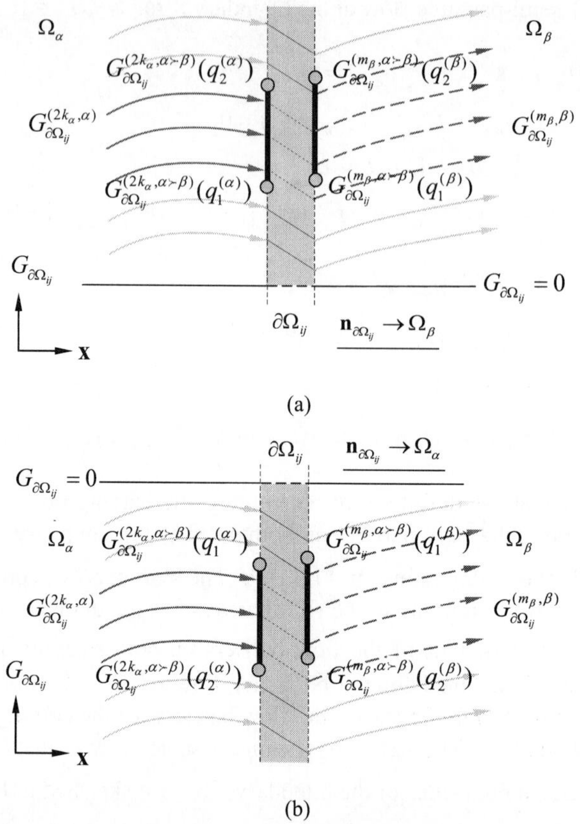

Fig. 4.9 G-functions for the coming and leaving flow barriers on both sides of the boundary in the $(2k_\alpha : m_\beta)$ - semi-passable flow: **(a)** $\mathbf{n}_{\partial\Omega_{ij}} \to \Omega_\beta$ and **(b)** $\mathbf{n}_{\partial\Omega_{ij}} \to \Omega_\alpha$. The red curves are the G-functions relative to the flow barriers. The thick lines are the G-function of the flow barriers at both α and β -side boundaries of the boundary $\partial\Omega_{ij}$. $G_{\partial\Omega_{ij}}^{(2k_\alpha,\alpha\succ\beta)}(q_1^{(\alpha)})$ and $G_{\partial\Omega_{ij}}^{(2k_\alpha,\alpha\succ\beta)}(q_2^{(\alpha)})$ are for *lower* and *upper* limits of the coming flow barrier on the α -side and, $G_{\partial\Omega_{ij}}^{(m_\beta,\alpha\succ\beta)}(q_1^{(\beta)})$ and $G_{\partial\Omega_{ij}}^{(m_\beta,\alpha\succ\beta)}(q_2^{(\beta)})$ are for *lower* and *upper* limits of the leaving flow barrier on the β -side. ($k_\alpha, m_\beta \in \{0,1,2,\cdots\}$, $\alpha = i,j$)

$$G_{\partial\Omega_{ij}}^{(\alpha\succ\beta)}(\mathbf{x}_m, q^{(\beta)}) \in \left[\hbar_\alpha G_{\partial\Omega_{ij}}^{(\alpha\succ\beta)}(\mathbf{x}_m, q_1^{(\beta)}), \hbar_\alpha G_{\partial\Omega_{ij}}^{(\alpha\succ\beta)}(\mathbf{x}_m, q_2^{(\beta)})\right]$$
$$\subset [0, +\infty). \tag{4.75}$$

The coming and leaving flows in the semi-passable flow satisfy

$$\hbar_\alpha G_{\partial\Omega_{ij}}^{(\alpha)}(\mathbf{x}_m, t_{m-}) > 0 \text{ and } \hbar_\alpha G_{\partial\Omega_{ij}}^{(\beta)}(\mathbf{x}_m, t_{m+}) > 0. \tag{4.76}$$

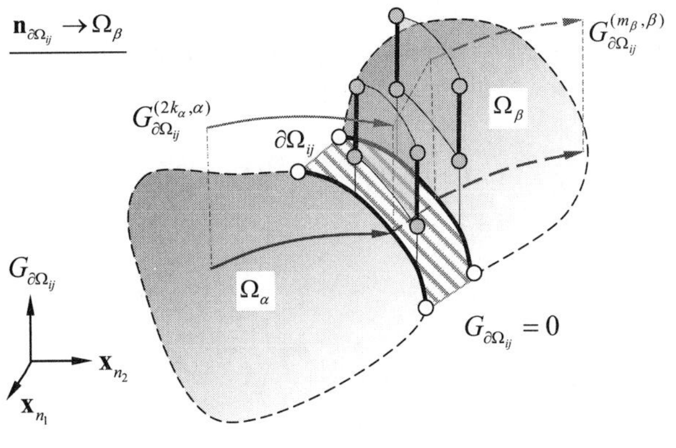

(a)

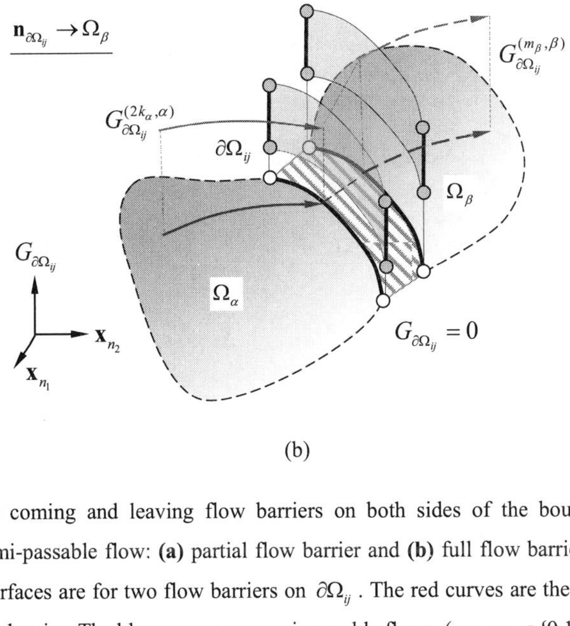

(b)

Fig. 4.10 The coming and leaving flow barriers on both sides of the boundary $\partial\Omega_{ij}$ in the $(2k_\alpha : m_\beta)$-semi-passable flow: **(a)** partial flow barrier and **(b)** full flow barrier. The green and blue shaded surfaces are for two flow barriers on $\partial\Omega_{ij}$. The red curves are the G-functions relative to the flow barrier. The blue curves are semi-passable flows. ($m_\alpha, m_\beta \in \{0,1,2,\cdots\}$)

(i) *The coming flow of* $\mathbf{x}^{(\alpha)}$ *cannot be switched to the leaving flow of* $\mathbf{x}^{(\beta)}$ *to form a semi-passable flow at the boundary if and only if*

$$\textit{either}\quad \hbar_\alpha G^{(\alpha)}_{\partial\Omega_{ij}}(\mathbf{x}_m,t_{m-}) \in \left(\hbar_\alpha G^{(\alpha\succ\beta)}_{\partial\Omega_{ij}}(\mathbf{x}_m,q_1^{(\alpha)}), \hbar_\alpha G^{(\alpha\succ\beta)}_{\partial\Omega_{ij}}(\mathbf{x}_m,q_2^{(\alpha)})\right), \tag{4.77}$$

$$or \quad \hbar_\alpha G^{(\beta)}_{\partial\Omega_{ij}}(\mathbf{x}_m, t_{m+}) \in \left(\hbar_\alpha G^{(\alpha\succ\beta)}_{\partial\Omega_{ij}}(\mathbf{x}_m, q_1^{(\beta)}), \hbar_\alpha G^{(\alpha\succ\beta)}_{\partial\Omega_{ij}}(\mathbf{x}_m, q_2^{(\beta)})\right). \tag{4.78}$$

(ii) *The coming flow of* $\mathbf{x}^{(\alpha)}$ *cannot be switched to the leaving flow of* $\mathbf{x}^{(\beta)}$ *to form a semi-passable flow at the boundary if and only if for* $\sigma_\alpha, \sigma_\beta \in \{1,2\}$

$$either \quad \left.\begin{array}{l} \hbar_\alpha G^{(s_\alpha,\alpha)}_{\partial\Omega_{ij}}(\mathbf{x}_m, t_{m-}) = \hbar_\alpha G^{(s_\alpha,\alpha\succ\beta)}_{\partial\Omega_{ij}}(\mathbf{x}_m, q^{(\alpha)}_{\sigma_\alpha}) \in (0,\infty) \\ \quad for\ s_\alpha = 0,1,\cdots,l_\alpha - 1, \\ (-1)^{\sigma_\alpha}\hbar_\alpha\left[G^{(l_\alpha,\alpha)}_{\partial\Omega_{ij}}(\mathbf{x}_m, t_{m-}) - G^{(l_\alpha,\alpha\succ\beta)}_{\partial\Omega_{ij}}(\mathbf{x}_m, q^{(\alpha)}_{\sigma_\alpha})\right] < 0; \end{array}\right\} \tag{4.79}$$

$$or \quad \left.\begin{array}{l} \hbar_\alpha G^{(s_\beta,\beta)}_{\partial\Omega_{ij}}(\mathbf{x}_m, t_{m+}) = \hbar_\alpha G^{(s_\beta,\alpha\succ\beta)}_{\partial\Omega_{ij}}(\mathbf{x}_m, q^{(\beta)}_{\sigma_\beta}) \in (0,\infty) \\ \quad for\ s_\beta = 0,1,\cdots,l_\beta - 1, \\ (-1)^{\sigma_\beta}\hbar_\alpha\left[G^{(l_\beta,\beta)}_{\partial\Omega_{ij}}(\mathbf{x}_m, t_{m+}) - G^{(l_\alpha,\alpha\succ\beta)}_{\partial\Omega_{ij}}(\mathbf{x}_m, q^{(\beta)}_{\sigma_\beta})\right] < 0. \end{array}\right\} \tag{4.80}$$

(iii) *The coming flow of* $\mathbf{x}^{(\alpha)}$ *is switched to the leaving flow of* $\mathbf{x}^{(\beta)}$ *to form a semi-passable flow at the boundary if and only if for* $\sigma_\alpha, \sigma_\beta \in \{1,2\}$

$$both \quad \left.\begin{array}{l} \hbar_\alpha G^{(s_\alpha,\alpha)}_{\partial\Omega_{ij}}(\mathbf{x}_m, t_{m-}) = \hbar_\alpha G^{(s_\alpha,\alpha\succ\beta)}_{\partial\Omega_{ij}}(\mathbf{x}_m, q^{(\alpha)}_{\sigma_\alpha}) \in (0,\infty) \\ \quad for\ s_\alpha = 0,1,\cdots,l_\alpha - 1, \\ (-1)^{\sigma_\alpha}\hbar_\alpha\left[G^{(l_\alpha,\alpha)}_{\partial\Omega_{ij}}(\mathbf{x}_m, t_{m-}) - G^{(l_\alpha,\alpha\succ\beta)}_{\partial\Omega_{ij}}(\mathbf{x}_m, q^{(\alpha)}_{\sigma_\alpha})\right] > 0; \end{array}\right\} \tag{4.81}$$

$$and \quad \left.\begin{array}{l} \hbar_\alpha G^{(s_\beta,\beta)}_{\partial\Omega_{ij}}(\mathbf{x}_m, t_{m+}) = \hbar_\alpha G^{(s_\beta,\alpha\succ\beta)}_{\partial\Omega_{ij}}(\mathbf{x}_m, q^{(\beta)}_{\sigma_\beta}) \in (0,\infty) \\ \quad for\ s_\beta = 0,1,\cdots,l_\beta - 1, \\ (-1)^{\sigma_\beta}\hbar_\alpha\left[G^{(l_\beta,\beta)}_{\partial\Omega_{ij}}(\mathbf{x}_m, t_{m+}) - G^{(l_\alpha,\alpha\succ\beta)}_{\partial\Omega_{ij}}(\mathbf{x}_m, q^{(\beta)}_{\sigma_\beta})\right] > 0. \end{array}\right\} \tag{4.82}$$

(iv) *The coming flow of* $\mathbf{x}^{(\alpha)}$ *is switched to the leaving flow of* $\mathbf{x}^{(\beta)}$ *to form a semi-passable flow at the boundary if and only if*

$$both \quad \hbar_\alpha G^{(\alpha)}_{\partial\Omega_{ij}}(\mathbf{x}_m, t_{m-}) \notin \left[\hbar_\alpha G^{(\alpha\succ\beta)}_{\partial\Omega_{ij}}(\mathbf{x}_m, q_1^{(\alpha)}), \hbar_\alpha G^{(\alpha\succ\beta)}_{\partial\Omega_{ij}}(\mathbf{x}_m, q_2^{(\alpha)})\right], \tag{4.83}$$

$$and \quad \hbar_\alpha G^{(\beta)}_{\partial\Omega_{ij}}(\mathbf{x}_m, t_{m+}) \notin \left[\hbar_\alpha G^{(\alpha\succ\beta)}_{\partial\Omega_{ij}}(\mathbf{x}_m, q_1^{(\beta)}), \hbar_\alpha G^{(\alpha\succ\beta)}_{\partial\Omega_{ij}}(\mathbf{x}_m, q_2^{(\beta)})\right]. \tag{4.84}$$

Proof. Similar to the proof of Theorem 4.1, this theorem can be proved. ∎

Theorem 4.6. *For a discontinuous dynamical system in Eq.* (4.1)*, there is a point* $\mathbf{x}^{(0)}(t_m) \equiv \mathbf{x}_m \in \partial\Omega_{ij}$ *at time* t_m *between two adjacent domains* Ω_α $(\alpha = i, j)$*. Sup-*

pose a flow barrier $\mathbf{F}^{(\alpha\succ\beta)}(\mathbf{x}^{(\alpha)},t,\boldsymbol{\pi}_\alpha,q^{(\alpha)})$ *at* $q^{(\alpha)}\in[q_1^{(\alpha)},q_2^{(\alpha)}]$ *exists for a coming flow of* $\mathbf{x}^{(\alpha)}$ *on the* α*-side of boundary* $\partial\Omega_{ij}$ *with*

$$\begin{aligned}
&G_{\partial\Omega_{ij}}^{(s_\alpha,\alpha\succ\beta)}(\mathbf{x}_m,q^{(\alpha)})=0 \text{ for } s_\alpha=0,1,\cdots,2k_\alpha-1,\\
&G_{\partial\Omega_{ij}}^{(2k_\alpha,\alpha\succ\beta)}(\mathbf{x}_m,q^{(\alpha)})\in\left[\hbar_\alpha G_{\partial\Omega_{ij}}^{(2k_\alpha,\alpha\succ\beta)}(\mathbf{x}_m,q_1^{(\alpha)}),\hbar_\alpha G_{\partial\Omega_{ij}}^{(2k_\alpha,\alpha\succ\beta)}(\mathbf{x}_m,q_2^{(\alpha)})\right]\\
&\qquad\subset[0,+\infty),
\end{aligned}\tag{4.85}$$

and also there is a flow barrier $\mathbf{F}^{(\alpha\mapsto\beta)}(\mathbf{x}^{(\beta)},t,\boldsymbol{\pi}_\beta,q^{(\beta)})$ *at* $q^{(\beta)}\in[q_1^{(\beta)},q_2^{(\beta)}]$ *for the leaving flow of* $\mathbf{x}^{(\beta)}$ *on the* β*-side of boundary* $\partial\Omega_{ij}$ *with*

$$\begin{aligned}
&G_{\partial\Omega_{ij}}^{(s_\beta,\alpha\succ\beta)}(\mathbf{x}_m,q^{(\beta)})=0 \text{ for } s_\beta=0,1,\cdots,m_\beta-1,\\
&G_{\partial\Omega_{ij}}^{(m_\beta,\alpha\succ\beta)}(\mathbf{x}_m,q^{(\beta)})\in\left[\hbar_\alpha G_{\partial\Omega_{ij}}^{(m_\beta,\alpha\succ\beta)}(\mathbf{x}_m,q_1^{(\beta)}),\hbar_\alpha G_{\partial\Omega_{ij}}^{(m_\beta,\alpha\succ\beta)}(\mathbf{x}_m,q_2^{(\beta)})\right]\\
&\qquad\subset[0,+\infty)
\end{aligned}\tag{4.86}$$

($\alpha,\beta\in\{i,j\}$ and $\alpha\neq\beta$). *The coming and leaving flows in the* $(2k_\alpha:m_\beta)$ *semi-passable flow satisfy*

$$\begin{aligned}
&G_{\partial\Omega_{ij}}^{(s_\alpha,\alpha)}(\mathbf{x}_m,t_{m-})=0 \text{ for } s_\alpha=0,1,\cdots,2k_\alpha-1,\\
&G_{\partial\Omega_{ij}}^{(s_\beta,\beta)}(\mathbf{x}_m,t_{m+})=0 \text{ for } s_\beta=0,1,\cdots,m_\beta-1,\\
&\hbar_\alpha G_{\partial\Omega_{ij}}^{(2k_\alpha,\alpha)}(\mathbf{x}_m,t_{m-})>0, \text{ and } \hbar_\alpha G_{\partial\Omega_{ij}}^{(m_\beta,\beta)}(\mathbf{x}_m,t_{m+})>0.
\end{aligned}\tag{4.87}$$

(i) *The coming flow of* $\mathbf{x}^{(\alpha)}$ *cannot be switched to the leaving flow of* $\mathbf{x}^{(\beta)}$ *to form the* $(2k_\alpha:m_\beta)$*-semi-passable flow if and only if*

$$\text{either } \hbar_\alpha G_{\partial\Omega_{ij}}^{(2k_\alpha,\alpha)}(\mathbf{x}_m,t_{m-})\in\left(\hbar_\alpha G_{\partial\Omega_{ij}}^{(2k_\alpha,\alpha\succ\beta)}(\mathbf{x}_m,q_1^{(\alpha)}),\ \hbar_\alpha G_{\partial\Omega_{ij}}^{(2k_\alpha,\alpha\succ\beta)}(\mathbf{x}_m,q_2^{(\alpha)})\right),\tag{4.88}$$

$$\text{or }\ \hbar_\alpha G_{\partial\Omega_{ij}}^{(m_\beta,\beta)}(\mathbf{x}_m,t_{m+})\in\left(\hbar_\alpha G_{\partial\Omega_{ij}}^{(m_\beta,\alpha\succ\beta)}(\mathbf{x}_m,q_1^{(\beta)}),\ \hbar_\alpha G_{\partial\Omega_{ij}}^{(m_\beta,\alpha\succ\beta)}(\mathbf{x}_m,q_2^{(\beta)})\right).\tag{4.89}$$

(ii) *The coming flow of* $\mathbf{x}^{(\alpha)}$ *cannot switched to the leaving flow* $\mathbf{x}^{(\beta)}$ *to form the* $(2k_\alpha:m_\beta)$*-semi-passable flow at the boundary if and only if for* $\sigma_\alpha,\sigma_\beta\in\{1,2\}$

$$\text{either}\quad\left.\begin{aligned}
&G_{\partial\Omega_{ij}}^{(s_\alpha,\alpha)}(\mathbf{x}_m,t_{m-})=G_{\partial\Omega_{ij}}^{(s_\alpha,\alpha\succ\beta)}(\mathbf{x}_m,q_{\sigma_\alpha}^{(\alpha)})\neq0\\
&\text{for } s_\alpha=2k_\alpha,2k_\alpha+1,\cdots,l_\alpha-1;\\
&(-1)^{\sigma_\alpha}\hbar_\alpha\left[G_{\partial\Omega_{ij}}^{(l_\alpha,\alpha)}(\mathbf{x}_m,t_{m-})-G_{\partial\Omega_{ij}}^{(l_\alpha,\alpha\succ\beta)}(\mathbf{x}_m,q_{\sigma_\alpha}^{(\alpha)})\right]<0
\end{aligned}\right\}\tag{4.90}$$

$$\left.\begin{array}{ll} & G_{\partial\Omega_{ij}}^{(s_\beta,\beta)}(\mathbf{x}_m,t_{m+}) = G_{\partial\Omega_{ij}}^{(s_\beta,\alpha\succ\beta)}(\mathbf{x}_m,q_{\sigma_\beta}^{(\beta)}) \neq 0 \\ or & \quad for\ s_\alpha = m_\beta, m_\beta+1,\cdots,l_\beta-1; \\ & (-1)^{\sigma_\beta}\hbar_\alpha\left[G_{\partial\Omega_{ij}}^{(l_\alpha,\beta)}(\mathbf{x}_m,t_{m+}) - G_{\partial\Omega_{ij}}^{(l_\alpha,\alpha\succ\beta)}(\mathbf{x}_m,q_{\sigma_\beta}^{(\beta)})\right] < 0. \end{array}\right\} \tag{4.91}$$

(iii) *The coming flow of* $\mathbf{x}^{(\alpha)}$ *is switched to the leaving flow* $\mathbf{x}^{(\beta)}$ *to form the* $(2k_\alpha : m_\beta)$*-semi-passable flow at the boundary if and only if for* $\sigma_\alpha \in \{1,2\}$ *and* $\sigma_\beta \in \{1,2\}$

$$\left.\begin{array}{ll} & G_{\partial\Omega_{ij}}^{(s_\alpha,\alpha)}(\mathbf{x}_m,t_{m-}) = G_{\partial\Omega_{ij}}^{(s_\alpha,\alpha\succ\beta)}(\mathbf{x}_m,q_{\sigma_\alpha}^{(\alpha)}) \neq 0 \\ both & \quad for\ s_\alpha = 2k_\alpha, 2k_\alpha+1,\cdots,l_\alpha-1, \\ & (-1)^{\sigma_\alpha}\hbar_\alpha\left[G_{\partial\Omega_{ij}}^{(l_\alpha,\alpha)}(\mathbf{x}_m,t_{m-}) - G_{\partial\Omega_{ij}}^{(l_\alpha,\alpha\succ\beta)}(\mathbf{x}_m,q_{\sigma_\alpha}^{(\alpha)})\right] > 0; \end{array}\right\} \tag{4.92}$$

$$\left.\begin{array}{ll} & G_{\partial\Omega_{ij}}^{(s_\beta,\beta)}(\mathbf{x}_m,t_{m+}) = G_{\partial\Omega_{ij}}^{(s_\beta,\alpha\succ\beta)}(\mathbf{x}_m,q_{\sigma_\beta}^{(\beta)}) \neq 0 \\ and & \quad for\ s_\alpha = m_\beta, m_\beta+1,\cdots,l_\beta-1, \\ & (-1)^{\sigma_\beta}\hbar_\alpha\left[G_{\partial\Omega_{ij}}^{(l_\alpha,\beta)}(\mathbf{x}_m,t_{m+}) - G_{\partial\Omega_{ij}}^{(l_\alpha,\alpha\succ\beta)}(\mathbf{x}_m,q_{\sigma_\beta}^{(\beta)})\right] > 0. \end{array}\right\} \tag{4.93}$$

(iv) *The coming flow of* $\mathbf{x}^{(\alpha)}$ *is switched to the leaving flow of* $\mathbf{x}^{(\beta)}$ *to form the* $(2k_\alpha : m_\beta)$*-semi-passable flow at the boundary if and only if*

$$\left.\begin{array}{ll} & G_{\partial\Omega_{ij}}^{(2k_\alpha,\alpha)}(\mathbf{x}_m,t_{m-}) \\ both & \notin\left[\hbar_\alpha G_{\partial\Omega_{ij}}^{(2k_\alpha,\alpha\succ\beta)}(\mathbf{x}_m,q_1^{(\alpha)}),\ \hbar_\alpha G_{\partial\Omega_{ij}}^{(2k_\alpha,\alpha\succ\beta)}(\mathbf{x}_m,q_2^{(\alpha)})\right], \end{array}\right\} \tag{4.94}$$

$$\left.\begin{array}{ll} & G_{\partial\Omega_{ij}}^{(m_\beta,\beta)}(\mathbf{x}_m,t_{m+}) \\ and & \notin\left[\hbar_\alpha G_{\partial\Omega_{ij}}^{(m_\beta,\alpha\succ\beta)}(\mathbf{x}_m,q_1^{(\beta)}),\ \hbar_\alpha G_{\partial\Omega_{ij}}^{(m_\beta,\alpha\succ\beta)}(\mathbf{x}_m,q_2^{(\beta)})\right]. \end{array}\right\} \tag{4.95}$$

Proof. As in the proof of Theorem 4.2, this theorem can be proved. ■

The forgoing two theorems presented that the coming and leaving flows on the boundary can pass over the corresponding flow barriers to form a semi-passable flow at the boundary. Theorems 4.5 and 4.6 give all the possible conditions for the coming flow to be switched into the leaving flow. The conditions for the leaving flow to pass over the corresponding flow barriers are presented in Fig. 4.11. The leaving flow at the critical point passes over the flow barrier and enters the domain Ω_β. Similarly, the other cases can be illustrated for a better understanding of the flow barriers in the semi-passable flow to the separation boundary.

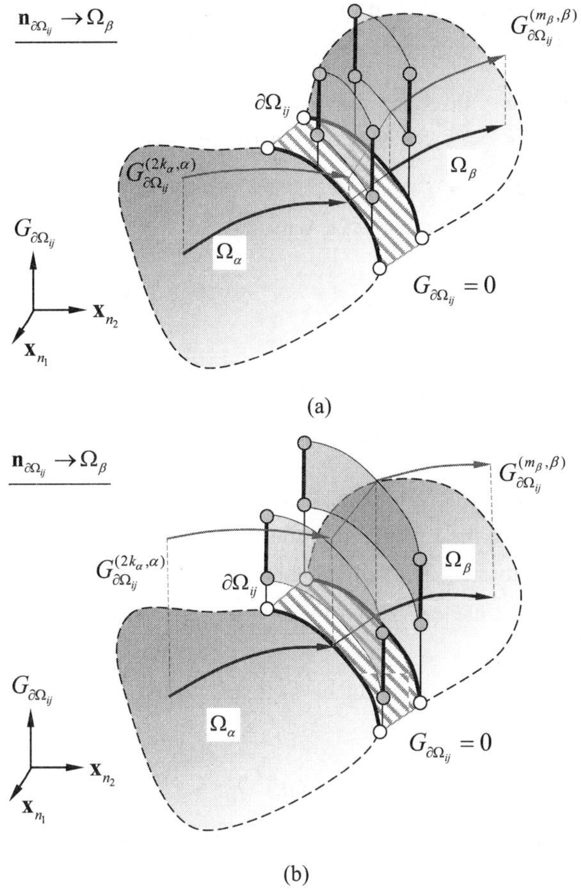

Fig. 4.11 The flow passing over the leaving flow barriers on the boundary $\partial\Omega_{ij}$: **(a)** partial flow barrier and **(b)** full flow barrier. The two vertical surfaces are for the flow barriers on the α -side and the β -side of the boundary $\partial\Omega_{ij}$, respectively. The two vertical surfaces are for two flow barriers on $\partial\Omega_{ij}$. The upper curves with arrows are the G-functions relative to the flow barrier. The lower curves with arrows are semi-passable flows. ($m_\alpha, m_\beta \in \{0,1,2,\cdots\}$).

4.2. Flow barriers for sink flows

Without any flow barriers, from Theorem 3.3, the necessary and sufficient conditions of a sink flow moving along the boundary in discontinuous dynamical systems are

$$\hbar_\alpha G^{(\alpha)}_{\partial\Omega_{ij}}(\mathbf{x}_m, t_{m-}, \mathbf{p}_\alpha, \boldsymbol{\lambda}) > 0 \text{ and } \hbar_\alpha G^{(\beta)}_{\partial\Omega_{ij}}(\mathbf{x}_m, t_{m-}, \mathbf{p}_\beta, \boldsymbol{\lambda}) < 0. \tag{4.96}$$

To investigate the sink flow property to the boundary with flow barriers, the sink flow barriers on the boundary will be discussed in this section.

Definition 4.15. For a discontinuous dynamical system in Eq. (4.1), there is a point $\mathbf{x}^{(0)}(t_m) \equiv \mathbf{x}_m \in \partial\Omega_{ij}$ at time t_m between two adjacent domains Ω_α ($\alpha = i, j$). Suppose there is a vector field $\mathbf{F}^{(\alpha\succ 0)}(\mathbf{x}^{(\alpha)}, t, \boldsymbol{\pi}_\alpha, q^{(\alpha)})$ for $q^{(\alpha)} \in [q_1^{(\alpha)}, q_2^{(\alpha)}]$ on the boundary $\partial\Omega_{ij}$ with

$$\begin{aligned}\hbar_\alpha G_{\partial\Omega_{ij}}^{(\alpha\succ 0)}(\mathbf{x}_m, q^{(\alpha)}) &\in \left[\hbar_\alpha G_{\partial\Omega_{ij}}^{(\alpha\succ 0)}(\mathbf{x}_m, q_1^{(\alpha)}), \hbar_\alpha G_{\partial\Omega_{ij}}^{(\alpha\succ 0)}(\mathbf{x}_m, q_2^{(\alpha)})\right] \\ &\subset [0, \infty).\end{aligned} \tag{4.97}$$

The two possible coming flows in the sink flow satisfy

$$\hbar_\alpha G_{\partial\Omega_{ij}}^{(\alpha)}(\mathbf{x}_m, t_{m-}) > 0 \text{ and } \hbar_\alpha G_{\partial\Omega_{ij}}^{(\beta)}(\mathbf{x}_m, t_{m-}) < 0 \tag{4.98}$$

($\alpha, \beta \in \{i, j\}$ and $\alpha \neq \beta$). The vector field of $\mathbf{F}^{(\alpha\succ 0)}(\mathbf{x}^{(\alpha)}, t, \boldsymbol{\pi}_\alpha, q^{(\alpha)})$ is called the *coming flow barrier* in the sink flow on the α -side of the boundary if the following conditions are satisfied. The critical values of $\mathbf{F}^{(\alpha\succ 0)}(\mathbf{x}^{(\alpha)}, t, \boldsymbol{\pi}_\alpha, q_\sigma^{(\alpha)})$ ($\sigma = 1, 2$) on the boundary $\partial\Omega_{ij}$ are called the *lower and upper limits* of the coming flow barriers on the α- side.

(i) The coming flow of $\mathbf{x}^{(\alpha)}$ cannot be switched to the boundary flow of $\mathbf{x}^{(0)}$ at $q^{(\alpha)} \in (q_1^{(\alpha)}, q_2^{(\alpha)})$ if

$$\begin{aligned}&\mathbf{x}^{(\alpha)}(t_{m-}) = \mathbf{x}^{(\alpha\succ 0)}(t_{m\pm}, q^{(\alpha)}) = \mathbf{x}_m, \\ &\hbar_\alpha G_{\partial\Omega_{ij}}^{(\alpha)}(\mathbf{x}_m, t_{m-}) \in \left(\hbar_\alpha G_{\partial\Omega_{ij}}^{(\alpha\succ 0)}(\mathbf{x}_m, q_1^{(\alpha)}), \hbar_\alpha G_{\partial\Omega_{ij}}^{(\alpha\succ 0)}(\mathbf{x}_m, q_2^{(\alpha)})\right).\end{aligned} \tag{4.99}$$

(ii) The coming flow of $\mathbf{x}^{(\alpha)}$ cannot be switched to the boundary flow of $\mathbf{x}^{(0)}$ at the critical points of the flow barrier (i.e., $q^{(\alpha)} = q_\sigma^{(\alpha)}$, $\sigma \in \{1, 2\}$) if

$$\begin{aligned}&\mathbf{x}^{(\alpha)}(t_{m-}) = \mathbf{x}^{(\alpha\succ 0)}(t_{m\pm}, q_\sigma^{(\alpha)}) = \mathbf{x}_m, \\ &G_{\partial\Omega_{ij}}^{(s_\alpha, \alpha)}(\mathbf{x}_m, t_{m-}) = G_{\partial\Omega_{ij}}^{(s_\alpha, \alpha\succ 0)}(\mathbf{x}_m, q_\sigma^{(\alpha)}) \neq 0 \text{ for } s_\alpha = 0, 1, 2, \cdots, l_\alpha - 1; \\ &(-1)^\sigma \hbar_\alpha \mathbf{n}_{\partial\Omega_{ij}}^{\mathrm{T}}(\mathbf{x}^{(0)}(t_{m+\varepsilon})) \cdot \left[\mathbf{x}^{(\alpha)}(t_{m+\varepsilon}) - \mathbf{x}^{(\alpha\succ 0)}(t_{m+\varepsilon}, q_\sigma^{(\alpha)})\right] < 0.\end{aligned} \tag{4.100}$$

(iii) The coming flow of $\mathbf{x}^{(\alpha)}$ is switched to the boundary flow of $\mathbf{x}^{(0)}$ at the critical points of the flow barrier (i.e., $q^{(\alpha)} = q_\sigma^{(\alpha)}$, $\sigma \in \{1, 2\}$) if

$$\begin{aligned}&\mathbf{x}^{(\alpha)}(t_{m-}) = \mathbf{x}^{(\alpha\succ 0)}(t_{m\pm}, q_\sigma^{(\alpha)}) = \mathbf{x}_m, \\ &G_{\partial\Omega_{ij}}^{(s_\alpha, \alpha)}(\mathbf{x}_m, t_{m-}) = G_{\partial\Omega_{ij}}^{(s_\alpha, \alpha\succ 0)}(\mathbf{x}_m, q_\sigma^{(\alpha)}) \neq 0 \text{ for } s_\alpha = 0, 1, 2, \cdots, l_\alpha - 1; \\ &(-1)^\sigma \hbar_\alpha \mathbf{n}_{\partial\Omega_{ij}}^{\mathrm{T}}(\mathbf{x}^{(0)}(t_{m+\varepsilon})) \cdot \left[\mathbf{x}^{(\alpha)}(t_{m+\varepsilon}) - \mathbf{x}^{(\alpha\succ 0)}(t_{m+\varepsilon}, q_\sigma^{(\alpha)})\right] > 0.\end{aligned} \tag{4.101}$$

Definition 4.16. For a discontinuous dynamical system in Eq. (4.1), there is a point $\mathbf{x}^{(0)}(t_m) \equiv \mathbf{x}_m \in \partial\Omega_{ij}$ at time t_m between two adjacent domains Ω_α ($\alpha = i, j$). There is a vector field $\mathbf{F}^{(\alpha \succ 0)}(\mathbf{x}^{(\alpha)}, t, \boldsymbol{\pi}_\alpha, q^{(\alpha)})$ for $q^{(\alpha)} \in [q_1^{(\alpha)}, q_2^{(\alpha)}]$ on the boundary $\partial\Omega_{ij}$ with the G-functions

$$
\begin{aligned}
G_{\partial\Omega_{ij}}^{(s_\alpha,\alpha\succ 0)}(\mathbf{x}_m, q^{(\alpha)}) = 0 \text{ for } s_\alpha = 0,1,\cdots,2k_\alpha - 1;\\
G_{\partial\Omega_{ij}}^{(2k_\alpha,\alpha\succ 0)}(\mathbf{x}_m, q^{(\alpha)}) \in \left[\hbar_\alpha G_{\partial\Omega_{ij}}^{(2k_\alpha,\alpha\succ 0)}(\mathbf{x}_m, q_1^{(\alpha)}), \hbar_\alpha G_{\partial\Omega_{ij}}^{(2k_\alpha,\alpha\succ 0)}(\mathbf{x}_m, q_2^{(\alpha)})\right]\\
\subset [0,\infty).
\end{aligned} \tag{4.102}
$$

The two possible coming flows in the $(2k_\alpha : 2k_\beta)$-sink flow satisfy

$$
\begin{aligned}
G_{\partial\Omega_{ij}}^{(s_\alpha,\alpha)}(\mathbf{x}_m, t_{m-}) = 0 \text{ for } s_\alpha = 0,1,\cdots,2k_\alpha - 1;\\
G_{\partial\Omega_{ij}}^{(s_\beta,\beta)}(\mathbf{x}_m, t_{m-}) = 0 \text{ for } s_\beta = 0,1,\cdots,2k_\beta - 1;\\
\hbar_\alpha G_{\partial\Omega_{ij}}^{(2k_\alpha,\alpha)}(\mathbf{x}_m, t_{m-}) > 0 \text{ and } \hbar_\alpha G_{\partial\Omega_{ij}}^{(2k_\beta,\beta)}(\mathbf{x}_m, t_{m-}) < 0
\end{aligned} \tag{4.103}
$$

($\alpha, \beta \in \{i, j\}$ and $\alpha \neq \beta$). The vector field of $\mathbf{F}^{(\alpha \succ 0)}(\mathbf{x}^{(\alpha)}, t, \boldsymbol{\pi}_\alpha, q^{(\alpha)})$ is called the *coming flow barrier* in the $(2k_\alpha : 2k_\beta)$-sink flow on the α-side if the following conditions are satisfied. The critical values of $\mathbf{F}^{(\alpha \succ 0)}(\mathbf{x}^{(\alpha)}, t, \boldsymbol{\pi}_\alpha, q_\sigma^{(\alpha)})$ ($\sigma = 1, 2$) at the boundary $\partial\Omega_{ij}$ are called the *lower and upper limits* of the coming flow barrier on the α - side.

(i) The coming flow of $\mathbf{x}^{(\alpha)}$ in the $(2k_\alpha : 2k_\beta)$-sink flow cannot be switched to the boundary flow of $\mathbf{x}^{(0)}$ if

$$
\hbar_\alpha G_{\partial\Omega_{ij}}^{(2k_\alpha,\alpha)}(\mathbf{x}_m, t_{m-}) \in \left(\hbar_\alpha G_{\partial\Omega_{ij}}^{(2k_\alpha,\alpha\succ 0)}(\mathbf{x}_m, q_1^{(\alpha)}), \hbar_\alpha G_{\partial\Omega_{ij}}^{(2k_\alpha,\alpha\succ 0)}(\mathbf{x}_m, q_2^{(\alpha)})\right). \tag{4.104}
$$

(ii) The coming flow of $\mathbf{x}^{(\alpha)}$ cannot be switched to the boundary flow of $\mathbf{x}^{(0)}$ at the critical points of the flow barrier (i.e., $q^{(\alpha)} = q_\sigma^{(\alpha)}$, $\sigma \in \{1, 2\}$) if

$$
\begin{aligned}
\mathbf{x}^{(\alpha)}(t_{m-}) = \mathbf{x}^{(\alpha\succ 0)}(t_{m\pm}, q_\sigma^{(\alpha)}) = \mathbf{x}_m,\\
G_{\partial\Omega_{ij}}^{(s_\alpha,\alpha)}(\mathbf{x}_m, t_{m-}) = G_{\partial\Omega_{ij}}^{(s_\alpha,\alpha\succ 0)}(\mathbf{x}_m, q_\sigma^{(\alpha)}) \neq 0\\
\text{for } s_\alpha = 2k_\alpha, 2k_\alpha + 1, \cdots, l_\alpha - 1;\\
(-1)^\sigma \hbar_\alpha \mathbf{n}_{\partial\Omega_{ij}}^{\mathrm{T}}(\mathbf{x}^{(0)}(t_{m+\varepsilon})) \cdot \left[\mathbf{x}^{(\alpha)}(t_{m+\varepsilon}) - \mathbf{x}^{(\alpha\succ 0)}(t_{m+\varepsilon}, q_\sigma^{(\alpha)})\right] < 0.
\end{aligned} \tag{4.105}
$$

(iii) The coming flow of $\mathbf{x}^{(\alpha)}$ is switched to the boundary flow of $\mathbf{x}^{(0)}$ at the critical points of the flow barrier (i.e., $q^{(\alpha)} = q_\sigma^{(\alpha)}$, $\sigma \in \{1, 2\}$) if

$$
\begin{aligned}
\mathbf{x}^{(\alpha)}(t_{m-}) = \mathbf{x}^{(\alpha\succ 0)}(t_{m\pm}, q_\sigma^{(\alpha)}) = \mathbf{x}_m,\\
G_{\partial\Omega_{ij}}^{(s_\alpha,\alpha)}(\mathbf{x}_m, t_{m-}) = G_{\partial\Omega_{ij}}^{(s_\alpha,\alpha\succ 0)}(\mathbf{x}_m, q_\sigma^{(\alpha)}) \neq 0\\
\text{for } s_\alpha = 2k_\alpha, 2k_\alpha + 1, \cdots, l_\alpha - 1;\\
(-1)^\sigma \hbar_\alpha \mathbf{n}_{\partial\Omega_{ij}}^{\mathrm{T}}(\mathbf{x}^{(0)}(t_{m+\varepsilon})) \cdot \left[\mathbf{x}^{(\alpha)}(t_{m+\varepsilon}) - \mathbf{x}^{(\alpha\succ 0)}(t_{m+\varepsilon}, q_\sigma^{(\alpha)})\right] > 0.
\end{aligned} \tag{4.106}
$$

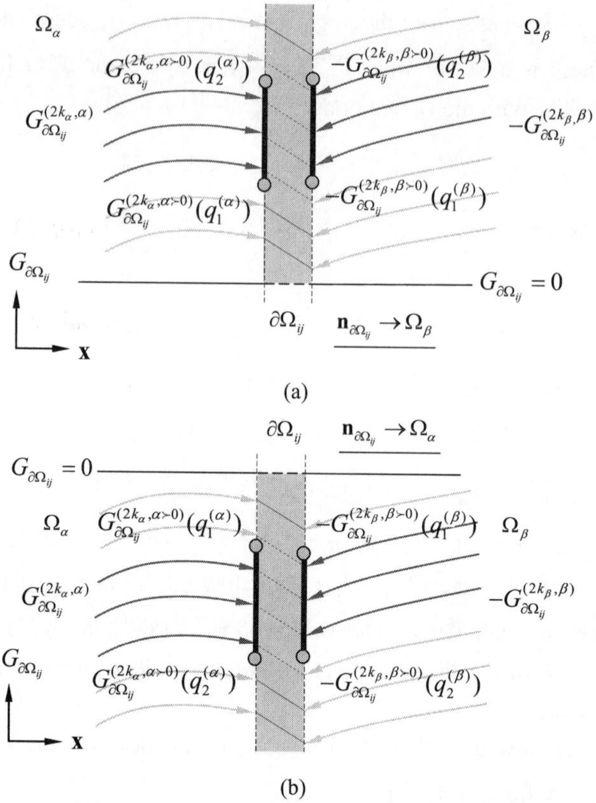

Fig. 4.12 *G*-functions for flow barriers in the $(2k_\alpha : 2k_\beta)$ -sink flow on the boundary: **(a)** $\mathbf{n}_{\partial\Omega_{ij}} \to \Omega_\beta$ and **(b)** $\mathbf{n}_{\partial\Omega_{ij}} \to \Omega_\alpha$. The red curves are the *G*-functions relative to the flow barrier. The thick lines are the *G*-functions of the flow barriers at both α and β -side boundaries of the boundary $\partial\Omega_{ij}$. $G^{(2k_\alpha,\alpha\succ 0)}_{\partial\Omega_{ij}}(q_1^{(\alpha)})$ and $G^{(2k_\alpha,\alpha\succ 0)}_{\partial\Omega_{ij}}(q_2^{(\alpha)})$ are for *lower* and *upper* barrier limits ($k_\alpha \in \{0,1,2,\cdots\}$, $\alpha = i, j$) and similarly for the β -side.

To explain the coming flow barrier in a sink flow on the boundary, the sink flow barriers on both sides of the boundary $\partial\Omega_{ij}$ are presented through the *G*-functions in Fig. 4.12. The red curves represent the *G*-function of flows pertaining to the flow barriers in each domain Ω_α ($\alpha = i, j$). The thick lines denote the *G*-functions of flow barriers on both sides of the boundary. The gray curves are the *G*-functions without any flow barriers, and the solid thin lines are used to connect the corresponding *G*-functions. To show the flow barriers, consider the *G*-function on the α -side of the boundary $\partial\Omega_{ij}$ as a reference. So the *G*-function on the β -side of the boundary $\partial\Omega_{ij}$ is presented through $-G^{(\beta)}_{\partial\Omega_{ij}}$ because the sink flow requires the *G*-functions of flows on both sides of the boundary should be with opposite signs. The hatched area is zoomed for the boundary flow of $\mathbf{x}^{(0)}$.

Definition 4.17. For a discontinuous dynamical system in Eq. (4.1), there is a point $\mathbf{x}^{(0)}(t_m) \equiv \mathbf{x}_m \in \partial\Omega_{ij}$ at time t_m between two adjacent domains Ω_α ($\alpha = i, j$). Suppose a coming flow barrier of $\mathbf{F}^{(\alpha \succ 0)}(\mathbf{x}^{(\alpha)}, t, \boldsymbol{\pi}_\alpha, q)$ for $q^{(\alpha)} \in [q_1^{(\alpha)}, q_2^{(\alpha)}]$ exists in the $(2k_\alpha : 2k_\beta)$-sink flow on the α-side of the boundary $\partial\Omega_{ij}$ ($k_\alpha, k_\beta = 0, 1, 2, \cdots$).

(i) The coming flow barrier in the $(2k_\alpha : 2k_\beta)$-sink flow is *partial* on the α-side if $\mathbf{x}_m \in S \subset \partial\Omega_{ij}$.

(ii) The coming flow barrier in the $(2k_\alpha : 2k_\beta)$-sink flow is *full* on the α-side if $\mathbf{x}_m \in S = \partial\Omega_{ij}$.

In a similar fashion, the partial and full $(2k_\alpha : 2k_\beta)$-sink flow barriers on both sides of the boundary $\partial\Omega_{ij}$ are sketched in Fig. 4.13 for $\mathbf{x}_m \in S \subseteq \partial\Omega_{ij}$ through the two different colored surfaces on the α and β-sides of the boundary $\partial\Omega_{ij}$. To clearly show the flow barriers, the G-function on the α-side of the boundary $\partial\Omega_{ij}$ is considered as a reference, and the G-function on the β-side of the boundary $\partial\Omega_{ij}$ is presented by $-G_{\partial\Omega_{ij}}^{(\beta)}$. The infinity $(2k_\alpha : 2k_\beta)$-sink flow barriers are presented in Fig. 4.14.

Definition 4.18. For a discontinuous dynamical system in Eq. (4.1), there is a point $\mathbf{x}^{(0)}(t_m) \equiv \mathbf{x}_m \in \partial\Omega_{ij}$ at time t_m between two adjacent domains Ω_α ($\alpha = i, j$). There is a coming barrier of $\mathbf{F}^{(\alpha \succ 0)}(\mathbf{x}^{(\alpha)}, t, \boldsymbol{\pi}_\alpha, q^{(\alpha)})$ for $q^{(\alpha)} \in [q_1^{(\alpha)}, q_2^{(\alpha)}]$ on the α-side of the boundary ($k_\alpha, k_\beta \in \{0, 1, 2, \cdots\}$) in the $(2k_\alpha : 2k_\beta)$-sink flow.

(i) The coming flow barrier in the $(2k_\alpha : 2k_\beta)$-sink flow is *with an upper limit* if for $\mathbf{x}_m \in S \subseteq \partial\Omega_{ij}$

$$\hbar_\alpha G_{\partial\Omega_{ij}}^{(2k_\alpha, \alpha \succ 0)}(\mathbf{x}_m, q_1^{(\alpha)}) = 0_- \text{ and } G_{\partial\Omega_{ij}}^{(2k_\alpha, \alpha \succ 0)}(\mathbf{x}_m, q_2^{(\alpha)}) \neq \infty. \tag{4.107}$$

(ii) The coming flow barrier in the $(2k_\alpha : 2k_\beta)$-sink flow is *with a lower limit* if for $\mathbf{x}_m \in S \subseteq \partial\Omega_{ij}$

$$\hbar_\alpha G_{\partial\Omega_{ij}}^{(2k_\alpha, \alpha \succ 0)}(\mathbf{x}_m, q_1^{(\alpha)}) \neq 0 \text{ and } \hbar_\alpha G_{\partial\Omega_{ij}}^{(2k_\alpha, \sigma \succ 0)}(\mathbf{x}_m, q_2^{(\alpha)}) \to +\infty. \tag{4.108}$$

(iii) The coming flow barrier in the $(2k_\alpha : 2k_\beta)$-sink flow is *absolute* if for $\mathbf{x}_m \in S \subseteq \partial\Omega_{ij}$

$$\hbar_\alpha G_{\partial\Omega_{ij}}^{(2k_\alpha, \alpha \succ 0)}(\mathbf{x}_m, q_1^{(\alpha)}) = 0_- \text{ and } \hbar_\alpha G_{\partial\Omega_{ij}}^{(2k_\alpha, \alpha \succ 0)}(\mathbf{x}_m, q_2^{(\alpha)}) \to +\infty. \tag{4.109}$$

(iv) The coming flow barrier in the $(2k_\alpha : 2k_\beta)$-sink flow is a *flow barrier wall* on the α-side if the absolute flow barrier exists on $\mathbf{x}_m \in S = \partial\Omega_{ij}$.

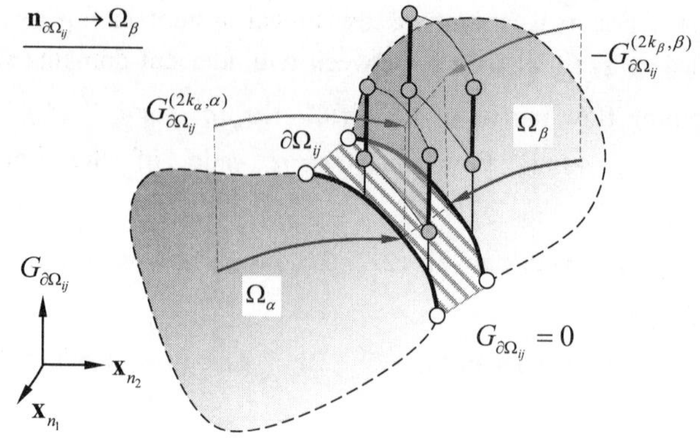

(a)

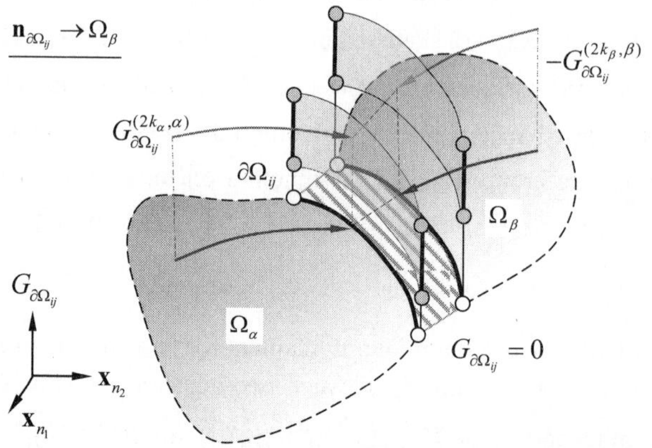

(b)

Fig. 4.13 The G-functions of the flow barriers in the $(2k_\alpha : 2k_\beta)$-sink flow on the boundary $\partial\Omega_{ij}$: **(a)** partial flow barrier and **(b)** full flow barrier. The upper curves with arrows are the G-functions relative to the flow barrier. The two vertical surfaces are the flow barrier surfaces. The hatched area is for the zoomed boundary. The lower curves with arrows are coming flows. ($k_\alpha, k_\beta \in \{0,1,2,\cdots\}$)

(v) The *coming* flow barrier in the $(2k_\alpha : 2k_\beta)$-sink flow is a *flow barrier fence* on the α-side if the flow barriers exist on $S_{k_1} \subset \partial\Omega_{ij}$ and no flow barriers on $S_{k_2} \subset \partial\Omega_{ij}$ ($k_1, k_2 \in \{1,2,\cdots\}$) for $S_{k_1} \cap S_{k_2} = \varnothing$.

Definition 4.19. For a discontinuous dynamical system in Eq. (4.1), there is a point $\mathbf{x}^{(0)}(t_m) \equiv \mathbf{x}_m \in \partial\Omega_{ij}$ at time t_m between two adjacent domains Ω_α ($\alpha = i, j$). The two possible coming flows in the sink flow satisfy Eq. (4.103). Suppose there is a coming flow barrier $\mathbf{F}^{(\alpha \succ 0)}(\mathbf{x}^{(\alpha)}, t, \boldsymbol{\pi}_\alpha, q^{(\alpha)})$ on the α-side in the $(2k_\alpha : 2k_\beta)$-sink

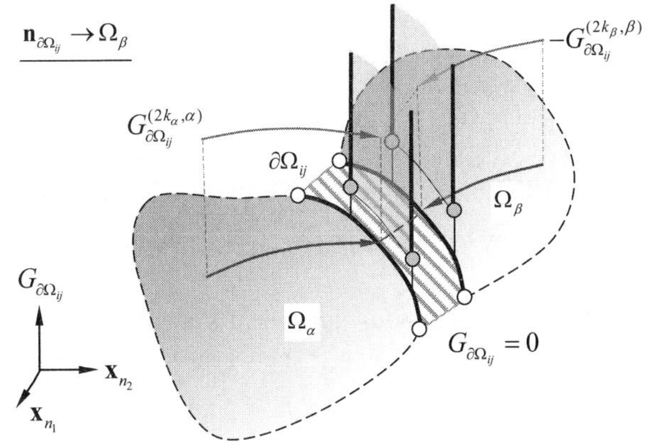

(a)

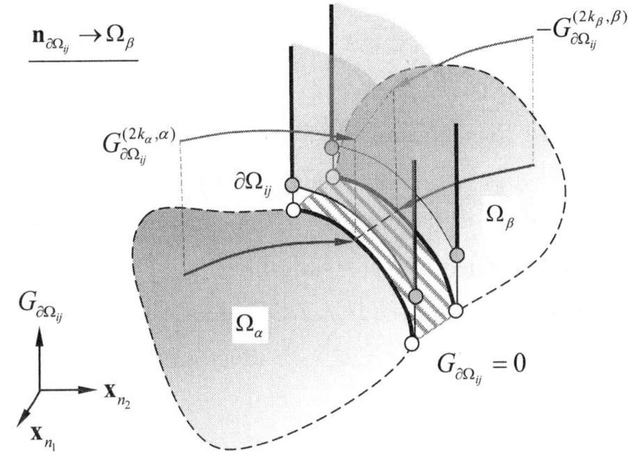

(b)

Fig. 4.14 The G-functions for the infinity $(2k_\alpha : 2k_\beta)$-sink flow barrier with lower boundary on $\partial\Omega_{ij}$: **(a)** partial flow barrier and **(b)** full flow barrier. The upper curves with arrows are the G-functions relative to the flow barrier. The two vertical surfaces are the flow barrier surfaces. The hatched area is for the zoomed boundary. The lower curves with arrows are coming flows. ($k_\alpha, k_\beta \in \{0,1,2,\cdots\}$).

flow ($k_\alpha, k_\beta \in \{0,1,2,\cdots\}$) for $\mathbf{x}_m \in S \subseteq \partial\Omega_{ij}$ and $q^{(\alpha)} \in [q_{2n-1}^{(\alpha)}, q_{2n}^{(\alpha)}]$ ($n = 1,2,\cdots$) with

$$\begin{aligned} &\hbar_\alpha G_{\partial\Omega_{ij}}^{(s_\alpha,\alpha\succ 0)}(\mathbf{x}_m, q^{(\alpha)}) = 0 \text{ for } s_\alpha = 0,1,2,\cdots,2k_\alpha - 1; \\ &\hbar_\alpha G_{\partial\Omega_{ij}}^{(2k_\alpha,\alpha\succ 0)}(\mathbf{x}_m, q^{(\alpha)}) \in \left[\hbar_\alpha G_{\partial\Omega_{ij}}^{(2k_\alpha,\alpha\succ 0)}(\mathbf{x}_m, q_{2n-1}^{(\alpha)}), \hbar_\alpha G_{\partial\Omega_{ij}}^{(2k_\alpha,\alpha\succ 0)}(\mathbf{x}_m, q_{2n}^{(\alpha)})\right]. \end{aligned} \tag{4.110}$$

For $q^{(\alpha)} \in (q_{2n}^{(\alpha)}, q_{2n+1}^{(\alpha)})$, no flow barriers are defined at $\mathbf{x}_m \in S \subseteq \partial\Omega_{ij}$. Thus, the

α- flow of $\mathbf{x}^{(\alpha)}$ can switch the flow of $\mathbf{x}^{(0)}$ for $\sigma = 2n, 2n+1$

$$\begin{aligned} &\hbar_\alpha G_{\partial\Omega_{ij}}^{(s_\alpha,\alpha)}(\mathbf{x}_m, q_\sigma^{(\alpha)}) = 0 \text{ for } s_\alpha = 0,1,2,\cdots,2k_\alpha - 1; \\ &\hbar_\alpha G_{\partial\Omega_{ij}}^{(2k_\alpha,\alpha)}(\mathbf{x}_m, t_{m-}) \in \left(\hbar_\alpha G_{\partial\Omega_{ij}}^{(2k_\alpha,\alpha\succ 0)}(\mathbf{x}_m, q_{2n}^{(\alpha)}), \hbar_\alpha G_{\partial\Omega_{ij}}^{(2k_\alpha,\alpha\succ 0)}(\mathbf{x}_m, q_{2n+1}^{(\alpha)})\right). \end{aligned} \tag{4.111}$$

The G-function intervals for all $\mathbf{x}_m \in S \subseteq \partial\Omega_{ij}$ with $q^{(\alpha)} \in (q_{2n}^{(\alpha)}, q_{2n+1}^{(\alpha)})$ are called the window of coming flow barrier in the $(2k_\alpha : 2k_\beta)$-sink flow on the α-side.

Definition 4.20. For a discontinuous dynamical system in Eq. (4.1), there is a point $\mathbf{x}^{(0)}(t_m) \equiv \mathbf{x}_m \in \partial\Omega_{ij}$ at time t_m between two adjacent domains Ω_α ($\alpha = i, j$). Suppose there is a coming flow barrier of $\mathbf{F}^{(\alpha\succ 0)}(\mathbf{x}^{(\alpha)}, t, \boldsymbol{\pi}_\alpha, q^{(\alpha)})$ for $q^{(\alpha)} \in [q_1^{(\alpha)}, q_2^{(\alpha)}]$ in the $(2k_\alpha : 2k_\beta)$-sink flow on the α-side of the boundary $\partial\Omega_{ij}$ ($k_\alpha, k_\beta \in \{0,1,2,\cdots\}$).

(i) The coming flow barrier in the $(2k_\alpha : 2k_\beta)$-sink flow is *permanent* on the α-side if the flow barrier is *independent* of time $t \in [0,\infty)$.

(ii) The coming flow barrier in the $(2k_\alpha : 2k_\beta)$-sink flow is *instantaneous* on the α-side if the flow barrier is continuously *dependent* on time $t \in [0,\infty)$.

(iii) The coming flow barrier in the $(2k_\alpha : 2k_\beta)$-sink flow is *intermittent* on the α-side if the flow barrier exists for time $t \in [t_k, t_{k+1}]$ with $k \in \mathbb{Z}$.

(iv) The coming flow barrier in the $(2k_\alpha : 2k_\beta)$-sink flow is *intermittent and static* on the α-side if the flow barrier is *independent* of time $t \in [t_k, t_{k+1}]$ with $k \in \mathbb{Z}$.

Similarly, the permanent and instantaneous windows of the sink flow barrier on the boundary can be discussed. Further, the concept of the door for the sink flow barrier wall on the boundary can also be described.

Definition 4.21. For a discontinuous dynamical system in Eq. (4.1), there is a point $\mathbf{x}^{(0)}(t_m) \equiv \mathbf{x}_m \in \partial\Omega_{ij}$ at time t_m between two adjacent domains Ω_α ($\alpha = i, j$). There is a coming flow barrier of $\mathbf{F}^{(\alpha\succ 0)}(\mathbf{x}^{(\alpha)}, t, \boldsymbol{\pi}_\alpha, q^{(\alpha)})$ for $q^{(\alpha)} \in [q_{2n-1}^{(\alpha)}, q_{2n}^{(\alpha)}]$ on the α-side of the boundary ($k_\alpha, k_\beta \in \{0,1,2,\cdots\}$, $n = 1,2,\cdots$). Suppose there is a window of the flow barrier for $S \subset \partial\Omega_{ij}$ and $q^{(\alpha)} \in [q_{2n}^{(\alpha)}, q_{2n+1}^{(\alpha)}]$.

(i) The window of the flow barrier in the $(2k_\alpha : 2k_\beta)$-sink flow is *permanent* on the α-side if the window is *independent* of t time $t \in [0,+\infty)$.

(ii) The window of the flow barrier in the $(2k_\alpha : 2k_\beta)$-sink flow is *instantaneous* on the α-side if the window is continuously *dependent* on time $t \in [0,+\infty)$.

(iii) The window of the flow barrier in the $(2k_\alpha : 2k_\beta)$-sink flow is *intermittent* on the α-side if the window exists for time $t \in [t_k, t_{k+1}]$ with $k \in \mathbb{Z}$.

(iv) The window of flow barrier in the $(2k_\alpha : 2k_\beta)$-sink flow is *intermittent and static* on the α-side if the window is *independent* of time $t \in [t_k, t_{k+1}]$ with $k \in \mathbb{Z}$.

Definition 4.22. For a discontinuous dynamical system in Eq. (4.1), there is a point $\mathbf{x}^{(0)}(t_m) \equiv \mathbf{x}_m \in \partial\Omega_{ij}$ at time t_m between two adjacent domains Ω_α ($\alpha = i, j$). There is a coming flow barrier of $\mathbf{F}^{(\alpha \succ 0)}(\mathbf{x}^{(\alpha)}, t, \boldsymbol{\pi}_\alpha, q^{(\alpha)})$ for $q^{(\alpha)} \in [q_1^{(\alpha)}, q_2^{(\alpha)}]$ on the α-side of the boundary ($k_\alpha, k_\beta \in \{0,1,2,\cdots\}$) in the $(2k_\alpha : 2k_\beta)$-sink flow. Suppose there is a *flow barrier wall* on the α-side for $S = \partial\Omega_{ij}$ and there is an *intermittent*, static window of the coming flow barrier on $S \subset \partial\Omega_{ij}$ for $q^{(\alpha)} \in [q_1^{(\alpha)}, q_2^{(\alpha)}]$ and $t \in [t_k, t_{k+1}]$ with $k \in \mathbb{Z}$.

(i) The window of the flow barrier in the $(2k_\alpha : 2k_\beta)$-sink flow is termed a *door* of the *flow barrier wall* on the α-side if the window and flow barriers exist alternatively.

(ii) The door of the coming flow barrier in the $(2k_\alpha : 2k_\beta)$-sink flow is *open* on the α-side if the window exists for time $t \in [t_k, t_{k+1}]$ with $k \in \mathbb{Z}$.

(iii) The door of the coming flow barrier in the $(2k_\alpha : 2k_\beta)$-sink flow is closed on the α-side if the flow barrier exists for time $t \in [t_{k+1}, t_{k+2}]$ with $k \in \mathbb{Z}$.

(iv) The door of the coming flow barrier in the $(2k_\alpha : 2k_\beta)$-sink flow is *permanently* open on the α-side if the window exists for $t \in [t_k, \infty)$.

(v) The door of the coming flow barrier in the $(2k_\alpha : 2k_\beta)$-sink flow is *permanently* closed on the α-side if the flow barrier exists for $t \in [t_{k+1}, \infty)$.

From the previous definition, the window of the coming flow barrier in the sink flow is sketched in Fig. 4.15. On the window area, the flow of $\mathbf{x}^{(\alpha)}$ can be switched to the boundary flow of $\mathbf{x}^{(0)}$. The door for the flow barrier wall in the $(2k_\alpha : 2k_\beta)$-sink flow on the boundary $\partial\Omega_{ij}$ is sketched in Fig. 4.16. In Fig. 4.16(a), the door of the sink flow barrier is open, and the coming flow of $\mathbf{x}^{(\alpha)}$-flow (or $\mathbf{x}^{(\beta)}$-flow) can be switched into the $\mathbf{x}^{(0)}$-flow. However, in Fig. 4.16(b), the door of the sink flow barrier is closed. No any flows can be switched into the $\mathbf{x}^{(0)}$-flow on the α-side through this door.

For a sink flow on the boundary $\partial\Omega_{ij}$, if the sink flow barrier exists on the α-side of such a boundary with a subset $S \subseteq \partial\Omega_{ij}$, the sink flow cannot be formed on the boundary under the following conditions for $\mathbf{x}_m \in S \subseteq \partial\Omega_{ij}$

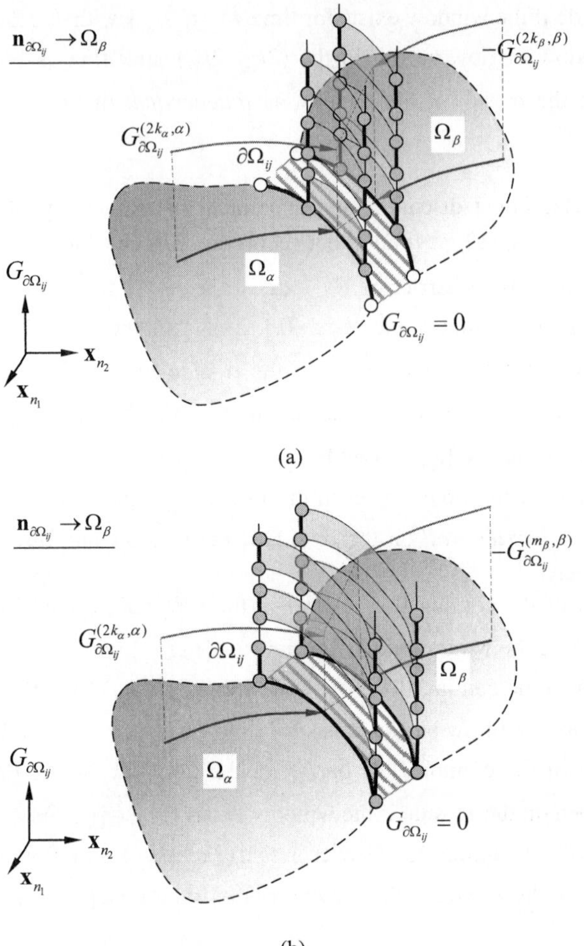

Fig. 4.15 The $(2k_\alpha : 2k_\beta)$-flow barrier windows on $\partial\Omega_{ij}$: **(a)** partial flow barrier and **(b)** full flow barrier. The upper curves with arrows are the *G*-functions relative to the flow barrier. The two vertical surfaces are the flow barrier surfaces. The hatched area is for the zoomed boundary. The lower curves with arrows are coming flows. ($k_\alpha, k_\beta \in \{0,1,2,\cdots\}$).

$$\hbar_\alpha G_{\partial\Omega_{ij}}^{(2k_\alpha,\alpha)}(\mathbf{x}_m) \in \left(\hbar_\alpha G_{\partial\Omega_{ij}}^{(2k_\alpha,\alpha\succ 0)}(\mathbf{x}_m, q_1^{(\alpha)}), \hbar_\alpha G_{\partial\Omega_{ij}}^{(2k_\alpha,\alpha\succ 0)}(\mathbf{x}_m, q_2^{(\alpha)})\right). \tag{4.112}$$

The dynamical system will be constrained on the boundary, given by Eq. (4.23) because the coming flow barrier exists on the α-side of the boundary $\partial\Omega_{ij}$, and the coming flow will be along the boundary until the condition in Eq. (4.112) cannot be satisfied. The following gives the corresponding sufficient and necessary conditions for the flow switchability under the flow barriers.

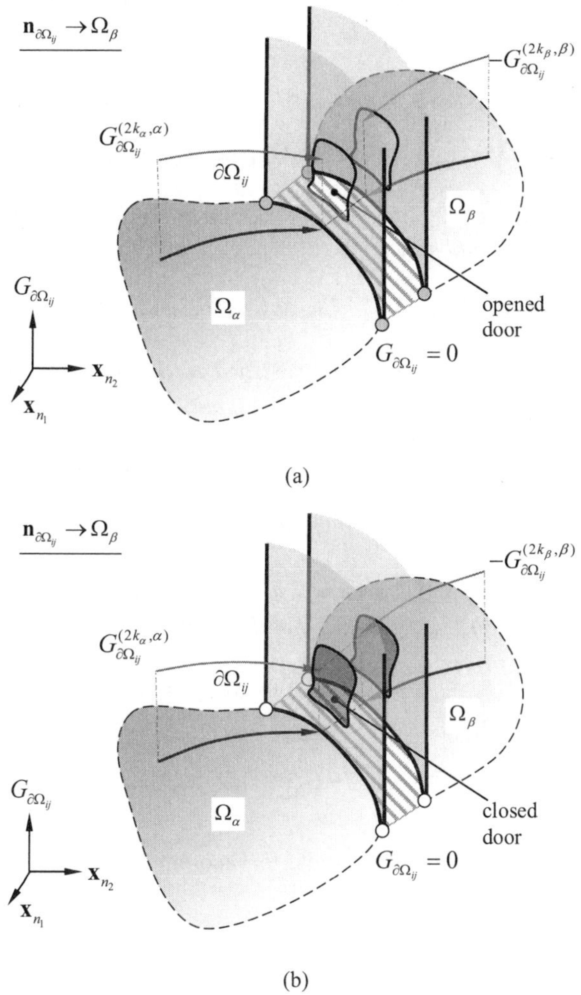

Fig. 4.16 The door of the absolute $(2k_\alpha : 2k_\beta)$-semi-passable flow barrier on $\partial\Omega_{ij}$: **(a)** opened door **(b)** closed door. The upper curves arrows are the *G*-functions relative to the flow barrier. The two vertical surfaces are the flow barrier surfaces. The hatched area is for the zoomed boundary. The lower curves arrows are coming flows. ($k_\alpha, k_\beta \in \{0,1,2,\cdots\}$).

Theorem 4.7. *For a discontinuous dynamical system in Eq.* (4.1)*, there is a point* $\mathbf{x}^{(0)}(t_m) \equiv \mathbf{x}_m \in \partial\Omega_{ij}$ *at time* t_m *between two adjacent domains* Ω_α ($\alpha = i, j$). *For* $\mathbf{x}_m \in S \subseteq \partial\Omega_{ij}$, *there is a flow barrier* $\mathbf{F}^{(\alpha\succ 0)}(\mathbf{x}^{(\alpha)}, t, \boldsymbol{\pi}_\alpha, q^{(\alpha)})$ *at* $q^{(\alpha)} \in [q_1^{(\alpha)}, q_2^{(\alpha)}]$ *on the* α *-side of the boundary* $\partial\Omega_{ij}$ *with*

$$\hbar_\alpha G^{(\alpha\succ 0)}_{\partial\Omega_{ij}}(\mathbf{x}_m,q^{(\alpha)})\in\left[\hbar_\alpha G^{(\alpha\succ 0)}_{\partial\Omega_{ij}}(\mathbf{x}_m,q_1^{(\alpha)}),\hbar_\alpha G^{(\alpha\succ 0)}_{\partial\Omega_{ij}}(\mathbf{x}_m,q_2^{(\alpha)})\right]\subset[0,\infty). \tag{4.113}$$

The two possible coming flows in the sink flow on the boundary satisfy

$$\hbar_\alpha G^{(\alpha)}_{\partial\Omega_{ij}}(\mathbf{x}_m,t_{m-})>0 \text{ and } \hbar_\alpha G^{(\beta)}_{\partial\Omega_{ij}}(\mathbf{x}_m,t_{m-})<0. \tag{4.114}$$

(i) *The coming flows of* $\mathbf{x}^{(\alpha)}$ *cannot be switched to the boundary flow of* $\mathbf{x}^{(0)}$ *to form a sink flow if and only if*

$$\hbar_\alpha G^{(\alpha)}_{\partial\Omega_{ij}}(\mathbf{x}_m,t_{m-})\in\left(\hbar_\alpha G^{(\alpha\succ 0)}_{\partial\Omega_{ij}}(\mathbf{x}_m,q_1),\hbar_\alpha G^{(\alpha\succ 0)}_{\partial\Omega_{ij}}(\mathbf{x}_m,q_2)\right). \tag{4.115}$$

(ii) *The coming flow of* $\mathbf{x}^{(\alpha)}$ *cannot be switched to the boundary flow of* $\mathbf{x}^{(0)}$ *to form a sink flow at* $q^{(\alpha)}=q_\sigma^{(\alpha)}$ ($\sigma\in\{1,2\}$) *if and only if*

$$\begin{aligned}&\hbar_\alpha G^{(s_\alpha,\alpha)}_{\partial\Omega_{ij}}(\mathbf{x}_m,t_{m-})=\hbar_\alpha G^{(s_\alpha,\alpha\succ 0)}_{\partial\Omega_{ij}}(\mathbf{x}_m,q_\sigma^{(\alpha)})\neq 0 \text{ for } s_\alpha=0,1,\cdots,l_\alpha-1;\\&(-1)^\sigma\hbar_\alpha\left[G^{(l_\alpha,\alpha)}_{\partial\Omega_{ij}}(\mathbf{x}_m,t_{m-})-G^{(l_\alpha,\alpha\succ 0)}_{\partial\Omega_{ij}}(\mathbf{x}_m,q_\sigma^{(\alpha)})\right]<0\end{aligned} \tag{4.116}$$

(iii) *The coming flow of* $\mathbf{x}^{(\alpha)}$ *is switched to the boundary flow of* $\mathbf{x}^{(0)}$ *to form a sink flow at* $q^{(\alpha)}=q_\sigma^{(\alpha)}$ ($\sigma\in\{1,2\}$) *if and only if*

$$\begin{aligned}&\hbar_\alpha G^{(s_\alpha,\alpha)}_{\partial\Omega_{ij}}(\mathbf{x}_m,t_{m-})=\hbar_\alpha G^{(s_\alpha,\alpha\succ 0)}_{\partial\Omega_{ij}}(\mathbf{x}_m,q_\sigma^{(\alpha)})\in(0,\infty)\\&\text{for } s_\alpha=0,1,\cdots,l_\alpha-1;\\&(-1)^\sigma\hbar_\alpha\left[G^{(l_\alpha,\alpha)}_{\partial\Omega_{ij}}(\mathbf{x}_m,t_{m-})-G^{(l_\alpha,\alpha\succ 0)}_{\partial\Omega_{ij}}(\mathbf{x}_m,q_\sigma^{(\alpha)})\right]>0.\end{aligned} \tag{4.117}$$

Proof. The proof of this theorem is similar to Theorem 4.1. ■

Theorem 4.8. *For a discontinuous dynamical system in Eq. (4.1), there is a point* $\mathbf{x}^{(0)}(t_m)\equiv\mathbf{x}_m\in\partial\Omega_{ij}$ *at time* t_m *between two adjacent domains* Ω_α ($\alpha=i,j$). *Suppose a* $(2k_\alpha:2k_\beta)$*-sink flow barrier* $\mathbf{F}^{(\alpha\succ 0)}(\mathbf{x}^{(\alpha)},t,\boldsymbol{\pi}_\alpha,q^{(\alpha)})$ *for* $q^{(\alpha)}\in[q_1^{(\alpha)},q_2^{(\alpha)}]$ *exists on the* α*-side of the boundary* $\partial\Omega_{ij}$ *with*

$$\begin{aligned}&G^{(s_\alpha,\alpha\succ 0)}_{\partial\Omega_{ij}}(\mathbf{x}_m,q^{(\alpha)})=0 \text{ for } s_\alpha=0,1,\cdots,2k_\alpha-1;\\&\hbar_\alpha G^{(2k_\alpha,\alpha\succ 0)}_{\partial\Omega_{ij}}(\mathbf{x}_m,q^{(\alpha)})\in\left[\hbar_\alpha G^{(2k_\alpha,\alpha\succ 0)}_{\partial\Omega_{ij}}(\mathbf{x}_m,q_1^{(\alpha)}),\hbar_\alpha G^{(2k_\alpha,\alpha\succ 0)}_{\partial\Omega_{ij}}(\mathbf{x}_m,q_2^{(\alpha)})\right]\\&\subset[0,\infty).\end{aligned} \tag{4.118}$$

The coming and leaving flows in the $(2k_\alpha:2k_\beta)$*-sink flow satisfy*

$$G^{(s_\alpha,\alpha)}_{\partial\Omega_{ij}}(\mathbf{x}_m,t_{m-},\mathbf{p}_\alpha,\boldsymbol{\lambda})=0 \text{ for } s_\alpha=0,1,\cdots,2k_\alpha-1;$$

$$\begin{aligned}&G_{\partial\Omega_{ij}}^{(s_\beta,\beta)}(\mathbf{x}_m,t_{m-},\mathbf{p}_\beta,\boldsymbol{\lambda})=0 \text{ for } s_\beta=0,1,\cdots,2k_\beta-1;\\&\hbar_\alpha G_{\partial\Omega_{ij}}^{(2k_\alpha,\alpha)}(\mathbf{x}_m,t_{m-})>0 \text{ and } \hbar_\alpha G_{\partial\Omega_{ij}}^{(2k_\beta,\beta)}(\mathbf{x}_m,t_{m-})<0.\end{aligned}\tag{4.119}$$

(i) *A coming flow of* $\mathbf{x}^{(\alpha)}$ *cannot be switched to the boundary flow of* $\mathbf{x}^{(0)}$ *to form a* $(2k_\alpha:2k_\beta)$ *-sink flow if and only if*

$$\hbar_\alpha G_{\partial\Omega_{ij}}^{(2k_\alpha,\alpha)}(\mathbf{x}_m,t_{m-})\in\left(\hbar_\alpha G_{\partial\Omega_{ij}}^{(2k_\alpha,\alpha\succ 0)}(\mathbf{x}_m,q_1^{(\alpha)}),\hbar_\alpha G_{\partial\Omega_{ij}}^{(2k_\alpha,\alpha\succ 0)}(\mathbf{x}_m,q_2^{(\alpha)})\right).\tag{4.120}$$

(ii) *The coming flow of* $\mathbf{x}^{(\alpha)}$ *cannot be switched to the boundary flow of* $\mathbf{x}^{(0)}$ *to form a* $(2k_\alpha:2k_\beta)$ *-sink flow at* $q^{(\alpha)}=q_\sigma^{(\alpha)}$ ($\sigma\in\{1,2\}$) *if and only if*

$$\begin{aligned}&\hbar_\alpha G_{\partial\Omega_{ij}}^{(s_\alpha,\alpha)}(\mathbf{x}_m,t_{m-})=\hbar_\alpha G_{\partial\Omega_{ij}}^{(s_\alpha,\alpha\succ 0)}(\mathbf{x}_m,q_\sigma^{(\alpha)})\in(0,\infty)\\&\quad \text{for } s_\alpha=2k_\alpha,2k_\alpha+1,\cdots,l_\alpha-1;\\&(-1)^\sigma\hbar_\alpha\left[G_{\partial\Omega_{ij}}^{(l_\alpha,\alpha)}(\mathbf{x}_m,t_{m-})-G_{\partial\Omega_{ij}}^{(l_\alpha,\alpha\succ 0)}(\mathbf{x}_m,q_\sigma^{(\alpha)})\right]<0.\end{aligned}\tag{4.121}$$

(iii) *The coming flow of* $\mathbf{x}^{(\alpha)}$ *is switched to the boundary flow of* $\mathbf{x}^{(0)}$ *to form a* $(2k_\alpha:2k_\beta)$ *-sink flow at* $q^{(\alpha)}=q_\sigma^{(\alpha)}$ ($\sigma\in\{1,2\}$) *if and only if*

$$\begin{aligned}&\hbar_\alpha G_{\partial\Omega_{ij}}^{(s_\alpha,\alpha)}(\mathbf{x}_m,t_{m-})=\hbar_\alpha G_{\partial\Omega_{ij}}^{(s_\alpha,\alpha\succ 0)}(\mathbf{x}_m,q_\sigma^{(\alpha)})\in(0,\infty)\\&\quad \text{for } s_\alpha=2k_\alpha,2k_\alpha+1,\cdots,l_\alpha-1;\\&(-1)^\sigma\hbar_\alpha\left[G_{\partial\Omega_{ij}}^{(l_\alpha,\alpha)}(\mathbf{x}_m,t_{m-})-G_{\partial\Omega_{ij}}^{(l_\alpha,\alpha\succ 0)}(\mathbf{x}_m,q_\sigma^{(\alpha)})\right]>0.\end{aligned}\tag{4.122}$$

Proof. The proof of this theorem is similar to the proof of Theorem 4.2. ■

4.3. Flow barriers for source flows

Without any flow barriers, from Theorem 3.5, the necessary and sufficient conditions for a source flow leaving the boundary are

$$\hbar_\alpha G_{\partial\Omega_{ij}}^{(\alpha)}(\mathbf{x}_m,t_{m+},\mathbf{p}_\alpha,\boldsymbol{\lambda})<0 \text{ and } \hbar_\alpha G_{\partial\Omega_{ij}}^{(\beta)}(\mathbf{x}_m,t_{m+},\mathbf{p}_\beta,\boldsymbol{\lambda})>0.\tag{4.123}$$

For the source flow, three is a boundary flow of $\mathbf{x}^{(0)}$ on the boundary, governed by

$$\dot{\mathbf{x}}^{(0)}=\mathbf{F}^{(0)}(\mathbf{x}^{(0)},t,\boldsymbol{\lambda}) \quad \text{with } \varphi_{ij}(\mathbf{x}^{(0)},t,\boldsymbol{\lambda})=0 \text{ on } \partial\Omega_{ij}.\tag{4.124}$$

The G-function for the boundary flow of $\mathbf{x}^{(0)}$ on $\partial\Omega_{ij}$ is already zero, i.e.,

$$G_{\partial\Omega_{ij}}(\mathbf{x}_m,t_m)\equiv 0 \text{ on } \partial\Omega_{ij}.\tag{4.125}$$

4.3.1. Boundary flow barriers

To avoid the boundary flow leaving the boundary to form a source flow, the flow barrier for the boundary flow should be exerted.

Definition 4.23. For a discontinuous dynamical system in Eq. (4.1), there is a point $\mathbf{x}^{(0)}(t_m) \equiv \mathbf{x}_m \in \partial\Omega_{ij}$ at time t_m between two adjacent domains Ω_α ($\alpha = i, j$). Suppose there is a vector field $\mathbf{F}^{(0 \succ 0_\alpha)}(\mathbf{x}^{(\alpha)}, t, \boldsymbol{\pi}_\alpha, q^{(\alpha)})$ for $q \in [q_1, q_2]$ on the boundary $\partial\Omega_{ij}$ with

$$0 \in \left[\hbar_\alpha G_{\partial\Omega_{ij}}^{(0 \succ 0_\alpha)}(\mathbf{x}_m, q_2^{(\alpha)}), \hbar_\alpha G_{\partial\Omega_{ij}}^{(0 \succ 0_\alpha)}(\mathbf{x}_m, q_1^{(\alpha)}) \right] \subset \mathscr{R}. \tag{4.126}$$

The two possible leaving flows in the source flow satisfy

$$\hbar_\alpha G_{\partial\Omega_{ij}}^{(\alpha)}(\mathbf{x}_m, t_{m+}) < 0 \text{ and } \hbar_\alpha G_{\partial\Omega_{ij}}^{(\beta)}(\mathbf{x}_m, t_{m+}) > 0 \tag{4.127}$$

($\alpha, \beta \in \{i, j\}$ and $\alpha \neq \beta$). The vector field of $\mathbf{F}^{(0 \succ 0_\alpha)}(\mathbf{x}^{(\alpha)}, t, \boldsymbol{\pi}_\alpha, q^{(\alpha)})$ is called the *flow barrier* of the boundary flow in the source flow on the α-side if the following conditions are satisfied. The two critical values of $\mathbf{F}^{(0 \succ 0_\alpha)}(\mathbf{x}^{(\alpha)}, t, \boldsymbol{\pi}_\alpha, q_\sigma^{(\alpha)})$ for $\sigma = 1, 2$ are called the *lower and upper limits* of the boundary flow barriers on the α-side.

(i) The boundary flow of $\mathbf{x}^{(0)}$ cannot be switched to the leaving flow of $\mathbf{x}^{(\alpha)}$ on the α-side if

$$\begin{aligned} &\mathbf{x}^{(0)}(t_m) = \mathbf{x}^{(0 \succ 0_\alpha)}(t_{m\pm}, q_\sigma^{(\alpha)}) = \mathbf{x}_m \quad \text{for } \sigma=1,2; \\ &\hbar_\alpha G_{\partial\Omega_{ij}}^{(0 \succ 0_\alpha)}(\mathbf{x}_m, q_1^{(\alpha)}) > 0 \text{ and } \hbar_\alpha G_{\partial\Omega_{ij}}^{(0 \succ 0_\alpha)}(\mathbf{x}_m, q_2^{(\alpha)}) < 0. \end{aligned} \tag{4.128}$$

(ii) The boundary flow of $\mathbf{x}^{(0)}$ cannot be switched to the leaving flow of $\mathbf{x}^{(\alpha)}$ on the α-side if

$$\begin{aligned} &\mathbf{x}^{(0)}(t_m) = \mathbf{x}^{(0 \succ 0_\alpha)}(t_{m\pm}, q_\sigma^{(\alpha)}) = \mathbf{x}_m, \\ &\hbar_\alpha G_{\partial\Omega_{ij}}^{(0 \succ 0_\alpha)}(\mathbf{x}_m, q_2^{(\alpha)}) < 0 \text{ and} \\ &\hbar_\alpha G_{\partial\Omega_{ij}}^{(s_\alpha, 0 \succ 0_\alpha)}(\mathbf{x}_m, q_1^{(\alpha)}) = 0 \ \text{ for } s_\alpha = 0, 1, 2, \cdots, l_\alpha - 1; \\ &\hbar_\alpha \mathbf{n}_{\partial\Omega_{ij}}^{\mathrm{T}}(\mathbf{x}^{(0)}(t_{m+\varepsilon})) \cdot \left[\mathbf{x}^{(0 \succ 0_\alpha)}(t_{m+\varepsilon}, q_1^{(\alpha)}) - \mathbf{x}^{(0)}(t_{m+\varepsilon}) \right] > 0. \end{aligned} \tag{4.129}$$

(iii) The boundary flow of $\mathbf{x}^{(0)}$ can be switched to the leaving flow of $\mathbf{x}^{(\alpha)}$ on the α-side if

$$\begin{aligned} &\mathbf{x}^{(0)}(t_m) = \mathbf{x}^{(0 \succ 0_\alpha)}(t_{m\pm}, q_\sigma^{(\alpha)}) = \mathbf{x}_m \ \text{ for } \sigma=1,2; \\ &\hbar_\alpha G_{\partial\Omega_{ij}}^{(0 \succ 0_\alpha)}(\mathbf{x}_m, q_2^{(\alpha)}) < 0 \text{ and} \end{aligned}$$

$$\begin{aligned}&\hbar_\alpha G_{\partial\Omega_{ij}}^{(s_\alpha,0\succ\mathbf{0}_\alpha)}(\mathbf{x}_m,q_1^{(\alpha)})=0\ \text{ for } s_\alpha=0,1,2,\cdots,l_\alpha-1;\\&\hbar_\alpha\mathbf{n}_{\partial\Omega_{ij}}^{\mathrm{T}}(\mathbf{x}^{(0)}(t_{m+\varepsilon}))\cdot\left[\mathbf{x}^{(0\succ0_\alpha)}(t_{m+\varepsilon},q_1^{(\alpha)})-\mathbf{x}^{(0)}(t_{m+\varepsilon})\right]<0.\end{aligned}\tag{4.130}$$

Definition 4.24. For a discontinuous dynamical system in Eq. (4.1), there is a point $\mathbf{x}^{(0)}(t_m)\equiv\mathbf{x}_m\in\partial\Omega_{ij}$ at time t_m between two adjacent domains Ω_α ($\alpha=i,j$). Suppose there is a vector field $\mathbf{F}^{(0\succ0_\alpha)}(\mathbf{x}^{(\alpha)},t,\boldsymbol{\pi}_\alpha,q^{(\alpha)})$ for $q^{(\alpha)}\in[q_1^{(\alpha)},q_2^{(\alpha)}]$ on the boundary $\partial\Omega_{ij}$ with its *G*-function

$$\begin{aligned}&G_{\partial\Omega_{ij}}^{(s_\alpha,0\succ0_\alpha)}(\mathbf{x}_m,q^{(\alpha)})=0\ \text{for } s_\alpha=0,1,\cdots,m_\alpha-1;\\&0\in\left[\hbar_\alpha G_{\partial\Omega_{ij}}^{(m_\alpha,0\succ0_\alpha)}(\mathbf{x}_m,q_2^{(\alpha)}),\hbar_\alpha G_{\partial\Omega_{ij}}^{(m_\alpha,0\succ0_\alpha)}(\mathbf{x}_m,q_1^{(\alpha)})\right]\subset\mathscr{R}.\end{aligned}\tag{4.131}$$

The two possible leaving flows in the (m_α,m_β)-source flow satisfy

$$\begin{aligned}&G_{\partial\Omega_{ij}}^{(s_\alpha,\alpha)}(\mathbf{x}_m,t_{m+})=0\ \text{for } s_\alpha=0,1,\cdots,m_\alpha-1;\\&G_{\partial\Omega_{ij}}^{(s_\beta,\beta)}(\mathbf{x}_m,t_{m+})=0\ \text{for } s_\beta=0,1,\cdots,m_\beta-1;\\&\hbar_\alpha G_{\partial\Omega_{ij}}^{(m_\alpha,\alpha)}(\mathbf{x}_m,t_{m+})<0 \text{ and } \hbar_\alpha G_{\partial\Omega_{ij}}^{(m_\beta,\beta)}(\mathbf{x}_m,t_{m+})>0.\end{aligned}\tag{4.132}$$

($\alpha,\beta\in\{i,j\}$ and $\alpha\neq\beta$). The vector field of $\mathbf{F}^{(0\succ0_\alpha)}(\mathbf{x}^{(\alpha)},t,\boldsymbol{\pi}_\alpha,q^{(\alpha)})$ is called the *flow barrier* of the boundary flow in the source flow on the α-side if the following conditions are satisfied. The two critical values of $\mathbf{F}^{(0\succ0_\alpha)}(\mathbf{x}^{(\alpha)},t,\boldsymbol{\pi}_\alpha,q_\sigma^{(\alpha)})$ for $\sigma=1,2$ are called the *lower and upper limits* of the boundary flow barriers on the α-side.

(i) The boundary flow of $\mathbf{x}^{(0)}$ cannot be switched to the (m_α)th-order leaving flow $\mathbf{x}^{(\alpha)}$ if

$$\begin{aligned}&\mathbf{x}^{(0)}(t_m)=\mathbf{x}^{(0\succ0_\alpha)}(t_{m\pm},q_\sigma^{(\alpha)})=\mathbf{x}_m\ \text{ for } \sigma=1,2;\\&\hbar_\alpha G_{\partial\Omega_{ij}}^{(m_\alpha,0\succ0_\alpha)}(\mathbf{x}_m,q_1^{(\alpha)})>0 \text{ and } \hbar_\alpha G_{\partial\Omega_{ij}}^{(m_\alpha,0\succ0_\alpha)}(\mathbf{x}_m,q_2^{(\alpha)})<0.\end{aligned}\tag{4.133}$$

(ii) The boundary flow of $\mathbf{x}^{(0)}$ cannot be switched to the (m_α)th-order leaving flow $\mathbf{x}^{(\alpha)}$ if

$$\begin{aligned}&\mathbf{x}^{(0)}(t_m)=\mathbf{x}^{(0\succ0_\alpha)}(t_{m\pm},q_\sigma^{(\alpha)})=\mathbf{x}_m\ \text{ for } \sigma=1,2;\\&\hbar_\alpha G_{\partial\Omega_{ij}}^{(m_\alpha,0\succ0_\alpha)}(\mathbf{x}_m,q_2^{(\alpha)})<0 \text{ and}\\&\hbar_\alpha G_{\partial\Omega_{ij}}^{(s_\alpha,0\succ0_\alpha)}(\mathbf{x}_m,q_1^{(\alpha)})=0\ \text{ for } s_\alpha=m_\alpha,m_\alpha+1,\cdots,l_\alpha-1;\\&\hbar_\alpha\mathbf{n}_{\partial\Omega_{ij}}^{\mathrm{T}}(\mathbf{x}^{(0)}(t_{m+\varepsilon}))\cdot\left[\mathbf{x}^{(0\succ0_\alpha)}(t_{m+\varepsilon},q_1^{(\alpha)})-\mathbf{x}^{(0)}(t_{m+\varepsilon})\right]>0.\end{aligned}\tag{4.134}$$

(iii) The boundary flow of $\mathbf{x}^{(0)}$ can be switched to the (m_α)th-order leaving flow $\mathbf{x}^{(\alpha)}$ if

$$\begin{aligned}
&\mathbf{x}^{(0)}(t_m) = \mathbf{x}^{(0 \succ 0_\alpha)}(t_{m\pm}, q_\sigma^{(\alpha)}) = \mathbf{x}_m \text{ for } \sigma = 1,2;\\
&\hbar_\alpha G_{\partial\Omega_{ij}}^{(m_\alpha, 0 \succ 0_\alpha)}(\mathbf{x}_m, q_2^{(\alpha)}) < 0 \text{ and}\\
&\hbar_\alpha G_{\partial\Omega_{ij}}^{(s_\alpha, 0 \succ 0_\alpha)}(\mathbf{x}_m, q_1^{(\alpha)}) = 0 \text{ for } s_\alpha = m_\alpha, m_\alpha + 1, \cdots, l_\alpha - 1;\\
&\hbar_\alpha \mathbf{n}_{\partial\Omega_{ij}}^{\mathrm{T}}(\mathbf{x}^{(0)}(t_{m+\varepsilon})) \cdot \left[\mathbf{x}^{(0 \succ 0_\alpha)}(t_{m+\varepsilon}, q_1^{(\alpha)}) - \mathbf{x}^{(0)}(t_{m+\varepsilon})\right] < 0.
\end{aligned} \tag{4.135}$$

To explain the boundary flow barrier, the G-functions of the flow barriers on both sides of the boundary $\partial\Omega_{ij}$ are presented in Fig. 4.17. The red dashed curves represent the G-function of the flows pertaining to the boundary flow barriers in each domain Ω_α. The thick lines denote the G-functions of the flow barriers on both sides of the boundary. To show the boundary flow barriers, consider the G-function on the β-side of the boundary $\partial\Omega_{ij}$ as a reference. The shade area is also zoomed for the boundary flow of $\mathbf{x}^{(0)}$. Because the G-function for the boundary flow is zero (i.e., $G_{\partial\Omega_{ij}}^{(0 \succ 0)} = 0$), the *lower* and *upper* limits of the boundary flow barriers should be negative and positive, respectively. Such flow barriers are independent of the flow of $\mathbf{x}^{(\alpha)}$ ($\alpha \in \{i, j\}$). However, once the flow barriers disappear, the boundary flow of $\mathbf{x}^{(0)}$ will be switched to the leaving flow of $\mathbf{x}^{(\alpha)}$ controlled by $G_{\partial\Omega_{ij}}^{(m_\alpha, \alpha)}$. For $\mathbf{n}_{\partial\Omega_{ij}} \to \Omega_\beta$, one obtains $\hbar_\alpha = 1$. The G-function $G_{\partial\Omega_{ij}}^{(m_\alpha, \alpha)}$ for the leaving flow of $\mathbf{x}^{(\alpha)}$ should be negative (i.e., $G_{\partial\Omega_{ij}}^{(m_\alpha, \alpha)} < 0$) from Eq. (4.132). If the G-function is positive (i.e., $G_{\partial\Omega_{ij}}^{(m_\alpha, \alpha)} > 0$), the leaving flow will become a coming flow in domain Ω_α. The leaving flow of $\mathbf{x}^{(\alpha)}$ on the α-side cannot be formed. In a similar fashion, the G-function $G_{\partial\Omega_{ij}}^{(m_\beta, \beta)}$ for the leaving flow of $\mathbf{x}^{(\beta)}$ should be positive (i.e., $G_{\partial\Omega_{ij}}^{(m_\beta, \beta)} > 0$) from Eq. (4.132) to form the source flow on the β-side. Otherwise, the leaving flow in domain Ω_β cannot be achieved. Such characteristics of the boundary flow barrier are sketched in Fig. 4.17(a). For $\mathbf{n}_{\partial\Omega_{ij}} \to \Omega_\alpha$, one obtains $\hbar_\alpha = -1$. The G-function for the leaving flow of $\mathbf{x}^{(\alpha)}$ should be positive ($G_{\partial\Omega_{ij}}^{(m_\alpha, \alpha)} > 0$), and the G-function for the leaving flow of $\mathbf{x}^{(\beta)}$ becomes negative ($G_{\partial\Omega_{ij}}^{(m_\beta, \beta)} < 0$) in order to form a source flow to the boundary. For this case, the G-functions of the flow barriers are sketched in 4.17(b).

Definition 4.25. For a discontinuous dynamical system in Eq. (4.1), there is a point $\mathbf{x}^{(0)}(t_m) \equiv \mathbf{x}_m \in \partial\Omega_{ij}$ at time t_m between two adjacent domains Ω_α ($\alpha = i, j$). Suppose there is a boundary flow barrier $\mathbf{F}^{(0 \succ 0_\alpha)}(\mathbf{x}^{(\alpha)}, t, \boldsymbol{\pi}_\alpha, q^{(\alpha)})$ for $q^{(\alpha)} \in [q_1^{(\alpha)}, q_2^{(\alpha)}]$

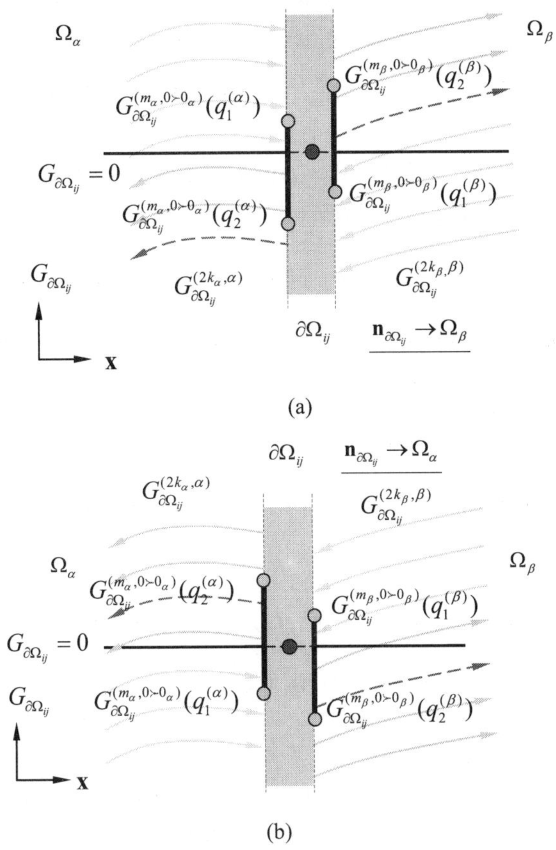

Fig. 4.17 The G-functions for the boundary flow barriers in the $(m_\alpha : m_\beta)$-source flow: **(a)** $\mathbf{n}_{\partial\Omega_{ij}} \to \Omega_\beta$ and **(b)** $\mathbf{n}_{\partial\Omega_{ij}} \to \Omega_\alpha$. The red dashed curves are the G-functions relative to the flow barrier. The thick lines are the G-functions of the flow barriers at both α and β-side boundaries of the boundary $\partial\Omega_{ij}$. $G^{(m_\alpha,0\succ 0_\alpha)}_{\partial\Omega_{ij}}(q_1^{(\alpha)})$ and $G^{(m_\alpha,0\succ 0_\alpha)}_{\partial\Omega_{ij}}(q_2^{(\alpha)})$ are for *lower* and *upper* barrier limits ($m_\alpha \in \{0,1,2,\cdots\}$, $\alpha = i,j$) and similarly for the β-side.

in the $(m_\alpha : m_\beta)$-source flow on the α-side of the boundary $\partial\Omega_{ij}$ ($\alpha \neq \beta$, m_α, $m_\beta \in \{0,1,2,\cdots\}$).

(i) The boundary flow barrier in the $(m_\alpha : m_\beta)$-source flow is *partial* on the α-side if $\mathbf{x}_m \in S \subset \partial\Omega_{ij}$.

(ii) The boundary flow barrier in the $(m_\alpha : m_\beta)$-source flow is *full* on the α-side if $\mathbf{x}_m \in S = \partial\Omega_{ij}$.

Definition 4.26. For a discontinuous dynamical system in Eq. (4.1), there is a point

$\mathbf{x}^{(0)}(t_m) \equiv \mathbf{x}_m \in \partial\Omega_{ij}$ at time t_m between two adjacent domains Ω_α ($\alpha = i, j$). Suppose there is a boundary flow barrier of $\mathbf{F}^{(0 \succ 0_\alpha)}(\mathbf{x}^{(\alpha)}, t, \boldsymbol{\pi}_\alpha, q^{(\alpha)})$ in the $(m_\alpha : m_\beta)$-source flow for $q^{(\alpha)} \in [q_1^{(\alpha)}, q_2^{(\alpha)}]$ on the α-side of the boundary $\partial\Omega_{ij}$ ($\alpha \neq \beta$, $m_\alpha, m_\beta \in \{0,1,2,\cdots\}$).

(i) The boundary flow barrier in the $(m_\alpha : m_\beta)$-source flow is with a lower limit on the α-side if for $\mathbf{x}_m \in S \subseteq \partial\Omega_{ij}$

$$\hbar_\alpha G_{\partial\Omega_{ij}}^{(m_\alpha, 0 \succ 0_\alpha)}(\mathbf{x}_m, q_1^{(\alpha)}) \to +\infty \text{ and } \hbar_\alpha G_{\partial\Omega_{ij}}^{(m_\alpha, 0 \succ 0_\alpha)}(\mathbf{x}_m, q_2^{(\alpha)}) < 0. \tag{4.136}$$

(ii) The boundary flow barrier in the $(m_\alpha : m_\beta)$-source flow is with an upper limit on the α-side if for $\mathbf{x}_m \in S \subseteq \partial\Omega_{ij}$

$$\hbar_\alpha G_{\partial\Omega_{ij}}^{(m_\alpha, 0 \succ 0_\alpha)}(\mathbf{x}_m, q_1^{(\alpha)}) > 0 \text{ and } \hbar_\alpha G_{\partial\Omega_{ij}}^{(m_\alpha, 0 \succ 0_\alpha)}(\mathbf{x}_m, q_2^{(\alpha)}) \to -\infty. \tag{4.137}$$

(iii) The boundary flow barrier in the $(m_\alpha : m_\beta)$-source flow is *absolute* on the α-side if for $\mathbf{x}_m \in S \subseteq \partial\Omega_{ij}$

$$G_{\partial\Omega_{ij}}^{(m_\alpha, 0 \succ 0_\alpha)}(\mathbf{x}_m, q_1^{(\alpha)}) \to +\infty \text{ and } \hbar_\alpha G_{\partial\Omega_{ij}}^{(m_\alpha, 0 \succ 0_\alpha)}(\mathbf{x}_m, q_2^{(\alpha)}) \to -\infty. \tag{4.138}$$

(iv) The boundary flow barrier is called the *complete boundary flow barrier wall* in the $(m_\alpha : m_\beta)$-source flow on the α-side if the absolute boundary flow barrier exists for $\mathbf{x}_m \in S = \partial\Omega_{ij}$.

(v) The boundary flow barrier is called the *boundary flow barrier fence* in the $(m_\alpha : m_\beta)$-source flow on the α-side if there are many partial flow barriers and many non-flow barriers on the α-side of the boundary $\partial\Omega_{ij}$.

In a similar fashion, the partial and full boundary flow barriers on both sides of $\partial\Omega_{ij}$ in the $(m_\alpha : m_\beta)$-source flow are sketched in Fig. 4.18 for $\mathbf{x}_m \in S \subseteq \partial\Omega_{ij}$ through the two different color surfaces on the α and β-sides of the boundary $\partial\Omega_{ij}$. To clearly present the boundary flow barriers, the *G*-function on the β-side of the boundary $\partial\Omega_{ij}$ is considered as a reference, and the *G*-function on the α-side of the boundary $\partial\Omega_{ij}$ is presented by $-G_{\partial\Omega_{ij}}^{(\alpha)}$. The infinity boundary flow barriers in the $(m_\alpha : m_\beta)$-source flow are sketched in Fig. 4.19.

Definition 4.27. For a discontinuous dynamical system in Eq. (4.1), there is a point $\mathbf{x}^{(0)}(t_m) \equiv \mathbf{x}_m \in \partial\Omega_{ij}$ at time t_m between two adjacent domains Ω_α ($\alpha = i, j$). The two possible leaving flows in the $(m_\alpha : m_\beta)$-source flow satisfy Eq. (4.132). There

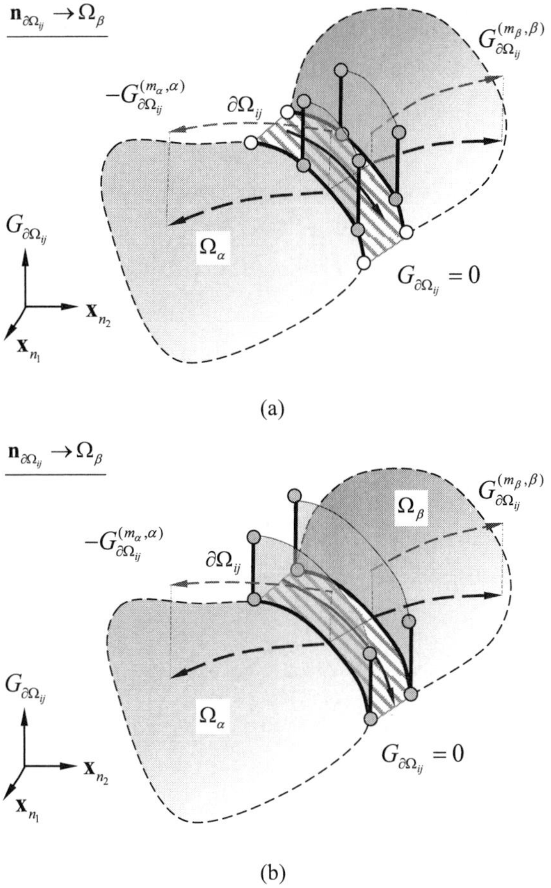

Fig. 4.18 The boundary flow barriers in the $(m_\alpha : m_\beta)$ -source flow on both sides of the boundary $\partial\Omega_{ij}$: **(a)** partial flow barrier and **(b)** full flow barrier. The red curves are the G-functions relative to the flow barrier. The green and blue surfaces are the flow barrier surfaces. The hatched area is for the zoomed boundary. The dark blue curves are coming flows. ($m_\alpha, m_\beta \in \{0,1,2,\cdots\}$)

is a boundary flow barrier $\mathbf{F}^{(0\succ 0_\alpha)}(\mathbf{x}^{(\alpha)}, t, \boldsymbol{\pi}_\alpha, q^{(\alpha)})$ in the $(m_\alpha : m_\beta)$ -source flow on the α -side of the boundary ($m_\alpha, m_\beta \in \{0,1,2,\cdots\}$) with

$$\begin{aligned} &\hbar_\alpha G_{\partial\Omega_{ij}}^{(s_\alpha, 0\succ 0_\alpha)}(\mathbf{x}_m, q^{(\alpha)}) = 0, \text{ for } s_\alpha = 0,1,2,\cdots, m_\alpha - 1; \\ &0 \in \left[\hbar_\alpha G_{\partial\Omega_{ij}}^{(m_\alpha, 0\succ 0_\alpha)}(\mathbf{x}_m, q_{2n-1}^{(\alpha)}), \hbar_\alpha G_{\partial\Omega_{ij}}^{(m_\alpha, 0\succ 0_\alpha)}(\mathbf{x}_m, q_{2n}^{(\alpha)}) \right] \subset \mathscr{R} \end{aligned} \tag{4.139}$$

for $\mathbf{x}_m \in S \subseteq \partial\Omega_{ij}$ and $q^{(\alpha)} \in [q_{2n-1}^{(\alpha)}, q_{2n}^{(\alpha)}]$ ($n = 1,2,\cdots$). For $q^{(\alpha)} \in (q_{2n}^{(\alpha)}, q_{2n+1}^{(\alpha)})$, no flow barriers are defined at $\mathbf{x}_m \in S \subseteq \partial\Omega_{ij}$. Thus, the boundary flow of $\mathbf{x}^{(0)}$ can be

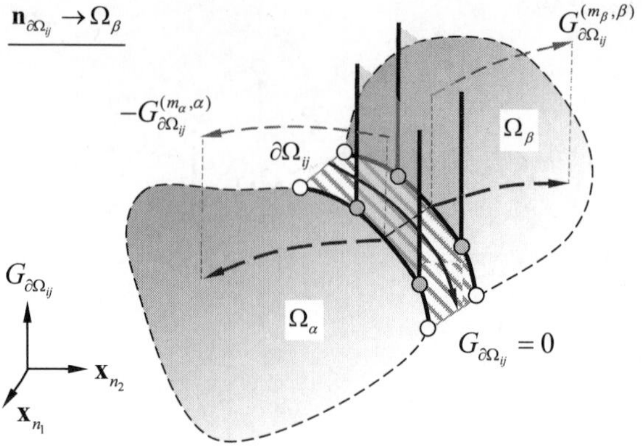

(a)

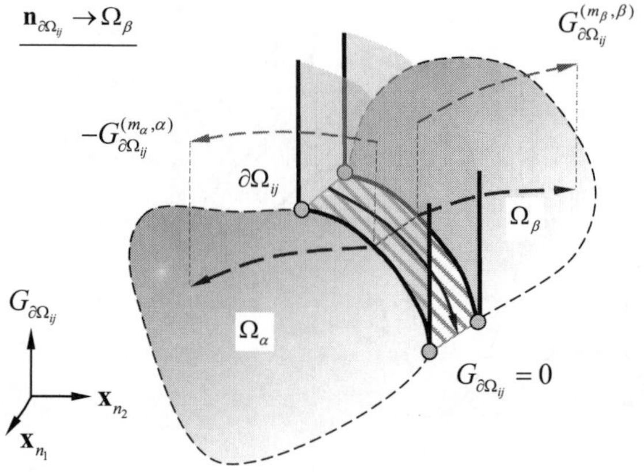

(b)

Fig. 4.19 The infinity boundary flow barrier with lower boundary on $\partial\Omega_{ij}$: **(a)** partial flow barrier and **(b)** full flow barrier. The upper curves arrows are the G-functions relative to the flow barrier. The two vertical surfaces are the flow barrier surfaces. The hatched area is for the zoomed boundary. The lower curves arrows are coming flows.

switched to the leaving flow of $\mathbf{x}^{(\alpha)}$ for

$$0 \in \left(\hbar G^{(m_\alpha,0\succ 0_\alpha)}_{\partial\Omega_{ij}}(\mathbf{x}_m, q^{(\alpha)}_{2n}), \hbar G^{(m_\alpha,0\succ 0_\alpha)}_{\partial\Omega_{ij}}(\mathbf{x}_m, q^{(\alpha)}_{2n+1})\right) \subset \mathscr{R}. \tag{4.140}$$

The G-function intervals for all $\mathbf{x}_m \in S \subseteq \partial\Omega_{ij}$ with $q^{(\alpha)} \in (q^{(\alpha)}_{2n}, q^{(\alpha)}_{2n+1})$ are called *the window* of the boundary flow barrier in the $(m_\alpha : m_\beta)$ -source flow on the α -side of the boundary.

Definition 4.28. For a discontinuous dynamical system in Eq. (4.1), there is a point

$\mathbf{x}^{(0)}(t_m) \equiv \mathbf{x}_m \in \partial\Omega_{ij}$ at time t_m between two adjacent domains Ω_α $(\alpha = i, j)$. The two leaving flows in the (m_α, m_β)-source flow satisfy Eq. (4.132). Suppose there is a boundary flow barrier of $\mathbf{F}^{(0 \succ 0_\alpha)}(\mathbf{x}^{(\alpha)}, t, \boldsymbol{\pi}_\alpha, q^{(\alpha)})$ for $q^{(\alpha)} \in [q_1^{(\alpha)}, q_2^{(\alpha)}]$ in the $(m_\alpha : m_\beta)$-source flow on the α-side of the boundary ($m_\alpha, m_\beta \in \{0,1,2,\cdots\}$).

(i) The boundary flow barrier in the (m_α, m_β)-source flow is *permanent* on the α-side if the flow barrier is *independent* of time $t \in [0, \infty)$.
(ii) The boundary flow barrier in the (m_α, m_β)-source flow is *instantaneous* on the α-side if the flow barrier is continuously *dependent* on time $t \in [0, \infty)$.
(iii) The boundary flow barrier in the (m_α, m_β)-source flow is *intermittent* on the α-side if the flow barrier exists for time $t \in [t_k, t_{k+1}]$ with $k \in \mathbb{Z}$.
(iv) The boundary flow barrier in the (m_α, m_β)-source flow is *intermittent and static* on the α-side if the flow barrier is *independent* of time $t \in [t_k, t_{k+1}]$ with $k \in \mathbb{Z}$.

Similarly, the permanent and instantaneous windows of the boundary flow barrier in the $(m_\alpha : m_\beta)$-source flow on the boundary can also be discussed. The door for the boundary flow barrier wall on the boundary can be presented.

Definition 4.29. For a discontinuous dynamical system in Eq. (4.1), there is a point $\mathbf{x}^{(0)}(t_m) \equiv \mathbf{x}_m \in \partial\Omega_{ij}$ at time t_m between two adjacent domains Ω_α $(\alpha = i, j)$. The two leaving flows in the (m_α, m_β)-source flow satisfy Eq. (4.132). There is a boundary flow barrier of $\mathbf{F}^{(0 \succ 0_\alpha)}(\mathbf{x}^{(\alpha)}, t, \boldsymbol{\pi}_\alpha, q^{(\alpha)})$ for $q^{(\alpha)} \in [q_{2n-1}^{(\alpha)}, q_{2n}^{(\alpha)}]$ in the $(m_\alpha : m_\beta)$-source flow on the α-side of the boundary ($m_\alpha, m_\beta \in \{0,1,2,\cdots\}$, $n = 1, 2, \cdots$). Suppose there is a window of the flow barrier for $S \subset \partial\Omega_{ij}$ and $q^{(\alpha)} \in [q_{2n}^{(\alpha)}, q_{2n+1}^{(\alpha)}]$.

(i) The window of boundary flow barrier in the $(m_\alpha : m_\beta)$-source flow is *permanent* on the α-side if the window is *independent* of time for $t \in [0, +\infty)$.
(ii) The window of boundary flow barrier in the $(m_\alpha : m_\beta)$-source flow is *instantaneous* on the α-side if the window is continuously *dependent* on time $t \in [0, +\infty)$.
(iii) The window of boundary flow barrier in the $(m_\alpha : m_\beta)$-source flow is *intermittent* on the α-side if the window exists for time $t \in [t_k, t_{k+1}]$ with $k \in \mathbb{Z}$.
(iv) The window of boundary flow barrier in the $(m_\alpha : m_\beta)$-source flow is *intermittent and static* on the α-side if the window is *independent* of time $t \in [t_k, t_{k+1}]$ with $k \in \mathbb{Z}$.

Definition 4.30. For a discontinuous dynamical system in Eq. (4.1), there is a point $\mathbf{x}^{(0)}(t_m) \equiv \mathbf{x}_m \in \partial\Omega_{ij}$ at time t_m between two adjacent domains Ω_α $(\alpha = i, j)$. There is a boundary flow barrier of $\mathbf{F}^{(0 \succ 0_\alpha)}(\mathbf{x}^{(\alpha)}, t, \boldsymbol{\pi}_\alpha, q^{(\alpha)})$ for $q^{(\alpha)} \in [q_1^{(\alpha)}, q_2^{(\alpha)}]$ on the α-side of the boundary ($m_\alpha, m_\beta \in \{0,1,2,\cdots\}$) in the $(m_\alpha : m_\beta)$-source flow. Suppose a boundary flow barrier wall exists on the α-side of the entire boundary $\partial\Omega_{ij}$ and there is an *intermittent* and static window of the boundary flow barrier on $S \subset \partial\Omega_{ij}$ for $q^{(\alpha)} \in [q_1^{(\alpha)}, q_2^{(\alpha)}]$ and $t \in [t_k, t_{k+1}]$ with $k \in \mathbb{Z}$.

(i) The window of the boundary flow barrier is termed the *door* of *the boundary flow barrier wall* on the α-side of the boundary $\partial\Omega_{ij}$ if the window and the flow barrier exists alternatively.

(ii) The door of the boundary flow barrier in the $(m_\alpha : m_\beta)$-source flow is open on the α-side if the window exists for time $t \in [t_k, t_{k+1}]$ with $k \in \mathbb{Z}$.

(iii) The door of the boundary flow barrier in the $(m_\alpha : m_\beta)$-source flow is closed on the α-side if the flow barrier exists for time $t \in [t_{k+1}, t_{k+2}]$ with $k \in \mathbb{Z}$.

(iv) The door of the boundary flow barrier in the $(m_\alpha : m_\beta)$-source flow is *permanently* open on the α-side if the window exists for time $t \in [t_k, \infty)$

(v) The door of the boundary flow barrier in the $(m_\alpha : m_\beta)$-source flow is *permanently* closed on the α-side if the flow barrier exists for time $t \in [t_{k+1}, \infty)$.

From the previous definition, the window of the boundary flow barrier in the $(m_\alpha : m_\beta)$-source flow is sketched in Fig. 4.20. On the window area, the source flow should satisfy the conditions stated in Chapters 2 and 3. The door for the boundary flow barrier wall on the boundary $\partial\Omega_{ij}$ is sketched in Fig. 4.21. In Fig. 4.21(a), the door of the boundary flow barrier wall is open, which implies the boundary flow of $\mathbf{x}^{(0)}$ can be switched to one of the $\mathbf{x}^{(\alpha)}$-flow. However, in Fig. 4.21(b), the door of the source flow barrier is closed, and the boundary flow of $\mathbf{x}^{(0)}$ cannot be switched into domain Ω_α from the α-side of the boundary.

For a source flow on the boundary $\partial\Omega_{ij}$, if the boundary flow barrier exists in the source flow on the α-side of $\partial\Omega_{ij}$ with a subset $S \subseteq \partial\Omega_{ij}$, the boundary flow $\mathbf{x}^{(0)}$ cannot leave the boundary from the α-side ($\alpha = i, j$) for $\mathbf{x}_m \in S \subseteq \partial\Omega_{ij}$

$$0 \in \left(\hbar_\alpha G_{\partial\Omega_{ij}}^{(m_\alpha, 0 \succ 0_\alpha)}(\mathbf{x}_m, q_2^{(\alpha)}), \hbar_\alpha G_{\partial\Omega_{ij}}^{(m_\alpha, 0 \succ 0_\alpha)}(\mathbf{x}_m, q_1^{(\alpha)})\right). \tag{4.141}$$

If there are two boundary flow barriers on both sides of the boundary to satisfy Eq. (4.141), the dynamical system will be constrained on the boundary, governed by Eq. (4.124).

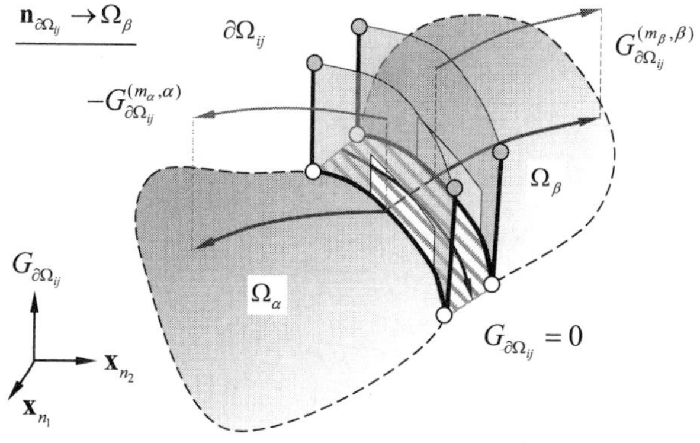

(a)

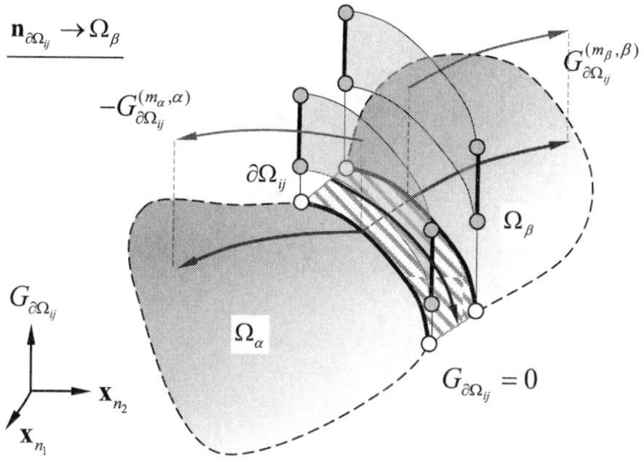

(b)

Fig. 4.20 The boundary flow barrier windows in the $(m_\alpha : m_\beta)$ -source flow on both sides of $\partial\Omega_{ij}$: **(a)** partial flow barrier and **(b)** full flow barrier. The upper curves arrows are the *G*-functions relative to the flow barrier. The two vertical surfaces are the flow barrier surfaces. The hatched area is for the zoomed boundary. The lower curves arrows are coming flows. ($m_\alpha, m_\beta \in \{0,1,2,\cdots\}$)

Theorem 4.9. *For a discontinuous dynamical system in Eq.* (4.1), *there is a point* $\mathbf{x}^{(0)}(t_m) \equiv \mathbf{x}_m \in \partial\Omega_{ij}$ *at time* t_m *between two adjacent domains* Ω_α ($\alpha = i, j$). *For* $\mathbf{x}_m \in S \subseteq \partial\Omega_{ij}$, *there is a source flow barrier* $\mathbf{F}^{(0 \succ 0_\alpha)}(\mathbf{x}^{(\alpha)}, t, \boldsymbol{\pi}_\alpha, q^{(\alpha)})$ *for* $q^{(\alpha)} \in [q_1^{(\alpha)}, q_2^{(\alpha)}]$ *on the* α *-side of the boundary* $\partial\Omega_{ij}$ *with*

$$0 \in \left(\hbar_\alpha G^{(0 \succ 0_\alpha)}_{\partial\Omega_{ij}}(\mathbf{x}_m, q_2^{(\alpha)}), \hbar_\alpha G^{(0 \succ 0_\alpha)}_{\partial\Omega_{ij}}(\mathbf{x}_m, q_1^{(\alpha)})\right) \subset \mathscr{R}. \tag{4.142}$$

The two leaving flows in the source flow satisfy

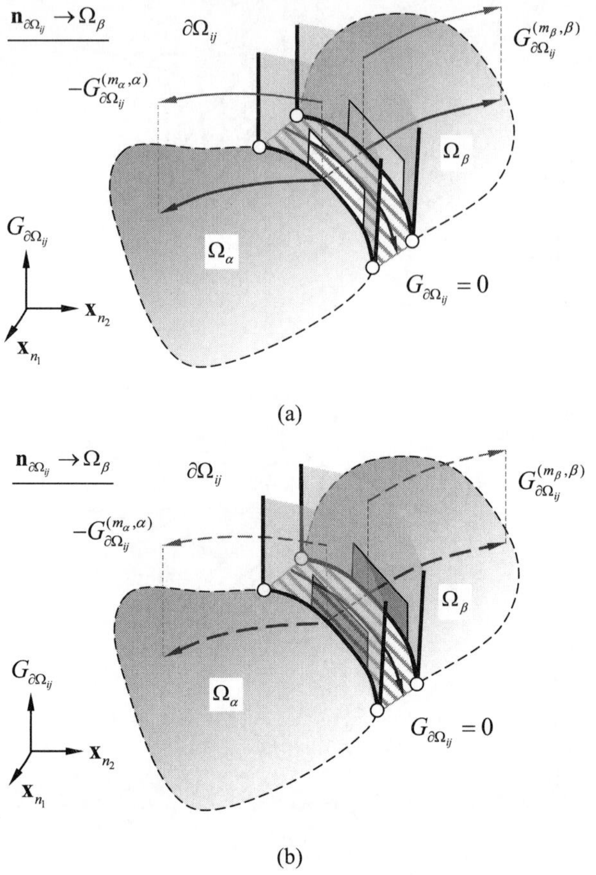

Fig. 4.21 The door of the absolute boundary flow barrier in the $(m_\alpha : m_\beta)$ -source flow on $\partial\Omega_{ij}$: **(a)** opened door **(b)** closed door. The upper curves with arrows are the *G*-functions relative to the flow barrier. The two vertical surfaces are the flow barrier surfaces. The hatched area is for the zoomed boundary. The lower curves with arrows are coming flows. ($m_\alpha, m_\beta \in \{0,1,2,\cdots\}$)

$$\hbar_\alpha G^{(\alpha)}_{\partial\Omega_{ij}}(\mathbf{x}_m, t_{m+}) < 0 \text{ and } \hbar_\alpha G^{(\beta)}_{\partial\Omega_{ij}}(\mathbf{x}_m, t_{m+}) > 0. \tag{4.143}$$

(i) *A boundary flow of* $\mathbf{x}^{(0)}$ *cannot be switched into the leaving flow of* $\mathbf{x}^{(\alpha)}$ *in a source flow on the* α *-side if and only if*

$$\hbar_\alpha G^{(0\succ 0_\alpha)}_{\partial\Omega_{ij}}(\mathbf{x}_m, q_1^{(\alpha)}) > 0 \text{ and } \hbar_\alpha G^{(0\succ 0_\alpha)}_{\partial\Omega_{ij}}(\mathbf{x}_m, q_2^{(\alpha)}) < 0. \tag{4.144}$$

(ii) *A boundary flow of* $\mathbf{x}^{(0)}$ *cannot be switched into the leaving flow of* $\mathbf{x}^{(\alpha)}$ *in a source flow on the* α *-side if and only if*

$$\begin{aligned}&\hbar_{\alpha} G_{\partial\Omega_{ij}}^{(s_{\alpha},0\succ 0_{\alpha})}(\mathbf{x}_m, q_1^{(\alpha)}) = 0 \;\; for\;\; s_{\alpha} = 0,1,2,\cdots,l_{\alpha}-1\\ &\hbar_{\alpha} G_{\partial\Omega_{ij}}^{(l_{\alpha},0\succ 0_{\alpha})}(\mathbf{x}_m, q_1^{(\alpha)}) > 0 \;\; and \;\; \hbar_{\alpha} G_{\partial\Omega_{ij}}^{(0\succ 0_{\alpha})}(\mathbf{x}_m, q_2^{(\alpha)}) < 0.\end{aligned} \tag{4.145}$$

(iii) *A boundary flow of* $\mathbf{x}^{(0)}$ *is switched into the leaving flow of* $\mathbf{x}^{(\alpha)}$ *in a source flow on the* α *-side if and only if*

$$\begin{aligned}&\hbar_{\alpha} G_{\partial\Omega_{ij}}^{(s_{\alpha},0\succ 0_{\alpha})}(\mathbf{x}_m, q_1^{(\alpha)}) = 0 \;\; for\;\; s_{\alpha} = 0,1,2,\cdots,l_{\alpha}-1;\\ &\hbar_{\alpha} G_{\partial\Omega_{ij}}^{(l_{\alpha},0\succ 0_{\alpha})}(\mathbf{x}_m, q_1^{(\alpha)}) < 0 \;\; and \;\; \hbar_{\alpha} G_{\partial\Omega_{ij}}^{(0\succ 0_{\alpha})}(\mathbf{x}_m, q_2^{(\alpha)}) < 0.\end{aligned} \tag{4.146}$$

Proof. (i) From Definition 4.23, the necessary and conditions in Eq. (4.144) are obtained.

(ii) An auxiliary flow of the boundary flow barrier is introduced as a fictitious flow of $\mathbf{x}^{(0\succ 0_{\alpha})}(t)$. For $\mathbf{x}^{(0\succ 0_{\alpha})}(t_{m\pm}) = \mathbf{x}^{(0)}(t_{m\pm})$ and $\mathbf{x}^{(\alpha)}(t_{m\pm}) = \mathbf{x}^{(0)}(t_{m\pm})$, the G-function definition gives

$$\begin{aligned}&\mathbf{n}_{\partial\Omega_{ij}}^{\mathrm{T}}(\mathbf{x}^{(0)}(t_{m+\varepsilon}))\cdot[\mathbf{x}^{(0\succ 0_{\alpha})}(t_{m+\varepsilon}, q_1^{(\alpha)}) - \mathbf{x}^{(0)}(t_{m+\varepsilon}, q_1^{(\alpha)})]\\ &= \sum\nolimits_{s_{\alpha}=0}^{l_{\alpha}-1} G_{\partial\Omega_{ij}}^{(s_{\alpha},0\succ 0_{\alpha})}(\mathbf{x}_m, t_{m+}, q_1^{(\alpha)})\varepsilon^{s_{\alpha}+1}\\ &\quad + G_{\partial\Omega_{ij}}^{(l_{\alpha},0\succ 0_{\alpha})}(\mathbf{x}_m, t_{m+}, q_1^{(\alpha)})\varepsilon^{l_{\alpha}+1} + o(\varepsilon^{l_{\alpha}+1}),\end{aligned}$$

Because of

$$\begin{aligned}&G_{\partial\Omega_{ij}}^{(s_{\alpha},0\succ 0_{\alpha})}(\mathbf{x}_m, t_{m\pm}, q_1^{(\alpha)}) = 0 \text{ for } s_{\alpha} = 0,1,2,\cdots,l_{\alpha}-1\\ &G_{\partial\Omega_{ij}}^{(l_{\alpha},0\succ 0_{\alpha})}(\mathbf{x}_m, t_{m\pm}, q_1^{(\alpha)}) \neq 0,\end{aligned}$$

one achieves

$$\begin{aligned}&\mathbf{n}_{\partial\Omega_{ij}}^{\mathrm{T}}(\mathbf{x}^{(0)}(t_{m+\varepsilon}))\cdot[\mathbf{x}^{(0)}(t_{m+\varepsilon}) - \mathbf{x}^{(0\succ 0_{\alpha})}(t_{m+\varepsilon}, q_1^{(\alpha)})]\\ &= -G_{\partial\Omega_{ij}}^{(l_{\alpha},0\succ 0_{\alpha})}(\mathbf{x}_m, t_{m+}, q_1^{(\alpha)})]\varepsilon^{l_{\alpha}+1}.\end{aligned}$$

From definition, the conditions for which the boundary flow of $\mathbf{x}^{(0)}$ cannot pass through the boundary flow barrier requires

$$\begin{aligned}&\mathbf{n}_{\partial\Omega_{ij}}^{\mathrm{T}}(\mathbf{x}^{(0)}(t_{m+\varepsilon}))\cdot[\mathbf{x}^{(0\succ 0_{\alpha})}(t_{m+\varepsilon}, q_1^{(\alpha)}) - \mathbf{x}^{(0)}(t_{m+\varepsilon})] > 0 \text{ for } \mathbf{n}_{\partial\Omega_{ij}} \to \Omega_{\beta},\\ &\mathbf{n}_{\partial\Omega_{ij}}^{\mathrm{T}}(\mathbf{x}^{(0)}(t_{m+\varepsilon}))\cdot[\mathbf{x}^{(0\succ 0_{\alpha})}(t_{m+\varepsilon}, q_1^{(\alpha)}) - \mathbf{x}^{(0)}(t_{m+\varepsilon})] < 0 \text{ for } \mathbf{n}_{\partial\Omega_{ij}} \to \Omega_{\alpha}.\end{aligned}$$

From the definition of $\hbar_{\alpha}$,

$$\hbar_{\alpha} = 1 \text{ for } \mathbf{n}_{\partial\Omega_{ij}} \to \Omega_{\beta} \text{ and } \hbar_{\alpha} = -1 \text{ for } \mathbf{n}_{\partial\Omega_{ij}} \to \Omega_{\alpha}$$

If a boundary flow of $\mathbf{x}^{(0)}$ cannot be switched into the leaving flow of $\mathbf{x}^{(\alpha)}$ in the source flow on the α -side, the conditions in Eq. (4.145) can be obtained, vice versa.

(iii) In a similar fashion, from definition, the conditions for which the boundary flow of $\mathbf{x}^{(0)}$ passes through the boundary flow barrier requires

$$\mathbf{n}^{\mathrm{T}}_{\partial\Omega_{ij}}(\mathbf{x}^{(0)}(t_{m+\varepsilon}))\cdot[\mathbf{x}^{(0\succ 0_\alpha)}(t_{m+\varepsilon},q_1^{(\alpha)})-\mathbf{x}^{(0)}(t_{m+\varepsilon})]<0 \text{ for } \mathbf{n}_{\partial\Omega_{ij}}\to\Omega_\beta,$$
$$\mathbf{n}^{\mathrm{T}}_{\partial\Omega_{ij}}(\mathbf{x}^{(0)}(t_{m+\varepsilon}))\cdot[\mathbf{x}^{(0\succ 0_\alpha)}(t_{m+\varepsilon},q_1^{(\alpha)})-\mathbf{x}^{(0)}(t_{m+\varepsilon})]>0 \text{ for } \mathbf{n}_{\partial\Omega_{ij}}\to\Omega_\alpha.$$

If a boundary flow of $\mathbf{x}^{(0)}$ is switched into the leaving flow of $\mathbf{x}^{(\alpha)}$ in the source flow on the α -side, the conditions in Eq. (4.146) can be obtained, vice versa. This theorem is proved. ■

For boundary flow barriers with the higher-order singularity, the switchability conditions of the boundary flow are presented as follows.

Theorem 4.10. *For a discontinuous dynamical system in Eq.* (4.1)*, there is a point* $\mathbf{x}^{(0)}(t_m)\equiv\mathbf{x}_m\in\partial\Omega_{ij}$ *at time* t_m *between two adjacent domains* Ω_α $(\alpha=i,j)$. *Suppose a boundary flow barrier* $\mathbf{F}^{(0\succ 0_\alpha)}(\mathbf{x}^{(\alpha)},t,\boldsymbol{\pi}_\alpha,q^{(\alpha)})$ *for* $q^{(\alpha)}\in[q_1^{(\alpha)},q_2^{(\alpha)}]$ *exists on the* α *-side of the boundary* $\partial\Omega_{ij}$ *in the* $(m_\alpha:m_\beta)$ *-source flow with*

$$\begin{aligned}&G^{(s_\alpha,0\succ 0_\alpha)}_{\partial\Omega_{ij}}(\mathbf{x}_m,q_\sigma^{(\alpha)})=0 \text{ for } s_\alpha=0,1,\cdots,m_\alpha-1;\\&0\in\left(\hbar_\alpha G^{(m_\alpha,0\succ 0_\alpha)}_{\partial\Omega_{ij}}(\mathbf{x}_m,q_2^{(\alpha)}),\hbar_\alpha G^{(m_\alpha,0\succ 0_\alpha)}_{\partial\Omega_{ij}}(\mathbf{x}_m,q_1^{(\alpha)})\right)\subset\mathscr{R}.\end{aligned}\tag{4.147}$$

The two possible leaving flows in the $(m_\alpha:m_\beta)$ *-source flow satisfy*

$$\begin{aligned}&G^{(s_\alpha,\alpha)}_{\partial\Omega_{ij}}(\mathbf{x}_m,t_{m+})=0 \text{ for } s_\alpha=0,1,\cdots,m_\alpha-1;\\&G^{(s_\beta,\beta)}_{\partial\Omega_{ij}}(\mathbf{x}_m,t_{m+})=0 \text{ for } s_\beta=0,1,\cdots,m_\beta-1;\\&\hbar_\alpha G^{(m_\alpha,\alpha)}_{\partial\Omega_{ij}}(\mathbf{x}_m,t_{m+})<0 \text{ and } \hbar_\alpha G^{(m_\beta,\beta)}_{\partial\Omega_{ij}}(\mathbf{x}_m,t_{m+})>0.\end{aligned}\tag{4.148}$$

(i) *A boundary flow of* $\mathbf{x}^{(0)}$ *cannot be switched to the leaving flow of* $\mathbf{x}^{(\alpha)}$ *on the* α *-side in the* $(m_\alpha:m_\beta)$ *-source flow if and only if*

$$\hbar_\alpha G^{(m_\alpha,0\succ 0_\alpha)}_{\partial\Omega_{ij}}(\mathbf{x}_m,q_1^{(\alpha)})>0 \text{ and } \hbar_\alpha G^{(m_\alpha,0\succ 0_\alpha)}_{\partial\Omega_{ij}}(\mathbf{x}_m,q_2^{(\alpha)})<0.\tag{4.149}$$

(ii) *A boundary flow of* $\mathbf{x}^{(0)}$ *cannot be switched to the leaving flow of* $\mathbf{x}^{(\alpha)}$ *on the* α *-side in the* $(m_\alpha:m_\beta)$ *source flow if and only if*

$$\begin{aligned}&G^{(s_\alpha,0\succ 0_\alpha)}_{\partial\Omega_{ij}}(\mathbf{x}_m,q_1^{(\alpha)})=0 \text{ for } s_\alpha=m_\alpha,m_\alpha+1,\cdots,l_\alpha-1;\\&\hbar_\alpha G^{(l_\alpha,0\succ 0_\alpha)}_{\partial\Omega_{ij}}(\mathbf{x}_m,q_1^{(\alpha)})>0 \text{ and } \hbar_\alpha G^{(m_\alpha,0\succ 0_\alpha)}_{\partial\Omega_{ij}}(\mathbf{x}_m,q_2^{(\alpha)})<0.\end{aligned}\tag{4.150}$$

(iii) *A boundary flow of* $\mathbf{x}^{(0)}$ *can be switched into the leaving flow of* $\mathbf{x}^{(\alpha)}$ *on the*

α -side in the $(m_\alpha : m_\beta)$ -source flow if and only if

$$\begin{aligned} &G_{\partial\Omega_{ij}}^{(s_\alpha,0\succ 0_\alpha)}(\mathbf{x}_m, q_1^{(\alpha)}) = 0 \ \mathit{for}\ s_\alpha = m_\alpha, m_\alpha+1,\cdots,l_\alpha-1; \\ &\hbar_\alpha G_{\partial\Omega_{ij}}^{(l_\alpha,0\succ 0_\alpha)}(\mathbf{x}_m, q_1^{(\alpha)}) < 0 \ \mathit{and}\ \hbar_\alpha G_{\partial\Omega_{ij}}^{(m_\alpha,0\succ 0_\alpha)}(\mathbf{x}_m, q_2^{(\alpha)}) < 0. \end{aligned} \tag{4.151}$$

Proof. (i) From Definition 4.24, the necessary and conditions in Eq. (4.148) are obtained.

(ii) An auxiliary flow of the boundary flow barrier is introduced as a fictitious flow of $\mathbf{x}^{(0\succ 0_\alpha)}(t)$. Since $\mathbf{x}^{(0\succ 0_\alpha)}(t_{m\pm}) = \mathbf{x}^{(0)}(t_{m\pm})$, the G-function definition gives

$$\begin{aligned} &\mathbf{n}_{\partial\Omega_{ij}}^{\mathrm{T}}(\mathbf{x}^{(0)}(t_{m+\varepsilon}))\cdot[\mathbf{x}^{(0\succ 0_\alpha)}(t_{m+\varepsilon}) - \mathbf{x}^{(0)}(t_{m+\varepsilon})] \\ &= \sum_{s_\alpha=0}^{m_\alpha-1} G_{\partial\Omega_{ij}}^{(s_\alpha,0\succ 0_\alpha)}(\mathbf{x}_m, t_{m+}, q_1^{(\alpha)})\varepsilon^{s_\alpha+1} \\ &\quad + \sum_{s_\alpha=m_\alpha}^{l_\alpha-1} G_{\partial\Omega_{ij}}^{(s_\alpha,0\succ 0_\alpha)}(\mathbf{x}_m, t_{m+}, q_1^{(\alpha)})\varepsilon^{s_\alpha+1} \\ &\quad + G_{\partial\Omega_{ij}}^{(l_\alpha,0\succ 0_\alpha)}(\mathbf{x}_m, t_{m+}, q_1^{(\alpha)})\varepsilon^{l_\alpha+1} + o(\varepsilon^{l_\alpha+1}). \end{aligned}$$

Because of

$$\begin{aligned} &G_{\partial\Omega_{ij}}^{(s_\alpha,0\succ 0_\alpha)}(\mathbf{x}_m, t_{m\pm}, q_1^{(\alpha)}) = 0 \ \text{ for } s_\alpha = 0,1,\cdots,m_\alpha-1; \\ &G_{\partial\Omega_{ij}}^{(s_\alpha,0\succ 0_\alpha)}(\mathbf{x}_m, q_1^{(\alpha)}) = 0 \ \text{ for } s_\alpha = m_\alpha, m_\alpha+1,\cdots,l_\alpha-1, \end{aligned}$$

one obtains

$$\begin{aligned} &\mathbf{n}_{\partial\Omega_{ij}}^{\mathrm{T}}(\mathbf{x}^{(0)}(t_{m+}))\cdot[\mathbf{x}^{(0)}(t_{m+\varepsilon}) - \mathbf{x}^{(0\succ 0_\alpha)}(t_{m+\varepsilon})] \\ &= -G_{\partial\Omega_{ij}}^{(l_\alpha,0\succ 0_\alpha)}(\mathbf{x}_m, q_1^{(\alpha)})\varepsilon^{l_\alpha+1}. \end{aligned}$$

From definition, the boundary flow of $\mathbf{x}^{(0)}(t)$ not passing over the boundary flow barrier gives

$$\begin{aligned} &\mathbf{n}_{\partial\Omega_{ij}}^{\mathrm{T}}(\mathbf{x}^{(0)}(t_{m+\varepsilon}))\cdot[\mathbf{x}^{(0)}(t_{m+\varepsilon}) - \mathbf{x}^{(0\succ 0_\alpha)}(t_{m+\varepsilon})] < 0 \text{ for } \mathbf{n}_{\partial\Omega_{ij}} \to \Omega_\beta, \\ &\mathbf{n}_{\partial\Omega_{ij}}^{\mathrm{T}}(\mathbf{x}^{(0)}(t_{m+\varepsilon}))\cdot[\mathbf{x}^{(0)}(t_{m+\varepsilon}) - \mathbf{x}^{(0\succ 0_\alpha)}(t_{m+\varepsilon})] > 0 \text{ for } \mathbf{n}_{\partial\Omega_{ij}} \to \Omega_\alpha. \end{aligned}$$

From the definition of $\hbar_\alpha$,

$$\hbar_\alpha = 1 \text{ for } \mathbf{n}_{\partial\Omega_{ij}} \to \Omega_\beta \text{ and } \hbar_\alpha = -1 \text{ for } \mathbf{n}_{\partial\Omega_{ij}} \to \Omega_\alpha.$$

Finally, one obtains Eq. (4.150). On the other hand, under Eq. (4.150), a boundary flow of $\mathbf{x}^{(0)}$ cannot be switched into the leaving flow of $\mathbf{x}^{(\alpha)}$ in the $(m_\alpha : m_\beta)$ source flow on the α -side.

(iii) In a similar fashion, from definition, the boundary flow of $\mathbf{x}^{(0)}(t)$ passing over the boundary flow barrier gives

$$\mathbf{n}_{\partial\Omega_{ij}}^{\mathrm{T}}(\mathbf{x}^{(0)}(t_{m+\varepsilon}))\cdot[\mathbf{x}^{(0)}(t_{m+\varepsilon})-\mathbf{x}^{(0\succ 0_\alpha)}(t_{m+\varepsilon})]>0 \text{ for } \mathbf{n}_{\partial\Omega_{ij}}\to\Omega_\beta,$$
$$\mathbf{n}_{\partial\Omega_{ij}}^{\mathrm{T}}(\mathbf{x}^{(0)}(t_{m+\varepsilon}))\cdot[\mathbf{x}^{(0)}(t_{m+\varepsilon})-\mathbf{x}^{(0\succ 0_\alpha)}(t_{m+\varepsilon})]<0 \text{ for } \mathbf{n}_{\partial\Omega_{ij}}\to\Omega_\alpha.$$

If a boundary flow $\mathbf{x}^{(0)}$ is switched into the leaving flow of $\mathbf{x}^{(\alpha)}$ in the $(m_\alpha : m_\beta)$ source flow on the α -side, the conditions in Eq. (4.151) are obtained, vice versa. This theorem is proved. ■

4.3.2. Leaving flows barriers

As similar to the coming flow barriers in the $(2k_\alpha : 2k_\beta)$-sink flow, the leaving flow barrier in the $(m_\alpha : m_\beta)$ -source flow can be described. For convenience, the corresponding discussion of the leaving flow barriers in the source flow will be given as follows.

Definition 4.31. For a discontinuous dynamical system in Eq. (4.1), there is a point $\mathbf{x}^{(0)}(t_m)\equiv\mathbf{x}_m\in\partial\Omega_{ij}$ at time t_m between two adjacent domains Ω_α $(\alpha=i,j)$. There is a vector field of $\mathbf{F}^{(0\succ\alpha)}(\mathbf{x}^{(\alpha)},t,\boldsymbol{\pi}_\alpha,q^{(\alpha)})$ for $q^{(\alpha)}\in[q_1^{(\alpha)},q_2^{(\alpha)}]$) on the boundary $\partial\Omega_{ij}$ with

$$\hbar_\alpha G_{\partial\Omega_{ij}}^{(0\succ\alpha)}(\mathbf{x}_m,q^{(\alpha)})\in\left[\hbar_\alpha G_{\partial\Omega_{ij}}^{(0\succ\alpha)}(\mathbf{x}_m,q_2^{(\alpha)}),\hbar_\alpha G_{\partial\Omega_{ij}}^{(0\succ\alpha)}(\mathbf{x}_m,q_1^{(\alpha)})\right]$$
$$\subset(-\infty,0] \tag{4.152}$$

The two possible leaving flows in the source flow satisfy

$$\hbar_\alpha G_{\partial\Omega_{ij}}^{(\alpha)}(\mathbf{x}_m,t_{m+})<0 \text{ and } \hbar_\alpha G_{\partial\Omega_{ij}}^{(\beta)}(\mathbf{x}_m,t_{m+})>0 \tag{4.153}$$

($\alpha,\beta\in\{i,j\}$ and $\alpha\neq\beta$). The vector field of $\mathbf{F}^{(0\succ\alpha)}(\mathbf{x}^{(\alpha)},t,\boldsymbol{\pi}_\alpha,q^{(\alpha)})$ is called the *flow barrier of the leaving flow* of $\mathbf{x}^{(\alpha)}$ in the source flow on the α -side if the following conditions are satisfied. $\mathbf{F}^{(0\succ\alpha)}(\mathbf{x}^{(\alpha)},t,\boldsymbol{\pi}_\alpha,q_\sigma^{(\alpha)})$ for $\sigma=1,2$ are called the *lower and upper limits* of the leaving flow barriers on the α- side.

(i) The leaving flow of $\mathbf{x}^{(\alpha)}$ cannot leave the boundary on the α -side if

$$\begin{aligned}&\mathbf{x}^{(\alpha)}(t_{m+})=\mathbf{x}^{(0\succ\alpha)}(t_{m\pm},q^{(\alpha)})=\mathbf{x}_m,\\&\hbar_\alpha G_{\partial\Omega_{ij}}^{(\alpha)}(\mathbf{x}_m,t_{m+})\in\left(\hbar_\alpha G_{\partial\Omega_{ij}}^{(0\succ\alpha)}(\mathbf{x}_m,q_2^{(\alpha)}),\hbar_\alpha G_{\partial\Omega_{ij}}^{(0\succ\alpha)}(\mathbf{x}_m,q_1^{(\alpha)})\right).\end{aligned} \tag{4.154}$$

(ii) The leaving flow of $\mathbf{x}^{(\alpha)}$ cannot leave the boundary on the α -side at the critical points (i.e., $q^{(\alpha)}=q_\sigma^{(\alpha)}$, $\sigma\in\{1,2\}$) if

$$\begin{aligned}&\mathbf{x}^{(\alpha)}(t_{m+})=\mathbf{x}^{(0\succ\alpha)}(t_{m\pm},q_{\sigma}^{(\alpha)})=\mathbf{x}_{m},\\&G_{\partial\Omega_{ij}}^{(s_\alpha,\alpha)}(\mathbf{x}_{m},t_{m+})=G_{\partial\Omega_{ij}}^{(s_\alpha,0\succ\alpha)}(\mathbf{x}_{m},q_{\sigma}^{(\alpha)})\neq 0 \text{ for } s_\alpha=0,1,2,\cdots,l_\alpha;\\&(-1)^{\sigma}\hbar_{\alpha}\mathbf{n}_{\partial\Omega_{ij}}^{\mathrm{T}}(\mathbf{x}^{(0)}(t_{m+\varepsilon}))\cdot\left[\mathbf{x}^{(\alpha)}(t_{m+\varepsilon})-\mathbf{x}^{(0\succ\alpha)}(t_{m+\varepsilon},q_{\sigma}^{(\alpha)})\right]>0.\end{aligned}\tag{4.155}$$

(iii) The leaving flow of $\mathbf{x}^{(\alpha)}$ leaves the boundary on the α -side at the critical points (i.e., $q^{(\alpha)}=q_{\sigma}^{(\alpha)}$, $\sigma\in\{1,2\}$) if

$$\begin{aligned}&\mathbf{x}^{(\alpha)}(t_{m+})=\mathbf{x}^{(0\succ\alpha)}(t_{m\pm},q_{\sigma}^{(\alpha)})=\mathbf{x}_{m},\\&G_{\partial\Omega_{ij}}^{(s_\alpha,\alpha)}(\mathbf{x}_{m},t_{m+})=G_{\partial\Omega_{ij}}^{(s_\alpha,0\succ\alpha)}(\mathbf{x}_{m},q_{\sigma}^{(\alpha)})\neq 0 \text{ for } s_\alpha=0,1,2,\cdots,l_\alpha;\\&(-1)^{\sigma}\hbar_{\alpha}\mathbf{n}_{\partial\Omega_{ij}}^{\mathrm{T}}(\mathbf{x}^{(0)}(t_{m+\varepsilon}))\cdot\left[\mathbf{x}^{(\alpha)}(t_{m+\varepsilon})-\mathbf{x}^{(0\succ\alpha)}(t_{m+\varepsilon},q_{\sigma}^{(\alpha)})\right]<0.\end{aligned}\tag{4.156}$$

Definition 4.32. For a discontinuous dynamical system in Eq. (4.1), there is a point $\mathbf{x}^{(0)}(t_m)\equiv\mathbf{x}_m\in\partial\Omega_{ij}$ at time t_m between two adjacent domains Ω_α ($\alpha=i,j$). A vector field $\mathbf{F}^{(0\succ\alpha)}(\mathbf{x}^{(\alpha)},t,\boldsymbol{\pi}_\alpha,q^{(\alpha)})$ for $q^{(\alpha)}\in[q_1^{(\alpha)},q_2^{(\alpha)}]$ exists on the boundary $\partial\Omega_{ij}$ with

$$\begin{aligned}&G_{\partial\Omega_{ij}}^{(s_\alpha,0\succ\alpha)}(\mathbf{x}_{m},q^{(\alpha)})=0 \text{ for } s_\alpha=0,1,\cdots,m_\alpha-1;\\&G_{\partial\Omega_{ij}}^{(m_\alpha,0\succ\alpha)}(\mathbf{x}_{m},q^{(\alpha)})\in\left[\hbar_\alpha G_{\partial\Omega_{ij}}^{(m_\alpha,0\succ\alpha)}(\mathbf{x}_{m},q_2^{(\alpha)}),\hbar_\alpha G_{\partial\Omega_{ij}}^{(m_\alpha,0\succ\alpha)}(\mathbf{x}_{m},q_1^{(\alpha)})\right]\\&\qquad\subset(-\infty,0]\end{aligned}\tag{4.157}$$

($\alpha,\beta\in\{i,j\}$ and $\alpha\neq\beta$). The leaving flows in the $(m_\alpha:m_\beta)$ -source flow satisfy

$$\begin{aligned}&G_{\partial\Omega_{ij}}^{(s_\alpha,\alpha)}(\mathbf{x}_{m},t_{m+})=0 \text{ for } s_\alpha=0,1,\cdots,m_\alpha-1;\\&G_{\partial\Omega_{ij}}^{(s_\beta,\beta)}(\mathbf{x}_{m},t_{m+})=0 \text{ for } s_\beta=0,1,\cdots,m_\beta-1;\\&\hbar_\alpha G_{\partial\Omega_{ij}}^{(m_\alpha,\alpha)}(\mathbf{x}_{m},t_{m+})<0 \text{ and } \hbar_\alpha G_{\partial\Omega_{ij}}^{(m_\beta,\beta)}(\mathbf{x}_{m},t_{m+})>0.\end{aligned}\tag{4.158}$$

The vector field of $\mathbf{F}^{(0\succ\alpha)}(\mathbf{x}^{(\alpha)},t,\boldsymbol{\pi}_\alpha,q^{(\alpha)})$ is called the *leaving flow barrier* in the $(m_\alpha:m_\beta)$ -source flow on the α -side if the following conditions are satisfied. The critical values of $\mathbf{F}^{(0\succ\alpha)}(\mathbf{x}^{(\alpha)},t,\boldsymbol{\pi}_\alpha,q_\sigma^{(\alpha)})$ ($\sigma=1,2$) at the boundary $\partial\Omega_{ij}$ are called the *lower and upper limits* of the leaving flow barriers in the $(m_\alpha:m_\beta)$ -source flow on the α - side.

(i) The leaving flow of $\mathbf{x}^{(\alpha)}$ in the $(m_\alpha:m_\beta)$ source flow cannot leave the boundary on the α -side if

$$
\begin{aligned}
&\mathbf{x}^{(\alpha)}(t_{m+}) = \mathbf{x}^{(0\succ\alpha)}(t_{m\pm}, q^{(\alpha)}) = \mathbf{x}_m, \\
&G^{(m_\alpha,\alpha)}_{\partial\Omega_{ij}}(\mathbf{x}_m, t_{m+}) \in \left(\hbar_\alpha G^{(m_\alpha,0\succ\alpha)}_{\partial\Omega_{ij}}(\mathbf{x}_m, q_2^{(\alpha)}),\ \hbar_\alpha G^{(m_\alpha,0\succ\alpha)}_{\partial\Omega_{ij}}(\mathbf{x}_m, q_1^{(\alpha)})\right).
\end{aligned}
\tag{4.159}
$$

(ii) The leaving flow of $\mathbf{x}^{(\alpha)}$ in the $(m_\alpha : m_\beta)$ source flow cannot leave the boundary on the α -side at the critical points (i.e., $q^{(\alpha)} = q_\sigma^{(\alpha)}$, $\sigma \in \{1,2\}$) if

$$
\begin{aligned}
&\mathbf{x}^{(\alpha)}(t_{m+}) = \mathbf{x}^{(0\succ\alpha)}(t_{m\pm}, q_\sigma^{(\alpha)}) = \mathbf{x}_m, \\
&G^{(s_\alpha,\alpha)}_{\partial\Omega_{ij}}(\mathbf{x}_m, t_{m+}) = G^{(s_\alpha,0\succ\alpha)}_{\partial\Omega_{ij}}(\mathbf{x}_m, q_\sigma^{(\alpha)}) \neq 0 \text{ for } s_\alpha = m_\alpha, m_\alpha+1, \cdots, l_\alpha; \\
&(-1)^\sigma \hbar_\alpha \mathbf{n}^{\mathrm{T}}_{\partial\Omega_{ij}}(\mathbf{x}^{(0)}(t_{m+\varepsilon})) \cdot \left[\mathbf{x}^{(\alpha)}(t_{m+\varepsilon}) - \mathbf{x}^{(0\succ\alpha)}(t_{m+\varepsilon}, q_\sigma^{(\alpha)})\right] > 0.
\end{aligned}
\tag{4.160}
$$

(iii) The leaving flow of $\mathbf{x}^{(\alpha)}$ in the $(m_\alpha : m_\beta)$ source flow leaves the boundary on the α -side at the critical points (i.e., $q^{(\alpha)} = q_\sigma^{(\alpha)}$, $\sigma \in \{1,2\}$) if

$$
\begin{aligned}
&\mathbf{x}^{(\alpha)}(t_{m+}) = \mathbf{x}^{(0\succ\alpha)}(t_{m\pm}, q_\sigma^{(\alpha)}) = \mathbf{x}_m, \\
&G^{(s_\alpha,\alpha)}_{\partial\Omega_{ij}}(\mathbf{x}_m, t_{m+}) = G^{(s_\alpha,0\succ\alpha)}_{\partial\Omega_{ij}}(\mathbf{x}_m, q_\sigma^{(\alpha)}) \neq 0 \text{ for } s_\alpha = m_\alpha, m_\alpha+1, \cdots, l_\alpha; \\
&(-1)^\sigma \hbar_\alpha \mathbf{n}^{\mathrm{T}}_{\partial\Omega_{ij}}(\mathbf{x}^{(0)}(t_{m+\varepsilon})) \cdot \left[\mathbf{x}^{(\alpha)}(t_{m+\varepsilon}) - \mathbf{x}^{(0\succ\alpha)}(t_{m+\varepsilon}, q_\sigma^{(\alpha)})\right] < 0.
\end{aligned}
\tag{4.161}
$$

To explain the leaving flow barrier in the source flow, the leaving flow barriers on both sides of the boundary $\partial\Omega_{ij}$ are also presented through the *G*-functions in Fig. 4.22. The red dashed curves represent the *G*-function of the flows pertaining to the flow barriers in each domain Ω_α ($\alpha \in \{i,j\}$). The thick lines denote the *G*-functions of the flow barriers on both sides of the boundary. To show the flow barriers, consider the *G*-function on the β -side of the boundary $\partial\Omega_{ij}$ as a reference ($\beta \in \{i,j\}, \beta \neq \alpha$). So the *G*-function on the α -side of the boundary $\partial\Omega_{ij}$ is presented through $-G^{(\alpha)}_{\partial\Omega_{ij}}$ because the source flow requires the *G*-functions of flow on both sides of the boundary should be opposite. The hatched area is zoomed for the boundary flow of $\mathbf{x}^{(0)}$.

Definition 4.33. For a discontinuous dynamical system in Eq. (4.1), there is a point $\mathbf{x}^{(0)}(t_m) \equiv \mathbf{x}_m \in \partial\Omega_{ij}$ at time t_m between two adjacent domains Ω_α ($\alpha = i,j$). Suppose there is a leaving flow barrier $\mathbf{F}^{(0\succ\alpha)}(\mathbf{x}^{(\alpha)}, t, \boldsymbol{\pi}_\alpha, q)$ for $q^{(\alpha)} \in [q_1^{(\alpha)}, q_2^{(\alpha)}]$ on the α -side of the boundary in the $(m_\alpha : m_\beta)$ -source flow ($m_\alpha, m_\beta = 0,1,\cdots$).

(i) The leaving flow barrier the $(m_\alpha : m_\beta)$ -source flow is *partial* on the α -side if $\mathbf{x}_m \in S \subset \partial\Omega_{ij}$.

(ii) The leaving flow barrier the $(m_\alpha : m_\beta)$ -source flow is *full* on the α -side if $\mathbf{x}_m \in S = \partial\Omega_{ij}$.

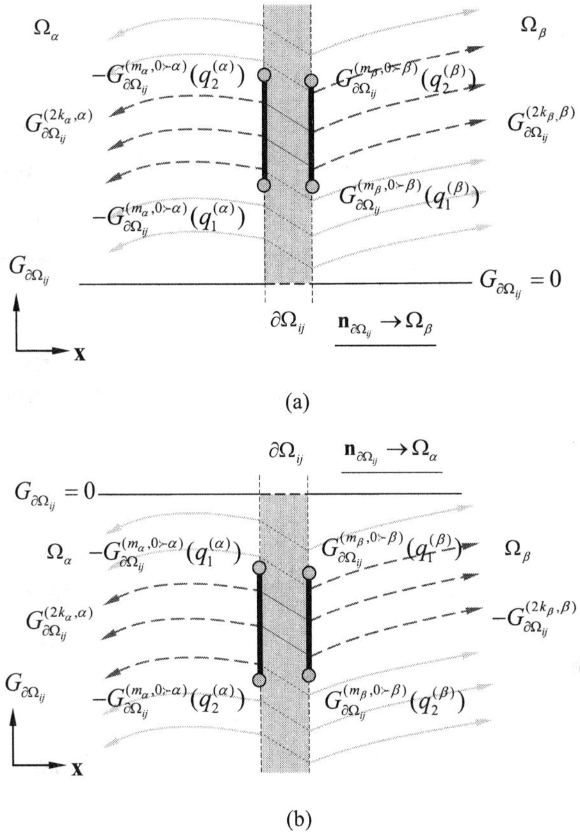

Fig. 4.22 The G-functions for leaving flow barriers in the $(m_\alpha : m_\beta)$-source flow: **(a)** $\mathbf{n}_{\partial\Omega_{ij}} \to \Omega_\beta$ and **(b)** $\mathbf{n}_{\partial\Omega_{ij}} \to \Omega_\alpha$. The red dashed curves are the G-functions relative to the flow barrier. The thick lines are the G-functions of the flow barriers at both α and β-side of the boundary $\partial\Omega_{ij}$. $G_{\partial\Omega_{ij}}^{(m_\alpha,0\succ\alpha)}(q_1^{(\alpha)})$ and $G_{\partial\Omega_{ij}}^{(m_\alpha,0\succ\alpha)}(q_2^{(\alpha)})$ are for *lower* and *upper* barrier limits ($m_\alpha \in \{0,1,2,\cdots\}$, $\alpha = i,j$) and similarly for the β-side.

Definition 4.34. For a discontinuous dynamical system in Eq. (4.1), there is a point $\mathbf{x}^{(0)}(t_m) \equiv \mathbf{x}_m \in \partial\Omega_{ij}$ at time t_m between two adjacent domains Ω_α ($\alpha = i,j$). There is a leaving flow barrier of $\mathbf{F}^{(0\succ\alpha)}(\mathbf{x}^{(\alpha)},t,\boldsymbol{\pi}_\alpha,q^{(\alpha)})$ for $q^{(\alpha)} \in [q_1^{(\alpha)},q_2^{(\alpha)}]$ in the $(m_\alpha : m_\beta)$-source flow on the α-side of the boundary ($m_\alpha, m_\beta \in \{0,1,2,\cdots\}$).

(i) The leaving flow barrier in the $(m_\alpha : m_\beta)$-source flow is with an *upper limit* on the α-side if for $\mathbf{x}_m \in S \subseteq \partial\Omega_{ij}$

$$\hbar_\alpha G_{\partial\Omega_{ij}}^{(2k_\alpha,0\succ\alpha)}(\mathbf{x}_m,q_1^{(\alpha)}) = 0_+ \text{ and } \hbar_\alpha G_{\partial\Omega_{ij}}^{(2k_\alpha,0\succ\alpha)}(\mathbf{x}_m,q_2^{(\alpha)}) \neq -\infty \tag{4.162}$$

(ii) The leaving flow barrier in the $(m_\alpha : m_\beta)$-source flow is with a *lower limit* on the α-side if for $\mathbf{x}_m \in S \subseteq \partial\Omega_{ij}$

$$\hbar_\alpha G_{\partial\Omega_{ij}}^{(m_\alpha,0\succ\alpha)}(\mathbf{x}_m, q_1^{(\alpha)}) < 0 \text{ and } \hbar_\alpha G_{\partial\Omega_{ij}}^{(m_\alpha,0\succ\alpha)}(\mathbf{x}_m, q_2^{(\alpha)}) \to -\infty. \tag{4.163}$$

(iii) The leaving flow barrier in the $(m_\alpha : m_\beta)$-source flow is *absolute* on the α-side if for $\mathbf{x}_m \in S \subseteq \partial\Omega_{ij}$

$$\hbar_\alpha G_{\partial\Omega_{ij}}^{(m_\alpha,0\succ\alpha)}(\mathbf{x}_m, q_1^{(\alpha)}) = 0_+ \text{ and } \hbar_\alpha G_{\partial\Omega_{ij}}^{(m_\alpha,0\succ\alpha)}(\mathbf{x}_m, q_2^{(\alpha)}) \to -\infty. \tag{4.164}$$

(iv) The leaving flow barrier is called the *leaving flow barrier wall* in the $(m_\alpha : m_\beta)$-source flow on the α-side if the absolute leaving flow barrier exists for $\mathbf{x}_m \in S = \partial\Omega_{ij}$.

(v) The flow barrier is called the *leaving flow barrier fence* in the $(m_\alpha : m_\beta)$-source flow on the α-side if there are many partial flow barriers and many non-barriers on the α-side of the boundary $\partial\Omega_{ij}$.

Definition 4.35. For a discontinuous dynamical system in Eq. (4.1), there is a point $\mathbf{x}^{(0)}(t_m) \equiv \mathbf{x}_m \in \partial\Omega_{ij}$ at time t_m between two adjacent domains Ω_α $(\alpha = i, j)$. The leaving flows in the $(m_\alpha : m_\beta)$-source flow satisfy Eq. (4.158). Suppose there is a leaving flow barrier of $\mathbf{F}^{(0\succ\alpha)}(\mathbf{x}^{(\alpha)}, t, \boldsymbol{\pi}_\alpha, q^{(\alpha)})$ in the $(m_\alpha : m_\beta)$-source flow on the α-side of the boundary ($m_\alpha, m_\beta \in \{0,1,2,\cdots\}$) with

$$\begin{aligned} &\hbar_\alpha G_{\partial\Omega_{ij}}^{(s_\alpha,0\succ\alpha)}(\mathbf{x}_m, q^{(\alpha)}) = 0 \text{ for } s_\alpha = 0,1,2,\cdots; \\ &\hbar_\alpha G_{\partial\Omega_{ij}}^{(m_\alpha,0\succ\alpha)}(\mathbf{x}_m, q^{(\alpha)}) \in \left[\hbar_\alpha G_{\partial\Omega_{ij}}^{(m_\alpha,0\succ\alpha)}(\mathbf{x}_m, q_{2n-1}^{(\alpha)}), \hbar_\alpha G_{\partial\Omega_{ij}}^{(m_\alpha,0\succ\alpha)}(\mathbf{x}_m, q_{2n}^{(\alpha)})\right] \end{aligned} \tag{4.165}$$

for $\mathbf{x}_m \in S \subseteq \partial\Omega_{ij}$ and $q^{(\alpha)} \in [q_{2n-1}^{(\alpha)}, q_{2n}^{(\alpha)}]$ $(n = 1,2,\cdots)$. For $q^{(\alpha)} \in (q_{2n}^{(\alpha)}, q_{2n+1}^{(\alpha)})$, no leaving flow barrier is defined at $\mathbf{x}_m \in S \subseteq \partial\Omega_{ij}$. Thus, the leaving flow of $\mathbf{x}^{(\alpha)}$ on the α-side can directly leave the boundary if for $\sigma = 2n, 2n+1$

$$\begin{aligned} &\hbar_\alpha G_{\partial\Omega_{ij}}^{(s_\alpha,0\succ\alpha)}(\mathbf{x}_m, q_\sigma^{(\alpha)}) = 0 \text{ for } s_\alpha = 0,1,2,\cdots; \\ &\hbar G_{\partial\Omega_{ij}}^{(m_\alpha,\alpha)}(\mathbf{x}_m, t_{m+}) \in \left(\hbar_\alpha G_{\partial\Omega_{ij}}^{(m_\alpha,0\succ\alpha)}(\mathbf{x}_m, q_{2n}^{(\alpha)}), \hbar_\alpha G_{\partial\Omega_{ij}}^{(m_\alpha,0\succ\alpha)}(\mathbf{x}_m, q_{2n+1}^{(\alpha)})\right). \end{aligned} \tag{4.166}$$

The G-function intervals for all $\mathbf{x}_m \in S \subseteq \partial\Omega_{ij}$ with $q^{(\alpha)} \in (q_{2n}^{(\alpha)}, q_{2n+1}^{(\alpha)})$ are called the *window* of the leaving flow barrier in the $(m_\alpha : m_\beta)$-source flow on the α-side of the boundary.

In a similar fashion, the partial and full leaving flow barrier in a $(m_\alpha : m_\beta)$-source flow on both sides of the boundary $\partial\Omega_{ij}$ are sketched in Fig. 4.23 for

$\mathbf{x}_m \in S \subseteq \partial\Omega_{ij}$ through the two different colored surfaces at the α and β-sides of the boundary $\partial\Omega_{ij}$. The two leaving flow barriers are different. To clearly show the leaving flow barriers, the G-function on the β-side of the boundary $\partial\Omega_{ij}$ is considered as a reference, and the G-function on the α-side of the boundary $\partial\Omega_{ij}$ is presented by $-G^{(\alpha)}_{\partial\Omega_{ij}}$. The infinity leaving flow barriers in the $(m_\alpha : m_\beta)$-source flow are sketched in Fig. 4.24.

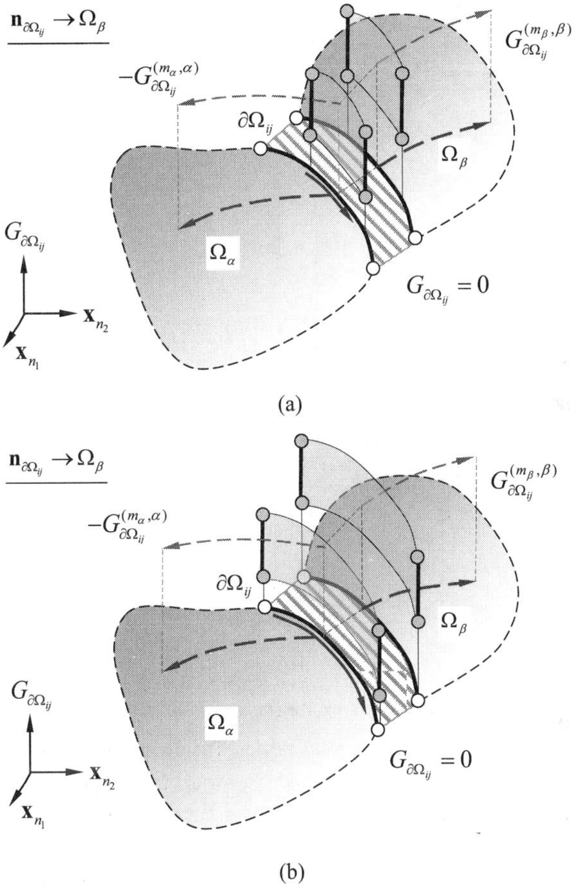

(a)

(b)

Fig. 4.23 The leaving flow barriers in the $(m_\alpha : m_\beta)$-source flow on both sides of the boundary $\partial\Omega_{ij}$: **(a)** partial flow barrier and **(b)** full flow barrier. The upper curves with arrows are the G-functions relative to the flow barrier. The two vertical surfaces are the flow barrier surfaces. The hatched area is for the zoomed boundary. The lower curves with arrows are coming flows. ($m_\alpha, m_\beta \in \{0,1,2,\cdots\}$).

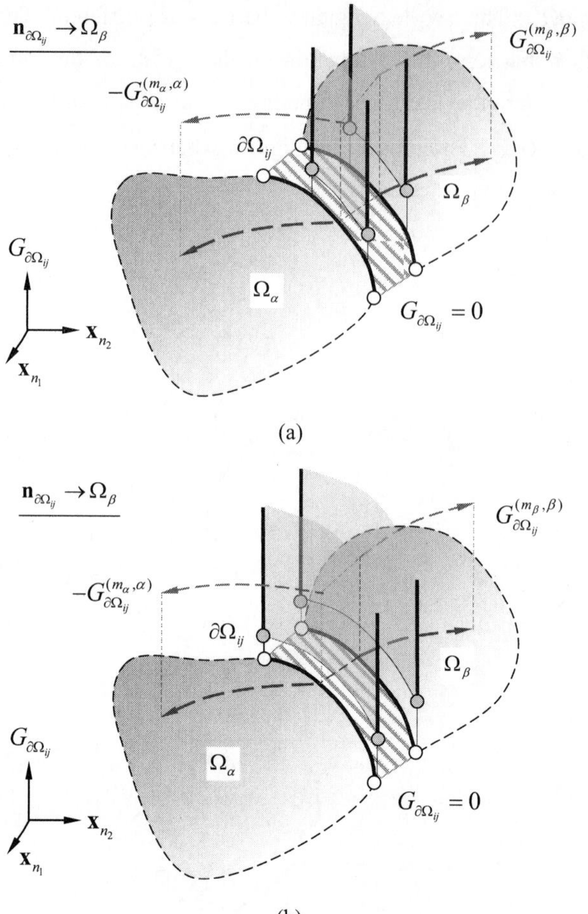

(b)

Fig. 4.24 The infinity leaving flow barrier in the $(m_\alpha : m_\beta)$ -source flow with lower boundary on $\partial\Omega_{ij}$: **(a)** partial flow barrier and **(b)** full flow barrier. The upper curves with arrows are the G-functions relative to the flow barrier. The two vertical surfaces are the flow barrier surfaces. The hatched area is for the zoomed boundary. The lower curves with arrows are coming flows. ($m_\alpha, m_\beta \in \{0,1,2,\cdots\}$).

Definition 4.36. For a discontinuous dynamical system in Eq. (4.1), there is a point $\mathbf{x}^{(0)}(t_m) \equiv \mathbf{x}_m \in \partial\Omega_{ij}$ at time t_m between two adjacent domains Ω_α ($\alpha = i, j$). The leaving flows in the $(m_\alpha : m_\beta)$ -source flow satisfy Eq. (4.158). Suppose there is a leaving flow barrier $\mathbf{F}^{(0\succ\alpha)}(\mathbf{x}^{(\alpha)}, t, \boldsymbol{\pi}_\alpha, q^{(\alpha)})$ for $q^{(\alpha)} \in [q_1^{(\alpha)}, q_2^{(\alpha)}]$ in the $(m_\alpha : m_\beta)$ -source flow on the α -side of the boundary ($m_\alpha, m_\beta \in \{0,1,2,\cdots\}$).

(i) The leaving flow barrier in the $(m_\alpha : m_\beta)$-source flow is *permanent* on the α-side if the flow barrier is *independent* of time $t \in [0, \infty)$.

(ii) The leaving flow barrier in the $(m_\alpha : m_\beta)$-source flow is *instantaneous* on the α-side if the flow barrier is continuously *dependent* on time $t \in [0, \infty)$.

(iii) The leaving flow barrier in the $(m_\alpha : m_\beta)$-source flow is *intermittent* on the α-side if the flow barrier exists for time $t \in [t_k, t_{k+1}]$ with $k \in \mathbb{Z}$.

(iv) The leaving flow barrier in the $(m_\alpha : m_\beta)$-source flow is *intermittent and static* on the α-side if the flow barrier is *independent* of time $t \in [t_k, t_{k+1}]$ with $k \in \mathbb{Z}$.

Similarly, the permanent and instantaneous windows of the leaving flow barrier in the source flow on the boundary can be also discussed. Further, the concept of the door of the leaving flow barrier wall on the boundary can be defined.

Definition 4.37. For a discontinuous dynamical system in Eq. (4.1), there is a point $\mathbf{x}^{(0)}(t_m) \equiv \mathbf{x}_m \in \partial\Omega_{ij}$ at time t_m between two adjacent domains Ω_α ($\alpha = i, j$). The leaving flows in the $(m_\alpha : m_\beta)$-source flow satisfy Eq. (4.158). Suppose there is a leaving flow barrier of $\mathbf{F}^{(0 \succ \alpha)}(\mathbf{x}^{(\alpha)}, t, \boldsymbol{\pi}_\alpha, q^{(\alpha)})$ in the $(m_\alpha : m_\beta)$-source flow for $q^{(\alpha)} \in [q_{2n-1}^{(\alpha)}, q_{2n}^{(\alpha)}]$ on the α-side ($m_\alpha, m_\beta \in \{0, 1, 2, \cdots\}$, $n = 1, 2, \cdots$), and there is a window of the leaving flow barrier for $S \subset \partial\Omega_{ij}$ and $q^{(\alpha)} \in [q_{2n}^{(\alpha)}, q_{2n+1}^{(\alpha)}]$.

(i) The window of the leaving flow barrier in the $(m_\alpha : m_\beta)$-source flow is *permanent* on the α-side if the window is *independent* of time $t \in [0, +\infty)$.

(ii) The window of the leaving flow barrier in the $(m_\alpha : m_\beta)$-source flow is *instantaneous* on the α-side if the window is continuously *dependent* of time $t \in [0, +\infty)$.

(iii) The window of the leaving flow barrier in the $(m_\alpha : m_\beta)$-source flow is *intermittent* on the α-side if the window exists for time $t \in [t_k, t_{k+1}]$ with $k \in \mathbb{Z}$.

(iv) The window of the leaving flow barrier in the $(m_\alpha : m_\beta)$-source flow is *intermittent and static* on the α-side if the window is *independent* of time $t \in [t_k, t_{k+1}]$ with $k \in \mathbb{Z}$.

Definition 4.38. For a discontinuous dynamical system in Eq. (4.1), there is a point $\mathbf{x}^{(0)}(t_m) \equiv \mathbf{x}_m \in \partial\Omega_{ij}$ at time t_m between two adjacent domains Ω_α ($\alpha = i, j$). There is a leaving flow barrier of $\mathbf{F}^{(0 \succ \alpha)}(\mathbf{x}^{(\alpha)}, t, \boldsymbol{\pi}_\alpha, q^{(\alpha)})$ for $q^{(\alpha)} \in [q_1^{(\alpha)}, q_2^{(\alpha)}]$ in the $(m_\alpha : m_\beta)$-source flow on the α-side of the boundary ($m_\alpha, m_\beta \in \{0, 1, 2, \cdots\}$). Suppose a leaving flow barrier wall in the source flow exists on the α-side, but there is an *intermittent* and static window of the leaving flow barrier on $S \subset \partial\Omega_{ij}$

for $q^{(\alpha)} \in [q_1^{(\alpha)}, q_2^{(\alpha)}]$ and $t \in [t_k, t_{k+1}]$ with $k \in \mathbb{Z}$.

(i) The window of the leaving flow barrier is termed the *door* of the leaving flow barrier wall in the $(m_\alpha : m_\beta)$-source flow on the α-side if the window and the barrier exist alternatively.

(ii) The door of the leaving flow barrier wall in the $(m_\alpha : m_\beta)$-source flow is *open* on the α-side if the window exists for time $t \in [t_k, t_{k+1}]$ with $k \in \mathbb{Z}$.

(iii) The door of the leaving flow barrier wall in the $(m_\alpha : m_\beta)$-source flow is *closed* on the α-side if the leaving flow barrier exists for time $t \in [t_{k+1}, t_{k+2}]$ with $k \in \mathbb{Z}$.

(iv) The door of the leaving flow barrier wall in the $(m_\alpha : m_\beta)$-source flow is *permanently open* on the α-side if the window exists for time $t \in [t_k, \infty)$.

(v) The door of the leaving flow barrier wall in the $(m_\alpha : m_\beta)$-source flow is *permanently closed* on the α-side if the leaving flow barrier exists for time $t \in [t_{k+1}, \infty)$.

From the previous definition, the window of the leaving flow barrier in the source flow is sketched in Fig. 4.25. On the window area, the source flow should satisfy the conditions in Chapters 2 and 3. The door for the leaving flow barrier wall in the $(m_\alpha : m_\beta)$-source flow on the boundary $\partial\Omega_{ij}$ is sketched in Fig. 4.26. In Fig. 4.26(a), the door of the flow barrier wall is open, so the leaving flow of $\mathbf{x}^{(\alpha)}$ can leave the boundary. However, in Fig. 4.26(b), the door of the leaving flow barrier is closed, and the leaving flow of $\mathbf{x}^{(\alpha)}$ cannot leave the boundary.

If the leaving flow barrier in the source flow exists on the α-side of such a boundary with a subset $S \subseteq \partial\Omega_{ij}$, the leaving flow of $\mathbf{x}^{(\alpha)}$ cannot leave the boundary if for $\mathbf{x}_m \in S \subseteq \partial\Omega_{ij}$

$$\hbar_\alpha G_{\partial\Omega_{ij}}^{(m_\alpha,\alpha)}(\mathbf{x}_m) \in \left(\hbar_\alpha G_{\partial\Omega_{ij}}^{(m_\alpha,0\succ\alpha)}(\mathbf{x}_m, q_2^{(\alpha)}), \hbar_\alpha G_{\partial\Omega_{ij}}^{(m_\alpha,0\succ\alpha)}(\mathbf{x}_m, q_1^{(\alpha)})\right). \tag{4.167}$$

For this case, the dynamical system on the α-side of the boundary will be constrained by the boundary $\partial\Omega_{ij}$, i.e.,

$$\begin{aligned} &\dot{\mathbf{x}}^{(\alpha)} = \mathbf{F}^{(\alpha)}(\mathbf{x}^{(\alpha)}, t, \boldsymbol{\mu}_\alpha) \text{ in } \Omega_\alpha \ (\alpha = i, j), \\ &\text{with } \varphi_{ij}(\mathbf{x}^{(\alpha)}, t, \boldsymbol{\lambda}) = 0 \text{ on } \partial\Omega_{ij}. \end{aligned} \tag{4.168}$$

Theorem 4.11. *For a discontinuous dynamical system in Eq. (4.1), there is a point* $\mathbf{x}^{(0)}(t_m) \equiv \mathbf{x}_m \in \partial\Omega_{ij}$ *at time* t_m *between two adjacent domains* Ω_α ($\alpha = i, j$)*. For* $\mathbf{x}_m \in S \subseteq \partial\Omega_{ij}$, *there is a leaving flow barrier* $\mathbf{F}^{(0\succ\alpha)}(\mathbf{x}^{(\alpha)}, t, \boldsymbol{\pi}_\alpha, q^{(\alpha)})$ *for* $q^{(\alpha)} \in [q_1^{(\alpha)}, q_2^{(\alpha)}]$ *in the source flow on the* α*-side of the boundary* $\partial\Omega_{ij}$ *with*

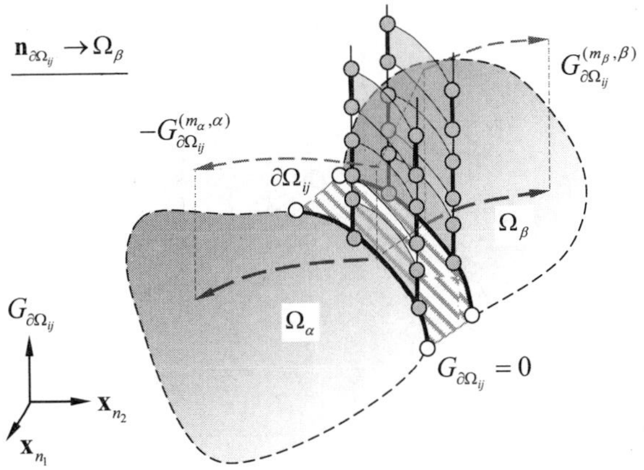

(a)

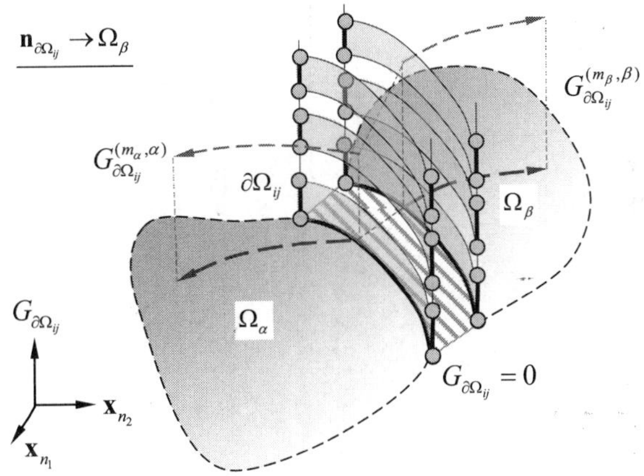

(b)

Fig. 4.25 The leaving flow barrier windows in the $(m_\alpha : m_\beta)$-source flow on $\partial\Omega_{ij}$: **(a)** partial flow barrier and **(b)** full flow barrier. The upper curves with arrows are the G-functions relative to the flow barrier. The two vertical surfaces are the flow barrier surfaces. The hatched area is for the zoomed boundary. The lower curves with arrows are coming flows. ($m_\alpha, m_\beta \in \{0,1,2,\cdots\}$).

$$\hbar_\alpha G^{(0\succ\alpha)}_{\partial\Omega_{ij}}(\mathbf{x}_m, q^{(\alpha)}) \in \left[\hbar_\alpha G^{(0\succ\alpha)}_{\partial\Omega_{ij}}(\mathbf{x}_m, q_2^{(\alpha)}), \hbar_\alpha G^{(0\succ\alpha)}_{\partial\Omega_{ij}}(\mathbf{x}_m, q_1^{(\alpha)})\right] \subset (-\infty, 0]. \tag{4.169}$$

The two possible leaving flows in the source flow satisfy

$$\hbar_\alpha G^{(\alpha)}_{\partial\Omega_{ij}}(\mathbf{x}_m, t_{m+}) < 0 \;\; and \;\; \hbar_\alpha G^{(\beta)}_{\partial\Omega_{ij}}(\mathbf{x}_m, t_{m+}) > 0 \tag{4.170}$$

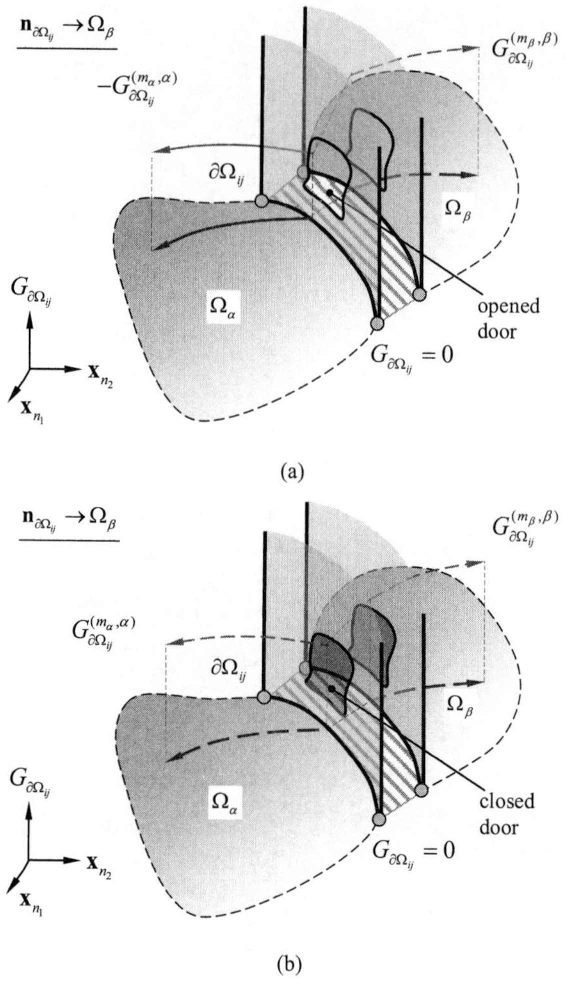

Fig. 4.26 The door of the leaving flow barrier in the absolute $(m_\alpha : m_\beta)$ -source flow on $\partial\Omega_{ij}$: **(a)** opened door **(b)** closed door. The upper curves with arrows are the G-functions relative to the flow barrier. The two vertical surfaces are the flow barrier surfaces. The hatched area is for the zoomed boundary. The lower curves with arrows are coming flows. ($m_\alpha, m_\beta \in \{0,1,2,\cdots\}$).

(i) *A leaving flow of* $\mathbf{x}^{(\alpha)}$ *in the domain* Ω_α *cannot leave the boundary* $\partial\Omega_{ij}$ *if and only if*

$$\hbar_\alpha G_{\partial\Omega_{ij}}^{(\alpha)}(\mathbf{x}_m, t_{m+}) \in \left(\hbar_\alpha G_{\partial\Omega_{ij}}^{(0\succ\alpha)}(\mathbf{x}_m, q_2), \hbar_\alpha G_{\partial\Omega_{ij}}^{(0\succ\alpha)}(\mathbf{x}_m, q_1)\right). \tag{4.171}$$

(ii) *A leaving flow of* $\mathbf{x}^{(\alpha)}$ *in the domain* Ω_α *cannot leave the boundary* $\partial\Omega_{ij}$ *at the critical points of* $q_\sigma^{(\alpha)}$ ($\sigma \in \{1,2\}$) *if and only if*

$$\begin{aligned}&G_{\partial\Omega_{ij}}^{(s_\alpha,\alpha)}(\mathbf{x}_m,t_{m+})=G_{\partial\Omega_{ij}}^{(s_\alpha,0\succ\alpha)}(\mathbf{x}_m,t_{m+},q_\sigma^{(\alpha)})\neq 0,\ s_\alpha=0,1,\cdots,l_\alpha-1;\\&(-1)^\sigma\hbar_\alpha[G_{\partial\Omega_{ij}}^{(l_\alpha,\alpha)}(\mathbf{x}_m,t_{m+})-G_{\partial\Omega_{ij}}^{(l_\alpha,0\succ\alpha)}(\mathbf{x}_m,t_{m+},q_\sigma^{(\alpha)})]>0.\end{aligned}\tag{4.172}$$

(iii) *A leaving flow of* $\mathbf{x}^{(\alpha)}$ *in the domain leaves the boundary at the critical points of* $q_\sigma^{(\alpha)}$ ($\sigma\in\{1,2\}$) *if and only if*

$$\begin{aligned}&G_{\partial\Omega_{ij}}^{(s_\alpha,\alpha)}(\mathbf{x}_m,t_{m+})=G_{\partial\Omega_{ij}}^{(s_\alpha,0\succ\alpha)}(\mathbf{x}_m,t_{m+},q_\sigma^{(\alpha)})\neq 0,\ s_\alpha=0,1,\cdots,l_\alpha-1;\\&(-1)^\sigma\hbar_\alpha[G_{\partial\Omega_{ij}}^{(l_\alpha,\alpha)}(\mathbf{x}_m,t_{m+})-G_{\partial\Omega_{ij}}^{(l_\alpha,0\succ\alpha)}(\mathbf{x}_m,t_{m+},q_\sigma^{(\alpha)})]<0.\end{aligned}\tag{4.173}$$

Proof. The proof of this theorem is similar to Theorem 4.1. ■

Theorem 4.12. *For a discontinuous dynamical system in Eq.* (4.1)*, there is a point* $\mathbf{x}^{(0)}(t_m)\equiv\mathbf{x}_m\in\partial\Omega_{ij}$ *at time* t_m *between two adjacent domains* Ω_α ($\alpha=i,j$). *Suppose a leaving barrier* $\mathbf{F}^{(0\succ\alpha)}(\mathbf{x}^{(\alpha)},t,\boldsymbol{\pi}_\alpha,q^{(\alpha)})$ *in the* $(m_\alpha:m_\beta)$*-source flow for* $q^{(\alpha)}\in[q_1^{(\alpha)},q_2^{(\alpha)}]$ *exists on the* α*-side of the boundary* $\partial\Omega_{ij}$ *with*

$$G_{\partial\Omega_{ij}}^{(s_\alpha,0\succ\alpha)}(\mathbf{x}_m,q^{(\alpha)})=0\ \ for\ s_\alpha=0,1,\cdots,m_\alpha-1;\tag{4.174}$$

$$\begin{aligned}\hbar_\alpha G_{\partial\Omega_{ij}}^{(m_\alpha,0\succ\alpha)}(\mathbf{x}_m,q^{(\alpha)})&\in\left[\hbar_\alpha G_{\partial\Omega_{ij}}^{(m_\alpha,0\succ\alpha)}(\mathbf{x}_m,q_2^{(\alpha)}),\hbar_\alpha G_{\partial\Omega_{ij}}^{(m_\alpha,0\succ\alpha)}(\mathbf{x}_m,q_1^{(\alpha)})\right]\\&\subset(-\infty,0].\end{aligned}\tag{4.175}$$

The leaving flows in the $(m_\alpha:m_\beta)$*-source flow satisfy*

$$\begin{aligned}&G_{\partial\Omega_{ij}}^{(s_\alpha,\alpha)}(\mathbf{x}_m,t_{m+},\mathbf{p}_\alpha,\boldsymbol{\lambda})=0\ \ for\ s_\alpha=0,1,\cdots,m_\alpha-1;\\&G_{\partial\Omega_{ij}}^{(s_\beta,\beta)}(\mathbf{x}_m,t_{m+},\mathbf{p}_\beta,\boldsymbol{\lambda})=0\ \ for\ s_\beta=0,1,\cdots,m_\beta-1;\\&\hbar_\alpha G_{\partial\Omega_{ij}}^{(m_\alpha,\alpha)}(\mathbf{x}_m,t_{m+})<0\ and\ \hbar_\alpha G_{\partial\Omega_{ij}}^{(m_\beta,\beta)}(\mathbf{x}_m,t_{m+})>0.\end{aligned}\tag{4.176}$$

(i) *The leaving flow of* $\mathbf{x}^{(\alpha)}$ *in the domain* Ω_α *cannot leave the boundary* $\partial\Omega_{ij}$ *at the critical points of* $q_\sigma^{(\alpha)}$ ($\sigma\in\{1,2\}$) *if and only if*

$$\hbar_\alpha G_{\partial\Omega_{ij}}^{(m_\alpha,\alpha)}(\mathbf{x}_m,t_{m+})\in\left(\hbar_\alpha G_{\partial\Omega_{ij}}^{(m_\alpha,0\succ\alpha)}(\mathbf{x}_m,q_2^{(\alpha)}),\hbar_\alpha G_{\partial\Omega_{ij}}^{(m_\alpha,0\succ\alpha)}(\mathbf{x}_m,q_1^{(\alpha)})\right).\tag{4.177}$$

(ii) *The leaving flow of* $\mathbf{x}^{(\alpha)}$ *in the domain* Ω_α *cannot leave the boundary* $\partial\Omega_{ij}$ *if and only if*

$$\begin{aligned}&G_{\partial\Omega_{ij}}^{(s_\alpha,\alpha)}(\mathbf{x}_m,t_{m+})=G_{\partial\Omega_{ij}}^{(s_\alpha,0\succ\alpha)}(\mathbf{x}_m,q_\sigma^{(\alpha)})\neq 0\\&for\ s_\alpha=m_\alpha,m_\alpha+1,\cdots,l_\alpha-1;\\&(-1)^\sigma\hbar_\alpha[G_{\partial\Omega_{ij}}^{(l_\alpha,\alpha)}(\mathbf{x}_m,t_{m+})-G_{\partial\Omega_{ij}}^{(l_\alpha,0\succ\alpha)}(\mathbf{x}_m,q_\sigma^{(\alpha)})]>0.\end{aligned}\tag{4.178}$$

(iii) *The leaving flow of* $\mathbf{x}^{(\alpha)}$ *in the domain* Ω_α *leaves the boundary* $\partial\Omega_{ij}$ *at the critical points of* $q_\sigma^{(\alpha)}$ ($\sigma \in \{1,2\}$) *if and only if*

$$\begin{aligned} &G_{\partial\Omega_{ij}}^{(s_\alpha,\alpha)}(\mathbf{x}_m, t_{m+}) = G_{\partial\Omega_{ij}}^{(s_\alpha,0\succ\alpha)}(\mathbf{x}_m, q_\sigma^{(\alpha)}) \neq 0 \\ &\mathit{for}\ s_\alpha = m_\alpha, m_\alpha+1,\cdots,l_\alpha-1; \\ &(-1)^\sigma \hbar_\alpha [G_{\partial\Omega_{ij}}^{(l_\alpha,\alpha)}(\mathbf{x}_m, t_{m+}) - G_{\partial\Omega_{ij}}^{(l_\alpha,0\succ\alpha)}(\mathbf{x}_m, q_\sigma^{(\alpha)})] < 0. \end{aligned} \tag{4.179}$$

Proof. The proof of this theorem is similar to Theorem 4.2. ■

4.4. Sink flows with flow barriers

Without flow barriers, the sink flow will be formed under conditions in Eq. (4.96). If the coming flow barrier in the sink flow exists, it will block the coming flow to form the sink flow, and the corresponding flow switchability was discussed in Section 4.2. Once the sink flow is formed on the boundary, the boundary flow can be controlled by Eq. (4.124) if Eq. (4.96) is satisfied. If the boundary flow disappears from the boundary in the source flow, the conditions in Eq. (4.123) should be satisfied. However, if the boundary flow barriers exist, the boundary flow in the source flow is independent of the conditions in Eq. (4.123). The boundary flow barrier will control the existence of the boundary flow. The existence of the sink with flow barriers will be further discussed.

Theorem 4.13. *For a discontinuous dynamical system in Eq.* (4.1), *there is a point* $\mathbf{x}^{(0)}(t_m) \equiv \mathbf{x}_m \in \partial\Omega_{ij}$ *at time* t_m *between two adjacent domains* Ω_α ($\alpha = i, j$). *For* $\mathbf{x}_m \in S \subseteq \partial\Omega_{ij}$, *there is a flow barrier* $\mathbf{F}^{(\alpha\succ 0)}(\mathbf{x}^{(\alpha)}, t, \boldsymbol{\pi}_\alpha, q^{(\alpha)})$ *on the* α*-side of the boundary* $\partial\Omega_{ij}$ *at* $q^{(\alpha)} \in [q_1^{(\alpha)}, q_2^{(\alpha)}]$ *with the G-function*

$$\begin{aligned} \hbar_\alpha G_{\partial\Omega_{ij}}^{(\alpha\succ 0)}(\mathbf{x}_m, q^{(\alpha)}) &\in \left[\hbar_\alpha G_{\partial\Omega_{ij}}^{(\alpha\succ 0)}(\mathbf{x}_m, q_1^{(\alpha)}), \hbar_\alpha G_{\partial\Omega_{ij}}^{(\alpha\succ 0)}(\mathbf{x}_m, q_2^{(\alpha)})\right] \\ &\subset [0, +\infty); \end{aligned} \tag{4.180}$$

and also there is a flow barrier $\mathbf{F}^{(\beta\succ 0)}(\mathbf{x}^{(\beta)}, t, \boldsymbol{\pi}_\alpha, q^{(\beta)})$ *at* $q^{(\beta)} \in [q_1^{(\beta)}, q_2^{(\beta)}]$ *on the* β*-side of the boundary* $\partial\Omega_{ij}$ *with the G-function*

$$\begin{aligned} \hbar_\alpha G_{\partial\Omega_{ij}}^{(\beta\succ 0)}(\mathbf{x}_m, q^{(\beta)}) &\in \left[\hbar_\alpha G_{\partial\Omega_{ij}}^{(\beta\succ 0)}(\mathbf{x}_m, q_2^{(\beta)}), \hbar_\alpha G_{\partial\Omega_{ij}}^{(\beta\succ 0)}(\mathbf{x}_m, q_1^{(\beta)})\right] \\ &\subset (-\infty, 0]. \end{aligned} \tag{4.181}$$

The two possible coming flows in the sink flow satisfy

$$\hbar_\alpha G^{(\alpha)}_{\partial\Omega_{ij}}(\mathbf{x}_m, t_{m-}) > 0 \textit{ and } \hbar_\alpha G^{(\beta)}_{\partial\Omega_{ij}}(\mathbf{x}_m, t_{m-}) < 0. \tag{4.182}$$

A coming flow of $\mathbf{x}^{(\alpha)}$ *switches to the boundary flow of* $\mathbf{x}^{(0)}$ *to form a sink flow on the boundary* $\partial\Omega_{ij}$ *if and only if for* $\sigma_\alpha, \sigma_\beta \in \{1,2\}$

$$\left.\begin{array}{ll} \textit{either} & \hbar_\alpha G^{(\alpha)}_{\partial\Omega_{ij}}(\mathbf{x}_m, t_{m-}) \notin \left[\hbar_\alpha G^{(\alpha\succ 0)}_{\partial\Omega_{ij}}(\mathbf{x}_m, q_1^{(\alpha)}), \hbar_\alpha G^{(\alpha\succ 0)}_{\partial\Omega_{ij}}(\mathbf{x}_m, q_2^{(\alpha)})\right], \\ \textit{or} & G^{(s_\alpha,\alpha)}_{\partial\Omega_{ij}}(\mathbf{x}_m, t_{m-}) = G^{(s_\alpha,\alpha\succ 0)}_{\partial\Omega_{ij}}(\mathbf{x}_m, q_{\sigma_\alpha}^{(\alpha)}) \neq 0 \text{ for } s_\alpha = 0,1,\cdots,l_\alpha - 1 \\ & \textit{with } (-1)^{\sigma_\alpha}\left[\hbar_\alpha G^{(l_\alpha,\alpha)}_{\partial\Omega_{ij}}(\mathbf{x}_m, t_{m-}) - \hbar_\alpha G^{(l_\alpha,\alpha\succ 0)}_{\partial\Omega_{ij}}(\mathbf{x}_m, q_{\sigma_\alpha}^{(\alpha)})\right] > 0; \end{array}\right\} \tag{4.183}$$

$$\left.\begin{array}{ll} \textit{either} & \hbar_\alpha G^{(\beta)}_{\partial\Omega_{ij}}(\mathbf{x}_m, t_{m-}) \notin \left[\hbar_\alpha G^{(\beta\succ 0)}_{\partial\Omega_{ij}}(\mathbf{x}_m, q_2^{(\beta)}), \hbar_\alpha G^{(\beta\succ 0)}_{\partial\Omega_{ij}}(\mathbf{x}_m, q_1^{(\beta)})\right], \\ \textit{or} & G^{(s_\beta,\beta)}_{\partial\Omega_{ij}}(\mathbf{x}_m, t_{m-}) = G^{(s_\beta,\beta\succ 0)}_{\partial\Omega_{ij}}(\mathbf{x}_m, q_{\sigma_\beta}^{(\beta)}) \neq 0 \text{ for } s_\alpha = 0,1,\cdots,l_\beta - 1 \\ & \textit{with } (-1)^{\sigma_\beta}\left[\hbar_\alpha G^{(l_\beta,\beta)}_{\partial\Omega_{ij}}(\mathbf{x}_m, t_{m-}) - \hbar_\alpha G^{(l_\beta,\beta\succ 0)}_{\partial\Omega_{ij}}(\mathbf{x}_m, q_{\sigma_\beta}^{(\beta)})\right] < 0. \end{array}\right\} \tag{4.184}$$

Proof. The proof of theorem is completed from the definition. ■

Theorem 4.14. *For a discontinuous dynamical system in Eq.* (4.1), *there is a point* $\mathbf{x}^{(0)}(t_m) \equiv \mathbf{x}_m \in \partial\Omega_{ij}$ *at time* t_m *between two adjacent domains* Ω_α $(\alpha = i, j)$. *For* $\mathbf{x}_m \in S \subseteq \partial\Omega_{ij}$, *there is a flow barrier* $\mathbf{F}^{(\alpha\succ 0)}(\mathbf{x}^{(\alpha)}, t, \boldsymbol{\pi}_\alpha, q^{(\alpha)})$ *on the* α*-side of the boundary* $\partial\Omega_{ij}$ *at* $q^{(\alpha)} \in [q_1^{(\alpha)}, q_2^{(\alpha)}]$ *with the G-function*

$$\begin{aligned} & G^{(s_\alpha,\alpha\succ 0)}_{\partial\Omega_{ij}}(\mathbf{x}_m, q_\sigma^{(\alpha)}) = 0 \text{ for } s_\alpha = 0,1,2,\cdots,2k_\alpha - 1; \\ & \hbar_\alpha G^{(2k_\alpha,\alpha\succ 0)}_{\partial\Omega_{ij}}(\mathbf{x}_m, q^{(\alpha)}) \in \left[\hbar_\alpha G^{(2k_\alpha,\alpha\succ 0)}_{\partial\Omega_{ij}}(\mathbf{x}_m, q_1^{(\alpha)}), \hbar_\alpha G^{(2k_\alpha,\alpha\succ 0)}_{\partial\Omega_{ij}}(\mathbf{x}_m, q_2^{(\alpha)})\right] \\ & \qquad \subset [0, +\infty); \end{aligned} \tag{4.185}$$

and also there is a flow barrier $\mathbf{F}^{(\beta\succ 0)}(\mathbf{x}^{(\beta)}, t, \boldsymbol{\pi}_\alpha, q^{(\beta)})$ *at* $q^{(\beta)} \in [q_1^{(\beta)}, q_2^{(\beta)}]$ *on the* β*-side of the boundary* $\partial\Omega_{ij}$ *with the G-function*

$$\begin{aligned} & G^{(s_\beta,\beta\succ 0)}_{\partial\Omega_{ij}}(\mathbf{x}_m, q_\sigma^{(\beta)}) = 0 \textit{ for } s_\beta = 0,1,2,\cdots,2k_\beta - 1; \\ & \hbar_\alpha G^{(2k_\beta,\beta\succ 0)}_{\partial\Omega_{ij}}(\mathbf{x}_m, q^{(\beta)}) \in \left[\hbar_\alpha G^{(2k_\beta,\beta\succ 0)}_{\partial\Omega_{ij}}(\mathbf{x}_m, q_2^{(\beta)}), \hbar_\alpha G^{(2k_\beta,\beta\succ 0)}_{\partial\Omega_{ij}}(\mathbf{x}_m, q_1^{(\beta)})\right] \\ & \qquad \subset (-\infty, 0]. \end{aligned} \tag{4.186}$$

The coming flows in the $(2k_\alpha : 2k_\beta)$*-sink flow satisfy*

$$\begin{aligned} & G^{(s_\alpha,\alpha)}_{\partial\Omega_{ij}}(\mathbf{x}_m, t_{m-}) = 0 \textit{ for } s_\alpha = 0,1,2,\cdots,2k_\alpha - 1; \\ & G^{(s_\beta,\beta)}_{\partial\Omega_{ij}}(\mathbf{x}_m, t_{m-}) = 0 \textit{ for } s_\beta = 0,1,2,\cdots,2k_\beta - 1; \end{aligned}$$

$$\hbar_\alpha G^{(2k_\alpha,\alpha)}_{\partial\Omega_{ij}}(\mathbf{x}_m,t_{m-})>0 \text{ and } \hbar_\alpha G^{(2k_\beta,\beta)}_{\partial\Omega_{ij}}(\mathbf{x}_m,t_{m-})<0. \tag{4.187}$$

A coming flow of $\mathbf{x}^{(\alpha)}$ *switches to the boundary flow of* $\mathbf{x}^{(0)}$ *to form a* $(2k_\alpha:2k_\beta)$*-sink flow on the boundary* $\partial\Omega_{ij}$ *if and only if for* $\sigma_\alpha,\sigma_\beta\in\{1,2\}$

$$\left.\begin{aligned}
&\text{either } \hbar_\alpha G^{(2k_\alpha,\alpha)}_{\partial\Omega_{ij}}(\mathbf{x}_m,t_{m-})\\
&\qquad\notin\left[\hbar_\alpha G^{(2k_\alpha,\alpha\succ 0)}_{\partial\Omega_{ij}}(\mathbf{x}_m,q_1^{(\alpha)}),\hbar_\alpha G^{(2k_\alpha,\alpha\succ 0)}_{\partial\Omega_{ij}}(\mathbf{x}_m,q_2^{(\alpha)})\right];\\
&\text{or}\quad G^{(s_\alpha,\alpha)}_{\partial\Omega_{ij}}(\mathbf{x}_m,t_{m-})=G^{(s_\alpha,\alpha\succ 0)}_{\partial\Omega_{ij}}(\mathbf{x}_m,q_{\sigma_\alpha}^{(\alpha)})\neq 0\\
&\qquad\text{for } s_\alpha=2k_\alpha,2k_\alpha+1,\cdots,l_\alpha-1;\\
&\qquad(-1)^{\sigma_\alpha}\left[\hbar_\alpha G^{(l_\alpha,\alpha)}_{\partial\Omega_{ij}}(\mathbf{x}_m,t_{m-})-\hbar_\alpha G^{(l_\alpha,\alpha\succ 0)}_{\partial\Omega_{ij}}(\mathbf{x}_m,q_{\sigma_\alpha}^{(\alpha)})\right]>0.
\end{aligned}\right\} \tag{4.188}$$

$$\left.\begin{aligned}
&\text{either } \hbar_\alpha G^{(2k_\beta,\beta)}_{\partial\Omega_{ij}}(\mathbf{x}_m,t_{m-})\\
&\qquad\notin\left[\hbar_\alpha G^{(2k_\beta,\beta\succ 0)}_{\partial\Omega_{ij}}(\mathbf{x}_m,q_2^{(\beta)}),\hbar_\alpha G^{(2k_\beta,\beta\succ 0)}_{\partial\Omega_{ij}}(\mathbf{x}_m,q_1^{(\beta)})\right];\\
&\text{or}\quad G^{(s_\beta,\beta)}_{\partial\Omega_{ij}}(\mathbf{x}_m,t_{m-})=G^{(s_\beta,\beta\succ 0)}_{\partial\Omega_{ij}}(\mathbf{x}_m,q_{\sigma_\beta}^{(\beta)})\neq 0\\
&\qquad\text{for } s_\beta=2k_\beta,2k_\beta+1,\cdots,l_\beta-1;\\
&\qquad(-1)^{\sigma_\beta}\left[\hbar_\alpha G^{(l_\beta,\beta)}_{\partial\Omega_{ij}}(\mathbf{x}_m,t_{m-})-\hbar_\alpha G^{(l_\beta,\beta\succ 0)}_{\partial\Omega_{ij}}(\mathbf{x}_m,q_{\sigma_\beta}^{(\beta)})\right]<0.
\end{aligned}\right\} \tag{4.189}$$

Proof. The proof of theorem is completed from the definition of higher-order singular sink flow barriers. ■

Once the sink flow is formed, the vanishing of the boundary flow in the source flow on the boundary is of great interest herein. Without the flow barriers, the vanishing conditions of the boundary flow in the sink flow were presented in Chapters 2 and 3. The following will give the vanishing condition for the boundary flow from the boundary with flow barriers.

Theorem 4.15. *For a discontinuous dynamical system in Eq.* (4.1)*, there is a point* $\mathbf{x}^{(0)}(t_m)\equiv\mathbf{x}_m\in\partial\Omega_{ij}$ *at time* t_m *between two adjacent domains* Ω_α $(\alpha=i,j)$*. Suppose the boundary flow in the sink flow on the boundary is formed under certain conditions. There is a boundary flow barrier of* $\mathbf{F}^{(0\succ 0_\alpha)}(\mathbf{x}^{(\alpha)},t,\boldsymbol{\pi}_\alpha,q^{(\alpha)})$ *at* $q^{(\alpha)}\in[q_1^{(\alpha)},q_2^{(\alpha)}]$ *on the* α*-side of the boundary* $\partial\Omega_{ij}$ *for* $\mathbf{x}_m\in S\subseteq\partial\Omega_{ij}$*, with the G-function*

$$0\in\left[\hbar_\alpha G^{(0\succ 0_\alpha)}_{\partial\Omega_{ij}}(\mathbf{x}_m,q_2^{(\alpha)}),\hbar_\alpha G^{(0\succ 0_\alpha)}_{\partial\Omega_{ij}}(\mathbf{x}_m,q_1^{(\alpha)})\right]\subset\mathscr{R} \tag{4.190}$$

and there is a boundary flow barrier of $\mathbf{F}^{(0\succ 0_\beta)}(\mathbf{x}^{(\beta)},t,\boldsymbol{\pi}_\beta,q^{(\beta)})$ *at* $q^{(\beta)}\in[q_1^{(\beta)},q_2^{(\beta)}]$ *with G-function*

$$0\in\left[\hbar_\alpha G_{\partial\Omega_{ij}}^{(0\succ 0_\beta)}(\mathbf{x}_m,q_1^{(\beta)}),\hbar_\alpha G_{\partial\Omega_{ij}}^{(0\succ 0_\beta)}(\mathbf{x}_m,q_2^{(\beta)})\right]\subset\mathscr{R} \tag{4.191}$$

on the β*-side of boundary* $\partial\Omega_{ij}$ ($\alpha,\beta\in\{i,j\}$ *and* $\alpha\neq\beta$). *The boundary flow of* $\mathbf{x}^{(0)}$ *in the sink flow disappears on the* α*-side of* $\partial\Omega_{ij}$ *if and only if*

$$\left.\begin{aligned}&\text{both } \hbar_\alpha G_{\partial\Omega_{ij}}^{(\alpha)}(\mathbf{x}_m,t_{m+})<0,\\&\text{and } \left.\begin{aligned}&\hbar_\alpha G_{\partial\Omega_{ij}}^{(s_\alpha,0\succ 0_\alpha)}(\mathbf{x}_m,q_1^{(\alpha)})=0 \text{ for } s_\alpha=0,1,2,\cdots,l_\alpha-1,\\&\hbar_\alpha G_{\partial\Omega_{ij}}^{(l_\alpha,0\succ 0_\alpha)}(\mathbf{x}_m,q_1^{(\alpha)})<0,\end{aligned}\right\}\end{aligned}\right. \tag{4.192}$$

on the α*-side;*

$$\left.\begin{aligned}&\text{either } \hbar_\alpha G_{\partial\Omega_{ij}}^{(0\succ 0_\beta)}(\mathbf{x}_m,q_1^{(\beta)})>0 \text{ but } \hbar_\alpha G_{\partial\Omega_{ij}}^{(\beta)}(\mathbf{x}_m,t_{m+})<0,\\&\text{or } \quad \hbar_\alpha G_{\partial\Omega_{ij}}^{(0\succ 0_\beta)}(\mathbf{x}_m,q_1^{(\beta)})<0 \text{ or } \hbar_\alpha G_{\partial\Omega_{ij}}^{(s_\beta,0\succ 0_\beta)}(\mathbf{x}_m,q_1^{(\beta)})=0,\\&\quad\text{for } s_\beta=0,1,2,\cdots,l_\beta-1 \text{ and } \hbar_\alpha G_{\partial\Omega_{ij}}^{(l_\beta,0\succ 0_\beta)}(\mathbf{x}_m,q_1^{(\beta)})<0,\end{aligned}\right\} \tag{4.193}$$

on the β*-side.*

Proof. This theorem can be proved from the boundary flow barrier in the *source* flow. ■

Theorem 4.16. *For a discontinuous dynamical system in Eq.* (4.1)*, there is a point* $\mathbf{x}^{(0)}(t_m)\equiv\mathbf{x}_m\in\partial\Omega_{ij}$ *at time* t_m *between two adjacent domains* Ω_α ($\alpha=i,j$). *Suppose the boundary flow in the* $(2k_\alpha:2k_\beta)$*-sink flow on the boundary is formed under certain conditions. There is a boundary flow barrier* $\mathbf{F}^{(0\succ 0_\alpha)}(\mathbf{x}^{(\alpha)},t,\boldsymbol{\pi}_\alpha,q^{(\alpha)})$ *at* $q^{(\alpha)}\in[q_1^{(\alpha)},q_2^{(\alpha)}]$ *with G-function*

$$\left.\begin{aligned}&G_{\partial\Omega_{ij}}^{(s_\alpha,0\succ 0_\alpha)}(\mathbf{x}_m,q_\sigma^{(\alpha)})=0 \text{ for } s_\alpha=0,1,2,\cdots,m_\alpha-1,\\&0\in\left[\hbar_\alpha G_{\partial\Omega_{ij}}^{(m_\alpha,0\succ 0_\alpha)}(\mathbf{x}_m,q_2^{(\alpha)}),\hbar_\alpha G_{\partial\Omega_{ij}}^{(m_\alpha,0\succ 0_\alpha)}(\mathbf{x}_m,q_1^{(\alpha)})\right]\subset\mathscr{R}\end{aligned}\right. \tag{4.194}$$

for the leaving flow of $\mathbf{x}^{(\alpha)}$ *on the* α*-side and also there is a flow barrier* $\mathbf{F}^{(0\succ 0_\beta)}(\mathbf{x}^{(\beta)},t,\boldsymbol{\pi}_\beta,q^{(\beta)})$ *at* $q^{(\beta)}\in[q_1^{(\beta)},q_2^{(\beta)}]$ *with G-function*

$$G_{\partial\Omega_{ij}}^{(s_\beta,0\succ 0_\beta)}(\mathbf{x}_m,q_\sigma^{(\beta)})=0 \text{ for } s_\beta=0,1,2,\cdots,m_\beta-1,$$

$$0\in\left[\hbar_\alpha G_{\partial\Omega_{ij}}^{(m_\beta,0\succ 0_\beta)}(\mathbf{x}_m,q_1^{(\beta)}),\hbar_\alpha G_{\partial\Omega_{ij}}^{(m_\beta,0\succ 0_\beta)}(\mathbf{x}_m,q_2^{(\beta)})\right]\subset\mathscr{R} \tag{4.195}$$

for the leaving flow of $\mathbf{x}^{(\beta)}$ *on the* β*-side* ($\alpha, \beta \in \{i, j\}$ *and* $\alpha \neq \beta$).

$$\left.\begin{aligned} &G_{\partial\Omega_{ij}}^{(s_\alpha,\alpha)}(\mathbf{x}_m, t_{m+}) = 0 \;\; \text{for } s_\alpha = 0,1,2,\cdots,m_\alpha - 1; \\ &G_{\partial\Omega_{ij}}^{(s_\beta,\beta)}(\mathbf{x}_m, t_{m+}) = 0 \;\; \text{for } s_\beta = 0,1,2,\cdots,m_\beta - 1. \end{aligned}\right. \tag{4.196}$$

The boundary flow of $\mathbf{x}^{(0)}$ *disappears on the* α*-side if and only if*

$$\begin{aligned} &\text{both } \hbar_\alpha G_{\partial\Omega_{ij}}^{(m_\alpha,\alpha)}(\mathbf{x}_m, t_{m+}) < 0, \\ &\text{and } \left.\begin{aligned} &\hbar_\alpha G_{\partial\Omega_{ij}}^{(s_\alpha,0\succ 0_\alpha)}(\mathbf{x}_m, q_1^{(\alpha)}) = 0 \text{ for } s_\alpha = m_\alpha, m_\alpha + 1, \cdots l_\alpha - 1, \\ &\hbar_\alpha G_{\partial\Omega_{ij}}^{(l_\alpha,0\succ 0_\alpha)}(\mathbf{x}_m, q_1^{(\alpha)}) < 0, \end{aligned}\right\} \\ &\text{on the } \alpha\text{-side;} \end{aligned} \tag{4.197}$$

$$\begin{aligned} &\text{either } \hbar_\alpha G_{\partial\Omega_{ij}}^{(m_\beta,0\succ 0_\beta)}(\mathbf{x}_m, q_1^{(\beta)}) > 0 \text{ but } \hbar_\alpha G_{\partial\Omega_{ij}}^{(m_\beta,\beta)}(\mathbf{x}_m, t_{m+}) < 0, \\ &\text{or } \left.\begin{aligned} &\hbar_\alpha G_{\partial\Omega_{ij}}^{(m_\beta,0\succ 0_\beta)}(\mathbf{x}_m, q_1^{(\beta)}) < 0 \;\text{ or } \hbar_\alpha G_{\partial\Omega_{ij}}^{(s_\beta,0\succ 0_\beta)}(\mathbf{x}_m, q_1^{(\beta)}) = 0, \\ &\text{for } s_\beta = m_\beta, m_\beta + 1, \cdots l_\beta - 1;\; \hbar_\alpha G_{\partial\Omega_{ij}}^{(l_\beta,0\succ 0_\beta)}(\mathbf{x}_m, q_1^{(\beta)}) > 0, \end{aligned}\right\} \\ &\text{on the } \beta\text{-side.} \end{aligned} \tag{4.198}$$

Proof. This theorem can be proved from the Theorems for the boundary flow barriers in the source flows with higher-order singularity. ■

4.5. An application

As in Luo (2006, 2007), consider a periodically-forced, friction-induced oscillator in Fig. 4.27(a). The dynamic system consists of a mass m, a spring of stiffness k, and a damper of viscous damping coefficient r. The moving mass rests on the horizontal belt surface traveling with a constant speed V. The coordinate system (x,t) is absolute with displacement x and time t. The periodic excitation force $Q_0 \cos\Omega t$ is exerted on the mass, where Q_0 and Ω are the excitation strength and frequency, respectively. The nonlinear friction is approximated by a piecewise linear model, as shown in Fig. 4.27(b). The parameters (μ_s, μ_k and F_N) are static and kinetic friction coefficients and a normal force to the contact surface, respectively. The coefficients μ_j (j=1,2,3,4) are the slope for friction force with velocity. The nonlinear friction force is expressed by

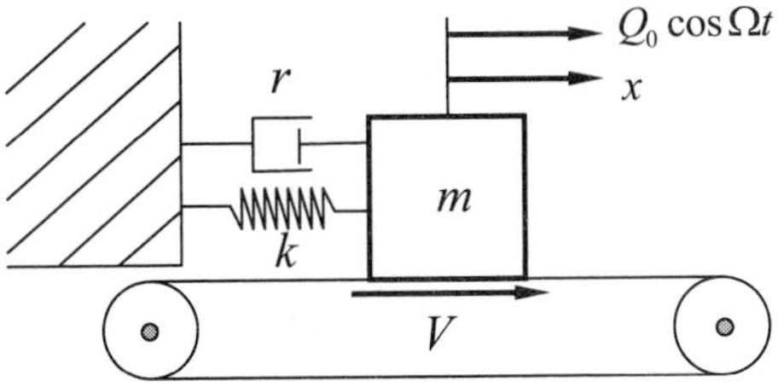

(a)

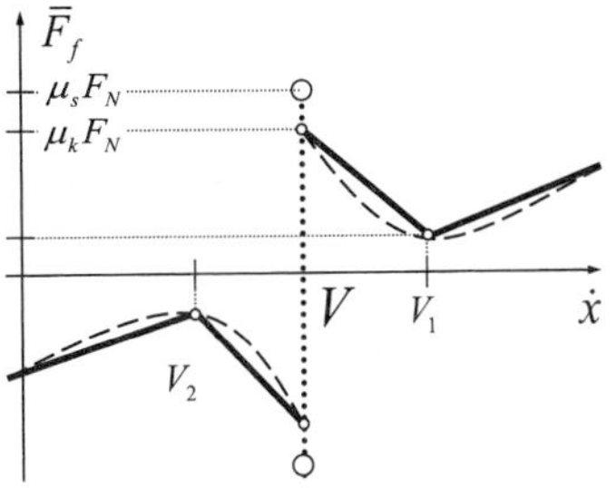

(b)

Fig. 4.27 (a) The friction-induced oscillator and **(b)** piecewise linear friction force model.

$$\bar{F}_f(\dot{x}) \begin{cases} = \mu_1(\dot{x}-V_1) - \mu_2(V_1 - V) + F_N\mu_k, & \dot{x} \in [V_1, \infty) \\ = -\mu_2(\dot{x}-V) + F_N\mu_k, & \dot{x} \in (V, V_1) \\ \in [-\mu_s F_N, \mu_s F_N], & \dot{x} = V \\ = -\mu_3(\dot{x}-V) - F_N\mu_k, & \dot{x} \in (V_2, V) \\ = \mu_4(\dot{x}-V_2) - \mu_3(V_2 - V) - F_N\mu_k, & \dot{x} \in (-\infty, V_2] \end{cases} \tag{4.199}$$

where $\dot{x} \triangleq dx/dt$.

For this problem, the normal force is $F_N = mg$ where g is the gravitational acceleration. The static friction force is in the interval of $[-\mu_s F_N, \mu_s F_N]$. The amplitude of the static friction force is $\mu_s F_N$. The dynamic friction forces just for the beginning of the relative motion are $\pm\mu_k F_N$. Two boundaries for the piecewise continuity of the friction force are at $\dot{x} = V_1$ and $\dot{x} = V_2$. The third boundary is at $\dot{x} = V$. For this boundary, the dynamical friction force for the passable motion is discontinuous, which can be referred to Luo (2006) (also see, Luo and Gegg, 2006a,b). Once the mass and the translation belt stick together, the relative motion does not exist between the mass and the belt. Only when the non-friction force is greater than the static friction force, the relative motion between the mass and belt can start. Three separation boundaries give four velocity regions on which the vector fields are different. In the aforementioned model, the normal force (F_N)

can be exerted externally and arbitrarily instead of $F_N = mg$. In addition, the non-friction forces per unit mass in the x-direction is determined by

$$F_s = A_0 \cos\Omega t - 2dV - cx, \quad \text{for } \dot{x} = V \tag{4.200}$$

where $A_0 = Q_0/m$ $d = r/2m$ and $c = k/m$. For stick motions, the non-friction force per unit mass is less than the static friction force amplitude per unit mass F_{f_s} (i.e., $|F_s| \le F_{f_s}$ and $F_{f_s} = \mu_s F_N/m$). The mass does not have any relative motion to the belt. Therefore, no acceleration exists because the belt speed is constant, i.e.,

$$\ddot{x} = 0, \qquad \text{for } \dot{x} = V. \tag{4.201}$$

If the non-friction force per unit mass is greater than the static friction force per unit mass (i.e., $|F_s| > F_{f_s}$), the non-stick motion occurs. For the non-stick motion, the total force per unit mass is

$$F = A_0 \cos\Omega t - F_f \operatorname{sgn}(\dot{x} - V) - 2d\dot{x} - cx, \quad \text{for } \dot{x} \ne V, \tag{4.202}$$

where the friction force per unit mass $F_f = \bar{F}_f/m$ for $\dot{x} \ne V$. Thus, equation of the non-stick motion for this dynamical system with a piecewise linear friction is

$$\ddot{x} + 2d\dot{x} + cx = A_0 \cos\Omega t - F_f \operatorname{sgn}(\dot{x} - V), \quad \text{for } \dot{x} \ne V. \tag{4.203}$$

4.5.1. Switchability conditions

For simplicity, the vectors for the flow and the corresponding vector field for such a system are introduced by

$$\mathbf{x} \triangleq (x, \dot{x})^{\mathrm{T}} \equiv (x, y)^{\mathrm{T}} \text{ and } \mathbf{F} \triangleq (y, F)^{\mathrm{T}}. \tag{4.204}$$

The discontinuities in this dynamical system are caused by the jumping from static to dynamic friction forces and piecewise linear dynamical friction model. As discussed before, there are four velocity regions caused by the three velocity boundaries. Therefore, the phase space can be partitioned into four sub-domains by the three velocity boundaries. Such a phase space partition is sketched in Fig. 4.28. Among three velocity boundaries, the friction force jumping is as a main discontinuity at $\dot{x} = V$. So the naming of the sub-domains in phase space starts from the domain near the main discontinuous boundary $\dot{x} = V$. In fact, the sub-domains can be named arbitrarily. From the direction of trajectories of mass motion in phase space, the corresponding boundaries are also named, as shown in Fig. 4.28(a). The boundary with the friction force jumping is represented by a dotted line. The rest boundaries are depicted by two dashed lines, respectively. The named domains and the oriented boundaries are expressed by

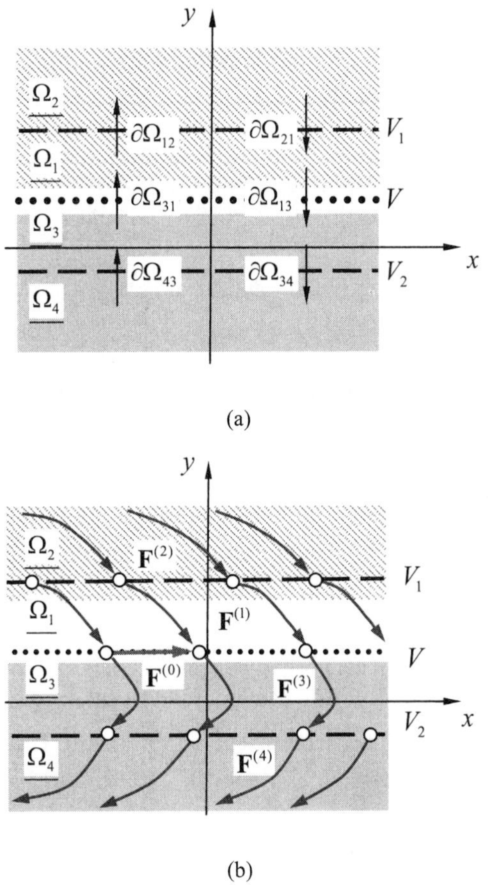

Fig. 4.28 **(a)** Phase plane partition and oriented boundaries **(b)** schematic vector fields in each domain.

$$\Omega_1 = \{(x,y) \mid y \in (V, V_1)\},\ \ \Omega_2 = \{(x,y) \mid y \in (V_1, \infty)\}, \qquad (4.205)$$
$$\Omega_3 = \{(x,y) \mid y \in (V_2, V)\},\ \ \Omega_4 = \{(x,y) \mid y \in (-\infty, V_2)\};$$

$$\partial\Omega_{\alpha\beta} = \{(x,y) \mid \varphi_{\alpha\beta}(x,y) \equiv y - V_\rho = 0\}, \qquad (4.206)$$

where $\rho = 1$ if $\alpha, \beta \in \{1,2\}$; $\rho = 0$ if $\alpha, \beta \in \{1,3\}$ and $\rho = 2$ if $\alpha, \beta \in \{3,4\}$. $V_0 \triangleq V$. The subscripts $\partial\Omega_{\alpha\beta}$ defines the boundary from Ω_α to Ω_β. The domains are accessible with a specific vector field. On the boundary $\partial\Omega_{13}$ or $\partial\Omega_{31}$, the vector fields are C^0-discontinuous, but on the boundaries $\partial\Omega_{12}$ and $\partial\Omega_{34}$, the vector fields are C^0-continuous. The vector fields for all the sub-domains are also sketched in Fig. 4.28(b) and labeled by $\mathbf{F}^{(\alpha)}$ for the sub-domain Ω_α ($\alpha = 1,\cdots,4$) and $\mathbf{F}^{(0)}$ is the sliding vector field on the boundary $\partial\Omega_{13}$.

Based on such a vector field in each sub-domain, the corresponding equations of motion in Eqs. (4.201) and (4.203) are rewritten as

$$\dot{\mathbf{x}} = \mathbf{F}^{(j)}(\mathbf{x},t), j \in \{0,\alpha\} \text{ and } \alpha \in \{1,2,3,4\} \tag{4.207}$$

where

$$\begin{aligned} &\mathbf{F}^{(0)}(\mathbf{x},t) = (V,0)^{\mathrm{T}} \quad \text{on } \partial\Omega_{13} \text{ or } \partial\Omega_{31}, \\ &\mathbf{F}^{(\alpha)}(\mathbf{x},t) = (y, F^{(\alpha)}(\mathbf{x},t))^{\mathrm{T}} \text{ in } \Omega_\alpha, \end{aligned} \tag{4.208}$$

$$F^{(\alpha)}(\mathbf{x},t) = A_0 \cos\Omega t - F_f^{(\alpha)}(\mathbf{x},t) - 2d_\alpha y - c_\alpha x. \tag{4.209}$$

From Eq. (4.199), the dynamical friction forces per unit mass can be expressed by

$$\begin{aligned} &F_f^{(2)}(\mathbf{x},t) = \nu_1(y - V_1) - \nu_2(V_1 - V) + F_N\nu_k, && y \in [V_1,\infty) \\ &F_f^{(1)}(\mathbf{x},t) = -\nu_2(y - V) + F_N\nu_k, && y \in (V,V_1) \\ &F_f^{(3)}(\mathbf{x},t) = -\nu_3(y - V) - F_N\nu_k, && y \in (V_2,V) \\ &F_f^{(4)}(\mathbf{x},t) = \nu_4(y - V_2) - \nu_3(V_2 - V) - F_N\nu_k, && y \in (-\infty,V_2] \end{aligned} \tag{4.210}$$

where $\nu_i = \mu_i / m$ ($i = 1,\cdots,4$) and $\nu_k = \mu_k / m$ are the slope coefficients of friction forces and dynamic friction coefficient per unit mass. The two force boundaries relative to $V_{1,2}$ (i.e., $y = V_1$ or V_2) are C^0-continuous. However, the boundary relative to the velocity V is a discontinuous force boundary. If the coming and leaving flow barriers does not exist, the coming and leaving flow vector fields for $\mathbf{x}_m \in \partial\Omega_{\alpha\beta}$ $(\alpha,\beta \in \{1,3\})$ are

$$\mathbf{F}^{(\alpha)}(\mathbf{x}_m,t_{m\pm}) = (y_m, F^{(\alpha)}(\mathbf{x}_m,t_{m\pm}))^{\mathrm{T}}. \tag{4.211}$$

The time t_m represents the moment for the motion just on the separation boundary and the time $t_{m\pm} = t_m \pm 0$ reflects the flows in the regions instead of the separation boundary. However, the boundary flow barriers on the boundary $\partial\Omega_{13}$ are for $\mathbf{x}_m \in \partial\Omega_{\alpha\beta}$ $(\alpha,\beta \in \{1,3\})$

$$\begin{aligned} &\mathbf{F}^{(0\succ 0_\alpha)}(\mathbf{x}_m,t_m,q^{(\alpha)}) = (y_m, F^{(0\succ 0_\alpha)}(\mathbf{x}_m,t_m,q^{(\alpha)}))^{\mathrm{T}}, \\ &F^{(0\succ 0_\alpha)}(\mathbf{x}_m,t_m,q^{(\alpha)}) = A_0 \cos\Omega t_m - F_{f_k}^{(\alpha)}(q^{(\alpha)}) - 2d_\alpha y_m - c_\alpha x_m. \end{aligned} \tag{4.212}$$

From Eq. (4.199), the static friction forces per unit mass on the boundary $\partial\Omega_{13}$ are

$$F_{f_s}^{(1)}(q^{(1)}) \in (-\infty, F_N\nu_s] \text{ and } F_{f_s}^{(3)}(q^{(3)}) \in [-F_N\nu_s, +\infty). \tag{4.213}$$

The boundary flow barrier on the α-side of the boundary $\partial\Omega_{13}$ are for

$\mathbf{x}_m \in \partial\Omega_{\alpha\beta}\quad (\alpha,\beta\in\{1,3\})$

$$
\begin{aligned}
&\mathbf{F}^{(0\succ 0_\alpha)}(\mathbf{x}_m,t_{m+},q_1^{(\alpha)}) = (y_m, F^{(0\succ 0_\alpha)}(\mathbf{x}_m,t_{m+},q_1^{(\alpha)}))^{\mathrm{T}},\\
&F^{(0\succ 0_\alpha)}(\mathbf{x}_m,t_{m+},q_1^{(\alpha)}) \equiv A_0\cos\Omega t_m - F_{f_s}^{(\alpha)}(q_1^{(\alpha)}) - 2d_\alpha y_m - c_\alpha x_m;\\
&\mathbf{F}^{(0\succ 0_1)}(\mathbf{x}_m,t_{m+},q_2^{(1)}) = (y_m,+\infty)^{\mathrm{T}},\ \mathbf{F}^{(0\succ 0_3)}(\mathbf{x}_m,t_{m+},q_2^{(3)}) = (y_m,-\infty)^{\mathrm{T}}.
\end{aligned}
\tag{4.214}
$$

Before discussion of the analytical conditions, the G-function can be reduced for the special boundary. The normal vector of the boundary $\partial\Omega_{\alpha\beta}$ is

$$
\mathbf{n}_{\partial\Omega_{\alpha\beta}} = \nabla\varphi_{\alpha\beta} = \left(\partial\varphi_{\alpha\beta}/\partial x, \partial\varphi_{\alpha\beta}/\partial y\right)^{\mathrm{T}}_{(x_m,y_m)},
\tag{4.215}
$$

where $\nabla = (\partial/\partial x, \partial/\partial y)^{\mathrm{T}}$ is the Hamilton operator. From Eq. (4.206), the boundaries are straight lines in phase space, which implies that the normal vectors are constant vectors. Furthermore, one obtains $D\mathbf{n}_{\partial\Omega_{\alpha\beta}} = 0$. Thus

$$
\begin{aligned}
&G_{\partial\Omega_{\alpha\beta}}^{(\alpha)}(\mathbf{x}^{(\alpha)},t) = \mathbf{n}_{\partial\Omega_{\alpha\beta}}^{\mathrm{T}}\cdot\mathbf{F}^{(\alpha)}(\mathbf{x}^{(\alpha)},t),\\
&G_{\partial\Omega_{\alpha\beta}}^{(1,\alpha)}(\mathbf{x}^{(\alpha)},t) = \mathbf{n}_{\partial\Omega_{\alpha\beta}}^{\mathrm{T}}\cdot D\mathbf{F}^{(\alpha)}(\mathbf{x}^{(\alpha)},t);\\
&G_{\partial\Omega_{\alpha\beta}}^{(0\succ 0_\alpha)}(\mathbf{x}^{(\alpha)},t) = \mathbf{n}_{\partial\Omega_{\alpha\beta}}^{\mathrm{T}}\cdot\mathbf{F}^{(0\succ 0_\alpha)}(\mathbf{x}^{(\alpha)},t),\\
&G_{\partial\Omega_{\alpha\beta}}^{(1,0\succ 0_\alpha)}(\mathbf{x}^{(\alpha)},t) = \mathbf{n}_{\partial\Omega_{\alpha\beta}}^{\mathrm{T}}\cdot D\mathbf{F}^{(0\succ 0_\alpha)}(\mathbf{x}^{(\alpha)},t)
\end{aligned}
\tag{4.216}
$$

where

$$
\begin{aligned}
&D\mathbf{F}^{(\alpha)}(\mathbf{x},t) = (F^{(\alpha)}(\mathbf{x},t), DF^{(\alpha)}(\mathbf{x},t))^{\mathrm{T}},\\
&DF^{(\alpha)}(\mathbf{x},t) \equiv \nabla F^{(\alpha)}(\mathbf{x},t)\cdot\mathbf{F}^{(\alpha)}(\mathbf{x},t) + \partial_t F^{(\alpha)}(\mathbf{x},t);\\
&D\mathbf{F}^{(0\succ 0_\alpha)}(\mathbf{x},t) = (F^{(0\succ 0_\alpha)}(\mathbf{x},t), DF^{(0\succ 0_\alpha)}(\mathbf{x},t))^{\mathrm{T}},\\
&DF^{(0\succ 0_\alpha)}(\mathbf{x},t) \equiv \nabla F^{(0\succ 0_\alpha)}(\mathbf{x},t)\cdot\mathbf{F}^{(0\succ 0_\alpha)}(\mathbf{x},t) + \partial_t F^{(0\succ 0_\alpha)}(\mathbf{x},t).
\end{aligned}
\tag{4.217}
$$

Note that $\partial_t(\cdot) = \partial(\cdot)/\partial t$. From Theorem 3.3 in Chapter 3 (also see, Luo, 2005, 2006), the existence condition of the stick motion (or sliding flow in mathematics) between oscillator and the translation belt on the boundary $\partial\Omega_{13}$ is

$$
\begin{aligned}
&\hbar_\alpha G_{\partial\Omega_{\alpha\beta}}^{(\alpha)}(\mathbf{x}_m,t_{m-}) = \hbar_\alpha\mathbf{n}_{\partial\Omega_{\alpha\beta}}^{\mathrm{T}}\cdot\mathbf{F}^{(\alpha)}(\mathbf{x}_m,t_{m-}) > 0;\\
&\hbar_\alpha G_{\partial\Omega_{\alpha\beta}}^{(\beta)}(\mathbf{x}_m,t_{m-}) = \hbar_\alpha\mathbf{n}_{\partial\Omega_{\alpha\beta}}^{\mathrm{T}}\cdot\mathbf{F}^{(\beta)}(\mathbf{x}_m,t_{m-}) < 0.
\end{aligned}
\tag{4.218}
$$

From Theorem 3.1, the necessary and sufficient conditions for the non-stick motion (or passable motion) are on the boundary $\partial\Omega_{\alpha\beta}$

$$
\begin{aligned}
&\hbar_\alpha G_{\partial\Omega_{\alpha\beta}}^{(\alpha)}(\mathbf{x}_m,t_{m-}) = \hbar_\alpha\mathbf{n}_{\partial\Omega_{\alpha\beta}}^{\mathrm{T}}\cdot\mathbf{F}^{(\alpha)}(\mathbf{x}_m,t_{m-}) > 0,\\
&\hbar_\alpha G_{\partial\Omega_{\alpha\beta}}^{(\beta)}(\mathbf{x}_m,t_{m+}) = \hbar_\alpha\mathbf{n}_{\partial\Omega_{\alpha\beta}}^{\mathrm{T}}\cdot\mathbf{F}^{(\beta)}(\mathbf{x}_m,t_{m+}) > 0.
\end{aligned}
\tag{4.219}
$$

The foregoing equations give the necessary and sufficient conditions for the motion without sliding. It indicates that the friction oscillator on the boundary will not stick with the translation belt together. From Theorem 3.15, the switching bifurcation from the non-stick motion to the stick motion is

$$\begin{aligned}
&\hbar_{\alpha} G_{\partial\Omega_{\alpha\beta}}^{(\alpha)}(\mathbf{x}_m, t_{m-}) = \hbar_{\alpha} \mathbf{n}_{\partial\Omega_{\alpha\beta}}^{\mathrm{T}} \cdot \mathbf{F}^{(\alpha)}(\mathbf{x}_m, t_{m-}) > 0,\\
&\hbar_{\alpha} G_{\partial\Omega_{\alpha\beta}}^{(\beta)}(\mathbf{x}_m, t_{m\pm}) = \hbar_{\alpha} \mathbf{n}_{\partial\Omega_{\alpha\beta}}^{\mathrm{T}} \cdot \mathbf{F}^{(\beta)}(\mathbf{x}_m, t_{m\pm}) = 0,\\
&\hbar_{\alpha} G_{\partial\Omega_{\alpha\beta}}^{(1,\beta)}(\mathbf{x}_m, t_{m\pm}) = \hbar_{\alpha} \mathbf{n}_{\partial\Omega_{\alpha\beta}}^{\mathrm{T}} \cdot D\mathbf{F}^{(\beta)}(\mathbf{x}_m, t_{m\pm}) < 0.
\end{aligned} \tag{4.220}$$

Once the stick motion (or the sink flow) is formed under Eq. (4.218), the boundary flow will control the motion on the boundary, which are independent of the vector fields in the two domains. For this problem, the boundary flow on the boundary possesses a boundary flow barrier caused by the static friction force. To obtain a new non-stick motion on the belt, the non-friction force must be greater than the static friction force. From Theorem 4.13, the necessary and sufficient conditions for the vanishing the sink flow (or sliding flow) on the α-side are

$$\left.\begin{aligned}
&\text{both } \hbar_{\alpha} G_{\partial\Omega_{\alpha\beta}}^{(\alpha)}(\mathbf{x}_m, t_{m+}) = \hbar_{\alpha} \mathbf{n}_{\partial\Omega_{\alpha\beta}}^{\mathrm{T}} \cdot \mathbf{F}^{(\alpha)}(\mathbf{x}_m, t_{m+}) < 0,\\
&\text{and } \hbar_{\alpha} G_{\partial\Omega_{\alpha\beta}}^{(0\succ 0_{\alpha})}(\mathbf{x}_m, q_1^{(\alpha)}) = \hbar_{\alpha} \mathbf{n}_{\partial\Omega_{\alpha\beta}}^{\mathrm{T}} \cdot \mathbf{F}^{(0\succ 0_{\alpha})}(\mathbf{x}_m, q_1^{(\alpha)}) = 0,\\
&\quad\text{with } \hbar_{\alpha} G_{\partial\Omega_{\alpha\beta}}^{(1,0\succ 0_{\alpha})}(\mathbf{x}_m, q_1^{(\alpha)}) = \hbar_{\alpha} \mathbf{n}_{\partial\Omega_{\alpha\beta}}^{\mathrm{T}} \cdot D\mathbf{F}^{(0\succ 0_{\alpha})}(\mathbf{x}_m, t_{m\pm}, q_1^{(\alpha)}) < 0;
\end{aligned}\right\} \tag{4.221}$$

$$\left.\begin{aligned}
&\text{either } \hbar_{\alpha} G_{\partial\Omega_{\alpha\beta}}^{(0\succ 0_{\beta})}(\mathbf{x}_m, q_1^{(\beta)}) = \hbar_{\alpha} \mathbf{n}_{\partial\Omega_{\alpha\beta}}^{\mathrm{T}} \cdot \mathbf{F}^{(0\succ 0_{\beta})}(\mathbf{x}_m, q_1^{(\beta)}) > 0,\\
&\quad\text{but } \hbar_{\alpha} G_{\partial\Omega_{\alpha\beta}}^{(\beta)}(\mathbf{x}_m, t_{m+}) = \hbar_{\alpha} \mathbf{n}_{\partial\Omega_{\alpha\beta}}^{\mathrm{T}} \cdot \mathbf{F}^{(\beta)}(\mathbf{x}_m, t_{m+}) < 0,\\
&\text{or } \hbar_{\alpha} G_{\partial\Omega_{\alpha\beta}}^{(0\succ 0_{\beta})}(\mathbf{x}_m, q_1^{(\beta)}) = \hbar_{\alpha} \mathbf{n}_{\partial\Omega_{\alpha\beta}}^{\mathrm{T}} \cdot \mathbf{F}^{(0\succ 0_{\beta})}(\mathbf{x}_m, q_1^{(\beta)}) < 0,\\
&\text{or } \hbar_{\alpha} G_{\partial\Omega_{\alpha\beta}}^{(0\succ 0_{\beta})}(\mathbf{x}_m, q_1^{(\beta)}) = \hbar_{\alpha} \mathbf{n}_{\partial\Omega_{\alpha\beta}}^{\mathrm{T}} \cdot \mathbf{F}^{(0\succ 0_{\beta})}(\mathbf{x}_m, q_1^{(\beta)}) = 0,\\
&\hbar_{\alpha} G_{\partial\Omega_{\alpha\beta}}^{(1,0\succ 0_{\beta})}(\mathbf{x}_m, q_1^{(\beta)}) = \hbar_{\alpha} \mathbf{n}_{\partial\Omega_{\alpha\beta}}^{\mathrm{T}} \cdot D\mathbf{F}^{(0\succ 0_{\beta})}(\mathbf{x}_m, q_1^{(\beta)}) < 0.
\end{aligned}\right\} \tag{4.222}$$

From Chapter 3, the necessary and sufficient conditions for grazing motion to the boundary in Eq. (4.207) are

$$\begin{aligned}
&G_{\partial\Omega_{\alpha\beta}}^{(\alpha)}(\mathbf{x}_m, t_{m+}) = \mathbf{n}_{\partial\Omega_{\alpha\beta}}^{\mathrm{T}} \cdot \mathbf{F}^{(\alpha)}(\mathbf{x}_m, t_{m\pm}) = 0 \text{ for } \alpha \neq \beta;\\
&G_{\partial\Omega_{\alpha\beta}}^{(1,\alpha)}(\mathbf{x}_m, t_{m+}) = \mathbf{n}_{\partial\Omega_{\alpha\beta}}^{\mathrm{T}} \cdot D\mathbf{F}^{(\alpha)}(\mathbf{x}_m, t_{m\pm}) > 0\\
&\quad \text{for } \alpha = 2,1,3 \text{ on } \partial\Omega_{\alpha\beta} \in \{\partial\Omega_{21}, \partial\Omega_{13}, \text{ and } \partial\Omega_{34}\},\\
&G_{\partial\Omega_{\alpha\beta}}^{(1,\alpha)}(\mathbf{x}_m, t_{m+}) = \mathbf{n}_{\partial\Omega_{\alpha\beta}}^{\mathrm{T}} \cdot D\mathbf{F}^{(\alpha)}(\mathbf{x}_m, t_{m\pm}) < 0\\
&\quad \text{for } \alpha = 1,3,4 \text{ on } \partial\Omega_{\alpha\beta} \in \{\partial\Omega_{12}, \partial\Omega_{31}, \text{ and } \partial\Omega_{43}\}.
\end{aligned} \tag{4.223}$$

Using Eq. (4.216), the normal vector of boundary $\partial\Omega_{\alpha\beta}$ with $\alpha, \beta \in \{1,2,3,4\}$ is

$$\mathbf{n}_{\partial\Omega_{\alpha\beta}} = \mathbf{n}_{\partial\Omega_{\beta\alpha}} = (0,1)^{\mathrm{T}}. \tag{4.224}$$

The normal vectors of the boundaries ($\partial\Omega_{12}$ and $\partial\Omega_{21}$), ($\partial\Omega_{13}$ and $\partial\Omega_{31}$) and ($\partial\Omega_{34}$ and $\partial\Omega_{43}$) point to the domains Ω_2, Ω_1 and Ω_3, respectively. Therefore, for $\alpha \in \{i,j\}$, one obtains

$$\begin{aligned}
&\mathbf{n}^{\mathrm{T}}_{\partial\Omega_{\alpha\beta}} \bullet \mathbf{F}^{(\alpha)}(\mathbf{x},t) = F^{(\alpha)}(\mathbf{x},t),\\
&\mathbf{n}^{\mathrm{T}}_{\partial\Omega_{\alpha\beta}} \bullet D\mathbf{F}^{(\alpha)}(\mathbf{x},t) = DF^{(\alpha)}(\mathbf{x},t);\\
&\mathbf{n}^{\mathrm{T}}_{\partial\Omega_{\alpha\beta}} \bullet \mathbf{F}^{(0\succ 0_\alpha)}(\mathbf{x},t,q^{(\alpha)}) = F^{(0\succ 0_\alpha)}(\mathbf{x},t,q^{(\alpha)}),\\
&\mathbf{n}^{\mathrm{T}}_{\partial\Omega_{\alpha\beta}} \bullet D\mathbf{F}^{(0\succ 0_\alpha)}(\mathbf{x},t,q^{(\alpha)}) = DF^{(0\succ 0_\alpha)}(\mathbf{x},t,q^{(\alpha)}).
\end{aligned} \tag{4.225}$$

With Eq. (4.225), the conditions in Eqs. (4.219) and (4.220) to form sink and passable motions to the boundary give the force conditions:

$$\begin{aligned}
&F^{(1)}(\mathbf{x}_m,t_{m-}) < 0 \text{ and } F^{(3)}(\mathbf{x}_m,t_{m-}) > 0 \text{ on } \partial\Omega_{13};\\
&\left.\begin{aligned}
&F^{(1)}(\mathbf{x}_m,t_{m-}) < 0 \text{ and } F^{(3)}(\mathbf{x}_m,t_{m+}) < 0 \text{ for } \Omega_1 \to \Omega_3,\\
&F^{(1)}(\mathbf{x}_m,t_{m+}) > 0 \text{ and } F^{(3)}(\mathbf{x}_m,t_{m-}) > 0 \text{ for } \Omega_3 \to \Omega_1.
\end{aligned}\right\}
\end{aligned} \tag{4.226}$$

The force condition for onset of the sink motion on $\partial\Omega_{13}$ is

$$\begin{aligned}
&\left.\begin{aligned}
&F^{(1)}(\mathbf{x}_m,t_{m-}) < 0 \text{ and } F^{(3)}(\mathbf{x}_m,t_{m\pm}) = 0\\
&\text{with } DF^{(3)}(\mathbf{x}_m,t_{m\pm}) < 0
\end{aligned}\right\} \text{for } \Omega_1 \to \partial\Omega_{13},\\
&\left.\begin{aligned}
&F^{(3)}(\mathbf{x}_m,t_{m-}) > 0 \text{ and } F^{(1)}(\mathbf{x}_m,t_{m+}) = 0\\
&\text{with } DF^{(1)}(\mathbf{x}_m,t_{m+}) > 0
\end{aligned}\right\} \text{for } \Omega_3 \to \partial\Omega_{13}.
\end{aligned} \tag{4.227}$$

Under the flow barrier, the force conditions for vanishing of the sink motion are

$$\left.\begin{aligned}
&\text{either } F^{(0\succ 0_1)}(\mathbf{x}_m,q_1^{(1)}) > 0 \text{ but } F^{(1)}(\mathbf{x}_m,t_{m-}) < 0\\
&\text{or } \quad F^{(0\succ 0_1)}(\mathbf{x}_m,q_1^{(1)}) < 0 \text{ or } F^{(0\succ 0_1)}(\mathbf{x}_m,q_1^{(1)}) = 0\\
&\qquad \text{with } DF^{(0\succ 0_1)}(\mathbf{x}_m,q_1^{(1)}) < 0;
\end{aligned}\right\} \tag{4.228}$$

$$\begin{aligned}
&\text{both } F^{(3)}(\mathbf{x}_m,t_{m-}) < 0\\
&\text{and } F^{(0\succ 0_3)}(\mathbf{x}_m,q_1^{(3)}) = 0 \text{ with } DF^{(0\succ 0_3)}(\mathbf{x}_m,q_1^{(3)}) < 0
\end{aligned} \tag{4.229}$$

from $\partial\Omega_{13} \to \Omega_3$, and

$$\left.\begin{aligned}
&\text{either } F^{(0\succ 0_3)}(\mathbf{x}_m,q_1^{(3)}) < 0 \text{ but } F^{(3)}(\mathbf{x}_m,t_{m-}) > 0\\
&\text{or } \quad F^{(0\succ 0_3)}(\mathbf{x}_m,q_1^{(3)}) > 0 \text{ or } F^{(0\succ 0_3)}(\mathbf{x}_m,q_1^{(3)}) = 0\\
&\qquad \text{with } DF^{(0\succ 0_3)}(\mathbf{x}_m,q_1^{(3)}) > 0;
\end{aligned}\right\} \tag{4.230}$$

$$\begin{aligned}&\text{both } F^{(1)}(\mathbf{x}_m,t_{m-})>0\\&\text{and } F^{(0\succ 0_1)}(\mathbf{x}_m,q_1^{(1)})=0 \text{ with } DF^{(0\succ 0_1)}(\mathbf{x}_m,q_1^{(1)})>0\end{aligned}\tag{4.231}$$

from $\partial\Omega_{13}\to\Omega_1$.

The force conditions for passable motions on the boundary $\partial\Omega_{\alpha\beta}$ are

$$\left.\begin{aligned}&F^{(\alpha)}(\mathbf{x}_m,t_{m-})<0 \text{ and } F^{(\beta)}(\mathbf{x}_m,t_{m+})<0\\&\quad\text{for } (\alpha,\beta)\in\{(2,1),(1,3),(3,4)\};\\&F^{(\alpha)}(\mathbf{x}_m,t_{m-})>0 \text{ and } F^{(\beta)}(\mathbf{x}_m,t_{m+})>0\\&\quad\text{for } (\alpha,\beta)=\{(1,2),(3,1),(4,3)\}.\end{aligned}\right\}\tag{4.232}$$

The force conditions for grazing motions are

$$\begin{aligned}&F^{(\alpha)}(\mathbf{x}_m,t_{m\pm})=0\\&\text{and}\left.\begin{aligned}&DF^{(\alpha)}(\mathbf{x}_m,t_{m\pm})>0 \text{ for } \alpha=2,1,3\\&\text{at } \partial\Omega_{\alpha\beta}\in\{\partial\Omega_{21},\partial\Omega_{13}, \text{ and } \partial\Omega_{34}\},\\&DF^{(\alpha)}(\mathbf{x}_m,t_{m\pm})<0 \text{ for } \alpha=1,3,4\\&\text{at } \partial\Omega_{\alpha\beta}\in\{\partial\Omega_{12},\partial\Omega_{31}, \text{ and } \partial\Omega_{43}\}.\end{aligned}\right\}\end{aligned}\tag{4.233}$$

4.5.2. Illustrations

To illustrate the motion with flow barriers in non-smooth dynamic systems, the basic mappings are introduced as in Luo and Zwiegart (2008). The mappings are determined by the close-form solution of the differential equation in the corresponding domain. With an initial condition (t_k,x_k,V), the direct integration of Eq. (4.201) yields

$$x=V\times(t-t_k)+x_k.\tag{4.234}$$

Substitution of Eq. (4.234) into (4.209) produces the forces for the very small-neighborhood of the stick motion in the domains Ω_j ($j\in\{1,3\}$). Because of the static friction jumping, the forces at $(\mathbf{x}_m,t_m)$ for the coming and leaving flows and the boundary flow barriers on the boundary $\partial\Omega_{13}$ are:

$$F^{(j)}(\mathbf{x}_m,t_{m\pm})=-2d_jV-c_jx_m+A_0\cos\Omega t_{m\pm}-a_j,\tag{4.235}$$

$$F^{(0\succ 0_j)}(\mathbf{x}_m,q_1^{(j)})=-2d_jV-c_jx_m+A_0\cos\Omega t_{m\pm}-a_j^{(0\succ 0_j)},\tag{4.236}$$

where $a_1=-a_3=\nu_kF_N$ and $a_1^{(0\succ 0_1)}=-a_3^{(0\succ 0_3)}=\nu_sF_N$.

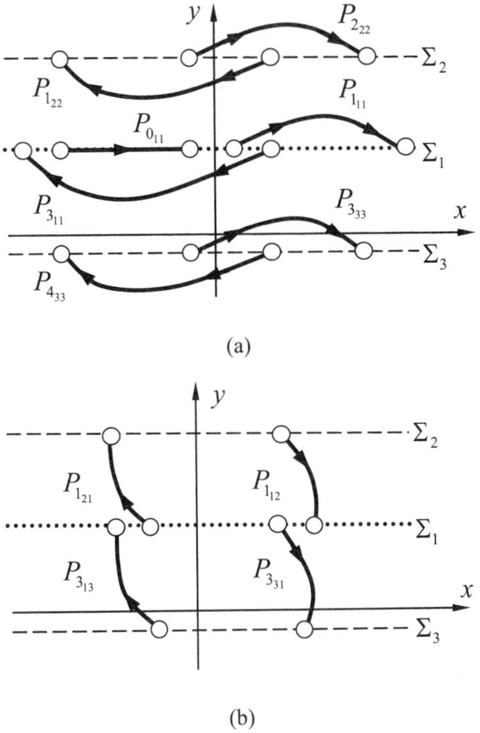

Fig. 4.29 Regular and stick mappings: **(a)** local and stick mappings and **(b)** global mappings.

To label the motion in this discontinuous dynamical system, the generic mappings are introduced. The switching sets on the boundaries should be numbered first. The switching set for the discontinuous force boundary is represented by Σ_1, and the other separation boundaries are Σ_2 and Σ_3. The switching sets for the three boundaries are

$$\Sigma_\alpha = \Sigma_\alpha^0 \cup \Sigma_\alpha^+ \cup \Sigma_\alpha^- \quad \text{for } \alpha = 1,2,3. \tag{4.237}$$

The corresponding, switching subsets are defined as

$$\Sigma_\alpha^0 = \left\{(x_k, \Omega t_k) \mid \dot{x}_k = V_\rho\right\} \text{ and } \Sigma_\alpha^\pm = \left\{(x_k, \Omega t_k \mid \dot{x}_k = V_\rho^{\,\pm}\right\}; \tag{4.238}$$

where $V_\sigma^{\,\pm} = \lim_{\delta \to 0}(V_\sigma \pm \delta)$ for an arbitrarily small $\delta > 0$ and $\rho = \{0,1,2\}$ for $\alpha = 1,2,3$. In phase space, the trajectories in Ω_j starting and ending at the separation boundaries are sketched in Fig. 4.29. The starting and ending points for mappings $P_{j_{\beta\alpha}}$ in Ω_j are $(x_k, \dot{x}_k, t_k)$ on Σ_α and $(x_{k+1}, \dot{x}_{k+1}, t_{k+1})$ on Σ_β, respectively. The indices $j = 1,2,3,4$ and $\alpha, \beta = 1,2,3$ are for domains and boundaries, respectively. The stick mapping is $P_{0_{11}}$. Thus, the mappings are defined as

$$\left.\begin{array}{l} P_{1_{11}}:\Sigma_1^+ \xrightarrow{\Omega_1} \Sigma_1^+,\ P_{3_{11}}:\Sigma_1^- \xrightarrow{\Omega_3} \Sigma_1^-, \\ P_{2_{22}}:\Sigma_2^+ \xrightarrow{\Omega_2} \Sigma_2^+,\ P_{1_{22}}:\Sigma_2^- \xrightarrow{\Omega_1} \Sigma_2^-, \\ P_{4_{33}}:\Sigma_3^- \xrightarrow{\Omega_4} \Sigma_3^-,\ P_{3_{33}}:\Sigma_3^+ \xrightarrow{\Omega_3} \Sigma_3^+ \end{array}\right\} \tag{4.239}$$

for the local mappings

$$\left.\begin{array}{l} P_{1_{21}}:\Sigma_1^+ \xrightarrow{\Omega_1} \Sigma_2^-,\ P_{1_{12}}:\Sigma_2^- \xrightarrow{\Omega_1} \Sigma_1^+, \\ P_{3_{31}}:\Sigma_1^- \xrightarrow{\Omega_3} \Sigma_3^+,\ P_{3_{13}}:\Sigma_3^+ \xrightarrow{\Omega_3} \Sigma_1^- \end{array}\right\} \tag{4.240}$$

for the global mappings and

$$P_{0_{11}}:\Sigma_1^0 \xrightarrow{\partial\Omega_{13}} \Sigma_1^0 \tag{4.241}$$

for the stick mapping.

The governing equations of $P_{0_{11}}$ for a sink flow leaving for Ω_j ($j \in \{1,3\}$) are

$$\left.\begin{array}{l} -x_{k+1} + V\times(t_{k+1}-t_k) + x_k = 0, \\ 2d_j V + c_j[V\times(t_{k+1}-t_k)+x_k] - A_0\cos\Omega t_{k+1} + a_j^{(0\succ 0_j)} = 0. \end{array}\right\} \tag{4.242}$$

Since the differential equation in each domain is linear, the closed form solution for such a linear differential equation can be obtained (e.g., Luo and Zwiegart, 2008). For the non-stick motion, the governing equations of mapping $P_{j_{\beta\alpha}}$ ($j = 1,2,3,4$ and $\alpha,\beta = 1,2,3$) are:

$$f_1^{(j_{\beta\alpha})}(x_k,\Omega t_k,x_{k+1},\Omega t_{k+1}) = 0 \text{ and } f_2^{(j_{\beta\alpha})}(x_k,\Omega t_k,x_{k+1},\Omega t_{k+1}) = 0. \tag{4.243}$$

From the foregoing relations, the periodic motions for such a periodically forced, frictional oscillator can be obtained. The details can be referred to Luo and Zwiegart (2008).The parameters ($m = 5$, $d_{1,2,3,4} = 0.1$, $c_{1,2,3,4} = 30$, $V = 3$, $V_1 = 4.5$, $V_2 = 1.5$, $\mu_s = 0.5$, $\mu_k = 0.4$, $\mu_{1,3} = 0.1$, $\mu_{2,4} = 0.5$ and $g = 9.8$) are adopted for numerical illustrations. The non-stick periodic motion of $P_{4_{33}3_{33}} = P_{4_{33}} \circ P_{3_{33}}$ is considered first. Using the above conditions, the periodic motions in such an oscillator can be obtained as in Luo and Zwiegart (2008). The phase plane, force distributions, and the responses of displacement, velocity and acceleration are presented respectively in Fig. 4.30(a)-(f) for the periodic motion of mapping $P_{4_{33}3_{33}}$ with $\Omega = 5$, $Q_0 = 70$ and the initial condition $(\Omega t_k, x_k, \dot{x}_k) \approx (0.0458,\ 3.0183,\ 1.50)$. The responses in Ω_3 and Ω_4 are depicted through the thin and dark curves, accordingly. The circles are switching points, and the gray filled cycle is the starting point of the periodic motion. The arrows give the directions of the periodic motion. In addition, the corresponding mappings are labeled in plots. In Fig. 4.30(a), the periodic trajectory in phase plane is clearly shown and the periodic motion does not have any intersection with the boundary $\partial\Omega_{13}$. This periodic motion only intersects with the boundary $\partial\Omega_{34}$. The force in domain Ω_α ($\alpha = 3,4$) are

$$\begin{aligned} F^{(3)} &\equiv F^{(3)}(\mathbf{x},t) \\ &= -2d_3\dot{x} - c_3 x + A_0\cos\Omega t + \nu_3(\dot{x} - V) + \mu_k g, \\ F^{(4)} &\equiv F^{(4)}(\mathbf{x},t) \\ &= -2d_4\dot{x} - c_4 x + A_0\cos\Omega t - \nu_4(\dot{x} - V_2) + \nu_3(V_2 - V) + \mu_k g. \end{aligned} \tag{4.244}$$

Therefore, with $\dot{x}_m = V_2$, the force conditions on the boundary $\partial\Omega_{34}$ from Ω_3 to Ω_4 are from Eq. (4.231) at time t_{m-} and t_{m+}

$$\begin{aligned} F_-^{(3)} &\equiv F^{(3)}(\mathbf{x}_m, t_{m-}) \\ &= -2d_3V_2 - c_3x_m + A_0\cos\Omega t_{m-} + \nu_3(V_2 - V) + \mu_k g < 0, \\ F_+^{(4)} &\equiv F^{(4)}(\mathbf{x}_m, t_{m+}) \\ &= -2d_4V_2 - c_4x_m + A_0\cos\Omega t_{m+} + \nu_3(V_2 - V) + \mu_k g < 0; \end{aligned} \tag{4.245}$$

and the force conditions on $\partial\Omega_{43}$ from Ω_4 to Ω_3 are at time t_{m-} and t_{m+}

$$\begin{aligned} F_-^{(4)} &\equiv F^{(4)}(\mathbf{x}_m, t_{m-}) \\ &= -2d_4V_2 - c_4x_m + A_0\cos\Omega t_{m-} + \nu_3(V_2 - V) + \mu_k g > 0, \\ F_+^{(3)} &\equiv F^{(3)}(\mathbf{x}_m, t_{m+}) \\ &= -2d_3V_2 - c_3x_m + A_0\cos\Omega t_{m+} + \nu_3(V_2 - V) + \mu_k g > 0. \end{aligned} \tag{4.246}$$

Because $c_3 = c_4$ and $d_3 = d_4$, one obtains $F_\pm^{(3)} = F_\mp^{(4)}$. From Eqs. (4.245) and (4.246), the total force on the boundary $\partial\Omega_{34}$ is continuous, but from Eq. (4.244), the derivative of the forces (i.e., $F_\pm^{(3)}$ and $F_\mp^{(4)}$) is discontinuous. Such force characteristics of the periodic flow can be observed in Figs. 4.30(b) and (c). The forces at the switching points on the boundary $\partial\Omega_{34}$ and $\partial\Omega_{43}$ are labeled by $F_\pm^{(3)}$ and $F_\mp^{(4)}$. The force distributions in domain Ω_3 and Ω_4 are presented through the thin and dark curves, respectively. In addition, such force distributions in domain Ω_3 and Ω_4 are labeled by $F^{(3)}$ and $F^{(4)}$. Note that the forces in Figs. 4.30(b) and (c) are the total force acting on the mass instead of the total force per unit mass. The displacement and velocity responses in Figs. 4.30(d)-(e) are very smooth owing to the continuity of the forces at the boundary. However, the non-smoothness of the acceleration is observed in Fig. 4.30(f). If the initial point is selected at the another switching point, the mapping structure of the periodic motion becomes $P_{3_{33}4_{33}} = P_{3_{33}} \circ P_{4_{33}}$. However, the two mapping structures present the same periodic motion except for the different initial conditions.

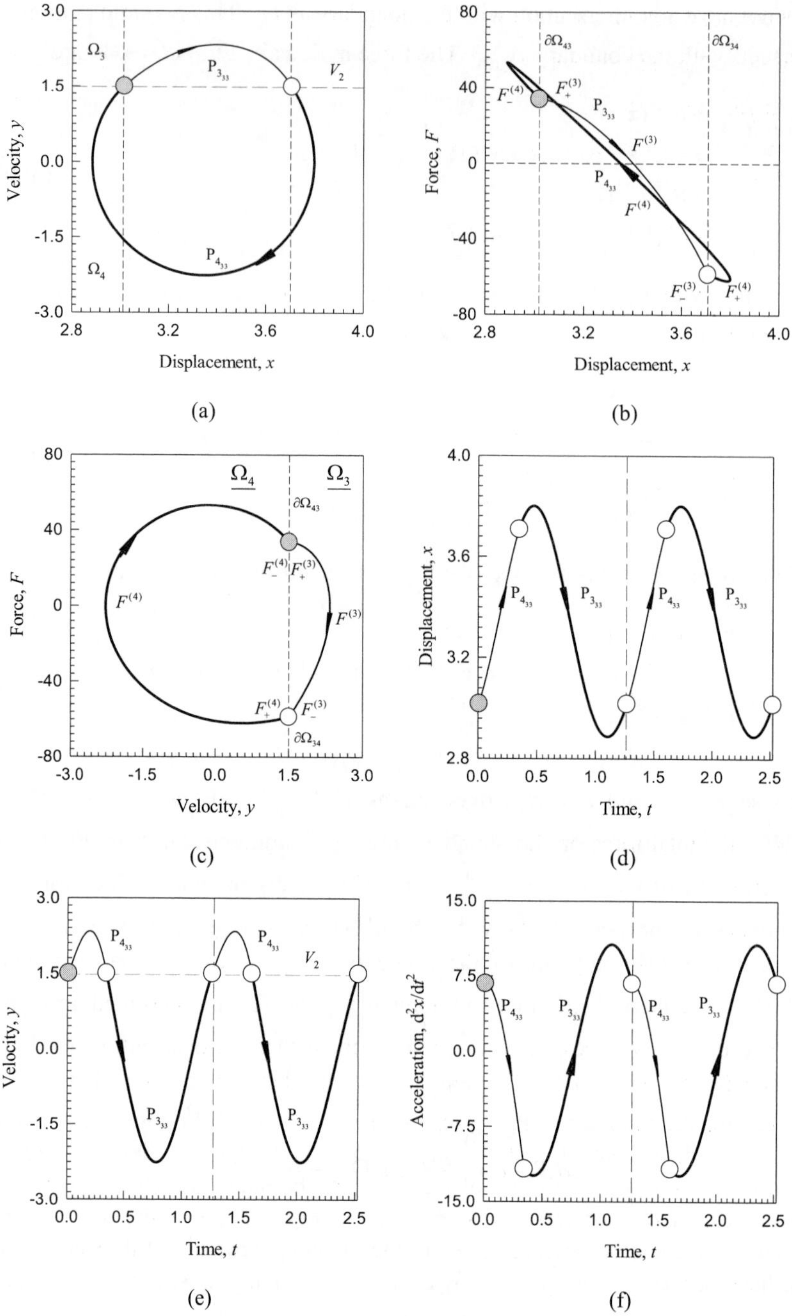

Fig. 4.30 Periodic responses of mapping $P_{4_{33}} \circ P_{3_{33}}$: **(a)** phase plane, **(b)** force distribution along displacement, **(c)** force distribution along velocity, **(d)** displacement, **(e)** velocity and **(f)** acceleration for $\Omega = 5$ and $Q_0 = 70$ with $(\Omega t_k, x_k, \dot{x}_k) \approx (0.0458, 3.0183, 1.50)$.

Consider a stick periodic motion relative to mapping $P_{3_{31}0_{11}3_{13}4_{33}}$ for the excitation frequency ($\Omega = 1$) and excitation amplitude ($Q_0 = 70$) with the initial conditions $(\Omega t_k, x_k, \dot{x}_k) \approx (0.6672, 6.5814, 1.50)$. The other parameters are the same as in the first example. The phase plane, force distributions, and displacement, velocity and acceleration responses for the periodic motion of $P_{3_{31}0_{11}3_{13}4_{33}}$ are shown in Figs. 4.31(a)-(f),respectively. In Fig. 4.31(a), the stick motion in phase plane is a straight line along the discontinuous boundary. The trajectory of this periodic motion exists in domains Ω_3 and Ω_4. The force description in domain Ω_3 and Ω_4 is given in Eq. (4.244). The forces on the switching points on the boundary $\partial\Omega_{34}$ or $\partial\Omega_{43}$ can be determined by Eqs. (4.245) and (4.246). Since the friction force on $\partial\Omega_{13}$ is C^0-discontinuous, such a force discontinuity causes the existence of the sliding (stick) motion along the boundary $\partial\Omega_{13}$. From Eq. (4.223), the condition for the stick motion appearing on $\partial\Omega_{13}$ is

$$\begin{aligned} F_{-}^{(1)} &\equiv F^{(1)}(\mathbf{x}_m, t_{m-}) \\ &= -2d_1V - c_1x_m + A_0\cos\Omega t_{m-} - \mu_k g < 0, \\ F_{-}^{(3)} &\equiv F^{(3)}(\mathbf{x}_m, t_{m-}) \\ &= -2d_3V - c_3x_m + A_0\cos\Omega t_{m-} + \mu_k g > 0. \end{aligned} \tag{4.247}$$

Because the static and kinetic fraction forces are different, the flow barriers exist in this dynamical system. Therefore, once the stick appears between the mass and the translation belt, the stick motion disappearance requires

$$\left.\begin{aligned} F^{(0\succ 0_1)} &\equiv F^{(0\succ 0_1)}(\mathbf{x}_m, q_1^{(1)}) \\ &= -2d_1V - c_1x_m + A_0\cos\Omega t_m - \mu_s g < 0 \\ F^{(0\succ 0_3)} &\equiv F^{(0\succ 0_3)}(\mathbf{x}_m, q_1^{(3)}) \\ &= -2d_3V - c_3x_m + A_0\cos\Omega t_{m\pm} + \mu_s g = 0 \\ DF^{(0\succ 0_3)} &\equiv DF^{(0\succ 0_3)}(\mathbf{x}_m, q_1^{(3)}) = -c_3V_m - A_0\Omega\sin\Omega t_{m\pm} < 0 \end{aligned}\right\} \tag{4.248}$$

for $\partial\Omega_{13} \to \Omega_3$;

$$\left.\begin{aligned} F^{(0\succ 0_3)} &\equiv F^{(0\succ 0_3)}(\mathbf{x}_m, q_1^{(3)}) \\ &= -2d_3V - c_3x_m + A_0\cos\Omega t_m + \mu_s g > 0 \\ F^{(0\succ 0_1)} &\equiv F^{(0\succ 0_1)}(\mathbf{x}_m, q_1^{(1)}) \\ &= -2d_1V - c_1x_m + A_0\cos\Omega t_{m\pm} - \mu_s g = 0 \\ DF^{(0\succ 0_1)} &\equiv DF^{(0\succ 0_1)}(\mathbf{x}_m, q_1^{(1)}) = -c_1V - \Omega A_0\sin\Omega t_{m\pm} > 0 \end{aligned}\right\} \tag{4.249}$$

for $\partial\Omega_{13} \to \Omega_1$.

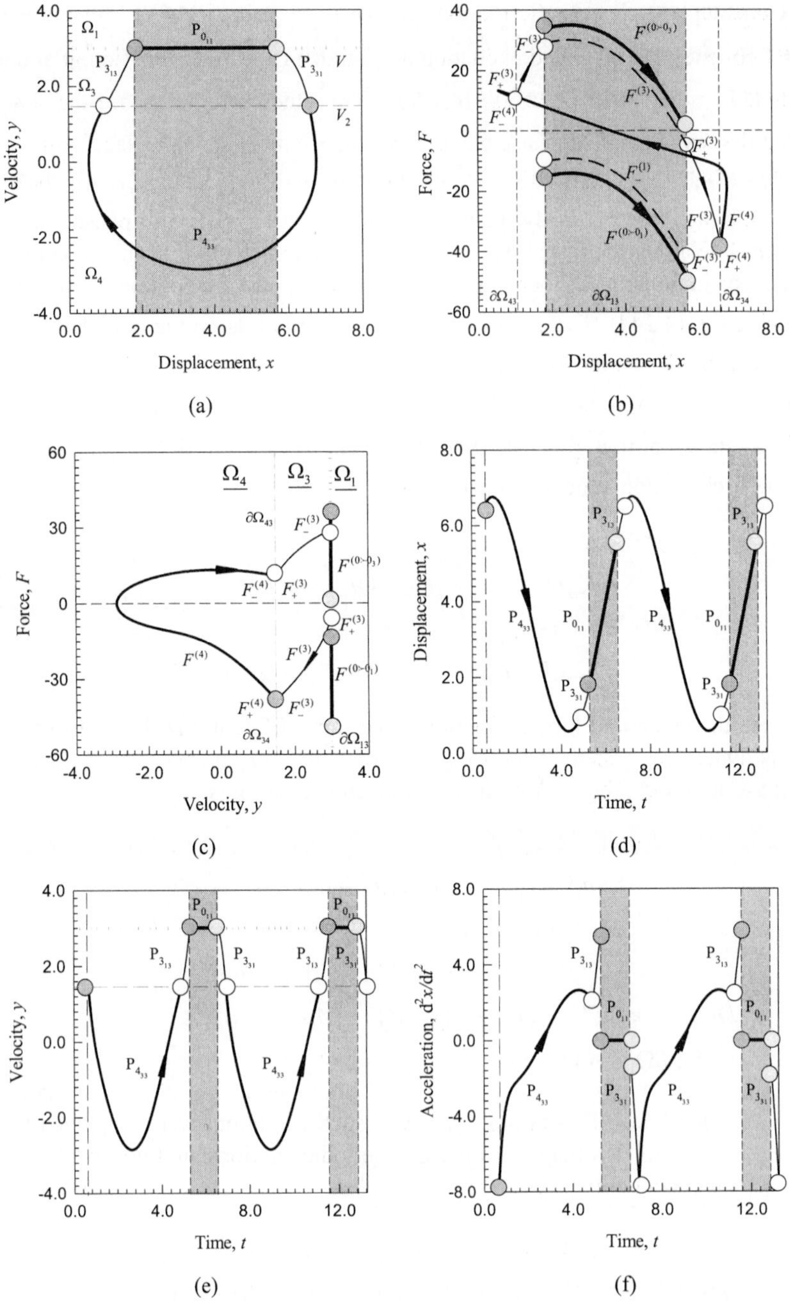

Fig. 4.31 Periodic responses of mapping $P_{3_{31}} \circ P_{0_{11}} \circ P_{3_{13}} \circ P_{4_{33}}$: **(a)** phase plane, **(b)** force distribution along displacement, **(c)** force distribution along velocity, **(d)** displacement, **(e)** velocity and **(f)** acceleration for $\Omega = 1$ and $Q_0 = 70$ with $(\Omega t_k, x_k, \dot{x}_k) \approx (0.6672, 6.5814, 1.50)$.

For simplicity, $F^{(0\succ 0_\alpha)} \triangleq F^{(0\succ 0_\alpha)}(\mathbf{x}_m, t_m)$ is depicted to observe the force criteria for the disappearance of the stick motion. In Fig. 4.31(b), the force $F_-^{(\alpha)}$ for the coming flow ($\alpha = 1,3$) is presented by the dashed curves. Since $F_-^{(3)} > 0$ and $F_-^{(1)} < 0$, the stick motion appears on the boundary $\partial\Omega_{13}$ because the conditions in Eq. (4.247) are satisfied. The force of the boundary flow barrier $F^{(0\succ 0_\alpha)}$ is plotted by the thick-solid curves. The stick motion disappears at $F^{(0\succ 0_3)} = 0$ and $DF^{(0\succ 0_3)} < 0$, which satisfies the conditions in Eq. (4.248), and such a condition indicates that the non-friction force must be greater than the static friction force (i.e., flow barriers). Furthermore, the oscillator will relatively oscillate on the moving belt. Once the relative motion starts between the oscillator and the belt, the kinetic friction force will control the motion in domain Ω_3. So the corresponding force jumps from zero to the negative one (i.e., $F_+^{(3)} < 0$), which is observed in Fig. 4.31(b). The non-smoothness of the forces on the boundary $\partial\Omega_{34}$ is also observed in Fig. 4.31(b) because of the piecewise continuity of the forces at the boundary. The forces at the switching points on the boundary $\partial\Omega_{34}$ and $\partial\Omega_{43}$ are labeled by $F_\pm^{(3)}$ and $F_\mp^{(4)}$. The force distributions in domain Ω_3 and Ω_4 are presented through the thin and thick curves, respectively. In addition, such force distributions in domain Ω_3 and Ω_4 are labeled by $F^{(3)}$ and $F^{(4)}$. The force discontinuity on the boundary $\partial\Omega_{34}$ between the two domains Ω_3 and Ω_4 is clearly observed. Such force characteristics of stick motion and domain switching are presented in Fig. 4.31(c) as well. In Fig. 4.31(d), the displacement is continuous because the velocity is C^0-continuous. The non-smoothness of the velocity response is observed because the force is C^0-discontinuous. The stick motion in the velocity response is clearly observed as well. Because the belt possesses a constant speed, the corresponding acceleration for the stick motion is zero. Therefore, in Fig. 4.31(f), the acceleration is zero for the stick motion on the boundary $\partial\Omega_{13}$ and non-smooth at $\partial\Omega_{34}$. The grazing phenomenon is independent of the flow barriers in discontinuous dynamical systems, such phenomena for this problem is identical to the friction oscillator without flow barriers, which can be found in references (e.g., Luo and Gegg, 2006c, 2007). The grazing conditions in Eq. (4.234) are not employed for numerical illustrations. However, such conditions were employed to determine the parameter map for different periodic motions in Luo and Zwiegart (2008). The grazing bifurcations were also discussed in Chapter 2.

4.6. Concluding remarks

In this chapter, a theory for flow barriers in discontinuous dynamical systems was presented. Because the flow barriers in the passable flow exist, the leaving and coming flows cannot reach the boundary but will slide along the boundary in the domain. The coming and leaving flow barriers in passable flows were discussed. If a coming flow in the sink flow is blocked by its flow barrier, the coming flow will slide along or stand at the boundary in the corresponding domain. Thus, the flow switchability of a sink flow with flow barriers was discussed. However, once the sink flow is formed, the boundary flow exists on the boundary. Once the boundary flow leaves the boundary, the boundary flow barrier in the source flow may exist, and the leaving flow barriers in the source flow may also exist. Both the boundary and leaving flow barriers in the source flow were presented. A practical discontinuous system with the flow barrier of the boundary flow was presented for a better understanding of the flow barrier theory in the discontinuous dynamical systems. The flow barrier theory in discontinuous dynamical systems is a new theory, which will provide a useful tool for one to design desired dynamical systems to satisfy engineering-oriented complex systems. The flow barrier theory will provide a theoretic base for stabilized dynamical systems in control theory.

References

Luo, A.C.J., 2006, *Singularity and Dynamics on Discontinuous Vector Fields*, Amsterdam: Elsevier.

Luo, A.C.J., 2007, Flow switching bifurcations on the separation boundary in discontinuous dynamical systems with flow barriers, IMeChe *Part K: Journal of Multibody Dynamics*, **221**, 475-495.

Luo, A.C.J. and Gegg, B.C., 2006a, On the mechanism of stick and non-stick, periodic motions in a forced linear oscillator including dry friction, *ASME Journal of Vibration and Acoustics*, **128**, 97-105.

Luo, A.C.J. and Gegg, B.C., 2006b, Stick and non-stick, periodic motions of a periodically forced, linear oscillator with dry friction, *Journal of Sound and Vibration*, **291**, 132-168.

Luo, A.C.J. and Gegg, B.C., 2006c, Grazing phenomena in a periodically forced, friction-induced, linear oscillator, *Communications in Nonlinear Science and Numerical Simulation*, **11**, 777-802.

Luo, A.C.J. and Gegg, B.C., 2007, An analytical prediction of sliding motions along discontinuous boundary in non-smooth dynamical systems, *Nonlinear Dynamics*, **40**, 401-424.

Luo, A.C.J. and Zwiegart, P., Jr., 2008, Existence and analytical prediction of periodic motions in a periodically forced, nonlinear friction oscillator, *Journal of Sound and Vibration*, **309**, 129-149.

Chapter 5
Transport Laws and Multi-valued Vector Fields

In this Chapter, a classification of discontinuity in discontinuous dynamical systems will be discussed first. To discuss the singularity to the boundary, the grazing and inflexional singular sets on the boundary will be presented, and the real and imaginary singular sets will be also discussed. With permanent flow barriers, the forbidden boundary and the boundary channel will be presented. The forbidden boundary will not allow any flows passing through the boundary, and the boundary channel will not allow any boundary flows getting into the corresponding domains. Further, the domain and boundary classification will be addressed. Sink and source domains will be discussed. Similarly, the sink and source boundary will be also presented. To make flows continue in discontinuous dynamical systems, transport laws are needed for C^0-discontinuity, flow barriers, isolated domains and boundary channels. Multi-valued vector fields in a single domain will be introduced. With the simplest transport law (i.e., the switching rule), bouncing flows on the boundary will be presented, and extendable flows will be discussed as well. A controlled piecewise linear system will be presented as an application, and vector fields on both sides of the boundary will be switched at the boundary. Bouncing flows will be illustrated in such a controlled piecewise linear system.

5.1. Discontinuity classification

Consider three dynamical systems

$$\begin{aligned} \dot{\mathbf{x}}^{(\alpha)} &= \mathbf{F}^{(\alpha)}(\mathbf{x}^{(\alpha)}, t, \mathbf{p}_\alpha) \ \text{ in } \Omega_\alpha \ (\alpha = i, j), \\ \dot{\mathbf{x}}^{(0)} &= \mathbf{F}^{(0)}(\mathbf{x}^{(0)}, t, \boldsymbol{\lambda}) \quad \text{with} \ \ \varphi_{ij}(\mathbf{x}^{(0)}, t, \boldsymbol{\lambda}) = 0 \ \text{ on } \partial\Omega_{ij}. \end{aligned} \tag{5.1}$$

The flow $\boldsymbol{\Phi}^{(\alpha)}(\mathbf{x}^{(\alpha)}, t)$ of Eq. (5.1) is C^1-discontinuous (or C^0- continuous) on a boundary $\partial\Omega_{ij}$ (or hyper surface), as shown in Fig. 5.1. In domain Ω_i, there is a flow $\mathbf{x}^{(i)}(t)$ coming to the ending point p_m with $\mathbf{x}_{m-}^{(i)}$. In domain Ω_j, there is a flow $\mathbf{x}^{(j)}(t)$ leaving from the starting point p_m with $\mathbf{x}_{m+}^{(j)}$. For a C^1-discontinuous flow $\mathbf{\Phi}^{(\alpha)}(\mathbf{x}^{(\alpha)},t) \equiv \mathbf{\Phi}^{(\alpha)}(\mathbf{x}_0^{(\alpha)},t,t_0)$ ($\alpha = i,j$), the points $\mathbf{x}^{(i)}$ and $\mathbf{x}^{(j)}$ relative to domains Ω_i and Ω_j on boundary $\partial\Omega_{ij}$ are identical (i.e., $\mathbf{x}^{(i)} = \mathbf{x}^{(j)}$) for the same time t_m. However, the vector fields of flows $\mathbf{x}^{(i)}(t)$ and $\mathbf{x}^{(j)}(t)$ on the boundary $\partial\Omega_{ij}$ are distinguished (i.e., $\mathbf{F}^{(i)}(\mathbf{x}^{(i)},t_{m-},\mathbf{p}_i) \neq \mathbf{F}^{(j)}(\mathbf{x}^{(j)},t_{m+},\mathbf{p}_j)$). Thus, a flow $\mathbf{x}^{(i)} \cup \mathbf{x}^{(j)}$ is C^1-discontinuous at $\mathbf{x}_m \in \partial\Omega_{ij}$. If the lowest order discontinuity of flow $\mathbf{x}^{(i)} \cup \mathbf{x}^{(j)}$ to such a boundary $\partial\Omega_{ij}$ is C^1-discontinuous, the corresponding system is called a C^1-discontinuous dynamical system at boundary $\partial\Omega_{ij}$, which is also called a non-smooth dynamical system at boundary $\partial\Omega_{ij}$ because the trajectory is continuous. The mathematical definition is given as follows:

Definition 5.1. For a discontinuous system in Eq. (5.1), there are a coming flow $\mathbf{x}^{(\alpha)}(t)$ ($\alpha \in \{i,j\}$) and a leaving flow $\mathbf{x}^{(\beta)}(t)$ ($\alpha,\beta \in \{i,j\}$, $\alpha \neq \beta$) at $\mathbf{x}_m \in \partial\Omega_{ij}$ for time t_m. The flow $\mathbf{x}^{(\alpha)}(t) \cup \mathbf{x}^{(\beta)}(t)$ at $\mathbf{x}_m \in \partial\Omega_{ij}$ is C^1-discontinuous if

$$\mathbf{x}^{(\alpha)}(t_{m-}) = \mathbf{x}_m = \mathbf{x}^{(\beta)}(t_{m+}), \tag{5.2}$$

$$\mathbf{F}^{(\alpha)}(\mathbf{x}^{(\alpha)},t_{m-},\mathbf{p}_\alpha) \neq \mathbf{F}^{(\beta)}(\mathbf{x}^{(\beta)},t_{m+},\mathbf{p}_\beta). \tag{5.3}$$

The dynamical system in Eq. (5.1) is called a C^1*-discontinuous* (or C^1*-non-smooth*) system to boundary $\partial\Omega_{ij}$.

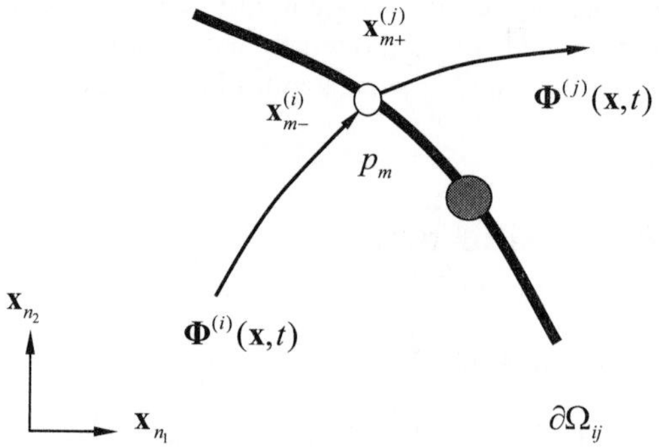

Fig. 5.1 The C^1-discontinuous flow $\mathbf{\Phi}(\mathbf{x},t)$ on the boundary $\partial\Omega_{ij}$ (or hyper-surface). The largest circle symbol is the set with tangential singularity.

Definition 5.2. For a discontinuous system in Eq. (5.1), there are a coming flow $\mathbf{x}^{(\alpha)}(t)$ ($\alpha \in \{i,j\}$) and a leaving flow $\mathbf{x}^{(\beta)}(t)$ ($\alpha,\beta \in \{i,j\}$, $\alpha \neq \beta$) at $\mathbf{x}_m \in \partial\Omega_{ij}$ for time t_m. The flow $\mathbf{x}^{(\alpha)}(t) \cup \mathbf{x}^{(\beta)}(t)$ at $\mathbf{x}_m \in \partial\Omega_{ij}$ is C^k-discontinuous if

$$\mathbf{x}^{(\alpha)}(t_{m-}) = \mathbf{x}_m = \mathbf{x}^{(\beta)}(t_{m+}), \tag{5.4}$$

$$\begin{gathered} D^{(r)}\mathbf{F}^{(\alpha)}(\mathbf{x}^{(\alpha)}, t_{m-}, \mathbf{p}_\alpha) = D^{(r)}\mathbf{F}^{(\beta)}(\mathbf{x}^{(\beta)}, t_{m+}, \mathbf{p}_\beta) \\ \text{for } r = 0,1,2,\cdots,k-2; \end{gathered} \tag{5.5}$$

$$D^{(k-1)}\mathbf{F}^{(\alpha)}(\mathbf{x}^{(\alpha)}, t_{m\mp}, \mathbf{p}_\alpha) \neq D^{(k-1)}\mathbf{F}^{(\beta)}(\mathbf{x}^{(\beta)}, t_{m\pm}, \mathbf{p}_\beta). \tag{5.6}$$

The dynamical system in Eq. (5.1) is called a C^k*-discontinuous* (or C^{k-1}*-non-smooth*) dynamical system to boundary $\partial\Omega_{ij}$.

Definition 5.3. For a discontinuous system in Eq. (5.1), there are a coming flow $\mathbf{x}^{(\alpha)}(t)$ ($\alpha \in \{i,j\}$) at $\mathbf{x}_m^{(\alpha)} \in \partial\Omega_{ij}$ and a leaving flow $\mathbf{x}^{(\beta)}(t)$ ($\alpha,\beta \in \{i,j\}$, $\alpha \neq \beta$) at $\mathbf{x}_m^{(\beta)} \in \partial\Omega_{ij}$ for time t_m. The flow $\mathbf{x}^{(\alpha)}(t) \cup \mathbf{x}^{(\beta)}(t)$ to boundary $\partial\Omega_{ij}$ is C^0-discontinuous if

$$\mathbf{x}^{(\alpha)}(t_{m-}) \neq \mathbf{x}^{(\beta)}(t_{m+}). \tag{5.7}$$

The dynamical system in Eq. (5.1) to boundary $\partial\Omega_{ij}$ is called a C^0*-discontinuous* system.

(i) The C^0-discontinuous dynamical system is called a *self-jumping* system to the boundary $\partial\Omega_{ij}$ if for $\mathbf{x}^{(\alpha)}(t_{m-}) = \mathbf{x}_{m-} \in \partial\Omega_{ij}$ and $\mathbf{x}^{(\alpha)}(t_{m+}) = \mathbf{x}_{m+} \in \partial\Omega_{ij}$.

$$\begin{gathered} \mathbf{x}_{m-} \neq \mathbf{x}_{m+}, \\ \hbar_\alpha G^{(\alpha)}(\mathbf{x}_{m-}, t_{m-}, \mathbf{p}_\alpha) > 0 \text{ and } \hbar_\alpha G^{(\alpha)}(\mathbf{x}_{m+}, t_{m+}, \mathbf{p}_\alpha) < 0. \end{gathered} \tag{5.8}$$

(ii) The C^0-discontinuous dynamical system is called a *jumping-switching* system to the boundary $\partial\Omega_{ij}$ if $\mathbf{x}^{(\alpha)}(t_{m-}) = \mathbf{x}_{m-} \in \partial\Omega_{ij}$ and $\mathbf{x}^{(\beta)}(t_{m+}) = \mathbf{x}_{m+} \in \partial\Omega_{ij}$.

$$\begin{gathered} \mathbf{x}_{m-} \neq \mathbf{x}_{m+}, \\ \hbar_\alpha G^{(\alpha)}(\mathbf{x}_{m-}, t_{m-}, \mathbf{p}_\alpha) > 0 \text{ and } \hbar_\alpha G^{(\beta)}(\mathbf{x}_{m+}, t_{m+}, \mathbf{p}_\beta) > 0. \end{gathered} \tag{5.9}$$

Consider an imaginary dynamical system as

$$\begin{aligned} &\dot{\mathbf{x}}_\alpha^{(\beta)} = \mathbf{F}^{(\beta)}(\mathbf{x}_\alpha^{(\beta)}, t, \mathbf{p}_\beta) \ \text{ in } \Omega_\alpha \ (\alpha,\beta = i,j \text{ and } \alpha \neq \beta), \\ &\dot{\mathbf{x}}^{(0)} = \mathbf{F}^{(0)}(\mathbf{x}^{(0)}, t, \boldsymbol{\lambda}) \quad \text{with } \varphi_{ij}(\mathbf{x}^{(0)}, t, \boldsymbol{\lambda}) = 0 \ \text{ on } \partial\Omega_{ij}. \end{aligned} \tag{5.10}$$

The discontinuity of imaginary flows $\mathbf{x}_\alpha^{(\beta)}$ and $\mathbf{x}_\beta^{(\alpha)}$ to the boundary in Eq. (5.10) is the same as the discontinuity of real flows $\mathbf{x}^{(\alpha)}$ and $\mathbf{x}^{(\beta)}$ in Eq. (5.1). The discontinuity of discontinuous systems relative to time will be discussed in sequel.

5.2. Singular sets on boundary

In Chapters 2 and 3, the switching bifurcation between the two kinds of flows was discussed. Once the switching bifurcation occurs, there is a kind of flow singularity to the boundary. To better understand the properties of flows to the boundary, the concept of singular set should be introduced. There are two classes of singular sets: grazing singularity and inflexional singularity sets. Consider the singular sets on the boundary.

Definition 5.4. For a discontinuous dynamical system of Eq. (5.1), the *grazing singular set* on the α-side of boundary $\partial\Omega_{ij}$ for $\alpha,\beta\in\{i,j\}$ and $\alpha\neq\beta$ is defined as

$$\mathcal{S}_{ij}^{(1,\alpha)}=\left\{(\mathbf{x}_m,t_m)\left|\begin{array}{l}\hbar_\alpha G_{\partial\Omega_{ij}}^{(\alpha)}(\mathbf{x}_m,t_{m\pm})=0,\\ \hbar_\alpha G_{\partial\Omega_{ij}}^{(1,\alpha)}(\mathbf{x}_m,t_{m\pm})<0\end{array}\right.\right\}\subset\partial\Omega_{ij}. \tag{5.11}$$

The *double grazing singular set* on both sides of boundary $\partial\Omega_{ij}$ is defined as

$$\begin{aligned}\mathcal{S}_{ij}^{(1:1)}&=\mathcal{S}_{ij}^{(1,\alpha)}\cup\mathcal{S}_{ij}^{(1,\beta)}\\ &=\left\{(\mathbf{x}_m,t_m)\left|\begin{array}{l}\hbar_\alpha G_{\partial\Omega_{ij}}^{(\alpha)}(\mathbf{x}_m,t_{m\pm})=0,\\ \hbar_\alpha G_{\partial\Omega_{ij}}^{(1,\alpha)}(\mathbf{x}_m,t_{m\pm})<0,\\ \hbar_\alpha G_{\partial\Omega_{ij}}^{(\beta)}(\mathbf{x}_m,t_{m\pm})=0,\\ \hbar_\alpha G_{\partial\Omega_{ij}}^{(1,\beta)}(\mathbf{x}_m,t_{m\pm})>0\end{array}\right.\right\}\subset\partial\Omega_{ij}.\end{aligned} \tag{5.12}$$

Definition 5.5. For a discontinuous dynamical system of Eq. (5.1), the *inflexional singular set* on the α-side of boundary $\partial\Omega_{ij}$ for $\alpha,\beta\in\{i,j\}$ and $\alpha\neq\beta$ is defined as

$$\mathcal{Q}_{ij}^{(2,\alpha)}=\left\{(\mathbf{x}_m,t_m)\left|\begin{array}{l}\hbar_\alpha G_{\partial\Omega_{ij}}^{(s_\alpha,\alpha)}(\mathbf{x}_m,t_{m\pm})=0 \text{ for } s_\alpha=0,1,\\ \hbar_\alpha G_{\partial\Omega_{ij}}^{(2,\alpha)}(\mathbf{x}_m,t_{m\pm})>0\end{array}\right.\right\}\subset\partial\Omega_{ij}. \tag{5.13}$$

The *double inflexional singular set* on both sides of $\partial\Omega_{ij}$ is defined as

$$\mathscr{Q}_{ij}^{(2:2)} = \mathscr{Q}_{ij}^{(2,\alpha)} \cup \mathscr{Q}_{ij}^{(2,\beta)}$$
$$= \left\{ (\mathbf{x}_m, t_m) \left| \begin{array}{l} \hbar_\alpha G_{\partial\Omega_{ij}}^{(s_\alpha,\alpha)}(\mathbf{x}_m, t_{m\pm}) = 0 \text{ for } s_\alpha = 0,1, \\ \hbar_\alpha G_{\partial\Omega_{ij}}^{(2,\alpha)}(\mathbf{x}_m, t_{m\pm}) > 0; \\ \hbar_\alpha G_{\partial\Omega_{ij}}^{(s_\beta,\beta)}(\mathbf{x}_m, t_{m\pm}) = 0 \text{ for } s_\beta = 0,1, \\ \hbar_\alpha G_{\partial\Omega_{ij}}^{(2,\beta)}(\mathbf{x}_m, t_{m\pm}) < 0 \end{array} \right. \right\} \subset \partial\Omega_{ij}. \tag{5.14}$$

Definition 5.6. For a discontinuous dynamical system of Eq. (5.1), the $(2k_\alpha - 1)$ th-*order, grazing singular set* on the α -side of boundary $\partial\Omega_{ij}$ for $\alpha, \beta \in \{i, j\}$ and $\alpha \neq \beta$ is defined as

$$\mathscr{S}_{ij}^{(2k_\alpha+1,\alpha)} = \left\{ (\mathbf{x}_m, t_m) \left| \begin{array}{l} \hbar_\alpha G_{\partial\Omega_{ij}}^{(s_\alpha,\alpha)}(\mathbf{x}_m, t_{m\pm}) = 0 \\ \text{for } s_\alpha = 0,1,2,\cdots,2k_\alpha, \\ \hbar_\alpha G_{\partial\Omega_{ij}}^{(2k_\alpha+1,\alpha)}(\mathbf{x}_m, t_{m\pm}) < 0 \end{array} \right. \right\} \subset \partial\Omega_{ij}. \tag{5.15}$$

The $(2k_\alpha + 1 : 2k_\beta + 1)$ *-double grazing singular set* on both sides of boundary $\partial\Omega_{ij}$ is defined as

$$\mathscr{S}_{ij}^{(2k_\alpha+1:2k_\beta+1)} = \mathscr{S}_{ij}^{(2k_\alpha+1,\alpha)} \cup \mathscr{S}_{ij}^{(2k_\beta+1,\beta)}$$
$$= \left\{ (\mathbf{x}_m, t_m) \left| \begin{array}{l} \hbar_\alpha G_{\partial\Omega_{ij}}^{(s_\alpha,\alpha)}(\mathbf{x}_m, t_{m\pm}) = 0 \text{ for } s_\alpha = 0,1,\cdots,2k_\alpha, \\ \hbar_\alpha G_{ij}^{(2k_\alpha+1,\alpha)}(\mathbf{x}_m, t_{m\pm}) < 0; \\ \hbar_\alpha G_{\partial\Omega_{ij}}^{(s_\beta,\beta)}(\mathbf{x}_m, t_{m\pm}) = 0 \text{ for } s_\beta = 0,1,\cdots,2k_\beta, \\ \hbar_\alpha G_{\partial\Omega_{ij}}^{(2k_\beta+1,\beta)}(\mathbf{x}_m, t_{m\pm}) > 0 \end{array} \right. \right\}$$
$$\subset \partial\Omega_{ij}. \tag{5.16}$$

Definition 5.7. For a discontinuous dynamical system of Eq. (5.1), the $(2k_\alpha)$ th-*order, inflexional singular set* on the α -side of boundary $\partial\Omega_{ij}$ for $\alpha, \beta \in \{i, j\}$ and $\alpha \neq \beta$ is defined as

$$\mathscr{Q}_{ij}^{(2k_\alpha,\alpha)} = \left\{ (\mathbf{x}_m, t_m) \left| \begin{array}{l} \hbar_\alpha G_{\partial\Omega_{ij}}^{(s_\alpha,\alpha)}(\mathbf{x}_m, t_{m\pm}) = 0 \\ \text{for } s_\alpha = 0,1,\cdots,2k_\alpha - 1, \\ \hbar_\alpha G_{\partial\Omega_{ij}}^{(2k_\alpha,\alpha)}(\mathbf{x}_m, t_{m\pm}) > 0 \end{array} \right. \right\} \subset \partial\Omega_{ij}. \tag{5.17}$$

The $(2k_\alpha : 2k_\beta)$ *-double inflexional singular set* on both sides of boundary $\partial\Omega_{ij}$ is defined as

$$\begin{aligned}\mathscr{Q}_{ij}^{(2k_\alpha:2k_\beta)} &= \mathscr{Q}_{ij}^{(2k_\alpha,\alpha)} \cup \mathscr{Q}_{ij}^{(2k_\beta,\beta)} \\ &= \left\{ (\mathbf{x}_m,t_m) \left| \begin{array}{l} \hbar_\alpha G_{\partial\Omega_{ij}}^{(s_\alpha,\alpha)}(\mathbf{x}_m,t_{m\pm}) = 0 \text{ for } s_\alpha = 0,1,\cdots,2k_\alpha-1, \\ \hbar_\alpha G_{\partial\Omega_{ij}}^{(2k_\alpha,\alpha)}(\mathbf{x}_m,t_{m\pm}) > 0; \\ \hbar_\alpha G_{\partial\Omega_{ij}}^{(s_\beta,\beta)}(\mathbf{x}_m,t_{m\pm}) = 0 \text{ for } s_\beta = 0,1,\cdots,2k_\beta-1, \\ \hbar_\alpha G_{\partial\Omega_{ij}}^{(2k_\beta,\beta)}(\mathbf{x}_m,t_{m\pm}) < 0 \end{array} \right. \right\} \\ &\subset \partial\Omega_{ij}. \end{aligned} \tag{5.18}$$

The above definitions of singular sets are given for real flows. Similarly, the singular sets for imaginary flows in the imaginary discontinuous dynamical system in Eq. (5.10) can be discussed.

Definition 5.8. For a discontinuous dynamical system of Eq. (5.10), the *imaginary grazing singular set* on the α-side of boundary $\partial\Omega_{ij}$ for $\alpha,\beta \in \{i,j\}$ and $\alpha \neq \beta$ is defined as

$$\mathscr{S}_{ij}^{(1,\alpha)} = \left\{ (\mathbf{x}_m,t_m) \left| \begin{array}{l} \hbar_\alpha G_{\partial\Omega_{ij}}^{(\beta)}(\mathbf{x}_\alpha^{(\beta)},t_{m\pm}) = 0, \\ \hbar_\alpha G_{\partial\Omega_{ij}}^{(1,\beta)}(\mathbf{x}_\alpha^{(\beta)},t_{m\pm}) < 0 \end{array} \right. \right\} \subset \partial\Omega_{ij}. \tag{5.19}$$

The *double imaginary grazing singular set* on both sides of boundary $\partial\Omega_{ij}$ is defined as

$$\begin{aligned}\mathscr{S}_{ij}^{(1:1)} &= \mathscr{S}_{ij}^{(1,\alpha)} \cup \mathscr{S}_{ij}^{(1,\beta)} \\ &= \left\{ (\mathbf{x}_m,t_m) \left| \begin{array}{l} \hbar_\alpha G_{\partial\Omega_{ij}}^{(\beta)}(\mathbf{x}_\alpha^{(\beta)},t_{m\pm}) = 0, \\ \hbar_\alpha G_{\partial\Omega_{ij}}^{(1,\beta)}(\mathbf{x}_\alpha^{(\beta)},t_{m\pm}) < 0, \\ \hbar_\alpha G_{\partial\Omega_{ij}}^{(\alpha)}(\mathbf{x}_\beta^{(\alpha)},t_{m\pm}) = 0, \\ \hbar_\alpha G_{\partial\Omega_{ij}}^{(1,\alpha)}(\mathbf{x}_\beta^{(\alpha)},t_{m\pm}) > 0 \end{array} \right. \right\} \subset \partial\Omega_{ij}. \end{aligned} \tag{5.20}$$

Definition 5.9. For a discontinuous dynamical system of Eq. (5.10), the *imaginary inflexional singular set* on the α-side of boundary $\partial\Omega_{ij}$ for $\alpha,\beta \in \{i,j\}$ and $\alpha \neq \beta$ is defined as

$$\mathscr{Q}_{ij}^{(2,\alpha)} = \left\{ (\mathbf{x}_m,t_m) \left| \begin{array}{l} \hbar_\alpha G_{\partial\Omega_{ij}}^{(s_\beta,\beta)}(\mathbf{x}_\alpha^{(\beta)},t_{m\pm}) = 0 \text{ for } s_\beta = 0,1, \\ \hbar_\alpha G_{\partial\Omega_{ij}}^{(2,\beta)}(\mathbf{x}_\alpha^{(\beta)},t_{m\pm}) > 0 \end{array} \right. \right\} \subset \partial\Omega_{ij}. \tag{5.21}$$

The *double imaginary, inflexional singular set* on both sides of boundary $\partial\Omega_{ij}$ is defined as

$$\mathscr{Q}_{ij}^{(2:2)}=\mathscr{Q}_{ij}^{(2,\alpha)}\cup\mathscr{Q}_{ij}^{(2,\beta)}$$
$$=\left\{(\mathbf{x}_m,t_m)\left|\begin{array}{l}\hbar_\alpha G_{\partial\Omega_{ij}}^{(s_\beta,\beta)}(\mathbf{x}_\alpha^{(\beta)},t_{m\pm})=0\text{ for }s_\beta=0,1,\\ \hbar_\alpha G_{\partial\Omega_{ij}}^{(2,\beta)}(\mathbf{x}_\alpha^{(\beta)},t_{m\pm})>0;\\ \hbar_\alpha G_{\partial\Omega_{ij}}^{(s_\alpha,\alpha)}(\mathbf{x}_\beta^{(\alpha)},t_{m\pm})=0\text{ for }s_\alpha=0,1,\\ \hbar_\alpha G_{\partial\Omega_{ij}}^{(2,\alpha)}(\mathbf{x}_\beta^{(\alpha)},t_{m\pm})<0\end{array}\right.\right\}\subset\partial\Omega_{ij}. \tag{5.22}$$

Definition 5.10. For a discontinuous dynamical system of Eq. (5.10), the $(2k_\beta+1)$ th-*order, imaginary grazing singular set* on the α -side of boundary $\partial\Omega_{ij}$ for $\alpha,\beta\in\{i,j\}$ and $\alpha\neq\beta$ is defined as

$$\mathscr{S}_{ij}^{(2k_\beta+1,\alpha)}=\left\{(\mathbf{x}_m,t_m)\left|\begin{array}{l}\hbar_\alpha G_{\partial\Omega_{ij}}^{(s_\beta,\beta)}(\mathbf{x}_\alpha^{(\beta)},t_{m\pm})=0\\ \text{for }s_\beta=0,1,2,\cdots,2k_\beta\\ \hbar_\alpha G_{\partial\Omega_{ij}}^{(2k_\beta+1,\beta)}(\mathbf{x}_\alpha^{(\beta)},t_{m\pm})>0\end{array}\right.\right\}\subset\partial\Omega_{ij}. \tag{5.23}$$

The $(2k_\alpha+1:2k_\beta+1)$ *-double imaginary grazing singular set* on both sides of boundary $\partial\Omega_{ij}$ is defined as

$$\mathscr{S}_{ij}^{(2k_\beta+1:2k_\alpha+1)}=\mathscr{S}_{ij}^{(2k_\beta+1,\alpha)}\cup\mathscr{S}_{ij}^{(2k_\alpha+1,\beta)}$$
$$=\left\{(\mathbf{x}_m,t_m)\left|\begin{array}{l}\hbar_\alpha G_{\partial\Omega_{ij}}^{(s_\beta,\beta)}(\mathbf{x}_\alpha^{(\beta)},t_{m\pm})=0\text{ for }s_\beta=0,1,\cdots,2k_\beta,\\ \hbar_\alpha G_{\partial\Omega_{ij}}^{(2k_\beta+1,\beta)}(\mathbf{x}_\alpha^{(\beta)},t_{m\pm})<0;\\ \hbar_\alpha G_{\partial\Omega_{ij}}^{(s_\alpha,\alpha)}(\mathbf{x}_\beta^{(\alpha)},t_{m\pm})=0\text{ for }s_\alpha=0,1,\cdots,2k_\alpha,\\ \hbar_\alpha G_{\partial\Omega_{ij}}^{(2k_\alpha+1,\alpha)}(\mathbf{x}_\beta^{(\alpha)},t_{m\pm})>0\end{array}\right.\right\}$$
$$\subset\partial\Omega_{ij}. \tag{5.24}$$

Definition 5.11. For a discontinuous dynamical system of Eq. (5.10), the $(2k_\alpha)$ th-*order, imaginary inflexional singular set* on the α -side of boundary $\partial\Omega_{ij}$ for $\alpha,\beta\in\{i,j\}$ and $\alpha\neq\beta$ is defined as

$$\mathscr{Q}_{ij}^{(2k_\beta,\alpha)}=\left\{(\mathbf{x}_m,t_m)\left|\begin{array}{l}\hbar_\alpha G_{\partial\Omega_{ij}}^{(s_\beta,\beta)}(\mathbf{x}_\alpha^{(\beta)},t_{m\pm})=0\\ \text{for }s_\alpha=0,1,\cdots,2k_\alpha-1\\ \hbar_\alpha G_{\partial\Omega_{ij}}^{(2k_\beta,\beta)}(\mathbf{x}_m,t_{m\pm})<0\end{array}\right.\right\}\subset\partial\Omega_{ij}. \tag{5.25}$$

The $(2k_\alpha:2k_\beta)$ *-double imaginary inflexional singular set* on both sides of boundary $\partial\Omega_{ij}$ is defined as

$$\begin{aligned}\boldsymbol{Q}_{ij}^{(2k_\alpha:2k_\beta)} &= \boldsymbol{Q}_{ij}^{(2k_\beta,\alpha)} \cup \boldsymbol{Q}_{ij}^{(2k_\alpha,\beta)}\\ &= \left\{ (\mathbf{x}_m, t_m) \left| \begin{array}{l} \hbar_\alpha G_{\partial\Omega_{ij}}^{(s_\beta,\beta)}(\mathbf{x}_\alpha^{(\beta)}, t_{m\pm}) = 0 \text{ for } s_\beta = 0,1,\cdots,2k_\beta - 1,\\ \hbar_\alpha G_{\partial\Omega_{ij}}^{(2k_\beta,\beta)}(\mathbf{x}_\alpha^{(\beta)}, t_{m\pm}) > 0;\\ \hbar_\alpha G_{\partial\Omega_{ij}}^{(s_\alpha,\alpha)}(\mathbf{x}_\beta^{(\alpha)}, t_{m\pm}) = 0 \text{ for } s_\alpha = 0,1,\cdots,2k_\alpha - 1,\\ \hbar_\alpha G_{\partial\Omega_{ij}}^{(2k_\alpha,\alpha)}(\mathbf{x}_\beta^{(\alpha)}, t_{m\pm}) < 0 \end{array} \right. \right\} \\ &\subset \partial\Omega_{ij}. \end{aligned} \tag{5.26}$$

Suppose there are two open subsets of the boundary (i.e., $S_I \subset \partial\Omega_{ij}, I = 1,2$) and a singular sets (i.e., $\Gamma_{12} = \bar{S}_1 \cap \bar{S}_2 \subset \partial\Omega_{ij}$). Thus, $\partial\Omega_{ij} = S_1 \cup S_2 \cup \Gamma_{12}$. If two passable flows passing through the two open subsets satisfy

$$\hbar_\alpha G_{\partial\Omega_{ij}}^{(\alpha)}(\mathbf{x}_m, t_{m-}) > 0 \text{ and } \hbar_\alpha G_{\partial\Omega_{ij}}^{(\beta)}(\mathbf{x}_m, t_{m+}) > 0 \tag{5.27}$$

for $\mathbf{x}_m \in S_1 \subset \partial\Omega_{ij}$ and

$$\hbar_\alpha G_{\partial\Omega_{ij}}^{(\beta)}(\mathbf{x}_m, t_{m-}) < 0 \text{ and } \hbar_\alpha G_{\partial\Omega_{ij}}^{(\alpha)}(\mathbf{x}_m, t_{m+}) < 0 \tag{5.28}$$

for $\mathbf{x}_m \in S_2 \subset \partial\Omega_{ij}$, then there is a singular set for $(\mathbf{x}_m, t_m) \in \Gamma_{12} = \boldsymbol{S}_{ij}^{(1:1)}$ with

$$\begin{aligned} &\hbar_\alpha G_{\partial\Omega_{ij}}^{(\alpha)}(\mathbf{x}_m, t_{m\mp}) = 0 \text{ and } \hbar_\alpha G_{\partial\Omega_{ij}}^{(1,\alpha)}(\mathbf{x}_m, t_{m\mp}) < 0,\\ &\hbar_\alpha G_{\partial\Omega_{ij}}^{(\beta)}(\mathbf{x}_m, t_{m\pm}) = 0 \text{ and } \hbar_\alpha G_{\partial\Omega_{ij}}^{(1,\beta)}(\mathbf{x}_m, t_{m\pm}) > 0. \end{aligned} \tag{5.29}$$

For a general case, if two passable flows passing through the two open sets are of the $(2k_\alpha : 2k_\beta)$-order, i.e.,

$$\begin{aligned} &\hbar_\alpha G_{\partial\Omega_{ij}}^{(s_\alpha,\alpha)}(\mathbf{x}_m, t_{m\pm}) = 0 \text{ for } s_\alpha = 0,1,\cdots,2k_\alpha,\\ &\hbar_\alpha G_{\partial\Omega_{ij}}^{(2k_\alpha,\alpha)}(\mathbf{x}_m, t_{m-}) > 0,\\ &\hbar_\alpha G_{\partial\Omega_{ij}}^{(s_\beta,\beta)}(\mathbf{x}_m, t_{m\pm}) = 0 \text{ for } s_\beta = 0,1,\cdots,2k_\beta,\\ &\hbar_\alpha G_{\partial\Omega_{ij}}^{(2k_\beta,\beta)}(\mathbf{x}_m, t_{m+}) > 0 \end{aligned} \tag{5.30}$$

for $\mathbf{x}_m \in S_1 \subset \partial\Omega_{ij}$ and

$$\begin{aligned} &\hbar_\alpha G_{\partial\Omega_{ij}}^{(s_\beta,\beta)}(\mathbf{x}_m, t_{m\pm}) = 0 \text{ for } s_\beta = 0,1,\cdots,2k_\beta,\\ &\hbar_\alpha G_{\partial\Omega_{ij}}^{(2k_\beta,\beta)}(\mathbf{x}_m, t_{m-}) < 0,\\ &\hbar_\alpha G_{\partial\Omega_{ij}}^{(s_\alpha,\alpha)}(\mathbf{x}_m, t_{m\pm}) = 0 \text{ for } s_\alpha = 0,1,\cdots,2k_\alpha, \end{aligned} \tag{5.31a}$$

$$\hbar_{\alpha} G_{\partial\Omega_{ij}}^{(2k_{\alpha},\alpha)}(\mathbf{x}_m, t_{m+}) < 0 \tag{5.31b}$$

for $\mathbf{x}_m \in S_2 \subset \partial\Omega_{ij}$, then there is a $(2k_{\alpha}+1:2k_{\beta}+1)$-singular set $\Gamma_{12} = \mathcal{S}_{ij}^{(2k_{\alpha}+1:2k_{\beta}+1)}$ with

$$\left.\begin{aligned}
&\hbar_{\alpha} G_{\partial\Omega_{ij}}^{(s_{\alpha},\alpha)}(\mathbf{x}_m, t_{m\mp}) = 0 \text{ for } s_{\alpha} = 0,1,\cdots,2k_{\alpha},\\
&\hbar_{\alpha} G_{\partial\Omega_{ij}}^{(2k_{\alpha}+1,\alpha)}(\mathbf{x}_m, t_{m\mp}) < 0;\\
&\hbar_{\alpha} G_{\partial\Omega_{ij}}^{(s_{\beta},\beta)}(\mathbf{x}_m, t_{m\pm}) = 0 \text{ for } s_{\beta} = 0,1,\cdots,2k_{\beta},\\
&\hbar_{\alpha} G_{\partial\Omega_{ij}}^{(2k_{\beta}+1,\beta)}(\mathbf{x}_m, t_{m\pm}) > 0.
\end{aligned}\right\} \tag{5.32}$$

The general case is illustrated in Fig. 5.2 for a better understanding of the singular set. For $\mathbf{x}_{m-}^{(i)}, \mathbf{x}_{m+}^{(j)} \in S_1$, the coming and leaving flows satisfy $G_{\partial\Omega_{ij}}^{(2k_i,i)} > 0$ and $G_{\partial\Omega_{ij}}^{(2k_j,j)} > 0$, respectively. For $\mathbf{x}_{m+}^{(i)}, \mathbf{x}_{m-}^{(j)} \in S_2$, $G_{\partial\Omega_{ij}}^{(2k_i,i)} < 0$ and $G_{\partial\Omega_{ij}}^{(2k_j,j)} < 0$ hold. The grazing singular set on the boundary should be double grazing singular sets, i.e., $\Gamma_{12} = \mathcal{S}_{ij}^{(2k_{\alpha}+1:2k_{\beta}+1)}$. In Fig. 5.2 (a), two passable flows relative to the subsets S_1 and S_2 are sketched by solid curves, and the passable directions of the two passable flows on the boundary are opposite. In Fig. 5.2(b), it is shown that the two grazing flows occur at the singular set. The $(2k_i:2k_j)$-passable flows and $(2k_i:2k_j)$-sink flows on the two subsets of the boundary with a singular set of $\Gamma_{12} = \mathcal{S}_{ij}^{(2k_j+1,j)}$ is sketched in Fig. 5.3 for $k_i, k_j \in \{0,1,2,\cdots\}$. The $(2k_i:2k_j)$-passable and $(2k_i:2k_j)$-source flows on the two subsets with a singular set of $\Gamma_{12} = \mathcal{S}_{ij}^{(2k_i+1,i)}$ is sketched in Fig. 5.4. The $(2k_i:2k_j)$-sink and $(2k_i:2k_j)$-source flows on the two subsets with a singular set of $\Gamma_{12} = \mathcal{S}_{ij}^{(2k_i+1:2k_j+1)}$ is sketched in Fig. 5.5. The discussions can be done as the two passable flows on the two subsets. Such illustrations give an intuitive explanation of the singular sets.

The afore-discussion is on the singularity among the $(2k_{\alpha}:2k_{\beta})$-passable, sink and source flows. The singularity among the $(2k_{\alpha}:2k_{\beta}-1)$-passable, sink and source flows involved with imaginary flows is also discussed herein. Consider the there are two open subsets of the boundary (i.e., $S_I \subset \partial\Omega_{ij}$, $I = 1,2$) and a singular sets (i.e., $\Gamma_{12} = \bar{S}_1 \cap \bar{S}_2 \subset \partial\Omega_{ij}$). If passable and source flows intersected with the two open sets are of the $(2k_{\alpha}:2k_{\beta}-1)$-order, i.e.,

$$\left.\begin{aligned}
&\hbar_{\alpha} G_{\partial\Omega_{ij}}^{(s_{\alpha},\alpha)}(\mathbf{x}_m, t_{m\mp}) = 0 \text{ for } s_{\alpha} = 0,1,\cdots,2k_{\alpha}-1\\
&\hbar_{\alpha} G_{\partial\Omega_{ij}}^{(2k_{\alpha},\alpha)}(\mathbf{x}_m, t_{m-}) > 0
\end{aligned}\right\} \text{for } \mathbf{x}_m \in S_1 \subset \partial\Omega_{ij},$$

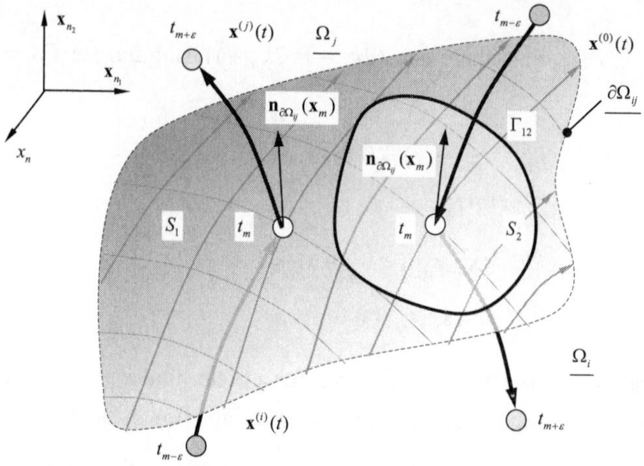

(a)

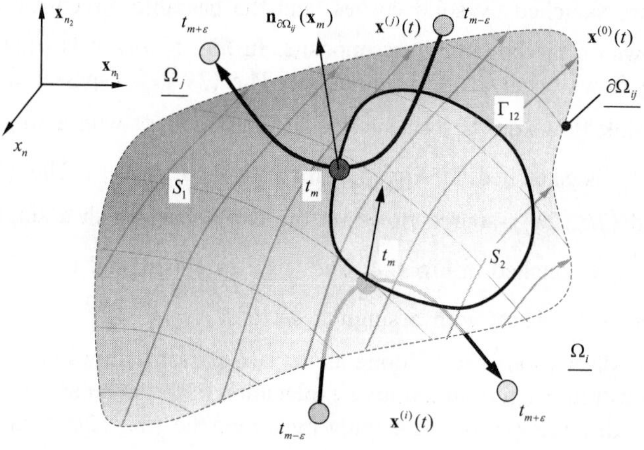

(b)

Fig. 5.2 Double grazing singular sets between two subsets on the boundary for the $(2k_i : 2k_j)$ and $(2k_j : 2k_i)$ passable flows: **(a)** on subsets and **(b)** singular sets of $\mathcal{S}_{ij}^{(2k_i+1:2k_j+1)}$. The flows $\mathbf{x}^{(i)}(t)$ and $\mathbf{x}^{(j)}(t)$ in domains Ω_i and Ω_j are depicted by solid curves, respectively. The flows on the boundary are described by $\mathbf{x}^{(0)}(t)$. The normal vector $\mathbf{n}_{\partial\Omega_{ij}}$ of the boundary is depicted with $\mathbf{n}_{\partial\Omega_{ij}} \to \Omega_j$. The hollow circular symbols are switching points on the subset of the boundary. The filled dark cycles are points on singular sets. The gray filled circular symbols are the points in the domains. ($n_1 + n_2 + 1 = n$)

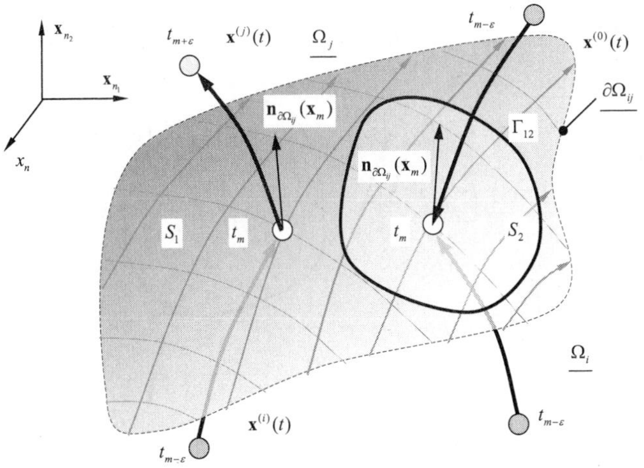

(a)

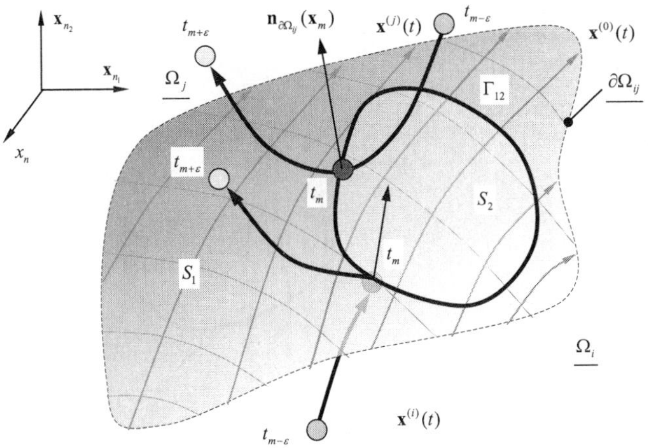

(b)

Fig. 5.3 Single grazing singular sets between the subsets on the boundary for the $(2k_i:2k_j)$-passable and $(2k_i:2k_j)$-sink flows: **(a)** on subsets and **(b)** on the singular set of $\mathcal{S}_{ij}^{(2k_j+1,j)}$. The flows $\mathbf{x}^{(i)}(t)$ and $\mathbf{x}^{(j)}(t)$ in domains Ω_i and Ω_j are depicted by solid curves, respectively. The flows on the boundary are described by $\mathbf{x}^{(0)}(t)$. The normal vector $\mathbf{n}_{\partial\Omega_{ij}}$ of the boundary is depicted with $\mathbf{n}_{\partial\Omega_{ij}} \to \Omega_j$. The hollow circular symbols are switching points on the subset of the boundary. The filled dark cycles are points on singular sets. The gray filled circular symbols are the points in the domains. ($n_1+n_2+1=n$)

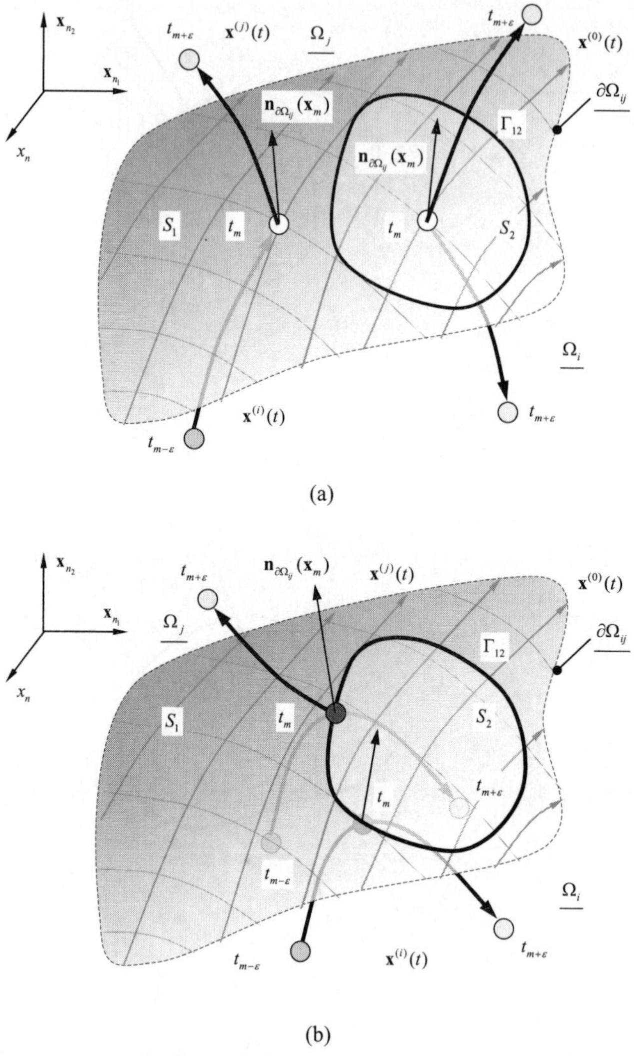

Fig. 5.4 Single grazing singular sets between the subsets on the boundary relative to the $(2k_i : 2k_j)$-passable and $(2k_i : 2k_j)$-source flows: **(a)** on subsets and **(b)** on the singular set of $\mathcal{S}_{ij}^{(2k_i+1,i)}$. The flows $\mathbf{x}^{(i)}(t)$ and $\mathbf{x}^{(j)}(t)$ in domains Ω_i and Ω_j are depicted by solid curves, respectively. The flows on the boundary are described by $\mathbf{x}^{(0)}(t)$. The normal vector $\mathbf{n}_{\partial\Omega_{ij}}$ of the boundary is depicted with $\mathbf{n}_{\partial\Omega_{ij}} \to \Omega_j$. The hollow circular symbols are switching points on the subset of the boundary. The filled dark cycles are points on singular sets. The gray filled circular symbols are the points in the domains. ($n_1 + n_2 + 1 = n$)

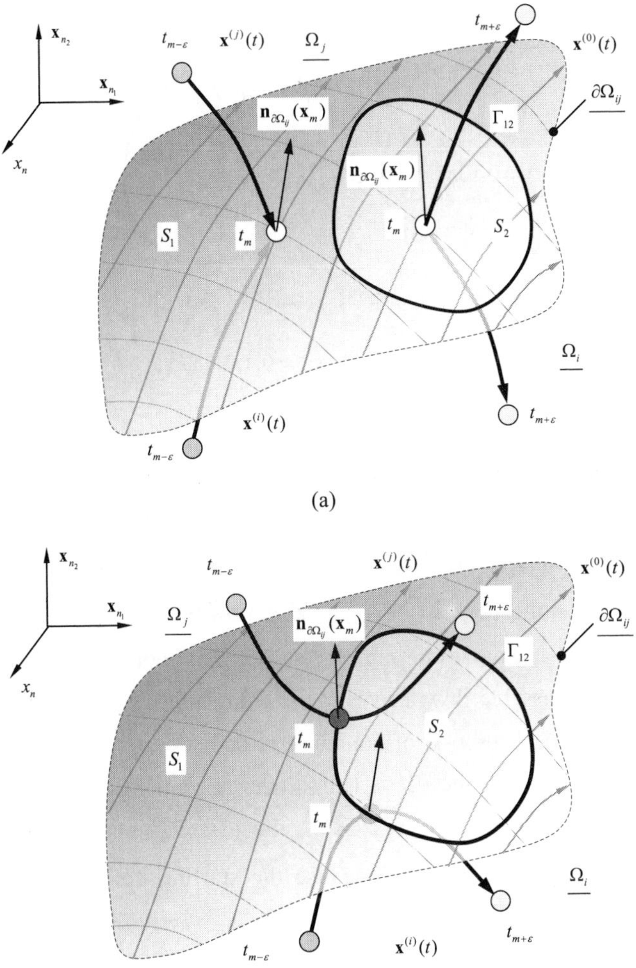

Fig. 5.5 Double grazing singular sets between two subsets on the boundary for the $(2k_i : 2k_j)$-sink and $(2k_i : 2k_j)$-source flows: **(a)** on subsets and **(b)** on the singular set of $\mathcal{S}_{ij}^{(2k_i+1:2k_j+1)}$. The flows $\mathbf{x}^{(i)}(t)$ and $\mathbf{x}^{(j)}(t)$ in domains Ω_i and Ω_j are depicted by solid curves, respectively. The flows on the boundary are described by $\mathbf{x}^{(0)}(t)$. The normal vector $\mathbf{n}_{\partial\Omega_{ij}}$ of the boundary is depicted with $\mathbf{n}_{\partial\Omega_{ij}} \to \Omega_j$. The hollow circular symbols are switching points on the subset of the boundary. The filled dark cycles are points on singular sets. The gray filled circular symbols are the points in the domains. ($n_1 + n_2 + 1 = n$)

$$\left.\begin{array}{l}\hbar_\alpha G_{\partial\Omega_{ij}}^{(s_\beta,\beta)}(\mathbf{x}_m,t_{m\mp})=0 \text{ for } s_\beta=0,1,\cdots,2k_\beta\\ \hbar_\alpha G_{\partial\Omega_{ij}}^{(2k_\beta+1,\beta)}(\mathbf{x}_m,t_{m+})>0\end{array}\right\}\text{for } \mathbf{x}_m\in S_1\subset\partial\Omega_{ij}, \tag{5.33}$$

$$\left.\begin{array}{l}\hbar_\alpha G_{\partial\Omega_{ij}}^{(s_\alpha,\alpha)}(\mathbf{x}_m,t_{m\mp})=0 \text{ for } s_\alpha=0,1,\cdots,2k_\alpha-1\\ \hbar_\alpha G_{\partial\Omega_{ij}}^{(2k_\alpha,\alpha)}(\mathbf{x}_m,t_{m+})<0,\\ \hbar_\alpha G_{\partial\Omega_{ij}}^{(s_\beta,\beta)}(\mathbf{x}_m,t_{m\mp})=0 \text{ for } s_\beta=0,1,\cdots,2k_\beta\\ \hbar_\alpha G_{\partial\Omega_{ij}}^{(2k_\beta+1,\beta)}(\mathbf{x}_m,t_{m+})>0\end{array}\right\}\text{for } \mathbf{x}_m\in S_2\subset\partial\Omega_{ij}; \tag{5.34}$$

then there is a $(2k_\alpha-1)$th -singular set $\Gamma_{12}=\mathcal{S}_{ij}^{(2k_\alpha+1,\alpha)}$ with

$$\begin{array}{l}\hbar_\alpha G_{\partial\Omega_{ij}}^{(s_\alpha,\alpha)}(\mathbf{x}_m,t_{m\mp})=0 \text{ for } s_\alpha=0,1,\cdots,2k_\alpha,\\ \hbar_\alpha G_{\partial\Omega_{ij}}^{(2k_\alpha+1,\alpha)}(\mathbf{x}_m,t_{m\mp})<0.\end{array} \tag{5.35}$$

Such a singular set between two subsets on the boundary for the $(2k_i:2k_j-1)$-passable flows and $(2k_i:2k_j-1)$-source flows is sketched in Fig. 5.6. Consider a flow relative to a subset $S_1\subset\partial\Omega_{ij}$. Suppose a starting point is on $\mathbf{x}^{(i)}(t)$ at time t_{m-s} in domain Ω_i, the $(2k_i:2k_j-1)$-passable flow with $G_{\partial\Omega_{ij}}^{(2k_i,i)}>0$ and $G_{\partial\Omega_{ij}}^{(2k_j-1,j)}>0$ can be formed and passes through the subset S_1 of boundary. In other words, the $(2k_i:2k_j-1)$-passable flow arrives to the boundary by the flow of $\mathbf{x}^{(i)}(t)$ and tangentially leave the subset S_1 of boundary at point $(\mathbf{x}_m,t_m)\in S_1$ because a flow of $\mathbf{x}^{(j)}(t)$ in domain Ω_j is the $(2k_j-1)$th -order, tangential flow, as shown in Fig. 5.6(a). Similarly, in subset of $S_2\subset\partial\Omega_{ij}$, the $(2k_i:2k_j-1)$-source flow with $G_{\partial\Omega_{ij}}^{(2k_i,i)}<0$ and $G_{\partial\Omega_{ij}}^{(2k_j-1,j)}>0$ exists because the starting point on the boundary $(\mathbf{x}_m,t_m)\in S_2$. If the starting point $(\mathbf{x}_m,t_m)\in S_2$ is exactly on the boundary, the flow will be along boundary flow $\mathbf{x}^{(0)}(t)$. If a starting point is $(\mathbf{x}_{m+},t_{m+})\in\Omega_1$, the $(2k_i)$th -order flow of $\mathbf{x}^{(i)}(t)$ with $G_{\partial\Omega_{ij}}^{(2k_i,i)}<0$ will leave the boundary in domain Ω_i. If a starting point is $(\mathbf{x}_{m+},t_{m+})\in\Omega_2$, the $(2k_j-1)$th -order, tangential flow of $\mathbf{x}^{(i)}(t)$ with $G_{\partial\Omega_{ij}}^{(2k_j-1,j)}>0$ will tangentially leave the boundary in domain Ω_j. On the singular set $\Gamma_{12}=\mathcal{S}_{ij}^{(2k_i+1,i)}$, a flow on the Ω_i-side of boundary is the $(2k_i+1)$th -order tangential flow. On the Ω_j-side, the $(2k_j-1)$th -order tangential source flow exists. If a starting point for the $(2k_j-1)$th -order tangential source flow is at the filled circular symbol with a dashed edge, the complete $(2k_j-1)$th -order tangential flow to the boundary can be observed. The dotted curves are real

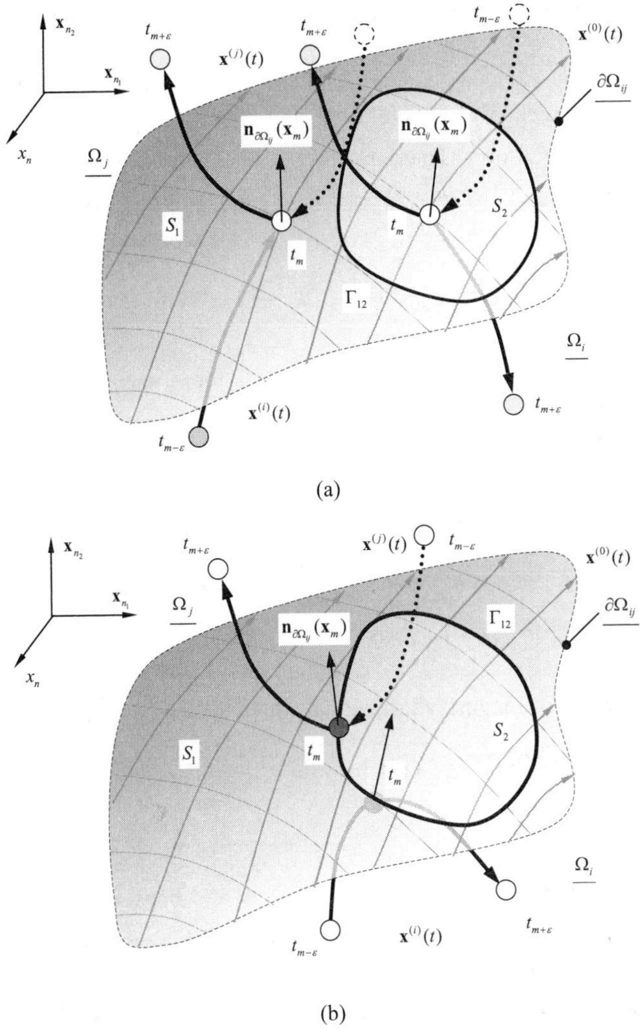

Fig. 5.6 Grazing singular set between two subsets on the boundary for $(2k_i : 2k_j - 1)$ - passable flows and $(2k_i : 2k_j - 1)$ -source flows: **(a)** on subsets and **(b)** on the singular set of $\mathcal{S}_{ij}^{(2k_i+1,i)}$. The flows $\mathbf{x}^{(i)}(t)$ and $\mathbf{x}^{(j)}(t)$ in domains Ω_i and Ω_j are depicted by solid curves, respectively. The flows on the boundary are described by $\mathbf{x}^{(0)}(t)$. The normal vector $\mathbf{n}_{\partial\Omega_{ij}}$ of the boundary is depicted with $\mathbf{n}_{\partial\Omega_{ij}} \to \Omega_j$. The dotted curves are real flows but the different initial condition does not start from there. The filled circular symbols with dashed edge are not real starting points. The hollow circular symbols are switching points on the subset of the boundary. The filled dark circular symbols are points on singular sets. The gray filled circular symbols are points in the domains. ($n_1 + n_2 + 1 = n$)

flows but a different initial condition does not start from time $t_{m-\varepsilon}$. However, the tangential source flow starts from a point $(\mathbf{x}_m, t_m)$ on the boundary, as shown in Fig. 5.6(b).

Consider a single grazing singular set between two subsets of boundary for the $(2k_\alpha : 2k_\beta + 1)$-tangential passable flow and the $(2k_\alpha : 2k_\beta + 1)$-tangential sink flow (or a half-sink flow). The $(2k_\alpha : 2k_\beta + 1)$-tangential passable flow satisfies Eq. (5.33). However, the $(2k_\alpha : 2k_\beta + 1)$-tangential sink flow (or a half-sink flow) satisfies

$$\left.\begin{aligned}
&\hbar_\alpha G_{\partial\Omega_{ij}}^{(s_\alpha,\alpha)}(\mathbf{x}_m, t_{m\mp}) = 0 \text{ for } s_\alpha = 0,1,\cdots,2k_\alpha - 1\\
&\hbar_\alpha G_{\partial\Omega_{ij}}^{(2k_\alpha,\alpha)}(\mathbf{x}_m, t_{m+}) < 0,\\
&\hbar_\alpha G_{\partial\Omega_{ij}}^{(s_\beta,\beta)}(\mathbf{x}_\alpha^{(\beta)}, t_{m\mp}) = 0 \text{ for } s_\beta = 0,1,\cdots,2k_\beta\\
&\hbar_\alpha G_{\partial\Omega_{ij}}^{(2k_\beta+1,\beta)}(\mathbf{x}_\alpha^{(\beta)}, t_{m+}) < 0
\end{aligned}\right\} \text{ for } \mathbf{x}_m \in S_2 \subset \partial\Omega_{ij}; \tag{5.36}$$

then the $(2k_\beta)$ th-order inflexional singular set $\Gamma_{12} = \mathscr{Q}_{ij}^{(2k_\beta,\beta)}$ exists with

$$\begin{aligned}
&\hbar_\alpha G_{\partial\Omega_{ij}}^{(s_\beta,\beta)}(\mathbf{x}_m, t_{m\mp}) = 0 \text{ for } s_\beta = 0,1,\cdots,2k_\beta - 1,\\
&\hbar_\alpha G_{\partial\Omega_{ij}}^{(2k_\beta,\beta)}(\mathbf{x}_m, t_{m\mp}) < 0.
\end{aligned} \tag{5.37}$$

or the $(2k_\beta)$ th-order imaginary inflexional singular set $\Gamma_{12} = \boldsymbol{Q}_{ij}^{(2k_\beta,\alpha)}$ exists with

$$\begin{aligned}
&\hbar_\alpha G_{\partial\Omega_{ij}}^{(s_\beta,\beta)}(\mathbf{x}_\alpha^{(\beta)}, t_{m\mp}) = 0 \text{ for } s_\beta = 0,1,\cdots,2k_\beta - 1,\\
&\hbar_\alpha G_{\partial\Omega_{ij}}^{(2k_\beta,\beta)}(\mathbf{x}_\alpha^{(\beta)}, t_{m\mp}) > 0.
\end{aligned} \tag{5.38}$$

The inflexional singular set $\mathscr{Q}_{ij}^{(2k_j,j)}$ of a real flow $\mathbf{x}^{(j)}$ between the $(2k_i : 2k_j - 1)$-passable tangential flow and imaginary tangential sink flow (or a half-sink flow) on subsets is sketched in Fig. 5.7. For $S_1 \subset \partial\Omega_{ij}$, the $(2k_i)$th -order, real flow of $\mathbf{x}^{(i)}$ exists in domain Ω_i and the $(2k_j - 1)$th -order, real tangential flow of $\mathbf{x}^{(j)}$ exists in domain Ω_j. Both of them form the $(2k_i : 2k_j - 1)$-passable tangential flow from domain Ω_i to Ω_j. For $S_2 \subset \partial\Omega_{ij}$, the $(2k_i)$th -order, real flow of $\mathbf{x}^{(i)}$ in domain Ω_i remains, but the $(2k_j - 1)$th -order, imaginary tangential flow of $\mathbf{x}_i^{(j)}$ exists in domain Ω_i rather than a real flow in domain Ω_j. Such a flow to the boundary is called the $(2k_i : 2k_j - 1)$-imaginary tangential sink flow (a half-sink flow in domain Ω_i). For a starting point in domain Ω_i, the imaginary flow $\mathbf{x}_i^{(j)}$ gives an imaginary tangential source flow. The flows in two subsets are depicted in Fig. 5.7(a). On the single inflexional singular set $\Gamma_{12} = \mathscr{Q}_{ij}^{(2k_j,j)}$, a real flow $\mathbf{x}^{(j)}$

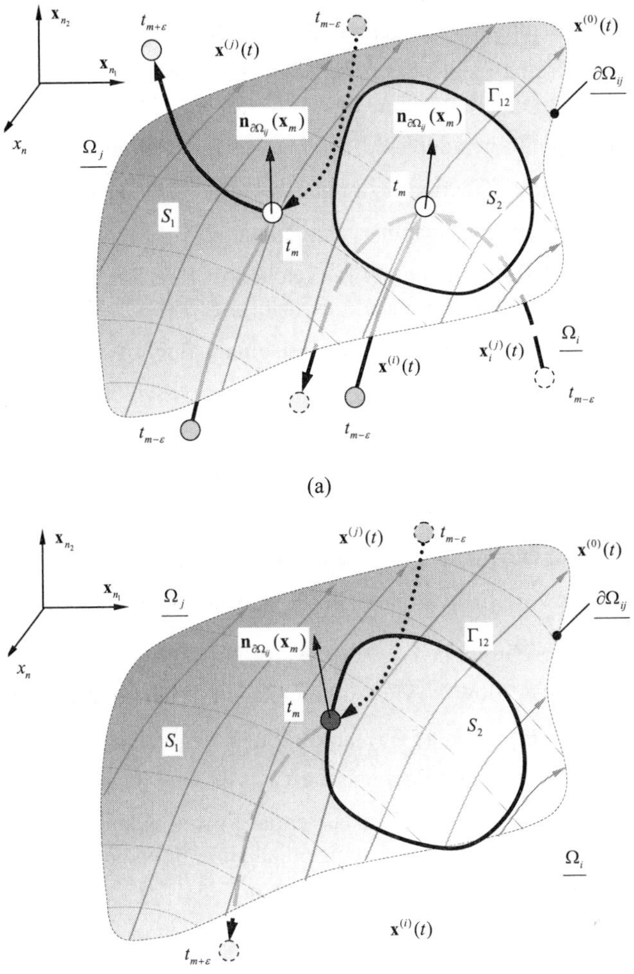

Fig. 5.7 Inflexional singular set of real flow $\mathbf{x}^{(j)}$ between two subsets on the boundary for the $(2k_i : 2k_j - 1)$ -passable tangential flow and the $(2k_i : 2k_j - 1)$ -imaginary tangential-sink (or half-sink flow) flows: **(a)** on subsets and **(b)** on singular sets of $\mathscr{Q}_{ij}^{(2k_j,j)}$. The flows $\mathbf{x}^{(i)}(t)$ and $\mathbf{x}^{(j)}(t)$ in domains Ω_i and Ω_j are depicted by solid curves, respectively. The flows on the boundary are described by $\mathbf{x}^{(0)}(t)$. The normal vector $\mathbf{n}_{\partial\Omega_{ij}}$ of the boundary is depicted with $\mathbf{n}_{\partial\Omega_{ij}} \to \Omega_j$. The dotted curves are real flows but the different initial condition does not start from there. The filled circular symbols with dashed edge are not real starting points. The hollow circular symbols are switching points on the subset of the boundary. The filled dark circular symbols are points on singular sets. The gray filled circular symbols are the points in the domains. ($n_1 + n_2 + 1 = n$)

starting from domain Ω_j extends to an imaginary flow of $\mathbf{x}_i^{(j)}$ with the $(2k_j)$th-order inflexion singularity, as sketched in Fig. 5.7(b). The switching from the $(2k_i : 2k_j - 1)$ passable-post tangential flow to the $(2k_i : 2k_j - 1)$ imaginary tangential-sink flow (a half sink) has been discussed. However, a switching from the $(2k_i : 2k_j - 1)$ imaginary tangential-sink flow to the $(2k_i : 2k_j - 1)$ passable-post tangential flow should be addressed herein. The flows on the two subsets remain same. The imaginary inflexional singular set $\boldsymbol{\mathcal{Q}}_{ij}^{(2k_j,i)}$ of the imaginary flow $\mathbf{x}_i^{(j)}$ between the $(2k_i : 2k_j - 1)$-imaginary tangential sink flow (half-sink flow) and passable tangential flow on subsets is sketched in Fig. 5.8. For comparison, the illustration in Fig. 5.8(a) is the same as in Fig. 5.7(a). On the single imaginary inflexional singular set $\Gamma_{12} = \boldsymbol{\mathcal{Q}}_{ij}^{(2k_j,i)}$, the imaginary flow of $\mathbf{x}_i^{(j)}$ starting from domain Ω_i extended to the real flow of $\mathbf{x}^{(j)}$ with the $(2k_j)$th-order inflexion singularity, as illustrated in Fig. 5.8(b).

For $S_1 \subset \partial\Omega_{ij}$, a coming flow of $\mathbf{x}^{(i)}$ with the $(2k_i)$th-order singularity in domain Ω_i and the $(2k_j - 1)$th-order tangential flow in domain Ω_j exist. However, the leaving flow to the boundary is one branch in domain Ω_j. For $S_2 \subset \partial\Omega_{ij}$, the coming flow of $\mathbf{x}^{(j)}$ remains in domain Ω_j and leaves the boundary tangentially as a leaving flow because of the $(2k_j - 1)$th-order tangency of the flow $\mathbf{x}^{(j)}$ to the boundary. The leaving flow of $\mathbf{x}^{(i)}$ is of the $(2k_i)$th-order. On the grazing singular set $\Gamma_{12} = \boldsymbol{\mathcal{S}}^{(2k_i+1,i)}$, Equation (5.35) holds. Such a grazing singular set for the $(2k_i : 2k_j - 1)$-tangential sink flow and the $(2k_i : 2k_j - 1)$-tangential source flow is presented in Fig. 5.9 (a) and (b). The tangential flow is independent of the flow in domain Ω_i. So the $(2k_i : 2k_j - 1)$-tangential sink flow possesses two coming flows and one leaving tangential flow, and the $(2k_i : 2k_j - 1)$-tangential source flow possesses one coming tangential flow and two leaving flows.

In a similar fashion, an inflexional singular set for the $(2k_i : 2k_j - 1)$-tangential sink flow and the $(2k_i : 2k_j - 1)$-imaginary tangential sink flow (half-sink flow) is presented in Fig. 5.10 (a) and (b). The single inflexional singular set $\Gamma_{12} = \boldsymbol{\mathcal{Q}}_{ij}^{(2k_j,j)}$ is with a starting point in domain Ω_j. An imaginary inflexional singular set for the $(2k_i : 2k_j - 1)$-*tangential* and *imaginary tangential* sink (half-sink flow) flows is presented in Fig. 5.11 (a) and (b). The single imaginary inflexional singular set is $\Gamma_{12} = \boldsymbol{\mathcal{Q}}_{ij}^{(2k_j,i)}$ with the starting point in domain Ω_i.

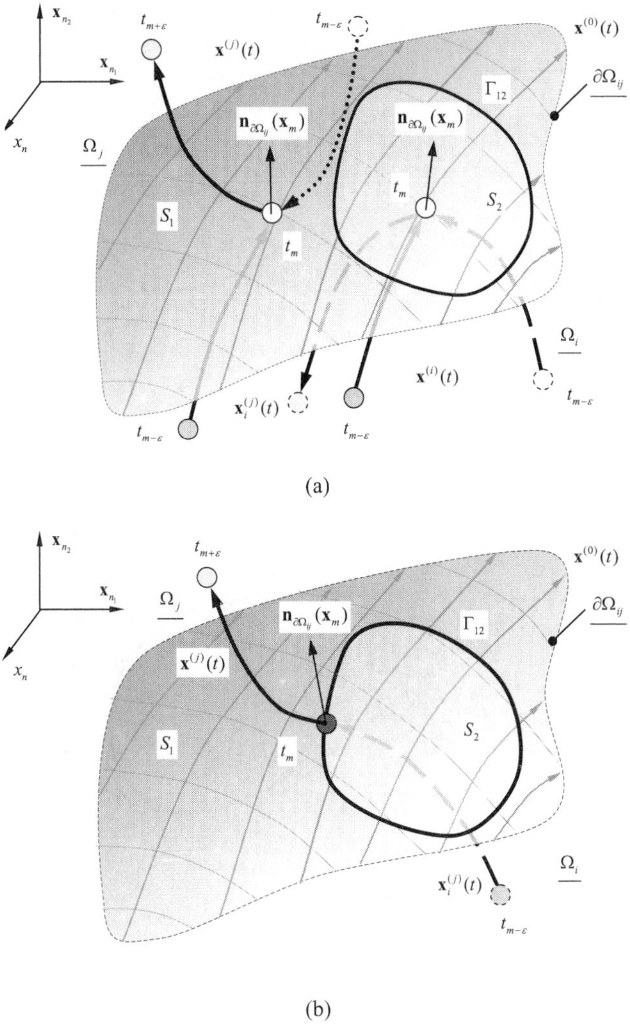

Fig. 5.8 Single grazing singular set between two subsets on the boundary for the $(2k_i : 2k_j - 1)$ -passable tangential flow and the $(2k_i : 2k_j - 1)$ -imaginary tangential-sink flow (or half-sink flow): **(a)** on subsets and **(b)** on singular set of $\mathscr{Q}_{ij}^{(2k_j,i)}$. The flows $\mathbf{x}^{(i)}(t)$ and $\mathbf{x}^{(j)}(t)$ in domains Ω_i and Ω_j are depicted by solid curves, respectively. The flows on the boundary are described by $\mathbf{x}^{(0)}(t)$. The normal vector $\mathbf{n}_{\partial\Omega_{ij}}$ of the boundary is depicted with $\mathbf{n}_{\partial\Omega_{ij}} \to \Omega_j$. The dotted curves are real flows but the different initial condition does not start from there. The filled circular symbols with dashed edge are not real starting points. The hollow circular symbols are switching points on the subset of the boundary. The filled dark circular symbols are points on singular sets. The gray filled circular symbols are the points in the domains. ($n_1 + n_2 + 1 = n$)

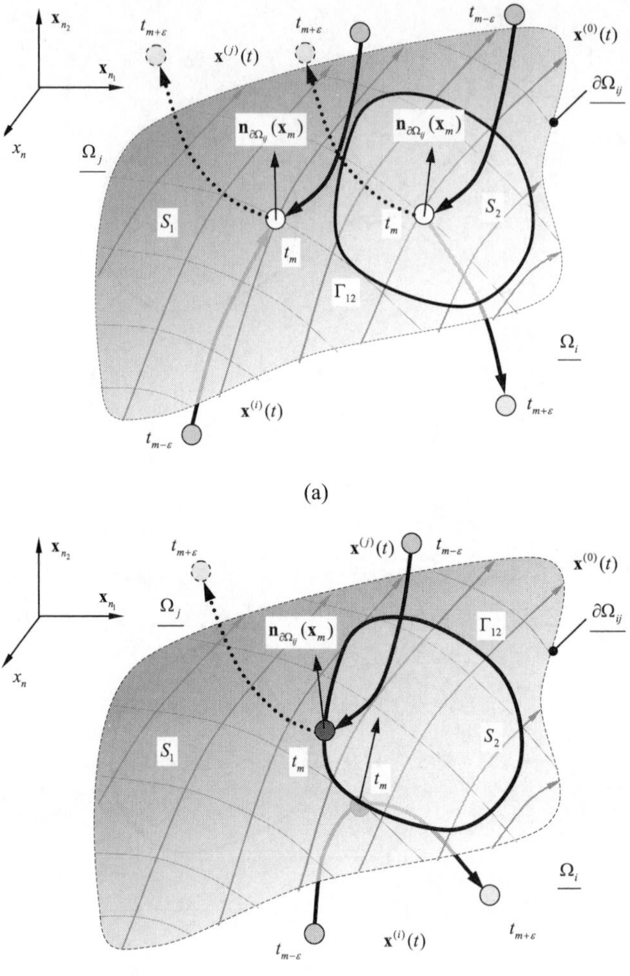

Fig. 5.9 Singular grazing set between two subsets on the $(2k_i : 2k_j - 1)$ -tangential sink flow and the $(2k_i : 2k_j - 1)$ - tangential-source flows on the boundary: **(a)** on subsets and **(b)** on the singular set of $\mathscr{C}_{ij}^{(2k_i,i)}$. The flows $\mathbf{x}^{(i)}(t)$ and $\mathbf{x}^{(j)}(t)$ in domains Ω_i and Ω_j are depicted by solid curves, respectively. The flows on the boundary are described by $\mathbf{x}^{(0)}(t)$. The normal vector $\mathbf{n}_{\partial\Omega_{ij}}$ of the boundary is depicted with $\mathbf{n}_{\partial\Omega_{ij}} \to \Omega_j$. The dotted curves are real flows. The filled circular symbols with dashed edge are not real starting points. The hollow circular symbols are switching points on the subset of the boundary. The filled dark circular symbols are points on singular sets. The gray filled circular symbols are the points in the domains. ($n_1 + n_2 + 1 = n$)

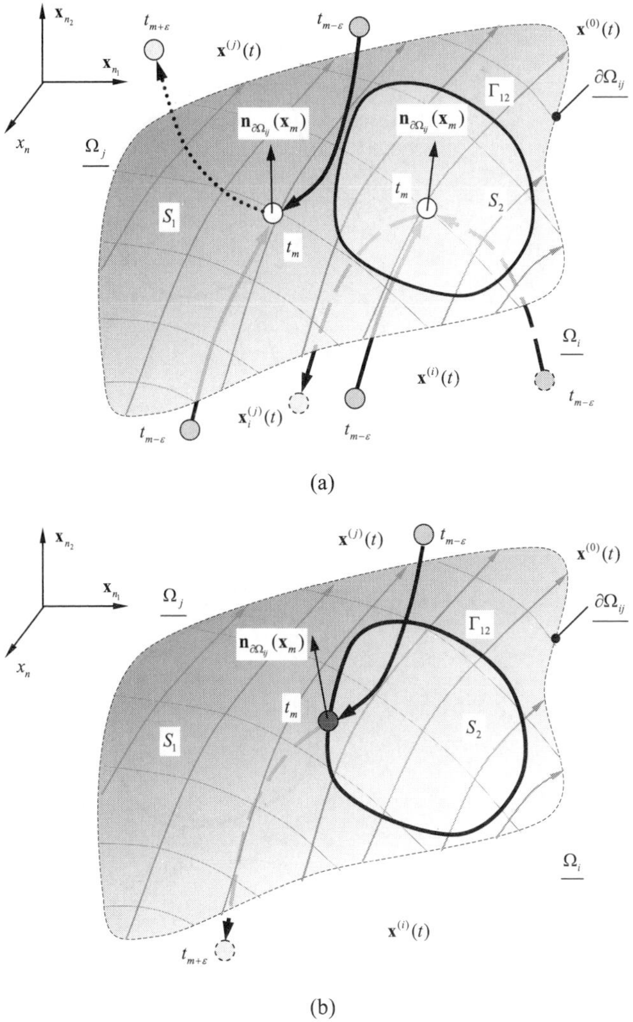

Fig. 5.10 Single inflexional singular set between two subsets on the boundary for the $(2k_i : 2k_j - 1)$ -tangential-sink and the $(2k_i : 2k_j - 1)$ -imaginary tangential-sink flow (or half-sink flow): **(a)** on subsets and **(b)** on the singular set of $\mathscr{C}_{ij}^{(2k_j,j)}$. The flows on the boundary are described by $\mathbf{x}^{(0)}(t)$. The normal vector $\mathbf{n}_{\partial\Omega_{ij}}$ of the boundary is depicted with $\mathbf{n}_{\partial\Omega_{ij}} \to \Omega_j$. The dotted curves are real flows. The filled circular symbols with dashed edge are not real starting points. The hollow circular symbols are switching points on the subset of the boundary. The filled dark circular symbols are points on singular sets. The gray filled circular symbols are the points in the domains. ($n_1 + n_2 + 1 = n$)

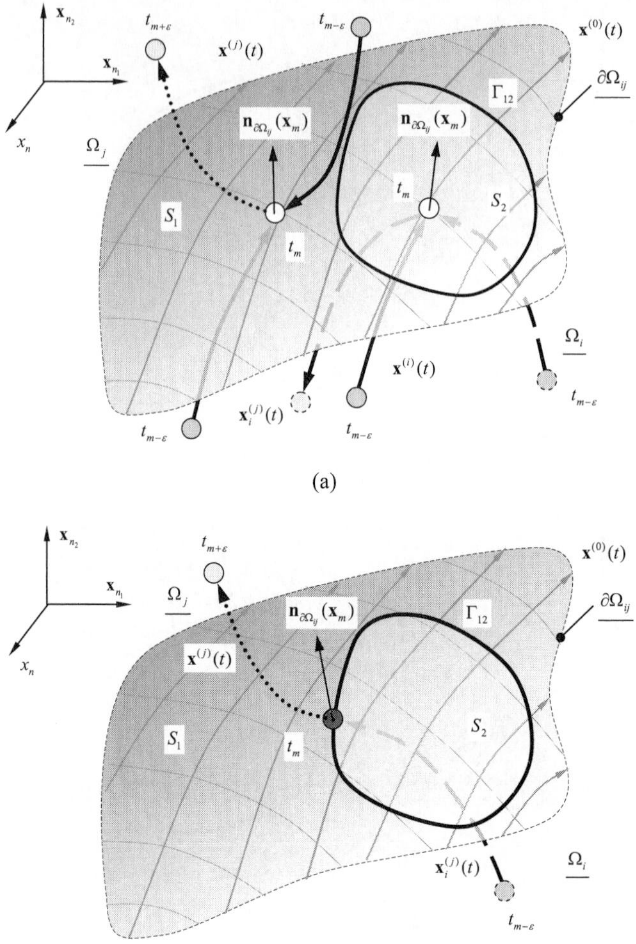

Fig. 5.11 Single imaginary inflexional singular set between two subset on the boundary for the $(2k_i : 2k_j - 1)$ -tangential-sink flow and the $(2k_i : 2k_j - 1)$ -imaginary tangential-sink flow (or half-sink flow): **(a)** on subsets and **(b)** on the singular set of $\boldsymbol{\wp}_{ij}^{(2k_j,i)}$. The flows on the boundary are described by $\mathbf{x}^{(0)}(t)$. The normal vector $\mathbf{n}_{\partial\Omega_{ij}}$ of the boundary is depicted with $\mathbf{n}_{\partial\Omega_{ij}} \to \Omega_j$. The dotted curves are real flows. The filled circular symbols with dashed edge are not real starting points. The hollow circular symbols are switching points on the subset of the boundary. The filled dark circular symbols are points on singular sets. The gray filled circular symbols are the points in the domains. ($n_1 + n_2 + 1 = n$)

Consider a single grazing singular set between two subsets of the boundary for the $(2k_\alpha : 2k_\beta - 1)$-tangential source flow and the $(2k_\alpha : 2k_\beta - 1)$-imaginary tangential source flow (or half-source flow), i.e.,

$$\left.\begin{aligned} &\hbar_\alpha G^{(s_\alpha,\alpha)}_{\partial\Omega_{ij}}(\mathbf{x}_m, t_{m\mp}) = 0 \text{ for } s_\alpha = 0,1,\cdots,2k_\alpha - 1 \\ &\hbar_\alpha G^{(2k_\alpha,\alpha)}_{\partial\Omega_{ij}}(\mathbf{x}_m, t_{m\mp}) < 0 \\ &\hbar_\alpha G^{(s_\beta,\beta)}_{\partial\Omega_{ij}}(\mathbf{x}_m, t_{m+}) = 0 \text{ for } s_\beta = 0,1,\cdots,2k_\beta - 2 \\ &\hbar_\alpha G^{(2k_\beta-1,\beta)}_{\partial\Omega_{ij}}(\mathbf{x}_m, t_{m+}) > 0 \end{aligned}\right\} \text{for } \mathbf{x}_m \in S_1 \subset \partial\Omega_{ij}, \tag{5.39}$$

$$\left.\begin{aligned} &\hbar_\alpha G^{(s_\alpha,\alpha)}_{\partial\Omega_{ij}}(\mathbf{x}^{(\alpha)}_\beta, t_{m\mp}) = 0 \text{ for } s_\alpha = 0,1,\cdots,2k_\alpha - 1 \\ &\hbar_\alpha G^{(2k_\alpha,\alpha)}_{\partial\Omega_{ij}}(\mathbf{x}^{(\alpha)}_\beta, t_{m\mp}) > 0 \\ &\hbar_\alpha G^{(s_\beta,\beta)}_{\partial\Omega_{ij}}(\mathbf{x}_m, t_{m+}) = 0 \text{ for } s_\beta = 0,1,\cdots,2k_\beta - 2 \\ &\hbar_\alpha G^{(2k_\beta-1,\beta)}_{\partial\Omega_{ij}}(\mathbf{x}_m, t_{m+}) > 0 \end{aligned}\right\} \text{for } \mathbf{x}_m \in S_2 \subset \partial\Omega_{ij}, \tag{5.40}$$

then there is a $(2k_\beta)$th -inflexional singular set $\Gamma_{12} = \mathscr{Q}^{(2k_\alpha:\alpha)}_{ij}$ with

$$\begin{aligned} &\hbar_\alpha G^{(s_\beta,\beta)}_{\partial\Omega_{ij}}(\mathbf{x}_m, t_{m\mp}) = 0 \text{ for } s_\beta = 0,1,\cdots,2k_\beta - 1, \\ &\hbar_\alpha G^{(2k_\beta,\beta)}_{\partial\Omega_{ij}}(\mathbf{x}_m, t_{m\mp}) > 0. \end{aligned} \tag{5.41}$$

or there is a $(2k_\beta)$th - imaginary inflexional singular set $\Gamma_{12} = \boldsymbol{\mathcal{Q}}^{(2k_\beta:\alpha)}_{ij}$ with

$$\begin{aligned} &\hbar_\alpha G^{(s_\beta,\beta)}_{\partial\Omega_{ij}}(\mathbf{x}^{(\beta)}_\alpha, t_{m\mp}) = 0 \text{ for } s_\beta = 0,1,\cdots,2k_\beta - 1, \\ &\hbar_\alpha G^{(2k_\beta,\beta)}_{\partial\Omega_{ij}}(\mathbf{x}^{(\beta)}_\alpha, t_{m\mp}) > 0. \end{aligned} \tag{5.42}$$

An inflexional singular set for the $(2k_i : 2k_j - 1)$-tangential source flow and the $(2k_i : 2k_j - 1)$-imaginary tangential source flow (half-sink flow) is presented in Fig. 5.12 (a) and (b). The single inflexional singular set is $\Gamma_{12} = \mathscr{Q}^{(2k_j,j)}_{ij}$ with the starting point in domain Ω_j. An imaginary inflexional singular set for the $(2k_i : 2k_j - 1)$-*tangential* and *imaginary tangential* source (half-source flow) flows is presented in Fig. 5.13 (a) and (b). The single imaginary inflexional singular set is $\Gamma_{12} = \boldsymbol{\mathcal{Q}}^{(2k_j,i)}_{ij}$ with the starting point in domain Ω_i.

Consider the singular set between the double $(2k_i - 1 : 2k_j - 1)$-tangential flows and the $(2k_i - 1 : 2k_j - 1)$-imaginary and real tangential flows, i.e.,

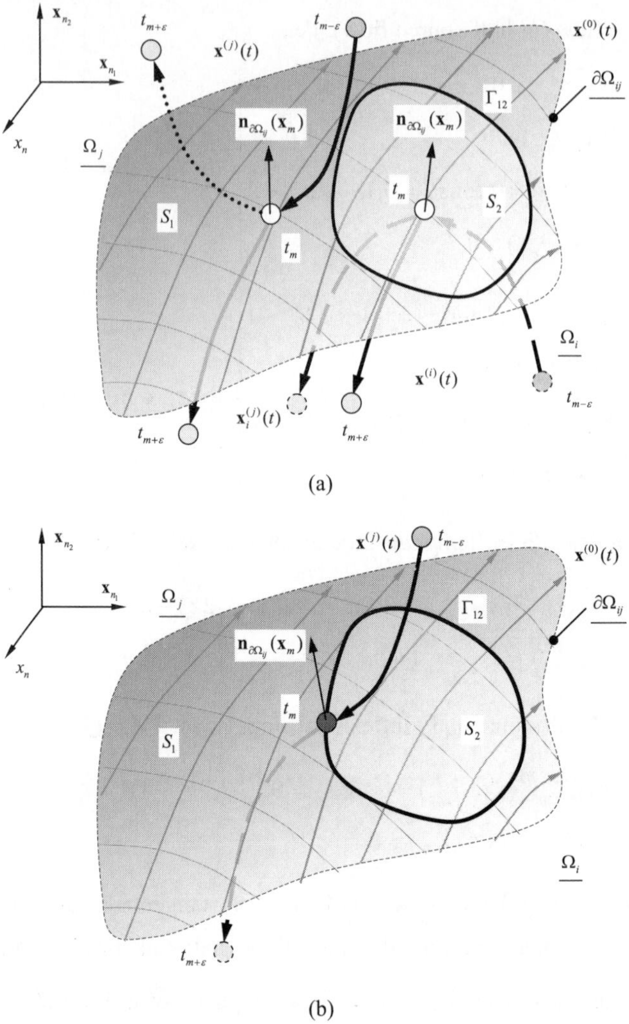

Fig. 5.12 Single real inflexional singular set between two subsets on the boundary for the $(2k_i : 2k_j - 1)$ -tangential-source flow and the $(2k_i : 2k_j - 1)$ -imaginary tangential-source flow (or half-source flow): **(a)** on subsets and **(b)** on the singular set of $\mathscr{Q}_{ij}^{(2k_j,j)}$. The flows on the boundary are described by $\mathbf{x}^{(0)}(t)$. The normal vector $\mathbf{n}_{\partial\Omega_{ij}}$ of the boundary is depicted with $\mathbf{n}_{\partial\Omega_{ij}} \to \Omega_j$. The dotted curves are real flows. The filled circular symbols with dashed edge are not real starting points. The hollow circular symbols are switching points on the subset of the boundary. The filled dark circular symbols are points on singular sets. The gray filled circular symbols are the points in the domains. ($n_1 + n_2 + 1 = n$)

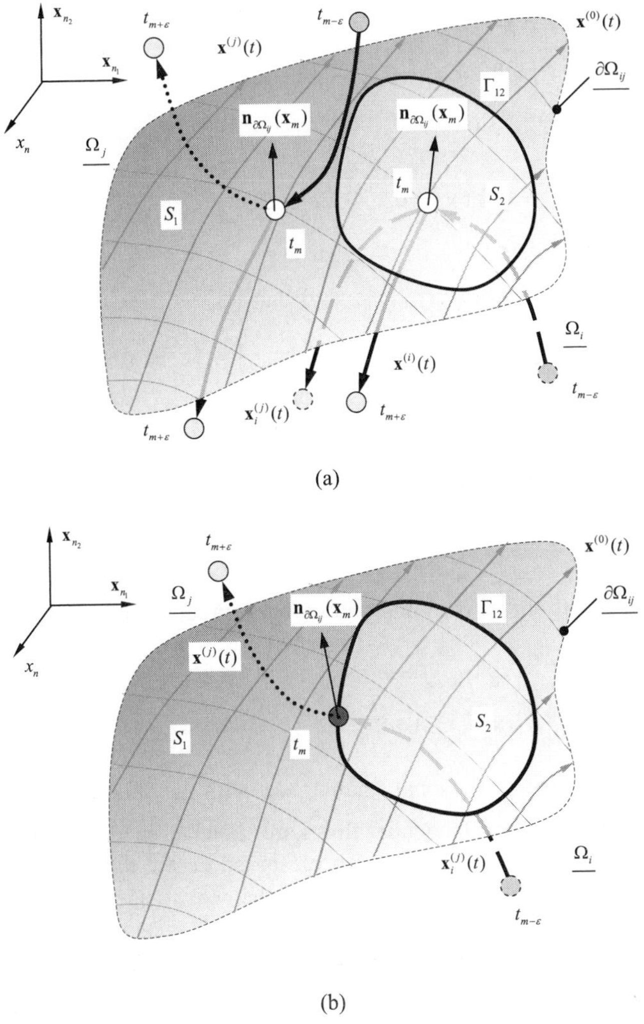

Fig. 5.13 Single imaginary inflexional singular set between two subsets on the boundary for the $(2k_i : 2k_j - 1)$ -tangential-sink flow and the $(2k_i : 2k_j - 1)$ -imaginary tangential-sink flow (or half-sink flow): **(a)** on subsets and **(b)** on the singular set of $\mathscr{Q}_{ij}^{(2k_j,i)}$. The flows on the boundary are described by $\mathbf{x}^{(0)}(t)$. The normal vector $\mathbf{n}_{\partial\Omega_{ij}}$ of the boundary is depicted with $\mathbf{n}_{\partial\Omega_{ij}} \to \Omega_j$. The dotted curves are real flows. The filled circular symbols with dashed edge are not real starting points. The hollow circular symbols are switching points on the subset of the boundary. The filled dark circular symbols are points on singular sets. The gray filled circular symbols are the points in the domains. ($n_1 + n_2 + 1 = n$)

$$\left.\begin{array}{l}\hbar_\alpha G_{\partial\Omega_{ij}}^{(s_\alpha,\alpha)}(\mathbf{x}_m,t_{m\mp})=0 \text{ for } s_\alpha=0,1,\cdots,2k_\alpha-2\\ \hbar_\alpha G_{\partial\Omega_{ij}}^{(2k_\alpha-1,\alpha)}(\mathbf{x}_m,t_{m\mp})<0\\ \hbar_\alpha G_{\partial\Omega_{ij}}^{(s_\beta,\beta)}(\mathbf{x}_m,t_{m+})=0 \text{ for } s_\beta=0,1,\cdots,2k_\beta-2\\ \hbar_\alpha G_{\partial\Omega_{ij}}^{(2k_\beta-1,\beta)}(\mathbf{x}_m,t_{m+})>0\end{array}\right\} \text{for } \mathbf{x}_m\in S_1\subset\partial\Omega_{ij}, \tag{5.43}$$

$$\left.\begin{array}{l}\hbar_\alpha G_{\partial\Omega_{ij}}^{(s_\alpha,\alpha)}(\mathbf{x}_m,t_{m\mp})=0 \text{ for } s_\alpha=0,1,\cdots,2k_\alpha-2\\ \hbar_\alpha G_{\partial\Omega_{ij}}^{(2k_\alpha-1,\alpha)}(\mathbf{x}_m,t_{m\mp})<0\\ \hbar_\alpha G_{\partial\Omega_{ij}}^{(s_\beta,\beta)}(\mathbf{x}_\alpha^{(\beta)},t_{m\mp})=0 \text{ for } s_\beta=0,1,\cdots,2k_\beta-2\\ \hbar_\alpha G_{\partial\Omega_{ij}}^{(2k_\beta-1,\beta)}(\mathbf{x}_\alpha^{(\beta)},t_{m\mp})<0\end{array}\right\} \text{for } \mathbf{x}_m\in S_2\subset\partial\Omega_{ij}, \tag{5.44}$$

then there is a $(2k_\beta)$th -order inflexional singular set $\Gamma_{12}=\mathscr{Q}_{ij}^{(2k_\beta:\beta)}$ with

$$\begin{array}{l}\hbar_\alpha G_{\partial\Omega_{ij}}^{(s_\beta,\beta)}(\mathbf{x}_m,t_{m\mp})=0 \text{ for } s_\beta=0,1,\cdots,2k_\beta-1,\\ \hbar_\alpha G_{\partial\Omega_{ij}}^{(2k_\beta,\beta)}(\mathbf{x}_m,t_{m\mp})>0.\end{array} \tag{5.45}$$

or there is a $(2k_\beta)$th -order imaginary inflexional singular set $\Gamma_{12}=\boldsymbol{\mathscr{Q}}_{ij}^{(2k_\beta:\alpha)}$ with

$$\begin{array}{l}\hbar_\alpha G_{\partial\Omega_{ij}}^{(s_\beta,\beta)}(\mathbf{x}_\alpha^{(\beta)},t_{m\mp})=0 \text{ for } s_\beta=0,1,\cdots,2k_\beta-1,\\ \hbar_\alpha G_{\partial\Omega_{ij}}^{(2k_\beta,\beta)}(\mathbf{x}_\alpha^{(\beta)},t_{m\mp})>0.\end{array} \tag{5.46}$$

The two $(2k_\beta)$th -order inflexional singular sets with the double -tangential flows and the imaginary and real tangential flows are sketched in Fig. 5.14 (a) and (b). The two real and imaginary singular sets are based on the different initial conditions. One flow on the singular set is from the real to imaginary portion, but another flow on the singular set is from the imaginary to real portion.

Consider the singular set between the double $(2k_i-1:2k_j-1)$ -imaginary tangential flows and the $(2k_i-1:2k_j-1)$ - imaginary and real tangential flows, i.e.,

$$\left.\begin{array}{l}\hbar_\alpha G_{\partial\Omega_{ij}}^{(s_\alpha,\alpha)}(\mathbf{x}_\beta^{(\alpha)},t_{m\mp})=0 \text{ for } s_\alpha=0,1,\cdots,2k_\alpha-2\\ \hbar_\alpha G_{\partial\Omega_{ij}}^{(2k_\alpha-1,\alpha)}(\mathbf{x}_\beta^{(\alpha)},t_{m\mp})<0\\ \hbar_\alpha G_{\partial\Omega_{ij}}^{(s_\beta,\beta)}(\mathbf{x}_\alpha^{(\beta)},t_{m+})=0 \text{ for } s_\beta=0,1,\cdots,2k_\beta-2\\ \hbar_\alpha G_{\partial\Omega_{ij}}^{(2k_\beta-1,\beta)}(\mathbf{x}_\alpha^{(\beta)},t_{m+})<0\end{array}\right\} \text{for } \mathbf{x}_m\in S_1\subset\partial\Omega_{ij}, \tag{5.47}$$

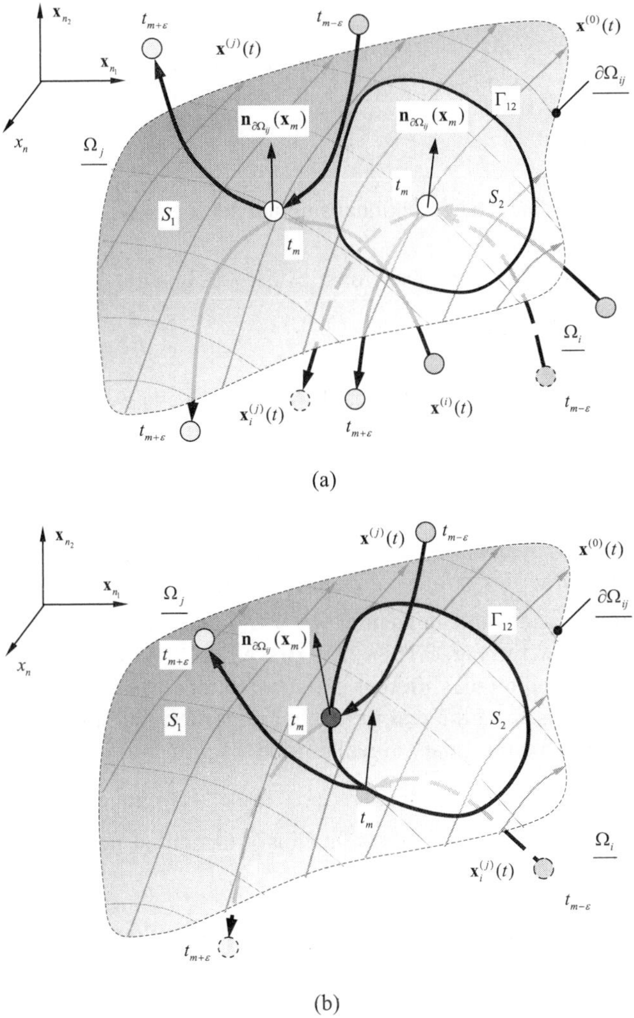

Fig. 5.14 Single real and imaginary inflexional singular sets between two subsets on the boundary for the $(2k_i - 1 : 2k_j - 1)$-doubling tangential flow and the $(2k_i - 1 : 2k_j - 1)$-imaginary and real tangential flows (or single tangential flow): **(a)** on subsets and **(b)** on the singular set of $\mathscr{Q}_{ij}^{(2k_j, j)}$ or $\mathbf{Q}_{ij}^{(2k_j, i)}$. The flows on the boundary are described by $\mathbf{x}^{(0)}(t)$. The normal vector $\mathbf{n}_{\partial\Omega_{ij}}$ of the boundary is depicted with $\mathbf{n}_{\partial\Omega_{ij}} \to \Omega_j$. The dotted curves are real flows. The filled circular symbols with dashed edge are not real starting points. The hollow circular symbols are switching points on the subset of the boundary. The filled dark circular symbols are points on singular sets. The gray filled circular symbols are the points in the domains. ($n_1 + n_2 + 1 = n$)

$$\left.\begin{array}{l}\hbar_{\alpha}G_{\partial\Omega_{ij}}^{(s_{\alpha},\alpha)}(\mathbf{x}_{m},t_{m\mp})=0\text{ for }s_{\alpha}=0,1,\cdots,2k_{\alpha}-2\\ \hbar_{\alpha}G_{\partial\Omega_{ij}}^{(2k_{\alpha}-1,\alpha)}(\mathbf{x}_{m},t_{m\mp})<0\\ \hbar_{\alpha}G_{\partial\Omega_{ij}}^{(s_{\beta},\beta)}(\mathbf{x}_{\alpha}^{(\beta)},t_{m\mp})=0\text{ for }s_{\beta}=0,1,\cdots,2k_{\beta}-2\\ \hbar_{\alpha}G_{\partial\Omega_{ij}}^{(2k_{\beta}-1,\beta)}(\mathbf{x}_{\alpha}^{(\beta)},t_{m\mp})<0\end{array}\right\}\text{ for }\mathbf{x}_{m}\in S_{2}\subset\partial\Omega_{ij},\tag{5.48}$$

then the $(2k_{\beta})$th - imaginary inflexional singular set $\Gamma_{12}=\boldsymbol{\mathscr{Q}}_{ij}^{(2k_{\alpha}:\beta)}$ exists with

$$\begin{array}{l}\hbar_{\alpha}G_{\partial\Omega_{ij}}^{(s_{\alpha},\alpha)}(\mathbf{x}_{\beta}^{(\alpha)},t_{m\mp})=0\text{ for }s_{\alpha}=0,1,\cdots,2k_{\alpha}-1,\\ \hbar_{\alpha}G_{\partial\Omega_{ij}}^{(2k_{\alpha},\alpha)}(\mathbf{x}_{\beta}^{(\alpha)},t_{m\mp})>0.\end{array}\tag{5.49}$$

or the $(2k_{\beta})$th - inflexional singular set $\Gamma_{12}=\mathscr{Q}_{ij}^{(2k_{\alpha}:\alpha)}$ exists with

$$\begin{array}{l}\hbar_{\alpha}G_{\partial\Omega_{ij}}^{(s_{\alpha},\alpha)}(\mathbf{x}_{m},t_{m\mp})=0\text{ for }s_{\alpha}=0,1,\cdots,2k_{\alpha}-1,\\ \hbar_{\alpha}G_{\partial\Omega_{ij}}^{(2k_{\alpha},\alpha)}(\mathbf{x}_{m},t_{m\mp})>0.\end{array}\tag{5.50}$$

The two $(2k_{\alpha})$th -order inflexional singular sets with the double imaginary tangential flow (double inaccessible tangential flow) and the imaginary and real tangential flows are sketched in Fig. 5.15 (a) and (b). Again, the two real and imaginary singular sets are based on the different initial conditions. As in Fig. 5.14, one flow on the singular set is from the real to imaginary portion, but another flow on the singular set is from the imaginary to real portion.

The singular set from the double $(2k_{i}-1:2k_{j}-1)$ -tangential flows in Eq. (5.43) to the double $(2k_{i}-1:2k_{j}-1)$ - imaginary tangential flows in Eq. (5.47) is defined by the double inflexional singular set of $\Gamma_{12}=\mathscr{Q}_{ij}^{(2k_{\alpha}:2k_{\beta})}$. However, the singular set from the double $(2k_{i}-1:2k_{j}-1)$ -imaginary tangential flows to the double $(2k_{i}-1:2k_{j}-1)$ - tangential flows is given by the double imaginary inflexional singular set $\Gamma_{12}=\boldsymbol{\mathscr{Q}}_{ij}^{(2k_{\alpha}:2k_{\beta})}$. Both of them are sketched in Fig. 5.16 and 5.17, respectively.

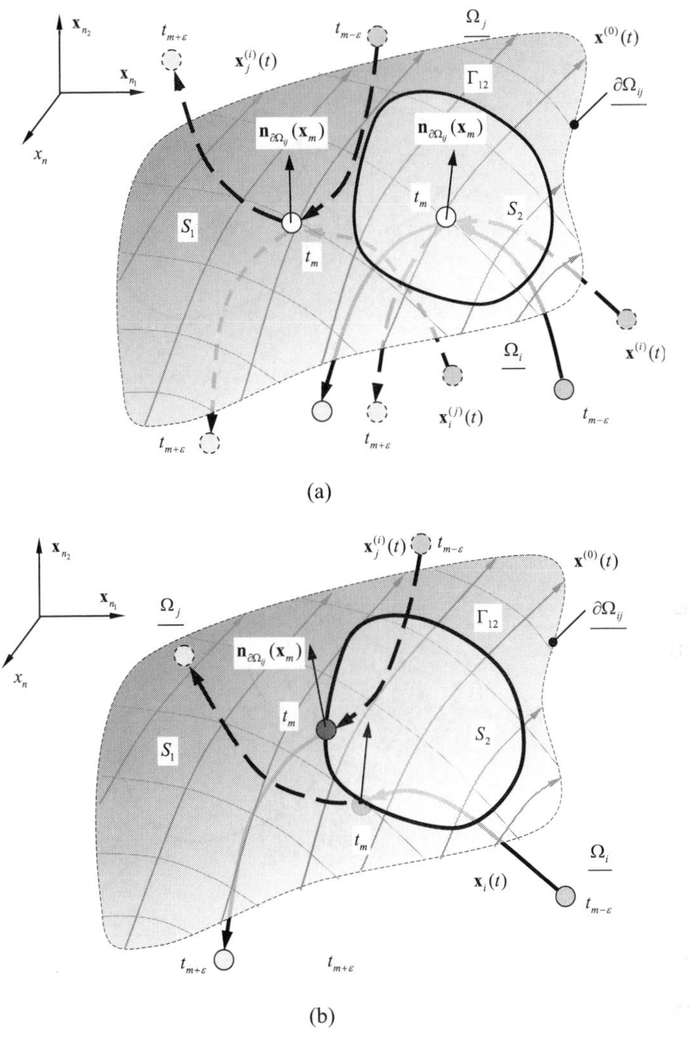

Fig. 5.15 Single real and imaginary inflexional singular sets between two subsets on the boundary for the $(2k_i-1:2k_j-1)$-double imaginary tangential flow and the $(2k_i-1:2k_j-1)$-imaginary and real tangential flows (or single tangential flow): **(a)** on subsets and **(b)** on singular set of $\mathscr{Q}_{ij}^{(2k_i,i)}$ or $\mathscr{Q}_{ij}^{(2k_i,j)}$. The flows on the boundary are described by $\mathbf{x}^{(0)}(t)$. The normal vector $\mathbf{n}_{\partial\Omega_{ij}}$ of the boundary is depicted with $\mathbf{n}_{\partial\Omega_{ij}} \to \Omega_j$. The dotted curves are real flows. The filled circular symbols with dashed edge are not real starting points. The hollow circular symbols are switching points on the subset of the boundary. The filled dark circular symbols are points on singular sets. The gray filled circular symbols are the points in the domains. ($n_1+n_2+1=n$)

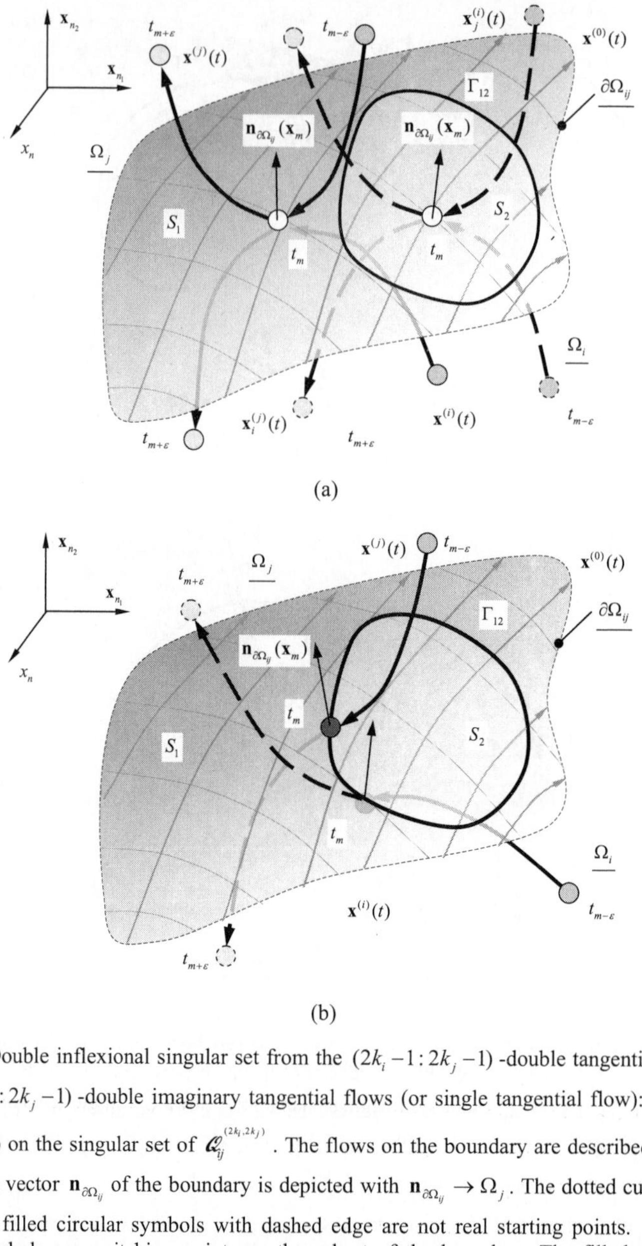

Fig. 5.16 Double inflexional singular set from the $(2k_i - 1 : 2k_j - 1)$ -double tangential flows and the $(2k_i - 1 : 2k_j - 1)$ -double imaginary tangential flows (or single tangential flow): **(a)** on subsets and **(b)** on the singular set of $\mathscr{C}_{ij}^{(2k_i,2k_j)}$. The flows on the boundary are described by $\mathbf{x}^{(0)}(t)$. The normal vector $\mathbf{n}_{\partial\Omega_{ij}}$ of the boundary is depicted with $\mathbf{n}_{\partial\Omega_{ij}} \to \Omega_j$. The dotted curves are real flows. The filled circular symbols with dashed edge are not real starting points. The hollow circular symbols are switching points on the subset of the boundary. The filled dark circular symbols are points on singular sets. The gray filled circular symbols are the points in the domains. ($n_1 + n_2 + 1 = n$)

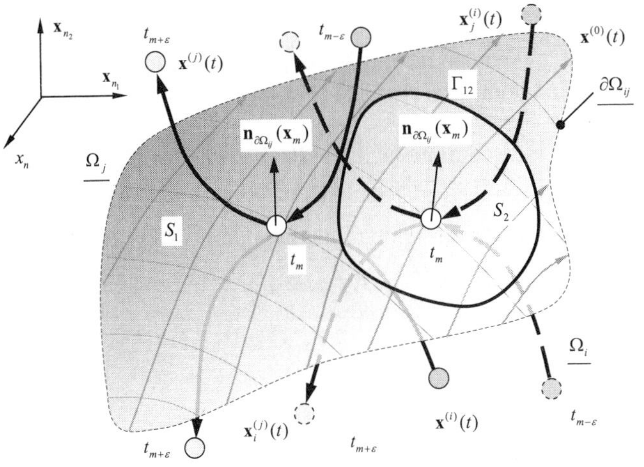

(a)

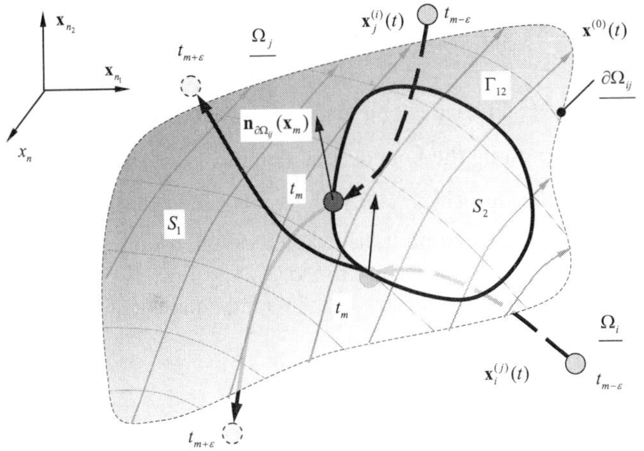

(b)

Fig. 5.17 Double imaginary inflexional singular set from the $(2k_i-1:2k_j-1)$-double tangential flows to the $(2k_i-1:2k_j-1)$-double imaginary tangential flows (or single tangential flow): **(a)** on subsets and **(b)** on the singular set of $\mathscr{Q}_{ij}^{(2k_i,2k_j)}$. The flows on the boundary are described by $\mathbf{x}^{(0)}(t)$. The normal vector $\mathbf{n}_{\partial\Omega_{ij}}$ of the boundary is depicted with $\mathbf{n}_{\partial\Omega_{ij}} \to \Omega_j$. The dotted curves are real flows. The filled circular symbols with dashed edge are not real starting points. The hollow circular symbols are switching points on the subset of the boundary. The filled dark circular symbols are points on singular sets. The gray filled circular symbols are the points in the domains. ($n_1+n_2+1=n$)

5.3. Forbidden boundary and boundary channels

The inner and outer boundaries of a boundary are introduced for discussion forbidden boundary and boundary channel.

Definition 5.12. In a discontinuous dynamical system of Eq. (5.1), the α-side of the boundary $\partial\Omega_{ij}$ ($\alpha \in \{i,j\}$) is called the *inner* boundary of $\partial\Omega_{ij}$ for a flow $\mathbf{x}^{(\alpha)}$ in domain Ω_{α}, and the β-side of the boundary $\partial\Omega_{ij}$ ($\beta \in \{i,j\}$ and $\beta \neq \alpha$) is called the *outer boundary* of $\partial\Omega_{ij}$ for a flow $\mathbf{x}^{(\alpha)}$ in domain Ω_{α}.

5.3.1. Forbidden boundary

From Chapter 4, flow barriers on the boundary were discussed. However, if the flow barriers exist on the inner and/or outer boundaries of a domain, the switchability of a flow to the boundary is different. For instance, if there is a permanent and absolute flow barrier $\mathbf{F}^{(\alpha\succ\beta)}(\mathbf{x}^{(\alpha)},t,\boldsymbol{\pi}_{\alpha},q^{(\alpha)})$ on the boundary $\partial\Omega_{ij}$, a coming flow $\mathbf{x}^{(\alpha)}(t)$ of Eq. (5.1) to the boundary $\partial\Omega_{ij}$ cannot pass over to the domain Ω_{β}, but it will move along the boundary in domain Ω_{α}. The coming flow $\mathbf{x}^{(\alpha)}(t)$ can return back to the domain Ω_{α} until the vector field changes the direction. In addition, the coming flow can be transported to another domain or boundary by a certain transport law. Such a boundary to the coming flow is called the permanently non-passable boundary on the α-side. A mathematical description of such a boundary is given as follows:

Definition 5.13. For a coming flow $\mathbf{x}^{(\alpha)}(t)$ ($\alpha \in \{i,j\}$) in Eq. (5.1) for $\mathbf{x}_m \in \partial\Omega_{ij}$ at time t_m, there is a coming flow barrier of $\mathbf{F}^{(\alpha\succ\beta)}(\mathbf{x}^{(\alpha)},t,\boldsymbol{\pi}_{\alpha},q^{(\alpha)})$ on the α-side of the boundary for $q^{(\alpha)} \in [q_1^{(\alpha)},q_2^{(\alpha)}]$ ($\beta \in \{0,i,j\}$ and $\beta \neq \alpha$) in domain Ω_{α}. The boundary $\partial\Omega_{ij}$ for the coming flow $\mathbf{x}^{(\alpha)}(t)$ is *permanently non-passable* if the flow barrier on the α-side is permanent and absolute, i.e., for all $\mathbf{x}_m \in S = \partial\Omega_{ij}$ and t_m,

$$\begin{aligned}&\hbar_{\alpha}G_{\partial\Omega_{ij}}^{(\alpha)}(\mathbf{x}_m,t_m) \in \left(\hbar_{\alpha}G_{\partial\Omega_{ij}}^{(\alpha\succ\beta)}(\mathbf{x}_m,q_1^{(\alpha)}),\hbar_{\alpha}G_{\partial\Omega_{ij}}^{(\alpha\succ\beta)}(\mathbf{x}_m,q_2^{(\alpha)})\right)\\&\text{with } \hbar_{\alpha}G_{\partial\Omega_{ij}}^{(\alpha\succ\beta)}(\mathbf{x}_m,q_1^{(\alpha)}) = 0_{-} \text{ and } \hbar_{\alpha}G_{\partial\Omega_{ij}}^{(\alpha\succ\beta)}(\mathbf{x}_m,q_2^{(\alpha)}) \to +\infty.\end{aligned} \tag{5.51}$$

Definition 5.14. For a coming flow $\mathbf{x}^{(\alpha)}(t)$ ($\alpha \in \{i,j\}$) in Eq. (5.1) for $\mathbf{x}_m \in \partial\Omega_{ij}$

at time t_m, there is a coming flow barrier of $\mathbf{F}^{(\alpha\succ\beta)}(\mathbf{x}^{(\alpha)},t,\boldsymbol{\pi}_\alpha,q^{(\alpha)})$ on the α-side of the boundary for $q^{(\alpha)}\in[q_1^{(\alpha)},q_2^{(\alpha)}]$ ($\alpha,\beta\in\{0,i,j\}$ and $\beta\neq\alpha$) in domain Ω_α. The boundary $\partial\Omega_{ij}$ for the coming flow $\mathbf{x}^{(\alpha)}(t)$ with the $(2k_\alpha)$th-order singularity is *permanently non-passable* if the flow barrier with the $(2k_\alpha)$th-order singularity on the α-side is permanent and absolute, i.e., for all $\mathbf{x}_m\in S=\partial\Omega_{ij}$ and t_m,

$$\begin{aligned}&G_{\partial\Omega_{ij}}^{(s_\alpha,\alpha)}(\mathbf{x}_m,t_m)=G_{\partial\Omega_{ij}}^{(s_\alpha,\alpha\succ\beta)}(\mathbf{x}_m,q_\sigma^{(\alpha)})=0\\&\text{for } \sigma=1,2 \text{ and } s_\alpha=0,1,2,\cdots,2k_\alpha-1;\end{aligned}\tag{5.52}$$

$$\begin{aligned}&\hbar_\alpha G_{\partial\Omega_{ij}}^{(2k_\alpha,\alpha)}(\mathbf{x}_m,t_m)\in\left(\hbar_\alpha G_{\partial\Omega_{ij}}^{(2k_\alpha,\alpha\succ\beta)}(\mathbf{x}_m,q_1^{(\alpha)}),\hbar_\alpha G_{\partial\Omega_{ij}}^{(2k_\alpha,\alpha\succ\beta)}(\mathbf{x}_m,q_2^{(\alpha)})\right)\\&\text{with } \hbar_\alpha G_{\partial\Omega_{ij}}^{(2k_\alpha,\alpha\succ\beta)}(\mathbf{x}_m,q_1^{(\alpha)})=0_- \text{ and } \hbar_\alpha G_{\partial\Omega_{ij}}^{(2k_\alpha,\alpha\succ\beta)}(\mathbf{x}_m,q_2^{(\alpha)})\to\infty.\end{aligned}\tag{5.53}$$

Note that if $\beta=0$, the flow barrier of $\mathbf{F}^{(\alpha\succ\beta)}(\mathbf{x}^{(\alpha)},t,\boldsymbol{\pi}_\alpha,q^{(\alpha)})$ gives the coming flow barrier of $\mathbf{F}^{(\alpha\succ0)}(\mathbf{x}^{(\alpha)},t,\boldsymbol{\pi}_\alpha,q^{(\alpha)})$ in the sink flow. From the permanently non-passable boundary, the forbidden boundary can be formed, which means that the flows in discontinuous dynamical systems cannot access the boundary from both domains. The flows only can stay in domains rather than the boundary.

Definition 5.15. For a coming flow $\mathbf{x}^{(\alpha)}(t)$ ($\alpha\in\{i,j\}$) in Eq. (5.1) for $\mathbf{x}_m\in\partial\Omega_{ij}$ at time t_m, there is a coming flow barrier of $\mathbf{F}^{(\alpha\succ\beta)}(\mathbf{x}^{(\alpha)},t,\boldsymbol{\pi}_\alpha,q^{(\alpha)})$ for $q^{(\alpha)}\in[q_1^{(\alpha)},q_2^{(\alpha)}]$ ($\beta\in\{0,i,j\}$ and $\beta\neq\alpha$) on the α-side of the boundary in domain Ω_α. The boundary $\partial\Omega_{ij}$ for the two coming flows $\mathbf{x}^{(\alpha)}(t)$ ($\alpha=i,j$) is *forbidden* if the boundary $\partial\Omega_{ij}$ for both coming flows are permanently non-passable.

Definition 5.16. For a coming flow $\mathbf{x}^{(\alpha)}(t)$ ($\alpha\in\{i,j\}$) in Eq. (5.1) for $\mathbf{x}_m\in\partial\Omega_{ij}$ at time t_m, there is a coming flow barrier of $\mathbf{F}^{(\alpha\succ\beta)}(\mathbf{x}^{(\alpha)},t,\boldsymbol{\pi}_\alpha,q^{(\alpha)})$ for $q^{(\alpha)}\in[q_1^{(\alpha)},q_2^{(\alpha)}]$ ($\beta\in\{0,i,j\}$ and $\alpha\neq\beta$) on the α-side of the boundary in domain Ω_α. The boundary $\partial\Omega_{ij}$ for the two coming flows $\mathbf{x}^{(\alpha)}(t)$ ($\alpha=i,j$) with the $(2k_\alpha)$th-order singularity is *forbidden* if the boundary $\partial\Omega_{ij}$ for both coming flows with the $(2k_i:2k_j)$ singularity is permanently non-passable.

Similar to the forbidden boundary, the accessible boundary to coming flows on both sides is defined from sink flows in discontinuous dynamical systems.

Definition 5.17. For two coming flows $\mathbf{x}^{(\alpha)}(t)$ ($\alpha=i,j$) in Eq. (5.1) for all $\mathbf{x}_m\in\partial\Omega_{ij}$ at time t_m, the boundary $\partial\Omega_{ij}$ ($\alpha,\beta=i,j$ but $\beta\neq\alpha$) is *accessible* if

$$\hbar_\alpha G_{\partial\Omega_{ij}}^{(\alpha)}(\mathbf{x}_m,t_{m-})>0 \text{ and } \hbar_\alpha G_{\partial\Omega_{ij}}^{(\beta)}(\mathbf{x}_m,t_{m-})<0,\tag{5.54}$$

without any coming flow barriers.

Definition 5.18. For two coming flows $\mathbf{x}^{(\alpha)}(t)$ $(\alpha = i, j)$ with the $(2k_\alpha)$th -order singularity in Eq. (5.1) for $\mathbf{x}_m \in \partial\Omega_{ij}$ at time t_m, the boundary $\partial\Omega_{ij}$ with the $(2k_i : 2k_j)$ -order singularity ($\alpha, \beta = i, j$ but $\beta \neq \alpha$) is *accessible* if

$$\begin{aligned}
&\hbar_\alpha G_{\partial\Omega_{ij}}^{(s_\alpha,\alpha)}(\mathbf{x}_m, t_{m-}) = 0 \text{ for } s_\alpha = 0,1,2,\cdots,2k_\alpha - 1,\\
&\hbar_\alpha G_{\partial\Omega_{ij}}^{(s_\beta,\beta)}(\mathbf{x}_m, t_{m-}) = 0 \text{ for } s_\beta = 0,1,2,\cdots,2k_\beta - 1,\\
&\hbar_\alpha G_{\partial\Omega_{ij}}^{(2k_\alpha,\alpha)}(\mathbf{x}_m, t_{m-}) > 0 \text{ and } \hbar_\alpha G_{\partial\Omega_{ij}}^{(2k_\beta,\beta)}(\mathbf{x}_m, t_{m-}) < 0,
\end{aligned} \tag{5.55}$$

without any coming flow barriers.

5.3.2. Boundary channels

Similarly, the boundary flow barriers on both sides of the boundary can form a boundary flow channel.

Definition 5.19. For a boundary flow $\mathbf{x}^{(0)}(t)$ in Eq. (5.1) for $\mathbf{x}_m \in \partial\Omega_{ij}$ at time t_m, there is a boundary flow barrier $\mathbf{F}^{(0\succ 0_\alpha)}(\mathbf{x}^{(\alpha)}, t, \boldsymbol{\pi}_\alpha, q^{(\alpha)})$ on the α -side of boundary $\partial\Omega_{ij}$ for $q^{(\alpha)} \in [q_1^{(\alpha)}, q_2^{(\alpha)}]$ ($\alpha \in \{i, j\}$). The boundary $\partial\Omega_{ij}$ for the boundary flow $\mathbf{x}^{(0)}(t)$ is *permanently non-leavable* on the α -side if the flow barrier on the α -side of the boundary is permanent and absolute, i.e., for all $\mathbf{x}_m \in S = \partial\Omega_{ij}$ and t_m,

$$\begin{aligned}
&0 \in \left(\hbar_\alpha G_{\partial\Omega_{ij}}^{(0\succ 0_\alpha)}(\mathbf{x}_m, q_2^{(\alpha)}), \hbar_\alpha G_{\partial\Omega_{ij}}^{(0\succ 0_\alpha)}(\mathbf{x}_m, q_1^{(\alpha)})\right)\\
&\text{with } \hbar_\alpha G_{\partial\Omega_{ij}}^{(0\succ 0_\alpha)}(\mathbf{x}_m, q_2^{(\alpha)}) \to -\infty \text{ and } \hbar_\alpha G_{\partial\Omega_{ij}}^{(0\succ 0_\alpha)}(\mathbf{x}_m, q_1^{(\alpha)}) \to \infty.
\end{aligned} \tag{5.56}$$

Definition 5.20. For a boundary flow $\mathbf{x}^{(0)}(t)$ in Eq. (5.1) for $\mathbf{x}_m \in \partial\Omega_{ij}$ at time t_m, there is a boundary flow barrier $\mathbf{F}^{(0\succ 0_\alpha)}(\mathbf{x}^{(\alpha)}, t, \boldsymbol{\pi}_\alpha, q^{(\alpha)})$ on the α -side of boundary $\partial\Omega_{ij}$ for $q^{(\alpha)} \in [q_1^{(\alpha)}, q_2^{(\alpha)}]$ ($\alpha \in \{i, j\}$). The boundary $\partial\Omega_{ij}$ for the boundary flow $\mathbf{x}^{(0)}(t)$ is *permanently non-leavable* with the m_αth -order singularity on the α -side if the flow barrier with the (m_α)th -order singularity on the α -side of the boundary is permanent and absolute, i.e., for all $\mathbf{x}_m \in S = \partial\Omega_{ij}$ and t_m,

$$G_{\partial\Omega_{ij}}^{(s_\alpha, 0\succ 0_\alpha)}(\mathbf{x}_m, q_\sigma^{(\alpha)}) = 0 \text{ for } \sigma = 1,2 \text{ and } s_\alpha = 0,1,2,\cdots,m_\alpha - 1; \tag{5.57}$$

$$0 \in \left[\hbar_{\alpha} G_{\partial\Omega_{ij}}^{(m_{\alpha},0\succ 0_{\alpha})}(\mathbf{x}_m, q_2^{(\alpha)}), \hbar_{\alpha} G_{\partial\Omega_{ij}}^{(m_{\alpha},0\succ 0_{\alpha})}(\mathbf{x}_m, q_1^{(\alpha)}) \right]$$
$$\text{with } \hbar_{\alpha} G_{\partial\Omega_{ij}}^{(m_{\alpha},0\succ 0_{\alpha})}(\mathbf{x}_m, q_2^{(\alpha)}) \to -\infty \text{ and } \hbar_{\alpha} G_{\partial\Omega_{ij}}^{(m_{\alpha},0\succ 0_{\alpha})}(\mathbf{x}_m, q_1^{(\alpha)}) \to \infty. \tag{5.58}$$

From the non-leavable boundary on the α-side, the boundary channel for the boundary flow can be formed, which means that the boundary flow in such a discontinuous dynamical system only slides on the non-leavable boundary.

Definition 5.21. For a boundary flow $\mathbf{x}^{(0)}(t)$ in Eq. (5.1) for $\mathbf{x}_m \in \partial\Omega_{ij}$ at time t_m, there is a boundary flow barrier $\mathbf{F}^{(0\succ 0_{\alpha})}(\mathbf{x}^{(\alpha)}, t, \boldsymbol{\pi}_{\alpha}, q^{(\alpha)})$ for $q^{(\alpha)} \in [q_1^{(\alpha)}, q_2^{(\alpha)}]$ ($\alpha = i, j$) on the α-side of boundary $\partial\Omega_{ij}$ in domain Ω_{α}. The boundary $\partial\Omega_{ij}$ for the boundary flow $\mathbf{x}^{(0)}(t)$ is a *boundary channel* if the boundary $\partial\Omega_{ij}$ for the boundary flow $\mathbf{x}^{(0)}(t)$ is permanently non-leavable on both sides.

Definition 5.22. For a boundary flow $\mathbf{x}^{(0)}(t)$ in Eq. (5.1) for $\mathbf{x}_m \in \partial\Omega_{ij}$ at time t_m, there is a boundary flow barrier of $\mathbf{F}^{(0\succ 0_{\alpha})}(\mathbf{x}^{(\alpha)}, t, \boldsymbol{\pi}_{\alpha}, q^{(\alpha)})$ for $q^{(\alpha)} \in [q_1^{(\alpha)}, q_2^{(\alpha)}]$ ($\alpha = i, j$) on the α-side of boundary $\partial\Omega_{ij}$ in domain Ω_{α}. The boundary $\partial\Omega_{ij}$ for the boundary flow $\mathbf{x}^{(0)}(t)$ is a *boundary channel* with the $(m_{\alpha} : m_{\beta})$ singularity if the boundary $\partial\Omega_{ij}$ for the boundary flow $\mathbf{x}^{(0)}(t)$ is non-leavable with the m_{α}th-order singularity ($\alpha = i, j$) on both sides.

On the other hand, the leavable boundary for the boundary flow in the source flow on both sides is defined from source flows as follows.

Definition 5.23. For the two possible leaving flows $\mathbf{x}^{(\alpha)}(t)$ and $\mathbf{x}^{(\beta)}(t)$ ($\alpha, \beta \in \{i, j\}$ and $\alpha \neq \beta$) in Eq. (5.1) for all $\mathbf{x}_m \in \partial\Omega_{ij}$ at time t_m, under the following conditions

$$\hbar_{\alpha} G_{\partial\Omega_{ij}}^{(\alpha)}(\mathbf{x}_m, t_{m+}) < 0 \text{ and } \hbar_{\alpha} G_{\partial\Omega_{ij}}^{(\beta)}(\mathbf{x}_m, t_{m+}) > 0. \tag{5.59}$$

(i) The boundary $\partial\Omega_{ij}$ for the boundary flow is *leavable on the* α*-side* if the non-leavable boundary does not exist on the α-side.
(ii) The boundary $\partial\Omega_{ij}$ for the boundary flow is *leavable* if the non-leavable boundary does not exist on both sides.

Definition 5.24. For two possible leaving flows of $\mathbf{x}^{(\alpha)}(t)$ and $\mathbf{x}^{(\beta)}(t)$ ($\alpha, \beta \in \{i, j\}$ and $\alpha \neq \beta$) with the $(m_{\alpha}{:}m_{\beta})$-singularity in Eq. (5.1) for $\mathbf{x}_m \in \partial\Omega_{ij}$ at time t_m, under the following conditions

$$\begin{aligned}&\hbar_\alpha G_{\partial\Omega_{ij}}^{(s_\alpha,\alpha)}(\mathbf{x}_m,t_{m+})=0 \text{ for } s_\alpha=0,1,2,\cdots,m_\alpha-1;\\&\hbar_\alpha G_{\partial\Omega_{ij}}^{(s_\beta,\beta)}(\mathbf{x}_m,t_{m+})=0 \text{ for } s_\beta=0,1,2,\cdots,m_\beta-1;\\&\hbar_\alpha G_{\partial\Omega_{ij}}^{(m_\alpha,\alpha)}(\mathbf{x}_m,t_{m+})<0 \text{ and } \hbar_\alpha G_{\partial\Omega_{ij}}^{(m_\beta,\beta)}(\mathbf{x}_m,t_{m+})>0.\end{aligned}\tag{5.60}$$

(i) The boundary $\partial\Omega_{ij}$ for the boundary flow is *leavable with the* m_αth *-order singularity* on the α -side if the non-leavable boundary does not exist on the α -side.

(ii) The boundary $\partial\Omega_{ij}$ for the boundary flow is *leavable with the* $(m_\alpha : m_\beta)$ *singularity* if the non-leavable boundary does not exist on both sides.

The accessible and forbidden boundaries for a coming flow to the boundary are sketched in Fig. 5.18. The accessible boundary does not have any flow barrier existence on both sides. Once a coming flow comes from one of the two domains, the boundary flow is obtained to form a sink flow because of $\hbar_\alpha G_{\partial\Omega_{ij}}^{(2k_\beta,\beta)} \le G_{\partial\Omega_{ij}}^{(2k_0,0)} \le \hbar_\alpha G_{\partial\Omega_{ij}}^{(2k_\alpha,\alpha)}$ and $G_{\partial\Omega_{ij}}^{(s_0,0)} \equiv 0$ ($s_0=0,1,2,\cdots$) for a boundary flow. The forbidden boundary possesses two permanent non-passable boundaries on both sides. Any coming flows from domains cannot assess the boundary. The two coming flow barriers are depicted by gray and blue surfaces. For comparison, an accessible boundary of coming flow is arranged in Fig. 5.18(a), and a forbidden boundary is expressed through a forbidden zone, as shown in Fig. 5.18(b). On the forbidden boundary, no any coming flows can be switched into a boundary flow. The width of the fictional forbidden zone is zero. To cross over the forbidden zone, a transport law should be exerted. Red curves are *G*-functions relative to coming flows. Dark and blue surfaces are for flow barrier surfaces. The hatched area is for the zoomed boundary. The dark blue curves are coming flows. For intuitive illustrations, light green surfaces represent "no flow barrier", which are used just for showing the *G*-function values at the boundary.

Similarly, a leavable boundary on both sides and a boundary channel for the boundary flow are illustrated through the *G*-function, as sketched in Fig. 5.19. The leavable boundary on both sides implies that the boundary flow can leave the boundary once initial conditions are in the neighborhood of the boundary. However, the channel possesses boundary flow barriers on both sides. The boundary flow is not allowed to get into the two domains. The boundary flow only can move along the boundary surface. This is why this channel is called the boundary channel for the boundary flow. For comparison, the leavable boundary on both sides for the boundary flow is presented in Fig. 5.19(a). Due to no flow barriers on the leavable boundary, the light green surface is used to present the *G*-function for "no flow barrier". The boundary channel is presented in Fig. 5.19(b). The boundary flow barriers are depicted by the dark and blue surfaces. The width of the fictitious boundary channel on the boundary is zero.

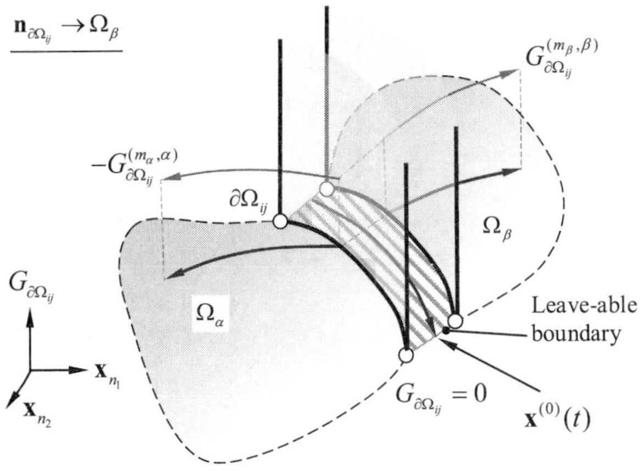

(a)

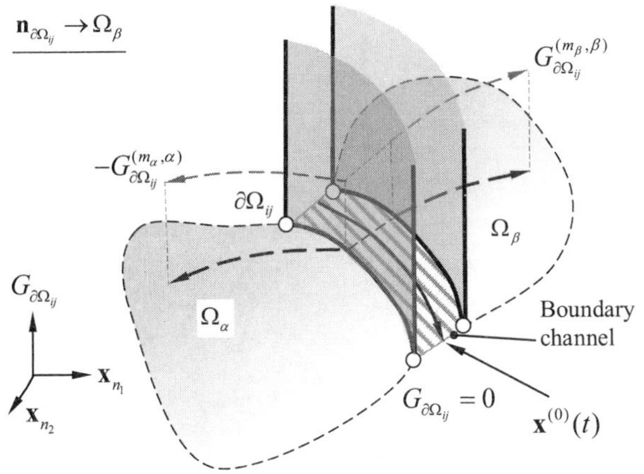

(b)

Fig. 5.18 **(a)** An accessible boundary and **(b)** a forbidden boundary. The upper curves with arrows are the *G*-functions relative to the coming flows. The two vertical surfaces are the flow barrier surfaces. The hatched area is for the zoomed boundary. The lower curves with arrows are coming flows. The light green surfaces represent "no flow barrier". ($k_\alpha, k_\beta \in \{0,1,2,\cdots\}$)

For a boundary flow leaving the boundary channel, a transport law should be exerted. For a boundary flow, it is not necessary to have the boundary flow barriers on both sides. If a boundary flow barrier exists on one side of the boundary, the other side of the boundary for the boundary flow will be leavable. In other words, the boundary flow on the leavable boundary can leave the boundary once initial conditions are chosen in neighborhood of a leavable boundary. A leavable

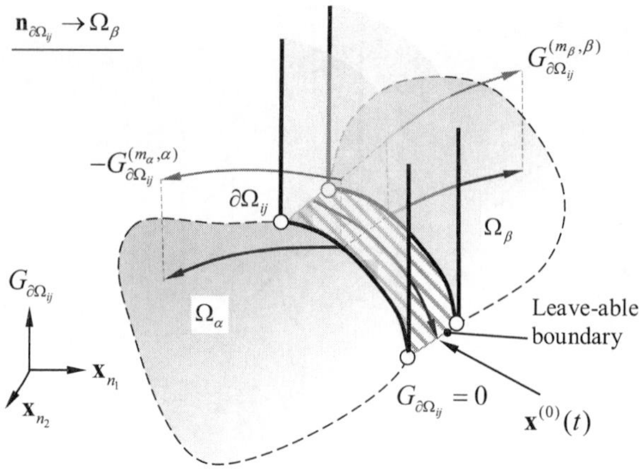

(a)

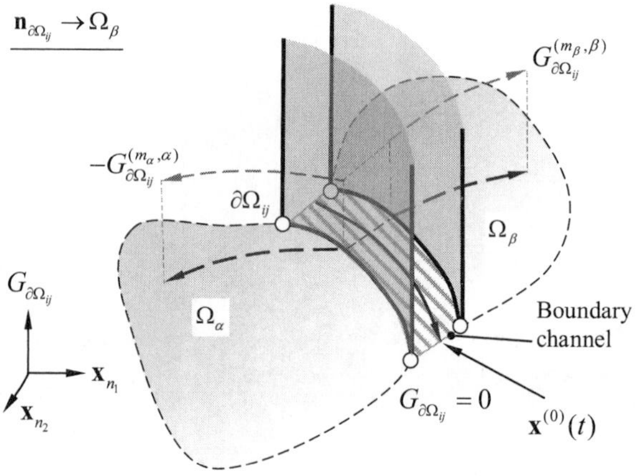

(b)

Fig. 5.19 **(a)** A leavable boundary on both sides, and **(b)** a boundary channel of the boundary flow of $\mathbf{x}^{(0)}$. The upper dashed and solid curves with arrows are the G-functions of the flows $\mathbf{x}^{(\alpha)}$ and $\mathbf{x}^{(\beta)}$ in the corresponding domains. The two vertical surfaces of (b) are the boundary flow barrier surfaces. The hatched area is for the zoomed boundary. The lower solid and dashed curves with arrows are boundary flows and leaving flows, respectively. The two vertical surfaces of (a) represent "no boundary flow barrier". ($m_\alpha, m_\beta \in \{0,1,2,\cdots\}$)

boundary on one side for the boundary flow is sketched in Fig. 5.20 (a) and (b). Once $\hbar_\alpha G^{(s_\alpha,\alpha)}_{\partial\Omega_{ij}} = 0$ ($s_\alpha = 0,1,2,\cdots,m_\alpha - 1$) and $\hbar_\alpha G^{(m_\alpha,\alpha)}_{\partial\Omega_{ij}} < 0$, a boundary flow can leave the boundary on the α-side ($\alpha \in \{i,j\}$). It should be mentioned, once a boundary flow is formed, its disappearance from the α-side is independent of the vector field in the β-domain. Thus, without conditions in Eq. (5.60), a boundary

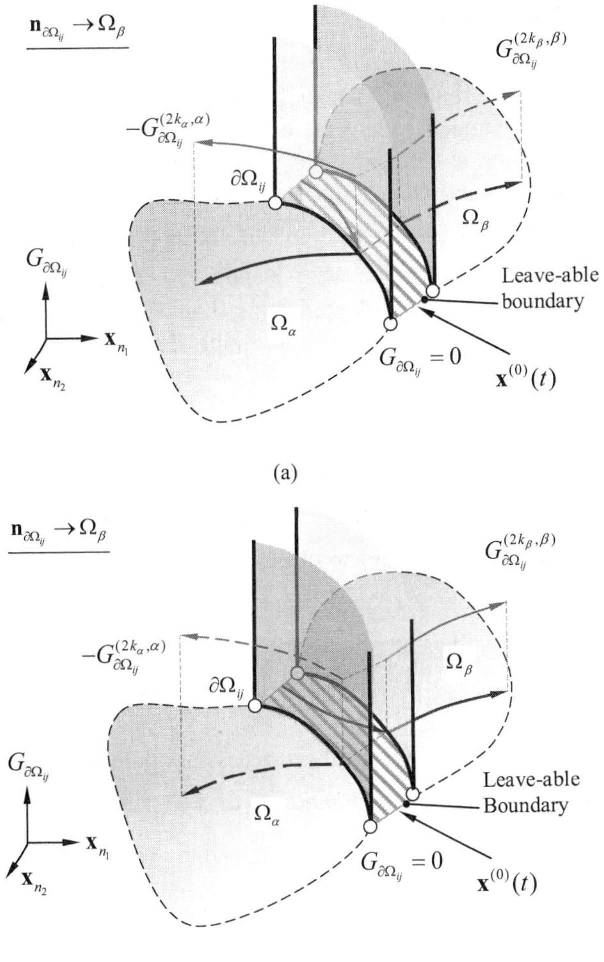

Fig. 5.20 The leavable boundary on: **(a)** the α -side and **(b)** the β -side. The dashed and solid upper curves with arrows are the G-functions of the flows of $\mathbf{x}^{(\alpha)}$ and $\mathbf{x}^{(\beta)}$ in the corresponding domains. The gray and dark vertical surfaces are the boundary flow barrier surface. The hatched area is for the zoomed boundary. The lower solid and dashed curves with arrows are boundary flows and leaving flows, respectively. The two bright vertical surfaces represent "no boundary flow barrier". ($m_\alpha, m_\beta \in \{0,1,2,\cdots\}$)

flow can leave the boundary under the conditions of $\hbar_\alpha G_{\partial\Omega_{ij}}^{(s_\alpha,\alpha)} = 0$ ($s_\alpha = 0,1,\cdots,$ $2k_\alpha - 1$) and $\hbar_\alpha G_{\partial\Omega_{ij}}^{(2k_\alpha,\alpha)} < 0$, as in Chapter 3. Such a case will not be interested. However, an accessible boundary to a boundary flow in the sink flow should be discussed. It is assumed that a boundary flow on the boundary cannot be formed if the coming flow barrier exists on one side of the boundary.

5.4. Domain and boundary classifications

After discussed the forbidden boundary, sink and source domains should be described for a better understanding the complexity of flows in discontinuous dynamical systems. Before doing so, the special boundary should be discussed. The boundary between an accessible domain Ω_α and an inaccessible domain $\Omega_\beta \subseteq \Omega_0$ should be defined, and the boundary of an accessible domain Ω_α is intersected with the universal domain boundary $\partial\mho$ should be described. The opened and closed boundaries of the accessible domains should be defined. The geometrical illustrations for such boundaries of the accessible domains are given in Figs. 5.21 and 5.22. The mathematical definitions are given as follows.

Definition 5.25. For a discontinuous dynamical system in Eq. (5.1), between an inaccessible domain $\Omega_\beta \subseteq \Omega_0$ and an accessible domain Ω_α, there is a boundary $\partial\Omega_{\alpha\beta} = \bar{\Omega}_\alpha \cap \bar{\Omega}_\beta$. The outer boundary of $\partial\Omega_{\alpha\beta}$ for the accessible domain is in the inaccessible domain Ω_β and suppose there is a fictitious flow of $\mathbf{x}^{(\beta)}$.

(i) The accessible domain Ω_α to the inaccessible domain is *closed* if there is a permanent wall on the outer boundary for a *fictitious* flow of $\mathbf{x}^{(\beta)}$ with

$$h_\alpha G^{(\beta)}_{\partial\Omega_{\alpha\beta}}(\mathbf{x}_m, t_{m-}) < 0 \text{ for all } \mathbf{x}_m \in S = \partial\Omega_{\alpha\beta}. \tag{5.61}$$

(ii) The accessible domain Ω_α to the inaccessible domain is *open* if the outer boundary is leavable for a *fictitious* flow of $\mathbf{x}^{(\beta)}$ with

$$h_\alpha G^{(\beta)}_{\partial\Omega_{\alpha\beta}}(\mathbf{x}_m, t_{m+}) > 0 \text{ for all } \mathbf{x}_m \in S \subseteq \partial\Omega_{\alpha\beta}. \tag{5.62}$$

(iii) The accessible domain Ω_α to the inaccessible domain is *uncertain* if the outer boundary is determined by a *fictitious* flow of $\mathbf{x}^{(\beta)}$ with

$$h_\alpha G^{(\beta)}_{\partial\Omega_{\alpha\beta}}(\mathbf{x}_m, t_{m+}) = 0 \text{ for all } \mathbf{x}_m \in S \subseteq \partial\Omega_{\alpha\beta}. \tag{5.63}$$

Definition 5.26. For a discontinuous dynamical system in Eq. (5.1), an accessible domain Ω_α and a bounded universal domain $\mho$ possess a common boundary $\partial\Omega_{\alpha\infty} = \bar{\Omega}_\alpha \cap \partial\mho$. The outer boundary of $\partial\Omega_{\alpha\infty}$ for the accessible domain is in the outside of the universal domain (expressed by Ω_∞), and suppose there is a fictitious flow of $\mathbf{x}^{(\infty)}$.

(i) The accessible domain Ω_α at the universal boundary is *closed* if there is a permanent wall on the outer boundary for a *fictitious* flow of $\mathbf{x}^{(\infty)}$ with

$$h_\alpha G^{(\infty)}_{\partial\Omega_{\alpha\infty}}(\mathbf{x}_m, t_{m-}) < 0 \text{ for all } \mathbf{x}_m \in S = \partial\Omega_{\alpha\infty}. \tag{5.64}$$

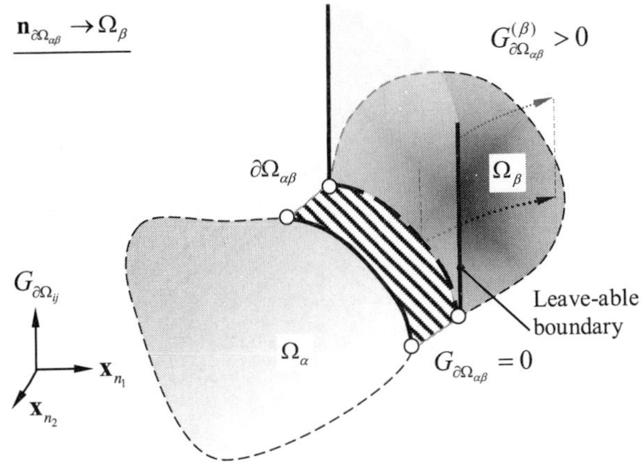

(a)

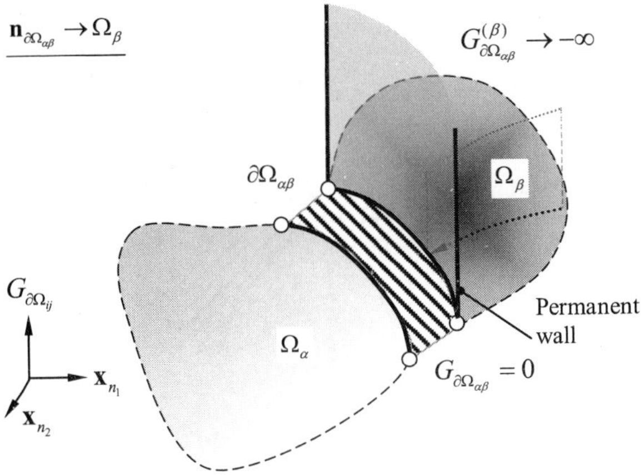

(b)

Fig. 5.21 (a) An opened boundary to the inaccessible domain and **(b)** a closed boundary to the inaccessible domain. The upper and lower curves are the fictitious flows $\mathbf{x}^{(\beta)}$ and the *G*-function in the inaccessible domain. The gray vertical surface is the permanent wall the fictitious flow in the inaccessible domain. The hatched area is for the zoomed boundary. The bright vertical surface represents the leavable outer boundary for the boundary flow to the inaccessible domain. Domains $\Omega_\beta, \Omega_\alpha$ represent the inaccessible and accessible domains, respectively.

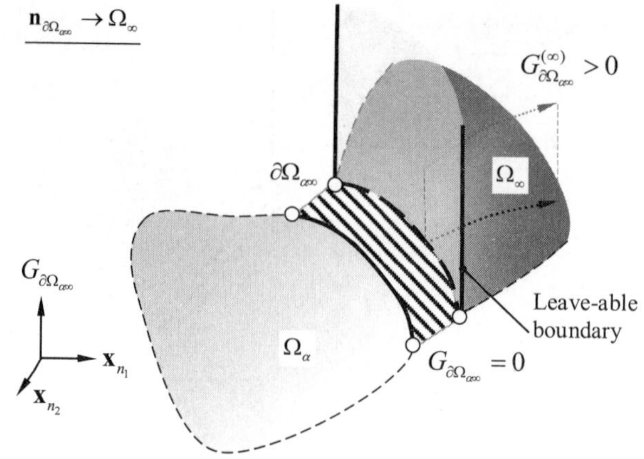

(a)

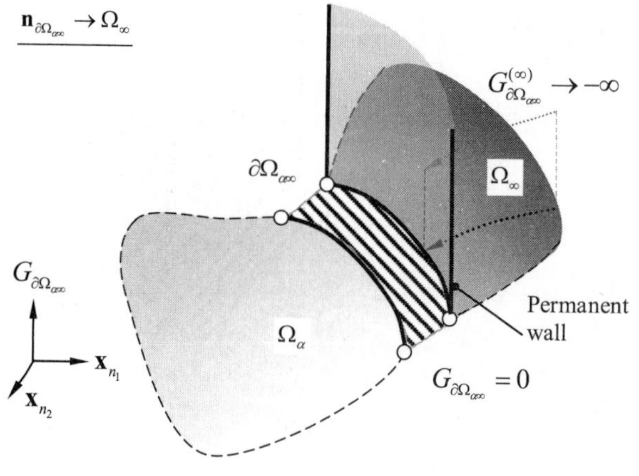

(b)

Fig. 5.22 (a) An opened boundary to the outside of the universal domain and **(b)** a closed boundary to the outside of the universal domain. The upper and lower curves are the G-function and the fictitious flows $\mathbf{x}^{(\infty)}$ in the inaccessible domain. The gray vertical surface is the permanent wall the fictitious flow in the outside of the universal domain. The hatched area is for the zoomed boundary. The bright vertical surface represents the leavable outer boundary for the boundary flow to the outside of the universal domain. Domains $\Omega_{\beta}, \Omega_{\alpha}$ represent the inaccessible and accessible domains, respectively.

(ii) The accessible domain Ω_α is *open at* the boundary of the universal domain if the outer boundary is leavable for a *fictitious* flow of $\mathbf{x}^{(\infty)}$ with

$$h_\alpha G^{(\infty)}_{\partial\Omega_{\alpha\beta}}(\mathbf{x}_m, t_{m+}) > 0 \text{ for } \mathbf{x}_m \in S \subseteq \partial\Omega_{\alpha\infty}. \tag{5.65}$$

(iii) The accessible domain Ω_α is *uncertain* at the boundary of the universal domain if the outer boundary is determined by a fictitious flow of $\mathbf{x}^{(\infty)}$ with

$$h_\alpha G^{(\infty)}_{\partial\Omega_{\alpha\beta}}(\mathbf{x}_m, t_{m+}) = 0 \text{ for all } \mathbf{x}_m \in S \subseteq \partial\Omega_{\alpha\infty}. \tag{5.66}$$

5.4.1. Domain classification

Definition 5.27. For a discontinuous dynamical system in Eq. (5.1), an accessible domain Ω_α and a unbounded universal domain $\mho$ possess a common boundary $\partial\Omega_{\alpha\infty} = \bar{\Omega}_\alpha \cap \partial\mho$ at the infinity. The accessible domain Ω_α is called an *unbounded, accessible* domain.

From the above definitions, the three cases imply:

(i) For a closed boundary between accessible and inaccessible domains, a boundary flow can be formed once a flow in the accessible domain comes to the boundary.
(ii) For an open boundary between accessible and inaccessible domains, a boundary flow cannot be formed once a flow in the accessible domain comes to the boundary. The coming flow will disappear at the boundary. Otherwise, a transport law should be exerted.
(iii) For an uncertain boundary between accessible and inaccessible domains, the higher order G-function should be exerted as a special assumption to determine whether the common boundary is opened or closed.

Similarly, using the higher-order G-functions of fictitious flows in inaccessible domains, one can define whether the common boundary between the accessible domain and the inaccessible domain is open or closed.

Definition 5.28. An accessible domain $\Omega_\alpha \subset \mho$ in a discontinuous dynamical system in Eq. (5.1) is termed a *sink* domain if

$$\left.\begin{aligned} &\hbar_\alpha G^{(\beta)}_{\partial\Omega_{\alpha\beta}}(\mathbf{x}_m, t_{m-}) < 0, \\ \text{either } &\hbar_\alpha G^{(\alpha)}_{\partial\Omega_{\alpha\beta}}(\mathbf{x}_m, t_{m+}) < 0, \\ \text{or } &\hbar_\alpha G^{(\alpha)}_{\partial\Omega_{\alpha\beta}}(\mathbf{x}_m, t_{m\pm}) = 0 \text{ and } \hbar_\alpha G^{(1,\alpha)}_{\partial\Omega_{\alpha\beta}}(\mathbf{x}_m, t_{m\pm}) < 0 \end{aligned}\right\} \tag{5.67}$$

for all $\mathbf{x}_m \in \partial\Omega_{\alpha\beta}$ and all the adjacent, accessible domains Ω_β of domain Ω_α without any flow barriers on the boundary $\partial\Omega_{\alpha\beta}$.

Definition 5.29. An accessible domain $\Omega_\alpha \subset \mho$ in a discontinuous dynamical system in Eq. (5.1) is termed a *sink* domain of the $(2k_\beta : m_\alpha)$-type if

$$\begin{aligned}
&\hbar_\alpha G_{\partial\Omega_{\alpha\beta}}^{(s_\alpha,\alpha)}(\mathbf{x}_m, t_{m+}) = 0 \text{ for } s_\alpha = 0,1,2,\cdots,m_\alpha - 1;\\
&\hbar_\alpha G_{\partial\Omega_{\alpha\beta}}^{(s_\beta,\beta)}(\mathbf{x}_m, t_{m-}) = 0 \text{ for } s_\beta = 0,1,2,\cdots,m_\beta - 1;\\
&\hbar_\alpha G_{\partial\Omega_{\alpha\beta}}^{(m_\alpha,\alpha)}(\mathbf{x}_m, t_{m+}) < 0 \text{ and } (-1)^{m_\beta}\,\hbar_\alpha G_{\partial\Omega_{\alpha\beta}}^{(m_\beta,\beta)}(\mathbf{x}_m, t_{m-}) < 0
\end{aligned} \tag{5.68}$$

for all $\mathbf{x}_m \in \partial\Omega_{\alpha\beta}$ and all the adjacent, accessible domains Ω_β of domain Ω_α without any flow barriers on the boundary $\partial\Omega_{\alpha\beta}$.

Similarly, the source domains are defined as follows.

Definition 5.30. An accessible domain $\Omega_\alpha \subset \mho$ in a discontinuous dynamical system in Eq. (5.1) is termed a *source* domain if

$$\begin{aligned}
&\hbar_\alpha G_{\partial\Omega_{\alpha\beta}}^{(\alpha)}(\mathbf{x}_m, t_{m-}) > 0,\\
&\left.\begin{aligned}
&\text{either } \hbar_\alpha G_{\partial\Omega_{\alpha\beta}}^{(\beta)}(\mathbf{x}_m, t_{m+}) > 0,\\
&\text{or} \quad \hbar_\alpha G_{\partial\Omega_{\alpha\beta}}^{(\beta)}(\mathbf{x}_m, t_{m\pm}) = 0 \text{ and } \hbar_\alpha G_{\partial\Omega_{\alpha\beta}}^{(1,\beta)}(\mathbf{x}_m, t_{m\pm}) > 0
\end{aligned}\right\}
\end{aligned} \tag{5.69}$$

for all $\mathbf{x}_m \in \partial\Omega_{\alpha\beta}$ and all the adjacent, accessible domains Ω_β of the domain Ω_α without any flow barriers on the boundary $\partial\Omega_{\alpha\beta}$.

Definition 5.31. An accessible domain $\Omega_\alpha \subset \mho$ in a discontinuous dynamical system in Eq. (5.1) is termed a *source* domain of the $(2k_\alpha : m_\beta)$-type if

$$\begin{aligned}
&\hbar_\alpha G_{\partial\Omega_{\alpha\beta}}^{(s_\alpha,\alpha)}(\mathbf{x}_m, t_{m-}) = 0 \text{ for } s_\alpha = 0,1,2,\cdots,2k_\alpha - 1;\\
&\hbar_\alpha G_{\partial\Omega_{\alpha\beta}}^{(s_\beta,\beta)}(\mathbf{x}_m, t_{m+}) = 0 \text{ for } s_\beta = 0,1,2,\cdots,m_\beta - 1;\\
&\hbar_\alpha G_{\partial\Omega_{\alpha\beta}}^{(2k_\alpha,\alpha)}(\mathbf{x}_m, t_{m-}) > 0 \text{ and } \hbar_\alpha G_{\partial\Omega_{\alpha\beta}}^{(m_\beta,\beta)}(\mathbf{x}_m, t_{m+}) > 0
\end{aligned} \tag{5.70}$$

for all $\mathbf{x}_m \in \partial\Omega_{\alpha\beta}$ and all the adjacent, accessible domains Ω_β of domain Ω_α without any flow barriers on the boundary $\partial\Omega_{\alpha\beta}$.

Definition 5.32 An accessible domain $\Omega_\alpha \subset \mho$ for discontinuous dynamical systems in Eq. (5.1) is termed an *isolated* domain if all the adjacent domains Ω_β of domain Ω_α are inaccessible or a permanently-non-passable boundary exists for either flow $\mathbf{x}^{(\alpha)}$ in domain Ω_α or flow $\mathbf{x}^{(\beta)}$ in the adjacent, accessible domain Ω_β.

The sink and source domains are sketched through the shaded area in Fig. 5.23. The neighbored sub-domains include the accessible and inaccessible sub-domains, which are depicted through the white and gray colors, respectively. The boundary between the accessible and inaccessible domains is permanently-non-passable. For the sink domain of Ω_i, a coming flow to the passable boundary is toward the inside of the domain, as shown in Fig. 5.23(a). However, for the source domain of Ω_i, a leaving flow to the boundary is toward the outside of the domain, as shown in Fig. 5.23(b). If the boundary between the accessible and inaccessible domains is opened, a transport law should be applied.

If all the neighbored domains of the accessible domain Ω_i are accessible and all the corresponding boundaries of Ω_i are semi-passable only in one direction, then the accessible domain Ω_i is sink or source. Consider a closed accessible domain Ω_i as an example for illustration. The sink and source domains are sketched in Fig.5.24(a) and (b). The hatched zone is the boundary of the closed domain Ω_i. The arrows represent flows going into the sink domain and getting out the source domain, respectively. This concept already extended the source and sink of equilibrium. If all the neighbored domains of the accessible domain Ω_i are inaccessible, the corresponding boundaries of Ω_i should be permanently-non-passable. The accessible domain is isolated, and this isolated domain can possess a closed or opened boundary. This isolated domain can also be a sink or source domain, as shown in Figs. 5.25 and 5.26. If the flows in the domain Ω_i pass over the boundary and into another accessible domain, at least one transport law should be exerted.

On an inaccessible domain, no any vector fields can be defined for either the real flow or imaginary flow. However, on an accessible domain, a vector field not possessing any real flow can be defined, but the imaginary flows exist in the other domains. Although the two domains did not have any real flows but they are distinguishing. Based on this reason, the sink domain without input and the source domain without output are defined through the imaginary flows.

Definition 5.33. An accessible domain $\Omega_\alpha \subset \mho$ in a discontinuous dynamical system in Eq. (5.1) is termed a *sink* domain without input if

$$\begin{aligned} &\hbar_\alpha G^{(\beta)}_{\partial\Omega_{\alpha\beta}}(\mathbf{x}_m, t_{m\pm}) = 0 \text{ and } \hbar_\alpha G^{(1,\beta)}_{\partial\Omega_{\alpha\beta}}(\mathbf{x}_m, t_{m\pm}) < 0; \\ &\hbar_\alpha G^{(\alpha)}_{\partial\Omega_{\alpha\beta}}(\mathbf{x}_m, t_{m+}) < 0 \end{aligned} \tag{5.71}$$

for all $\mathbf{x}_m \in \partial\Omega_{\alpha\beta}$, and no any real flows exist in all the adjacent, accessible domains Ω_β except its imaginary flows exist in domain Ω_α.

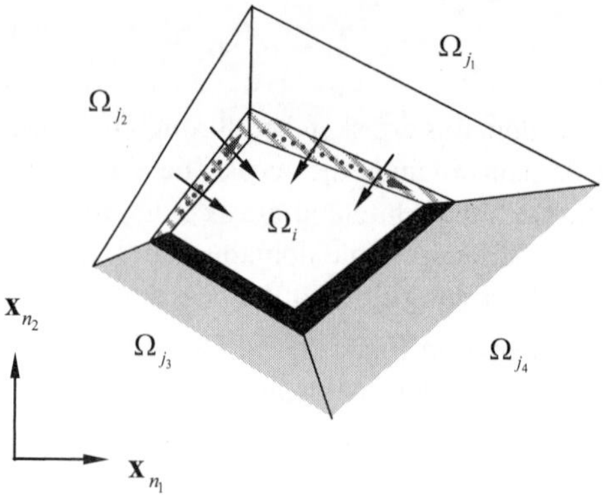

(a)

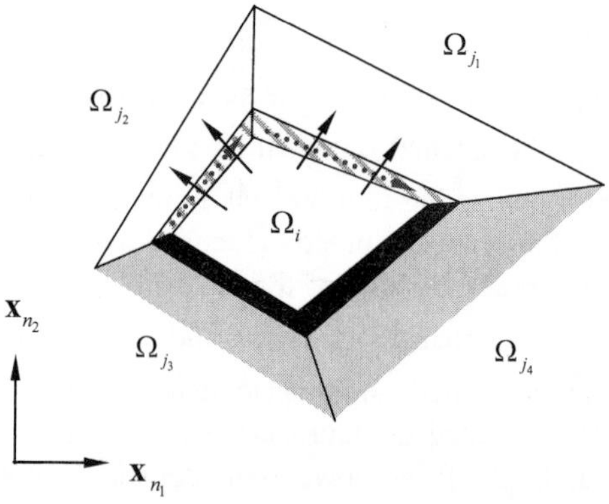

(b)

Fig. 5.23 (a) Sink domain and **(b)** source domain of Ω_i . The white sub-domains are accessible. The gray sub-domains are inaccessible domain. The hatched boundaries are for passable flows, and the dark boundaries are between the accessible and inaccessible boundaries.

Definition 5.34. An accessible domain $\Omega_\alpha \subset \mho$ in a discontinuous dynamical system in Eq. (5.1) is termed a *sink* domain of the $(2k_\beta+1:2k_\alpha)$-type without input if

$$\begin{aligned} &\hbar_\alpha G_{\partial\Omega_{\alpha\beta}}^{(s_\alpha,\alpha)}(\mathbf{x}_m,t_{m+})=0 \text{ for } s_\alpha=0,1,2,\cdots,m_\alpha-1;\\ &\hbar_\alpha G_{\partial\Omega_{\alpha\beta}}^{(s_\beta,\beta)}(\mathbf{x}_m,t_{m\pm})=0 \text{ for } s_\beta=0,1,2,\cdots,2k_\beta;\\ &\hbar_\alpha G_{\partial\Omega_{\alpha\beta}}^{(m_\alpha,\alpha)}(\mathbf{x}_m,t_{m+})<0 \text{ and } \hbar_\alpha G_{\partial\Omega_{\alpha\beta}}^{(2k_\beta+1,\beta)}(\mathbf{x}_m,t_{m\pm})<0 \end{aligned} \tag{5.72}$$

for all $\mathbf{x}_m \in \partial\Omega_{\alpha\beta}$ and no any real flows exist in all the adjacent, accessible domains Ω_β except its imaginary flows exist in domain Ω_α .

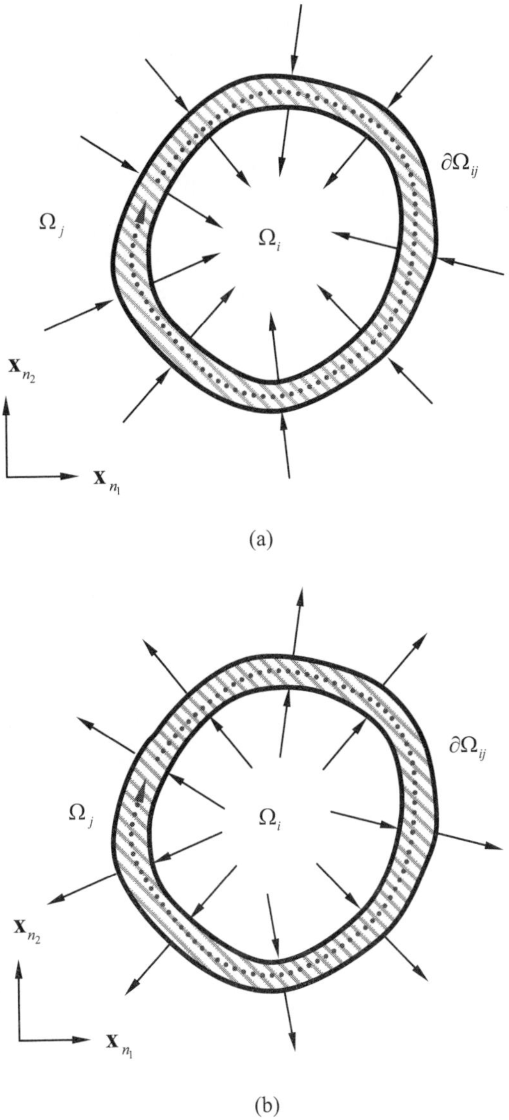

Fig. 5.24 Two closed accessible domains: **(a)** sink and **(b)** source. The hatched zone is the boundary of the closed accessible domain Ω_i. The solid closed curves on the hatched zone are the inner and outer boundaries of the closed accessible domain Ω_i. The inward and outward arrows to domain Ω_i are the flows going into the sink domain and going out the source domains, respectively. The dotted curves are the boundary flows for the case of passable flows only.

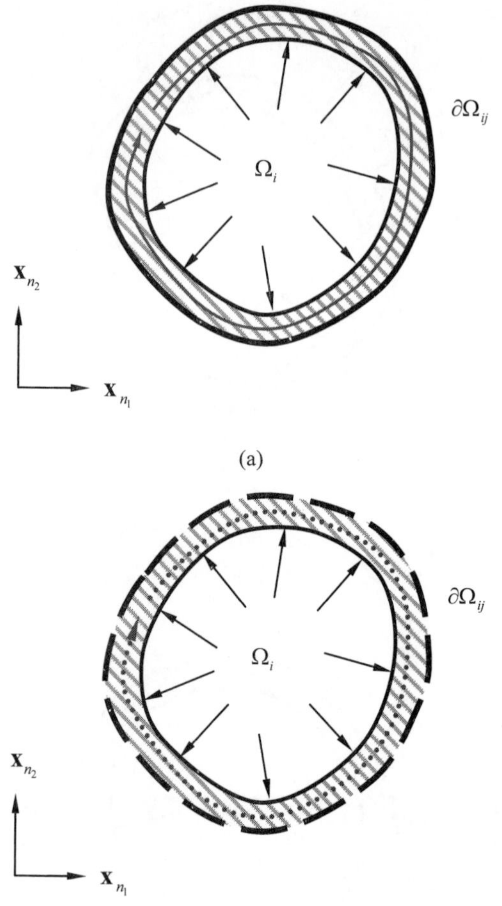

(b)

Fig. 5.25 Two source domains with: **(a)** a closed boundary and **(b)** an opened boundary. The hatched zone is the boundary of the closed accessible domain Ω_i. The solid and dashed curves on the boundary zone represent the closed and opened boundaries, respectively. The arrows depict the flows going out the source domain. The boundary flow can be formed on the closed boundary in the source domains, but no boundary flow can be formed on the opened boundary. If the initial conditions are on the boundary, the boundary flow can exist, expressed by the dotted curve.

From the above definitions, no real flows exist in the accessible domain of Ω_β, and the vector field in such a domain is defined indeed, but such a vector field can exist on domain Ω_α only as an imaginary vector field relative to domain Ω_β. For the inaccessible domain, no any vector field can be defined. Similarly, source domains without output are defined as follows.

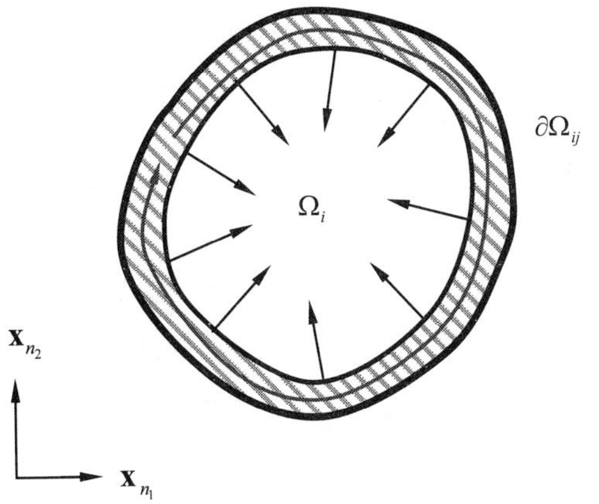

(a)

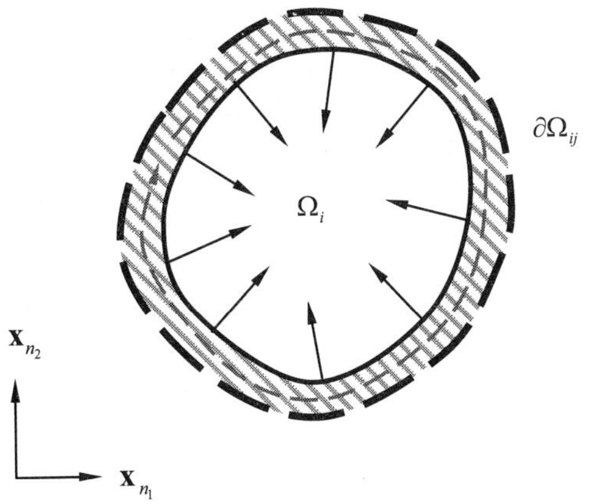

(b)

Fig. 5.26 Two closed sink domains with: **(a)** closed boundary and **(b)** open boundary. The hatched zone is the boundary of the closed accessible domain Ω_i. The arrows depict the flows going into the sink domain. The solid and dashed curves on the outer of the hatched zone represent the closed and opened boundaries, respectively. The boundary flow on the closed boundary can only get into in the sink domain, but the boundary flow on the opened boundary may enter the sink domain or disappear.

Definition 5.35. An accessible domain $\Omega_\alpha \subset \mho$ in a discontinuous dynamical system in Eq. (5.1) is termed a *source* domain without output if

$$\begin{gathered}\hbar_\alpha G^{(\alpha)}_{\partial\Omega_{\alpha\beta}}(\mathbf{x}_m, t_{m-}) > 0,\\ \hbar_\alpha G^{(\beta)}_{\partial\Omega_{\alpha\beta}}(\mathbf{x}_m, t_{m\pm}) = 0 \text{ and } \hbar_\alpha G^{(1,\beta)}_{\partial\Omega_{\alpha\beta}}(\mathbf{x}_m, t_{m\pm}) < 0\end{gathered} \tag{5.73}$$

for all $\mathbf{x}_m \in \partial\Omega_{\alpha\beta}$ and no any real flows exist in all the adjacent, accessible domains Ω_β except for its imaginary flows exist in domain Ω_α.

Definition 5.36. An accessible domain $\Omega_\alpha \subset \mho$ in a discontinuous dynamical system in Eq. (5.1) is termed a *source* domain of the $(2k_\alpha : 2k_\beta + 1)$-type without output if

$$\begin{gathered}\hbar_\alpha G_{\partial\Omega_{\alpha\beta}}^{(s_\alpha,\alpha)}(\mathbf{x}_m, t_{m-}) = 0 \text{ for } s_\alpha = 0,1,2,\cdots,2k_\alpha - 1;\\ \hbar_\alpha G_{\partial\Omega_{\alpha\beta}}^{(s_\beta,\beta)}(\mathbf{x}_m, t_{m\mp}) = 0 \text{ for } s_\beta = 0,1,2,\cdots,2k_\beta;\\ \hbar_\alpha G_{\partial\Omega_{\alpha\beta}}^{(2k_\alpha,\alpha)}(\mathbf{x}_m, t_{m-}) > 0 \text{ and } \hbar_\alpha G_{\partial\Omega_{\alpha\beta}}^{(2k_\beta+1,\beta)}(\mathbf{x}_m, t_{m\mp}) < 0\end{gathered} \tag{5.74}$$

for all $\mathbf{x}_m \in \partial\Omega_{\alpha\beta}$ and no any real flows exist in all the adjacent, accessible domains Ω_β except for its imaginary flows exist in domain Ω_α.

A sink domain without input and a source domain without output are formed by the half-sink and half source flows, as sketched in Fig. 5.27. Dashed curves are the vector fields of imaginary flows, which are defined in the accessible domain Ω_j and the flow of $\mathbf{x}_i^{(j)}$ exist in domain Ω_i. The domain Ω_i is a sink or source domain. In such sink and source domains, a flow $\mathbf{x}^{(i)}$ at boundary $\partial\Omega_{ij}$ is with the $(2k_i)$th-order singularity ($k_i \in \{0,1,2,\cdots\}$). If a flow $\mathbf{x}^{(i)}$ in domain Ω_i is with the $(2k_i+1)$th-order singularity at the entire boundary ($k_i \in \{1,2,\cdots\}$), the flow $\mathbf{x}^{(i)}$ will not get out the domain, and such a domain is called the *self-closed* domain. More strictly, the definition of the self-closed domain can be defined.

Definition 5.37. An accessible domain $\Omega_\alpha \subset \mho$ in a discontinuous dynamical system in Eq. (5.1) is termed a *self-closed* domain if a flow $\mathbf{x}^{(\alpha)}$ to points $\mathbf{x}_m \in \partial\Omega_{\alpha\beta}$ for all the adjacent, accessible domains Ω_β of domain Ω_α satisfy

$$\hbar_\alpha G_{\partial\Omega_{\alpha\beta}}^{(\alpha)}(\mathbf{x}_m, t_{m\pm}) = 0 \text{ and } \hbar_\alpha G_{\partial\Omega_{\alpha\beta}}^{(1,\alpha)}(\mathbf{x}_m, t_{m\pm}) < 0. \tag{5.75}$$

(i) The self-closed domain Ω_α is with *input* if a real flow $\mathbf{x}^{(\beta)}$ satisfies

$$\hbar_\alpha G_{\partial\Omega_{\alpha\beta}}^{(\beta)}(\mathbf{x}_m, t_{m-}) < 0. \tag{5.76}$$

(ii) The self-closed domain Ω_α is with *output* if a real flow $\mathbf{x}^{(\beta)}$ satisfies

$$\hbar_\alpha G_{\partial\Omega_{\alpha\beta}}^{(\beta)}(\mathbf{x}_m, t_{m+}) > 0. \tag{5.77}$$

(iii) The self-closed domain Ω_α is *independent* if a real flow $\mathbf{x}^{(\beta)}$ satisfies

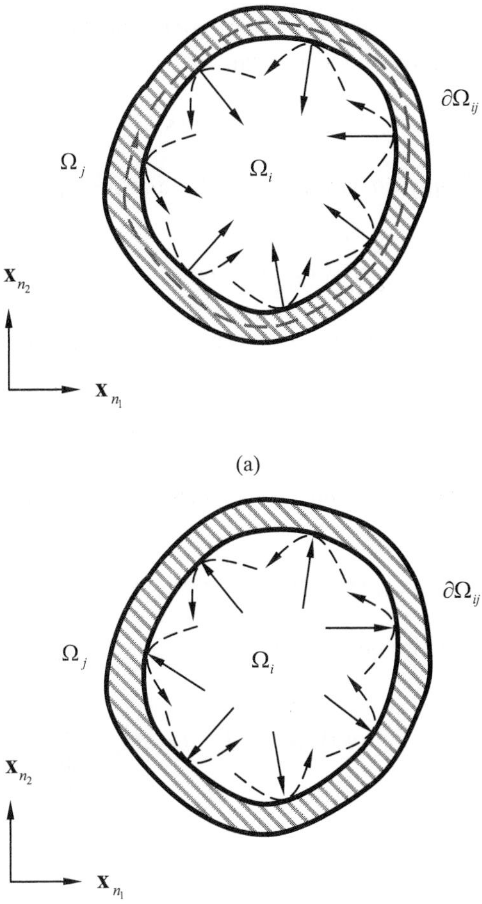

(b)

Fig. 5.27 (a) Sink domain without input and **(b)** source domain without output. The hatched zone is the boundary of the closed accessible domain Ω_i. The solid closed curves on the hatched zone are the inner and outer boundaries of the closed accessible domain Ω_i. The inward and outward arrows to the domain Ω_i are the flows going into the sink domain and going out the source domains, respectively. The dashed curves with arrows are the imaginary grazing flows.

$$\hbar_\alpha G^{(\beta)}_{\partial\Omega_{\alpha\beta}}(\mathbf{x}_m, t_{m\pm}) = 0 \text{ and } \hbar_\alpha G^{(1,\beta)}_{\partial\Omega_{\alpha\beta}}(\mathbf{x}_m, t_{m\pm}) > 0. \tag{5.78}$$

(iv) The self-closed domain Ω_α is *isolated* if an imaginary flow $\mathbf{x}^{(\beta)}_\alpha$ satisfies

$$\hbar_\alpha G^{(\beta)}_{\partial\Omega_{\alpha\beta}}(\mathbf{x}_m, t_{m\pm}) = 0 \text{ and } \hbar_\alpha G^{(1,\beta)}_{\partial\Omega_{\alpha\beta}}(\mathbf{x}_m, t_{m\pm}) < 0. \tag{5.79}$$

Definition 5.38. An accessible domain $\Omega_\alpha \subset \mho$ in a discontinuous dynamical system in Eq. (5.1) is termed a *self-closed empty* domain if an imaginary flow $\mathbf{x}_\beta^{(\alpha)}$ to point $\mathbf{x}_m \in \partial\Omega_{\alpha\beta}$ for all the adjacent domains Ω_β of domain Ω_α satisfy

$$\hbar_\alpha G^{(\alpha)}_{\partial\Omega_{\alpha\beta}}(\mathbf{x}_m, t_{m\pm}) = 0 \text{ and } \hbar_\alpha G^{(1,\alpha)}_{\partial\Omega_{\alpha\beta}}(\mathbf{x}_m, t_{m\pm}) > 0. \tag{5.80}$$

(i) The self-closed empty domain Ω_α is with input if the real flow of $\mathbf{x}^{(\beta)}$ satisfies

$$\hbar_\alpha G^{(\beta)}_{\partial\Omega_{\alpha\beta}}(\mathbf{x}_m, t_{m-}) < 0. \tag{5.81}$$

(ii) The self-closed empty domain Ω_α is with output if the real flow of $\mathbf{x}^{(\beta)}$ satisfies

$$\hbar_\alpha G^{(\beta)}_{\partial\Omega_{\alpha\beta}}(\mathbf{x}_m, t_{m+}) > 0. \tag{5.82}$$

(iii) The self-closed empty domain Ω_α is independent if the real flow of $\mathbf{x}^{(\beta)}$ satisfies

$$\hbar_\alpha G^{(\beta)}_{\partial\Omega_{\alpha\beta}}(\mathbf{x}_m, t_{m\pm}) = 0 \text{ and } \hbar_\alpha G^{(1,\beta)}_{\partial\Omega_{\alpha\beta}}(\mathbf{x}_m, t_{m\pm}) > 0. \tag{5.83}$$

(iv) The self-closed empty domain Ω_α is isolated if the imaginary flow of $\mathbf{x}_\alpha^{(\beta)}$ satisfies

$$\hbar_\alpha G^{(\beta)}_{\partial\Omega_{\alpha\beta}}(\mathbf{x}_m, t_{m\pm}) = 0 \text{ and } \hbar_\alpha G^{(1,\beta)}_{\partial\Omega_{\alpha\beta}}(\mathbf{x}_m, t_{m\pm}) < 0. \tag{5.84}$$

With higher-order singularity, the self-closed domain and self-closed-empty domains are discussed as follows.

Definition 5.39. An accessible domain $\Omega_\alpha \subset \mho$ in a discontinuous dynamical system in Eq. (5.1) is termed a *self-closed* domain with the $(2k_\alpha + 1)$th-order singularity if a flow $\mathbf{x}^{(\alpha)}$ to point $\mathbf{x}_m \in \partial\Omega_{\alpha\beta}$ for all the adjacent, accessible domains Ω_β of domain Ω_α satisfies

$$\left.\begin{aligned} &\hbar_\alpha G^{(s_\alpha,\alpha)}_{\partial\Omega_{\alpha\beta}}(\mathbf{x}_m, t_{m\pm}) = 0 \ \text{ for } s_\alpha = 0,1,2,\cdots,2k_\alpha, \\ &\hbar_\alpha G^{(2k_\alpha+1,\alpha)}_{\partial\Omega_{\alpha\beta}}(\mathbf{x}_m, t_{m\pm}) < 0. \end{aligned}\right\} \tag{5.85}$$

(i) The $(2k_\alpha + 1 : 2k_\beta)$ self-closed domain Ω_α is with the $(2k_\beta)$th-order flow input if a real flow $\mathbf{x}^{(\beta)}$ satisfies

$$\left.\begin{aligned} &\hbar_\alpha G^{(s_\beta,\beta)}_{\partial\Omega_{\alpha\beta}}(\mathbf{x}_m, t_{m-}) = 0 \ \text{ for } s_\beta = 0,1,2,\cdots,2k_\beta - 1, \\ &\hbar_\alpha G^{(2k_\beta,\beta)}_{\partial\Omega_{\alpha\beta}}(\mathbf{x}_m, t_{m-}) < 0. \end{aligned}\right\} \tag{5.86}$$

(ii) The $(2k_\alpha+1:2k_\beta)$ self-closed domain Ω_α is with the $(2k_\beta)$ th-order flow output if a real flow $\mathbf{x}^{(\beta)}$ satisfies

$$\begin{aligned}&\hbar_\alpha G^{(s_\beta,\beta)}_{\partial\Omega_{\alpha\beta}}(\mathbf{x}_m,t_{m+})=0 \ \text{ for } s_\beta=0,1,2,\cdots,2k_\beta-1,\\&\hbar_\alpha G^{(2k_\beta,\beta)}_{\partial\Omega_{\alpha\beta}}(\mathbf{x}_m,t_{m+})>0.\end{aligned}\tag{5.87}$$

(iii) The $(2k_\alpha+1:2k_\beta+1)$ self-closed domain Ω_α is independent if the $(2k_\beta+1)$ th-order real flow $\mathbf{x}^{(\beta)}$ satisfies

$$\begin{aligned}&\hbar_\alpha G^{(s_\beta,\beta)}_{\partial\Omega_{\alpha\beta}}(\mathbf{x}_m,t_{m\mp})=0 \ \text{ for } s_\beta=0,1,2,\cdots,2k_\beta,\\&\hbar_\alpha G^{(2k_\beta+1,\beta)}_{\partial\Omega_{\alpha\beta}}(\mathbf{x}_m,t_{m\mp})>0.\end{aligned}\tag{5.88}$$

(iv) The $(2k_\alpha+1:2k_\beta+1)$ self-closed domain Ω_α is isolated if the $(2k_\beta+1)$th - order imaginary flow $\mathbf{x}^{(\beta)}_\alpha$ satisfies

$$\begin{aligned}&\hbar_\alpha G^{(s_\beta,\beta)}_{\partial\Omega_{\alpha\beta}}(\mathbf{x}_m,t_{m\mp})=0 \ \text{ for } s_\beta=0,1,2,\cdots,2k_\beta,\\&\hbar_\alpha G^{(2k_\beta+1,\beta)}_{\partial\Omega_{\alpha\beta}}(\mathbf{x}_m,t_{m\mp})<0.\end{aligned}\tag{5.89}$$

Definition 5.40. An accessible domain $\Omega_\alpha \subset \mho$ in a discontinuous dynamical system in Eq. (5.1) is termed a *self-closed empty* domain with the $(2k_\alpha+1)$th - order singularity if an imaginary flow $\mathbf{x}^{(\alpha)}_\beta$ to point $\mathbf{x}_m \in \partial\Omega_{\alpha\beta}$ for all the adjacent domains Ω_β of domain Ω_α satisfies

$$\begin{aligned}&\hbar_\alpha G^{(s_\alpha,\alpha)}_{\partial\Omega_{\alpha\beta}}(\mathbf{x}_m,t_{m\pm})=0 \ \text{ for } s_\alpha=0,1,2,\cdots,2k_\alpha,\\&\hbar_\alpha G^{(2k_\alpha+1,\alpha)}_{\partial\Omega_{\alpha\beta}}(\mathbf{x}_m,t_{m\pm})>0.\end{aligned}\tag{5.90}$$

(i) The $(2k_\alpha+1:2k_\beta)$ self-closed empty domain Ω_α is with the $(2k_\beta)$th -order flow input if the real flow of $\mathbf{x}^{(\beta)}$ satisfies

$$\begin{aligned}&\hbar_\alpha G^{(s_\beta,\beta)}_{\partial\Omega_{\alpha\beta}}(\mathbf{x}_m,t_{m-})=0 \ \text{ for } s_\beta=0,1,2,\cdots,2k_\beta-1,\\&\hbar_\alpha G^{(2k_\beta,\beta)}_{\partial\Omega_{\alpha\beta}}(\mathbf{x}_m,t_{m-})<0.\end{aligned}\tag{5.91}$$

(ii) The $(2k_\alpha+1:2k_\beta)$ self-closed empty domain Ω_α is with the $(2k_\beta)$th -order flow output if a real flow $\mathbf{x}^{(\beta)}$ satisfies

$$\begin{aligned}&\hbar_\alpha G^{(s_\beta,\beta)}_{\partial\Omega_{\alpha\beta}}(\mathbf{x}_m,t_{m+})=0 \ \text{ for } s_\beta=0,1,2,\cdots,2k_\beta-1,\\&\hbar_\alpha G^{(2k_\beta,\beta)}_{\partial\Omega_{\alpha\beta}}(\mathbf{x}_m,t_{m+})>0.\end{aligned}\tag{5.92}$$

(iii) The $(2k_\alpha+1:2k_\beta+1)$ self-closed empty domain Ω_α is independent if the $(2k_\beta+1)$ th-order real flow $\mathbf{x}^{(\beta)}$ satisfies

$$\begin{aligned}&\hbar_\alpha G^{(s_\beta,\beta)}_{\partial\Omega_{\alpha\beta}}(\mathbf{x}_m,t_{m\mp})=0 \text{ for } s_\beta=0,1,2,\cdots,2k_\beta,\\&\hbar_\alpha G^{(2k_\beta+1,\beta)}_{\partial\Omega_{\alpha\beta}}(\mathbf{x}_m,t_{m\mp})>0.\end{aligned} \tag{5.93}$$

(iv) The $(2k_\alpha+1:2k_\beta+1)$ self-closed empty domain Ω_α is isolated if the $(2k_\beta+1)$th -order imaginary flow $\mathbf{x}^{(\beta)}_\alpha$ satisfies

$$\begin{aligned}&\hbar_\alpha G^{(s_\beta,\beta)}_{\partial\Omega_{\alpha\beta}}(\mathbf{x}_m,t_{m\mp})=0 \text{ for } s_\beta=0,1,2,\cdots,2k_\beta,\\&\hbar_\alpha G^{(2k_\beta+1,\beta)}_{\partial\Omega_{\alpha\beta}}(\mathbf{x}_m,t_{m\mp})<0.\end{aligned} \tag{5.94}$$

The self-closed domains and self-closed empty domains are sketched in Figs. 5.28-5.33. In Fig. 5.28, self-closed domains with input and output are presented. In a self-closed domain Ω_i, all the flows graze to boundary $\partial\Omega_{ij}$. On the outside of domain Ω_i (i.e., Ω_j), all flows to the entire boundary are either sink or source. In Fig. 5.29, an independent self-closed domain Ω_i and an isolated self-closed empty domain Ω_i are presented. The self-closed domain Ω_i is independent of its adjacent domains because no any flows in such domains are either sink or source. If a self-closed empty domain is isolated, no any real flow $\mathbf{x}^{(i)}$ exists. However, only imaginary flows $\mathbf{x}^{(j)}_i$ exist in the self-closed empty domain and such the imaginary flows are grazing to the boundary. In Fig. 5.30, an isolated self-closed domain and an independent, self-closed empty domain are presented. In an isolated self-closed domain Ω_i, the real flow $\mathbf{x}^{(i)}$ and the imaginary flow $\mathbf{x}^{(j)}_i$ are grazing to the boundary $\partial\Omega_{ij}$. In an independent self-closed empty domain, no any real and imaginary flows exist, and the real and imaginary flows exist outside domain Ω_i only. In Fig. 5.31, no any real flows $\mathbf{x}^{(i)}$ exist in a self-closed empty domain. The imaginary flow $\mathbf{x}^{(i)}_j$ exist outside domain Ω_i, and the corresponding real flows in domain Ω_j are either sink flow or source flow. For self-closed empty domains, real flows are half-sink or half-source flows. If the outside of self-closed domains are inaccessible, then there are two self-closed domains with closed and opened boundaries. The real flow in the self-closed domain is grazing to the boundary. In a self-closed empty domain, no any real flows exist except for the imaginary flow of $\mathbf{x}^{(j)}_i$ on the outside of domain Ω_i. Such self-closed domains are presented in Figs. 5.32 and 5.33.

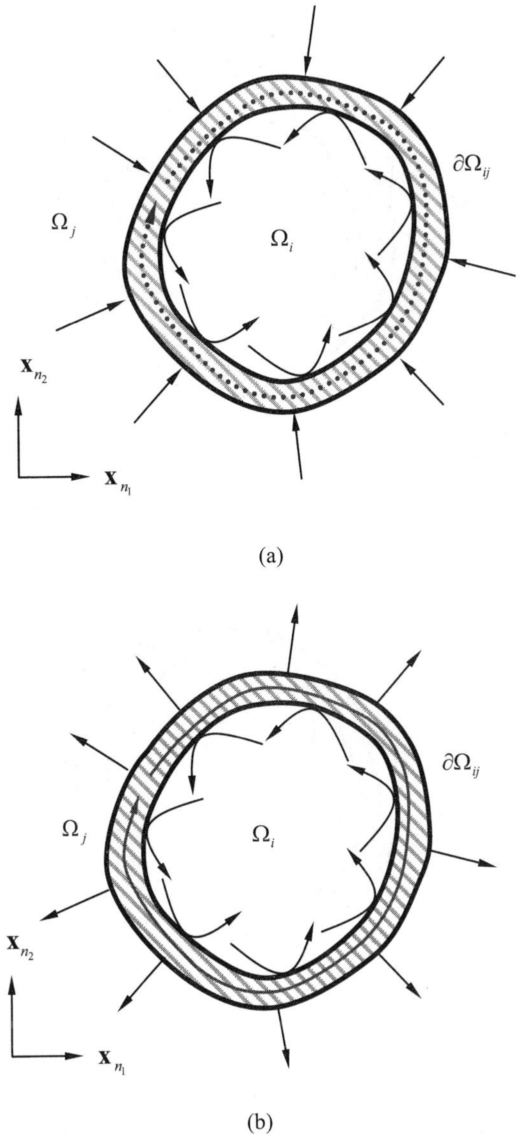

Fig. 5.28 Two self-closed domains with: **(a)** input and **(b)** output. The hatched zone is the boundary of the closed accessible domain Ω_i. The solid closed curves on the hatched zone are the inner and outer boundaries of the closed accessible domain Ω_i. The inward and outward arrows to domain Ω_i are the flows going into the sink domain and going out the source domains, respectively. The solid curves with arrows are grazing real flows.

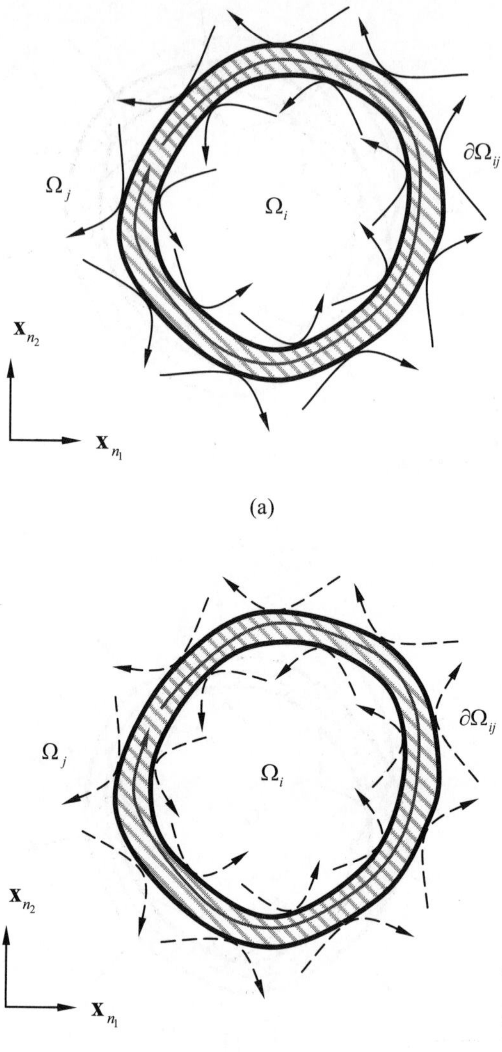

Fig. 5.29 Two self-closed domains: **(a)** independent self-closed domain and **(b)** isolated self-closed empty domain. The hatched zone is the boundary of the closed accessible domain Ω_i. The solid closed curves on the hatched zone are the inner and outer boundaries of the closed accessible domain Ω_i. The dashed and solid curves with arrows are grazing real and imaginary flows, respectively.

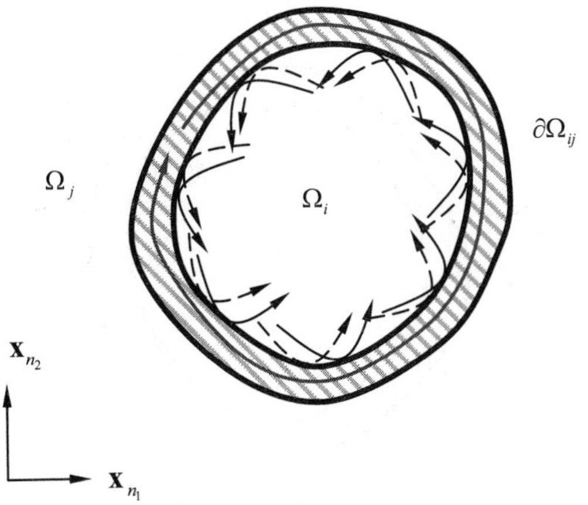

(a)

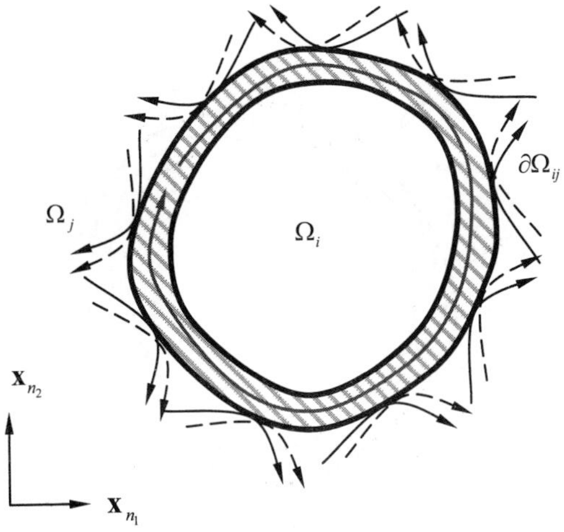

(b)

Fig. 5.30 Two self-closed domains of Ω_i: **(a)** isolated self-closed domain and **(b)** independent self-closed empty domain. The hatched zone is the boundary of the closed accessible domain Ω_i. The solid closed curves on the hatched zone are the inner and outer boundaries of the closed accessible domain Ω_i. The dashed and solid curves with arrows are grazing real and imaginary flows, respectively.

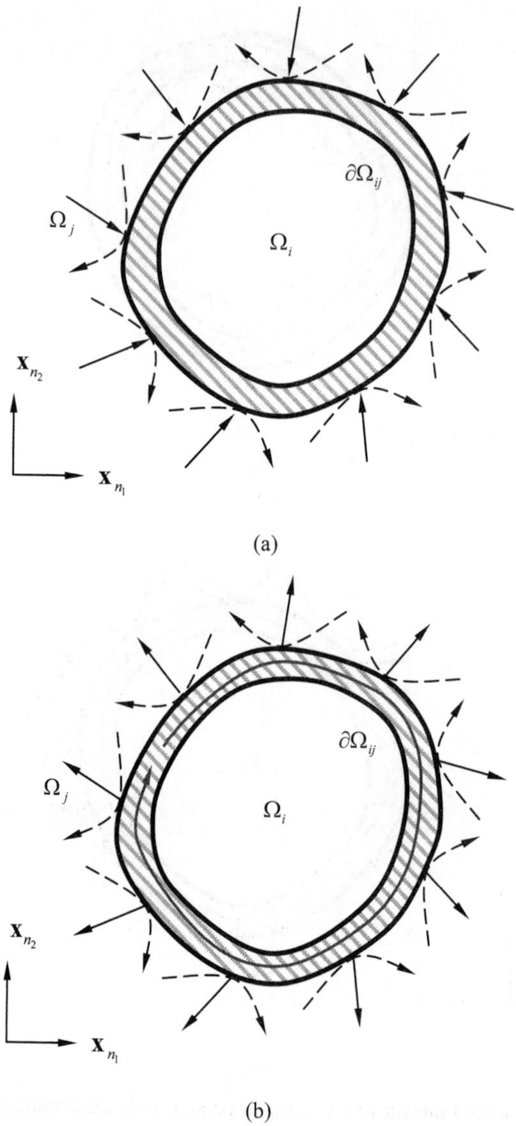

Fig. 5.31 Two self-closed domains: **(a)** self-closed empty domain with half sink and **(b)** self-closed empty domain with half-source. The hatched zone is the boundary of the closed accessible domain Ω_i. The solid closed curves on the hatched zone are the inner and outer boundaries of the closed accessible domain Ω_i. The inward and outward arrows to the domain Ω_i are the flows going into the sink domain and going out the source domains, respectively. The dashed curves with arrows are imaginary grazing flows.

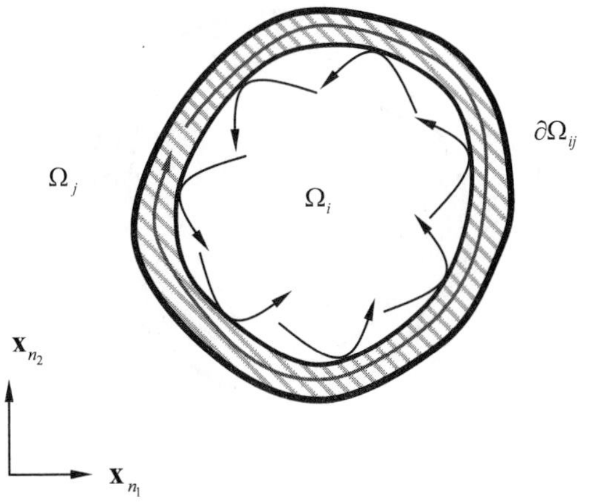

(a)

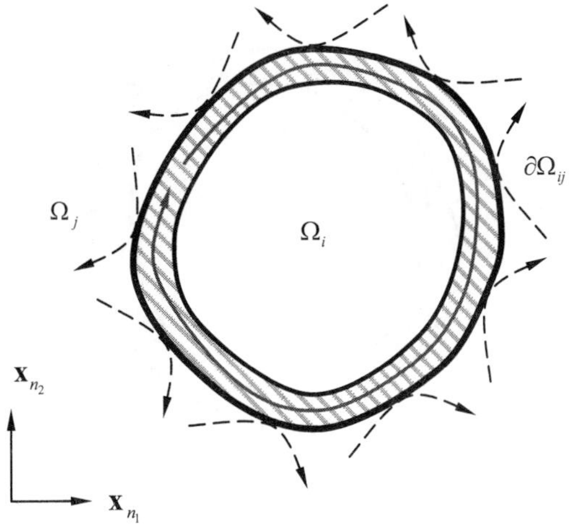

(b)

Fig. 5.32 Two self-closed domains of Ω_i rounded by an inaccessible domain: **(a)** isolated self-closed isolated domain with a closed boundary and **(b)** self-closed empty domain with a closed boundary. The hatched zone is the boundary of the closed accessible domain Ω_i . The solid closed curves on the hatched zone are the inner and outer boundaries of the closed accessible domain Ω_i . The dashed and solid curves with arrows are real and imaginary grazing flows, respectively.

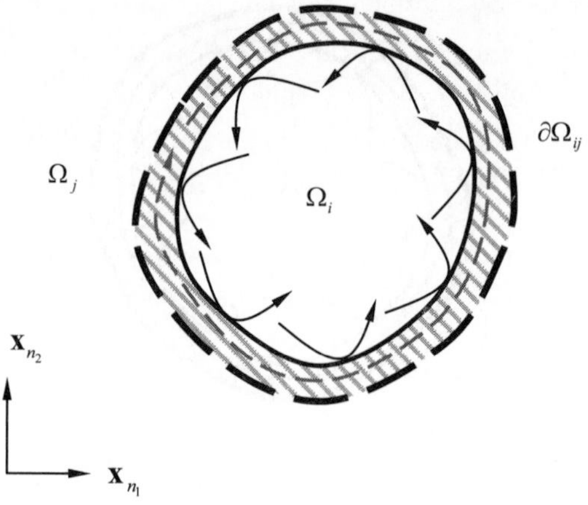

(a)

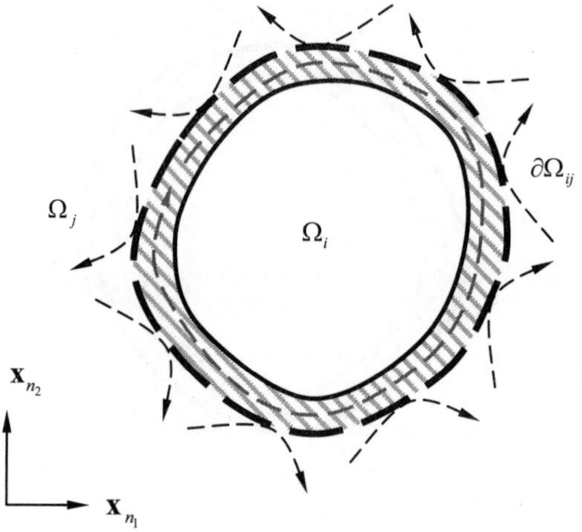

(b)

Fig. 5.33 Two self-closed domains of Ω_i: **(a)** isolated self-closed domain and **(b)** self-closed empty domain. The hatched zone is the boundary of the closed accessible domain Ω_i. The solid and dashed, closed curves on the hatched zone are the inner and outer boundaries of domain Ω_i. The dashed and solid curves with arrows are imaginary and real grazing flows, respectively.

5.4.2. Boundary classifications

As mentioned in Chapter 2, for boundary $\partial\Omega_{\alpha\beta}$, if flows in domains Ω_α and Ω_β at the such a boundary are in the same direction, such a boundary is called the passable boundary. However, all flows in domains Ω_α and Ω_β possess the opposite directions at such a boundary, such a boundary is called sink or source boundary. Herein, such issues will be further discussed.

Definition 5.41. A boundary between two adjacent, accessible domains of Ω_α and Ω_β in a discontinuous dynamical system in Eq. (5.1) is termed a *sink* boundary if

$$\hbar_\alpha G^{(\alpha)}_{\partial\Omega_{\alpha\beta}}(\mathbf{x}_m, t_{m-}) < 0 \text{ and } \hbar_\alpha G^{(\beta)}_{\partial\Omega_{\alpha\beta}}(\mathbf{x}_m, t_{m-}) < 0 \tag{5.95}$$

for point $\mathbf{x}_m \in \partial\Omega_{\alpha\beta}$.

Definition 5.42. A boundary between two adjacent, accessible domains of Ω_α and Ω_β in a discontinuous dynamical system in Eq. (5.1) is termed a *sink* boundary of the $(2k_\alpha : 2k_\beta)$-type if

$$\begin{aligned}
&\hbar_\alpha G^{(s_\alpha,\alpha)}_{\partial\Omega_{\alpha\beta}}(\mathbf{x}_m, t_{m-}) = 0 \text{ for } s_\alpha = 0,1,2,\cdots,2k_\alpha - 1;\\
&\hbar_\alpha G^{(s_\beta,\beta)}_{\partial\Omega_{\alpha\beta}}(\mathbf{x}_m, t_{m-}) = 0 \text{ for } s_\beta = 0,1,2,\cdots,2k_\beta - 1;\\
&\hbar_\alpha G^{(2k_\alpha,\alpha)}_{\partial\Omega_{\alpha\beta}}(\mathbf{x}_m, t_{m-}) < 0 \text{ and } \hbar_\alpha G^{(2k_\beta,\beta)}_{\partial\Omega_{\alpha\beta}}(\mathbf{x}_m, t_{m-}) < 0
\end{aligned} \tag{5.96}$$

for point $\mathbf{x}_m \in \partial\Omega_{\alpha\beta}$.

Similarly, the source boundary of the accessible domain is defined as follows.

Definition 5.43. A boundary between two adjacent, accessible domains of Ω_α and Ω_β in a discontinuous dynamical system in Eq. (5.1) is termed a *source* boundary if

$$\hbar_\alpha G^{(\alpha)}_{\partial\Omega_{\alpha\beta}}(\mathbf{x}_m, t_{m+}) > 0 \text{ and } \hbar_\alpha G^{(\beta)}_{\partial\Omega_{\alpha\beta}}(\mathbf{x}_m, t_{m+}) > 0 \tag{5.97}$$

for point $\mathbf{x}_m \in \partial\Omega_{\alpha\beta}$.

Definition 5.44. A boundary between two adjacent, accessible domains of Ω_α and Ω_β in a discontinuous dynamical system in Eq. (5.1) is termed a *source* boundary of the $(m_\alpha : m_\beta)$-type if

$$
\begin{aligned}
&\hbar_{\alpha} G_{\partial\Omega_{\alpha\beta}}^{(s_{\alpha},\alpha)}(\mathbf{x}_m, t_{m+}) = 0 \text{ for } s_{\alpha} = 0,1,2,\cdots,m_{\alpha}-1;\\
&\hbar_{\alpha} G_{\partial\Omega_{\alpha\beta}}^{(s_{\beta},\beta)}(\mathbf{x}_m, t_{m+}) = 0 \text{ for } s_{\beta} = 0,1,2,\cdots,m_{\beta}-1;\\
&\hbar_{\alpha} G_{\partial\Omega_{\alpha\beta}}^{(m_{\alpha},\alpha)}(\mathbf{x}_m, t_{m+}) > 0 \text{ and } \hbar_{\alpha} G_{\partial\Omega_{\alpha\beta}}^{(m_{\beta},\beta)}(\mathbf{x}_m, t_{m+}) > 0
\end{aligned}
\tag{5.98}
$$

for point $\mathbf{x}_m \in \partial\Omega_{\alpha\beta}$.

All the boundaries of domain Ω_i with its adjacent, accessible domains are sink and source boundaries, sketched in Fig. 5.34 (a) and (b), respectively. Without any inaccessible domains, the *full* sink and source boundaries of domain Ω_i are presented in Fig. 5.35 (a) and (b). For the *full* sink boundary of domain Ω_i, the boundary flow can be formed on the boundary. For a *full* source boundary, the boundary flow with a small perturbation in its two domains will disappear. If all the adjacent domains of Ω_i are inaccessible, then a closed boundary can be a sink boundary with a source flow or a closed boundary can be a source boundary with a sink flow. However, an opened boundary of domain Ω_i cannot form any sink boundary with a source flow in the domain. For this case, the transport law will be added. Otherwise, the flow will disappear from the boundary. On the other hand, the open boundary of domain Ω_i can form a source boundary with a sink flow.

Definition 5.45. A boundary between two adjacent, accessible domains of Ω_{α} and Ω_{β} in a discontinuous dynamical system in Eq. (5.1) is termed a *semi-passable* boundary from Ω_{α} to Ω_{β} if

$$
\hbar_{\alpha} G_{\partial\Omega_{\alpha\beta}}^{(\alpha)}(\mathbf{x}_m, t_{m-}) > 0 \text{ and } \hbar_{\alpha} G_{\partial\Omega_{\alpha\beta}}^{(\beta)}(\mathbf{x}_m, t_{m+}) > 0
\tag{5.99}
$$

for all $\mathbf{x}_m \in \partial\Omega_{\alpha\beta}$.

Definition 5.46. A boundary between two adjacent, accessible domains of Ω_{α} and Ω_{β} in a discontinuous dynamical system in Eq. (5.1) is termed a *semi-passable* boundary of the $(2k_{\alpha} : m_{\beta})$-type from Ω_{α} to Ω_{β} if

$$
\begin{aligned}
&\hbar_{\alpha} G_{\partial\Omega_{\alpha\beta}}^{(s_{\alpha},\alpha)}(\mathbf{x}_m, t_{m-}) = 0 \text{ for } s_{\alpha} = 0,1,2,\cdots,2k_{\alpha}-1;\\
&\hbar_{\alpha} G_{\partial\Omega_{\alpha\beta}}^{(s_{\beta},\beta)}(\mathbf{x}_m, t_{m+}) = 0 \text{ for } s_{\beta} = 0,1,2,\cdots,m_{\beta}-1;\\
&\hbar_{\alpha} G_{\partial\Omega_{\alpha\beta}}^{(2k_{\alpha},\alpha)}(\mathbf{x}_m, t_{m-}) > 0 \text{ and } \hbar_{\alpha} G_{\partial\Omega_{\alpha\beta}}^{(m_{\beta},\beta)}(\mathbf{x}_m, t_{m+}) > 0
\end{aligned}
\tag{5.100}
$$

for all $\mathbf{x}_m \in \partial\Omega_{\alpha\beta}$.

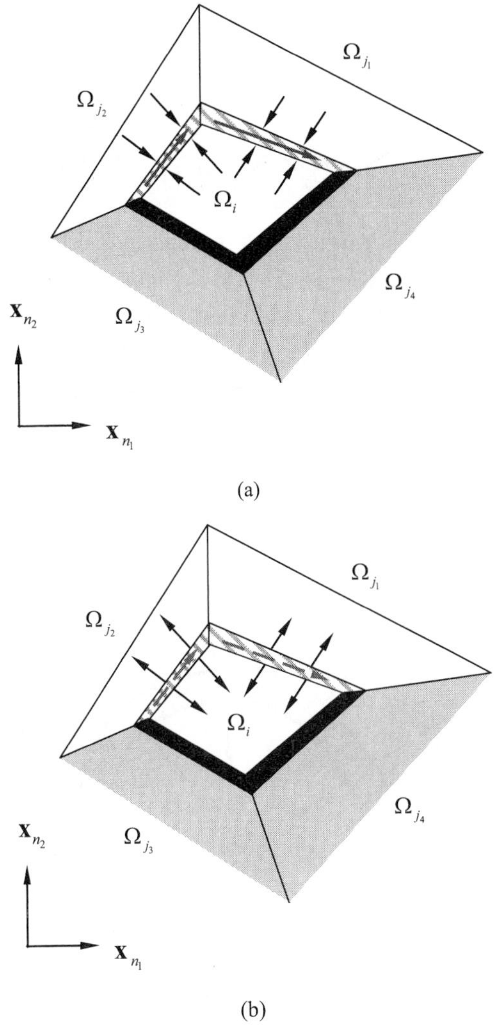

Fig. 5.34 (a) sink boundary and **(b)** source boundary of $\partial\Omega_{ij}$. The white sub-domains are accessible. The gray sub-domains are inaccessible. The hatched boundaries are for non-passable flows, and the dark boundaries are between the accessible and inaccessible boundaries.

If all the boundaries in a discontinuous dynamical system are the sink boundaries, the sink boundary network can be formed, and the boundary flows in such a network are stable. If all the boundaries in a discontinuous dynamical system are the source boundaries, the source boundary network can be formed, and the boundary flows in such a network are unstable to the domains. The boundary network with a mixing of the sink and source can also be formed. The flow complexity in discontinuous dynamical systems can be discussed.

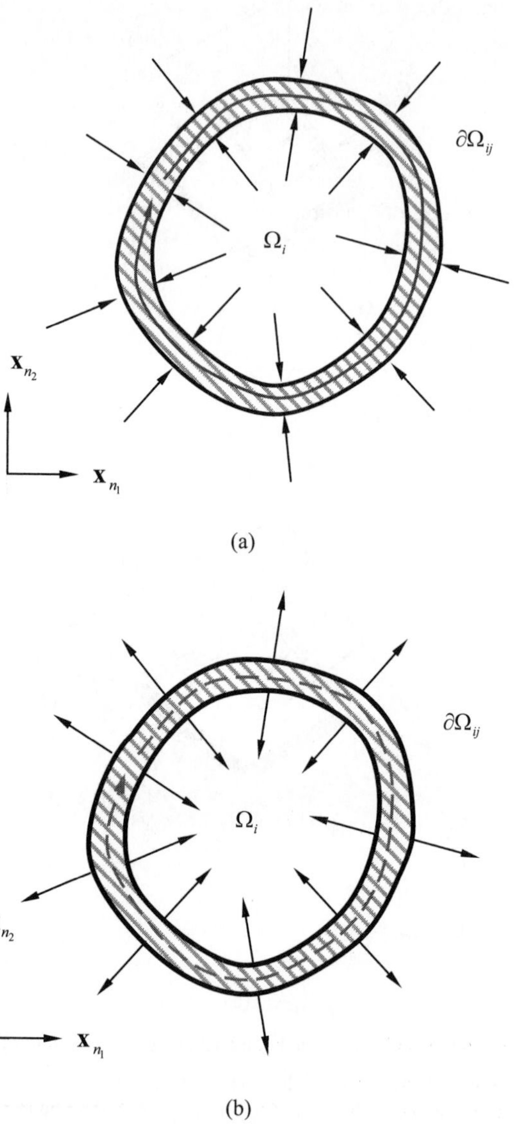

Fig. 5.35 Two boundaries: **(a)** sink and **(b)** source. The solid closed curves is the outer and inner boundary of the closed accessible domain Ω_i. The arrows are the flows pointing to the boundary or leaving from the boundary. The solid and dashed curves are the boundary flows for the sink and source boundary, respectively. The boundary flow on the sink boundary can be formed. However, the boundary flow on the source boundary cannot be formed. If the initial conditions on the boundary, the boundary flow will exist. With a small perturbation to the domains, the boundary flow will get into one of the two domains.

In a boundary network in n-dimensional phase space, there is at least an $(n-2)$-dimensional edge to connect two different surfaces. Otherwise, the two boundaries are absolute independent, and the boundary network cannot be formed. The sink and source boundary networks are presented in Fig. 5.36 (a) and (b). Only one $(n-2)$-dimensional edge is sketched which connect three braches of boundary network, which is depicted by a large filled circle. In Fig. 5.36(a), flows in the three corresponding domains are sink flows to the boundary. The three boundary flows can be formed on a boundary network, presented by solid curves. The direction of boundary flows is dependent on dynamical systems on the boundary, which will determine the property of edges. The dynamical behaviors of edge will be discussed later. In Fig. 5.36(b), flows in the three corresponding domains are source flows to the boundary. The boundary flows are expressed by dashed curves because the boundary flow can easily leave the boundary once a small perturbation of the boundary to one of domains exists. In a boundary network, sink and source boundary branches can be mixed together, which presented in Fig. 5.37(a). In three branches, there are two sink boundaries and one source boundary. In addition, two grazing singular sets of the flows at the boundary are just at the singular edge, depicted by two hollow circles. In Fig. 5.37(b), the boundary network formed by the sink, source and passable boundary branches. For the semi-passable boundary, the boundary follow cannot be formed. However, if the boundary flow exists on the boundary, such a flow can be perturbed to one of two domains from the leaving flow. Such a boundary flow is represented by a dotted curve.

5.5. Transport laws

From discussions in the previous sections, because the separation boundary, inaccessible domains and flow barriers on the boundary exist in discontinuous dynamical systems, it is necessary to introduce physical laws or rules to continue the flow. Before discussing transport laws on the discontinuous boundary $\partial\Omega_{ij}$, the *coming* and *leaving* boundaries of the flow are introduced first.

Definition 5.47. For a discontinuous system in Eq. (5.1), a boundary $\partial\Omega_{ij}$ to which only a coming flow in domain Ω_α ($\alpha \in \{i, j\}$) arrives is termed the α*-side coming* boundary $\partial\Omega_{\alpha\beta}^{\downarrow\alpha}$ ($\beta \in \{i, j\}$ and $\alpha \neq \beta$).

Definition 5.48. For a discontinuous system in Eq. (5.1), a boundary $\partial\Omega_{ij}$ where only a leaving flow leaves for the sub-domain Ω_α ($\alpha \in \{i, j\}$) is termed the α*-side leaving* boundary $\partial\Omega_{\beta\alpha}^{\uparrow\alpha}$ ($\beta \in \{i, j\}$ and $\alpha \neq \beta$).

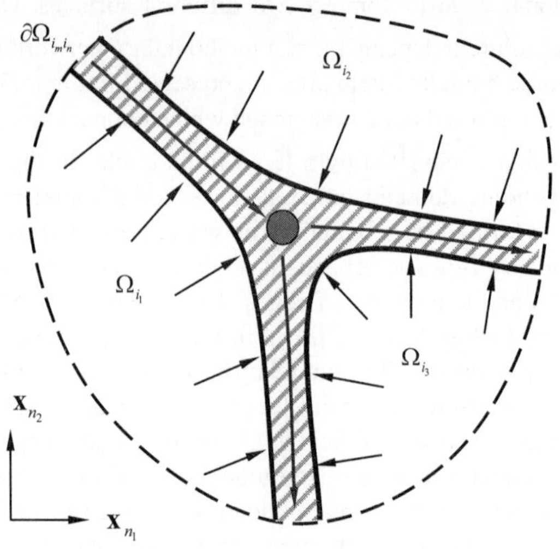

(a)

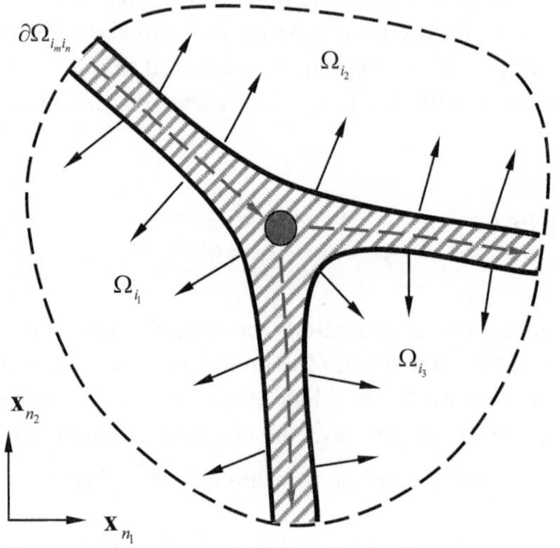

(b)

Fig. 5.36 Boundary flow network on the boundary $\partial\Omega_{i_m i_n}$: **(a)** sink and **(b)** source. Solid closed curves are the outer and inner boundaries of the accessible domain Ω_{i_m} . The arrows are the flows coming to the boundary or leaving the boundary. Boundary flows on the sink and source boundaries are presented by solid and dashed curves with arrows, respectively. The filled circle is the singular edge.

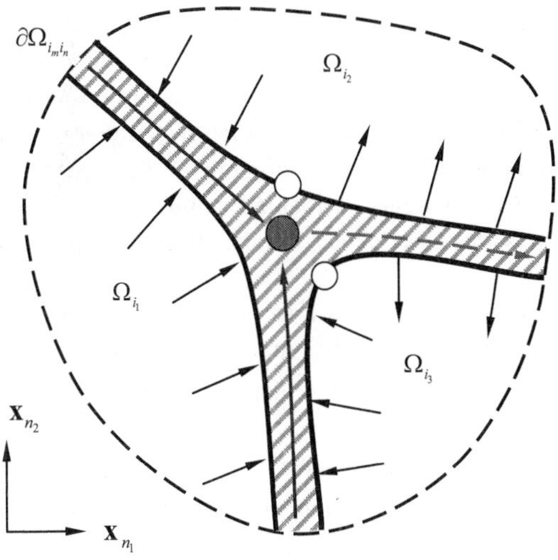

(a)

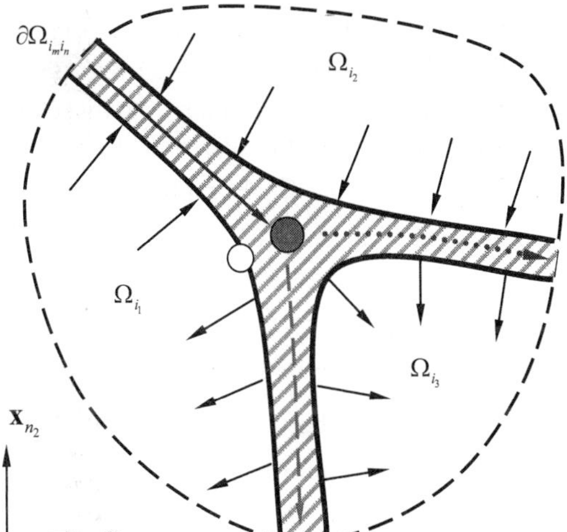

(b)

Fig. 5.37 Boundary flow network on the boundary $\partial\Omega_{i_m i_n}$: **(a)** sink and **(b)** source. Solid closed curves are the outer and inner boundaries of a closed accessible domain Ω_{i_m} . The arrows are the flows coming to the boundary or leaving the boundary. Boundary flows on sink, source and passable boundaries are presented by the solid, dashed and dotted curves with arrows, respectively. The filled circle is the singular edge. The hollow cycles are the grazing singular sets.

The above two definitions are applicable to semi-passable boundary (i.e., $\overrightarrow{\partial\Omega_{ij}}$) and non-passable boundary (i.e., $\overline{\partial\Omega_{ij}}$). If a flow in a domain comes to its boundary, this boundary is termed the coming boundary of a flow. On the other hand, if a flow leaves the boundary and gets into a domain, the boundary is called the leaving boundary. To restrict our discussion, the investigation focuses on a flow at the boundary with C^0-discontinuity. Consider such a boundary to be the part of the boundary for the unbounded, universal domain $\mho$ first. Since the universal domain is unbounded, the boundary is located at the infinity. For such a case, no transport rule or law is required. Hence, a more strict description is given, i.e.,

Definition 5.49. Under Assumptions (H2.1)-(H2.3), if a discontinuous dynamic system of Eq. (5.1) on an unbounded domain $\Omega_\alpha \subset \mho$ possesses a bounded flow, there are *no any laws* exerted on the boundary $\partial\Omega_{\alpha\infty} = \Omega_\alpha \cap \partial\mho \neq \varnothing$.

Definition 5.50. Under Assumptions (H2.1)-(H2.3), if a flow of a discontinuous dynamic system of Eq. (5.1) on an bounded domain $\Omega_\alpha \subset \mho$ arrives to the boundary of $\partial\Omega_{\alpha\infty} = \Omega_\alpha \cap \partial\mho \neq \varnothing$, there is *at least a transport law* on such a boundary.

To show the above definition intuitively, the boundaries of domain Ω_α intersected with unbounded and bounded, universal domains are sketched in Fig. 5.38 (a) and (b) through dotted and solid curves, respectively. The domain Ω_α is accessible. From Assumption (H2.2-H2.3) in Chapter 2, to make physical flows exist in the discontinuous system, a flow in the unbounded sub-domain Ω_α must be bounded because of $\partial\Omega_{\alpha\infty} = \Omega_\alpha \cap \partial\mho \neq \varnothing$. For a bounded, universal domain $\mho$, there is a common boundary $\partial\Omega_{\alpha\infty} = \Omega_\alpha \cap \partial\mho \neq \varnothing$ between domain $\mho$ and Ω_α is non-passable.

On a common boundary $\partial\Omega_{\alpha\infty}$, there is a grazing singular point dividing the boundary the coming and leaving sub-boundaries on the α-side of $\partial\Omega_{\alpha\infty}$ (i.e., $\partial\Omega_{\alpha\infty}^{\downarrow\alpha}$ and $\partial\Omega_{\alpha\infty}^{\uparrow\alpha}$). On a non-passable boundary, there is at least a transport law to map a flow at the coming boundary into the leaving boundary. If the boundary $\partial\Omega_{i\infty}$ is a coming or leaving boundary only, a transport law on the boundary must exist to map a *coming* boundary of $\partial\Omega_{\alpha\infty}^{\downarrow\alpha}$ into a leaving boundary of $\partial\Omega_{\beta\infty}^{\uparrow\beta}$. Otherwise, the coming boundary of $\partial\Omega_{\alpha\infty}^{\downarrow\alpha}$ may become an *attractive* boundary set (i.e., "blackhole"). In other words, if a bounded boundary $\partial\Omega_{\alpha\infty}$ is closed, a coming flow will be switched to a boundary flow moving along the boundary $\partial\Omega_{\alpha\infty}$. If the bounded boundary $\partial\Omega_{\alpha\infty}$ is open, a coming flow in Ω_α will disappear without any transport law.

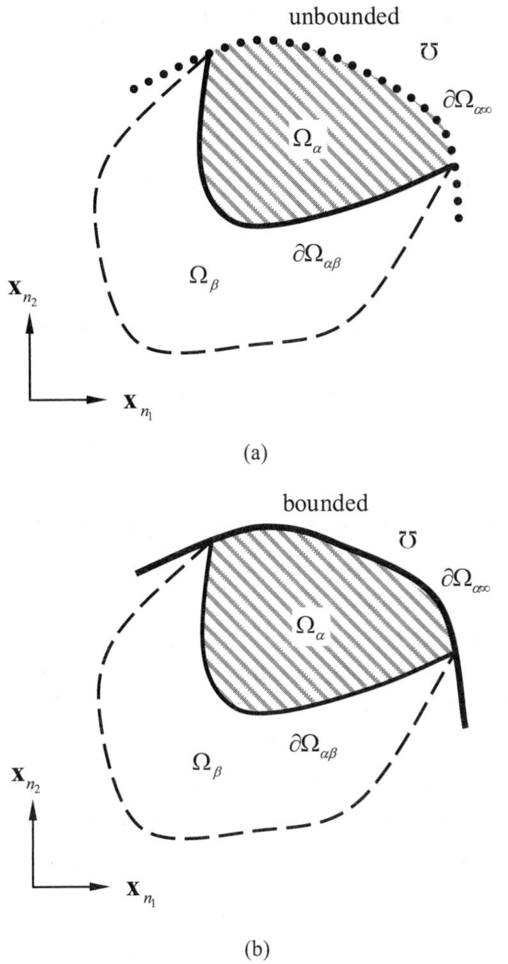

Fig. 5.38 (a) Unbounded and **(b)** bounded domains of Ω_i. The hatched domain is a specific accessible domain.

On the other hand, the leaving boundary $\partial\Omega_{\alpha\infty}$ may become *source* to repel flows into the corresponding domain. If a leaving boundary $\partial\Omega_{\beta\infty}^{\uparrow\beta}$ is closed, a boundary flow on such a leaving boundary is source. If a leaving boundary of $\partial\Omega_{\beta\infty}^{\uparrow\beta}$ is open, the boundary flow on such a leaving boundary may disappear on the outside of the bounded universal domain. This investigation will not make too much sense.

Definition 5.51. Under Assumptions (H2.1)-(H2.3), for a discontinuous system in

Eq. (5.1), there is a flow $\mathbf{x}^{(\alpha)} \in \Omega_\alpha$ at a common boundary

$$\partial\Omega_{\alpha\infty} = \partial\Omega_{\alpha\infty}^{\downarrow\alpha} \cup \partial\Omega_{\alpha\infty}^{\uparrow\alpha} \cup \mathcal{S}_{\alpha\infty}. \tag{5.101}$$

A map $T_{\alpha\infty}$ transporting a point $\mathbf{x}_{m-}^{(\alpha)} \in \partial\Omega_{i\infty}^{\downarrow\alpha}$ to a point $\mathbf{x}_{m+}^{(\alpha)} \in \partial\Omega_{\alpha\infty}^{\uparrow\alpha}$ for time t_m without time consumption is termed the *transport law* on the common boundary $\partial\Omega_{\alpha\infty}$, i.e.,

$$T_{\alpha\infty} : \mathbf{x}_{m-}^{(\alpha)} \to \mathbf{x}_{m+}^{(\alpha)}, \tag{5.102}$$

which is determined by a one-to-one vector function $\mathbf{g}_{\alpha\infty} \in \mathscr{R}^{n-1}$ as

$$\mathbf{g}_{\alpha\infty}(\mathbf{x}_{m-}^{(\alpha)}, \mathbf{x}_{m+}^{(\alpha)}, \mathbf{p}^{(\alpha\infty)}) = 0 \quad \text{on } \partial\Omega_{\alpha\infty}, \tag{5.103}$$

with a parameter vector $\mathbf{p}^{(\alpha\infty)} \in \mathscr{R}^{k_\alpha}$ relative to the boundary $\partial\Omega_{\alpha\infty}$.

The transport law on the boundary of $\partial\Omega_{\alpha\infty}$ can be explained geometrically in Fig. 5.39 through the dashed curve. The white circular points are for a transportation from the coming sub-boundary to the leaving sub-boundary of $\partial\Omega_{\alpha\infty}$. The transport law can be exerted to discontinuous dynamic systems, or can exist generically in discontinuous dynamical systems. The impact law in physics is a transport law to map the motion on the coming boundary onto the leaving boundary.

Consider $\mathbf{x}_{m-}^{(\alpha)} \in \partial\Omega_{\alpha\infty}^{\downarrow\alpha} \subset \mathscr{R}^{n-1}$ on a α-side coming boundary of $\varphi_{\alpha\infty}(\mathbf{x}_{m-}^{(\alpha)}, t_{m-}) = 0$. Because a function for the separation boundary is unique, one of n-components of $\mathbf{x}_{m-}^{(\alpha)}$ can be expressed by the $(n-1)$-components. Similarly, for $\mathbf{x}_{m+}^{(\alpha)} \in \partial\Omega_{\alpha\infty}^{\uparrow\alpha} \subset \mathscr{R}^{n-1}$ on the α-side leaving boundary, a function of $\varphi_{\alpha\infty}(\mathbf{x}_{m+}^{(\alpha)}, t_{m+}) = 0$ should be satisfied. One of n-components of $\mathbf{x}_{m+}^{(\alpha)}$ can also expressed by the $(n-1)$-components. Therefore, a transport law for two mapping points of $\mathbf{x}_{m-}^{(\alpha)}$ and $\mathbf{x}_{m+}^{(\alpha)}$ should give a one-to-one, $(n-1)$-dimensional vector function. In Definition 5.51, the transport law is defined on the same domain with the coming and leaving boundaries. A transport law between the boundaries of two separated, accessible sub-domains is defined as follows.

Definition 5.52. Under Assumptions (H2.1)-(H2.3), for a discontinuous system in Eq. (5.1), two accessible domains Ω_α $(\alpha = i, j)$ exist. For two flows $\mathbf{x}^{(\alpha)} \in \Omega_\alpha$, a map T_{ij}^{∞} to transport a point $\mathbf{x}_{m-}^{(i)} \in \partial\Omega_{i\infty}^{\downarrow i}$ to the point $\mathbf{x}_{m+}^{(j)} \in \partial\Omega_{j\infty}^{\uparrow j}$ for time t_m without time consumption is termed the *transport law* from $\partial\Omega_{i\infty}^{\downarrow i}$ to $\partial\Omega_{j\infty}^{\uparrow j}$, i.e.,

$$T_{ij}^{\infty} : \mathbf{x}_{m-}^{(i)} \to \mathbf{x}_{m+}^{(j)}, \tag{5.104}$$

which is determined by a one-to-one vector function $\mathbf{g}_{ij}^{\infty} \in \mathscr{R}^{n-1}$ as

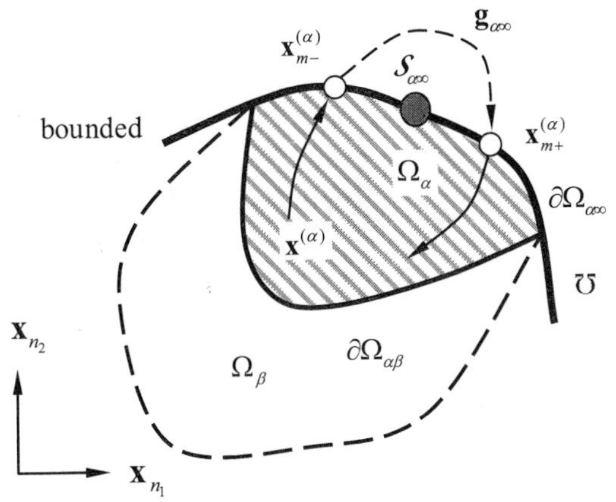

Fig. 5.39 Transport law on the boundary of $\partial\Omega_{\alpha\infty}$. The arrows represent the input and output boundaries on $\partial\Omega_{i\infty}$ for the flow. The dark circular symbols are connecting sets, and the white circular points are for flow transporting.

$$\mathbf{g}_{ij}^{\infty}(\mathbf{x}_{m-}^{(i)}, \mathbf{x}_{m+}^{(j)}, \mathbf{p}_{ij}^{\infty}) = 0, \quad \text{for } \mathbf{x}_{m-}^{(i)} \in \partial\Omega_{i\infty}^{\downarrow i} \text{ and } \mathbf{x}_{m+}^{(j)} \in \partial\Omega_{j\infty}^{\uparrow j}, \tag{5.105}$$

with a parameter vector $\mathbf{p}_{ij}^{\infty} \in \mathscr{R}^{k_{ij}^{\infty}}$.

This transport law between two boundaries of $\partial\Omega_{i\infty}$ and $\partial\Omega_{j\infty}$ must be exerted for flow continuity. However, for two separated, accessible sub-domains, a transport law mapping a flow from a coming boundary of $\partial\Omega_{i\infty}^{\downarrow i}$ to a leaving boundary of $\partial\Omega_{j\infty}^{\downarrow j}$ is presented in Fig. 5.40. If the coming and leaving boundaries of two separated domains are closed, a transport law will transport a flow from a *source* domain into a *sink* domain. The afore-discussed transport law is relative to the common boundary between a accessible domain and its universal domain. The transport law on the boundary between the two accessible sub-domains will be discussed, and then a generalized transport law on boundaries between accessible and/or inaccessible domains will be discussed.

Definition 5.53. Under Assumptions (H.2.1)-(H2.3), for a discontinuous system in Eq. (5.1), two adjacent accessible domains Ω_{α} $(\alpha = i, j)$ exist. A map $T_{i\beta}$ transporting $\mathbf{x}_{m-}^{(i)} \in \partial\Omega_{ij}^{\downarrow i}$ to $\mathbf{x}_{m+}^{(\beta)} \in \partial\Omega_{\alpha\beta}^{\uparrow\beta}$ $(\alpha, \beta \in \{i, j\}$ and $\beta \neq \alpha)$ with $\mathbf{x}_{m-}^{(i)} \neq \mathbf{x}_{m+}^{(\beta)}$ is termed a *transport law* from $\partial\Omega_{ij}^{\downarrow i}$ to $\partial\Omega_{\alpha\beta}^{\uparrow\beta}$, i.e.,

$$T_{i\beta} : \mathbf{x}_{m-}^{(i)} \to \mathbf{x}_{m+}^{(\beta)}, \tag{5.106}$$

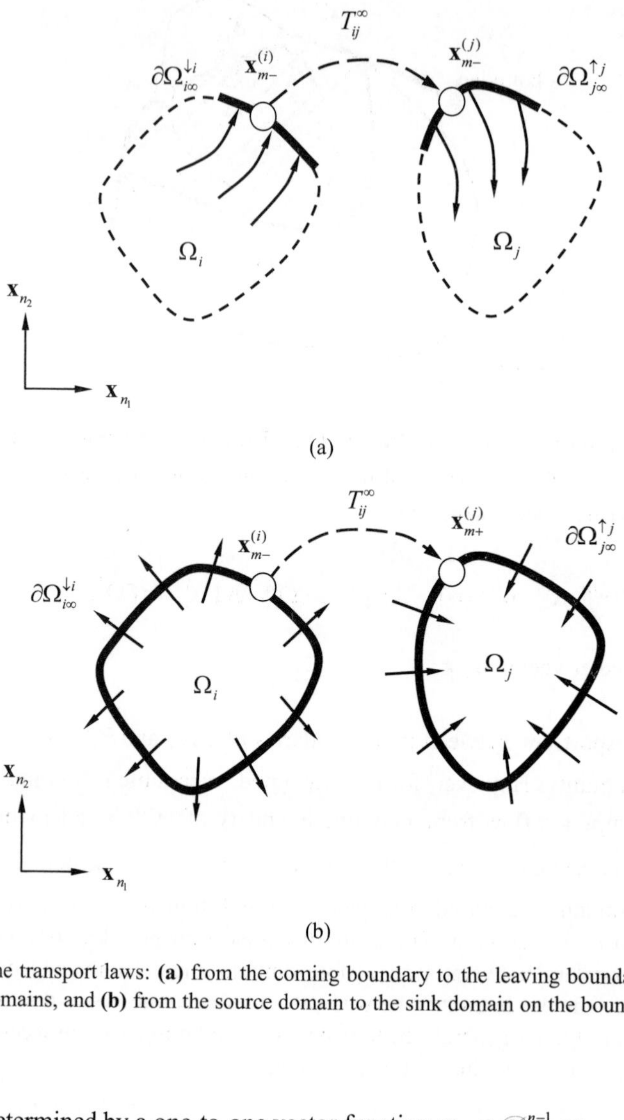

Fig. 5.40 The transport laws: **(a)** from the coming boundary to the leaving boundary to the two separated domains, and **(b)** from the source domain to the sink domain on the boundary of $\partial\Omega_{\alpha\infty}$ ($\alpha = i, j$).

which is determined by a one-to-one vector function $\mathbf{g}_{i\beta} \in \mathscr{R}^{n-1}$ as

$$g_{i\beta}(\mathbf{x}_{m-}^{(i)}, \mathbf{x}_{m+}^{(\beta)}, \mathbf{p}_{i\beta}) = 0, \quad \text{on } \partial\Omega_{ij}^{\downarrow i} \text{ and } \partial\Omega_{\alpha\beta}^{\uparrow\beta}, \tag{5.107}$$

with a parameter vector $\mathbf{p}_{i\beta} \in \mathscr{R}^{k_{i\beta}}$.

The transport laws for the C^0-discontinuous flow $\mathbf{\Phi}(\mathbf{x}, t)$ on the boundary $\partial\Omega_{ij}$ are sketched in Fig. 5.41 for the above definition. Note that the subscripts "-"

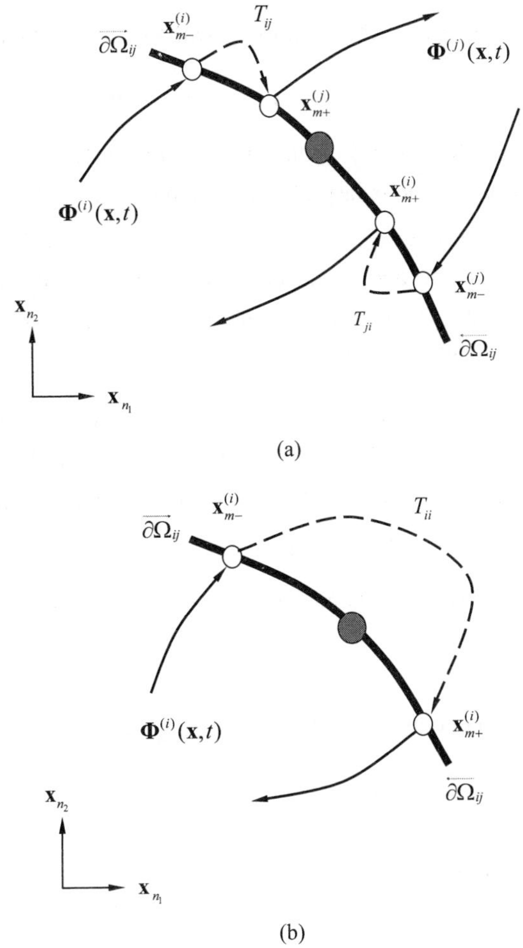

Fig. 5.41 Transport laws for the C^0-discontinuous flow of $\mathbf{\Phi}(\mathbf{x},t)$ on the boundary $\partial\Omega_{ij}$: **(a)** from Ω_α to Ω_β on $\partial\Omega_{\alpha\beta}^{\downarrow\alpha}$ ($\alpha,\beta \in \{i,j\}$ and $\alpha \neq \beta$), and **(b)** from $\partial\Omega_{ij}^{\downarrow i}$ to $\partial\Omega_{ji}^{\uparrow i}$.

and "+" represent before and after the transport on the boundary. The dashed curves represent a transport law on the boundary. Solid thin curves represent the sub-domains. In Fig. 5.41(a), transport laws are given by the following equation,

$$T_{\alpha\beta} : \mathbf{x}_{m-}^{(\alpha)} \to \mathbf{x}_{m+}^{(\beta)}, \quad \text{for } \mathbf{x}_{m-}^{(\alpha)}, \mathbf{x}_{m+}^{(\beta)} \in \overrightarrow{\partial\Omega}_{\alpha\beta} \text{ and } \alpha,\beta \in \{i,j\} \;\; \beta \neq \alpha. \tag{5.108}$$

The two transport laws transport a flow on the same boundary with different domains. For a transport law T_{ij}, $\partial\Omega_{ij}^{\downarrow i}, \partial\Omega_{ij}^{\uparrow j} \subset \overrightarrow{\partial\Omega}_{ij}$ are for the C^0-discontinuous system on the semi-passable boundary $\overrightarrow{\partial\Omega}_{ij}$. However, for another transport law

T_{ji}, $\partial\Omega_{ij}^{\downarrow j}, \partial\Omega_{ij}^{\uparrow i} \subset \overleftarrow{\partial\Omega_{ij}}$. In Fig. 5.41(b), the transport law maps a point $\mathbf{x}_{m-}^{(i)} \in \partial\Omega_{ij}^{\downarrow i} \subset \overrightarrow{\partial\Omega_{ij}}$ to a point $\mathbf{x}_{m+}^{(i)} \in \partial\Omega_{ij}^{\uparrow i} \subset \overleftarrow{\partial\Omega_{ij}}$ at time t_m without time consumption, i.e.,

$$T_{ii} : \mathbf{x}_{m-}^{(i)} \to \mathbf{x}_{m+}^{(i)}, \quad \text{for } \mathbf{x}_{m-}^{(i)} \in \overrightarrow{\partial\Omega_{ij}} \text{ and } \mathbf{x}_{m+}^{(i)} \in \overleftarrow{\partial\Omega_{ij}}. \tag{5.109}$$

After transported, the flow is still on the side of domain Ω_i. It indicates that a flow cannot pass over the boundary $\partial\Omega_{ij}$ into domain Ω_j. This is because the transport law brings the flow from $\overrightarrow{\partial\Omega_{ij}}$ to $\overleftarrow{\partial\Omega_{ij}}$ and avoids a flow getting into domain Ω_j. To extend the above concept, a transport law can map a coming boundary of an accessible domain to a leaving boundary of another domain. In a similar fashion, the transport law for this case is described as follows.

Definition 5.54. Under Assumptions (H2.1)-(H2.3), for a discontinuous system in Eq. (5.1), there are two accessible domains of Ω_{α_1} and Ω_{β_2} with the α_1-side coming boundary of $\partial\Omega_{\alpha_1\beta_1}^{\downarrow\alpha_1}$ and the β_2-side leaving boundaries and $\partial\Omega_{\alpha_2\beta_2}^{\uparrow\beta_2}$. A map $T_{\alpha_1\beta_2}$ transporting $\mathbf{x}_{m-}^{(\alpha_1)} \in \partial\Omega_{\alpha_1\beta_1}^{\downarrow\alpha_1}$ to $\mathbf{x}_{m+}^{(\beta_2)} \in \partial\Omega_{\alpha_2\beta_2}^{\uparrow\beta_2}$ at time t_m without the time consumption is termed *a transport law* from $\partial\Omega_{\alpha_1\beta_1}^{\downarrow\alpha_1}$ to $\partial\Omega_{\alpha_2\beta_2}^{\uparrow\beta_2}$, i.e.,

$$T_{\alpha_1\beta_2} : \mathbf{x}_{m-}^{(\alpha_1)} \to \mathbf{x}_{m+}^{(\beta_2)}, \tag{5.110}$$

which is governed by a one-to-one vector function $\mathbf{g}_{\alpha_1\beta_2} \in \mathscr{R}^{n-1}$ as

$$g_{\alpha_1\beta_2}(\mathbf{x}_{m-}^{(\alpha_1)}, \mathbf{x}_{m+}^{(\beta_2)}, \mathbf{p}_{\alpha_1\beta_2}) = 0 \text{ for } \mathbf{x}_{m-}^{(\alpha_1)} \in \partial\Omega_{\alpha_1\beta_1}^{\downarrow\alpha_1} \text{ and } \mathbf{x}_{m+}^{(\beta_2)} \in \partial\Omega_{\alpha_2\beta_2}^{\uparrow\beta_2}, \tag{5.111}$$

with a parameter vector $\mathbf{p}_{\alpha_1\beta_2} \in \mathscr{R}^{k_{\alpha_1\beta_2}}$.

From this definition, it is not necessary to require the domains of Ω_{β_1} and Ω_{α_1} to be accessible. They can be accessible, or inaccessible or universal domain. Such a definition is illustrated in Fig. 5.42. In addition, the boundary can be permanently non-passable. To continue the flow in the entire discontinuous system, the transport law should be exerted. The above transport laws are time-independent.

In fact, the transport law gives the two states relations via nonlinear algebraic equations. Thus, transport laws can be time-dependent, which can be similar to regular dynamical systems to determine the relation of the two states. Therefore, the time-dependent transport law is introduced herein.

Definition 5.55. For a discontinuous system in Eq. (5.1), under Assumptions (H2.1)-(H2.3), there are two accessible domains of Ω_{α_1} and Ω_{β_2} with the α_1-side coming boundary of $\partial\Omega_{\alpha_1\beta_1}^{\downarrow\alpha_1}$ and the β_2-side leaving boundaries and $\partial\Omega_{\alpha_2\beta_2}^{\uparrow\beta_2}$. A map $T_{\alpha_1\beta_2}$ transporting from $\mathbf{x}_{m-}^{(\alpha_1)} \in \partial\Omega_{\alpha_1\beta_1}^{\downarrow\alpha_1}$ to $\mathbf{x}_{(m+1)+}^{(\beta_2)} \in \partial\Omega_{\alpha_2\beta_2}^{\uparrow\beta_2}$ with a time difference

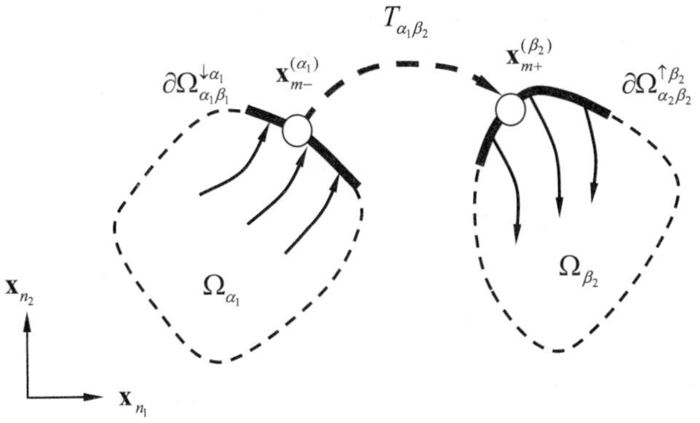

Fig. 5.42 A generalized transport laws from $\partial\Omega^{\downarrow\alpha_1}_{\alpha_1\beta_1}$ to $\partial\Omega^{\downarrow\beta_2}_{\alpha_2\beta_2}$ for the C^0-discontinuous flow.

$\Delta t = t_{m+1} - t_m$ is termed a *transport law* from $\partial\Omega^{\downarrow\alpha_1}_{\alpha_1\beta_1}$ to $\partial\Omega^{\uparrow\beta_2}_{\alpha_2\beta_2}$, i.e.,

$$T_{\alpha_1\beta_2} : \mathbf{x}^{(\alpha_1)}_{m-} \to \mathbf{x}^{(\beta_2)}_{(m+1)+}, \tag{5.112}$$

which is determined by a one-to-one vector function as

$$\begin{gathered} g_{\alpha_1\beta_2}(\mathbf{x}^{(\alpha_1)}_{m-}, t_m, \mathbf{x}^{(\beta_2)}_{(m+1)+}, t_{m+1}, \mathbf{p}_{\alpha_1\beta_2}) = 0, \\ \text{for } \mathbf{x}^{(\alpha_1)}_{m-} \in \partial\Omega^{\downarrow\alpha_1}_{\alpha_1\beta_1} \text{ and } \mathbf{x}^{(\beta_2)}_{(m+1)+} \in \partial\Omega^{\uparrow\beta_2}_{\alpha_2\beta_2}, \end{gathered} \tag{5.113}$$

with a parameter vector $\mathbf{p}_{\alpha_1\beta_2} \in \mathscr{R}^{k_{\alpha_1\beta_2}}$.

The afore-discussed transport law provides a connection between the two states on the boundary of distinct domains in discontinuous dynamical systems. The transport law also exists in discontinuous systems at some time moment. Such a transport law will be discussed in sequel.

5.6. Multi-valued vector fields and bouncing flows

In the previous discussion, only at most one vector field is defined on each domain in discontinuous dynamical systems. On the different adjacent domains, the vector fields are distinguishing. Thus, the passability of flows in two adjacent domains at the boundary is very significant for a better understanding of the global behaviors of flows in discontinuous dynamical system, which was discussed in Chapters 2 and 3. If flow barriers on the boundary exist, the flow passability of flows in two adjacent domains at the boundary is totally different. In Chapter 4, the comprehensive discussion on such flow passability was given. Before multi-valued vector fields on each accessible domain are discussed, two leaving flows at the separation boundary will be discussed first once a coming flow arrives to the boundary. The imaginary flow in Chapter 3 is recalled. If a vector field defined on a domain can be extended to other domains continuously, such an extended vector field generates an imaginary flow in the corresponding domains. Equation (5.1) can be rewritten for real, imaginary and boundary flows, i.e.,

$$\begin{aligned} &\dot{\mathbf{x}}_{\alpha}^{(\alpha)} = \mathbf{F}^{(\alpha)}(\mathbf{x}_{\alpha}^{(\alpha)},t,\mathbf{p}_{\alpha}) \text{ in } \Omega_{\alpha}, \\ &\dot{\mathbf{x}}_{\alpha}^{(\beta)} = \mathbf{F}^{(\beta)}(\mathbf{x}_{\alpha}^{(\beta)},t,\mathbf{p}_{\beta}) \text{ in } \Omega_{\alpha} \text{ from } \Omega_{\beta} \ (\beta = \beta_1,\beta_2,\cdots,\beta_k), \\ &\dot{\mathbf{x}}^{(0)} = \mathbf{F}^{(0)}(\mathbf{x}^{(0)},t,\boldsymbol{\lambda}) \quad \text{with } \varphi_{\alpha\beta}(\mathbf{x}^{(0)},t,\boldsymbol{\lambda}) = 0 \text{ on } \partial\Omega_{\alpha\beta}, \end{aligned} \tag{5.114}$$

where $\mathbf{x}_{\alpha}^{(\alpha)}$, $\mathbf{x}_{\alpha}^{(\beta)}$ and $\mathbf{x}^{(0)}$ are real, imaginary and boundary flows, respectively. $\mathbf{x}_{\alpha}^{(\alpha)} \equiv \mathbf{x}^{(\alpha)}$. The imaginary flow $\mathbf{x}_{\alpha}^{(\beta)}$ is a continuous extension of $\mathbf{x}_{\beta}^{(\beta)}$ from domain Ω_{β} via the boundary of $\partial\Omega_{\alpha\beta}$. In other words, they have the same vector field in different domains. For a better description of multi-valued vector fields in discontinuous dynamical systems, the passable and sink flows are sketched in Fig. 5.43. The solid and dashed curves represent real and imaginary flows in the domains, respectively. If the imaginary flows in all domains become real flows, then multi-valued, vector fields in each domain of discontinuous dynamical systems will be obtained. Once a real flow in a domain arrives to its boundary, the flow will continue because this real flow can extend to the adjacent domain. For example, a flow of $\mathbf{x}^{(\alpha)}$ in domain Ω_{α} arrives to the boundary of $\partial\Omega_{\alpha\beta}$, and continue in domain Ω_{β} through the flow of $\mathbf{x}_{\beta}^{(\alpha)}$, as in Fig. 5.43(a). This is because the flow can follow the "continuation axiom" in discontinuous dynamical systems (i.e., $\mathbf{x}_{\alpha}^{(\alpha)}(t_{m-}) = \mathbf{x}_{\beta}^{(\alpha)}(t_{m+})$). To make this flow be switched at the boundary, a transport law should be involved. Consider a switching rule as the simplest transport law, i.e.,

$$\mathbf{x}_{\alpha}^{(\alpha)}(t_{m-}) = \mathbf{x}_{\beta}^{(\beta)}(t_{m+}) \text{ or } \mathbf{x}_{\alpha}^{(\alpha)}(t_{m-}) = \mathbf{x}_{\alpha}^{(\beta)}(t_{m-}). \tag{5.115}$$

Consider a passable flow at the boundary. If the switching location is on the side of domain Ω_{α} or Ω_{β}, the flow $\mathbf{x}^{(\alpha)}$ in domain Ω_{α} will be switched to a flow $\mathbf{x}^{(\beta)}$ in domain Ω_{β}. This case is the same as for the single-valued vector field. However, for a sink flow to the boundary, the switching location will change the flow passability. In Fig. 5.43(b), real flows $\mathbf{x}^{(\alpha)}$ in domain Ω_{α} and $\mathbf{x}^{(\beta)}$ in domain Ω_{β} will come to the boundary at the same time. If imaginary flows $\mathbf{x}_{\alpha}^{(\beta)}$ in domain Ω_{α} and $\mathbf{x}_{\beta}^{(\alpha)}$ in domain Ω_{β} become real flows. For the coming flow $\mathbf{x}^{(\alpha)}$, without switching, the flow will extend to $\mathbf{x}_{\beta}^{(\alpha)}$ in domain Ω_{β}. If the switching rule is considered, there are two possibilities. If the switching occurs at $\mathbf{x}_{\alpha}^{(\alpha)}(t_{m-}) = \mathbf{x}_{\beta}^{(\beta)}(t_{m-})$, such a flow will be switched to the boundary flow $\mathbf{x}^{(0)}$. This

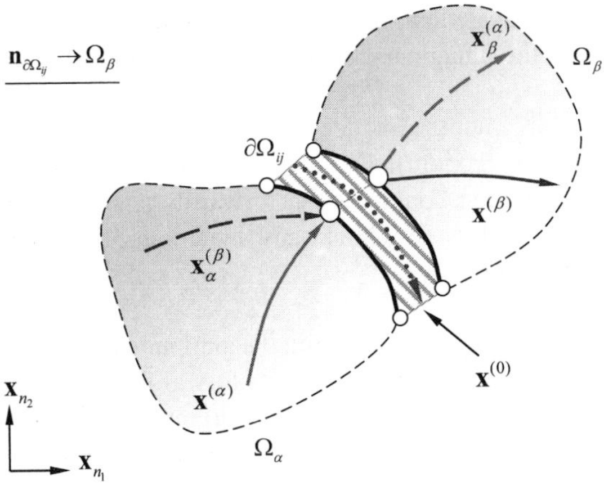

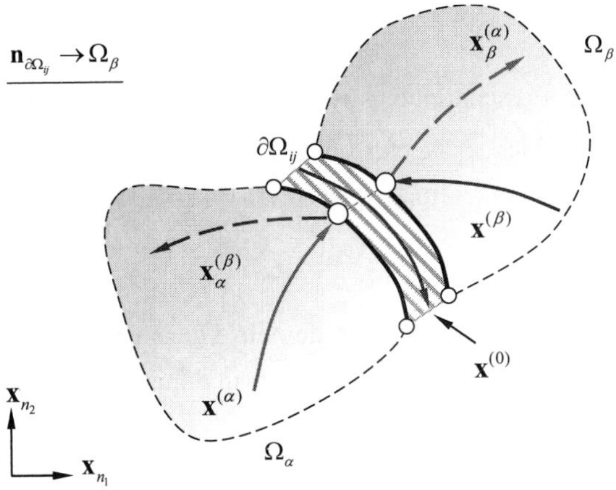

Fig. 5.43 Real and imaginary flows: **(a)** semi-passable flows and **(b)** sink flows. The hatched area with two thick solid curves is the boundary. The boundary flows for the sink and passable flows are presented by the solid and dotted curves with arrows, respectively. The solid and dashed curves in the domains are the real and imaginary flows. (color plot in the book end)

is because the flow switching happens on the side of domain Ω_β. This case is also the same as for the single-valued vector field. For a discontinuous dynamical system with the single-valued vector-field in each domain, the switching rule is assumed to be satisfied automatically. If the switching is at $\mathbf{x}_\alpha^{(\alpha)}(t_{m-}) = \mathbf{x}_\alpha^{(\beta)}(t_{m+})$, the coming flow $\mathbf{x}^{(\alpha)}$ will be switched to the flow $\mathbf{x}_\alpha^{(\beta)}$ in domain Ω_α. Such a switching shows that the flow is still in the same domain. The flow switching is like a flow bouncing. Similarly, a source flow at the boundary can be discussed, and the coming flow will be an imaginary flow. From the previous discussion, if the imaginary flow becomes a real flow, there are two possibilities for the coming flow to switch. To make the imaginary flow exist, the following axioms will be adopted.

Axiom 5.1 (Continuation axiom). For a discontinuous dynamical system, any flows coming to a separation boundary can forwardly and/or backwardly, extend to the adjacent accessible domain without any switching rules, transport laws and any flow barriers at the boundary.

Axiom 5.2 (Non-extendibility axiom). For a discontinuous dynamical system, any flows coming to a separation boundary cannot forwardly and backwardly extend to any inaccessible domain. Even if such flows are extendible to the inaccessible domain, then the corresponding extended flows are fictitious only.

The multi-valued vector fields on an accessible sub-domain in discontinuous dynamical systems will be introduced, and the bouncing flow will be discussed in the same domain.

Definition 5.56. For a discontinuous system, there is an accessible domain Ω_α ($\alpha \in \{1,2,\cdots,N\}$) on which k_α -vector fields $\mathbf{F}^{(\alpha_k)}(\mathbf{x}^{(\alpha_k)},t,\mathbf{p}_{\alpha_k})$ ($k=1,2,\cdots,k_\alpha$) are defined, and the corresponding dynamical system is given by

$$\dot{\mathbf{x}}^{(\alpha_k)} = \mathbf{F}^{(\alpha_k)}(\mathbf{x}^{(\alpha_k)},t,\mathbf{p}_{\alpha_k}). \tag{5.116}$$

The discontinuous dynamical system on domain Ω_α is called to be *of* the *multi-valued vector fields*. A set of dynamical systems in domain Ω_α is defined as

$$\begin{aligned} \mathscr{D}_\alpha &= \cup_{k=1}^{k_\alpha} \mathscr{D}_{\alpha_k} \\ \mathscr{D}_{\alpha_k} &= \left\{ \dot{\mathbf{x}}^{(\alpha_k)} = \mathbf{F}^{(\alpha_k)}(\mathbf{x}^{(\alpha_k)},t,\mathbf{p}_{\alpha_k}) \middle| k \in \{1,2,\cdots,k_\alpha\} \right\} \end{aligned} \tag{5.117}$$

A total set of dynamical systems in the discontinuous dynamical system is

$$\mathscr{D} = \cup_{\alpha=1}^{N} \mathscr{D}_\alpha. \tag{5.118}$$

5.6.1. Bouncing flows

Definition 5.57. For a discontinuous system with multi-valued vector fields in Eq. (5.118), there is a point $\mathbf{x}^{(0)}(t_m) \equiv \mathbf{x}_m \in \partial\Omega_{\alpha\beta}$ at time t_m between two adjacent domains Ω_α and Ω_β. For an arbitrarily small $\varepsilon > 0$, there are two time intervals $[t_{m-\varepsilon}, t_m)$ and $(t_m, t_{m+\varepsilon}]$. If a coming flow of $\mathbf{x}^{(\alpha_k)}$ arrives to boundary $\partial\Omega_{\alpha\beta}$, there is a switching rule between the two flows of $\mathbf{x}^{(\alpha_k)}$ and $\mathbf{x}^{(\alpha_l)}$ with $\mathbf{x}^{(\alpha_k)}(t_{m-}) = \mathbf{x}_m = \mathbf{x}^{(\alpha_l)}(t_{m+})$. The flows of $\mathbf{x}^{(\alpha_k)}(t)$ and $\mathbf{x}^{(\alpha_l)}(t)$ to boundary $\partial\Omega_{ij}$ are called a *bouncing* flow in Ω_α if

$$\left.\begin{aligned} \hbar_\alpha \mathbf{n}^{\mathrm{T}}_{\partial\Omega_{ij}}(\mathbf{x}^{(0)}_{m-\varepsilon})\cdot[\mathbf{x}^{(0)}_{m-\varepsilon} - \mathbf{x}^{(\alpha_k)}_{m-\varepsilon}] > 0, \\ \hbar_\alpha \mathbf{n}^{\mathrm{T}}_{\partial\Omega_{ij}}(\mathbf{x}^{(0)}_{m+\varepsilon})\cdot[\mathbf{x}^{(\alpha_l)}_{m+\varepsilon} - \mathbf{x}^{(0)}_{m+\varepsilon}] < 0. \end{aligned}\right\} \tag{5.119}$$

Theorem 5.1. *For a discontinuous system with multi-valued vector fields in Eq. (5.118), there is a point $\mathbf{x}^{(0)}(t_m) \equiv \mathbf{x}_m \in \partial\Omega_{\alpha\beta}$ at time t_m between two adjacent domains Ω_α and Ω_β. For an arbitrarily small $\varepsilon > 0$, there are two time intervals $[t_{m-\varepsilon}, t_m)$ and $(t_m, t_{m+\varepsilon}]$. If a coming flow of $\mathbf{x}^{(\alpha_1)}$ arrives to the boundary of $\partial\Omega_{\alpha\beta}$, there is a switching rule between the two flows of $\mathbf{x}^{(\alpha_k)}$ and $\mathbf{x}^{(\alpha_l)}$ with $\mathbf{x}^{(\alpha_k)}(t_{m-}) = \mathbf{x}_m = \mathbf{x}^{(\alpha_l)}(t_{m+})$. A flow $\mathbf{x}^{(\alpha_k)}(t)$ is $C^{r_{\alpha_k}}_{[t_{m-\varepsilon}, t_m)}$-continuous ($r_{\alpha_k} \geq 2$) for time t with $\| d^{r_{\alpha_k}+1}\mathbf{x}^{(\alpha)} / dt^{r_{\alpha_k}+1} \| < \infty$, and a flow $\mathbf{x}^{(\alpha_l)}(t)$ is $C^{r_{\alpha_l}}_{(t_m, t_{m+\varepsilon}]}$-continuous ($r_{\alpha_k} \geq 2$) for time t with $\| d^{r_{\alpha_l}+1}\mathbf{x}^{(\alpha)} / dt^{r_{\alpha_l}+1} \| < \infty$. The flows of $\mathbf{x}^{(\alpha_k)}(t)$ and $\mathbf{x}^{(\alpha_l)}(t)$ to the boundary $\partial\Omega_{ij}$ form a bouncing flow in Ω_α if and only if*

$$\hbar_\alpha G^{(\alpha_k)}_{\partial\Omega_{\alpha\beta}}(\mathbf{x}_m, t_{m-}) > 0 \textit{ and } \hbar_\alpha G^{(\alpha_l)}_{\partial\Omega_{\alpha\beta}}(\mathbf{x}_m, t_{m+}) < 0. \tag{5.120}$$

Proof. Following the proof of Theorem 3.1, the proof of the theorem is proved. ■

Using Eqs. (3.14) and (3.15), the G-function in Eq. (5.120) becomes

$$\begin{aligned} G^{(\alpha_k)}_{\partial\Omega_{\alpha\beta}}(\mathbf{x}_m, t_{m-}) = \mathbf{n}^{\mathrm{T}}_{\partial\Omega_{\alpha\beta}}\cdot\dot{\mathbf{x}}^{(\alpha_k)} = \mathbf{n}^{\mathrm{T}}_{\partial\Omega_{\alpha\beta}}\cdot\mathbf{F}^{(\alpha_k)}, \\ G^{(\alpha_l)}_{\partial\Omega_{\alpha\beta}}(\mathbf{x}_m, t_{m+}) = \mathbf{n}^{\mathrm{T}}_{\partial\Omega_{\alpha\beta}}\cdot\dot{\mathbf{x}}^{(\alpha_l)} = \mathbf{n}^{\mathrm{T}}_{\partial\Omega_{\alpha\beta}}\cdot\mathbf{F}^{(\alpha_l)}. \end{aligned} \tag{5.121}$$

If the two vector fields are same, then $\dot{\mathbf{x}}^{(\alpha_k)}(t_{m-}) = \dot{\mathbf{x}}^{(\alpha_l)}_\alpha(t_{m+})$ and from Eq. (5.105),

$$G^{(\alpha_k)}_{\partial\Omega_{\alpha\beta}}(\mathbf{x}_m, t_{m-}) = G^{(\alpha_l)}_{\partial\Omega_{\alpha\beta}}(\mathbf{x}_m, t_{m+}) = 0. \tag{5.122}$$

Theorem 3.9 gives

$$\hbar_\alpha G^{(1,\alpha_k)}_{\partial\Omega_{\alpha\beta}}(\mathbf{x}_m, t_{m-}) = \hbar_\alpha G^{(1,\alpha_l)}_{\partial\Omega_{\alpha\beta}}(\mathbf{x}_m, t_{m+}) < 0. \tag{5.123}$$

The grazing flow to the boundary is a special case of the bouncing flow. The conditions in Eq. (5.120) are different from Eqs. (5.122) and (5.123). This is because the bouncing flow possesses two different vector fields with the switching rule. However, the grazing flow to the boundary keeps a same vector field without any switching rule because of the continuation of flow. The bouncing flow is an extension of the grazing flow to make flows exist in the same domain. The bouncing flows to the boundary are independent of the boundary flow as grazing flows. So the discontinuity at the boundary can exist in the same domain. Without the switching rule, the grazing flow to the boundary can be observed. However, the bouncing flow to the boundary needs a switching rule to be exerted.

Definition 5.58. For a discontinuous system with multi-valued vector fields in Eq. (5.118), there is a point $\mathbf{x}^{(0)}(t_m) \equiv \mathbf{x}_m \in \partial\Omega_{\alpha\beta}$ at time t_m between two adjacent domains Ω_α and Ω_β. For an arbitrarily small $\varepsilon > 0$, there are two time intervals $[t_{m-\varepsilon}, t_m)$ and $(t_m, t_{m+\varepsilon}]$. If a coming flow $\mathbf{x}^{(\alpha_k)}$ arrives to the boundary $\partial\Omega_{\alpha\beta}$, there is a switching rule between two flows $\mathbf{x}^{(\alpha_k)}$ and $\mathbf{x}^{(\alpha_l)}$ with $\mathbf{x}^{(\alpha_k)}(t_{m-}) = \mathbf{x}_m =$ $\mathbf{x}^{(\alpha_l)}(t_{m+})$. A flow $\mathbf{x}^{(\alpha_k)}(t)$ is $C^{r_{\alpha_k}}_{[t_{m-\varepsilon}, t_m)}$-continuous ($r_{\alpha_k} \geq m_{\alpha_k} + 1$) for time t with $\| d^{r_{\alpha_k}+1}\mathbf{x}^{(\alpha)} / dt^{r_{\alpha_k}+1} \| < \infty$, and a flow $\mathbf{x}^{(\alpha_l)}(t)$ is $C^{r_{\alpha_l}}_{(t_m, t_{m+\varepsilon}]}$-continuous ($r_{\alpha_l} \geq m_{\alpha_l} + 1$) for time t with $\| d^{r_{\alpha_l}+1}\mathbf{x}^{(\alpha)} / dt^{r_{\alpha_l}+1} \| < \infty$. The flows of $\mathbf{x}^{(\alpha_k)}(t)$ and $\mathbf{x}^{(\alpha_l)}(t)$ to boundary $\partial\Omega_{ij}$ are called a $(m_{\alpha_k} : m_{\alpha_l})$*-bouncing flow* in domain Ω_α if

$$
\begin{aligned}
&\hbar_\alpha G^{(s_{\alpha_k}, \alpha_k)}_{\partial\Omega_{\alpha\beta}}(\mathbf{x}_m, t_{m-}) = 0 \ \text{ for } s_{\alpha_k} = 0, 1, 2, \cdots, m_{\alpha_k} - 1, \\
&\hbar_\alpha G^{(s_{\alpha_l}, \alpha_l)}_{\partial\Omega_{\alpha\beta}}(\mathbf{x}_m, t_{m+}) = 0 \ \text{ for } s_{\alpha_l} = 0, 1, 2, \cdots, m_{\alpha_l} - 1;
\end{aligned}
\tag{5.124}
$$

$$
\left.
\begin{aligned}
&\hbar_\alpha \mathbf{n}^{\mathrm{T}}_{\partial\Omega_{ij}}(\mathbf{x}^{(0)}_{m-\varepsilon}) \cdot [\mathbf{x}^{(0)}_{m-\varepsilon} - \mathbf{x}^{(\alpha_k)}_{m-\varepsilon}] > 0, \\
&\hbar_\alpha \mathbf{n}^{\mathrm{T}}_{\partial\Omega_{ij}}(\mathbf{x}^{(0)}_{m+\varepsilon}) \cdot [\mathbf{x}^{(\alpha_l)}_{m+\varepsilon} - \mathbf{x}^{(0)}_{m+\varepsilon}] < 0.
\end{aligned}
\right\}
\tag{5.125}
$$

Theorem 5.2. *For a discontinuous system with multi-valued vector fields in Eq. (5.118), there is a point* $\mathbf{x}^{(0)}(t_m) \equiv \mathbf{x}_m \in \partial\Omega_{\alpha\beta}$ *at time* t_m *between two adjacent domains* Ω_α *and* Ω_β. *For an arbitrarily small* $\varepsilon > 0$, *there are two time intervals* $[t_{m-\varepsilon}, t_m)$ *and* $(t_m, t_{m+\varepsilon}]$. *If a coming flow* $\mathbf{x}^{(\alpha_k)}$ *arrives to the boundary* $\partial\Omega_{\alpha\beta}$, *there is a switching rule between two flows* $\mathbf{x}^{(\alpha_k)}$ *and* $\mathbf{x}^{(\alpha_l)}$ *with* $\mathbf{x}^{(\alpha_k)}(t_{m-}) = \mathbf{x}_m =$ $\mathbf{x}^{(\alpha_l)}(t_{m+})$. *A flow* $\mathbf{x}^{(\alpha_k)}(t)$ *is* $C^{r_{\alpha_k}}_{[t_{m-\varepsilon}, t_m)}$*-continuous* ($r_{\alpha_k} \geq m_{\alpha_k} + 1$) *for time t with* $\| d^{r_{\alpha_k}+1}\mathbf{x}^{(\alpha)} / dt^{r_{\alpha_k}+1} \| < \infty$, *and a flow* $\mathbf{x}^{(\alpha_l)}(t)$ *is* $C^{r_{\alpha_l}}_{(t_m, t_{m+\varepsilon}]}$*-continuous* ($r_{\alpha_l} \geq m_{\alpha_l} + 1$) *for time t with* $\| d^{r_{\alpha_l}+1}\mathbf{x}^{(\alpha)} / dt^{r_{\alpha_l}+1} \| < \infty$. *The flows of* $\mathbf{x}^{(\alpha_k)}(t)$ *and* $\mathbf{x}^{(\alpha_l)}(t)$ *to the boundary* $\partial\Omega_{ij}$ *form an* $(m_{\alpha_k} : m_{\alpha_l})$*-bouncing flow in domain* Ω_α *if and only if*

$$\begin{aligned}&\hbar_{\alpha}G_{\partial\Omega_{\alpha\beta}}^{(s_{\alpha_k},\alpha_k)}(\mathbf{x}_m,t_{m-})=0 \ \ for\ s_{\alpha_k}=0,1,2,\cdots,m_{\alpha_k}-1,\\&\hbar_{\alpha}G_{\partial\Omega_{\alpha\beta}}^{(s_{\alpha_l},\alpha_l)}(\mathbf{x}_m,t_{m+})=0 \ \ for\ s_{\alpha_l}=0,1,2,\cdots,m_{\alpha_l}-1;\end{aligned} \tag{5.126}$$

$$\hbar_{\alpha}G_{\partial\Omega_{\alpha\beta}}^{(m_{\alpha_k},\alpha_k)}(\mathbf{x}_m,t_{m-})>0 \ and\ \hbar_{\alpha}G_{\partial\Omega_{\alpha\beta}}^{(m_{\alpha_l},\alpha_l)}(\mathbf{x}_m,t_{m+})<0. \tag{5.127}$$

Proof. Following the proof of Theorem 3.2, the proof of the theorem is proved. ■

A bouncing flow with two different vector fields in domain Ω_α switches at the boundary of $\partial\Omega_{\alpha\beta}$, including four bouncing flows: (i) the $(2k_{\alpha_k}:2k_{\alpha_l})$-bouncing flow, (ii) the $(2k_{\alpha_k}:2k_{\alpha_l}+1)$-bouncing flow, (iii) the $(2k_{\alpha_k}+1:2k_{\alpha_l})$-bouncing flows, and (iv) the $(2k_{\alpha_k}+1:2k_{\alpha_l}+1)$-bouncing flow. The $(2k_{\alpha_k}:2k_{\alpha_l})$-bouncing flow in domain Ω_α and $(2k_{\alpha_k}:2k_{\alpha_l})$-sink flow are shown Fig. 5.44 (a) and (b), respectively. Solid curves in domain Ω_α depict real flows. If a bouncing flow with two vector fields exists in domain Ω_α, then an corresponding imaginary bouncing flow will exist in domain Ω_β. Dashed curves in domain Ω_β give imaginary bouncing flows. The $(2k_{\alpha_k}:2k_{\alpha_l})$-bouncing flow in domain Ω_α are inflexionally singular. A sink flow to the boundary also is sketched in Fig. 5.44(b). Compared with the bouncing flow in Fig. 5.44 (a), the real and imaginary flows are swapped. The source flows can be converted to the bouncing flow if the imaginary flow becomes a real coming flow.

The $(2k_{\alpha_k}+1:2k_{\alpha_l}+1)$-bouncing flow and the $(2k_\alpha+1:2k_\beta+1)$-real and imaginary flows in domain Ω_α are presented in Fig. 5.45(a) and (b), respectively. Dotted curves are the forward and backward continuations of real flows without any switching rule. The $(2k_{\alpha_k}+1:2k_{\alpha_l}+1)$-bouncing flow in domain Ω_α possesses the grazing singularity fully. If the $(2k_\beta+1)$-order imaginary flow $\mathbf{x}_\alpha^{(\beta)}$ in domain Ω_α become a real flow, with a switching rule, both the coming flow $\mathbf{x}^{(\alpha)}$ and the leaving flow $\mathbf{x}_\alpha^{(\beta)}$ form a bouncing flow. The imaginary flow is presented by a dashed curve in domain Ω_α. The $(2k_\alpha+1:2k_\beta+1)$ double tangential flow and the $(2k_\alpha+1:2k_\beta+1)$ inaccessible tangential flow do not have the corresponding bouncing flows. The two $(2k_{\alpha_1}+1:2k_{\alpha_2})$ and $(2k_{\alpha_1}:2k_{\alpha_2}+1)$ bouncing flows are sketched in Fig. 5.46. The two bouncing flows can be swapped from the half-inflexional singularity and the grazing singularity with a switching rule, similar to the sink flow and the real and imaginary tangential flows.

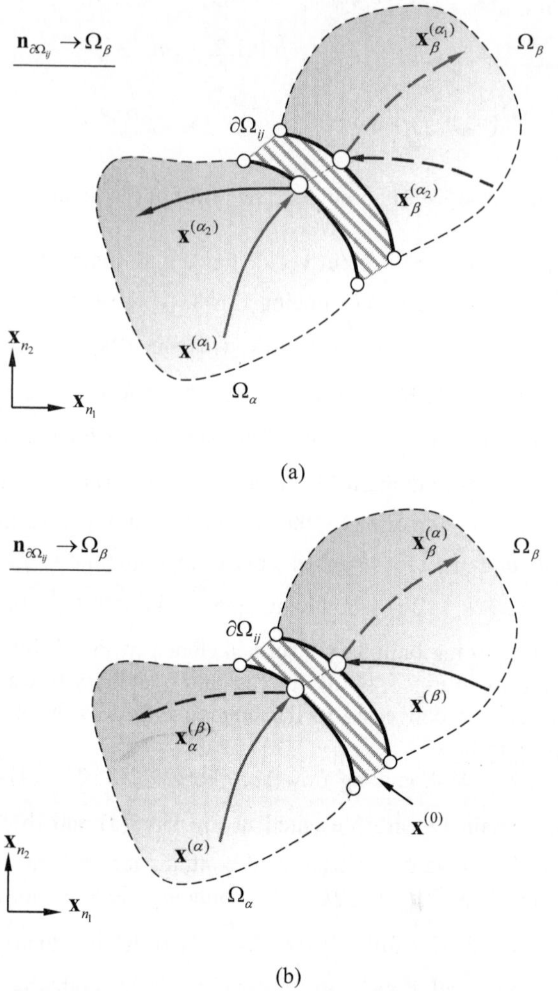

Fig. 5.44 Bouncing flows: **(a)** $(2k_{\alpha_1} : 2k_{\alpha_2})$-bouncing flows and **(b)** $(2k_\alpha : 2k_\beta)$-sink flows. The hatched area with two thick solid curves is the boundary. The solid and dashed curves are the real and imaginary flows in the domains. The dotted curves are the forward and backward extensions of the real flows without switching. (color plot in the book end)

5.6.2. Extended passable flows

Consider the switching rule in domain Ω_α only. After the switching, a flow in the domain Ω_α can extend to domain Ω_β with the continuation axiom.

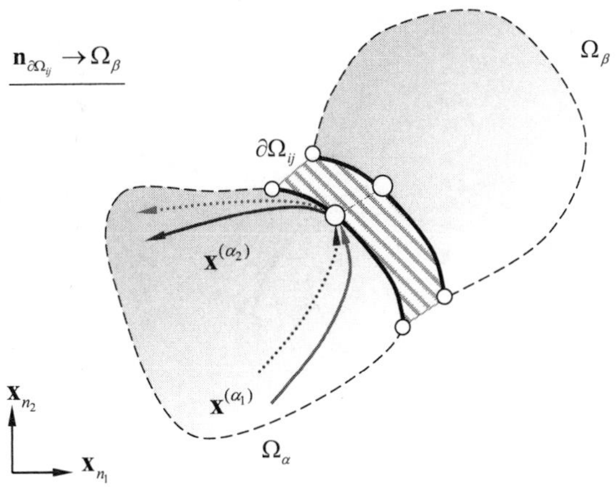

(a)

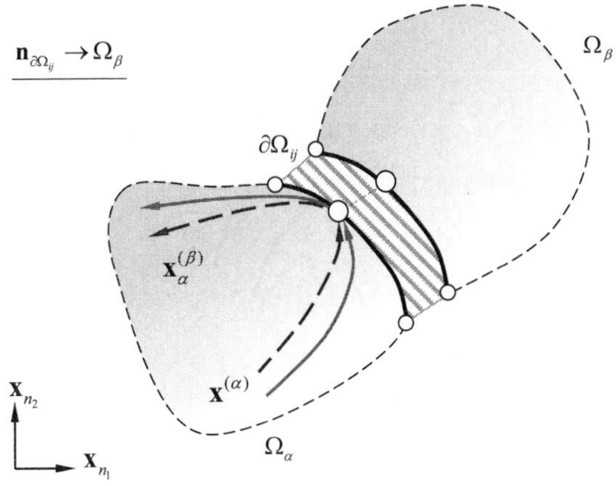

(b)

Fig. 5.45 (a) $(2k_{\alpha_1}+1:2k_{\alpha_2}+1)$ -bouncing flow and **(b)** $(2k_\alpha+1:2k_\beta+1)$ -real and imaginary flow. The hatched area with two thick solid curves is the boundary. The solid and dashed curves are the real and imaginary flows in the domains. The dotted curves are the forward and backward extensions of the real flows without switching.

Definition 5.59. For a discontinuous system with multi-valued vector fields in Eq. (5.118), there is a point $\mathbf{x}^{(0)}(t_m) \equiv \mathbf{x}_m \in \partial\Omega_{\alpha\beta}$ at time t_m between two adjacent domains Ω_α and Ω_β. For an arbitrarily small $\varepsilon > 0$, there are two time intervals $[t_{m-\varepsilon}, t_m)$ and $(t_m, t_{m+\varepsilon}]$. If a coming flow of $\mathbf{x}^{(\alpha_k)}$ arrives to the boundary $\partial\Omega_{\alpha\beta}$, there is a switching rule between two flows $\mathbf{x}^{(\alpha_k)}$ and $\mathbf{x}^{(\alpha_l)}$ with $\mathbf{x}^{(\alpha_k)}(t_{m-}) = \mathbf{x}_m =$

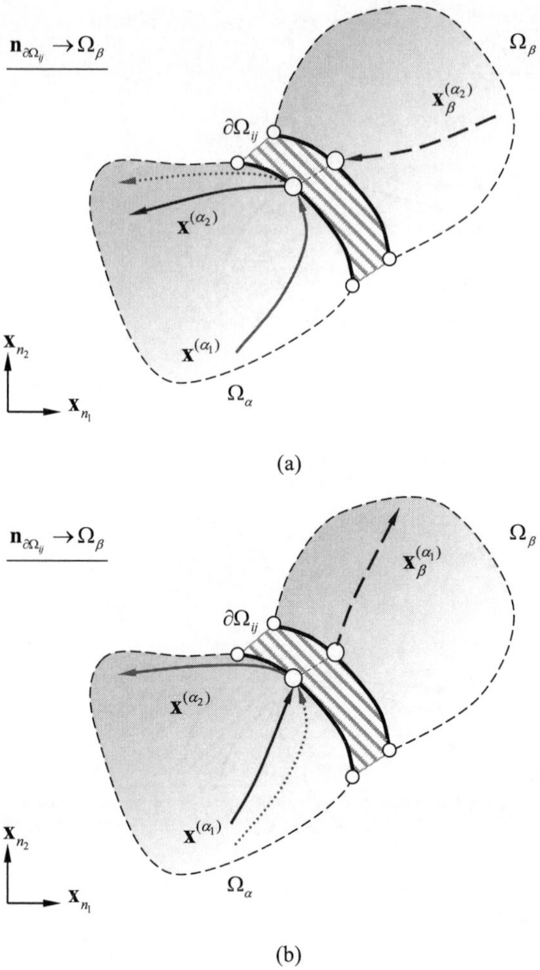

Fig. 5.46 (a) $(2k_{\alpha_1}+1:2k_{\alpha_2})$ -bouncing flows and **(b)** $(2k_\alpha:2k_\beta+1)$ -bouncing flows. The hatched area with two thick solid curves is the boundary. The solid and dashed curves are the real and imaginary flows in the domains. The dotted curves are the forward and backward extensions of the real flows without switching.

$\mathbf{x}^{(\alpha_l)}(t_{m+})$. The flows of $\mathbf{x}^{(\alpha_k)}(t)$ and $\mathbf{x}^{(\alpha_l)}(t)$ to boundary $\partial\Omega_{ij}$ are called an *extended passable* flow from Ω_α to Ω_β if

$$\left.\begin{aligned} \hbar_\alpha \mathbf{n}_{\partial\Omega_{ij}}^{\mathrm{T}}(\mathbf{x}_{m-\varepsilon}^{(0)})\cdot[\mathbf{x}_{m-\varepsilon}^{(0)} - \mathbf{x}_{m-\varepsilon}^{(\alpha_k)}] > 0, \\ \hbar_\alpha \mathbf{n}_{\partial\Omega_{ij}}^{\mathrm{T}}(\mathbf{x}_{m+\varepsilon}^{(0)})\cdot[\mathbf{x}_{m+\varepsilon}^{(\alpha_l)} - \mathbf{x}_{m+\varepsilon}^{(0)}] > 0. \end{aligned}\right\} \tag{5.128}$$

Theorem 5.3. *For a discontinuous system with multi-valued vector fields in*

Eq. (5.118), *there is a point* $\mathbf{x}^{(0)}(t_m) \equiv \mathbf{x}_m \in \partial\Omega_{\alpha\beta}$ *at time* t_m *between two adjacent domains* Ω_α *and* Ω_β. *For an arbitrarily small* $\varepsilon > 0$, *there are two time intervals* $[t_{m-\varepsilon}, t_m)$ *and* $(t_m, t_{m+\varepsilon}]$. *If a coming flow* $\mathbf{x}^{(\alpha_k)}$ *arrives to the boundary* $\partial\Omega_{\alpha\beta}$, *there is a switching rule between two flows* $\mathbf{x}^{(\alpha_k)}$ *and* $\mathbf{x}^{(\alpha_l)}$ *with* $\mathbf{x}^{(\alpha_k)}(t_{m-}) = \mathbf{x}_m = \mathbf{x}^{(\alpha_l)}(t_{m+})$. *A flow* $\mathbf{x}^{(\alpha_k)}(t)$ *is* $C^{r_{\alpha_k}}_{[t_{m-\varepsilon}, t_m)}$*-continuous* ($r_{\alpha_k} \geq 2$) *for time* t *with* $\| d^{r_{\alpha_k}+1}\mathbf{x}^{(\alpha)} / dt^{r_{\alpha_k}+1} \| < \infty$, *and a flow* $\mathbf{x}^{(\alpha_l)}(t)$ *is* $C^{r_{\alpha_l}}_{(t_m, t_{m+\varepsilon}]}$*-continuous* ($r_{\alpha_k} \geq 2$) *for time* t *with* $\| d^{r_{\alpha_l}+1}\mathbf{x}^{(\alpha)} / dt^{r_{\alpha_l}+1} \| < \infty$. *The flows of* $\mathbf{x}^{(\alpha_k)}(t)$ *and* $\mathbf{x}^{(\alpha_l)}(t)$ *to boundary* $\partial\Omega_{ij}$ *form an extended passable flow from* Ω_α *to* Ω_β *if and only if*

$$\hbar_\alpha G^{(\alpha_k)}_{\partial\Omega_{\alpha\beta}}(\mathbf{x}_m, t_{m-}) > 0 \text{ and } \hbar_\alpha G^{(\alpha_l)}_{\partial\Omega_{\alpha\beta}}(\mathbf{x}_m, t_{m+}) > 0. \tag{5.129}$$

Proof. Following the proof of Theorem 3.1, the proof of the theorem is proved. ■

Definition 5.60. For a discontinuous system with multi-valued vector fields in Eq. (5.118), there is a point $\mathbf{x}^{(0)}(t_m) \equiv \mathbf{x}_m \in \partial\Omega_{\alpha\beta}$ at time t_m between two adjacent domains Ω_α and Ω_β. For an arbitrarily small $\varepsilon > 0$, there are two time intervals $[t_{m-\varepsilon}, t_m)$ and $(t_m, t_{m+\varepsilon}]$. If a coming flow of $\mathbf{x}^{(\alpha_k)}$ arrives to boundary $\partial\Omega_{\alpha\beta}$, there is a switching rule between two flows $\mathbf{x}^{(\alpha_k)}$ and $\mathbf{x}^{(\alpha_l)}$ with $\mathbf{x}^{(\alpha_k)}(t_{m-}) = \mathbf{x}_m = \mathbf{x}^{(\alpha_2)}(t_{m+})$. A flow $\mathbf{x}^{(\alpha_k)}(t)$ is $C^{r_{\alpha_k}}_{[t_{m-\varepsilon}, t_m)}$-continuous ($r_{\alpha_k} \geq m_{\alpha_k} + 1$) for time t with $\| d^{r_{\alpha_k}+1}\mathbf{x}^{(\alpha)} / dt^{r_{\alpha_k}+1} \| < \infty$, and a flow $\mathbf{x}^{(\alpha_l)}(t)$ is $C^{r_{\alpha_l}}_{(t_m, t_{m+\varepsilon}]}$-continuous ($r_{\alpha_l} \geq m_{\alpha_l} + 1$) for time t with $\| d^{r_{\alpha_l}+1}\mathbf{x}^{(\alpha)} / dt^{r_{\alpha_l}+1} \| < \infty$. The flows of $\mathbf{x}^{(\alpha_k)}(t)$ and $\mathbf{x}^{(\alpha_l)}(t)$ to boundary $\partial\Omega_{ij}$ are called an $(m_{\alpha_k} : m_{\alpha_l})$ - extended passable flow from Ω_α to Ω_β if

$$\begin{aligned} &\hbar_\alpha G^{(s_{\alpha_k}, \alpha_k)}_{\partial\Omega_{\alpha\beta}}(\mathbf{x}_m, t_{m-}) = 0 \ \text{ for } s_{\alpha_k} = 0, 1, 2, \cdots, m_{\alpha_k} - 1, \\ &\hbar_\alpha G^{(s_{\alpha_l}, \alpha_l)}_{\partial\Omega_{\alpha\beta}}(\mathbf{x}_m, t_{m+}) = 0 \ \text{ for } s_{\alpha_l} = 0, 1, 2, \cdots, m_{\alpha_l} - 1; \end{aligned} \tag{5.130}$$

$$\left.\begin{aligned} &\hbar_\alpha \mathbf{n}^{\mathrm{T}}_{\partial\Omega_{ij}}(\mathbf{x}^{(0)}_{m-\varepsilon}) \cdot [\mathbf{x}^{(0)}_{m-\varepsilon} - \mathbf{x}^{(\alpha_k)}_{m-\varepsilon}] > 0, \\ &\hbar_\alpha \mathbf{n}^{\mathrm{T}}_{\partial\Omega_{ij}}(\mathbf{x}^{(0)}_{m+\varepsilon}) \cdot [\mathbf{x}^{(\alpha_l)}_{m+\varepsilon} - \mathbf{x}^{(0)}_{m+\varepsilon}] > 0. \end{aligned}\right\} \tag{5.131}$$

Theorem 5.4. *For a discontinuous system with multi-valued vector fields in Eq.* (5.118), *there is a point* $\mathbf{x}^{(0)}(t_m) \equiv \mathbf{x}_m \in \partial\Omega_{\alpha\beta}$ *at time* t_m *between two adjacent domains* Ω_α *and* Ω_β. *For an arbitrarily small* $\varepsilon > 0$, *there are two time inter-*

vals $[t_{m-\varepsilon},t_m)$ *and* $(t_m,t_{m+\varepsilon}]$. *If a coming flow of* $\mathbf{x}^{(\alpha_k)}$ *arrives to boundary* $\partial\Omega_{\alpha\beta}$, *there is a switching rule between two flows* $\mathbf{x}^{(\alpha_k)}$ *and* $\mathbf{x}^{(\alpha_l)}$ *with* $\mathbf{x}^{(\alpha_k)}(t_{m-})=\mathbf{x}_m=\mathbf{x}^{(\alpha_2)}(t_{m+})$. *A flow* $\mathbf{x}^{(\alpha_k)}(t)$ *is* $C^{r_{\alpha_k}}_{[t_{m-\varepsilon},t_m)}$*-continuous* ($r_{\alpha_k}\geq m_{\alpha_k}+1$) *for time t with* $\| d^{r_{\alpha_k}+1}\mathbf{x}^{(\alpha)}/dt^{r_{\alpha_k}+1}\|<\infty$, *and a flow* $\mathbf{x}^{(\alpha_l)}(t)$ *is* $C^{r_{\alpha_l}}_{(t_m,t_{m+\varepsilon}]}$*-continuous* ($r_{\alpha_l}\geq m_{\alpha_l}+1$) *for time t with* $\| d^{r_{\alpha_l}+1}\mathbf{x}^{(\alpha)}/dt^{r_{\alpha_l}+1}\|<\infty$. *The flows of* $\mathbf{x}^{(\alpha_k)}(t)$ *and* $\mathbf{x}^{(\alpha_l)}(t)$ *to boundary* $\partial\Omega_{ij}$ *form an* $(m_{\alpha_k}:m_{\alpha_l})$*-extended passable flow from* Ω_α *to* Ω_β *if and only if*

$$\left.\begin{array}{l}\hbar_\alpha G^{(s_{\alpha_k},\alpha_k)}_{\partial\Omega_{\alpha\beta}}(\mathbf{x}_m,t_{m-})=0 \ \ for\ s_{\alpha_k}=0,1,2,\cdots,m_{\alpha_k}-1,\\ \hbar_\alpha G^{(s_{\alpha_l},\alpha_l)}_{\partial\Omega_{\alpha\beta}}(\mathbf{x}_m,t_{m+})=0 \ \ for\ s_{\alpha_l}=0,1,2,\cdots,m_{\alpha_l}-1;\end{array}\right. \tag{5.132}$$

$$\hbar_\alpha G^{(m_{\alpha_k},\alpha_k)}_{\partial\Omega_{\alpha\beta}}(\mathbf{x}_m,t_{m-})>0 \ and\ \hbar_\alpha G^{(m_{\alpha_l},\alpha_l)}_{\partial\Omega_{\alpha\beta}}(\mathbf{x}_m,t_{m+})>0. \tag{5.133}$$

Proof. Following the proof of Theorem 3.2, the proof of the theorem is proved. ■

Consider switching rules at the boundary on domains Ω_α and Ω_β. After the switching in Ω_α, if a bouncing flow in Ω_α is formed, the switching at the boundary on domain Ω_β will not be used. After the switching in Ω_α, if the extended passable flow from domain Ω_α to Ω_β is formed, the switching rule at the boundary on domain Ω_β is very significant. If such a switching rule exists, a passable flow in domain Ω_β can be switched rather than an extended passable flow from domain Ω_α to Ω_β if

$$\left.\begin{array}{l}\hbar_\alpha G^{(s_{\alpha_k},\alpha_k)}_{\partial\Omega_{\alpha\beta}}(\mathbf{x}_m,t_{m-})=0 \ \ for\ s_{\alpha_k}=0,1,2,\cdots,m_{\alpha_k}-1,\\ \hbar_\alpha G^{(s_{\beta_l},\beta_l)}_{\partial\Omega_{\alpha\beta}}(\mathbf{x}_m,t_{m+})=0 \ \ for\ s_{\beta_l}=0,1,2,\cdots,m_{\beta_l}-1;\end{array}\right. \tag{5.134}$$

$$\left.\begin{array}{l}\hbar_\alpha \mathbf{n}^{\mathrm{T}}_{\partial\Omega_{ij}}(\mathbf{x}^{(0)}_{m-\varepsilon})\cdot[\mathbf{x}^{(0)}_{m-\varepsilon}-\mathbf{x}^{(\alpha_k)}_{m-\varepsilon}]>0,\\ \hbar_\alpha \mathbf{n}^{\mathrm{T}}_{\partial\Omega_{ij}}(\mathbf{x}^{(0)}_{m+\varepsilon})\cdot[\mathbf{x}^{(\beta_l)}_{m+\varepsilon}-\mathbf{x}^{(0)}_{m+\varepsilon}]>0.\end{array}\right\} \tag{5.135}$$

For this case, the direct switching at the boundary in domain Ω_β can be completed. Such a case was the passable flows, discussed in Chapters 2 and 3. If the switching rule is applied at the boundary in domain Ω_β only, the tangential flow in domain Ω_α can exist. If the two switching rules occur at the boundary, no tangential flow exists in Ω_α, but the sink flow can be formed, which will not discussed herein.

5.7 A controlled piecewise linear system

Consider a following piecewise linear dynamical system with a control law in two domains as

$$\ddot{x}+2d_{\alpha}\dot{x}+c_{\alpha}x=b_{\alpha}+Q_0\cos\Omega t, \tag{5.136}$$

where $\alpha\in\{1,2\}$ and the two domains are separated by the following equation

$$ax+b\dot{x}=c. \tag{5.137}$$

Once a coming flow in domain Ω_{α} arrives to boundary $\partial\Omega_{12}$ ($\alpha,\beta\in\{1,2\}$, $\beta\neq\alpha$) at moment t_m, the control force will be exerted in the corresponding domain Ω_{α}. The controlling force on the boundary is given through the constant force varying with the switching location at the boundary. This is a switching rule, which requires the vector fields must be switched with the constant force

$$b_{\alpha_k}=c_{\alpha}x(t_m)\ for\ \alpha\in\{1,2\}\text{ and }k\in\{1,2,3,\cdots\}. \tag{5.138}$$

5.7.1. Passable and bouncing conditions

In phase plane, the vectors are introduced by

$$\mathbf{x}\triangleq(x,\dot{x})^{\mathrm{T}}\equiv(x,y)^{\mathrm{T}}\text{ and }\mathbf{F}\triangleq(y,F)^{\mathrm{T}}. \tag{5.139}$$

From Eq. (5.137), this control logic generates a discontinuous boundary in the system. To analyze dynamics of the system, two domains are defined as

$$\Omega_1=\{(x,y)\mid ax+by>c\}\text{ and }\Omega_2=\{(x,y)\mid ax+by<c\}. \tag{5.140}$$

The separation boundary $\partial\Omega_{\alpha\beta}=\bar{\Omega}_{\alpha}\cap\bar{\Omega}_{\beta}$ ($\alpha,\beta=1,2$) is defined as

$$\begin{aligned}\partial\Omega_{12}&=\partial\Omega_{21}=\bar{\Omega}_1\cap\bar{\Omega}_2\\&=\{(x,y)\mid\varphi_{12}(x,y)\equiv ax+by-c=0\}.\end{aligned} \tag{5.141}$$

The domains and boundary are sketched in Fig. 5.47. The boundary is depicted by a dotted straight line, governed by Eq. (5.137). The two domains are shaded. The arrows crossing the boundary indicate the possible directions of flows to the boundary. If a flow of motion in phase space is in domain Ω_{α} ($\alpha=1,2$), the vector fields in such a domain are continuous. However, if a flow switches from domain Ω_{α} to domain Ω_{β} ($\alpha,\beta\in\{1,2\}$, $\beta\neq\alpha$) through boundary $\partial\Omega_{\alpha\beta}$, the vector field in domain Ω_{α} will be changed in domain Ω_{β} ($\alpha,\beta\in\{1,2\}$ accordingly.

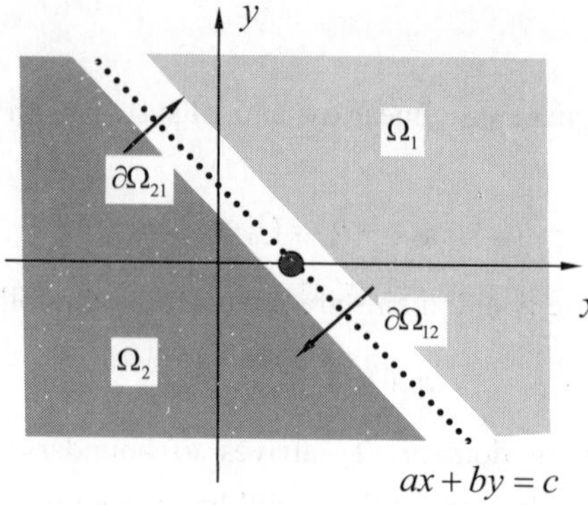

Fig. 5.47 Sub-domains and boundary of the switched dynamical system ($a>0$ and $b>0$).

Because of discontinuity, the sliding motion along the boundary may exist. As discussed in Luo (2006), the sliding motion on the boundary has an equilibrium point $(E,0)$ where $E=c/a$. For the equilibrium point, the parabolicity and hyperbolicity in its vicinity can be discussed. From Eq. (5.137), with initial condition $(x_i^{(0)},\dot{x}_i^{(0)})$ on the boundary, displacement and velocity for the sliding motion are given by

$$\left.\begin{aligned} x^{(0)} &= \frac{c}{a}+\frac{1}{a}(ax_i^{(0)}-c)\exp[-\frac{a}{b}(t-t_i)],\\ y^{(0)} &= -\frac{1}{b}(ax_i^{(0)}-c)\exp[-\frac{a}{b}(t-t_i)]. \end{aligned}\right\} \tag{5.142}$$

With Eq. (5.139), equations of motion are described as

$$\dot{\mathbf{x}}=\mathbf{F}^{(\lambda)}(\mathbf{x},t)\ \ for\ \lambda\in\{0,\alpha_k\} \tag{5.143}$$

where

$$\left.\begin{aligned} &\mathbf{F}^{(\alpha_k)}(\mathbf{x},t)=(y,F_{\alpha_k}(\mathbf{x},t))^{\mathrm{T}} \ \text{ in } \Omega_\alpha \ (\alpha\in\{1,2\},\ k,l=1,2,3\cdots),\\ &\mathbf{F}^{(0)}(\mathbf{x},t)=(y,-\frac{a}{b}y)^{\mathrm{T}} \quad \text{for sliding on } \partial\Omega_{\alpha\beta}(\alpha,\beta\in\{1,2\}),\\ &\mathbf{F}^{(0)}(\mathbf{x},t)=[\mathbf{F}^{(\alpha_k)}(\mathbf{x},t),\mathbf{F}^{(\beta_l)}(\mathbf{x},t)] \ \text{ for non-sliding on } \partial\Omega_{\alpha\beta}, \end{aligned}\right\} \tag{5.144}$$

$$F_{\alpha_k}(\mathbf{x},t)=-2d_{\alpha_k}y-c_{\alpha_k}x+A_0\cos(\Omega t+\phi)+b_{\alpha_k}. \tag{5.145}$$

From the theory of discontinuous dynamical systems in Chapters 2 and 3 (e.g., Luo, 2005, 2006, 2008), for a passable motion with the corresponding normal vector $\mathbf{n}_{\partial\Omega_{\alpha\beta}}$ pointing to domain Ω_α (i.e., $\mathbf{n}_{\partial\Omega_{\alpha\beta}}\to\Omega_\beta$), the necessary and sufficient conditions of a passable motion to boundary $\partial\Omega_{\alpha\beta}$ are

$$\left.\begin{array}{l}\hbar_\alpha G^{(0,\alpha_k)}(\mathbf{x}_m,t_{m-})=\hbar_\alpha \mathbf{n}^{\mathrm{T}}_{\partial\Omega_{\alpha\beta}}\bullet\mathbf{F}^{(\alpha_k)}(\mathbf{x}_m,t_{m-})>0\\ \hbar_\alpha G^{(0,\beta_l)}(\mathbf{x}_m,t_{m+})=\hbar_\alpha \mathbf{n}^{\mathrm{T}}_{\partial\Omega_{\alpha\beta}}\bullet\mathbf{F}^{(\beta_l)}(\mathbf{x}_m,t_{m+})>0\end{array}\right\}\text{from }\Omega_\alpha\to\Omega_\beta \tag{5.146}$$

where $\alpha,\beta\in\{1,2\}$, $\alpha\neq\beta$ and $k,l\in\{1,2,3,\cdots\}$ with

$$\mathbf{n}_{\partial\Omega_{\alpha\beta}}=\nabla\varphi_{\alpha\beta}=(\partial_x\varphi_{\alpha\beta},\partial_y\varphi_{\alpha\beta})^{\mathrm{T}}_{(x_m,y_m)}. \tag{5.147}$$

$\nabla=(\partial_x,\partial_y)^{\mathrm{T}}=(\partial/\partial x,\partial/\partial y)^{\mathrm{T}}$ is the Hamilton operator. Note that t_m is switching time for the motion to the switching boundary and $t_{m\pm}=t_m\pm 0$ reflects the responses in domains rather than on the boundary.

The necessary and sufficient conditions of a $(0:1)$-passable motion to boundary $\partial\Omega_{\alpha\beta}$ are

$$\left.\begin{array}{l}\hbar_\alpha G^{(0,\alpha_k)}(\mathbf{x}_m,t_{m-})=\hbar_\alpha \mathbf{n}^{\mathrm{T}}_{\partial\Omega_{\alpha\beta}}\cdot\mathbf{F}^{(\alpha_k)}(\mathbf{x}_m,t_{m-})>0\\ \hbar_\alpha G^{(0,\beta_l)}(\mathbf{x}_m,t_{m+})=\hbar_\alpha \mathbf{n}^{\mathrm{T}}_{\partial\Omega_{\alpha\beta}}\cdot\mathbf{F}^{(\beta_l)}(\mathbf{x}_m,t_{m+})=0\\ \hbar_\alpha G^{(1,\beta_l)}(\mathbf{x}_m,t_{m+})=\hbar_\alpha \mathbf{n}^{\mathrm{T}}_{\partial\Omega_{\alpha\beta}}\cdot D\mathbf{F}^{(\beta_l)}(\mathbf{x}_m,t_{m+})>0\end{array}\right\}\text{from }\Omega_\alpha\to\Omega_\beta \tag{5.148}$$

where

$$D\mathbf{F}^{(\beta_l)}(\mathbf{x},t)=(F_{\beta_l}(\mathbf{x},t),\nabla F_{\beta_l}(\mathbf{x},t)\cdot\mathbf{F}^{(\beta_l)}(\mathbf{x},t)+\partial_t F_{\beta_l}(\mathbf{x},t))^{\mathrm{T}}. \tag{5.149}$$

The $(0:1)$-passable motion is also called the *post-passable tangential motion* (or the *post-passable grazing motion*).

The necessary and sufficient conditions for sliding motions on the separation boundary with $\mathbf{n}_{\partial\Omega_{\alpha\beta}}\to\Omega_\beta$ are from Chapter 2 (e.g., Luo, 2005, 2006, 2008)

$$\left.\begin{array}{l}\hbar_\alpha G^{(0,\alpha_k)}(\mathbf{x}_m,t_{m-})=\hbar_\alpha \mathbf{n}^{\mathrm{T}}_{\partial\Omega_{\alpha\beta}}\cdot\mathbf{F}^{(\alpha_k)}(\mathbf{x}_m,t_{m-})>0,\\ \hbar_\alpha G^{(0,\beta_l)}(\mathbf{x}_m,t_{m-})=\hbar_\alpha \mathbf{n}^{\mathrm{T}}_{\partial\Omega_{\alpha\beta}}\cdot\mathbf{F}^{(\beta_l)}(\mathbf{x}_m,t_{m-})<0\end{array}\right\}\text{ on }\partial\Omega_{\alpha\beta}. \tag{5.150}$$

The necessary and sufficient conditions for vanishing of the sliding motions on the separation boundary with $\mathbf{n}_{\partial\Omega_{\alpha\beta}}\to\Omega_\beta$ are from Chapters 2 (e.g., Luo, 2005, 2006, 2008), i.e.,

$$\left.\begin{array}{l}\hbar_\alpha G^{(0,\alpha_k)}(\mathbf{x}_m,t_{m\mp})=\hbar_\alpha \mathbf{n}^{\mathrm{T}}_{\partial\Omega_{\alpha\beta}}\cdot\mathbf{F}^{(\alpha_k)}(\mathbf{x}_m,t_{m\mp})=0,\\ \hbar_\alpha G^{(1,\alpha_k)}(\mathbf{x}_m,t_{m\mp})=\hbar_\alpha \mathbf{n}^{\mathrm{T}}_{\partial\Omega_{\alpha\beta}}\cdot D\mathbf{F}^{(\alpha_k)}(\mathbf{x}_m,t_{m\mp})<0,\\ \hbar_\alpha G^{(0,\beta_l)}(\mathbf{x}_m,t_{m-})=\hbar_\alpha \mathbf{n}^{\mathrm{T}}_{\partial\Omega_{\alpha\beta}}\cdot\mathbf{F}^{(\beta_l)}(\mathbf{x}_m,t_{m-})<0\end{array}\right\} \tag{5.151}$$

where

$$D\mathbf{F}^{(\alpha_k)}(\mathbf{x},t)=(F_{\alpha_k}(\mathbf{x},t),\nabla F_{\alpha_k}(\mathbf{x},t)\cdot\mathbf{F}^{(\alpha_k)}(\mathbf{x},t)+\partial_t F_{\alpha_k}(\mathbf{x},t))^{\mathrm{T}}. \tag{5.152}$$

Similarly, the necessary and sufficient conditions for onset of the sliding motion on the separation boundary with $\mathbf{n}_{\partial\Omega_{\alpha\beta}} \to \Omega_\beta$ are from Chapter 2 (e.g., Luo, 2005, 2006, 2008), i.e.,

$$\left.\begin{aligned} \hbar_\alpha G^{(0,\alpha_k)}(\mathbf{x}_m, t_{m\pm}) &= \hbar_\alpha \mathbf{n}^{\mathrm{T}}_{\partial\Omega_{\alpha\beta}} \cdot \mathbf{F}^{(\alpha_k)}(\mathbf{x}_m, t_{m\pm}) = 0, \\ \hbar_\alpha G^{(1,\alpha_k)}(\mathbf{x}_m, t_{m\pm}) &= \hbar_\alpha \mathbf{n}^{\mathrm{T}}_{\partial\Omega_{\alpha\beta}} \cdot D\mathbf{F}^{(\alpha_k)}(\mathbf{x}_m, t_{m\pm}) > 0, \\ \hbar_\alpha G^{(0,\beta_l)}(\mathbf{x}_m, t_{m-}) &= \hbar_\alpha \mathbf{n}^{\mathrm{T}}_{\partial\Omega_{\alpha\beta}} \cdot \mathbf{F}^{(\beta_l)}(\mathbf{x}_m, t_{m-}) < 0. \end{aligned}\right\} \tag{5.153}$$

The necessary and sufficient conditions for a grazing motion to the separation boundary $\partial\Omega_{\alpha\beta}$ are from Luo (2005, 2006, 2008), i.e.,

$$\left.\begin{aligned} \hbar_\alpha G^{(0,\alpha_k)}(\mathbf{x}_m, t_{m\pm}) &= \hbar_\alpha \mathbf{n}^{\mathrm{T}}_{\partial\Omega_{\alpha\beta}} \cdot \mathbf{F}^{(\alpha_k)}(\mathbf{x}_m, t_{m\pm}) = 0 \\ \hbar_\alpha G^{(1,\alpha_k)}(\mathbf{x}_m, t_{m\pm}) &= \hbar_\alpha \mathbf{n}^{\mathrm{T}}_{\partial\Omega_{12}} \cdot D\mathbf{F}^{(\alpha_k)}(\mathbf{x}_m, t_{m\pm}) < 0 \end{aligned}\right\} \tag{5.154}$$

for ($\alpha \in \{1,2\}$ and $k \in \{1,2,3,\cdots\}$). Because the switching rule exists, the grazing flow of a coming flow cannot be formed in its own domain. Once the flow arrives to the boundary, the vector field must be switched. Only if the starting and ending points of the coming flow on the same boundary is at the same point, the tangential flow can be observed. Thus, the grazing flow flows to the boundary will not be discussed.

The necessary and sufficient conditions for a bouncing motion to the boundary $\partial\Omega_{\alpha\beta}$ are for $\alpha \in \{1,2\}$ and $k \in \{1,2,3,\cdots\}$

$$\left.\begin{aligned} \hbar_\alpha G^{(0,\alpha_k)}(\mathbf{x}_m, t_{m-}) &= \hbar_\alpha \mathbf{n}^{\mathrm{T}}_{\partial\Omega_{\alpha\beta}} \bullet \mathbf{F}^{(\alpha_k)}(\mathbf{x}_m, t_{m-}) > 0 \\ \hbar_\alpha G^{(0,\alpha_l)}(\mathbf{x}_m, t_{m+}) &= \hbar_\alpha \mathbf{n}^{\mathrm{T}}_{\partial\Omega_{\alpha\beta}} \bullet \mathbf{F}^{(\alpha_l)}(\mathbf{x}_m, t_{m+}) < 0 \end{aligned}\right\} \text{in } \Omega_\alpha. \tag{5.155}$$

The necessary and sufficient conditions for a $(0:1)$-bouncing motion to the boundary $\partial\Omega_{\alpha\beta}$ are for $\alpha \in \{1,2\}$ and $k, l \in \{1,2,3,\cdots\}$

$$\left.\begin{aligned} \hbar_\alpha G^{(0,\alpha_k)}(\mathbf{x}_m, t_{m-}) &= \hbar_\alpha \mathbf{n}^{\mathrm{T}}_{\partial\Omega_{\alpha\beta}} \bullet \mathbf{F}^{(\alpha_k)}(\mathbf{x}_m, t_{m-}) > 0 \\ \hbar_\alpha G^{(0,\alpha_l)}(\mathbf{x}_m, t_{m+}) &= \hbar_\alpha \mathbf{n}^{\mathrm{T}}_{\partial\Omega_{\alpha\beta}} \bullet \mathbf{F}^{(\alpha_l)}(\mathbf{x}_m, t_{m+}) = 0 \\ \hbar_\alpha G^{(1,\alpha_l)}(\mathbf{x}_m, t_{m+}) &= \hbar_\alpha \mathbf{n}^{\mathrm{T}}_{\partial\Omega_{\alpha\beta}} \bullet D\mathbf{F}^{(\alpha_l)}(\mathbf{x}_m, t_{m+}) < 0 \end{aligned}\right\} \text{in } \Omega_\alpha. \tag{5.156}$$

$$D\mathbf{F}^{(\alpha_k)}(\mathbf{x}, t) = (F_{\alpha_k}(\mathbf{x}, t), \nabla F_{\alpha_k}(\mathbf{x}, t) \cdot \mathbf{F}^{(\alpha_k)}(\mathbf{x}, t) + \partial_t F_{\alpha_k}(\mathbf{x}, t))^{\mathrm{T}}. \tag{5.157}$$

The necessary and sufficient conditions for a $(1:0)$-bouncing motion to the boundary $\partial\Omega_{\alpha\beta}$ are for $\alpha \in \{1,2\}$ and $k \in \{1,2,3,\cdots\}$

$$\left.\begin{aligned}
&\hbar_{\alpha} G^{(0,\alpha_k)}(\mathbf{x}_m, t_{m-}) = \hbar_{\alpha} \mathbf{n}^{\mathrm{T}}_{\partial\Omega_{\alpha\beta}} \bullet \mathbf{F}^{(\alpha_k)}(\mathbf{x}_m, t_{m-}) = 0 \\
&\hbar_{\alpha} G^{(1,\alpha_k)}(\mathbf{x}_m, t_{m-}) = \hbar_{\alpha} \mathbf{n}^{\mathrm{T}}_{\partial\Omega_{\alpha\beta}} \bullet D\mathbf{F}^{(\alpha_k)}(\mathbf{x}_m, t_{m+}) > 0 \\
&\hbar_{\alpha} G^{(0,\alpha_{k+1})}(\mathbf{x}_m, t_{m+}) = \hbar_{\alpha} \mathbf{n}^{\mathrm{T}}_{\partial\Omega_{\alpha\beta}} \bullet \mathbf{F}^{(\alpha_{k+1})}(\mathbf{x}_m, t_{m+}) < 0
\end{aligned}\right\} \text{in } \Omega_{\alpha}. \tag{5.158}$$

The necessary and sufficient conditions for a $(1:1)$-bouncing motion to the boundary $\partial\Omega_{\alpha\beta}$ are for $\alpha \in \{1,2\}$ and $k,l \in \{1,2,3,\cdots\}$

$$\left.\begin{aligned}
&\hbar_{\alpha} G^{(0,\alpha_k)}(\mathbf{x}_m, t_{m-}) = \hbar_{\alpha} \mathbf{n}^{\mathrm{T}}_{\partial\Omega_{\alpha\beta}} \bullet \mathbf{F}^{(\alpha_k)}(\mathbf{x}_m, t_{m-}) = 0 \\
&\hbar_{\alpha} G^{(1,\alpha_k)}(\mathbf{x}_m, t_{m-}) = \hbar_{\alpha} \mathbf{n}^{\mathrm{T}}_{\partial\Omega_{\alpha\beta}} \bullet D\mathbf{F}^{(\alpha_k)}(\mathbf{x}_m, t_{m-}) > 0 \\
&\hbar_{\alpha} G^{(0,\alpha_l)}(\mathbf{x}_m, t_{m+}) = \hbar_{\alpha} \mathbf{n}^{\mathrm{T}}_{\partial\Omega_{\alpha\beta}} \bullet \mathbf{F}^{(\alpha_l)}(\mathbf{x}_m, t_{m+}) = 0 \\
&\hbar_{\alpha} G^{(1,\alpha_l)}(\mathbf{x}_m, t_{m+}) = \hbar_{\alpha} \mathbf{n}^{\mathrm{T}}_{\partial\Omega_{\alpha\beta}} \bullet D\mathbf{F}^{(\alpha_l)}(\mathbf{x}_m, t_{m+}) < 0
\end{aligned}\right\} \text{in } \Omega_{\alpha}. \tag{5.159}$$

The conditions for the bouncing flow with higher-order singularity can be similarly obtained. Notice all the conditions for the curve boundary are different, and those conditions can be developed from the G-function definitions.

Substitution of Eq. (5.140) into Eq. (5.147) gives

$$\mathbf{n}_{\partial\Omega_{12}} = \mathbf{n}_{\partial\Omega_{21}} = (a,b)^{\mathrm{T}}. \tag{5.160}$$

From the forgoing equation, the normal vector always points to domain Ω_1 (i.e., $\mathbf{n}_{\partial\Omega_{12}} \to \Omega_1$). Therefore,

$$G^{(0,\alpha_k)}(\mathbf{x}_m, t_m) = \mathbf{n}^{\mathrm{T}}_{\partial\Omega_{\alpha\beta}} \cdot \mathbf{F}^{(\alpha_k)}(\mathbf{x}_m, t_m) = a y_m + b F_{\alpha_k}(\mathbf{x}_m, t_m), \tag{5.161}$$

$$\begin{aligned}
&G^{(1,\alpha_k)}(\mathbf{x}_m, t_m) \\
&= \mathbf{n}^{\mathrm{T}}_{\partial\Omega_{\alpha\beta}} \bullet D\mathbf{F}^{(\alpha_k)}(\mathbf{x}_m, t_m) \\
&= a F_{\alpha_k}(\mathbf{x}_m, t_m) + b\left[\nabla F_{\alpha_k}(\mathbf{x},t) \cdot \mathbf{F}^{(\alpha_k)}(\mathbf{x},t) + \partial_t F_{\alpha_k}(\mathbf{x},t)\right]_{(\mathbf{x}_m, t_m)}.
\end{aligned} \tag{5.162}$$

From Eqs. (5.146) and (5.160), the conditions for passable motions on the boundary are for $\alpha = 1, \beta = 2$ and $k,l \in \{1,2,3,\cdots\}$:

$$\left.\begin{aligned}
&G^{(0,\alpha_k)}(\mathbf{x}_m, t_{m-}) < 0 \text{ and } G^{(0,\beta_l)}(\mathbf{x}_m, t_{m+}) < 0, \text{ from } \Omega_{\alpha} \to \Omega_{\beta}; \\
&G^{(0,\beta_k)}(\mathbf{x}_m, t_{m-}) > 0 \text{ and } G^{(0,\alpha_l)}(\mathbf{x}_m, t_{m+}) > 0, \text{ from } \Omega_{\beta} \to \Omega_{\alpha}.
\end{aligned}\right\} \tag{5.163}$$

From Eqs. (5.148) and (5.160), the conditions of a $(0:1)$-passable tangential motion to boundary $\partial\Omega_{\alpha\beta}$ are for $\alpha, \beta \in \{1,2\}$, $\alpha \neq \beta$ and $k,l \in \{1,2,3,\cdots\}$

$$\left.\begin{array}{l}(-1)^{\alpha} G^{(0,\alpha_k)}(\mathbf{x}_m, t_{m-}) > 0; \\ G^{(0,\beta_l)}(\mathbf{x}_m, t_{m+}) = 0 \text{ and } (-1)^{\alpha}\ G^{(1,\beta_l)}(\mathbf{x}_m, t_{m+}) < 0\end{array}\right\} \text{from } \Omega_\alpha \to \Omega_\beta \tag{5.164}$$

From Eqs. (5.150) and (5.160), the conditions for sliding motion on the separation boundary are for $\alpha = 1, \beta = 2$ and $k, l \in \{1, 2, 3, \cdots\}$

$$G^{(0,\alpha_k)}(\mathbf{x}_m, t_{m-}) < 0 \text{ and } G^{(0,\beta_l)}(\mathbf{x}_m, t_{m-}) > 0. \tag{5.165}$$

From Eq. (5.151), the vanishing conditions for sliding motions on the separation boundary are for $\alpha, \beta \in \{1, 2\}$, $\alpha \neq \beta$ and $k, l \in \{1, 2, 3, \cdots\}$

$$\left.\begin{array}{l}(-1)^{\alpha} G^{(0,\alpha_k)}(\mathbf{x}_m, t_{m-}) > 0 \text{ and } G^{(0,\beta_l)}(\mathbf{x}_m, t_{m\mp}) = 0 \text{ with} \\ (-1)^{\beta} G^{(1,\beta_k)}(\mathbf{x}_m, t_{m\mp}) < 0 \text{ from } \partial\Omega_{12} \to \Omega_\beta\ (\alpha, \beta \in \{1, 2\});\end{array}\right\} \tag{5.166}$$

From Eq. (5.153), the onset condition of the sliding motion on the switching boundary is for $\alpha, \beta \in \{1, 2\}$, $\alpha \neq \beta$ and $k, l \in \{1, 2, 3, \cdots\}$

$$\left.\begin{array}{l}(-1)^{\alpha} G^{(0,\alpha_k)}(\mathbf{x}_m, t_{m-}) > 0 \text{ and } G^{(0,\beta_l)}(\mathbf{x}_m, t_{m\pm}) = 0 \text{ with} \\ (-1)^{\beta} G^{(1,\beta_l)}(\mathbf{x}_m, t_{m\pm}) < 0 \text{ from } \Omega_\alpha \to \partial\Omega_{12}\ (\alpha, \beta \in \{1, 2\}).\end{array}\right\} \tag{5.167}$$

From Eq. (5.155), the conditions for a bouncing motion to the boundary $\partial\Omega_{\alpha\beta}$ are for $\alpha \in \{1, 2\}$ and $k, l \in \{1, 2, 3, \cdots\}$

$$(-1)^{\alpha} G^{(0,\alpha_k)}(\mathbf{x}_m, t_{m-}) < 0 \text{ and } (-1)^{\alpha} G^{(0,\alpha_l)}(\mathbf{x}_m, t_{m+}) > 0 \text{ in } \Omega_\alpha. \tag{5.168}$$

From Eq. (5.156), the conditions for the $(0:1)$-bouncing motion to the boundary $\partial\Omega_{\alpha\beta}$ are for $\alpha \in \{1, 2\}$ and $k, l \in \{1, 2, 3, \cdots\}$

$$\left.\begin{array}{l}(-1)^{\alpha} G^{(0,\alpha_k)}(\mathbf{x}_m, t_{m-}) < 0, \\ G^{(0,\alpha_l)}(\mathbf{x}_m, t_{m+}) = 0 \text{ and } (-1)^{\alpha} G^{(1,\alpha_l)}(\mathbf{x}_m, t_{m+}) > 0\end{array}\right\} \text{in } \Omega_\alpha. \tag{5.169}$$

From Eq. (5.157), the conditions for the $(1:0)$-bouncing motion to the boundary $\partial\Omega_{\alpha\beta}$ are for $\alpha \in \{1, 2\}$ and $k, l \in \{1, 2, 3, \cdots\}$

$$\left.\begin{array}{l}G^{(0,\alpha_k)}(\mathbf{x}_m, t_{m-}) = 0 \text{ and } (-1)^{\alpha} G^{(1,\alpha_k)}(\mathbf{x}_m, t_{m-}) < 0, \\ (-1)^{\alpha} G^{(0,\alpha_l)}(\mathbf{x}_m, t_{m+}) < 0\end{array}\right\} \text{in } \Omega_\alpha. \tag{5.170}$$

From Eq. (5.159), the conditions for the $(1:1)$-bouncing motion to the boundary $\partial\Omega_{\alpha\beta}$ are for $\alpha \in \{1, 2\}$ and $k, l \in \{1, 2, 3, \cdots\}$

$$\left.\begin{array}{l}G^{(0,\alpha_k)}(\mathbf{x}_m, t_{m-}) = 0 \text{ and } (-1)^{\alpha} G^{(1,\alpha_k)}(\mathbf{x}_m, t_{m-}) < 0, \\ G^{(0,\alpha_l)}(\mathbf{x}_m, t_{m+}) = 0 \text{ and } (-1)^{\alpha} G^{(1,\alpha_l)}(\mathbf{x}_m, t_{m+}) > 0\end{array}\right\} \text{in } \Omega_\alpha. \tag{5.171}$$

5.7.2. Illustrations

For each domain Ω_α ($\alpha \in \{1,2\}$), consider the initial conditions on the boundary, the closed-form solutions in Eq. (5.130) can be given. Through the closed form solutions for discontinuous dynamical systems in Eq. (5.144), the bouncing flows to the separation boundary are presented in Figs. 5.48-5.50. The parameters ($a_1 = a_2 = 1$, $c_1 = 2$, $c_2 = 6$, $d_1 = d_2 = 0.01$, $Q_0 = 40$) are used. Select an initial condition on the boundary $(\Omega t_i, x_i, y_i) = (4.8520,\ -0.2150,\ 1.2150)$ with $\Omega = 1.6$. The phase trajectory and the G-functions of a bouncing flow in domain Ω_2 are illustrated in Fig. 5.48 ($\alpha = 1$ and $\beta = 2$). In Fig. 5.48(a), the arrow direction is the direction of flow. Circular symbols represent the flow switching on the boundary. The bouncing phenomenon of the flow in domain Ω_2 is observed, which is labeled "Bouncing". In Fig. 5.48(b), the G-function to the boundary is presented. Hollow and filled circular symbols are relative to vector fields in domain Ω_1 and Ω_2, respectively. For the initial point, $G_{\partial\Omega_{12}}^{(0,\alpha_1)}(t_{m+}) > 0$ and $G_{\partial\Omega_{12}}^{(0,\beta_1)}(t_{m+}) > 0$, the flow $\mathbf{x}^{(\alpha_1)}$ will be in domain Ω_1 and the G-function relative to $\mathbf{x}^{(\alpha_1)}$ is depicted by a solid curve. However, based on the value of $\mathbf{x}^{(\alpha_1)}$, the G-function relative to the vector fields of $\mathbf{x}^{(\beta_1)}$ is depicted by a dashed curve because this G-function is fictitious. Once the flow $\mathbf{x}^{(\alpha_1)}$ hits the boundary, $G_{\partial\Omega_{12}}^{(0,\alpha_1)}(t_{m-}) < 0$ and $G_{\partial\Omega_{12}}^{(0,\beta_1)}(t_{m-}) < 0$. The flow must pass from domain Ω_1 to Ω_2. Under the control law in Eq. (5.124), the two vectors are switched into the two new vector fields on the boundary. The G-functions for the two new vectors are $G_{\partial\Omega_{12}}^{(0,\alpha_2)}(t_{m+}) > 0$ and $G_{\partial\Omega_{12}}^{(0,\beta_2)}(t_{m+}) < 0$; and the flow is already in domain Ω_2. Thus, the flow of $\mathbf{x}^{(\beta_2)}$ will exist, and the G-function to the boundary becomes real, depicted by the solid curve. The G-function relative to the vector field of $\mathbf{x}^{(\alpha_2)}$ becomes fictitious and is represented by the dashed curve. Once the flow of $\mathbf{x}^{(\beta_2)}$ hits the boundary, $G_{\partial\Omega_{12}}^{(0,\alpha_2)}(t_{m-}) > 0$ and $G_{\partial\Omega_{12}}^{(0,\beta_2)}(t_{m-}) > 0$. Without switching, the flow will pass through the boundary and will enter the domain Ω_1. However, under the control law in Eq. (5.138), two vector fields at the boundary are switched to two new vector fields, and the G-functions are $G_{\partial\Omega_{12}}^{(0,\alpha_3)}(t_{m+}) < 0$ and $G_{\partial\Omega_{12}}^{(0,\beta_3)}(t_{m+}) < 0$. The bouncing flow in domain Ω_2 is formed. A new flow $\mathbf{x}^{(\beta_3)}$ in domain Ω_2 appears. Once the flow $\mathbf{x}^{(\beta_3)}$ hits the boundary, the G-functions are $G_{\partial\Omega_{12}}^{(0,\alpha_3)}(t_{m-}) > 0$ and $G_{\partial\Omega_{12}}^{(0,\beta_3)}(t_{m-}) > 0$. The flow $\mathbf{x}^{(\beta_3)}$ will pass through the boundary from Ω_2 to Ω_1. Under the control law, the switched vector fields possess $G_{\partial\Omega_{12}}^{(0,\alpha_4)}(t_{m+}) > 0$ and $G_{\partial\Omega_{12}}^{(0,\beta_4)}(t_{m+}) > 0$, and such vector fields are the same as the initial vector fields to form a periodic motion.

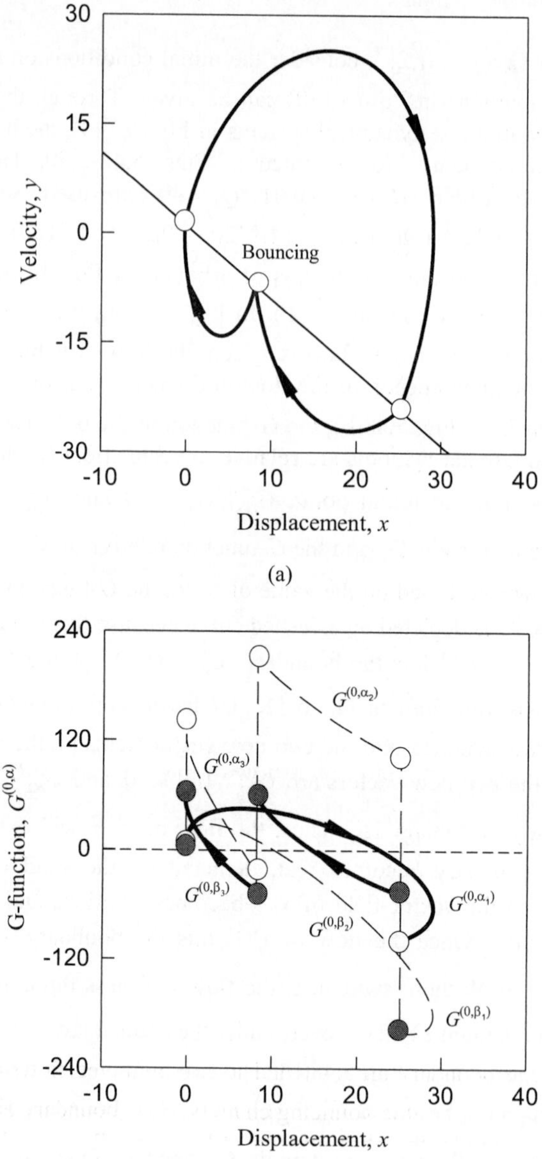

Fig. 5.48 The bouncing flow on the boundary in the lower domain ($\Omega = 1.60$): **(a)** phase trajectory and **(b)** G-functions to the boundary $\partial\Omega_{12}$. The initial condition is ($\Omega t_i, x_i, y_i$)= (4.8520, -0.2150, 1.2150). ($a_1 = a_2 = 1$, $c_1 = 2$, $c_2 = 6$, $d_1 = d_2 = 0.01$, $Q_0 = 40$)

In Fig. 5.49, a periodic motion with a bouncing flow and the $(0:1)$ -passable tangential flow to the boundary is presented with $\Omega = 1.49$. The initial condition is $(\Omega t_i, x_i, y_i) = (4.6458,\ -1.8672,\ 2.8672)$. Again, the arrow direction is the direction of flow. Circular symbols represent flow switching on the boundary. The bouncing phenomenon of flow in domain Ω_2 is observed, which is labeled by "Bouncing". The acronym "$(0:1)$ -passable" represents the $(0:1)$ -passable tangential point. Through the zoomed area, it is clearly observed that the flow is tangential to the boundary just after such a flow passes over the boundary. The G-functions for the bouncing and passable flows are presented in Fig. 5.49(b). Hollow and filled circular symbols are relative to vector fields in domains Ω_1 and Ω_2, respectively. However, $G_{\partial\Omega_{12}}^{(0,\alpha_1)}(t_{m+}) = 0$ and $G_{\partial\Omega_{12}}^{(0,\beta_1)}(t_{m+}) = 0$ at the initial point. However, $G_{\partial\Omega_{12}}^{(0,\alpha_1)}(t_{m+\varepsilon}) > 0$ and $G_{\partial\Omega_{12}}^{(0,\beta_1)}(t_{m+\varepsilon}) > 0$, the flow $\mathbf{x}^{(\alpha_1)}$ will be tangential to the boundary in domain Ω_1 and the G-functions relative to $\mathbf{x}^{(\alpha_1)}$ is depicted by a solid curve. For bouncing and passable flows at the boundary, the G-functions of the corresponding vector fields are the same as in Fig. 5.48. Once the flow $\mathbf{x}^{(\beta_3)}$ hits the boundary, the corresponding G-functions are $G_{\partial\Omega_{12}}^{(0,\alpha_3)}(t_{m-}) > 0$ and $G_{\partial\Omega_{12}}^{(0,\beta_3)}(t_{m-}) > 0$. The flow $\mathbf{x}^{(\beta_3)}$ will pass through the boundary from domain Ω_2 to Ω_1. Under the control law, the switched vector fields possess the G-functions of $G_{\partial\Omega_{12}}^{(0,\alpha_4)}(t_{m+}) = 0$ and $G_{\partial\Omega_{12}}^{(0,\beta_4)}(t_{m+}) = 0$ with $G_{\partial\Omega_{12}}^{(0,\alpha_4)}(t_{m+\varepsilon}) > 0$ and $G_{\partial\Omega_{12}}^{(0,\beta_4)}(t_{m+\varepsilon}) > 0$, which implies that $G_{\partial\Omega_{12}}^{(1,\alpha_4)}(t_{m+}) > 0$ and $G_{\partial\Omega_{12}}^{(1,\beta_4)}(t_{m+}) > 0$. The passable tangential flow is a $(0:1)$ -passable, tangential flow to the boundary. Such vector fields are the same as the initial vector fields to form a periodic motion.

The periodic motion with multiple bouncing flows is presented in Fig. 5.50 for $\Omega = 1.37$. The initial condition is $(\Omega t_i, x_i, y_i) = (5.1197, -1.2459,\ 2.2459)$. The trajectory in phase plane is plotted in Fig. 5.50(a). The acronym "B" represents bouncing point. The G-function for the periodic motion is shown in Fig. 5.50(b). For the initial condition, $G_{\partial\Omega_{12}}^{(0,\alpha_1)}(t_{m+}) > 0$ and $G_{\partial\Omega_{12}}^{(0,\beta_1)}(t_{m+}) > 0$ in domain Ω_1. The flow of $\mathbf{x}^{(\alpha_1)}$ starts in domain Ω_1 and continues. The corresponding discussion of G-function is the same as in Fig. 5.48(b). The two bouncing flows in domain Ω_2 are observed, which implies the bouncing chatters to the boundary exist.

Through this example, the phenomenon of bouncing flows to the boundary is presented. If there are many vector fields in a single, bounded domain, then, under switching rules, bouncing flows to its boundary will form periodic flows and chaos. In addition, the bounding charters can be observed. Such a system will form a class of generalized billiard systems with arbitrary vector fields, which will allow people to develop new switching systems and controls. The complexity of such a class of systems should be further discussed.

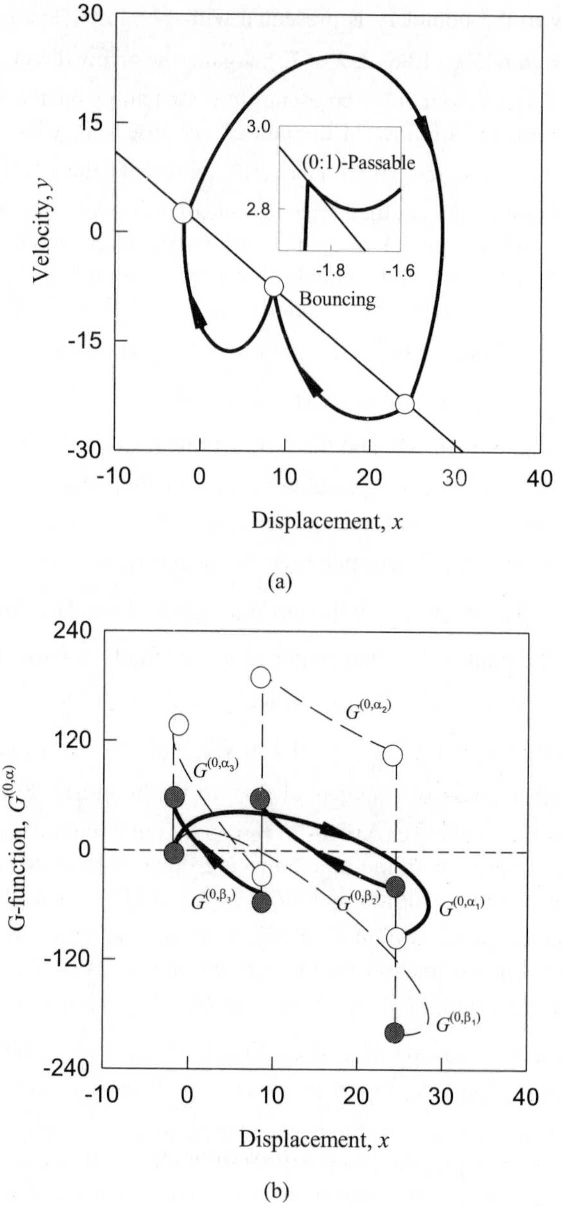

Fig. 5.49 The (0:1)-passable tangential flow ($\Omega = 1.49$): **(a)** phase trajectory and **(b)** *G*-functions to the boundary $\partial\Omega_{12}$. The initial condition is $(\Omega t_i, x_i, y_i) = (4.6458, -1.8672,\ 2.8672)$. ($a_1 = a_2 = 1,\ c_1 = 2,\ c_2 = 6$, $d_1 = d_2 = 0.01$, $Q_0 = 40$). The acronym "(0:1)-passable" represents the (0:1)-passable tangential flow.

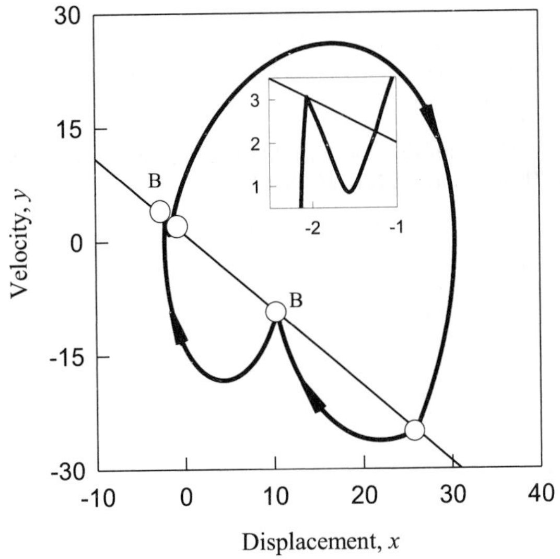

(a)

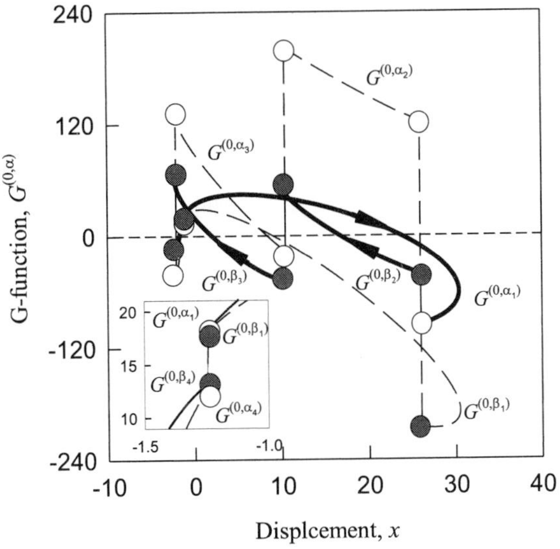

(b)

Fig. 5.50 The bouncing flows on the boundary ($\Omega = 1.37$): **(a)** phase trajectory and **(b)** the G-functions to the boundary $\partial\Omega_{12}$. The initial condition is $(\Omega t_i, x_i, y_i) = (4.3675, -2.0605, 3.0605)$ ($a_1 = a_2 = 1,\ c_1 = 2,\ c_2 = 6,\ d_1 = d_2 = 0.01,\ Q_0 = 40$). The acronym "B" represents bouncing flows.

References

Luo, A.C.J., 2005, A theory for non-smooth dynamic systems on the connectable domains, *Communications in Nonlinear Science and Numerical Simulation*, **10**, 1-55.

Luo, A.C.J., 2006, *Singularity and Dynamics on Discontinuous Vector Fields*, Amsterdam: Elsevier.

Luo, A.C.J., 2008, A theory for flow switchability in discontinuous dynamical systems, *Nonlinear Analysis: Hybrid Systems*, **2**(4), 1030-1061.

Chapter 6
Switchability and Attractivity of Domain Flows

In this chapter, the switchability and attractivity of domain flows to edges of domains will be discussed. The classification and definition of edges will be presented, and the corresponding dynamical systems on domains, boundaries, edges, and vertexes will be defined. The coming, leaving and tangency of a domain flow to a specific edge will be discussed through the corresponding boundaries. The switchability and passability of a flow from an accessible domain to another accessible domain will be presented with a switching rule. The convex and concave edges are introduced for discontinuous dynamical systems. In addition, the mirror domains will also be introduced through the extension of boundaries at the convex edge. The transversally grazing passability of a flow to the concave edges will be presented. The equi-measuring surface will be introduced, and the attractivity of a domain flow to the boundary will be discussed. Further, an equi-measuring edge in domain will be presented, and the attractivity of a domain flow to a specific edge will be discussed. The bouncing domain flows to a specific edge will be discussed for multi-valued vector fields.

6.1. Dynamical systems on edges

In Chapter 2, the singular sets of boundary intersection include edges and vertexes. Consider a sub-domain $\Omega_\alpha \subset \mathscr{R}^n$ ($\alpha \in \mathscr{N}_{\mathscr{D}} = \{1,2,\cdots,N_d\}$) with the corresponding boundary $\partial\Omega_{\alpha i_\alpha} \subset \mathscr{R}^{n-1}$ ($i_\alpha \in \mathscr{N}_{\alpha\mathscr{B}} \subset \mathscr{N}_{\mathscr{B}} = \{1,2,\cdots,N_b\}$). The intersection of $(n-r)$ linearly independent boundaries relative to domain Ω_α forms a r-dimensional edge (e.g., Luo, 2005, 2006)

$$\mathscr{E}_{\sigma_r}^{(r)} \equiv \cap_{i_\alpha \in \mathscr{N}_{\alpha\mathscr{B}}} \partial\Omega_{\alpha i_\alpha} \subset \mathscr{R}^r,\ (r = 0,1,\cdots,n-2). \tag{6.1}$$

For an $(n-2)$-dimensional edge, at least there are two boundary surfaces to form two or three sub-domains. For linearly dependent boundaries, the edge will connect with many domains and boundaries. Consider four boundaries with a common edge, as sketched in Fig. 6.1. The common edge $\mathscr{E}^{(n-2)}$ is an intersection of the boundaries $\partial\Omega_{\alpha\beta}$ ($\alpha,\beta \in \{i,j,k,l\}$ and $\alpha \neq \beta$), depicted by a thick curve.

Definition 6.1. In a discontinuous dynamical system, if there are N_{n-1}-boundaries of $\partial\Omega_{\alpha\beta} \subset \mathscr{R}^{n-1}$ ($\alpha,\beta \in \{1,2,\cdots,N_d\}$, $N_{n-1} \geq 2$) with a common edge to form N_d-sub-domains, the common edge is called an $(n-2)$*-dimensional* edge if

$$\mathscr{E}^{(n-2)} = \cap_{\sigma_{n-1}=1}^{N_{n-1}} \mathscr{E}_{\sigma_{n-1}}^{(n-1)} = \cap_{\alpha=1}^{N_d} \cap_{\beta=1}^{N_d} \partial\Omega_{\alpha\beta} = \cap_{\alpha=1}^{N_d} \bar{\Omega}_{\alpha} \subset \mathscr{R}^{n-2}. \tag{6.2}$$

To extend the above definition of the $(n-2)$-dimensional edge to the r-dimensional edge ($r = 0,1,\cdots,n-2$), the generalized edge concepts is introduced.

Definition 6.2. In a discontinuous dynamical system, if there are N_{r+1}-edges of $\mathscr{E}_{\sigma_{r+1}}^{(r+1)}(\sigma_r) \subset \mathscr{R}^{r+1}$ ($\sigma_{r+1} = 1,2,\cdots,N_{r+1}$ and $N_{r+1} \geq n-(r+1)$) with a common edge of $\mathscr{E}_{\sigma_r}^{(r)}$, the common edge is called the r*-dimensional* edge if

$$\mathscr{E}_{\sigma_r}^{(r)} = \cap_{\sigma_{r+1}=1}^{N_{r+1}} \mathscr{E}_{\sigma_{r+1}}^{(r+1)}(\sigma_r) \subset \mathscr{R}^{r} \tag{6.3}$$

for $r = 0,1,2,\cdots,n-2,n-1$. There are three special cases.
(i) The r-dimensional edge is called the *sub-domain* $\mathscr{D} \equiv \mathscr{E}^{(n)}$ for $r = n$.
(ii) The r-dimensional edge is called the *boundary* $\mathscr{B} \equiv \mathscr{E}^{(n-1)}$ for $r = n-1$.
(iii) The r-dimensional edge is called the *vertex* $\mathscr{V} \equiv \mathscr{E}^{(0)}$ for $r = 0$.

Consider five edges of $\mathscr{E}_{\sigma_{n-2}}^{(n-2)}$ ($\sigma_{n-2} = 1,2,\cdots,5$) to have a common edge $\mathscr{E}^{(n-3)}$, as sketched in Fig. 6.2 through dark curves. The filled circular symbol depicts the common edge $\mathscr{E}^{(n-3)}$, which is an intersection of five edges $\mathscr{E}_{\sigma_{n-2}}^{(n-2)}$. Since a universal domain in $\mathscr{R}^n$ is partitioned into many sub-domains by separation boundaries. On accessible domains, dynamical systems can be defined, and dynamical systems on the separation boundaries can be defined as well. The intersections of boundaries will form edges and vertexes. On edges, the corresponding dynamical systems can also be defined. At vertexes, dynamical systems are independent of state variables. Thus the vertexes can be either stationary points or points varying with time.

To describe dynamical systems, the corresponding flows in domains, boundary and edges are called the *domain*, *boundary* and *edge* flows, respectively. The flows $\mathbf{x}^{(\alpha;\mathscr{D})}$ in domain Ω_α, $\mathbf{x}^{(i_\alpha;\mathscr{B})}$ on boundary $\partial\Omega_{\alpha i_\alpha}$ and $\mathbf{x}^{(\sigma_r;\mathscr{E})}$ on edge $\mathscr{E}_{\sigma_r}^{(r)}$. The stationary vertexes are denoted by $\mathbf{x}^{(\sigma_0;\mathscr{V})} \equiv \mathbf{x}^{(\sigma_0;\mathscr{E})}$ in phase space. Such flows in domains, boundaries and edges are sketched in Figs. 6.3 and 6.4. In Fig. 6.3,

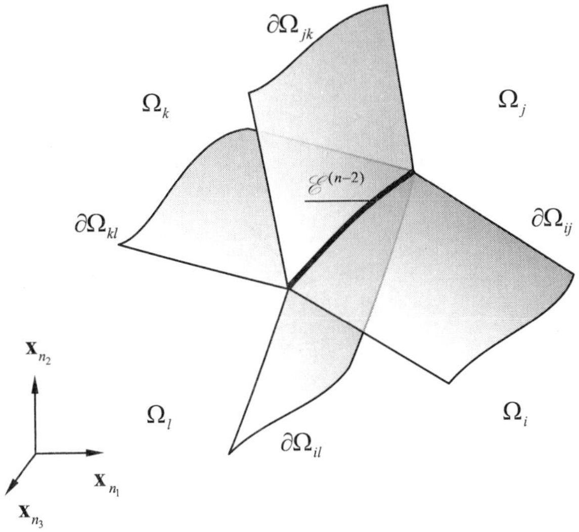

Fig. 6.1 An $(n-2)$-dimensional edge of four boundaries with four domains. ($n_1 + n_2 + n_3 = n$)

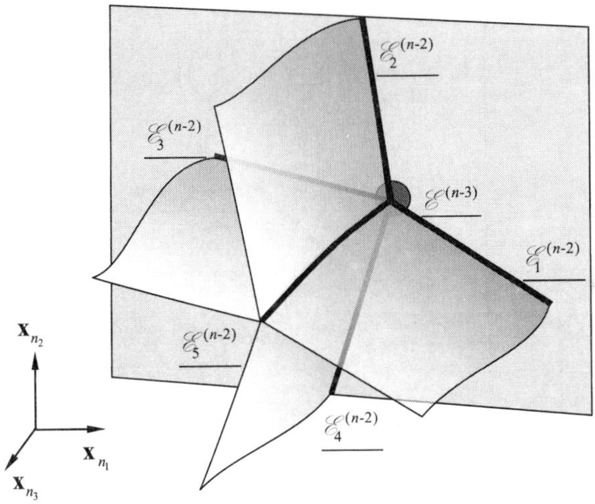

Fig. 6.2 An $(n-3)$-dimensional edge of five $(n-2)$-dimensional edges with five boundaries and five domains. The point is an $(n-3)$-dimensional edge. On the vertical wall, the $(n-3)$-dimensional edge is specified. ($n_1 + n_2 + n_3 = n$)

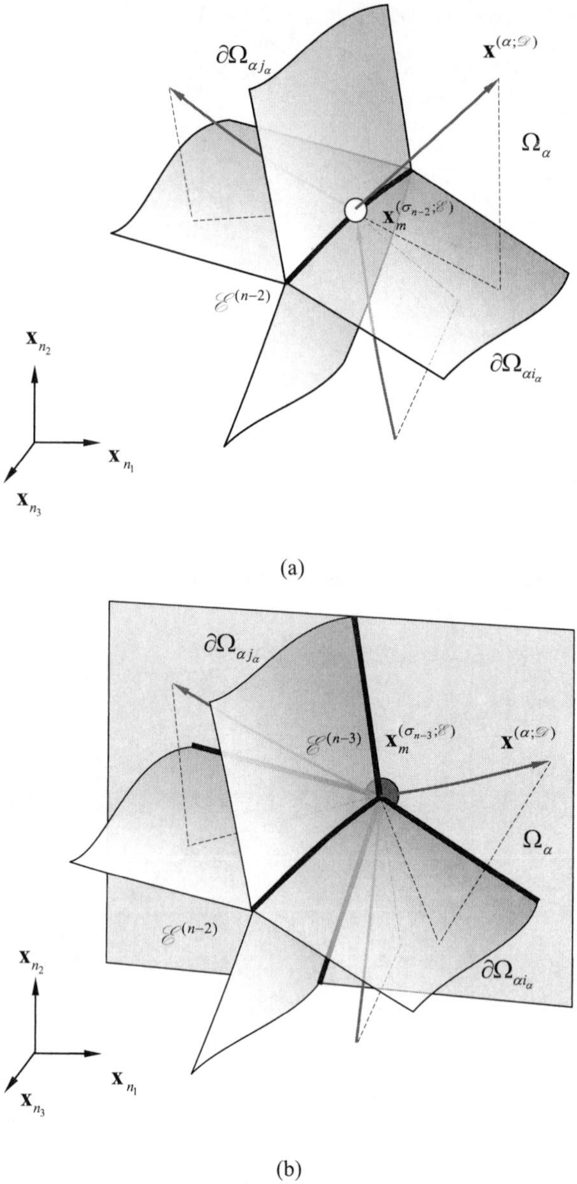

Fig. 6.3 The domain flows $\mathbf{x}^{(\alpha;\mathscr{D})}$ in accessible domains to edges: **(a)** $\mathscr{E}^{(n-2)}$ and **(b)** $\mathscr{E}^{(n-3)}$. The solid curves with arrows represent domain flows in accessible domains. On the vertical wall, the $(n-3)$-dimensional edge is specified. ($n_1 + n_2 + n_3 = n$) (color plot in the book end)

domain flows $\mathbf{x}^{(\alpha;\mathscr{D})}$ to the edges of $\mathscr{E}^{(n-2)}$ and $\mathscr{E}^{(n-3)}$ are presented by curves with arrows. In Fig. 6.4, boundary flows $\mathbf{x}^{(i_\alpha;\mathscr{B})}$ on the boundary of $\partial\Omega_{\alpha i_\alpha}$ to the edge of $\mathscr{E}^{(n-2)}$, and edge flows $\mathbf{x}^{(\sigma_{n-2};\mathscr{E})}$ on the edge of $\mathscr{E}^{(n-2)}_{\sigma_{n-2}}$ to the edge of $\mathscr{E}^{(n-3)}$ are sketched accordingly.

Definition 6.3. In a discontinuous dynamical system in $\mathscr{R}^n$, phase space partitioned by boundaries is composed of

(C$_1$) N_d-sub-domains of $\Omega_\alpha \subset \mathscr{R}^n$ ($\alpha \in \mathscr{N}_{\mathscr{D}} = \{1,2,\cdots,N_d\}$),

(C$_2$) N_b-boundaries of $S_{i_\alpha} = \partial\Omega_{\alpha i_\alpha} \subset \mathscr{R}^{n-1}$ ($i_\alpha \in \mathscr{N}_{\mathscr{B}} = \{1,2,\cdots,N_b\}$),

(C$_3$) N_r-edges of $\mathscr{E}^{(r)}_{\sigma_r} \subset \mathscr{R}^r$ ($\sigma_r \in \mathscr{N}^r_{\mathscr{E}} = \{1,2,\cdots,N_r\}$, $r \in \{1,2,\cdots,n-2\}$), and

(C$_4$) N_0-vertexes of $\mathscr{V}_{\sigma_0} \equiv \mathscr{E}^{(0)}_{\sigma_0} \subset \mathscr{R}^0$ ($\sigma_0 \in \mathscr{N}_{\mathscr{V}} = \{1,2,\cdots,N_0\}$).

(i) The *dynamical system on an accessible domain* relative to the edge of $\mathscr{E}^{(r)}_{\sigma_r}$ is defined as

$$\dot{\mathbf{x}}^{(\alpha;\mathscr{D})} = \mathbf{F}^{(\alpha)}(\mathbf{x}^{(\alpha;\mathscr{D})}, t, \mathbf{p}_\alpha) \text{ on } \Omega_\alpha \; (\alpha \in \mathscr{N}_{\mathscr{D}}). \tag{6.4}$$

(ii) The *dynamical system on a boundary* relative to the edge of $\mathscr{E}^{(r)}_{\sigma_r}$ is defined as

$$\begin{aligned} &\dot{\mathbf{x}}^{(i_\alpha;\mathscr{B})} = \mathbf{F}^{(i_\alpha)}(\mathbf{x}^{(i_\alpha;\mathscr{B})}, t, \boldsymbol{\lambda}_{i_\alpha}) \\ &\text{with } \varphi_{\alpha i_\alpha}(\mathbf{x}^{(i_\alpha;\mathscr{B})}, t, \boldsymbol{\lambda}_{i_\alpha}) = 0 \text{ on } \partial\Omega_{\alpha i_\alpha} \\ &(\alpha \in \mathscr{N}_{\mathscr{D}}, i_\alpha \in \mathscr{N}_{\mathscr{B}}). \end{aligned} \tag{6.5}$$

(iii) The *dynamical system on an edge of* $\mathscr{E}^{(s)}_{\sigma_s}(\sigma_r)$ relative to the edge of $\mathscr{E}^{(r)}_{\sigma_r}$ ($s > r$) is defined as

$$\begin{aligned} &\dot{\mathbf{x}}^{(\sigma_s;\mathscr{E})} = \mathbf{F}^{(\sigma_s)}(\mathbf{x}^{(\sigma_s;\mathscr{E})}, t, \boldsymbol{\pi}_{\sigma_s}) \\ &\text{with } \varphi_{\alpha i_\alpha}(\mathbf{x}^{(\sigma_s;\mathscr{E})}, t, \boldsymbol{\lambda}_{i_\alpha}) = 0 \text{ on } \mathscr{E}^{(s)}_{\sigma_s}(\sigma_r) \\ &(\alpha \in \mathscr{N}_{\mathscr{D}}(\sigma_s) \subset \mathscr{N}_{\mathscr{D}}, i_\alpha \in \mathscr{N}_{\alpha\mathscr{B}} \subset \mathscr{N}_{\mathscr{B}}; \\ &\sigma_s \in \mathscr{N}^s_{\mathscr{E}}; s \in \{r+1, r+2,\cdots,n-2\}). \end{aligned} \tag{6.6}$$

(iv) The *dynamical system on an edge of* $\mathscr{E}^{(r)}_{\sigma_r}$ is defined as

$$\begin{aligned} &\dot{\mathbf{x}}^{(\sigma_r;\mathscr{E})} = \mathbf{F}^{(\sigma_r)}(\mathbf{x}^{(\sigma_r;\mathscr{E})}, t, \boldsymbol{\pi}_{\sigma_r}) \\ &\text{with } \varphi_{\alpha i_\alpha}(\mathbf{x}^{(\sigma_r;\mathscr{E})}, t, \boldsymbol{\lambda}_{i_\alpha}) = 0 \text{ on } \mathscr{E}^{(r)}_{\sigma_r} \\ &(\alpha \in \mathscr{N}_{\mathscr{D}}(\sigma_r) \subset \mathscr{N}_{\mathscr{D}}, i_\alpha \in \mathscr{N}_{\alpha\mathscr{B}} \subset \mathscr{N}_{\mathscr{B}}; \\ &\sigma_r \in \mathscr{N}^r_{\mathscr{E}}; r \in \{1,2,\cdots,n-2\}). \end{aligned} \tag{6.7}$$

(v) The *dynamical system on a vertex of* $\mathscr{V}_{\sigma_0} \equiv \mathscr{E}^{(0)}_{\sigma_0}$ is defined as

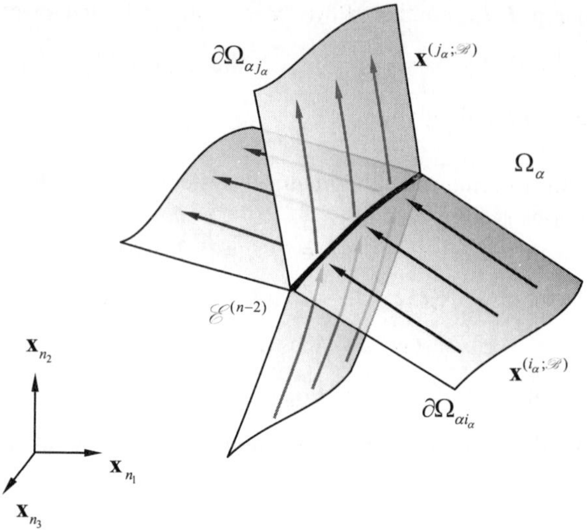

(a)

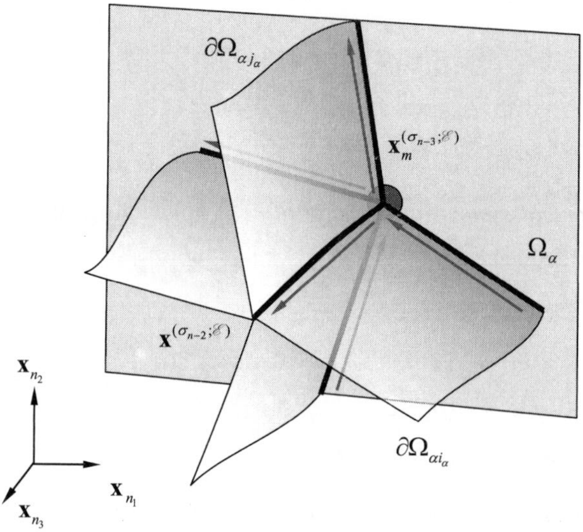

(b)

Fig. 6.4 (a) The boundary flows $\mathbf{x}^{(i_\alpha;\mathscr{B})}$ on the boundaries of $\partial\Omega_{\alpha i_\alpha}$ to the edges of $\mathscr{E}^{(n-2)}$, and **(b)** the edge flows $\mathbf{x}^{(\sigma_{n-2};\mathscr{E})}$ on the edges of $\mathscr{E}^{(n-2)}$ to an edge of $\mathscr{E}^{(n-3)}$. The curves with arrows represent the boundary and edge flows. On the vertical wall, the $(n-3)$-dimensional edge is specified. ($n_1+n_2+n_3=n$) (color plot in the book end)

$$
\begin{aligned}
&\dot{\mathbf{x}}^{(\sigma_0;\mathscr{V})} = \mathbf{F}^{(\sigma_0)}(t,\boldsymbol{\pi}_{\sigma_0}) \\
&\text{with } \varphi_{\alpha i_\alpha}(\mathbf{x}^{(\sigma_0;\mathscr{V})},t,\boldsymbol{\lambda}_{i_\alpha}) = 0 \text{ on } \mathscr{V}_{\sigma_0} \\
&(\alpha \in \mathscr{N}_{\mathscr{D}}(\sigma_0) \subset \mathscr{N}_{\mathscr{D}}, i_\alpha \in \mathscr{N}_{\alpha\mathscr{B}} \subset \mathscr{N}_{\mathscr{B}}, \sigma_0 \in \mathscr{N}_{\mathscr{V}}).
\end{aligned}
\tag{6.8}
$$

Notice that $\varphi_{\alpha i_\alpha}(\mathbf{x}^{(i_\alpha;\mathscr{B})},t,\boldsymbol{\lambda}_{i_\alpha}) \equiv \varphi_{\alpha i_\alpha}(\mathbf{x}^{(\alpha i_\alpha)},t,\boldsymbol{\lambda}_{\alpha i_\alpha})$ for simplicity. Without labeling domain and boundary, the corresponding dynamical systems on domains and boundaries are expressed as before. The dynamical systems of edge flows are

$$
\begin{aligned}
&\dot{\mathbf{x}}^{(\sigma_r)} = \mathbf{F}^{(\sigma_r)}(\mathbf{x}^{(\sigma_r)},t,\boldsymbol{\pi}_{\sigma_r}) \\
&\text{with } \varphi_{\alpha i_\alpha}(\mathbf{x}^{(\sigma_r)},t,\boldsymbol{\lambda}_{i_\alpha}) = 0 \text{ on } \mathscr{E}_{\sigma_r}^{(r)} \\
&(\sigma_r = 1,2,\cdots,N_r; r = 1,2,\cdots,n-2;\ \alpha, i_\alpha \in \{1,2,\cdots,N\}, i_\alpha \neq \alpha).
\end{aligned}
\tag{6.9}
$$

6.2. Edge classification and mirror domains

The edge classification will be discussed through the angle between two boundaries, and the concept for imaginary boundaries and edges will be introduced through the extension of boundaries and edges.

Definition 6.4. For a discontinuous dynamical system in Eqs. (6.4)-(6.8), there is a point $\mathbf{x}^{(\sigma_r;\mathscr{E})}(t_m) \equiv \mathbf{x}_m \in \mathscr{E}_{\sigma_r}^{(r)}$ ($\sigma_r \in \mathscr{N}_{\mathscr{E}}^{r} = \{1,2,\cdots,N_r\}$ and $r \in \{0,1,2,\cdots,n-2\}$) at time t_m with the related boundaries $\partial\Omega_{\alpha i_\alpha}(\sigma_r)$ and related domains $\Omega_\alpha(\sigma_r)$ ($\rho \in \mathscr{N}_{\mathscr{D}}(\sigma_r) \subset \mathscr{N}_{\mathscr{D}} = \{1,2,\cdots,N_d\}$ and $i_\alpha \in \mathscr{N}_{\alpha\mathscr{B}}(\sigma_r) \subset \mathscr{N}_{\mathscr{B}} = \{1,2,\cdots,N_b\}$), where the boundary index subset of $\mathscr{N}_{\alpha\mathscr{B}}(\sigma_r)$ has n_{σ_r}-elements ($n_{\sigma_r} \geq (n-r)$). Suppose there are two adjacent boundaries $\partial\Omega_{\alpha i_\alpha}$ and $\partial\Omega_{\alpha j_\alpha}$ to form an $(n-2)$-dimensional edge for the specific edge of $\mathscr{E}_{\sigma_r}^{(r)}$. An *angle* between the two boundaries at the point $\mathbf{x}_m \in \mathscr{E}_{\sigma_r}^{(r)}$ in domain Ω_α is determined by

$$
\cos\theta_{i_\alpha j_\alpha} = -\frac{\mathbf{n}_{\partial\Omega_\alpha i_\alpha}^{\mathrm{T}} \cdot \mathbf{n}_{\partial\Omega_\alpha j_\alpha}}{|\mathbf{n}_{\partial\Omega_\alpha i_\alpha}| \times |\mathbf{n}_{\partial\Omega_\alpha j_\alpha}|}.
\tag{6.10}
$$

The total set of the $(n-2)$-dimensional edges for the r-dimensional edge of domain Ω_α is defined as

$$
\Theta_\alpha = \Theta_\alpha^{\mathscr{V}} \cup \Theta_\alpha^{\mathscr{C}} \text{ and } \Theta_\alpha^{\mathscr{V}} \cap \Theta_\alpha^{\mathscr{C}} = \varnothing,
\tag{6.11}
$$

where the convex and concave sets of the $(n-2)$-dimensional edge are defined as

$$\left.\begin{aligned}\Theta_{\alpha}^{\mathscr{V}} &= \{\theta_{i_\alpha j_\alpha} \,|\, \theta_{i_\alpha j_\alpha} \in (0,\pi) \quad \text{for } i_\alpha, j_\alpha \in \mathscr{N}_{\alpha\mathscr{B}}(\sigma_r)\},\\ \Theta_{\alpha}^{\mathscr{C}} &= \{\theta_{i_\alpha j_\alpha} \,|\, \theta_{i_\alpha j_\alpha} \in (\pi,2\pi) \ \text{for } i_\alpha, j_\alpha \in \mathscr{N}_{\alpha\mathscr{B}}(\sigma_r)\}.\end{aligned}\right\} \tag{6.12}$$

(i) The $(n-2)$-dimensional edge between two boundaries $\partial\Omega_{\alpha i_\alpha}$ and $\partial\Omega_{\alpha j_\alpha}$ is called a *folded edge* at point $\mathbf{x}_m \in \mathscr{E}_{\sigma_{n-2}}^{(n-2)}$ if

$$\theta_{i_\alpha j_\alpha} = 0 \quad \text{for } i_\alpha, j_\alpha \in \mathscr{N}_{\alpha\mathscr{B}}(\sigma_r). \tag{6.13}$$

(ii) The $(n-2)$-dimensional edge between two boundaries $\partial\Omega_{\alpha i_\alpha}$ and $\partial\Omega_{\alpha j_\alpha}$ is a *flattening edge* at point $\mathbf{x}_m \in \mathscr{E}_{\sigma_{n-2}}^{(n-2)}$ if

$$\theta_{i_\alpha j_\alpha} = \pi \quad \text{for } i_\alpha, j_\alpha \in \mathscr{N}_{\alpha\mathscr{B}}(\sigma_r). \tag{6.14}$$

(iii) The $(n-2)$-dimensional edge between two boundaries $\partial\Omega_{\alpha i_\alpha}$ and $\partial\Omega_{\alpha j_\alpha}$ is called a *convex edge* at point $\mathbf{x}_m \in \mathscr{E}_{\sigma_{n-2}}^{(n-2)}$ if

$$\theta_{i_\alpha j_\alpha} \in (0,\pi) \quad \text{for } i_\alpha, j_\alpha \in \mathscr{N}_{\alpha\mathscr{B}}(\sigma_r). \tag{6.15}$$

(iv) The $(n-2)$-dimensional edge between two boundaries $\partial\Omega_{\alpha i_\alpha}$ and $\partial\Omega_{\alpha j_\alpha}$ is called a *concave edge* at point $\mathbf{x}_m \in \mathscr{E}_{\sigma_{n-2}}^{(n-2)}$ if

$$\theta_{i_\alpha j_\alpha} \in (\pi,2\pi) \quad \text{for } i_\alpha, j_\alpha \in \mathscr{N}_{\alpha\mathscr{B}}(\sigma_r). \tag{6.16}$$

(v) The r-dimensional edge of domain Ω_α is called a *convex edge* if

$$\theta_{i_\alpha j_\alpha} \in \Theta_{\alpha}^{\mathscr{V}} \quad \text{for all } i_\alpha, j_\alpha \in \mathscr{N}_{\alpha\mathscr{B}}(\sigma_r). \tag{6.17}$$

(vi) The r-dimensional edge of domain Ω_α is called a *concave edge* if

$$\theta_{i_\alpha j_\alpha} \in \Theta_{\alpha}^{\mathscr{C}} \quad \text{for all } i_\alpha, j_\alpha \in \mathscr{N}_{\alpha\mathscr{B}}(\sigma_r). \tag{6.18}$$

(vii) The r-dimensional edge of domain Ω_α is called a *convex-concave edge* if

$$\begin{aligned}&\theta_{i_\alpha j_\alpha} \in \Theta_{\alpha}^{\mathscr{V}} \cup \Theta_{\alpha}^{\mathscr{C}},\ \ \Theta_{\alpha}^{\mathscr{V}} \neq \varnothing \text{ and } \Theta_{\alpha}^{\mathscr{C}} \neq \varnothing\\ &\text{for all } i_\alpha, j_\alpha \in \mathscr{N}_{\alpha\mathscr{B}}(\sigma_r).\end{aligned} \tag{6.19}$$

(viii) A domain is *convex* at the r-dimensional edge if the edge is convex.
(ix) A domain is *concave* at the r-dimensional edge if the edge is concave.
(x) A domain is *convex-concave* at the r-dimensional edge if the edge is convex-concave.

To explain the foregoing definition, an $(n-2)$-dimensional edge of two $(n-1)$-dimensional boundaries and an $(n-3)$-dimensional edge are sketched in Fig. 6.5(a) and (b), respectively. The edge angle $\theta_{i_\alpha j_\alpha}$ of two $(n-1)$-dimensional boundaries $\partial\Omega_{\alpha i_\alpha}$ and $\partial\Omega_{\alpha j_\alpha}$ at $\mathbf{x}_m \in \mathscr{E}^{(n-2)}$ is defined through the normal vectors of the two boundaries. If $\theta_{i_\alpha j_\alpha}=0$ with a common edge, then the two boundaries will be foldable at the edge. If $\theta_{i_\alpha j_\alpha}=0$ without a common edge, then the two boundaries should be paralleled inversely. If $\theta_{i_\alpha j_\alpha}=\pi$ with a common edge, then the two boundaries should be edgeless at $\mathbf{x}_m \in \mathscr{E}^{(n-2)}$. If $\theta_{i_\alpha j_\alpha}=\pi$ without a common edge, then the two boundaries should be paralleled. If $\theta_{i_\alpha j_\alpha}\in(0,\pi)$, the edge to the domain is a convex edge, as shown in Fig. 6.5(a). However, if the edge angle lies in $\theta_{i_\alpha j_\alpha}\in(\pi,2\pi)$, the edge to the domain is a concave edge. In other words, the edge in Fig. 6.5(a) is a concave edge for the complimentary domains of Ω_α. In Fig. 6.5(b), an $(n-3)$-dimensional convex edge of domain Ω_α is an intersection of three $(n-2)$-dimensional convex edges. The $(n-3)$-dimensional edge is a concave edge to the complementary domain of domain Ω_α. The three angles of the $(n-2)$-dimensional convex edges are depicted. Two convex edges with one concave edge form an $(n-3)$-dimensional convex and concave edge, as shown in Fig. 6.6. Three $(n-1)$-dimensional boundaries are shaded. The two edge angles are $\theta_{k_\alpha j_\alpha},\theta_{k_\alpha i_\alpha}\in(0,\pi)$ and the edge angle is $\theta_{i_\alpha j_\alpha}\in(\pi,2\pi)$. However, for the complimentary domain, the three edges are two concave edges and one convex edge.

Definition 6.5. For a discontinuous dynamical system in Eqs. (6.4)-(6.8), there is a point $\mathbf{x}^{(\sigma_r;\mathscr{E})}(t_m)\equiv\mathbf{x}_m\in\mathscr{E}_{\sigma_r}^{(r)}$ ($\sigma_r\in\mathscr{N}_{\mathscr{E}}^{r}=\{1,2,\cdots,N_r\}$ and $r\in\{0,1,2,\cdots,n-2\}$) at time t_m with the related boundaries $\partial\Omega_{\alpha i_\alpha}(\sigma_r)$ and related domains $\Omega_\alpha(\sigma_r)$ ($\alpha\in\mathscr{N}_{\mathscr{D}}(\sigma_r)\subset\mathscr{N}_{\mathscr{D}}=\{1,2,\cdots,N_d\}$ and $i_\alpha\in\mathscr{N}_{\alpha\mathscr{B}}(\sigma_r)\subset\mathscr{N}_{\mathscr{B}}=\{1,2,\cdots,N_b\}$), where the boundary index subset of $\mathscr{N}_{\alpha\mathscr{B}}(\sigma_r)$ has n_{σ_r}-elements ($n_{\sigma_r}\geq(n-r)$).

(i) The extension of a finite boundary $\partial\Omega_{\alpha i_\alpha}$ of an convex edge σ_r in domain Ω_α is called a *fictitious boundary* $\partial\bar{\Omega}_{\alpha i_\alpha}$.

(ii) The extension of an edge $\mathscr{E}_{\sigma_s}^{(s)}$ ($s=r+1,r+2,\cdots,n-2$) of a convex edge σ_r in domain Ω_α is called a *fictitious edge* $\mathscr{E}_{\bar{\sigma}_s}^{(s)}$.

(iii) The extension of a domain at the convex edge σ_r in the corresponding concave domain Ω_β is called a *full mirror imaginary domain* of domain Ω_α at the convex edge σ_r (i.e., $\Omega_\alpha^{\mathscr{FM}}$).

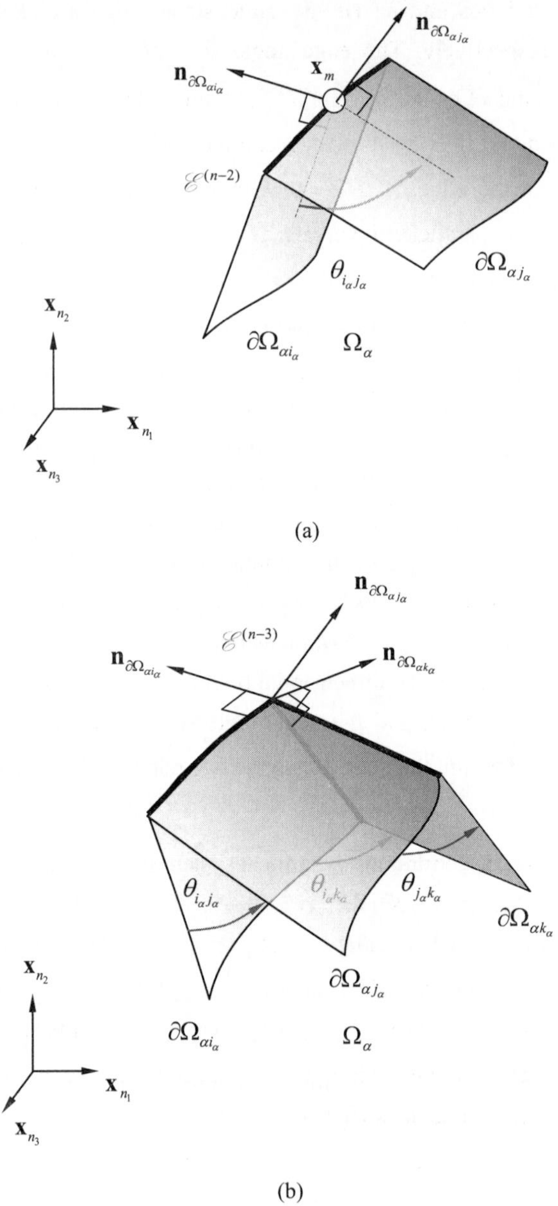

Fig. 6.5 Edge angles and convex edges in domain Ω_α : **(a)** an $(n-2)$ -dimensional edge, and **(b)** an $(n-3)$ -dimensional edge. The angle $\theta_{i_\alpha j_\alpha}$ of the two boundaries $\partial\Omega_{\alpha i_\alpha}$ and $\partial\Omega_{\alpha j_\alpha}$ is depicted. The $(n-1)$ -dimensional boundaries are shaded. The $(n-2)$ -dimensional edges are represented by thick curves. The $(n-3)$ -edge is shown by a corner. (color plot in the book end)

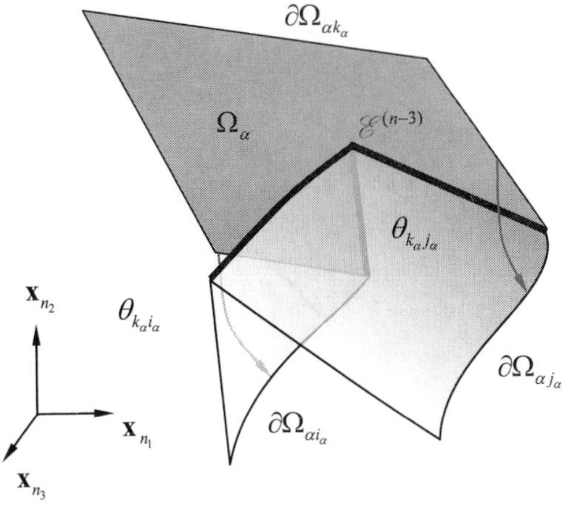

Fig. 6.6 A convex and concave edge. The angle $\theta_{i_\alpha j_\alpha}$ of two boundaries $\partial\Omega_{\alpha i_\alpha}$ and $\partial\Omega_{\alpha j_\alpha}$ is depicted. The $(n-1)$ -dimensional boundaries are shaded. The $(n-2)$ -dimensional edges are represented by thick curves. The $(n-3)$ -edge is formed by two convex edges and one concave edge.

(iv) The complement domain of the full mirror imaginary domain of domain Ω_α in the concave domain Ω_β is called the *self-mirror imaginary domain* at the convex edge σ_r (i.e., $\Omega_\beta^{\mathscr{SM}}$). $\Omega_\beta = \Omega_\beta^{\mathscr{SM}} \cup \Omega_\alpha^{\mathscr{FM}}$.

To explain the above definitions, consider extensions of boundaries for an $(n-2)$ -dimensional convex edge and an $(n-3)$ -dimensional convex edge in Fig. 6.7(a) and (b), respectively. The corresponding mirror imaginary domains are presented. In Fig. 6.7(a), the boundaries of $\partial\Omega_{\alpha i_\alpha}$ and $\partial\Omega_{\alpha j_\alpha}$ of the convex domain Ω_α are extended into the concave domain Ω_β. The extended boundaries given through the dashed curves are fictitious boundaries, denoted by $\partial\bar{\Omega}_{\alpha i_\alpha}$ and $\partial\bar{\Omega}_{\alpha j_\alpha}$. The two boundaries in the concave domain Ω_β forms a full mirror domain of domain Ω_α, denoted by $\Omega_\alpha^{\mathscr{FM}}$. The concave domain Ω_β is divided into three portions. The rest two portions are self–mirror domains. A flow in the minor domain arrives to the edge and must be switched. However, a flow in a self-mirror domain in Ω_β can pass over to the corresponding self-mirror domain without any switching laws. In Fig. 6.7(b), a mirror imaginary domain for an $(n-3)$ -dimensional convex edge is presented through dashed curves. The $(n-3)$ -dimensional convex edge of the three convex $(n-2)$ -dimensional edges is depicted by a corner point. The mirror imaginary domain lies in the corresponding concave domain. A flow in the mirror imaginary domain must be switched at the $(n-3)$ -dimensional convex

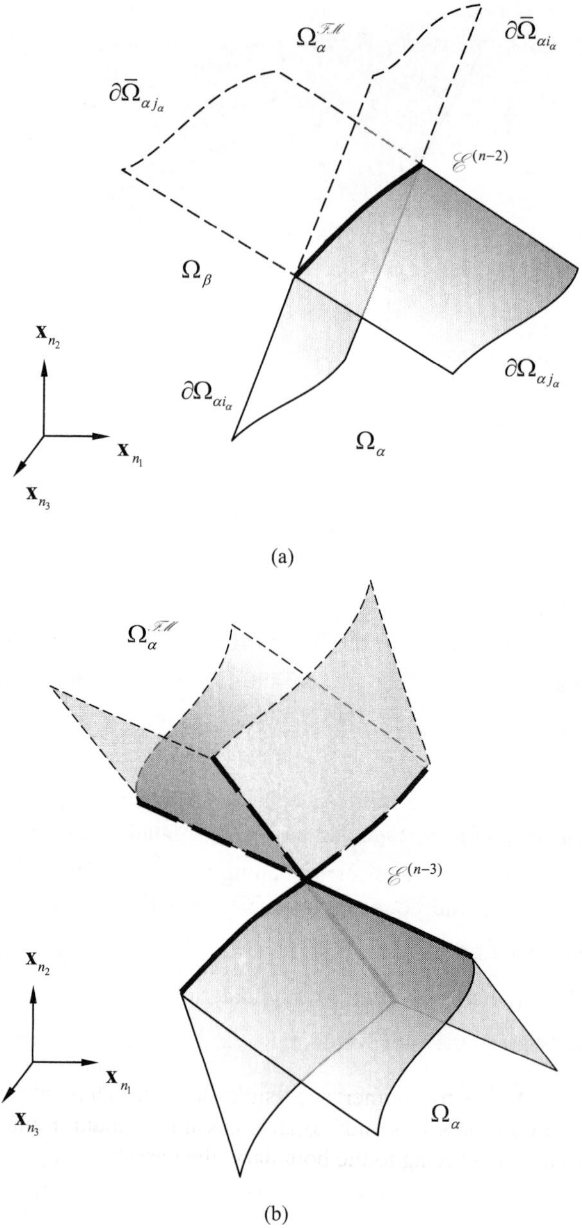

Fig. 6.7 Imaginary extensions of the boundaries and edges: **(a)** $(n-2)$ -dimensional convex edge, and **(b)** $(n-3)$ -dimensional convex edge. The boundaries are shaded, and the $(n-2)$ -dimensional edges are represented by thick curves. The $(n-3)$ -dimensional edge is presented by a corner point. Dashed curves are for minor imaginary domains.

edge. However, a flow in the rest of the concave domain can pass over each other through the $(n-3)$-dimensional convex edge without any switching laws.

In addition to the convex edge, there is a mixing of the concave-convex edge, the corresponding mirror domains are presented in Fig. 6.8. In Fig. 6.8(a), two domains Ω_α and Ω_β with three $(n-2)$-dimensional edges and an $(n-3)$ dimensional edge are sketched. The $(n-3)$-dimensional edge is formed by two convex edges and one concave edge for domain Ω_α. The mirrored domain of the convex sub-domain of domain Ω_α is in domain Ω_α. Suppose a domain is partitioned into sub-domains by the extensions of boundaries, (i.e., $\Omega_\alpha = \cup_{\alpha_1} \Omega_{\alpha_1}$). In Fig. 6.8(a), suppose a domain Ω_{α_1} is a unique convex domain, which has its mirror domain in domain Ω_α (i.e., $\Omega_{\alpha_1}^{\mathscr{F}\mathscr{M}}$). The mirror domain of the rest portion of domain Ω_α is the domain of Ω_β. In other words, the mirror domain of Ω_β is in domain Ω_α. In Fig. 6.8(b), the complicated mirror imaginary domains are presented through the $(n-3)$-dimensional convex-concave edge. The mirror domains scatters in two domains of Ω_α and Ω_β. If the mirror sub-domain with its original sub-domains are in different domains, a flow in the mirror or original domain should be switched at the edge. If a mirror sub-domain with its original sub-domain is in the same domain, a flow in the mirror or original domain can pass over at the edge.

6.3. Domain flow properties to convex edges

In the previous section, edges in discontinuous dynamical systems can be classified into convex, concave, and convex-concave mixing edges. The coming, leaving and tangential domain flows to convex edges will be discussed first. Further, the passability of domain flows to concave edges will be discussed later. Because many sub-domains are around edges, once a domain flow in an accessible sub-domain arrives to an edge, such a domain flow will be more chance to switch from one accessible domain to another accessible domain. Thus, a switching rule is needed for a coming flow to switch to an accessible domain. From the passability conditions for a flow arriving to the boundary, the passability conditions for a flow hitting a specific edge can be developed. Before the passability of a domain flow to an edge is discussed, coming, leaving and tangential flows to an edge will be introduced. The coming and leaving flows of $\mathbf{x}^{(\alpha;\mathscr{D})}$ to the edges $\mathscr{E}^{(n-2)}$ and $\mathscr{E}^{(n-3)}$ are sketched in Figs. 6.9 and 6.10, respectively. Curves with arrows represent coming and leaving domain flows. In addition, a tangential, domain flow to the edge of $\mathscr{E}^{(n-2)}$ is presented in Fig. 6.11. This tangential flow is also called the grazing domain flow to such an edge. The existence of the grazing domain flow requires the domain flows graze all the boundaries associated with the edge.

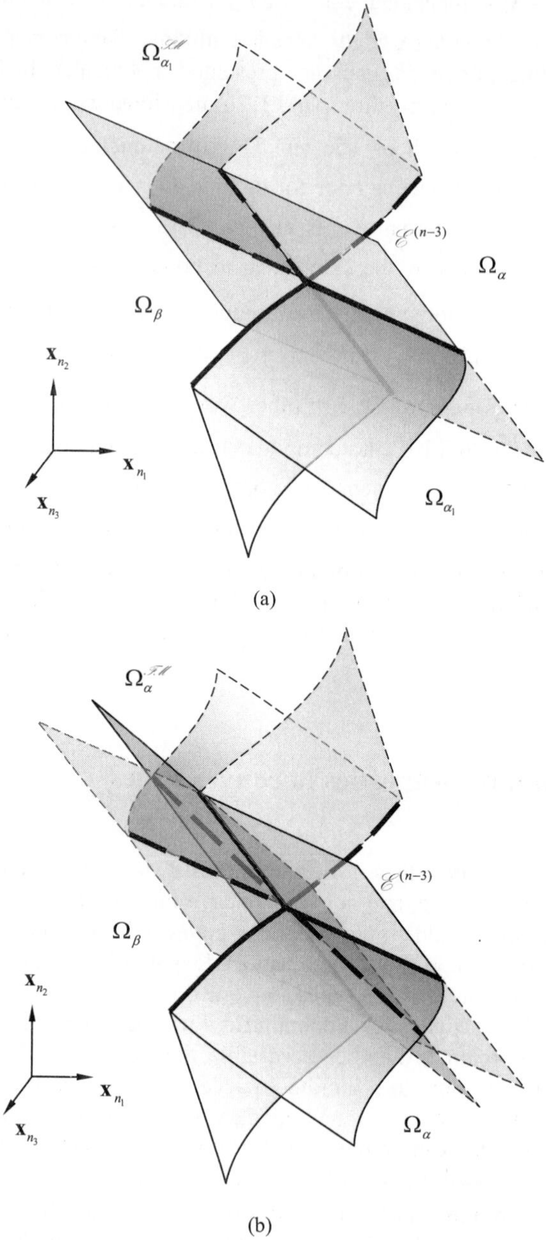

Fig. 6.8 An $(n-3)$-dimensional convex-concave edge: **(a)** self-mirror imaginary domain in its own domain, and **(b)** scaring mirror imaginary domains. The boundaries are shaded, and the $(n-2)$-dimensional edges are depicted by thick curves. The $(n-3)$-dimensional edge is presented by a corner point. Dashed curves are for minor imaginary domains. (color plot in the book end)

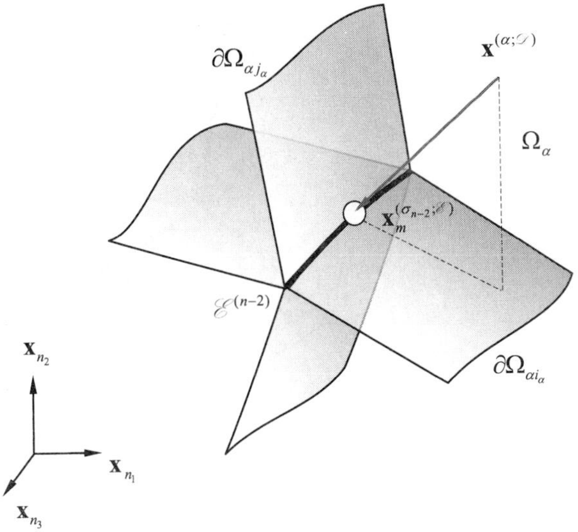

(a)

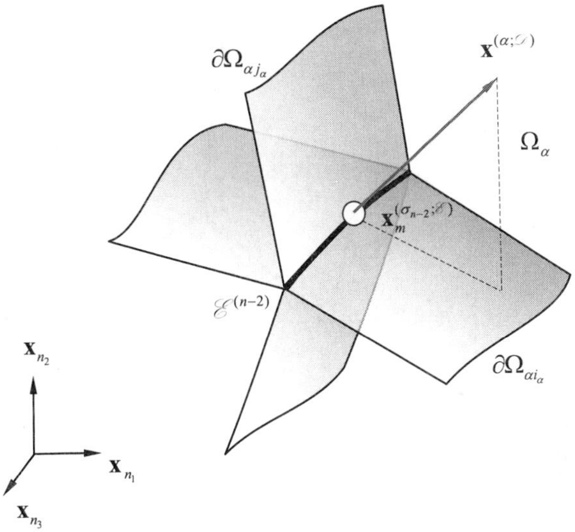

(b)

Fig. 6.9 (a) An incoming domain flow and **(b)** a leaving domain flow of $\mathbf{x}^{(\alpha;\mathscr{D})}$ in accessible domains to an edge of $\mathscr{E}^{(n-2)}$. The red curves with arrows represent incoming and leaving domain flows in the accessible domain. ($n_1 + n_2 + n_3 = n$)

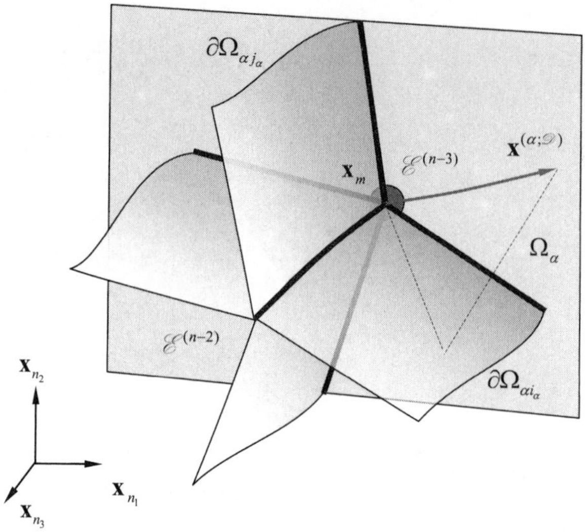

(a)

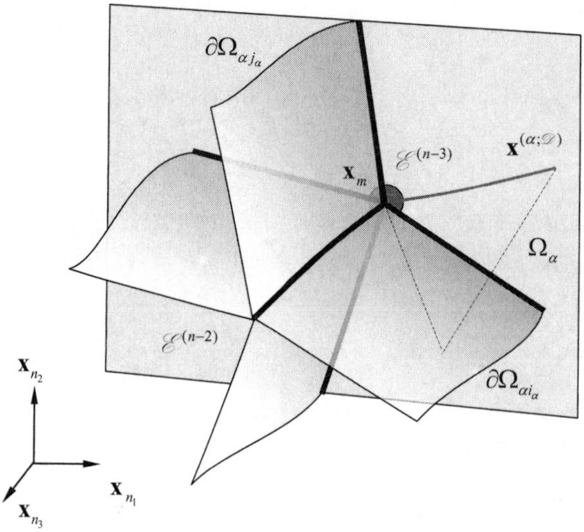

(b)

Fig. 6.10 (a) A coming domain flow and **(b)** a leaving domain flow of $\mathbf{x}^{(\alpha;\mathscr{D})}$ in an accessible domain to an edge of $\mathscr{E}^{(n-3)}$. The solid curves with arrows represent domain flows in the accessible domain. On the vertical wall, the $(n-3)$-dimensional edge is specified. ($n_1+n_2+n_3=n$)

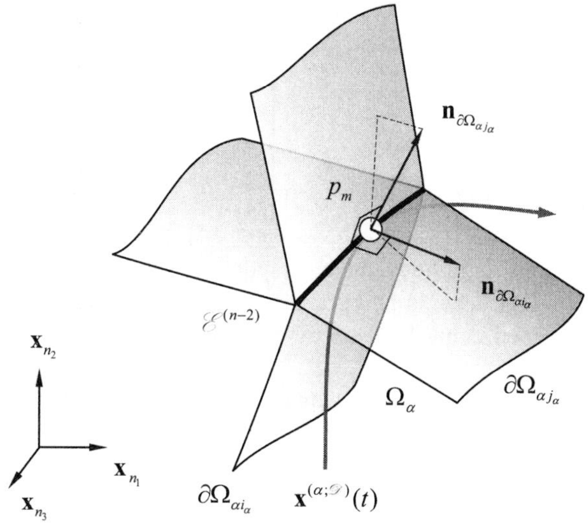

Fig. 6.11 A domain of $\mathbf{x}^{(\alpha;\mathscr{D})}$ in accessible domains grazing an edge of $\mathscr{E}^{(n-2)}$. The gray curve with arrow represents the grazing domain flow in an accessible domain. ($n_1+n_2+n_3=n$)

Definition 6.6. For a discontinuous dynamical system in Eqs. (6.4)-(6.8), there is a point $\mathbf{x}^{(\sigma_r;\mathscr{E})}(t_m)\equiv\mathbf{x}_m\in\mathscr{E}_{\sigma_r}^{(r)}$ ($\sigma_r\in\mathscr{N}_{\mathscr{E}}^{r}=\{1,2,\cdots,N_r\}$ and $r\in\{0,1,2,\cdots,n-2\}$) at time t_m with the related boundaries $\partial\Omega_{\alpha i_\alpha}(\sigma_r)$ and related domains $\Omega_\alpha(\sigma_r)$ ($\alpha\in\mathscr{N}_{\mathscr{D}}(\sigma_r)\subset\mathscr{N}_{\mathscr{D}}=\{1,2,\cdots,N_d\}$ and $i_\alpha\in\mathscr{N}_{\alpha\mathscr{B}}(\sigma_r)\subset\mathscr{N}_{\mathscr{B}}=\{1,2,\cdots,N_b\}$), where the boundary index subset of $\mathscr{N}_{\alpha\mathscr{B}}(\sigma_r)$ has n_{σ_r}-elements ($n_{\sigma_r}\geq(n-r)$). For an arbitrarily small $\varepsilon>0$, there are two time intervals $[t_{m-\varepsilon},t_m)$ and $(t_m,t_{m+\varepsilon}]$. A flow $\mathbf{x}^{(\alpha;\mathscr{D})}(t)$ is $C_{[t_{m-\varepsilon},t_m)}^{r_\alpha}$ and/or $C_{(t_m,t_{m+\varepsilon}]}^{r_\alpha}$-continuous *(* $r_\alpha\geq 0$ or $r_\alpha\geq 1$*)* for time t and $\| d^{r_\alpha+1}\mathbf{x}^{(\alpha;\mathscr{D})}/dt^{r_\alpha+1}\|<\infty$.

(i) The domain flow $\mathbf{x}^{(\alpha;\mathscr{D})}(t)$ in domain Ω_α to the edge $\mathscr{E}_{\sigma_r}^{(r)}$ is a *coming flow* at point $\mathbf{x}_m\in\mathscr{E}_{\sigma_r}^{(r)}$ if

$$\hbar_{i_\alpha}\mathbf{n}_{\partial\Omega_{\alpha i_\alpha}}^{\mathrm{T}}(\mathbf{x}_{m-\varepsilon}^{(i_\alpha;\mathscr{B})})\cdot[\mathbf{x}_{m-\varepsilon}^{(i_\alpha;\mathscr{B})}-\mathbf{x}_{m-\varepsilon}^{(\alpha;\mathscr{D})}]<0\ \text{ for all } i_\alpha\in\mathscr{N}_{\alpha\mathscr{B}}(\sigma_r). \tag{6.20}$$

(ii) The domain flow $\mathbf{x}^{(\alpha;\mathscr{D})}(t)$ in domain Ω_α to the edge $\mathscr{E}_{\sigma_r}^{(r)}$ is a *leaving flow* at point $\mathbf{x}_m\in\mathscr{E}_{\sigma_r}^{(r)}$ if

$$\hbar_{i_\alpha}\mathbf{n}_{\partial\Omega_{\alpha i_\alpha}}^{\mathrm{T}}(\mathbf{x}_{m+\varepsilon}^{(i_\alpha;\mathscr{B})})\cdot[\mathbf{x}_{m+\varepsilon}^{(\alpha;\mathscr{D})}-\mathbf{x}_{m+\varepsilon}^{(i_\alpha;\mathscr{B})}]>0\ \text{ for all } i_\alpha\in\mathscr{N}_{\alpha\mathscr{B}}(\sigma_r). \tag{6.21}$$

(iii) The domain flow $\mathbf{x}^{(\alpha;\mathscr{D})}(t)$ in domain Ω_α to the edge $\mathscr{E}_{\sigma_r}^{(r)}$ is a *grazing flow* at point $\mathbf{x}_m \in \mathscr{E}_{\sigma_r}^{(r)}$ if

$$\left.\begin{array}{l} \hbar_{i_\alpha} G_{\partial\Omega_{\alpha i_\alpha}}^{(\alpha;\mathscr{D})}(\mathbf{x}_m, t_{m\pm}, \mathbf{p}_\alpha, \boldsymbol{\lambda}_{i_\alpha}) = 0; \\ \hbar_{i_\alpha} \mathbf{n}_{\partial\Omega_{\alpha i_\alpha}}^{\mathrm{T}} (\mathbf{x}_{m-\varepsilon}^{(i_\alpha;\mathscr{B})}) \cdot [\mathbf{x}_{m-\varepsilon}^{(i_\alpha;\mathscr{B})} - \mathbf{x}_{m-\varepsilon}^{(\alpha;\mathscr{D})}] < 0, \\ \hbar_{i_\alpha} \mathbf{n}_{\partial\Omega_{\alpha i_\alpha}}^{\mathrm{T}} (\mathbf{x}_{m+\varepsilon}^{(i_\alpha;\mathscr{B})}) \cdot [\mathbf{x}_{m+\varepsilon}^{(\alpha;\mathscr{D})} - \mathbf{x}_{m+\varepsilon}^{(i_\alpha;\mathscr{B})}] > 0 \end{array}\right\} \text{ for all } i_\alpha \in \mathscr{N}_{\alpha\mathscr{B}}(\sigma_r). \tag{6.22}$$

From the previous definitions, the corresponding sufficient and necessary conditions for the coming, leaving and grazing flows to the boundaries of a specific edge will be developed as in Chapter 3 for each boundary.

Theorem 6.1. *For a discontinuous dynamical system in Eqs.* (6.4)-(6.8)*, there is a point* $\mathbf{x}^{(\sigma_r;\mathscr{E})}(t_m) \equiv \mathbf{x}_m \in \mathscr{E}_{\sigma_r}^{(r)}$ ($\sigma_r \in \mathscr{N}_{\mathscr{E}}^{r} = \{1,2,\cdots,N_r\}$ *and* $r \in \{0,1,2,\cdots,n-2\}$) *at time* t_m *with the related boundaries* $\partial\Omega_{\alpha i_\alpha}(\sigma_r)$ *and related domains* $\Omega_\alpha(\sigma_r)$ ($\alpha \in \mathscr{N}_{\mathscr{D}}(\sigma_r) \subset \mathscr{N}_{\mathscr{D}} = \{1,2,\cdots,N_d\}$ *and* $i_\alpha \in \mathscr{N}_{\alpha\mathscr{B}}(\sigma_r) \subset \mathscr{N}_{\mathscr{B}} = \{1,2,\cdots,N_b\}$), *where the boundary index subset of* $\mathscr{N}_{\alpha\mathscr{B}}(\sigma_r)$ *has* n_{σ_r}*-elements* ($n_{\sigma_r} \geq (n-r)$). *For an arbitrarily small* $\varepsilon > 0$, *there are two time intervals* $[t_{m-\varepsilon}, t_m)$ *and* $(t_m, t_{m+\varepsilon}]$. *A flow* $\mathbf{x}^{(\alpha;\mathscr{D})}(t)$ *is* $C_{[t_{m-\varepsilon}, t_m)}^{r_\alpha}$ *and/or* $C_{(t_m, t_{m+\varepsilon}]}^{r_\alpha}$*-continuous* ($r_\alpha \geq 1$ or $r_\alpha \geq 2$) *for time* t *and* $\| d^{r_\alpha+1}\mathbf{x}^{(\alpha;\mathscr{D})} / dt^{r_\alpha+1} \| < \infty$.

(i) *The domain flow* $\mathbf{x}^{(\alpha;\mathscr{D})}(t)$ *in domain* Ω_α *to the edge* $\mathscr{E}_{\sigma_r}^{(r)}$ *is a coming flow if and only if*

$$\hbar_{i_\alpha} G_{\partial\Omega_{\alpha i_\alpha}}^{(\alpha;\mathscr{D})}(\mathbf{x}_m, t_{m-}, \mathbf{p}_\alpha, \boldsymbol{\lambda}_{i_\alpha}) < 0 \text{ for all } i_\alpha \in \mathscr{N}_{\alpha\mathscr{B}}(\sigma_r). \tag{6.23}$$

(ii) *A domain flow* $\mathbf{x}^{(\alpha;\mathscr{D})}(t)$ *in domain* Ω_α *to the edge* $\mathscr{E}_{\sigma_r}^{(r)}$ *is a leaving flow if and only if*

$$\hbar_{i_\alpha} G_{\partial\Omega_{\alpha i_\alpha}}^{(\alpha;\mathscr{D})}(\mathbf{x}_m, t_{m+}, \mathbf{p}_\alpha, \boldsymbol{\lambda}_{i_\alpha}) > 0 \text{ for all } i_\alpha \in \mathscr{N}_{\alpha\mathscr{B}}(\sigma_r). \tag{6.24}$$

(iii) *A domain flow* $\mathbf{x}^{(\alpha;\mathscr{D})}(t)$ *in domain* Ω_α at $\mathbf{x}_m \in \mathscr{E}_{\sigma_r}^{(r)}$ ($r \neq 0$) *is a grazing flow if and only if*

$$\left.\begin{array}{l} \hbar_{i_\alpha} G_{\partial\Omega_{\alpha i_\alpha}}^{(\alpha;\mathscr{D})}(\mathbf{x}_m, t_{m\pm}, \mathbf{p}_\alpha, \boldsymbol{\lambda}_{i_\alpha}) = 0 \\ \hbar_{i_\alpha} G_{\partial\Omega_{\alpha i_\alpha}}^{(1,\alpha;\mathscr{D})}(\mathbf{x}_m, t_{m\pm}, \mathbf{p}_\alpha, \boldsymbol{\lambda}_{i_\alpha}) > 0 \end{array}\right\} \text{for all } i_\alpha \in \mathscr{N}_{\alpha\mathscr{B}}(\sigma_r). \tag{6.25}$$

Proof. The proof is the same as in Chapter 3 for each boundary. ■

With the higher-order singularity, the coming, leaving and grazing domain flows to the boundaries of the edge will be discussed herein. Since the boundary index subset of $\mathscr{N}_{\alpha\mathscr{B}}(\sigma_r)$ has $(n-r)$-elements, the vectors of the order of singularity are defined as

$$\begin{aligned}&\mathbf{m}_\alpha \equiv (m_1, m_2, \cdots, m_{i_\alpha}, \cdots, m_{n-r}),\ 2\mathbf{k}_\alpha \equiv (2k_1, 2k_2, \cdots, 2k_{i_\alpha}, \cdots, 2k_{n-r}),\\ &2\mathbf{k}_\alpha + 1 \equiv (2k_1+1, 2k_2+1, \cdots, 2k_{i_\alpha}+1, \cdots, 2k_{n-r}+1).\end{aligned} \tag{6.26}$$

Definition 6.7. For a discontinuous dynamical system in Eqs. (6.4)-(6.8), there is a point $\mathbf{x}^{(\sigma_r;\mathscr{E})}(t_m) \equiv \mathbf{x}_m \in \mathscr{E}_{\sigma_r}^{(r)}$ $\sigma_r \in \mathscr{N}_{\mathscr{E}}^{r} = \{1, 2, \cdots, N_r\}$ and $r \in \{0, 1, 2, \cdots, n-2\}$) at time t_m with the related boundaries $\partial\Omega_{\alpha i_\alpha}(\sigma_r)$ and related domains $\Omega_\alpha(\sigma_r)$ ($\alpha \in \mathscr{N}_{\mathscr{D}}(\sigma_r) \subset \mathscr{N}_{\mathscr{D}} = \{1, 2, \cdots, N_d\}$ and $i_\alpha \in \mathscr{N}_{\alpha\mathscr{B}}(\sigma_r) \subset \mathscr{N}_{\mathscr{B}} = \{1, 2, \cdots, N_b\}$), where the boundary index subset of $\mathscr{N}_{\alpha\mathscr{B}}(\sigma_r)$ has n_{σ_r}-elements ($n_{\sigma_r} \geq n-r$). For an arbitrarily small $\varepsilon > 0$, there are two time intervals $[t_{m-\varepsilon}, t_m)$ and $(t_m, t_{m+\varepsilon}]$. A flow $\mathbf{x}^{(\alpha;\mathscr{D})}(t)$ is $C^{r_\alpha}_{[t_{m-\varepsilon}, t_m)}$ and/or $C^{r_\alpha}_{(t_m, t_{m+\varepsilon}]}$-continuous for time t and $\| d^{r_\alpha+1}\mathbf{x}^{(\alpha;\mathscr{D})} / dt^{r_\alpha+1} \| < \infty$ ($r_\alpha \geq \max_{i_\alpha \in \mathscr{N}_{\alpha\mathscr{B}}(\sigma_r)} \{m_{i_\alpha}\}$). As $r = 0$, $\mathbf{m}_\rho \neq 2\mathbf{k}_\rho + 1$.

(i) The domain flow $\mathbf{x}^{(\alpha;\mathscr{D})}(t)$ in domain Ω_α at point $\mathbf{x}_m \in \mathscr{E}_{\sigma_r}^{(r)}$ is a *coming flow* with the $(\mathbf{m}_\alpha;\mathscr{D})$-singularity if

$$\left.\begin{aligned}&\hbar_{i_\alpha} G^{(s_{i_\alpha},\alpha;\mathscr{D})}_{\partial\Omega_{\alpha i_\alpha}}(\mathbf{x}_m, t_{m-}, \mathbf{p}_\alpha, \boldsymbol{\lambda}_{i_\alpha}) = 0\\ &\text{for } s_{i_\alpha} = 0, 1, 2, \cdots, m_{i_\alpha} - 1;\\ &\hbar_{i_\alpha} \mathbf{n}^{\mathrm{T}}_{\partial\Omega_{\alpha i_\alpha}}(\mathbf{x}^{(i_\alpha;\mathscr{B})}_{m-\varepsilon}) \cdot [\mathbf{x}^{(i_\alpha;\mathscr{B})}_{m-\varepsilon} - \mathbf{x}^{(\alpha;\mathscr{D})}_{m-\varepsilon}] < 0\end{aligned}\right\} \text{for all } i_\alpha \in \mathscr{N}_{\alpha\mathscr{B}}(\sigma_r). \tag{6.27}$$

(ii) The domain flow $\mathbf{x}^{(\alpha;\mathscr{D})}(t)$ in domain Ω_α at point $\mathbf{x}_m \in \mathscr{E}_{\sigma_r}^{(r)}$ is a *leaving flow* with the $(\mathbf{m}_\alpha;\mathscr{D})$-singularity if

$$\left.\begin{aligned}&\hbar_{i_\alpha} G^{(s_{i_\alpha},\alpha;\mathscr{D})}_{\partial\Omega_{\alpha i_\alpha}}(\mathbf{x}_m, t_{m+}, \mathbf{p}_\alpha, \boldsymbol{\lambda}_{i_\alpha}) = 0\\ &\text{for } s_{i_\alpha} = 0, 1, 2, \cdots, m_{i_\alpha} - 1;\\ &\hbar_{i_\alpha} \mathbf{n}^{\mathrm{T}}_{\partial\Omega_{\alpha i_\alpha}}(\mathbf{x}^{(i_\alpha;\mathscr{B})}_{m+\varepsilon}) \cdot [\mathbf{x}^{(\alpha;\mathscr{D})}_{m+\varepsilon} - \mathbf{x}^{(i_\alpha;\mathscr{B})}_{m+\varepsilon}] > 0\end{aligned}\right\} \text{for all } i_\alpha \in \mathscr{N}_{\alpha\mathscr{B}}(\sigma_r). \tag{6.28}$$

(iii) The domain flow $\mathbf{x}^{(\alpha;\mathscr{D})}(t)$ in domain Ω_α at point $\mathbf{x}_m \in \mathscr{E}_{\sigma_r}^{(r)}$ ($r \neq 0$) is a *grazing flow* with the $(2\mathbf{k}_\alpha+1;\mathscr{D})$-singularity if

$$\left.\begin{aligned}&\hbar_{i_\alpha} G^{(s_{i_\alpha},\alpha;\mathscr{D})}_{\partial\Omega_{\alpha i_\alpha}}(\mathbf{x}_m, t_{m\pm}, \mathbf{p}_\alpha, \boldsymbol{\lambda}_{i_\alpha}) = 0\\ &\text{for } s_{i_\alpha} = 0, 1, 2, \cdots, 2k_{i_\alpha};\\ &\hbar_{i_\alpha} \mathbf{n}^{\mathrm{T}}_{\partial\Omega_{\alpha i_\alpha}}(\mathbf{x}^{(i_\alpha;\mathscr{B})}_{m-\varepsilon}) \cdot [\mathbf{x}^{(i_\alpha;\mathscr{B})}_{m-\varepsilon} - \mathbf{x}^{(\alpha;\mathscr{D})}_{m-\varepsilon}] < 0\\ &\hbar_{i_\alpha} \mathbf{n}^{\mathrm{T}}_{\partial\Omega_{\alpha i_\alpha}}(\mathbf{x}^{(i_\alpha;\mathscr{B})}_{m+\varepsilon}) \cdot [\mathbf{x}^{(\alpha;\mathscr{D})}_{m+\varepsilon} - \mathbf{x}^{(i_\alpha;\mathscr{B})}_{m+\varepsilon}] > 0\end{aligned}\right\} \text{for all } i_\alpha \in \mathscr{N}_{\alpha\mathscr{B}}(\sigma_r). \tag{6.29}$$

The sufficient and necessary conditions for a domain flow to a specific edge with the higher-order singularity will be presented.

Theorem 6.2. *For a discontinuous dynamical system in Eqs.* (6.4)-(6.8)*, there is a point* $\mathbf{x}^{(\sigma_r;\mathscr{E})}(t_m)\equiv\mathbf{x}_m\in\mathscr{E}_{\sigma_r}^{(r)}$ ($\sigma_r\in\mathscr{N}_{\mathscr{E}}^{r}=\{1,2,\cdots,N_r\}$ *and* $r\in\{0,1,2,\cdots,n-2\}$) *at time* t_m *with the related boundaries* $\partial\Omega_{\alpha i_\alpha}(\sigma_r)$ *and related domains* $\Omega_\alpha(\sigma_r)$ ($\alpha\in\mathscr{N}_{\mathscr{D}}(\sigma_r)\subset\mathscr{N}_{\mathscr{D}}=\{1,2,\cdots,N_d\}$ *and* $i_\alpha\in\mathscr{N}_{\alpha\mathscr{B}}(\sigma_r)\subset\mathscr{N}_{\mathscr{B}}=\{1,2,\cdots,N_b\}$), *where the boundary index subset of* $\mathscr{N}_{\alpha\mathscr{B}}(\sigma_r)$ *has* n_{σ_r} -elements ($n_{\sigma_r}\geq(n-r)$). *For an arbitrarily small* $\varepsilon>0$, *there are two time intervals* $[t_{m-\varepsilon},t_m)$ *and* $(t_m,t_{m+\varepsilon}]$. *A flow* $\mathbf{x}^{(\alpha;\mathscr{D})}(t)$ *is* $C^{r_\alpha}_{[t_{m-\varepsilon},t_m)}$ *and/or* $C^{r_\alpha}_{(t_m,t_{m+\varepsilon}]}$*-continuous for time* t *and* $\|d^{r_\alpha+1}\mathbf{x}^{(\alpha;\mathscr{D})}/dt^{r_\alpha+1}\|<\infty$ ($r_\alpha\geq\max_{i_\alpha\in\mathscr{N}_{\alpha\mathscr{B}}(\sigma_r)}\{m_{i_\alpha}\}$). *As* $r=0$, $\mathbf{m}_\alpha\neq 2\mathbf{k}_\alpha+1$.

(i) *The domain flow* $\mathbf{x}^{(\alpha;\mathscr{D})}(t)$ *in domain* Ω_α *at point* $\mathbf{x}_m\in\mathscr{E}_{\sigma_r}^{(r)}$ *is a coming flow with the* $(\mathbf{m}_\alpha;\mathscr{D})$*-singularity if and only if*

$$\left.\begin{array}{l}\hbar_{i_\alpha}G^{(s_{i_\alpha},\alpha;\mathscr{D})}_{\partial\Omega_{\alpha i_\alpha}}(\mathbf{x}_m,t_{m-},\mathbf{p}_\alpha,\boldsymbol{\lambda}_{i_\alpha})=0\\ \text{for } s_{i_\alpha}=0,1,2,\cdots,m_{i_\alpha}-1;\\ (-1)^{m_{i_\alpha}}\hbar_{i_\alpha}G^{(2k_{i_\alpha},\alpha;\mathscr{D})}_{\partial\Omega_{\alpha i_\alpha}}(\mathbf{x}_m,t_{m-},\mathbf{p}_\alpha,\boldsymbol{\lambda}_{i_\alpha})<0\end{array}\right\}\textit{for all } i_\alpha\in\mathscr{N}_{\alpha\mathscr{B}}(\sigma_r). \tag{6.30}$$

(ii) *The domain flow* $\mathbf{x}^{(\alpha;\mathscr{D})}(t)$ *in domain* Ω_α *at point* $\mathbf{x}_m\in\mathscr{E}_{\sigma_r}^{(r)}$ *is a leaving flow with the* $(\mathbf{m}_\alpha;\mathscr{D})$*-singularity if and only if*

$$\left.\begin{array}{l}\hbar_{i_\alpha}G^{(s_{i_\alpha},\alpha;\mathscr{D})}_{\partial\Omega_{\alpha i_\alpha}}(\mathbf{x}_m,t_{m+},\mathbf{p}_\alpha,\boldsymbol{\lambda}_{i_\alpha})=0\\ \text{for } s_{i_\alpha}=0,1,2,\cdots,m_{i_\alpha}-1;\\ \hbar_{i_\alpha}G^{(m_{i_\alpha},\alpha;\mathscr{D})}_{\partial\Omega_{\alpha i_\alpha}}(\mathbf{x}_m,t_{m+},\mathbf{p}_\alpha,\boldsymbol{\lambda}_{i_\alpha})>0\end{array}\right\}\textit{for all } i_\alpha\in\mathscr{N}_{\alpha\mathscr{B}}(\sigma_r). \tag{6.31}$$

(iii) *The domain flow* $\mathbf{x}^{(\alpha;\mathscr{D})}(t)$ *in domain* Ω_α *at point* $\mathbf{x}_m\in\mathscr{E}_{\sigma_r}^{(r)}$ ($r\neq 0$) *is a grazing flow with the* $(2\mathbf{k}_\alpha+1;\mathscr{D})$*-singularity if and only if*

$$\left.\begin{array}{l}\hbar_{i_\alpha}G^{(s_{i_\alpha},\alpha;\mathscr{D})}_{\partial\Omega_{\alpha i_\alpha}}(\mathbf{x}_m,t_{m\pm},\mathbf{p}_\alpha,\boldsymbol{\lambda}_{i_\alpha})=0\\ \text{for } s_{i_\alpha}=0,1,2,\cdots,2k_{i_\alpha};\\ \hbar_{i_\alpha}G^{(2k_{i_\alpha}+1,\alpha;\mathscr{D})}_{\partial\Omega_{\alpha i_\alpha}}(\mathbf{x}_m,t_{m\pm},\mathbf{p}_\alpha,\boldsymbol{\lambda}_{i_\alpha})>0\end{array}\right\}\textit{for all } i_\alpha\in\mathscr{N}_{\alpha\mathscr{B}}(\sigma_r). \tag{6.32}$$

Proof. The proof is the same as in Chapter 3 for each boundary. ■

6.4. Domain flow switchability to convex edges

Because many sub-domains exist, once a flow in a sub-domain arrives to a specific edge, the switching of such a flow from one domain to another domain has more possibility. From the passability conditions of a flow arriving to the boundary, passability conditions for a flow to the specific edge can be developed.

Definition 6.8. For a discontinuous dynamical system in Eqs. (6.4)-(6.8), there is a point $\mathbf{x}^{(\sigma_r;\mathscr{E})}(t_m)\equiv\mathbf{x}_m\in\mathscr{E}_{\sigma_r}^{(r)}$ ($\sigma_r\in\mathscr{N}_{\mathscr{E}}^{r}=\{1,2,\cdots,N_r\}$ and $r\in\{0,1,2,\cdots,n-2\}$) at time t_m with the related boundaries $\partial\Omega_{\rho i_\rho}(\sigma_r)$ and related domains $\Omega_\rho(\sigma_r)$ ($\rho\in\mathscr{N}_{\mathscr{D}}(\sigma_r)\subset\mathscr{N}_{\mathscr{D}}=\{1,2,\cdots,N_d\}$ and $i_\rho\in\mathscr{N}_{\rho\mathscr{B}}(\sigma_r)\subset\mathscr{N}_{\mathscr{B}}=\{1,2,\cdots,N_b\}$), where the boundary index subset of $\mathscr{N}_{\rho\mathscr{B}}(\sigma_r)$ has n_{σ_r}-elements ($\rho=\alpha,\beta$ and $n_{\sigma_r}\geq n-r$). For an arbitrarily small $\varepsilon>0$, there are two time intervals $[t_{m-\varepsilon},t_m)$ and $(t_m,t_{m+\varepsilon}]$. The switching rule of $\mathbf{x}^{(\alpha;\mathscr{D})}(t_{m\pm})=\mathbf{x}_m=\mathbf{x}^{(\beta;\mathscr{D})}(t_{m\pm})$ exists.

(i) Two flows of $\mathbf{x}^{(\alpha;\mathscr{D})}(t)$ and $\mathbf{x}^{(\beta;\mathscr{D})}(t)$ at point $\mathbf{x}_m\in\mathscr{E}_{\sigma_r}^{(r)}$ are *switchable* from domain Ω_α to domain Ω_β if

$$\begin{aligned}&\hbar_{i_\alpha}\mathbf{n}_{\partial\Omega_{\alpha i_\alpha}}^{\mathrm{T}}(\mathbf{x}_{m-\varepsilon}^{(i_\alpha;\mathscr{B})})\cdot[\mathbf{x}_{m-\varepsilon}^{(i_\alpha;\mathscr{B})}-\mathbf{x}_{m-\varepsilon}^{(\alpha;\mathscr{D})}]<0\ \text{ for all } i_\alpha\in\mathscr{N}_{\alpha\mathscr{B}}(\sigma_r),\\&\hbar_{j_\beta}\mathbf{n}_{\partial\Omega_{\beta j_\beta}}^{\mathrm{T}}(\mathbf{x}_{m+\varepsilon}^{(j_\beta;\mathscr{B})})\cdot[\mathbf{x}_{m+\varepsilon}^{(\beta;\mathscr{D})}-\mathbf{x}_{m+\varepsilon}^{(j_\beta;\mathscr{B})}]>0\ \text{ for all } j_\beta\in\mathscr{N}_{\beta\mathscr{B}}(\sigma_r).\end{aligned}\tag{6.33}$$

(ii) Two flows of $\mathbf{x}^{(\alpha;\mathscr{D})}(t)$ and $\mathbf{x}^{(\beta;\mathscr{D})}(t)$ at point $\mathbf{x}_m\in\mathscr{E}_{\sigma_r}^{(r)}$ are *non-switchable of the first kind* if

$$\begin{aligned}&\hbar_{i_\alpha}\mathbf{n}_{\partial\Omega_{\alpha i_\alpha}}^{\mathrm{T}}(\mathbf{x}_{m-\varepsilon}^{(i_\alpha;\mathscr{B})})\cdot[\mathbf{x}_{m-\varepsilon}^{(i_\alpha;\mathscr{B})}-\mathbf{x}_{m-\varepsilon}^{(\alpha;\mathscr{D})}]<0\ \text{ for all } i_\alpha\in\mathscr{N}_{\alpha\mathscr{B}}(\sigma_r),\\&\hbar_{j_\beta}\mathbf{n}_{\partial\Omega_{\beta j_\beta}}^{\mathrm{T}}(\mathbf{x}_{m-\varepsilon}^{(j_\beta;\mathscr{B})})\cdot[\mathbf{x}_{m-\varepsilon}^{(j_\beta;\mathscr{B})}-\mathbf{x}_{m-\varepsilon}^{(\beta;\mathscr{D})}]<0\ \text{ for all } j_\beta\in\mathscr{N}_{\beta\mathscr{B}}(\sigma_r).\end{aligned}\tag{6.34}$$

(iii) Two flows of $\mathbf{x}^{(\alpha;\mathscr{D})}(t)$ and $\mathbf{x}^{(\beta;\mathscr{D})}(t)$ at point $\mathbf{x}_m\in\mathscr{E}_{\sigma_r}^{(r)}$ are *non-switchable of the second kind* if

$$\begin{aligned}&\hbar_{i_\alpha}\mathbf{n}_{\partial\Omega_{\alpha i_\alpha}}^{\mathrm{T}}(\mathbf{x}_{m+\varepsilon}^{(i_\alpha;\mathscr{B})})\cdot[\mathbf{x}_{m+\varepsilon}^{(\alpha;\mathscr{D})}-\mathbf{x}_{m+\varepsilon}^{(i_\alpha;\mathscr{B})}]>0\ \text{ for all } i_\alpha\in\mathscr{N}_{\alpha\mathscr{B}}(\sigma_r),\\&\hbar_{j_\beta}\mathbf{n}_{\partial\Omega_{\beta j_\beta}}^{\mathrm{T}}(\mathbf{x}_{m+\varepsilon}^{(j_\beta;\mathscr{B})})\cdot[\mathbf{x}_{m+\varepsilon}^{(\beta;\mathscr{D})}-\mathbf{x}_{m+\varepsilon}^{(j_\beta;\mathscr{B})}]>0\ \text{ for all } j_\beta\in\mathscr{N}_{\beta\mathscr{B}}(\sigma_r).\end{aligned}\tag{6.35}$$

Theorem 6.3. *For a discontinuous dynamical system in Eqs.* (6.4)-(6.8), *there is a point* $\mathbf{x}^{(\sigma_r;\mathscr{E})}(t_m)\equiv\mathbf{x}_m\in\mathscr{E}_{\sigma_r}^{(r)}$ ($\sigma_r\in\mathscr{N}_{\mathscr{E}}^{r}=\{1,2,\cdots,N_r\}$ *and* $r\in\{0,1,2,\cdots,n-2\}$) *at time* t_m *with the related boundaries* $\partial\Omega_{\rho i_\rho}(\sigma_r)$ *and related domains* $\Omega_\rho(\sigma_r)$ ($\rho\in\mathscr{N}_{\mathscr{D}}(\sigma_r)\subset\mathscr{N}_{\mathscr{D}}=\{1,2,\cdots,N_d\}$ *and* $i_\rho\in\mathscr{N}_{\rho\mathscr{B}}(\sigma_r)\subset\mathscr{N}_{\mathscr{B}}=\{1,2,\cdots,N_b\}$), *where the boundary index subset of* $\mathscr{N}_{\rho\mathscr{B}}(\sigma_r)$ *has* n_{σ_r}*-elements (*$\rho=\alpha,\beta$ *and* $n_{\sigma_r}\geq(n-r)$*). For an arbitrarily small* $\varepsilon>0$*, there are two time intervals* $[t_{m-\varepsilon}, t_m)$ *and* $(t_m, t_{m+\varepsilon}]$*. A flow* $\mathbf{x}^{(\rho;\mathscr{D})}(t)$ *is* $C^{r_\rho}_{[t_{m-\varepsilon},t_m)}$ *and/or* $C^{r_\rho}_{(t_m,t_{m+\varepsilon}]}$ *continuous (*$r_\rho\geq 1$*) for time t and* $\| d^{r_\rho+1}\mathbf{x}^{(\rho;\mathscr{D})}/dt^{r_\rho+1}\|<\infty$*. Suppose there is a switching rule of* $\mathbf{x}^{(\alpha;\mathscr{D})}(t_{m\pm})=\mathbf{x}_m=\mathbf{x}^{(\beta;\mathscr{D})}(t_{m\pm})$*.*

(i) *Two flows of* $\mathbf{x}^{(\alpha;\mathscr{D})}(t)$ *and* $\mathbf{x}^{(\beta;\mathscr{D})}(t)$ *at* $\mathbf{x}_m\in\mathscr{E}^{(r)}_{\sigma_r}$ *are switchable from* Ω_α *to* Ω_β *if and only if*

$$\begin{aligned}&\hbar_{i_\alpha} G^{(\alpha;\mathscr{D})}_{\partial\Omega_{\alpha i_\alpha}}(\mathbf{x}_m,t_{m-},\mathbf{p}_\alpha,\boldsymbol{\lambda}_{i_\alpha})<0 \text{ for all } i_\alpha\in\mathscr{N}_{\alpha\mathscr{B}}(\sigma_r),\\ &\hbar_{j_\beta} G^{(\beta;\mathscr{D})}_{\partial\Omega_{\beta j_\beta}}(\mathbf{x}_m,t_{m+},\mathbf{p}_\beta,\boldsymbol{\lambda}_{j_\beta})>0 \text{ for all } j_\beta\in\mathscr{N}_{\beta\mathscr{B}}(\sigma_r).\end{aligned}\tag{6.36}$$

(ii) *Two flows of* $\mathbf{x}^{(\alpha;\mathscr{D})}(t)$ *and* $\mathbf{x}^{(\beta;\mathscr{D})}(t)$ *at* $\mathbf{x}_m\in\mathscr{E}^{(r)}_{\sigma_r}$ *are non-switchable of the first kind if and only if*

$$\begin{aligned}&\hbar_{i_\alpha} G^{(\alpha;\mathscr{D})}_{\partial\Omega_{\alpha i_\alpha}}(\mathbf{x}_m,t_{m-},\mathbf{p}_\alpha,\boldsymbol{\lambda}_{i_\alpha})<0 \text{ for all } i_\alpha\in\mathscr{N}_{\alpha\mathscr{B}}(\sigma_r),\\ &\hbar_{j_\beta} G^{(\beta;\mathscr{D})}_{\partial\Omega_{\beta j_\beta}}(\mathbf{x}_m,t_{m-},\mathbf{p}_\beta,\boldsymbol{\lambda}_{j_\beta})<0 \text{ for all } j_\beta\in\mathscr{N}_{\beta\mathscr{B}}(\sigma_r).\end{aligned}\tag{6.37}$$

(iii) *Two flows of* $\mathbf{x}^{(\alpha;\mathscr{D})}(t)$ *and* $\mathbf{x}^{(\beta;\mathscr{D})}(t)$ *at* $\mathbf{x}_m\in\mathscr{E}^{(r)}_{\sigma_r}$ *are non-switchable of the second kind if and only if*

$$\begin{aligned}&\hbar_{i_\alpha} G^{(\alpha;\mathscr{D})}_{\partial\Omega_{\alpha i_\alpha}}(\mathbf{x}_m,t_{m+},\mathbf{p}_\alpha,\boldsymbol{\lambda}_{i_\alpha})>0 \text{ for all } i_\alpha\in\mathscr{N}_{\alpha\mathscr{B}}(\sigma_r),\\ &\hbar_{j_\beta} G^{(\beta;\mathscr{D})}_{\partial\Omega_{\beta j_\beta}}(\mathbf{x}_m,t_{m+},\mathbf{p}_\beta,\boldsymbol{\lambda}_{j_\beta})>0 \text{ for all } j_\beta\in\mathscr{N}_{\beta\mathscr{B}}(\sigma_r).\end{aligned}\tag{6.38}$$

Proof. The proof is the same as in Chapter 3 for each boundary. ■

To explain the switchability of domain flows to edges in the foregoing definition, two switchable domain flows of $\mathbf{x}^{(\alpha;\mathscr{D})}$ and $\mathbf{x}^{(\beta;\mathscr{D})}$ to edges $\mathscr{E}^{(n-2)}$ and $\mathscr{E}^{(n-3)}$ are sketched in Fig. 6.12. Curves with arrows represent domain flows and points $\mathbf{x}_m\in\mathscr{E}^{(n-2)}$ and $\mathbf{x}_m\in\mathscr{E}^{(n-3)}$ are on edges. Once a domain flow arrives to a specific edge, there are many possibilities for a coming flow to switch. Under a switching rule, a leaving flow in an accessible sub-domain will be for the coming flow to switch. Without the switching rule, it is very difficult for such a coming flow to switch. Thus, the switching rule is very important.

For flow switchability with higher-order singularity, under the switching rule, a coming domain flow tangential to the edge can be switched into another domain. Otherwise, the coming flow with the grazing singularity will form a tangential flow to the edge. Thus, the coming grazing flow cannot be switched into another domain flow.

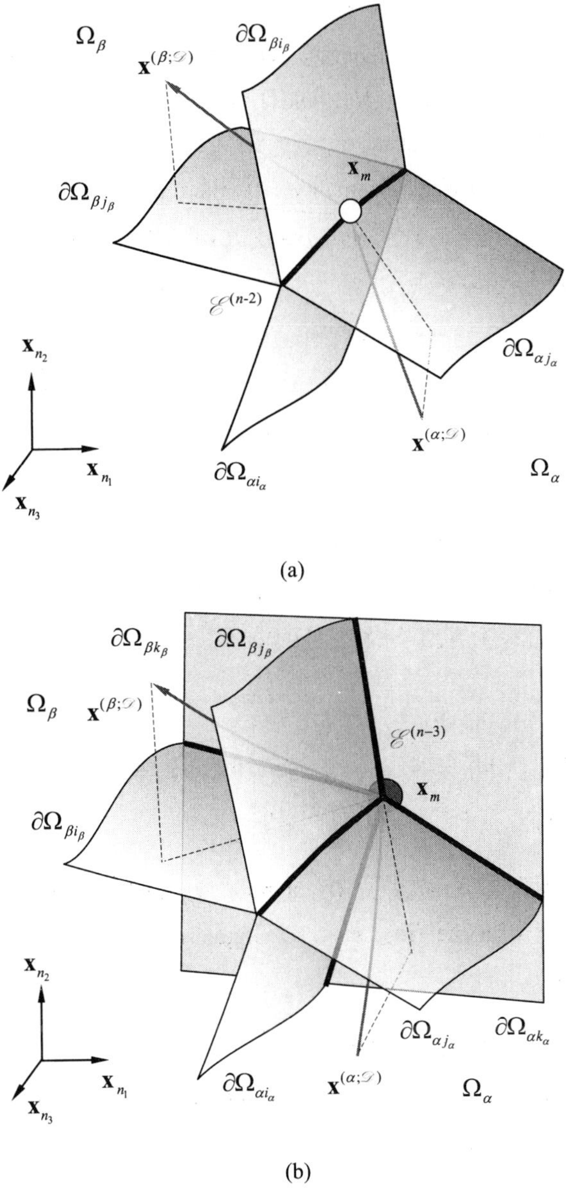

Fig. 6.12 Two switchable domains flows of $\mathbf{x}^{(\alpha;\mathscr{D})}$ and $\mathbf{x}^{(\beta;\mathscr{D})}$ to edges of: **(a)** $\mathscr{E}^{(n-2)}$ and **(b)** $\mathscr{E}^{(n-3)}$. Curves with arrows represent the domain flows ($n_1 + n_2 + n_3 = n$). The filled point is two $(n-3)$ -dimensional edge.

Definition 6.9. For a discontinuous dynamical system in Eqs. (6.4)-(6.8), there is a point $\mathbf{x}^{(\sigma_r;\mathscr{E})}(t_m) \equiv \mathbf{x}_m \in \mathscr{E}_{\sigma_r}^{(r)}$ ($\sigma_r \in \mathscr{N}_{\mathscr{E}}^r = \{1,2,\cdots,N_r\}$ and $r \in \{0,1,2,\cdots,n-2\}$) at time t_m with the related boundaries $\partial\Omega_{\rho i_\rho}(\sigma_r)$ and related domains $\Omega_\rho(\sigma_r)$ ($\rho \in \mathscr{N}_{\mathscr{D}}(\sigma_r) \subset \mathscr{N}_{\mathscr{D}} = \{1,2,\cdots,N_d\}$ and $i_\rho \in \mathscr{N}_{\rho\mathscr{B}}(\sigma_r) \subset \mathscr{N}_{\mathscr{B}} = \{1,2,\cdots,N_b\}$), where the boundary index subset of $\mathscr{N}_{\rho\mathscr{B}}(\sigma_r)$ has n_{σ_r} -elements ($\rho = \alpha, \beta$ and $n_{\sigma_r} \geq (n-r)$). For an arbitrarily small $\varepsilon > 0$, there are two time intervals $[t_{m-\varepsilon}, t_m)$ and $(t_m, t_{m+\varepsilon}]$. A flow $\mathbf{x}^{(\rho;\mathscr{D})}(t)$ ($\rho = \alpha, \beta$) is $C^{r_\rho}_{[t_{m-\varepsilon},t_m)}$ or $C^{r_\rho}_{(t_m,t_{m+\varepsilon}]}$ continuous for time t and $\| d^{r_\rho+1}\mathbf{x}^{(\rho;\mathscr{D})}(t)/dt^{r_\rho+1} \| < \infty$ ($r_\rho \geq \max_{i_\rho \in \mathscr{N}_{\rho\mathscr{B}}} \{m_{i_\rho}\}$). Suppose a switching rule $\mathbf{x}^{(\alpha;\mathscr{D})}(t_{m\pm}) = \mathbf{x}_m = \mathbf{x}^{(\beta;\mathscr{D})}(t_{m\pm})$ exists. As $r = 0$, $\mathbf{m}_\rho \neq 2\mathbf{k}_\rho + 1$.

(i) With the switching rule, two flows $\mathbf{x}^{(\alpha;\mathscr{D})}(t)$ and $\mathbf{x}^{(\beta;\mathscr{D})}(t)$ at point $\mathbf{x}_m \in \mathscr{E}_{\sigma_r}^{(r)}$ are *switchable* with the $(\mathbf{m}_\alpha : \mathbf{m}_\beta; \mathscr{D})$ singularity from domain Ω_α to domain Ω_β if

$$\left.\begin{array}{l} \hbar_{i_\alpha} G^{(s_{i_\alpha},\alpha;\mathscr{D})}_{\partial\Omega_{\alpha i_\alpha}}(\mathbf{x}_m, t_{m-}, \mathbf{p}_\alpha, \boldsymbol{\lambda}_{i_\alpha}) = 0 \\ \text{for } s_{i_\alpha} = 0,1,\cdots,m_{i_\alpha}-1; \\ \hbar_{i_\alpha} \mathbf{n}^{\mathrm{T}}_{\partial\Omega_{\alpha i_\alpha}}(\mathbf{x}^{(i_\alpha,\mathscr{B})}_{m-\varepsilon}) \cdot [\mathbf{x}^{(i_\alpha;\mathscr{B})}_{m-\varepsilon} - \mathbf{x}^{(\alpha;\mathscr{D})}_{m-\varepsilon}] < 0 \end{array}\right\} \text{for all } i_\alpha \in \mathscr{N}_{\alpha\mathscr{B}}(\sigma_r), \tag{6.39}$$

$$\left.\begin{array}{l} \hbar_{j_\beta} G^{(s_{j_\beta},\beta;\mathscr{D})}_{\partial\Omega_{\beta j_\beta}}(\mathbf{x}_m, t_{m+}, \mathbf{p}_\beta, \boldsymbol{\lambda}_{j_\beta}) = 0 \\ \text{for } s_{j_\beta} = 0,1,\cdots,m_{j_\beta}-1; \\ \hbar_{j_\beta} \mathbf{n}^{\mathrm{T}}_{\partial\Omega_{\beta j_\beta}}(\mathbf{x}^{(j_\beta,\mathscr{B})}_{m+\varepsilon}) \cdot [\mathbf{x}^{(\beta;\mathscr{D})}_{m+\varepsilon} - \mathbf{x}^{(j_\beta;\mathscr{B})}_{m+\varepsilon}] > 0 \end{array}\right\} \text{for all } j_\beta \in \mathscr{N}_{\beta\mathscr{B}}(\sigma_r). \tag{6.40}$$

(ii) Two flows $\mathbf{x}^{(\alpha;\mathscr{D})}(t)$ and $\mathbf{x}^{(\beta;\mathscr{D})}(t)$ at point $\mathbf{x}_m \in \mathscr{E}_{\sigma_r}^{(r)}$ are *non-switchable of the first kind* with the $(\mathbf{m}_\alpha : \mathbf{m}_\beta; \mathscr{D})$ singularity ($\mathbf{m}_\alpha \neq 2\mathbf{k}_\alpha + 1$ and $\mathbf{m}_\beta \neq 2\mathbf{k}_\beta + 1$) if

$$\left.\begin{array}{l} \hbar_{i_\alpha} G^{(s_{i_\alpha},\alpha;\mathscr{D})}_{\partial\Omega_{\alpha i_\alpha}}(\mathbf{x}_m, t_{m-}, \mathbf{p}_\alpha, \boldsymbol{\lambda}_{i_\alpha}) = 0 \\ \text{for } s_{i_\alpha} = 0,1,\cdots,m_{i_\alpha}-1; \\ \hbar_{i_\alpha} \mathbf{n}^{\mathrm{T}}_{\partial\Omega_{\alpha i_\alpha}}(\mathbf{x}^{(i_\alpha,\mathscr{B})}_{m-\varepsilon}) \cdot [\mathbf{x}^{(i_\alpha;\mathscr{B})}_{m-\varepsilon} - \mathbf{x}^{(\alpha;\mathscr{D})}_{m-\varepsilon}] < 0 \end{array}\right\} \text{for all } i_\alpha \in \mathscr{N}_{\alpha\mathscr{B}}(\sigma_r), \tag{6.41}$$

$$\left.\begin{array}{l} \hbar_{j_\beta} G^{(s_{j_\beta},\beta;\mathscr{D})}_{\partial\Omega_{\beta j_\beta}}(\mathbf{x}_m, t_{m-}, \mathbf{p}_\beta, \boldsymbol{\lambda}_{j_\beta}) = 0 \\ \text{for } s_{j_\beta} = 0,1,\cdots,m_{j_\beta}-1; \\ \hbar_{j_\beta} \mathbf{n}^{\mathrm{T}}_{\partial\Omega_{\beta j_\beta}}(\mathbf{x}^{(j_\beta,\mathscr{B})}_{m-\varepsilon}) \cdot [\mathbf{x}^{(j_\beta;\mathscr{B})}_{m-\varepsilon} - \mathbf{x}^{(\beta;\mathscr{D})}_{m-\varepsilon}] < 0 \end{array}\right\} \text{for all } j_\beta \in \mathscr{N}_{\beta\mathscr{B}}(\sigma_r). \tag{6.42}$$

(iii) Two flows $\mathbf{x}^{(\alpha;\mathscr{D})}(t)$ and $\mathbf{x}^{(\beta;\mathscr{D})}(t)$ at point $\mathbf{x}_m \in \mathscr{E}_{\sigma_r}^{(r)}$ are *non-switchable of the second kind with the* $(\mathbf{m}_\alpha : \mathbf{m}_\beta; \mathscr{D})$ *singularity* if

$$\left.\begin{array}{l} \hbar_{i_\alpha} G_{\partial\Omega_{\alpha i_\alpha}}^{(s_{i_\alpha},\alpha;\mathscr{D})}(\mathbf{x}_m, t_{m+}, \mathbf{p}_\alpha, \boldsymbol{\lambda}_{i_\alpha}) = 0 \\ \text{for } s_{i_\alpha} = 0,1,\cdots,m_{i_\alpha}-1; \\ \hbar_{i_\alpha} \mathbf{n}_{\partial\Omega_{\alpha i_\alpha}}^{\mathrm{T}}(\mathbf{x}_{m+\varepsilon}^{(i_\alpha,\mathscr{B})}) \cdot [\mathbf{x}_{m+\varepsilon}^{(\alpha;\mathscr{D})} - \mathbf{x}_{m+\varepsilon}^{(i_\alpha;\mathscr{B})}] > 0 \end{array}\right\} \text{for all } i_\alpha \in \mathscr{N}_{\alpha\mathscr{B}}(\sigma_r), \tag{6.43}$$

$$\left.\begin{array}{l} \hbar_{j_\beta} G_{\partial\Omega_{\beta j_\beta}}^{(s_{j_\beta},\beta;\mathscr{D})}(\mathbf{x}_m, t_{m+}, \mathbf{p}_\beta, \boldsymbol{\lambda}_{j_\beta}) = 0 \\ \text{for } s_{j_\beta} = 0,1,\cdots,m_{j_\beta}-1; \\ \hbar_{j_\beta} \mathbf{n}_{\partial\Omega_{\beta j_\beta}}^{\mathrm{T}}(\mathbf{x}_{m+\varepsilon}^{(j_\beta,\mathscr{B})}) \cdot [\mathbf{x}_{m+\varepsilon}^{(\beta;\mathscr{D})} - \mathbf{x}_{m+\varepsilon}^{(j_\beta;\mathscr{B})}] > 0 \end{array}\right\} \text{for all } j_\beta \in \mathscr{N}_{\beta\mathscr{B}}(\sigma_r). \tag{6.44}$$

(iv) Without any switching rules, two flows $\mathbf{x}^{(\alpha;\mathscr{D})}(t)$ and $\mathbf{x}^{(\beta;\mathscr{D})}(t)$ at point $\mathbf{x}_m \in \mathscr{E}_{\sigma_r}^{(r)}$ are *potentially switchable with the* $(\mathbf{m}_\alpha : \mathbf{m}_\beta; \mathscr{D})$ *singularity* from domain Ω_α to domain Ω_β ($\mathbf{m}_\alpha \neq 2\mathbf{k}_\alpha + 1$) if Eqs. (6.39) and (6.40) hold.

(v) Without any switching rules, two flows $\mathbf{x}^{(\alpha;\mathscr{D})}(t)$ and $\mathbf{x}^{(\beta;\mathscr{D})}(t)$ at point $\mathbf{x}_m \in \mathscr{E}_{\sigma_r}^{(r)}$ are *non-switchable with the* $(2\mathbf{k}_\alpha + 1 : \mathbf{m}_\beta; \mathscr{D})$ *singularity* from domain Ω_α to domain Ω_β if Eqs. (6.39) and (6.40) hold ($\mathbf{m}_\alpha = 2\mathbf{k}_\alpha + 1$).

(vi) With the switching rule, two flows $\mathbf{x}^{(\alpha;\mathscr{D})}(t)$ and $\mathbf{x}^{(\beta;\mathscr{D})}(t)$ grazing to the edge at point $\mathbf{x}_m \in \mathscr{E}_{\sigma_r}^{(r)}$ are *switchable of the* $((2\mathbf{k}_\alpha + 1) : (2\mathbf{k}_\beta + 1); \mathscr{D})$ *singularity* if

$$\left.\begin{array}{l} \hbar_{i_\alpha} G_{\partial\Omega_{\alpha i_\alpha}}^{(s_{i_\alpha},\alpha;\mathscr{D})}(\mathbf{x}_m, t_{m\pm}, \mathbf{p}_\alpha, \boldsymbol{\lambda}_{i_\alpha}) = 0 \\ \text{for } s_{i_\alpha} = 0,1,\cdots,2k_{i_\alpha}; \\ \hbar_{i_\alpha} \mathbf{n}_{\partial\Omega_{\alpha i_\alpha}}^{\mathrm{T}}(\mathbf{x}_{m-\varepsilon}^{(i_\alpha,\mathscr{B})}) \cdot [\mathbf{x}_{m-\varepsilon}^{(i_\alpha;\mathscr{B})} - \mathbf{x}_{m-\varepsilon}^{(\alpha;\mathscr{D})}] < 0, \\ \hbar_{i_\alpha} \mathbf{n}_{\partial\Omega_{\alpha i_\alpha}}^{\mathrm{T}}(\mathbf{x}_{m+\varepsilon}^{(i_\alpha,\mathscr{B})}) \cdot [\mathbf{x}_{m+\varepsilon}^{(\alpha;\mathscr{D})} - \mathbf{x}_{m+\varepsilon}^{(i_\alpha;\mathscr{B})}] > 0 \end{array}\right\} \text{for all } i_\alpha \in \mathscr{N}_{\alpha\mathscr{B}}(\sigma_r), \tag{6.45}$$

$$\left.\begin{array}{l} \hbar_{j_\beta} G_{\partial\Omega_{\beta j_\beta}}^{(s_{j_\beta},\beta;\mathscr{D})}(\mathbf{x}_m, t_{m\pm}, \mathbf{p}_\beta, \boldsymbol{\lambda}_{j_\beta}) = 0 \\ \text{for } s_{j_\beta} = 0,1,\cdots,2k_{j_\beta}; \\ \hbar_{j_\beta} \mathbf{n}_{\partial\Omega_{\beta j_\beta}}^{\mathrm{T}}(\mathbf{x}_{m-\varepsilon}^{(j_\beta,\mathscr{B})}) \cdot [\mathbf{x}_{m-\varepsilon}^{(j_\beta;\mathscr{B})} - \mathbf{x}_{m-\varepsilon}^{(\beta;\mathscr{D})}] < 0, \\ \hbar_{j_\beta} \mathbf{n}_{\partial\Omega_{\beta j_\beta}}^{\mathrm{T}}(\mathbf{x}_{m+\varepsilon}^{(j_\beta,\mathscr{B})}) \cdot [\mathbf{x}_{m+\varepsilon}^{(\beta;\mathscr{D})} - \mathbf{x}_{m+\varepsilon}^{(j_\beta;\mathscr{B})}] > 0 \end{array}\right\} \text{for all } j_\beta \in \mathscr{N}_{\beta\mathscr{B}}(\sigma_r). \tag{6.46}$$

(vii) Without any switching rules, two grazing flows $\mathbf{x}^{(\alpha;\mathscr{D})}(t)$ and $\mathbf{x}^{(\beta;\mathscr{D})}(t)$ to the edge at point $\mathbf{x}_m \in \mathscr{E}_{\sigma_r}^{(r)}$ ($r \neq 0$) are the *double, tangential flows* of the

$((2\mathbf{k}_\alpha+1):(2\mathbf{k}_\beta+1);\mathscr{D})$-singularity if Eqs. (6.45) and (6.46) holds.

From the above definition, the corresponding theorem is given for the sufficient and necessary conditions for the switchability of a domain flow to a specific edge with the higher-order singularity.

Theorem 6.4. *For a discontinuous dynamical system in Eqs. (6.4)-(6.8), there is a point* $\mathbf{x}^{(\sigma_r;\mathscr{E})}(t_m)\equiv\mathbf{x}_m\in\mathscr{E}_{\sigma_r}^{(r)}$ ($\sigma_r\in\mathscr{N}_{\mathscr{E}}^{r}=\{1,2,\cdots,N_r\}$ *and* $r\in\{0,1,2,\cdots,n-2\}$) *at time* t_m *with the related boundaries* $\partial\Omega_{\rho i_\rho}(\sigma_r)$ *and related domains* $\Omega_\rho(\sigma_r)$ ($\rho\in\mathscr{N}_{\mathscr{D}}(\sigma_r)\subset\mathscr{N}_{\mathscr{D}}=\{1,2,\cdots,N_d\}$ *and* $i_\rho\in\mathscr{N}_{\rho\mathscr{B}}(\sigma_r)\subset\mathscr{N}_{\mathscr{B}}=\{1,2,\cdots,N_b\}$), *where the boundary index subset of* $\mathscr{N}_{\rho\mathscr{B}}(\sigma_r)$ *has* n_{σ_r}*-elements* ($\rho=\alpha,\beta$ *and* $n_{\sigma_r}\geq(n-r)$). *For an arbitrarily small* $\varepsilon>0$, *there are two time intervals* $[t_{m-\varepsilon},t_m)$ *and* $(t_m,t_{m+\varepsilon}]$. *A flow* $\mathbf{x}^{(\rho;\mathscr{D})}(t)$ ($\rho=\alpha,\beta$) *are* $C^{r_\rho}_{[t_{m-\varepsilon},t_m)}$ *or* $C^{r_\rho}_{(t_m,t_{m+\varepsilon}]}$*-continuous for time t and* $\|d^{r_\rho+1}\mathbf{x}^{(\rho;\mathscr{D})}/dt^{r_\rho+1}\|<\infty$ ($r_\rho\geq\max_{i_\rho\in\mathscr{N}_{\rho\mathscr{B}}}\{m_{i_\rho}\}$). *Suppose* $\mathbf{x}^{(\alpha;\mathscr{D})}(t_{m\pm})=\mathbf{x}_m=\mathbf{x}^{(\beta;\mathscr{D})}(t_{m\pm})$ *is a switching rule. As* $r=0$, $\mathbf{m}_\rho\neq 2\mathbf{k}_\rho+1$.

(i) *With the switching rule, two flows* $\mathbf{x}^{(\alpha;\mathscr{D})}(t)$ *and* $\mathbf{x}^{(\beta;\mathscr{D})}(t)$ *at point* $\mathbf{x}_m\in\mathscr{E}_{\sigma_r}^{(r)}$ *are switchable with the* $(\mathbf{m}_\alpha:\mathbf{m}_\beta;\mathscr{D})$*- singularity from domain* Ω_α *to domain* Ω_β *if and only if*

$$\left.\begin{array}{l}\hbar_{i_\alpha}G_{\partial\Omega_{\alpha i_\alpha}}^{(s_{i_\alpha},\alpha;\mathscr{D})}(\mathbf{x}_m,t_{m-},\mathbf{p}_\alpha,\boldsymbol{\lambda}_{i_\alpha})=0\\ \text{for } s_{i_\alpha}=0,1,\cdots,m_{i_\alpha}-1;\\ (-1)^{m_{i_\alpha}}\hbar_{i_\alpha}G_{\partial\Omega_{\alpha i_\alpha}}^{(m_{i_\alpha},\alpha;\mathscr{D})}(\mathbf{x}_m,t_{m-},\mathbf{p}_\alpha,\boldsymbol{\lambda}_{i_\alpha})<0\end{array}\right\}\textit{for all } i_\alpha\in\mathscr{N}_{\alpha\mathscr{B}}(\sigma_r), \tag{6.47}$$

$$\left.\begin{array}{l}\hbar_{j_\beta}G_{\partial\Omega_{\beta j_\beta}}^{(s_{j_\beta},\beta;\mathscr{D})}(\mathbf{x}_m,t_{m+},\mathbf{p}_\beta,\boldsymbol{\lambda}_{j_\beta})=0\\ \text{for } s_{j_\beta}=0,1,\cdots,m_{j_\beta}-1;\\ \hbar_{j_\beta}G_{\partial\Omega_{\beta j_\beta}}^{(m_{j_\beta},\beta;\mathscr{D})}(\mathbf{x}_m,t_{m+},\mathbf{p}_\beta,\boldsymbol{\lambda}_{j_\beta})>0\end{array}\right\}\textit{for all } j_\beta\in\mathscr{N}_{\beta\mathscr{B}}(\sigma_r). \tag{6.48}$$

(ii) *Two flows* $\mathbf{x}^{(\alpha;\mathscr{D})}(t)$ *and* $\mathbf{x}^{(\beta;\mathscr{D})}(t)$ *at point* $\mathbf{x}_m\in\mathscr{E}_{\sigma_r}^{(r)}$ *are non-switchable of the first kind with the* $(\mathbf{m}_\alpha:\mathbf{m}_\beta;\mathscr{D})$ *singularity* ($\mathbf{m}_\alpha\neq 2\mathbf{k}_\alpha+1$ *and* $\mathbf{m}_\beta\neq 2\mathbf{k}_\beta+1$) *if and only if*

$$\left.\begin{array}{l}\hbar_{i_\alpha}G_{\partial\Omega_{\alpha i_\alpha}}^{(s_{i_\alpha},\alpha;\mathscr{D})}(\mathbf{x}_m,t_{m-},\mathbf{p}_\alpha,\boldsymbol{\lambda}_{i_\alpha})=0\\ \text{for } s_{i_\alpha}=0,1,\cdots,m_{i_\alpha}-1;\\ (-1)^{m_{i_\alpha}}\hbar_{i_\alpha}G_{\partial\Omega_{\alpha i_\alpha}}^{(m_{i_\alpha},\alpha;\mathscr{D})}(\mathbf{x}_m,t_{m-},\mathbf{p}_\alpha,\boldsymbol{\lambda}_{i_\alpha})<0\end{array}\right\}\textit{for all } i_\alpha\in\mathscr{N}_{\alpha\mathscr{B}}(\sigma_r), \tag{6.49}$$

$$\left.\begin{array}{l}\hbar_{j_\beta} G^{(s_{j_\beta},\beta;\mathscr{D})}_{\partial\Omega_{\beta j_\beta}}(\mathbf{x}_m,t_{m-},\mathbf{p}_\beta,\boldsymbol{\lambda}_{j_\beta})=0\\ \text{for } s_{j_\beta}=0,1,\cdots,m_{j_\beta}-1;\\ (-1)^{m_{j_\alpha}}\hbar_{j_\beta} G^{(m_{j_\beta},\beta;\mathscr{D})}_{\partial\Omega_{\beta j_\beta}}(\mathbf{x}_m,t_{m-},\mathbf{p}_\beta,\boldsymbol{\lambda}_{j_\beta})<0\end{array}\right\} \textit{for all } j_\beta\in\mathscr{N}_{\beta\mathscr{B}}(\sigma_r). \tag{6.50}$$

(iii) *Two flows* $\mathbf{x}^{(\alpha;\mathscr{D})}(t)$ *and* $\mathbf{x}^{(\beta;\mathscr{D})}(t)$ *at point* $\mathbf{x}_m\in\mathscr{E}^{(r)}_{\sigma_r}$ *are non-switchable of the second kind with the* $(\mathbf{m}_\alpha:\mathbf{m}_\beta;\mathscr{D})$ *singularity if and only if*

$$\left.\begin{array}{l}\hbar_{i_\alpha} G^{(s_{i_\alpha},\alpha;\mathscr{D})}_{\partial\Omega_{\alpha i_\alpha}}(\mathbf{x}_m,t_{m+},\mathbf{p}_\alpha,\boldsymbol{\lambda}_{i_\alpha})=0\\ \text{for } s_{i_\alpha}=0,1,\cdots,m_{i_\alpha}-1;\\ \hbar_{i_\alpha} G^{(m_{i_\alpha},\alpha;\mathscr{D})}_{\partial\Omega_{\alpha i_\alpha}}(\mathbf{x}_m,t_{m+},\mathbf{p}_\alpha,\boldsymbol{\lambda}_{i_\alpha})>0\end{array}\right\} \textit{for all } i_\alpha\in\mathscr{N}_{\alpha\mathscr{B}}(\sigma_r), \tag{6.51}$$

$$\left.\begin{array}{l}\hbar_{j_\beta} G^{(s_{j_\beta},\beta;\mathscr{D})}_{\partial\Omega_{\beta j_\beta}}(\mathbf{x}_m,t_{m+},\mathbf{p}_\beta,\boldsymbol{\lambda}_{j_\beta})=0\\ \text{for } s_{j_\beta}=0,1,\cdots,m_{j_\beta}-1;\\ \hbar_{j_\beta} G^{(m_{j_\beta},\beta;\mathscr{D})}_{\partial\Omega_{\beta j_\beta}}(\mathbf{x}_m,t_{m+},\mathbf{p}_\beta,\boldsymbol{\lambda}_{j_\beta})>0\end{array}\right\} \textit{for all } j_\beta\in\mathscr{N}_{\beta\mathscr{B}}(\sigma_r). \tag{6.52}$$

(iv) *Without any switching rules, two flows* $\mathbf{x}^{(\alpha;\mathscr{D})}(t)$ *and* $\mathbf{x}^{(\beta;\mathscr{D})}(t)$ *at point* $\mathbf{x}_m\in\mathscr{E}^{(r)}_{\sigma_r}$ *are potentially switchable with the* $(\mathbf{m}_\alpha:\mathbf{m}_\beta;\mathscr{D})$*-singularity from domain* Ω_α *to domain* Ω_β *if and only if Eqs. (6.47) and (6.48) holds with* $\mathbf{m}_\alpha\neq 2\mathbf{k}_\alpha+1$.

(v) *Without any switching rules, two flows* $\mathbf{x}^{(\alpha;\mathscr{D})}(t)$ *and* $\mathbf{x}^{(\beta;\mathscr{D})}(t)$ *at point* $\mathbf{x}_m\in\mathscr{E}^{(r)}_{\sigma_r}$ *are non-switchable with the* $(2\mathbf{k}_\alpha+1:\mathbf{m}_\beta;\mathscr{D})$ *-singularity from domain* Ω_α *to domain* Ω_β *if and only if Eqs. (6.47) and (6.48) holds with* $\mathbf{m}_\alpha=2\mathbf{k}_\alpha+1$.

(vi) *With the switching rule, two flows* $\mathbf{x}^{(\alpha;\mathscr{D})}(t)$ *and* $\mathbf{x}^{(\beta;\mathscr{D})}(t)$ *grazing to the edge at point* $\mathbf{x}_m\in\mathscr{E}^{(r)}_{\sigma_r}$ *are switchable of the* $((2\mathbf{k}_\alpha+1):(2\mathbf{k}_\beta+1);\mathscr{D})$*-singularity if and only if*

$$\left.\begin{array}{l}\hbar_{i_\alpha} G^{(s_{i_\alpha},\alpha;\mathscr{D})}_{\partial\Omega_{\alpha i_\alpha}}(\mathbf{x}_m,t_{m\pm},\mathbf{p}_\alpha,\boldsymbol{\lambda}_{i_\alpha})=0\\ \text{for } s_{i_\alpha}=0,1,\cdots,2k_{i_\alpha};\\ \hbar_{i_\alpha} G^{(2k_{i_\alpha}+1,\alpha;\mathscr{D})}_{\partial\Omega_{\alpha i_\alpha}}(\mathbf{x}_m,t_{m\pm},\mathbf{p}_\alpha,\boldsymbol{\lambda}_{i_\alpha})>0\end{array}\right\} \textit{for all } i_\alpha\in\mathscr{N}_{\alpha\mathscr{B}}(\sigma_r), \tag{6.53}$$

$$\hbar_{j_\beta} G^{(s_{j_\beta},\beta;\mathscr{D})}_{\partial\Omega_{\beta j_\beta}}(\mathbf{x}_m,t_{m\pm},\mathbf{p}_\beta,\boldsymbol{\lambda}_{j_\beta})=0 \;\; \textit{for } s_{j_\beta}=0,1,\cdots,2k_{j_\beta};$$

$$\hbar_{j_\beta} G^{(2k_{j_\beta}+1,\beta;\mathscr{D})}_{\partial\Omega_{\beta j_\beta}}(\mathbf{x}_m, t_{m\pm}, \mathbf{p}_\beta, \boldsymbol{\lambda}_{j_\beta}) > 0 \ \textit{for all } j_\beta \in \mathscr{N}_{\beta\mathscr{B}}(\sigma_r). \tag{6.54}$$

(vii) *Without any switching rules, two grazing flows* $\mathbf{x}^{(\alpha;\mathscr{D})}(t)$ *and* $\mathbf{x}^{(\beta;\mathscr{D})}(t)$ *to the edge at point* $\mathbf{x}_m \in \mathscr{E}^{(r)}_{\sigma_r}$ *are the double, tangential flows of the* $((2\mathbf{k}_\alpha+1):(2\mathbf{k}_\beta+1);\mathscr{D})$*-singularity if and only if Eqs.* (6.53) *and* (6.54) *hold.*

Proof. The proof is the same as in Chapter 3 for each domain and boundary. ■

Definition 6.10. For a discontinuous dynamical system in Eqs. (6.4)-(6.8), there is a point $\mathbf{x}^{(\sigma_r;\mathscr{E})}(t_m) \equiv \mathbf{x}_m \in \mathscr{E}^{(r)}_{\sigma_r}$ ($\sigma_r \in \mathscr{N}^r_{\mathscr{E}} = \{1,2,\cdots,N_r\}$ and $r \in \{0,1,2,\cdots,n-2\}$) at time t_m with the related boundaries $\partial\Omega_{\rho i_\rho}(\sigma_r)$ and related domains $\Omega_\rho(\sigma_r)$ ($\rho \in \mathscr{N}_{\mathscr{D}}(\sigma_r) \subset \mathscr{N}_{\mathscr{D}} = \{1,2,\cdots,N_d\}$ and $i_\rho \in \mathscr{N}_{\rho\mathscr{B}}(\sigma_r) \subset \mathscr{N}_{\mathscr{B}} = \{1,2,\cdots,N_b\}$), where the boundary index subset of $\mathscr{N}_{\rho\mathscr{B}}(\sigma_r)$ has n_{σ_r}-elements ($\rho = \alpha,\beta$ and $n_{\sigma_r} \ge (n-r)$). For an arbitrarily small $\varepsilon > 0$, there are two time intervals $[t_{m-\varepsilon}, t_m)$ and $(t_m, t_{m+\varepsilon}]$. A flow $\mathbf{x}^{(\rho;\mathscr{D})}(t)$ is $C^{r_\rho}_{[t_{m-\varepsilon},t_m)}$ and/or $C^{r_\rho}_{(t_m,t_{m+\varepsilon}]}$-continuous and $\| d^{r_\rho+1}\mathbf{x}^{(\rho;\mathscr{D})}/dt^{r_\rho+1} \| < \infty$ ($r_\rho \ge 1$) for time t. Suppose a switching rule $\mathbf{x}^{(\alpha;\mathscr{D})}(t_{m\pm}) = \mathbf{x}_m = \mathbf{x}^{(\beta;\mathscr{D})}(t_{m\pm})$ exists. $\mathscr{N}_{\mathscr{D}}(\sigma_r) = \cup_{j\in\mathscr{J}} \mathscr{N}^j_{\mathscr{D}}(\sigma_r)$ ($\mathscr{J} = \{C, L, I\}$) and $\cap_{j\in\mathscr{J}} \mathscr{N}^j_{\mathscr{D}}(\sigma_r) = \varnothing$.

(i) With the switching rule, the r-dimensional edge $\mathscr{E}^{(r)}_{\sigma_r}$ for all coming domain flows $\mathbf{x}^{(\alpha;\mathscr{D})}(t)$ and all leaving domain flows $\mathbf{x}^{(\beta;\mathscr{D})}(t)$ at point $\mathbf{x}_m \in \mathscr{E}^{(r)}_{\sigma_r}$ is $(\mathscr{N}^C_{\mathscr{D}} : \mathscr{N}^L_{\mathscr{D}})$ *switchable* if for all $\alpha \in \mathscr{N}^C_{\mathscr{D}}(\sigma_r)$ and $\beta \in \mathscr{N}^L_{\mathscr{D}}(\sigma_r)$

$$\begin{aligned} &\hbar_{i_\alpha} \mathbf{n}^{\mathrm{T}}_{\partial\Omega_{\alpha i_\alpha}}(\mathbf{x}^{(i_\alpha;\mathscr{B})}_{m-\varepsilon}) \cdot [\mathbf{x}^{(i_\alpha;\mathscr{B})}_{m-\varepsilon} - \mathbf{x}^{(\alpha;\mathscr{D})}_{m-\varepsilon}] < 0 \ \text{for all } i_\alpha \in \mathscr{N}_{\alpha\mathscr{B}}(\sigma_r), \\ &\hbar_{j_\beta} \mathbf{n}^{\mathrm{T}}_{\partial\Omega_{\beta j_\beta}}(\mathbf{x}^{(j_\beta;\mathscr{B})}_{m+\varepsilon}) \cdot [\mathbf{x}^{(\beta;\mathscr{D})}_{m+\varepsilon} - \mathbf{x}^{(j_\beta;\mathscr{B})}_{m+\varepsilon}] > 0 \ \text{for all } j_\beta \in \mathscr{N}_{\beta\mathscr{B}}(\sigma_r). \end{aligned} \tag{6.55}$$

(ii) The r-dimensional edge $\mathscr{E}^{(r)}_{\sigma_r}$ for all coming domain flows $\mathbf{x}^{(\alpha;\mathscr{D})}(t)$ at point $\mathbf{x}_m \in \mathscr{E}^{(r)}_{\sigma_r}$ is $(\mathscr{N}^C_{\mathscr{D}};\mathscr{D})$ *non-switchable of the first kind* if for $\mathscr{N}^L_{\mathscr{D}}(\sigma_r) = \varnothing$, and all $\alpha \in \mathscr{N}^C_{\mathscr{D}}(\sigma_r)$

$$\hbar_{i_\alpha} \mathbf{n}^{\mathrm{T}}_{\partial\Omega_{\alpha i_\alpha}}(\mathbf{x}^{(i_\alpha;\mathscr{B})}_{m-\varepsilon}) \cdot [\mathbf{x}^{(i_\alpha;\mathscr{B})}_{m-\varepsilon} - \mathbf{x}^{(\alpha;\mathscr{D})}_{m-\varepsilon}] < 0 \ \text{for all } i_\alpha \in \mathscr{N}_{\alpha\mathscr{B}}(\sigma_r). \tag{6.56}$$

(iii) The r-dimensional edge $\mathscr{E}^{(r)}_{\sigma_r}$ for all leaving domain flows $\mathbf{x}^{(\beta;\mathscr{D})}(t)$ at point $\mathbf{x}_m \in \mathscr{E}^{(r)}_{\sigma_r}$ is $(\mathscr{N}^L_{\mathscr{D}};\mathscr{D})$ *non-switchable of the second kind* if for $\mathscr{N}^C_{\mathscr{D}}(\sigma_r) = \varnothing$ and all $\beta \in \mathscr{N}^L_{\mathscr{D}}(\sigma_r)$

$$\hbar_{j_\beta}\mathbf{n}^{\mathrm{T}}_{\partial\Omega_{\beta j_\beta}}(\mathbf{x}^{(j_\beta;\mathscr{B})}_{m+\varepsilon})\cdot[\mathbf{x}^{(\beta;\mathscr{D})}_{m+\varepsilon}-\mathbf{x}^{(j_\beta;\mathscr{B})}_{m+\varepsilon}]>0 \text{ for all } j_\beta\in\mathscr{N}_{\beta\mathscr{B}}(\sigma_r). \tag{6.57}$$

Theorem 6.5. *For a discontinuous dynamical system in Eqs.* (6.4)-(6.8), *there is a point* $\mathbf{x}^{(\sigma_r;\mathscr{E})}(t_m)\equiv\mathbf{x}_m\in\mathscr{E}^{(r)}_{\sigma_r}$ ($\sigma_r\in\mathscr{N}^r_{\mathscr{E}}=\{1,2,\cdots,N_r\}$ *and* $r\in\{0,1,2,\cdots,n-2\}$) *at time* t_m *with the related boundaries* $\partial\Omega_{\rho i_\rho}(\sigma_r)$ *and related domains* $\Omega_\rho(\sigma_r)$ ($\rho\in\mathscr{N}_{\mathscr{D}}(\sigma_r)\subset\mathscr{N}_{\mathscr{D}}=\{1,2,\cdots,N_d\}$ *and* $i_\rho\in\mathscr{N}_{\rho\mathscr{B}}(\sigma_r)\subset\mathscr{N}_{\mathscr{B}}=\{1,2,\cdots,N_b\}$), *where the boundary index subset of* $\mathscr{N}_{\rho\mathscr{B}}(\sigma_r)$ *has* n_{σ_r}*-elements* ($\rho=\alpha,\beta$ *and* $n_{\sigma_r}\ge(n-r)$). *For an arbitrarily small* $\varepsilon>0$, *there are two time intervals* $[t_{m-\varepsilon},t_m)$ *and* $(t_m,t_{m+\varepsilon}]$. *A flow* $\mathbf{x}^{(\rho;\mathscr{D})}(t)$ *is* $C^{r_\rho}_{[t_{m-\varepsilon},t_m)}$ *and/or* $C^{r_\rho}_{(t_m,t_{m+\varepsilon}]}$*-continuous and* $\|d^{r_\rho+1}\mathbf{x}^{(\rho;\mathscr{D})}/dt^{r_\rho+1}\|<\infty$ ($r_\rho\ge1$) *for time* t . *Suppose a switching rule* $\mathbf{x}^{(\alpha;\mathscr{D})}(t_{m\pm})=\mathbf{x}_m=\mathbf{x}^{(\beta;\mathscr{D})}(t_{m\pm})$ *exists.* $\mathscr{N}_{\mathscr{D}}(\sigma_r)=\cup_{j\in\mathscr{J}}\mathscr{N}^j_{\mathscr{D}}(\sigma_r)$ ($\mathscr{J}=\{C,L,I\}$) *and* $\cap_{j\in\mathscr{J}}\mathscr{N}^j_{\mathscr{D}}(\sigma_r)=\varnothing$.

(i) *With the switching rule, the r-dimensional edge* $\mathscr{E}^{(r)}_{\sigma_r}$ *for all coming domain flows* $\mathbf{x}^{(\alpha;\mathscr{D})}(t)$ *and all leaving domain flows* $\mathbf{x}^{(\beta;\mathscr{D})}(t)$ *at point* $\mathbf{x}_m\in\mathscr{E}^{(r)}_{\sigma_r}$ *is* ($\mathscr{N}^C_{\mathscr{D}}:\mathscr{N}^L_{\mathscr{D}}$) *switchable if and only if for all* $\alpha\in\mathscr{N}^C_{\mathscr{D}}(\sigma_r)$ *and* $\beta\in\mathscr{N}^L_{\mathscr{D}}(\sigma_r)$

$$\begin{aligned}&\hbar_{i_\alpha}G^{(\alpha;\mathscr{D})}_{\partial\Omega_{\alpha i_\alpha}}(\mathbf{x}_m,t_{m-},\mathbf{p}_\alpha,\boldsymbol{\lambda}_{i_\alpha})<0 \text{ for all } i_\alpha\in\mathscr{N}_{\alpha\mathscr{B}}(\sigma_r),\\&\hbar_{j_\beta}G^{(\beta;\mathscr{D})}_{\partial\Omega_{\beta j_\beta}}(\mathbf{x}_m,t_{m+},\mathbf{p}_\beta,\boldsymbol{\lambda}_{j_\beta})>0 \text{ for all } j_\beta\in\mathscr{N}_{\beta\mathscr{B}}(\sigma_r).\end{aligned} \tag{6.58}$$

(ii) *The r-dimensional edge* $\mathscr{E}^{(r)}_{\sigma_r}$ *for all coming domain flows* $\mathbf{x}^{(\alpha;\mathscr{D})}(t)$ *at point* $\mathbf{x}_m\in\mathscr{E}^{(r)}_{\sigma_r}$ *is* ($\mathscr{N}^C_{\mathscr{D}};\mathscr{D}$) *non-switchable of the first kind if and only if for* $\mathscr{N}^L_{\mathscr{D}}(\sigma_r)=\varnothing$ *and all* $\alpha\in\mathscr{N}^C_{\mathscr{D}}(\sigma_r)$

$$\hbar_{i_\alpha}G^{(\alpha;\mathscr{D})}_{\partial\Omega_{\alpha i_\alpha}}(\mathbf{x}_m,t_{m-},\mathbf{p}_\alpha,\boldsymbol{\lambda}_{i_\alpha})<0 \text{ for all } i_\alpha\in\mathscr{N}_{\alpha\mathscr{B}}(\sigma_r). \tag{6.59}$$

(iii) *The r-dimensional edge* $\mathscr{E}^{(r)}_{\sigma_r}$ *for all leaving domain flows* $\mathbf{x}^{(\beta;\mathscr{D})}(t)$ *at point* $\mathbf{x}_m\in\mathscr{E}^{(r)}_{\sigma_r}$ *is* ($\mathscr{N}^L_{\mathscr{D}};\mathscr{D}$) *non-switchable of the second kind if and only if for* $\mathscr{N}^C_{\mathscr{D}}(\sigma_r)=\varnothing$ *and all* $\beta\in\mathscr{N}^L_{\mathscr{D}}(\sigma_r)$

$$\hbar_{j_\beta}G^{(\beta;\mathscr{D})}_{\partial\Omega_{\beta j_\beta}}(\mathbf{x}_m,t_{m+},\mathbf{p}_\beta,\boldsymbol{\lambda}_{j_\beta})>0 \text{ for all } j_\beta\in\mathscr{N}_{\beta\mathscr{B}}(\sigma_r). \tag{6.60}$$

Proof. The proof is the same as in Chapter 3 for each boundary. ∎

Definition 6.11. For a discontinuous dynamical system in Eqs. (6.4)-(6.8), there is a point $\mathbf{x}^{(\sigma_r;\mathscr{E})}(t_m)\equiv\mathbf{x}_m\in\mathscr{E}^{(r)}_{\sigma_r}$ ($\sigma_r\in\mathscr{N}^r_{\mathscr{E}}=\{1,2,\cdots,N_r\}$ and $r\in\{0,1,2,\cdots,n-2\}$)

at time t_m with the related boundaries $\partial\Omega_{\rho i_\rho}(\sigma_r)$ and related domains $\Omega_\rho(\sigma_r)$ ($\rho \in \mathscr{N}_{\mathscr{D}}(\sigma_r) \subset \mathscr{N}_{\mathscr{D}} = \{1,2,\cdots,N_d\}$ and $i_\rho \in \mathscr{N}_{\rho\mathscr{B}}(\sigma_r) \subset \mathscr{N}_{\mathscr{B}} = \{1,2,\cdots,N_b\}$), where the boundary index subset of $\mathscr{N}_{\rho\mathscr{B}}(\sigma_r)$ has n_{σ_r}-elements ($\rho = \alpha, \beta$ and $n_{\sigma_r} \geq (n-r)$). For an arbitrarily small $\varepsilon > 0$, there are two time intervals $[t_{m-\varepsilon}, t_m)$ and $(t_m, t_{m+\varepsilon}]$. A flow $\mathbf{x}^{(\rho;\mathscr{D})}(t)$ is $C^{r_\rho}_{[t_{m-\varepsilon},t_m)}$ and/or $C^{r_\rho}_{(t_m,t_{m+\varepsilon}]}$-continuous and $\| d^{r_\rho+1}\mathbf{x}^{(\rho;\mathscr{D})}/dt^{r_\rho+1} \| < \infty$ ($r_\rho \geq \max_{i_\rho \in \mathscr{N}_{\rho\mathscr{B}}}\{m_{i_\rho}\}$) for time t. Suppose a switching rule $\mathbf{x}^{(\alpha;\mathscr{D})}(t_{m\pm}) = \mathbf{x}_m = \mathbf{x}^{(\beta;\mathscr{D})}(t_{m\pm})$ exists. $\mathscr{N}_{\mathscr{D}}(\sigma_r) = \cup_{j\in\mathscr{J}} \mathscr{N}^j_{\mathscr{D}}(\sigma_r)$ and $\cap_{j\in\mathscr{J}} \mathscr{N}^j_{\mathscr{D}}(\sigma_r) = \varnothing$ ($\mathscr{J} = \{C, L, I\}$). As $r = 0$, $\mathbf{m}_\rho \neq 2\mathbf{k}_\rho + 1$.

(i) With the switching rule, the r-dimensional edge $\mathscr{E}^{(r)}_{\sigma_r}$ for all coming domain flows $\mathbf{x}^{(\alpha;\mathscr{D})}(t)$ and all leaving domain flows $\mathbf{x}^{(\beta;\mathscr{D})}(t)$ at point $\mathbf{x}_m \in \mathscr{E}^{(r)}_{\sigma_r}$ is *switchable with the* $(\cup_{\alpha\in\mathscr{N}^C_{\mathscr{D}}} \mathbf{m}_\alpha : \cup_{\beta\in\mathscr{N}^L_{\mathscr{D}}} \mathbf{m}_\beta; \mathscr{D})$ *singularity* if for all $\alpha \in \mathscr{N}^C_{\mathscr{D}}(\sigma_r)$ and $\beta \in \mathscr{N}^L_{\mathscr{D}}(\sigma_r)$

$$\left.\begin{array}{l} \hbar_{i_\alpha} G^{(s_{i_\alpha},\alpha;\mathscr{D})}_{\partial\Omega_{\alpha i_\alpha}}(\mathbf{x}_m, t_{m-}, \mathbf{p}_\alpha, \boldsymbol{\lambda}_{i_\alpha}) = 0 \\ \text{for } s_{i_\alpha} = 0,1,\cdots,m_{i_\alpha}-1; \\ \hbar_{i_\alpha} \mathbf{n}^{\mathrm{T}}_{\partial\Omega_{\alpha i_\alpha}}(\mathbf{x}^{(i_\alpha,\mathscr{B})}_{m-\varepsilon}) \cdot [\mathbf{x}^{(i_\alpha;\mathscr{B})}_{m-\varepsilon} - \mathbf{x}^{(\alpha;\mathscr{D})}_{m-\varepsilon}] < 0 \end{array}\right\} \text{for all } i_\alpha \in \mathscr{N}_{\alpha\mathscr{B}}(\sigma_r), \tag{6.61}$$

$$\left.\begin{array}{l} \hbar_{j_\beta} G^{(s_{j_\beta},\beta;\mathscr{D})}_{\partial\Omega_{\beta j_\beta}}(\mathbf{x}_m, t_{m+}, \mathbf{p}_\beta, \boldsymbol{\lambda}_{j_\beta}) = 0 \\ \text{for } s_{j_\beta} = 0,1,\cdots,m_{j_\beta}-1; \\ \hbar_{j_\beta} \mathbf{n}^{\mathrm{T}}_{\partial\Omega_{\beta j_\beta}}(\mathbf{x}^{(j_\beta,\mathscr{B})}_{m+\varepsilon}) \cdot [\mathbf{x}^{(\beta;\mathscr{D})}_{m+\varepsilon} - \mathbf{x}^{(j_\beta;\mathscr{B})}_{m+\varepsilon}] > 0 \end{array}\right\} \text{for all } j_\beta \in \mathscr{N}_{\beta\mathscr{B}}(\sigma_r). \tag{6.62}$$

(ii) The r-dimensional edge $\mathscr{E}^{(r)}_{\sigma_r}$ for all coming domain flows $\mathbf{x}^{(\alpha;\mathscr{D})}(t)$ at point $\mathbf{x}_m \in \mathscr{E}^{(r)}_{\sigma_r}$ is *non-switchable of the first kind with the* $(\cup_{\alpha\in\mathscr{N}^C_{\mathscr{D}}} \mathbf{m}_\alpha; \mathscr{D})$ *singularity* ($\mathbf{m}_\alpha \neq 2\mathbf{k}_\alpha + 1$) if for $\mathscr{N}^L_{\mathscr{D}}(\sigma_r) = \varnothing$ and all $\alpha \in \mathscr{N}^C_{\mathscr{D}}(\sigma_r)$

$$\left.\begin{array}{l} \hbar_{i_\alpha} G^{(s_{i_\alpha},\alpha;\mathscr{D})}_{\partial\Omega_{\alpha i_\alpha}}(\mathbf{x}_m, t_{m-}, \mathbf{p}_\alpha, \boldsymbol{\lambda}_{i_\alpha}) = 0 \\ \text{for } s_{i_\alpha} = 0,1,\cdots,m_{i_\alpha}-1; \\ \hbar_{i_\alpha} \mathbf{n}^{\mathrm{T}}_{\partial\Omega_{\alpha i_\alpha}}(\mathbf{x}^{(i_\alpha,\mathscr{B})}_{m-\varepsilon}) \cdot [\mathbf{x}^{(i_\alpha;\mathscr{B})}_{m-\varepsilon} - \mathbf{x}^{(\alpha;\mathscr{D})}_{m-\varepsilon}] < 0 \end{array}\right\} \text{for all } i_\alpha \in \mathscr{N}_{\alpha\mathscr{B}}(\sigma_r). \tag{6.63}$$

(iii) The r-dimensional edge $\mathscr{E}^{(r)}_{\sigma_r}$ for all leaving domain flows $\mathbf{x}^{(\beta;\mathscr{D})}(t)$ at point $\mathbf{x}_m \in \mathscr{E}^{(r)}_{\sigma_r}$ is *non-switchable with the* $(\cup_{\beta\in\mathscr{N}^L_{\mathscr{D}}} \mathbf{m}_\beta; \mathscr{D})$ *singularity* if for $\mathscr{N}^C_{\mathscr{D}}(\sigma_r) = \varnothing$ and all $\beta \in \mathscr{N}^L_{\mathscr{D}}(\sigma_r)$

$$\left.\begin{array}{l} \hbar_{j_\beta} G^{(s_{j_\beta},\beta;\mathscr{D})}_{\partial\Omega_{\beta j_\beta}}(\mathbf{x}_m,t_{m+},\mathbf{p}_\beta,\boldsymbol{\lambda}_{j_\beta})=0 \\ \text{for } s_{j_\beta}=0,1,\cdots,m_{j_\beta}-1; \\ \hbar_{j_\beta}\mathbf{n}^{\mathrm{T}}_{\partial\Omega_{\beta j_\beta}}(\mathbf{x}^{(j_\beta,\mathscr{B})}_{m+\varepsilon})\cdot[\mathbf{x}^{(\beta;\mathscr{D})}_{m+\varepsilon}-\mathbf{x}^{(j_\beta;\mathscr{B})}_{m+\varepsilon}]>0 \end{array}\right\} \text{for all } j_\beta\in\mathscr{N}_{\beta\mathscr{B}}(\sigma_r). \tag{6.64}$$

(iv) Without the switching rule, the r-dimensional edge $\mathscr{E}^{(r)}_{\sigma_r}$ for all coming domain flows $\mathbf{x}^{(\alpha;\mathscr{D})}(t)$ and all leaving domain flows $\mathbf{x}^{(\beta;\mathscr{D})}(t)$ at point $\mathbf{x}_m\in\mathscr{E}^{(r)}_{\sigma_r}$ is *potentially switchable with the* $(\cup_{\alpha\in\mathscr{N}^C_{\mathscr{D}}}\mathbf{m}_\alpha:\cup_{\beta\in\mathscr{N}^L_{\mathscr{D}}}\mathbf{m}_\beta;\mathscr{D})$ *singularity* if Eqs. (6.61) and (6.62) hold with $\mathbf{m}_\alpha\neq 2\mathbf{k}_\alpha+1$ for all $\alpha\in\mathscr{N}^C_{\mathscr{D}}(\sigma_r)$ and $\beta\in\mathscr{N}^L_{\mathscr{D}}(\sigma_r)$.

(v) Without any switching rules, the r-dimensional edge $\mathscr{E}^{(r)}_{\sigma_r}$ for all coming domain flows $\mathbf{x}^{(\alpha;\mathscr{D})}(t)$ and all leaving domain flows $\mathbf{x}^{(\beta;\mathscr{D})}(t)$ at point $\mathbf{x}_m\in\mathscr{E}^{(r)}_{\sigma_r}$ is *non-switchable with the* $(\cup_{\alpha\in\mathscr{N}^C_{\mathscr{D}}}\mathbf{m}_\alpha:\cup_{\beta\in\mathscr{N}^L_{\mathscr{D}}}\mathbf{m}_\beta;\mathscr{D})$ *singularity* if Eqs. (6.61) and (6.62) hold for all $\alpha\in\mathscr{N}^C_{\mathscr{D}}(\sigma_r)$ and $\beta\in\mathscr{N}^L_{\mathscr{D}}(\sigma_r)$ with $\mathbf{m}_\alpha=2\mathbf{k}_\alpha+1$ for at least one of coming flows.

(vi) With the switching rule, the r-dimensional edge $\mathscr{E}^{(r)}_{\sigma_r}$ at point $\mathbf{x}_m\in\mathscr{E}^{(r)}_{\sigma_r}$ is *switchable with the* $(\cup_{\alpha\in\mathscr{N}^C_{\mathscr{D}}}(2\mathbf{k}_\alpha+1):\cup_{\alpha\in\mathscr{N}^L_{\mathscr{D}}}(2\mathbf{k}_\alpha+1);\mathscr{D})$ *singularity* for all the grazing domain flows in domains Ω_α $(\alpha\in\mathscr{N}_{\mathscr{D}}(\sigma_r))$ if

$$\left.\begin{array}{l} \hbar_{i_\alpha} G^{(s_{i_\alpha},\alpha;\mathscr{D})}_{\partial\Omega_{\alpha i_\alpha}}(\mathbf{x}_m,t_{m\pm},\mathbf{p}_\alpha,\boldsymbol{\lambda}_{i_\alpha})=0 \\ \text{for } s_{i_\alpha}=0,1,\cdots,2k_{i_\alpha}; \\ \hbar_{i_\alpha}\mathbf{n}^{\mathrm{T}}_{\partial\Omega_{\alpha i_\alpha}}(\mathbf{x}^{(i_\alpha,\mathscr{B})}_{m-\varepsilon})\cdot[\mathbf{x}^{(i_\alpha;\mathscr{B})}_{m-\varepsilon}-\mathbf{x}^{(\alpha;\mathscr{D})}_{m-\varepsilon}]<0, \\ \hbar_{i_\alpha}\mathbf{n}^{\mathrm{T}}_{\partial\Omega_{\alpha i_\alpha}}(\mathbf{x}^{(i_\alpha,\mathscr{B})}_{m+\varepsilon})\cdot[\mathbf{x}^{(\alpha;\mathscr{D})}_{m+\varepsilon}-\mathbf{x}^{(i_\alpha;\mathscr{B})}_{m+\varepsilon}]>0 \end{array}\right\} \text{for all } i_\alpha\in\mathscr{N}_{\alpha\mathscr{B}}(\sigma_r). \tag{6.65}$$

(vii) Without any switching rules, the r-dimensional edge at point $\mathbf{x}_m\in\mathscr{E}^{(r)}_{\sigma_r}$ is a *grazing edge with the* $(\cup_{\alpha\in\mathscr{N}^C_{\mathscr{D}}}(2\mathbf{k}_\alpha+1):\cup_{\alpha\in\mathscr{N}^L_{\mathscr{D}}}(2\mathbf{k}_\alpha+1);\mathscr{D})$ *singularity* for all the domain flows in domains Ω_α ($\alpha\in\mathscr{N}_{\mathscr{D}}(\sigma_r)$) if Eq. (6.65) holds.

Theorem 6.6. *For a discontinuous dynamical system in Eqs.* (6.4)-(6.8), *there is a point* $\mathbf{x}^{(\sigma_r;\mathscr{E})}(t_m)\equiv\mathbf{x}_m\in\mathscr{E}^{(r)}_{\sigma_r}$ ($\sigma_r\in\mathscr{N}^r_{\mathscr{E}}=\{1,2,\cdots,N_r\}$ *and* $r\in\{0,1,2,\cdots,n-2\}$) *at time* t_m *with the related boundaries* $\partial\Omega_{\rho i_\rho}(\sigma_r)$ *and related domains* $\Omega_\rho(\sigma_r)$ ($\rho\in\mathscr{N}_{\mathscr{D}}(\sigma_r)\subset\mathscr{N}_{\mathscr{D}}=\{1,2,\cdots,N_d\}$ *and* $i_\rho\in\mathscr{N}_{\rho\mathscr{B}}(\sigma_r)\subset\mathscr{N}_{\mathscr{B}}=\{1,2,\cdots,N_b\}$), *where the boundary index subset of* $\mathscr{N}_{\rho\mathscr{B}}(\sigma_r)$ *has* n_{σ_r}*-elements* ($\rho=\alpha,\beta$ *and* $n_{\sigma_r}\geq(n-r)$). *For an arbitrarily small* $\varepsilon>0$, *there are two time intervals*

$[t_{m-\varepsilon}, t_m)$ *and* $(t_m, t_{m+\varepsilon}]$. *A flow* $\mathbf{x}^{(\rho;\mathscr{D})}(t)$ *is* $C^{r_\rho}_{[t_{m-\varepsilon},t_m)}$ *and/or* $C^{r_\rho}_{(t_m,t_{m+\varepsilon}]}$*-continuous and* $\| d^{r_\rho+1}\mathbf{x}^{(\rho;\mathscr{D})}/dt^{r_\rho+1} \| < \infty$ ($r_\rho \geq \max_{i_\rho \in \mathscr{N}_{\rho\mathscr{B}}}\{m_{i_\rho}\}+1$) *for time* t. *Suppose a switching rule* $\mathbf{x}^{(\alpha;\mathscr{D})}(t_{m\pm}) = \mathbf{x}_m = \mathbf{x}^{(\beta;\mathscr{D})}(t_{m\pm})$ *exists.* $\mathscr{N}_{\mathscr{D}}(\sigma_r) = \cup_{j\in\mathscr{J}}\mathscr{N}^j_{\mathscr{D}}(\sigma_r)$ *and* $\cap_{j\in\mathscr{J}}\mathscr{N}^j_{\mathscr{D}}(\sigma_r) = \varnothing$ ($\mathscr{J} = \{C, L, I\}$). *As* $r = 0$, $\mathbf{m}_\rho \neq 2\mathbf{k}_\rho + 1$.

(i) *With the switching rule, the r-dimensional edge* $\mathscr{E}^{(r)}_{\sigma_r}$ *for all coming domain flows* $\mathbf{x}^{(\alpha;\mathscr{D})}(t)$ *and all leaving domain flows* $\mathbf{x}^{(\beta;\mathscr{D})}(t)$ *at point* $\mathbf{x}_m \in \mathscr{E}^{(r)}_{\sigma_r}$ *is switchable with the* $(\cup_{\alpha\in\mathscr{N}^C_{\mathscr{D}}}\mathbf{m}_\alpha : \cup_{\beta\in\mathscr{N}^L_{\mathscr{D}}}\mathbf{m}_\beta;\mathscr{D})$ *singularity if and only if for all* $\alpha \in \mathscr{N}^C_{\mathscr{D}}(\sigma_r)$ *and* $\beta \in \mathscr{N}^L_{\mathscr{D}}(\sigma_r)$

$$\left.\begin{array}{l} \hbar_{i_\alpha} G^{(s_{i_\alpha},\alpha;\mathscr{D})}_{\partial\Omega_{\alpha i_\alpha}}(\mathbf{x}_m, t_{m-}, \mathbf{p}_\alpha, \boldsymbol{\lambda}_{i_\alpha}) = 0 \\ \text{for } s_{i_\alpha} = 0,1,\cdots,m_{i_\alpha}-1; \\ (-1)^{m_{i_\alpha}}\hbar_{i_\alpha} G^{(m_{i_\alpha},\alpha;\mathscr{D})}_{\partial\Omega_{\alpha i_\alpha}}(\mathbf{x}_m, t_{m-}, \mathbf{p}_\alpha, \boldsymbol{\lambda}_{i_\alpha}) < 0 \end{array}\right\} \text{for all } i_\alpha \in \mathscr{N}_{\alpha\mathscr{B}}(\sigma_r), \tag{6.66}$$

$$\left.\begin{array}{l} \hbar_{j_\beta} G^{(s_{j_\beta},\beta;\mathscr{D})}_{\partial\Omega_{\beta j_\beta}}(\mathbf{x}_m, t_{m+}, \mathbf{p}_\beta, \boldsymbol{\lambda}_{j_\beta}) = 0 \\ \text{for } s_{j_\beta} = 0,1,\cdots,m_{j_\beta}-1; \\ \hbar_{j_\beta} G^{(m_{j_\beta},\beta;\mathscr{D})}_{\partial\Omega_{\beta j_\beta}}(\mathbf{x}_m, t_{m+}, \mathbf{p}_\beta, \boldsymbol{\lambda}_{j_\beta}) > 0 \end{array}\right\} \text{for all } j_\beta \in \mathscr{N}_{\beta\mathscr{B}}(\sigma_r). \tag{6.67}$$

(ii) *The r-dimensional edge* $\mathscr{E}^{(r)}_{\sigma_r}$ *for all coming domain flows* $\mathbf{x}^{(\alpha;\mathscr{D})}(t)$ *at point* $\mathbf{x}_m \in \mathscr{E}^{(r)}_{\sigma_r}$ *is non-switchable of the first kind with the* $(\cup_{\alpha\in\mathscr{N}^C_{\mathscr{D}}}\mathbf{m}_\alpha;\mathscr{D})$ *singularity* ($\mathbf{m}_\alpha \neq 2\mathbf{k}_\alpha + 1$) *if and only if for* $\mathscr{N}^L_{\mathscr{D}}(\sigma_r) = \varnothing$ *and all* $\alpha \in \mathscr{N}^C_{\mathscr{D}}(\sigma_r)$

$$\left.\begin{array}{l} \hbar_{i_\alpha} G^{(s_{i_\alpha},\alpha;\mathscr{D})}_{\partial\Omega_{\alpha i_\alpha}}(\mathbf{x}_m, t_{m-}, \mathbf{p}_\alpha, \boldsymbol{\lambda}_{i_\alpha}) = 0 \\ \text{for } s_{i_\alpha} = 0,1,\cdots,m_{i_\alpha}-1; \\ (-1)^{m_{i_\alpha}}\hbar_{i_\alpha} G^{(m_{i_\alpha},\alpha;\mathscr{D})}_{\partial\Omega_{\alpha i_\alpha}}(\mathbf{x}_m, t_{m-}, \mathbf{p}_\alpha, \boldsymbol{\lambda}_{i_\alpha}) < 0 \end{array}\right\} \text{for all } i_\alpha \in \mathscr{N}_{\alpha\mathscr{B}}(\sigma_r). \tag{6.68}$$

(iii) *The r-dimensional edge* $\mathscr{E}^{(r)}_{\sigma_r}$ *for all leaving domain flows* $\mathbf{x}^{(\beta;\mathscr{D})}(t)$ *at point* $\mathbf{x}_m \in \mathscr{E}^{(r)}_{\sigma_r}$ *is non-switchable with the* $(\cup_{\beta\in\mathscr{N}^L_{\mathscr{D}}}\mathbf{m}_\beta;\mathscr{D})$ *singularity if for* $\mathscr{N}^C_{\mathscr{D}}(\sigma_r) = \varnothing$ *and all* $\beta \in \mathscr{N}^L_{\mathscr{D}}(\sigma_r)$

$$\left.\begin{array}{l} \hbar_{j_\beta} G^{(s_{j_\beta},\beta;\mathscr{D})}_{\partial\Omega_{\beta j_\beta}}(\mathbf{x}_m, t_{m+}, \mathbf{p}_\beta, \boldsymbol{\lambda}_{j_\beta}) = 0 \\ \text{for } s_{j_\beta} = 0,1,\cdots,m_{j_\beta}-1; \\ \hbar_{j_\beta} G^{(m_{j_\beta},\beta;\mathscr{D})}_{\partial\Omega_{\beta j_\beta}}(\mathbf{x}_m, t_{m+}, \mathbf{p}_\beta, \boldsymbol{\lambda}_{j_\beta}) > 0 \end{array}\right\} \text{for all } j_\beta \in \mathscr{N}_{\beta\mathscr{B}}(\sigma_r). \tag{6.69}$$

(iv) *Without the switching rule, the r-dimensional edge* $\mathscr{E}_{\sigma_r}^{(r)}$ *for all coming domain flows* $\mathbf{x}^{(\alpha;\mathscr{D})}(t)$ *and all leaving domain flows* $\mathbf{x}^{(\beta;\mathscr{D})}(t)$ *at point* $\mathbf{x}_m \in \mathscr{E}_{\sigma_r}^{(r)}$ *is potentially switchable with the* $(\cup_{\alpha\in\mathscr{N}_{\mathscr{D}}^C}\mathbf{m}_\alpha : \cup_{\beta\in\mathscr{N}_{\mathscr{D}}^L}\mathbf{m}_\beta;\mathscr{D})$ *singularity if and only if Eqs.* (6.66) *and* (6.67) *hold with* $\mathbf{m}_\alpha \neq 2\mathbf{k}_\alpha + 1$ *for all* $\alpha \in \mathscr{N}_{\mathscr{D}}^C(\sigma_r)$ *and* $\beta \in \mathscr{N}_{\mathscr{D}}^L(\sigma_r)$.

(v) *Without any switching rules, the r-dimensional edge* $\mathscr{E}_{\sigma_r}^{(r)}$ *for all coming domain flows* $\mathbf{x}^{(\alpha;\mathscr{D})}(t)$ *and all leaving domain flows* $\mathbf{x}^{(\beta;\mathscr{D})}(t)$ *at point* $\mathbf{x}_m \in \mathscr{E}_{\sigma_r}^{(r)}$ *is non-switchable with the* $(\cup_{\alpha\in\mathscr{N}_{\mathscr{D}}^C}\mathbf{m}_\alpha : \cup_{\beta\in\mathscr{N}_{\mathscr{D}}^L}\mathbf{m}_\beta;\mathscr{D})$ *singularity if Eqs.* (6.66) *and* (6.67) *hold for all* $\alpha \in \mathscr{N}_{\mathscr{D}}^C(\sigma_r)$ *and* $\beta \in \mathscr{N}_{\mathscr{D}}^L(\sigma_r)$ *with* $\mathbf{m}_\alpha = 2\mathbf{k}_\alpha + 1$ *for at least one of coming flows.*

(vi) *With the switching rule, the r-dimensional edge at point* $\mathbf{x}_m \in \mathscr{E}_{\sigma_r}^{(r)}$ *is switchable with the* $(\cup_{\alpha\in\mathscr{N}_{\mathscr{D}}^C}(2\mathbf{k}_\alpha + 1) : \cup_{\alpha\in\mathscr{N}_{\mathscr{D}}^L}(2\mathbf{k}_\alpha + 1);\mathscr{D})$ *singularity for all the grazing domain flows in domains* Ω_α $(\alpha \in \mathscr{N}_{\mathscr{D}}(\sigma_r))$ *if and only if*

$$\left.\begin{array}{l} \hbar_{i_\alpha} G_{\partial\Omega_{\alpha i_\alpha}}^{(s_{i_\alpha},\alpha;\mathscr{D})}(\mathbf{x}_m, t_{m\pm}, \mathbf{p}_\alpha, \boldsymbol{\lambda}_{i_\alpha}) = 0 \\ \text{for } s_{i_\alpha} = 0,1,\cdots,2k_{i_\alpha}; \\ \hbar_{i_\alpha} G_{\partial\Omega_{\alpha i_\alpha}}^{(2k_{i_\alpha}+1,\alpha;\mathscr{D})}(\mathbf{x}_m, t_{m\pm}, \mathbf{p}_\alpha, \boldsymbol{\lambda}_{i_\alpha}) > 0 \end{array}\right\} \text{for all } i_\alpha \in \mathscr{N}_{\alpha\mathscr{B}}(\sigma_r). \tag{6.70}$$

(vii) *Without any switching rules, the r-dimensional edge* $\mathscr{E}_{\sigma_r}^{(r)}$ *at point* $\mathbf{x}_m \in \mathscr{E}_{\sigma_r}^{(r)}$ *is a grazing edge with the* $(\cup_{\alpha\in\mathscr{N}_{\mathscr{D}}^C}(2\mathbf{k}_\alpha + 1) : \cup_{\alpha\in\mathscr{N}_{\mathscr{D}}^L}(2\mathbf{k}_\alpha + 1);\mathscr{D})$ *singularity for all the domain flows in domains* Ω_α $(\alpha \in \mathscr{N}_{\mathscr{D}}(\sigma_r))$ *if and only if Eq.* (6.70) *holds.*

Proof. The proof is the same as in Chapter 3 for each domain and boundary. ■

6.5. Transverse grazing passability to concave edges

To discuss the passability of a domain flow at the concave edge, the corresponding definition should be given. $\mathbf{n}_{\partial\bar{\Omega}_{\alpha i_\alpha}} = \mathbf{n}_{\partial\Omega_{\alpha i_\alpha}}$ because the imaginary extension of the boundary has the same function as itself from the boundary extension.

Definition 6.12. For a discontinuous dynamical system in Eqs. (6.4)-(6.8), there is a point $\mathbf{x}^{(\sigma_r;\mathscr{E})}(t_m) \equiv \mathbf{x}_m \in \mathscr{E}_{\sigma_r}^{(r)}$ $(\sigma_r \in \mathscr{N}_{\mathscr{E}}^r = \{1,2,\cdots,N_r\}$ and $r \in \{0,1,2,\cdots,n-2\})$

at time t_m with the related boundaries $\partial\Omega_{\rho i_\rho}(\sigma_r)$ and related domains $\Omega_\rho(\sigma_r)$ ($\rho \in \mathscr{N}_{\mathscr{D}}(\sigma_r) \subset \mathscr{N}_{\mathscr{D}} = \{1,2,\cdots,N_d\}$ and $i_\rho \in \mathscr{N}_{\rho\mathscr{B}}(\sigma_r) \subset \mathscr{N}_{\mathscr{B}} = \{1,2,\cdots,N_b\}$), where the boundary index subset of $\mathscr{N}_{\rho\mathscr{B}}(\sigma_r)$ has n_{σ_r}-elements ($\rho = \alpha,\beta$ and $n_{\sigma_r} \ge n-r$). For an arbitrarily small $\varepsilon > 0$, there are two time intervals $[t_{m-\varepsilon}, t_m)$ and $(t_m, t_{m+\varepsilon}]$. A flow $\mathbf{x}^{(\rho;\mathscr{D})}(t)$ is $C^{r_\rho}_{[t_{m-\varepsilon},t_m)}$ and/or $C^{r_\rho}_{(t_m,t_{m+\varepsilon}]}$-continuous and $\| d^{r_\rho+1}\mathbf{x}^{(\rho;\mathscr{D})}/dt^{r_\rho+1} \| < \infty$ for time t. $\cap_{j\in\mathscr{J}} \mathscr{N}^j_{\rho\mathscr{B}}(\sigma_r) = \varnothing$ and $\mathscr{N}_{\rho\mathscr{B}}(\sigma_r) = \cup_{j\in\mathscr{J}} \mathscr{N}^j_{\rho\mathscr{B}}(\sigma_r)$ ($\mathscr{J} = \{C,L\}$). $\lambda_{i_\alpha} = 1,-1$ are for real and imaginary boundaries in the self-mirror domains, respectively.

(i) A domain flow $\mathbf{x}^{(\alpha;\mathscr{D})}$ in the self-mirror domain $\Omega_\alpha^{\mathscr{SM}}$ is called to be *transversely grazing* to the concave edge $\mathscr{E}^{(r)}_{\sigma_r}$ at point $\mathbf{x}_m \in \mathscr{E}^{(r)}_{\sigma_r}$ if

$$
\left.
\begin{aligned}
&\lambda_{i_\alpha}\hbar_{i_\alpha}\mathbf{n}^{\mathrm{T}}_{\partial\Omega_{\alpha i_\alpha}}(\mathbf{x}^{(i_\alpha;\mathscr{B})}_{m-\varepsilon})\cdot[\mathbf{x}^{(i_\alpha;\mathscr{B})}_{m-\varepsilon} - \mathbf{x}^{(\alpha;\mathscr{D})}_{m-\varepsilon}] < 0\\
&\text{for all } i_\alpha \in \mathscr{N}^C_{\alpha\mathscr{B}} \subset \mathscr{N}_{\alpha\mathscr{B}}(\sigma_r,\mathscr{SM}),\\
&\lambda_{j_\alpha}\hbar_{j_\alpha}\mathbf{n}^{\mathrm{T}}_{\partial\Omega_{\alpha j_\alpha}}(\mathbf{x}^{(j_\alpha;\mathscr{B})}_{m+\varepsilon})\cdot[\mathbf{x}^{(\alpha;\mathscr{D})}_{m+\varepsilon} - \mathbf{x}^{(j_\alpha;\mathscr{B})}_{m+\varepsilon}] > 0\\
&\text{for all } j_\alpha \in \mathscr{N}^L_{\alpha\mathscr{B}} \subset \mathscr{N}_{\alpha\mathscr{B}}(\sigma_r,\mathscr{SM}).
\end{aligned}
\right. \tag{6.71}
$$

(ii) A domain flow $\mathbf{x}^{(\alpha;\mathscr{D})}$ tangential to the imaginary extended edges $\mathscr{E}^{(s)}_{\bar{\sigma}_s}$ ($s = r+1, r+2,\cdots,n-1$) with the $(2\mathbf{k}_{n-s}+1)$ singularity in a self-mirror domain $\Omega_\alpha^{\mathscr{SM}}$ at $\mathbf{x}_m \in \mathscr{E}^{(r)}_{\sigma_r}$ does *not transversely pass* over the concave edge of $\mathscr{E}^{(r)}_{\sigma_r}$ if

$$
\left.
\begin{aligned}
&\hbar_{i_\alpha} G^{(s_{i_\alpha},\alpha;\mathscr{D})}_{\partial\bar{\Omega}_{\alpha i_\alpha}}(\mathbf{x}_m, t_{m-}, \mathbf{p}_\alpha, \boldsymbol{\lambda}_{i_\alpha}) = 0\\
&\text{for } s_{i_\alpha} = 0,1,2,\cdots,2k_{i_\alpha};\\
&\hbar_{i_\alpha}\mathbf{n}^{\mathrm{T}}_{\partial\bar{\Omega}_{\alpha i_\alpha}}(\mathbf{x}^{(i_\alpha;\mathscr{B})}_{m-\varepsilon})\cdot[\mathbf{x}^{(i_\alpha;\mathscr{B})}_{m-\varepsilon} - \mathbf{x}^{(\alpha;\mathscr{D})}_{m-\varepsilon}] > 0
\end{aligned}
\right\}
\text{ for all } i_\alpha \in \mathscr{N}^C_{\alpha\mathscr{B}}(\bar{\sigma}_s) \subset \mathscr{N}_{\alpha\mathscr{B}}(\sigma_r,\mathscr{SM}). \tag{6.72}
$$

(iii) A domain flow $\mathbf{x}^{(\alpha;\mathscr{D})}$ inflexional to the imaginary extended edges $\mathscr{E}^{(s)}_{\bar{\sigma}_s}$ ($s = r+1, r+2,\cdots,n-1$) with the $(2\mathbf{k}_{n-s})$th-order singularity in a self-mirror domain $\Omega_\alpha^{\mathscr{SM}}$ at $\mathbf{x}_m \in \mathscr{E}^{(r)}_{\sigma_r}$ *transversely passes* over the concave edge of $\mathscr{E}^{(r)}_{\sigma_r}$ if

$$
\left.
\begin{aligned}
&\hbar_{i_\alpha} G^{(s_{i_\alpha},\alpha;\mathscr{D})}_{\partial\bar{\Omega}_{\alpha i_\alpha}}(\mathbf{x}_m, t_{m-}, \mathbf{p}_\alpha, \boldsymbol{\lambda}_{i_\alpha}) = 0\\
&\text{for } s_{i_\alpha} = 0,1,2,\cdots,2k_{i_\alpha}-1;\\
&\hbar_{i_\alpha}\mathbf{n}^{\mathrm{T}}_{\partial\bar{\Omega}_{\alpha i_\alpha}}(\mathbf{x}^{(i_\alpha;\mathscr{B})}_{m-\varepsilon})\cdot[\mathbf{x}^{(i_\alpha;\mathscr{B})}_{m-\varepsilon} - \mathbf{x}^{(\alpha;\mathscr{D})}_{m-\varepsilon}] > 0
\end{aligned}
\right\}
\text{ for all } i_\alpha \in \mathscr{N}^C_{\alpha\mathscr{B}}(\sigma_{\bar{s}}) \subset \mathscr{N}_{\alpha\mathscr{B}}(\sigma_r,\mathscr{SM}), \tag{6.73}
$$

$$\left.\begin{array}{l} \hbar_{i_\alpha} G^{(s_{i_\alpha},\alpha;\mathscr{D})}_{\partial\Omega_{\alpha i_\alpha}}(\mathbf{x}_m, t_{m+}, \mathbf{p}_\alpha, \boldsymbol{\lambda}_{i_\alpha}) = 0 \\ \text{for } s_{i_\alpha} = 0,1,2,\cdots,2k_{i_\alpha}-1; \\ \hbar_{i_\alpha} \mathbf{n}^{\mathrm{T}}_{\partial\Omega_{\alpha i_\alpha}}(\mathbf{x}^{(i_\alpha;\mathscr{B})}_{m+\varepsilon}) \cdot [\mathbf{x}^{(\alpha;\mathscr{D})}_{m+\varepsilon} - \mathbf{x}^{(i_\alpha;\mathscr{B})}_{m+\varepsilon}] > 0 \end{array}\right\} \tag{6.74}$$
$$\text{for all } i_\alpha \in \mathscr{N}^{L}_{\alpha\mathscr{B}}(\sigma_s) \subset \mathscr{N}_{\alpha\mathscr{B}}(\sigma_r, \mathscr{SM}).$$

(iv) A domain flow $\mathbf{x}^{(\alpha;\mathscr{D})}$ tangential to the real edges $\mathscr{E}^{(s)}_{\sigma_s}$ ($s = r+1, r+2, \cdots, n-1$) with the $(2\mathbf{k}_{n-s}+1)$th -order singularity in a self-mirror domain $\Omega^{\mathscr{SM}}_\alpha$ at $\mathbf{x}_m \in \mathscr{E}^{(r)}_{\sigma_r}$ *transversely passes* over the concave edge of $\mathscr{E}^{(r)}_{\sigma_r}$ if

$$\left.\begin{array}{l} \hbar_{i_\alpha} G^{(s_{i_\alpha},\alpha;\mathscr{D})}_{\partial\Omega_{\alpha i_\alpha}}(\mathbf{x}_m, t_{m-}, \mathbf{p}_\alpha, \boldsymbol{\lambda}_{i_\alpha}) = 0 \\ \text{for } s_{i_\alpha} = 0,1,2,\cdots,2k_{i_\alpha}; \\ \hbar_{i_\alpha} \mathbf{n}^{\mathrm{T}}_{\partial\Omega_{\alpha i_\alpha}}(\mathbf{x}^{(i_\alpha;\mathscr{B})}_{m-\varepsilon}) \cdot [\mathbf{x}^{(i_\alpha;\mathscr{B})}_{m-\varepsilon} - \mathbf{x}^{(\alpha;\mathscr{D})}_{m-\varepsilon}] < 0 \end{array}\right\} \tag{6.75}$$
$$\text{for all } i_\alpha \in \mathscr{N}^{C}_{\alpha\mathscr{B}}(\sigma_s) \subset \mathscr{N}_{\alpha\mathscr{B}}(\sigma_r, \mathscr{SM})$$

$$\left.\begin{array}{l} \hbar_{i_\alpha} G^{(s_{i_\alpha},\alpha;\mathscr{D})}_{\partial\bar{\Omega}_{\alpha i_\alpha}}(\mathbf{x}_m, t_{m+}, \mathbf{p}_\alpha, \boldsymbol{\lambda}_{i_\alpha}) = 0 \\ \text{for } s_{i_\alpha} = 0,1,2,\cdots,2k_{i_\alpha}; \\ \hbar_{i_\alpha} \mathbf{n}^{\mathrm{T}}_{\partial\bar{\Omega}_{\alpha i_\alpha}}(\mathbf{x}^{(i_\alpha;\mathscr{B})}_{m+\varepsilon}) \cdot [\mathbf{x}^{(\alpha;\mathscr{D})}_{m+\varepsilon} - \mathbf{x}^{(i_\alpha;\mathscr{B})}_{m+\varepsilon}] > 0 \end{array}\right\} \tag{6.76}$$
$$\text{for all } i_\alpha \in \mathscr{N}^{L}_{\alpha\mathscr{B}}(\bar{\sigma}_s) \subset \mathscr{N}_{\alpha\mathscr{B}}(\sigma_r, \mathscr{FM}).$$

(v) A domain flow $\mathbf{x}^{(\alpha;\mathscr{D})}$ inflexional to the real edges $\mathscr{E}^{(s)}_{\sigma_s}$ ($s = r+1, r+2, \cdots, n-1$) with the $(2\mathbf{k}_{n-s})$th -order singularity in a self-mirror domain $\Omega^{\mathscr{SM}}_\alpha$ at $\mathbf{x}_m \in \mathscr{E}^{(r)}_{\sigma_r}$ *transversely passes* over the concave edge $\mathscr{E}^{(r)}_{\sigma_r}$ if

$$\left.\begin{array}{l} \hbar_{i_\alpha} G^{(s_{i_\alpha},\alpha;\mathscr{D})}_{\partial\Omega_{\alpha i_\alpha}}(\mathbf{x}_m, t_{m-}, \mathbf{p}_\alpha, \boldsymbol{\lambda}_{i_\alpha}) = 0 \\ \text{for } s_{i_\alpha} = 0,1,2,\cdots,2k_{i_\alpha}-1; \\ \hbar_{i_\alpha} \mathbf{n}^{\mathrm{T}}_{\partial\Omega_{\alpha i_\alpha}}(\mathbf{x}^{(i_\alpha;\mathscr{B})}_{m-\varepsilon}) \cdot [\mathbf{x}^{(i_\alpha;\mathscr{B})}_{m-\varepsilon} - \mathbf{x}^{(\alpha;\mathscr{D})}_{m-\varepsilon}] < 0 \end{array}\right\} \tag{6.77}$$
$$\text{for all } i_\alpha \in \mathscr{N}^{C}_{\alpha\mathscr{B}}(\sigma_s) \subset \mathscr{N}_{\alpha\mathscr{B}}(\sigma_r, \mathscr{SM}),$$

$$\left.\begin{array}{l} \hbar_{i_\alpha} G^{(s_{i_\alpha},\alpha;\mathscr{D})}_{\partial\bar{\Omega}_{\alpha i_\alpha}}(\mathbf{x}_m, t_{m+}, \mathbf{p}_\alpha, \boldsymbol{\lambda}_{i_\alpha}) = 0 \\ \text{for } s_{i_\alpha} = 0,1,2,\cdots,2k_{i_\alpha}-1; \\ \hbar_{i_\alpha} \mathbf{n}^{\mathrm{T}}_{\partial\bar{\Omega}_{\alpha i_\alpha}}(\mathbf{x}^{(i_\alpha;\mathscr{B})}_{m+\varepsilon}) \cdot [\mathbf{x}^{(\alpha;\mathscr{D})}_{m+\varepsilon} - \mathbf{x}^{(i_\alpha;\mathscr{B})}_{m+\varepsilon}] < 0 \end{array}\right\} \tag{6.78}$$
$$\text{for all } i_\alpha \in \mathscr{N}^{L}_{\alpha\mathscr{B}}(\sigma_{\bar{s}}) \subset \mathscr{N}_{\alpha\mathscr{B}}(\sigma_r, \mathscr{SM}).$$

For the full mirror imaginary domain of a convex domain in the corresponding concave domain, domain flow properties to edges are presented.

Definition 6.13. For a discontinuous dynamical system in Eqs. (6.4)-(6.8), there is a point $\mathbf{x}^{(\sigma_r;\mathscr{E})}(t_m) \equiv \mathbf{x}_m \in \mathscr{E}_{\sigma_r}^{(r)}$ ($\sigma_r \in \mathscr{N}_{\mathscr{E}}^r = \{1,2,\cdots,N_r\}$ and $r \in \{0,1,2,\cdots,n-2\}$) at time t_m with the related boundaries $\partial\Omega_{\rho i_\rho}(\sigma_r)$ and related domains $\Omega_\rho(\sigma_r)$ ($\rho \in \mathscr{N}_{\mathscr{D}}(\sigma_r) \subset \mathscr{N}_{\mathscr{D}} = \{1,2,\cdots,N_d\}$ and $i_\rho \in \mathscr{N}_{\rho\mathscr{B}}(\sigma_r) \subset \mathscr{N}_{\mathscr{B}} = \{1,2,\cdots,N_b\}$), where the boundary index subset of $\mathscr{N}_{\rho\mathscr{B}}(\sigma_r)$ has n_{σ_r}-elements ($\rho = \alpha, \beta$ and $n_{\sigma_r} \geq n-r$). For an arbitrarily small $\varepsilon > 0$, there are two time intervals $[t_{m-\varepsilon}, t_m)$ and $(t_m, t_{m+\varepsilon}]$. A flow $\mathbf{x}^{(\rho;\mathscr{D})}(t)$ is $C^{r_\rho}_{[t_{m-\varepsilon},t_m)}$ and/or $C^{r_\rho}_{(t_m,t_{m+\varepsilon}]}$-continuous for time t and $\| d^{r_\rho+1}\mathbf{x}^{(\rho;\mathscr{D})}/dt^{r_\rho+1} \| < \infty$. $\mathscr{N}_{\rho\mathscr{B}}(\sigma_r) = \cup_{j\in\mathscr{J}} \mathscr{N}_{\rho\mathscr{B}}^{j}(\sigma_r)$ where $\mathscr{J} = \{C, L\}$ and $\cap_{j\in\mathscr{J}} \mathscr{N}_{\rho\mathscr{B}}^{j}(\sigma_r) = \varnothing$.

(i) A domain flow $\mathbf{x}^{(\alpha;\mathscr{D})}$ in a full-mirror domain $\Omega_\alpha^{\mathscr{FM}}$ at $\mathbf{x}_m \in \mathscr{E}_{\sigma_r}^{(r)}$ is called a *coming flow* to the concave edge $\mathscr{E}_{\sigma_r}^{(r)}$ if

$$\hbar_{i_\alpha} \mathbf{n}^{\mathrm{T}}_{\partial\bar{\Omega}_{\alpha i_\alpha}}(\mathbf{x}^{(i_\alpha;\mathscr{B})}_{m-\varepsilon}) \cdot [\mathbf{x}^{(i_\alpha;\mathscr{B})}_{m-\varepsilon} - \mathbf{x}^{(\alpha;\mathscr{D})}_{m-\varepsilon}] < 0 \text{ for all } i_\alpha \in \mathscr{N}_{\alpha\mathscr{B}}^{\mathscr{FM}}(\sigma_r), \tag{6.79}$$

(ii) A domain flow $\mathbf{x}^{(\alpha;\mathscr{D})}$ in a full-mirror domain $\Omega_\alpha^{\mathscr{FM}}$ at point $\mathbf{x}_m \in \mathscr{E}_{\sigma_r}^{(r)}$ is called a *leaving flow* to the concave edge $\mathscr{E}_{\sigma_r}^{(r)}$ if

$$\hbar_{i_\alpha} \mathbf{n}^{\mathrm{T}}_{\partial\bar{\Omega}_{\alpha i_\alpha}}(\mathbf{x}^{(i_\alpha;\mathscr{B})}_{m+\varepsilon}) \cdot [\mathbf{x}^{(\alpha;\mathscr{D})}_{m+\varepsilon} - \mathbf{x}^{(i_\alpha;\mathscr{B})}_{m+\varepsilon}] > 0 \text{ for all } i_\alpha \in \mathscr{N}_{\alpha\mathscr{B}}^{\mathscr{FM}}(\sigma_r), \tag{6.80}$$

(iii) A domain flow $\mathbf{x}^{(\alpha;\mathscr{D})}$ tangential to the imaginary extended edges $\mathscr{E}_{\bar{\sigma}_s}^{(s)}$ ($s = r+1, r+2, \cdots, n-1$) with the $(2\mathbf{k}_{n-s}+1)$ singularity in the maximum full-mirror domain $\Omega_\alpha^{\mathscr{FM}}$ at point $\mathbf{x}_m \in \mathscr{E}_{\sigma_r}^{(r)}$ *transversely passes over* the concave edge $\mathscr{E}_{\sigma_r}^{(r)}$ if

$$\left.\begin{aligned}
&\hbar_{i_\alpha} G^{(s_{i_\alpha},\alpha)}_{\partial\bar{\Omega}_{\alpha i_\alpha}}(\mathbf{x}_m, t_{m-}, \mathbf{p}_\alpha, \boldsymbol{\lambda}_{i_\alpha}) = 0\\
&\text{for } s_{i_\alpha} = 0,1,2,\cdots,2k_{i_\alpha};\\
&\hbar_{i_\alpha} \mathbf{n}^{\mathrm{T}}_{\partial\bar{\Omega}_{\alpha i_\alpha}}(\mathbf{x}^{(i_\alpha;\mathscr{B})}_{m-\varepsilon}) \cdot [\mathbf{x}^{(i_\alpha;\mathscr{B})}_{m-\varepsilon} - \mathbf{x}^{(\alpha;\mathscr{D})}_{m-\varepsilon}] < 0
\end{aligned}\right\} \tag{6.81}$$

$$\text{for all } i_\alpha \in \mathscr{N}^{C}_{\alpha\mathscr{B}}(\sigma_{\bar{s}}) \subset \mathscr{N}_{\alpha\mathscr{B}}(\sigma_r, \mathscr{FM}),$$

$$\left.\begin{aligned}
&\hbar_{i_\alpha} G^{(s_{i_\alpha},\alpha)}_{\partial\bar{\Omega}_{\alpha i_\alpha}}(\mathbf{x}_m, t_{m+}, \mathbf{p}_\alpha, \boldsymbol{\lambda}_{i_\alpha}) = 0\\
&\text{for } s_{i_\alpha} = 0,1,2,\cdots,2k_{i_\alpha};\\
&\hbar_{i_\alpha} \mathbf{n}^{\mathrm{T}}_{\partial\bar{\Omega}_{\alpha i_\alpha}}(\mathbf{x}^{(i_\alpha;\mathscr{B})}_{m+\varepsilon}) \cdot [\mathbf{x}^{(\alpha;\mathscr{D})}_{m+\varepsilon} - \mathbf{x}^{(i_\alpha;\mathscr{B})}_{m+\varepsilon}] > 0
\end{aligned}\right\}$$

$$\text{for all } i_\alpha \in \mathscr{N}^{L}_{\alpha\mathscr{B}} \subset \mathscr{N}_{\alpha\mathscr{B}}(\sigma_s, \mathscr{SM}). \tag{6.82}$$

(iv) A domain flow $\mathbf{x}^{(\alpha;\mathscr{D})}$ inflexional to the imaginary extended edges $\mathscr{E}^{(s)}_{\sigma_s}$ ($s = r+1, r+2, \cdots, n-1$) with the $(2\mathbf{k}_{n-s})$-singularity in the maximum full-mirror domain $\Omega^{\mathscr{FM}}_\alpha$ at $\mathbf{x}_m \in \mathscr{E}^{(r)}_{\sigma_r}$ does not *transversely pass* over the edge $\mathscr{E}^{(r)}_{\sigma_r}$ if

$$\left.\begin{array}{l} \hbar_{i_\alpha} G^{(s_{i_\alpha},\alpha)}_{\partial\bar{\Omega}_{\alpha i_\alpha}}(\mathbf{x}_m, t_{m-}, \mathbf{p}_\alpha, \boldsymbol{\lambda}_{i_\alpha}) = 0 \\ \text{for } s_{i_\alpha} = 0,1,2,\cdots,2k_{i_\alpha}-1; \\ \hbar_{i_\alpha} \mathbf{n}^{\mathrm{T}}_{\partial\bar{\Omega}_{\alpha i_\alpha}}(\mathbf{x}^{(i_\alpha;\mathscr{B})}_{m-\varepsilon}) \cdot [\mathbf{x}^{(i_\alpha;\mathscr{B})}_{m-\varepsilon} - \mathbf{x}^{(\alpha;\mathscr{D})}_{m-\varepsilon}] < 0 \\ \text{for all } i_\alpha \in \mathscr{N}^{C}_{\alpha\mathscr{B}}(\sigma_{\bar{s}}) \subset \mathscr{N}_{\alpha\mathscr{B}}(\sigma_r, \mathscr{FM}). \end{array}\right\} \tag{6.83}$$

In the foregoing two definitions, a domain flow in the concave domain was discussed. For comparison, the properties of a flow to the edge in a convex domain will be discussed, and the corresponding definition is presented as follows.

Definition 6.14. For a discontinuous dynamical system in Eqs. (6.4)-(6.8), there is a point $\mathbf{x}^{(\sigma_r;\mathscr{E})}(t_m) \equiv \mathbf{x}_m \in \mathscr{E}^{(r)}_{\sigma_r}$ ($\sigma_r \in \mathscr{N}^{r}_{\mathscr{E}} = \{1,2,\cdots,N_r\}$ and $r \in \{0,1,2,\cdots,n-2\}$) at time t_m with the related boundaries $\partial\Omega_{\rho i_\rho}(\sigma_r)$ and related domains $\Omega_\rho(\sigma_r)$ ($\rho \in \mathscr{N}_{\mathscr{D}}(\sigma_r) \subset \mathscr{N}_{\mathscr{D}} = \{1,2,\cdots,N_d\}$ and $i_\rho \in \mathscr{N}_{\rho\mathscr{B}}(\sigma_r) \subset \mathscr{N}_{\mathscr{B}} = \{1,2,\cdots,N_b\}$), where the boundary index subset of $\mathscr{N}_{\rho\mathscr{B}}(\sigma_r)$ has n_{σ_r}-elements ($\rho = \alpha, \beta$ and $n_{\sigma_r} \geq (n-r)$). For an arbitrarily small $\varepsilon > 0$, there are two time intervals $[t_{m-\varepsilon}, t_m)$ and $(t_m, t_{m+\varepsilon}]$. A flow $\mathbf{x}^{(\rho;\mathscr{D})}(t)$ is $C^{r_\rho}_{[t_{m-\varepsilon},t_m)}$ and/or $C^{r_\rho}_{(t_m,t_{m+\varepsilon}]}$-continuous for time t and $\| d^{r_\rho+1}\mathbf{x}^{(\rho;\mathscr{D})}/dt^{r_\rho+1} \| < \infty$. $\mathscr{N}_{\rho\mathscr{B}}(\sigma_r) = \cup_{j\in\mathscr{J}} \mathscr{N}^{j}_{\rho\mathscr{B}}(\sigma_r)$ where $\mathscr{J} = \{C, L\}$ and $\cap_{j\in\mathscr{J}} \mathscr{N}^{j}_{\rho\mathscr{B}}(\sigma_r) = \varnothing$.

(i) A domain flow $\mathbf{x}^{(\alpha;\mathscr{D})}$ in domain Ω_α at point $\mathbf{x}_m \in \mathscr{E}^{(r)}_{\sigma_r}$ is called a *coming* flow to the convex edge $\mathscr{E}^{(r)}_{\sigma_r}$ if

$$\hbar_{i_\alpha} \mathbf{n}^{\mathrm{T}}_{\partial\Omega_{\alpha i_\alpha}}(\mathbf{x}^{(i_\alpha;\mathscr{B})}_{m-\varepsilon}) \cdot [\mathbf{x}^{(i_\alpha;\mathscr{B})}_{m-\varepsilon} - \mathbf{x}^{(\alpha;\mathscr{D})}_{m-\varepsilon}] < 0 \text{ for all } i_\alpha \in \mathscr{N}_{\alpha\mathscr{B}}(\sigma_r). \tag{6.84}$$

(ii) A domain flow $\mathbf{x}^{(\alpha;\mathscr{D})}$ in domain Ω_α at $\mathbf{x}_m \in \mathscr{E}^{(r)}_{\sigma_r}$ is called a *leaving* flow to the convex edge $\mathscr{E}^{(r)}_{\sigma_r}$ if

$$\hbar_{i_\alpha} \mathbf{n}^{\mathrm{T}}_{\partial\Omega_{\alpha i_\alpha}}(\mathbf{x}^{(i_\alpha;\mathscr{B})}_{m+\varepsilon}) \cdot [\mathbf{x}^{(\alpha;\mathscr{D})}_{m+\varepsilon} - \mathbf{x}^{(i_\alpha;\mathscr{B})}_{m+\varepsilon}] > 0 \text{ for all } i_\alpha \in \mathscr{N}_{\alpha\mathscr{B}}(\sigma_r). \tag{6.85}$$

(iii) A domain flow $\mathbf{x}^{(\alpha;\mathscr{D})}$ in domain Ω_α at $\mathbf{x}_m \in \mathscr{E}^{(r)}_{\sigma_r}$ is called a *tangential com-*

ing flow to the convex edges of $\mathscr{E}_{\sigma_s}^{(s)}$ ($s=r+1,r+2,\cdots,n-1$) with the $(2\mathbf{k}_{n-s}+1)$ singularity if

$$\left.\begin{array}{l}\hbar_{i_\alpha} G_{\partial\Omega_{\alpha i_\alpha}}^{(s_{i_\alpha},\alpha;\mathscr{D})}(\mathbf{x}_m,t_{m-},\mathbf{p}_\alpha,\boldsymbol{\lambda}_{i_\alpha})=0\\ \text{for } s_{i_\alpha}=0,1,2,\cdots,2k_{i_\alpha};\\ \hbar_{i_\alpha}\mathbf{n}_{\partial\Omega_{\alpha i_\alpha}}^{\mathrm{T}}(\mathbf{x}_{m-\varepsilon}^{(i_\alpha;\mathscr{B})})\cdot[\mathbf{x}_{m-\varepsilon}^{(i_\alpha;\mathscr{B})}-\mathbf{x}_{m-\varepsilon}^{(\alpha;\mathscr{D})}]<0\end{array}\right\}\text{ for all } i_\alpha\in\mathscr{N}_{\alpha\mathscr{B}}(\sigma_s). \tag{6.86}$$

(iv) A domain flow $\mathbf{x}^{(\alpha;\mathscr{D})}$ in domain Ω_α at $\mathbf{x}_m\in\mathscr{E}_{\sigma_r}^{(r)}$ is called a *tangential leaving* flow to the convex edges of $\mathscr{E}_{\sigma_s}^{(s)}$ ($s=r+1,r+2,\cdots,n-1$) with the $(2\mathbf{k}_{n-s}+1)$ -singularity if

$$\left.\begin{array}{l}\hbar_{i_\alpha} G_{\partial\Omega_{\alpha i_\alpha}}^{(s_{i_\alpha},\alpha;\mathscr{D})}(\mathbf{x}_m,t_{m+},\mathbf{p}_\alpha,\boldsymbol{\lambda}_{i_\alpha})=0\\ \text{for } s_{i_\alpha}=0,1,2,\cdots,2k_{i_\alpha};\\ \hbar_{i_\alpha}\mathbf{n}_{\partial\Omega_{\alpha i_\alpha}}^{\mathrm{T}}(\mathbf{x}_{m+\varepsilon}^{(i_\alpha;\mathscr{B})})\cdot[\mathbf{x}_{m+\varepsilon}^{(\alpha;\mathscr{D})}-\mathbf{x}_{m+\varepsilon}^{(i_\alpha;\mathscr{B})}]>0\end{array}\right\}\text{ for all } i_\alpha\in\mathscr{N}_{\alpha\mathscr{B}}(\sigma_s). \tag{6.87}$$

(v) A domain flow $\mathbf{x}^{(\alpha;\mathscr{D})}$ in domain Ω_α at $\mathbf{x}_m\in\mathscr{E}_{\sigma_r}^{(r)}$ is called an *inflexional coming* flow to the convex edges $\mathscr{E}_{\sigma_s}^{(s)}$ ($s=r+1,r+2,\cdots,n-1$) with the $(2\mathbf{k}_{n-s})$ -singularity if

$$\left.\begin{array}{l}\hbar_{i_\alpha} G_{\partial\Omega_{\alpha i_\alpha}}^{(s_{i_\alpha},\alpha;\mathscr{D})}(\mathbf{x}_m,t_{m-},\mathbf{p}_\alpha,\boldsymbol{\lambda}_{i_\alpha})=0\\ \text{for } s_{i_\alpha}=0,1,2,\cdots,2k_{i_\alpha}-1;\\ \hbar_{i_\alpha}\mathbf{n}_{\partial\Omega_{\alpha i_\alpha}}^{\mathrm{T}}(\mathbf{x}_{m-\varepsilon}^{(i_\alpha;\mathscr{B})})\cdot[\mathbf{x}_{m-\varepsilon}^{(i_\alpha;\mathscr{B})}-\mathbf{x}_{m-\varepsilon}^{(\alpha;\mathscr{D})}]<0\end{array}\right\}\text{ for all } i_\alpha\in\mathscr{N}_{\alpha\mathscr{B}}(\sigma_s). \tag{6.88}$$

(vi) A domain flow $\mathbf{x}^{(\alpha;\mathscr{D})}$ in domain Ω_α at $\mathbf{x}_m\in\mathscr{E}_{\sigma_r}^{(r)}$ is called an *inflexional leaving* flow to the convex edges $\mathscr{E}_{\sigma_s}^{(s)}$ ($s=r+1,r+2,\cdots,n-1$) with the $2\mathbf{k}_{n-s}$ -singularity if

$$\left.\begin{array}{l}\hbar_{i_\alpha} G_{\partial\Omega_{\alpha i_\alpha}}^{(s_{i_\alpha},\alpha;\mathscr{D})}(\mathbf{x}_m,t_{m+},\mathbf{p}_\alpha,\boldsymbol{\lambda}_{i_\alpha})=0\\ \text{for } s_{i_\alpha}=0,1,2,\cdots,2k_{i_\alpha}-1;\\ \hbar_{i_\alpha}\mathbf{n}_{\partial\Omega_{\alpha i_\alpha}}^{\mathrm{T}}(\mathbf{x}_{m+\varepsilon}^{(i_\alpha;\mathscr{B})})\cdot[\mathbf{x}_{m+\varepsilon}^{(\alpha;\mathscr{D})}-\mathbf{x}_{m+\varepsilon}^{(i_\alpha;\mathscr{B})}]>0\end{array}\right\}\text{ for all } i_\alpha\in\mathscr{N}_{\alpha\mathscr{B}}(\sigma_s). \tag{6.89}$$

To interpret the above definitions, consider domain flows in the self-mirror and full-mirror domains via an $(n-2)$ -dimensional edge in Fig. 6.13. In Fig. 6.13(a), a domain flow of $\mathbf{x}^{(\alpha;\mathscr{D})}$ in a sub-domain of domain Ω_α arrives to the edge $\mathscr{E}^{(n-2)}$ at $\mathbf{x}_m\in\mathscr{E}^{(n-2)}$. The mirror domain of the sub-domain through the edge $\mathscr{E}^{(n-2)}$ is in domain Ω_α, which is called a self-mirror domain. Since the coming flow to the edge can be extended with the extension theorem of a solution of differential equation, without any switching rules, the coming flow will transversely graze to the edge to the self-mirror domain $\Omega_\alpha^{\mathscr{LM}}$. In Fig. 6.13(b), a flow of $\mathbf{x}^{(\alpha;\mathscr{D})}$ in a sub-domain of domain Ω_α arrives to $\mathbf{x}_m \in \mathscr{E}^{(n-2)}$, and this sub-domain is the mirror domain of convex domain Ω_β. If a domain flow $\mathbf{x}^{(\beta;\mathscr{D})}$ in domain Ω_β at $\mathbf{x}_m \in \mathscr{E}^{(n-2)}$ is a coming flow, the coming flow cannot be switched without the switching rule. Thus the sliding flow on the edge $\mathscr{E}^{(n-2)}$ will be formed as discussed on the boundary.

The tangential domain flow of $\mathbf{x}^{(\alpha;\mathscr{D})}$ to one of the boundaries of the edge $\mathscr{E}^{(n-2)}$ will be presented in Fig. 6.14. In Fig. 6.14(a), a domain flow $\mathbf{x}^{(\alpha;\mathscr{D})}$ tangential to the extended imaginary boundary in the full-mirror domain will be in the domain Ω_α without the switching rule. However, in the self-mirror domain, a coming domain flow tangential to the extended imaginary boundary of the edge in the self-mirror domain will stop at the edge without any switching rules, as shown in Fig. 6.14(b). However, a domain flow $\mathbf{x}^{(\alpha;\mathscr{D})}$ tangential to the real boundary in the self-mirror domain will keep in the domain Ω_α without any switching rules, which can be found from 6.14(a). For a better understanding of lower-order edges, consider a domain flow $\mathbf{x}^{(\alpha;\mathscr{D})}$ to the $(n-3)$-dimensional edge in Fig. 6.15. A domain flow $\mathbf{x}^{(\alpha;\mathscr{D})}$ in the self-mirror domain transversely pass over the $(n-3)$-dimensional edge via $\mathbf{x}_m \in \mathscr{E}^{(n-3)}$, as shown in Fig. 6.15(a). A domain flow $\mathbf{x}^{(\alpha;\mathscr{D})}$ in the full-mirror domain of the convex domain Ω_β stops at $\mathbf{x}_m \in \mathscr{E}^{(n-3)}$ without the switching rule, as shown in Fig. 6.15(b). From the previous definitions, the necessary and sufficient conditions for domain flow singularity to the edges and boundaries in the concave and convex domains will be presented. The corresponding proofs can be given as in Chapter 3.

Theorem 6.7. *For a discontinuous dynamical system in Eqs.* (6.4)-(6.8), *there is a point* $\mathbf{x}^{(\sigma_r;\mathscr{E})}(t_m) \equiv \mathbf{x}_m \in \mathscr{E}_{\sigma_r}^{(r)}$ ($\sigma_r \in \mathscr{N}_{\mathscr{E}}^r = \{1,2,\cdots,N_r\}$ *and* $r \in \{0,1,2,\cdots,n-2\}$) *at time* t_m *with the related boundaries* $\partial\Omega_{\rho i_\rho}(\sigma_r)$ *and related domains* $\Omega_\rho(\sigma_r)$ ($\rho \in \mathscr{N}_{\mathscr{D}}(\sigma_r) \subset \mathscr{N}_{\mathscr{D}} = \{1,2,\cdots,N_d\}$ *and* $i_\rho \in \mathscr{N}_{\rho\mathscr{B}}(\sigma_r) \subset \mathscr{N}_{\mathscr{B}} = \{1,2,\cdots,N_b\}$), *where the boundary index subset of* $\mathscr{N}_{\rho\mathscr{B}}(\sigma_r)$ *has* n_{σ_r}*-elements* ($\rho = \alpha,\beta$ *and* $n_{\sigma_r} \geq (n-r)$). *For an arbitrarily small* $\varepsilon > 0$, *there are two time intervals* $[t_{m-\varepsilon}, t_m)$ *and* $(t_m, t_{m+\varepsilon}]$. *A flow* $\mathbf{x}^{(\rho;\mathscr{D})}(t)$ *is* $C^{r_\rho}_{[t_{m-\varepsilon},t_m)}$ *and/or* $C^{r_\rho}_{(t_m,t_{m+\varepsilon}]}$ *-continuous for time* t *and* $\| d^{r_\rho+1}\mathbf{x}^{(\rho;\mathscr{D})} / dt^{r_\rho+1} \| < \infty$. $\mathscr{N}_{\rho\mathscr{B}}(\sigma_r) = \cup_{j\in\mathscr{J}} \mathscr{N}_{\rho\mathscr{B}}^j(\sigma_r)$ *where* $\mathscr{J} = \{C, L\}$ *and* $\cap_{j\in\mathscr{J}} \mathscr{N}_{\rho\mathscr{B}}^j(\sigma_r) = \varnothing$. $\lambda_{i_\alpha} = 1,-1$ *are for real and imaginary boundaries in the self-mirror domains, respectively.*

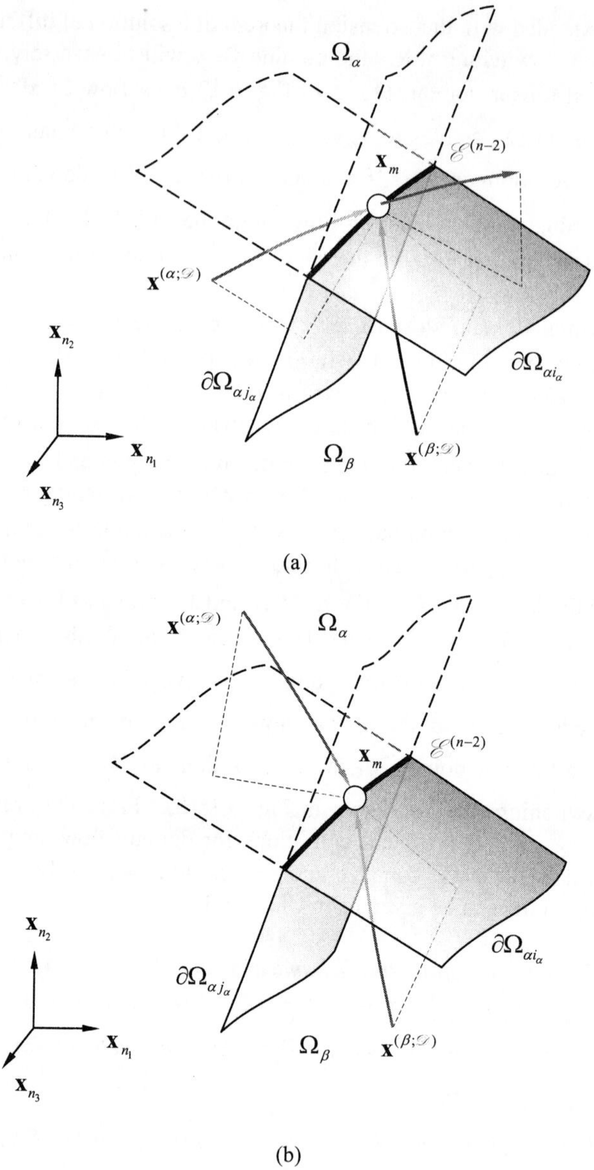

Fig. 6.13 A coming domain flow to an $(n-2)$-dimensional convex-concave edge in: **(a)** self-mirror imaginary domain, and **(b)** full-mirror domains. The boundaries are shaded, and the $(n-2)$-dimensional edges are depicted by thick curves. Dashed curves are for extended boundaries.

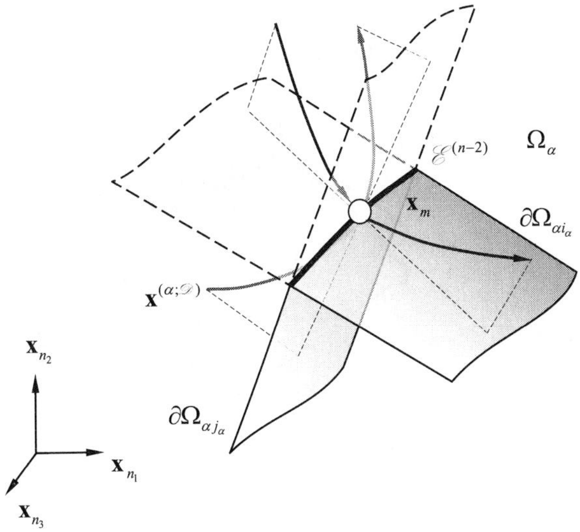

(a)

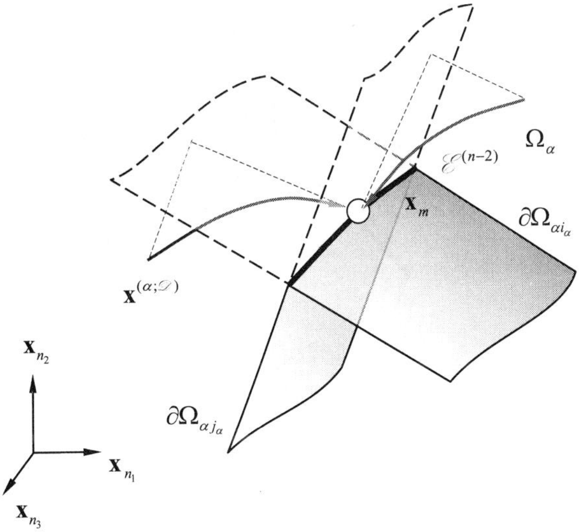

(b)

Fig. 6.14 Tangential flows at an $(n-2)$-dimensional convex-concave edge to: **(a)** the extended boundary in full-mirror domain, and **(b)** the extended boundary in the self-mirror domain. The boundaries are shaded, and the $(n-2)$-dimensional edges are depicted by thick curves. Dashed curves are for extended boundaries.

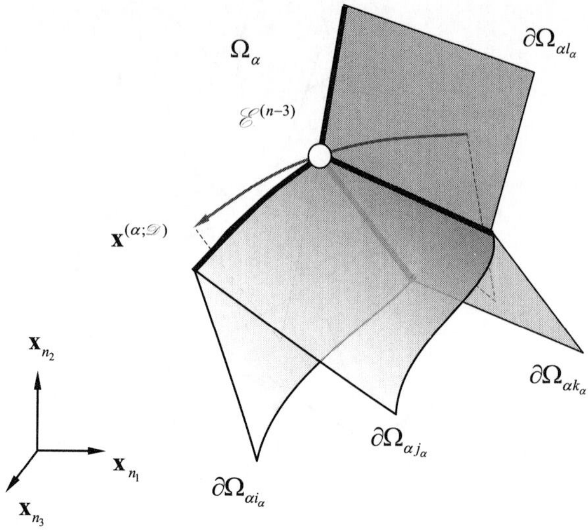

(a)

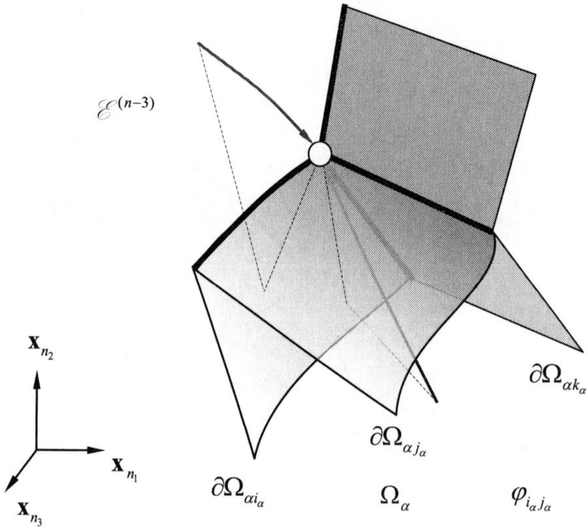

(b)

Fig. 6.15 An $(n-3)$ -dimensional convex-concave edge: **(a)** self-mirror imaginary domain in its own domain, and **(b)** full mirror imaginary domains. The boundaries are shaded, and the $(n-2)$ -dimensional edges are depicted by thick curves. The $(n-3)$ -dimensional edge is presented by a corner point. Dashed curves are for minor imaginary domains.

(i) *A domain flow* $\mathbf{x}^{(\alpha;\mathscr{D})}$ *in the self-mirror domain* $\Omega_{\alpha}^{\mathscr{SM}}$ *transversely grazes to the concave edge* $\mathscr{E}_{\sigma_r}^{(r)}$ *at* $\mathbf{x}_m \in \mathscr{E}_{\sigma_r}^{(r)}$ *if and only if*

$$\begin{aligned}
&\hbar_{i_\alpha} \hbar_{i_\alpha} G_{\partial\Omega_{\alpha i_\alpha}}^{(\alpha;\mathscr{D})}(\mathbf{x}_m, t_{m-}, \mathbf{p}_\alpha, \boldsymbol{\lambda}_{i_\alpha}) < 0 \\
&\text{for all } i_\alpha \in \mathscr{N}_{\alpha\mathscr{B}}^{C} \subset \mathscr{N}_{\alpha\mathscr{B}}(\sigma_r, \mathscr{SM}), \\
&\hbar_{i_\alpha} \hbar_{j_\alpha} G_{\partial\Omega_{\alpha j_\alpha}}^{(\alpha;\mathscr{D})}(\mathbf{x}_m, t_{m+}, \mathbf{p}_\alpha, \boldsymbol{\lambda}_{j_\alpha}) > 0 \\
&\text{for all } j_\alpha \in \mathscr{N}_{\alpha\mathscr{B}}^{L} \subset \mathscr{N}_{\alpha\mathscr{B}}(\sigma_r, \mathscr{SM}).
\end{aligned} \tag{6.90}$$

(ii) *A domain flow* $\mathbf{x}^{(\alpha;\mathscr{D})}$ *tangential to the imaginary extended edges* $\mathscr{E}_{\bar{\sigma}_s}^{(s)}$ $(s = r+1, r+2, \cdots, n-1)$ *with the* $(2\mathbf{k}_{n-s}+\mathbf{1})$ *singularity in a self-mirror domain* $\Omega_{\alpha}^{\mathscr{SM}}$ *at* $\mathbf{x}_m \in \mathscr{E}_{\sigma_r}^{(r)}$ *does not transversely pass over the concave edge* $\mathscr{E}_{\sigma_r}^{(r)}$ *if and only if*

$$\left.\begin{aligned}
&\hbar_{i_\alpha} G_{\partial\bar{\Omega}_{\alpha i_\alpha}}^{(s_{i_\alpha},\alpha;\mathscr{D})}(\mathbf{x}_m, t_{m-}, \mathbf{p}_\alpha, \boldsymbol{\lambda}_{i_\alpha}) = 0 \\
&\text{for } s_{i_\alpha} = 0, 1, 2, \cdots, 2k_{i_\alpha}; \\
&\hbar_{i_\alpha} G_{\partial\bar{\Omega}_{\alpha i_\alpha}}^{(2k_{i_\alpha}+1,\alpha;\mathscr{D})}(\mathbf{x}_m, t_{m-}, \mathbf{p}_\alpha, \boldsymbol{\lambda}_{i_\alpha}) < 0
\end{aligned}\right\} \tag{6.91}$$
$$\text{for all } i_\alpha \in \mathscr{N}_{\alpha\mathscr{B}}^{C}(\sigma_{\bar{s}}) \subset \mathscr{N}_{\alpha\mathscr{B}}(\sigma_r, \mathscr{SM}).$$

(iii) *A domain flow* $\mathbf{x}^{(\alpha;\mathscr{D})}$ *inflexional to the imaginary extended edges* $\mathscr{E}_{\bar{\sigma}_s}^{(s)}$ $(s = r+1, r+2, \cdots, n-1)$ *with the* $(2\mathbf{k}_{n-s})$th *-order singularity in a self-mirror domain* $\Omega_{\alpha}^{\mathscr{SM}}$ *at* $\mathbf{x}_m \in \mathscr{E}_{\sigma_r}^{(r)}$ *transversely passes over the concave edge* $\mathscr{E}_{\sigma_r}^{(r)}$ *if and only if*

$$\left.\begin{aligned}
&\hbar_{i_\alpha} G_{\partial\bar{\Omega}_{\alpha i_\alpha}}^{(s_{i_\alpha},\alpha;\mathscr{D})}(\mathbf{x}_m, t_{m-}, \mathbf{p}_\alpha, \boldsymbol{\lambda}_{i_\alpha}) = 0 \\
&\text{for } s_{i_\alpha} = 0, 1, 2, \cdots, 2k_{i_\alpha}-1; \\
&\hbar_{i_\alpha} G_{\partial\bar{\Omega}_{\alpha i_\alpha}}^{(2k_{i_\alpha},\alpha;\mathscr{D})}(\mathbf{x}_m, t_{m-}, \mathbf{p}_\alpha, \boldsymbol{\lambda}_{i_\alpha}) > 0
\end{aligned}\right\} \tag{6.92}$$
$$\text{for all } i_\alpha \in \mathscr{N}_{\alpha\mathscr{B}}^{C}(\sigma_{\bar{s}}) \subset \mathscr{N}_{\alpha\mathscr{B}}(\sigma_r, \mathscr{SM})$$

$$\left.\begin{aligned}
&\hbar_{i_\alpha} G_{\partial\Omega_{\alpha i_\alpha}}^{(s_{i_\alpha},\alpha;\mathscr{D})}(\mathbf{x}_m, t_{m+}, \mathbf{p}_\alpha, \boldsymbol{\lambda}_{i_\alpha}) = 0 \\
&\text{for } s_{i_\alpha} = 0, 1, 2, \cdots, 2k_{i_\alpha}-1; \\
&\hbar_{i_\alpha} G_{\partial\Omega_{\alpha i_\alpha}}^{(2k_{i_\alpha},\alpha;\mathscr{D})}(\mathbf{x}_m, t_{m+}, \mathbf{p}_\alpha, \boldsymbol{\lambda}_{i_\alpha}) > 0
\end{aligned}\right\} \tag{6.93}$$
$$\text{for all } i_\alpha \in \mathscr{N}_{\alpha\mathscr{B}}^{L}(\sigma_s) \subset \mathscr{N}_{\alpha\mathscr{B}}(\sigma_r, \mathscr{SM}).$$

(iv) *A domain flow* $\mathbf{x}^{(\alpha;\mathscr{D})}$ *tangential to the real edges* $\mathscr{E}_{\sigma_s}^{(s)}$ $(s = r+1, r+2, \cdots,$

$n-1$) *with the* $(2\mathbf{k}_{n-s}+\mathbf{1})$th *-order singularity in a self-mirror domain* $\Omega_{\alpha}^{\mathscr{SM}}$ *at* $\mathbf{x}_m \in \mathscr{E}_{\sigma_r}^{(r)}$ *transversely passes over the concave edge* $\mathscr{E}_{\sigma_r}^{(r)}$ *if and only if*

$$\left.\begin{aligned}&\hbar_{i_\alpha} G_{\partial\Omega_{\alpha i_\alpha}}^{(s_{i_\alpha},\alpha;\mathscr{D})}(\mathbf{x}_m,t_{m-},\mathbf{p}_\alpha,\boldsymbol{\lambda}_{i_\alpha})=0\\&\text{for } s_{i_\alpha}=0,1,2,\cdots,2k_{i_\alpha};\\&\hbar_{i_\alpha} G_{\partial\Omega_{\alpha i_\alpha}}^{(2k_{i_\alpha}+1,\alpha;\mathscr{D})}(\mathbf{x}_m,t_{m-},\mathbf{p}_\alpha,\boldsymbol{\lambda}_{i_\alpha})>0\end{aligned}\right\}\tag{6.94}$$

for all $i_\alpha \in \mathscr{N}_{\alpha\mathscr{B}}^{C}(\sigma_s)\subset \mathscr{N}_{\alpha\mathscr{B}}(\sigma_r,\mathscr{SM})$,

$$\left.\begin{aligned}&\hbar_{i_\alpha} G_{\partial\bar{\Omega}_{\alpha i_\alpha}}^{(s_{i_\alpha},\alpha;\mathscr{D})}(\mathbf{x}_m,t_{m+},\mathbf{p}_\alpha,\boldsymbol{\lambda}_{i_\alpha})=0\\&\text{for } s_{i_\alpha}=0,1,2,\cdots,2k_{i_\alpha};\\&\hbar_{i_\alpha} G_{\partial\bar{\Omega}_{\alpha i_\alpha}}^{(2k_{i_\alpha}+1,\alpha;\mathscr{D})}(\mathbf{x}_m,t_{m+},\mathbf{p}_\alpha,\boldsymbol{\lambda}_{i_\alpha})>0\end{aligned}\right\}\tag{6.95}$$

for all $i_\alpha \in \mathscr{N}_{\alpha\mathscr{B}}^{L}(\sigma_{\bar{s}})\subset \mathscr{N}_{\alpha\mathscr{B}}(\sigma_r,\mathscr{FM})$.

(v) *A domain flow* $\mathbf{x}^{(\alpha;\mathscr{D})}$ *inflexional to the real edges* $\mathscr{E}_{\sigma_s}^{(s)}$ ($s=r+1,r+2,\cdots,n-1$) *with the* $(2\mathbf{k}_{n-s})$th *-order singularity in a self-mirror domain* $\Omega_{\alpha}^{\mathscr{SM}}$ *at* $\mathbf{x}_m \in \mathscr{E}_{\sigma_r}^{(r)}$ *transversely passes over the concave edge of* $\mathscr{E}_{\sigma_r}^{(r)}$ *if and only if*

$$\left.\begin{aligned}&\hbar_{i_\alpha} G_{\partial\Omega_{\alpha i_\alpha}}^{(s_{i_\alpha},\alpha;\mathscr{D})}(\mathbf{x}_m,t_{m-},\mathbf{p}_\alpha,\boldsymbol{\lambda}_{i_\alpha})=0\\&\text{for } s_{i_\alpha}=0,1,2,\cdots,2k_{i_\alpha}-1;\\&\hbar_{i_\alpha} G_{\partial\Omega_{\alpha i_\alpha}}^{(2k_{i_\alpha},\alpha;\mathscr{D})}(\mathbf{x}_m,t_{m-},\mathbf{p}_\alpha,\boldsymbol{\lambda}_{i_\alpha})<0\end{aligned}\right\}\tag{6.96}$$

for all $i_\alpha \in \mathscr{N}_{\alpha\mathscr{B}}^{C}(\sigma_s)\subset \mathscr{N}_{\alpha\mathscr{B}}(\sigma_r,\mathscr{SM})$,

$$\left.\begin{aligned}&\hbar_{i_\alpha} G_{\partial\bar{\Omega}_{\alpha i_\alpha}}^{(s_{i_\alpha},\alpha;\mathscr{D})}(\mathbf{x}_m,t_{m+},\mathbf{p}_\alpha,\boldsymbol{\lambda}_{i_\alpha})=0\\&\text{for } s_{i_\alpha}=0,1,2,\cdots,2k_{i_\alpha}-1;\\&\hbar_{i_\alpha} G_{\partial\bar{\Omega}_{\alpha i_\alpha}}^{(2k_{i_\alpha},\alpha;\mathscr{D})}(\mathbf{x}_m,t_{m+},\mathbf{p}_\alpha,\boldsymbol{\lambda}_{i_\alpha})<0\end{aligned}\right\}\tag{6.97}$$

for all $i_\alpha \in \mathscr{N}_{\alpha\mathscr{B}}^{L}(\sigma_{\bar{s}})\subset \mathscr{N}_{\alpha\mathscr{B}}(\sigma_r,\mathscr{SM})$.

Proof. This theorem can be proved as in Chapter 3. ■

Theorem 6.8. *For a discontinuous dynamical system in Eqs.* (6.4)-(6.8), *there is a point* $\mathbf{x}^{(\sigma_r;\mathscr{E})}(t_m)\equiv \mathbf{x}_m \in \mathscr{E}_{\sigma_r}^{(r)}$ ($\sigma_r \in \mathscr{N}_{\mathscr{E}}^{r}=\{1,2,\cdots,N_r\}$ *and* $r\in\{0,1,2,\cdots,n-2\}$) *at time* t_m *with the related boundaries* $\partial\Omega_{\rho i_\rho}(\sigma_r)$ *and related domains* $\Omega_\rho(\sigma_r)$ ($\rho \in \mathscr{N}_{\mathscr{D}}(\sigma_r)\subset \mathscr{N}_{\mathscr{D}}=\{1,2,\cdots,N_d\}$ *and* $i_\rho \in \mathscr{N}_{\rho\mathscr{B}}(\sigma_r)\subset \mathscr{N}_{\mathscr{B}}=\{1,2,\cdots,N_b\}$), *where the boundary index subset of* $\mathscr{N}_{\rho\mathscr{B}}(\sigma_r)$ *has* n_{σ_r}*-elements* ($\rho=\alpha,\beta$ *and* $n_{\sigma_r}\geq(n-r)$). *For an arbitrarily small* $\varepsilon>0$, *there are two time intervals* $[t_{m-\varepsilon},t_m)$ *and* $(t_m,t_{m+\varepsilon}]$. *A flow* $\mathbf{x}^{(\rho;\mathscr{D})}(t)$ *is* $C^{r_\rho}_{[t_{m-\varepsilon},t_m)}$ *and/or* $C^{r_\rho}_{(t_m,t_{m+\varepsilon}]}$*-continuous for time* t *and* $\| d^{r_\rho+1}\mathbf{x}^{(\rho;\mathscr{D})}/dt^{r_\rho+1}\|<\infty$. $\mathscr{N}_{\rho\mathscr{B}}(\sigma_r)=\cup_{j\in\mathscr{J}}\mathscr{N}^{j}_{\rho\mathscr{B}}(\sigma_r)$ *where* $\mathscr{J}=\{C,L\}$ *and* $\cap_{j\in\mathscr{J}}\mathscr{N}^{j}_{\rho\mathscr{B}}(\sigma_r)=\varnothing$.

(i) *A domain flow* $\mathbf{x}^{(\alpha;\mathscr{D})}$ *in a full-mirror domain* $\Omega_\alpha^{\mathscr{FM}}$ *at* $\mathbf{x}_m\in\mathscr{E}^{(r)}_{\sigma_r}$ *is a coming flow to the concave edge* $\mathscr{E}^{(r)}_{\sigma_r}$ *if and only if*

$$\hbar_{i_\alpha}G^{(\alpha)}_{\partial\bar{\Omega}_{\alpha i_\alpha}}(\mathbf{x}_m,t_{m-},\mathbf{p}_\alpha,\boldsymbol{\lambda}_{i_\alpha})<0 \text{ for all } i_\alpha\in\mathscr{N}^{\mathscr{FM}}_{\alpha\mathscr{B}}(\sigma_r), \tag{6.98}$$

(ii) *A domain flow* $\mathbf{x}^{(\alpha;\mathscr{D})}$ *in a full-mirror domain* $\Omega_\alpha^{\mathscr{FM}}$ *at* $\mathbf{x}_m\in\mathscr{E}^{(r)}_{\sigma_r}$ *is a leaving flow to the concave edge* $\mathscr{E}^{(r)}_{\sigma_r}$ *if and only if*

$$\hbar_{i_\alpha}G^{(\alpha;\mathscr{D})}_{\partial\bar{\Omega}_{\alpha i_\alpha}}(\mathbf{x}_m,t_{m+},\mathbf{p}_\alpha,\boldsymbol{\lambda}_{i_\alpha})>0 \text{ for all } i_\alpha\in\mathscr{N}^{\mathscr{FM}}_{\alpha\mathscr{B}}(\sigma_r), \tag{6.99}$$

(iii) *A domain flow* $\mathbf{x}^{(\alpha;\mathscr{D})}$ *tangential to the imaginary extended edges* $\mathscr{E}^{(s)}_{\bar{\sigma}_s}$ ($s=r+1,r+2,\cdots,n-1$) *with the* $(2\mathbf{k}_{n-s}+1)$ *th-order singularity in the maximum full-mirror domain* $\Omega_\alpha^{\mathscr{FM}}$ *at* $\mathbf{x}_m\in\mathscr{E}^{(r)}_{\sigma_r}$ *transversely passes over the concave edge* $\mathscr{E}^{(r)}_{\sigma_r}$ *if and only if*

$$\left.\begin{array}{l}\hbar_{i_\alpha}G^{(s_{i_\alpha},\alpha;\mathscr{D})}_{\partial\bar{\Omega}_{\alpha i_\alpha}}(\mathbf{x}_m,t_{m-},\mathbf{p}_\alpha,\boldsymbol{\lambda}_{i_\alpha})=0\\ \text{for } s_{i_\alpha}=0,1,2,\cdots,2k_{i_\alpha};\\ \hbar_{i_\alpha}G^{(2k_{i_\alpha}+1,\alpha;\mathscr{D})}_{\partial\bar{\Omega}_{\alpha i_\alpha}}(\mathbf{x}_m,t_{m-},\mathbf{p}_\alpha,\boldsymbol{\lambda}_{i_\alpha})>0\end{array}\right\} \tag{6.100}$$
$$\text{for all } i_\alpha\in\mathscr{N}^{C}_{\alpha\mathscr{B}}(\sigma_{\bar{s}})\subset\mathscr{N}_{\alpha\mathscr{B}}(\sigma_r,\mathscr{FM}),$$

$$\left.\begin{array}{l}\hbar_{i_\alpha}G^{(s_{i_\alpha},\alpha;\mathscr{D})}_{\partial\Omega_{\alpha i_\alpha}}(\mathbf{x}_m,t_{m+},\mathbf{p}_\alpha,\boldsymbol{\lambda}_{i_\alpha})=0\\ \text{for } s_{i_\alpha}=0,1,2,\cdots,2k_{i_\alpha};\\ \hbar_{i_\alpha}G^{(2k_{i_\alpha}+1,\alpha;\mathscr{D})}_{\partial\Omega_{\alpha i_\alpha}}(\mathbf{x}_m,t_{m+},\mathbf{p}_\alpha,\boldsymbol{\lambda}_{i_\alpha})>0\end{array}\right\} \tag{6.101}$$
$$\text{for all } i_\alpha\in\mathscr{N}^{L}_{\alpha\mathscr{B}}\subset\mathscr{N}_{\alpha\mathscr{B}}(\sigma_r,\mathscr{SM}).$$

(iv) *A domain flow* $\mathbf{x}^{(\alpha;\mathscr{D})}$ *inflexional to the imaginary extended edges* $\mathscr{E}^{(s)}_{\sigma_s}$ ($s=r+1,r+2,\cdots,n-1$) *with the* $(2\mathbf{k}_{n-s})$*-singularity in the maximum full-mirror domain* $\Omega_\alpha^{\mathscr{FM}}$ *at* $\mathbf{x}_m\in\mathscr{E}^{(r)}_{\sigma_r}$ *does not transversely pass over the concave edge* $\mathscr{E}^{(r)}_{\sigma_r}$ *if and only if*

$$\left.\begin{array}{l}\hbar_{i_\alpha} G_{\partial\Omega_{\alpha i_\alpha}}^{(s_{i_\alpha},\alpha;\mathscr{D})}(\mathbf{x}_m, t_{m-}, \mathbf{p}_\alpha, \boldsymbol{\lambda}_{i_\alpha}) = 0 \\ for\ s_{i_\alpha} = 0,1,2,\cdots,2k_{i_\alpha}-1; \\ \hbar_{i_\alpha} G_{\partial\Omega_{\alpha i_\alpha}}^{(2k_{i_\alpha},\alpha;\mathscr{D})}(\mathbf{x}_m, t_{m-}, \mathbf{p}_\alpha, \boldsymbol{\lambda}_{i_\alpha}) < 0 \end{array}\right\} \tag{6.102}$$
$$for\ all\ i_\alpha \in \mathscr{N}_{\alpha\mathscr{B}}^{C}(\sigma_{\bar{s}}) \subset \mathscr{N}_{\alpha\mathscr{B}}(\sigma_r, \mathscr{FM}).$$

Proof. This theorem can be proved as in Chapter 3. ■

Theorem 6.9. *For a discontinuous dynamical system in Eqs. (6.4)-(6.8), there is a point* $\mathbf{x}^{(\sigma_r;\mathscr{E})}(t_m) \equiv \mathbf{x}_m \in \mathscr{E}_{\sigma_r}^{(r)}$ ($\sigma_r \in \mathscr{N}_{\mathscr{E}}^{r} = \{1,2,\cdots,N_r\}$ *and* $r \in \{0,1,2,\cdots,n-2\}$) *at time* t_m *with the related boundaries* $\partial\Omega_{\rho i_\rho}(\sigma_r)$ *and related domains* $\Omega_\rho(\sigma_r)$ ($\rho \in \mathscr{N}_{\mathscr{D}}(\sigma_r) \subset \mathscr{N}_{\mathscr{D}} = \{1,2,\cdots,N_d\}$ *and* $i_\rho \in \mathscr{N}_{\rho\mathscr{B}}(\sigma_r) \subset \mathscr{N}_{\mathscr{B}} = \{1,2,\cdots,N_b\}$), *where the boundary index subset of* $\mathscr{N}_{\rho\mathscr{B}}(\sigma_r)$ *has* n_{σ_r} *-elements* ($\rho = \alpha,\beta$ *and* $n_{\sigma_r} \geq (n-r)$). *For an arbitrarily small* $\varepsilon > 0$, *there are two time intervals* $[t_{m-\varepsilon}, t_m)$ *and* $(t_m, t_{m+\varepsilon}]$. *A flow* $\mathbf{x}^{(\rho;\mathscr{D})}(t)$ *is* $C_{[t_{m-\varepsilon},t_m)}^{r_\rho}$ *and/or* $C_{(t_m,t_{m+\varepsilon}]}^{r_\rho}$ *-continuous for time* t *and* $\| d^{r_\rho+1}\mathbf{x}^{(\rho;\mathscr{D})}/dt^{r_\rho+1} \| < \infty$. $\mathscr{N}_{\rho\mathscr{B}}(\sigma_r) = \cup_{j\in\mathscr{J}} \mathscr{N}_{\rho\mathscr{B}}^{j}(\sigma_r)$ *where* $\mathscr{J} = \{C, L\}$ *and* $\cap_{j\in\mathscr{J}} \mathscr{N}_{\rho\mathscr{B}}^{j}(\sigma_r) = \varnothing$.

(i) *A domain flow* $\mathbf{x}^{(\alpha;\mathscr{D})}$ *in domain* Ω_α *at* $\mathbf{x}_m \in \mathscr{E}_{\sigma_r}^{(r)}$ *is a coming flow to the convex edge* $\mathscr{E}_{\sigma_r}^{(r)}$ *if and only if*

$$\hbar_{i_\alpha} G_{\partial\Omega_{\alpha i_\alpha}}^{(\alpha;\mathscr{D})}(\mathbf{x}_m, t_{m-}, \mathbf{p}_\alpha, \boldsymbol{\lambda}_{i_\alpha}) < 0\ \ for\ all\ i_\alpha \in \mathscr{N}_{\alpha\mathscr{B}}(\sigma_r). \tag{6.103}$$

(ii) *A domain flow* $\mathbf{x}^{(\alpha;\mathscr{D})}$ *in domain* Ω_α *at* $\mathbf{x}_m \in \mathscr{E}_{\sigma_r}^{(r)}$ *is a leaving flow to the convex edge* $\mathscr{E}_{\sigma_r}^{(r)}$ *if and only if*

$$\hbar_{i_\alpha} G_{\partial\Omega_{\alpha i_\alpha}}^{(\alpha;\mathscr{D})}(\mathbf{x}_m, t_{m+}, \mathbf{p}_\alpha, \boldsymbol{\lambda}_{i_\alpha}) > 0\ \ for\ all\ i_\alpha \in \mathscr{N}_{\alpha\mathscr{B}}(\sigma_r). \tag{6.104}$$

(iii) *A domain flow* $\mathbf{x}^{(\alpha;\mathscr{D})}$ *in domain* Ω_α *at* $\mathbf{x}_m \in \mathscr{E}_{\sigma_r}^{(r)}$ *is a tangential coming flow to the convex edges* $\mathscr{E}_{\sigma_s}^{(s)}$ ($s = r+1, r+2, \cdots, n-1$) *with the* $(2\mathbf{k}_{n-s}+1)$ *-singularity if and only if*

$$\left.\begin{array}{l}\hbar_{i_\alpha} G_{\partial\Omega_{\alpha i_\alpha}}^{(s_{i_\alpha},\alpha;\mathscr{D})}(\mathbf{x}_m, t_{m-}, \mathbf{p}_\alpha, \boldsymbol{\lambda}_{i_\alpha}) = 0 \\ for\ s_{i_\alpha} = 0,1,2,\cdots,2k_{i_\alpha}; \\ \hbar_{i_\alpha} G_{\partial\Omega_{\alpha i_\alpha}}^{(2k_{i_\alpha}+1,\alpha;\mathscr{D})}(\mathbf{x}_m, t_{m-}, \mathbf{p}_\alpha, \boldsymbol{\lambda}_{i_\alpha}) > 0 \end{array}\right\}\ for\ all\ i_\alpha \in \mathscr{N}_{\alpha\mathscr{B}}(\sigma_s). \tag{6.105}$$

(iv) *A domain flow* $\mathbf{x}^{(\alpha;\mathscr{D})}$ *in domain* Ω_α *at* $\mathbf{x}_m \in \mathscr{E}_{\sigma_r}^{(r)}$ *is a tangential leaving flow to the convex edges* $\mathscr{E}_{\sigma_s}^{(s)}$ ($s = r+1, r+2, \cdots, n-1$) *with the* $(2\mathbf{k}_{n-s}+1)$ *-singularity if and only if*

$$\left.\begin{array}{l} \hbar_{i_\alpha} G_{\partial\Omega_{\alpha i_\alpha}}^{(s_{i_\alpha},\alpha;\mathscr{D})}(\mathbf{x}_m, t_{m+}, \mathbf{p}_\alpha, \boldsymbol{\lambda}_{i_\alpha}) = 0 \\ \text{for } s_{i_\alpha} = 0,1,2,\cdots,2k_{i_\alpha}; \\ \hbar_{i_\alpha} G_{\partial\Omega_{\alpha i_\alpha}}^{(2k_{i_\alpha}+1,\alpha;\mathscr{D})}(\mathbf{x}_m, t_{m+}, \mathbf{p}_\alpha, \boldsymbol{\lambda}_{i_\alpha}) > 0 \end{array}\right\} \text{for all } i_\alpha \in \mathscr{N}_{\alpha\mathscr{B}}(\sigma_s). \tag{6.106}$$

(v) *A domain flow* $\mathbf{x}^{(\alpha;\mathscr{D})}$ *in domain* Ω_α *at* $\mathbf{x}_m \in \mathscr{E}_{\sigma_r}^{(r)}$ *is an inflexional coming flow to the convex edges* $\mathscr{E}_{\sigma_s}^{(s)}$ $(s = r+1, r+2, \cdots, n-1)$ *with the* $(2\mathbf{k}_{n-s})$ *singularity if and only if*

$$\left.\begin{array}{l} \hbar_{i_\alpha} G_{\partial\Omega_{\alpha i_\alpha}}^{(s_{i_\alpha},\alpha;\mathscr{D})}(\mathbf{x}_m, t_{m-}, \mathbf{p}_\alpha, \boldsymbol{\lambda}_{i_\alpha}) = 0 \\ \text{for } s_{i_\alpha} = 0,1,2,\cdots,2k_{i_\alpha}-1; \\ \hbar_{i_\alpha} G_{\partial\Omega_{\alpha i_\alpha}}^{(2k_{i_\alpha},\alpha;\mathscr{D})}(\mathbf{x}_m, t_{m-}, \mathbf{p}_\alpha, \boldsymbol{\lambda}_{i_\alpha}) < 0 \end{array}\right\} \text{for all } i_\alpha \in \mathscr{N}_{\alpha\mathscr{B}}(\sigma_s). \tag{6.107}$$

(vi) *A domain flow* $\mathbf{x}^{(\alpha;\mathscr{D})}$ *in domain* Ω_α *at* $\mathbf{x}_m \in \mathscr{E}_{\sigma_r}^{(r)}$ *is an inflexional leaving flow to the convex edges* $\mathscr{E}_{\sigma_s}^{(s)}$ $(s = r+1, r+2, \cdots, n-1)$ *with the* $2\mathbf{k}_{n-s}$ *singularity if and only if*

$$\left.\begin{array}{l} \hbar_{i_\alpha} G_{\partial\Omega_{\alpha i_\alpha}}^{(s_{i_\alpha},\alpha;\mathscr{D})}(\mathbf{x}_m, t_{m+}, \mathbf{p}_\alpha, \boldsymbol{\lambda}_{i_\alpha}) = 0 \\ \text{for } s_{i_\alpha} = 0,1,2,\cdots,2k_{i_\alpha}-1; \\ \hbar_{i_\alpha} G_{\partial\Omega_{\alpha i_\alpha}}^{(2k_{i_\alpha},\alpha;\mathscr{D})}(\mathbf{x}_m, t_{m+}, \mathbf{p}_\alpha, \boldsymbol{\lambda}_{i_\alpha}) > 0 \end{array}\right\} \text{for all } i_\alpha \in \mathscr{N}_{\alpha\mathscr{B}}(\sigma_s). \tag{6.108}$$

Proof. This theorem can be proved as in Chapter 3. ■

6.6. Domains flow attractivity

The switchability of a domain flow to the edge was discussed in the previous section. In this section, the attractivity of a domain flow to edges will be discussed. Before such a discussion, the attractivity of a flow to separation boundary will be discussed first. Further, the attractivity of a flow to a specific edge can be discussed.

6.6.1. Attractivity to boundary

Consider a discontinuous dynamical system with three vectors fields to form three dynamical systems

$$\begin{aligned}&\dot{\mathbf{x}}^{(\alpha;\mathscr{D})}=\mathbf{F}^{(\alpha;\mathscr{D})}(\mathbf{x}^{(\alpha;\mathscr{D})},t,\mathbf{p}_\alpha)\ \text{in}\ \Omega_\alpha\ (\alpha=i,j),\\&\dot{\mathbf{x}}^{(i_\alpha;\mathscr{B})}=\mathbf{F}^{(i_\alpha;\mathscr{B})}(\mathbf{x}^{(i_\alpha;\mathscr{B})},t,\boldsymbol{\lambda}_{i_\alpha})\\&\text{with}\ \varphi_{\alpha i_\alpha}(\mathbf{x}^{(i_\alpha;\mathscr{B})},t,\boldsymbol{\lambda}_{i_\alpha})=0\ \text{on}\ \partial\Omega_{\alpha i_\alpha}.\end{aligned}\tag{6.109}$$

In Chapters 2 and 3, the passability of a domain flow to a specific boundary is dependent on the vector fields on both sides of boundary. In Chapter 4, flow barriers on a boundary and the corresponding switchability were discussed. Since the flow barriers and multiple vector fields exist, the transport laws were discussed in Chapter 5. To discuss the attractivity of a flow to the separaration boundary (or separtrix), the *G*-function of a flow in the domain will be introduced. Thus a separatrix surface family will be introduced. The separatrix surface family will be formed by many measuring surfaces possessing the same function of the separatrix surface with different nonzero constants in right-hand side.

Definition 6.15. In a discontinuous dynamical system, suppose there is a boundary of $\partial\Omega_{\alpha i_\alpha}$ with $\varphi_{\alpha i_\alpha}(\mathbf{x}^{(i_\alpha;\mathscr{B})},t,\boldsymbol{\lambda}_{i_\alpha})=0$ which divide the phase space into two domains Ω_i and Ω_j. The normal vector of the boundary is

$$\mathbf{n}_{\partial\Omega_{\alpha i_\alpha}}=\nabla\varphi_{\alpha i_\alpha}=(\frac{\partial\varphi_{\alpha i_\alpha}}{\partial x_1},\cdots,\frac{\partial\varphi_{\alpha i_\alpha}}{\partial x_n})^{\mathrm{T}}.\tag{6.110}$$

Suppose there is a family of constants C_σ monotonically increasing with $\sigma\in\mathbb{Z}$ for $\mathbf{n}_{\partial\Omega_{\alpha i_\alpha}}\to\Omega_\alpha$, the *corresponding family of separatrix measuring surfaces* in two domains (Ω_α and Ω_β) separated by the boundary $\partial\Omega_{\alpha\beta}$ are defined as

$$\begin{aligned}&S_\sigma^{(i_\alpha,\alpha)}=\left\{\mathbf{x}_\sigma^{(i_\alpha,\alpha)}\left|\begin{array}{l}\varphi_{\alpha i_\alpha}(\mathbf{x}_\sigma^{(i_\alpha,\alpha)},t,\boldsymbol{\lambda}_{i_\alpha})=C_\sigma^{(i_\alpha,\alpha)}\in(0,\infty)\\\text{monotonically increaing with }\sigma\in\mathbb{Z}_+\\\text{for }\mathbf{n}_{\partial\Omega_{\alpha i_\alpha}}\to\Omega_\alpha\end{array}\right.\right\}\subset\Omega_\alpha,\\&S_0^{(i_\alpha,\alpha)}=\left\{\mathbf{x}^{(i_\alpha;\mathscr{B})}\left|\begin{array}{l}\varphi_{\alpha i_\alpha}(\mathbf{x}^{(i_\alpha;\mathscr{B})},t,\boldsymbol{\lambda}_{i_\alpha})=C_0^{(i_\alpha,\alpha)}=0\\\text{for }\mathbf{n}_{\partial\Omega_{\alpha i_\alpha}}\to\Omega_\alpha\end{array}\right.\right\}=\partial\Omega_{\alpha i_\alpha},\\&S_\sigma^{(i_\alpha,\beta)}=\left\{(\mathbf{x}_\sigma^{(i_\alpha,\beta)}\left|\begin{array}{l}\varphi_{\beta i_\alpha}(\mathbf{x}_\sigma^{(i_\alpha,\beta)},t,\boldsymbol{\lambda}_{i_\alpha})=C_\sigma^{(i_\alpha,\beta)}\in(-\infty,0)\\\text{monotonically inceasing with }\sigma\in\mathbb{Z}_-\\\text{for }\mathbf{n}_{\partial\Omega_{\alpha i_\alpha}}=\mathbf{n}_{\partial\Omega_{\alpha\beta}}\to\Omega_\alpha\end{array}\right.\right\}\subset\Omega_\beta.\end{aligned}\tag{6.111}$$

From the previous definition, with different non-zero constants, the family of separatrix surfaces is sketched in Fig. 6.16. The dark surface is a separatrix. The gray surfaces are a family of separatrix measuring surfaces with different non-zero constants. The boundary is determined by $\varphi_{\alpha i_\alpha}(\mathbf{x}^{(i_\alpha;\mathscr{B})},t,\boldsymbol{\lambda}_{i_\alpha})=0$, but the separatrix measuring surface in domain is given by $\varphi_{\alpha i_\alpha}(\mathbf{x}_\sigma^{(i_\alpha,\alpha)},t,\boldsymbol{\lambda}_{i_\alpha})=C_\sigma^{(i_\alpha,\alpha)}$. If $\mathbf{n}_{\partial\Omega_{ij}}\to\Omega_i$, the non-zero constants $C_\sigma^{(i)}$ and $C_\sigma^{(j)}$ should be greater than zero in domain Ω_i and less then zero in domain Ω_j, respectively. To determine the attractivity of a flow to the separation boundary, the G-function of a flow to the separatrix measuring surface should be introduced as the G-function of a flow to the boundary.

Consider a separatrix measuring surface in domain Ω_i in Fig. 6.17. This surface of $\varphi_{\alpha i_\alpha}(\mathbf{x}_\sigma^{(i_\alpha,\alpha)},t,\boldsymbol{\lambda}_{i_\alpha})=C_\sigma^{(i_\alpha,\alpha)}$ is an equi-constant surface with the same property of the separation boundary of $\varphi_{\alpha i_\alpha}(\mathbf{x}^{(i_\alpha;\mathscr{B})},t,\boldsymbol{\lambda}_{i_\alpha})=0$. Suppose there is a flow intersected with such a surface, and the corresponding vector fields are also expressed. Thus the corresponding G-function of the flow to the separatrix measuring surface can be defined from Luo (2008a,b). Through the G-function, the increasing and decreasing of the constant $C_\sigma^{(i)}$ can be measured.

Definition 6.16. For a discontinuous dynamical system in Eq. (6.109), there is a point $\mathbf{x}_\sigma^{(i_\alpha,\alpha)}(t)\equiv\mathbf{x}\in S_\sigma^{(i_\alpha,\alpha)}$ ($\sigma\in\mathbb{N}$) at time t in domains Ω_α ($\alpha=i,j$) to satisfy

$$\begin{aligned}&\dot{\mathbf{x}}_\sigma^{(i_\alpha,\alpha)}=\mathbf{F}^{(i_\alpha,\alpha)}(\mathbf{x}_\sigma^{(i_\alpha,\alpha)},t,\boldsymbol{\lambda}_{i_\alpha})\\ &\text{with }\varphi_{\alpha i_\alpha}(\mathbf{x}_\sigma^{(i_\alpha,\alpha)},t,\boldsymbol{\lambda}_{i_\alpha})=C_\sigma^{(i_\alpha,\alpha)}\ \text{ on } S_\sigma^{(i_\alpha,\alpha)},\end{aligned}\tag{6.112}$$

where $C_\sigma^{(i_\alpha,\alpha)}$ is a non-zero constant.

The G-functions of a flow to the surface $S_\sigma^{(i_\alpha,\alpha)}$ are defined as follows.

Definition 6.17. For a discontinuous dynamical system in Eq. (6.109), there is a point $\mathbf{x}_\sigma^{(i_\alpha,\alpha)}(t)\equiv\mathbf{x}\in S_\sigma^{(i_\alpha,\alpha)}$ ($\sigma\in\mathbb{N}$) at time t in domains Ω_α ($\alpha=i,j$) to satisfy Eq. (6.112). For a point $\mathbf{x}^{(\alpha;\mathscr{S})}(t)=\mathbf{x}$ on the flow, the G-function of the vector field $\mathbf{F}^{(\alpha;\mathscr{S})}(\mathbf{x}^{(\alpha;\mathscr{S})},t,\mathbf{p}_\alpha)$ for the flow $\mathbf{x}^{(\alpha;\mathscr{S})}(t)$ is defined as

$$\begin{aligned}&G_{S_\sigma^{(i_\alpha,\alpha)}}^{(\alpha;\mathscr{S})}(\mathbf{x},t,\mathbf{p}_\alpha,\boldsymbol{\lambda}_{i_\alpha})\\ &=\mathbf{n}_{S_\sigma^{(i_\alpha,\alpha)}}^{\mathrm{T}}(\mathbf{x}_\sigma^{(i_\alpha,\alpha)},t,\boldsymbol{\lambda}_{i_\alpha})\cdot[\mathbf{F}^{(\alpha;\mathscr{S})}(\mathbf{x}^{(\alpha;\mathscr{S})},t,\mathbf{p}_\alpha)\\ &\quad-\mathbf{F}^{(i_\alpha,\alpha)}(\mathbf{x}_\sigma^{(i_\alpha,\alpha)},t,\boldsymbol{\lambda}_{i_\alpha})]\Big|_{(\mathbf{x}^{(\alpha;\mathscr{S})}=\mathbf{x},\mathbf{x}_\sigma^{(i_\alpha,\alpha)}=\mathbf{x},t)}.\end{aligned}\tag{6.113}$$

The *higher-order G-function* of the vector field $\mathbf{F}^{(\alpha;\mathscr{S})}(\mathbf{x}^{(\alpha;\mathscr{S})},t,\mathbf{p}_\alpha)$ is defined as

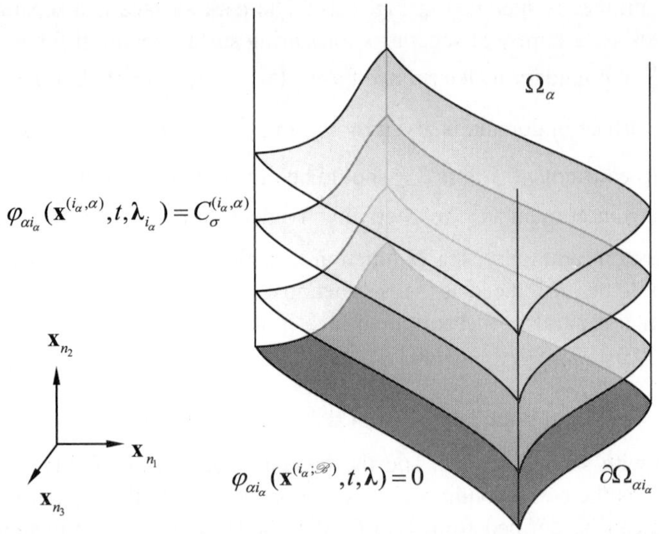

Fig. 6.16 A family of separatrix measuring surfaces in domain Ω_i . The dark surface is the separation boundary. The gray surfaces are a family of separatrix measuring surfaces with different non-zero constants $C_\sigma^{(i_\alpha,\alpha)}$ ($\sigma=1,2,\cdots$).

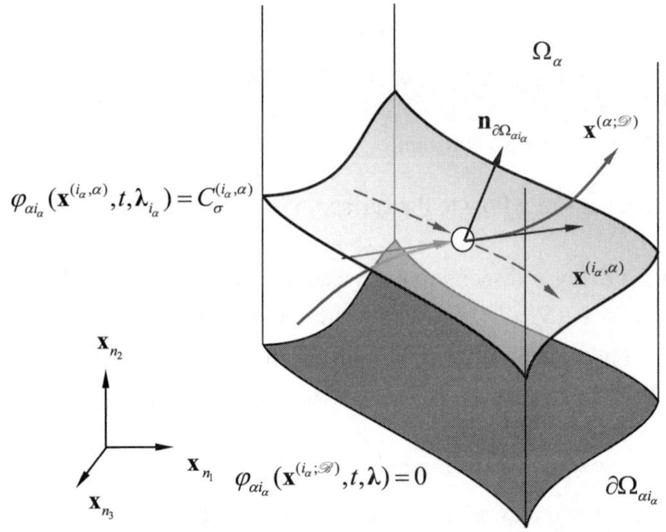

Fig. 6.17 A flow intersected with a separatrix measuring surface in domain Ω_i . The dark surface is the separation boundary. The gray surface is the separatrix measuring surface with a non-zero constant $C_\sigma^{(i_\alpha,\alpha)}$ ($\sigma=1,2,\cdots$). The solid and dashed curves are a domain flow $\mathbf{x}^{(\alpha;\mathscr{D})}$ and the flow on measuring surface $\mathbf{x}_\sigma^{(i_\alpha,\alpha)}$, respectively.

$$
\begin{aligned}
&G_{S_\sigma^{(i_\alpha,\alpha)}}^{(k_\alpha,\alpha;\mathscr{D})}(\mathbf{x},t,\mathbf{p}_\alpha,\boldsymbol{\lambda}_{i_\alpha}) \\
&=\sum_{r=1}^{k_\alpha+1} C_{k_\alpha+1}^{r} D_\sigma^{k_\alpha+1-r}\mathbf{n}_{S_\sigma^{(i_\alpha,\alpha)}}^{\mathrm{T}}(\mathbf{x}_\sigma^{(i_\alpha,\alpha)},t,\boldsymbol{\lambda}_{i_\alpha})\cdot[D_\alpha^{r-1}\mathbf{F}^{(\alpha;\mathscr{D})}(\mathbf{x}^{(\alpha;\mathscr{D})},t,\mathbf{p}_\alpha) \\
&\quad - D_\sigma^{r-1}\mathbf{F}^{(i_\alpha,\alpha)}(\mathbf{x}_\sigma^{(i_\alpha,\alpha)},t,\boldsymbol{\lambda}_{i_\alpha})]\Big|_{(\mathbf{x}^{(\alpha;\mathscr{D})}=\mathbf{x},\mathbf{x}_\sigma^{(i_\alpha,\alpha)}=\mathbf{x},t)}.
\end{aligned}
\tag{6.114}
$$

for $k_\alpha = 0,1,2,\cdots$ where $D_\alpha = \partial/\partial\mathbf{x}^{(\alpha;\mathscr{D})} + \partial/\partial t$ and $D_\sigma = \partial/\partial\mathbf{x}_\sigma^{(i_\alpha,\alpha)} + \partial/\partial t$.

Since the boundary of $\varphi_{\alpha i_\alpha}(\mathbf{x}^{(i_\alpha;\mathscr{B})},t,\boldsymbol{\lambda}_{i_\alpha})=0$ and the measuring surface of $\varphi_{\alpha i_\alpha}(\mathbf{x}_\sigma^{(i_\alpha,\alpha)},t,\boldsymbol{\lambda}_{i_\alpha})=C_\sigma^{(i_\alpha,\alpha)}$ has the same function with different constant, the normal vector expressions of the boundary and measuring surfaces are identical. i.e.

$$
\mathbf{n}_{S_\sigma^{(i_\alpha,\alpha)}}^{\mathrm{T}}(\mathbf{x}_\sigma^{(i_\alpha,\alpha)},t,\boldsymbol{\lambda}_{i_\alpha}) = \mathbf{n}_{\partial\Omega_{\alpha i_\alpha}}^{\mathrm{T}}(\mathbf{x}_\sigma^{(i_\alpha,\alpha)},t,\boldsymbol{\lambda}_{i_\alpha}).
\tag{6.115}
$$

The boundary and the measuring surface do not have any intersection.

Definition 6.18. For a discontinuous dynamical system in Eq. (6.109), there is a flow $\mathbf{x}^{(\alpha;\mathscr{D})}(t)$ in domains Ω_α ($\alpha = i,j$). At time t_m, $\mathbf{x}_m^{(i_\alpha,\alpha)} = \mathbf{x}_\sigma^{(i_\alpha,\alpha)}(t_m) \in S_{\sigma_m}^{(i_\alpha,\alpha)}$ ($\sigma_m \in \mathbb{N}$) satisfying Eq. (6.112). For an arbitrarily small $\varepsilon > 0$, there is a time interval $[t_{m-\varepsilon}, t_{m+\varepsilon}]$. A flow $\mathbf{x}^{(\alpha;\mathscr{D})}(t)$ is $C_{[t_{m-\varepsilon},t_{m+\varepsilon}]}^{r_\alpha}$-continuous ($r_\alpha \geq 1$) for time t and $\| d^{r_\alpha+1}\mathbf{x}^{(\alpha;\mathscr{D})}/dt^{r_\alpha+1} \| < \infty$. $\mathbf{x}_{m\pm\varepsilon}^{(\alpha;\mathscr{D})} = \mathbf{x}^{(\alpha;\mathscr{D})}(t_{m\pm\varepsilon}) \in S_{\sigma_{m\pm\varepsilon}}^{(i_\alpha,\alpha)}$ and $\mathbf{x}_{\sigma(m\pm\varepsilon)}^{(i_\alpha,\alpha)} \in S_{\sigma_m}^{(i_\alpha,\alpha)}$.

(i) The flow $\mathbf{x}^{(\alpha;\mathscr{D})}(t)$ at time t_m to the boundary of $\partial\Omega_{\alpha i_\alpha}$ is *attractive* if

$$
\begin{aligned}
&\hbar_{i_\alpha}\mathbf{n}_{\partial\Omega_{\alpha i_\alpha}}^{\mathrm{T}}(\mathbf{x}_{\sigma(m-\varepsilon)}^{(i_\alpha,\alpha)})\cdot[\mathbf{x}_{\sigma(m-\varepsilon)}^{(i_\alpha,\alpha)} - \mathbf{x}_{m-\varepsilon}^{(\alpha;\mathscr{D})}] < 0, \\
&\hbar_{i_\alpha}\mathbf{n}_{\partial\Omega_{\alpha i_\alpha}}^{\mathrm{T}}(\mathbf{x}_{\sigma(m+\varepsilon)}^{(i_\alpha,\alpha)})\cdot[\mathbf{x}_{m+\varepsilon}^{(\alpha;\mathscr{D})} - \mathbf{x}_{\sigma(m+\varepsilon)}^{(i_\alpha,\alpha)}] < 0, \\
&\hbar_{i_\alpha}C_{m-\varepsilon}^{(i_\alpha,\alpha)} > \hbar_{i_\alpha}C_{m}^{(i_\alpha,\alpha)} > \hbar_{i_\alpha}C_{m+\varepsilon}^{(i_\alpha,\alpha)}.
\end{aligned}
\tag{6.116}
$$

(ii) The flow $\mathbf{x}^{(\alpha;\mathscr{D})}(t)$ at time t_m to the boundary of $\partial\Omega_{\alpha i_\alpha}$ is *repulsive* if

$$
\begin{aligned}
&\hbar_{i_\alpha}\mathbf{n}_{\partial\Omega_{\alpha i_\alpha}}^{\mathrm{T}}(\mathbf{x}_{\sigma(m-\varepsilon)}^{(i_\alpha,\alpha)})\cdot[\mathbf{x}_{\sigma(m-\varepsilon)}^{(i_\alpha,\alpha)} - \mathbf{x}_{m-\varepsilon}^{(\alpha;\mathscr{D})}] > 0, \\
&\hbar_{i_\alpha}\mathbf{n}_{\partial\Omega_{\alpha i_\alpha}}^{\mathrm{T}}(\mathbf{x}_{\sigma(m+\varepsilon)}^{(i_\alpha,\alpha)})\cdot[\mathbf{x}_{m+\varepsilon}^{(\alpha;\mathscr{D})} - \mathbf{x}_{\sigma(m+\varepsilon)}^{(i_\alpha,\alpha)}] > 0, \\
&\hbar_{i_\alpha}C_{m-\varepsilon}^{(i_\alpha,\alpha)} < \hbar_{i_\alpha}C_{m}^{(i_\alpha,\alpha)} < \hbar_{i_\alpha}C_{m+\varepsilon}^{(i_\alpha,\alpha)}.
\end{aligned}
\tag{6.117}
$$

(iii) The flow $\mathbf{x}^{(\alpha;\mathscr{D})}(t)$ at time t_m to the boundary of $\partial\Omega_{\alpha i_\alpha}$ is *from the attractive to repulsive* state if

$$
\hbar_{i_\alpha}\mathbf{n}_{\partial\Omega_{\alpha i_\alpha}}^{\mathrm{T}}(\mathbf{x}_{\sigma(m-\varepsilon)}^{(i_\alpha,\alpha)})\cdot[\mathbf{x}_{\sigma(m-\varepsilon)}^{(i_\alpha,\alpha)} - \mathbf{x}_{m-\varepsilon}^{(\alpha;\mathscr{D})}] < 0,
$$

$$\begin{aligned}&\hbar_{i_\alpha}\mathbf{n}^{\mathrm{T}}_{\partial\Omega_{\alpha i_\alpha}}(\mathbf{x}^{(i_\alpha,\alpha)}_{\sigma(m+\varepsilon)})\cdot[\mathbf{x}^{(\alpha;\mathscr{D})}_{m+\varepsilon}-\mathbf{x}^{(i_\alpha,\alpha)}_{\sigma(m+\varepsilon)}]>0,\\&\hbar_{i_\alpha}C^{(i_\alpha,\alpha)}_{m}<\hbar_{i_\alpha}C^{(i_\alpha,\alpha)}_{m-\varepsilon}\text{ and }\hbar_{i_\alpha}C^{(i_\alpha,\alpha)}_{m}<\hbar_{i_\alpha}C^{(i_\alpha,\alpha)}_{m+\varepsilon}.\end{aligned}\tag{6.118}$$

(iv) The flow $\mathbf{x}^{(\alpha;\mathscr{D})}(t)$ at time t_m to the boundary of $\partial\Omega_{\alpha i_\alpha}$ is *from the repulsive to attractive* state if

$$\begin{aligned}&\hbar_{i_\alpha}\mathbf{n}^{\mathrm{T}}_{\partial\Omega_{\alpha i_\alpha}}(\mathbf{x}^{(i_\alpha,\alpha)}_{\sigma(m-\varepsilon)})\cdot[\mathbf{x}^{(i_\alpha,\alpha)}_{\sigma(m-\varepsilon)}-\mathbf{x}^{(\alpha;\mathscr{D})}_{m-\varepsilon}]>0,\\&\hbar_{i_\alpha}\mathbf{n}^{\mathrm{T}}_{\partial\Omega_{\alpha i_\alpha}}(\mathbf{x}^{(i_\alpha,\alpha)}_{\sigma(m+\varepsilon)})\cdot[\mathbf{x}^{(\alpha;\mathscr{D})}_{m+\varepsilon}-\mathbf{x}^{(i_\alpha,\alpha)}_{\sigma(m+\varepsilon)}]<0,\\&\hbar_{i_\alpha}C^{(i_\alpha,\alpha)}_{m}>\hbar_{i_\alpha}C^{(i_\alpha,\alpha)}_{m-\varepsilon}\text{ and }\hbar_{i_\alpha}C^{(i_\alpha,\alpha)}_{m}>\hbar_{i_\alpha}C^{(i_\alpha,\alpha)}_{m+\varepsilon}.\end{aligned}\tag{6.119}$$

(v) The flow $\mathbf{x}^{(\alpha;\mathscr{D})}(t)$ at time t_m to the boundary of $\partial\Omega_{\alpha i_\alpha}$ is *invariant* with an equi-measuring quantity of $C^{(i_\alpha,\alpha)}_{m}$ if

$$\begin{aligned}&\mathbf{x}^{(i_\alpha,\alpha)}_{\sigma(m-\varepsilon)}=\mathbf{x}^{(\alpha;\mathscr{D})}_{m-\varepsilon}\text{ and }\mathbf{x}^{(\alpha;\mathscr{D})}_{m+\varepsilon}=\mathbf{x}^{(i_\alpha,\alpha)}_{\sigma(m+\varepsilon)},\\&\hbar_{i_\alpha}C^{(i_\alpha,\alpha)}_{m-\varepsilon}=\hbar_{i_\alpha}C^{(i_\alpha,\alpha)}_{m}=\hbar_{i_\alpha}C^{(i_\alpha,\alpha)}_{m+\varepsilon}.\end{aligned}\tag{6.120}$$

From the above definition, the corresponding theorem is given as follows.

Theorem 6.10. *For a discontinuous dynamical system in Eq.* (6.109), *there is a flow* $\mathbf{x}^{(\alpha;\mathscr{D})}(t)$ *in domains* Ω_α $(\alpha=i,j)$. *At time* t_m, $\mathbf{x}^{(i_\alpha,\alpha)}_{m}=\mathbf{x}^{(i_\alpha,\alpha)}_{\sigma}(t_m)\in S^{(i_\alpha,\alpha)}_{\sigma_m}$ $(\sigma_m\in\mathbb{N})$ *satisfying Eq.* (6.112). *For an arbitrarily small* $\varepsilon>0$, *there is a time interval* $[t_{m-\varepsilon},t_{m+\varepsilon}]$. *A flow* $\mathbf{x}^{(\alpha;\mathscr{D})}(t)$ *is* $C^{r_\alpha}_{[t_{m-\varepsilon},t_{m+\varepsilon}]}$ *-continuous* $(r_\alpha\ge 2)$ *for time t and* $\|d^{r_\alpha+1}\mathbf{x}^{(\alpha;\mathscr{D})}/dt^{r_\alpha+1}\|<\infty$. $\mathbf{x}^{(\alpha;\mathscr{D})}_{m\pm\varepsilon}=\mathbf{x}^{(\alpha;\mathscr{D})}(t_{m\pm\varepsilon})\in S^{(i_\alpha,\alpha)}_{\sigma_{m\pm\varepsilon}}$ *and* $\mathbf{x}^{(i_\alpha,\alpha)}_{\sigma(m\pm\varepsilon)}\in S^{(i_\alpha,\alpha)}_{\sigma_m}$.

(i) *The flow* $\mathbf{x}^{(\alpha;\mathscr{D})}(t)$ *at time* t_m *is attractive to the boundary of* $\partial\Omega_{ij}$ *if and only if*

$$\hbar_{i_\alpha}G^{(\alpha;\mathscr{D})}_{\partial\Omega_{ij}}(\mathbf{x}^{(i_\alpha,\alpha)}_{m},t_m,\mathbf{p}_\alpha,\boldsymbol{\lambda}_{i_\alpha})<0.\tag{6.121}$$

(ii) *The flow* $\mathbf{x}^{(\alpha;\mathscr{D})}(t)$ *at time* t_m *is repulsive to the boundary of* $\partial\Omega_{ij}$ *if and only if*

$$\hbar_{i_\alpha}G^{(\alpha;\mathscr{D})}_{\partial\Omega_{ij}}(\mathbf{x}^{(i_\alpha,\alpha)}_{m},t_m,\mathbf{p}_\alpha,\boldsymbol{\lambda}_{i_\alpha})>0.\tag{6.122}$$

(iii) *The flow* $\mathbf{x}^{(\alpha;\mathscr{D})}(t)$ *at time* t_m *is from the attractive to repulsive state to the boundary of* $\partial\Omega_{ij}$ *if and only if*

$$\begin{aligned}&\hbar_{i_\alpha}G^{(\alpha;\mathscr{D})}_{\partial\Omega_{ij}}(\mathbf{x}^{(i_\alpha,\alpha)}_{m},t_m,\mathbf{p}_\alpha,\boldsymbol{\lambda}_{i_\alpha})=0,\\&\hbar_{i_\alpha}G^{(1,\alpha;\mathscr{D})}_{\partial\Omega_{ij}}(\mathbf{x}^{(i_\alpha,\alpha)}_{m},t_m,\mathbf{p}_\alpha,\boldsymbol{\lambda}_{i_\alpha})>0.\end{aligned}\tag{6.123}$$

(iv) *The flow* $\mathbf{x}^{(\alpha;\mathscr{D})}(t)$ *at time* t_m *is from the repulsive to attractive state for the* *boundary of $\partial\Omega_{ij}$ if and only if*

$$\begin{aligned} &\hbar_{i_\alpha} G_{\partial\Omega_{ij}}^{(\alpha;\mathscr{D})}(\mathbf{x}_m^{(i_\alpha,\alpha)}, t_m, \mathbf{p}_\alpha, \boldsymbol{\lambda}_{i_\alpha}) = 0, \\ &\hbar_{i_\alpha} G_{\partial\Omega_{ij}}^{(1,\alpha;\mathscr{D})}(\mathbf{x}_m^{(i_\alpha,\alpha)}, t_m, \mathbf{p}_\alpha, \boldsymbol{\lambda}_{i_\alpha}) < 0. \end{aligned} \tag{6.124}$$

Proof. The surface of $\varphi_{\alpha i_\alpha}(\mathbf{x}_m^{(i_\alpha,\alpha)}, t_m, \boldsymbol{\lambda}_{i_\alpha}) = C_m^{(i_\alpha,\alpha)}$ is as a boundary. As in Chapter 3, this theorem can be proved. ■

Theorem 6.11. *For a discontinuous dynamical system in Eq.* (6.109), *there is a flow* $\mathbf{x}^{(\alpha;\mathscr{D})}(t)$ *in domains* Ω_α ($\alpha = i, j$). *At time* t_m, $\mathbf{x}_m^{(i_\alpha,\alpha)} = \mathbf{x}_\sigma^{(i_\alpha,\alpha)}(t_m) \in S_{\sigma_m}^{(i_\alpha,\alpha)}$ ($\sigma_m \in \mathbb{N}$) *satisfying Eq.* (6.112). *For an arbitrarily small* $\varepsilon > 0$, *there is a time interval* $[t_{m-\varepsilon}, t_{m+\varepsilon}]$. *A flow* $\mathbf{x}^{(\alpha;\mathscr{D})}(t)$ *is* $C_{[t_{m-\varepsilon}, t_{m+\varepsilon}]}^{r_\alpha}$*-continuous* ($r_\alpha < \infty$) *for time* t *and* $\| d^{r_\alpha+1}\mathbf{x}^{(\alpha;\mathscr{D})} / dt^{r_\alpha+1} \| < \infty$. $\mathbf{x}_{m\pm\varepsilon}^{(\alpha;\mathscr{D})} = \mathbf{x}^{(\alpha;\mathscr{D})}(t_{m\pm\varepsilon}) \in S_{\sigma_{m\pm\varepsilon}}^{(i_\alpha,\alpha)}$ *and* $\mathbf{x}_{\sigma(m\pm\varepsilon)}^{(i_\alpha,\alpha)} \in S_{\sigma_m}^{(i_\alpha,\alpha)}$. *The flow* $\mathbf{x}^{(\alpha;\mathscr{D})}(t)$ *at time* t_m *to the boundary of* $\partial\Omega_{ij}$ *is invariant with an equi-measuring quantity of* $C_m^{(i_\alpha,\alpha)}$ *if and only if*

$$G_{\partial\Omega_{ij}}^{(s_\alpha,\alpha;\mathscr{D})}(\mathbf{x}_m^{(i_\alpha,\alpha)}, t_m, \mathbf{p}_\alpha, \boldsymbol{\lambda}_{i_\alpha}) = 0 \ \textit{for}\ s_\alpha = 0,1,2,\cdots. \tag{6.125}$$

Proof. The theorem is obvious. In other words, if the flow $\mathbf{x}^{(\alpha;\mathscr{D})}(t)$ is on the surface of $\varphi_{\alpha i_\alpha}(\mathbf{x}_m^{(i_\alpha,\alpha)}, t_m, \boldsymbol{\lambda}_{i_\alpha}) = C_m^{(i_\alpha,\alpha)}$, then Eq. (6.125) holds always. If Eq. (6.125) holds, the flow $\mathbf{x}^{(\alpha;\mathscr{D})}(t)$ contacts with the surface of $\varphi_{\alpha i_\alpha}(\mathbf{x}_m^{(i_\alpha,\alpha)}, t_m, \boldsymbol{\lambda}_{i_\alpha}) = C_m^{(i_\alpha,\alpha)}$, which implies the flow lies in such a surface. Thus, this theorem can be proved. ■

If a flow in the domains possesses the higher order singularity to the measuring surface, the corresponding attractivity of the flow to the corresponding edge is also presented herein.

Definition 6.19. For a discontinuous dynamical system in Eq. (6.109), there is a flow $\mathbf{x}^{(\alpha;\mathscr{D})}(t)$ in domains Ω_α ($\alpha = i, j$). At time t_m, $\mathbf{x}_m^{(i_\alpha,\alpha)} = \mathbf{x}_\sigma^{(i_\alpha,\alpha)}(t_m) \in S_{\sigma_m}^{(i_\alpha,\alpha)}$ ($\sigma_m \in \mathbb{N}$) satisfying Eq. (6.112). For an arbitrarily small $\varepsilon > 0$, there is a time interval $[t_{m-\varepsilon}, t_{m+\varepsilon}]$. A flow $\mathbf{x}^{(\alpha;\mathscr{D})}(t)$ is $C_{[t_{m-\varepsilon}, t_{m+\varepsilon}]}^{r_\alpha}$-continuous ($r_\alpha \geq 2k_\alpha$) for time t and $\| d^{r_\alpha+1}\mathbf{x}^{(\alpha;\mathscr{D})} / dt^{r_\alpha+1} \| < \infty$. $\mathbf{x}_{m\pm\varepsilon}^{(\alpha;\mathscr{D})} = \mathbf{x}^{(\alpha;\mathscr{D})}(t_{m\pm\varepsilon}) \in S_{\sigma_{m\pm\varepsilon}}^{(i_\alpha,\alpha)}$ and $\mathbf{x}_{\sigma(m\pm\varepsilon)}^{(i_\alpha,\alpha)} \in S_{\sigma_m}^{(i_\alpha,\alpha)}$.

(i) The flow $\mathbf{x}^{(\alpha;\mathscr{D})}(t)$ at time t_m is *attractive* of the $(2k_\alpha)$th-order to the boundary of $\partial\Omega_{ij}$ if

$$\left.\begin{array}{l} G_{\partial\Omega_{ij}}^{(s_\alpha,\alpha;\mathscr{D})}(\mathbf{x}_m^{(i_\alpha,\alpha)},t_m,\mathbf{p}_\alpha,\boldsymbol{\lambda}_{i_\alpha})=0 \\ \text{for } s_\alpha=0,1,2,\cdots,2k_\alpha-1; \\ \hbar_{i_\alpha}\mathbf{n}_{\partial\Omega_{\alpha i_\alpha}}^{\mathrm{T}}(\mathbf{x}_{\sigma(m-\varepsilon)}^{(i_\alpha,\alpha)})\cdot[\mathbf{x}_{\sigma(m-\varepsilon)}^{(i_\alpha,\alpha)}-\mathbf{x}_{m-\varepsilon}^{(\alpha;\mathscr{D})}]<0, \\ \hbar_{i_\alpha}\mathbf{n}_{\partial\Omega_{\alpha i_\alpha}}^{\mathrm{T}}(\mathbf{x}_{\sigma(m+\varepsilon)}^{(i_\alpha,\alpha)})\cdot[\mathbf{x}_{m+\varepsilon}^{(\alpha;\mathscr{D})}-\mathbf{x}_{\sigma(m+\varepsilon)}^{(i_\alpha,\alpha)}]<0, \\ \hbar_{i_\alpha}C_{m-\varepsilon}^{(i_\alpha,\alpha)}>\hbar_{i_\alpha}C_{m}^{(i_\alpha,\alpha)}>\hbar_{i_\alpha}C_{m+\varepsilon}^{(i_\alpha,\alpha)}. \end{array}\right\} \tag{6.126}$$

(ii) The flow $\mathbf{x}^{(\alpha;\mathscr{D})}(t)$ at time t_m is *repulsive* of the $(2k_\alpha)$th-order to the boundary of $\partial\Omega_{ij}$ if

$$\left.\begin{array}{l} G_{\partial\Omega_{ij}}^{(s_\alpha,\alpha;\mathscr{D})}(\mathbf{x}_m^{(i_\alpha,\alpha)},t_m,\mathbf{p}_\alpha,\boldsymbol{\lambda}_{i_\alpha})=0 \\ \text{for } s_\alpha=0,1,2,\cdots,2k_\alpha-1; \\ \hbar_{i_\alpha}\mathbf{n}_{\partial\Omega_{\alpha i_\alpha}}^{\mathrm{T}}(\mathbf{x}_{\sigma(m-\varepsilon)}^{(i_\alpha,\alpha)})\cdot[\mathbf{x}_{\sigma(m-\varepsilon)}^{(i_\alpha,\alpha)}-\mathbf{x}_{m-\varepsilon}^{(\alpha;\mathscr{D})}]>0, \\ \hbar_{i_\alpha}\mathbf{n}_{\partial\Omega_{\alpha i_\alpha}}^{\mathrm{T}}(\mathbf{x}_{\sigma(m+\varepsilon)}^{(i_\alpha,\alpha)})\cdot[\mathbf{x}_{m+\varepsilon}^{(\alpha;\mathscr{D})}-\mathbf{x}_{\sigma(m+\varepsilon)}^{(i_\alpha,\alpha)}]>0, \\ \hbar_{i_\alpha}C_{m-\varepsilon}^{(i_\alpha,\alpha)}<\hbar_{i_\alpha}C_{m}^{(i_\alpha,\alpha)}<\hbar_{i_\alpha}C_{m+\varepsilon}^{(i_\alpha,\alpha)}. \end{array}\right\} \tag{6.127}$$

(iii) The flow $\mathbf{x}^{(\alpha;\mathscr{D})}(t)$ at time t_m to the boundary of $\partial\Omega_{ij}$ is *from the attractive to repulsive* state with the $(2k_\alpha+1)$th-order singularity if

$$\left.\begin{array}{l} G_{\partial\Omega_{ij}}^{(s_\alpha,\alpha;\mathscr{D})}(\mathbf{x}_m^{(i_\alpha,\alpha)},t_m,\mathbf{p}_\alpha,\boldsymbol{\lambda}_{i_\alpha})=0 \\ \text{for } s_\alpha=0,1,2,\cdots,2k_\alpha-1; \\ \hbar_{i_\alpha}\mathbf{n}_{\partial\Omega_{\alpha i_\alpha}}^{\mathrm{T}}(\mathbf{x}_{\sigma(m-\varepsilon)}^{(i_\alpha,\alpha)})\cdot[\mathbf{x}_{\sigma(m-\varepsilon)}^{(i_\alpha,\alpha)}-\mathbf{x}_{m-\varepsilon}^{(\alpha;\mathscr{D})}]<0, \\ \hbar_{i_\alpha}\mathbf{n}_{\partial\Omega_{\alpha i_\alpha}}^{\mathrm{T}}(\mathbf{x}_{\sigma(m+\varepsilon)}^{(i_\alpha,\alpha)})\cdot[\mathbf{x}_{m+\varepsilon}^{(\alpha;\mathscr{D})}-\mathbf{x}_{\sigma(m+\varepsilon)}^{(i_\alpha,\alpha)}]>0, \\ \hbar_{i_\alpha}C_{m}^{(i_\alpha,\alpha)}<\hbar_{i_\alpha}C_{m-\varepsilon}^{(i_\alpha,\alpha)} \text{ and } \hbar_{i_\alpha}C_{m}^{(i_\alpha,\alpha)}<\hbar_{i_\alpha}C_{m+\varepsilon}^{(i_\alpha,\alpha)}. \end{array}\right\} \tag{6.128}$$

(iv) The flow $\mathbf{x}^{(\alpha;\mathscr{D})}(t)$ at time t_m to the boundary of $\partial\Omega_{ij}$ is *from the repulsive to attractive* state with the $(2k_\alpha+1)$th-order singularity if

$$\left.\begin{array}{l} G_{\partial\Omega_{ij}}^{(s_\alpha,\alpha;\mathscr{D})}(\mathbf{x}_m^{(i_\alpha,\alpha)},t_m,\mathbf{p}_\alpha,\boldsymbol{\lambda}_{i_\alpha})=0 \\ \text{for } s_\alpha=0,1,2,\cdots,2k_\alpha; \\ \hbar_{i_\alpha}\mathbf{n}_{\partial\Omega_{\alpha i_\alpha}}^{\mathrm{T}}(\mathbf{x}_{\sigma(m-\varepsilon)}^{(i_\alpha,\alpha)})\cdot[\mathbf{x}_{\sigma(m-\varepsilon)}^{(i_\alpha,\alpha)}-\mathbf{x}_{m-\varepsilon}^{(\alpha;\mathscr{D})}]>0, \\ \hbar_{i_\alpha}\mathbf{n}_{\partial\Omega_{\alpha i_\alpha}}^{\mathrm{T}}(\mathbf{x}_{\sigma(m+\varepsilon)}^{(i_\alpha,\alpha)})\cdot[\mathbf{x}_{m+\varepsilon}^{(\alpha;\mathscr{D})}-\mathbf{x}_{\sigma(m+\varepsilon)}^{(i_\alpha,\alpha)}]<0, \\ \hbar_{i_\alpha}C_{m}^{(i_\alpha,\alpha)}>\hbar_{i_\alpha}C_{m-\varepsilon}^{(i_\alpha,\alpha)} \text{ and } \hbar_{i_\alpha}C_{m}^{(i_\alpha,\alpha)}>\hbar_{i_\alpha}C_{m+\varepsilon}^{(i_\alpha,\alpha)}. \end{array}\right\} \tag{6.129}$$

Theorem 6.12. *For a discontinuous dynamical system in Eq.* (6.109), *there is a* *flow* $\mathbf{x}^{(\alpha;\mathscr{D})}(t)$ *in domains* Ω_α ($\alpha = i, j$). *At time* t_m , $\mathbf{x}_m^{(i_\alpha,\alpha)} = \mathbf{x}_\sigma^{(i_\alpha,\alpha)}(t_m) \in S_{\sigma_m}^{(i_\alpha,\alpha)}$ ($\sigma_m \in \mathbb{N}$) *satisfying Eq.* (6.112). *For an arbitrarily small* $\varepsilon > 0$, *there is a time interval* $[t_{m-\varepsilon}, t_{m+\varepsilon}]$. *A flow* $\mathbf{x}^{(\alpha;\mathscr{D})}(t)$ *is* $C^{r_\alpha}_{[t_{m-\varepsilon}, t_{m+\varepsilon}]}$ *-continuous* ($r_\alpha \geq 2k_\alpha + 1$) *for time* t *and* $\| d^{r_\alpha+1}\mathbf{x}^{(\alpha;\mathscr{D})} / dt^{r_\alpha+1} \| < \infty$. *Suppose* $\mathbf{x}_{m\pm\varepsilon}^{(\alpha;\mathscr{D})} = \mathbf{x}^{(\alpha;\mathscr{D})}(t_{m\pm\varepsilon}) \in S_{\sigma_{m\pm\varepsilon}}^{(i_\alpha,\alpha)}$ *and* $\mathbf{x}_{\sigma(m\pm\varepsilon)}^{(i_\alpha,\alpha)} \in S_{\sigma_m}^{(i_\alpha,\alpha)}$.

(i) *The flow* $\mathbf{x}^{(\alpha;\mathscr{D})}(t)$ *at time* t_m *is attractive of the* $(2k_\alpha)$ *th-order to the boundary of* $\partial\Omega_{ij}$ *if and only if*

$$\left.\begin{aligned} & G_{\partial\Omega_{ij}}^{(s_\alpha,\alpha;\mathscr{D})}(\mathbf{x}_m^{(i_\alpha,\alpha)}, t_m, \mathbf{p}_\alpha, \boldsymbol{\lambda}_{i_\alpha}) = 0 \\ & \text{for } s_\alpha = 0,1,2,\cdots,2k_\alpha - 1; \\ & \hbar_{i_\alpha} G_{\partial\Omega_{ij}}^{(2k_\alpha,\alpha;\mathscr{D})}(\mathbf{x}_m^{(i_\alpha,\alpha)}, t_m, \mathbf{p}_\alpha, \boldsymbol{\lambda}_{i_\alpha}) < 0. \end{aligned}\right\} \tag{6.130}$$

(ii) *The flow* $\mathbf{x}^{(\alpha;\mathscr{D})}(t)$ *at time* t_m *is repulsive of the* $(2k_\alpha)$ *th-order to the boundary of* $\partial\Omega_{ij}$ *if and only if*

$$\left.\begin{aligned} & G_{\partial\Omega_{ij}}^{(s_\alpha,\alpha;\mathscr{D})}(\mathbf{x}_m^{(i_\alpha,\alpha)}, t_m, \mathbf{p}_\alpha, \boldsymbol{\lambda}_{i_\alpha}) = 0 \\ & \text{for } s_\alpha = 0,1,2,\cdots,2k_\alpha - 1; \\ & \hbar_{i_\alpha} G_{\partial\Omega_{ij}}^{(2k_\alpha,\alpha;\mathscr{D})}(\mathbf{x}_m^{(i_\alpha,\alpha)}, t_m, \mathbf{p}_\alpha, \boldsymbol{\lambda}_{i_\alpha}) > 0. \end{aligned}\right\} \tag{6.131}$$

(iii) *The flow* $\mathbf{x}^{(\alpha;\mathscr{D})}(t)$ *at time* t_m *to the boundary of* $\partial\Omega_{ij}$ *is from the attractive to repulsive state with the* $(2k_\alpha + 1)$ *th-order singularity if and only if*

$$\left.\begin{aligned} & G_{\partial\Omega_{ij}}^{(s_\alpha,\alpha;\mathscr{D})}(\mathbf{x}_m^{(i_\alpha,\alpha)}, t_m, \mathbf{p}_\alpha, \boldsymbol{\lambda}_{i_\alpha}) = 0 \\ & \text{for } s_\alpha = 0,1,2,\cdots,2k_\alpha; \\ & \hbar_{i_\alpha} G_{\partial\Omega_{ij}}^{(2k_\alpha+1,\alpha;\mathscr{D})}(\mathbf{x}_m^{(i_\alpha,\alpha)}, t_m, \mathbf{p}_\alpha, \boldsymbol{\lambda}_{i_\alpha}) > 0. \end{aligned}\right\} \tag{6.132}$$

(iv) *The flow* $\mathbf{x}^{(\alpha;\mathscr{D})}(t)$ *at time* t_m *to the boundary of* $\partial\Omega_{ij}$ *is from the repulsive to attractive state with the* $(2k_\alpha + 1)$ *th-order singularity if and only if*

$$\left.\begin{aligned} & G_{\partial\Omega_{ij}}^{(s_\alpha,\alpha;\mathscr{D})}(\mathbf{x}_m^{(i_\alpha,\alpha)}, t_m, \mathbf{p}_\alpha, \boldsymbol{\lambda}_{i_\alpha}) = 0 \\ & \text{for } s_\alpha = 0,1,2,\cdots,2k_\alpha; \\ & \hbar_{i_\alpha} G_{\partial\Omega_{ij}}^{(2k_\alpha+1,\alpha;\mathscr{D})}(\mathbf{x}_m^{(i_\alpha,\alpha)}, t_m, \mathbf{p}_\alpha, \boldsymbol{\lambda}_{i_\alpha}) < 0. \end{aligned}\right\} \tag{6.133}$$

Proof. The surface of $\varphi_{ij}(\mathbf{x}_m^{(\alpha)}, t_m, \boldsymbol{\lambda}) = C_m^{(\alpha)}$ is as a boundary. As in Chapter 3, this theorem can be proved. ■

6.6.2. Attractivity to edge

To discuss the attractivity of a flow to an edge, a new edge in the domain should be considered, and the new edge will not intersected with the edge $\mathscr{E}^{(n-2)}$. For illustration, consider an edge between two measuring surfaces relative to two boundaries $\partial\Omega_{\alpha i_\alpha}$ and $\partial\Omega_{\alpha j_\alpha}$ in domain Ω_α, as sketched in Fig. 6.18. The two measuring surfaces are two equi-constant surfaces of boundaries, respectively. To determine the attractivity of a flow to a specific edge, a flow $\mathbf{x}^{(\alpha;\mathscr{D})}$ intersecting with such an edge should be discussed.

Definition 6.20. For a discontinuous dynamical system in Eqs. (6.4)-(6.8), there is an edge $\mathscr{E}_{\sigma_r}^{(r)}$ ($\sigma_r \in \{1,2,\cdots,N_r\}$ and $r \in \{0,1,2,\cdots,n-2\}$) formed by the intersection of $(n-r)$-boundaries $\partial\Omega_{\alpha i_\alpha}(\sigma_r)$ ($\alpha \in \mathscr{N}_{\mathscr{D}}(\sigma_r) \subset \mathscr{N}_{\mathscr{D}} = \{1,2,\cdots,N_d\}$ and $i_\alpha \in \mathscr{N}_{\alpha\mathscr{B}}(\sigma_r) \subset \mathscr{N}_{\mathscr{B}} = \{1,2,\cdots,N_b\}$) in domain $\Omega_\alpha(\sigma_r)$. Suppose there is a flow $\mathbf{x}^{(i_\alpha;\alpha)}(t)$ on the measuring surface $S_\sigma^{(i_\alpha,\alpha)}$ in domain $\Omega_\alpha(\sigma_r)$. At time t_m, $\mathbf{x}_m^{(\alpha;\mathscr{D})} = \mathbf{x}_\sigma^{(i_\alpha,\alpha)}(t_m) \in S_\sigma^{(i_\alpha,\alpha)}$ ($\sigma \in \mathbb{N}$). The corresponding dynamical system is

$$
\begin{aligned}
&\dot{\mathbf{x}}_\sigma^{(i_\alpha,\alpha)} = \mathbf{F}^{(\sigma)}(\mathbf{x}_\sigma^{(i_\alpha,\alpha)}, t, \boldsymbol{\lambda}_{i_\alpha}) \\
&\text{with } \varphi_{\alpha i_\alpha}(\mathbf{x}_\sigma^{(i_\alpha,\alpha)}, t, \boldsymbol{\lambda}_{i_\alpha}) = \varphi_{\alpha i_\alpha}(\mathbf{x}_m^{(i_\alpha,\alpha)}, t_m, \boldsymbol{\lambda}_{i_\alpha}) \equiv C_m^{(i_\alpha,\alpha)} \text{ on } S_\sigma^{(i_\alpha,\alpha)}
\end{aligned}
\tag{6.134}
$$

where $C_m^{(i_\alpha,\alpha)}$ is a non-zero constant.

From discussions in the previous section, there are five types of attractivity of a flow in domain to the boundary. For simplicity, consider the attractivity for a flow to all boundaries relative to the edge of $\mathscr{E}_{\sigma_r}^{(r)}$ to be same first.

Definition 6.21. For a discontinuous dynamical system in Eqs. (6.4)-(6.8), there is an edge $\mathscr{E}_{\sigma_r}^{(r)}$ ($\sigma_r \in \{1,2,\cdots,N_r\}$ and $r \in \{0,1,2,\cdots,n-2\}$) formed by the intersection of $(n-r)$-boundaries $\partial\Omega_{\alpha i_\alpha}(\sigma_r)$ ($\alpha \in \mathscr{N}_{\mathscr{D}}(\sigma_r) \subset \mathscr{N}_{\mathscr{D}} = \{1,2,\cdots,N_d\}$ and $i_\alpha \in \mathscr{N}_{\alpha\mathscr{B}}(\sigma_r) \subset \mathscr{N}_{\mathscr{B}} = \{1,2,\cdots,N_b\}$) in domain $\Omega_\alpha(\sigma_r)$. Suppose there is a measuring edge $\mathscr{E}_{\sigma_r}^{(r,\alpha)}$ of a flow $\mathbf{x}^{(\alpha;\mathscr{D})}(t)$ in domain $\Omega_\alpha(\sigma_r)$ with an equi-constant vector of the edge $\mathscr{E}_{\sigma_r}^{(r)}$, and the measuring edge is formed by an intersection of all measuring surfaces $S_{\sigma_m}^{(i_\alpha,\alpha)}$ for $i_\alpha \in \mathscr{N}_{\alpha\mathscr{B}}(\sigma_r)$. At time t_m, $\mathbf{x}_m^{(\alpha;\mathscr{D})} = \mathbf{x}_\sigma^{(i_\alpha,\alpha)}(t_m) \in S_\sigma^{(i_\alpha,\alpha)}$ ($\sigma \in \mathbb{N}$) satisfies Eq. (6.134). For an arbitrarily small $\varepsilon > 0$, there is a time interval $[t_{m-\varepsilon}, t_{m+\varepsilon}]$. Suppose $\mathbf{x}_{m\pm\varepsilon}^{(\alpha;\mathscr{D})} = \mathbf{x}^{(\alpha;\mathscr{D})}(t_{m\pm\varepsilon}) \in S_{\sigma_{m\pm\varepsilon}}^{(i_\alpha,\alpha)}$ and $\mathbf{x}_{\sigma(m\pm\varepsilon)}^{(i_\alpha,\alpha)} \in S_{\sigma_m}^{(i_\alpha,\alpha)}$.

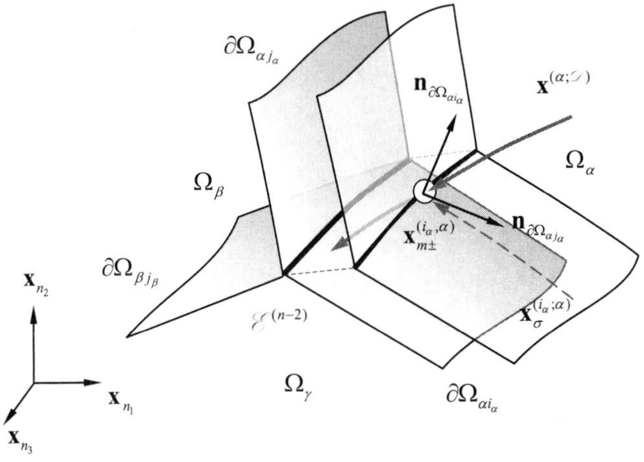

Fig. 6.18 A flow intersected with an edge between two separatrix measuring surfaces in domain Ω_α. The two measuring surfaces are two equi-constant surfaces relative to the two separation boundaries of $\partial\Omega_{\alpha i_\alpha}$ and $\partial\Omega_{\alpha j_\alpha}$. The common edge $\mathscr{E}^{(n\text{-}2)}$ is the $(n-2)$-dimensional, intersection surface of three separation boundaries. The corresponding normal vectors are presented. (color plot in the book end)

(i) The flow $\mathbf{x}^{(\alpha;\mathscr{D})}(t)$ at time t_m is *attractive to* the edge of $\mathscr{E}^{(r)}_{\sigma_r}$ if for all $i_\alpha \in \mathscr{N}_{\alpha\mathscr{B}}(\sigma_r)$

$$\begin{aligned}
&{}_{i_\alpha}\mathbf{n}^{\mathrm{T}}_{\partial\Omega_{\alpha i_\alpha}}(\mathbf{x}^{(i_\alpha,\alpha)}_{\sigma(m-\varepsilon)})\cdot[\mathbf{x}^{(i_\alpha,\alpha)}_{\sigma(m-\varepsilon)}-\mathbf{x}^{(\alpha;\mathscr{D})}_{m-\varepsilon}]<0,\\
&{}_{i_\alpha}\mathbf{n}^{\mathrm{T}}_{\partial\Omega_{\alpha i_\alpha}}(\mathbf{x}^{(i_\alpha,\alpha)}_{\sigma(m+\varepsilon)})\cdot[\mathbf{x}^{(\alpha;\mathscr{D})}_{m+\varepsilon}-\mathbf{x}^{(i_\alpha,\alpha)}_{\sigma(m+\varepsilon)}]<0,\\
&{}_{i_\alpha}C^{(i_\alpha,\alpha)}_{m-\varepsilon} > {}_{i_\alpha}C^{(i_\alpha,\alpha)}_{m} > {}_{i_\alpha}C^{(i_\alpha,\alpha)}_{m+\varepsilon}.
\end{aligned} \tag{6.135}$$

(ii) The flow $\mathbf{x}^{(\alpha;\mathscr{D})}(t)$ at time t_m is *repulsive* to the edge of $\mathscr{E}^{(r)}_{\sigma_r}$ if for all $i_\alpha \in \mathscr{N}_{\alpha\mathscr{B}}(\sigma_r)$

$$\begin{aligned}
&{}_{i_\alpha}\mathbf{n}^{\mathrm{T}}_{\partial\Omega_{\alpha i_\alpha}}(\mathbf{x}^{(i_\alpha,\alpha)}_{\sigma(m-\varepsilon)})\cdot[\mathbf{x}^{(i_\alpha,\alpha)}_{\sigma(m-\varepsilon)}-\mathbf{x}^{(\alpha;\mathscr{D})}_{m-\varepsilon}]>0,\\
&{}_{i_\alpha}\mathbf{n}^{\mathrm{T}}_{\partial\Omega_{\alpha i_\alpha}}(\mathbf{x}^{(i_\alpha,\alpha)}_{\sigma(m+\varepsilon)})\cdot[\mathbf{x}^{(\alpha;\mathscr{D})}_{m+\varepsilon}-\mathbf{x}^{(i_\alpha,\alpha)}_{\sigma(m+\varepsilon)}]>0,\\
&{}_{i_\alpha}C^{(i_\alpha,\alpha)}_{m-\varepsilon} < {}_{i_\alpha}C^{(i_\alpha,\alpha)}_{m} < {}_{i_\alpha}C^{(i_\alpha,\alpha)}_{m+\varepsilon}.
\end{aligned} \tag{6.136}$$

(iii) The flow $\mathbf{x}^{(\alpha;\mathscr{D})}(t)$ at time t_m to the edge of $\mathscr{E}^{(r)}_{\sigma_r}$ is *from the attractive to repulsive* state if for all $i_\alpha \in \mathscr{N}_{\alpha\mathscr{B}}(\sigma_r)$

$$\begin{aligned}
&{}_{i_\alpha}\mathbf{n}^{\mathrm{T}}_{\partial\Omega_{\alpha i_\alpha}}(\mathbf{x}^{(i_\alpha,\alpha)}_{\sigma(m-\varepsilon)})\cdot[\mathbf{x}^{(i_\alpha,\alpha)}_{\sigma(m-\varepsilon)}-\mathbf{x}^{(\alpha;\mathscr{D})}_{m-\varepsilon}]<0,\\
&{}_{i_\alpha}\mathbf{n}^{\mathrm{T}}_{\partial\Omega_{\alpha i_\alpha}}(\mathbf{x}^{(i_\alpha,\alpha)}_{\sigma(m+\varepsilon)})\cdot[\mathbf{x}^{(\alpha;\mathscr{D})}_{m+\varepsilon}-\mathbf{x}^{(i_\alpha,\alpha)}_{\sigma(m+\varepsilon)}]>0,
\end{aligned}$$

$$\hbar_{i_\alpha} C_m^{(i_\alpha,\alpha)} < \hbar_{i_\alpha} C_{m-\varepsilon}^{(i_\alpha,\alpha)} \text{ and } \hbar_{i_\alpha} C_m^{(i_\alpha,\alpha)} < \hbar_{i_\alpha} C_{m+\varepsilon}^{(i_\alpha,\alpha)}. \tag{6.137}$$

(iv) The flow $\mathbf{x}^{(\alpha;\mathscr{D})}(t)$ at time t_m to the edge of $\mathscr{E}_{\sigma_r}^{(r)}$ is *from the repulsive to attractive* state if for all $i_\alpha \in \mathscr{N}_{\alpha\mathscr{B}}(\sigma_r)$

$$\begin{aligned} &\hbar_{i_\alpha} \mathbf{n}_{\partial\Omega_{\alpha i_\alpha}}^{\mathrm{T}} (\mathbf{x}_{\sigma(m-\varepsilon)}^{(i_\alpha,\alpha)}) \cdot [\mathbf{x}_{\sigma(m-\varepsilon)}^{(i_\alpha,\alpha)} - \mathbf{x}_{m-\varepsilon}^{(\alpha;\mathscr{D})}] > 0, \\ &\hbar_{i_\alpha} \mathbf{n}_{\partial\Omega_{\alpha i_\alpha}}^{\mathrm{T}} (\mathbf{x}_{\sigma(m+\varepsilon)}^{(i_\alpha,\alpha)}) \cdot [\mathbf{x}_{m+\varepsilon}^{(\alpha;\mathscr{D})} - \mathbf{x}_{\sigma(m+\varepsilon)}^{(i_\alpha,\alpha)}] < 0, \\ &\hbar_{i_\alpha} C_m^{(i_\alpha,\alpha)} > \hbar_{i_\alpha} C_{m-\varepsilon}^{(i_\alpha,\alpha)} \text{ and } \hbar_{i_\alpha} C_m^{(i_\alpha,\alpha)} > \hbar_{i_\alpha} C_{m+\varepsilon}^{(i_\alpha,\alpha)}. \end{aligned} \tag{6.138}$$

(v) The flow $\mathbf{x}^{(\alpha;\mathscr{D})}(t)$ at time t_m to the edge of $\mathscr{E}_{\sigma_r}^{(r)}$ is *invariant* with equi-measuring quantities of $C_m^{(i_\alpha,\alpha)}$ if for all $i_\alpha \in \mathscr{N}_{\alpha\mathscr{B}}(\sigma_r)$

$$\begin{aligned} &\mathbf{x}_{m-\varepsilon}^{(\alpha;\mathscr{D})} = \mathbf{x}_{\sigma(m-\varepsilon)}^{(i_\alpha,\alpha)} \text{ and } \mathbf{x}_{m+\varepsilon}^{(\alpha;\mathscr{D})} = \mathbf{x}_{\sigma(m+\varepsilon)}^{(i_\alpha,\alpha)}, \\ &C_{m-\varepsilon}^{(i_\alpha,\alpha)} = C_m^{(i_\alpha,\alpha)} = C_{m+\varepsilon}^{(i_\alpha,\alpha)}. \end{aligned} \tag{6.139}$$

From the definition, the corresponding conditions for such attractivity can be obtained through the following theorem.

Theorem 6.13. *For a discontinuous dynamical system in Eqs.* (6.4)-(6.8), *there is an edge* $\mathscr{E}_{\sigma_r}^{(r)}$ ($\sigma_r \in \{1,2,\cdots,N_r\}$ *and* $r \in \{0,1,2,\cdots,n-2\}$) *formed by the intersection of* $(n-r)$*-boundaries* $\partial\Omega_{\alpha i_\alpha}(\sigma_r)$ ($\alpha \in \mathscr{N}_{\mathscr{D}}(\sigma_r) \subset \mathscr{N}_{\mathscr{D}} = \{1,2,\cdots,N_d\}$ *and* $i_\alpha \in \mathscr{N}_{\alpha\mathscr{B}}(\sigma_r) \subset \mathscr{N}_{\mathscr{B}} = \{1,2,\cdots,N_b\}$) *in domain* $\Omega_\alpha(\sigma_r)$. *Suppose there is a measuring edge* $\mathscr{E}_{\sigma_r}^{(r,\alpha)}$ *of a flow* $\mathbf{x}^{(\alpha;\mathscr{D})}(t)$ *in domain* $\Omega_\alpha(\sigma_r)$ *with an equi-constant vector of the edge* $\mathscr{E}_{\sigma_r}^{(r)}$, *and the measuring edge is formed by an intersection of all measuring surfaces* $S_{\sigma_m}^{(i_\alpha,\alpha)}$ *for* $i_\alpha \in \mathscr{N}_{\alpha\mathscr{B}}(\sigma_r)$. *At time* t_m, $\mathbf{x}_m^{(\alpha;\mathscr{D})} = \mathbf{x}_\sigma^{(i_\alpha,\alpha)}(t_m) \in S_\sigma^{(i_\alpha,\alpha)}$ ($\sigma \in \mathbb{N}$) *satisfies Eq.* (6.134). *For an arbitrarily small* $\varepsilon > 0$, *there is a time interval* $[t_{m-\varepsilon}, t_{m+\varepsilon}]$. *A flow* $\mathbf{x}^{(\alpha;\mathscr{D})}(t)$ *is* $C_{[t_{m-\varepsilon},t_{m+\varepsilon}]}^{r_\alpha}$*-continuous* ($r_\alpha \geq 1$) *for time* t *and* $\| d^{r_\alpha+1}\mathbf{x}^{(\alpha;\mathscr{D})}/dt^{r_\alpha+1} \| < \infty$. *Suppose* $\mathbf{x}_{m\pm\varepsilon}^{(\alpha;\mathscr{D})} = \mathbf{x}^{(\alpha;\mathscr{D})}(t_{m\pm\varepsilon}) \in S_{\sigma_{m\pm\varepsilon}}^{(i_\alpha,\alpha)}$ *and* $\mathbf{x}_{\sigma(m\pm\varepsilon)}^{(i_\alpha,\alpha)} \in S_{\sigma_m}^{(i_\alpha,\alpha)}$.

(i) *The flow* $\mathbf{x}^{(\alpha;\mathscr{D})}(t)$ *at time* t_m *is attractive to the edge of* $\mathscr{E}_{\sigma_r}^{(r)}$ *if and only if for all* $i_\alpha \in \mathscr{N}_{\alpha\mathscr{B}}(\sigma_r)$

$$\hbar_{i_\alpha} G_{\partial\Omega_{\alpha i_\alpha}}^{(\alpha;\mathscr{D})}(\mathbf{x}_m^{(\alpha;\mathscr{D})}, t_m, \mathbf{p}_\alpha, \boldsymbol{\lambda}_{i_\alpha}) < 0. \tag{6.140}$$

(ii) *The flow* $\mathbf{x}^{(\alpha;\mathscr{D})}(t)$ *at time* t_m *is repulsive to the edge of* $\mathscr{E}_{\sigma_r}^{(r)}$ *if and only if for all* $i_\alpha \in \mathscr{N}_{\alpha\mathscr{B}}(\sigma_r)$

$$\hbar_{i_\alpha} G^{(\alpha;\mathscr{D})}_{\partial\Omega_{\alpha i_\alpha}}(\mathbf{x}^{(\alpha:\mathscr{D})}_m, t_m, \mathbf{p}_\alpha, \boldsymbol{\lambda}_{i_\alpha}) > 0. \tag{6.141}$$

(iii) *The flow* $\mathbf{x}^{(\alpha;\mathscr{D})}(t)$ *at time* t_m *to the edge of* $\mathscr{E}^{(r)}_{\sigma_r}$ *is from the attractive to repulsive state if and only if for all* $i_\alpha \in \mathscr{N}_{\alpha\mathscr{B}}(\sigma_r)$

$$\begin{aligned} &\hbar_{i_\alpha} G^{(\alpha;\mathscr{D})}_{\partial\Omega_{\alpha i_\alpha}}(\mathbf{x}^{(\alpha:\mathscr{D})}_m, t_m, \mathbf{p}_\alpha, \boldsymbol{\lambda}_{i_\alpha}) = 0, \\ &\hbar_{i_\alpha} G^{(1,\alpha;\mathscr{D})}_{\partial\Omega_{\alpha i_\alpha}}(\mathbf{x}^{(\alpha:\mathscr{D})}_m, t_m, \mathbf{p}_\alpha, \boldsymbol{\lambda}_{i_\alpha}) > 0. \end{aligned} \tag{6.142}$$

(iv) *The flow* $\mathbf{x}^{(\alpha;\mathscr{D})}(t)$ *at time* t_m *to the edge of* $\mathscr{E}^{(r)}_{\sigma_r}$ *is from the repulsive to attractive state if and only if for all* $i_\alpha \in \mathscr{N}_{\alpha\mathscr{B}}(\sigma_r)$

$$\begin{aligned} &\hbar_{i_\alpha} G^{(\alpha;\mathscr{D})}_{\partial\Omega_{\alpha i_\alpha}}(\mathbf{x}^{(\alpha:\mathscr{D})}_m, t_m, \mathbf{p}_\alpha, \boldsymbol{\lambda}_{i_\alpha}) = 0, \\ &\hbar_{i_\alpha} G^{(1,\alpha;\mathscr{D})}_{\partial\Omega_{\alpha i_\alpha}}(\mathbf{x}^{(\alpha:\mathscr{D})}_m, t_m, \mathbf{p}_\alpha, \boldsymbol{\lambda}_{i_\alpha}) < 0. \end{aligned} \tag{6.143}$$

(v) *The flow* $\mathbf{x}^{(\alpha;\mathscr{D})}(t)$ *at time* t_m *to the edge of* $\mathscr{E}^{(r)}_{\sigma_r}$ *is invariant with equi-measuring quantities of* $C^{(i_\alpha,\alpha)}_m$ *if and only if for all* $i_\alpha \in \mathscr{N}_{\alpha\mathscr{B}}(\sigma_r)$

$$\hbar_{i_\alpha} G^{(s_\alpha,\alpha;\mathscr{D})}_{\partial\Omega_{\alpha i_\alpha}}(\mathbf{x}^{(\alpha:\mathscr{D})}_m, t_m, \mathbf{p}_\alpha, \boldsymbol{\lambda}_{i_\alpha}) = 0 \text{ for } s_\alpha = 0,1,2,\cdots. \tag{6.144}$$

Similarly, the flow attractitivity to the edge with the higher-order singularity is described as follows.

Definition 6.22. For a discontinuous dynamical system in Eqs. (6.4)-(6.8), there is an edge $\mathscr{E}^{(r)}_{\sigma_r}$ ($\sigma_r \in \{1,2,\cdots,N_r\}$ and $r \in \{0,1,2,\cdots,n-2\}$) formed by the intersection of $(n-r)$-boundaries $\partial\Omega_{\alpha i_\alpha}(\sigma_r)$ ($\alpha \in \mathscr{N}_{\mathscr{D}}(\sigma_r) \subset \mathscr{N}_{\mathscr{D}} = \{1,2,\cdots,N_d\}$ and $i_\alpha \in \mathscr{N}_{\alpha\mathscr{B}}(\sigma_r) \subset \mathscr{N}_{\mathscr{B}} = \{1,2,\cdots,N_b\}$) in domain $\Omega_\alpha(\sigma_r)$. Suppose there is a measuring edge $\mathscr{E}^{(r,\alpha)}_{\sigma_r}$ of a flow $\mathbf{x}^{(\alpha;\mathscr{D})}(t)$ in domain $\Omega_\alpha(\sigma_r)$ with an equi-constant vector of the edge $\mathscr{E}^{(r)}_{\sigma_r}$, and the measuring edge is formed by an intersection of all measuring surfaces $S^{(i_\alpha,\alpha)}_{\sigma_m}$ for $i_\alpha \in \mathscr{N}_{\alpha\mathscr{B}}(\sigma_r)$. At time t_m, $\mathbf{x}^{(\alpha;\mathscr{D})}_m = \mathbf{x}^{(i_\alpha,\alpha)}_\sigma(t_m) \in S^{(i_\alpha,\alpha)}_\sigma$ ($\sigma \in \mathbb{N}$) satisfies Eq. (6.134). For an arbitrarily small $\varepsilon > 0$, there is a time interval $[t_{m-\varepsilon}, t_{m+\varepsilon}]$. A flow $\mathbf{x}^{(\alpha;\mathscr{D})}(t)$ is $C^{r_\alpha}_{[t_{m-\varepsilon}, t_{m+\varepsilon}]}$-continuous for time t and $\| d^{r_\alpha+1}\mathbf{x}^{(\alpha;\mathscr{D})}(t)/dt^{r_\alpha+1} \| < \infty$ ($r_\alpha \geq \max_{i_\alpha \in \mathscr{N}_{\alpha\mathscr{B}}}(2k_{i_\alpha}+1)$). Suppose $\mathbf{x}^{(i_\alpha,\alpha)}_{\sigma(m\pm\varepsilon)} \in S^{(i_\alpha,\alpha)}_{\sigma_m}$ and $\mathbf{x}^{(\alpha;\mathscr{D})}_{m\pm\varepsilon} = \mathbf{x}^{(\alpha;\mathscr{D})}(t_{m\pm\varepsilon}) \in S^{(i_\alpha,\alpha)}_{\sigma_{m\pm\varepsilon}}$.

(i) The flow $\mathbf{x}^{(\alpha;\mathscr{D})}(t)$ at time t_m to the edge of $\mathscr{E}^{(r)}_{\sigma_r}$ is *attractive* with the $2\mathbf{k}_{\mathscr{E}}$-

order singularity if for all $i_\alpha \in \mathscr{N}_{\alpha\mathscr{B}}(\sigma_r)$

$$\left.\begin{array}{l} G^{(s_\alpha,\alpha;\mathscr{D})}_{\partial\Omega_{\alpha i_\alpha}}(\mathbf{x}^{(\alpha;\mathscr{D})}_m, t_m, \mathbf{p}_\alpha, \boldsymbol{\lambda}_{i_\alpha}) = 0 \\ \text{for } s_\alpha = 0,1,2,\cdots,2k_{i_\alpha}-1; \\ \hbar_{i_\alpha}\mathbf{n}^{\mathrm{T}}_{\partial\Omega_{\alpha i_\alpha}}(\mathbf{x}^{(i_\alpha,\alpha)}_{\sigma(m-\varepsilon)})\cdot[\mathbf{x}^{(i_\alpha,\alpha)}_{\sigma(m-\varepsilon)} - \mathbf{x}^{(\alpha;\mathscr{D})}_{m-\varepsilon}] < 0, \\ \hbar_{i_\alpha}\mathbf{n}^{\mathrm{T}}_{\partial\Omega_{\alpha i_\alpha}}(\mathbf{x}^{(i_\alpha,\alpha)}_{\sigma(m+\varepsilon)})\cdot[\mathbf{x}^{(\alpha;\mathscr{D})}_{m+\varepsilon} - \mathbf{x}^{(i_\alpha,\alpha)}_{\sigma(m+\varepsilon)}] < 0, \\ \hbar_{i_\alpha}C^{(i_\alpha,\alpha)}_{m-\varepsilon} > \hbar_{i_\alpha}C^{(i_\alpha,\alpha)}_{m} > \hbar_{i_\alpha}C^{(i_\alpha,\alpha)}_{m+\varepsilon}. \end{array}\right\} \tag{6.145}$$

(ii) The flow $\mathbf{x}^{(\alpha;\mathscr{D})}(t)$ at time t_m to the edge of $\mathscr{E}^{(r)}_{\sigma_r}$ is *repulsive* with the $2\mathbf{k}_{\mathscr{E}}$-order singularity if for all $i_\alpha \in \mathscr{N}_{\alpha\mathscr{B}}(\sigma_r)$

$$\left.\begin{array}{l} G^{(s_\alpha,\alpha;\mathscr{D})}_{\partial\Omega_{\alpha i_\alpha}}(\mathbf{x}^{(\alpha;\mathscr{D})}_m, t_m, \mathbf{p}_\alpha, \boldsymbol{\lambda}_{i_\alpha}) = 0 \\ \text{for } s_\alpha = 0,1,2,\cdots,2k_{i_\alpha}-1; \\ \hbar_{i_\alpha}\mathbf{n}^{\mathrm{T}}_{\partial\Omega_{\alpha i_\alpha}}(\mathbf{x}^{(i_\alpha,\alpha)}_{\sigma(m-\varepsilon)})\cdot[\mathbf{x}^{(i_\alpha,\alpha)}_{\sigma(m-\varepsilon)} - \mathbf{x}^{(\alpha;\mathscr{D})}_{m-\varepsilon}] > 0, \\ \hbar_{i_\alpha}\mathbf{n}^{\mathrm{T}}_{\partial\Omega_{\alpha i_\alpha}}(\mathbf{x}^{(i_\alpha,\alpha)}_{\sigma(m+\varepsilon)})\cdot[\mathbf{x}^{(\alpha;\mathscr{D})}_{m+\varepsilon} - \mathbf{x}^{(i_\alpha,\alpha)}_{\sigma(m+\varepsilon)}] > 0, \\ \hbar_{i_\alpha}C^{(i_\alpha,\alpha)}_{m-\varepsilon} < \hbar_{i_\alpha}C^{(i_\alpha,\alpha)}_{m} < \hbar_{i_\alpha}C^{(i_\alpha,\alpha)}_{m+\varepsilon}. \end{array}\right\} \tag{6.146}$$

(iii) The flow $\mathbf{x}^{(\alpha;\mathscr{D})}(t)$ at time t_m to the edge of $\mathscr{E}^{(r)}_{\sigma_r}$ is *from the attractive to repulsive* state with the $(2\mathbf{k}_{\mathscr{E}}+1)$-order singularity if for all $i_\alpha \in \mathscr{N}_{\alpha\mathscr{B}}(\sigma_r)$

$$\left.\begin{array}{l} G^{(s_\alpha,\alpha;\mathscr{D})}_{\partial\Omega_{\alpha i_\alpha}}(\mathbf{x}^{(\alpha;\mathscr{D})}_m, t_m, \mathbf{p}_\alpha, \boldsymbol{\lambda}_{i_\alpha}) = 0 \\ \text{for } s_\alpha = 0,1,2,\cdots,2k_{i_\alpha}; \\ \hbar_{i_\alpha}\mathbf{n}^{\mathrm{T}}_{\partial\Omega_{\alpha i_\alpha}}(\mathbf{x}^{(i_\alpha,\alpha)}_{\sigma(m-\varepsilon)})\cdot[\mathbf{x}^{(i_\alpha,\alpha)}_{\sigma(m-\varepsilon)} - \mathbf{x}^{(\alpha;\mathscr{D})}_{m-\varepsilon}] < 0, \\ \hbar_{i_\alpha}\mathbf{n}^{\mathrm{T}}_{\partial\Omega_{\alpha i_\alpha}}(\mathbf{x}^{(i_\alpha,\alpha)}_{\sigma(m+\varepsilon)})\cdot[\mathbf{x}^{(\alpha;\mathscr{D})}_{m+\varepsilon} - \mathbf{x}^{(i_\alpha,\alpha)}_{\sigma(m+\varepsilon)}] > 0, \\ \hbar_{i_\alpha}C^{(i_\alpha,\alpha)}_{m} < \hbar_{i_\alpha}C^{(i_\alpha,\alpha)}_{m-\varepsilon} \text{ and } \hbar_{i_\alpha}C^{(i_\alpha,\alpha)}_{m} < \hbar_{i_\alpha}C^{(i_\alpha,\alpha)}_{m+\varepsilon}. \end{array}\right\} \tag{6.147}$$

(iv) The flow $\mathbf{x}^{(\alpha;\mathscr{D})}(t)$ at time t_m to the edge of $\mathscr{E}^{(r)}_{\sigma_r}$ is *from the repulsive to attractive* state with the $(2\mathbf{k}_{\mathscr{E}}+1)$-order singularity if for all $i_\alpha \in \mathscr{N}_{\alpha\mathscr{B}}(\sigma_r)$

$$\left.\begin{array}{l} G^{(s_\alpha,\alpha;\mathscr{D})}_{\partial\Omega_{\alpha i_\alpha}}(\mathbf{x}^{(\alpha;\mathscr{D})}_m, t_m, \mathbf{p}_\alpha, \boldsymbol{\lambda}_{i_\alpha}) = 0 \\ \text{for } s_\alpha = 0,1,2,\cdots,2k_{i_\alpha}; \\ \hbar_{i_\alpha}\mathbf{n}^{\mathrm{T}}_{\partial\Omega_{\alpha i_\alpha}}(\mathbf{x}^{(i_\alpha,\alpha)}_{\sigma(m-\varepsilon)})\cdot[\mathbf{x}^{(i_\alpha,\alpha)}_{\sigma(m-\varepsilon)} - \mathbf{x}^{(\alpha;\mathscr{D})}_{m-\varepsilon}] > 0, \\ \hbar_{i_\alpha}\mathbf{n}^{\mathrm{T}}_{\partial\Omega_{\alpha i_\alpha}}(\mathbf{x}^{(i_\alpha,\alpha)}_{\sigma(m+\varepsilon)})\cdot[\mathbf{x}^{(\alpha;\mathscr{D})}_{m+\varepsilon} - \mathbf{x}^{(i_\alpha,\alpha)}_{\sigma(m+\varepsilon)}] < 0, \\ \hbar_{i_\alpha}C^{(i_\alpha,\alpha)}_{m} > \hbar_{i_\alpha}C^{(i_\alpha,\alpha)}_{m-\varepsilon} \text{ and } \hbar_{i_\alpha}C^{(i_\alpha,\alpha)}_{m} > \hbar_{i_\alpha}C^{(i_\alpha,\alpha)}_{m+\varepsilon}. \end{array}\right\} \tag{6.148}$$

The conditions for the attractivity of a flow to the edge are stated as follows.

Theorem 6.14. *For a discontinuous dynamical system in Eqs.* (6.4)-(6.8), *there is an edge* $\mathscr{E}_{\sigma_r}^{(r)}$ ($\sigma_r \in \{1,2,\cdots,N_r\}$ *and* $r \in \{0,1,2,\cdots,n-2\}$) *formed by the intersection of* $(n-r)$ *-boundaries* $\partial\Omega_{\alpha i_\alpha}(\sigma_r)$ ($\alpha \in \mathscr{N}_{\mathscr{D}}(\sigma_r) \subset \mathscr{N}_{\mathscr{D}} = \{1,2,\cdots,N_d\}$ *and* $i_\alpha \in \mathscr{N}_{\alpha\mathscr{B}}(\sigma_r) \subset \mathscr{N}_{\mathscr{B}} = \{1,2,\cdots,N_b\}$) *in domain* $\Omega_\alpha(\sigma_r)$. *Suppose there is a measuring edge* $\mathscr{E}_{\sigma_r}^{(r,\alpha)}$ *of a flow* $\mathbf{x}^{(\alpha;\mathscr{D})}(t)$ *in domain* $\Omega_\alpha(\sigma_r)$ *with an equi-constant vector of the edge* $\mathscr{E}_{\sigma_r}^{(r)}$, *and the measuring edge is formed by an intersection of all measuring surfaces* $S_{\sigma_m}^{(i_\alpha,\alpha)}$ *for* $i_\alpha \in \mathscr{N}_{\alpha\mathscr{B}}(\sigma_r)$. *At time* t_m , $\mathbf{x}_m^{(\alpha;\mathscr{D})} = \mathbf{x}_\sigma^{(i_\alpha,\alpha)}(t_m) \in S_\sigma^{(i_\alpha,\alpha)}$ ($\sigma \in \mathrm{N}$) *satisfies Eq.* (6.134). *For an arbitrarily small* $\varepsilon > 0$, *there is a time interval* $[t_{m-\varepsilon}, t_{m+\varepsilon}]$. *A flow* $\mathbf{x}^{(\alpha;\mathscr{D})}(t)$ *is* $C_{[t_{m-\varepsilon},t_{m+\varepsilon}]}^{r_\alpha}$ *-continuous for time* t *and* $\| d^{r_\alpha+1}\mathbf{x}^{(\alpha;\mathscr{D})}(t)/dt^{r_\alpha+1} \| < \infty$ ($r_\alpha \geq \max_{i_\alpha \in \mathscr{N}_{\alpha\mathscr{B}}}(2k_{i_\alpha}+2)$). *Suppose* $\mathbf{x}_{\sigma(m\pm\varepsilon)}^{(i_\alpha,\alpha)} \in S_{\sigma_m}^{(i_\alpha,\alpha)}$ *and* $\mathbf{x}_{m\pm\varepsilon}^{(\alpha;\mathscr{D})} = \mathbf{x}^{(\alpha;\mathscr{D})}(t_{m\pm\varepsilon}) \in S_{\sigma_{m\pm\varepsilon}}^{(i_\alpha,\alpha)}$.

(i) *The flow* $\mathbf{x}^{(\alpha;\mathscr{D})}(t)$ *at time* t_m *to the edge of* $\mathscr{E}_{\sigma_r}^{(r)}$ *is attractive with the* $2\mathbf{k}_{\mathscr{E}}$ *-order singularity if and only if for all* $i_\alpha \in \mathscr{N}_{\alpha\mathscr{B}}(\sigma_r)$

$$\left.\begin{array}{l} G_{\partial\Omega_{\alpha i_\alpha}}^{(s_{i_\alpha},\alpha;\mathscr{D})}(\mathbf{x}_m^{(\alpha;\mathscr{D})},t_m,\mathbf{p}_\alpha,\boldsymbol{\lambda}_{i_\alpha})=0 \\ \mathit{for}\ s_{i_\alpha}=0,1,2,\cdots,2k_{i_\alpha}-1; \\ \hbar_{i_\alpha} G_{\partial\Omega_{\alpha i_\alpha}}^{(2k_{i_\alpha},\alpha;\mathscr{D})}(\mathbf{x}_m^{(\alpha;\mathscr{D})},t_m,\mathbf{p}_\alpha,\boldsymbol{\lambda}_{i_\alpha})<0. \end{array}\right\} \tag{6.149}$$

(ii) *The flow* $\mathbf{x}^{(\alpha;\mathscr{D})}(t)$ *at time* t_m *to the edge of* $\mathscr{E}_{\sigma_r}^{(r)}$ *is repulsive with the* $2\mathbf{k}_{\mathscr{E}}$ *-order singularity if and only if for all* $i_\alpha \in \mathscr{N}_{\alpha\mathscr{B}}(\sigma_r)$

$$\left.\begin{array}{l} G_{\partial\Omega_{\alpha i_\alpha}}^{(s_{i_\alpha},\alpha;\mathscr{D})}(\mathbf{x}_m^{(\alpha;\mathscr{D})},t_m,\mathbf{p}_\alpha,\boldsymbol{\lambda}_{i_\alpha})=0 \\ \mathit{for}\ s_{i_\alpha}=0,1,2,\cdots,2k_{i_\alpha}-1; \\ \hbar_{i_\alpha} G_{\partial\Omega_{\alpha i_\alpha}}^{(2k_{i_\alpha},\alpha;\mathscr{D})}(\mathbf{x}_m^{(\alpha;\mathscr{D})},t_m,\mathbf{p}_\alpha,\boldsymbol{\lambda}_{i_\alpha})>0. \end{array}\right\} \tag{6.150}$$

(iii) *The flow* $\mathbf{x}^{(\alpha;\mathscr{D})}(t)$ *at time* t_m *to the edge of* $\mathscr{E}_{\sigma_r}^{(r)}$ *is from the attractive to repulsive state with the* $(2\mathbf{k}_{\mathscr{E}}+1)$ *-order singularity if for all* $i_\alpha \in \mathscr{N}_{\alpha\mathscr{B}}(\sigma_r)$

$$\left.\begin{array}{l} G_{\partial\Omega_{\alpha i_\alpha}}^{(s_{i_\alpha},\alpha;\mathscr{D})}(\mathbf{x}_m^{(\alpha;\mathscr{D})},t_m,\mathbf{p}_\alpha,\boldsymbol{\lambda}_{i_\alpha})=0 \\ \mathit{for}\ s_{i_\alpha}=0,1,2,\cdots,2k_{i_\alpha}; \\ \hbar_{i_\alpha} G_{\partial\Omega_{\alpha i_\alpha}}^{(2k_{i_\alpha}+1,\alpha;\mathscr{D})}(\mathbf{x}_m^{(\alpha;\mathscr{D})},t_m,\mathbf{p}_\alpha,\boldsymbol{\lambda}_{i_\alpha})>0. \end{array}\right\} \tag{6.151}$$

(iv) *The flow* $\mathbf{x}^{(\alpha;\mathscr{D})}(t)$ *at time* t_m *to the edge of* $\mathscr{E}_{\sigma_r}^{(r)}$ *is from the repulsive to attractive state with the* $(2\mathbf{k}_{\mathscr{E}}+1)$ *-order singularity if and only if for all* $i_\alpha \in \mathscr{N}_{\alpha\mathscr{B}}(\sigma_r)$

$$\left.\begin{aligned} &G_{\partial\Omega_{\alpha i_\alpha}}^{(s_{i_\alpha},\alpha;\mathscr{D})}(\mathbf{x}_m^{(\alpha;\mathscr{D})},t_m,\mathbf{p}_\alpha,\boldsymbol{\lambda}_{i_\alpha})=0\\ &\textit{for } s_{i_\alpha}=0,1,2,\cdots,2k_{i_\alpha};\\ &\hbar_{i_\alpha}G_{\partial\Omega_{\alpha i_\alpha}}^{(2k_{i_\alpha}+1,\alpha;\mathscr{D})}(\mathbf{x}_m^{(\alpha;\mathscr{D})},t_m,\mathbf{p}_\alpha,\boldsymbol{\lambda}_{i_\alpha})<0. \end{aligned}\right\} \tag{6.152}$$

It is very difficult to make a flow to all boundaries possess the same attractivity at time. Thus consider five types of flow attractivity to the edge divided into five groups, and the general description for the flow attractivity to the edge is given as follows.

Definition 6.23. For a discontinuous dynamical system in Eqs. (6.4)-(6.8), there is an edge $\mathscr{E}_{\sigma_r}^{(r)}$ ($\sigma_r \in \{1,2,\cdots,N_r\}$ and $r\in\{0,1,2,\cdots,n-2\}$) formed by the intersection of $(n-r)$-boundaries $\partial\Omega_{\alpha i_\alpha}(\sigma_r)$ ($\alpha \in \mathscr{N}_{\mathscr{D}}(\sigma_r)\subset\mathscr{N}_{\mathscr{D}}=\{1,2,\cdots,N_d\}$ and $i_\alpha \in \mathscr{N}_{\alpha\mathscr{B}}(\sigma_r)\subset\mathscr{N}_{\mathscr{B}}=\{1,2,\cdots,N_b\}$) in domain $\Omega_\alpha(\sigma_r)$. Suppose there is a measuring edge $\mathscr{E}_{\sigma_r}^{(r,\alpha)}$ of a flow $\mathbf{x}^{(\alpha;\mathscr{D})}(t)$ in domain $\Omega_\alpha(\sigma_r)$ with an equi-constant vector of the edge $\mathscr{E}_{\sigma_r}^{(r)}$, and the measuring edge is formed by an intersection of all measuring surfaces $S_{\sigma_m}^{(i_\alpha,\alpha)}$ for $i_\alpha\in\mathscr{N}_{\alpha\mathscr{B}}(\sigma_r)$. At time t_m, $\mathbf{x}_m^{(\alpha;\mathscr{D})}=\mathbf{x}_\sigma^{(i_\alpha,\alpha)}(t_m)\in S_\sigma^{(i_\alpha,\alpha)}$ ($\sigma\in\mathrm{N}$) satisfies Eq. (6.134). For an arbitrarily small $\varepsilon>0$, there is a time interval $[t_{m-\varepsilon},t_{m+\varepsilon}]$. A flow $\mathbf{x}^{(\alpha;\mathscr{D})}(t)$ is $C_{[t_{m-\varepsilon},t_{m+\varepsilon}]}^{r_\alpha}$-continuous for time t and $\|d^{r_\alpha+1}\mathbf{x}^{(\alpha;\mathscr{D})}(t)/dt^{r_\alpha+1}\|<\infty$ ($r_\alpha\ge 1$). Suppose $\mathbf{x}_{m\pm\varepsilon}^{(\alpha;\mathscr{D})}=\mathbf{x}^{(\alpha;\mathscr{D})}(t_{m\pm\varepsilon})\in S_{\sigma_{m\pm\varepsilon}}^{(i_\alpha,\alpha)}$ and $\mathbf{x}_{\sigma(m\pm\varepsilon)}^{(i_\alpha,\alpha)}\in S_{\sigma_m}^{(i_\alpha,\alpha)}$. $\mathscr{N}_{\alpha\mathscr{B}}(\sigma_r)=\cup_{k=1}^{5}\mathscr{N}_{\alpha\mathscr{B}}^{(k,\sigma_r)}$ and $\cap_{k=1}^{5}\mathscr{N}_{\alpha\mathscr{B}}^{(k,\sigma_r)}=\varnothing$.

(i) The flow $\mathbf{x}^{(\alpha;\mathscr{D})}(t)$ at time t_m is *partially attractive* to the edge of $\mathscr{E}_{\sigma_r}^{(r)}$ for all the boundaries of $\partial\Omega_{\alpha i_\alpha}$ ($i_\alpha\in\mathscr{N}_{\alpha\mathscr{B}}^{(1,\sigma_r)}$) if

$$\left.\begin{aligned} &\hbar_{i_\alpha}\mathbf{n}_{\partial\Omega_{\alpha i_\alpha}}^{\mathrm{T}}(\mathbf{x}_{\sigma(m-\varepsilon)}^{(i_\alpha,\alpha)})\cdot[\mathbf{x}_{\sigma(m-\varepsilon)}^{(i_\alpha,\alpha)}-\mathbf{x}_{m-\varepsilon}^{(\alpha;\mathscr{D})}]<0,\\ &\hbar_{i_\alpha}\mathbf{n}_{\partial\Omega_{\alpha i_\alpha}}^{\mathrm{T}}(\mathbf{x}_{\sigma(m+\varepsilon)}^{(i_\alpha,\alpha)})\cdot[\mathbf{x}_{m+\varepsilon}^{(\alpha;\mathscr{D})}-\mathbf{x}_{\sigma(m+\varepsilon)}^{(i_\alpha,\alpha)}]<0,\\ &\hbar_{i_\alpha}C_{m-\varepsilon}^{(i_\alpha,\alpha)}>\hbar_{i_\alpha}C_{m}^{(i_\alpha,\alpha)}>\hbar_{i_\alpha}C_{m+\varepsilon}^{(i_\alpha,\alpha)}. \end{aligned}\right\} \tag{6.153}$$

(ii) The flow $\mathbf{x}^{(\alpha;\mathscr{D})}(t)$ at time t_m is *partially repulsive* to the edge of $\mathscr{E}_{\sigma_r}^{(r)}$ for all the boundaries of $\partial\Omega_{\alpha i_\alpha}$ ($i_\alpha\in\mathscr{N}_{\alpha\mathscr{B}}^{(2,\sigma_r)}$) if

$$\left.\begin{array}{l}\hbar_{i_\alpha}\mathbf{n}^{\mathrm{T}}_{\partial\Omega_{\alpha i_\alpha}}(\mathbf{x}^{(i_\alpha,\alpha)}_{\sigma(m-\varepsilon)})\cdot[\mathbf{x}^{(i_\alpha,\alpha)}_{\sigma(m-\varepsilon)}-\mathbf{x}^{(\alpha;\mathscr{D})}_{m-\varepsilon}]>0,\\ \hbar_{i_\alpha}\mathbf{n}^{\mathrm{T}}_{\partial\Omega_{\alpha i_\alpha}}(\mathbf{x}^{(i_\alpha,\alpha)}_{\sigma(m+\varepsilon)})\cdot[\mathbf{x}^{(\alpha;\mathscr{D})}_{m+\varepsilon}-\mathbf{x}^{(i_\alpha,\alpha)}_{\sigma(m+\varepsilon)}]>0,\\ \hbar_{i_\alpha}C^{(i_\alpha,\alpha)}_{m-\varepsilon}<\hbar_{i_\alpha}C^{(i_\alpha,\alpha)}_{m}<\hbar_{i_\alpha}C^{(i_\alpha,\alpha)}_{m+\varepsilon}.\end{array}\right\}\tag{6.154}$$

(iii) The flow $\mathbf{x}^{(\alpha;\mathscr{D})}(t)$ at time t_m to the edge of $\mathscr{E}^{(r)}_{\sigma_r}$ for all the boundaries of $\partial\Omega_{\alpha i_\alpha}$ $(i_\alpha\in\mathscr{N}^{(3,\sigma_r)}_{\alpha\mathscr{B}})$ is *partially from the attractive to repulsive* state if

$$\left.\begin{array}{l}\hbar_{i_\alpha}\mathbf{n}^{\mathrm{T}}_{\partial\Omega_{\alpha i_\alpha}}(\mathbf{x}^{(i_\alpha,\alpha)}_{\sigma(m-\varepsilon)})\cdot[\mathbf{x}^{(i_\alpha,\alpha)}_{\sigma(m-\varepsilon)}-\mathbf{x}^{(\alpha;\mathscr{D})}_{m-\varepsilon}]<0,\\ \hbar_{i_\alpha}\mathbf{n}^{\mathrm{T}}_{\partial\Omega_{\alpha i_\alpha}}(\mathbf{x}^{(i_\alpha,\alpha)}_{\sigma(m+\varepsilon)})\cdot[\mathbf{x}^{(\alpha;\mathscr{D})}_{m+\varepsilon}-\mathbf{x}^{(i_\alpha,\alpha)}_{\sigma(m+\varepsilon)}]>0,\\ \hbar_{i_\alpha}C^{(i_\alpha,\alpha)}_{m}<\hbar_{i_\alpha}C^{(i_\alpha,\alpha)}_{m-\varepsilon}\text{ and }\hbar_{i_\alpha}C^{(i_\alpha,\alpha)}_{m}<\hbar_{i_\alpha}C^{(i_\alpha,\alpha)}_{m+\varepsilon}.\end{array}\right\}\tag{6.155}$$

(iv) The flow $\mathbf{x}^{(\alpha;\mathscr{D})}(t)$ at time t_m to the edge of $\mathscr{E}^{(r)}_{\sigma_r}$ for all the boundaries of $\partial\Omega_{\alpha i_\alpha}$ $(i_\alpha\in\mathscr{N}^{(4,\sigma_r)}_{\alpha\mathscr{B}})$ is *partially from the repulsive to attractive* state if

$$\left.\begin{array}{l}\hbar_{i_\alpha}\mathbf{n}^{\mathrm{T}}_{\partial\Omega_{\alpha i_\alpha}}(\mathbf{x}^{(i_\alpha,\alpha)}_{\sigma(m-\varepsilon)})\cdot[\mathbf{x}^{(i_\alpha,\alpha)}_{\sigma(m-\varepsilon)}-\mathbf{x}^{(\alpha;\mathscr{D})}_{m-\varepsilon}]>0,\\ \hbar_{i_\alpha}\mathbf{n}^{\mathrm{T}}_{\partial\Omega_{\alpha i_\alpha}}(\mathbf{x}^{(i_\alpha,\alpha)}_{\sigma(m+\varepsilon)})\cdot[\mathbf{x}^{(\alpha;\mathscr{D})}_{m+\varepsilon}-\mathbf{x}^{(i_\alpha,\alpha)}_{\sigma(m+\varepsilon)}]<0,\\ \hbar_{i_\alpha}C^{(i_\alpha,\alpha)}_{m}>\hbar_{i_\alpha}C^{(i_\alpha,\alpha)}_{m-\varepsilon}\text{ and }\hbar_{i_\alpha}C^{(i_\alpha,\alpha)}_{m}>\hbar_{i_\alpha}C^{(i_\alpha,\alpha)}_{m+\varepsilon}.\end{array}\right\}\tag{6.156}$$

(v) The flow $\mathbf{x}^{(\alpha;\mathscr{D})}(t)$ at time t_m to the edge of $\mathscr{E}^{(r)}_{\sigma_r}$ for all the boundaries of $\partial\Omega_{\alpha i_\alpha}$ $(i_\alpha\in\mathscr{N}^{(5,\sigma_r)}_{\alpha\mathscr{B}})$ is *partially invariant* with the equi-measuring quantities of $C^{(i_\alpha,\alpha)}_{m}$ if

$$\begin{array}{l}\mathbf{x}^{(i_\alpha,\alpha)}_{\sigma(m-\varepsilon)}=\mathbf{x}^{(\alpha;\mathscr{D})}_{m-\varepsilon}\text{ and }\mathbf{x}^{(\alpha;\mathscr{D})}_{m+\varepsilon}=\mathbf{x}^{(i_\alpha,\alpha)}_{\sigma(m+\varepsilon)},\\ C^{(i_\alpha,\alpha)}_{m-\varepsilon}=C^{(i_\alpha,\alpha)}_{m}=C^{(i_\alpha,\alpha)}_{m+\varepsilon}.\end{array}\tag{6.157}$$

Similar to the Theorem 6.14, the conditions for the flow attractivity to the edge can be stated by the following theorem for convenience.

Theorem 6.15. *For a discontinuous dynamical system in Eqs.* (6.4)-(6.8), *there is an edge* $\mathscr{E}^{(r)}_{\sigma_r}$ $(\sigma_r\in\{1,2,\cdots,N_r\}$ *and* $r\in\{0,1,2,\cdots,n-2\})$ *formed by the intersection of* $(n-r)$ *-boundaries* $\partial\Omega_{\alpha i_\alpha}(\sigma_r)$ ($\alpha\in\mathscr{N}_{\mathscr{D}}(\sigma_r)\subset\mathscr{N}_{\mathscr{D}}=\{1,2,\cdots,N_d\}$ *and* $i_\alpha\in\mathscr{N}_{\alpha\mathscr{B}}(\sigma_r)\subset\mathscr{N}_{\mathscr{B}}=\{1,2,\cdots,N_b\}$) *in domain* $\Omega_\alpha(\sigma_r)$. *Suppose there is a measuring edge* $\mathscr{E}^{(r,\alpha)}_{\sigma_r}$ *of a flow* $\mathbf{x}^{(\alpha;\mathscr{D})}(t)$ *in domain* $\Omega_\alpha(\sigma_r)$ *with an equi-constant vector of the edge* $\mathscr{E}^{(r)}_{\sigma_r}$, *and the measuring edge is formed by an inter-*

section of all measuring surfaces $S_{\sigma_m}^{(i_\alpha,\alpha)}$ *for* $i_\alpha \in \mathscr{N}_{\alpha\mathscr{B}}(\sigma_r)$. *At time* t_m, $\mathbf{x}_m^{(\alpha;\mathscr{D})} = \mathbf{x}_\sigma^{(i_\alpha,\alpha)}(t_m) \in S_\sigma^{(i_\alpha,\alpha)}$ $(\sigma \in \mathbb{N})$ *satisfies Eq.* (6.134). *For an arbitrarily small* $\varepsilon > 0$, *there is a time interval* $[t_{m-\varepsilon}, t_{m+\varepsilon}]$. *A flow* $\mathbf{x}^{(\alpha;\mathscr{D})}(t)$ *is* $C_{[t_{m-\varepsilon},t_{m+\varepsilon}]}^{r_\alpha}$*-continuous for time* t *and* $\| d^{r_\alpha+1}\mathbf{x}^{(\alpha;\mathscr{D})}(t)/dt^{r_\alpha+1} \| < \infty$ ($r_\alpha \geq 1$). *Suppose* $\mathbf{x}_{m\pm\varepsilon}^{(\alpha;\mathscr{D})} = \mathbf{x}^{(\alpha;\mathscr{D})}(t_{m\pm\varepsilon}) \in S_{\sigma_{m\pm\varepsilon}}^{(i_\alpha,\alpha)}$ *and* $\mathbf{x}_{\sigma(m\pm\varepsilon)}^{(i_\alpha,\alpha)} \in S_{\sigma_m}^{(i_\alpha,\alpha)}$. $\mathscr{N}_{\alpha\mathscr{B}}(\sigma_r) = \cup_{k=1}^{5} \mathscr{N}_{\alpha\mathscr{B}}^{(k,\sigma_r)}$ *and* $\cap_{k=1}^{5} \mathscr{N}_{\alpha\mathscr{B}}^{(k,\sigma_r)} = \varnothing$.

(i) *The flow* $\mathbf{x}^{(\alpha;\mathscr{D})}(t)$ *at time* t_m *is partially attractive to the edge of* $\mathscr{E}_{\sigma_r}^{(r)}$ *for all the boundaries of* $\partial\Omega_{\alpha i_\alpha}$ ($i_\alpha \in \mathscr{N}_{\alpha\mathscr{B}}^{(1,\sigma_r)}$) *if and only if*

$$\hbar_{i_\alpha} G_{\partial\Omega_{\alpha i_\alpha}}^{(\alpha;\mathscr{D})}(\mathbf{x}_m^{(\alpha;\mathscr{D})}, t_m, \mathbf{p}_\alpha, \boldsymbol{\lambda}_{i_\alpha}) < 0. \tag{6.158}$$

(ii) *The flow* $\mathbf{x}^{(\alpha;\mathscr{D})}(t)$ *at time* t_m *is partially repulsive to the edge of* $\mathscr{E}_{\sigma_r}^{(r)}$ *for all the boundaries of* $\partial\Omega_{\alpha i_\alpha}$ ($i_\alpha \in \mathscr{N}_{\alpha\mathscr{B}}^{(2,\sigma_r)}$) *if and only if*

$$\hbar_{i_\alpha} G_{\partial\Omega_{\alpha i_\alpha}}^{(\alpha;\mathscr{D})}(\mathbf{x}_m^{(\alpha;\mathscr{D})}, t_m, \mathbf{p}_\alpha, \boldsymbol{\lambda}_{i_\alpha}) > 0. \tag{6.159}$$

(iii) *The flow* $\mathbf{x}^{(\alpha;\mathscr{D})}(t)$ *at time* t_m *to the edge of* $\mathscr{E}_{\sigma_r}^{(r)}$ *for all the boundaries of* $\partial\Omega_{\alpha i_\alpha}$ ($i_\alpha \in \mathscr{N}_{\alpha\mathscr{B}}^{(3,\sigma_r)}$) *is partially from the attractive to repulsive state if and only if*

$$\begin{aligned} &\hbar_{i_\alpha} G_{\partial\Omega_{\alpha i_\alpha}}^{(\alpha;\mathscr{D})}(\mathbf{x}_m^{(\alpha;\mathscr{D})}, t_m, \mathbf{p}_\alpha, \boldsymbol{\lambda}_{i_\alpha}) = 0,\\ &\hbar_{i_\alpha} G_{\partial\Omega_{\alpha i_\alpha}}^{(1,\alpha;\mathscr{D})}(\mathbf{x}_m^{(\alpha;\mathscr{D})}, t_m, \mathbf{p}_\alpha, \boldsymbol{\lambda}_{i_\alpha}) > 0. \end{aligned} \tag{6.160}$$

(iv) *The flow* $\mathbf{x}^{(\alpha;\mathscr{D})}(t)$ *at time* t_m *to the edge of* $\mathscr{E}_{\sigma_r}^{(r)}$ *for all the boundaries of* $\partial\Omega_{\alpha i_\alpha}$ ($i_\alpha \in \mathscr{N}_{\alpha\mathscr{B}}^{(4,\sigma_r)}$) *is from the repulsive to attractive state if and only if*

$$\begin{aligned} &\hbar_{i_\alpha} G_{\partial\Omega_{\alpha i_\alpha}}^{(\alpha;\mathscr{D})}(\mathbf{x}_m^{(\alpha;\mathscr{D})}, t_m, \mathbf{p}_\alpha, \boldsymbol{\lambda}_{i_\alpha}) = 0,\\ &\hbar_{i_\alpha} G_{\partial\Omega_{\alpha i_\alpha}}^{(1,\alpha;\mathscr{D})}(\mathbf{x}_m^{(\alpha;\mathscr{D})}, t_m, \mathbf{p}_\alpha, \boldsymbol{\lambda}_{i_\alpha}) < 0. \end{aligned} \tag{6.161}$$

(v) *The flow* $\mathbf{x}^{(\alpha;\mathscr{D})}(t)$ *at time* t_m *to the edge of* $\mathscr{E}_{\sigma_r}^{(r)}$ *for all the boundaries of* $\partial\Omega_{\alpha i_\alpha}$ ($i_\alpha \in \mathscr{N}_{\alpha\mathscr{B}}^{(5,\sigma_r)}$) *is partially invariant with the equi-measuring quantities of* $C_m^{(i_\alpha,\alpha)}$ *if and only if*

$$G_{\partial\Omega_{\alpha i_\alpha}}^{(s_{i_\alpha},\alpha;\mathscr{D})}(\mathbf{x}_m^{(\alpha;\mathscr{D})}, t_m, \mathbf{p}_\alpha, \boldsymbol{\lambda}_{i_\alpha}) = 0 \text{ for } s_{i_\alpha} = 0,1,2,\cdots. \tag{6.162}$$

Proof. The surface of $\varphi_{\alpha i_\alpha}(\mathbf{x}_m^{(i_\alpha,\alpha)}, t_m, \boldsymbol{\lambda}_{i_\alpha}) = C_m^{(i_\alpha,\alpha)}$ is as a boundary. As in Chapter 3, this theorem can be proved. ■

As in the previous subsection, the flow attractivity to the edge with the higher order singularity is also introduced herein.

Definition 6.24. For a discontinuous dynamical system in Eqs. (6.4)-(6.8), there is an edge $\mathscr{E}_{\sigma_r}^{(r)}$ ($\sigma_r \in \{1,2,\cdots,N_r\}$ and $r \in \{0,1,2,\cdots,n-2\}$) formed by the intersection of $(n-r)$-boundaries $\partial\Omega_{\alpha i_\alpha}(\sigma_r)$ ($\alpha \in \mathscr{N}_{\mathscr{D}}(\sigma_r) \subset \mathscr{N}_{\mathscr{D}} = \{1,2,\cdots,N_d\}$ and $i_\alpha \in \mathscr{N}_{\alpha\mathscr{B}}(\sigma_r) \subset \mathscr{N}_{\mathscr{B}} = \{1,2,\cdots,N_b\}$) in domain $\Omega_\alpha(\sigma_r)$. Suppose there is a measuring edge $\mathscr{E}_{\sigma_r}^{(r,\alpha)}$ of a flow $\mathbf{x}^{(\alpha;\mathscr{D})}(t)$ in domain $\Omega_\alpha(\sigma_r)$ with an equi-constant vector of the edge $\mathscr{E}_{\sigma_r}^{(r)}$, and the measuring edge is formed by an intersection of all measuring surfaces $S_{\sigma_m}^{(i_\alpha,\alpha)}$ for $i_\alpha \in \mathscr{N}_{\alpha\mathscr{B}}(\sigma_r)$. At time t_m, $\mathbf{x}_m^{(\alpha;\mathscr{D})} = \mathbf{x}_\sigma^{(i_\alpha,\alpha)}(t_m) \in S_\sigma^{(i_\alpha,\alpha)}$ ($\sigma \in \mathbb{N}$) satisfies Eq. (6.134). For an arbitrarily small $\varepsilon > 0$, there is a time interval $[t_{m-\varepsilon}, t_{m+\varepsilon}]$. A flow $\mathbf{x}^{(\alpha;\mathscr{D})}(t)$ is $C_{[t_{m-\varepsilon},t_{m+\varepsilon}]}^{r_\alpha}$-continuous for time t and $\| d^{r_\alpha+1}\mathbf{x}^{(\alpha;\mathscr{D})}(t)/dt^{r_\alpha+1} \| < \infty$ ($r_\alpha \geq \max_{i_\alpha \in \mathscr{N}_{\sigma_r}^{\alpha}}(2k_{i_\alpha}+1)$). Suppose $\mathbf{x}_{\sigma(m\pm\varepsilon)}^{(i_\alpha,\alpha)} \in S_{\sigma_m}^{(i_\alpha,\alpha)}$ and $\mathbf{x}_{m\pm\varepsilon}^{(\alpha;\mathscr{D})} = \mathbf{x}^{(\alpha;\mathscr{D})}(t_{m\pm\varepsilon}) \in S_{\sigma_{m\pm\varepsilon}}^{(i_\alpha,\alpha)}$. $\mathscr{N}_{\alpha\mathscr{B}}(\sigma_r) = \cup_{k=1}^{5} \mathscr{N}_{\alpha\mathscr{B}}^{(k,\sigma_r)}$ and $\cap_{k=1}^{5} \mathscr{N}_{\alpha\mathscr{B}}^{(k,\sigma_r)} = \varnothing$.

(i) The flow $\mathbf{x}^{(\alpha;\mathscr{D})}(t)$ at time t_m is *partially attractive* of the $2\mathbf{k}_{\mathscr{E}}^{(1)}$-order to the edge of $\mathscr{E}_{\sigma_r}^{(r)}$ for all the boundaries of $\partial\Omega_{\alpha i_\alpha}$ ($i_\alpha \in \mathscr{N}_{\alpha\mathscr{B}}^{(1,\sigma_r)}$) if

$$
\begin{aligned}
&G_{\partial\Omega_{\alpha i_\alpha}}^{(s_{i_\alpha},\alpha;\mathscr{D})}(\mathbf{x}_m^{(\alpha;\mathscr{D})}, t_m, \mathbf{p}_\alpha, \boldsymbol{\lambda}_{i_\alpha}) = 0 \\
&\text{for } s_{i_\alpha} = 0,1,2,\cdots,2k_{i_\alpha}-1; \\
&\hbar_{i_\alpha}\mathbf{n}_{\partial\Omega_{\alpha i_\alpha}}^{\mathrm{T}}(\mathbf{x}_{\sigma(m-\varepsilon)}^{(i_\alpha,\alpha)})\cdot[\mathbf{x}_{\sigma(m-\varepsilon)}^{(i_\alpha,\alpha)} - \mathbf{x}_{m-\varepsilon}^{(\alpha;\mathscr{D})}] < 0, \\
&\hbar_{i_\alpha}\mathbf{n}_{\partial\Omega_{\alpha i_\alpha}}^{\mathrm{T}}(\mathbf{x}_{\sigma(m+\varepsilon)}^{(i_\alpha,\alpha)})\cdot[\mathbf{x}_{m+\varepsilon}^{(\alpha;\mathscr{D})} - \mathbf{x}_{\sigma(m+\varepsilon)}^{(i_\alpha,\alpha)}] < 0, \\
&\hbar_{i_\alpha}C_{m-\varepsilon}^{(i_\alpha,\alpha)} > \hbar_{i_\alpha}C_m^{(i_\alpha,\alpha)} > \hbar_{i_\alpha}C_{m+\varepsilon}^{(i_\alpha,\alpha)}.
\end{aligned}
\tag{6.163}
$$

(ii) The flow $\mathbf{x}^{(\alpha;\mathscr{D})}(t)$ at time t_m is *partially repulsive* of the $2\mathbf{k}_{\mathscr{E}}^{(2)}$-order to the edge of $\mathscr{E}_{\sigma_r}^{(r)}$ for all the boundaries of $\partial\Omega_{\alpha i_\alpha}$ ($i_\alpha \in \mathscr{N}_{\alpha\mathscr{B}}^{(2,\sigma_r)}$) if

$$
\begin{aligned}
&G_{\partial\Omega_{\alpha i_\alpha}}^{(s_{i_\alpha},\alpha;\mathscr{D})}(\mathbf{x}_m^{(\alpha;\mathscr{D})}, t_m, \mathbf{p}_\alpha, \boldsymbol{\lambda}_{i_\alpha}) = 0 \\
&\text{for } s_{i_\alpha} = 0,1,2,\cdots,2k_{i_\alpha}-1; \\
&\hbar_{i_\alpha}\mathbf{n}_{\partial\Omega_{\alpha i_\alpha}}^{\mathrm{T}}(\mathbf{x}_{\sigma(m-\varepsilon)}^{(i_\alpha,\alpha)})\cdot[\mathbf{x}_{\sigma(m-\varepsilon)}^{(i_\alpha,\alpha)} - \mathbf{x}_{m-\varepsilon}^{(\alpha;\mathscr{D})}] > 0, \\
&\hbar_{i_\alpha}\mathbf{n}_{\partial\Omega_{\alpha i_\alpha}}^{\mathrm{T}}(\mathbf{x}_{\sigma(m+\varepsilon)}^{(i_\alpha,\alpha)})\cdot[\mathbf{x}_{m+\varepsilon}^{(\alpha;\mathscr{D})} - \mathbf{x}_{\sigma(m+\varepsilon)}^{(i_\alpha,\alpha)}] > 0, \\
&\hbar_{i_\alpha}C_{m-\varepsilon}^{(i_\alpha,\alpha)} > \hbar_{i_\alpha}C_m^{(i_\alpha,\alpha)} > \hbar_{i_\alpha}C_{m+\varepsilon}^{(i_\alpha,\alpha)}.
\end{aligned}
\tag{6.164}
$$

(iii) The flow $\mathbf{x}^{(\alpha;\mathscr{D})}(t)$ at time t_m to the edge of $\mathscr{E}_{\sigma_r}^{(r)}$ for all the boundaries of

$\partial\Omega_{\alpha i_\alpha}$ ($i_\alpha \in \mathscr{N}_{\alpha\mathscr{B}}^{(3,\sigma_r)}$) is *partially from the attractive to repulsive* state with the $(2\mathbf{k}_{\mathscr{E}}^{(3)}+1)$-order singularity if

$$\begin{aligned}
&G_{\partial\Omega_{\alpha i_\alpha}}^{(s_{i_\alpha},\alpha;\mathscr{D})}(\mathbf{x}_m^{(\alpha;\mathscr{D})},t_m,\mathbf{p}_\alpha,\boldsymbol{\lambda}_{i_\alpha})=0\\
&\text{for } s_{i_\alpha}=0,1,2,\cdots,2k_{i_\alpha};\\
&\hbar_{i_\alpha}\mathbf{n}_{\partial\Omega_{\alpha i_\alpha}}^{\mathrm{T}}(\mathbf{x}_{\sigma(m-\varepsilon)}^{(i_\alpha,\alpha)})\cdot[\mathbf{x}_{\sigma(m-\varepsilon)}^{(i_\alpha,\alpha)}-\mathbf{x}_{m-\varepsilon}^{(\alpha;\mathscr{D})}]<0,\\
&\hbar_{i_\alpha}\mathbf{n}_{\partial\Omega_{\alpha i_\alpha}}^{\mathrm{T}}(\mathbf{x}_{\sigma(m+\varepsilon)}^{(i_\alpha,\alpha)})\cdot[\mathbf{x}_{m+\varepsilon}^{(\alpha;\mathscr{D})}-\mathbf{x}_{\sigma(m+\varepsilon)}^{(i_\alpha,\alpha)}]>0,\\
&\hbar_{i_\alpha}C_m^{(i_\alpha,\alpha)}<\hbar_{i_\alpha}C_{m-\varepsilon}^{(i_\alpha,\alpha)}\text{ and }\hbar_{i_\alpha}C_m^{(i_\alpha,\alpha)}<\hbar_{i_\alpha}C_{m+\varepsilon}^{(i_\alpha,\alpha)}.
\end{aligned}\tag{6.165}$$

(iv) The flow $\mathbf{x}^{(\alpha;\mathscr{D})}(t)$ at time t_m to the edge of $\mathscr{E}_{\sigma_r}^{(r)}$ for all the boundaries of $\partial\Omega_{\alpha i_\alpha}$ ($i_\alpha \in \mathscr{N}_{\alpha\mathscr{B}}^{(4,\sigma_r)}$) is *partially from the repulsive to attractive* state with the $(2\mathbf{k}_{\mathscr{E}}^{(4)}+1)$-order singularity if

$$\begin{aligned}
&G_{\partial\Omega_{\alpha i_\alpha}}^{(s_{i_\alpha},\alpha;\mathscr{D})}(\mathbf{x}_m^{(\alpha;\mathscr{D})},t_m,\mathbf{p}_\alpha,\boldsymbol{\lambda}_{i_\alpha})=0\\
&\text{for } s_{i_\alpha}=0,1,2,\cdots,2k_{i_\alpha};\\
&\hbar_{i_\alpha}\mathbf{n}_{\partial\Omega_{\alpha i_\alpha}}^{\mathrm{T}}(\mathbf{x}_{\sigma(m-\varepsilon)}^{(i_\alpha,\alpha)})\cdot[\mathbf{x}_{\sigma(m-\varepsilon)}^{(i_\alpha,\alpha)}-\mathbf{x}_{m-\varepsilon}^{(\alpha;\mathscr{D})}]>0,\\
&\hbar_{i_\alpha}\mathbf{n}_{\partial\Omega_{\alpha i_\alpha}}^{\mathrm{T}}(\mathbf{x}_{\sigma(m+\varepsilon)}^{(i_\alpha,\alpha)})\cdot[\mathbf{x}_{m+\varepsilon}^{(\alpha;\mathscr{D})}-\mathbf{x}_{\sigma(m+\varepsilon)}^{(i_\alpha,\alpha)}]<0,\\
&\hbar_{i_\alpha}C_m^{(i_\alpha,\alpha)}>\hbar_{i_\alpha}C_{m-\varepsilon}^{(i_\alpha,\alpha)}\text{ and }\hbar_{i_\alpha}C_m^{(i_\alpha,\alpha)}>\hbar_{i_\alpha}C_{m+\varepsilon}^{(i_\alpha,\alpha)}.
\end{aligned}\tag{6.166}$$

From the definition, the theorem for the flow attractivity to the edge with the higher order singularity is presented as follows.

Theorem 6.16. *For a discontinuous dynamical system in Eqs.* (6.4)-(6.8), *there is an edge* $\mathscr{E}_{\sigma_r}^{(r)}$ ($\sigma_r \in \{1,2,\cdots,N_r\}$ *and* $r \in \{0,1,2,\cdots,n-2\}$) *formed by the intersection of* $(n-r)$*-boundaries* $\partial\Omega_{\alpha i_\alpha}(\sigma_r)$ ($\alpha \in \mathscr{N}_{\mathscr{D}}(\sigma_r) \subset \mathscr{N}_{\mathscr{D}} = \{1,2,\cdots,N_d\}$ *and* $i_\alpha \in \mathscr{N}_{\alpha\mathscr{B}}(\sigma_r) \subset \mathscr{N}_{\mathscr{B}} = \{1,2,\cdots,N_b\}$) *in domain* $\Omega_\alpha(\sigma_r)$. *Suppose there is a measuring edge* $\mathscr{E}_{\sigma_r}^{(r,\alpha)}$ *of a flow* $\mathbf{x}^{(\alpha;\mathscr{D})}(t)$ *in domain* $\Omega_\alpha(\sigma_r)$ *with an equi-constant vector of the edge* $\mathscr{E}_{\sigma_r}^{(r)}$, *and the measuring edge is formed by an intersection of all measuring surfaces* $S_m^{(i_\alpha,\alpha)}$ *for* $i_\alpha \in \mathscr{N}_{\alpha\mathscr{B}}(\sigma_r)$. *At time* t_m, $\mathbf{x}_m^{(\alpha;\mathscr{D})} = \mathbf{x}^{(i_\alpha,\alpha)}(t_m) \in S_m^{(i_\alpha,\alpha)}$ ($\sigma \in \mathrm{N}$) *satisfies Eq.* (6.134). *For an arbitrarily small* $\varepsilon > 0$, *there is a time interval* $[t_{m-\varepsilon}, t_{m+\varepsilon}]$. *A flow* $\mathbf{x}^{(\alpha;\mathscr{D})}(t)$ *is* $C_{[t_{m-\varepsilon},t_{m+\varepsilon}]}^{r_\alpha}$*-continuous for time* t *and* $\| d^{r_\alpha+1}\mathbf{x}^{(\alpha;\mathscr{D})}(t)/dt^{r_\alpha+1} \| < \infty$ ($r_\alpha \geq \max_{i_\alpha \in \mathscr{N}_{\sigma_r}^{\alpha}}(2k_{i_\alpha}+2)$). *Suppose* $\mathbf{x}_{m\pm\varepsilon}^{(i_\alpha,\alpha)} \in S_m^{(i_\alpha,\alpha)}$ *and* $\mathbf{x}_{m\pm\varepsilon}^{(\alpha;\mathscr{D})} = \mathbf{x}^{(\alpha;\mathscr{D})}(t_{m\pm\varepsilon}) \in S_{m\pm\varepsilon}^{(i_\alpha,\alpha)}$. $\mathscr{N}_{\alpha\mathscr{B}}(\sigma_r) = \cup_{k=1}^{5}\mathscr{N}_{\alpha\mathscr{B}}^{(k,\sigma_r)}$ *and* $\cap_{k=1}^{5}\mathscr{N}_{\alpha\mathscr{B}}^{(k,\sigma_r)} = \varnothing$.

(i) *The flow* $\mathbf{x}^{(\alpha;\mathscr{D})}(t)$ *at time* t_m *is partially attractive of the* $2\mathbf{k}_{\mathscr{E}}^{(1)}$ *-order to the edge of* $\mathscr{E}_{\sigma_r}^{(r)}$ *for all the boundaries of* $\partial\Omega_{\alpha i_\alpha}$ $(i_\alpha \in \mathscr{N}_{\alpha\mathscr{B}}^{(1,\sigma_r)})$ *if and only if*

$$\begin{aligned} & G_{\partial\Omega_{\alpha i_\alpha}}^{(s_{i_\alpha},\alpha;\mathscr{D})}(\mathbf{x}_m^{(\alpha;\mathscr{D})},t_m,\mathbf{p}_\alpha,\boldsymbol{\lambda}_{i_\alpha})=0 \\ & \text{for } s_{i_\alpha}=0,1,2,\cdots,2k_{i_\alpha}-1; \\ & \hbar_{i_\alpha} G_{\partial\Omega_{\alpha i_\alpha}}^{(2k_{i_\alpha},\alpha;\mathscr{D})}(\mathbf{x}_m^{(\alpha;\mathscr{D})},t_m,\mathbf{p}_\alpha,\boldsymbol{\lambda}_{i_\alpha})<0. \end{aligned} \tag{6.167}$$

(ii) *The flow* $\mathbf{x}^{(\alpha;\mathscr{D})}(t)$ *at time* t_m *is partially repulsive of the* $2\mathbf{k}_{\mathscr{E}}^{(2)}$ *-order to the edge of* $\mathscr{E}_{\sigma_r}^{(r)}$ *for all the boundaries of* $\partial\Omega_{\alpha i_\alpha}$ $(i_\alpha \in \mathscr{N}_{\alpha\mathscr{B}}^{(2,\sigma_r)})$ *if and only if*

$$\begin{aligned} & G_{\partial\Omega_{\alpha i_\alpha}}^{(s_{i_\alpha},\alpha;\mathscr{D})}(\mathbf{x}_m^{(\alpha;\mathscr{D})},t_m,\mathbf{p}_\alpha,\boldsymbol{\lambda}_{i_\alpha})=0 \\ & \text{for } s_{i_\alpha}=0,1,2,\cdots,2k_{i_\alpha}-1; \\ & \hbar_{i_\alpha} G_{\partial\Omega_{\alpha i_\alpha}}^{(2k_{i_\alpha},\alpha;\mathscr{D})}(\mathbf{x}_m^{(\alpha;\mathscr{D})},t_m,\mathbf{p}_\alpha,\boldsymbol{\lambda}_{i_\alpha})>0. \end{aligned} \tag{6.168}$$

(iii) *The flow* $\mathbf{x}^{(\alpha;\mathscr{D})}(t)$ *at time* t_m *to the edge of* $\mathscr{E}_{\sigma_r}^{(r)}$ *for all the boundaries of* $\partial\Omega_{\alpha i_\alpha}$ $(i_\alpha \in \mathscr{N}_{\alpha\mathscr{B}}^{(3,\sigma_r)})$ *is partially from the attractive to repulsive state with the* $(2\mathbf{k}_{\mathscr{E}}^{(3)}+1)$ *-order singularity if and only if*

$$\begin{aligned} & G_{\partial\Omega_{\alpha i_\alpha}}^{(s_{i_\alpha},\alpha;\mathscr{D})}(\mathbf{x}_m^{(\alpha;\mathscr{D})},t_m,\mathbf{p}_\alpha,\boldsymbol{\lambda}_{i_\alpha})=0 \\ & \text{for } s_{i_\alpha}=0,1,2,\cdots,2k_{i_\alpha}; \\ & \hbar_{i_\alpha} G_{\partial\Omega_{\alpha i_\alpha}}^{(2k_{i_\alpha}+1,\alpha;\mathscr{D})}(\mathbf{x}_m^{(\alpha;\mathscr{D})},t_m,\mathbf{p}_\alpha,\boldsymbol{\lambda}_{i_\alpha})>0. \end{aligned} \tag{6.169}$$

(iv) *The flow* $\mathbf{x}^{(\alpha;\mathscr{D})}(t)$ *at time* t_m *to the edge of* $\mathscr{E}_{\sigma_r}^{(r)}$ *for all the boundaries of* $\partial\Omega_{\alpha i_\alpha}$ $(i_\alpha \in \mathscr{N}_{\alpha\mathscr{B}}^{(4,\sigma_r)})$ *is partially form the repulsive to attractive state with the* $(2\mathbf{k}_{\mathscr{E}}^{(4)}+1)$ *-order singularity if and only if*

$$\begin{aligned} & G_{\partial\Omega_{\alpha i_\alpha}}^{(s_{i_\alpha},\alpha;\mathscr{D})}(\mathbf{x}_m^{(\alpha;\mathscr{D})},t_m,\mathbf{p}_\alpha,\boldsymbol{\lambda}_{i_\alpha})=0 \\ & \text{for } s_{i_\alpha}=0,1,2,\cdots,2k_{i_\alpha}; \\ & \hbar_{i_\alpha} G_{\partial\Omega_{\alpha j_\alpha}}^{(2k_{i_\alpha}+1,\alpha;\mathscr{D})}(\mathbf{x}_m^{(\alpha;\mathscr{D})},t_m,\mathbf{p}_\alpha,\boldsymbol{\lambda}_{i_\alpha})<0. \end{aligned} \tag{6.170}$$

Proof. The surface of $\varphi_{\alpha i_\alpha}(\mathbf{x}_m^{(i_\alpha,\alpha)},t_m,\boldsymbol{\lambda}_{i_\alpha})=C_m^{(i_\alpha,\alpha)}$ is as a boundary. As in Chapter 3, this theorem can be proved. ■

The previous discussion on the attractivity of a flow to the common edge is based on the attractivity of a flow to each $(n-1)$-dimensional boundary. The subset $\mathscr{N}_{\alpha\mathscr{B}}$ of $\mathscr{N}_{\mathscr{B}}$ includes all $(n-r)$-indices of the $(n-r)$-boundaries of the edge of $\mathscr{E}_{\sigma_r}^{(r)}$ ($\sigma_r \in \{1,2,\cdots,N_r\}$. If $\mathscr{N}_{\alpha\mathscr{B}}^{(1,\sigma_r)} = \mathscr{N}_{\alpha\mathscr{B}}$ and $\mathscr{N}_{\alpha\mathscr{B}}^{(j,\sigma_r)} = \varnothing$ ($j = 2,3,4,5$), then the flow to the edge $\mathscr{E}_{\sigma_r}^{(r)}$ in domain Ω_α is attractive. If $\mathscr{N}_{\alpha\mathscr{B}}^{(2,\sigma_r)} = \mathscr{N}_{\alpha\mathscr{B}}$ and $\mathscr{N}_{\alpha\mathscr{B}}^{(j,\sigma_r)} = \varnothing$ ($j = 1,3,4,5$), then the flow to the edge $\mathscr{E}_{\sigma_r}^{(r)}$ in domain Ω_α is repulsive. If $\mathscr{N}_{\alpha\mathscr{B}}^{(3,\sigma_r)} = \mathscr{N}_{\alpha\mathscr{B}}$ and $\mathscr{N}_{\alpha\mathscr{B}}^{(j,\sigma_r)} = \varnothing$ ($j = 1,2,4,5$), then the flow to the edge $\mathscr{E}_{\sigma_r}^{(r)}$ in domain Ω_α is from the attractive to repulsive state. If $\mathscr{N}_{\alpha\mathscr{B}}^{(4,\sigma_r)} = \mathscr{N}_{\alpha\mathscr{B}}$ and $\mathscr{N}_{\alpha\mathscr{B}}^{(j,\sigma_r)} = \varnothing$ ($j = 1,2,3,5$), then the flow to the edge $\mathscr{E}_{\sigma_r}^{(r)}$ in domain Ω_α is from the repulsive to attractive state. If $\mathscr{N}_{\alpha\mathscr{B}}^{(5,\sigma_r)} = \mathscr{N}_{\alpha\mathscr{B}}$ and $\mathscr{N}_{\alpha\mathscr{B}}^{(j,\sigma_r)} = \varnothing$ ($j = 1,2,3,4$), then the flow to the edge of $\mathscr{E}_{\sigma_r}^{(r)}$ in domain Ω_α is invariant. If $\mathscr{N}_{\alpha\mathscr{B}}^{(1,\sigma_r)} \cup \mathscr{N}_{\alpha\mathscr{B}}^{(2,\sigma_r)} = \mathscr{N}_{\sigma_r}^{\alpha}$ and $\mathscr{N}_{\sigma_r}^{(j,\alpha)} = \varnothing$ ($j = 3,4,5$), is the flow to the edge of $\mathscr{E}_{\sigma_r}^{(r)}$ in domain Ω_α attracted or not? If there is an r_1-dimensional edge ($r_1 \le r+1$) for $\mathscr{N}_{\alpha\mathscr{B}}^{(5,\sigma_r)} \ne \varnothing$, then the edge of $\mathscr{E}_{\sigma_r}^{(r)}$ to the r_1-dimensional edge should not be intersected.

To describe the measuring edge varying with time in domain, consider the measuring edge which should not be intersected with the common edge. For instance, each measuring edge $\mathscr{E}^{(\rho,n-2)}$ is the intersection of two separatrix measuring surfaces in a domain. The two measuring surfaces are two equi-constant surfaces relative to the two corresponding separation boundaries. The common edge $\mathscr{E}^{(n-2)}$ is the $(n-2)$-dimensional, intersection surface of three separation boundaries, as sketched in Fig. 6.19. With different time, the flow will move to different location, and the corresponding edge formed by the separatrix measuring surfaces in domain is different. To describe the location of the measuring edge, the non-zero constant for each equi-constant separatrix measuring surface is replaced by a variable. For a specific time, the variables should be treated as constants to form the separatrix measuring surfaces. Thus, for a new set of constants, the new measuring surfaces will give the new measuring edge, which can be expressed by a vector of non-zero constant variables. The definition is given as follows.

Definition 6.25. For a discontinuous dynamical system in Eqs. (6.4)-(6.8), there is an edge $\mathscr{E}_{\sigma_r}^{(r)}$ ($\sigma_r \in \{1,2,\cdots,N_r\}$ and $r \in \{0,1,2,\cdots,n-2\}$) formed by the intersection of $(n-r)$-boundaries $\partial\Omega_{\alpha i_\alpha}(\sigma_r)$ ($\alpha \in \mathscr{N}_{\mathscr{D}}(\sigma_r) \subset \mathscr{N}_{\mathscr{D}} = \{1,2,\cdots,N_d\}$ and $i_\alpha \in \mathscr{N}_{\alpha\mathscr{B}}(\sigma_r) \subset \mathscr{N}_{\mathscr{B}} = \{1,2,\cdots,N_b\}$) in domain $\Omega_\alpha(\sigma_r)$.

(i) The *unit normal vector* of the boundary $\partial\Omega_{\alpha i_\alpha}(\sigma_r)$ is defined by

$$\mathbf{e}_{i_\alpha} = \frac{\mathbf{n}_{\partial\Omega_{\alpha i_\alpha}}}{|\mathbf{n}_{\partial\Omega_{\alpha i_\alpha}}|} = \frac{1}{\sqrt{g_{i_\alpha i_\alpha}}}\frac{\partial\varphi_{\alpha i_\alpha}}{\partial\mathbf{x}} = \frac{1}{\sqrt{g_{i_\alpha i_\alpha}}}\left(\frac{\partial\varphi_{\alpha i_\alpha}}{\partial x_1},\frac{\partial\varphi_{\alpha i_\alpha}}{\partial x_2},\cdots,\frac{\partial\varphi_{\alpha i_\alpha}}{\partial x_n}\right)^{\mathrm{T}} \tag{6.171}$$

where

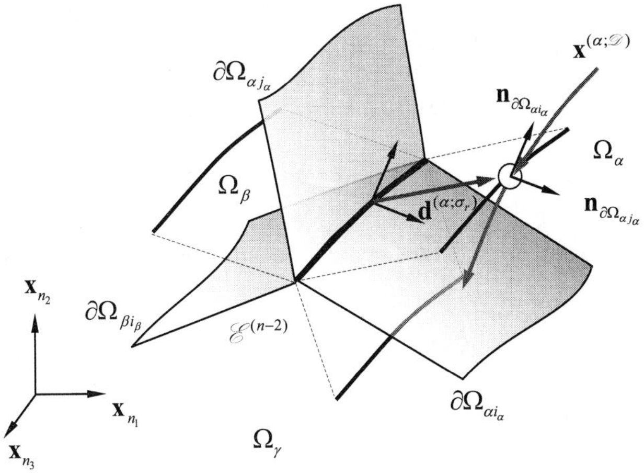

Fig. 6.19 Three measuring edges paralleled with the common edge $\mathscr{E}^{(n-2)}$ in three different domains. Each of the three measuring edges is the intersection of two separatrix measuring surfaces in each domain. The two measuring surfaces are two equi-constant surfaces relative to the two corresponding separation boundaries. The common edge $\mathscr{E}^{(n-2)}$ is the $(n-2)$-dimensional, intersection surface of three separation boundaries. The corresponding normal vectors are presented. ($n_1+n_2+n_3=n$) (color plot in the book end)

$$g_{i_\alpha i_\alpha}=(\frac{\partial\varphi_{\alpha i_\alpha}}{\partial x_1})^2+(\frac{\partial\varphi_{\alpha i_\alpha}}{\partial x_2})^2+\cdots+(\frac{\partial\varphi_{\alpha i_\alpha}}{\partial x_n})^2. \tag{6.172}$$

(ii) The *measuring vector* from the common edge to the measuring edge in domain Ω_α is defined as

$$\mathbf{d}^{(\alpha;\sigma_r)}=\sum\nolimits_{i_\alpha=1}^{n-r}z^{i_\alpha}\mathbf{e}_{i_\alpha} \tag{6.173}$$

where the measuring variable z^{i_α} relative to $\partial\Omega_{\alpha i_\alpha}$ is

$$z^{i_\alpha}=\varphi_{\alpha i_\alpha}(\mathbf{x}^{(\alpha;\mathscr{D})},t,\boldsymbol{\lambda}_{i_\alpha}). \tag{6.174}$$

(iii) The *measuring function* is defined as

$$\mathscr{D}^{(\alpha;\sigma_r)}=\|\mathbf{d}^{(\alpha;\sigma_r)}\|^2=\sum\nolimits_{j_\alpha=1}^{n-r}\sum\nolimits_{i_\alpha=1}^{n-r}z^{i_\alpha}z^{j_\alpha}e_{i_\alpha j_\alpha} \tag{6.175}$$

where the metric tensor is

$$e_{i_\alpha j_\alpha}=\mathbf{e}_{i_\alpha}\cdot\mathbf{e}_{j_\alpha}=\frac{1}{\sqrt{g_{i_\alpha i_\alpha}g_{j_\alpha j_\alpha}}}\sum\nolimits_{k=1}^{n}\frac{\partial\varphi_{\alpha i_\alpha}}{\partial x_k}\frac{\partial\varphi_{\alpha j_\alpha}}{\partial x_k}. \tag{6.176}$$

(iv) If $\mathscr{D}^{(\alpha;\sigma_r)}=\mathscr{C}^{(\alpha;\sigma_r)}$ is constant, the *measuring surface* $\mathscr{M}^{(\alpha;\sigma_r)}$ is defined as

$$\sum_{j_\alpha=1}^{n-r}\sum_{i_\alpha=1}^{n-r} z^{i_\alpha} z^{j_\alpha} e_{i_\alpha j_\alpha} = \mathscr{C}^{(\alpha;\sigma_r)}. \tag{6.177}$$

(v) The *normal vector* of the measuring surface is defined as

$$\mathbf{n}_{\mathscr{M}^{(\alpha;\sigma_r)}} = \frac{\partial}{\partial \mathbf{x}} \sum_{j_\alpha=1}^{n-r}\sum_{i_\alpha=1}^{n-r} z^{i_\alpha} z^{j_\alpha} e_{i_\alpha j_\alpha}. \tag{6.178}$$

(vi) If $\mathscr{D}^{(\alpha;\sigma_r)}$ varies with time, the *derivative* of the measuring function is defined as

$$\begin{aligned} \frac{D}{Dt}\mathscr{D}^{(\alpha;\sigma_r)} &= \dot{\mathscr{D}}^{(\alpha;\sigma_r)} = \frac{\partial \mathscr{D}^{(\alpha;\sigma_r)}}{\partial \mathbf{x}}\dot{\mathbf{x}} + \frac{\partial \mathscr{D}^{(\alpha;\sigma_r)}}{\partial t} \\ &= \sum_{j_\alpha=1}^{n-r}\sum_{i_\alpha=1}^{n-r} (\dot{z}^{i_\alpha} z^{j_\alpha} e_{i_\alpha j_\alpha} + z^{i_\alpha} \dot{z}^{j_\alpha} e_{i_\alpha j_\alpha} + z^{i_\alpha} z^{j_\alpha} \dot{e}_{i_\alpha j_\alpha}). \end{aligned} \tag{6.179}$$

From Luo (2008a,b), the G-functions for such a measuring surface can be defined similarly as in Eqs. (6.113) and (6.114), i.e., $G_{\partial\Omega_{ij}}^{(\alpha;\mathscr{D})} = G_{\mathscr{M}^{(\alpha,\sigma_r)}}^{(\alpha;\mathscr{M})}$ and $G_{\partial\Omega_{ij}}^{(s_\alpha,\alpha;\mathscr{D})} = G_{\mathscr{M}^{(\alpha,\sigma_r)}}^{(s_\alpha,\alpha;\mathscr{M})}$. The measuring vector can be decomposed by many measuring sub-vectors. The corresponding definition is given as follows.

Definition 6.26. For a discontinuous dynamical system in Eqs. (6.4)-(6.8), there is an edge $\mathscr{E}_{\sigma_r}^{(r)}$ ($\sigma_r \in \{1,2,\cdots,N_r\}$ and $r \in \{0,1,2,\cdots,n-2\}$) formed by the intersection of $(n-r)$-boundaries $\partial\Omega_{\alpha i_\alpha}(\sigma_r)$ ($\alpha \in \mathscr{N}_{\mathscr{D}}(\sigma_r) \subset \mathscr{N}_{\mathscr{D}} = \{1,2,\cdots,N_d\}$ and $i_\alpha \in \mathscr{N}_{\alpha\mathscr{B}}(\sigma_r) \subset \mathscr{N}_{\mathscr{B}} = \{1,2,\cdots,N_b\}$) in domain $\Omega_\alpha(\sigma_r)$. The *measuring vector* from the common edge to the measuring edge in domain Ω_α in Eq. (6.173) is composed of the measuring sub-vectors, i.e.,

$$\begin{aligned} \mathbf{d}^{(\alpha;\sigma_r)} &= \sum_{i_\alpha=1}^{n-r} z^{i_\alpha}\mathbf{e}_{i_\alpha} = \sum_{j=1}^{5} \mathbf{d}_j^{(\alpha;\sigma_r)}, \\ \mathbf{d}_j^{(\alpha;\sigma_r)} &= \sum_{i_\alpha=1}^{n-r} z_{(j)}^{i_\alpha}\mathbf{e}_{i_\alpha}^{(j)} \text{ for } i_\alpha \in \mathscr{N}_{\alpha\mathscr{B}}^{(j,\sigma_r)} \end{aligned} \tag{6.180}$$

where $\mathscr{N}_{\alpha\mathscr{B}}(\sigma_r) = \cup_{j=1}^{5}\mathscr{N}_{\alpha\mathscr{B}}^{(j,\sigma_r)}$ and $\cap_{j=1}^{5}\mathscr{N}_{\alpha\mathscr{B}}^{(j,\sigma_r)} = \varnothing$.

Definition 6.27. For a discontinuous dynamical system in Eqs.(6.4)-(6.8), there is an edge $\mathscr{E}_{\sigma_r}^{(r)}$ ($\sigma_r \in \{1,2,\cdots,N_r\}$ and $r \in \{0,1,2,\cdots,n-2\}$) formed by the intersection of $(n-r)$-boundaries $\partial\Omega_{\alpha i_\alpha}(\sigma_r)$ ($\alpha \in \mathscr{N}_{\mathscr{D}}(\sigma_r) \subset \mathscr{N}_{\mathscr{D}} = \{1,2,\cdots,N_d\}$ and $i_\alpha \in \mathscr{N}_{\alpha\mathscr{B}}(\sigma_r) \subset \mathscr{N}_{\mathscr{B}} = \{1,2,\cdots,N_b\}$) in domain $\Omega_\alpha(\sigma_r)$. Suppose there is a measuring edge $\mathscr{E}_{\sigma_r}^{(r,\alpha)}$ of a flow $\mathbf{x}^{(\alpha;\mathscr{D})}(t)$ in domain $\Omega_\alpha(\sigma_r)$ with an equi-constant vector of the edge $\mathscr{E}_{\sigma_r}^{(r)}$ and the measuring edge is an intersection of $S_\sigma^{(i_\alpha,\alpha)}$

$$\mathscr{E}_{\sigma_r}^{(r,\alpha)} = \bigcap\nolimits_{i_\alpha=1}^{n-r} S_{\sigma}^{(i_\alpha,\alpha)}. \tag{6.181}$$

which lies in a measuring surface $\mathscr{M}^{(\alpha;\sigma_r)}$ given by

$$\mathscr{M}^{(\alpha;\sigma_r)} = \left\{ \mathbf{x}^{(\alpha;\mathscr{D})} \left| \begin{array}{l} \sum\nolimits_{j_\alpha=1}^{n-r} \sum\nolimits_{i_\alpha=1}^{n-r} z^{i_\alpha} z^{j_\alpha} e_{i_\alpha j_\alpha} = \mathscr{C}^{(\alpha;\sigma_r)} = \text{const} \\ \text{where } z^{i_\alpha} = \varphi_{\alpha i_\alpha}(\mathbf{x}^{(\alpha;\mathscr{D})}, t, \boldsymbol{\lambda}_{i_\alpha}) \end{array} \right. \right\}. \tag{6.182}$$

From the description of measuring vector and measuring surface, the flow attractivity to the edge can be described.

Definition 6.28. For a discontinuous dynamical system in Eqs. (6.4)-(6.8), there is an edge $\mathscr{E}_{\sigma_r}^{(r)}$ ($\sigma_r \in \{1,2,\cdots,N_r\}$ and $r \in \{0,1,2,\cdots,n-2\}$) formed by the intersection of $(n-r)$-boundaries $\partial\Omega_{\alpha i_\alpha}(\sigma_r)$ ($\alpha \in \mathscr{N}_{\mathscr{D}}(\sigma_r) \subset \mathscr{N}_{\mathscr{D}} = \{1,2,\cdots,N_d\}$ and $i_\alpha \in \mathscr{N}_{\alpha\mathscr{B}}(\sigma_r) \subset \mathscr{N}_{\mathscr{B}} = \{1,2,\cdots,N_b\}$) in domain $\Omega_\alpha(\sigma_r)$. Suppose there is a measuring edge $\mathscr{E}_{\sigma_r}^{(r,\alpha)}$ of a flow $\mathbf{x}^{(\alpha;\mathscr{D})}(t)$ in domain $\Omega_\alpha(\sigma_r)$ with an equi-constant vector of the edge $\mathscr{E}_{\sigma_r}^{(r)}$, and the measuring edge and an equi-constant measuring surface $\mathscr{M}^{(\alpha;\sigma_r)}$ are determined by Eqs. (6.171) and (6.182), respectively. At time t_m, $\mathbf{x}_m^{(\alpha;\mathscr{M})} = \mathbf{x}_\sigma^{(\alpha;\mathscr{M})}(t_m) \in \mathscr{M}_m^{(\alpha,\sigma_r)}$ satisfies Eq. (6.134). For an arbitrarily small $\varepsilon > 0$, there is a time interval $[t_{m-\varepsilon}, t_{m+\varepsilon}]$. The flow $\mathbf{x}^{(\alpha;\mathscr{D})}(t)$ is $C_{[t_{m-\varepsilon},t_{m+\varepsilon}]}^{r_\alpha}$-continuous for time t and $\| d^{r_\alpha+1}\mathbf{x}^{(\alpha;\mathscr{D})}/dt^{r_\alpha+1} \| < \infty$ ($r_\alpha \ge 1$). Suppose $\mathbf{x}_{m\pm\varepsilon}^{(\alpha;\mathscr{M})} \in \mathscr{M}_m^{(\alpha,\sigma_r)}$ and $\mathbf{x}_{m\pm\varepsilon}^{(\alpha;\mathscr{D})} = \mathbf{x}^{(\alpha;\mathscr{D})}(t_{m\pm\varepsilon}) \in \mathscr{M}_{m\pm\varepsilon}^{(\alpha,\sigma_r)}$. $\mathscr{D}^{(\alpha;\sigma_r)}(t_m) \equiv \mathscr{C}_m^{(\alpha;\sigma_r)}$ and $\mathscr{D}^{(\alpha;\sigma_r)}(t_{m\pm\varepsilon}) \equiv \mathscr{C}_{m\pm\varepsilon}^{(\alpha;\sigma_r)}$.

(i) The flow $\mathbf{x}^{(\alpha;\mathscr{D})}(t)$ at time t_m to the edge of $\mathscr{E}_{\sigma_r}^{(r)}$ is *attractive* in sense of the measuring surface of $\mathscr{M}^{(\alpha,\sigma_r)}$ if

$$\left.\begin{array}{l} \mathbf{n}^{\mathrm{T}}_{\mathscr{M}_m^{(\alpha;\sigma_r)}}(\mathbf{x}_{m-\varepsilon}^{(\alpha;\mathscr{M})}) \cdot [\mathbf{x}_{m-\varepsilon}^{(\alpha;\mathscr{M})} - \mathbf{x}_{m-\varepsilon}^{(\alpha;\mathscr{D})}] < 0, \\ \mathbf{n}^{\mathrm{T}}_{\mathscr{M}_m^{(\alpha;\sigma_r)}}(\mathbf{x}_{m+\varepsilon}^{(\alpha;\mathscr{M})}) \cdot [\mathbf{x}_{m+\varepsilon}^{(\alpha;\mathscr{D})} - \mathbf{x}_{m+\varepsilon}^{(\alpha;\mathscr{M})}] < 0, \\ \mathscr{C}_{m-\varepsilon}^{(\alpha,\sigma_r)} > \mathscr{C}_m^{(\alpha,\sigma_r)} > \mathscr{C}_{m+\varepsilon}^{(\alpha,\sigma_r)} > 0. \end{array}\right\} \tag{6.183}$$

(ii) The flow $\mathbf{x}^{(\alpha;\mathscr{D})}(t)$ at time t_m to the edge of $\mathscr{E}_{\sigma_r}^{(r)}$ is *repulsive* in sense of the measuring surface of $\mathscr{M}^{(\alpha,\sigma_r)}$

$$\left.\begin{array}{l} \mathbf{n}^{\mathrm{T}}_{\mathscr{M}_m^{(\alpha;\sigma_r)}}(\mathbf{x}_{m-\varepsilon}^{(\alpha;\mathscr{M})}) \cdot [\mathbf{x}_{m-\varepsilon}^{(\alpha;\mathscr{M})} - \mathbf{x}_{m-\varepsilon}^{(\alpha;\mathscr{D})}] > 0, \\ \mathbf{n}^{\mathrm{T}}_{\mathscr{M}_m^{(\alpha;\sigma_r)}}(\mathbf{x}_{m+\varepsilon}^{(\alpha;\mathscr{M})}) \cdot [\mathbf{x}_{m+\varepsilon}^{(\alpha;\mathscr{D})} - \mathbf{x}_{m+\varepsilon}^{(\alpha;\mathscr{M})}] > 0, \\ 0 < \mathscr{C}_{m-\varepsilon}^{(\alpha,\sigma_r)} < \mathscr{C}_m^{(\alpha,\sigma_r)} < \mathscr{C}_{m+\varepsilon}^{(\alpha,\sigma_r)}. \end{array}\right\} \tag{6.184}$$

(iii) The flow $\mathbf{x}^{(\alpha;\mathscr{D})}(t)$ at time t_m to the edge of $\mathscr{E}_{\sigma_r}^{(r)}$ is *from the attractive to repulsive* state in sense of the measuring surface of $\mathscr{M}^{(\alpha,\sigma_r)}$ if

$$\left.\begin{aligned}&\mathbf{n}^{\mathrm{T}}_{\mathscr{M}_m^{(\alpha;\sigma_r)}}(\mathbf{x}_{m-\varepsilon}^{(\alpha;\mathscr{M})})\cdot[\mathbf{x}_{m-\varepsilon}^{(\alpha;\mathscr{M})}-\mathbf{x}_{m-\varepsilon}^{(\alpha;\mathscr{D})}]<0,\\&\mathbf{n}^{\mathrm{T}}_{\mathscr{M}_m^{(\alpha;\sigma_r)}}(\mathbf{x}_{m+\varepsilon}^{(\alpha;\mathscr{M})})\cdot[\mathbf{x}_{m+\varepsilon}^{(\alpha;\mathscr{D})}-\mathbf{x}_{m+\varepsilon}^{(\alpha;\mathscr{M})}]>0,\\&0<\mathscr{C}_m^{(\alpha)}<\mathscr{C}_{m-\varepsilon}^{(\alpha)}\text{ and }0<\mathscr{C}_m^{(\alpha)}<\mathscr{C}_{m+\varepsilon}^{(\alpha)}.\end{aligned}\right\}\tag{6.185}$$

(iv) The flow $\mathbf{x}^{(\alpha;\mathscr{D})}(t)$ at time t_m to the edge of $\mathscr{E}_{\sigma_r}^{(r)}$ is *from the repulsive to attractive* state in sense of the measuring surface of $\mathscr{M}^{(\alpha,\sigma_r)}$ if

$$\left.\begin{aligned}&\mathbf{n}^{\mathrm{T}}_{\mathscr{M}_m^{(\alpha;\sigma_r)}}(\mathbf{x}_{m-\varepsilon}^{(\alpha;\mathscr{M})})\cdot[\mathbf{x}_{m-\varepsilon}^{(\alpha;\mathscr{M})}-\mathbf{x}_{m-\varepsilon}^{(\alpha;\mathscr{D})}]>0,\\&\mathbf{n}^{\mathrm{T}}_{\mathscr{M}_m^{(\alpha;\sigma_r)}}(\mathbf{x}_{m+\varepsilon}^{(\alpha;\mathscr{M})})\cdot[\mathbf{x}_{m+\varepsilon}^{(\alpha;\mathscr{D})}-\mathbf{x}_{m+\varepsilon}^{(\alpha;\mathscr{M})}]<0,\\&\mathscr{C}_m^{(\alpha,\sigma_r)}>\mathscr{C}_{m-\varepsilon}^{(\alpha,\sigma_r)}>0\text{ and }\mathscr{C}_m^{(\alpha,\sigma_r)}>\mathscr{C}_{m+\varepsilon}^{(\alpha,\sigma_r)}>0.\end{aligned}\right\}\tag{6.186}$$

(v) The flow $\mathbf{x}^{(\alpha;\mathscr{D})}(t)$ at time t_m to the edge of $\mathscr{E}_{\sigma_r}^{(r)}$ is *invariant* in sense of the measuring surface of $\mathscr{M}^{(\alpha,\sigma_r)}$ if

$$\begin{aligned}&\mathbf{x}_{m-\varepsilon}^{(\alpha;\mathscr{M})}=\mathbf{x}_{m-\varepsilon}^{(\alpha;\mathscr{D})}\text{ and }\mathbf{x}_{m+\varepsilon}^{(\alpha;\mathscr{D})}=\mathbf{x}_{m+\varepsilon}^{(\alpha;\mathscr{M})}\\&\mathscr{C}_{m-\varepsilon}^{(\alpha,\sigma_r)}=\mathscr{C}_m^{(\alpha,\sigma_r)}=\mathscr{C}_{m+\varepsilon}^{(\alpha,\sigma_r)}.\end{aligned}\tag{6.187}$$

Theorem 6.17. *For a discontinuous dynamical system in Eqs. (6.4)-(6.8), there is an edge* $\mathscr{E}_{\sigma_r}^{(r)}$ ($\sigma_r\in\{1,2,\cdots,N_r\}$ *and* $r\in\{0,1,2,\cdots,n-2\}$) *formed by the intersection of* $(n-r)$*-boundaries* $\partial\Omega_{\alpha i_\alpha}(\sigma_r)$ ($\alpha\in\mathscr{N}_{\mathscr{D}}(\sigma_r)\subset\mathscr{N}_{\mathscr{D}}=\{1,2,\cdots,N_d\}$ *and* $i_\alpha\in\mathscr{N}_{\alpha\mathscr{B}}(\sigma_r)\subset\mathscr{N}_{\mathscr{B}}=\{1,2,\cdots,N_b\}$) *in domain* $\Omega_\alpha(\sigma_r)$. *Suppose there is a measuring edge* $\mathscr{E}_{\sigma_r}^{(r,\alpha)}$ *of a flow* $\mathbf{x}^{(\alpha;\mathscr{D})}(t)$ *in domain* $\Omega_\alpha(\sigma_r)$ *with an equi-constant vector of the edge* $\mathscr{E}_{\sigma_r}^{(r)}$, *and the measuring edge and an equi-constant measuring surface* $\mathscr{M}^{(\alpha;\sigma_r)}$ *are determined by Eqs. (6.173) and (6.182), respectively. At time* t_m, $\mathbf{x}_m^{(\alpha;\mathscr{M})}=\mathbf{x}_\sigma^{(\alpha;\mathscr{M})}(t_m)\in\mathscr{M}_m^{(\alpha,\sigma_r)}$ *satisfies Eq. (6.134). For an arbitrarily small* $\varepsilon>0$, *there is a time interval* $[t_{m-\varepsilon},t_{m+\varepsilon}]$. *The flow* $\mathbf{x}^{(\alpha;\mathscr{D})}(t)$ *is* $C^{r_\alpha}_{[t_{m-\varepsilon},t_{m+\varepsilon}]}$*-continuous for time* t *and* $\|d^{r_\alpha+1}\mathbf{x}^{(\alpha;\mathscr{D})}/dt^{r_\alpha+1}\|<\infty$ ($r_\alpha\geq 2$). *Suppose* $\mathbf{x}_{m\pm\varepsilon}^{(\alpha;\mathscr{M})}\in\mathscr{M}_m^{(\alpha,\sigma_r)}$ *and* $\mathbf{x}_{m\pm\varepsilon}^{(\alpha;\mathscr{D})}=\mathbf{x}^{(\alpha;\mathscr{D})}(t_{m\pm\varepsilon})\in\mathscr{M}_{m\pm\varepsilon}^{(\alpha,\sigma_r)}$. $\mathscr{D}^{(\alpha;\sigma_r)}(t_m)\equiv\mathscr{C}_m^{(\alpha;\sigma_r)}$ *and* $\mathscr{D}^{(\alpha;\sigma_r)}(t_{m\pm\varepsilon})\equiv\mathscr{C}_{m\pm\varepsilon}^{(\alpha;\sigma_r)}$.

(i) *The flow* $\mathbf{x}^{(\alpha;\mathscr{D})}(t)$ *at time* t_m *to the edge of* $\mathscr{E}_{\sigma_r}^{(r)}$ *is attractive in sense of the measuring surface of* $\mathscr{M}^{(\alpha,\sigma_r)}$ *if and only if*

$$G^{(\alpha;\mathscr{D})}_{\mathscr{M}^{(\alpha;\sigma_r)}}(\mathbf{x}_m^{(\alpha;\mathscr{M})},t_m,\mathbf{p}_\alpha,\boldsymbol{\lambda})<0.\tag{6.188}$$

(ii) *The flow* $\mathbf{x}^{(\alpha;\mathscr{D})}(t)$ *at time* t_m *to the edge of* $\mathscr{E}_{\sigma_r}^{(r)}$ *is repulsive in sense of the measuring surface of* $\mathscr{M}^{(\alpha,\sigma_r)}$ *if and only if*

$$G_{\mathscr{M}^{(\alpha;\sigma_r)}}^{(\alpha;\mathscr{D})}(\mathbf{x}_m^{(\alpha;\mathscr{M})},t_m,\mathbf{p}_\alpha,\boldsymbol{\lambda})>0. \tag{6.189}$$

(iii) *The flow* $\mathbf{x}^{(\alpha;\mathscr{D})}(t)$ *at time* t_m *to the edge of* $\mathscr{E}_{\sigma_r}^{(r)}$ *is from the attractive to repulsive state in sense of the measuring surface of* $\mathscr{M}^{(\alpha,\sigma_r)}$ *if and only if*

$$\begin{aligned}&G_{\mathscr{M}^{(\alpha;\sigma_r)}}^{(\alpha;\mathscr{D})}(\mathbf{x}_m^{(\alpha;\mathscr{M})},t_m,\mathbf{p}_\alpha,\boldsymbol{\lambda})=0,\\&G_{\mathscr{M}^{(\alpha;\sigma_r)}}^{(1,\alpha;\mathscr{D})}(\mathbf{x}_m^{(\alpha;\mathscr{M})},t_m,\mathbf{p}_\alpha,\boldsymbol{\lambda})>0.\end{aligned} \tag{6.190}$$

(iv) *The flow* $\mathbf{x}^{(\alpha;\mathscr{D})}(t)$ *at time* t_m *to the edge of* $\mathscr{E}_{\sigma_r}^{(r)}$ *is from the repulsive to attractive state in sense of the measuring surface of* $\mathscr{M}^{(\alpha,\sigma_r)}$ *if and only if*

$$\begin{aligned}&G_{\mathscr{M}^{(\alpha;\sigma_r)}}^{(\alpha;\mathscr{D})}(\mathbf{x}_m^{(\alpha;\mathscr{M})},t_m,\mathbf{p}_\alpha,\boldsymbol{\lambda})=0,\\&G_{\mathscr{M}^{(\alpha;\sigma_r)}}^{(1,\alpha;\mathscr{D})}(\mathbf{x}_m^{(\alpha;\mathscr{M})},t_m,\mathbf{p}_\alpha,\boldsymbol{\lambda})<0.\end{aligned} \tag{6.191}$$

(v) *The flow* $\mathbf{x}^{(\alpha;\mathscr{D})}(t)$ *at time* t_m *to the edge of* $\mathscr{E}_{\sigma_r}^{(r)}$ *is invariant in sense of the measuring surface of* $\mathscr{M}^{(\alpha,\sigma_r)}$ *if and only if*

$$G_{\mathscr{M}^{(\alpha;\sigma_r)}}^{(s_\alpha,\alpha;\mathscr{D})}(\mathbf{x}_m^{(\alpha;\mathscr{M})},t_m,\mathbf{p}_\alpha,\boldsymbol{\lambda})=0 \;\; for\; s_\alpha=1,2,\cdots. \tag{6.192}$$

Proof. The measuring surface of $\mathscr{M}^{(\alpha,\sigma_r)}$ is as a boundary. As in Chapter 3, this theorem can be proved. ■

Definition 6.29. For a discontinuous dynamical system in Eqs. (6.4)-(6.8), there is an edge $\mathscr{E}_{\sigma_r}^{(r)}$ ($\sigma_r\in\{1,2,\cdots,N_r\}$ and $r\in\{0,1,2,\cdots,n-2\}$) formed by the intersection of $(n-r)$-boundaries $\partial\Omega_{\alpha i_\alpha}(\sigma_r)$ ($\alpha\in\mathscr{N}_{\mathscr{D}}(\sigma_r)\subset\mathscr{N}_{\mathscr{D}}=\{1,2,\cdots,N_d\}$ and $i_\alpha\in\mathscr{N}_{\alpha\mathscr{B}}(\sigma_r)\subset\mathscr{N}_{\mathscr{B}}=\{1,2,\cdots,N_b\}$) in domain $\Omega_\alpha(\sigma_r)$. Suppose there is a measuring edge $\mathscr{E}_{\sigma_r}^{(r,\alpha)}$ of a flow $\mathbf{x}^{(\alpha;\mathscr{D})}(t)$ in domain $\Omega_\alpha(\sigma_r)$ with an equi-constant vector of the edge $\mathscr{E}_{\sigma_r}^{(r)}$. The measuring edge and an equi-constant measuring surface $\mathscr{M}^{(\alpha;\sigma_r)}$ are determined by Eqs.(6.173) and (6.182), respectively. At time t_m, $\mathbf{x}_m^{(\alpha;\mathscr{M})}=\mathbf{x}_\sigma^{(\alpha;\mathscr{M})}(t_m)\in\mathscr{M}_m^{(\alpha,\sigma_r)}$ satisfies Eq.(6.134). For an arbitrarily small $\varepsilon>0$, there is a time interval $[t_{m-\varepsilon},t_{m+\varepsilon}]$. A flow $\mathbf{x}^{(\alpha;\mathscr{D})}(t)$ is $C_{[t_{m-\varepsilon},t_{m+\varepsilon}]}^{r_\alpha}$-continuous for time t and $\|d^{r_\alpha+1}\mathbf{x}^{(\alpha;\mathscr{D})}/dt^{r_\alpha+1}\|<\infty$ ($r_\alpha\geq 2k_\alpha-1\text{ o }2k_\alpha$). Suppose $\mathbf{x}_{m\pm\varepsilon}^{(\alpha;\mathscr{M})}\in\mathscr{M}_m^{(\alpha,\sigma_r)}$ and $\mathbf{x}_{m\pm\varepsilon}^{(\alpha;\mathscr{D})}\in\mathscr{M}_{m\pm\varepsilon}^{(\alpha,\sigma_r)}$. $\mathscr{D}^{(\alpha;\sigma_r)}(t_{m\pm\varepsilon})\equiv\mathscr{C}_{m\pm\varepsilon}^{(\alpha;\sigma_r)}$ and $\mathscr{C}_m^{(\alpha;\sigma_r)}\equiv\mathscr{D}^{(\alpha;\sigma_r)}(t_m)$.

(i) The flow $\mathbf{x}^{(\alpha;\mathscr{D})}(t)$ at time t_m to the edge of $\mathscr{E}_{\sigma_r}^{(r)}$ is *attractive* in sense of the measuring surface of $\mathscr{M}^{(\alpha,\sigma_r)}$ with the $2k_\alpha$-singularity if

$$\left.\begin{array}{l} G^{(s_\alpha,\alpha;\mathscr{D})}_{\mathscr{M}^{(\alpha;\sigma_r)}}(\mathbf{x}^{(\alpha;\mathscr{M})}_m, t_m, \mathbf{p}_\alpha, \boldsymbol{\lambda}) = 0 \\ \text{for } s_\alpha = 1,2,\cdots 2k_\alpha - 1; \\ \mathbf{n}^{\mathrm{T}}_{\mathscr{M}_m^{(\alpha;\sigma_r)}}(\mathbf{x}^{(\alpha;\mathscr{M})}_{m-\varepsilon})\cdot[\mathbf{x}^{(\alpha;\mathscr{M})}_{m-\varepsilon} - \mathbf{x}^{(\alpha;\mathscr{D})}_{m-\varepsilon}] < 0, \\ \mathbf{n}^{\mathrm{T}}_{\mathscr{M}_m^{(\alpha;\sigma_r)}}(\mathbf{x}^{(\alpha;\mathscr{M})}_{m+\varepsilon})\cdot[\mathbf{x}^{(\alpha;\mathscr{D})}_{m+\varepsilon} - \mathbf{x}^{(\alpha;\mathscr{M})}_{m+\varepsilon}] < 0, \\ \mathscr{C}^{(\alpha,\sigma_r)}_{m-\varepsilon} > \mathscr{C}^{(\alpha,\sigma_r)}_{m} > \mathscr{C}^{(\alpha,\sigma_r)}_{m+\varepsilon} > 0. \end{array}\right\} \tag{6.193}$$

(ii) The flow $\mathbf{x}^{(\alpha;\mathscr{D})}(t)$ at time t_m to the edge of $\mathscr{E}^{(r)}_{\sigma_r}$ is *repulsive* in sense of the measuring surface of $\mathscr{M}^{(\alpha,\sigma_r)}$ with the $2k_\alpha$-singularity if

$$\left.\begin{array}{l} G^{(s_\alpha,\alpha;\mathscr{D})}_{\mathscr{M}^{(\alpha;\sigma_r)}}(\mathbf{x}^{(\alpha;\mathscr{M})}_m, t_m, \mathbf{p}_\alpha, \boldsymbol{\lambda}) = 0 \\ \text{for } s_\alpha = 1,2,\cdots 2k_\alpha - 1; \\ \mathbf{n}^{\mathrm{T}}_{\mathscr{M}_m^{(\alpha;\sigma_r)}}(\mathbf{x}^{(\alpha;\mathscr{M})}_{m-\varepsilon})\cdot[\mathbf{x}^{(\alpha;\mathscr{M})}_{m-\varepsilon} - \mathbf{x}^{(\alpha;\mathscr{D})}_{m-\varepsilon}] > 0, \\ \mathbf{n}^{\mathrm{T}}_{\mathscr{M}_m^{(\alpha;\sigma_r)}}(\mathbf{x}^{(\alpha;\mathscr{M})}_{m+\varepsilon})\cdot[\mathbf{x}^{(\alpha;\mathscr{D})}_{m+\varepsilon} - \mathbf{x}^{(\alpha;\mathscr{M})}_{m+\varepsilon}] > 0, \\ 0 < \mathscr{C}^{(\alpha,\sigma_r)}_{m-\varepsilon} < \mathscr{C}^{(\alpha,\sigma_r)}_{m} < \mathscr{C}^{(\alpha,\sigma_r)}_{m+\varepsilon}. \end{array}\right\} \tag{6.194}$$

(iii) The flow $\mathbf{x}^{(\alpha;\mathscr{D})}(t)$ at time t_m to the edge of $\mathscr{E}^{(r)}_{\sigma_r}$ is *from the attractive to repulsive* state in sense of the measuring surface of $\mathscr{M}^{(\alpha,\sigma_r)}$ with the $(2k_\alpha+1)$-singularity if

$$\left.\begin{array}{l} G^{(s_\alpha,\alpha;\mathscr{D})}_{\mathscr{M}^{(\alpha;\sigma_r)}}(\mathbf{x}^{(\alpha;\mathscr{M})}_m, t_m, \mathbf{p}_\alpha, \boldsymbol{\lambda}) = 0 \text{ for } s_\alpha = 1,2,\cdots,2k_\alpha; \\ \mathbf{n}^{\mathrm{T}}_{\mathscr{M}_m^{(\alpha;\sigma_r)}}(\mathbf{x}^{(\alpha;\mathscr{M})}_{m-\varepsilon})\cdot[\mathbf{x}^{(\alpha;\mathscr{M})}_{m-\varepsilon} - \mathbf{x}^{(\alpha;\mathscr{D})}_{m-\varepsilon}] < 0, \\ \mathbf{n}^{\mathrm{T}}_{\mathscr{M}_m^{(\alpha;\sigma_r)}}(\mathbf{x}^{(\alpha;\mathscr{M})}_{m+\varepsilon})\cdot[\mathbf{x}^{(\alpha;\mathscr{D})}_{m+\varepsilon} - \mathbf{x}^{(\alpha;\mathscr{M})}_{m+\varepsilon}] > 0, \\ 0 < \mathscr{C}^{(\alpha,\sigma_r)}_{m} < \mathscr{C}^{(\alpha,\sigma_r)}_{m-\varepsilon} \text{ and } 0 < \mathscr{C}^{(\alpha,\sigma_r)}_{m} < \mathscr{C}^{(\alpha,\sigma_r)}_{m+\varepsilon}. \end{array}\right\} \tag{6.195}$$

(iv) The flow $\mathbf{x}^{(\alpha;\mathscr{D})}(t)$ at time t_m to the edge of $\mathscr{E}^{(r)}_{\sigma_r}$ is *from the repulsive to attractive* state in sense of the measuring surface of $\mathscr{M}^{(\alpha,\sigma_r)}$ with the $(2k_\alpha+1)$-singularity if

$$\left.\begin{array}{l} G^{(s_\alpha,\alpha;\mathscr{D})}_{\mathscr{M}^{(\alpha;\sigma_r)}}(\mathbf{x}^{(\alpha;\mathscr{M})}_m, t_m, \mathbf{p}_\alpha, \boldsymbol{\lambda}) = 0 \text{ for } s_\alpha = 1,2,\cdots,2k_\alpha; \\ \mathbf{n}^{\mathrm{T}}_{\mathscr{M}_m^{(\alpha;\sigma_r)}}(\mathbf{x}^{(\alpha;\mathscr{M})}_{m-\varepsilon})\cdot[\mathbf{x}^{(\alpha;\mathscr{M})}_{m-\varepsilon} - \mathbf{x}^{(\alpha;\mathscr{D})}_{m-\varepsilon}] > 0, \\ \mathbf{n}^{\mathrm{T}}_{\mathscr{M}_m^{(\alpha;\sigma_r)}}(\mathbf{x}^{(\alpha;\mathscr{M})}_{m+\varepsilon})\cdot[\mathbf{x}^{(\alpha;\mathscr{D})}_{m+\varepsilon} - \mathbf{x}^{(\alpha;\mathscr{M})}_{m+\varepsilon}] < 0, \\ \mathscr{C}^{(\alpha,\sigma_r)}_{m} > \mathscr{C}^{(\alpha,\sigma_r)}_{m-\varepsilon} > 0 \text{ and } \mathscr{C}^{(\alpha,\sigma_r)}_{m} > \mathscr{C}^{(\alpha,\sigma_r)}_{m+\varepsilon} > 0. \end{array}\right\} \tag{6.196}$$

Theorem 6.18. *For a discontinuous dynamical system in Eqs.* (6.4)-(6.8), *there is* *an edge* $\mathscr{E}_{\sigma_r}^{(r)}$ ($\sigma_r \in \{1,2,\cdots,N_r\}$ *and* $r \in \{0,1,2,\cdots,n-2\}$) *formed by the intersection of* $(n-r)$*-boundaries* $\partial\Omega_{\alpha i_\alpha}(\sigma_r)$ ($\alpha \in \mathscr{N}_{\mathscr{D}}(\sigma_r) \subset \mathscr{N}_{\mathscr{D}} = \{1,2,\cdots,N_d\}$ *and* $i_\alpha \in \mathscr{N}_{\alpha\mathscr{B}}(\sigma_r) \subset \mathscr{N}_{\mathscr{B}} = \{1,2,\cdots,N_b\}$) *in domain* $\Omega_\alpha(\sigma_r)$. *Suppose there is a measuring edge* $\mathscr{E}_{\sigma_r}^{(r,\alpha)}$ *of a flow* $\mathbf{x}^{(\alpha;\mathscr{D})}(t)$ *in domain* $\Omega_\alpha(\sigma_r)$ *with an equi-constant vector of the edge* $\mathscr{E}_{\sigma_r}^{(r)}$, *and the measuring edge and an equi-constant measuring surface* $\mathscr{M}^{(\alpha;\sigma_r)}$ *are determined by Eqs.* (6.173) *and* (6.182), *respectively. At time* t_m, $\mathbf{x}_m^{(\alpha;\mathscr{M})} = \mathbf{x}_\sigma^{(\alpha;\mathscr{M})}(t_m) \in \mathscr{M}_m^{(\alpha,\sigma_r)}$ *satisfies Eq.* (6.134). *For an arbitrarily small* $\varepsilon > 0$, *there is a time interval* $[t_{m-\varepsilon}, t_{m+\varepsilon}]$. *A flow* $\mathbf{x}^{(\alpha;\mathscr{D})}(t)$ *is* $C_{[t_{m-\varepsilon},t_{m+\varepsilon}]}^{r_\alpha}$*-continuous for time* t *and* $\| d^{r_\alpha+1}\mathbf{x}^{(\alpha;\mathscr{D})}/dt^{r_\alpha+1} \| < \infty$ ($r_\alpha \geq 2k_\alpha$ *or* $2k_\alpha+1$). *Suppose* $\mathbf{x}_{m\pm\varepsilon}^{(\alpha;\mathscr{M})} \in \mathscr{M}_m^{(\alpha,\sigma_r)}$ *and* $\mathbf{x}_{m\pm\varepsilon}^{(\alpha;\mathscr{D})} = \mathbf{x}^{(\alpha;\mathscr{D})}(t_{m\pm\varepsilon}) \in \mathscr{M}_{m\pm\varepsilon}^{(\alpha,\sigma_r)}$. $\mathscr{D}^{(\alpha;\sigma_r)}(t_{m\pm\varepsilon}) \equiv \mathscr{C}_{m\pm\varepsilon}^{(\alpha;\sigma_r)}$ *and* $\mathscr{C}_m^{(\alpha;\sigma_r)} \equiv \mathscr{D}^{(\alpha;\sigma_r)}(t_m)$.

(i) *The flow* $\mathbf{x}^{(\alpha;\mathscr{D})}(t)$ *at time* t_m *to the edge of* $\mathscr{E}_{\sigma_r}^{(r)}$ *is attractive in sense of the measuring surface of* $\mathscr{M}^{(\alpha,\sigma_r)}$ *with the* $2k_\alpha$*-singularity if and only if*

$$\left.\begin{array}{l} G_{\mathscr{M}^{(\alpha;\sigma_r)}}^{(s_\alpha,\alpha;\mathscr{D})}(\mathbf{x}_m^{(\alpha;\mathscr{M})}, t_m, \mathbf{p}_\alpha, \boldsymbol{\lambda}) = 0 \\ \text{for } s_\alpha = 1,2,\cdots,2k_\alpha-1; \\ G_{\mathscr{M}^{(\alpha;\sigma_r)}}^{(2k_\alpha,\alpha;\mathscr{D})}(\mathbf{x}_m^{(\alpha;\mathscr{M})}, t_m, \mathbf{p}_\alpha, \boldsymbol{\lambda}) < 0. \end{array}\right\} \tag{6.197}$$

(ii) *The flow* $\mathbf{x}^{(\alpha;\mathscr{D})}(t)$ *at time* t_m *to the edge of* $\mathscr{E}_{\sigma_r}^{(r)}$ *is repulsive in sense of the measuring surface of* $\mathscr{M}^{(\alpha,\sigma_r)}$ *with the* $2k_\alpha$*-singularity if and only if*

$$\left.\begin{array}{l} G_{\mathscr{M}^{(\alpha;\sigma_r)}}^{(s_\alpha,\alpha;\mathscr{D})}(\mathbf{x}_m^{(\alpha;\mathscr{M})}, t_m, \mathbf{p}_\alpha, \boldsymbol{\lambda}) = 0 \\ \text{for } s_\alpha = 1,2,\cdots,2k_\alpha-1; \\ G_{\mathscr{M}^{(\alpha;\sigma_r)}}^{(2k_\alpha,\alpha;\mathscr{D})}(\mathbf{x}_m^{(\alpha;\mathscr{M})}, t_m, \mathbf{p}_\alpha, \boldsymbol{\lambda}) > 0. \end{array}\right\} \tag{6.198}$$

(iii) *The flow* $\mathbf{x}^{(\alpha;\mathscr{D})}(t)$ *at time* t_m *to the edge of* $\mathscr{E}_{\sigma_r}^{(r)}$ *is from the attractive to repulsive state in sense of the measuring surface of* $\mathscr{M}^{(\alpha,\sigma_r)}$ *with the* $(2k_\alpha+1)$*-singularity if and only if*

$$\left.\begin{array}{l} G_{\mathscr{M}^{(\alpha;\sigma_r)}}^{(s_\alpha,\alpha;\mathscr{D})}(\mathbf{x}_m^{(\alpha;\mathscr{M})}, t_m, \mathbf{p}_\alpha, \boldsymbol{\lambda}) = 0 \\ \text{for } s_\alpha = 1,2,\cdots,2k_\alpha; \\ G_{\mathscr{M}^{(\alpha;\sigma_r)}}^{(2k_\alpha+1,\alpha;\mathscr{D})}(\mathbf{x}_m^{(\alpha;\mathscr{M})}, t_m, \mathbf{p}_\alpha, \boldsymbol{\lambda}) > 0. \end{array}\right\} \tag{6.199}$$

(iv) *The flow* $\mathbf{x}^{(\alpha;\mathscr{D})}(t)$ *at time* t_m *to the edge of* $\mathscr{E}_{\sigma_r}^{(r)}$ *is from the repulsive to attractive state in sense of the measuring surface of* $\mathscr{M}^{(\alpha,\sigma_r)}$ *with the* $(2k_\alpha+1)$*-singularity if and only if*

$$\left.\begin{aligned}&G^{(s_\alpha,\alpha;\mathscr{D})}_{\mathscr{M}(\alpha;\sigma_r)}(\mathbf{x}^{(\alpha;\mathscr{M})}_m,t_m,\mathbf{p}_\alpha,\boldsymbol{\lambda})=0\\&for\ s_\alpha=1,2,\cdots,2k_\alpha;\\&G^{(2k_\alpha+1,\alpha;\mathscr{D})}_{\mathscr{M}(\alpha;\sigma_r)}(\mathbf{x}^{(\alpha;\mathscr{M})}_m,t_m,\mathbf{p}_\alpha,\boldsymbol{\lambda})<0.\end{aligned}\right\}\tag{6.200}$$

Proof. The measuring surface of $\mathscr{D}^{(\alpha)}$ is as a boundary. As in Chapter 3, this theorem can be proved. ■

6.7. Multi-valued vector fields switching at edges

The multi-valued vector fields on domains to edges can be defined as in Section 5.6. If a vector field defined on a domain can be extended continuously to other domains via the edge instead of the boundary, such an extended vector field generates an imaginary flow in the corresponding domains. Equation (5.114) can be rewritten for real and imaginary domain flows via edges, i.e.,

$$\begin{aligned}&\dot{\mathbf{x}}^{(\alpha;\mathscr{D})}_\alpha=\mathbf{F}^{(\alpha;\mathscr{D})}(\mathbf{x}^{(\alpha;\mathscr{D})}_\alpha,t,\mathbf{p}_\alpha)\ \text{ on }\Omega_\alpha,\\&\dot{\mathbf{x}}^{(\beta;\mathscr{D})}_\alpha=\mathbf{F}^{(\beta;\mathscr{D})}(\mathbf{x}^{(\beta;\mathscr{D})}_\alpha,t,\mathbf{p}_\beta)\ \text{ on }\Omega_\alpha\\&\text{from }\Omega_\beta\ (\beta=\beta_1,\beta_2,\cdots,\beta_k)\end{aligned}\tag{6.201}$$

where $\mathbf{x}^{(\alpha;\mathscr{D})}_\alpha$ and $\mathbf{x}^{(\beta;\mathscr{D})}_\alpha$ are real and imaginary domain flows to edges, respectively. In Fig. 6.20, imaginary domain flows to the $(n-2)$-dimensional edge are extensions of the real domain flows. The black curve is the $(n-2)$-dimensional edge. The domain flows are depicted by thick curves with arrows in the domains.

Consider the switching rule as the simplest transport law, i.e.,

$$\mathbf{x}^{(\alpha;\mathscr{D})}_\alpha(t_{m-})=\mathbf{x}^{(\beta;\mathscr{D})}_\beta(t_{m+})\ \text{or}\ \mathbf{x}^{(\alpha;\mathscr{D})}_\alpha(t_{m-})=\mathbf{x}^{(\beta;\mathscr{D})}_\alpha(t_{m-}).\tag{6.202}$$

To make the imaginary flow exist, Axioms 5.1 and 5.2 can be extended to the domain flows through edges and vertexes, i.e.,

Axiom 6.1(Continuation axiom). For a discontinuous dynamical system, any domain flows coming to (or leaving from) a convex edge can be, forwardly (or backwardly) extendible to the adjacent accessible domain without any switching rules, transport laws and any flow barriers at the edge.

Axiom 6.2(Non-extendibility axiom). For a discontinuous dynamical system, any domain flows coming to (or leaving from) a convex edge cannot forwardly (or backwardly) extend to any inaccessible domain or edges. Even if such domain flows are extendible to the inaccessible domains, then the corresponding extended flows are fictitious only.

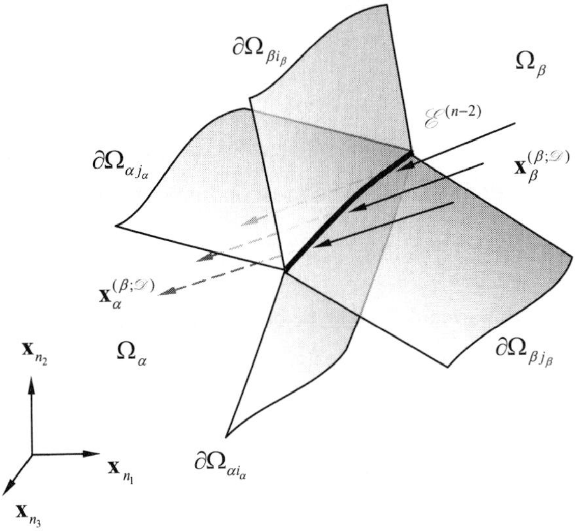

(a)

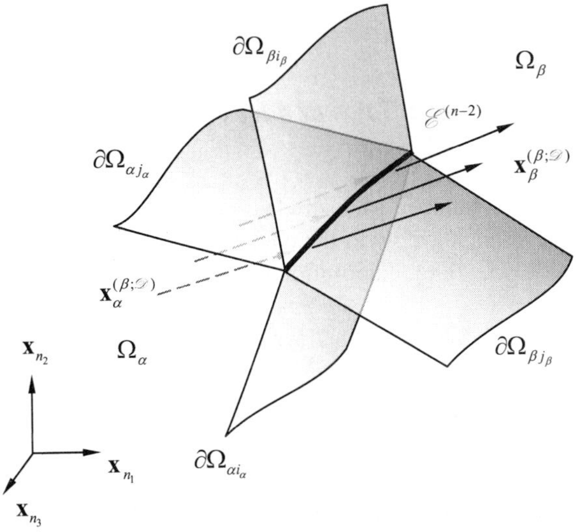

(b)

Fig. 6.20 Real and imaginary domain flows to the $(n-2)$-dimensional edge : **(a)** coming real flow, **(b)** leaving real flow. The black curve represents the $(n-2)$-dimensional edge. The domain flows are represented by the thick curves with arrows in the domain. ($n_1+n_2+n_3=n$)

As in Definition 5.53, the multi-valued vector fields on an accessible sub-domain in discontinuous dynamical systems can be re-stated, and the bouncing flow can be discussed.

Definition 6.30. For a discontinuous system, there is an accessible domain Ω_α ($\alpha \in \mathscr{N}_{\mathscr{D}} = \{1,2,\cdots,N_d\}$) on which k_α-vector fields of $\mathbf{F}^{(\alpha_k;\mathscr{D})}(\mathbf{x}^{(\alpha_k;\mathscr{D})}, t, \mathbf{p}_{\alpha_k})$ ($k = 1,2,\cdots,k_\alpha$) are defined, and the corresponding dynamical system is given by

$$\dot{\mathbf{x}}^{(\alpha_k;\mathscr{D})} = \mathbf{F}^{(\alpha_k;\mathscr{D})}(\mathbf{x}^{(\alpha_k;\mathscr{D})}, t, \mathbf{p}_{\alpha_k}). \tag{6.203}$$

The discontinuous dynamical system in domain Ω_α is called to be *of the multi-valued vector fields*. A set of dynamical systems in domain Ω_α is defined as

$$\begin{aligned} &\mathscr{D}_\alpha = \cup_{k=1}^{k_\alpha} \mathscr{D}_{\alpha_k}, \\ &\mathscr{D}_{\alpha_k} = \left\{ \dot{\mathbf{x}}^{(\alpha_k;\mathscr{D})} = \mathbf{F}^{(\alpha_k;\mathscr{D})}(\mathbf{x}^{(\alpha_k;\mathscr{D})}, t, \mathbf{p}_{\alpha_k}) \big| k \in \{1,2,\cdots,k_\alpha\} \right\}. \end{aligned} \tag{6.204}$$

A total set of dynamical systems in the discontinuous dynamical system is

$$\mathscr{D} = \cup_{\alpha \in \mathscr{N}_{\mathscr{D}}} \mathscr{D}_\alpha. \tag{6.205}$$

6.7.1. Bouncing domain flows at edges

If multi-valued vector fields in each single domain exist, with a switching rule or transport law, the bouncing flow to the edge will exist as for the boundary. As in found from Luo (2005, 2006), a definition for the bouncing flow is given herein.

Definition 6.31. For a discontinuous system with multi-valued vector fields in Eq. (6.205), there is a point $\mathbf{x}^{(\sigma_r;\mathscr{E})}(t_m) \equiv \mathbf{x}_m \in \mathscr{E}_{\sigma_r}^{(r)}$ ($\sigma_r \in \mathscr{N}_{\mathscr{E}}^r = \{1,2,\cdots,N_r\}$ and $r \in \{0,1,2,\cdots,n-1\}$) at time t_m with the related boundaries $\partial\Omega_{\alpha i_\alpha}(\sigma_r)$ and related domains $\Omega_\alpha(\sigma_r)$ ($\alpha \in \mathscr{N}_{\mathscr{D}}(\sigma_r) \subset \mathscr{N}_{\mathscr{D}} = \{1,2,\cdots,N_d\}$ and $i_\alpha \in \mathscr{N}_{\alpha\mathscr{B}}(\sigma_r) \subset \mathscr{N}_{\mathscr{B}} = \{1,2,\cdots,N_b\}$), where the boundary index subset of $\mathscr{N}_{\alpha\mathscr{B}}(\sigma_r)$ has n_{σ_r}-elements ($n_{\sigma_r} \geq n-r$). For an arbitrarily small $\varepsilon > 0$, there are two time intervals $[t_{m-\varepsilon}, t_m)$ and $(t_m, t_{m+\varepsilon}]$. If a coming flow of $\mathbf{x}^{(\alpha_k;\mathscr{D})}$ arrives to the edge of $\mathscr{E}_{\sigma_r}^{(r)}$, there is a switching rule between two flows $\mathbf{x}^{(\alpha_k;\mathscr{D})}$ and $\mathbf{x}^{(\alpha_l;\mathscr{D})}$ with $\mathbf{x}^{(\alpha_k;\mathscr{D})}(t_{m-}) = \mathbf{x}_m = \mathbf{x}^{(\alpha_l;\mathscr{D})}(t_{m+})$. The flows of $\mathbf{x}^{(\alpha_k;\mathscr{D})}$ and $\mathbf{x}^{(\alpha_l;\mathscr{D})}$ to the edge of $\mathscr{E}_{\sigma_r}^{(r)}$ are called a *bouncing* flow in Ω_α if

$$\left.\begin{aligned}&\hbar_{i_\alpha}\mathbf{n}^{\mathrm{T}}_{\partial\Omega_{\alpha i_\alpha}}(\mathbf{x}^{(i_\alpha;\mathscr{B})}_{m-\varepsilon})\cdot[\mathbf{x}^{(i_\alpha;\mathscr{B})}_{m-\varepsilon}-\mathbf{x}^{(\alpha_k;\mathscr{D})}_{m-\varepsilon}]<0\\&\hbar_{i_\alpha}\mathbf{n}^{\mathrm{T}}_{\partial\Omega_{\alpha i_\alpha}}(\mathbf{x}^{(i_\alpha;\mathscr{B})}_{m+\varepsilon})\cdot[\mathbf{x}^{(\alpha_l;\mathscr{D})}_{m+\varepsilon}-\mathbf{x}^{(i_\alpha;\mathscr{B})}_{m+\varepsilon}]>0\end{aligned}\right\}\tag{6.206}$$

for all $i_\alpha\in\mathscr{N}_{\alpha\mathscr{B}}(\sigma_r)$.

Using the G-function to the boundary, the foregoing definition of the bouncing flow to the edge yields the corresponding theorem.

Theorem 6.19. *For a discontinuous system with multi-valued vector fields in Eq. (6.205), there is a point* $\mathbf{x}^{(\sigma_r;\mathscr{E})}(t_m)\equiv\mathbf{x}_m\in\mathscr{E}^{(r)}_{\sigma_r}$ ($\sigma_r\in\mathscr{N}^{\,r}_{\mathscr{E}}=\{1,2,\cdots,N_r\}$ *and* $r\in\{0,1,2,\cdots,n-1\}$) *at time* t_m *with the related boundaries* $\partial\Omega_{\alpha i_\alpha}(\sigma_r)$ *and related domains* $\Omega_\alpha(\sigma_r)(\alpha\in\mathscr{N}_{\mathscr{D}}(\sigma_r)\subset\mathscr{N}_{\mathscr{D}}=\{1,2,\cdots,N_d\}$ *and* $i_\alpha\in\mathscr{N}_{\alpha\mathscr{B}}(\sigma_r)\subset\mathscr{N}_{\mathscr{B}}=\{1,2,\cdots,N_b\}$), *where the boundary index subset of* $\mathscr{N}_{\alpha\mathscr{B}}(\sigma_r)$ *has* n_{σ_r} *elements* ($n_{\sigma_r}\geq n-r$). *For an arbitrarily small* $\varepsilon>0$, *there are two time intervals* $[t_{m-\varepsilon},t_m)$ *and* $(t_m,t_{m+\varepsilon}]$. *If a coming flow of* $\mathbf{x}^{(\alpha_k;\mathscr{D})}$ *arrives to the edge of* $\mathscr{E}^{(r)}_{\sigma_r}$, *there is a switching rule between two flows* $\mathbf{x}^{(\alpha_k;\mathscr{D})}$ *and* $\mathbf{x}^{(\alpha_l;\mathscr{D})}$ *with* $\mathbf{x}^{(\alpha_k;\mathscr{D})}(t_{m-})=\mathbf{x}_m=\mathbf{x}^{(\alpha_l;\mathscr{D})}(t_{m+})$. *A flow* $\mathbf{x}^{(\alpha_k;\mathscr{D})}(t)$ *is* $C^{r_{\alpha_k}}_{[t_{m-\varepsilon},t_m)}$*-continuous* ($r_{\alpha_k}\geq 2$) *for time* t *with* $\|d^{r_{\alpha_k}+1}\mathbf{x}^{(\alpha_k;\mathscr{D})}/dt^{r_{\alpha_k}+1}\|<\infty$, *and a flow* $\mathbf{x}^{(\alpha_l;\mathscr{D})}(t)$ *is* $C^{r_{\alpha_l}}_{(t_m,t_{m+\varepsilon}]}$ *-continuous* ($r_{\alpha_k}\geq 2$) *for time* t *with* $\|d^{r_{\alpha_l}+1}\mathbf{x}^{(\alpha_l;\mathscr{D})}/dt^{r_{\alpha_l}+1}\|<\infty$. *The flows of* $\mathbf{x}^{(\alpha_k;\mathscr{D})}$ *and* $\mathbf{x}^{(\alpha_l;\mathscr{D})}$ *to the edge* $\mathscr{E}^{(r)}_{\sigma_r}$ *form a bouncing flow in* Ω_α *if and only if*

$$\left.\begin{aligned}&\hbar_{i_\alpha}G^{(\alpha_k;\mathscr{D})}_{\partial\Omega_{\alpha i_\alpha}}(\mathbf{x}_m,t_{m-},\mathbf{p}_{\alpha_k},\boldsymbol{\lambda}_{i_\alpha})<0,\\&\hbar_{i_\alpha}G^{(\alpha_l;\mathscr{D})}_{\partial\Omega_{\alpha i_\alpha}}(\mathbf{x}_m,t_{m+},\mathbf{p}_{\alpha_l},\boldsymbol{\lambda}_{i_\alpha})>0.\end{aligned}\right\}\tag{6.207}$$

Proof. Following the proof of Theorem 3.1, the theorem can be proved. ■

The bouncing flow with singularity will be stated as follows.

Definition 6.32. For a discontinuous system with multi-valued vector fields in Eq. (6.205), there is a point $\mathbf{x}^{(\sigma_r;\mathscr{E})}(t_m)\equiv\mathbf{x}_m\in\mathscr{E}^{(r)}_{\sigma_r}$ ($\sigma_r\in\mathscr{N}^{\,r}_{\mathscr{E}}=\{1,2,\cdots,N_r\}$ and $r\in\{0,1,2,\cdots,n-1\}$) at time t_m with the related boundaries $\partial\Omega_{\alpha i_\alpha}(\sigma_r)$ and related domains $\Omega_\alpha(\sigma_r)$ ($\alpha\in\mathscr{N}_{\mathscr{D}}(\sigma_r)\subset\mathscr{N}_{\mathscr{D}}=\{1,2,\cdots,N_d\}$ and $i_\alpha\in\mathscr{N}_{\alpha\mathscr{B}}(\sigma_r)\subset\mathscr{N}_{\mathscr{B}}=\{1,2,\cdots,N_b\}$), where the boundary index subset of $\mathscr{N}_{\alpha\mathscr{B}}(\sigma_r)$ has n_{σ_r} -elements ($n_{\sigma_r}\geq(n-r)$). For an arbitrarily small $\varepsilon>0$, there are two time intervals $[t_{m-\varepsilon},t_m)$ and $(t_m,t_{m+\varepsilon}]$. If a coming flow of $\mathbf{x}^{(\alpha_k;\mathscr{D})}$ arrives to the edge of

$\mathscr{E}_{\sigma_r}^{(r)}$, there is a switching rule between two flows $\mathbf{x}^{(\alpha_k;\mathscr{D})}$ and $\mathbf{x}^{(\alpha_l;\mathscr{D})}$ with $\mathbf{x}^{(\alpha_k;\mathscr{D})}(t_{m-}) = \mathbf{x}_m = \mathbf{x}^{(\alpha_l;\mathscr{D})}(t_{m+})$. A flow $\mathbf{x}^{(\alpha_k;\mathscr{D})}(t)$ is $C_{[t_{m-\varepsilon},t_m)}^{r_{\alpha_k}}$-continuous ($r_{\alpha_k} \geq m_{\alpha_k}+1$) for time t with $\| d^{r_{\alpha_k}+1}\mathbf{x}^{(\alpha_k;\mathscr{D})}/dt^{r_{\alpha_k}+1}\| < \infty$, and a flow $\mathbf{x}^{(\alpha_l;\mathscr{D})}(t)$ is $C_{(t_m,t_{m+\varepsilon}]}^{r_{\alpha_l}}$ continuous ($r_{\alpha_k} \geq m_{\alpha_l}+1$) for time t with $\| d^{r_{\alpha_l}+1}\mathbf{x}^{(\alpha_l;\mathscr{D})}/dt^{r_{\alpha_l}+1}\| < \infty$. The flows of $\mathbf{x}^{(\alpha_k;\mathscr{D})}(t)$ and $\mathbf{x}^{(\alpha_l;\mathscr{D})}(t)$ to the edge of $\mathscr{E}_{\sigma_r}^{(r)}$ are called an $(\mathbf{m}_{\alpha_k}:\mathbf{m}_{\alpha_l})$-*bouncing flow* in Ω_α if

$$\left.\begin{aligned}
&\hbar_{i_\alpha} G_{\partial\Omega_{\alpha\beta}}^{(s_{\alpha_k},\alpha_k;\mathscr{D})}(\mathbf{x}_m,t_{m-},\mathbf{p}_{\alpha_k},\boldsymbol{\lambda}_{i_\alpha}) = 0\\
&\text{for } s_{\alpha_k} = 0,1,2,\cdots,m_{\alpha_k}-1,\\
&\hbar_{i_\alpha}\mathbf{n}_{\partial\Omega_{\alpha i_\alpha}}^{\mathrm{T}}(\mathbf{x}_{m-\varepsilon}^{(i_\alpha;\mathscr{B})})\cdot[\mathbf{x}_{m-\varepsilon}^{(i_\alpha;\mathscr{B})} - \mathbf{x}_{m-\varepsilon}^{(\alpha_k;\mathscr{D})}] < 0;
\end{aligned}\right\} \tag{6.208}$$

$$\left.\begin{aligned}
&\hbar_{i_\alpha} G_{\partial\Omega_{\alpha i_\alpha}}^{(s_{\alpha_l},\alpha_l;\mathscr{D})}(\mathbf{x}_m,t_{m+},\mathbf{p}_{\alpha_l},\boldsymbol{\lambda}_{i_\alpha}) = 0\\
&\text{for } s_{\alpha_l} = 0,1,2,\cdots,m_{\alpha_l}-1,\\
&\hbar_{i_\alpha}\mathbf{n}_{\partial\Omega_{\alpha i_\alpha}}^{\mathrm{T}}(\mathbf{x}_{m+\varepsilon}^{(i_\alpha;\mathscr{B})})\cdot[\mathbf{x}_{m+\varepsilon}^{(\alpha_k;\mathscr{D})} - \mathbf{x}_{m+\varepsilon}^{(i_\alpha;\mathscr{B})}] > 0
\end{aligned}\right\} \tag{6.209}$$

$$\text{for all } i_\alpha \in \mathscr{N}_{\alpha\mathscr{B}}(\sigma_r).$$

Theorem 6.20. *For a discontinuous system with multi-valued vector fields in Eq. (6.205), there is a point* $\mathbf{x}^{(\sigma_r;\mathscr{E})}(t_m) \equiv \mathbf{x}_m \in \mathscr{E}_{\sigma_r}^{(r)}$ ($\sigma_r \in \mathscr{N}_{\mathscr{E}}^{r} = \{1,2,\cdots,N_r\}$ *and* $r \in \{0,1,2,\cdots,n-1\}$) *at time* t_m *with the related boundaries* $\partial\Omega_{\alpha i_\alpha}(\sigma_r)$ *and related domains* $\Omega_\alpha(\sigma_r)$ ($\alpha \in \mathscr{N}_{\mathscr{D}}(\sigma_r) \subset \mathscr{N}_{\mathscr{D}} = \{1,2,\cdots,N_d\}$ *and* $i_\alpha \in \mathscr{N}_{\alpha\mathscr{B}}(\sigma_r) \subset \mathscr{N}_{\mathscr{B}} = \{1,2,\cdots,N_b\}$), *where the boundary index subset of* $\mathscr{N}_{\alpha\mathscr{B}}(\sigma_r)$ *has* n_{σ_r} *elements* ($n_{\sigma_r} \geq n-r$). *For an arbitrarily small* $\varepsilon > 0$, *there are two time intervals* $[t_{m-\varepsilon},t_m)$ *and* $(t_m,t_{m+\varepsilon}]$. *If a coming flow of* $\mathbf{x}^{(\alpha_k;\mathscr{D})}$ *arrives to edge of* $\mathscr{E}_{\sigma_r}^{(r)}$, *there is a switching rule between two flows* $\mathbf{x}^{(\alpha_k;\mathscr{D})}$ *and* $\mathbf{x}^{(\alpha_l;\mathscr{D})}$ *with* $\mathbf{x}^{(\alpha_k;\mathscr{D})}(t_{m-}) = \mathbf{x}_m = \mathbf{x}^{(\alpha_l;\mathscr{D})}(t_{m+})$. *A flow* $\mathbf{x}^{(\alpha_k;\mathscr{D})}(t)$ *is* $C_{[t_{m-\varepsilon},t_m)}^{r_{\alpha_k}}$*-continuous* ($r_{\alpha_k} \geq m_{\alpha_k}+1$) *for time* t *with* $\| d^{r_{\alpha_k}+1}\mathbf{x}^{(\alpha_k;\mathscr{D})}/dt^{r_{\alpha_k}+1}\| < \infty$, *and a flow* $\mathbf{x}^{(\alpha_l;\mathscr{D})}(t)$ *is* $C_{(t_m,t_{m+\varepsilon}]}^{r_{\alpha_l}}$ *continuous* ($r_{\alpha_k} \geq m_{\alpha_l}+1$) *for time* t *with* $\| d^{r_{\alpha_l}+1}\mathbf{x}^{(\alpha_l;\mathscr{D})}/dt^{r_{\alpha_l}+1}\| < \infty$. *The flows of* $\mathbf{x}^{(\alpha_k;\mathscr{D})}(t)$ *and* $\mathbf{x}^{(\alpha_l;\mathscr{D})}(t)$ *to edge* $\mathscr{E}_{\sigma_r}^{(r)}$ *form an* $(\mathbf{m}_{\alpha_k}:\mathbf{m}_{\alpha_l})$*-bouncing flow in* Ω_α *if and only if*

$$\left.\begin{aligned} &\hbar_{i_\alpha} G_{\partial\Omega_{\alpha\beta}}^{(s_{\alpha_k},\alpha_k;\mathscr{D})}(\mathbf{x}_m, t_{m-}, \mathbf{p}_{\alpha_k}, \boldsymbol{\lambda}_{i_\alpha}) = 0 \\ &for\ s_{\alpha_k} = 0,1,2,\cdots,m_{\alpha_k}-1, \\ &(-1)^{m_{\alpha_k}} \hbar_{i_\alpha} G_{\partial\Omega_{\alpha\beta}}^{(m_{\alpha_k},\alpha_k;\mathscr{D})}(\mathbf{x}_m, t_{m-}, \mathbf{p}_{\alpha_k}, \boldsymbol{\lambda}_{i_\alpha}) < 0; \end{aligned}\right\} \tag{6.210}$$

$$\left.\begin{aligned} &\hbar_{i_\alpha} G_{\partial\Omega_{\alpha\beta}}^{(s_{\alpha_l},\alpha_l;\mathscr{D})}(\mathbf{x}_m, t_{m+}, \mathbf{p}_{\alpha_l}, \boldsymbol{\lambda}_{i_\alpha}) = 0 \\ &for\ s_{\alpha_l} = 0,1,2,\cdots,m_{\alpha_l}-1, \\ &\hbar_{i_\alpha} G_{\partial\Omega_{\alpha\beta}}^{(m_{\alpha_l},\alpha_l;\mathscr{D})}(\mathbf{x}_m, t_{m+}, \mathbf{p}_{\alpha_l}, \boldsymbol{\lambda}_{i_\alpha}) > 0 \end{aligned}\right\} \tag{6.211}$$

$$for\ all\ i_\alpha \in \mathscr{N}_{\alpha\mathscr{B}}(\sigma_r).$$

Proof. Following the proof of Theorem 3.2, the theorem can be proved. ■

The bouncing flow with two different vector fields in domain Ω_α switches at the edge of $\mathscr{E}^{(n\text{-}2)}$, including nine types bouncing flows: (i) the $(2\mathbf{k}_{\alpha_k}:2\mathbf{k}_{\alpha_l})$ bouncing flow, (ii) the $(2\mathbf{k}_{\alpha_k}:\mathbf{m}_{\alpha_l})$ bouncing flow ($\mathbf{m}_{\alpha_l} \neq 2\mathbf{k}_{\alpha_l}+1$), (iii) the $(2\mathbf{k}_{\alpha_k}:2\mathbf{k}_{\alpha_l}+1)$ bouncing flow; (iv) the $(\mathbf{m}_{\alpha_k}:2\mathbf{k}_{\alpha_l})$ bouncing flow ($\mathbf{m}_{\alpha_l} \neq 2\mathbf{k}_{\alpha_l}+1$), (v) the $(\mathbf{m}_{\alpha_k}:2\mathbf{k}_{\alpha_l}+1)$-bouncing flow ($\mathbf{m}_{\alpha_k} \neq 2\mathbf{k}_{\alpha_k}+1$), (vi) the $(\mathbf{m}_{\alpha_k}:\mathbf{m}_{\alpha_l})$ bouncing flow ($\mathbf{m}_{\alpha_k} \neq 2\mathbf{k}_{\alpha_k}+1$ and $\mathbf{m}_{\alpha_l} \neq 2\mathbf{k}_{\alpha_l}+1$); (vii) the $(2\mathbf{k}_{\alpha_k}+1:2\mathbf{k}_{\alpha_l})$ bouncing flow, (viii) the $(2\mathbf{k}_{\alpha_k}+1:\mathbf{m}_{\alpha_l})$ bouncing flow ($\mathbf{m}_{\alpha_k} \neq 2\mathbf{k}_{\alpha_k}+1$), (ix) the $(2\mathbf{k}_{\alpha_k}+1:2\mathbf{k}_{\alpha_l}+1)$ bouncing flow. The $(2\mathbf{k}_{\alpha_k}:2\mathbf{k}_{\alpha_l})$ bouncing flow in domain Ω_α and $(2\mathbf{k}_{\alpha_k}+1:2\mathbf{k}_{\alpha_l}+1)$ bouncing flow are shown Fig. 6.21(a) and (b), respectively. The solid curves in domain Ω_α depict the real flow. If a bouncing flow with two vector fields exists in domain Ω_α, then the two corresponding imaginary flows will exist in domain Ω_β and Ω_γ, which are depicted by the dashed curves. The $(2\mathbf{k}_{\alpha_k}:2\mathbf{k}_{\alpha_l})$ bouncing flow in domain Ω_α is inflexionally singular, and the $(2\mathbf{k}_{\alpha_k}+1:2\mathbf{k}_{\alpha_l}+1)$ bouncing flow in domain Ω_α is tangentially singular. The other cases are presented in Figs. 6.22-6.24. The bouncing flows with the $(2\mathbf{k}_{\alpha_k}:\mathbf{m}_{\alpha_l})$ singularity ($\mathbf{m}_{\alpha_l} \neq 2\mathbf{k}_{\alpha_l}+1$) and the $(\mathbf{m}_{\alpha_k}:2\mathbf{k}_{\alpha_l})$ singularity ($\mathbf{m}_{\alpha_k} \neq 2\mathbf{k}_{\alpha_k}+1$) are sketched in Fig. 6.22(a) and (b), respectively. The bouncing flow with the $(2\mathbf{k}_{\alpha_k}+1:2\mathbf{k}_{\alpha_l})$ singularity and the $(2\mathbf{k}_{\alpha_k}:2\mathbf{k}_{\alpha_l}+1)$ singularity are sketched in Fig. 6.23(a) and (b), respectively. The bouncing flow with the $(2\mathbf{k}_{\alpha_k}+1:\mathbf{m}_{\alpha_l})$ singularity ($\mathbf{m}_{\alpha_l} \neq 2\mathbf{k}_{\alpha_l}+1$) and the $(\mathbf{m}_{\alpha_k}:2\mathbf{k}_{\alpha_l}+1)$ singularity ($\mathbf{m}_{\alpha_k} \neq 2\mathbf{k}_{\alpha_k}+1$) are sketched in Fig. 6.24(a) and (b), respectively. The bouncing flow with the $(\mathbf{m}_{\alpha_k}:\mathbf{m}_{\alpha_l})$-singularity ($\mathbf{m}_{\alpha_k} \neq 2\mathbf{k}_{\alpha_k}+1$ and $\mathbf{m}_{\alpha_l} \neq 2\mathbf{k}_{\alpha_l}+1$) is presented in Fig. 6.25.

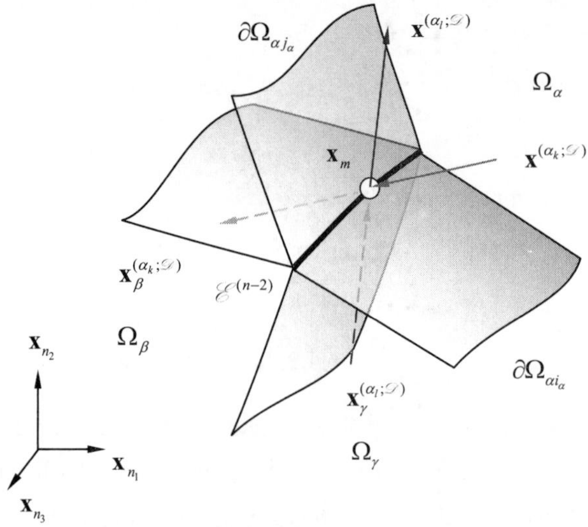

(a)

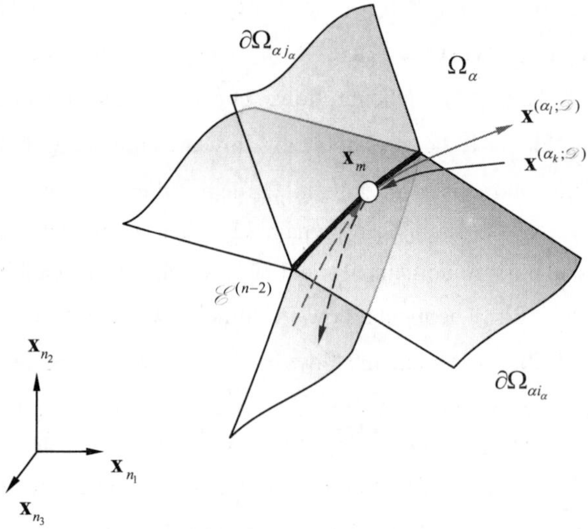

(b)

Fig. 6.21 Bouncing flows to the edge: **(a)** the $(2\mathbf{k}_{\alpha_k} : 2\mathbf{k}_{\alpha_l})$ singularity, **(b)** the $(2\mathbf{k}_{\alpha_k} + 1 : 2\mathbf{k}_{\alpha_l} + 1)$ singularity. The black curve represents the $(n-2)$ -dimensional edge. The domain flows are represented by the thick curves with arrows in the domain. The dashed curves are the extensions of the real flows. ($n_1 + n_2 + n_3 = n$)

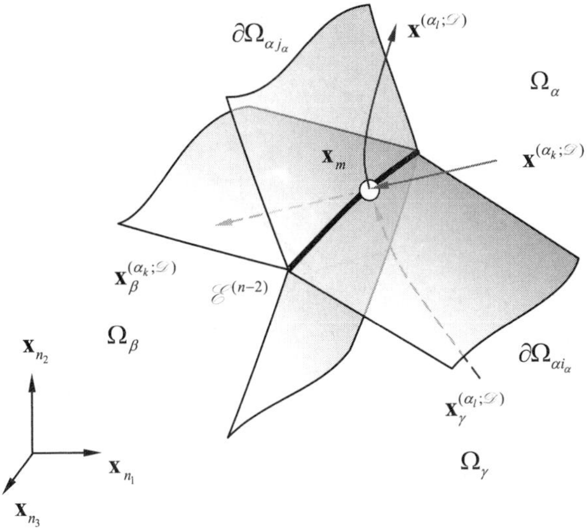

(a)

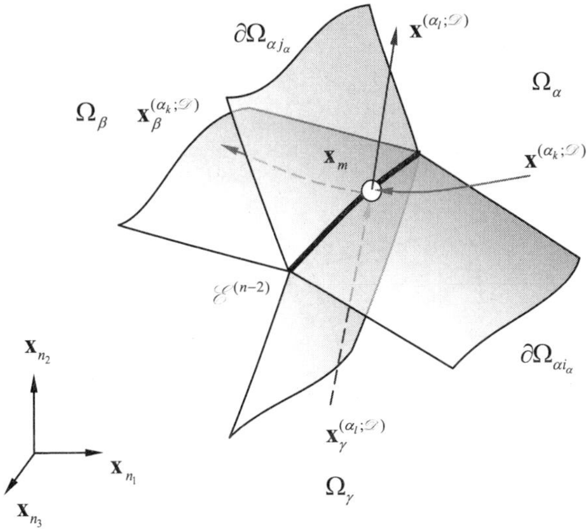

(b)

Fig. 6.22 Bouncing flows to the edge: **(a)** the $(2\mathbf{k}_{\alpha_k}:\mathbf{m}_{\alpha_l})$ - singularity ($\mathbf{m}_{\alpha_l}\neq 2\mathbf{k}_{\alpha_l}+1$), **(b)** the $(\mathbf{m}_{\alpha_k}:2\mathbf{k}_{\alpha_l})$ - singularity ($\mathbf{m}_{\alpha_k}\neq 2\mathbf{k}_{\alpha_k}+1$). The black curve represents the $(n-2)$ -dimensional edge. The domain flows are represented by the thick curves with arrows in the domain. The dashed curves are the extended imaginary flow of real. ($n_1+n_2+n_3=n$)

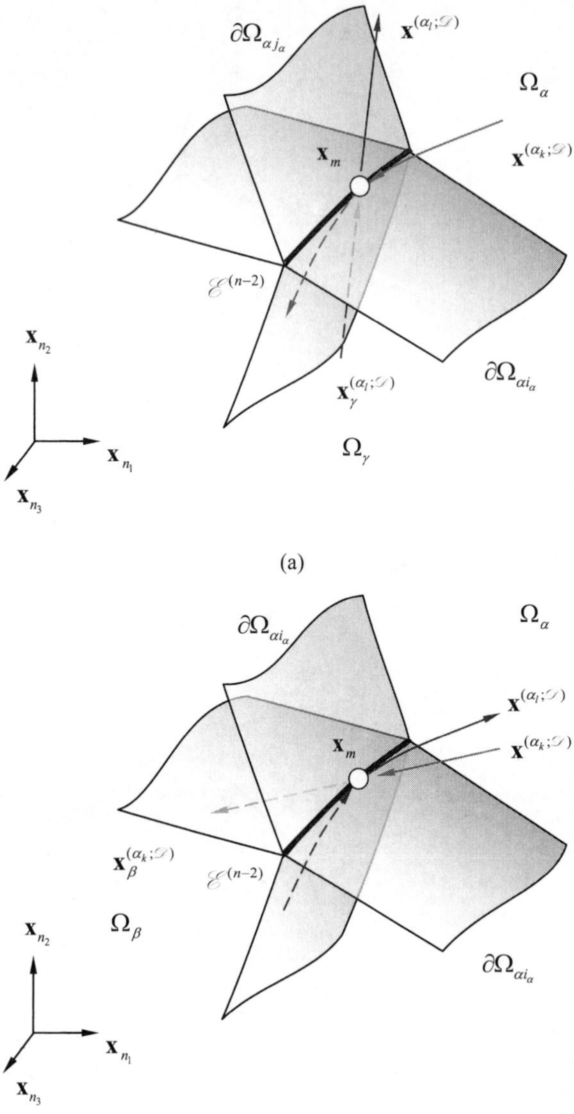

(b)

Fig. 6.23 Bouncing flows to the edge: **(a)** the $(2\mathbf{k}_{\alpha_k}+1:2\mathbf{k}_{\alpha_l})$ -singularity, **(b)** the $(2\mathbf{k}_{\alpha_k}:2\mathbf{k}_{\alpha_l}+1)$ -singularity. The black curve represents the $(n-2)$ -dimensional edge. The domain flows are represented by the thick curves with arrows in the domain. The dashed curves are the backward or forward extensions of real flows. ($n_1+n_2+n_3=n$)

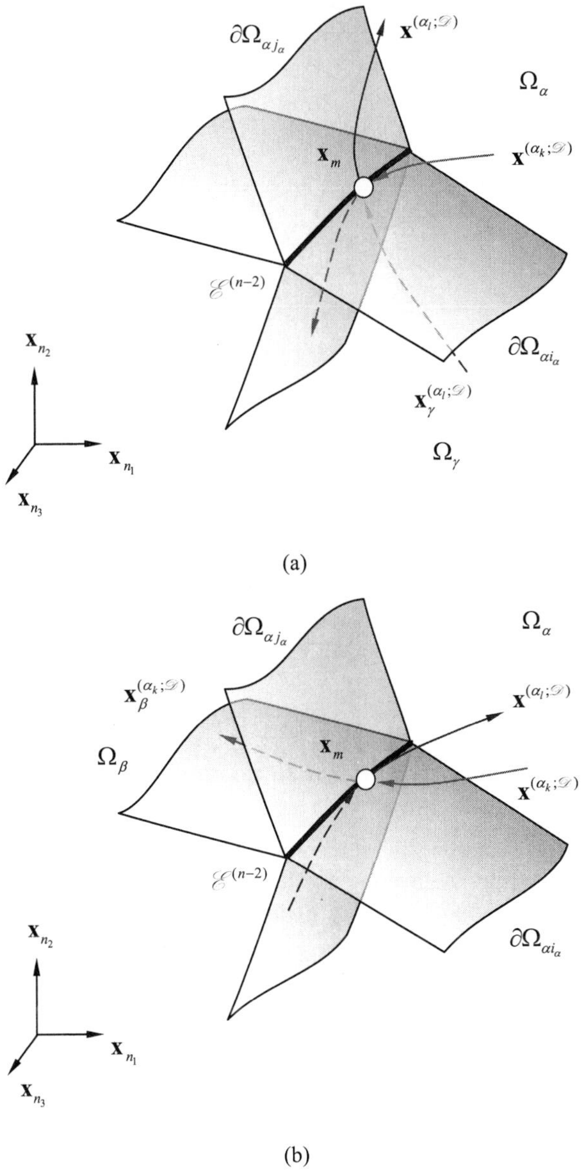

Fig. 6.24 Bouncing flows to the edge: **(a)** the $(2\mathbf{k}_{\alpha_k}+1:\mathbf{m}_{\alpha_l})$ -singularity ($\mathbf{m}_{\alpha_l}\neq 2\mathbf{k}_{\alpha_l}+1$), **(b)** the $(\mathbf{m}_{\alpha_k}:2\mathbf{k}_{\alpha_l}+1)$ -singularity ($\mathbf{m}_{\alpha_k}\neq 2\mathbf{k}_{\alpha_k}+1$). The black curve represents the $(n-2)$ dimensional edge. The domain flows are represented by the thick curves with arrows in the domain. The dashed curves are the backward or forward extensions of real flows. ($n_1+n_2+n_3=n$)

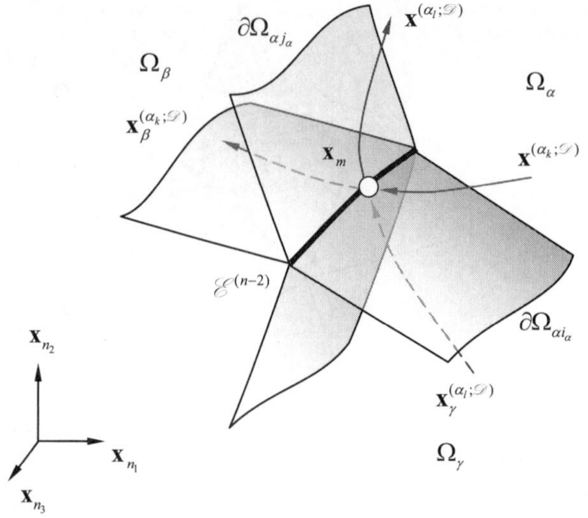

Fig. 6.25 Bouncing flows to the edge for the $(\mathbf{m}_{\alpha_k}:\mathbf{m}_{\alpha_l})$-singularity ($\mathbf{m}_{\alpha_k}\neq 2\mathbf{k}_{\alpha_k}+1$ and $\mathbf{m}_{\alpha_l}\neq 2\mathbf{k}_{\alpha_l}+1$). The black curve represents the $(n-2)$-dimensional edge. The domain flows are represented by the thick curves with arrows in the domain. The dashed curves are the extended imaginary flows of real flows. ($n_1+n_2+n_3=n$)

6.7.2. Extended passable domain flows to edges

The bouncing flow requires at least a coming flow and a leaving flow at the edge. With the switching rule, the coming and leaving domain flows will form a bouncing flow. If two coming domain flows in domain Ω_α at the same time arrive at the same location of the edge, after switching, the flow in the domain Ω_α can extend to domain Ω_β with the Continuation axiom, as sketched in Fig. 6.26. The definitions and theorems for such extended flows are presented as follows.

Definition 6.33. For a discontinuous system with multi-valued vector fields in Eq. (6.205), there is a point $\mathbf{x}^{(\sigma_r;\mathscr{E})}(t_m)\equiv\mathbf{x}_m\in\mathscr{E}_{\sigma_r}^{(r)}$ ($\sigma_r\in\mathscr{N}_{\mathscr{E}}^{r}=\{1,2,\cdots,N_r\}$ and $r\in\{0,1,2,\cdots,n-1\}$) at time t_m with the related boundaries $\partial\Omega_{\alpha i_\alpha}(\sigma_r)$ and related domains $\Omega_\alpha(\sigma_r)$ ($\alpha\in\mathscr{N}_{\mathscr{D}}(\sigma_r)\subset\mathscr{N}_{\mathscr{D}}=\{1,2,\cdots,N_d\}$ and $i_\alpha\in\mathscr{N}_{\alpha\mathscr{B}}(\sigma_r)\subset\mathscr{N}_{\mathscr{B}}=\{1,2,\cdots,N_b\}$), where the boundary index subset of $\mathscr{N}_{\alpha\mathscr{B}}(\sigma_r)$ has n_{σ_r}-elements ($n_{\sigma_r}\geq(n-r)$). For an arbitrarily small $\varepsilon>0$, there are two time intervals $[t_{m-\varepsilon},t_m)$ and $(t_m,t_{m+\varepsilon}]$. A flow $\mathbf{x}^{(\alpha_k)}(t)$ is $C_{[t_{m-\varepsilon},t_m)}^{r_{\alpha_k}}$-continuous ($r_{\alpha_k}\geq 1$) for time t with $\| d^{r_{\alpha_k}+1}\mathbf{x}^{(\alpha_k;\mathscr{D})}/dt^{r_{\alpha_k}+1}\| < \infty$, and a flow $\mathbf{x}^{(\alpha_l)}(t)$ is $C^{r_{\alpha_l}}_{(t_m,t_{m+\varepsilon}]}$-continuous ($r_{\alpha_k} \geq 1$) for time t with $\| d^{r_{\alpha_l}+1}\mathbf{x}^{(\alpha_l;\mathscr{D})}/dt^{r_{\alpha_l}+1}\| < \infty$. If a coming flow of $\mathbf{x}^{(\alpha_k;\mathscr{D})}$ arrives to the edge of $\mathscr{E}^{(r)}_{\sigma_r}$, there is a switching rule between the two flows of $\mathbf{x}^{(\alpha_k;\mathscr{D})}$ and $\mathbf{x}^{(\alpha_l;\mathscr{D})}$ with $\mathbf{x}^{(\alpha_k;\mathscr{D})}(t_{m-}) = \mathbf{x}_m = \mathbf{x}^{(\alpha_l;\mathscr{D})}(t_{m+})$. The flows of $\mathbf{x}^{(\alpha_k)}(t)$ and $\mathbf{x}^{(\alpha_l)}(t)$ to edge $\mathscr{E}^{(r)}_{\sigma_r}$ are called an *extended passable flow* from Ω_α to Ω_β if

$$\left.\begin{aligned}&\hbar_{i_\alpha}\mathbf{n}^{\mathrm{T}}_{\partial\Omega_{\alpha i_\alpha}}(\mathbf{x}^{(i_\alpha;\mathscr{B})}_{m-\varepsilon})\cdot[\mathbf{x}^{(i_\alpha;\mathscr{B})}_{m-\varepsilon} - \mathbf{x}^{(\alpha_k;\mathscr{D})}_{m-\varepsilon}] < 0\\ &\hbar_{i_\alpha}\mathbf{n}^{\mathrm{T}}_{\partial\Omega_{\alpha i_\alpha}}(\mathbf{x}^{(i_\alpha;\mathscr{B})}_{m+\varepsilon})\cdot[\mathbf{x}^{(\alpha_k;\mathscr{D})}_{m+\varepsilon} - \mathbf{x}^{(i_\alpha;\mathscr{B})}_{m+\varepsilon}] < 0\end{aligned}\right\} \tag{6.212}$$
$$\text{for all } i_\alpha \in \mathscr{N}_{\alpha\mathscr{B}}(\sigma_r).$$

An extended, passable flow from one domain to another one is described in the foregoing definition, and the following theorem will give the sufficient and necessary conditions.

Theorem 6.21. *For a discontinuous system with multi-valued vector fields in Eq.* (6.205)*, there is a point* $\mathbf{x}^{(\sigma_r;\mathscr{E})}(t_m) \equiv \mathbf{x}_m \in \mathscr{E}^{(r)}_{\sigma_r}$ ($\sigma_r \in \mathscr{N}^r_{\mathscr{E}} = \{1,2,\cdots,N_r\}$ *and* $r \in \{0,1,2,\cdots,n-1\}$) *at time* t_m *with the related boundaries* $\partial\Omega_{\alpha i_\alpha}(\sigma_r)$ *and related domains* $\Omega_\alpha(\sigma_r)$ ($\alpha \in \mathscr{N}_{\mathscr{D}}(\sigma_r) \subset \mathscr{N}_{\mathscr{D}} = \{1,2,\cdots,N_d\}$ *and* $i_\alpha \in \mathscr{N}_{\alpha\mathscr{B}}(\sigma_r) \subset \mathscr{N}_{\mathscr{B}} = \{1,2,\cdots,N_b\}$)*, where the boundary index subset of* $\mathscr{N}_{\alpha\mathscr{B}}(\sigma_r)$ *has* n_{σ_r}*-elements* ($n_{\sigma_r} \geq n-r$)*. For an arbitrarily small* $\varepsilon > 0$*, there are two time intervals* $[t_{m-\varepsilon}, t_m)$ *and* $(t_m, t_{m+\varepsilon}]$*. A flow* $\mathbf{x}^{(\alpha_k)}(t)$ *is* $C^{r_{\alpha_k}}_{[t_{m-\varepsilon},t_m)}$*-continuous* ($r_{\alpha_k} \geq 2$) *for time* t *with* $\| d^{r_{\alpha_k}+1}\mathbf{x}^{(\alpha_k;\mathscr{D})}/dt^{r_{\alpha_k}+1}\| < \infty$*, and a flow* $\mathbf{x}^{(\alpha_l)}(t)$ *is* $C^{r_{\alpha_l}}_{(t_m,t_{m+\varepsilon}]}$*-continuous for time* t *with* $\| d^{r_{\alpha_l}+1}\mathbf{x}^{(\alpha_l;\mathscr{D})}/dt^{r_{\alpha_l}+1}\| < \infty$ ($r_{\alpha_k} \geq 2$)*. If a coming flow of* $\mathbf{x}^{(\alpha_k;\mathscr{D})}$ *arrives to the edge of* $\mathscr{E}^{(r)}_{\sigma_r}$*, there is a switching rule between two flows* $\mathbf{x}^{(\alpha_k;\mathscr{D})}$ *and* $\mathbf{x}^{(\alpha_l;\mathscr{D})}$ *with* $\mathbf{x}^{(\alpha_k;\mathscr{D})}(t_{m-}) = \mathbf{x}_m = \mathbf{x}^{(\alpha_l;\mathscr{D})}(t_{m+})$*. The flows* $\mathbf{x}^{(\alpha_k;\mathscr{D})}(t)$ *and* $\mathbf{x}^{(\alpha_l;\mathscr{D})}(t)$ *to edge* $\mathscr{E}^{(r)}_{\sigma_r}$ *form an extended passable flow from* Ω_α *to* Ω_β *if and only if*

$$\left.\begin{aligned}&\hbar_{i_\alpha}G^{(\alpha_k;\mathscr{D})}_{\partial\Omega_{\alpha i_\alpha}}(\mathbf{x}_m, t_{m-}, \mathbf{p}_{\alpha_k}, \boldsymbol{\lambda}_{i_\alpha}) < 0\\ &\hbar_{i_\alpha}G^{(\alpha_l;\mathscr{D})}_{\partial\Omega_{\alpha i_\alpha}}(\mathbf{x}_m, t_{m+}, \mathbf{p}_{\alpha_l}, \boldsymbol{\lambda}_{i_\alpha}) < 0\end{aligned}\right\} \text{for all } i_\alpha \in \mathscr{N}_{\alpha\mathscr{B}}(\sigma_r). \tag{6.213}$$

Proof. Consider a flow to all boundaries relative to the edge. Following the proof of Theorem 3.1, the theorem can be proved. ■

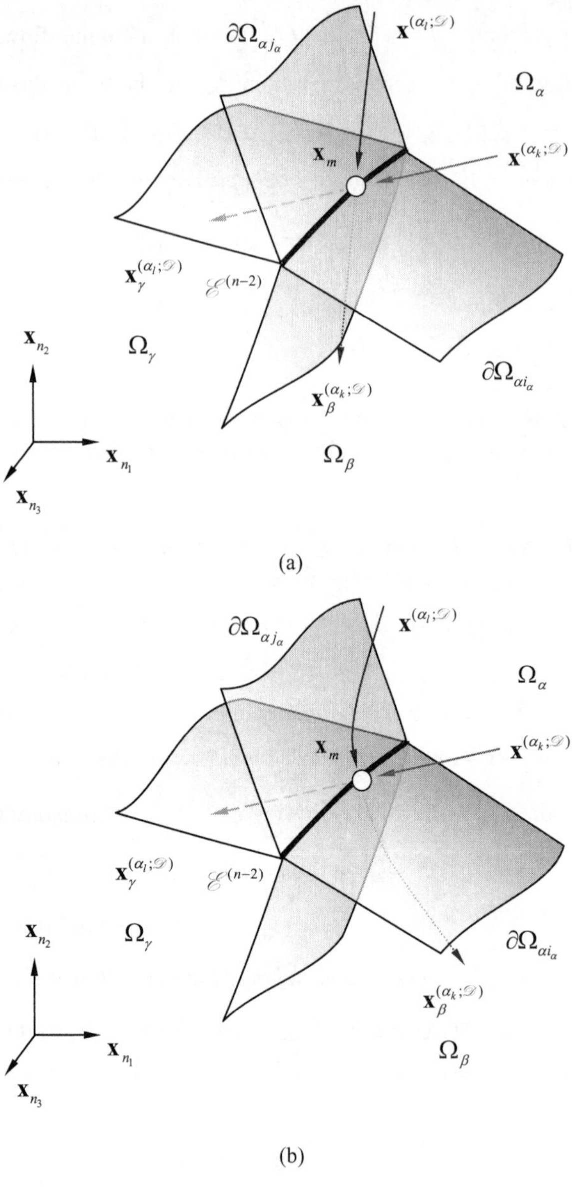

(a)

(b)

Fig. 6.26 Extended passable flows to the edge: **(a)** the $(2\mathbf{k}_{\alpha_k}:2\mathbf{k}_{\alpha_l})$ -singularity, **(b)** the $(2\mathbf{k}_{\alpha_k}:\mathbf{m}_{\alpha_l})$ -singularity. The black curve represents the $(n-2)$ -dimensional edge. The domain flows are represented by the thick curves with arrows in the domain. The dashed curves are the extended imaginary flow of real. The dotted curve is an extended passable flow. ($n_1+n_2+n_3=n$)

Definition 6.34. For a discontinuous system with multi-valued vector fields in Eq. (6.205), there is a point $\mathbf{x}^{(\sigma_r;\mathscr{E})}(t_m) \equiv \mathbf{x}_m \in \mathscr{E}_{\sigma_r}^{(r)}$ ($\sigma_r \in \mathscr{N}_{\mathscr{E}}^{r} = \{1,2,\cdots,N_r\}$ and $r \in \{0,1,2,\cdots,n-1\}$) at time t_m with the related boundaries $\partial\Omega_{\alpha i_\alpha}(\sigma_r)$ and related domains $\Omega_\alpha(\sigma_r)$ ($\alpha \in \mathscr{N}_{\mathscr{D}}(\sigma_r) \subset \mathscr{N}_{\mathscr{D}} = \{1,2,\cdots,N_d\}$ and $i_\alpha \in \mathscr{N}_{\alpha\mathscr{B}}(\sigma_r) \subset \mathscr{N}_{\mathscr{B}} = \{1,2,\cdots,N_b\}$), where the boundary index subset of $\mathscr{N}_{\alpha\mathscr{B}}(\sigma_r)$ has n_{σ_r} -elements ($n_{\sigma_r} \geq n-r$). For an arbitrarily small $\varepsilon > 0$, there are two time intervals $[t_{m-\varepsilon}, t_m)$ and $(t_m, t_{m+\varepsilon}]$. If a coming flow of $\mathbf{x}^{(\alpha_k;\mathscr{D})}$ arrives to edge of $\mathscr{E}_{\sigma_r}^{(r)}$, there is a switching rule between two flows $\mathbf{x}^{(\alpha_k;\mathscr{D})}$ and $\mathbf{x}^{(\alpha_l;\mathscr{D})}$ with $\mathbf{x}^{(\alpha_k;\mathscr{D})}(t_{m-}) = \mathbf{x}_m = \mathbf{x}^{(\alpha_l;\mathscr{D})}(t_{m+})$. A flow $\mathbf{x}^{(\alpha_k;\mathscr{D})}(t)$ is $C_{[t_{m-\varepsilon},t_m)}^{r_{\alpha_k}}$-continuous ($r_{\alpha_k} \geq m_{\alpha_k}+1$) for time t with $\| d^{r_{\alpha_k}+1}\mathbf{x}^{(\alpha_k;\mathscr{D})}/dt^{r_{\alpha_k}+1} \| < \infty$, and a flow $\mathbf{x}^{(\alpha_l;\mathscr{D})}(t)$ is $C_{(t_m,t_{m+\varepsilon}]}^{r_{\alpha_l}}$-continuous ($r_{\alpha_k} \geq m_{\alpha_l}+1$) for time t with $\| d^{r_{\alpha_l}+1}\mathbf{x}^{(\alpha_l;\mathscr{D})}/dt^{r_{\alpha_l}+1} \| < \infty$. The flows of $\mathbf{x}^{(\alpha_k;\mathscr{D})}(t)$ and $\mathbf{x}^{(\alpha_l;\mathscr{D})}(t)$ to the edge $\mathscr{E}_{\sigma_r}^{(r)}$ are called an $(\mathbf{m}_{\alpha_k} : \mathbf{m}_{\alpha_l})$ *-extended passable flow* from Ω_α to Ω_β ($\mathbf{m}_{\alpha_l} \neq 2\mathbf{k}_{\alpha_l}+1$) if

$$\left.\begin{array}{l} \hbar_{i_\alpha} G_{\partial\Omega_{\alpha\beta}}^{(s_{\alpha_k},\alpha_k;\mathscr{D})}(\mathbf{x}_m, t_{m-}, \mathbf{p}_{\alpha_k}, \boldsymbol{\lambda}_{i_\alpha}) = 0 \\ \text{for } s_{\alpha_k} = 0,1,2,\cdots, m_{\alpha_k}-1, \\ \hbar_{i_\alpha} \mathbf{n}_{\partial\Omega_{\alpha i_\alpha}}^{\mathrm{T}}(\mathbf{x}_{m-\varepsilon}^{(i_\alpha;\mathscr{B})}) \cdot [\mathbf{x}_{m-\varepsilon}^{(i_\alpha;\mathscr{B})} - \mathbf{x}_{m-\varepsilon}^{(\alpha_k;\mathscr{D})}] < 0; \end{array}\right\} \tag{6.214}$$

$$\left.\begin{array}{l} \hbar_{i_\alpha} G_{\partial\Omega_{\alpha\beta}}^{(s_{\alpha_l},\alpha_l;\mathscr{D})}(\mathbf{x}_m, t_{m+}, \mathbf{p}_{\alpha_l}, \boldsymbol{\lambda}_{i_\alpha}) = 0 \\ \text{for } s_{\alpha_l} = 0,1,2,\cdots, m_{\alpha_l}-1, \\ \hbar_{i_\alpha} \mathbf{n}_{\partial\Omega_{\alpha i_\alpha}}^{\mathrm{T}}(\mathbf{x}_{m+\varepsilon}^{(i_\alpha;\mathscr{B})}) \cdot [\mathbf{x}_{m+\varepsilon}^{(\alpha_k;\mathscr{D})} - \mathbf{x}_{m+\varepsilon}^{(i_\alpha;\mathscr{B})}] < 0 \end{array}\right\} \tag{6.215}$$

for all $i_\alpha \in \mathscr{N}_{\alpha\mathscr{B}}(\sigma_r)$.

From the foregoing description of the $(\mathbf{m}_{\alpha_k} : \mathbf{m}_{\alpha_l})$ extended, passable flow from one domain to another one, the corresponding conditions can be given via the following theorem.

Theorem 6.22. *For a discontinuous system with multi-valued vector fields in Eq.* (6.205), *there is a point* $\mathbf{x}^{(\sigma_r;\mathscr{E})}(t_m) \equiv \mathbf{x}_m \in \mathscr{E}_{\sigma_r}^{(r)}$ ($\sigma_r \in \mathscr{N}_{\mathscr{E}}^{r} = \{1,2,\cdots,N_r\}$ *and* $r \in \{0,1,2,\cdots,n-1\}$) *at time* t_m *with the related boundaries* $\partial\Omega_{\alpha i_\alpha}(\sigma_r)$ *and related domains* $\Omega_\alpha(\sigma_r)$ ($\alpha \in \mathscr{N}_{\mathscr{D}}(\sigma_r) \subset \mathscr{N}_{\mathscr{D}} = \{1,2,\cdots,N_d\}$ *and* $i_\alpha \in \mathscr{N}_{\alpha\mathscr{B}}(\sigma_r) \subset \mathscr{N}_{\mathscr{B}} = \{1,2,\cdots,N_b\}$), *where the boundary index subset of* $\mathscr{N}_{\alpha\mathscr{B}}(\sigma_r)$ *has* n_{σ_r} -

elements ($n_{\sigma_r} \geq n-r$). *For an arbitrarily small* $\varepsilon > 0$, *there are two time intervals* $[t_{m-\varepsilon}, t_m)$ *and* $(t_m, t_{m+\varepsilon}]$. *If a coming flow of* $\mathbf{x}^{(\alpha_k;\mathscr{D})}$ *arrives to the edge of* $\mathscr{E}_{\sigma_r}^{(r)}$, *there is a switching rule between two flows* $\mathbf{x}^{(\alpha_k;\mathscr{D})}$ *and* $\mathbf{x}^{(\alpha_l;\mathscr{D})}$ *with* $\mathbf{x}^{(\alpha_k;\mathscr{D})}(t_{m-}) = \mathbf{x}_m = \mathbf{x}^{(\alpha_l;\mathscr{D})}(t_{m+})$. *A flow* $\mathbf{x}^{(\alpha_k;\mathscr{D})}(t)$ *is* $C_{[t_{m-\varepsilon}, t_m)}^{r_{\alpha_k}}$ *-continuous* ($r_{\alpha_k} \geq m_{\alpha_k} + 1$) *for time* t *with* $\| d^{r_{\alpha_k}+1}\mathbf{x}^{(\alpha_k;\mathscr{D})} / dt^{r_{\alpha_k}+1} \| < \infty$, *and a flow* $\mathbf{x}^{(\alpha_l;\mathscr{D})}(t)$ *is* $C_{(t_m, t_{m+\varepsilon}]}^{r_{\alpha_l}}$ *continuous* ($r_{\alpha_k} \geq m_{\alpha_l} + 1$) *for time* t *with* $\| d^{r_{\alpha_l}+1}\mathbf{x}^{(\alpha_l;\mathscr{D})} / dt^{r_{\alpha_l}+1} \| < \infty$. *The flows of* $\mathbf{x}^{(\alpha_k;\mathscr{D})}(t)$ *and* $\mathbf{x}^{(\alpha_l;\mathscr{D})}(t)$ *to the edge* $\mathscr{E}_{\sigma_r}^{(r)}$ *form an* $(\mathbf{m}_{\alpha_k} : \mathbf{m}_{\alpha_l})$ *extended passable flow from* Ω_α *to* Ω_β ($\mathbf{m}_{\alpha_l} \neq 2\mathbf{k}_{\alpha_l} + 1$) *if and only if*

$$\left.\begin{array}{l} \hbar_{i_\alpha} G_{\partial\Omega_{\alpha\beta}}^{(s_{\alpha_k}, \alpha_k;\mathscr{D})}(\mathbf{x}_m, t_{m-}, \mathbf{p}_{\alpha_k}, \boldsymbol{\lambda}_{i_\alpha}) = 0 \\ \text{for } s_{\alpha_k} = 0, 1, 2, \cdots, m_{\alpha_k} - 1, \\ (-1)^{m_{\alpha_k}} \hbar_{i_\alpha} G_{\partial\Omega_{\alpha\beta}}^{(s_{\alpha_k}, \alpha_k;\mathscr{D})}(\mathbf{x}_m, t_{m-}, \mathbf{p}_{\alpha_k}, \boldsymbol{\lambda}_{i_\alpha}) < 0; \end{array}\right\} \tag{6.216}$$

$$\left.\begin{array}{l} \hbar_{i_\alpha} G_{\partial\Omega_{\alpha\beta}}^{(s_{\alpha_l}, \alpha_l;\mathscr{D})}(\mathbf{x}_m, t_{m+}, \mathbf{p}_{\alpha_l}, \boldsymbol{\lambda}_{i_\alpha}) = 0 \\ \text{for } s_{\alpha_l} = 0, 1, 2, \cdots, m_{\alpha_l} - 1, \\ \hbar_{i_\alpha} G_{\partial\Omega_{\alpha\beta}}^{(s_{\alpha_l}, \alpha_l;\mathscr{D})}(\mathbf{x}_m, t_{m+}, \mathbf{p}_{\alpha_l}, \boldsymbol{\lambda}_{i_\alpha}) < 0 \end{array}\right\} \tag{6.217}$$

$$\text{for all } i_\alpha \in \mathscr{N}_{\alpha\mathscr{B}}(\sigma_r).$$

Proof. Consider a flow to all boundaries relative to the edge. Following the proof of Theorem 3.2, the theorem can be proved. ■

References

Luo, A.C.J., 2005, A theory for non-smooth dynamical systems on connectable domains, *Communication in Nonlinear Science and Numerical Simulation,* **10**, 1-55.

Luo, A.C.J., 2006, *Singularity and Dynamics on Discontinuous Vector Fields*, Amsterdam: Elsevier.

Luo, A.C.J., 2008a, A theory for flow switchability in discontinuous dynamical systems, *Nonlinear Analysis: Hybrid Systems*, **2**, 1030-1061.

Luo, A.C.J., 2008b, *Global Transversality, Resonance and Chaotic Dynamics*, Singapore: World Scientific.

Chapter 7
Dynamics and Singularity of Boundary Flows

In this chapter, the switchability and attractivity of boundary flows to $(n-2)$-dimensional edges will be discussed in order to understand edge and vertex dynamics in discontinuous dynamical systems. The basic properties of boundary flows to edges will be discussed first. The coming, leaving and tangency of a boundary flow to an $(n-2)$-dimensional edge will be presented. The switchability and passability of a boundary flow from an accessible boundary to another accessible boundary will be presented with a switching rule. In addition, the switchability and passability of boundary flows and domain flows will be discussed with a switching rule. The equi-measuring edge for boundary flows will be introduced, and the attractivity of a boundary flow to the $(n-2)$-dimensional edge will be presented. Finally, the bouncing characteristics of a boundary flow to the $(n-2)$-dimensional edge will be discussed.

7.1. Boundary flow properties

In Chapter 5, the boundary flow network was discussed once flows exist only on the boundaries. To investigate the boundary flow behaviors, the basic properties of a boundary flow to the edge will be discussed. Consider an $(n-2)$-dimensional edge to be formed by intersection of three $(n-1)$-dimensional boundaries, as shown in Fig. 7.1. The black thick curve denotes the $(n-2)$-dimensional edge. The investigated boundary flow is a black curve, and the referenced boundary flow is a red curve. The $(n-2)$-dimensional edge must be intersected by two $(n-1)$-dimensional boundary surfaces. So the other surfaces of the $(n-2)$-dimensional edge should be correlated to the aforementioned two boundary surface. Thus, only one of the rest boundary surfaces can be adopted for the attractivity of a boundary flow to the edge. Similarly, the coming, leaving and totally grazing boundary flows to the edge are also sketched in Figs. 7.2 and 7.3.

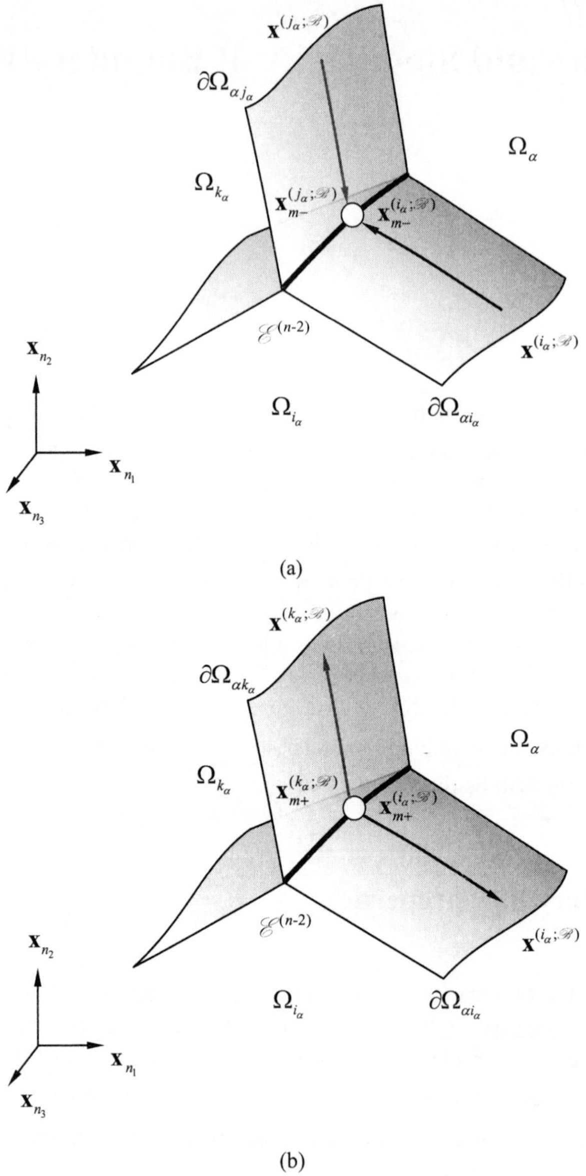

Fig. 7.1 Boundary flows on the three $(n-1)$-dimensional boundaries: **(a)** coming flow and **(b)** leaving flows. The dark thick curve represents the $(n-2)$-dimensional edge. The investigated boundary flow is a black curve with arrow on the boundary. The referenced boundary flow is a gray curve with arrow. ($n_1 + n_2 + n_3 = n$)

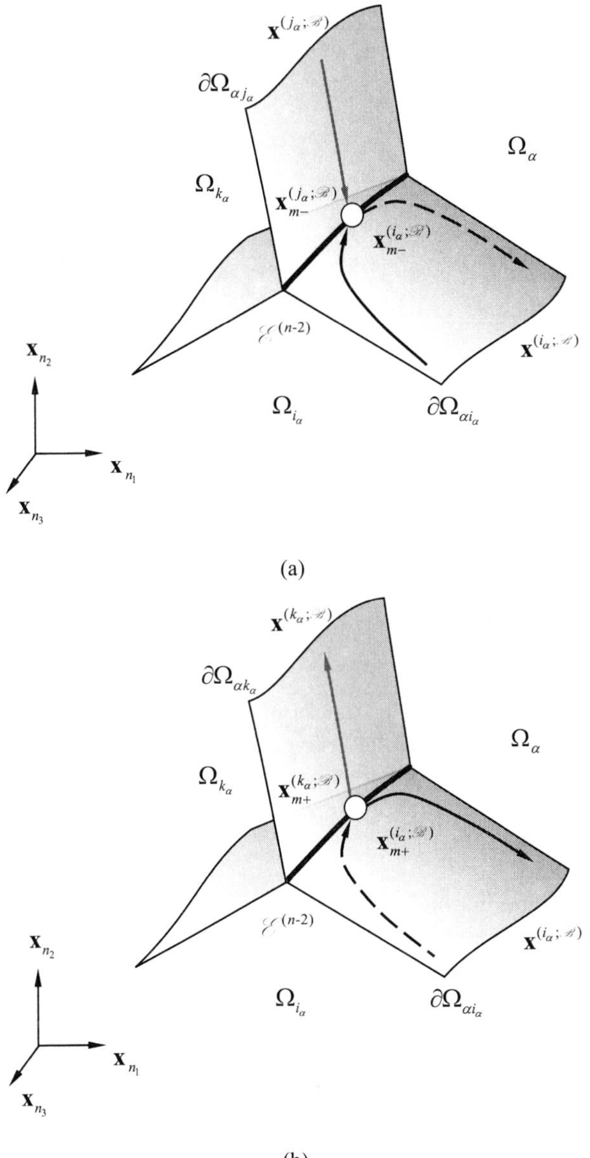

Fig. 7.2 Boundary flows on the three $(n-1)$-dimensional boundaries: **(a)** coming grazing flow and **(b)** leaving grazing flows. The dark thick curve represents the $(n-2)$-dimensional edge. The investigated boundary flow is a black curve on the boundary. The referenced boundary flow is a gray curve with arrow. ($n_1+n_2+n_3=n$)

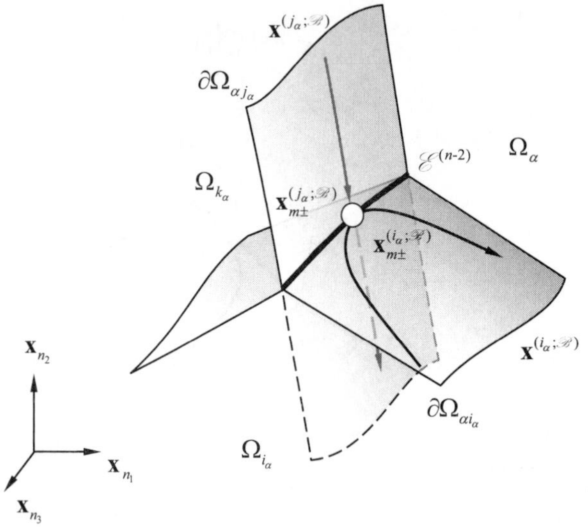

Fig. 7.3 Total grazing boundary flows on the three $(n-1)$-dimensional boundaries. The dark thick curve represents the $(n-2)$-dimensional edge. The investigated boundary flow is a dark curve with arrow on the boundary. The referenced boundary flow is a gray curve with arrow. ($n_1+n_2+n_3=n$)

From the previous geometrical description, the coming, leaving and grazing boundary flows to the edge can be defined. In addition, the concepts for the coming grazing, leaving grazing and total grazing flows to the edge will be presented. With the switching role, such coming and leaving boundary flows at specific edges are switchable.

Definition 7.1. For a discontinuous dynamical system in Eqs. (6.4)-(6.8), there is an $(n-2)$-dimensional edge $\mathscr{E}_{\sigma_{n-2}}^{(n-2)}$ ($\sigma_{n-2}\in\mathscr{N}_{\mathscr{E}}^{n-2}=\{1,2,\cdots,N_{n-2}\}$) formed by the intersection of two $(n-1)$-dimensional boundaries $\partial\Omega_{\alpha i_\alpha}(\sigma_{n-2})$ of domain Ω_α ($\alpha\in\mathscr{N}_{\mathscr{D}}(\sigma_{n-2})\subset\mathscr{N}_{\mathscr{D}}=\{1,2,\cdots,N_d\}$ and $i_\alpha\in\mathscr{N}_{\alpha\mathscr{B}}(\sigma_{n-2})\subset\mathscr{N}_{\mathscr{B}}=\{1,2,\cdots,N_b\}$). Suppose there is a boundary flow $\mathbf{x}^{(i_\alpha;\mathscr{B})}(t)$ on the boundary $\partial\Omega_{\alpha i_\alpha}(\sigma_{n-2})$. At time t_m, $\mathbf{x}_m=\mathbf{x}_m^{(i_\alpha;\mathscr{B})}=\mathbf{x}^{(i_\alpha;\mathscr{B})}(t_m)\in\mathscr{E}_{\sigma_{n-2}}^{(n-2)}$. For an arbitrarily small $\varepsilon>0$, there is a time interval $[t_{m-\varepsilon},t_{m+\varepsilon}]$.

(i) The boundary flow $\mathbf{x}^{(i_\alpha;\mathscr{B})}(t)$ to the edge of $\mathscr{E}_{\sigma_{n-2}}^{(n-2)}$ for time t_m is a *coming* flow if

$$\hbar_{j_\alpha}\mathbf{n}_{\partial\Omega_{\alpha j_\alpha}}^{\mathrm{T}}(\mathbf{x}_{m-\varepsilon}^{(j_\alpha;\mathscr{B})})\cdot[\mathbf{x}_{m-\varepsilon}^{(j_\alpha;\mathscr{B})}-\mathbf{x}_{m-\varepsilon}^{(i_\alpha;\mathscr{B})}]<0. \tag{7.1}$$

(ii) The boundary flow $\mathbf{x}^{(i_\alpha;\mathscr{B})}(t)$ to the edge of $\mathscr{E}_{\sigma_{n-2}}^{(n-2)}$ for time t_m is a *leaving* flow if

$$\hbar_{j_\alpha} \mathbf{n}^{\mathrm{T}}_{\partial\Omega_{\alpha j_\alpha}}(\mathbf{x}^{(j_\alpha;\mathscr{B})}_{m+\varepsilon})\cdot[\mathbf{x}^{(i_\alpha;\mathscr{B})}_{m+\varepsilon} - \mathbf{x}^{(j_\alpha;\mathscr{B})}_{m+\varepsilon}] > 0. \tag{7.2}$$

(iii) The boundary flow $\mathbf{x}^{(i_\alpha;\mathscr{B})}(t)$ to the edge of $\mathscr{E}^{(n-2)}_{\sigma_{n-2}}$ ($n \neq 2$) for time t_m is a *total grazing* flow if

$$\begin{aligned}
&\hbar_{j_\alpha} G^{(i_\alpha;\mathscr{B})}_{\partial\Omega_{\alpha j_\alpha}}(\mathbf{x}_m, t_{m\pm}, \boldsymbol{\lambda}_{i_\alpha}, \boldsymbol{\lambda}_{j_\alpha}) = 0,\\
&\hbar_{j_\alpha} \mathbf{n}^{\mathrm{T}}_{\partial\Omega_{\alpha j_\alpha}}(\mathbf{x}^{(j_\alpha;\mathscr{B})}_{m-\varepsilon})\cdot[\mathbf{x}^{(j_\alpha;\mathscr{B})}_{m-\varepsilon} - \mathbf{x}^{(i_\alpha;\mathscr{B})}_{m-\varepsilon}] < 0,\\
&\hbar_{j_\alpha} \mathbf{n}^{\mathrm{T}}_{\partial\Omega_{\alpha j_\alpha}}(\mathbf{x}^{(j_\alpha;\mathscr{B})}_{m-\varepsilon})\cdot[\mathbf{x}^{(i_\alpha;\mathscr{B})}_{m+\varepsilon} - \mathbf{x}^{(j_\alpha;\mathscr{B})}_{m+\varepsilon}] > 0.
\end{aligned} \tag{7.3}$$

(iv) The boundary flow $\mathbf{x}^{(i_\alpha;\mathscr{B})}(t)$ to the edge of $\mathscr{E}^{(n-2)}_{\sigma_{n-2}}$ ($n \neq 2$) for time t_m is a *coming grazing* flow if

$$\begin{aligned}
&\hbar_{j_\alpha} G^{(i_\alpha;\mathscr{B})}_{\partial\Omega_{\alpha j_\alpha}}(\mathbf{x}_m, t_{m-}, \boldsymbol{\lambda}_{i_\alpha}, \boldsymbol{\lambda}_{j_\alpha}) = 0,\\
&\hbar_{j_\alpha} \mathbf{n}^{\mathrm{T}}_{\partial\Omega_{\alpha j_\alpha}}(\mathbf{x}^{(j_\alpha;\mathscr{B})}_{m-\varepsilon})\cdot[\mathbf{x}^{(j_\alpha;\mathscr{B})}_{m-\varepsilon} - \mathbf{x}^{(i_\alpha;\mathscr{B})}_{m-\varepsilon}] < 0.
\end{aligned} \tag{7.4}$$

(v) The boundary flow $\mathbf{x}^{(i_\alpha;\mathscr{B})}(t)$ to the edge of $\mathscr{E}^{(n-2)}_{\sigma_{n-2}}$ ($n \neq 2$) for time t_m is a *leaving grazing* flow if

$$\begin{aligned}
&\hbar_{j_\alpha} G^{(i_\alpha;\mathscr{B})}_{\partial\Omega_{\alpha j_\alpha}}(\mathbf{x}_m, t_{m+}, \boldsymbol{\lambda}_{i_\alpha}, \boldsymbol{\lambda}_{j_\alpha}) = 0,\\
&\hbar_{j_\alpha} \mathbf{n}^{\mathrm{T}}_{\partial\Omega_{\alpha j_\alpha}}(\mathbf{x}^{(j_\alpha;\mathscr{B})}_{m-\varepsilon})\cdot[\mathbf{x}^{(i_\alpha;\mathscr{B})}_{m+\varepsilon} - \mathbf{x}^{(j_\alpha;\mathscr{B})}_{m+\varepsilon}] > 0.
\end{aligned} \tag{7.5}$$

From the foregoing definition, the corresponding necessary and sufficient conditions for the coming, leaving and grazing boundary flows to the edge can be presented as follows.

Theorem 7.1. *For a discontinuous dynamical system in Eqs.* (6.4)-(6.8), *there is an* $(n-2)$ *dimensional edge* $\mathscr{E}^{(n-2)}_{\sigma_{n-2}}$ ($\sigma_{n-2} \in \mathscr{N}^{n-2}_{\mathscr{E}} = \{1,2,\cdots,N_{n-2}\}$) *formed by the intersection of two* $(n-1)$ *-dimensional boundaries* $\partial\Omega_{\alpha i_\alpha}(\sigma_{n-2})$ *of domain* Ω_α ($\alpha \in \mathscr{N}_{\mathscr{D}}(\sigma_{n-2}) \subset \mathscr{N}_{\mathscr{D}} = \{1,2,\cdots,N_d\}$ *and* $i_\alpha \in \mathscr{N}_{\alpha\mathscr{B}}(\sigma_{n-2}) \subset \mathscr{N}_{\mathscr{B}} = \{1,2,\cdots, N_b\}$). *Suppose there is a boundary flow* $\mathbf{x}^{(i_\alpha;\mathscr{B})}(t)$ *on the boundary* $\partial\Omega_{\alpha i_\alpha}(\sigma_{n-2})$. *At time* t_m, $\mathbf{x}_m = \mathbf{x}^{(i_\alpha;\mathscr{B})}_m = \mathbf{x}^{(i_\alpha;\mathscr{B})}(t_m) \in \mathscr{E}^{(n-2)}_{\sigma_{n-2}}$. *For an arbitrarily small* $\varepsilon > 0$, *there is a time interval* $[t_{m-\varepsilon}, t_{m+\varepsilon}]$. *The flow* $\mathbf{x}^{(i_\alpha;\mathscr{B})}(t)$ *is* $C^{r_{i_\alpha}}_{[t_{m-\varepsilon},t_{m+\varepsilon}]}$*-continuous for time* t *and* $\| d^{r_{i_\alpha}+1}\mathbf{x}^{(i_\alpha;\mathscr{B})}/dt^{r_{i_\alpha}+1} \| < \infty$ ($r_{i_\alpha} \geq 2$).

(i) *The boundary flow* $\mathbf{x}^{(i_\alpha;\mathscr{B})}(t)$ *to the edge of* $\mathscr{E}^{(n-2)}_{\sigma_{n-2}}$ *at time* t_m *is a coming flow if and only if*

$$\hbar_{j_\alpha} G^{(i_\alpha;\mathscr{B})}_{\partial\Omega_{\alpha j_\alpha}}(\mathbf{x}_m, t_{m-}, \boldsymbol{\lambda}_{i_\alpha}, \boldsymbol{\lambda}_{j_\alpha}) < 0. \tag{7.6}$$

(ii) *The boundary flow* $\mathbf{x}^{(i_\alpha;\mathscr{B})}(t)$ *to the edge of* $\mathscr{E}^{(n-2)}_{\sigma_{n-2}}$ ($n \neq 2$) *at time* t_m *is a leaving flow if and only if*

$$\hbar_{j_\alpha} G^{(i_\alpha;\mathscr{B})}_{\partial\Omega_{\alpha j_\alpha}}(\mathbf{x}_m, t_{m+}, \boldsymbol{\lambda}_{i_\alpha}, \boldsymbol{\lambda}_{j_\alpha}) > 0. \tag{7.7}$$

(iii) *The boundary flow* $\mathbf{x}^{(i_\alpha;\mathscr{B})}(t)$ *to the edge of* $\mathscr{E}^{(n-2)}_{\sigma_{n-2}}$ ($n \neq 2$) *at time* t_m *is a total grazing flow if and only if*

$$\begin{aligned} &\hbar_{j_\alpha} G^{(i_\alpha;\mathscr{B})}_{\partial\Omega_{\alpha j_\alpha}}(\mathbf{x}_m, t_{m\pm}, \boldsymbol{\lambda}_{i_\alpha}, \boldsymbol{\lambda}_{j_\alpha}) = 0,\\ &\hbar_{j_\alpha} G^{(1,i_\alpha;\mathscr{B})}_{\partial\Omega_{\alpha j_\alpha}}(\mathbf{x}_m, t_{m\pm}, \boldsymbol{\lambda}_{i_\alpha}, \boldsymbol{\lambda}_{j_\alpha}) > 0. \end{aligned} \tag{7.8}$$

(iv) *The boundary flow* $\mathbf{x}^{(i_\alpha;\mathscr{B})}(t)$ *to the edge of* $\mathscr{E}^{(n-2)}_{\sigma_{n-2}}$ *at time* t_m *is a coming grazing flow if and only if*

$$\begin{aligned} &\hbar_{j_\alpha} G^{(i_\alpha;\mathscr{B})}_{\partial\Omega_{\alpha j_\alpha}}(\mathbf{x}_m, t_{m-}, \boldsymbol{\lambda}_{i_\alpha}, \boldsymbol{\lambda}_{j_\alpha}) = 0,\\ &\hbar_{j_\alpha} G^{(1,i_\alpha;\mathscr{B})}_{\partial\Omega_{\alpha j_\alpha}}(\mathbf{x}_m, t_{m-}, \boldsymbol{\lambda}_{i_\alpha}, \boldsymbol{\lambda}_{j_\alpha}) > 0. \end{aligned} \tag{7.9}$$

(v) *The boundary flow* $\mathbf{x}^{(i_\alpha;\mathscr{B})}(t)$ *to the edge of* $\mathscr{E}^{(n-2)}_{\sigma_{n-2}}$ ($n \neq 2$) *at time* t_m *is a leaving grazing flow if and only if*

$$\begin{aligned} &\hbar_{j_\alpha} G^{(i_\alpha;\mathscr{B})}_{\partial\Omega_{\alpha j_\alpha}}(\mathbf{x}_m, t_{m+}, \boldsymbol{\lambda}_{i_\alpha}, \boldsymbol{\lambda}_{j_\alpha}) = 0,\\ &\hbar_{j_\alpha} G^{(1,i_\alpha;\mathscr{B})}_{\partial\Omega_{\alpha j_\alpha}}(\mathbf{x}_m, t_{m+}, \boldsymbol{\lambda}_{i_\alpha}, \boldsymbol{\lambda}_{j_\alpha}) > 0. \end{aligned} \tag{7.10}$$

Proof. The boundary flow on the boundary can be treated as a special flow in domain to the other boundary of the edge. As in Chapter 3, this theorem can be proved. ■

For the higher-order singularity of the boundary flow to the edge, the corresponding definitions for the coming, leaving and grazing flows to the edge are given.

Definition 7.2. For a discontinuous dynamical system in Eqs. (6.4)-(6.8), there is an $(n-2)$ -dimensional edge $\mathscr{E}^{(n-2)}_{\sigma_{n-2}}$ ($\sigma_{n-2} \in \mathscr{N}^{n-2}_{\mathscr{E}} = \{1,2,\cdots,N_{n-2}\}$) formed by the intersection of two $(n-1)$ -dimensional boundaries $\partial\Omega_{\alpha i_\alpha}(\sigma_{n-2})$ of domain Ω_α ($\alpha \in \mathscr{N}_{\mathscr{D}}(\sigma_{n-2}) \subset \mathscr{N}_{\mathscr{D}} = \{1,2,\cdots,N_d\}$ and $i_\alpha \in \mathscr{N}_{\alpha\mathscr{B}}(\sigma_{n-2}) \subset \mathscr{N}_{\mathscr{B}} = \{1,2,\cdots, N_b\}$). Suppose there is a boundary flow $\mathbf{x}^{(i_\alpha,\mathscr{B})}(t)$ on the boundary $\partial\Omega_{\alpha i_\alpha}(\sigma_{n-2})$.

At time t_m, $\mathbf{x}_m = \mathbf{x}_m^{(i_\alpha;\mathscr{B})} = \mathbf{x}^{(i_\alpha;\mathscr{B})}(t_m) \in \mathscr{E}_{\sigma_{n-2}}^{(n-2)}$. For an arbitrarily small $\varepsilon > 0$, there is a time interval $[t_{m-\varepsilon}, t_{m+\varepsilon}]$. The flow $\mathbf{x}^{(i_\alpha;\mathscr{B})}(t)$ is $C_{[t_{m-\varepsilon},t_{m+\varepsilon}]}^{r_{i_\alpha}}$-continuous for time t and $\| d^{r_{i_\alpha}+1}\mathbf{x}^{(i_\alpha;\mathscr{B})} / dt^{r_{i_\alpha}+1} \| < \infty$ ($r_{i_\alpha} \geq 2k_{i_\alpha}$ or $2k_{i_\alpha}+1$).

(i) The boundary flow $\mathbf{x}^{(i_\alpha;\mathscr{B})}(t)$ to the edge of $\mathscr{E}_{\sigma_{n-2}}^{(n-2)}$ at time t_m is a *coming flow* of the $(2k_{i_\alpha})$th-order singularity if

$$
\begin{aligned}
&\hbar_{j_\alpha} G_{\partial\Omega_{\alpha j_\alpha}}^{(s_{i_\alpha}, i_\alpha;\mathscr{B})}(\mathbf{x}_m, t_{m-}, \boldsymbol{\lambda}_{i_\alpha}, \boldsymbol{\lambda}_{j_\alpha}) = 0 \;\; for\; s_{i_\alpha} = 0,1,2,\cdots,2k_{i_\alpha}-1;\\
&\hbar_{j_\alpha} \mathbf{n}_{\partial\Omega_{\alpha j_\alpha}}^{\mathrm{T}}(\mathbf{x}_{m-\varepsilon}^{(j_\alpha;\mathscr{B})}) \cdot [\mathbf{x}_{m-\varepsilon}^{(j_\alpha;\mathscr{B})} - \mathbf{x}_{m-\varepsilon}^{(i_\alpha;\mathscr{B})}] < 0.
\end{aligned}
\tag{7.11}
$$

(ii) The boundary flow $\mathbf{x}^{(i_\alpha;\mathscr{B})}(t)$ to the edge of $\mathscr{E}_{\sigma_{n-2}}^{(n-2)}$ at time t_m is a *leaving flow* of the $(2k_{i_\alpha})$th-order singularity if

$$
\begin{aligned}
&\hbar_{j_\alpha} G_{\partial\Omega_{\alpha j_\alpha}}^{(s_{i_\alpha}, i_\alpha;\mathscr{B})}(\mathbf{x}_m, t_{m+}, \boldsymbol{\lambda}_{i_\alpha}, \boldsymbol{\lambda}_{j_\alpha}) = 0 \;\; for\; s_{i_\alpha} = 0,1,2,\cdots,2k_{i_\alpha}-1;\\
&\hbar_{j_\alpha} \mathbf{n}_{\partial\Omega_{\alpha j_\alpha}}^{\mathrm{T}}(\mathbf{x}_{m-\varepsilon}^{(j_\alpha;\mathscr{B})}) \cdot [\mathbf{x}_{m+\varepsilon}^{(i_\alpha;\mathscr{B})} - \mathbf{x}_{m+\varepsilon}^{(j_\alpha;\mathscr{B})}] > 0.
\end{aligned}
\tag{7.12}
$$

(iii) The boundary flow $\mathbf{x}^{(i_\alpha;\mathscr{B})}(t)$ to the edge of $\mathscr{E}_{\sigma_{n-2}}^{(n-2)}$ ($n \neq 2$) at time t_m is a total *grazing flow* with the $(2k_{i_\alpha}+1)$th-order singularity if

$$
\begin{aligned}
&\hbar_{j_\alpha} G_{\partial\Omega_{\alpha j_\alpha}}^{(s_{i_\alpha}, i_\alpha;\mathscr{B})}(\mathbf{x}_m, t_{m\pm}, \boldsymbol{\lambda}_{i_\alpha}, \boldsymbol{\lambda}_{j_\alpha}) = 0 \;\; for\; s_{i_\alpha} = 0,1,2,\cdots,2k_{i_\alpha};\\
&\hbar_{j_\alpha} \mathbf{n}_{\partial\Omega_{\alpha j_\alpha}}^{\mathrm{T}}(\mathbf{x}_{m-\varepsilon}^{(j_\alpha;\mathscr{B})}) \cdot [\mathbf{x}_{m-\varepsilon}^{(j_\alpha;\mathscr{B})} - \mathbf{x}_{m-\varepsilon}^{(i_\alpha;\mathscr{B})}] < 0,\\
&\hbar_{j_\alpha} \mathbf{n}_{\partial\Omega_{\alpha j_\alpha}}^{\mathrm{T}}(\mathbf{x}_{m-\varepsilon}^{(j_\alpha;\mathscr{B})}) \cdot [\mathbf{x}_{m+\varepsilon}^{(i_\alpha;\mathscr{B})} - \mathbf{x}_{m+\varepsilon}^{(j_\alpha;\mathscr{B})}] > 0.
\end{aligned}
\tag{7.13}
$$

(iv) The boundary flow $\mathbf{x}^{(i_\alpha;\mathscr{B})}(t)$ to the edge of $\mathscr{E}_{\sigma_{n-2}}^{(n-2)}$ ($n \neq 2$) at time t_m is a *coming grazing flow* with the $(2k_{i_\alpha}+1)$th-order singularity if

$$
\begin{aligned}
&\hbar_{j_\alpha} G_{\partial\Omega_{\alpha j_\alpha}}^{(s_{i_\alpha}, i_\alpha;\mathscr{B})}(\mathbf{x}_m, t_{m-}, \boldsymbol{\lambda}_{i_\alpha}, \boldsymbol{\lambda}_{j_\alpha}) = 0 \;\; for\; s_{i_\alpha} = 0,1,2,\cdots,2k_{i_\alpha};\\
&\hbar_{j_\alpha} \mathbf{n}_{\partial\Omega_{\alpha j_\alpha}}^{\mathrm{T}}(\mathbf{x}_{m-\varepsilon}^{(j_\alpha;\mathscr{B})}) \cdot [\mathbf{x}_{m-\varepsilon}^{(j_\alpha;\mathscr{B})} - \mathbf{x}_{m-\varepsilon}^{(i_\alpha;\mathscr{B})}] < 0.
\end{aligned}
\tag{7.14}
$$

(v) The boundary flow $\mathbf{x}^{(i_\alpha;\mathscr{B})}(t)$ to the edge of $\mathscr{E}_{\sigma_{n-2}}^{(n-2)}$ ($n \neq 2$) at time t_m is a *leaving grazing flow* with the $(2k_{i_\alpha}+1)$th-order singularity if

$$
\begin{aligned}
&\hbar_{j_\alpha} G_{\partial\Omega_{\alpha j_\alpha}}^{(s_{i_\alpha}, i_\alpha;\mathscr{B})}(\mathbf{x}_m, t_{m+}, \boldsymbol{\lambda}_{i_\alpha}, \boldsymbol{\lambda}_{j_\alpha}) = 0 \;\; for\; s_{i_\alpha} = 0,1,2,\cdots,2k_{i_\alpha};\\
&\hbar_{j_\alpha} \mathbf{n}_{\partial\Omega_{\alpha j_\alpha}}^{\mathrm{T}}(\mathbf{x}_{m-\varepsilon}^{(j_\alpha;\mathscr{B})}) \cdot [\mathbf{x}_{m+\varepsilon}^{(i_\alpha;\mathscr{B})} - \mathbf{x}_{m+\varepsilon}^{(j_\alpha;\mathscr{B})}] > 0.
\end{aligned}
\tag{7.15}
$$

Similarly, the necessary and sufficient conditions for the coming, leaving and grazing boundary flows to the edge with the higher-order singularity will be presented.

Theorem 7.2. *For a discontinuous dynamical system in Eqs.* (6.4)-(6.8), *there is an* $(n-2)$ *-dimensional edge* $\mathscr{E}_{\sigma_{n-2}}^{(n-2)}$ ($\sigma_{n-2} \in \mathscr{N}_{\mathscr{E}}^{n-2} = \{1,2,\cdots,N_{n-2}\}$) *formed by the intersection of two* $(n-1)$ *-dimensional boundaries* $\partial\Omega_{\alpha i_\alpha}(\sigma_{n-2})$ *of domain* Ω_α ($\alpha \in \mathscr{N}_{\mathscr{D}}(\sigma_{n-2}) \subset \mathscr{N}_{\mathscr{D}} = \{1,2,\cdots,N_d\}$ *and* $i_\alpha \in \mathscr{N}_{\alpha\mathscr{B}}(\sigma_{n-2}) \subset \mathscr{N}_{\mathscr{B}} = \{1,2,\cdots,N_b\}$). *Suppose there is a boundary flow* $\mathbf{x}^{(i_\alpha;\mathscr{B})}(t)$ *on the boundary* $\partial\Omega_{\alpha i_\alpha}(\sigma_{n-2})$. *At time* t_m, $\mathbf{x}_m = \mathbf{x}_m^{(i_\alpha;\mathscr{B})} = \mathbf{x}^{(i_\alpha;\mathscr{B})}(t_m) \in \mathscr{E}_{\sigma_{n-2}}^{(n-2)}$. *For an arbitrarily small* $\varepsilon > 0$, *there is a time interval* $[t_{m-\varepsilon}, t_{m+\varepsilon}]$. *The flow* $\mathbf{x}^{(i_\alpha;\mathscr{B})}(t)$ *is* $C_{[t_{m-\varepsilon},t_{m+\varepsilon}]}^{r_{i_\alpha}}$*-continuous for time* t *and* $\| d^{r_{i_\alpha}+1}\mathbf{x}^{(i_\alpha;\mathscr{B})} / dt^{r_{i_\alpha}+1} \| < \infty$ ($r_{i_\alpha} \geq 2k_{i_\alpha}$ *or* $2k_{i_\alpha}+1$).

(i) *The boundary flow* $\mathbf{x}^{(i_\alpha;\mathscr{B})}(t)$ *to the edge of* $\mathscr{E}_{\sigma_{n-2}}^{(n-2)}$ *at time* t_m *is a coming flow of the* $(2k_{i_\alpha})$th-*order singularity if and only if*

$$
\begin{aligned}
&\hbar_{j_\alpha} G_{\partial\Omega_{\alpha j_\alpha}}^{(s_{i_\alpha}, i_\alpha;\mathscr{B})}(\mathbf{x}_m, t_{m-}, \boldsymbol{\lambda}_{i_\alpha}, \boldsymbol{\lambda}_{j_\alpha}) = 0 \text{ for } s_{i_\alpha} = 0,1,2,\cdots,2k_{i_\alpha}-1;\\
&\hbar_{j_\alpha} G_{\partial\Omega_{\alpha j_\alpha}}^{(2k_{i_\alpha}, i_\alpha;\mathscr{B})}(\mathbf{x}_m, t_{m-}, \boldsymbol{\lambda}_{i_\alpha}, \boldsymbol{\lambda}_{j_\alpha}) < 0.
\end{aligned}
\tag{7.16}
$$

(ii) *The boundary flow* $\mathbf{x}^{(i_\alpha;\mathscr{B})}(t)$ *to the edge of* $\mathscr{E}_{\sigma_{n-2}}^{(n-2)}$ *at time* t_m *is a leaving flow of the* $(2k_{i_\alpha})$th-*order singularity if and only if*

$$
\begin{aligned}
&\hbar_{j_\alpha} G_{\partial\Omega_{\alpha j_\alpha}}^{(s_{i_\alpha}, i_\alpha;\mathscr{B})}(\mathbf{x}_m, t_{m+}, \boldsymbol{\lambda}_{i_\alpha}, \boldsymbol{\lambda}_{j_\alpha}) = 0 \text{ for } s_{i_\alpha} = 0,1,2,\cdots,2k_{i_\alpha}-1;\\
&\hbar_{j_\alpha} G_{\partial\Omega_{\alpha j_\alpha}}^{(2k_{i_\alpha}, i_\alpha;\mathscr{B})}(\mathbf{x}_m, t_{m+}, \boldsymbol{\lambda}_{i_\alpha}, \boldsymbol{\lambda}_{j_\alpha}) > 0.
\end{aligned}
\tag{7.17}
$$

(iii) *The boundary flow* $\mathbf{x}^{(i_\alpha;\mathscr{B})}(t)$ *to the edge of* $\mathscr{E}_{\sigma_{n-2}}^{(n-2)}$ ($n \neq 2$) *at time* t_m *is a total grazing flow with the* $(2k_{i_\alpha}+1)$th-*order singularity if and only if*

$$
\begin{aligned}
&\hbar_{j_\alpha} G_{\partial\Omega_{\alpha j_\alpha}}^{(s_{i_\alpha}, i_\alpha;\mathscr{B})}(\mathbf{x}_m, t_{m\pm}, \boldsymbol{\lambda}_{i_\alpha}, \boldsymbol{\lambda}_{j_\alpha}) = 0 \text{ for } s_{i_\alpha} = 0,1,2,\cdots,2k_{i_\alpha};\\
&\hbar_{j_\alpha} G_{\partial\Omega_{\alpha j_\alpha}}^{(2k_{i_\alpha}+1, i_\alpha;\mathscr{B})}(\mathbf{x}_m, t_{m\pm}, \boldsymbol{\lambda}_{i_\alpha}, \boldsymbol{\lambda}_{j_\alpha}) > 0.
\end{aligned}
\tag{7.18}
$$

(iv) *The boundary flow* $\mathbf{x}^{(i_\alpha;\mathscr{B})}(t)$ *to the edge of* $\mathscr{E}_{\sigma_{n-2}}^{(n-2)}$ ($n \neq 2$) *at time* t_m *is a coming grazing flow with the* $(2k_{i_\alpha}+1)$th-*order singularity if and only if*

$$
\begin{aligned}
&\hbar_{j_\alpha} G_{\partial\Omega_{\alpha j_\alpha}}^{(s_{i_\alpha}, i_\alpha;\mathscr{B})}(\mathbf{x}_m, t_{m-}, \boldsymbol{\lambda}_{i_\alpha}, \boldsymbol{\lambda}_{j_\alpha}) = 0 \text{ for } s_{i_\alpha} = 0,1,2,\cdots,2k_{i_\alpha};\\
&\hbar_{j_\alpha} G_{\partial\Omega_{\alpha j_\alpha}}^{(2k_{i_\alpha}+1, i_\alpha;\mathscr{B})}(\mathbf{x}_m, t_{m-}, \boldsymbol{\lambda}_{i_\alpha}, \boldsymbol{\lambda}_{j_\alpha}) > 0.
\end{aligned}
\tag{7.19}
$$

(v) *The boundary flow* $\mathbf{x}^{(i_\alpha;\mathscr{B})}(t)$ *to the edge of* $\mathscr{E}^{(n-2)}_{\sigma_{n-2}}$ ($n \neq 2$) *at time* t_m *is a leaving grazing flow with the* $(2k_{i_\alpha}+1)$th-*order singularity if and only if*

$$\begin{aligned}
&\hbar_{j_\alpha} G^{(s_{i_\alpha}, i_\alpha; \mathscr{B})}_{\partial\Omega_{\alpha j_\alpha}}(\mathbf{x}_m, t_{m+}, \boldsymbol{\lambda}_{i_\alpha}, \boldsymbol{\lambda}_{j_\alpha}) = 0 \ \textit{for}\ s_{i_\alpha} = 0,1,2,\cdots,2k_{i_\alpha};\\
&\hbar_{j_\alpha} G^{(2k_{i_\alpha}+1, i_\alpha; \mathscr{B})}_{\partial\Omega_{\alpha j_\alpha}}(\mathbf{x}_m, t_{m+}, \boldsymbol{\lambda}_{i_\alpha}, \boldsymbol{\lambda}_{j_\alpha}) > 0.
\end{aligned} \tag{7.20}$$

Proof. The boundary flow on the boundary can be treated as a special flow in domain to the other boundary of the edge. As in Chapter 3, this theorem can be proved. ■

7.2. Boundary flow switchability

Consider boundary flows existing on boundary surfaces only. Once a boundary flow arrives to the edge, the switchability of two boundary flows via the edge will be discussed.

For an $(n-2)$-dimensional edge, two boundary flows switching is sketched in Figs. 7.4-7.9. The black thick curve represents the $(n-2)$-dimensional edge again. The boundary flows are denoted by curves with arrows on the boundaries. Thus, to determine the attractivity of a boundary flow to the $(n-2)$-dimensional edge, only one of the rest boundary surfaces can be adopted. In Fig. 7.4(a), the boundary flow switching from one boundary $\partial\Omega_{\alpha i_\alpha}$ to another boundary $\partial\Omega_{\beta i_\beta}$ is presented (i.e., from $\mathbf{x}^{(i_\alpha;\mathscr{B})}(t)$ to $\mathbf{x}^{(i_\beta;\mathscr{B})}(t)$). However, the inverse switching of boundary flows from $\mathbf{x}^{(i_\beta;\mathscr{B})}(t)$ to $\mathbf{x}^{(i_\alpha;\mathscr{B})}(t)$ is presented in Fig. 7.4(b). In Fig. 7.5, the non-switchable flows of the first and second kinds between the two boundary flows are presented. Since the switching role exists, the coming grazing flow can be switched to the inflexional and grazing leaving flows in different boundary flows. Without any switching rules, the coming grazing flow cannot be switched into the other boundary flows or domain flows. The grazing-inflexional boundary flow switching and the inflexional-grazing boundary flow switching at the edge are sketched in Fig. 7.6 (a) and (b), respectively. Because the switching rule exists, two grazing flows can be switched each other as in Fig. 7.7(a) and (b). The inflexional-grazing and grazing-grazing, non-switchable boundary flows of the *first* kind are presented in Figs. 7.8 (a) and (b), respectively. Similarly, the inflexional-grazing and grazing-grazing, non-switchable boundary flows of the *second* kind are presented in Figs. 7.9 (a) and (b), respectively. The mathematical descriptions for the switchability of two boundary flows to edges will be presented as in Luo (2008a, b).

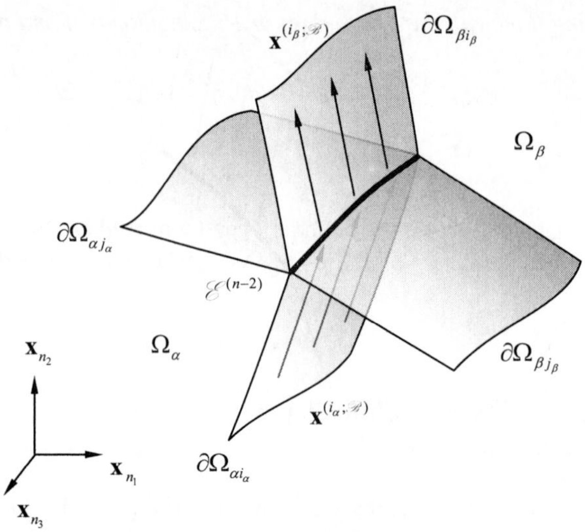

(a)

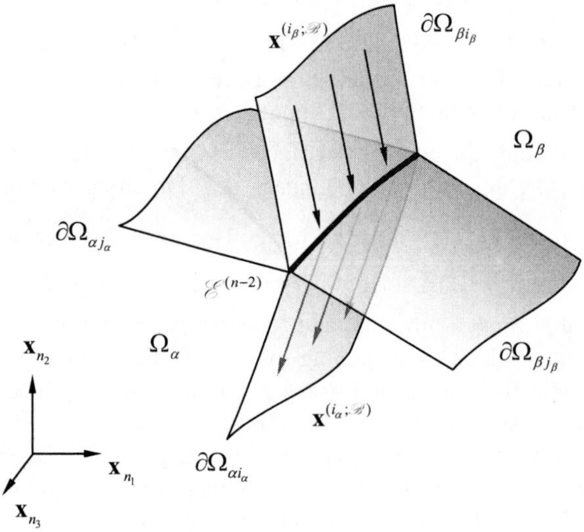

(b)

Fig. 7.4 The switchable boundary flow on the three $(n-1)$-dimensional boundaries to the edge: **(a)** from $\mathbf{x}^{(i_\alpha;\mathscr{B})}(t)$ to $\mathbf{x}^{(i_\beta;\mathscr{B})}(t)$; and **(b)** $\mathbf{x}^{(i_\beta;\mathscr{B})}(t)$ to $\mathbf{x}^{(i_\alpha;\mathscr{B})}(t)$; The thick curve represents the $(n-2)$-dimensional edge. The boundary flows are denoted by curves with arrows on the boundaries. ($n_1+n_2+n_3=n$)

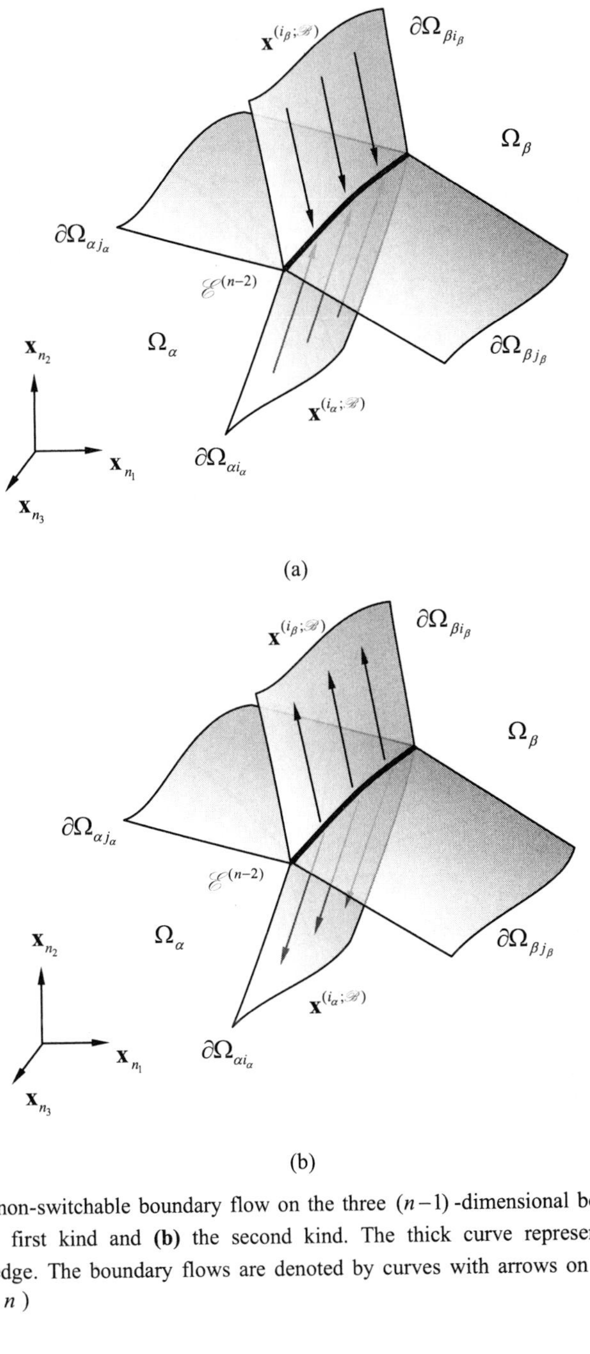

Fig. 7.5 The non-switchable boundary flow on the three $(n-1)$-dimensional boundaries to the edge: **(a)** the first kind and **(b)** the second kind. The thick curve represents the $(n-2)$-dimensional edge. The boundary flows are denoted by curves with arrows on the boundaries. ($n_1+n_2+n_3=n$)

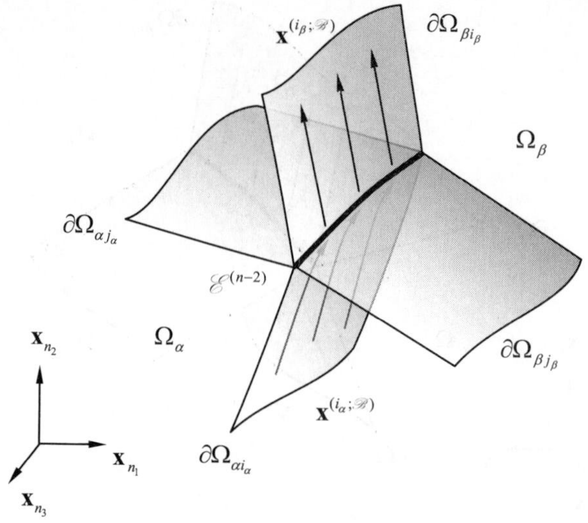

(a)

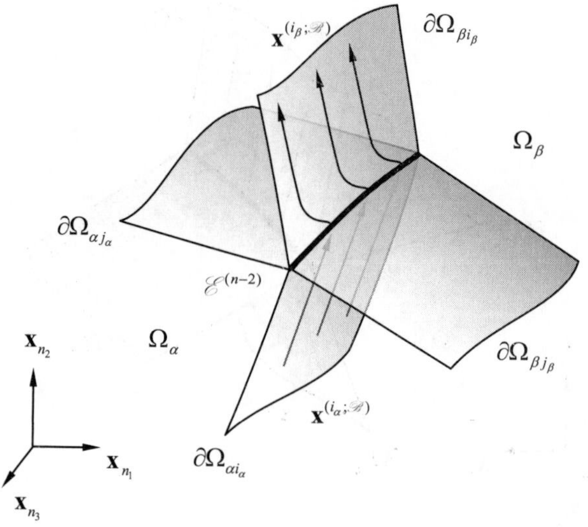

(b)

Fig. 7.6 The grazing-switchable boundary flow on the three $(n-1)$-dimensional boundaries to the edge: **(a)** the inflexion-grazing switching and **(b)** the grazing-inflexion switching. The thick curve represents the $(n-2)$-dimensional edge. The boundary flows are denoted by curves with arrows on the boundaries. ($n_1+n_2+n_3=n$)

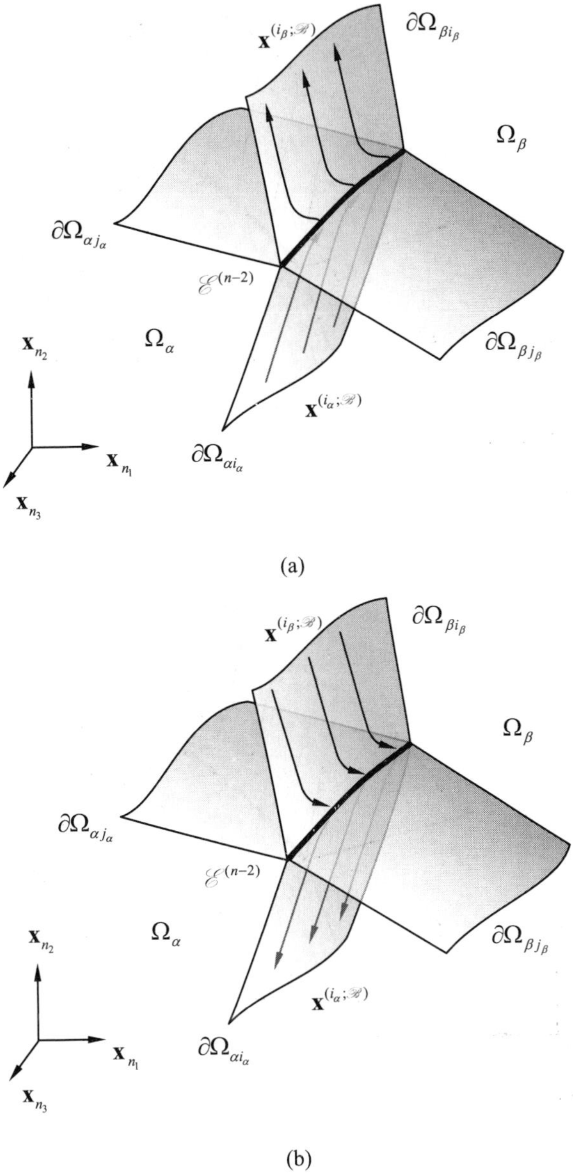

Fig. 7.7 The grazing switchable boundary flow on the three $(n-1)$-dimensional boundaries to the edge: **(a)** $\mathbf{x}^{(i_\alpha;\mathscr{B})}(t)$ to $\mathbf{x}^{(i_\beta;\mathscr{B})}(t)$; and **(b)** $\mathbf{x}^{(i_\beta;\mathscr{B})}(t)$ to $\mathbf{x}^{(i_\alpha;\mathscr{B})}(t)$. The thick curve represents the $(n-2)$-dimensional edge. The boundary flows are denoted by curves with arrows on the boundaries. ($n_1+n_2+n_3=n$)

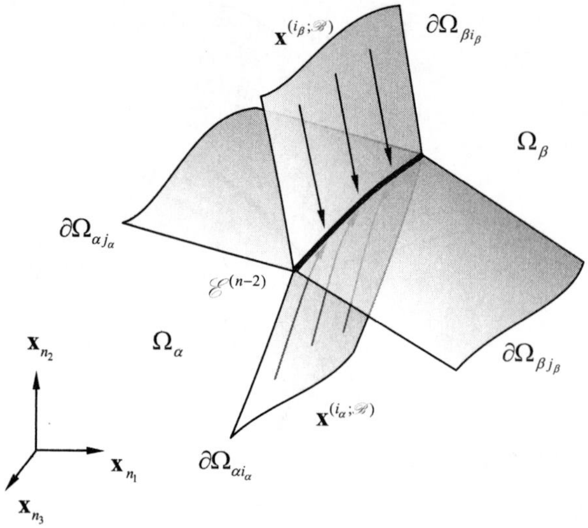

(a)

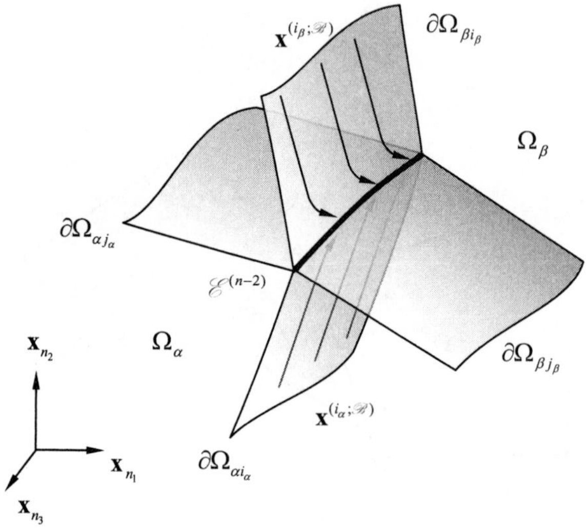

(b)

Fig. 7.8 The non-switchable boundary flow of the first kind on the three $(n-1)$-dimensional boundaries to the edge: **(a)** the inflexional-grazing non-switching and **(b)** the grazing-grazing non-switching. The thick curve represents the $(n-2)$-dimensional edge. The boundary flows are denoted by curves with arrows on the boundaries. ($n_1 + n_2 + n_3 = n$)

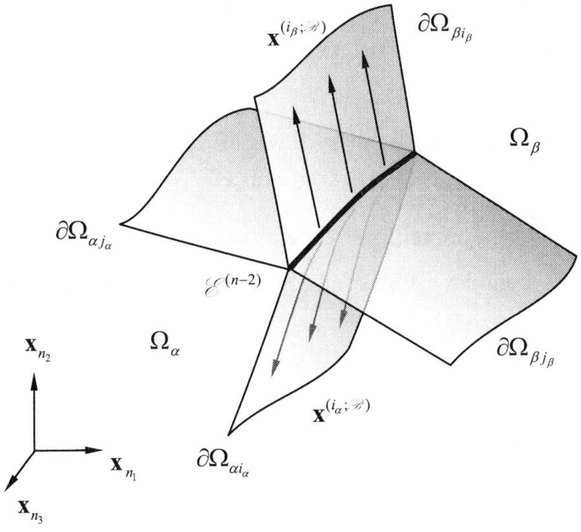

(a)

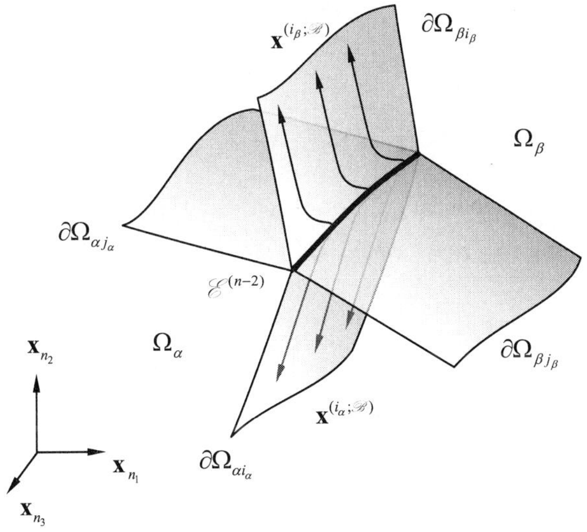

(b)

Fig. 7.9 The non-switchable boundary flow of the second kind on the three $(n-1)$-dimensional boundaries to the edge: **(a)** the inflexional-grazing non-switching and **(b)** the grazing-grazing non-switching. The thick curve represents the $(n-2)$-dimensional edge. The boundary flows are denoted by curves with arrows on the boundaries. ($n_1+n_2+n_3=n$)

Definition 7.3. For a discontinuous dynamical system in Eqs. (6.4)-(6.8), there is an $(n-2)$ -dimensional edge $\mathscr{E}_{\sigma_{n-2}}^{(n-2)}$ ($\sigma_{n-2} \in \mathscr{N}_{\mathscr{E}}^{n-2} = \{1,2,\cdots,N_{n-2}\}$) formed by the intersection of two $(n-1)$ -dimensional boundaries $\partial\Omega_{\alpha i_\alpha}(\sigma_{n-2})$ of domain Ω_α ($\alpha \in \mathscr{N}_{\mathscr{D}}(\sigma_{n-2}) \subset \mathscr{N}_{\mathscr{D}} = \{1,2,\cdots,N_d\}$ and $i_\alpha \in \mathscr{N}_{\alpha\mathscr{B}}(\sigma_{n-2}) \subset \mathscr{N}_{\mathscr{B}} = \{1,2,\cdots, N_b\}$). Suppose there are two boundary flows of $\mathbf{x}^{(i_\rho;\mathscr{B})}(t)$ ($\rho = \alpha,\beta$ and $i_\alpha \neq i_\beta$) on the boundary $\partial\Omega_{\rho i_\rho}$. $\mathbf{x}_m = \mathbf{x}_m^{(i_\alpha;\mathscr{B})} = \mathbf{x}^{(i_\alpha;\mathscr{B})}(t_m) \in \mathscr{E}_{\sigma_{n-2}}^{(n-2)}$ and $\mathbf{x}_m^{(i_\alpha;\mathscr{B})} = \mathbf{x}_m^{(i_\beta;\mathscr{B})}$ at time t_m. For an arbitrarily small $\varepsilon > 0$, there is a time interval $[t_{m-\varepsilon}, t_{m+\varepsilon}]$.

(i) The coming boundary flows of $\mathbf{x}^{(i_\alpha;\mathscr{B})}(t)$ to the leaving boundary flow $\mathbf{x}^{(i_\beta;\mathscr{B})}(t)$ at the edge of $\mathscr{E}_{\sigma_{n-2}}^{(n-2)}$ for time t_m is *switchable* from boundary $\partial\Omega_{\alpha i_\alpha}$ to $\partial\Omega_{\beta i_\beta}$ if

$$\begin{aligned} &\hbar_{j_\alpha} \mathbf{n}_{\partial\Omega_{\alpha j_\alpha}}^{\mathrm{T}}(\mathbf{x}_{m-\varepsilon}^{(j_\alpha;\mathscr{B})}) \cdot [\mathbf{x}_{m-\varepsilon}^{(j_\alpha;\mathscr{B})} - \mathbf{x}_{m-\varepsilon}^{(i_\alpha;\mathscr{B})}] < 0; \\ &\hbar_{j_\beta} \mathbf{n}_{\partial\Omega_{\beta j_\beta}}^{\mathrm{T}}(\mathbf{x}_{m+\varepsilon}^{(j_\beta;\mathscr{B})}) \cdot [\mathbf{x}_{m+\varepsilon}^{(i_\beta;\mathscr{B})} - \mathbf{x}_{m+\varepsilon}^{(j_\beta;\mathscr{B})}] > 0. \end{aligned} \tag{7.21}$$

(ii) The two coming boundary flows of $\mathbf{x}^{(i_\alpha;\mathscr{B})}(t)$ and $\mathbf{x}^{(i_\beta;\mathscr{B})}(t)$ at the edge of $\mathscr{E}_{\sigma_{n-2}}^{(n-2)}$ for time t_m is *non-switchable of the first kind* if

$$\begin{aligned} &\hbar_{j_\alpha} \mathbf{n}_{\partial\Omega_{\alpha j_\alpha}}^{\mathrm{T}}(\mathbf{x}_{m-\varepsilon}^{(j_\alpha;\mathscr{B})}) \cdot [\mathbf{x}_{m-\varepsilon}^{(j_\alpha;\mathscr{B})} - \mathbf{x}_{m-\varepsilon}^{(i_\alpha;\mathscr{B})}] < 0; \\ &\hbar_{j_\beta} \mathbf{n}_{\partial\Omega_{\beta j_\beta}}^{\mathrm{T}}(\mathbf{x}_{m-\varepsilon}^{(j_\beta;\mathscr{B})}) \cdot [\mathbf{x}_{m-\varepsilon}^{(j_\beta;\mathscr{B})} - \mathbf{x}_{m-\varepsilon}^{(i_\beta;\mathscr{B})}] < 0. \end{aligned} \tag{7.22}$$

(iii) The two leaving boundary flows of $\mathbf{x}^{(i_\alpha;\mathscr{B})}(t)$ and $\mathbf{x}^{(i_\beta;\mathscr{B})}(t)$ at the edge of $\mathscr{E}_{\sigma_{n-2}}^{(n-2)}$ for time t_m is *non-switchable of the second kind* if

$$\begin{aligned} &\hbar_{j_\alpha} \mathbf{n}_{\partial\Omega_{\alpha j_\alpha}}^{\mathrm{T}}(\mathbf{x}_{m+\varepsilon}^{(j_\alpha;\mathscr{B})}) \cdot [\mathbf{x}_{m+\varepsilon}^{(i_\alpha;\mathscr{B})} - \mathbf{x}_{m+\varepsilon}^{(j_\alpha;\mathscr{B})}] > 0; \\ &\hbar_{j_\beta} \mathbf{n}_{\partial\Omega_{\beta j_\beta}}^{\mathrm{T}}(\mathbf{x}_{m+\varepsilon}^{(j_\beta;\mathscr{B})}) \cdot [\mathbf{x}_{m+\varepsilon}^{(i_\beta;\mathscr{B})} - \mathbf{x}_{m+\varepsilon}^{(j_\beta;\mathscr{B})}] > 0. \end{aligned} \tag{7.23}$$

From the above definition, the conditions for the boundary flow to edges can be developed as for domain flows to the boundary in Chapters 3 and 6.

Theorem 7.3. *For a discontinuous dynamical system in Eqs.* (6.4)-(6.8), *there is an* $(n-2)$ *-dimensional edge* $\mathscr{E}_{\sigma_{n-2}}^{(n-2)}$ $(\sigma_{n-2} \in \mathscr{N}_{\mathscr{E}}^{n-2} = \{1,2,\cdots,N_{n-2}\})$ *formed by the intersection of two* $(n-1)$ *-dimensional boundaries* $\partial\Omega_{\alpha i_\alpha}(\sigma_{n-2})$ *of domain* Ω_α ($\alpha \in \mathscr{N}_{\mathscr{D}}(\sigma_{n-2}) \subset \mathscr{N}_{\mathscr{D}} = \{1,2,\cdots,N_d\}$ *and* $i_\alpha \in \mathscr{N}_{\alpha\mathscr{B}}(\sigma_{n-2}) \subset \mathscr{N}_{\mathscr{B}} = \{1,2,\cdots, N_b\}$). *Suppose there are two boundary flows of* $\mathbf{x}^{(i_\rho;\mathscr{B})}(t)$ $(\rho = \alpha,\beta$ *and* $i_\alpha \neq i_\beta)$ *on the boundary* $\partial\Omega_{\rho i_\rho}$. $\mathbf{x}_m = \mathbf{x}_m^{(i_\alpha;\mathscr{B})} = \mathbf{x}^{(i_\alpha;\mathscr{B})}(t_m) \in \mathscr{E}_{\sigma_{n-2}}^{(n-2)}$ *and* $\mathbf{x}_m^{(i_\alpha;\mathscr{B})} = \mathbf{x}_m^{(i_\beta;\mathscr{B})}$ *at* *time* t_m. *For an arbitrarily small* $\varepsilon > 0$, *there is a time interval* $[t_{m-\varepsilon}, t_{m+\varepsilon}]$. *The boundary flows of* $\mathbf{x}^{(i_\rho;\mathscr{B})}(t)$ *is* $C^{(r_{i_\rho};\mathscr{B})}_{[t_{m-\varepsilon},0)}$ *or* $C^{(r_{i_\rho};\mathscr{B})}_{(0,t_{m+\varepsilon}]}$*-continuous* ($r_{i_\rho} \geq 2$) *for time* t *and* $\| d^{r_{i_\rho}+1}\mathbf{x}^{(i_\rho;\mathscr{B})} / dt^{r_{i_\rho}+1} \| < \infty$.

(i) *The coming boundary flows of* $\mathbf{x}^{(i_\alpha;\mathscr{B})}(t)$ *to the leaving boundary flow* $\mathbf{x}^{(i_\beta;\mathscr{B})}(t)$ *at the edge of* $\mathscr{E}^{(n-2)}_{\sigma_{n-2}}$ *for time* t_m *is switchable from boundary* $\partial\Omega_{\alpha i_\alpha}$ *to* $\partial\Omega_{\beta i_\beta}$ *if and only if*

$$\begin{aligned} &\hbar_{j_\alpha} G^{(i_\alpha;\mathscr{B})}_{\partial\Omega_{\alpha j_\alpha}}(\mathbf{x}_m, t_{m-}, \boldsymbol{\lambda}_{i_\alpha}, \boldsymbol{\lambda}_{j_\alpha}) < 0;\\ &\hbar_{j_\beta} G^{(i_\beta;\mathscr{B})}_{\partial\Omega_{\beta j_\beta}}(\mathbf{x}_m, t_{m+}, \boldsymbol{\lambda}_{i_\beta}, \boldsymbol{\lambda}_{j_\beta}) > 0. \end{aligned} \tag{7.24}$$

(ii) *The two coming boundary flows of* $\mathbf{x}^{(i_\alpha;\mathscr{B})}(t)$ *and* $\mathbf{x}^{(i_\beta;\mathscr{B})}(t)$ *at the edge of* $\mathscr{E}^{(n-2)}_{\sigma_{n-2}}$ *for time* t_m *is non-switchable of the first kind if and only if*

$$\begin{aligned} &\hbar_{j_\alpha} G^{(i_\alpha;\mathscr{B})}_{\partial\Omega_{\alpha j_\alpha}}(\mathbf{x}_m, t_{m-}, \boldsymbol{\lambda}_{i_\alpha}, \boldsymbol{\lambda}_{j_\alpha}) < 0;\\ &\hbar_{j_\beta} G^{(i_\beta;\mathscr{B})}_{\partial\Omega_{\beta j_\beta}}(\mathbf{x}_m, t_{m-}, \boldsymbol{\lambda}_{i_\beta}, \boldsymbol{\lambda}_{j_\beta}) < 0. \end{aligned} \tag{7.25}$$

(iii) *The two leaving boundary flows of* $\mathbf{x}^{(i_\alpha;\mathscr{B})}(t)$ *and* $\mathbf{x}^{(i_\beta;\mathscr{B})}(t)$ *at the edge of* $\mathscr{E}^{(n-2)}_{\sigma_{n-2}}$ *for time* t_m *is non-switchable of the second kind if and only if*

$$\begin{aligned} &\hbar_{j_\alpha} G^{(i_\alpha;\mathscr{B})}_{\partial\Omega_{\alpha j_\alpha}}(\mathbf{x}_m, t_{m+}, \boldsymbol{\lambda}_{i_\alpha}, \boldsymbol{\lambda}_{j_\alpha}) > 0;\\ &\hbar_{j_\beta} G^{(i_\beta;\mathscr{B})}_{\partial\Omega_{\beta j_\beta}}(\mathbf{x}_m, t_{m+}, \boldsymbol{\lambda}_{i_\beta}, \boldsymbol{\lambda}_{j_\beta}) > 0. \end{aligned} \tag{7.26}$$

Proof. The boundary flow on the boundary can be treated as a special flow in domain to the other boundary of the edge. As in Chapter 3, this theorem can be proved. ■

The boundary flow switchability to edges with the higher-order singularity is described as follows.

Definition 7.4. For a discontinuous dynamical system in Eqs. (6.4)-(6.8), there is an $(n-2)$-dimensional edge $\mathscr{E}^{(n-2)}_{\sigma_{n-2}}$ ($\sigma_{n-2} \in \mathscr{N}^{n-2}_{\mathscr{E}} = \{1, 2, \cdots, N_{n-2}\}$) formed by the intersection of two $(n-1)$-dimensional boundaries $\partial\Omega_{\alpha i_\alpha}(\sigma_{n-2})$ of domain Ω_α ($\alpha \in \mathscr{N}_{\mathscr{D}}(\sigma_{n-2}) \subset \mathscr{N}_{\mathscr{D}} = \{1, 2, \cdots, N_d\}$ and $i_\alpha \in \mathscr{N}_{\alpha\mathscr{B}}(\sigma_{n-2}) \subset \mathscr{N}_{\mathscr{B}} = \{1, 2, \cdots, N_b\}$). Suppose there are two boundary flows of $\mathbf{x}^{(i_\rho;\mathscr{B})}(t)$ ($\rho = \alpha, \beta$ and $i_\alpha \neq i_\beta$) on the boundary $\partial\Omega_{\rho i_\rho}$. $\mathbf{x}_m = \mathbf{x}^{(i_\alpha;\mathscr{B})}_m = \mathbf{x}^{(i_\alpha;\mathscr{B})}(t_m) \in \mathscr{E}^{(n-2)}_{\sigma_{n-2}}$ and $\mathbf{x}^{(i_\alpha;\mathscr{B})}_m = \mathbf{x}^{(i_\beta;\mathscr{B})}_m$ at time t_m. For an arbitrarily small $\varepsilon > 0$, there is a time interval $[t_{m-\varepsilon}, t_{m+\varepsilon}]$. The boundary flows of $\mathbf{x}^{(i_\rho;\mathscr{B})}(t)$ is $C^{(r_{i_\rho};\mathscr{B})}_{[t_{m-\varepsilon},0)}$ or $C^{(r_{i_\rho};\mathscr{B})}_{(0,t_{m+\varepsilon}]}$-continuous ($r_{i_\rho} \geq m_{i_\rho}$) for time t and $\| d^{r_{i_\rho}+1}\mathbf{x}^{(i_\rho;\mathscr{B})}/dt^{r_{i_\rho}+1} \| < \infty$. (As $m_{i_\rho} = 2k_{i_\rho}+1$, $n \neq 2$).

(i) With a switching rule, the coming boundary flows of $\mathbf{x}^{(i_\alpha;\mathscr{B})}(t)$ to the leaving boundary flow $\mathbf{x}^{(i_\beta;\mathscr{B})}(t)$ at the edge of $\mathscr{E}^{(n-2)}_{\sigma_{n-2}}$ for time t_m is *switchable* with the $(m_{i_\alpha} : m_{i_\beta};\mathscr{B})$-singularity from boundary $\partial\Omega_{\alpha i_\alpha}$ to $\partial\Omega_{\beta i_\beta}$ if

$$\left.\begin{array}{l}
\hbar_{j_\alpha} G^{(s_{i_\alpha},i_\alpha;\mathscr{B})}_{\partial\Omega_{\alpha j_\alpha}}(\mathbf{x}_m,t_{m-},\boldsymbol{\lambda}_{i_\alpha},\boldsymbol{\lambda}_{j_\alpha}) = 0 \text{ for } s_{i_\alpha} = 0,1,2,\cdots,m_{i_\alpha}-1;\\
\hbar_{j_\beta} G^{(s_{i_\beta},i_\beta;\mathscr{B})}_{\partial\Omega_{\beta j_\beta}}(\mathbf{x}_m,t_{m+},\boldsymbol{\lambda}_{i_\beta},\boldsymbol{\lambda}_{j_\beta}) = 0 \text{ for } s_{i_\beta} = 0,1,2,\cdots,m_{i_\beta}-1;\\
\hbar_{j_\alpha} \mathbf{n}^{\mathrm{T}}_{\partial\Omega_{\alpha j_\alpha}}(\mathbf{x}^{(j_\alpha;\mathscr{B})}_{m-\varepsilon})\cdot[\mathbf{x}^{(j_\alpha;\mathscr{B})}_{m-\varepsilon} - \mathbf{x}^{(i_\alpha;\mathscr{B})}_{m-\varepsilon}] < 0;\\
\hbar_{j_\beta} \mathbf{n}^{\mathrm{T}}_{\partial\Omega_{\beta j_\beta}}(\mathbf{x}^{(j_\beta;\mathscr{B})}_{m+\varepsilon})\cdot[\mathbf{x}^{(i_\beta;\mathscr{B})}_{m+\varepsilon} - \mathbf{x}^{(j_\beta;\mathscr{B})}_{m+\varepsilon}] > 0.
\end{array}\right\} \tag{7.27}$$

(ii) The two coming boundary flows of $\mathbf{x}^{(i_\alpha;\mathscr{B})}(t)$ and $\mathbf{x}^{(i_\beta;\mathscr{B})}(t)$ at the edge of $\mathscr{E}^{(n-2)}_{\sigma_{n-2}}$ for time t_m is *non-switchable of the first kind* with the $(2k_{i_\alpha} : 2k_{i_\beta};\mathscr{B})$-singularity if

$$\left.\begin{array}{l}
\hbar_{j_\alpha} G^{(s_{i_\alpha},i_\alpha;\mathscr{B})}_{\partial\Omega_{\alpha j_\alpha}}(\mathbf{x}_m,t_{m-},\boldsymbol{\lambda}_{i_\alpha},\boldsymbol{\lambda}_{j_\alpha}) = 0 \text{ for } s_{i_\alpha} = 0,1,2,\cdots 2k_{i_\alpha}-1;\\
\hbar_{j_\beta} G^{(s_{i_\beta},i_\beta;\mathscr{B})}_{\partial\Omega_{\beta j_\beta}}(\mathbf{x}_m,t_{m-},\boldsymbol{\lambda}_{i_\beta},\boldsymbol{\lambda}_{j_\beta}) = 0 \text{ for } s_{i_\beta} = 0,1,2,\cdots 2k_{i_\beta}-1;\\
\hbar_{j_\alpha} \mathbf{n}^{\mathrm{T}}_{\partial\Omega_{\alpha j_\alpha}}(\mathbf{x}^{(j_\alpha;\mathscr{B})}_{m-\varepsilon})\cdot[\mathbf{x}^{(j_\alpha;\mathscr{B})}_{m-\varepsilon} - \mathbf{x}^{(i_\alpha;\mathscr{B})}_{m-\varepsilon}] < 0;\\
\hbar_{j_\beta} \mathbf{n}^{\mathrm{T}}_{\partial\Omega_{\beta j_\beta}}(\mathbf{x}^{(j_\beta;\mathscr{B})}_{m-\varepsilon})\cdot[\mathbf{x}^{(j_\beta;\mathscr{B})}_{m-\varepsilon} - \mathbf{x}^{(i_\beta;\mathscr{B})}_{m-\varepsilon}] < 0.
\end{array}\right\} \tag{7.28}$$

(iii) The two leaving boundary flows of $\mathbf{x}^{(i_\alpha;\mathscr{B})}(t)$ and $\mathbf{x}^{(i_\beta;\mathscr{B})}(t)$ at the edge of $\mathscr{E}^{(n-2)}_{\sigma_{n-2}}$ for time t_m is *non-switchable of the second kind* with the $(m_{i_\alpha} : m_{i_\beta};\mathscr{B})$-singularity if

$$\left.\begin{array}{l}
\hbar_{j_\alpha} G^{(s_{i_\alpha},i_\alpha;\mathscr{B})}_{\partial\Omega_{\alpha j_\alpha}}(\mathbf{x}_m,t_{m+},\boldsymbol{\lambda}_{i_\alpha},\boldsymbol{\lambda}_{j_\alpha}) = 0 \text{ for } s_{i_\alpha} = 0,1,2,\cdots,m_{i_\alpha}-1;\\
\hbar_{j_\beta} G^{(s_{i_\beta},i_\beta;\mathscr{B})}_{\partial\Omega_{\beta j_\beta}}(\mathbf{x}_m,t_{m+},\boldsymbol{\lambda}_{i_\beta},\boldsymbol{\lambda}_{j_\beta}) = 0 \text{ for } s_{i_\beta} = 0,1,2,\cdots,m_{i_\beta}-1;\\
\hbar_{j_\alpha} \mathbf{n}^{\mathrm{T}}_{\partial\Omega_{\alpha j_\alpha}}(\mathbf{x}^{(j_\alpha;\mathscr{B})}_{m+\varepsilon})\cdot[\mathbf{x}^{(i_\alpha;\mathscr{B})}_{m+\varepsilon} - \mathbf{x}^{(j_\alpha;\mathscr{B})}_{m+\varepsilon}] > 0;\\
\hbar_{j_\beta} \mathbf{n}^{\mathrm{T}}_{\partial\Omega_{\beta j_\beta}}(\mathbf{x}^{(j_\beta;\mathscr{B})}_{m+\varepsilon})\cdot[\mathbf{x}^{(i_\beta;\mathscr{B})}_{m+\varepsilon} - \mathbf{x}^{(j_\beta;\mathscr{B})}_{m+\varepsilon}] > 0.
\end{array}\right\} \tag{7.29}$$

(iv) Without any switching rules, the coming boundary flows of $\mathbf{x}^{(i_\alpha;\mathscr{B})}(t)$ to the leaving boundary flow $\mathbf{x}^{(i_\beta;\mathscr{B})}(t)$ at the edge of $\mathscr{E}^{(n-2)}_{\sigma_{n-2}}$ for time t_m is *potentially switchable* with the $(2k_{i_\alpha} : m_{i_\beta};\mathscr{B})$-singularity from boundary $\partial\Omega_{\alpha i_\alpha}$ to

$\partial\Omega_{\beta i_\beta}$ if Eq. (7.27) holds with $m_{i_\alpha} = 2k_{i_\alpha}$.

(v) Without any switching rules, the coming boundary flows of $\mathbf{x}^{(i_\alpha;\mathscr{B})}(t)$ to the leaving boundary flow $\mathbf{x}^{(i_\beta;\mathscr{B})}(t)$ at the edge of $\mathscr{E}^{(n-2)}_{\sigma_{n-2}}$ for time t_m is *non-switchable* with the $(2k_{i_\alpha}+1:m_{i_\beta};\mathscr{B})$-singularity from boundary $\partial\Omega_{\alpha i_\alpha}$ to $\partial\Omega_{\beta i_\beta}$ if Eq. (7.27) holds with $m_{i_\alpha} = 2k_{i_\alpha}+1$.

(vi) With any switching rules, two grazing boundary flows of $\mathbf{x}^{(i_\alpha;\mathscr{B})}(t)$ and $\mathbf{x}^{(i_\beta;\mathscr{B})}(t)$ at the edge of $\mathscr{E}^{(n-2)}_{\sigma_{n-2}}$ for time t_m are *switchable* at the edge with the $(2k_{i_\alpha}+1:2k_{i_\beta}+1;\mathscr{B})$-singularity if

$$\left.\begin{aligned}
&\hbar_{j_\alpha} G^{(s_{i_\alpha},i_\alpha;\mathscr{B})}_{\partial\Omega_{\alpha j_\alpha}}(\mathbf{x}_m,t_{m\pm},\boldsymbol{\lambda}_{i_\alpha},\boldsymbol{\lambda}_{j_\alpha}) = 0 \text{ for } s_{i_\alpha} = 0,1,2,\cdots,2k_{i_\alpha};\\
&\hbar_{j_\beta} G^{(s_{i_\beta},i_\beta;\mathscr{B})}_{\partial\Omega_{\beta j_\beta}}(\mathbf{x}_m,t_{m\pm},\boldsymbol{\lambda}_{i_\beta},\boldsymbol{\lambda}_{j_\beta}) = 0 \text{ for } s_{i_\beta} = 0,1,2,\cdots,2k_{i_\beta};\\
&\hbar_{j_\alpha}\mathbf{n}^{\mathrm{T}}_{\partial\Omega_{\alpha j_\alpha}}(\mathbf{x}^{(j_\alpha;\mathscr{B})}_{m-\varepsilon})\cdot[\mathbf{x}^{(j_\alpha;\mathscr{B})}_{m-\varepsilon}-\mathbf{x}^{(i_\alpha;\mathscr{B})}_{m-\varepsilon}]<0,\\
&\hbar_{j_\beta}\mathbf{n}^{\mathrm{T}}_{\partial\Omega_{\beta j_\beta}}(\mathbf{x}^{(j_\beta;\mathscr{B})}_{m-\varepsilon})\cdot[\mathbf{x}^{(j_\beta;\mathscr{B})}_{m-\varepsilon}-\mathbf{x}^{(i_\beta;\mathscr{B})}_{m-\varepsilon}]<0;\\
&\hbar_{j_\alpha}\mathbf{n}^{\mathrm{T}}_{\partial\Omega_{\alpha j_\alpha}}(\mathbf{x}^{(j_\alpha;\mathscr{B})}_{m+\varepsilon})\cdot[\mathbf{x}^{(i_\alpha;\mathscr{B})}_{m+\varepsilon}-\mathbf{x}^{(j_\alpha;\mathscr{B})}_{m+\varepsilon}]>0,\\
&\hbar_{j_\beta}\mathbf{n}^{\mathrm{T}}_{\partial\Omega_{\beta j_\beta}}(\mathbf{x}^{(j_\beta;\mathscr{B})}_{m+\varepsilon})\cdot[\mathbf{x}^{(i_\beta;\mathscr{B})}_{m+\varepsilon}-\mathbf{x}^{(j_\beta;\mathscr{B})}_{m+\varepsilon}]>0.
\end{aligned}\right\} \tag{7.30}$$

(vii) Without any switching rules, two grazing boundary flows of $\mathbf{x}^{(i_\alpha;\mathscr{B})}(t)$ and $\mathbf{x}^{(i_\beta;\mathscr{B})}(t)$ at the edge of $\mathscr{E}^{(n-2)}_{\sigma_{n-2}}$ for time t_m are the *double tangential flows* to the edge with the $(2k_{i_\alpha}+1:2k_{i_\beta}+1;\mathscr{B})$-singularity if Eq. (7.30) holds.

Theorem 7.4. *For a discontinuous dynamical system in Eqs.* (6.4)-(6.8)*, there is an* $(n-2)$*-dimensional edge* $\mathscr{E}^{(n-2)}_{\sigma_{n-2}}$ ($\sigma_{n-2}\in\mathscr{N}^{n-2}_{\mathscr{E}}=\{1,2,\cdots,N_{n-2}\}$) *formed by the intersection of two* $(n-1)$*-dimensional boundaries* $\partial\Omega_{\alpha i_\alpha}(\sigma_{n-2})$ *of domain* Ω_α ($\alpha\in\mathscr{N}_{\mathscr{D}}(\sigma_{n-2})\subset\mathscr{N}_{\mathscr{D}}=\{1,2,\cdots,N_{\mathscr{D}}\}$ *and* $i_\alpha\in\mathscr{N}_{\alpha\mathscr{B}}(\sigma_{n-2})\subset\mathscr{N}_{\mathscr{B}}=\{1,2,\cdots,N_b\}$)*. Suppose there are two boundary flows of* $\mathbf{x}^{(i_\rho;\mathscr{B})}(t)$ ($\rho=\alpha,\beta$ *and* $i_\alpha\neq i_\beta$) *on the boundary* $\partial\Omega_{\rho i_\rho}$. $\mathbf{x}_m=\mathbf{x}^{(i_\alpha;\mathscr{B})}_m=\mathbf{x}^{(i_\alpha;\mathscr{B})}(t_m)\in\mathscr{E}^{(n-2)}_{\sigma_{n-2}}$ *and* $\mathbf{x}^{(i_\alpha;\mathscr{B})}_m=\mathbf{x}^{(i_\beta;\mathscr{B})}_m$ *at time* t_m. *For an arbitrarily small* $\varepsilon>0$, *there is a time interval* $[t_{m-\varepsilon},t_{m+\varepsilon}]$. *The boundary flows of* $\mathbf{x}^{(i_\rho;\mathscr{B})}(t)$ *is* $C^{(r_{i_\rho};\mathscr{B})}_{[t_{m-\varepsilon},0)}$ *or* $C^{(r_{i_\rho};\mathscr{B})}_{(0,t_{m+\varepsilon}]}$*-continuous* ($r_{i_\rho}\geq m_{i_\rho}+1$) *for time* t *and* $\|d^{r_{i_\rho}+1}\mathbf{x}^{(i_\rho;\mathscr{B})}/dt^{r_{i_\rho}+1}\|<\infty$. (*As* $m_{i_\rho}=2k_{i_\rho}+1$, $n\neq 2$).

(i) *With a switching rule, the coming boundary flows of* $\mathbf{x}^{(i_\alpha;\mathscr{B})}(t)$ *to the leaving boundary flow* $\mathbf{x}^{(i_\beta;\mathscr{B})}(t)$ *at the edge of* $\mathscr{E}^{(n-2)}_{\sigma_{n-2}}$ *for time* t_m *is switchable with the* $(m_{i_\alpha}:m_{i_\beta};\mathscr{B})$*-singularity from boundary* $\partial\Omega_{\alpha i_\alpha}$ *to* $\partial\Omega_{\beta i_\beta}$ *if and only if*

$$\left.\begin{aligned}
&\hbar_{j_\alpha} G^{(s_{i_\alpha},i_\alpha;\mathscr{B})}_{\partial\Omega_{\alpha j_\alpha}}(\mathbf{x}_m,t_{m-},\boldsymbol{\lambda}_{i_\alpha},\boldsymbol{\lambda}_{j_\alpha})=0 \ \textit{for}\ s_{i_\alpha}=0,1,2,\cdots,m_{i_\alpha}-1;\\
&\hbar_{j_\beta} G^{(s_{i_\beta},i_\beta;\mathscr{B})}_{\partial\Omega_{\beta j_\beta}}(\mathbf{x}_m,t_{m+},\boldsymbol{\lambda}_{i_\beta},\boldsymbol{\lambda}_{j_\beta})=0 \ \textit{for}\ s_{i_\beta}=0,1,2,\cdots,m_{i_\beta}-1;\\
&(-1)^{m_{i_\alpha}}\hbar_{j_\alpha} G^{(m_{i_\alpha},i_\alpha;\mathscr{B})}_{\partial\Omega_{\alpha j_\alpha}}(\mathbf{x}_m,t_{m-},\boldsymbol{\lambda}_{i_\alpha},\boldsymbol{\lambda}_{j_\alpha})<0;\\
&\hbar_{j_\beta} G^{(m_{i_\beta},i_\beta;\mathscr{B})}_{\partial\Omega_{\beta j_\beta}}(\mathbf{x}_m,t_{m+},\boldsymbol{\lambda}_{i_\beta},\boldsymbol{\lambda}_{j_\beta})>0.
\end{aligned}\right\}\qquad(7.31)$$

(ii) *The two coming boundary flows of* $\mathbf{x}^{(i_\alpha;\mathscr{B})}(t)$ *and* $\mathbf{x}^{(i_\beta;\mathscr{B})}(t)$ *at the edge of* $\mathscr{E}^{(n-2)}_{\sigma_{n-2}}$ *for time* t_m *is non-switchable of the first kind with the* $(2k_{i_\alpha}:2k_{i_\beta};\mathscr{B})$ *-singularity if and only if*

$$\left.\begin{aligned}
&\hbar_{j_\alpha} G^{(s_{i_\alpha},i_\alpha;\mathscr{B})}_{\partial\Omega_{\alpha j_\alpha}}(\mathbf{x}_m,t_{m-},\boldsymbol{\lambda}_{i_\alpha},\boldsymbol{\lambda}_{j_\alpha})=0 \ \textit{for}\ s_{i_\alpha}=0,1,2,\cdots,2k_{i_\alpha}-1;\\
&\hbar_{j_\beta} G^{(s_{i_\beta},i_\beta;\mathscr{B})}_{\partial\Omega_{\beta j_\beta}}(\mathbf{x}_m,t_{m-},\boldsymbol{\lambda}_{i_\beta},\boldsymbol{\lambda}_{j_\beta})=0 \ \textit{for}\ s_{i_\beta}=0,1,2,\cdots,2k_{i_\beta}-1;\\
&\hbar_{j_\alpha} G^{(2k_{i_\alpha},i_\alpha;\mathscr{B})}_{\partial\Omega_{\alpha j_\alpha}}(\mathbf{x}_m,t_{m-},\boldsymbol{\lambda}_{i_\alpha},\boldsymbol{\lambda}_{j_\alpha})<0;\\
&\hbar_{j_\beta} G^{(2k_{i_\beta},i_\beta;\mathscr{B})}_{\partial\Omega_{\beta j_\beta}}(\mathbf{x}_m,t_{m-},\boldsymbol{\lambda}_{i_\beta},\boldsymbol{\lambda}_{j_\beta})<0.
\end{aligned}\right\}\qquad(7.32)$$

(iii) *The two leaving boundary flows of* $\mathbf{x}^{(i_\alpha;\mathscr{B})}(t)$ *and* $\mathbf{x}^{(i_\beta;\mathscr{B})}(t)$ *at the edge of* $\mathscr{E}^{(n-2)}_{\sigma_{n-2}}$ *for time* t_m *is non-switchable of the second kind with the* $(m_{i_\alpha}:m_{i_\beta};\mathscr{B})$ *-singularity if and only if*

$$\left.\begin{aligned}
&\hbar_{j_\alpha} G^{(s_{i_\alpha},i_\alpha;\mathscr{B})}_{\partial\Omega_{\alpha j_\alpha}}(\mathbf{x}_m,t_{m+},\boldsymbol{\lambda}_{i_\alpha},\boldsymbol{\lambda}_{j_\alpha})=0 \ \textit{for}\ s_{i_\alpha}=0,1,2,\cdots,m_{i_\alpha}-1;\\
&\hbar_{j_\beta} G^{(s_{i_\beta},i_\beta;\mathscr{B})}_{\partial\Omega_{\beta j_\beta}}(\mathbf{x}_m,t_{m+},\boldsymbol{\lambda}_{i_\beta},\boldsymbol{\lambda}_{j_\beta})=0 \ \textit{for}\ s_{i_\beta}=0,1,2,\cdots,m_{i_\beta}-1;\\
&\hbar_{j_\alpha} G^{(m_{i_\alpha},i_\alpha;\mathscr{B})}_{\partial\Omega_{\alpha j_\alpha}}(\mathbf{x}_m,t_{m+},\boldsymbol{\lambda}_{i_\alpha},\boldsymbol{\lambda}_{j_\alpha})>0;\\
&\hbar_{j_\beta} G^{(m_{i_\beta},i_\beta;\mathscr{B})}_{\partial\Omega_{\beta j_\beta}}(\mathbf{x}_m,t_{m+},\boldsymbol{\lambda}_{i_\beta},\boldsymbol{\lambda}_{j_\beta})>0.
\end{aligned}\right\}\qquad(7.33)$$

(iv) *Without any switching rules, the coming boundary flows of* $\mathbf{x}^{(i_\alpha;\mathscr{B})}(t)$ *to the leaving boundary flow* $\mathbf{x}^{(i_\beta;\mathscr{B})}(t)$ *at the edge of* $\mathscr{E}^{(n-2)}_{\sigma_{n-2}}$ *for time* t_m *is potentially switchable with the* $(2k_{i_\alpha}:m_{i_\beta};\mathscr{B})$*-singularity from boundary* $\partial\Omega_{\alpha i_\alpha}$ *to* $\partial\Omega_{\beta i_\beta}$ *if and only if Eq. (7.31) holds with* $m_{i_\alpha}=2k_{i_\alpha}$.

(v) *Without any switching rules, the coming boundary flows of* $\mathbf{x}^{(i_\alpha;\mathscr{B})}(t)$ *to the leaving boundary flow* $\mathbf{x}^{(i_\beta;\mathscr{B})}(t)$ *at the edge of* $\mathscr{E}^{(n-2)}_{\sigma_{n-2}}$ *for time* t_m *is non-switchable with the* $(2k_{i_\alpha}+1:m_{i_\beta};\mathscr{B})$ *-singularity from boundary* $\partial\Omega_{\alpha i_\alpha}$ *to*

$\partial\Omega_{\beta i_\beta}$ *if and only if Eq.* (7.31) *holds with* $m_{i_\alpha} = 2k_{i_\alpha} + 1$.

(vi) *With the switching rule, two grazing boundary flows of* $\mathbf{x}^{(i_\alpha;\mathscr{B})}(t)$ *and* $\mathbf{x}^{(i_\beta;\mathscr{B})}(t)$ *at the edge of* $\mathscr{E}^{(n-2)}_{\sigma_{n-2}}$ *for time* t_m *are switchable to the edge with the* $(2k_{i_\alpha}+1:2k_{i_\beta}+1;\mathscr{B})$*-singularity if and only if Eq.* (7.34) *holds.*

$$\left.\begin{aligned}
&\hbar_{j_\alpha} G^{(s_{i_\alpha},i_\alpha;\mathscr{B})}_{\partial\Omega_{\alpha j_\alpha}}(\mathbf{x}_m,t_{m\pm},\boldsymbol{\lambda}_{i_\alpha},\boldsymbol{\lambda}_{j_\alpha}) = 0 \ \ \textit{for}\ s_{i_\alpha} = 0,1,2,\cdots,2k_{i_\alpha};\\
&\hbar_{j_\beta} G^{(s_{i_\beta},i_\beta;\mathscr{B})}_{\partial\Omega_{\beta j_\beta}}(\mathbf{x}_m,t_{m\pm},\boldsymbol{\lambda}_{i_\beta},\boldsymbol{\lambda}_{j_\beta}) = 0 \ \ \textit{for}\ s_{i_\beta} = 0,1,2,\cdots,2k_{i_\beta};\\
&\hbar_{j_\alpha} G^{(2k_{i_\alpha}+1,i_\alpha;\mathscr{B})}_{\partial\Omega_{\alpha j_\alpha}}(\mathbf{x}_m,t_{m\pm},\boldsymbol{\lambda}_{i_\alpha},\boldsymbol{\lambda}_{j_\alpha}) > 0;\\
&\hbar_{j_\beta} G^{(2k_{i_\beta}+1,i_\beta;\mathscr{B})}_{\partial\Omega_{\beta j_\beta}}(\mathbf{x}_m,t_{m\pm},\boldsymbol{\lambda}_{i_\beta},\boldsymbol{\lambda}_{j_\beta}) > 0.
\end{aligned}\right\} \tag{7.34}$$

(vii) *Without any switching rules, two grazing boundary flows of* $\mathbf{x}^{(i_\alpha;\mathscr{B})}(t)$ *and* $\mathbf{x}^{(i_\beta;\mathscr{B})}(t)$ *at the edge of* $\mathscr{E}^{(n-2)}_{\sigma_{n-2}}$ *for time* t_m *are the double, tangential flows to the edge with the* $(2k_{i_\alpha}+1:2k_{i_\beta}+1;\mathscr{B})$*-singularity if and only if*

Proof. The boundary flow on the boundary can be treated as a special flow in domain to the other boundary of the edge. As in Chapter 3, this theorem can be proved. ■

The foregoing discussion on the switchability of a boundary flow to another boundary flow via their edge was given. In fact, a coming boundary flow can possess the similar switchability to several leaving boundary flows, or such a boundary flow can be switched into different leaving boundary at the same time. On the other hand, several coming flows can be switched into a leaving boundary flow. Consider a specific $(n-2)$-dimensional edge to be the intersection of m_b-boundary surfaces. Among the m_b-boundary surfaces, there are m_{b1}- boundary surfaces on which the coming flows exist and m_{b2}-boundary surface on which the leaving flows exist. Thus an index set for the boundary surfaces related to the edge of $\mathscr{E}^{(n-2)}_{\sigma_{n-2}}$ is $\mathscr{N}_{\mathscr{B}}(\sigma_{n-2}) = \{1,2,\cdots,m_b\}$, and the index set can be decomposed of two index subsets for the coming and leaving flow boundary surfaces, i.e., $\mathscr{N}_{\mathscr{B}}(\sigma_{n-2}) = \mathscr{N}^C_{\mathscr{B}}(\sigma_{n-2}) \cup \mathscr{N}^L_{\mathscr{B}}(\sigma_{n-2})$ and $\mathscr{N}^C_{\mathscr{B}}(\sigma_{n-2}) \cap \mathscr{N}^L_{\mathscr{B}}(\sigma_{n-2}) = \varnothing$.

Definition 7.5. For a discontinuous dynamical system in Eqs. (6.4)-(6.8), there is an $(n-2)$-dimensional edge $\mathscr{E}^{(n-2)}_{\sigma_{n-2}}$ ($\sigma_{n-2} \in \mathscr{N}^{n-2}_{\mathscr{E}} = \{1,2,\cdots,N_{n-2}\}$) to be formed by m_b-boundaries $\partial\Omega_{\alpha i_\alpha}(\sigma_{n-2})$ of $(n-1)$-dimensions ($\alpha, i_\alpha \in \mathscr{N}_{\mathscr{D}}(\sigma_{n-2}) \subset \mathscr{N}_{\mathscr{D}} \in \{1,2,\cdots,N_d\}$, and also $i_\alpha \in \mathscr{N}_{\mathscr{B}}(\sigma_{n-2}) \subset \mathscr{N}_{\mathscr{B}} = \{1,2,\cdots,N_b\}$ with $\alpha \neq i_\alpha$). $\mathscr{N}_{\mathscr{B}}(\sigma_{n-2}) = \mathscr{N}^C_{\mathscr{B}}(\sigma_{n-2}) \cup \mathscr{N}^L_{\mathscr{B}}(\sigma_{n-2})$ and $\mathscr{N}^C_{\mathscr{B}}(\sigma_{n-2}) \cap \mathscr{N}^L_{\mathscr{B}}(\sigma_{n-2}) = \varnothing$. Suppose the coming and leaving boundary flows are $\mathbf{x}^{(i_\alpha;\mathscr{B})}(t)$ ($i_\alpha \in \mathscr{N}_{\mathscr{B}}^C(\sigma_{n-2})$) and $\mathbf{x}^{(i_\beta;\mathscr{B})}(t)$ ($i_\beta \in \mathscr{N}_{\mathscr{B}}^L(\sigma_{n-2})$) on the boundary $\partial\Omega_{\alpha i_\alpha}$ and $\partial\Omega_{\beta i_\beta}$, respectively. At time t_m, $\mathbf{x}_m^{(i_\alpha;\mathscr{B})} = \mathbf{x}_m^{(i_\beta;\mathscr{B})}$ and $\mathbf{x}_m = \mathbf{x}_m^{(i_\alpha;\mathscr{B})} = \mathbf{x}^{(i_\alpha;\mathscr{B})}(t_m) \in \mathscr{E}_{\sigma_{n-2}}^{(n-2)}$. For an arbitrarily small $\varepsilon > 0$, there is a time interval $[t_{m-\varepsilon}, t_{m+\varepsilon}]$. The boundary flows of $\mathbf{x}^{(i_\rho;\mathscr{B})}(t)$ ($\rho = \alpha, \beta$) is $C_{[t_{m-\varepsilon},0)}^{(r_{i_\rho};\mathscr{B})}$ or $C_{(0,t_{m+\varepsilon}]}^{(r_{i_\rho};\mathscr{B})}$-continuous and $\| d^{r_{i_\rho}+1}\mathbf{x}^{(i_\rho;\mathscr{B})} / dt^{r_{i_\rho}+1} \| < \infty$ ($r_{i_\rho} \geq 1$) for time t.

(i) The coming boundary flows of $\mathbf{x}^{(i_\alpha;\mathscr{B})}(t)$ ($i_\alpha \in \mathscr{N}_{\mathscr{B}}^C(\sigma_{n-2})$) to the leaving boundary flows of $\mathbf{x}^{(i_\beta;\mathscr{B})}(t)$ ($i_\beta \in \mathscr{N}_{\mathscr{B}}^L(\sigma_{n-2})$) at the edge of $\mathscr{E}_{\sigma_{n-2}}^{(n-2)}$ for time t_m are ($\mathscr{N}_{\mathscr{B}}^C : \mathscr{N}_{\mathscr{B}}^L$)-*switchable* from boundary $\partial\Omega_{\alpha i_\alpha}$ to $\partial\Omega_{\beta i_\beta}$ if

$$\begin{aligned} &\hbar_{j_\alpha}\mathbf{n}_{\partial\Omega_{\alpha j_\alpha}}^{\mathrm{T}}(\mathbf{x}_{m-\varepsilon}^{(j_\alpha;\mathscr{B})})\cdot[\mathbf{x}_{m-\varepsilon}^{(j_\alpha;\mathscr{B})} - \mathbf{x}_{m-\varepsilon}^{(i_\alpha;\mathscr{B})}] < 0 \text{ for all } i_\alpha \in \mathscr{N}_{\mathscr{B}}^C(\sigma_{n-2});\\ &\hbar_{j_\beta}\mathbf{n}_{\partial\Omega_{\beta j_\beta}}^{\mathrm{T}}(\mathbf{x}_{m+\varepsilon}^{(j_\beta;\mathscr{B})})\cdot[\mathbf{x}_{m+\varepsilon}^{(i_\beta;\mathscr{B})} - \mathbf{x}_{m+\varepsilon}^{(j_\beta;\mathscr{B})}] > 0 \text{ for all } i_\beta \in \mathscr{N}_{\mathscr{B}}^L(\sigma_{n-2}). \end{aligned} \tag{7.35}$$

(ii) The edge of $\mathscr{E}_{\sigma_{n-2}}^{(n-2)}$ to the corresponding boundary flows of $\mathbf{x}^{(i_\alpha;\mathscr{B})}(t)$ is ($\mathscr{N}_{\mathscr{B}}^C;\mathscr{B}$)-*non-switchable of the first kind* if all the boundary flows of $\mathbf{x}^{(i_\alpha;\mathscr{B})}(t)$ to the edge of $\mathscr{E}_{\sigma_{n-2}}^{(n-2)}$ for time t_m are coming flows, i.e.,

$$\begin{aligned} &\hbar_{j_\alpha}\mathbf{n}_{\partial\Omega_{\alpha j_\alpha}}^{\mathrm{T}}(\mathbf{x}_{m-\varepsilon}^{(j_\alpha;\mathscr{B})})\cdot[\mathbf{x}_{m-\varepsilon}^{(j_\alpha;\mathscr{B})} - \mathbf{x}_{m-\varepsilon}^{(i_\alpha;\mathscr{B})}] < 0\\ &\text{for all } i_\alpha \in \mathscr{N}_{\mathscr{B}}^C(\sigma_{n-2}) \text{ and } \mathscr{N}_{\mathscr{B}}^L(\sigma_{n-2}) = \varnothing. \end{aligned} \tag{7.36}$$

(iii) The edge of $\mathscr{E}_{\sigma_{n-2}}^{(n-2)}$ to the corresponding boundary flows of $\mathbf{x}^{(i_\beta;\mathscr{B})}(t)$ is ($\mathscr{N}_{\mathscr{B}}^L;\mathscr{B}$)-*non-switchable of the second kind* if all the boundary flows of $\mathbf{x}^{(i_\beta;\mathscr{B})}(t)$ to the edge of $\mathscr{E}_{\sigma_{n-2}}^{(n-2)}$ for time t_m are leaving flows, i.e.,

$$\begin{aligned} &\hbar_{j_\beta}\mathbf{n}_{\partial\Omega_{\beta j_\beta}}^{\mathrm{T}}(\mathbf{x}_{m+\varepsilon}^{(j_\beta;\mathscr{B})})\cdot[\mathbf{x}_{m+\varepsilon}^{(i_\beta;\mathscr{B})} - \mathbf{x}_{m+\varepsilon}^{(j_\beta;\mathscr{B})}] > 0\\ &\text{if all } i_\alpha \in \mathscr{N}_{\mathscr{B}}^L(\sigma_{n-2}) \text{ and } \mathscr{N}_{\mathscr{B}}^C(\sigma_{n-2}) = \varnothing. \end{aligned} \tag{7.37}$$

From the proceeding definition of the boundary flow switchability to a specific edge, the corresponding conditions for such boundary flow switchability will be presented.

Theorem 7.5. *For a discontinuous dynamical system in Eqs.* (6.4)-(6.8)*, there is an* $(n-2)$*-dimensional edge* $\mathscr{E}_{\sigma_{n-2}}^{(n-2)}$ ($\sigma_{n-2} \in \mathscr{N}_{\mathscr{E}}^{n-2} = \{1, 2, \cdots, N_{n-2}\}$) *to be formed by* m_b*-boundaries* $\partial\Omega_{\alpha i_\alpha}(\sigma_{n-2})$ *of* $(n-1)$*-dimensions* ($\alpha, i_\alpha \in \mathscr{N}_{\mathscr{D}}(\sigma_{n-2}) \subset \mathscr{N}_{\mathscr{D}} \in \{1, 2, \cdots, N_d\}$, *and also* $i_\alpha \in \mathscr{N}_{\mathscr{B}}(\sigma_{n-2}) \subset \mathscr{N}_{\mathscr{B}} = \{1, 2, \cdots, N_b\}$ *with* $\alpha \neq i_\alpha$).

$\mathscr{N}_{\mathscr{B}}(\sigma_{n-2}) = \mathscr{N}_{\mathscr{B}}^{C}(\sigma_{n-2}) \cup \mathscr{N}_{\mathscr{B}}^{L}(\sigma_{n-2})$ *and* $\mathscr{N}_{\mathscr{B}}^{C}(\sigma_{n-2}) \cap \mathscr{N}_{\mathscr{B}}^{L}(\sigma_{n-2}) = \varnothing$. *Suppose the coming and leaving boundary flows are* $\mathbf{x}^{(i_\alpha;\mathscr{B})}(t)$ ($i_\alpha \in \mathscr{N}_{\mathscr{B}}^{C}(\sigma_{n-2})$) *and* $\mathbf{x}^{(i_\beta;\mathscr{B})}(t)$ ($i_\beta \in \mathscr{N}_{\mathscr{B}}^{L}(\sigma_{n-2})$) *on the boundary* $\partial\Omega_{\alpha i_\alpha}$ *and* $\partial\Omega_{\beta i_\beta}$, *respectively. At time* t_m, $\mathbf{x}_m^{(i_\alpha;\mathscr{B})} = \mathbf{x}_m^{(i_\beta;\mathscr{B})}$ *and* $\mathbf{x}_m = \mathbf{x}_m^{(i_\alpha;\mathscr{B})} = \mathbf{x}^{(i_\alpha;\mathscr{B})}(t_m) \in \mathscr{E}^{(n-2)}$. *For an arbitrarily small* $\varepsilon > 0$, *there is a time interval* $[t_{m-\varepsilon}, t_{m+\varepsilon}]$. *The boundary flows of* $\mathbf{x}^{(i_\rho;\mathscr{B})}(t)$ ($\rho = \alpha, \beta$) *is* $C_{[t_{m-\varepsilon},0)}^{(r_{i_\rho};\mathscr{B})}$ *or* $C_{(0,t_{m+\varepsilon}]}^{(r_{i_\rho};\mathscr{B})}$*-continuous and* $\| d^{r_{i_\rho}+1}\mathbf{x}^{(i_\sigma;\mathscr{B})} / dt^{r_{i_\rho}+1} \| < \infty$ ($r_{i_\rho} \geq 2$) *for time* t.

(i) *The coming boundary flows of* $\mathbf{x}^{(i_\alpha;\mathscr{B})}(t)$ ($i_\alpha \in \mathscr{N}_{\mathscr{B}}^{C}(\sigma_{n-2})$) *to the leaving boundary flows of* $\mathbf{x}^{(i_\beta;\mathscr{B})}(t)$ ($i_\beta \in \mathscr{N}_{\mathscr{B}}^{L}(\sigma_{n-2})$) *at the edge of* $\mathscr{E}_{\sigma_{n-2}}^{(n-2)}$ *for time* t_m *are* ($\mathscr{N}_{\mathscr{B}}^{C} : \mathscr{N}_{\mathscr{B}}^{L}$)*-switchable from boundary* $\partial\Omega_{\alpha i_\alpha}$ *to* $\partial\Omega_{\beta i_\beta}$ *if and only if*

$$\begin{aligned} &\hbar_{j_\alpha} G_{\partial\Omega_{\alpha j_\alpha}}^{(i_\alpha;\mathscr{B})}(\mathbf{x}_m, t_{m-}, \boldsymbol{\lambda}_{i_\alpha}, \boldsymbol{\lambda}_{j_\alpha}) < 0 \ \textit{for}\ \text{all}\ i_\alpha \in \mathscr{N}_{\mathscr{B}}^{C}(\sigma_{n-2}); \\ &\hbar_{j_\beta} G_{\partial\Omega_{\beta j_\beta}}^{(i_\beta;\mathscr{B})}(\mathbf{x}_m, t_{m+}, \boldsymbol{\lambda}_{i_\beta}, \boldsymbol{\lambda}_{j_\beta}) > 0 \ \textit{for}\ \text{all}\ i_\beta \in \mathscr{N}_{\mathscr{B}}^{L}(\sigma_{n-2}). \end{aligned} \tag{7.38}$$

(ii) *The edge of* $\mathscr{E}_{\sigma_{n-2}}^{(n-2)}$ *to the corresponding boundary flows of* $\mathbf{x}^{(i_\alpha;\mathscr{B})}(t)$ *is* ($\mathscr{N}_{\mathscr{B}}^{C};\mathscr{B}$)*-non-switchable of the first kind if all the boundary flows of* $\mathbf{x}^{(i_\alpha;\mathscr{B})}(t)$ *to the edge of* $\mathscr{E}_{\sigma_{n-2}}^{(n-2)}$ *for time* t_m *are coming flows, i.e.,*

$$\begin{aligned} &\hbar_{j_\alpha} G_{\partial\Omega_{\alpha j_\alpha}}^{(i_\alpha;\mathscr{B})}(\mathbf{x}_m, t_{m-}, \boldsymbol{\lambda}_{i_\alpha}, \boldsymbol{\lambda}_{j_\alpha}) < 0 \\ &\textit{for all}\ i_\alpha \in \mathscr{N}_{\mathscr{B}}^{C}(\sigma_{n-2}) \ \textit{and}\ \mathscr{N}_{\mathscr{B}}^{L}(\sigma_{n-2}) = \varnothing. \end{aligned} \tag{7.39}$$

(iii) *The edge of* $\mathscr{E}_{\sigma_{n-2}}^{(n-2)}$ *to the corresponding boundary flows of* $\mathbf{x}^{(i_\beta;\mathscr{B})}(t)$ *is* ($\mathscr{N}_{\mathscr{B}}^{L};\mathscr{B}$)*-non-switchable of the second kind if all the boundary flows of* $\mathbf{x}^{(i_\beta;\mathscr{B})}(t)$ *to the edge of* $\mathscr{E}_{\sigma_{n-2}}^{(n-2)}$ *for time* t_m *are leaving flows, i.e.,*

$$\begin{aligned} &\hbar_{j_\beta} G_{\partial\Omega_{\beta j_\beta}}^{(i_\beta;\mathscr{B})}(\mathbf{x}_m, t_{m+}, \boldsymbol{\lambda}_{i_\beta}, \boldsymbol{\lambda}_{j_\beta}) > 0 \\ &\textit{for all}\ i_\beta \in \mathscr{N}_{\mathscr{B}}^{L}(\sigma_{n-2}) \ \textit{and}\ \mathscr{N}_{\mathscr{B}}^{C}(\sigma_{n-2}) = \varnothing. \end{aligned} \tag{7.40}$$

Proof. The boundary flow on the boundary can be treated as a special flow in domain to the other boundary of the edge. As in Chapter 3, this theorem can be proved. ■

Definition 7.6. For a discontinuous dynamical system in Eqs. (6.4)-(6.8), there is an $(n-2)$-dimensional edge $\mathscr{E}_{\sigma_{n-2}}^{(n-2)}$ ($\sigma_{n-2} \in \mathscr{N}_{\mathscr{E}}^{n-2} = \{1,2,\cdots,N_{n-2}\}$) to be formed by m_b-boundaries $\partial\Omega_{\alpha i_\alpha}(\sigma_{n-2})$ of $(n-1)$-dimensions ($\alpha, i_\alpha \in \mathscr{N}_{\mathscr{D}}(\sigma_{n-2}) \subset \mathscr{N}_{\mathscr{D}} \in \{1,2,\cdots,N_d\}$, and also $i_\alpha \in \mathscr{N}_{\mathscr{B}}(\sigma_{n-2}) \subset \mathscr{N}_{\mathscr{B}} = \{1,2,\cdots,N_b\}$ with $\alpha \neq i_\alpha$). $\mathscr{N}_{\mathscr{B}}(\sigma_{n-2}) = \mathscr{N}_{\mathscr{B}}^{C} \cup \mathscr{N}_{\mathscr{B}}^{L}$. Suppose the coming and leaving boundary flows are $\mathbf{x}^{(i_\alpha;\mathscr{B})}(t)$ on the boundary $\partial\Omega_{\alpha i_\alpha}$ ($i_\alpha \in \mathscr{N}_{\mathscr{B}}^{C}(\sigma_{n-2})$) and $\mathbf{x}^{(i_\beta;\mathscr{B})}(t)$ on the boundary $\partial\Omega_{\beta i_\beta}$ ($i_\beta \in \mathscr{N}_{\mathscr{B}}^{L}(\sigma_{n-2})$), respectively. At time t_m, $\mathbf{x}^{(i_\alpha;\mathscr{B})}(t_m) = \mathbf{x}_m^{(i_\alpha;\mathscr{B})} = \mathbf{x}_m \in \mathscr{E}_{\sigma_{n-2}}^{(n-2)}$. Suppose a switching rule $\mathbf{x}_m^{(i_\beta;\mathscr{B})} = \mathbf{x}_m^{(i_\alpha;\mathscr{B})}$. For an arbitrarily small $\varepsilon > 0$, there is a time interval $[t_{m-\varepsilon}, t_{m+\varepsilon}]$. The boundary flows of $\mathbf{x}^{(i_\rho;\mathscr{B})}(t)$ ($\rho = \alpha, \beta$) is $C_{[t_{m-\varepsilon},0)}^{(r_{i_\rho};\mathscr{B})}$ or $C_{(0,t_{m+\varepsilon}]}^{(r_{i_\rho};\mathscr{B})}$-continuous for time t and $\| d^{r_{i_\rho}+1}\mathbf{x}^{(i_\rho;\mathscr{B})} / dt^{r_{i_\rho}+1} \| < \infty$ ($r_{i_\rho} \geq m_{i_\rho}$). As $n=2$, $m_{i_\rho} \neq 2k_{i_\rho}+1$.

(i) With the switching rule, the $(n-2)$-dimensional edge $\mathscr{E}_{\sigma_{n-2}}^{(n-2)}$ for all coming boundary flows $\mathbf{x}^{(i_\alpha;\mathscr{B})}(t)$ and all leaving boundary flows $\mathbf{x}^{(i_\beta;\mathscr{B})}(t)$ at point $\mathbf{x}_m \in \mathscr{E}_{\sigma_{n-2}}^{(n-2)}$ is *switchable* with the $(\cup_{i_\alpha \in \mathscr{N}_{\mathscr{B}}^{C}} m_{i_\alpha} : \cup_{j_\alpha \in \mathscr{N}_{\mathscr{B}}^{L}} m_{j_\alpha}; \mathscr{B})$ singularity if for all $i_\alpha \in \mathscr{N}_{\mathscr{B}}^{C}(\sigma_{n-2})$ and $i_\beta \in \mathscr{N}_{\mathscr{B}}^{L}(\sigma_{n-2})$

$$\left.\begin{array}{l}
\hbar_{j_\alpha} G_{\partial\Omega_{\alpha j_\alpha}}^{(s_{i_\alpha}, i_\alpha;\mathscr{B})}(\mathbf{x}_m, t_{m-}, \boldsymbol{\lambda}_{i_\alpha}, \boldsymbol{\lambda}_{j_\alpha}) = 0 \text{ for } s_{i_\alpha} = 0,1,2,\cdots,m_{i_\alpha}-1;\\
\hbar_{j_\beta} G_{\partial\Omega_{\beta j_\beta}}^{(s_{i_\beta}, i_\beta;\mathscr{B})}(\mathbf{x}_m, t_{m+}, \boldsymbol{\lambda}_{i_\beta}, \boldsymbol{\lambda}_{j_\beta}) = 0 \text{ for } s_{i_\beta} = 0,1,2,\cdots,m_{i_\beta}-1;\\
\hbar_{j_\alpha} \mathbf{n}_{\partial\Omega_{\alpha j_\alpha}}^{\mathrm{T}}(\mathbf{x}_{m-\varepsilon}^{(j_\alpha;\mathscr{B})}) \cdot [\mathbf{x}_{m-\varepsilon}^{(j_\alpha;\mathscr{B})} - \mathbf{x}_{m-\varepsilon}^{(i_\alpha;\mathscr{B})}] < 0;\\
\hbar_{j_\beta} \mathbf{n}_{\partial\Omega_{\beta j_\beta}}^{\mathrm{T}}(\mathbf{x}_{m+\varepsilon}^{(j_\beta;\mathscr{B})}) \cdot [\mathbf{x}_{m+\varepsilon}^{(i_\beta;\mathscr{B})} - \mathbf{x}_{m+\varepsilon}^{(j_\beta;\mathscr{B})}] > 0.
\end{array}\right\} \tag{7.41}$$

(ii) The $(n-2)$-dimensional edge $\mathscr{E}_{\sigma_{n-2}}^{(n-2)}$ at point $\mathbf{x}_m \in \mathscr{E}_{\sigma_{n-2}}^{(n-2)}$ for all the coming boundary flows $\mathbf{x}^{(i_\alpha;\mathscr{B})}(t)$ is *non-switchable of the first kind* with the $(\cup_{i_\alpha \in \mathscr{N}_{\mathscr{B}}^{C}} 2k_{i_\alpha}; \mathscr{B})$ singularity if $\mathscr{N}_{\mathscr{B}}^{L}(\sigma_{n-2}) = \varnothing$ and all $i_\alpha \in \mathscr{N}_{\mathscr{B}}^{C}(\sigma_{n-2})$

$$\begin{array}{l}
\hbar_{j_\alpha} G_{\partial\Omega_{\alpha j_\alpha}}^{(s_{i_\alpha}, i_\alpha;\mathscr{B})}(\mathbf{x}_m, t_{m-}, \boldsymbol{\lambda}_{i_\alpha}, \boldsymbol{\lambda}_{j_\alpha}) = 0 \text{ for } s_{i_\alpha} = 0,1,2,\cdots,2k_{i_\alpha}-1;\\
\hbar_{j_\alpha} \mathbf{n}_{\partial\Omega_{\alpha j_\alpha}}^{\mathrm{T}}(\mathbf{x}_{m-\varepsilon}^{(j_\alpha;\mathscr{B})}) \cdot [\mathbf{x}_{m-\varepsilon}^{(j_\alpha;\mathscr{B})} - \mathbf{x}_{m-\varepsilon}^{(i_\alpha;\mathscr{B})}] < 0.
\end{array} \tag{7.42}$$

(iii) The $(n-2)$-dimensional edge $\mathscr{E}_{\sigma_{n-2}}^{(n-2)}$ at point $\mathbf{x}_m \in \mathscr{E}_{\sigma_{n-2}}^{(n-2)}$ for all the leaving boundary flows $\mathbf{x}^{(i_\beta;\mathscr{B})}(t)$ is *non-switchable of the second kind* with the $(\cup_{i_\beta \in \mathscr{N}_{\mathscr{B}}} m_{i_\beta}; \mathscr{B})$ singularity if for $\mathscr{N}_{\mathscr{B}}^{C}(\sigma_{n-2}) = \varnothing$ and all $i_\alpha \in \mathscr{N}_{\mathscr{B}}^{L}(\sigma_{n-2})$

$$\hbar_{j_\beta} G_{\partial\Omega_{\beta j_\beta}}^{(s_{i_\beta}, i_\beta;\mathscr{B})}(\mathbf{x}_m, t_{m+}, \boldsymbol{\lambda}_{i_\beta}, \boldsymbol{\lambda}_{j_\beta}) = 0 \text{ for } s_{i_\beta} = 0,1,2,\cdots,m_{i_\beta}-1;$$

$$\hbar_{j_\beta}\mathbf{n}^{\mathrm{T}}_{\partial\Omega_{\beta j_\beta}}(\mathbf{x}^{(j_\beta;\mathscr{B})}_{m+\varepsilon})\cdot[\mathbf{x}^{(i_\beta;\mathscr{B})}_{m+\varepsilon}-\mathbf{x}^{(j_\beta;\mathscr{B})}_{m+\varepsilon}]>0. \quad (7.43)$$

(iv) Without any switching rules, the $(n-2)$-dimensional edge $\mathscr{E}^{(n-2)}_{\sigma_{n-2}}$ for all coming boundary flows $\mathbf{x}^{(i_\alpha;\mathscr{B})}(t)$ and all leaving boundary flows $\mathbf{x}^{(i_\beta;\mathscr{B})}(t)$ at point $\mathbf{x}_m\in\mathscr{E}^{(n-2)}_{\sigma_{n-2}}$ is *potentially switchable* with the $(\cup_{i_\alpha\in\mathscr{N}^C_{\mathscr{B}}}m_{i_\alpha}:\cup_{j_\alpha\in\mathscr{N}^L_{\mathscr{B}}}m_{j_\alpha};\mathscr{B})$ singularity if Eq. (7.41) holds with $m_{i_\alpha}=2k_{i_\alpha}$ for all $i_\alpha\in\mathscr{N}^C_{\mathscr{B}}(\sigma_{n-2})$ and $i_\beta\in\mathscr{N}^L_{\mathscr{B}}(\sigma_{n-2})$.

(v) Without any switching rules, the $(n-2)$-dimensional edge $\mathscr{E}^{(n-2)}_{\sigma_{n-2}}$ for all coming boundary flows $\mathbf{x}^{(i_\alpha;\mathscr{B})}(t)$ and all leaving boundary flows $\mathbf{x}^{(i_\beta;\mathscr{B})}(t)$ at point $\mathbf{x}_m\in\mathscr{E}^{(n-2)}_{\sigma_{n-2}}$ is n*on-switchable* with the $(\cup_{i_\alpha\in\mathscr{N}^C_{\mathscr{B}}}m_{i_\alpha}:\cup_{j_\alpha\in\mathscr{N}^L_{\mathscr{B}}}m_{j_\alpha};\mathscr{B})$ singularity if Eq. (7.41) holds for all $i_\alpha\in\mathscr{N}^C_{\mathscr{B}}(\sigma_{n-2})$ and $i_\beta\in\mathscr{N}^L_{\mathscr{B}}(\sigma_{n-2})$ with $m_{i_\alpha}=2k_{i_\alpha}+1$ for at least one of coming flows.

(vi) With the switching rule, the $(n-2)$-dimensional edge $\mathscr{E}^{(n-2)}_{\sigma_{n-2}}$ at point $\mathbf{x}_m\in\mathscr{E}^{(n-2)}_{\sigma_{n-2}}$ is *switchable* of the $(\cup_{i_\alpha\in\mathscr{N}^C_{\mathscr{B}}}(2k_{i_\alpha}+1):\cup_{i_\alpha\in\mathscr{N}^L_{\mathscr{B}}}(2k_{i_\alpha}+1);\mathscr{B})$ for all grazing boundary flows in boundary $\partial\Omega_{\alpha i_\alpha}$ ($i_\alpha\in\mathscr{N}_{\mathscr{B}}(\sigma_{n-2})$) if

$$\left.\begin{aligned}&\hbar_{j_\alpha}G^{(s_{i_\alpha},i_\alpha;\mathscr{B})}_{\partial\Omega_{\alpha j_\alpha}}(\mathbf{x}_m,t_{m\pm},\boldsymbol{\lambda}_{i_\alpha},\boldsymbol{\lambda}_{j_\alpha})=0 \text{ for } s_{i_\alpha}=0,1,2,\cdots,2k_{i_\alpha};\\&\hbar_{j_\alpha}\mathbf{n}^{\mathrm{T}}_{\partial\Omega_{\alpha j_\alpha}}(\mathbf{x}^{(j_\alpha;\mathscr{B})}_{m-\varepsilon})\cdot[\mathbf{x}^{(j_\alpha;\mathscr{B})}_{m-\varepsilon}-\mathbf{x}^{(i_\alpha;\mathscr{B})}_{m-\varepsilon}]<0,\\&\hbar_{j_\alpha}\mathbf{n}^{\mathrm{T}}_{\partial\Omega_{\alpha j_\alpha}}(\mathbf{x}^{(j_\alpha;\mathscr{B})}_{m+\varepsilon})\cdot[\mathbf{x}^{(i_\alpha;\mathscr{B})}_{m+\varepsilon}-\mathbf{x}^{(j_\alpha;\mathscr{B})}_{m+\varepsilon}]>0.\end{aligned}\right\} \quad (7.44)$$

(vii) Without any switching rules, the $(n-2)$-dimensional edge $\mathscr{E}^{(n-2)}_{\sigma_{n-2}}$ at point $\mathbf{x}_m\in\mathscr{E}^{(n-2)}_{\sigma_{n-2}}$ is a *grazing edge* with the $(\cup_{i_\alpha\in\mathscr{N}^C_{\mathscr{B}}}(2k_{i_\alpha}+1):\cup_{i_\alpha\in\mathscr{N}^L_{\mathscr{B}}}(2k_{i_\alpha}+1);\mathscr{B})$ singularity for all the boundary flows in $\partial\Omega_{\alpha i_\alpha}$ ($i_\alpha\in\mathscr{N}_{\mathscr{B}}(\sigma_{n-2})$) if Eq. (7.44) holds.

For all coming and leaving boundary flows to a specific edge, the non-switchable boundary flows of the first and second kinds can be sketched in Fig. 7.10(a) and (b). If all the boundary flows are coming (leaving) flows to the edge, the edge is called the non-switching edge of the first (second) kind. If the boundary flows are partially coming and leaving flows to the edge, the boundary flows to the edge is switchable. Such a switchable boundary flow to the edge is sketched in Fig. 7.11. In Fig. 7.11(a), the 3-D view is presented. To clearly show the boundary flow map, a 2-D in-plane view is given in Fig. 7.11(b). For a boundary flow to the edge with the higher-order inflexional singularity, the sink and source edges of the boundary flows can be shown in Fig. 7.10(a) and (b), and the switchable boundary flow to the edge is presented in Fig. 7.11 as well.

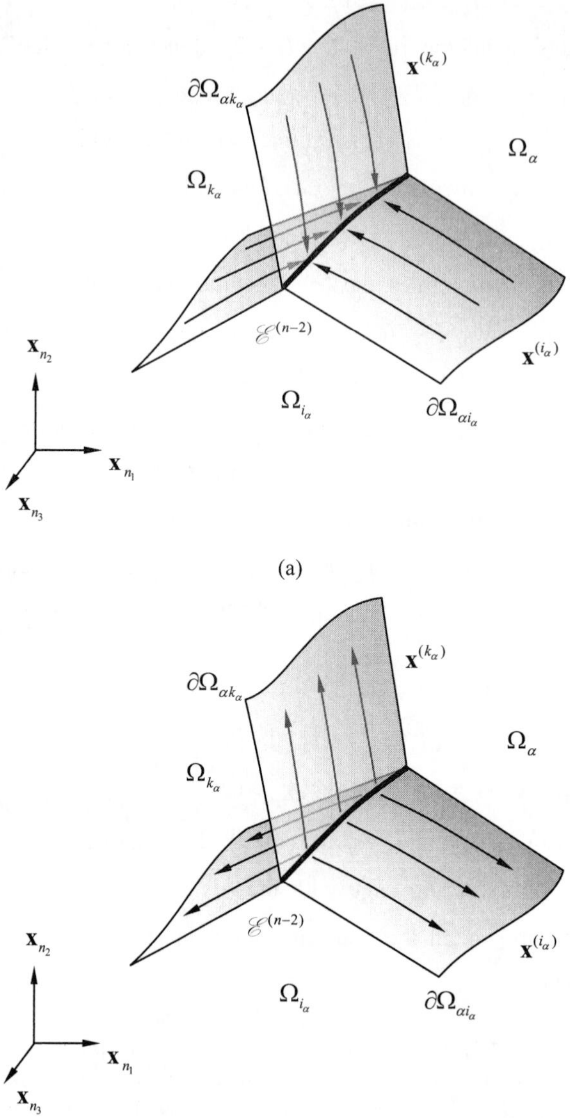

(b)

Fig. 7.10 (a) The inflexional sink boundary flows and **(b)** the inflexional source boundary flows on the three $(n-1)$ -dimensional boundaries to the $(n-2)$ -dimensional edge. The black curve represents the $(n-2)$ -dimensional edge. The boundary flows are denoted by curves with arrows on the boundaries. ($n_1 + n_2 + n_3 = n$)

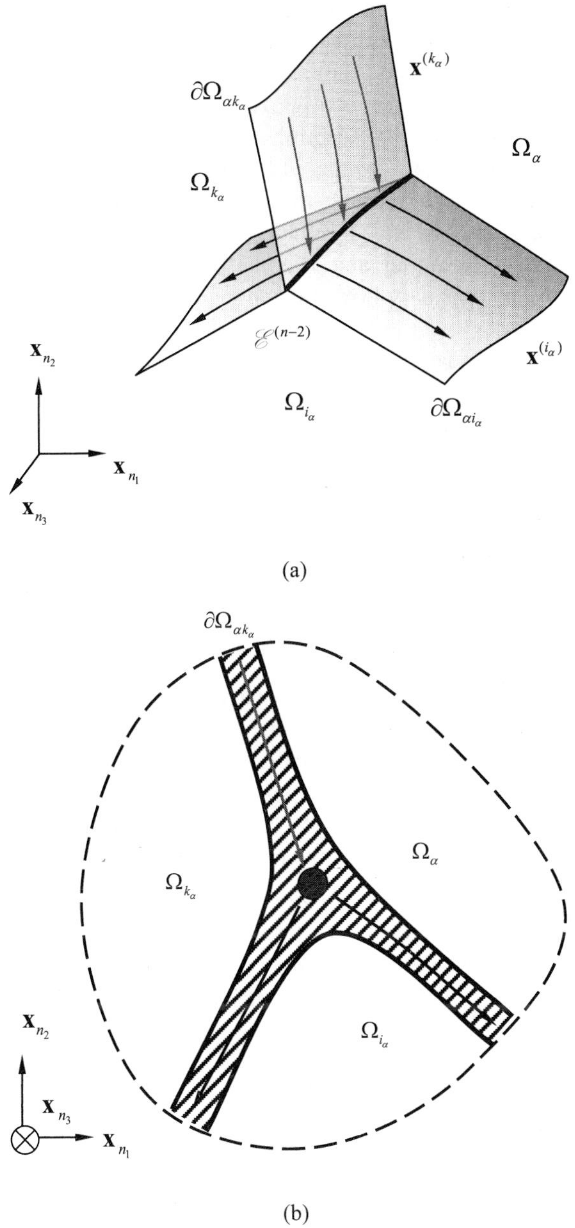

Fig. 7.11 (a) The inflexional switachable boundary flows on the three $(n-1)$-dimensional boundaries to the $(n-2)$-dimensional edge, and **(b)** the projection to the boundary flows on the three $(n-1)$-dimensional boundaries. The black curve represents the $(n-2)$-dimensional edge. The boundary flows are denoted by curves with arrows on the boundaries. ($n_1+n_2+n_3=n$)

To help one understand such singularity of the boundary flows to the specific edge, all the leaving boundary flows to the non-switchable edge of the second kind with the inflexional-grazing and fully grazing singularity are presented in Fig. 7.12(a) and (b), respectively. Similarly, with switching rules, the switchable flows to the edge with the inflexional-grazing and fully grazing singularity are sketched in Fig. 7.13(a) and (b), accordingly. Under switching rules, the fully grazing, coming boundary flows to the edge can be switched to the fully grazing, leaving flows. Without any switching rules, the fully grazing boundary flows to the edge will be tangential to the edge. Thus, no any boundary flow switching occurs. The full grazing, coming and leaving flow switching is shown in Fig. 7.14(a), and the all the boundary flows tangential to the edges are presented in Fig. 7.14(b).

Consider all the boundary flows to be coming flows. If the coming boundary flows experience the inflexional singularity, the edge to all the coming boundary flows will be a non-switchable edge of the first kind. However, if at least one of the coming boundary flows is a grazing flow to the edge, then there is a grazing leaving flow. The non-switchable boundary flows to the edge will disappear and a new switching flow will be formed. Such a bifurcation is a switching bifurcation between the switchable and non-switchable boundary flows. For a more general case, if many coming boundary flows are grazing flows to the edge, then the non-switchable boundary flows of the first kind will disappear and form an $(\cup_{i_\alpha \in \mathscr{N}_{\mathscr{B}}^C} m_{i_\alpha} : \cup_{j_\alpha \in \mathscr{N}_{\mathscr{B}}^L} m_{j_\alpha}; \mathscr{B})$ -switchable boundary flows. On the other hand, such a switchable boundary flows can be switched back to form a sink boundary flow. For the switchable flow, more than one leaving boundary, the switching rule should be added. The switching bifurcations from the first non-switchable to switchable boundary flows and from the switchable to first non-switchable boundary flow are sketched in Fig. 7.15 (a) and (b), respectively. If all the $(\cup_{i_\alpha \in \mathscr{N}_{\mathscr{B}}} 2k_{i_\alpha}; \mathscr{B})$-boundary flows to the non-switchable edge of the first kind are with the grazing singularity, then, without any switching rules, all the boundary flows to the non-switchable edge of the first kind will become fully grazing flows to the edge. However, with the switching rules, the $(\cup_{i_\alpha \in \mathscr{N}_{\mathscr{B}}^C} (2k_{i_\alpha}+1) : \cup_{i_\alpha \in \mathscr{N}_{\mathscr{B}}^L} (2k_{i_\alpha}+1); \mathscr{B})$ switchable flows will exist. From the foregoing discussion, the switching bifurcation between the non-switchable and switchable boundary flows is discussed. For a set of $(\cup_{i_\alpha \in \mathscr{N}_{\mathscr{B}}} 2k_{i_\alpha}; \mathscr{B})$ boundary flows, if one of boundary flows possesses the grazing singularity, a new set of $(\cup_{i_\alpha \in \mathscr{N}_{\mathscr{B}}^C} m_{i_\alpha} : m_{j_\alpha}; \mathscr{B})$-boundary flows will be formed. If all the leaving boundary flows in $(\cup_{i_\alpha \in \mathscr{N}_{\mathscr{B}}^C} m_{i_\alpha} : \cup_{j_\alpha \in \mathscr{N}_{\mathscr{B}}^L} m_{j_\alpha}; \mathscr{B})$ are switched to coming boundary flows, the $(\cup_{i_\alpha \in \mathscr{N}_{\mathscr{B}}^C} m_{i_\alpha} : \cup_{j_\alpha \in \mathscr{N}_{\mathscr{B}}^L} m_{j_\alpha}; \mathscr{B})$ flows will become the $(\cup_{i_\alpha \in \mathscr{N}_{\mathscr{B}}} 2k_{i_\alpha}; \mathscr{B})$ non-switchable flows of the first kind. Similarly, the switching bifurcations between the switchable flows and non-switchable flows of the second kind, and between switchable and switchable flows can be discussed as in Chapter 3.

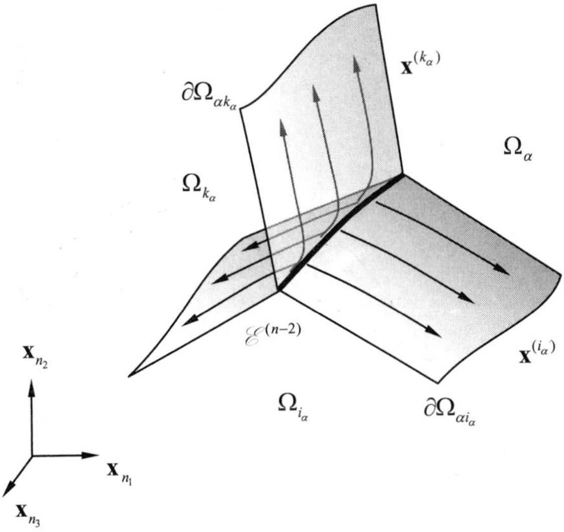

(a)

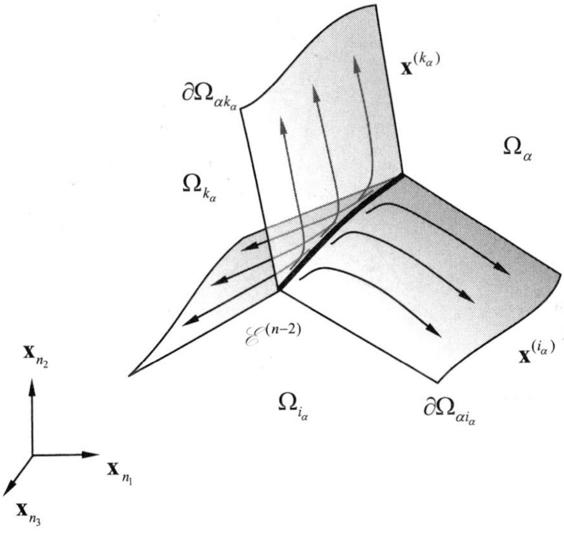

(b)

Fig. 7.12 (a) The inflexional-grazing source boundary flow and **(b)** grazing source boundary flow on the three $(n-1)$-dimensional boundaries to the $(n-2)$-dimensional edge. The black curve represents the $(n-2)$-dimensional edge. The boundary flows are denoted by curves with arrows on the boundaries. ($n_1+n_2+n_3=n$)

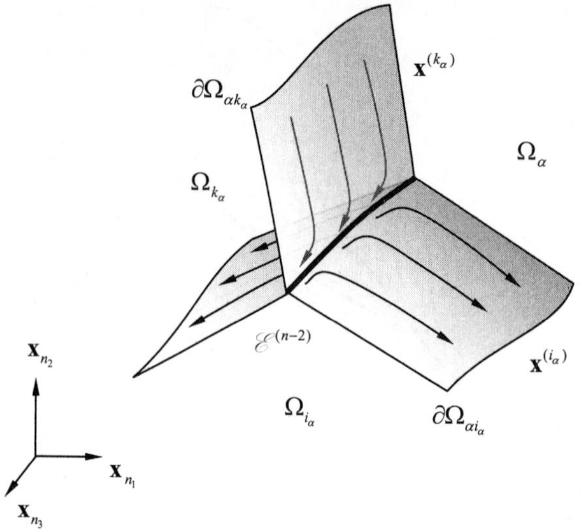

(a)

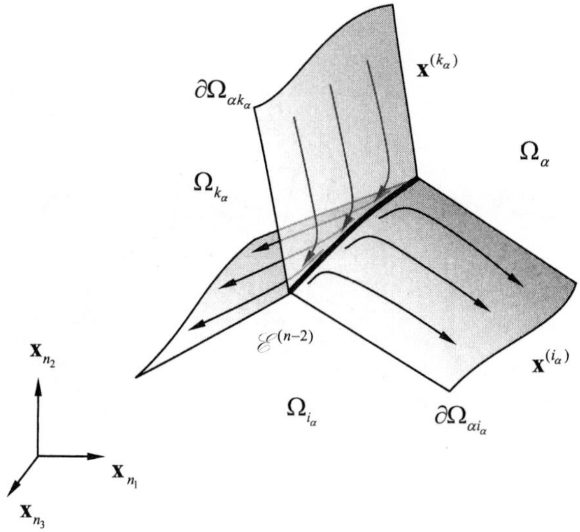

(b)

Fig. 7.13 (a) The inflexional-grazing switchable boundary flows and **(b)** fully-grazing switchable boundary flows on the three $(n-1)$ -dimensional boundaries to the $(n-2)$ -dimensional edge. The black curve represents the $(n-2)$ -dimensional edge. The boundary flows are denoted by curves with arrows on the boundaries. ($n_1+n_2+n_3=n$)

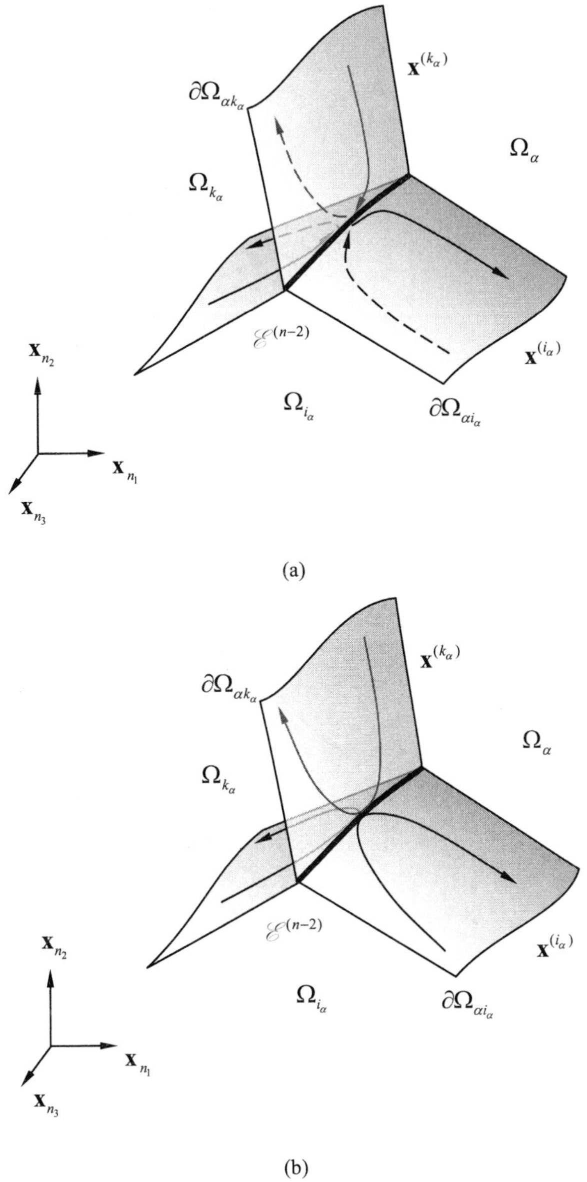

Fig. 7.14 **(a)** The switchable-grazing boundary flows with switching rules and **(b)** full tangential boundary flows without any switching rules on the three $(n-1)$-dimensional boundaries. The black curve represents the $(n-2)$-dimensional edge. The boundary flows are denoted by curves with arrows on the boundaries. ($n_1 + n_2 + n_3 = n$)

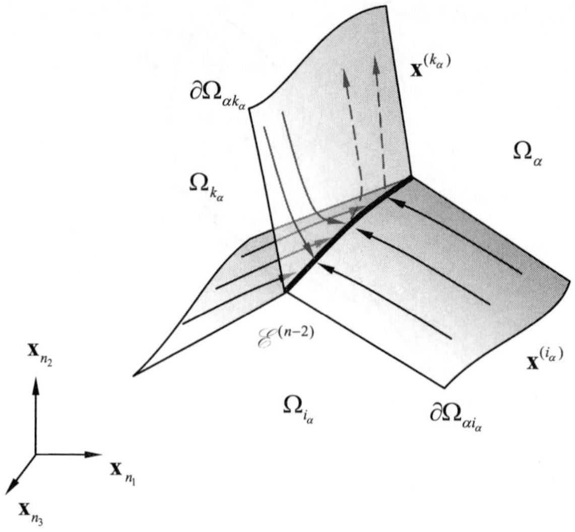

(a)

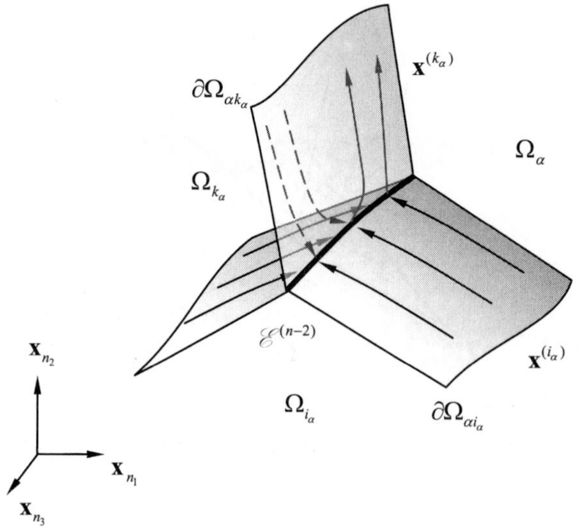

(b)

Fig. 7.15 Switching bifurcation from **(a)** the switchable to sink boundary flow and **(b)** the sink to switchable boundary flows on the three $(n-1)$ -dimensional boundaries to the $(n-2)$ -dimensional edge. The black curve represents the $(n-2)$ -dimensional edge. The boundary flows are denoted by solid and dashed curves with arrows on the boundaries. The solid and dashed curves are before and after switching bifurcation. ($n_1 + n_2 + n_3 = n$)

Theorem 7.6. *For a discontinuous dynamical system in Eqs.* (6.4)-(6.8), *there is an* $(n-2)$ *-dimensional edge* $\mathscr{E}_{\sigma_{n-2}}^{(n-2)}$ ($\sigma_{n-2} \in \mathscr{N}_{\mathscr{E}}^{n-2} = \{1,2,\cdots,N_{n-2}\}$) *to be formed by* m_b *-boundaries* $\partial\Omega_{\alpha i_\alpha}(\sigma_{n-2})$ *of* $(n-1)$ *-dimensions* ($\alpha, i_\alpha \in \mathscr{N}_{\mathscr{D}}(\sigma_{n-2}) \subset \mathscr{N}_{\mathscr{D}} \in \{1,2,\cdots,N_d\}$, *and also* $i_\alpha \in \mathscr{N}_{\mathscr{B}}(\sigma_{n-2}) \subset \mathscr{N}_{\mathscr{B}} = \{1,2,\cdots,N_b\}$ *with* $\alpha \neq i_\alpha$). $\mathscr{N}_{\mathscr{B}}(\sigma_{n-2}) = \mathscr{N}_{\mathscr{B}}^C \cup \mathscr{N}_{\mathscr{B}}^L$. *Suppose the coming and leaving boundary flows are* $\mathbf{x}^{(i_\alpha;\mathscr{B})}(t)$ *on the boundary* $\partial\Omega_{\alpha i_\alpha}$ ($i_\alpha \in \mathscr{N}_{\mathscr{B}}^C(\sigma_{n-2})$) *and* $\mathbf{x}^{(i_\beta;\mathscr{B})}(t)$ *on the boundary* $\partial\Omega_{\beta i_\beta}$ ($i_\beta \in \mathscr{N}_{\mathscr{B}}^L(\sigma_{n-2})$)*, respectively. At time* t_m , $\mathbf{x}^{(i_\alpha;\mathscr{B})}(t_m) = \mathbf{x}_m^{(i_\alpha;\mathscr{B})} = \mathbf{x}_m \in \mathscr{E}_{\sigma_{n-2}}^{(n-2)}$. *Suppose a switching rule* $\mathbf{x}_m^{(i_\beta;\mathscr{B})} = \mathbf{x}_m^{(i_\alpha;\mathscr{B})}$. *For an arbitrarily small* $\varepsilon > 0$, *there is a time interval* $[t_{m-\varepsilon}, t_{m+\varepsilon}]$. *The boundary flows of* $\mathbf{x}^{(i_\rho;\mathscr{B})}(t)$ ($\rho = \alpha, \beta$) *is* $C_{[t_{m-\varepsilon},0)}^{(r_{i_\rho};\mathscr{B})}$ *or* $C_{(0,t_{m+\varepsilon}]}^{(r_{i_\rho};\mathscr{B})}$ *-continuous for time* t *and* $\| d^{r_{i_\rho}+1}\mathbf{x}^{(i_\rho;\mathscr{B})} / dt^{r_{i_\rho}+1} \| < \infty$ ($r_{i_\rho} \geq m_{i_\rho} + 1$). *As* $n = 2$, $m_{i_\rho} \neq 2k_{i_\rho} + 1$.

(i) *With the switching rule, the* $(n-2)$ *-dimensional edge* $\mathscr{E}_{\sigma_{n-2}}^{(n-2)}$ *for all coming boundary flows* $\mathbf{x}^{(i_\alpha;\mathscr{B})}(t)$ *and all leaving boundary flows* $\mathbf{x}^{(i_\beta;\mathscr{B})}(t)$ *at point* $\mathbf{x}_m \in \mathscr{E}_{\sigma_{n-2}}^{(n-2)}$ *is switchable with the* $(\cup_{i_\alpha \in \mathscr{N}_{\mathscr{B}}^C} m_{i_\alpha} : \cup_{j_\alpha \in \mathscr{N}_{\mathscr{B}}^L} m_{j_\alpha}; \mathscr{B})$ *singularity if and only if for all* $i_\alpha \in \mathscr{N}_{\mathscr{B}}^C(\sigma_{n-2})$ *and* $i_\beta \in \mathscr{N}_{\mathscr{B}}^L(\sigma_{n-2})$

$$\left.\begin{aligned}
&\hbar_{j_\alpha} G_{\partial\Omega_{\alpha j_\alpha}}^{(s_{i_\alpha}, i_\alpha;\mathscr{B})}(\mathbf{x}_m, t_{m-}, \boldsymbol{\lambda}_{i_\alpha}, \boldsymbol{\lambda}_{j_\alpha}) = 0 \ \ for\ s_{i_\alpha} = 0,1,2,\cdots, m_{i_\alpha} - 1;\\
&\hbar_{j_\beta} G_{\partial\Omega_{\beta j_\beta}}^{(s_{i_\beta}, i_\beta;\mathscr{B})}(\mathbf{x}_m, t_{m+}, \boldsymbol{\lambda}_{i_\beta}, \boldsymbol{\lambda}_{j_\beta}) = 0 \ \ for\ s_{i_\beta} = 0,1,2,\cdots, m_{i_\beta} - 1;\\
&(-1)^{m_{i_\alpha}} \hbar_{j_\alpha} G_{\partial\Omega_{\alpha j_\alpha}}^{(m_{i_\alpha}, i_\alpha;\mathscr{B})}(\mathbf{x}_m, t_{m-}, \boldsymbol{\lambda}_{i_\alpha}, \boldsymbol{\lambda}_{j_\alpha}) < 0;\\
&\hbar_{j_\beta} G_{\partial\Omega_{\beta j_\beta}}^{(m_{i_\beta}, i_\beta;\mathscr{B})}(\mathbf{x}_m, t_{m+}, \boldsymbol{\lambda}_{i_\beta}, \boldsymbol{\lambda}_{j_\beta}) > 0.
\end{aligned}\right\} \tag{7.45}$$

(ii) *The* $(n-2)$ *-dimensional edge* $\mathscr{E}_{\sigma_{n-2}}^{(n-2)}$ *at point* $\mathbf{x}_m \in \mathscr{E}_{\sigma_{n-2}}^{(n-2)}$ *for all the coming boundary flows* $\mathbf{x}^{(i_\alpha;\mathscr{B})}(t)$ *is non-switchable of the first kind with the* $(\cup_{i_\alpha \in \mathscr{N}_{\mathscr{B}}^C} 2k_{i_\alpha}; \mathscr{B})$ *singularity if* $\mathscr{N}_{\mathscr{B}}^L(\sigma_{n-2}) = \varnothing$ *and all* $i_\alpha \in \mathscr{N}_{\mathscr{B}}^C(\sigma_{n-2})$

$$\begin{aligned}
&\hbar_{j_\alpha} G_{\partial\Omega_{\alpha j_\alpha}}^{(s_{i_\alpha}, i_\alpha;\mathscr{B})}(\mathbf{x}_m, t_{m-}, \boldsymbol{\lambda}_{i_\alpha}, \boldsymbol{\lambda}_{j_\alpha}) = 0 \ \ for\ s_{i_\alpha} = 0,1,2,\cdots, 2k_{i_\alpha} - 1;\\
&\hbar_{j_\alpha} G_{\partial\Omega_{\alpha j_\alpha}}^{(2k_{i_\alpha}, i_\alpha;\mathscr{B})}(\mathbf{x}_m, t_{m-}, \boldsymbol{\lambda}_{i_\alpha}, \boldsymbol{\lambda}_{j_\alpha}) < 0.
\end{aligned} \tag{7.46}$$

(iii) *The* $(n-2)$ *-dimensional edge* $\mathscr{E}_{\sigma_{n-2}}^{(n-2)}$ *at point* $\mathbf{x}_m \in \mathscr{E}_{\sigma_{n-2}}^{(n-2)}$ *for all the leaving boundary flows* $\mathbf{x}^{(i_\beta;\mathscr{B})}(t)$ *is non-switchable of the second kind with the* $(\cup_{i_\beta \in \mathscr{N}_{\mathscr{B}}} m_{i_\beta}; \mathscr{B})$ *singularity if for* $\mathscr{N}_{\mathscr{B}}^C(\sigma_{n-2}) = \varnothing$ *and all* $i_\alpha \in \mathscr{N}_{\mathscr{B}}^L(\sigma_{n-2})$

$$\hbar_{j_\beta} G_{\partial\Omega_{\beta j_\beta}}^{(s_{i_\beta}, i_\beta;\mathscr{B})}(\mathbf{x}_m, t_{m+}, \boldsymbol{\lambda}_{i_\beta}, \boldsymbol{\lambda}_{j_\beta}) = 0 \ \ for\ s_{i_\beta} = 0,1,2,\cdots, m_{i_\beta} - 1;$$

$$\hbar_{j_\beta} G_{\partial\Omega_{\beta j_\beta}}^{(m_{i_\beta}, i_\beta; \mathscr{B})}(\mathbf{x}_m, t_{m+}, \boldsymbol{\lambda}_{i_\beta}, \boldsymbol{\lambda}_{j_\beta}) > 0. \tag{7.47}$$

(iv) *Without any switching rules, the* $(n-2)$ *-dimensional edge* $\mathscr{E}_{\sigma_{n-2}}^{(n-2)}$ *for all coming boundary flows* $\mathbf{x}^{(i_\alpha;\mathscr{B})}(t)$ *and all leaving boundary flows* $\mathbf{x}^{(i_\beta;\mathscr{B})}(t)$ *at point* $\mathbf{x}_m \in \mathscr{E}_{\sigma_{n-2}}^{(n-2)}$ *is potentially switchable with the* $(\cup_{i_\alpha \in \mathscr{N}_{\mathscr{B}}^C} m_{i_\alpha} : \cup_{j_\alpha \in \mathscr{N}_{\mathscr{B}}^L} m_{j_\alpha}; \mathscr{B})$ *singularity if Eq.* (7.45) *holds with* $m_{i_\alpha} = 2k_{i_\alpha}$ *for all* $i_\alpha \in \mathscr{N}_{\mathscr{B}}^C(\sigma_{n-2})$ *and* $i_\beta \in \mathscr{N}_{\mathscr{B}}^L(\sigma_{n-2})$.

(v) *Without any switching rules, the* $(n-2)$ *-dimensional edge* $\mathscr{E}_{\sigma_{n-2}}^{(n-2)}$ *for all coming boundary flows* $\mathbf{x}^{(i_\alpha;\mathscr{B})}(t)$ *and all leaving boundary flows* $\mathbf{x}^{(i_\beta;\mathscr{B})}(t)$ *at point* $\mathbf{x}_m \in \mathscr{E}_{\sigma_{n-2}}^{(n-2)}$ *is non-switchable with the* $(\cup_{i_\alpha \in \mathscr{N}_{\mathscr{B}}^C} m_{i_\alpha} : \cup_{j_\alpha \in \mathscr{N}_{\mathscr{B}}^L} m_{j_\alpha}; \mathscr{B})$ *singularity if Eq.* (7.45) *holds for all* $i_\alpha \in \mathscr{N}_{\mathscr{B}}^C(\sigma_{n-2})$ *and* $i_\beta \in \mathscr{N}_{\mathscr{B}}^L(\sigma_{n-2})$ *with* $m_{i_\alpha} = 2k_{i_\alpha} + 1$ *for at least one of coming flows.*

(vi) *With the switching rule, the* $(n-2)$ *-dimensional edge* $\mathscr{E}_{\sigma_{n-2}}^{(n-2)}$ *at point* $\mathbf{x}_m \in \mathscr{E}_{\sigma_{n-2}}^{(n-2)}$ *is switchable of the* $(\cup_{i_\alpha \in \mathscr{N}_{\mathscr{B}}^C}(2k_{i_\alpha}+1) : \cup_{i_\alpha \in \mathscr{N}_{\mathscr{B}}^L}(2k_{i_\alpha}+1); \mathscr{B})$ *for all grazing boundary flows in boundary* $\partial\Omega_{\alpha i_\alpha}$ $(i_\alpha \in \mathscr{N}_{\mathscr{B}}(\sigma_{n-2}))$ *if*

$$\begin{gathered}\hbar_{j_\alpha} G_{\partial\Omega_{\alpha j_\alpha}}^{(s_{i_\alpha}, i_\alpha; \mathscr{B})}(\mathbf{x}_m, t_{m\pm}, \boldsymbol{\lambda}_{i_\alpha}, \boldsymbol{\lambda}_{j_\alpha}) = 0 \ \textit{for} \ s_{i_\alpha} = 0,1,2,\cdots,2k_{i_\alpha};\\ \hbar_{j_\alpha} G_{\partial\Omega_{\alpha j_\alpha}}^{(2k_{i_\alpha}+1, i_\alpha; \mathscr{B})}(\mathbf{x}_m, t_{m\pm}, \boldsymbol{\lambda}_{i_\alpha}, \boldsymbol{\lambda}_{j_\alpha}) > 0.\end{gathered} \tag{7.48}$$

(vii) *Without any switching rules, the* $(n-2)$ *-dimensional edge* $\mathscr{E}_{\sigma_{n-2}}^{(n-2)}$ *at point* $\mathbf{x}_m \in \mathscr{E}_{\sigma_{n-2}}^{(n-2)}$ *is a grazing edge with the* $(\cup_{i_\alpha \in \mathscr{N}_{\mathscr{B}}^C}(2k_{i_\alpha}+1) : \cup_{i_\alpha \in \mathscr{N}_{\mathscr{B}}^L}(2k_{i_\alpha}+1);$ $\mathscr{B})$ *singularity for all the boundary flows in* $\partial\Omega_{\alpha i_\alpha}$ $(i_\alpha \in \mathscr{N}_{\mathscr{B}}(\sigma_{n-2}))$ *if Eq.* (7.48) *holds.*

Proof. As in Chapter 3, this theorem can be proved. ■

7.3. Switchability of boundary and domain flows

In previous sections, the switchability of the domain and boundary flows to edges was discussed individually. In fact, the domain flow can be switched into the boundary flows through the edge, and vice versa.

Definition 7.7. For a discontinuous dynamical system in Eqs. (6.4)-(6.8), there is an $(n-2)$ -dimensional edge of $\mathscr{E}_{\sigma_{n-2}}^{(n-2)}$ to be formed by m_b -boundaries $\partial\Omega_{\alpha i_\alpha}$ of

$(n-1)$ -dimensions, with m_d -domains Ω_α ($\alpha \in \mathcal{N}_{\mathscr{D}}(\sigma_r) \subset \mathcal{N}_{\mathscr{D}} = \{1,2,\cdots,N_d\}$ and $i_\alpha \in \mathcal{N}_{\alpha\mathscr{B}}(\sigma_{n-2}) \subset \mathcal{N}_{\mathscr{B}} = \{1,2,\cdots,N_b\}$). Suppose the boundary and domain flows are $\mathbf{x}^{(i_\alpha;\mathscr{B})}(t)$ on the boundary $\partial\Omega_{\sigma i_\sigma}$ and $\mathbf{x}^{(i_\beta;\mathscr{D})}(t)$ in the domain Ω_β, respectively. At time t_m, $\mathbf{x}_m^{(\mathscr{E})} = \mathbf{x}^{(\mathscr{E})}(t_m) \in \mathscr{E}_{\sigma_{n-2}}^{(n-2)}$ and $\mathbf{x}_{m\pm}^{(i_\alpha;\mathscr{B})} = \mathbf{x}_{m\mp}^{(\beta;\mathscr{D})} = \mathbf{x}_m^{(\mathscr{E})}$. For an arbitrarily small $\varepsilon > 0$, there is a time interval $[t_{m-\varepsilon}, t_{m+\varepsilon}]$.

(i) The coming boundary flow $\mathbf{x}^{(i_\alpha;\mathscr{B})}(t)$ to the leaving domain flow $\mathbf{x}^{(\beta;\mathscr{D})}(t)$ at the edge $\mathscr{E}_{\sigma_{n-2}}^{(n-2)}$ for time t_m is *switchable* from $\partial\Omega_{\alpha i_\alpha}$ to Ω_β if

$$
\begin{aligned}
&\hbar_{j_\alpha} \mathbf{n}_{\partial\Omega_{\alpha j_\alpha}}^{\mathrm{T}}(\mathbf{x}_{m-\varepsilon}^{(j_\alpha;\mathscr{B})}) \cdot [\mathbf{x}_{m-\varepsilon}^{(j_\alpha;\mathscr{B})} - \mathbf{x}_{m-\varepsilon}^{(i_\alpha;\mathscr{B})}] < 0;\\
&\hbar_{j_\beta} \mathbf{n}_{\partial\Omega_{\beta j_\beta}}^{\mathrm{T}}(\mathbf{x}_{m+\varepsilon}^{(j_\beta;\mathscr{D})}) \cdot [\mathbf{x}_{m+\varepsilon}^{(\beta;\mathscr{D})} - \mathbf{x}_{m+\varepsilon}^{(j_\beta;\mathscr{D})}] > 0 \text{ for all } j_\beta \in \mathcal{N}_{\beta\mathscr{B}}(\sigma_{n-2}).
\end{aligned}
\tag{7.49}
$$

(ii) The coming domain flow $\mathbf{x}^{(\beta;\mathscr{D})}(t)$ to the leaving boundary flow $\mathbf{x}^{(i_\alpha;\mathscr{B})}(t)$ at the edge $\mathscr{E}_{\sigma_{n-2}}^{(n-2)}$ for time t_m is *switchable* from Ω_β to $\partial\Omega_{\alpha i_\alpha}$ if

$$
\begin{aligned}
&\hbar_{j_\beta} \mathbf{n}_{\partial\Omega_{\beta j_\beta}}^{\mathrm{T}}(\mathbf{x}_{m-\varepsilon}^{(j_\beta;\mathscr{B})}) \cdot [\mathbf{x}_{m-\varepsilon}^{(j_\beta;\mathscr{B})} - \mathbf{x}_{m-\varepsilon}^{(\beta;\mathscr{D})}] < 0 \text{ for } j_\beta \in \mathcal{N}_{\beta\mathscr{B}}(\sigma_{n-2});\\
&\hbar_{j_\alpha} \mathbf{n}_{\partial\Omega_{\alpha j_\alpha}}^{\mathrm{T}}(\mathbf{x}_{m+\varepsilon}^{(j_\alpha;\mathscr{B})}) \cdot [\mathbf{x}_{m+\varepsilon}^{(i_\alpha;\mathscr{B})} - \mathbf{x}_{m+\varepsilon}^{(j_\alpha;\mathscr{B})}] > 0.
\end{aligned}
\tag{7.50}
$$

(iii) The coming boundary and domain flows of $\mathbf{x}^{(i_\alpha;\mathscr{B})}(t)$ and $\mathbf{x}^{(\beta;\mathscr{D})}(t)$ at the edge $\mathscr{E}_{\sigma_{n-2}}^{(n-2)}$ for time t_m are *non-switchable of the first kind* if

$$
\begin{aligned}
&\hbar_{j_\alpha} \mathbf{n}_{\partial\Omega_{\alpha j_\alpha}}^{\mathrm{T}}(\mathbf{x}_{m-\varepsilon}^{(j_\alpha;\mathscr{B})}) \cdot [\mathbf{x}_{m-\varepsilon}^{(j_\alpha;\mathscr{B})} - \mathbf{x}_{m-\varepsilon}^{(i_\alpha;\mathscr{B})}] < 0;\\
&\hbar_{j_\beta} \mathbf{n}_{\partial\Omega_{\beta j_\beta}}^{\mathrm{T}}(\mathbf{x}_{m-\varepsilon}^{(j_\beta;\mathscr{B})}) \cdot [\mathbf{x}_{m-\varepsilon}^{(j_\beta;\mathscr{B})} - \mathbf{x}_{m-\varepsilon}^{(\beta;\mathscr{D})}] < 0 \text{ for all } j_\beta \in \mathcal{N}_{\beta\mathscr{B}}(\sigma_{n-2}).
\end{aligned}
\tag{7.51}
$$

(iv) The leaving boundary and domain flows of $\mathbf{x}^{(i_\alpha;\mathscr{B})}(t)$ and $\mathbf{x}^{(\beta;\mathscr{D})}(t)$ at the edge $\mathscr{E}_{\sigma_{n-2}}^{(n-2)}$ for time t_m are *non-switchable of the second kind* if

$$
\begin{aligned}
&\hbar_{j_\alpha} \mathbf{n}_{\partial\Omega_{\alpha j_\alpha}}^{\mathrm{T}}(\mathbf{x}_{m+\varepsilon}^{(j_\alpha;\mathscr{B})}) \cdot [\mathbf{x}_{m+\varepsilon}^{(i_\alpha;\mathscr{B})} - \mathbf{x}_{m+\varepsilon}^{(j_\alpha;\mathscr{B})}] > 0;\\
&\hbar_{j_\beta} \mathbf{n}_{\partial\Omega_{\beta j_\beta}}^{\mathrm{T}}(\mathbf{x}_{m+\varepsilon}^{(j_\beta;\mathscr{B})}) \cdot [\mathbf{x}_{m+\varepsilon}^{(\beta;\mathscr{D})} - \mathbf{x}_{m+\varepsilon}^{(j_\beta;\mathscr{B})}] > 0 \text{ for all } j_\beta \in \mathcal{N}_{\beta\mathscr{B}}(\sigma_{n-2}).
\end{aligned}
\tag{7.52}
$$

From the definition, the corresponding conditions can be developed as follows.

Theorem 7.7. *For a discontinuous dynamical system in Eqs.* (6.4)-(6.8), *there is an* $(n-2)$ *-dimensional edge of* $\mathscr{E}_{\sigma_{n-2}}^{(n-2)}$ *to be formed by* m_b *-boundaries* $\partial\Omega_{\alpha i_\alpha}$ *of* $(n-1)$ *-dimensions, with* m_d *-domains* Ω_α ($\alpha \in \mathcal{N}_{\mathscr{D}}(\sigma_r) \subset \mathcal{N}_{\mathscr{D}} = \{1,2,\cdots,N_d\}$ *and* $i_\alpha \in \mathcal{N}_{\alpha\mathscr{B}}(\sigma_{n-2}) \subset \mathcal{N}_{\mathscr{B}} = \{1,2,\cdots,N_b\}$). *Suppose the boundary and domain* *flows are* $\mathbf{x}^{(i_\alpha;\mathscr{B})}(t)$ *on the boundary* $\partial\Omega_{\sigma i_\sigma}$ *and* $\mathbf{x}^{(i_\beta;\mathscr{D})}(t)$ *in the domain* Ω_β, *respectively. At time* t_m, $\mathbf{x}_m^{(\mathscr{E})}=\mathbf{x}^{(\mathscr{E})}(t_m)\in\mathscr{E}_{\sigma_{n-2}}^{(n-2)}$ *and* $\mathbf{x}_{m\pm}^{(i_\alpha;\mathscr{B})}=\mathbf{x}_{m\mp}^{(\beta;\mathscr{D})}=\mathbf{x}_m$. *For an arbitrarily small* $\varepsilon>0$, *there is a time interval* $[t_{m-\varepsilon},t_{m+\varepsilon}]$. *The boundary flow of* $\mathbf{x}^{(i_\alpha;\mathscr{B})}(t)$ *is* $C_{[t_{m-\varepsilon},0)}^{(r_{i_\alpha};\mathscr{B})}$ *or* $C_{(0,t_{m+\varepsilon}]}^{(r_{i_\alpha};\mathscr{B})}$*-continuous for time* t *and* $\| d^{r_{i_\alpha}+1}\mathbf{x}^{(i_\alpha)}/dt^{r_{i_\alpha}+1}\|<\infty$ $(r_{i_\alpha}\geq 2)$. *The domain flow of* $\mathbf{x}^{(\beta;\mathscr{D})}(t)$ *is* $C_{[t_{m-\varepsilon},0)}^{(\beta;\mathscr{D})}$ *or* $C_{(0,t_{m+\varepsilon}]}^{(\beta;\mathscr{D})}$*-continuous for time* t *and* $\| d^{r_\beta+1}\mathbf{x}^{(\beta;\mathscr{D})}/dt^{r_\beta+1}\|<\infty$ $(r_\beta\geq 2)$.

(i) *The coming boundary flow* $\mathbf{x}^{(i_\alpha;\mathscr{B})}(t)$ *to the leaving domain flow* $\mathbf{x}^{(\beta;\mathscr{B})}(t)$ *at the edge* $\mathscr{E}_{\sigma_{n-2}}^{(n-2)}$ *for time* t_m *is switchable from* $\partial\Omega_{\alpha i_\alpha}$ *to* Ω_β *if and only if*

$$\begin{aligned}&\hbar_{j_\alpha} G_{\partial\Omega_{\alpha j_\alpha}}^{(i_\alpha;\mathscr{B})}(\mathbf{x}_m,t_{m-},\boldsymbol{\lambda}_{i_\alpha},\boldsymbol{\lambda}_{j_\alpha})<0;\\&\hbar_{j_\beta} G_{\partial\Omega_{\beta j_\beta}}^{(\beta;\mathscr{D})}(\mathbf{x}_m,t_{m+},\mathbf{p}_\beta,\boldsymbol{\lambda}_{j_\beta})>0 \textit{ for all } j_\beta\in\mathscr{N}_{\beta\mathscr{B}}(\sigma_{n-2}).\end{aligned}\tag{7.53}$$

(ii) *The coming domain flow of* $\mathbf{x}^{(\beta;\mathscr{D})}(t)$ *to the leaving boundary flow* $\mathbf{x}^{(i_\alpha;\mathscr{B})}(t)$ *at the edge* $\mathscr{E}_{\sigma_{n-2}}^{(n-2)}$ *for time* t_m *is switchable from* Ω_β *to* $\partial\Omega_{\alpha i_\alpha}$ *if and only if*

$$\begin{aligned}&\hbar_{j_\beta} G_{\partial\Omega_{\beta j_\beta}}^{(\beta;\mathscr{D})}(\mathbf{x}_m,t_{m-},\mathbf{p}_\beta,\boldsymbol{\lambda}_{j_\beta})<0 \textit{ for all } j_\beta\in\mathscr{N}_{\beta\mathscr{B}}(\sigma_{n-2});\\&\hbar_{j_\alpha} G_{\partial\Omega_{\alpha j_\alpha}}^{(i_\alpha;\mathscr{B})}(\mathbf{x}_m,t_{m+},\boldsymbol{\lambda}_{i_\alpha},\boldsymbol{\lambda}_{j_\alpha})>0.\end{aligned}\tag{7.54}$$

(iii) *The coming boundary and domain flows of* $\mathbf{x}^{(i_\alpha;\mathscr{B})}(t)$ *and* $\mathbf{x}^{(\beta;\mathscr{D})}(t)$ *at the edge* $\mathscr{E}_{\sigma_{n-2}}^{(n-2)}$ *for time* t_m *are non-switchable of the first kind if and only if*

$$\begin{aligned}&\hbar_{j_\alpha} G_{\partial\Omega_{\alpha j_\alpha}}^{(i_\alpha;\mathscr{B})}(\mathbf{x}_m,t_{m-},\boldsymbol{\lambda}_{i_\alpha},\boldsymbol{\lambda}_{j_\alpha})<0;\\&\hbar_{j_\beta} G_{\partial\Omega_{\beta j_\beta}}^{(\beta;\mathscr{D})}(\mathbf{x}_m,t_{m-},\mathbf{p}_\beta,\boldsymbol{\lambda}_{j_\beta})<0 \textit{ for all } j_\beta\in\mathscr{N}_{\beta\mathscr{B}}(\sigma_{n-2}).\end{aligned}\tag{7.55}$$

(iv) *The leaving boundary and domain flows of* $\mathbf{x}^{(i_\alpha;\mathscr{B})}(t)$ *and* $\mathbf{x}^{(\beta;\mathscr{D})}(t)$ *at the edge* $\mathscr{E}_{\sigma_{n-2}}^{(n-2)}$ *for time* t_m *are non-switchable of the second kind if and only if*

$$\begin{aligned}&\hbar_{j_\alpha} G_{\partial\Omega_{\alpha j_\alpha}}^{(i_\alpha;\mathscr{B})}(\mathbf{x}_m,t_{m+},\boldsymbol{\lambda}_{i_\alpha},\boldsymbol{\lambda}_{j_\alpha})>0;\\&\hbar_{j_\beta} G_{\partial\Omega_{\beta j_\beta}}^{(\beta;\mathscr{D})}(\mathbf{x}_m,t_{m+},\mathbf{p}_\beta,\boldsymbol{\lambda}_{j_\beta})>0 \textit{ for all } j_\beta\in\mathscr{N}_{\beta\mathscr{B}}(\sigma_{n-2}).\end{aligned}\tag{7.56}$$

Proof. From the domain flow and boundary flow properties, the theorem can be proved as in Chapter 3. ■

Similarly, the switchability between the boundary and domain flows with the higher-order singularity is presented as follows.

Definition 7.8. For a discontinuous dynamical system in Eqs. (6.4)-(6.8), there is an $(n-2)$-dimensional edge of $\mathscr{E}_{\sigma_{n-2}}^{(n-2)}$ to be formed by m_b-boundaries $\partial\Omega_{\alpha i_\alpha}$ of $(n-1)$-dimensions, with m_d-domains Ω_α ($\alpha, i_\alpha \in \mathscr{N}_{\mathscr{D}} = \{1,2,\cdots,N_d\}$ and $i_\alpha \in \mathscr{N}_{\mathscr{B}} = \{1,2,\cdots,N_b\}$). Suppose the boundary and domain flows are $\mathbf{x}^{(\beta;\mathscr{B})}(t)$ on the boundary $\partial\Omega_{\sigma i_\sigma}$ and $\mathbf{x}^{(\beta;\mathscr{D})}(t)$ in the domain Ω_β, respectively. At time t_m, $\mathbf{x}_m = \mathbf{x}^{(\mathscr{E})}(t_m) \in \mathscr{E}_{\sigma_{n-2}}^{(n-2)}$ and $\mathbf{x}_{m\pm}^{(i_\alpha;\mathscr{B})} = \mathbf{x}_{m\mp}^{(\beta;\mathscr{D})} = \mathbf{x}_m$. For an arbitrarily small $\varepsilon > 0$, there is a time interval $[t_{m-\varepsilon}, t_{m+\varepsilon}]$. The boundary flow of $\mathbf{x}^{(i_\alpha;\mathscr{B})}(t)$ is $C_{[t_{m-\varepsilon},0)}^{(r_{i_\alpha};\mathscr{B})}$ or $C_{(0,t_{m+\varepsilon}]}^{(r_{i_\alpha};\mathscr{B})}$-continuous and $\| d^{r_{i_\alpha}+1}\mathbf{x}^{(i_\alpha)}/dt^{r_{i_\alpha}+1} \| < \infty$ ($r_{i_\alpha} \geq m_{i_\alpha}$) for time t. The domain flow of $\mathbf{x}^{(\beta;\mathscr{D})}(t)$ is $C_{[t_{m-\varepsilon},0)}^{(\gamma_\beta;\mathscr{D})}$ or $C_{(0,t_{m+\varepsilon}]}^{(\gamma_\beta;\mathscr{D})}$-continuous and $\| d^{r_\beta+1}\mathbf{x}^{(\beta;\mathscr{D})}/dt^{r_\beta+1} \| < \infty$ ($r_\beta \geq m_\beta$) for time t. As $n=2$, $m_{i_\alpha} \neq 2k_{i_\alpha}+1$ and $\mathbf{m}_\beta \neq 2\mathbf{k}_\beta + 1$.

(i) With the switching rule, the coming boundary flow $\mathbf{x}^{(i_\alpha;\mathscr{B})}(t)$ to the leaving domain flow $\mathbf{x}^{(\beta;\mathscr{D})}(t)$ at the edge $\mathscr{E}_{\sigma_{n-2}}^{(n-2)}$ for time t_m is *switchable* with the $(m_{i_\alpha};\mathscr{B}:\mathbf{m}_\beta;\mathscr{D})$-singularity *from* $\partial\Omega_{\alpha i_\alpha}$ *to* Ω_β if

$$\left.\begin{aligned}
&\hbar_{j_\alpha} G_{\partial\Omega_{\alpha j_\alpha}}^{(s_{i_\alpha},i_\alpha;\mathscr{B})}(\mathbf{x}_m, t_{m-}, \boldsymbol{\lambda}_{i_\alpha}, \boldsymbol{\lambda}_{j_\alpha}) = 0\\
&\text{for } s_{i_\alpha} = 0,1,2,\cdots,m_{i_\alpha}-1,\\
&\hbar_{j_\alpha} \mathbf{n}_{\partial\Omega_{\alpha j_\alpha}}^{\mathrm{T}}(\mathbf{x}_{m-\varepsilon}^{(j_\alpha;\mathscr{B})})\cdot[\mathbf{x}_{m-\varepsilon}^{(j_\alpha;\mathscr{B})} - \mathbf{x}_{m-\varepsilon}^{(i_\alpha;\mathscr{B})}] < 0;
\end{aligned}\right\}$$

$$\left.\begin{aligned}
&\hbar_{j_\beta} G_{\partial\Omega_{\beta j_\beta}}^{(s_\beta,\beta;\mathscr{D})}(\mathbf{x}_m, t_{m+}, \mathbf{p}_\beta, \boldsymbol{\lambda}_{j_\beta}) = 0\\
&\text{for } s_\beta = 0,1,2,\cdots,m_\beta-1,\\
&\hbar_{j_\beta} \mathbf{n}_{\partial\Omega_{\beta j_\beta}}^{\mathrm{T}}(\mathbf{x}_{m+\varepsilon}^{(j_\beta;\mathscr{B})})\cdot[\mathbf{x}_{m+\varepsilon}^{(\beta;\mathscr{D})} - \mathbf{x}_{m+\varepsilon}^{(j_\beta;\mathscr{B})}] > 0
\end{aligned}\right\} \tag{7.57}$$

for all $j_\beta \in \mathscr{N}_{\beta\mathscr{B}}(\sigma_{n-2})$.

(ii) With the switching rule, the coming domain flow $\mathbf{x}^{(\beta;\mathscr{D})}(t)$ to the leaving boundary flow $\mathbf{x}^{(i_\alpha;\mathscr{B})}(t)$ at the edge of $\mathscr{E}_{\sigma_{n-2}}^{(n-2)}$ for time t_m is *switchable* with the $(\mathbf{m}_\beta;\mathscr{D}:m_{i_\alpha};\mathscr{B})$-singularity *from* Ω_β *to* $\partial\Omega_{\alpha i_\alpha}$ if

$$\left.\begin{aligned}
&\hbar_{j_\beta} G_{\partial\Omega_{\beta j_\beta}}^{(s_\beta,\beta;\mathscr{D})}(\mathbf{x}_m, t_{m-}, \mathbf{p}_\beta, \boldsymbol{\lambda}_{j_\beta}) = 0\\
&\text{for } s_\beta = 0,1,2,\cdots,m_\beta-1,\\
&\hbar_{j_\beta} \mathbf{n}_{\partial\Omega_{\beta j_\beta}}^{\mathrm{T}}(\mathbf{x}_{m-\varepsilon}^{(j_\beta;\mathscr{B})})\cdot[\mathbf{x}_{m-\varepsilon}^{(j_\beta;\mathscr{B})} - \mathbf{x}_{m-\varepsilon}^{(\beta;\mathscr{D})}] < 0
\end{aligned}\right\}$$

for all $j_\beta \in \mathscr{N}_{\beta\mathscr{B}}(\sigma_{n-2})$;

$$\left.\begin{array}{l}\hbar_{j_\alpha} G^{(s_{i_\alpha}, i_\alpha;\mathscr{B})}_{\partial\Omega_{\alpha j_\alpha}}(\mathbf{x}_m, t_{m+}, \boldsymbol{\lambda}_{i_\alpha}, \boldsymbol{\lambda}_{j_\alpha}) = 0 \\ \text{for } s_{i_\alpha} = 0,1,2,\cdots, m_{i_\alpha} - 1, \\ \hbar_{j_\alpha} \mathbf{n}^{\mathrm{T}}_{\partial\Omega_{\alpha j_\alpha}}(\mathbf{x}^{(j_\alpha;\mathscr{B})}_{m+\varepsilon}) \cdot [\mathbf{x}^{(i_\alpha;\mathscr{B})}_{m+\varepsilon} - \mathbf{x}^{(j_\alpha;\mathscr{B})}_{m+\varepsilon}] > 0. \end{array}\right\} \tag{7.58}$$

(iii) The coming boundary and domain flows of $\mathbf{x}^{(i_\alpha;\mathscr{B})}(t)$ and $\mathbf{x}^{(\beta;\mathscr{D})}(t)$ at the edge $\mathscr{E}^{(n-2)}_{\sigma_{n-2}}$ for time t_m are *non-switchable of the first kind* with the $(2k_{i_\alpha};\mathscr{B}:\mathbf{m}_\beta;\mathscr{D})$-singularity ($\mathbf{m}_\beta \neq 2\mathbf{k}_\beta + 1$) if

$$\left.\begin{array}{l}\left.\begin{array}{l}\hbar_{j_\alpha} G^{(s_{i_\alpha}, i_\alpha;\mathscr{B})}_{\partial\Omega_{\alpha j_\alpha}}(\mathbf{x}_m, t_{m-}, \boldsymbol{\lambda}_{i_\alpha}, \boldsymbol{\lambda}_{j_\alpha}) = 0 \\ \text{for } s_{i_\alpha} = 0,1,2,\cdots, 2k_{i_\alpha} - 1, \\ \hbar_{j_\alpha} \mathbf{n}^{\mathrm{T}}_{\partial\Omega_{\alpha j_\alpha}}(\mathbf{x}^{(j_\alpha;\mathscr{B})}_{m-\varepsilon}) \cdot [\mathbf{x}^{(j_\alpha;\mathscr{B})}_{m-\varepsilon} - \mathbf{x}^{(i_\alpha;\mathscr{B})}_{m-\varepsilon}] < 0; \end{array}\right\} \\ \left.\begin{array}{l}\hbar_{j_\beta} G^{(s_\beta,\beta;\mathscr{D})}_{\partial\Omega_{\beta j_\beta}}(\mathbf{x}_m, t_{m-}, \mathbf{p}_\beta, \boldsymbol{\lambda}_{j_\beta}) = 0 \\ \text{for } s_\beta = 0,1,2,\cdots, m_\beta - 1, \\ \hbar_{j_\beta} \mathbf{n}^{\mathrm{T}}_{\partial\Omega_{\beta j_\beta}}(\mathbf{x}^{(j_\beta;\mathscr{B})}_{m-\varepsilon}) \cdot [\mathbf{x}^{(j_\beta;\mathscr{B})}_{m-\varepsilon} - \mathbf{x}^{(\beta;\mathscr{D})}_{m-\varepsilon}] < 0 \end{array}\right\} \\ \text{for all } j_\beta \in \mathscr{N}_{\beta\mathscr{B}}(\sigma_{n-2}). \end{array}\right. \tag{7.59}$$

(iv) The leaving boundary and domain flows of $\mathbf{x}^{(i_\alpha;\mathscr{B})}(t)$ and $\mathbf{x}^{(\beta;\mathscr{D})}(t)$ at the edge of $\mathscr{E}^{(n-2)}_{\sigma_{n-2}}$ for time t_m are *non-switchable of the second kind* with the $(m_{i_\alpha};\mathscr{B}:\mathbf{m}_\beta;\mathscr{D})$-singularity if

$$\left.\begin{array}{l}\left.\begin{array}{l}\hbar_{j_\alpha} G^{(s_{i_\alpha}, i_\alpha;\mathscr{B})}_{\partial\Omega_{\alpha j_\alpha}}(\mathbf{x}_m, t_{m+}, \boldsymbol{\lambda}_{i_\alpha}, \boldsymbol{\lambda}_{j_\alpha}) = 0 \\ \text{for } s_{i_\alpha} = 0,1,2,\cdots, m_{i_\alpha} - 1, \\ \hbar_{j_\alpha} \mathbf{n}^{\mathrm{T}}_{\partial\Omega_{\alpha j_\alpha}}(\mathbf{x}^{(j_\alpha;\mathscr{B})}_{m+\varepsilon}) \cdot [\mathbf{x}^{(i_\alpha;\mathscr{B})}_{m+\varepsilon} - \mathbf{x}^{(j_\alpha;\mathscr{B})}_{m+\varepsilon}] > 0; \end{array}\right\} \\ \left.\begin{array}{l}\hbar_{j_\beta} G^{(s_\beta,\beta;\mathscr{D})}_{\partial\Omega_{\beta j_\beta}}(\mathbf{x}_m, t_{m+}, \mathbf{p}_\beta, \boldsymbol{\lambda}_{j_\beta}) = 0 \\ \text{for } s_\beta = 0,1,2,\cdots, m_\beta - 1, \\ \hbar_{j_\beta} \mathbf{n}^{\mathrm{T}}_{\partial\Omega_{\beta j_\beta}}(\mathbf{x}^{(j_\beta;\mathscr{B})}_{m+\varepsilon}) \cdot [\mathbf{x}^{(\beta;\mathscr{D})}_{m+\varepsilon} - \mathbf{x}^{(j_\beta;\mathscr{B})}_{m+\varepsilon}] > 0 \end{array}\right\} \\ \text{for all } j_\beta \in \mathscr{N}_{\beta\mathscr{B}}(\sigma_{n-2}). \end{array}\right. \tag{7.60}$$

(v) Without any switching rules, the coming boundary flow of $\mathbf{x}^{(i_\alpha;\mathscr{B})}(t)$ to the leaving domain flow $\mathbf{x}^{(\beta;\mathscr{D})}(t)$ at the edge $\mathscr{E}^{(n-2)}_{\sigma_{n-2}}$ for time t_m is *potentially switchable* with the $(2k_{i_\alpha};\mathscr{B}:\mathbf{m}_\beta;\mathscr{D})$-singularity from $\partial\Omega_{\alpha i_\alpha}$ to Ω_β if Eq. (7.57) holds with $m_{i_\alpha} = 2k_{i_\alpha}$.

(vi) Without any switching rules, the coming boundary flow of $\mathbf{x}^{(i_\alpha;\mathscr{B})}(t)$ to the leaving domain flow $\mathbf{x}^{(\beta;\mathscr{D})}(t)$ at the edge $\mathscr{E}_{\sigma_{n-2}}^{(n-2)}$ for time t_m is *non- switchable* with the $(2k_{i_\alpha}+1;\mathscr{B}:\mathbf{m}_\beta;\mathscr{D})$ -singularity from $\partial\Omega_{\alpha i_\alpha}$ to Ω_β if Eq. (7.57) holds with $m_{i_\alpha}=2k_{i_\alpha}+1$.

(vii) Without any switching rules, the coming domain flow $\mathbf{x}^{(\beta;\mathscr{D})}(t)$ to the leaving boundary flows of $\mathbf{x}^{(i_\alpha;\mathscr{B})}(t)$ at the edge $\mathscr{E}_{\sigma_{n-2}}^{(n-2)}$ for time t_m is *potentially switchable* with the $(\mathbf{m}_\beta;\mathscr{D}:m_{i_\alpha};\mathscr{B})$-singularity ($\mathbf{m}_\beta\neq 2\mathbf{k}_\beta+1$) from $\partial\Omega_{\alpha i_\alpha}$ to Ω_β if Eq. (7.58) holds with $\mathbf{m}_\beta\neq 2\mathbf{k}_\beta+1$.

(viii) Without any switching rules, the coming domain flow $\mathbf{x}^{(\beta;\mathscr{D})}(t)$ to the leaving boundary flows of $\mathbf{x}^{(i_\alpha;\mathscr{B})}(t)$ at the edge $\mathscr{E}_{\sigma_{n-2}}^{(n-2)}$ for time t_m is *non-switchable* with the $(2\mathbf{k}_\beta+1;\mathscr{D}:m_{i_\alpha};\mathscr{B})$ -singularity from $\partial\Omega_{\alpha i_\alpha}$ to Ω_β if Eq. (7.58) holds with $\mathbf{m}_\beta=2\mathbf{k}_\beta+1$.

(ix) With the switching rule, the domain and boundary flows of $\mathbf{x}^{(\beta;\mathscr{D})}(t)$ and $\mathbf{x}^{(i_\alpha;\mathscr{B})}(t)$ grazing to the edge $\mathscr{E}_{\sigma_{n-2}}^{(n-2)}$ for time t_m are *switchable* of the $(2\mathbf{k}_\beta+1;\mathscr{D}:2k_{i_\alpha}+1;\mathscr{B})$ -singularity if

$$\left.\begin{aligned}
&\left.\begin{aligned}
&\hbar_{j_\alpha} G_{\partial\Omega_{\alpha j_\alpha}}^{(s_{i_\alpha},i_\alpha;\mathscr{B})}(\mathbf{x}_m,t_{m\pm},\boldsymbol{\lambda}_{i_\alpha},\boldsymbol{\lambda}_{j_\alpha})=0\\
&\text{for } s_{i_\alpha}=0,1,2,\cdots,2k_{i_\alpha},\\
&\hbar_{j_\alpha}\mathbf{n}_{\partial\Omega_{\alpha j_\alpha}}^{\mathrm{T}}(\mathbf{x}_{m-\varepsilon}^{(j_\alpha;\mathscr{B})})\cdot[\mathbf{x}_{m-\varepsilon}^{(j_\alpha;\mathscr{B})}-\mathbf{x}_{m-\varepsilon}^{(i_\alpha;\mathscr{B})}]<0,\\
&\hbar_{j_\alpha}\mathbf{n}_{\partial\Omega_{\alpha j_\alpha}}^{\mathrm{T}}(\mathbf{x}_{m+\varepsilon}^{(j_\alpha;\mathscr{B})})\cdot[\mathbf{x}_{m+\varepsilon}^{(i_\alpha;\mathscr{B})}-\mathbf{x}_{m+\varepsilon}^{(j_\alpha;\mathscr{B})}]>0;
\end{aligned}\right\}\\
&\left.\begin{aligned}
&\hbar_{j_\beta} G_{\partial\Omega_{\beta j_\beta}}^{(s_\beta,\beta;\mathscr{D})}(\mathbf{x}_m,t_{m\pm},\mathbf{p}_\beta,\boldsymbol{\lambda}_{j_\beta})=0\\
&\text{for } s_\beta=0,1,2,\cdots,2k_\beta,\\
&\hbar_{j_\beta}\mathbf{n}_{\partial\Omega_{\beta j_\beta}}^{\mathrm{T}}(\mathbf{x}_{m-\varepsilon}^{(j_\beta;\mathscr{B})})\cdot[\mathbf{x}_{m-\varepsilon}^{(j_\beta;\mathscr{B})}-\mathbf{x}_{m-\varepsilon}^{(\beta;\mathscr{D})}]<0,\\
&\hbar_{j_\beta}\mathbf{n}_{\partial\Omega_{\beta j_\beta}}^{\mathrm{T}}(\mathbf{x}_{m+\varepsilon}^{(j_\beta;\mathscr{B})})\cdot[\mathbf{x}_{m+\varepsilon}^{(\beta;\mathscr{D})}-\mathbf{x}_{m+\varepsilon}^{(j_\beta;\mathscr{B})}]>0
\end{aligned}\right\}\\
&\text{for all } j_\beta\in\mathscr{N}_{\beta\mathscr{B}}(\sigma_{n-2}).
\end{aligned}\right. \tag{7.61}$$

(x) Without any switching rules, a domain flow of $\mathbf{x}^{(\beta;\mathscr{D})}(t)$ and a boundary flow $\mathbf{x}^{(i_\alpha;\mathscr{B})}(t)$ grazing to the edge of $\mathscr{E}_{\sigma_{n-2}}^{(n-2)}$ for time t_m are a *double tangential flow* of the $(2\mathbf{k}_\beta+1;\mathscr{D}:2k_{i_\alpha}+1;\mathscr{B})$ -singularity if Eq. (7.61) holds.

Theorem 7.8. *For a discontinuous dynamical system in Eqs.* (6.4)-(6.8), *there is an* $(n-2)$ *-dimensional edge* $\mathscr{E}_{\sigma_{n-2}}^{(n-2)}$ *to be formed by* m_b *-boundaries* $\partial\Omega_{\alpha i_\alpha}$ *of* $(n-1)$ -

dimensions, with m_d *-domains* Ω_α ($\alpha, i_\alpha \in \mathscr{N}_{\mathscr{D}} = \{1,2,\cdots,N_d\}$ *and* $\alpha \neq i_\alpha$ *, also* $i_\alpha \in \mathscr{N}_{\mathscr{B}} = \{1,2,\cdots,N_b\}$). *Suppose the boundary and domain flows are* $\mathbf{x}^{(i_\alpha;\mathscr{B})}(t)$ *on the boundary* $\partial\Omega_{\sigma i_\sigma}$ *and* $\mathbf{x}^{(\beta;\mathscr{D})}(t)$ *in the domain* Ω_β *. At time* t_m *,* $\mathbf{x}_m = \mathbf{x}^{(\mathscr{E})}(t_m) \in \mathscr{E}^{(n-2)}_{\sigma_{n-2}}$ *and* $\mathbf{x}^{(i_\alpha;\mathscr{B})}_{m\pm} = \mathbf{x}^{(\beta;\mathscr{D})}_{m\mp} = \mathbf{x}_m$ *. For an arbitrarily small* $\varepsilon > 0$ *, there is a time interval* $[t_{m-\varepsilon}, t_{m+\varepsilon}]$ *. The boundary flow of* $\mathbf{x}^{(i_\alpha;\mathscr{B})}(t)$ *is* $C^{(r_{i_\alpha};\mathscr{B})}_{[t_{m-\varepsilon},0)}$ *or* $C^{(r_{i_\alpha};\mathscr{B})}_{(0,t_{m+\varepsilon}]}$ *-continuous and* $\| d^{r_{i_\alpha}+1}\mathbf{x}^{(i_\alpha)}/dt^{r_{i_\alpha}+1} \| < \infty$ ($r_{i_\alpha} \geq m_{i_\alpha}$) *for time* t *. The domain flow of* $\mathbf{x}^{(\beta;\mathscr{D})}(t)$ *is* $C^{(\gamma_\beta;\mathscr{D})}_{[t_{m-\varepsilon},0)}$ *or* $C^{(\gamma_\beta;\mathscr{D})}_{(0,t_{m+\varepsilon}]}$ *-continuous and* $\| d^{r_\beta+1}\mathbf{x}^{(\beta;\mathscr{D})}/dt^{r_\beta+1} \| < \infty$ ($r_\beta \geq m_\beta$) *for time* t . As $n=2$, $m_{i_\alpha} \neq 2k_{i_\alpha}+1$ and $\mathbf{m}_\beta \neq 2\mathbf{k}_\beta + 1$.

(i) *With the switching rule, the coming boundary flows of* $\mathbf{x}^{(i_\alpha;\mathscr{B})}(t)$ *to the leaving domain flow* $\mathbf{x}^{(\beta;\mathscr{D})}(t)$ *at the edge* $\mathscr{E}^{(n-2)}_{\sigma_{n-2}}$ *for time* t_m *is switchable with the* $(m_{i_\alpha};\mathscr{B}:\mathbf{m}_\beta;\mathscr{D})$ *-singularity from* $\partial\Omega_{\alpha i_\alpha}$ *to* Ω_β *if and only if*

$$\left.\begin{aligned}
&\left.\begin{aligned}
&\hbar_{j_\alpha} G^{(s_{i_\alpha}, i_\alpha;\mathscr{B})}_{\partial\Omega_{\alpha j_\alpha}}(\mathbf{x}_m, t_{m-}, \boldsymbol{\lambda}_{i_\alpha}, \boldsymbol{\lambda}_{j_\alpha}) = 0\\
&\text{for } s_{i_\alpha} = 0,1,2,\cdots, m_{i_\alpha}-1,\\
&(-1)^{m_{i_\alpha}} \hbar_{j_\alpha} G^{(m_{i_\alpha}, i_\alpha;\mathscr{B})}_{\partial\Omega_{\alpha j_\alpha}}(\mathbf{x}_m, t_{m-}, \boldsymbol{\lambda}_{i_\alpha}, \boldsymbol{\lambda}_{j_\alpha}) < 0;
\end{aligned}\right\}\\
&\left.\begin{aligned}
&\hbar_{j_\beta} G^{(s_\beta,\beta;\mathscr{D})}_{\partial\Omega_{\beta j_\beta}}(\mathbf{x}_m, t_{m+}, \mathbf{p}_\beta, \boldsymbol{\lambda}_{j_\beta}) = 0\\
&\text{for } s_\beta = 0,1,2,\cdots, m_\beta - 1,\\
&\hbar_{j_\beta} G^{(m_\beta,\beta;\mathscr{D})}_{\partial\Omega_{\beta j_\beta}}(\mathbf{x}_m, t_{m+}, \mathbf{p}_\beta, \boldsymbol{\lambda}_{j_\beta}) > 0
\end{aligned}\right\} \text{for all } j_\beta \in \mathscr{N}_{\beta\mathscr{B}}(\sigma_{n-2}).
\end{aligned}\right. \tag{7.62}$$

(ii) *With the switching rule, the coming domain flow of* $\mathbf{x}^{(\beta;\mathscr{D})}(t)$ *to the leaving boundary flow* $\mathbf{x}^{(i_\alpha;\mathscr{B})}(t)$ *at the edge* $\mathscr{E}^{(n-2)}_{\sigma_{n-2}}$ *for time* t_m *is switchable* $(\mathbf{m}_\beta;\mathscr{D}:m_{i_\alpha};\mathscr{B})$ *-singularity from* Ω_β *to* $\partial\Omega_{\alpha i_\alpha}$ *if and only if*

$$\left.\begin{aligned}
&\left.\begin{aligned}
&\hbar_{j_\beta} G^{(s_\beta,\beta;\mathscr{D})}_{\partial\Omega_{\beta j_\beta}}(\mathbf{x}_m, t_{m-}, \mathbf{p}_\beta, \boldsymbol{\lambda}_{j_\beta}) = 0\\
&\text{for } s_\beta = 0,1,2,\cdots, m_\beta - 1,\\
&(-1)^{m_\beta} \hbar_{j_\beta} G^{(m_\beta,\beta;\mathscr{D})}_{\partial\Omega_{\beta j_\beta}}(\mathbf{x}_m, t_{m-}, \mathbf{p}_\beta, \boldsymbol{\lambda}_{j_\beta}) < 0
\end{aligned}\right\} \text{for all } j_\beta \in \mathscr{N}_{\beta\mathscr{B}}(\sigma_{n-2});\\
&\left.\begin{aligned}
&\hbar_{j_\alpha} G^{(s_{i_\alpha}, i_\alpha;\mathscr{B})}_{\partial\Omega_{\alpha j_\alpha}}(\mathbf{x}_m, t_{m+}, \boldsymbol{\lambda}_{i_\alpha}, \boldsymbol{\lambda}_{j_\alpha}) = 0\\
&\text{for } s_{i_\alpha} = 0,1,2,\cdots, m_{i_\alpha}-1,\\
&\hbar_{j_\alpha} G^{(m_{i_\alpha}, i_\alpha;\mathscr{B})}_{\partial\Omega_{\alpha j_\alpha}}(\mathbf{x}_m, t_{m+}, \boldsymbol{\lambda}_{i_\alpha}, \boldsymbol{\lambda}_{j_\alpha}) > 0.
\end{aligned}\right\}
\end{aligned}\right. \tag{7.63}$$

(iii) *The coming boundary and domain flows of* $\mathbf{x}^{(i_\alpha;\mathscr{B})}(t)$ *and* $\mathbf{x}^{(\beta;\mathscr{D})}(t)$ *at the edge of* $\mathscr{E}^{(n-2)}_{\sigma_{n-2}}$ *for time* t_m *are non-switchable of the first kind with the* $(2k_{i_\alpha};$ $\mathscr{B}:\mathbf{m}_\beta;\mathscr{D})$ -singularity ($\mathbf{m}_\beta \neq 2\mathbf{k}_\beta+1$) *if and only if*

$$\left.\begin{aligned}&\left.\begin{aligned}&\hbar_{j_\alpha} G^{(s_{i_\alpha},i_\alpha;\mathscr{B})}_{\partial\Omega_{\alpha j_\alpha}}(\mathbf{x}_m,t_{m-},\boldsymbol{\lambda}_{i_\alpha},\boldsymbol{\lambda}_{j_\alpha})=0\\&\text{for } s_{i_\alpha}=0,1,2,\cdots,2k_{i_\alpha}-1,\\&\hbar_{j_\alpha} G^{(2k_{i_\alpha},i_\alpha;\mathscr{B})}_{\partial\Omega_{\alpha j_\alpha}}(\mathbf{x}_m,t_{m-},\boldsymbol{\lambda}_{i_\alpha},\boldsymbol{\lambda}_{j_\alpha})<0;\end{aligned}\right\}\\&\left.\begin{aligned}&\hbar_{j_\beta} G^{(s_\beta,\beta;\mathscr{D})}_{\partial\Omega_{\beta j_\beta}}(\mathbf{x}_m,t_{m-},\mathbf{p}_\beta,\boldsymbol{\lambda}_{j_\beta})=0\\&\text{for } s_\beta=0,1,2,\cdots,m_\beta-1,\\&(-1)^{m_\beta}\hbar_{j_\beta} G^{(m_\beta,\beta;\mathscr{D})}_{\partial\Omega_{\beta j_\beta}}(\mathbf{x}_m,t_{m-},\mathbf{p}_\beta,\boldsymbol{\lambda}_{j_\beta})<0\end{aligned}\right\}\text{for all } j_\beta\in\mathscr{N}_{\beta\mathscr{B}}(\sigma_{n-2}).\end{aligned}\right. \tag{7.64}$$

(iv) *The leaving boundary and domain flows of* $\mathbf{x}^{(i_\alpha;\mathscr{B})}(t)$ *and* $\mathbf{x}^{(\beta;\mathscr{D})}(t)$ *at the edge of* $\mathscr{E}^{(n-2)}_{\sigma_{n-2}}$ *for time* t_m *are non-switchable of the second kind with the* $(m_{i_\alpha};\mathscr{B}:\mathbf{m}_\beta;\mathscr{D})$ *-singularity if and only if*

$$\left.\begin{aligned}&\left.\begin{aligned}&\hbar_{j_\alpha} G^{(s_{i_\alpha},i_\alpha;\mathscr{B})}_{\partial\Omega_{\alpha j_\alpha}}(\mathbf{x}_m,t_{m+},\boldsymbol{\lambda}_{i_\alpha},\boldsymbol{\lambda}_{j_\alpha})=0\\&\text{for } s_{i_\alpha}=0,1,2,\cdots,m_{i_\alpha}-1,\\&\hbar_{j_\alpha} G^{(m_{i_\alpha},i_\alpha;\mathscr{B})}_{\partial\Omega_{\alpha j_\alpha}}(\mathbf{x}_m,t_{m+},\boldsymbol{\lambda}_{i_\alpha},\boldsymbol{\lambda}_{j_\alpha})>0;\end{aligned}\right\}\\&\left.\begin{aligned}&\hbar_{j_\beta} G^{(s_\beta,\beta;\mathscr{D})}_{\partial\Omega_{\beta j_\beta}}(\mathbf{x}_m,t_{m+},\mathbf{p}_\beta,\boldsymbol{\lambda}_{j_\beta})=0\\&\text{for } s_\beta=0,1,2,\cdots,m_\beta-1,\\&\hbar_{j_\beta} G^{(m_\beta,\beta;\mathscr{D})}_{\partial\Omega_{\beta j_\beta}}(\mathbf{x}_m,t_{m+},\mathbf{p}_\beta,\boldsymbol{\lambda}_{j_\beta})>0\end{aligned}\right\}\text{for all } j_\beta\in\mathscr{N}_{\beta\mathscr{B}}(\sigma_{n-2}).\end{aligned}\right. \tag{7.65}$$

(v) *Without any switching rules, the coming boundary flow of* $\mathbf{x}^{(i_\alpha;\mathscr{B})}(t)$ *to the leaving domain flow* $\mathbf{x}^{(\beta;\mathscr{D})}(t)$ *at the edge of* $\mathscr{E}^{(n-2)}_{\sigma_{n-2}}$ *for time* t_m *is potentially switchable with the* $(2k_{i_\alpha};\mathscr{B}:\mathbf{m}_\beta;\mathscr{D})$ *-singularity from* $\partial\Omega_{\alpha i_\alpha}$ *to* Ω_β *if and only* if *Eq.* (7.61) *holds with* $m_{i_\alpha}=2k_{i_\alpha}$.

(vi) *Without any switching rules, the coming boundary flow of* $\mathbf{x}^{(i_\alpha;\mathscr{B})}(t)$ *to the leaving domain flow* $\mathbf{x}^{(\beta;\mathscr{D})}(t)$ *at the edge of* $\mathscr{E}^{(n-2)}_{\sigma_{n-2}}$ *for time* t_m *is non-switchable with the* $(2k_{i_\alpha}+1;\mathscr{B}:\mathbf{m}_\beta;\mathscr{D})$ *-singularity from* $\partial\Omega_{\alpha i_\alpha}$ *to* Ω_β *if and only if Eq.* (7.61) *holds with* $m_{i_\alpha}=2k_{i_\alpha}+1$.

(vii) *Without any switching rules, the coming domain flow of* $\mathbf{x}^{(\beta;\mathscr{D})}(t)$ *to the leaving boundary flow* $\mathbf{x}^{(i_\alpha;\mathscr{B})}(t)$ *at the edge of* $\mathscr{E}^{(n-2)}_{\sigma_{n-2}}$ *for time* t_m *is non-*

switchable $(2\mathbf{k}_\beta+1;\mathscr{D}:m_{i_\alpha};\mathscr{B})$*-singularity from* Ω_β *to* $\partial\Omega_{\alpha i_\alpha}$ *if and only if Eq.* (7.62) *holds with* $\mathbf{m}_\beta \neq 2\mathbf{k}_\beta+1$.

(viii) *Without any switching rules, the coming domain flow of* $\mathbf{x}^{(\beta;\mathscr{D})}(t)$ *to the leaving boundary flow* $\mathbf{x}^{(i_\alpha;\mathscr{B})}(t)$ *at the edge of* $\mathscr{E}^{(n-2)}_{\sigma_{n-2}}$ *for time* t_m *is non-switchable* $(\mathbf{m}_\beta;\mathscr{D}:m_{i_\alpha};\mathscr{B})$*-singularity from* Ω_β *to* $\partial\Omega_{\alpha i_\alpha}$ *if and only if Eq.* (7.62) *holds with* $\mathbf{m}_\beta = 2\mathbf{k}_\beta+1$.

(ix) *With the switching rule, a domain flow of* $\mathbf{x}^{(\beta;\mathscr{D})}(t)$ *and a boundary flow* $\mathbf{x}^{(i_\alpha;\mathscr{B})}(t)$ *grazing to the edge of* $\mathscr{E}^{(n-2)}_{\sigma_{n-2}}$ *for time* t_m *are switchable* $(2\mathbf{k}_\beta+1;\mathscr{D}:2k_{i_\alpha}+1;\mathscr{B})$*-singularity from the domain* Ω_β *to the boundary* $\partial\Omega_{\alpha i_\alpha}$ *if and only if*

$$\left.\begin{aligned}&\hbar_{j_\beta} G^{(s_\beta,\beta;\mathscr{D})}_{\partial\Omega_{\beta j_\beta}}(\mathbf{x}_m,t_{m\pm},\mathbf{p}_\beta,\boldsymbol{\lambda}_{j_\beta})=0\\&\text{for } s_\beta=0,1,2,\cdots,2k_\beta,\\&\hbar_{j_\beta} G^{(2k_\beta+1,\beta;\mathscr{D})}_{\partial\Omega_{\beta j_\beta}}(\mathbf{x}_m,t_{m\pm},\mathbf{p}_\beta,\boldsymbol{\lambda}_{j_\beta})>0\end{aligned}\right\}$$
$$\text{for all } j_\beta\in\mathscr{N}_{\beta\mathscr{B}}(\sigma_{n-2}); \tag{7.66}$$
$$\left.\begin{aligned}&\hbar_{j_\alpha} G^{(s_{i_\alpha},i_\alpha;\mathscr{B})}_{\partial\Omega_{\alpha j_\alpha}}(\mathbf{x}_m,t_{m\pm},\boldsymbol{\lambda}_{i_\alpha},\boldsymbol{\lambda}_{j_\alpha})=0\\&\text{for } s_{i_\alpha}=0,1,2,\cdots,2k_{i_\alpha},\\&\hbar_{j_\alpha} G^{(2k_{i_\alpha}+1,i_\alpha;\mathscr{B})}_{\partial\Omega_{\alpha j_\alpha}}(\mathbf{x}_m,t_{m\pm},\boldsymbol{\lambda}_{i_\alpha},\boldsymbol{\lambda}_{j_\alpha})>0.\end{aligned}\right\}$$

(x) *Without any switching rules, a domain flow of* $\mathbf{x}^{(\beta;\mathscr{D})}(t)$ *and a boundary flow* $\mathbf{x}^{(i_\alpha;\mathscr{B})}(t)$ *grazing to the edge of* $\mathscr{E}^{(n-2)}_{\sigma_{n-2}}$ *for time* t_m *are a double tangential flow of the* $(2\mathbf{k}_\beta+1;\mathscr{D}:2k_{i_\alpha}+1;\mathscr{B})$*-singularity from the domain* Ω_β *to the boundary* $\partial\Omega_{\alpha i_\alpha}$ *if and only if Eq.* (7.66) *hold.*

Proof. The theorem can be proved as in Chapter 3. ■

As in Figs. 7.4-7.9, from the foregoing definitions and theorems, the switchabilty of boundary and domain flows for an $(n-2)$-dimensional edge is sketched in Figs. 7.16 and 7.17. The gray and black curves with arrows represent the boundary and domain flows. The switching between boundary and domain flows is given in Fig. 7.16. Two types of non-switching for coming and leaving flows are presented in Fig. 7.17. Illustrations of the switchability with singularity can be given as well.

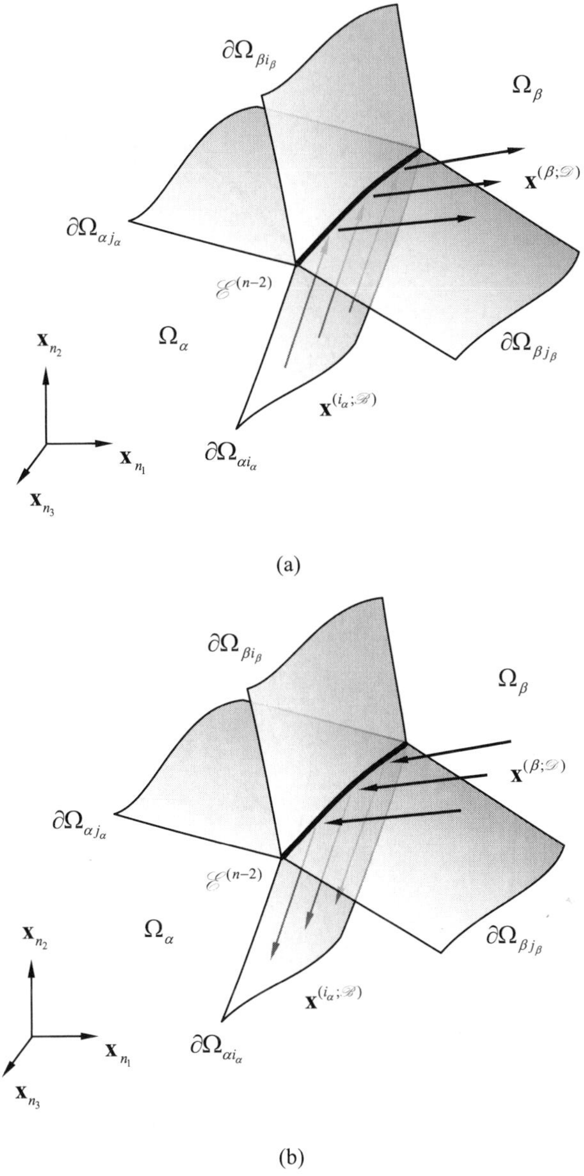

Fig. 7.16 The switchable flows: **(a)** from a boundary flow to a domain flow, and **(b)** from a domain flow to boundary flow. The black curve represents the $(n-2)$-dimensional edge. The boundary flows are denoted by thin curves with arrows on the boundaries. The domain flows are represented by the thick curves with arrows in the domain. ($n_1 + n_2 + n_3 = n$)

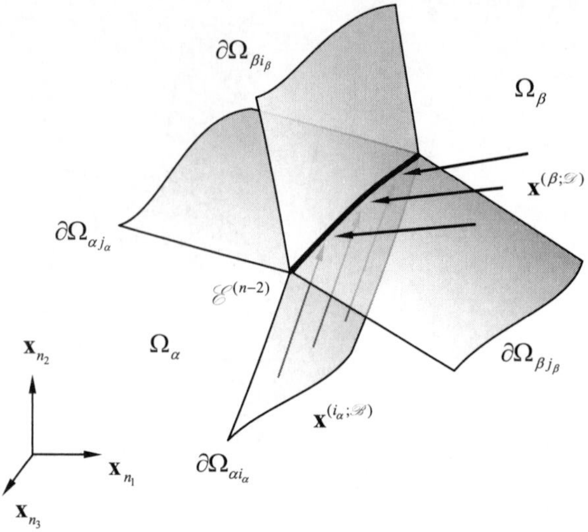

(a)

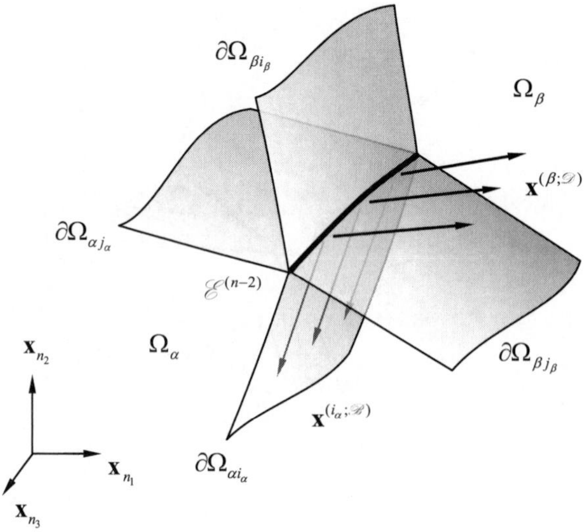

(b)

Fig. 7.17 The non-switchable flows between the boundary and domain flows: **(a)** the first kind, and **(b)** the second kind. The black curve represents the $(n-2)$ -dimensional edge. The boundary flows are denoted by thin curves with arrows on the boundaries. The domain flows are represented by the thick curves with arrows in the domain. ($n_1 + n_2 + n_3 = n$)

Definition 7.9. For a discontinuous dynamical system in Eqs. (6.4)-(6.8), there is an $(n-2)$-dimensional edge $\mathscr{E}_{\sigma_{n-2}}^{(n-2)}$ to be formed by m_b-boundaries $\partial\Omega_{\alpha i_\alpha}$ of $(n-1)$-dimensions with m_d-domains Ω_α ($\alpha, i_\alpha \in \mathscr{N}_{\mathscr{D}} = \{1,2,\cdots,m_d\}$ and $\alpha \neq i_\alpha$, also $i_\alpha \in \mathscr{N}_{\mathscr{B}} = \{1,2,\cdots,m_b\}$). $\mathscr{N}_{\mathscr{D}}(\sigma_{n-2}) = \mathscr{N}_{\mathscr{D}}^C(\sigma_{n-2}) \cup \mathscr{N}_{\mathscr{D}}^L(\sigma_{n-2}) \subset \mathscr{N}_{\mathscr{D}}$, and $\mathscr{N}_{\mathscr{D}}^C(\sigma_{n-2}) \cap \mathscr{N}_{\mathscr{D}}^L(\sigma_{n-2}) = \varnothing$. $\mathscr{N}_{\mathscr{B}}(\sigma_{n-2}) = \mathscr{N}_{\mathscr{B}}^C(\sigma_{n-2}) \cup \mathscr{N}_{\mathscr{B}}^L(\sigma_{n-2}) \subset \mathscr{N}_{\mathscr{B}}$ and $\mathscr{N}_{\mathscr{B}}^C(\sigma_{n-2}) \cap \mathscr{N}_{\mathscr{B}}^L(\sigma_{n-2}) = \varnothing$. Suppose the coming and leaving boundary flows are $\mathbf{x}^{(i_\alpha;\mathscr{B})}(t)$ ($i_\alpha \in \mathscr{N}_{\mathscr{B}}^C(\sigma_{n-2})$) and $\mathbf{x}^{(i_\beta;\mathscr{B})}(t)$ ($i_\beta \in \mathscr{N}_{\mathscr{B}}^L(\sigma_{n-2})$) on the boundaries $\partial\Omega_{\alpha i_\alpha}$ and $\partial\Omega_{\beta i_\beta}$, respectively. Suppose the coming and leaving domain flows are $\mathbf{x}^{(\mu;\mathscr{D})}(t)$ ($\mu \in \mathscr{N}_{\mathscr{D}}^C(\sigma_{n-2})$) and $\mathbf{x}^{(\nu;\mathscr{D})}(t)$ ($\nu \in \mathscr{N}_{\mathscr{D}}^L(\sigma_{n-2})$) in domains Ω_μ and Ω_ν, respectively. At time t_m, $\mathbf{x}_m = \mathbf{x}^{(\mathscr{E})}(t_m) \in \mathscr{E}_{\sigma_{n-2}}^{(n-2)}$, $\mathbf{x}_{m\pm}^{(i_\alpha;\mathscr{B})} = \mathbf{x}_{m\mp}^{(i_\beta;\mathscr{B})} = \mathbf{x}_m$ and $\mathbf{x}_{m\pm}^{(\mu;\mathscr{D})} = \mathbf{x}_{m\mp}^{(\nu;\mathscr{D})} = \mathbf{x}_m$. For an arbitrarily small $\varepsilon > 0$, there is a time interval $[t_{m-\varepsilon}, t_{m+\varepsilon}]$. The boundary flows of $\mathbf{x}^{(i_\rho;\mathscr{B})}(t)$ ($\rho = \alpha, \beta$) are $C_{[t_{m-\varepsilon},0)}^{(r_{i_\rho};\mathscr{B})}$ or $C_{(0,t_{m+\varepsilon}]}^{(r_{i_\rho};\mathscr{B})}$-continuous for time t and $\| d^{r_{i_\rho}+1}\mathbf{x}^{(i_\rho;\mathscr{B})} / dt^{r_{i_\rho}+1} \| < \infty$ ($r_{i_\rho} \geq 1$). The domain flows of $\mathbf{x}^{(\sigma;\mathscr{D})}(t)$ ($\sigma = \mu, \nu$) are $C_{[t_{m-\varepsilon},0)}^{(r_\sigma;\mathscr{D})}$ or $C_{(0,t_{m+\varepsilon}]}^{(r_\sigma;\mathscr{D})}$-continuous for time t and $\| d^{r_\sigma+1}\mathbf{x}^{(\sigma;\mathscr{D})} / dt^{r_\sigma+1} \| < \infty$ ($r_\sigma \geq 1$).

(i) The coming boundary and domain flows of $\mathbf{x}^{(i_\alpha;\mathscr{B})}(t)$ ($i_\alpha \in \mathscr{N}_{\mathscr{B}}^C(\sigma_{n-2})$) and $\mathbf{x}^{(\mu;\mathscr{D})}(t)$ ($\mu \in \mathscr{N}_{\mathscr{D}}^C(\sigma_{n-2})$) to the leaving boundary and domain flows of $\mathbf{x}^{(i_\beta;\mathscr{B})}(t)$ ($i_\beta \in \mathscr{N}_{\beta\mathscr{B}}^L(\sigma_{n-2})$) and $\mathbf{x}^{(\nu;\mathscr{D})}(t)$ ($\nu \in \mathscr{N}_{\mathscr{D}}^L(\sigma_{n-2})$) at the edge $\mathscr{E}_{\sigma_{n-2}}^{(n-2)}$ for time t_m are $(\mathscr{N}_{\mathscr{B}}^C(\sigma_{n-2}) \oplus \mathscr{N}_{\mathscr{D}}^C(\sigma_{n-2}) : \mathscr{N}_{\mathscr{B}}^L(\sigma_{n-2}) \oplus \mathscr{N}_{\mathscr{D}}^L(\sigma_{n-2}))$ *-switchable* from $\partial\Omega_{\alpha i_\alpha}$ and Ω_μ to $\partial\Omega_{\beta i_\beta}$ and Ω_ν if

$$
\left.
\begin{aligned}
&\hbar_{j_\alpha} \mathbf{n}_{\partial\Omega_{\alpha j_\alpha}}^{\mathrm{T}} (\mathbf{x}_{m-\varepsilon}^{(j_\alpha;\mathscr{B})}) \cdot [\mathbf{x}_{m-\varepsilon}^{(j_\alpha;\mathscr{B})} - \mathbf{x}_{m-\varepsilon}^{(i_\alpha;\mathscr{B})}] < 0 \\
&\text{for all } i_\alpha \in \mathscr{N}_{\mathscr{B}}^C(\sigma_{n-2}), \\
&\hbar_{j_\beta} \mathbf{n}_{\partial\Omega_{\beta j_\beta}}^{\mathrm{T}} (\mathbf{x}_{m+\varepsilon}^{(j_\beta;\mathscr{B})}) \cdot [\mathbf{x}_{m+\varepsilon}^{(i_\beta;\mathscr{B})} - \mathbf{x}_{m+\varepsilon}^{(j_\beta;\mathscr{B})}] > 0 \\
&\text{for all } i_\beta \in \mathscr{N}_{\mathscr{B}}^L(\sigma_{n-2}); \\
&\hbar_{i_\mu} \mathbf{n}_{\partial\Omega_{\mu i_\mu}}^{\mathrm{T}} (\mathbf{x}_{m-\varepsilon}^{(i_\mu;\mathscr{B})}) \cdot [\mathbf{x}_{m-\varepsilon}^{(i_\mu;\mathscr{B})} - \mathbf{x}_{m-\varepsilon}^{(\mu;\mathscr{D})}] < 0 \\
&\text{for all } \mu \in \mathscr{N}_{\mathscr{D}}^C(\sigma_{n-2}) \text{ and } i_\mu \in \mathscr{N}_{\mu\mathscr{B}}(\sigma_{n-2}), \\
&\hbar_{j_\nu} \mathbf{n}_{\partial\Omega_{\nu j_\nu}}^{\mathrm{T}} (\mathbf{x}_{m+\varepsilon}^{(j_\nu;\mathscr{B})}) \cdot [\mathbf{x}_{m+\varepsilon}^{(\nu;\mathscr{D})} - \mathbf{x}_{m+\varepsilon}^{(j_\nu;\mathscr{B})}] > 0 \\
&\text{for all } \nu \in \mathscr{N}_{\mathscr{D}}^L(\sigma_{n-2}) \text{ and all } j_\nu \in \mathscr{N}_{\nu\mathscr{B}}(\sigma_{n-2}).
\end{aligned}
\right\} \quad (7.67)
$$

(ii) The edge of $\mathscr{E}_{\sigma_{n-2}}^{(n-2)}$ to the boundary and domain flows of $\mathbf{x}^{(i_\alpha;\mathscr{B})}(t)$ ($i_\alpha \in \mathscr{N}_{\mathscr{B}}(\sigma_{n-2})$ with all $\alpha \in \mathscr{N}_{\mathscr{D}}(\sigma_{n-2})$) and $\mathbf{x}^{(\mu;\mathscr{D})}(t)$ ($\mu \in \mathscr{N}_{\mathscr{D}}(\sigma_{n-2})$) is

$(\mathscr{N}_{\mathscr{B}}^{C}(\sigma_{n-2})\oplus\mathscr{N}_{\mathscr{D}}^{C}(\sigma_{n-2}))$ *non-switchable of the first kind* (sink) if

$$
\begin{aligned}
&\hbar_{j_\alpha}\mathbf{n}^{\mathrm{T}}_{\partial\Omega_{\alpha j_\alpha}}(\mathbf{x}^{(j_\alpha;\mathscr{B})}_{m-\varepsilon})\cdot[\mathbf{x}^{(j_\alpha;\mathscr{B})}_{m-\varepsilon}-\mathbf{x}^{(i_\alpha;\mathscr{B})}_{m-\varepsilon}]<0\\
&\text{for all } i_\alpha\in\mathscr{N}_{\mathscr{B}}(\sigma_{n-2})=\mathscr{N}_{\mathscr{B}}^{C}(\sigma_{n-2});\\
&\hbar_{i_\mu}\mathbf{n}^{\mathrm{T}}_{\partial\Omega_{\mu i_\mu}}(\mathbf{x}^{(i_\mu;\mathscr{B})}_{m-\varepsilon})\cdot[\mathbf{x}^{(i_\mu;\mathscr{B})}_{m-\varepsilon}-\mathbf{x}^{(\mu;\mathscr{D})}_{m-\varepsilon}]<0\\
&\text{for all } \mu\in\mathscr{N}_{\mathscr{D}}(\sigma_{n-2})=\mathscr{N}_{\mathscr{D}}^{C}(\sigma_{n-2})\\
&\text{and } i_\mu\in\mathscr{N}_{\mu\mathscr{B}}(\sigma_{n-2}).
\end{aligned}
\qquad (7.68)
$$

(iii) The edge of $\mathscr{E}^{(n-2)}_{\sigma_{n-2}}$ to the boundary and domain flows of $\mathbf{x}^{(i_\beta;\mathscr{B})}(t)$ ($i_\beta\in\mathscr{N}_{\mathscr{B}}(\sigma_{n-2})$ with all $\beta\in\mathscr{N}_{\mathscr{D}}(\sigma_{n-2})$) and $\mathbf{x}^{(\nu;\mathscr{D})}(t)$ ($\nu\in\mathscr{N}_{\mathscr{D}}(\sigma_{n-2})$) is $(\mathscr{N}_{\mathscr{B}}^{L}(\sigma_{n-2})\oplus\mathscr{N}_{\mathscr{D}}^{L}(\sigma_{n-2}))$ *non-switchable of the second* (source) if

$$
\begin{aligned}
&\hbar_{j_\beta}\mathbf{n}^{\mathrm{T}}_{\partial\Omega_{\beta j_\beta}}(\mathbf{x}^{(j_\beta;\mathscr{B})}_{m+\varepsilon})\cdot[\mathbf{x}^{(i_\beta;\mathscr{B})}_{m+\varepsilon}-\mathbf{x}^{(j_\beta;\mathscr{B})}_{m+\varepsilon}]>0\\
&\text{for all } i_\beta\in\mathscr{N}_{\mathscr{B}}^{L}(\sigma_{n-2})=\mathscr{N}_{\mathscr{B}}(\sigma_{n-2});\\
&\hbar_{j_\nu}\mathbf{n}^{\mathrm{T}}_{\partial\Omega_{\nu j_\nu}}(\mathbf{x}^{(j_\nu;\mathscr{B})}_{m+\varepsilon})\cdot[\mathbf{x}^{(\nu;\mathscr{D})}_{m+\varepsilon}-\mathbf{x}^{(j_\nu;\mathscr{B})}_{m+\varepsilon}]>0\\
&\text{for all } \nu\in\mathscr{N}_{\mathscr{D}}^{L}(\sigma_{n-2})=\mathscr{N}_{\mathscr{D}}(\sigma_{n-2})\\
&\text{and } j_\nu\in\mathscr{N}_{\nu\mathscr{B}}(\sigma_{n-2}).
\end{aligned}
\qquad (7.69)
$$

The foregoing definition gives three basic types of the switchabilty among all the boundary and domains to the edge of $\mathscr{E}^{(n-2)}_{\sigma_{n-2}}$. The switcahability described in the definition can be sketched in Figs. 7.18-7.20. In Fig. 7.18(a), the edge of $\mathscr{E}^{(n-2)}$ is a complete sink. All the boundary and domain flows will come to such an edge. The edge flow will be formed in such an edge. As in Chapter 3, once the sink flow on the edge disappears, the passable flows on the edge will be formed. If all the domain and boundary flows to the edge are leaving flows, then the edge to all the boundary and domain flows is a complete source, as sketched in Fig. 7.18(b).When the boundary flows to the edge are coming (or leaving) flows, the domain flows to the edge are leaving (or coming) flows, as shown in Fig. 7.19 (a) and (b). For an edge of $\mathscr{E}^{(n-2)}$, partial boundary and domain flows are coming flows and the rest boundary and domain flows are leaving flows, which is sketched in Fig. 7.20(a). Thus, a complex switching of the boundary and boundary flows will exist. If all domain flows to the edge are the sink flows, the boundary flows can form a boundary flow channel to the edge, as shown in Fig. 7.20(b). With the flow barrier to the domain flow, the boundary flow channel will be reliable and robust. From such a definition, the corresponding theorem will be presented for the sufficient and necessary conditions as follows.

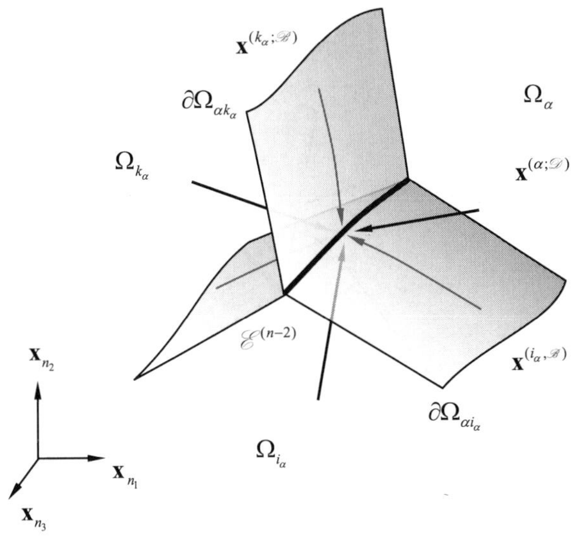

(a)

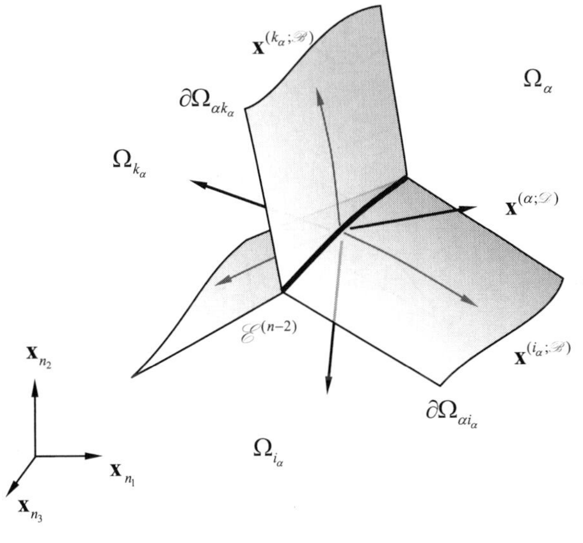

(b)

Fig. 7.18 (a) The complete sink edge and **(b)** complete source edge for domain and boundary flows. The black curve represents the $(n-2)$ -dimensional edge. The boundary flows are denoted by thin curves with arrows on the boundaries. The domain flows are represented by the thick curves with arrows in the domain. ($n_1 + n_2 + n_3 = n$) (color plot in the book end)

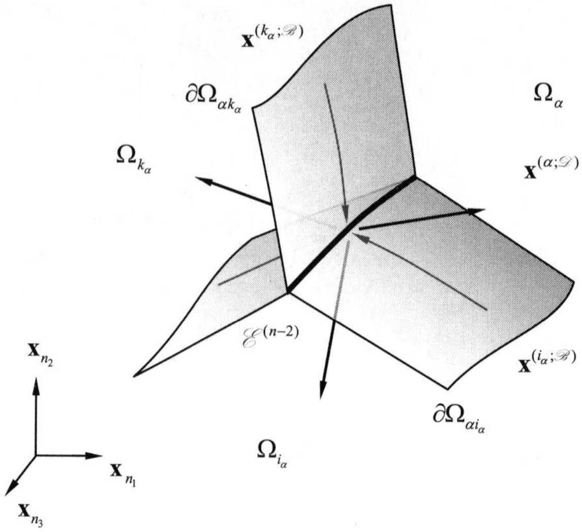

(a)

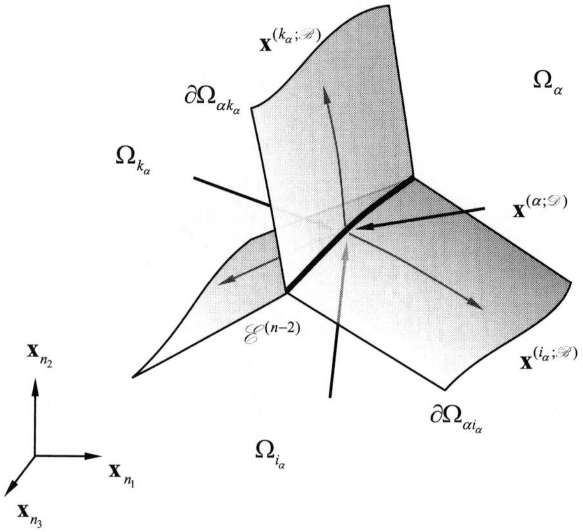

(b)

Fig. 7.19 (a) The sink boundary flows to source domain flows via the edge, and **(b)** the sink domain flows to source boundary flows via the edge. The black curve represents the $(n-2)$-dimensional edge. The two blue points are two $(n-3)$-dimensional edges. The boundary flows are denoted by thin curves with arrows on the boundaries. The domain flows are represented by the thick curves with arrows in the domain. ($n_1+n_2+n_3=n$) (color plot in the book end)

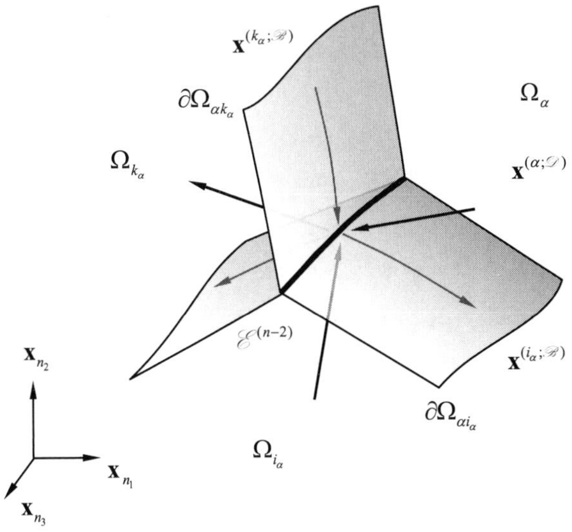

(a)

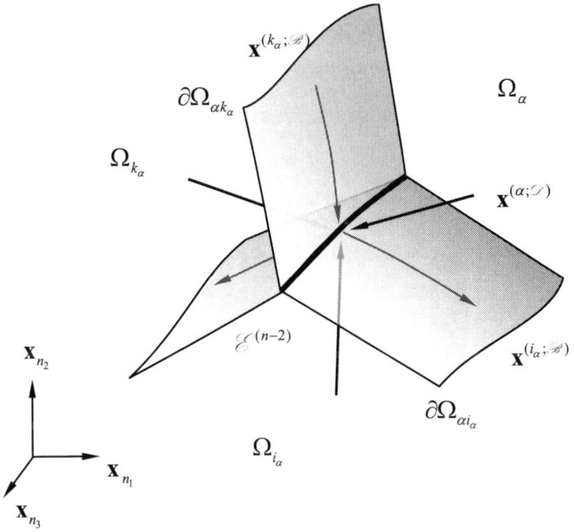

(b)

Fig. 7.20 (a) The coming and leaving mixing of boundary flows and domain flows to the edge, and **(b)** the sink domain flows with partially coming and partially leaving boundary flows via the edge. The black curve represents the $(n-2)$ -dimensional edge. The boundary flows are denoted by thin curves with arrows on the boundaries. The domain flows are represented by thick curves with arrows in the domain. ($n_1+n_2+n_3=n$)

Theorem 7.9. *For a discontinuous dynamical system in Eqs. (6.4)-(6.8), there is an* $(n-2)$ *-dimensional edge* $\mathscr{E}_{\sigma_{n-2}}^{(n-2)}$ *to be formed by* m_b *-boundaries* $\partial\Omega_{\alpha i_\alpha}$ *of* $(n-1)$ *-dimensions with* m_d *-domains* Ω_α $(\alpha, i_\alpha \in \mathscr{N}_{\mathscr{D}} = \{1,2,\cdots,m_d\}$ *and* $\alpha \neq i_\alpha$, *also* $i_\alpha \in \mathscr{N}_{\mathscr{B}} = \{1,2,\cdots,m_b\})$. $\mathscr{N}_{\mathscr{D}}(\sigma_{n-2}) = \mathscr{N}_{\mathscr{D}}^{C}(\sigma_{n-2}) \cup \mathscr{N}_{\mathscr{D}}^{L}(\sigma_{n-2}) \subset \mathscr{N}_{\mathscr{D}}$ *and* $\mathscr{N}_{\mathscr{D}}^{C}(\sigma_{n-2}) \cap \mathscr{N}_{\mathscr{D}}^{L}(\sigma_{n-2}) = \varnothing$. $\mathscr{N}_{\mathscr{B}}(\sigma_{n-2}) = \mathscr{N}_{\mathscr{B}}^{C}(\sigma_{n-2}) \cup \mathscr{N}_{\mathscr{B}}^{L}(\sigma_{n-2}) \subset \mathscr{N}_{\mathscr{B}}$ *and* $\mathscr{N}_{\mathscr{B}}^{C}(\sigma_{n-2}) \cap \mathscr{N}_{\mathscr{B}}^{L}(\sigma_{n-2}) = \varnothing$. *Suppose the coming and leaving boundary flows are* $\mathbf{x}^{(i_\alpha;\mathscr{B})}(t)$ $(i_\alpha \in \mathscr{N}_{\mathscr{B}}^{C}(\sigma_{n-2}))$ *and* $\mathbf{x}^{(i_\beta;\mathscr{B})}(t)$ $(i_\beta \in \mathscr{N}_{\mathscr{B}}^{L}(\sigma_{n-2}))$ *on the boundaries* $\partial\Omega_{\alpha i_\alpha}$ *and* $\partial\Omega_{\beta i_\beta}$, *respectively. Suppose the coming and leaving domain flows are* $\mathbf{x}^{(\mu;\mathscr{D})}(t)$ $(\mu \in \mathscr{N}_{\mathscr{D}}^{C}(\sigma_{n-2}))$ *and* $\mathbf{x}^{(\nu;\mathscr{D})}(t)$ $(\nu \in \mathscr{N}_{\mathscr{D}}^{L}(\sigma_{n-2}))$ *in domains* Ω_μ *and* Ω_ν, *respectively. At time* t_m, $\mathbf{x}_m = \mathbf{x}^{(\mathscr{E})}(t_m) \in \mathscr{E}_{\sigma_{n-2}}^{(n-2)}$, $\mathbf{x}_{m\pm}^{(i_\alpha;\mathscr{B})} = \mathbf{x}_{m\mp}^{(i_\beta;\mathscr{B})} = \mathbf{x}_m$ *and* $\mathbf{x}_{m\pm}^{(\mu;\mathscr{D})} = \mathbf{x}_{m\mp}^{(\nu;\mathscr{D})} = \mathbf{x}_m$. *For an arbitrarily small* $\varepsilon > 0$, *there is a time interval* $[t_{m-\varepsilon}, t_{m+\varepsilon}]$. *The boundary flows of* $\mathbf{x}^{(i_\rho;\mathscr{B})}(t)$ $(\rho = \alpha, \beta)$ *are* $C_{[t_{m-\varepsilon},0)}^{(r_{i_\rho};\mathscr{B})}$ *or* $C_{(0,t_{m+\varepsilon}]}^{(r_{i_\rho};\mathscr{B})}$ *-continuous for time t and* $\| d^{r_{i_\rho}+1}\mathbf{x}^{(i_\rho;\mathscr{B})} / dt^{r_{i_\rho}+1} \| < \infty$ $(r_{i_\rho} \geq 2)$. *The domain flows of* $\mathbf{x}^{(\sigma;\mathscr{D})}(t)$ $(\sigma = \mu, \nu)$ *are* $C_{[t_{m-\varepsilon},0)}^{(r_\sigma;\mathscr{D})}$ *or* $C_{(0,t_{m+\varepsilon}]}^{(r_\sigma;\mathscr{D})}$ *-continuous for time t and* $\| d^{r_\sigma+1}\mathbf{x}^{(\sigma;\mathscr{D})} / dt^{r_\sigma+1} \| < \infty$ $(r_\sigma \geq 2)$.

(i) *The coming boundary and domain flows of* $\mathbf{x}^{(i_\alpha;\mathscr{B})}(t)$ $(i_\alpha \in \mathscr{N}_{\alpha\mathscr{B}}^{C}(\sigma_{n-2}))$ *and* $\mathbf{x}^{(\mu;\mathscr{D})}(t)$ $(\mu \in \mathscr{N}_{\mathscr{D}}^{C}(\sigma_{n-2}))$ *to the leaving boundary and domain flows of* $\mathbf{x}^{(i_\beta;\mathscr{B})}(t)$ $(i_\beta \in \mathscr{N}_{\beta\mathscr{B}}^{L}(\sigma_{n-2}))$ *and* $\mathbf{x}^{(\nu;\mathscr{D})}(t)$ $(\nu \in \mathscr{N}_{\mathscr{D}}^{L}(\sigma_{n-2}))$ *at the edge* $\mathscr{E}_{\sigma_{n-2}}^{(n-2)}$ *for time* t_m *are* $(\mathscr{N}_{\mathscr{B}}^{C}(\sigma_{n-2}) \oplus \mathscr{N}_{\mathscr{D}}^{C}(\sigma_{n-2}) : \mathscr{N}_{\mathscr{B}}^{L}(\sigma_{n-2}) \oplus \mathscr{N}_{\mathscr{D}}^{L}(\sigma_{n-2}))$ *-switchable from* $\partial\Omega_{\alpha i_\alpha}$ *and* Ω_μ *to* $\partial\Omega_{\beta i_\beta}$ *and* Ω_ν *if and only if*

$$
\begin{aligned}
&\hbar_{j_\alpha} G_{\partial\Omega_{\alpha j_\alpha}}^{(i_\alpha;\mathscr{B})}(\mathbf{x}_m, t_{m-}, \boldsymbol{\lambda}_{i_\alpha}, \boldsymbol{\lambda}_{j_\alpha}) < 0 \\
&\text{for all } i_\alpha \in \mathscr{N}_{\alpha\mathscr{B}}^{C}(\sigma_{n-2}) \text{ and } \alpha \in \mathscr{N}_{\mathscr{D}}(\sigma_{n-2}), \\
&\hbar_{j_\beta} G_{\partial\Omega_{\beta j_\beta}}^{(i_\beta;\mathscr{B})}(\mathbf{x}_m, t_{m+}, \boldsymbol{\lambda}_{i_\beta}, \boldsymbol{\lambda}_{j_\beta}) > 0 \\
&\text{for all } i_\beta \in \mathscr{N}_{\beta\mathscr{B}}^{L}(\sigma_{n-2}) \text{ and } \beta \in \mathscr{N}_{\mathscr{D}}(\sigma_{n-2}); \\
&\hbar_{j_\mu} G_{\partial\Omega_{\mu j_\mu}}^{(\mu;\mathscr{D})}(\mathbf{x}_m, t_{m-}, \mathbf{p}_\mu, \boldsymbol{\lambda}_{j_\mu}) < 0 \\
&\text{for all } \mu \in \mathscr{N}_{\mathscr{D}}^{C}(\sigma_{n-2}) \text{ and } j_\mu \in \mathscr{N}_{\mu\mathscr{B}}(\sigma_{n-2}), \\
&\hbar_{j_\nu} G_{\partial\Omega_{\nu j_\nu}}^{(\nu;\mathscr{D})}(\mathbf{x}_m, t_{m+}, \mathbf{p}_\nu, \boldsymbol{\lambda}_{j_\nu}) > 0 \\
&\text{for all } \nu \in \mathscr{N}_{\mathscr{D}}^{L}(\sigma_{n-2}) \text{ and } j_\nu \in \mathscr{N}_{\nu\mathscr{B}}(\sigma_{n-2}).
\end{aligned}
\tag{7.70}
$$

(ii) *The edge of* $\mathscr{E}_{\sigma_{n-2}}^{(n-2)}$ *to the boundary and domain flows of* $\mathbf{x}^{(i_\alpha;\mathscr{B})}(t)$ $(i_\alpha \in \mathscr{N}_{\alpha\mathscr{B}}(\sigma_{n-2})$ *with all* $\alpha \in \mathscr{N}_{\mathscr{D}}(\sigma_{n-2}))$ *and* $\mathbf{x}^{(\mu;\mathscr{D})}(t)$ $(\mu \in \mathscr{N}_{\mathscr{D}}(\sigma_{n-2}))$ *is an*

$(\mathscr{N}_{\mathscr{B}}^{C}(\sigma_{n-2})\oplus\mathscr{N}_{\mathscr{D}}^{C}(\sigma_{n-2}))$ *non-switchable of the first kind (sink) if and only if*

$$\begin{aligned}
&\hbar_{j_\alpha} G_{\partial\Omega_{\alpha j_\alpha}}^{(i_\alpha;\mathscr{B})}(\mathbf{x}_m,t_{m-},\boldsymbol{\lambda}_{i_\alpha},\boldsymbol{\lambda}_{j_\alpha})<0\\
&\textit{for all } i_\alpha\in\mathscr{N}_{\alpha\mathscr{B}}(\sigma_{n-2})=\mathscr{N}_{\alpha\mathscr{B}}^{C}(\sigma_{n-2});\\
&\hbar_{j_\mu} G_{\partial\Omega_{\mu j_\mu}}^{(\beta;\mathscr{D})}(\mathbf{x}_m,t_{m-},\mathbf{p}_\mu,\boldsymbol{\lambda}_{j_\mu})<0\\
&\textit{for all } \mu\in\mathscr{N}_{\mathscr{D}}(\sigma_{n-2})=\mathscr{N}_{\mathscr{D}}^{C}(\sigma_{n-2})\\
&\textit{and } j_\mu\in\mathscr{N}_{\mu\mathscr{B}}(\sigma_{n-2}).
\end{aligned}\tag{7.71}$$

(iii) *The edge of* $\mathscr{E}_{\sigma_{n-2}}^{(n-2)}$ *to the boundary and domain flows of* $\mathbf{x}^{(i_\beta;\mathscr{B})}(t)$ $(i_\beta\in\mathscr{N}_{\mathscr{B}}(\sigma_{n-2})$ *with all* $\beta\in\mathscr{N}_{\mathscr{D}}(\sigma_{n-2})$) *and* $\mathbf{x}^{(\nu;\mathscr{D})}(t)$ ($\nu\in\mathscr{N}_{\mathscr{D}}(\sigma_{n-2})$) *is an* $(\mathscr{N}_{\mathscr{B}}^{L}(\sigma_{n-2})\oplus\mathscr{N}_{\mathscr{D}}^{L}(\sigma_{n-2}))$ *source if and only if*

$$\begin{aligned}
&\hbar_{j_\beta} G_{\partial\Omega_{\beta j_\beta}}^{(i_\beta;\mathscr{B})}(\mathbf{x}_m,t_{m+},\boldsymbol{\lambda}_{i_\beta},\boldsymbol{\lambda}_{j_\beta})>0\\
&\textit{for all } i_\beta\in\mathscr{N}_{\beta\mathscr{B}}^{L}(\sigma_{n-2})=\mathscr{N}_{\beta\mathscr{B}}(\sigma_{n-2});\\
&\hbar_{j_\mu} G_{\partial\Omega_{\mu j_\mu}}^{(\beta;\mathscr{D})}(\mathbf{x}_m,t_{m+},\mathbf{p}_\mu,\boldsymbol{\lambda}_{j_\mu})>0\\
&\textit{for all } \nu\in\mathscr{N}_{\mathscr{D}}^{L}(\sigma_{n-2})=\mathscr{N}_{\mathscr{D}}(\sigma_{n-2})\\
&\textit{and } j_\nu\in\mathscr{N}_{\nu\mathscr{B}}(\sigma_{n-2}).
\end{aligned}\tag{7.72}$$

Proof. The theorem can be proved as in Chapter 3. ■

If boundary and domain flows to the edge with higher order singularity, the corresponding description is also given.

Definition 7.10. For a discontinuous dynamical system in Eqs. (6.4)-(6.8), there is an $(n-2)$ -dimensional edge $\mathscr{E}_{\sigma_{n-2}}^{(n-2)}$ to be formed by m_b -boundaries $\partial\Omega_{\alpha i_\alpha}$ of $(n-1)$ -dimensions with m_d -domains Ω_α ($\alpha,i_\alpha\in\mathscr{N}_{\mathscr{D}}=\{1,2,\cdots,m_d\}$ and $\alpha\neq i_\alpha$, also $i_\alpha\in\mathscr{N}_{\mathscr{B}}=\{1,2,\cdots,m_b\}$). $\mathscr{N}_{\mathscr{D}}(\sigma_{n-2})=\mathscr{N}_{\mathscr{D}}^{C}(\sigma_{n-2})\cup\mathscr{N}_{\mathscr{D}}^{L}(\sigma_{n-2})\subset\mathscr{N}_{\mathscr{D}}$, and $\mathscr{N}_{\mathscr{D}}^{C}(\sigma_{n-2})\cap\mathscr{N}_{\mathscr{D}}^{L}(\sigma_{n-2})=\varnothing$. $\mathscr{N}_{\mathscr{B}}(\sigma_{n-2})=\mathscr{N}_{\mathscr{B}}^{C}(\sigma_{n-2})\cup\mathscr{N}_{\mathscr{B}}^{L}(\sigma_{n-2})\subset\mathscr{N}_{\mathscr{B}}$ and $\mathscr{N}_{\mathscr{B}}^{C}(\sigma_{n-2})\cap\mathscr{N}_{\mathscr{B}}^{L}(\sigma_{n-2})=\varnothing$. Suppose the coming and leaving boundary flows are $\mathbf{x}^{(i_\alpha;\mathscr{B})}(t)$ ($i_\alpha\in\mathscr{N}_{\mathscr{B}}^{C}(\sigma_{n-2})$) and $\mathbf{x}^{(i_\beta;\mathscr{B})}(t)$ ($i_\beta\in\mathscr{N}_{\mathscr{B}}^{L}(\sigma_{n-2})$) on the boundary $\partial\Omega_{\alpha i_\alpha}$ and $\partial\Omega_{\beta i_\beta}$, respectively. Suppose the coming and leaving domain flows are $\mathbf{x}^{(\mu;\mathscr{D})}(t)$ ($\mu\in\mathscr{N}_{\mathscr{D}}^{C}(\sigma_{n-2})$) and $\mathbf{x}^{(\nu;\mathscr{D})}(t)$ ($\nu\in\mathscr{N}_{\mathscr{D}}^{L}(\sigma_{n-2})$) in domains Ω_μ and Ω_ν , respectively. At time t_m , $\mathbf{x}_m=\mathbf{x}^{(\mathscr{E})}(t_m)\in\mathscr{E}_{\sigma_{n-2}}^{(n-2)}$, $\mathbf{x}_{m\pm}^{(i_\alpha;\mathscr{B})}=\mathbf{x}_{m\mp}^{(i_\beta;\mathscr{B})}=\mathbf{x}_m$ and $\mathbf{x}_{m\pm}^{(\mu;\mathscr{D})}=\mathbf{x}_{m\mp}^{(\nu;\mathscr{D})}=\mathbf{x}_m$. For an arbitrarily small $\varepsilon>0$, there is a time interval $[t_{m-\varepsilon},t_{m+\varepsilon}]$. The boundary flows of $\mathbf{x}^{(i_\rho;\mathscr{B})}(t)$ ($\rho=\alpha,\beta$) are $C_{[t_{m-\varepsilon},0)}^{(r_{i_\rho};\mathscr{B})}$ or $C_{(0,t_{m+\varepsilon}]}^{(r_{i_\rho};\mathscr{B})}$ -

continuous for time t and $\| d^{r_{i_\rho}+1}\mathbf{x}^{(i_\rho;\mathscr{B})}/dt^{r_{i_\rho}+1}\|<\infty$ ($r_{i_\rho} \geq m_{i_\rho}$). The domain flows of $\mathbf{x}^{(\sigma;\mathscr{D})}(t)$ ($\sigma=\mu,\nu$) are $C^{(r_\sigma;\mathscr{D})}_{[t_{m-\varepsilon},0)}$ or $C^{(r_\sigma;\mathscr{D})}_{(0,t_{m+\varepsilon}]}$-continuous for time t and $\| d^{r_\sigma+1}\mathbf{x}^{(\sigma;\mathscr{D})}/dt^{r_\sigma+1}\|<\infty$ ($r_\sigma \geq \mathbf{m}_\sigma$). As $n=2$, $m_{i_\rho}=2k_{i_\rho}+1$ and $\mathbf{m}_\sigma = 2\mathbf{k}_\sigma+1$.

(i) With the switching rule, the coming boundary and domain flows of $\mathbf{x}^{(i_\alpha;\mathscr{B})}(t)$ ($i_\alpha \in \mathscr{N}^{C}_{\mathscr{B}}(\sigma_{n-2})$) and $\mathbf{x}^{(\mu;\mathscr{D})}(t)$ ($\mu \in \mathscr{N}^{C}_{\mathscr{D}}(\sigma_{n-2})$) to the leaving boundary and domain flows of $\mathbf{x}^{(i_\beta;\mathscr{B})}(t)$ ($i_\beta \in \mathscr{N}^{L}_{\beta\mathscr{B}}(\sigma_{n-2})$) and $\mathbf{x}^{(\nu;\mathscr{D})}(t)$ ($\nu \in \mathscr{N}^{L}_{\mathscr{D}}(\sigma_{n-2})$) at the edge $\mathscr{E}^{(n-2)}_{\sigma_{n-2}}$ for time t_m are *switchable* with the $(\cup_{i_\alpha}(m_{i_\alpha};\mathscr{B})\oplus\cup_\mu(\mathbf{m}_\mu;\mathscr{D}):\cup_{i_\beta}(m_{i_\beta};\mathscr{B})\oplus\cup_\nu(\mathbf{m}_\nu;\mathscr{D}))$ singularity from $\partial\Omega_{\alpha i_\alpha}$ and Ω_μ to $\partial\Omega_{\beta i_\beta}$ and Ω_ν if

$$\left.\begin{array}{l}
\hbar_{j_\alpha} G^{(s_{i_\alpha},i_\alpha;\mathscr{B})}_{\partial\Omega_{\alpha j_\alpha}}(\mathbf{x}_m,t_{m-},\boldsymbol{\lambda}_{i_\alpha},\boldsymbol{\lambda}_{j_\alpha})=0\\
\text{for } s_{i_\alpha}=0,1,2,\cdots,m_{i_\alpha}-1,\\
\hbar_{j_\alpha}\mathbf{n}^{\mathrm{T}}_{\partial\Omega_{\alpha j_\alpha}}(\mathbf{x}^{(j_\alpha;\mathscr{B})}_{m-\varepsilon})\cdot[\mathbf{x}^{(j_\alpha;\mathscr{B})}_{m-\varepsilon}-\mathbf{x}^{(i_\alpha;\mathscr{B})}_{m-\varepsilon}]<0
\end{array}\right\} \tag{7.73}$$

for all $i_\alpha \in \mathscr{N}^{C}_{\alpha\mathscr{B}}(\sigma_{n-2})$ and $\alpha \in \mathscr{N}_{\mathscr{D}}(\sigma_{n-2})$;

$$\left.\begin{array}{l}
\hbar_{j_\beta} G^{(s_{i_\beta},i_\beta;\mathscr{B})}_{\partial\Omega_{\beta j_\beta}}(\mathbf{x}_m,t_{m+},\boldsymbol{\lambda}_{i_\beta},\boldsymbol{\lambda}_{j_\beta})=0\\
\text{for } s_{i_\beta}=0,1,2,\cdots,m_{i_\beta}-1,\\
\hbar_{j_\beta}\mathbf{n}^{\mathrm{T}}_{\partial\Omega_{\beta j_\beta}}(\mathbf{x}^{(j_\beta;\mathscr{B})}_{m+\varepsilon})\cdot[\mathbf{x}^{(i_\beta;\mathscr{B})}_{m+\varepsilon}-\mathbf{x}^{(j_\beta;\mathscr{B})}_{m+\varepsilon}]>0
\end{array}\right\} \tag{7.74}$$

for all $i_\beta \in \mathscr{N}^{L}_{\beta\mathscr{B}}(\sigma_{n-2})$ and $\beta \in \mathscr{N}_{\mathscr{D}}(\sigma_{n-2})$;

$$\left.\begin{array}{l}
\hbar_{j_\mu} G^{(s_\mu,\beta;\mathscr{D})}_{\partial\Omega_{\mu j_\mu}}(\mathbf{x}_m,t_{m-},\mathbf{p}_\mu,\boldsymbol{\lambda}_{j_\mu})=0\\
\text{for } s_\mu=0,1,2,\cdots,m_\mu-1,\\
\hbar_{i_\mu}\mathbf{n}^{\mathrm{T}}_{\partial\Omega_{\mu i_\mu}}(\mathbf{x}^{(i_\mu;\mathscr{B})}_{m-\varepsilon})\cdot[\mathbf{x}^{(i_\mu;\mathscr{B})}_{m-\varepsilon}-\mathbf{x}^{(\mu;\mathscr{D})}_{m-\varepsilon}]<0
\end{array}\right\} \tag{7.75}$$

for all $\mu \in \mathscr{N}^{C}_{\mathscr{D}}(\sigma_{n-2})$ and $j_\mu \in \mathscr{N}_{\mu\mathscr{B}}(\sigma_{n-2})$;

$$\left.\begin{array}{l}
\hbar_{j_\nu} G^{(s_\nu,\nu;\mathscr{D})}_{\partial\Omega_{\nu j_\nu}}(\mathbf{x}_m,t_{m+},\mathbf{p}_\nu,\boldsymbol{\lambda}_{j_\nu})=0\\
\text{for } s_\nu=0,1,2,\cdots,m_\nu-1,\\
\hbar_{j_\nu}\mathbf{n}^{\mathrm{T}}_{\partial\Omega_{\nu j_\nu}}(\mathbf{x}^{(j_\nu;\mathscr{B})}_{m+\varepsilon})\cdot[\mathbf{x}^{(\nu;\mathscr{D})}_{m+\varepsilon}-\mathbf{x}^{(j_\nu;\mathscr{B})}_{m+\varepsilon}]>0
\end{array}\right\} \tag{7.76}$$

for all $\nu \in \mathscr{N}^{L}_{\mathscr{D}}(\sigma_{n-2})$ and $j_\nu \in \mathscr{N}_{\nu\mathscr{B}}(\sigma_{n-2})$.

(ii) The edge of $\mathscr{E}^{(n-2)}_{\sigma_{n-2}}$ to the boundary and domain flows of $\mathbf{x}^{(i_\alpha;\mathscr{B})}(t)$ ($i_\alpha \in \mathscr{N}_{\mathscr{B}}(\sigma_{n-2})$ with all $\alpha \in \mathscr{N}_{\mathscr{D}}(\sigma_{n-2})$) and $\mathbf{x}^{(\mu;\mathscr{D})}(t)$ ($\mu \in \mathscr{N}_{\mathscr{D}}(\sigma_{n-2})$) is *non-*

switchable of the first kind (sink) with the $(\cup_{i_\alpha}(2k_{i_\alpha};\mathscr{B})\oplus(\cup_\mu \mathbf{m}_\mu;\mathscr{D})$ singularity ($\mathbf{m}_{\mu\mathscr{D}}^C \neq 2\mathbf{k}_{\mu\mathscr{D}}^C+\mathbf{1}$) if

$$\left.\begin{array}{l}\hbar_{j_\alpha} G_{\partial\Omega_{\alpha j_\alpha}}^{(s_{i_\alpha},i_\alpha;\mathscr{B})}(\mathbf{x}_m,t_{m-},\boldsymbol{\lambda}_{i_\alpha},\boldsymbol{\lambda}_{j_\alpha})=0\\ \text{for } s_{i_\alpha}=0,1,2,\cdots,2k_{i_\alpha}-1,\\ \hbar_{j_\alpha}\mathbf{n}_{\partial\Omega_{\alpha j_\alpha}}^{\mathrm{T}}(\mathbf{x}_{m-\varepsilon}^{(j_\alpha;\mathscr{B})})\cdot[\mathbf{x}_{m-\varepsilon}^{(j_\alpha;\mathscr{B})}-\mathbf{x}_{m-\varepsilon}^{(i_\alpha;\mathscr{B})}]<0\\ \text{for all } i_\alpha\in\mathscr{N}_{\alpha\mathscr{B}}^C(\sigma_{n-2}) \text{ and } \alpha\in\mathscr{N}_{\mathscr{D}}(\sigma_{n-2});\end{array}\right\} \tag{7.77}$$

$$\left.\begin{array}{l}\hbar_{j_\mu} G_{\partial\Omega_{\mu j_\mu}}^{(s_\mu,\beta;\mathscr{D})}(\mathbf{x}_m,t_{m-},\mathbf{p}_\mu,\boldsymbol{\lambda}_{j_\mu})=0\\ \text{for } s_\mu=0,1,2,\cdots,m_\mu-1,\\ \hbar_{i_\mu}\mathbf{n}_{\partial\Omega_{\mu i_\mu}}^{\mathrm{T}}(\mathbf{x}_{m-\varepsilon}^{(i_\mu;\mathscr{B})})\cdot[\mathbf{x}_{m-\varepsilon}^{(i_\mu;\mathscr{B})}-\mathbf{x}_{m-\varepsilon}^{(\mu;\mathscr{D})}]<0\\ \text{for all } \mu\in\mathscr{N}_{\mathscr{D}}^C(\sigma_{n-2}) \text{ and } j_\mu\in\mathscr{N}_{\mu\mathscr{B}}(\sigma_{n-2}).\end{array}\right\} \tag{7.78}$$

(iii) The edge of $\mathscr{E}_{\sigma_{n-2}}^{(n-2)}$ to the boundary and domain flows of $\mathbf{x}^{(i_\beta;\mathscr{B})}(t)$ ($i_\beta \in \mathscr{N}_{\mathscr{B}}(\sigma_{n-2})$ with all $\beta\in\mathscr{N}_{\mathscr{D}}(\sigma_{n-2})$) and $\mathbf{x}^{(\nu;\mathscr{D})}(t)$ ($\nu\in\mathscr{N}_{\mathscr{D}}(\sigma_{n-2})$) is *switchable of the second kind* (source) with the $(\cup_{i_\alpha}(2k_{i_\alpha};\mathscr{B})\oplus(\cup_\mu \mathbf{m}_\mu;\mathscr{D})$ singularity if

$$\left.\begin{array}{l}\hbar_{j_\beta} G_{\partial\Omega_{\beta j_\beta}}^{(s_{i_\beta},i_\beta;\mathscr{B})}(\mathbf{x}_m,t_{m+},\boldsymbol{\lambda}_{i_\beta},\boldsymbol{\lambda}_{j_\beta})=0\\ \text{for } s_{i_\beta}=0,1,2,\cdots,m_{i_\beta}-1,\\ \hbar_{j_\beta}\mathbf{n}_{\partial\Omega_{\beta j_\beta}}^{\mathrm{T}}(\mathbf{x}_{m+\varepsilon}^{(j_\beta;\mathscr{B})})\cdot[\mathbf{x}_{m+\varepsilon}^{(i_\beta;\mathscr{B})}-\mathbf{x}_{m+\varepsilon}^{(j_\beta;\mathscr{B})}]>0\\ \text{for all } i_\beta\in\mathscr{N}_{\beta\mathscr{B}}^L(\sigma_{n-2}), \text{and } \beta\in\mathscr{N}_{\mathscr{D}}(\sigma_{n-2});\end{array}\right\} \tag{7.79}$$

$$\left.\begin{array}{l}\hbar_{j_\nu} G_{\partial\Omega_{\nu j_\nu}}^{(s_\nu,\nu;\mathscr{D})}(\mathbf{x}_m,t_{m+},\mathbf{p}_\nu,\boldsymbol{\lambda}_{j_\nu})=0\\ \text{for } s_\nu=0,1,2,\cdots,m_\nu-1,\\ \hbar_{j_\nu}\mathbf{n}_{\partial\Omega_{\nu j_\nu}}^{\mathrm{T}}(\mathbf{x}_{m+\varepsilon}^{(j_\nu;\mathscr{B})})\cdot[\mathbf{x}_{m+\varepsilon}^{(\nu;\mathscr{D})}-\mathbf{x}_{m+\varepsilon}^{(j_\nu;\mathscr{B})}]>0\\ \text{for all } \nu\in\mathscr{N}_{\mathscr{D}}^L(\sigma_{n-2}) \text{ and } j_\nu\in\mathscr{N}_{\nu\mathscr{B}}(\sigma_{n-2}).\end{array}\right\} \tag{7.80}$$

(iv) Without the switching rule, the coming boundary and domain flows of $\mathbf{x}^{(i_\alpha;\mathscr{B})}(t)$ ($i_\alpha\in\mathscr{N}_{\mathscr{B}}^C(\sigma_{n-2})$) and $\mathbf{x}^{(\mu;\mathscr{D})}(t)$ ($\mu\in\mathscr{N}_{\mathscr{D}}^C(\sigma_{n-2})$) to the leaving boundary and domain flows of $\mathbf{x}^{(i_\beta;\mathscr{B})}(t)$ ($i_\beta\in\mathscr{N}_{\beta\mathscr{B}}^L(\sigma_{n-2})$) and $\mathbf{x}^{(\nu;\mathscr{D})}(t)$ ($\nu\in\mathscr{N}_{\mathscr{D}}^L(\sigma_{n-2})$) at the edge of $\mathscr{E}_{\sigma_{n-2}}^{(n-2)}$ for time t_m are *potentially switchable* with the $(\cup_{i_\alpha}(2k_{i_\alpha};\mathscr{B})\oplus\cup_\mu(\mathbf{m}_\mu;\mathscr{D}):\cup_{i_\beta}(m_{i_\beta};\mathscr{B})\oplus\cup_\nu(\mathbf{m}_\nu;\mathscr{D}))$ singularity if Eqs. (7.73)-(7.76) hold with $m_{i_\alpha}=2k_{i_\alpha}$ and $\mathbf{m}_\mu\neq 2\mathbf{k}_\mu+1$.

(v) Without the switching rule, the edge $\mathscr{E}^{(n-2)}_{\sigma_{n-2}}$ to the coming boundary and domain flows of $\mathbf{x}^{(i_\alpha;\mathscr{B})}(t)$ ($i_\alpha \in \mathscr{N}^C_{\mathscr{B}}(\sigma_{n-2})$) and $\mathbf{x}^{(\mu;\mathscr{D})}(t)$ ($\mu \in \mathscr{N}^C_{\mathscr{D}}(\sigma_{n-2})$) to the leaving boundary and domain flows of $\mathbf{x}^{(i_\beta;\mathscr{B})}(t)$ ($i_\beta \in \mathscr{N}^L_{\beta\mathscr{B}}(\sigma_{n-2})$) and $\mathbf{x}^{(\nu;\mathscr{D})}(t)$ ($\nu \in \mathscr{N}^L_{\mathscr{D}}(\sigma_{n-2})$) at the edge of $\mathscr{E}^{(n-2)}_{\sigma_{n-2}}$ for time t_m is *non-switchable* with the $(\cup_{i_\alpha}(m_{i_\alpha};\mathscr{B})\oplus\cup_\mu(\mathbf{m}_\mu;\mathscr{D}):\cup_{i_\beta}(m_{i_\beta};\mathscr{B})\oplus\cup_\nu(\mathbf{m}_\nu;\mathscr{D}))$ singularity if Eqs. (7.73)-(7.76) hold with (at least either $m_{i_\alpha}=2k_{i_\alpha}+1$ for specific i_α or $\mathbf{m}_\mu = 2\mathbf{k}_\mu+1$ for a specific μ).

(vi) With switching rules, the edge $\mathscr{E}^{(n-2)}_{\sigma_{n-2}}$ is switchable with the $(\cup_{i_\alpha}(2k_{i_\alpha}+1;\mathscr{B})\oplus\cup_\mu(2\mathbf{k}_\mu+1;\mathscr{D}):\cup_{i_\alpha}(2k_{i_\alpha}+1;\mathscr{B})\oplus\cup_\mu(2\mathbf{k}_\mu+1;\mathscr{D}))$ *singularity* for all the gazing boundary and domain flows of $\mathbf{x}^{(i_\alpha;\mathscr{B})}(t)$ ($i_\alpha \in \mathscr{N}_{\mathscr{B}}(\sigma_{n-2})$ with all $\alpha\in\mathscr{N}_{\mathscr{D}}(\sigma_{n-2})$) and $\mathbf{x}^{(\mu;\mathscr{D})}(t)$ ($\mu\in\mathscr{N}_{\mathscr{D}}(\sigma_{n-2})$) for time t_m if

$$\left.\begin{array}{l}
\hbar_{j_\alpha} G^{(s_{i_\alpha},i_\alpha;\mathscr{B})}_{\partial\Omega_{\alpha j_\alpha}}(\mathbf{x}_m,t_{m\pm},\boldsymbol{\lambda}_{i_\alpha},\boldsymbol{\lambda}_{j_\alpha})=0\\
\text{for } s_{i_\alpha}=0,1,2,\cdots,2k_{i_\alpha},\\
\hbar_{j_\alpha}\mathbf{n}^{\mathrm{T}}_{\partial\Omega_{\alpha j_\alpha}}(\mathbf{x}^{(j_\alpha;\mathscr{B})}_{m-\varepsilon})\cdot[\mathbf{x}^{(j_\alpha;\mathscr{B})}_{m-\varepsilon}-\mathbf{x}^{(i_\alpha;\mathscr{B})}_{m-\varepsilon}]<0,\\
\hbar_{j_\alpha}\mathbf{n}^{\mathrm{T}}_{\partial\Omega_{\alpha j_\alpha}}(\mathbf{x}^{(j_\alpha;\mathscr{B})}_{m+\varepsilon})\cdot[\mathbf{x}^{(i_\alpha;\mathscr{B})}_{m+\varepsilon}-\mathbf{x}^{(j_\alpha;\mathscr{B})}_{m+\varepsilon}]>0
\end{array}\right\}\quad(7.81)$$

for all $i_\alpha\in\mathscr{N}^C_{\alpha\mathscr{B}}(\sigma_{n-2})$ and $\alpha\in\mathscr{N}_{\mathscr{D}}(\sigma_{n-2})$;

$$\left.\begin{array}{l}
\hbar_{j_\mu} G^{(s_\mu,\beta;\mathscr{D})}_{\partial\Omega_{\mu j_\mu}}(\mathbf{x}_m,t_{m\pm},\mathbf{p}_\mu,\boldsymbol{\lambda}_{j_\mu})=0\\
\text{for } s_\mu=0,1,2,\cdots,2k_\mu,\\
\hbar_{i_\mu}\mathbf{n}^{\mathrm{T}}_{\partial\Omega_{\mu i_\mu}}(\mathbf{x}^{(i_\mu;\mathscr{B})}_{m-\varepsilon})\cdot[\mathbf{x}^{(i_\mu;\mathscr{B})}_{m-\varepsilon}-\mathbf{x}^{(\mu;\mathscr{D})}_{m-\varepsilon}]<0,\\
\hbar_{i_\mu}\mathbf{n}^{\mathrm{T}}_{\partial\Omega_{\mu i_\mu}}(\mathbf{x}^{(i_\mu;\mathscr{B})}_{m+\varepsilon})\cdot[\mathbf{x}^{(\mu;\mathscr{D})}_{m+\varepsilon}-\mathbf{x}^{(i_\mu;\mathscr{B})}_{m+\varepsilon}]>0
\end{array}\right\}\quad(7.82)$$

for all $\mu\in\mathscr{N}^C_{\mathscr{D}}(\sigma_{n-2})$ and $j_\mu\in\mathscr{N}_{\mu\mathscr{B}}(\sigma_{n-2})$.

(vii) Without switching rules, the edge $\mathscr{E}^{(n-2)}_{\sigma_{n-2}}$ is a grazing edge with the $(\cup_{i_\alpha}(2k_{i_\alpha}+1;\mathscr{B})\oplus\cup_\mu(2\mathbf{k}_\mu+1;\mathscr{D}):\cup_{i_\alpha}(2k_{i_\alpha}+1;\mathscr{B})\oplus\cup_\mu(2\mathbf{k}_\mu+1;\mathscr{D}))$ singularity for all the gazing boundary and domain flows of $\mathbf{x}^{(i_\alpha;\mathscr{B})}(t)$ ($i_\alpha\in\mathscr{N}_{\mathscr{B}}(\sigma_{n-2})$ with all $\alpha\in\mathscr{N}_{\mathscr{D}}(\sigma_{n-2})$) and $\mathbf{x}^{(\mu;\mathscr{D})}(t)$ ($\mu\in\mathscr{N}_{\mathscr{D}}(\sigma_{n-2})$) for time t_m if Eqs. (7.81) and (7.82) hold.

From definition, the theorem for the switchabilty among all boundary and domains to edge $\mathscr{E}^{(n-2)}_{\sigma_{n-2}}$ with higher order singularity will be presented as follows.

Theorem 7.10. *For a discontinuous dynamical system in Eqs.* (6.4)-(6.8), *there is an* $(n-2)$ *-dimensional edge* $\mathscr{E}_{\sigma_{n-2}}^{(n-2)}$ *to be formed by* m_b *-boundaries* $\partial\Omega_{\alpha i_\alpha}$ *of* $(n-1)$ *-dimensions with* m_d *-domains* Ω_α ($\alpha, i_\alpha \in \mathscr{N}_{\mathscr{D}} = \{1,2,\cdots,m_d\}$ *and* $\alpha \neq i_\alpha$, *also* $i_\alpha \in \mathscr{N}_{\mathscr{B}} = \{1,2,\cdots,m_b\}$). $\mathscr{N}_{\mathscr{D}}(\sigma_{n-2}) = \mathscr{N}_{\mathscr{D}}^{C}(\sigma_{n-2}) \cup \mathscr{N}_{\mathscr{D}}^{L}(\sigma_{n-2}) \subset \mathscr{N}_{\mathscr{D}}$, *and* $\mathscr{N}_{\mathscr{D}}^{C}(\sigma_{n-2}) \cap \mathscr{N}_{\mathscr{D}}^{L}(\sigma_{n-2}) = \varnothing$. $\mathscr{N}_{\mathscr{B}}(\sigma_{n-2}) = \mathscr{N}_{\mathscr{B}}^{C}(\sigma_{n-2}) \cup \mathscr{N}_{\mathscr{B}}^{L}(\sigma_{n-2}) \subset \mathscr{N}_{\mathscr{B}}$ *and* $\mathscr{N}_{\mathscr{B}}^{C}(\sigma_{n-2}) \cap \mathscr{N}_{\mathscr{B}}^{L}(\sigma_{n-2}) = \varnothing$. *Suppose the coming and leaving boundary flows are* $\mathbf{x}^{(i_\alpha;\mathscr{B})}(t)$ ($i_\alpha \in \mathscr{N}_{\mathscr{B}}^{C}(\sigma_{n-2})$) *and* $\mathbf{x}^{(i_\beta;\mathscr{B})}(t)$ ($i_\beta \in \mathscr{N}_{\mathscr{B}}^{L}(\sigma_{n-2})$) *on the boundary* $\partial\Omega_{\alpha i_\alpha}$ *and* $\partial\Omega_{\beta i_\beta}$, *respectively. Suppose the coming and leaving domain flows are* $\mathbf{x}^{(\mu;\mathscr{D})}(t)$ ($\mu \in \mathscr{N}_{\mathscr{D}}^{C}(\sigma_{n-2})$) *and* $\mathbf{x}^{(\nu;\mathscr{D})}(t)$ ($\nu \in \mathscr{N}_{\mathscr{D}}^{L}(\sigma_{n-2})$) *in domains* Ω_μ *and* Ω_ν, *respectively. At time* t_m, $\mathbf{x}_m = \mathbf{x}^{(\mathscr{E})}(t_m) \in \mathscr{E}_{\sigma_{n-2}}^{(n-2)}$, $\mathbf{x}_{m\pm}^{(i_\alpha;\mathscr{B})} = \mathbf{x}_{m\mp}^{(i_\beta;\mathscr{B})} = \mathbf{x}_m$ *and* $\mathbf{x}_{m\pm}^{(\mu;\mathscr{D})} = \mathbf{x}_{m\mp}^{(\nu;\mathscr{D})} = \mathbf{x}_m$. *For an arbitrarily small* $\varepsilon > 0$, *there is a time interval* $[t_{m-\varepsilon}, t_{m+\varepsilon}]$. *The boundary flows of* $\mathbf{x}^{(i_\rho;\mathscr{B})}(t)$ ($\rho = \alpha, \beta$) *are* $C_{[t_{m-\varepsilon},0)}^{(r_{i_\rho};\mathscr{B})}$ *or* $C_{(0,t_{m+\varepsilon}]}^{(r_{i_\rho};\mathscr{B})}$ *-continuous for time* t *and* $\| d^{r_{i_\rho}+1}\mathbf{x}^{(i_\rho;\mathscr{B})} / dt^{r_{i_\rho}+1} \| < \infty$ ($r_{i_\rho} \geq m_{i_\rho} + 1$). *The domain flows of* $\mathbf{x}^{(\sigma;\mathscr{D})}(t)$ ($\sigma = \mu, \nu$) *are* $C_{[t_{m-\varepsilon},0)}^{(r_\sigma;\mathscr{D})}$ *or* $C_{(0,t_{m+\varepsilon}]}^{(r_\sigma;\mathscr{N})}$ *-continuous for time* t *and* $\| d^{r_\sigma+1}\mathbf{x}^{(\sigma;\mathscr{D})} / dt^{r_\sigma+1} \| < \infty$ ($r_\sigma \geq m_\sigma + 1$). *As* $n = 2$, $m_{i_\rho} = 2k_{i_\rho} + 1$ *and* $\mathbf{m}_\sigma = 2\mathbf{k}_\sigma + 1$.

(i) *With the switching rule, the coming boundary and domain flows of* $\mathbf{x}^{(i_\alpha;\mathscr{B})}(t)$ ($i_\alpha \in \mathscr{N}_{\mathscr{B}}^{C}(\sigma_{n-2})$) *and* $\mathbf{x}^{(\mu;\mathscr{D})}(t)$ ($\mu \in \mathscr{N}_{\mathscr{D}}^{C}(\sigma_{n-2})$) *to the leaving boundary and domain flows of* $\mathbf{x}^{(i_\beta;\mathscr{B})}(t)$ ($i_\beta \in \mathscr{N}_{\beta\mathscr{B}}^{L}(\sigma_{n-2})$) *and* $\mathbf{x}^{(\nu;\mathscr{D})}(t)$ ($\nu \in \mathscr{N}_{\mathscr{D}}^{L}(\sigma_{n-2})$) *at the edge* $\mathscr{E}_{\sigma_{n-2}}^{(n-2)}$ *for time* t_m *are switchable with the* $(\cup_{i_\alpha}(m_{i_\alpha};\mathscr{B}) \oplus \cup_\mu(\mathbf{m}_\mu;\mathscr{D}) : \cup_{i_\beta}(m_{i_\beta};\mathscr{B}) \oplus \cup_\nu(\mathbf{m}_\nu;\mathscr{D}))$ *singularity from* Ω_μ *and* $\partial\Omega_{\alpha i_\alpha}$ *to* $\partial\Omega_{\beta i_\beta}$ *and* Ω_ν *if and only if*

$$\left.\begin{aligned} &\hbar_{j_\alpha} G_{\partial\Omega_{\alpha j_\alpha}}^{(s_{i_\alpha}, i_\alpha;\mathscr{B})}(\mathbf{x}_m, t_{m-}, \boldsymbol{\lambda}_{i_\alpha}, \boldsymbol{\lambda}_{j_\alpha}) = 0 \\ &\text{for } s_{i_\alpha} = 0,1,2,\cdots, m_{i_\alpha} - 1, \\ &(-1)^{m_{i_\alpha}} \hbar_{j_\alpha} G_{\partial\Omega_{\alpha j_\alpha}}^{(m_{i_\alpha}, i_\alpha;\mathscr{B})}(\mathbf{x}_m, t_{m-}, \boldsymbol{\lambda}_{i_\alpha}, \boldsymbol{\lambda}_{j_\alpha}) < 0 \end{aligned}\right\} \tag{7.83}$$

for all $i_\alpha \in \mathscr{N}_{\alpha\mathscr{B}}^{C}(\sigma_{n-2})$ *and* $\alpha \in \mathscr{N}_{\mathscr{D}}(\sigma_{n-2})$;

$$\left.\begin{aligned} &\hbar_{j_\beta} G_{\partial\Omega_{\beta j_\beta}}^{(s_{i_\beta}, i_\beta;\mathscr{B})}(\mathbf{x}_m, t_{m+}, \boldsymbol{\lambda}_{i_\beta}, \boldsymbol{\lambda}_{j_\beta}) = 0 \\ &\textit{for } s_{i_\beta} = 0,1,2,\cdots, m_{i_\beta} - 1, \\ &\hbar_{j_\beta} G_{\partial\Omega_{\beta j_\beta}}^{(m_{i_\beta}, i_\beta;\mathscr{B})}(\mathbf{x}_m, t_{m+}, \boldsymbol{\lambda}_{i_\beta}, \boldsymbol{\lambda}_{j_\beta}) > 0 \end{aligned}\right\} \tag{7.84}$$

$$\text{for all } i_\beta \in \mathscr{N}^{L}_{\beta\mathscr{B}}(\sigma_{n-2}) \text{ and } \beta \in \mathscr{N}_{\mathscr{D}}(\sigma_{n-2});$$

$$\left.\begin{array}{l} \hbar_{j_\mu} G^{(s_\mu,\beta;\mathscr{D})}_{\partial\Omega_{\mu j_\mu}}(\mathbf{x}_m, t_{m-}, \mathbf{p}_\mu, \boldsymbol{\lambda}_{j_\mu}) = 0 \\ \text{for } s_\mu = 0,1,2,\cdots,m_\mu - 1, \\ (-1)^{m_\mu} \hbar_{j_\mu} G^{(m_\mu,\beta;\mathscr{D})}_{\partial\Omega_{\mu j_\mu}}(\mathbf{x}_m, t_{m-}, \mathbf{p}_\mu, \boldsymbol{\lambda}_{j_\mu}) < 0 \end{array}\right\} \tag{7.85}$$

$$\text{for all } \mu \in \mathscr{N}^{C}_{\mathscr{D}}(\sigma_{n-2}) \text{ and } j_\mu \in \mathscr{N}_{\mu\mathscr{B}}(\sigma_{n-2});$$

$$\left.\begin{array}{l} \hbar_{j_\nu} G^{(s_\nu,\nu;\mathscr{D})}_{\partial\Omega_{\nu j_\nu}}(\mathbf{x}_m, t_{m+}, \mathbf{p}_\nu, \boldsymbol{\lambda}_{j_\nu}) = 0 \\ \text{for } s_\nu = 0,1,2,\cdots,m_\nu - 1, \\ \hbar_{j_\nu} G^{(m_\nu,\nu;\mathscr{D})}_{\partial\Omega_{\nu j_\nu}}(\mathbf{x}_m, t_{m+}, \mathbf{p}_\nu, \boldsymbol{\lambda}_{j_\nu}) > 0 \end{array}\right\} \tag{7.86}$$

$$\text{for all } \nu \in \mathscr{N}^{L}_{\mathscr{D}}(\sigma_{n-2}) \text{ and } j_\nu \in \mathscr{N}_{\nu\mathscr{B}}(\sigma_{n-2}).$$

(ii) *The edge of* $\mathscr{E}^{(n-2)}_{\sigma_{n-2}}$ *to the boundary and domain flows of* $\mathbf{x}^{(i_\alpha;\mathscr{B})}(t)$ ($i_\alpha \in \mathscr{N}_{\mathscr{B}}(\sigma_{n-2})$ *with all* $\alpha \in \mathscr{N}_{\mathscr{D}}(\sigma_{n-2})$) *and* $\mathbf{x}^{(\mu;\mathscr{D})}(t)$ ($\mu \in \mathscr{N}_{\mathscr{D}}(\sigma_{n-2})$) *is non-switchable of the first kind with the* $(\cup_{i_\alpha}(2k_{i_\alpha};\mathscr{B})\oplus(\cup_\mu \mathbf{m}_\mu;\mathscr{D})$ *singularity* ($\mathbf{m}^{C}_{\mu\mathscr{D}} \neq 2\mathbf{k}^{C}_{\mu\mathscr{D}} + 1$) *if and only if*

$$\left.\begin{array}{l} \hbar_{j_\alpha} G^{(s_{i_\alpha},i_\alpha;\mathscr{B})}_{\partial\Omega_{\alpha j_\alpha}}(\mathbf{x}_m, t_{m-}, \boldsymbol{\lambda}_{i_\alpha}, \boldsymbol{\lambda}_{j_\alpha}) = 0 \\ \text{for } s_{i_\alpha} = 0,1,2,\cdots,2k_{i_\alpha} - 1, \\ \hbar_{j_\alpha} G^{(2k_{i_\alpha},i_\alpha;\mathscr{B})}_{\partial\Omega_{\alpha j_\alpha}}(\mathbf{x}_m, t_{m-}, \boldsymbol{\lambda}_{i_\alpha}, \boldsymbol{\lambda}_{j_\alpha}) < 0 \end{array}\right\} \tag{7.87}$$

$$\text{for all } i_\alpha \in \mathscr{N}^{C}_{\alpha\mathscr{B}}(\sigma_{n-2}) \text{ and } \alpha \in \mathscr{N}_{\mathscr{D}}(\sigma_{n-2});$$

$$\left.\begin{array}{l} \hbar_{j_\mu} G^{(s_\mu,\beta;\mathscr{D})}_{\partial\Omega_{\mu j_\mu}}(\mathbf{x}_m, t_{m-}, \mathbf{p}_\mu, \boldsymbol{\lambda}_{j_\mu}) = 0 \\ \text{for } s_\mu = 0,1,2,\cdots,m_\mu - 1, \\ (-1)^{m_\mu} \hbar_{j_\mu} G^{(m_\mu,\beta;\mathscr{D})}_{\partial\Omega_{\mu j_\mu}}(\mathbf{x}_m, t_{m-}, \mathbf{p}_\mu, \boldsymbol{\lambda}_{j_\mu}) < 0 \end{array}\right\} \tag{7.88}$$

$$\text{for all } \mu \in \mathscr{N}^{C}_{\mathscr{D}}(\sigma_{n-2}) \text{ and } j_\mu \in \mathscr{N}_{\mu\mathscr{B}}(\sigma_{n-2}).$$

(iii) *The edge of* $\mathscr{E}^{(n-2)}_{\sigma_{n-2}}$ *to the boundary and domain flows of* $\mathbf{x}^{(i_\beta;\mathscr{B})}(t)$ ($i_\beta \in \mathscr{N}_{\mathscr{B}}(\sigma_{n-2})$ *with all* $\beta \in \mathscr{N}_{\mathscr{D}}(\sigma_{n-2})$) *and* $\mathbf{x}^{(\nu;\mathscr{D})}(t)$ ($\nu \in \mathscr{N}_{\mathscr{D}}(\sigma_{n-2})$) *is switchable of the second kind with the* $(\cup_{i_\alpha}(2k_{i_\alpha};\mathscr{B})\oplus(\cup_\mu \mathbf{m}_\mu;\mathscr{D})$ *singularity if and only if*

$$\hbar_{j_\beta} G^{(s_{i_\beta},i_\beta;\mathscr{B})}_{\partial\Omega_{\beta j_\beta}}(\mathbf{x}_m, t_{m+}, \boldsymbol{\lambda}_{i_\beta}, \boldsymbol{\lambda}_{j_\beta}) = 0 \text{ for } s_{i_\beta} = 0,1,2,\cdots,m_{i_\beta} - 1;$$

$$\begin{aligned}&\hbar_{j_\beta} G^{(m_{i_\beta},i_\beta;\mathscr{B})}_{\partial\Omega_{\beta j_\beta}}(\mathbf{x}_m,t_{m+},\boldsymbol{\lambda}_{i_\beta},\boldsymbol{\lambda}_{j_\beta})>0\\ &\textit{for all } i_\beta\in\mathscr{N}^{L}_{\beta\mathscr{B}}(\sigma_{n-2}) \textit{ and } \beta\in\mathscr{N}_{\mathscr{D}}(\sigma_{n-2});\end{aligned}\tag{7.89}$$

$$\left.\begin{aligned}&\hbar_{j_\nu} G^{(s_\nu,\nu;\mathscr{D})}_{\partial\Omega_{\nu j_\nu}}(\mathbf{x}_m,t_{m+},\mathbf{p}_\nu,\boldsymbol{\lambda}_{j_\nu})=0\\ &\textit{for } s_\nu=0,1,2,\cdots,m_\nu-1,\\ &\hbar_{j_\nu} G^{(m_\nu,\nu;\mathscr{D})}_{\partial\Omega_{\nu j_\nu}}(\mathbf{x}_m,t_{m+},\mathbf{p}_\nu,\boldsymbol{\lambda}_{j_\nu})>0\end{aligned}\right\}\tag{7.90}$$
$$\textit{for all } \nu\in\mathscr{N}^{L}_{\mathscr{D}}(\sigma_{n-2}) \textit{ and } j_\nu\in\mathscr{N}_{\nu\mathscr{B}}(\sigma_{n-2}).$$

(iv) *Without the switching rule, the coming boundary and domain flows of* $\mathbf{x}^{(i_\alpha;\mathscr{B})}(t)$ ($i_\alpha\in\mathscr{N}^{C}_{\mathscr{B}}(\sigma_{n-2})$) *and* $\mathbf{x}^{(\mu;\mathscr{D})}(t)$ ($\mu\in\mathscr{N}^{C}_{\mathscr{D}}(\sigma_{n-2})$) *to the leaving boundary and domain flows of* $\mathbf{x}^{(i_\beta;\mathscr{B})}(t)$ ($i_\beta\in\mathscr{N}^{L}_{\beta\mathscr{B}}(\sigma_{n-2})$) *and* $\mathbf{x}^{(\nu;\mathscr{D})}(t)$ ($\nu\in\mathscr{N}^{L}_{\mathscr{D}}(\sigma_{n-2})$) *at the edge of* $\mathscr{E}^{(n-2)}_{\sigma_{n-2}}$ *for time* t_m *are potentially switchable with the* $(\cup_{i_\alpha}(2k_{i_\alpha};\mathscr{B})\oplus\cup_\mu(\mathbf{m}_\mu;\mathscr{D}):\cup_{i_\beta}(m_{i_\beta};\mathscr{B})\oplus\cup_\nu(\mathbf{m}_\nu;\mathscr{D}))$ *singularity if and only if Eqs.* (7.83)-(7.86) *hold with* $m_{i_\alpha}=2k_{i_\alpha}$ *and* $\mathbf{m}_\mu\neq 2\mathbf{k}_\mu+1$.

(v) *Without the switching rule, the edge* $\mathscr{E}^{(n-2)}_{\sigma_{n-2}}$ *to the coming boundary and domain flows of* $\mathbf{x}^{(i_\alpha;\mathscr{B})}(t)$ ($i_\alpha\in\mathscr{N}^{C}_{\mathscr{B}}(\sigma_{n-2})$) *and* $\mathbf{x}^{(\mu;\mathscr{D})}(t)$ ($\mu\in\mathscr{N}^{C}_{\mathscr{D}}(\sigma_{n-2})$) *to the leaving boundary and domain flows of* $\mathbf{x}^{(i_\beta;\mathscr{B})}(t)$ ($i_\beta\in\mathscr{N}^{L}_{\beta\mathscr{B}}(\sigma_{n-2})$) *and* $\mathbf{x}^{(\nu;\mathscr{D})}(t)$ ($\nu\in\mathscr{N}^{L}_{\mathscr{D}}(\sigma_{n-2})$) *at the edge of* $\mathscr{E}^{(n-2)}_{\sigma_{n-2}}$ *for time* t_m *is non-switchable with the* $(\cup_{i_\alpha}(m_{i_\alpha};\mathscr{B})\oplus\cup_\mu(\mathbf{m}_\mu;\mathscr{D}):\cup_{i_\beta}(m_{i_\beta};\mathscr{B})\oplus\cup_\nu(\mathbf{m}_\nu;\mathscr{D}))$ *singularity if and only if Eqs.* (7.83)-(7.86) *hold with (at least either* m_{i_α} = $2k_{i_\alpha}+1$ *for specific* i_α *or* $\mathbf{m}_\mu=2\mathbf{k}_\mu+1$ *for a specific* μ).

(vi) *With switching rules, the edge* $\mathscr{E}^{(n-2)}_{\sigma_{n-2}}$ *is switchable with the* $(\cup_{i_\alpha}(2k_{i_\alpha}+1;\mathscr{B})\oplus\cup_\mu(2\mathbf{k}_\mu+1;\mathscr{D}):\cup_{i_\alpha}(2k_{i_\alpha}+1;\mathscr{B})\oplus\cup_\mu(2\mathbf{k}_\mu+1;\mathscr{D}))$ *singularity for all the gazing boundary and domain flows of* $\mathbf{x}^{(i_\alpha;\mathscr{B})}(t)$ ($i_\alpha\in\mathscr{N}_{\mathscr{B}}(\sigma_{n-2})$ *with all* $\alpha\in\mathscr{N}_{\mathscr{D}}(\sigma_{n-2})$) *and* $\mathbf{x}^{(\mu;\mathscr{D})}(t)$ ($\mu\in\mathscr{N}_{\mathscr{D}}(\sigma_{n-2})$) *for time* t_m *if and only if*

$$\left.\begin{aligned}&\hbar_{j_\alpha} G^{(s_{i_\alpha},i_\alpha;\mathscr{B})}_{\partial\Omega_{\alpha j_\alpha}}(\mathbf{x}_m,t_{m\pm},\boldsymbol{\lambda}_{i_\alpha},\boldsymbol{\lambda}_{j_\alpha})=0\\ &\textit{for } s_{i_\alpha}=0,1,2,\cdots,2k_{i_\alpha},\\ &\hbar_{j_\alpha} G^{(2k_{i_\alpha}+1,i_\alpha;\mathscr{B})}_{\partial\Omega_{\alpha j_\alpha}}(\mathbf{x}_m,t_{m\pm},\boldsymbol{\lambda}_{i_\alpha},\boldsymbol{\lambda}_{j_\alpha})>0\end{aligned}\right\}\tag{7.91}$$
$$\textit{for all } i_\alpha\in\mathscr{N}^{C}_{\alpha\mathscr{B}}(\sigma_{n-2}) \textit{ and } \alpha\in\mathscr{N}_{\mathscr{D}}(\sigma_{n-2});$$

$$\hbar_{j_\mu} G^{(s_\mu,\beta;\mathscr{D})}_{\partial\Omega_{\mu j_\mu}}(\mathbf{x}_m,t_{m\pm},\mathbf{p}_\mu,\boldsymbol{\lambda}_{j_\mu})=0 \textit{ for } s_\mu=0,1,2,\cdots,2k_\mu;$$

$$\begin{aligned}&\hbar_{j_\mu} G_{\partial\Omega_{\mu j_\mu}}^{(2k_\mu+1,\beta;\mathscr{D})}(\mathbf{x}_m, t_{m\pm}, \mathbf{p}_\mu, \boldsymbol{\lambda}_{j_\mu}) > 0\\ &\textit{for all } \mu \in \mathscr{N}_{\mathscr{D}}^{C}(\sigma_{n-2}) \textit{ and } j_\mu \in \mathscr{N}_{\mu\mathscr{B}}(\sigma_{n-2}).\end{aligned} \tag{7.92}$$

(vii) *Without switching rules, the edge* $\mathscr{E}_{\sigma_{n-2}}^{(n-2)}$ *is a grazing edge with the* $(\cup_{i_\alpha}(2k_{i_\alpha}+1;\mathscr{B})\oplus\cup_\mu(2\mathbf{k}_\mu+1;\mathscr{D}):\cup_{i_\alpha}(2k_{i_\alpha}+1;\mathscr{B})\oplus\cup_\mu(2\mathbf{k}_\mu+1;\mathscr{D}))$ *singularity for all the gazing boundary and domain flows of* $\mathbf{x}^{(i_\alpha;\mathscr{B})}(t)$ ($i_\alpha \in \mathscr{N}_{\mathscr{B}}(\sigma_{n-2})$ *with all* $\alpha \in \mathscr{N}_{\mathscr{D}}(\sigma_{n-2})$) *and* $\mathbf{x}^{(\mu;\mathscr{D})}(t)$ ($\mu \in \mathscr{N}_{\mathscr{D}}(\sigma_{n-2})$) *for time* t_m *if Eqs.* (7.91) *and* (7.92) *hold.*

Proof. The theorem can be proved as in Chapter 3. ■

7.4. Boundary flow attractivity

Definition 7.11. For a discontinuous dynamical system in Eqs. (6.4)-(6.8), there is an $(n-2)$-dimensional edge $\mathscr{E}_{\sigma_{n-2}}^{(n-2)}$ to be formed by m_b-boundaries $\partial\Omega_{\alpha i_\alpha}$ of $(n-1)$-dimensions with m_d-domains Ω_α ($\alpha, i_\alpha \in \mathscr{N}_{\mathscr{D}} = \{1,2,\cdots,m_d\}$ and $\alpha \neq i_\alpha$, also $i_\alpha \in \mathscr{N}_{\mathscr{B}} = \{1,2,\cdots,m_b\}$). There is a measuring $(n-2)$-dimensional edge $\mathscr{E}_{\sigma_{n-2}}^{(j_\alpha,i_\alpha)} = S_\sigma^{(j_\alpha,\alpha)} \cap \partial\Omega_{\alpha i_\alpha}$ and $C_{m+\varepsilon}^{(j_\alpha,i_\alpha)} \equiv C_{m+\varepsilon}^{(j_\alpha,\alpha)}$. Suppose there is a boundary flow $\mathbf{x}^{(i_\alpha;\mathscr{B})}(t)$ on the boundary $\partial\Omega_{\alpha i_\alpha}(\sigma_{n-2})$. At time t_m, $\mathbf{x}^{(j_\alpha,i_\alpha)}(t_m) = \mathbf{x}_m^{(j_\alpha,i_\alpha)} \in \mathscr{E}_{\sigma_{n-2}}^{(j_\alpha,i_\alpha)}$. For an arbitrarily small $\varepsilon > 0$, there is a time interval $[t_{m-\varepsilon}, t_{m+\varepsilon}]$.

(i) The flow $\mathbf{x}^{(i_\alpha;\mathscr{B})}(t)$ at time t_m to the edge of $\mathscr{E}_{\sigma_{n-2}}^{(n-2)}$ is *attractive* if

$$\left.\begin{aligned}&\hbar_{j_\alpha}\mathbf{n}_{\partial\Omega_{\alpha j_\alpha}}^{\mathrm{T}}(\mathbf{x}_{m-\varepsilon}^{(j_\alpha,i_\alpha)})\cdot[\mathbf{x}_{m-\varepsilon}^{(j_\alpha,i_\alpha)} - \mathbf{x}_{m-\varepsilon}^{(i_\alpha;\mathscr{B})}] < 0,\\ &\hbar_{j_\alpha}\mathbf{n}_{\partial\Omega_{\alpha j_\alpha}}^{\mathrm{T}}(\mathbf{x}_{m+\varepsilon}^{(j_\alpha,i_\alpha)})\cdot[\mathbf{x}_{m+\varepsilon}^{(i_\alpha;\mathscr{B})} - \mathbf{x}_{m+\varepsilon}^{(j_\alpha,i_\alpha)}] < 0;\\ &\hbar_{j_\alpha}C_{m-\varepsilon}^{(j_\alpha,i_\alpha)} > \hbar_{j_\alpha}C_{m}^{(j_\alpha,i_\alpha)} > \hbar_{j_\alpha}C_{m+\varepsilon}^{(j_\alpha,i_\alpha)}.\end{aligned}\right\} \tag{7.93}$$

(ii) The flow $\mathbf{x}^{(i_\alpha;\mathscr{B})}(t)$ at time t_m to the edge of $\mathscr{E}_{\sigma_{n-2}}^{(n-2)}$ is *repulsive* if

$$\left.\begin{aligned}&\hbar_{j_\alpha}\mathbf{n}_{\partial\Omega_{\alpha j_\alpha}}^{\mathrm{T}}(\mathbf{x}_{m-\varepsilon}^{(j_\alpha,i_\alpha)})\cdot[\mathbf{x}_{m-\varepsilon}^{(j_\alpha,i_\alpha)} - \mathbf{x}_{m-\varepsilon}^{(i_\alpha;\mathscr{B})}] > 0,\\ &\hbar_{j_\alpha}\mathbf{n}_{\partial\Omega_{\alpha j_\alpha}}^{\mathrm{T}}(\mathbf{x}_{m+\varepsilon}^{(j_\alpha,i_\alpha)})\cdot[\mathbf{x}_{m+\varepsilon}^{(i_\alpha;\mathscr{B})} - \mathbf{x}_{m+\varepsilon}^{(j_\alpha,i_\alpha)}] > 0;\\ &\hbar_{j_\alpha}C_{m-\varepsilon}^{(j_\alpha,i_\alpha)} < \hbar_{j_\alpha}C_{m}^{(j_\alpha,i_\alpha)} < \hbar_{j_\alpha}C_{m+\varepsilon}^{(j_\alpha,i_\alpha)}.\end{aligned}\right\} \tag{7.94}$$

(iii) The flow $\mathbf{x}^{(i_\alpha;\mathscr{B})}(t)$ at time t_m to the edge of $\mathscr{E}_{\sigma_{n-2}}^{(n-2)}$ is *from the attractive to repulsive* state if

$$\left.\begin{array}{l} \hbar_{j_\alpha}\mathbf{n}^{\mathrm{T}}_{\partial\Omega_{\alpha j_\alpha}}(\mathbf{x}^{(j_\alpha,i_\alpha)}_{m-\varepsilon})\cdot[\mathbf{x}^{(j_\alpha,i_\alpha)}_{m-\varepsilon}-\mathbf{x}^{(i_\alpha;\mathscr{B})}_{m-\varepsilon}]<0,\\ \hbar_{j_\alpha}\mathbf{n}^{\mathrm{T}}_{\partial\Omega_{\alpha j_\alpha}}(\mathbf{x}^{(j_\alpha,i_\alpha)}_{m+\varepsilon})\cdot[\mathbf{x}^{(i_\alpha;\mathscr{B})}_{m+\varepsilon}-\mathbf{x}^{(j_\alpha,i_\alpha)}_{m+\varepsilon}]>0;\\ \hbar_{j_\alpha}C^{(j_\alpha,i_\alpha)}_{m}<\hbar_{j_\alpha}C^{(j_\alpha,i_\alpha)}_{m-\varepsilon}\text{ and }\hbar_{j_\alpha}C^{(j_\alpha,i_\alpha)}_{m}<\hbar_{j_\alpha}C^{(j_\alpha,i_\alpha)}_{m+\varepsilon}. \end{array}\right\}\tag{7.95}$$

(iv) The flow $\mathbf{x}^{(i_\alpha;\mathscr{B})}(t)$ at time t_m to the edge of $\mathscr{E}^{(n-2)}_{\sigma_{n-2}}$ is *from the repulsive to attractive* state if

$$\left.\begin{array}{l} \hbar_{j_\alpha}\mathbf{n}^{\mathrm{T}}_{S^{(j_\alpha,\alpha)}_{\sigma}}(\mathbf{x}^{(j_\alpha,i_\alpha)}_{m-\varepsilon})\cdot[\mathbf{x}^{(j_\alpha,i_\alpha)}_{m-\varepsilon}-\mathbf{x}^{(i_\alpha;\mathscr{B})}_{m-\varepsilon}]>0,\\ \hbar_{j_\alpha}\mathbf{n}^{\mathrm{T}}_{S^{(j_\alpha,\alpha)}_{\sigma}}(\mathbf{x}^{(j_\alpha,i_\alpha)}_{m+\varepsilon})\cdot[\mathbf{x}^{(i_\alpha;\mathscr{B})}_{m+\varepsilon}-\mathbf{x}^{(j_\alpha,i_\alpha)}_{m+\varepsilon}]<0;\\ \hbar_{j_\alpha}C^{(j_\alpha,\alpha)}_{m}<\hbar_{j_\alpha}C^{(j_\alpha,\alpha)}_{m-\varepsilon}\text{ and }\hbar_{j_\alpha}C^{(j_\alpha,\alpha)}_{m}<\hbar_{j_\alpha}C^{(j_\alpha,\alpha)}_{m+\varepsilon}. \end{array}\right\}\tag{7.96}$$

(v) The flow $\mathbf{x}^{(i_\alpha;\mathscr{B})}(t)$ at time t_m to the edge of $\mathscr{E}^{(n-2)}_{\sigma_{n-2}}$ is *invariant* with an equi-measuring quantity $C^{(j_\alpha,\alpha)}_{m}$ if

$$\left.\begin{array}{l} \mathbf{x}^{(j_\alpha,i_\alpha)}_{m-\varepsilon}=\mathbf{x}^{(i_\alpha;\mathscr{B})}_{m-\varepsilon}\text{ and }\mathbf{x}^{(i_\alpha;\mathscr{B})}_{m+\varepsilon}=\mathbf{x}^{(j_\alpha,i_\alpha)}_{m+\varepsilon};\\ C^{(j_\alpha,\alpha)}_{m-\varepsilon}=C^{(j_\alpha,\alpha)}_{m}=C^{(j_\alpha,\alpha)}_{m+\varepsilon}. \end{array}\right.\tag{7.97}$$

From the above definition, the attractivity of a boundary flow to an edge can be investigated through the measuring edge $\mathscr{E}^{(j_\alpha,i_\alpha)}_{n-2}$ in the boundary of a boundary flow, as shown in Fig. 7.21.

Theorem 7.11. *For a discontinuous dynamical system in Eqs.* (6.4)-(6.8), *there is an* $(n-2)$ *-dimensional edge* $\mathscr{E}^{(n-2)}_{\sigma_{n-2}}$ *to be formed by* m_b *-boundaries* $\partial\Omega_{\alpha i_\alpha}$ *of* $(n-1)$ *-dimensions with* m_d *-domains* Ω_α ($\alpha,i_\alpha\in\mathscr{N}_{\mathscr{D}}=\{1,2,\cdots,m_d\}$ *and* $\alpha\neq i_\alpha$, *also* $i_\alpha\in\mathscr{N}_{\mathscr{B}}=\{1,2,\cdots,m_b\}$). *There is a measuring* $(n-2)$ *-dimensional edge* $\mathscr{E}^{(j_\alpha,i_\alpha)}_{\sigma_{n-2}}=S^{(j_\alpha,\alpha)}_{\sigma}\cap\partial\Omega_{\alpha i_\alpha}$ *and* $C^{(j_\alpha,i_\alpha)}_{m+\varepsilon}\equiv C^{(j_\alpha,\alpha)}_{m+\varepsilon}$. *Suppose there is a boundary flow* $\mathbf{x}^{(i_\alpha;\mathscr{B})}(t)$ *on the boundary* $\partial\Omega_{\alpha i_\alpha}(\sigma_{n-2})$. *At time* t_m, $\mathbf{x}^{(j_\alpha,i_\alpha)}(t_m)=\mathbf{x}^{(j_\alpha,i_\alpha)}_{m}\in\mathscr{E}^{(j_\alpha,i_\alpha)}_{\sigma_{n-2}}$. *For an arbitrarily small* $\varepsilon>0$, *there is a time interval* $[t_{m-\varepsilon},t_{m+\varepsilon}]$. *The flow* $\mathbf{x}^{(i_\alpha;\mathscr{B})}(t)$ *is* $C^{r_{i_\alpha}}_{[t_{m-\varepsilon},t_{m+\varepsilon}]}$ *-continuous for time* t *and* $\|d^{r_{i_\alpha}+1}\mathbf{x}^{(i_\alpha;\mathscr{B})}/dt^{r_{i_\alpha}+1}\|<\infty$ ($r_{i_\alpha}\geq 2$).

(i) *The flow* $\mathbf{x}^{(i_\alpha;\mathscr{B})}(t)$ *to the edge of* $\mathscr{E}^{(n-2)}_{\sigma_{n-2}}$ *at time* t_m *is attractive if and only if*

$$\hbar_{j_\alpha}G^{(i_\alpha;\mathscr{B})}_{\partial\Omega_{\alpha j_\alpha}}(\mathbf{x}^{(j_\alpha,i_\alpha)}_{m},t_m,\boldsymbol{\lambda}_{i_\alpha},\boldsymbol{\lambda}_{j_\alpha})<0.\tag{7.98}$$

(ii) *The flow* $\mathbf{x}^{(i_\alpha;\mathscr{B})}(t)$ *to the edge of* $\mathscr{E}^{(n-2)}_{\sigma_{n-2}}$ *at time* t_m *is repulsive if and only if*

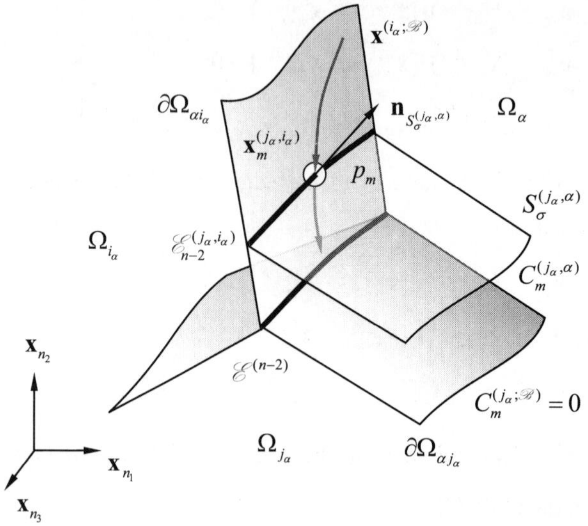

Fig. 7.21 Boundary flows on the three $(n-1)$ -dimensional boundaries. The black curve represents the $(n-2)$ -dimensional edge. The boundary flows are denoted by the three curves with arrows. ($n_1+n_2+n_3=n$).

$$\hbar_{j_\alpha} G_{\partial\Omega_{\alpha j_\alpha}}^{(i_\alpha;\mathscr{B})}(\mathbf{x}_m^{(j_\alpha,i_\alpha)},t_m,\boldsymbol{\lambda}_{i_\alpha},\boldsymbol{\lambda}_{j_\alpha})>0. \tag{7.99}$$

(iii) *The flow* $\mathbf{x}^{(i_\alpha;\mathscr{B})}(t)$ *at time* t_m *to the edge of* $\mathscr{E}_{\sigma_{n-2}}^{(n-2)}$ *is from the attractive to repulsive state if and only if*

$$\begin{aligned}&\hbar_{j_\alpha} G_{\partial\Omega_{\alpha j_\alpha}}^{(i_\alpha;\mathscr{B})}(\mathbf{x}_m^{(j_\alpha,i_\alpha)},t_m,\boldsymbol{\lambda}_{i_\alpha},\boldsymbol{\lambda}_{j_\alpha})=0,\\&\hbar_{j_\alpha} G_{\partial\Omega_{\alpha j_\alpha}}^{(1,i_\alpha;\mathscr{B})}(\mathbf{x}_m^{(j_\alpha,i_\alpha)},t_m,\boldsymbol{\lambda}_{i_\alpha},\boldsymbol{\lambda}_{j_\alpha})>0.\end{aligned} \tag{7.100}$$

(iv) *The flow* $\mathbf{x}^{(i_\alpha;\mathscr{B})}(t)$ *at time* t_m *to the edge of* $\mathscr{E}_{\sigma_{n-2}}^{(n-2)}$ *is from the repulsive to attractive state if and only if*

$$\begin{aligned}&\hbar_{j_\alpha} G_{\partial\Omega_{\alpha j_\alpha}}^{(i_\alpha;\mathscr{B})}(\mathbf{x}_m^{(j_\alpha,i_\alpha)},t_m,\boldsymbol{\lambda}_{i_\alpha},\boldsymbol{\lambda}_{j_\alpha})=0,\\&\hbar_{j_\alpha} G_{\partial\Omega_{\alpha j_\alpha}}^{(1,i_\alpha;\mathscr{B})}(\mathbf{x}_m^{(j_\alpha,i_\alpha)},t_m,\boldsymbol{\lambda}_{i_\alpha},\boldsymbol{\lambda}_{j_\alpha})<0.\end{aligned} \tag{7.101}$$

(v) *The flow* $\mathbf{x}^{(i_\alpha;\mathscr{B})}(t)$ *at time* t_m *to the edge of* $\mathscr{E}_{\sigma_{n-2}}^{(n-2)}$ *is invariant with an equi-measuring quantity* $C_m^{(j_\alpha,\alpha)}$ *if and only if*

$$G_{\partial\Omega_{\alpha j_\alpha}}^{(s_{i_\alpha},i_\alpha;\mathscr{B})}(\mathbf{x}_m^{(j_\alpha,i_\alpha)},t_m,\boldsymbol{\lambda}_{i_\alpha},\boldsymbol{\lambda}_{j_\alpha})=0,\ s_{i_\alpha}=0,1,2,\cdots. \tag{7.102}$$

Proof. The boundary flow can be treated as a special flow in domain to the other boundary of the edge. As in Chapter 3, this theorem can be proved.

Definition 7.12. For a discontinuous dynamical system in Eqs. (6.4)-(6.8), there is an $(n-2)$-dimensional edge $\mathscr{E}_{\sigma_{n-2}}^{(n-2)}$ to be formed by m_b-boundaries $\partial\Omega_{\alpha i_\alpha}$ of $(n-1)$-dimensions with m_d-domains Ω_α ($\alpha, i_\alpha \in \mathscr{N}_{\mathscr{D}} = \{1,2,\cdots,m_d\}$ and $\alpha \neq i_\alpha$, also $i_\alpha \in \mathscr{N}_{\mathscr{B}} = \{1,2,\cdots,m_b\}$). There is a measuring $(n-2)$-dimensional edge $\mathscr{E}_{\sigma_{n-2}}^{(j_\alpha,i_\alpha)} = S_\sigma^{(j_\alpha,\alpha)} \cap \partial\Omega_{\alpha i_\alpha}$ and $C_{m+\varepsilon}^{(j_\alpha,i_\alpha)} \equiv C_{m+\varepsilon}^{(j_\alpha,\alpha)}$. Suppose there is a boundary flow $\mathbf{x}^{(i_\alpha;\mathscr{B})}(t)$ on the boundary $\partial\Omega_{\alpha i_\alpha}(\sigma_{n-2})$. At time t_m, $\mathbf{x}^{(j_\alpha,i_\alpha)}(t_m) = \mathbf{x}_m^{(j_\alpha,i_\alpha)} \in \mathscr{E}_{\sigma_{n-2}}^{(j_\alpha,i_\alpha)}$. For an arbitrarily small $\varepsilon > 0$, there is a time interval $[t_{m-\varepsilon}, t_{m+\varepsilon}]$. The flow $\mathbf{x}^{(i_\alpha;\mathscr{B})}(t)$ is $C_{[t_{m-\varepsilon},t_{m+\varepsilon}]}^{r_{i_\alpha}}$-continuous for time t and $\| d^{r_{i_\alpha}+1}\mathbf{x}^{(i_\alpha;\mathscr{B})}/dt^{r_{i_\alpha}+1} \| < \infty$ ($r_{i_\alpha} \geq 2k_{i_\alpha}$ or $2k_{i_\alpha}+1$).

(i) The flow $\mathbf{x}^{(i_\alpha;\mathscr{B})}(t)$ to the edge of $\mathscr{E}_{\sigma_{n-2}}^{(n-2)}$ at time t_m is *attractive* with the $(2k_{i_\alpha})$th-order singularity if

$$\left.\begin{array}{l} \hbar_{j_\alpha} G_{\partial\Omega_{\alpha j_\alpha}}^{(s_{i_\alpha},i_\alpha;\mathscr{B})}(\mathbf{x}_m^{(j_\alpha,i_\alpha)}, \mathbf{x}_m^{(i_\alpha;\mathscr{B})}, \boldsymbol{\lambda}_{i_\alpha}, \boldsymbol{\lambda}_{j_\alpha}) = 0 \\ \text{for } s_{i_\alpha} = 0,1,2,\cdots,2k_{i_\alpha}-1; \\ \hbar_{j_\alpha} \mathbf{n}_{\partial\Omega_{\alpha j_\alpha}}^{\mathrm{T}}(\mathbf{x}_{m-\varepsilon}^{(j_\alpha,i_\alpha)}) \cdot [\mathbf{x}_{m-\varepsilon}^{(j_\alpha,i_\alpha)} - \mathbf{x}_{m-\varepsilon}^{(i_\alpha;\mathscr{B})}] < 0 \end{array}\right\} \tag{7.103}$$
$$\hbar_{j_\alpha} C_{m-\varepsilon}^{(j_\alpha,i_\alpha)} > \hbar_{j_\alpha} C_m^{(j_\alpha,i_\alpha)} > \hbar_{j_\alpha} C_{m+\varepsilon}^{(j_\alpha,i_\alpha)}.$$

(ii) The boundary flow $\mathbf{x}^{(i_\alpha;\mathscr{B})}(t)$ to the edge of $\mathscr{E}_{\sigma_{n-2}}^{(n-2)}$ at time t_m is *repulsive* with the $(2k_{i_\alpha})$th-order singularity if

$$\begin{array}{l} \hbar_{j_\alpha} G_{\partial\Omega_{\alpha j_\alpha}}^{(s_{i_\alpha},i_\alpha;\mathscr{B})}(\mathbf{x}_m^{(j_\alpha,i_\alpha)}, t_m, \boldsymbol{\lambda}_{i_\alpha}, \boldsymbol{\lambda}_{j_\alpha}) = 0 \\ \text{for } s_{i_\alpha} = 0,1,2,\cdots,2k_{i_\alpha}-1; \\ \hbar_{j_\alpha} \mathbf{n}_{\partial\Omega_{\alpha j_\alpha}}^{\mathrm{T}}(\mathbf{x}_{m+\varepsilon}^{(j_\alpha,i_\alpha)}) \cdot [\mathbf{x}_{m+\varepsilon}^{(i_\alpha;\mathscr{B})} - \mathbf{x}_{m+\varepsilon}^{(j_\alpha,i_\alpha)}] > 0 \\ \hbar_{j_\alpha} C_{m-\varepsilon}^{(j_\alpha,i_\alpha)} < \hbar_{j_\alpha} C_m^{(j_\alpha,i_\alpha)} < \hbar_{j_\alpha} C_{m+\varepsilon}^{(j_\alpha,i_\alpha)}. \end{array} \tag{7.104}$$

(iii) The flow $\mathbf{x}^{(i_\alpha;\mathscr{B})}(t)$ to the edge of $\mathscr{E}_{\sigma_{n-2}}^{(n-2)}$ at time t_m is *from the attractive to repulsive* state with the $(2k_{i_\alpha}+1)$th-order singularity if

$$\begin{array}{l} \hbar_{j_\alpha} G_{\partial\Omega_{\alpha j_\alpha}}^{(s_{i_\alpha},i_\alpha;\mathscr{B})}(\mathbf{x}_m^{(j_\alpha,i_\alpha)}, t_m, \boldsymbol{\lambda}_{i_\alpha}, \boldsymbol{\lambda}_{j_\alpha}) = 0 \\ \text{for } s_{i_\alpha} = 0,1,2,\cdots,2k_{i_\alpha}; \\ \hbar_{j_\alpha} \mathbf{n}_{\partial\Omega_{\alpha j_\alpha}}^{\mathrm{T}}(\mathbf{x}_{m-\varepsilon}^{(j_\alpha,i_\alpha)}) \cdot [\mathbf{x}_{m-\varepsilon}^{(j_\alpha,i_\alpha)} - \mathbf{x}_{m-\varepsilon}^{(i_\alpha;\mathscr{B})}] < 0, \\ \hbar_{j_\alpha} \mathbf{n}_{\partial\Omega_{\alpha j_\alpha}}^{\mathrm{T}}(\mathbf{x}_{m+\varepsilon}^{(j_\alpha,i_\alpha)}) \cdot [\mathbf{x}_{m+\varepsilon}^{(i_\alpha;\mathscr{B})} - \mathbf{x}_{m+\varepsilon}^{(j_\alpha,i_\alpha)}] > 0, \end{array}$$

$$\hbar_{j_\alpha} C_m^{(j_\alpha,i_\alpha)} < \hbar_{j_\alpha} C_{m-\varepsilon}^{(j_\alpha,i_\alpha)} \text{ and } \hbar_{j_\alpha} C_m^{(j_\alpha,i_\alpha)} < \hbar_{j_\alpha} C_{m+\varepsilon}^{(j_\alpha,i_\alpha)}. \tag{7.105}$$

(iv) The flow $\mathbf{x}^{(i_\alpha;\mathscr{B})}(t)$ to the edge of $\mathscr{E}_{\sigma_{n-2}}^{(n-2)}$ at time t_m is *from the repulsive to attractive* state with the $(2k_{i_\alpha}+1)$th -order singularity if

$$\left.\begin{array}{l}
\hbar_{j_\alpha} G_{\partial\Omega_{\alpha j_\alpha}}^{(s_{i_\alpha},i_\alpha;\mathscr{B})}(\mathbf{x}_m^{(j_\alpha,i_\alpha)},t_m,\boldsymbol{\lambda}_{i_\alpha},\boldsymbol{\lambda}_{j_\alpha})=0\\
\text{for } s_{i_\alpha}=0,1,2,\cdots,2k_{i_\alpha};\\
\hbar_{j_\alpha}\mathbf{n}_{\partial\Omega_{\alpha j_\alpha}}^{\mathrm{T}}(\mathbf{x}_{m-\varepsilon}^{(j_\alpha,i_\alpha)})\cdot[\mathbf{x}_{m-\varepsilon}^{(j_\alpha,i_\alpha)}-\mathbf{x}_{m-\varepsilon}^{(i_\alpha;\mathscr{B})}]>0,\\
\hbar_{j_\alpha}\mathbf{n}_{\partial\Omega_{\alpha j_\alpha}}^{\mathrm{T}}(\mathbf{x}_{m+\varepsilon}^{(j_\alpha,i_\alpha)})\cdot[\mathbf{x}_{m+\varepsilon}^{(i_\alpha;\mathscr{B})}-\mathbf{x}_{m+\varepsilon}^{(j_\alpha,i_\alpha)}]<0;\\
\hbar_{j_\alpha} C_m^{(j_\alpha,\alpha)} < \hbar_{j_\alpha} C_{m-\varepsilon}^{(j_\alpha,\alpha)} \text{ and } \hbar_{j_\alpha} C_m^{(j_\alpha,\alpha)} < \hbar_{j_\alpha} C_{m+\varepsilon}^{(j_\alpha,\alpha)}.
\end{array}\right\} \tag{7.106}$$

Theorem 7.12. *For a discontinuous dynamical system in Eqs.* (6.4)-(6.8), *there is an* $(n-2)$ *-dimensional edge* $\mathscr{E}_{\sigma_{n-2}}^{(n-2)}$ *to be formed by* m_b *-boundaries* $\partial\Omega_{\alpha i_\alpha}$ *of* $(n-1)$ *-dimensions with* m_d *-domains* Ω_α ($\alpha,i_\alpha\in\mathscr{N}_{\mathscr{D}}=\{1,2,\cdots,m_d\}$ *and* $\alpha\neq i_\alpha$, *also* $i_\alpha\in\mathscr{N}_{\mathscr{B}}=\{1,2,\cdots,m_b\}$). *There is a measuring* $(n-2)$ *-dimensional edge* $\mathscr{E}_{\sigma_{n-2}}^{(j_\alpha,i_\alpha)}=S_\sigma^{(j_\alpha,\alpha)}\cap\partial\Omega_{\alpha i_\alpha}$ *and* $C_{m+\varepsilon}^{(j_\alpha,i_\alpha)}\equiv C_{m+\varepsilon}^{(j_\alpha,\alpha)}$. *Suppose there is a boundary flow* $\mathbf{x}^{(i_\alpha;\mathscr{B})}(t)$ *on the boundary* $\partial\Omega_{\alpha i_\alpha}(\sigma_{n-2})$. *At time* t_m, $\mathbf{x}^{(j_\alpha,i_\alpha)}(t_m)=\mathbf{x}_m^{(j_\alpha,i_\alpha)}\in\mathscr{E}_{\sigma_{n-2}}^{(j_\alpha,i_\alpha)}$. *For an arbitrarily small* $\varepsilon>0$, *there is a time interval* $[t_{m-\varepsilon},t_{m+\varepsilon}]$. *The flow* $\mathbf{x}^{(i_\alpha;\mathscr{B})}(t)$ *is* $C_{[t_{m-\varepsilon},t_{m+\varepsilon}]}^{r_{i_\alpha}}$ *-continuous for time* t *and* $\|d^{r_{i_\alpha}+1}\mathbf{x}^{(i_\alpha;\mathscr{B})}/dt^{r_{i_\alpha}+1}\|<\infty$ ($r_{i_\alpha}\geq 2k_{i_\alpha}$ *or* $2k_{i_\alpha}+1$).

(i) *The flow* $\mathbf{x}^{(i_\alpha;\mathscr{B})}(t)$ *to the edge of* $\mathscr{E}_{\sigma_{n-2}}^{(n-2)}$ *at time* t_m *is attractive with the* $(2k_{i_\alpha})$th-*order singularity if and only if*

$$\left.\begin{array}{l}
\hbar_{j_\alpha} G_{\partial\Omega_{\alpha j_\alpha}}^{(s_{i_\alpha},i_\alpha;\mathscr{B})}(\mathbf{x}_m^{(j_\alpha,i_\alpha)},t_m,\boldsymbol{\lambda}_{i_\alpha},\boldsymbol{\lambda}_{j_\alpha})=0\\
\text{for } s_{i_\alpha}=0,1,2,\cdots,2k_{i_\alpha}-1;\\
\hbar_{j_\alpha} G_{\partial\Omega_{\alpha j_\alpha}}^{(2k_{i_\alpha},i_\alpha;\mathscr{B})}(\mathbf{x}_m^{(j_\alpha,i_\alpha)},t_m,\boldsymbol{\lambda}_{i_\alpha},\boldsymbol{\lambda}_{j_\alpha})<0.
\end{array}\right\} \tag{7.107}$$

(ii) *The boundary flow* $\mathbf{x}^{(i_\alpha;\mathscr{B})}(t)$ *to the edge of* $\mathscr{E}_{\sigma_{n-2}}^{(n-2)}$ *at time* t_m *is repulsive with the* $(2k_{i_\alpha})$th-*order singularity if and only if*

$$\left.\begin{array}{l}
\hbar_{j_\alpha} G_{\partial\Omega_{\alpha j_\alpha}}^{(s_{i_\alpha},i_\alpha;\mathscr{B})}(\mathbf{x}_m^{(j_\alpha,i_\alpha)},t_m,\boldsymbol{\lambda}_{i_\alpha},\boldsymbol{\lambda}_{j_\alpha})=0\\
\text{for } s_{i_\alpha}=0,1,2,\cdots,2k_{i_\alpha}-1;\\
\hbar_{j_\alpha} G_{\partial\Omega_{\alpha j_\alpha}}^{(2k_{i_\alpha},i_\alpha;\mathscr{B})}(\mathbf{x}_m^{(j_\alpha,i_\alpha)},t_m,\boldsymbol{\lambda}_{i_\alpha},\boldsymbol{\lambda}_{j_\alpha})>0.
\end{array}\right\} \tag{7.108}$$

(iii) *The flow* $\mathbf{x}^{(i_\alpha;\mathscr{B})}(t)$ *to the edge of* $\mathscr{E}_{\sigma_{n-2}}^{(n-2)}$ *at time* t_m *is from the attractive to repulsive state with the* $(2k_{i_\alpha}+1)$th-*order singularity if and only if*

$$\left.\begin{aligned}&\hbar_{j_\alpha} G_{\partial\Omega_{\alpha j_\alpha}}^{(s_{i_\alpha},i_\alpha;\mathscr{B})}(\mathbf{x}_m^{(j_\alpha,i_\alpha)},t_m,\boldsymbol{\lambda}_{i_\alpha},\boldsymbol{\lambda}_{j_\alpha})=0\\ &for\ s_{i_\alpha}=0,1,2,\cdots,2k_{i_\alpha};\\ &\hbar_{j_\alpha} G_{\partial\Omega_{\alpha j_\alpha}}^{(2k_{i_\alpha}+1,i_\alpha;\mathscr{B})}(\mathbf{x}_m^{(j_\alpha,i_\alpha)},t_m,\boldsymbol{\lambda}_{i_\alpha},\boldsymbol{\lambda}_{j_\alpha})>0.\end{aligned}\right\}\tag{7.109}$$

(iv) *The flow* $\mathbf{x}^{(i_\alpha;\mathscr{B})}(t)$ *at time* t_m *to the edge of* $\mathscr{E}_{\sigma_{n-2}}^{(n-2)}$ *is from the repulsive to attractive state with the* $(2k_{i_\alpha}+1)$th-*order singularity if and only if*

$$\left.\begin{aligned}&\hbar_{j_\alpha} G_{\partial\Omega_{\alpha j_\alpha}}^{(s_{i_\alpha},i_\alpha;\mathscr{B})}(\mathbf{x}_m^{(j_\alpha,i_\alpha)},t_m,\boldsymbol{\lambda}_{i_\alpha},\boldsymbol{\lambda}_{j_\alpha})=0\\ &for\ s_{i_\alpha}=0,1,2,\cdots,2k_{i_\alpha};\\ &\hbar_{j_\alpha} G_{\partial\Omega_{\alpha j_\alpha}}^{(2k_{i_\alpha}+1,i_\alpha;\mathscr{B})}(\mathbf{x}_m^{(j_\alpha,i_\alpha)},t_m,\boldsymbol{\lambda}_{i_\alpha},\boldsymbol{\lambda}_{j_\alpha})<0.\end{aligned}\right\}\tag{7.110}$$

Proof. The boundary flow can be treated as a special flow in domain to the other boundary of the edge. As in Chapter 3, this theorem can be proved.

7.5. Boundary dynamics with multi-valued vector fields

The multi-valued vector fields on a boundary can be discussed as in Section 5.6. If a vector field defined on the imaginary extension of a boundary in a domain can be extended continuously to the boundary via the edge, such an extended vector field generates an imaginary flow in the corresponding boundary. On the other hand, a boundary flow can be extended to the domain, and an imaginary domain flow on the imaginary extension of boundary in the domain will be formed. Equation (5.114) can be rewritten for real and imaginary boundary flows, i.e.,

$$\begin{aligned}&\dot{\mathbf{x}}_{i_\alpha}^{(i_\alpha;\mathscr{B})}=\mathbf{F}^{(i_\alpha;\mathscr{B})}(\mathbf{x}_{i_\alpha}^{(i_\alpha;\mathscr{B})},t,\boldsymbol{\lambda}_{i_\alpha})\ \text{on}\ \partial\Omega_{\alpha i_\alpha},\\ &\dot{\mathbf{x}}_{i_\alpha}^{(\beta;\mathscr{D})}=\mathbf{F}^{(\beta;\mathscr{D})}(\mathbf{x}_{i_\alpha}^{(\beta;\mathscr{D})},t,\mathbf{p}_\beta)\ \text{on}\ \partial\Omega_{\alpha i_\alpha}\ (\beta=\beta_1,\beta_2,\cdots,\beta_k),\\ &\dot{\mathbf{x}}_{\beta}^{(i_\alpha;\mathscr{B})}=\mathbf{F}^{(i_\alpha;\mathscr{B})}(\mathbf{x}_{\beta}^{(i_\alpha;\mathscr{B})},t,\boldsymbol{\lambda}_{i_\alpha})\ \text{on}\ \Omega_\beta\ (\beta=\beta_1,\beta_2,\cdots,\beta_k),\end{aligned}\tag{7.111}$$

where $\mathbf{x}_{i_\alpha}^{(i_\alpha;\mathscr{B})}$ is a real boundary flow, but $\mathbf{x}_{i_\alpha}^{(\beta;\mathscr{D})}$ and $\mathbf{x}_{\beta}^{(i_\alpha;\mathscr{B})}$ are imaginary boundary and domain flows, respectively. In Fig. 7.22, the imaginary domain flows $\mathbf{x}_{\beta}^{(i_\alpha;\mathscr{B})}$ to the $(n-2)$-dimensional edge are the extensions of the real boundary flows. The black curve represents the $(n-2)$-dimensional edge. The imaginary domain flows are represented by the dashed curves with arrows in the domains.

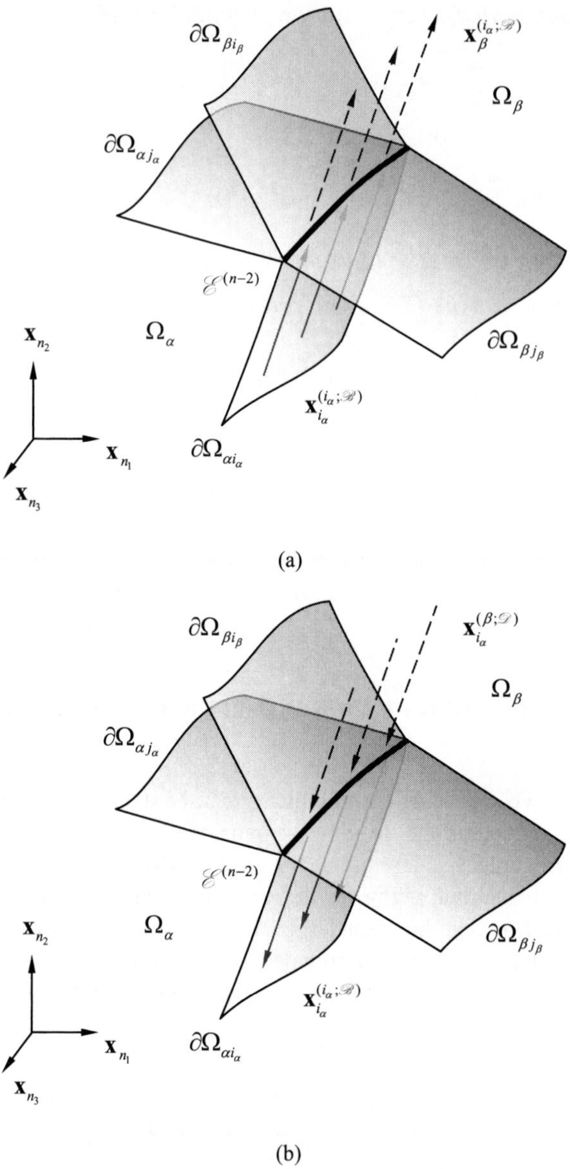

Fig. 7.22 Imaginary domain flows from boundary: **(a)** coming boundary flow extension and **(b)** leaving boundary flow extension. The black curve represents the $(n-2)$ -dimensional edge. The boundary flows are denoted by thin curves with arrows on the boundaries. The domain flows are represented by the dashed curves with arrows in the domain. ($n_1+n_2+n_3=n$)

Similarly, the imaginary boundary flow $\mathbf{x}_{i_\alpha}^{(\beta;\mathscr{D})}$ to the $(n-2)$-dimensional edge is the extension of the real domain flow in Fig. 7.23. Consider the switching rule as

$$\mathbf{x}_{i_\alpha}^{(i_\alpha;\mathscr{B})}(t_{m-}) = \mathbf{x}_{\beta}^{(\beta;\mathscr{D})}(t_{m+}) \text{ or } \mathbf{x}_{i_\alpha}^{(i_\alpha;\mathscr{B})}(t_{m-}) = \mathbf{x}_{i_\alpha}^{(\beta;\mathscr{D})}(t_{m-}). \tag{7.112}$$

To make the imaginary boundary flow exist, Axioms 6.1 and 6.2 can be extended to the edges and vertexes, i.e.,

Axiom 7.1 (Continuation axiom)**:** For a discontinuous dynamical system, any *boundary* flows coming to (or leaving from) an $(n-2)$-dimensional edge can be, forwardly (or backwardly) extended to the adjacent accessible domain without any switching rules, transport laws and any flow barriers at the edge.

Axiom 7.2 (Continuation axiom)**.** For a discontinuous dynamical system, a *domain* flow on the imaginary extension of a boundary, coming to (or leaving from) an $(n-2)$-dimensional edge, can be, forwardly (or backwardly) extended to the corresponding boundary without any switching rules, transport laws and any flow barriers at the edge.

Axiom 7.3 (Non-extendibility axiom)**.** For a discontinuous dynamical system, *any boundary* flows coming to (or leaving from) an $(n-2)$-dimensional edge cannot be, forwardly (or backwardly) extended to any inaccessible domain or edges. Even if such boundary flows are extendible to the inaccessible domains, then the corresponding extended flows are fictitious only.

Axiom 7.4 (Non-extendibility axiom)**.** For a discontinuous dynamical system, a *domain* flow on the imaginary extension of an inaccessible boundary, coming to (or leaving from) an $(n-2)$-dimensional edge, *cannot* be, forwardly (or backwardly) extended to the inaccessible boundary without any switching rules, transport laws and any flow barriers at the edge.

The multi-valued vector fields on an accessible boundary can be defined.

Definition 7.13. For a discontinuous system, there is an accessible boundary $\partial\Omega_{\alpha i_\alpha}$ ($\alpha \in \{1,2,\cdots,N\}$) on which k_α-vector fields of $\mathbf{F}^{(i_{\alpha_k};\mathscr{B})}(\mathbf{x}^{(i_{\alpha_k};\mathscr{B})},t,\boldsymbol{\lambda}_{i_{\alpha_k}})$ ($k = 1,2,\cdots,k_\alpha$) are defined, and the dynamical system on the boundary is defined as

$$\dot{\mathbf{x}}^{(i_{\alpha_k};\mathscr{B})} = \mathbf{F}^{(i_{\alpha_k};\mathscr{B})}(\mathbf{x}^{(i_{\alpha_k};\mathscr{B})},t,\boldsymbol{\lambda}_{i_{\alpha_k}}). \tag{7.113}$$

The dynamical system on the boundary $\partial\Omega_{\alpha i_\alpha}$ is called to be *of the multi-valued vector fields*. A set of dynamical systems on the boundary of $\partial\Omega_{\alpha i_\alpha}$ is defined as

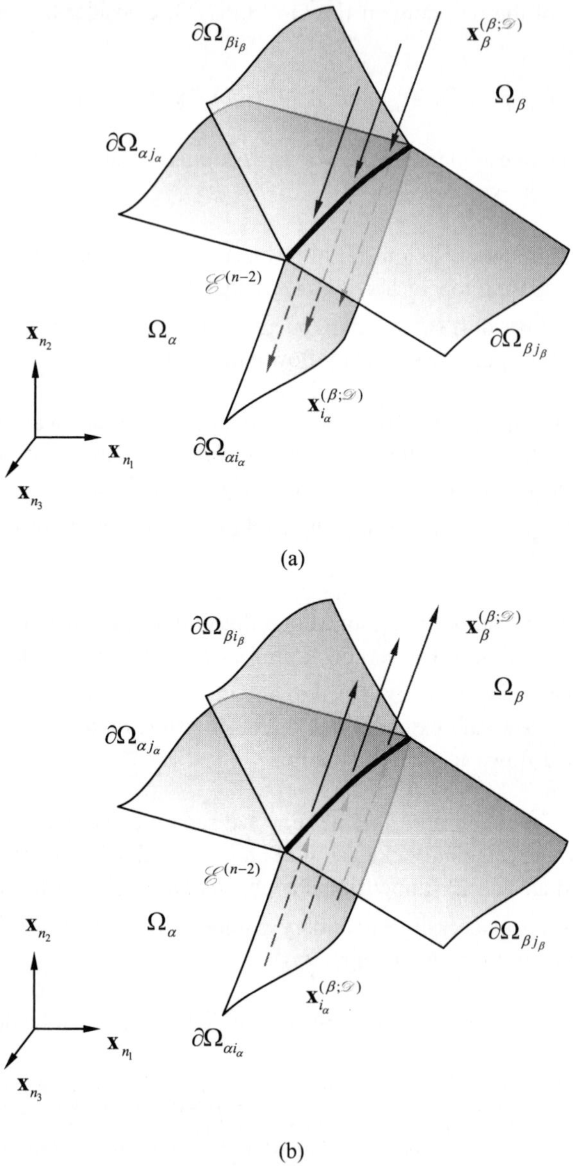

Fig. 7.23 Imaginary boundary flows from domain: **(a)** coming domain flow extension and **(b)** leaving domain flow extension. The black curve represents the $(n-2)$-dimensional edge. The boundary flows are denoted by thin curves with arrows on the boundaries. The domain flows are represented by the thick curves with arrows in the domain. ($n_1 + n_2 + n_3 = n$)

$$\begin{aligned} \mathscr{B}_{i_\alpha} &= \cup_{k=1}^{k_\alpha} \mathscr{B}_{i_{\alpha_k}}, \\ \mathscr{B}_{i_{\alpha_k}} &= \left\{ \dot{\mathbf{x}}^{(i_{\alpha_k};\mathscr{B})} = \mathbf{F}^{(i_{\alpha_k};\mathscr{B})}(\mathbf{x}^{(i_{\alpha_k};\mathscr{B})}, t, \boldsymbol{\lambda}_{i_{\alpha_k}}) \big| k \in \{1,2,\cdots,k_\alpha\} \right\}. \end{aligned} \tag{7.114}$$

A total set of dynamical systems on all the boundaries is

$$\mathscr{B} = \cup_{\alpha \in \mathscr{N}_{\mathscr{D}}} \cup_{i_\alpha \in \mathscr{N}_{\alpha\mathscr{B}}} \mathscr{B}_{i_\alpha}. \tag{7.115}$$

7.5.1. Bouncing boundary flows

The initial discussion on the bouncing flows can be found from Luo (2005, 2006). Herein, such an idea will be extended to the boundary flows.

Definition 7.14. For a discontinuous system with multi-valued vector fields on a boundary in Eq. (7.115), there is a point $\mathbf{x}^{(\sigma_{n-2};\mathscr{E})}(t_m) \equiv \mathbf{x}_m \in \mathscr{E}_{\sigma_{n-2}}^{(n-2)}$ ($\sigma_{n-2} \in \mathscr{N}_{\mathscr{E}}^{n-2} = \{1,2,\cdots,N_{n-2}\}$) at time t_m with the related boundaries $\partial\Omega_{\alpha i_\alpha}(\sigma_{n-2})$ and related domains $\Omega_\alpha(\sigma_{n-2})$ ($\alpha \in \mathscr{N}_{\mathscr{D}}(\sigma_{n-2}) \subset \mathscr{N}_{\mathscr{D}} = \{1,2,\cdots,N_d\}$ and $i_\alpha \in \mathscr{N}_{\alpha\mathscr{B}}(\sigma_{n-2}) \subset \mathscr{N}_{\mathscr{B}} = \{1,2,\cdots,N_b\}$). If a coming flow of $\mathbf{x}^{(i_{\alpha_k};\mathscr{B})}$ arrives to edge of $\mathscr{E}_{\sigma_{n-2}}^{(n-2)}$, there is a switching rule between the two flows of $\mathbf{x}^{(i_{\alpha_k};\mathscr{B})}$ and $\mathbf{x}^{(i_{\alpha_l};\mathscr{B})}$ with $\mathbf{x}^{(i_{\alpha_k};\mathscr{B})}(t_{m-}) = \mathbf{x}_m = \mathbf{x}^{(i_{\alpha_l};\mathscr{B})}(t_{m+})$. The flows of $\mathbf{x}^{(i_{\alpha_k};\mathscr{B})}$ and $\mathbf{x}^{(i_{\alpha_l};\mathscr{B})}$ to edge of $\mathscr{E}_{\sigma_r}^{(r)}$ are called a *bouncing* flow in $\partial\Omega_{\alpha i_\alpha}$ if

$$\begin{aligned} \hbar_{j_\alpha} \mathbf{n}_{\partial\Omega_{\alpha j_\alpha}}^{\mathrm{T}}(\mathbf{x}_{m-\varepsilon}^{(j_\alpha;\mathscr{B})}) \cdot [\mathbf{x}_{m-\varepsilon}^{(j_\alpha;\mathscr{B})} - \mathbf{x}_{m-\varepsilon}^{(i_{\alpha_k};\mathscr{B})}] < 0; \\ \hbar_{j_\alpha} \mathbf{n}_{\partial\Omega_{\alpha j_\alpha}}^{\mathrm{T}}(\mathbf{x}_{m+\varepsilon}^{(j_\alpha;\mathscr{B})}) \cdot [\mathbf{x}_{m+\varepsilon}^{(i_{\alpha_l};\mathscr{B})} - \mathbf{x}_{m+\varepsilon}^{(j_\alpha;\mathscr{B})}] > 0. \end{aligned} \tag{7.116}$$

Theorem 7.13. *For a discontinuous system with multi-valued vector fields on a boundary in Eq.* (7.115), *there is a point* $\mathbf{x}^{(\sigma_{n-2};\mathscr{E})}(t_m) \equiv \mathbf{x}_m \in \mathscr{E}_{\sigma_{n-2}}^{(n-2)}$ ($\sigma_{n-2} \in \mathscr{N}_{\mathscr{E}}^{n-2} = \{1,2,\cdots,N_{n-2}\}$) *at time* t_m *with the related boundaries* $\partial\Omega_{\alpha i_\alpha}(\sigma_{n-2})$ *and related domains* $\Omega_\alpha(\sigma_{n-2})$ ($\alpha \in \mathscr{N}_{\mathscr{D}}(\sigma_{n-2}) \subset \mathscr{N}_{\mathscr{D}} = \{1,2,\cdots,N_d\}$ *and* $i_\alpha \in \mathscr{N}_{\alpha\mathscr{B}}(\sigma_{n-2}) \subset \mathscr{N}_{\mathscr{B}} = \{1,2,\cdots,N_b\}$). *If a coming flow of* $\mathbf{x}^{(i_{\alpha_k};\mathscr{B})}$ *arrives to edge of* $\mathscr{E}_{\sigma_{n-2}}^{(n-2)}$, *there is a switching rule between the two flows of* $\mathbf{x}^{(i_{\alpha_k};\mathscr{B})}$ *and* $\mathbf{x}^{(i_{\alpha_l};\mathscr{B})}$ *with* $\mathbf{x}^{(i_{\alpha_k};\mathscr{B})}(t_{m-}) = \mathbf{x}_m = \mathbf{x}^{(i_{\alpha_l};\mathscr{B})}(t_{m+})$. *A flow* $\mathbf{x}^{(i_{\alpha_k};\mathscr{B})}(t)$ *is* $C_{[t_{m-\varepsilon},t_m)}^{r_{i_{\alpha_k}}}$*-continuous* ($r_{i_{\alpha_k}} \geq 2$) *for time* t *with* $\| d^{r_{i_{\alpha_k}}+1}\mathbf{x}^{(i_{\alpha_k};\mathscr{B})} / dt^{r_{i_{\alpha_k}}+1} \| < \infty$, *and a flow* $\mathbf{x}^{(i_{\alpha_l};\mathscr{B})}(t)$ *is* $C_{(t_m,t_{m+\varepsilon}]}^{r_{i_{\alpha_l}}}$-

continuous for time t with $\| d^{r_{i_{\alpha l}}+1} \mathbf{x}^{(i_{\alpha l};\mathscr{B})} / dt^{r_{i_{\alpha l}}+1} \| < \infty$ ($r_{i_{\alpha l}} \geq 2$). *The flows of* $\mathbf{x}^{(i_{\alpha k};\mathscr{B})}(t)$ *and* $\mathbf{x}^{(i_{\alpha l};\mathscr{B})}(t)$ *to edge* $\mathscr{E}_{\sigma_{n-2}}^{(n-2)}$ *form a bouncing flow in the boundary* $\partial\Omega_{\alpha i_\alpha}$ *if and only if*

$$\begin{aligned} &\hbar_{j_\alpha} G_{\partial\Omega_{\alpha\beta}}^{(i_{\alpha k};\mathscr{B})}(\mathbf{x}_m, t_{m-}, \boldsymbol{\lambda}_{i_{\alpha k}}, \boldsymbol{\lambda}_{j_\alpha}) < 0, \\ &\hbar_{j_\alpha} G_{\partial\Omega_{\alpha\beta}}^{(i_{\alpha l};\mathscr{B})}(\mathbf{x}_m, t_{m+}, \boldsymbol{\lambda}_{i_{\alpha l}}, \boldsymbol{\lambda}_{j_\alpha}) > 0. \end{aligned} \tag{7.117}$$

Proof. Following the proof of Theorem 3.1, the proof of the theorem is proved. ■

The definition of a boundary bouncing flow with the higher order singularity is given as follows.

Definition 7.15. For a discontinuous system with multi-valued vector fields on a boundary in Eq. (7.115), there is a point $\mathbf{x}^{(\sigma_{n-2};\mathscr{E})}(t_m) \equiv \mathbf{x}_m \in \mathscr{E}_{\sigma_{n-2}}^{(n-2)}$ ($\sigma_{n-2} \in \mathscr{N}_{\mathscr{E}}^{n-2} = \{1,2,\cdots,N_{n-2}\}$) at time t_m with the related boundaries $\partial\Omega_{\alpha i_\alpha}(\sigma_{n-2})$ and related domains $\Omega_\alpha(\sigma_{n-2})$ ($\alpha \in \mathscr{N}_{\mathscr{D}}(\sigma_{n-2}) \subset \mathscr{N}_{\mathscr{D}} = \{1,2,\cdots,N_d\}$ and $i_\alpha \in \mathscr{N}_{\alpha\mathscr{B}}(\sigma_{n-2}) \subset \mathscr{N}_{\mathscr{B}} = \{1,2,\cdots,N_b\}$). If a coming flow of $\mathbf{x}^{(i_{\alpha k};\mathscr{B})}$ arrives to edge of $\mathscr{E}_{\sigma_{n-2}}^{(n-2)}$, there is a switching rule between the two flows of $\mathbf{x}^{(i_{\alpha k};\mathscr{B})}$ and $\mathbf{x}^{(i_{\alpha l};\mathscr{B})}$ with $\mathbf{x}^{(i_{\alpha k};\mathscr{B})}(t_{m-}) = \mathbf{x}_m = \mathbf{x}^{(i_{\alpha l};\mathscr{B})}(t_{m+})$. A flow $\mathbf{x}^{(i_{\alpha k};\mathscr{B})}(t)$ is $C_{[t_{m-\varepsilon}, t_m)}^{r_{i_{\alpha k}}}$-continuous ($r_{i_{\alpha k}} \geq m_{i_{\alpha k}} + 1$) for time t with $\| d^{r_{i_{\alpha k}}+1} \mathbf{x}^{(i_{\alpha k};\mathscr{B})} / dt^{r_{i_{\alpha k}}+1} \| < \infty$, and a flow $\mathbf{x}^{(i_{\alpha l};\mathscr{B})}(t)$ is $C_{(t_m, t_{m+\varepsilon}]}^{r_{i_{\alpha l}}}$-continuous for time t with $\| d^{r_{i_{\alpha l}}+1} \mathbf{x}^{(i_{\alpha l};\mathscr{B})} / dt^{r_{i_{\alpha l}}+1} \| < \infty$ ($r_{i_{\alpha l}} \geq m_{i_{\alpha l}} + 1$). The flows of $\mathbf{x}^{(i_{\alpha k};\mathscr{B})}(t)$ and $\mathbf{x}^{(i_{\alpha l};\mathscr{B})}(t)$ to edge $\mathscr{E}_{\sigma_{n-2}}^{(n-2)}$ form an $(m_{i_{\alpha k}} : m_{i_{\alpha l}})$-*bouncing flow* in $\partial\Omega_{\alpha i_\alpha}$ if

$$\left.\begin{aligned} &\hbar_{j_\alpha} G_{\partial\Omega_{\alpha\beta}}^{(s_{i_{\alpha k}}, i_{\alpha k};\mathscr{B})}(\mathbf{x}_m, t_{m-}, \boldsymbol{\lambda}_{i_{\alpha k}}, \boldsymbol{\lambda}_{j_\alpha}) = 0 \\ &\text{for } s_{i_{\alpha k}} = 0,1,2,\cdots, m_{i_{\alpha k}} - 1, \\ &\hbar_{j_\alpha} \mathbf{n}_{\partial\Omega_{\alpha j_\alpha}}^{\mathrm{T}}(\mathbf{x}_{m-\varepsilon}^{(j_\alpha;\mathscr{B})}) \cdot [\mathbf{x}_{m-\varepsilon}^{(j_\alpha;\mathscr{B})} - \mathbf{x}_{m-\varepsilon}^{(i_{\alpha k};\mathscr{B})}] < 0; \end{aligned}\right\} \tag{7.118}$$

$$\left.\begin{aligned} &\hbar_{j_\alpha} G_{\partial\Omega_{\alpha\beta}}^{(s_{i_{\alpha l}}, i_{\alpha l};\mathscr{B})}(\mathbf{x}_m, t_{m+}, \boldsymbol{\lambda}_{i_{\alpha l}}, \boldsymbol{\lambda}_{j_\alpha}) = 0 \\ &\text{for } s_{i_{\alpha l}} = 0,1,2,\cdots, m_{i_{\alpha l}} - 1, \\ &\hbar_{j_\alpha} \mathbf{n}_{\partial\Omega_{\alpha j_\alpha}}^{\mathrm{T}}(\mathbf{x}_{m+\varepsilon}^{(j_\alpha;\mathscr{B})}) \cdot [\mathbf{x}_{m+\varepsilon}^{(i_{\alpha l};\mathscr{B})} - \mathbf{x}_{m+\varepsilon}^{(j_\alpha;\mathscr{B})}] > 0. \end{aligned}\right\} \tag{7.119}$$

Theorem 7.14. *For a discontinuous system with multi-valued vector fields on a boundary in Eq.* (7.115), *there is a point* $\mathbf{x}^{(\sigma_{n-2};\mathscr{E})}(t_m) \equiv \mathbf{x}_m \in \mathscr{E}^{(n-2)}_{\sigma_{n-2}}$ ($\sigma_{n-2} \in \mathscr{N}^{n-2}_{\mathscr{E}} = \{1,2,\cdots,N_{n-2}\}$) *at time* t_m *with the related boundaries* $\partial\Omega_{\alpha i_\alpha}(\sigma_{n-2})$ *and related domains* $\Omega_\alpha(\sigma_{n-2})$ ($\alpha \in \mathscr{N}_{\mathscr{D}}(\sigma_{n-2}) \subset \mathscr{N}_{\mathscr{D}} = \{1,2,\cdots,N_d\}$ *and* $i_\alpha \in \mathscr{N}_{\alpha\mathscr{B}}(\sigma_{n-2}) \subset \mathscr{N}_{\mathscr{B}} = \{1,2,\cdots,N_b\}$). *If a coming flow of* $\mathbf{x}^{(i_{\alpha k};\mathscr{B})}$ *arrives to the edge of* $\mathscr{E}^{(n-2)}_{\sigma_{n-2}}$, *there is a switching rule between the two flows of* $\mathbf{x}^{(i_{\alpha k};\mathscr{B})}$ *and* $\mathbf{x}^{(i_{\alpha l};\mathscr{B})}$ *with* $\mathbf{x}^{(i_{\alpha k};\mathscr{B})}(t_{m-}) = \mathbf{x}_m = \mathbf{x}^{(i_{\alpha l};\mathscr{B})}(t_{m+})$. *A flow* $\mathbf{x}^{(i_{\alpha k};\mathscr{B})}(t)$ *is* $C^{r_{i_{\alpha k}}}_{[t_{m-\varepsilon},t_m)}$*-continuous* ($r_{i_{\alpha k}} \ge m_{i_{\alpha k}} + 2$) *for time t with* $\| d^{r_{i_{\alpha k}}+1}\mathbf{x}^{(i_{\alpha k};\mathscr{B})}/dt^{r_{i_{\alpha k}}+1} \| < \infty$, *and a flow* $\mathbf{x}^{(i_{\alpha l};\mathscr{B})}(t)$ *is* $C^{r_{i_{\alpha l}}}_{(t_m,t_{m+\varepsilon}]}$*-continuous for time t with* $\| d^{r_{i_{\alpha l}}+1}\mathbf{x}^{(i_{\alpha l};\mathscr{B})}/dt^{r_{i_{\alpha l}}+1} \| < \infty$ ($r_{i_{\alpha l}} \ge m_{i_{\alpha l}} + 2$). *The flow* $\mathbf{x}^{(i_{\alpha k};\mathscr{B})}(t)$ *and* $\mathbf{x}^{(i_{\alpha l};\mathscr{B})}(t)$ *to the edge* $\mathscr{E}^{(n-2)}_{\sigma_{n-2}}$ *is an* $(m_{i_{\alpha k}} : m_{i_{\alpha l}})$*-bouncing flow in* $\partial\Omega_{\alpha i_\alpha}$ *if and only if*

$$\left.\begin{aligned}
&\hbar_{j_\alpha} G^{(s_{i_{\alpha k}}, i_{\alpha k};\mathscr{B})}_{\partial\Omega_{\alpha\beta}}(\mathbf{x}_m, t_{m-}, \boldsymbol{\lambda}_{i_{\alpha k}}, \boldsymbol{\lambda}_{j_\alpha}) = 0\\
&\text{for } s_{i_{\alpha k}} = 0,1,2,\cdots, m_{i_{\alpha k}} - 1,\\
&\hbar_{j_\alpha} G^{(m_{i_{\alpha k}}, i_{\alpha k};\mathscr{B})}_{\partial\Omega_{\alpha\beta}}(\mathbf{x}_m, t_{m-}, \boldsymbol{\lambda}_{i_{\alpha k}}, \boldsymbol{\lambda}_{j_\alpha}) < 0;
\end{aligned}\right\} \tag{7.120}$$

$$\left.\begin{aligned}
&\hbar_{j_\alpha} G^{(s_{i_{\alpha l}}, i_{\alpha l};\mathscr{B})}_{\partial\Omega_{\alpha\beta}}(\mathbf{x}_m, t_{m+}, \boldsymbol{\lambda}_{i_{\alpha l}}, \boldsymbol{\lambda}_{j_\alpha}) = 0\\
&\text{for } s_{i_{\alpha l}} = 0,1,2,\cdots, m_{i_{\alpha l}} - 1,\\
&\hbar_{j_\alpha} G^{(m_{i_{\alpha l}}, i_{\alpha l};\mathscr{B})}_{\partial\Omega_{\alpha\beta}}(\mathbf{x}_m, t_{m+}, \boldsymbol{\lambda}_{i_{\alpha l}}, \boldsymbol{\lambda}_{j_\alpha}) > 0.
\end{aligned}\right\} \tag{7.121}$$

Proof. Following the proof of Theorem 3.2, the theorem can be proved. ■

The bouncing flow with two different vector fields on the boundary of $\partial\Omega_{\alpha i_\alpha}$ switches at the edge of $\mathscr{E}^{(n-2)}_{\sigma_{n-2}}$, and there are four cases: (i) the $(2k_{i_{\alpha k}} : 2k_{i_{\alpha l}})$-bouncing flow, (ii) the $(2k_{i_{\alpha k}} : 2k_{i_{\alpha l}} + 1)$-bouncing flow, (iii) the $(2k_{i_{\alpha k}} + 1 : 2k_{i_{\alpha l}})$-bouncing flows, and (iv) the $(2k_{i_{\alpha k}} + 1 : 2k_{i_{\alpha l}} + 1)$-bouncing flow. The $(2k_{i_{\alpha k}} : 2k_{i_{\alpha l}})$-bouncing flow and $(2k_{i_{\alpha k}} + 1 : 2k_{i_{\alpha l}} + 1)$-bouncing flows are shown Fig. 7.24(a) and (b), respectively. The solid curves on the boundary $\partial\Omega_{\alpha i_\alpha}$ depict the real flow. The dashed curves give the backward or forward extension of real flows. In addition, The $(2k_{i_{\alpha k}} + 1 : 2k_{i_{\alpha l}})$-bouncing flow and $(2k_{i_{\alpha k}} : 2k_{i_{\alpha l}} + 1)$-bouncing flows are shown Fig. 7.25(a) and (b), respectively.

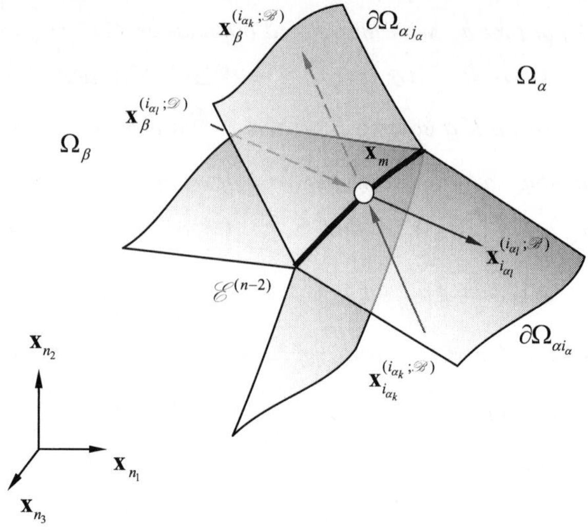

(a)

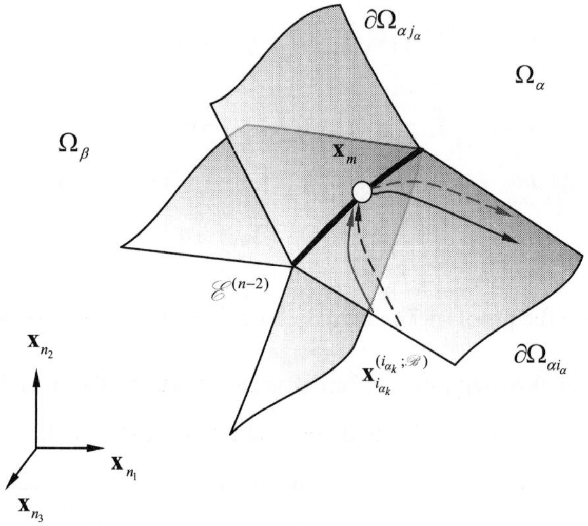

(b)

Fig. 7.24 Bouncing boundary flows to the edge: **(a)** the $(2k_{i_{\alpha_k}} : 2k_{i_{\alpha_l}})$ -singularity, **(b)** the $(2k_{i_{\alpha_k}} + 1 : 2k_{i_{\alpha_l}} + 1)$ -singularity. The black curve represents the $(n-2)$ -dimensional edge. The real boundary flows are denoted by solid curves with arrows on the boundaries. The extension flows are represented by the thick curves with arrows in the domain. ($n_1 + n_2 + n_3 = n$)

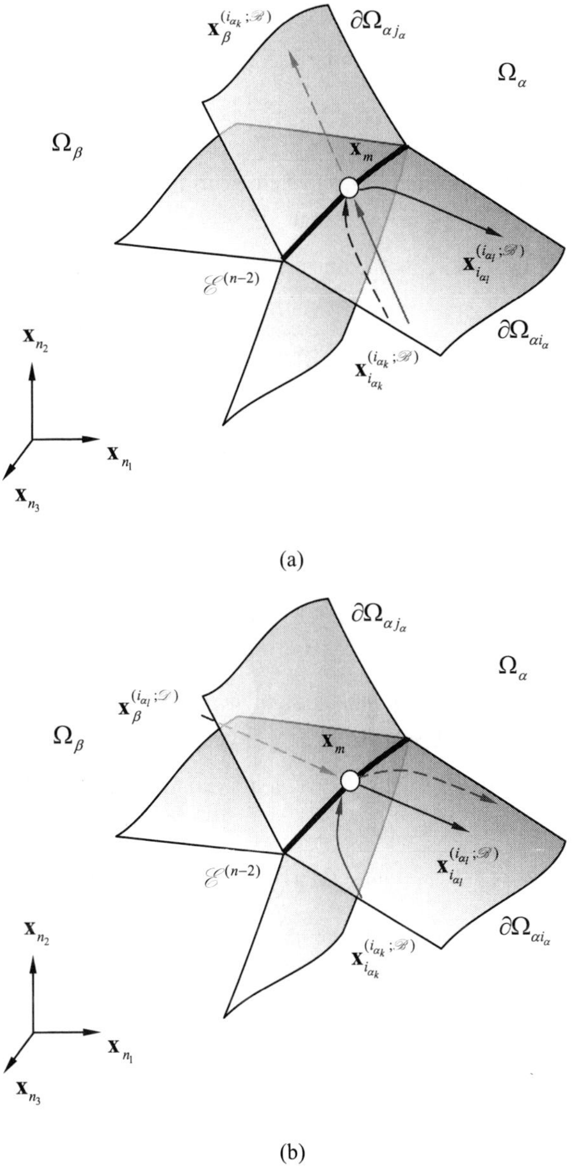

Fig. 7.25 Bouncing boundary flows to the edge: **(a)** the $(2k_{i_{\alpha k}} : 2k_{i_{\alpha l}} + 1)$-singularity, **(b)** the $(2k_{i_{\alpha k}} + 1 : 2k_{i_{\alpha l}})$-singularity. The black curve represents the $(n-2)$-dimensional edge. The real boundary flows are denoted by solid curves with arrows on the boundaries. The extension flows are represented by the thick curves with arrows in the domain. ($n_1 + n_2 + n_3 = n$)

7.5.2. Extended passable boundary flows

Consider the switching rule in boundary $\partial\Omega_{\alpha i_\alpha}$ only. After the switching, the flow in the boundary $\partial\Omega_{\alpha i_\alpha}$ can extend to domain Ω_β with the Continuation axiom.

Definition 7.16. For a discontinuous system with multi-valued vector fields on a boundary in Eq. (7.115), there is a point $\mathbf{x}^{(\sigma_{n-2};\mathscr{E})}(t_m) \equiv \mathbf{x}_m \in \mathscr{E}_{\sigma_{n-2}}^{(n-2)}$ ($\sigma_{n-2} \in \mathscr{N}_{\mathscr{E}}^{n-2} = \{1,2,\cdots,N_{n-2}\}$) at time t_m with the related boundaries $\partial\Omega_{\alpha i_\alpha}(\sigma_{n-2})$ and related domains $\Omega_\alpha(\sigma_{n-2})$ ($\alpha \in \mathscr{N}_{\mathscr{D}}(\sigma_{n-2}) \subset \mathscr{N}_{\mathscr{D}} = \{1,2,\cdots,N_d\}$ and $i_\alpha \in \mathscr{N}_{\alpha\mathscr{B}}(\sigma_{n-2}) \subset \mathscr{N}_{\mathscr{B}} = \{1,2,\cdots,N_b\}$). If a coming flow of $\mathbf{x}^{(i_{\alpha_k};\mathscr{B})}$ arrives to edge of $\mathscr{E}_{\sigma_{n-2}}^{(n-2)}$, there is a switching rule between the two flows of $\mathbf{x}^{(i_{\alpha_k};\mathscr{B})}$ and $\mathbf{x}^{(i_{\alpha_l};\mathscr{B})}$ with $\mathbf{x}_m = \mathbf{x}^{(i_{\alpha_k};\mathscr{B})}(t_{m-}) = \mathbf{x}^{(i_{\alpha_l};\mathscr{B})}(t_{m+})$. The flows of $\mathbf{x}^{(i_{\alpha_k};\mathscr{B})}$ and $\mathbf{x}^{(i_{\alpha_k};\mathscr{B})}$ to edge of $\mathscr{E}_{\sigma_{n-2}}^{(n-2)}$ are called an *extended passable* flow from $\partial\Omega_{\alpha i_\alpha}$ to Ω_β if

$$\left.\begin{aligned} &\hbar_{j_\alpha}\mathbf{n}_{\partial\Omega_{\alpha j_\alpha}}^{\mathrm{T}}(\mathbf{x}_{m-\varepsilon}^{(j_\alpha;\mathscr{B})})\cdot[\mathbf{x}_{m-\varepsilon}^{(j_\alpha;\mathscr{B})} - \mathbf{x}_{m-\varepsilon}^{(i_{\alpha_k};\mathscr{B})}] < 0;\\ &\hbar_{j_\alpha}\mathbf{n}_{\partial\Omega_{\alpha j_\alpha}}^{\mathrm{T}}(\mathbf{x}_{m+\varepsilon}^{(j_\alpha;\mathscr{B})})\cdot[\mathbf{x}_{m+\varepsilon}^{(i_{\alpha_l};\mathscr{B})} - \mathbf{x}_{m+\varepsilon}^{(j_\alpha;\mathscr{B})}] < 0. \end{aligned}\right\} \tag{7.122}$$

Theorem 7.15. *For a discontinuous system with multi-valued vector fields on a boundary in Eq.* (7.115), *there is a point* $\mathbf{x}^{(\sigma_{n-2};\mathscr{E})}(t_m) \equiv \mathbf{x}_m \in \mathscr{E}_{\sigma_{n-2}}^{(n-2)}$ ($\sigma_{n-2} \in \mathscr{N}_{\mathscr{E}}^{n-2} = \{1,2,\cdots,N_{n-2}\}$) *at time* t_m *with the related boundaries* $\partial\Omega_{\alpha i_\alpha}(\sigma_{n-2})$ *and related domains* $\Omega_\alpha(\sigma_{n-2})$ ($\alpha \in \mathscr{N}_{\mathscr{D}}(\sigma_{n-2}) \subset \mathscr{N}_{\mathscr{D}} = \{1,2,\cdots,N_d\}$ *and* $i_\alpha \in \mathscr{N}_{\alpha\mathscr{B}}(\sigma_{n-2}) \subset \mathscr{N}_{\mathscr{B}} = \{1,2,\cdots,N_b\}$). *If a coming flow of* $\mathbf{x}^{(i_{\alpha_k};\mathscr{B})}$ *arrives to edge of* $\mathscr{E}_{\sigma_{n-2}}^{(n-2)}$, *there is a switching rule between the two flows of* $\mathbf{x}^{(i_{\alpha_k};\mathscr{B})}$ *and* $\mathbf{x}^{(i_{\alpha_l};\mathscr{B})}$ *with* $\mathbf{x}_m = \mathbf{x}^{(i_{\alpha_k};\mathscr{B})}(t_{m-}) = \mathbf{x}^{(i_{\alpha_l};\mathscr{B})}(t_{m+})$. A flow $\mathbf{x}^{(i_{\alpha_k};\mathscr{B})}(t)$ is $C_{[t_{m-\varepsilon},t_m)}^{r_{i_{\alpha_k}}}$-continuous ($r_{i_{\alpha_k}} \geq 2$) for time t with $\| d^{r_{i_{\alpha_k}}+1}\mathbf{x}^{(i_{\alpha_k};\mathscr{B})}/dt^{r_{i_{\alpha_k}}+1}\| < \infty$, and a flow $\mathbf{x}^{(i_{\alpha_l};\mathscr{B})}(t)$ is $C_{(t_m,t_{m+\varepsilon}]}^{r_{i_{\alpha_l}}}$-continuous for time t with $\| d^{r_{i_{\alpha_l}}+1}\mathbf{x}^{(i_{\alpha_l};\mathscr{B})}/dt^{r_{i_{\alpha_l}}+1}\| < \infty$ ($r_{i_{\alpha_l}} \geq 2$). *The flows of* $\mathbf{x}^{(i_{\alpha_k};\mathscr{B})}$ *and* $\mathbf{x}^{(i_{\alpha_k};\mathscr{B})}$ *to edge of* $\mathscr{E}_{\sigma_{n-2}}^{(n-2)}$ *form an extended passable flow from* $\partial\Omega_{\alpha i_\alpha}$ *to* Ω_β *if and only if*

$$\left.\begin{aligned} &\hbar_{j_\alpha} G_{\partial\Omega_{\alpha\beta}}^{(i_{\alpha_k};\mathscr{B})}(\mathbf{x}_m, t_{m-}, \boldsymbol{\lambda}_{i_{\alpha_k}}, \boldsymbol{\lambda}_{j_\alpha}) < 0;\\ &\hbar_{j_\alpha} G_{\partial\Omega_{\alpha\beta}}^{(i_{\alpha_l};\mathscr{B})}(\mathbf{x}_m, t_{m+}, \boldsymbol{\lambda}_{i_{\alpha_l}}, \boldsymbol{\lambda}_{j_\alpha}) < 0. \end{aligned}\right\} \tag{7.123}$$

Proof. Following the proof of Theorem 3.1, the theorem can be proved. ■

Definition 7.17. For a discontinuous system with multi-valued vector fields on a boundary in Eq. (7.115), there is a point $\mathbf{x}^{(\sigma_{n-2};\mathscr{E})}(t_m) \equiv \mathbf{x}_m \in \mathscr{E}_{\sigma_{n-2}}^{(n-2)}$ ($\sigma_{n-2} \in \mathscr{N}_{\mathscr{E}}^{n-2} = \{1,2,\cdots,N_{n-2}\}$) at time t_m with the related boundaries $\partial\Omega_{\alpha i_\alpha}(\sigma_{n-2})$ and related domains $\Omega_\alpha(\sigma_{n-2})$ ($\alpha \in \mathscr{N}_{\mathscr{D}}(\sigma_{n-2}) \subset \mathscr{N}_{\mathscr{D}} = \{1,2,\cdots,N_d\}$ and $i_\alpha \in \mathscr{N}_{\alpha\mathscr{B}}(\sigma_{n-2}) \subset \mathscr{N}_{\mathscr{B}} = \{1,2,\cdots,N_b\}$). If a coming flow of $\mathbf{x}^{(i_{\alpha_k};\mathscr{B})}$ arrives to edge of $\mathscr{E}_{\sigma_{n-2}}^{(n-2)}$, there is a switching rule between the two flows of $\mathbf{x}^{(i_{\alpha_k};\mathscr{B})}$ and $\mathbf{x}^{(i_{\alpha_l};\mathscr{B})}$ with $\mathbf{x}_m = \mathbf{x}^{(i_{\alpha_k};\mathscr{B})}(t_{m-}) = \mathbf{x}^{(i_{\alpha_l};\mathscr{B})}(t_{m+})$. A flow $\mathbf{x}^{(i_{\alpha_k};\mathscr{B})}(t)$ is $C_{[t_{m-\varepsilon},t_m)}^{r_{i_{\alpha_k}}}$-continuous ($r_{i_{\alpha_k}} \geq m_{i_{\alpha_k}}+1$) for time t with $\| d^{r_{i_{\alpha_k}}+1}\mathbf{x}^{(i_{\alpha_k};\mathscr{B})}/dt^{r_{i_{\alpha_k}}+1} \| < \infty$, and a flow $\mathbf{x}^{(i_{\alpha_l};\mathscr{B})}(t)$ is $C_{(t_m,t_{m+\varepsilon}]}^{r_{i_{\alpha_l}}}$-continuous for time t with $\| d^{r_{i_{\alpha_l}}+1}\mathbf{x}^{(i_{\alpha_l};\mathscr{B})}/dt^{r_{i_{\alpha_l}}+1} \| < \infty$ ($r_{i_{\alpha_l}} \geq 2k_{i_{\alpha_l}}+1$). The flows of $\mathbf{x}^{(i_{\alpha_k};\mathscr{B})}$ and $\mathbf{x}^{(i_{\alpha_k};\mathscr{B})}$ to edge of $\mathscr{E}_{\sigma_{n-2}}^{(n-2)}$ are called an $(m_{i_{\alpha_k}} : 2k_{i_{\alpha_l}})$ extended passable flow from $\partial\Omega_{\alpha i_\alpha}$ to Ω_β if

$$\left.\begin{array}{l}
\hbar_{j_\alpha} G_{\partial\Omega_{\alpha\beta}}^{(s_{i_{\alpha_k}}, i_{\alpha_k};\mathscr{B})}(\mathbf{x}_m, t_{m-}, \boldsymbol{\lambda}_{i_{\alpha_k}}, \boldsymbol{\lambda}_{j_\alpha}) = 0 \\
\text{for } s_{i_{\alpha_k}} = 0,1,2,\cdots,m_{i_{\alpha_k}}-1, \\
\hbar_{j_\alpha} \mathbf{n}_{\partial\Omega_{\alpha j_\alpha}}^{\mathrm{T}}(\mathbf{x}_{m-\varepsilon}^{(j_\alpha;\mathscr{B})}) \cdot [\mathbf{x}_{m-\varepsilon}^{(j_\alpha;\mathscr{B})} - \mathbf{x}_{m-\varepsilon}^{(i_{\alpha_k};\mathscr{B})}] < 0;
\end{array}\right\} \quad (7.124)$$

$$\left.\begin{array}{l}
\hbar_{j_\alpha} G_{\partial\Omega_{\alpha\beta}}^{(s_{i_{\alpha_l}}, i_{\alpha_l};\mathscr{B})}(\mathbf{x}_m, t_{m+}, \boldsymbol{\lambda}_{i_{\alpha_l}}, \boldsymbol{\lambda}_{j_\alpha}) = 0 \\
\text{for } s_{i_{\alpha_l}} = 0,1,2,\cdots,2k_{i_{\alpha_l}}-1, \\
\hbar_{j_\alpha} \mathbf{n}_{\partial\Omega_{\alpha j_\alpha}}^{\mathrm{T}}(\mathbf{x}_{m+\varepsilon}^{(j_\alpha;\mathscr{B})}) \cdot [\mathbf{x}_{m+\varepsilon}^{(i_{\alpha_l};\mathscr{B})} - \mathbf{x}_{m+\varepsilon}^{(j_\alpha;\mathscr{B})}] < 0.
\end{array}\right\} \quad (7.125)$$

Theorem 7.16. *For a discontinuous system with multi-valued vector fields on a boundary in Eq.* (7.115), *there is a point* $\mathbf{x}^{(\sigma_{n-2};\mathscr{E})}(t_m) \equiv \mathbf{x}_m \in \mathscr{E}_{\sigma_{n-2}}^{(n-2)}$ ($\sigma_{n-2} \in \mathscr{N}_{\mathscr{E}}^{n-2} = \{1,2,\cdots,N_{n-2}\}$) *at time* t_m *with the related boundaries* $\partial\Omega_{\alpha i_\alpha}(\sigma_{n-2})$ *and related domains* $\Omega_\alpha(\sigma_{n-2})$ ($\alpha \in \mathscr{N}_{\mathscr{D}}(\sigma_{n-2}) \subset \mathscr{N}_{\mathscr{D}} = \{1,2,\cdots,N_d\}$ *and* $i_\alpha \in \mathscr{N}_{\alpha\mathscr{B}}(\sigma_{n-2}) \subset \mathscr{N}_{\mathscr{B}} = \{1,2,\cdots,N_b\}$). *If a coming flow of* $\mathbf{x}^{(i_{\alpha_k};\mathscr{B})}$ *arrives to edge of* $\mathscr{E}_{\sigma_{n-2}}^{(n-2)}$, *there is a switching rule between the two flows of* $\mathbf{x}^{(i_{\alpha_k};\mathscr{B})}$ *and* $\mathbf{x}^{(i_{\alpha_l};\mathscr{B})}$ *with* $\mathbf{x}_m = \mathbf{x}^{(i_{\alpha_k};\mathscr{B})}(t_{m-}) = \mathbf{x}^{(i_{\alpha_l};\mathscr{B})}(t_{m+})$. *A flow* $\mathbf{x}^{(i_{\alpha_k};\mathscr{B})}(t)$ *is* $C_{[t_{m-\varepsilon},t_m)}^{r_{i_{\alpha_k}}}$*-continuous* ($r_{i_{\alpha_k}} \geq m_{i_{\alpha_k}}+1$) *for time t with* $\| d^{r_{i_{\alpha_k}}+1}\mathbf{x}^{(i_{\alpha_k};\mathscr{B})}/dt^{r_{i_{\alpha_k}}+1} \| < \infty$, *and a flow* $\mathbf{x}^{(i_{\alpha_l};\mathscr{B})}(t)$ *is* $C_{(t_m,t_{m+\varepsilon}]}^{r_{i_{\alpha_l}}}$-

continuous for time t with $\| d^{r_{i_{\alpha_l}}+1}\mathbf{x}^{(i_{\alpha_l};\mathscr{B})}/dt^{r_{i_{\alpha_l}}+1}\| < \infty$ $(r_{i_{\alpha_l}} \geq 2k_{i_{\alpha_l}}+1)$. *The flows of* $\mathbf{x}^{(i_{\alpha_k};\mathscr{B})}$ *and* $\mathbf{x}^{(i_{\alpha_k};\mathscr{B})}$ *to edge of* $\mathscr{E}^{(n-2)}_{\sigma_{n-2}}$ *form an* $(m_{i_{\alpha_k}} : m_{i_{\alpha_l}})$ *extended passable flow from* $\partial\Omega_{\alpha i_\alpha}$ *to* Ω_β *if and only if*

$$\left.\begin{array}{l} \hbar_{j_\alpha} G^{(s_{i_{\alpha_k}}, i_{\alpha_k};\mathscr{B})}_{\partial\Omega_{\alpha\beta}}(\mathbf{x}_m, t_{m-}, \boldsymbol{\lambda}_{i_{\alpha_k}}, \boldsymbol{\lambda}_{j_\alpha}) = 0 \\ \textit{for } s_{i_{\alpha_k}} = 0,1,2,\cdots, m_{i_{\alpha_k}} - 1, \\ \hbar_{j_\alpha} G^{(m_{i_{\alpha_k}}, i_{\alpha_k};\mathscr{B})}_{\partial\Omega_{\alpha\beta}}(\mathbf{x}_m, t_{m-}, \boldsymbol{\lambda}_{i_{\alpha_k}}, \boldsymbol{\lambda}_{j_\alpha}) < 0; \end{array}\right\} \tag{7.126}$$

$$\left.\begin{array}{l} \hbar_{j_\alpha} G^{(s_{i_{\alpha_l}}, i_{\alpha_l};\mathscr{B})}_{\partial\Omega_{\alpha\beta}}(\mathbf{x}_m, t_{m+}, \boldsymbol{\lambda}_{i_{\alpha_l}}, \boldsymbol{\lambda}_{j_\alpha}) = 0 \\ \textit{for } s_{i_{\alpha_l}} = 0,1,2,\cdots, 2k_{i_{\alpha_l}} - 1, \\ \hbar_{j_\alpha} G^{(2k_{i_{\alpha_l}}, i_{\alpha_l};\mathscr{B})}_{\partial\Omega_{\alpha\beta}}(\mathbf{x}_m, t_{m+}, \boldsymbol{\lambda}_{i_{\alpha_l}}, \boldsymbol{\lambda}_{j_\alpha}) < 0. \end{array}\right\} \tag{7.127}$$

Proof. Following the proof of Theorem 3.2, the theorem is proved. ■

References

Luo, A.C.J., 2005, A theory for non-smooth dynamical systems on connectable domains, *Communication in Nonlinear Science and Numerical Simulation,* **10**, 1-55.

Luo, A.C.J., 2006, *Singularity and Dynamics on Discontinuous Vector Fields*, Amsterdam: Elsevier.

Luo, A.C.J., 2008a, A theory for flow switchability in discontinuous dynamical systems, *Nonlinear Analysis: Hybrid Systems*, **2**(4), 1030-1061.

Luo, A.C.J., 2008b, Global Transversality, Resonance and Chaotic Dynamics, Singapore: World Scientific.

Chapter 8
Edge Dynamics and Switching Complexity

In this chapter, the switchability and attractivity of edge flows to the lower-dimensional edges will be discussed with a generation of the theory of switchability and attractivity for domain and boundary flows. The basic properties of edge flows to a specific edge will be discussed first. The coming, leaving and tangency of an edge flow to a specific edge will be presented through the separation boundaries. The switchability and passability of an edge flow from an accessible edge to another accessible edge (or boundary or domain) will be discussed with a switching rule. Similarly, the equi-measuring edge for edge flows will be introduced, and the attractivity of an edge flow to the lower-dimensional edge will be presented. Finally, a bouncing edge flow to a specific lower-dimensional edge will be discussed as well. The switchability of a flow in a 2-DOF frictional oscillator will be discussed as a sample problem to illustrate edge dynamics.

8.1. Edge flows

In Chapter 7, the switchability and singularity of a boundary flow to an $(n-2)$-dimensional edge were discussed in n-dimensional, discontinuous dynamical systems. Herein, such an idea will be extended to an s-dimensional edge flow to an r-dimensional edge ($r < s \leq n$) in n-dimensional, discontinuous dynamical systems. The r-dimensional edge can be formed by the intersection of the s-dimensional edge and $(s-r)$ boundaries in domain Ω_α. Thus, from Definition 6.2, a r-dimensional edge can be expressed by

$$
\begin{aligned}
&\mathscr{E}_{\sigma_r}^{(r)} = \cup_\alpha \mathscr{E}_{\alpha\sigma_r}^{(r)} \subset \mathscr{R}^r,\\
&\mathscr{E}_{\alpha\sigma_r}^{(r)} = \cap_{i_\alpha \in \mathscr{I}_{\alpha\mathscr{B}}(\sigma_r)} \partial\Omega_{\alpha i_\alpha}(\sigma_r),\\
&\mathscr{E}_{\alpha\sigma_r}^{(r)} = \mathscr{E}_{\alpha\sigma_s}^{(s)} \cap \cap_{i_\alpha \in \bar{\mathscr{I}}_{\alpha\mathscr{B}}(\sigma_r,\sigma_s)} \partial\Omega_{\alpha i_\alpha}(\sigma_r) \subset \mathscr{R}^r;
\end{aligned}
\tag{8.1}
$$

where

$$\begin{aligned}&\mathscr{E}_{\alpha\sigma_s}^{(s)}(\sigma_r)=\cap_{i_\alpha\in\mathscr{N}_{\alpha\mathscr{B}}(\sigma_r,\sigma_s)}\partial\Omega_{\alpha i_\alpha}(\sigma_r),\\&\mathscr{N}_{\alpha\mathscr{B}}(\sigma_r)\subset\mathscr{N}_{\mathscr{B}}=\{1,2,\cdots,N_b\},\\&\mathscr{N}_{\alpha\mathscr{B}}(\sigma_r)=\mathscr{N}_{\alpha\mathscr{B}}(\sigma_r,\sigma_s)\cup\bar{\mathscr{N}}_{\alpha\mathscr{B}}(\sigma_r,\sigma_s).\end{aligned}\tag{8.2}$$

Consider a s-dimensional edge flow to a r-dimensional edge ($s>r$) in $\mathscr{R}^n$. One can treat an s-dimensional edge flow in $\mathscr{R}^n$ as a domain flow in $\mathscr{R}^s$ analogically, and the boundary flows discussed in Chapter 7 are as a special case of s-dimensional edge flows ($s=n-1$). The coming, leaving and grazing edge flows are introduced in order to discuss the switchability of edge flows to a specific edge.

Definition 8.1. For a discontinuous dynamical system in Eqs. (6.4)-(6.8), there is a point $\mathbf{x}^{(\sigma_r;\mathscr{E})}(t_m)\equiv\mathbf{x}_m\in\mathscr{E}_{\sigma_r}^{(r)}$ ($\sigma_r\in\mathscr{N}_{\mathscr{E}}^r=\{1,2,\cdots,N_r\}$ and $r\in\{0,1,2,\cdots,n-2\}$) at time t_m with the related boundaries $\partial\Omega_{\alpha i_\alpha}(\sigma_r)$ and related domains $\Omega_\alpha(\sigma_r)$ ($\alpha\in\mathscr{N}_{\mathscr{D}}(\sigma_r)\subset\mathscr{N}_{\mathscr{D}}=\{1,2,\cdots,N_d\}$ and $i_\alpha\in\mathscr{N}_{\alpha\mathscr{B}}(\sigma_r)\subset\mathscr{N}_{\mathscr{B}}=\{1,2,\cdots,N_b\}$), where the boundary index subset of $\mathscr{N}_{\alpha\mathscr{B}}(\sigma_r)$ has n_{σ_r}-elements ($n_{\sigma_r}\geq(n-r)$). The edge of $\mathscr{E}_{\sigma_r}^{(r)}$ is the intersection of edges (including domains and boundaries) $\mathscr{E}_{\sigma_s}^{(s)}(\sigma_r)$ ($\sigma_s\in\mathscr{N}_{\mathscr{E}}^s(\sigma_r)\subset\mathscr{N}_{\mathscr{E}}^s=\{1,2,\cdots,N_s\}$ and $s\in\{r+1,r+2,\cdots,n-1,n\}$). Suppose there is an s-dimensional edge $\mathscr{E}_{\sigma_s}^{(s)}(\sigma_r)$ with a boundary index set $\mathscr{N}_{\alpha\mathscr{B}}(\sigma_r,\sigma_s)$ relative to domain $\Omega_\alpha(\sigma_r)$. The complementary boundary index set is $\bar{\mathscr{N}}_{\alpha\mathscr{B}}(\sigma_r,\sigma_s)=\mathscr{N}_{\alpha\mathscr{B}}(\sigma_r)/\mathscr{N}_{\alpha\mathscr{B}}(\sigma_r,\sigma_s)$. For an arbitrarily small $\varepsilon>0$, there are two time intervals $[t_{m-\varepsilon},t_m)$ and $(t_m,t_{m+\varepsilon}]$. An edge flow $\mathbf{x}^{(\sigma_s;\mathscr{E})}(t)$ is $C_{[t_{m-\varepsilon},t_m)}^{r_{\sigma_s}}$ and/or $C_{(t_m,t_{m+\varepsilon}]}^{r_{\sigma_s}}$-continuous for time t and $\| d^{r_{\sigma_s}+1}\mathbf{x}^{(\sigma_s;\mathscr{E})}/dt^{r_{\sigma_s}+1}\|<\infty$ ($r_{\sigma_s}\geq1$).

(i) The s-dimensional edge flow $\mathbf{x}^{(\sigma_s;\mathscr{E})}(t)$ in an edge $\mathscr{E}_{\sigma_s}^{(s)}$ to the edge $\mathscr{E}_{\sigma_r}^{(r)}$ is a *coming edge flow* at point $\mathbf{x}_m\in\mathscr{E}_{\sigma_r}^{(r)}$ if

$$\hbar_{i_\alpha}\mathbf{n}_{\partial\Omega_{\alpha i_\alpha}}^{\mathrm{T}}(\mathbf{x}_{m-\varepsilon}^{(i_\alpha;\mathscr{B})})\cdot[\mathbf{x}_{m-\varepsilon}^{(i_\alpha;\mathscr{B})}-\mathbf{x}_{m-\varepsilon}^{(\sigma_s;\mathscr{E})}]<0 \text{ for all } i_\alpha\in\bar{\mathscr{N}}_{\alpha\mathscr{B}}(\sigma_r,\sigma_s).\tag{8.3}$$

(ii) The s-dimensional edge flow $\mathbf{x}^{(\sigma_s;\mathscr{E})}(t)$ in an edge $\mathscr{E}_{\sigma_s}^{(s)}$ to the edge $\mathscr{E}_{\sigma_r}^{(r)}$ is a *leaving edge flow* at point $\mathbf{x}_m\in\mathscr{E}_{\sigma_r}^{(r)}$ if

$$\hbar_{i_\alpha}\mathbf{n}_{\partial\Omega_{\alpha i_\alpha}}^{\mathrm{T}}(\mathbf{x}_{m+\varepsilon}^{(i_\alpha;\mathscr{B})})\cdot[\mathbf{x}_{m+\varepsilon}^{(\sigma_s;\mathscr{E})}-\mathbf{x}_{m+\varepsilon}^{(i_\alpha;\mathscr{B})}]>0 \text{ for all } i_\alpha\in\bar{\mathscr{N}}_{\alpha\mathscr{B}}(\sigma_r,\sigma_s).\tag{8.4}$$

(iii) The s -dimensional edge flow $\mathbf{x}^{(\sigma_s;\mathscr{E})}(t)$ in an edge $\mathscr{E}_{\sigma_s}^{(s)}$ to the edge $\mathscr{E}_{\sigma_r}^{(r)}$ is a grazing edge flow at point $\mathbf{x}_m \in \mathscr{E}_{\sigma_r}^{(r)}$ ($r \neq 0$) if

$$\left.\begin{array}{l} \hbar_{i_\alpha} G_{\partial\Omega_{\alpha i_\alpha}}^{(\sigma_s;\mathscr{E})}(\mathbf{x}_m, t_{m\pm}, \boldsymbol{\pi}_{\sigma_s}, \boldsymbol{\lambda}_{i_\alpha}) = 0 \\ \hbar_{i_\alpha} \mathbf{n}_{\partial\Omega_{\alpha i_\alpha}}^{\mathrm{T}}(\mathbf{x}_{m-\varepsilon}^{(i_\alpha;\mathscr{B})}) \cdot [\mathbf{x}_{m-\varepsilon}^{(i_\alpha;\mathscr{B})} - \mathbf{x}_{m-\varepsilon}^{(\sigma_s;\mathscr{E})}] < 0 \\ \hbar_{i_\alpha} \mathbf{n}_{\partial\Omega_{\alpha i_\alpha}}^{\mathrm{T}}(\mathbf{x}_{m+\varepsilon}^{(i_\alpha;\mathscr{B})}) \cdot [\mathbf{x}_{m+\varepsilon}^{(\sigma_s;\mathscr{E})} - \mathbf{x}_{m+\varepsilon}^{(i_\alpha;\mathscr{B})}] > 0 \end{array}\right\} \text{for all } i_\alpha \in \bar{\mathscr{N}}_{\alpha\mathscr{B}}(\sigma_r, \sigma_s). \tag{8.5}$$

The foregoing definition yields the corresponding theorem as follows.

Theorem 8.1. *For a discontinuous dynamical system in Eqs.* (6.4)-(6.8), *there is a point* $\mathbf{x}^{(\sigma_r;\mathscr{E})}(t_m) \equiv \mathbf{x}_m \in \mathscr{E}_{\sigma_r}^{(r)}$ ($\sigma_r \in \mathscr{N}_{\mathscr{E}}^{r} = \{1,2,\cdots,N_r\}$ *and* $r \in \{0,1,2,\cdots,n-2\}$) *at time* t_m *with the related boundaries* $\partial\Omega_{\alpha i_\alpha}(\sigma_r)$ *and related domains* $\Omega_\alpha(\sigma_r)$ ($\alpha \in \mathscr{N}_{\mathscr{D}}(\sigma_r) \subset \mathscr{N}_{\mathscr{D}} = \{1,2,\cdots,N_d\}$ *and* $i_\alpha \in \mathscr{N}_{\alpha\mathscr{B}}(\sigma_r) \subset \mathscr{N}_{\mathscr{B}} = \{1,2,\cdots,N_b\}$), *where the boundary index subset of* $\mathscr{N}_{\alpha\mathscr{B}}(\sigma_r)$ *has* n_{σ_r} *-elements* ($n_{\sigma_r} \geq (n-r)$). *The edge of* $\mathscr{E}_{\sigma_r}^{(r)}$ *is the intersection of edges* (*including domains and boundaries*) $\mathscr{E}_{\sigma_s}^{(s)}(\sigma_r)$ ($\sigma_s \in \mathscr{N}_{\mathscr{E}}^{s}(\sigma_r) \subset \mathscr{N}_{\mathscr{E}}^{s} = \{1,2,\cdots,N_s\}$ *and* $s \in \{r+1, r+2,\cdots,n-1,n\}$). *Suppose there is an* s *-dimensional edge* $\mathscr{E}_{\sigma_s}^{(s)}(\sigma_r)$ *with a boundary index set* $\mathscr{N}_{\alpha\mathscr{B}}(\sigma_r,\sigma_s)$ *relative to domain* $\Omega_\alpha(\sigma_r)$. *The complementary boundary index set is* $\bar{\mathscr{N}}_{\alpha\mathscr{B}}(\sigma_r,\sigma_s) = \mathscr{N}_{\alpha\mathscr{B}}(\sigma_r) / \mathscr{N}_{\alpha\mathscr{B}}(\sigma_r,\sigma_s)$. *For an arbitrarily small* $\varepsilon > 0$, *there are two time intervals* $[t_{m-\varepsilon}, t_m)$ *and* $(t_m, t_{m+\varepsilon}]$. *An edge flow* $\mathbf{x}^{(\sigma_s;\mathscr{E})}(t)$ *is* $C_{[t_{m-\varepsilon},t_m)}^{r_{\sigma_s}}$ *and/or* $C_{(t_m,t_{m+\varepsilon}]}^{r_{\sigma_s}}$*-continuous for time* t *and* $\| d^{r_{\sigma_s}+1}\mathbf{x}^{(\sigma_s;\mathscr{E})} / dt^{r_{\sigma_s}+1} \| < \infty$ ($r_{\sigma_s} \geq 2$).

(i) *The* s *-dimensional edge flow* $\mathbf{x}^{(\sigma_s;\mathscr{E})}(t)$ *in an edge* $\mathscr{E}_{\sigma_s}^{(s)}$ *to the edge* $\mathscr{E}_{\sigma_r}^{(r)}$ *is a coming flow at point* $\mathbf{x}_m \in \mathscr{E}_{\sigma_r}^{(r)}$ *if and only if*

$$\hbar_{i_\alpha} G_{\partial\Omega_{\alpha i_\alpha}}^{(\sigma_s;\mathscr{E})}(\mathbf{x}_m, t_{m-}, \boldsymbol{\pi}_{\sigma_s}, \boldsymbol{\lambda}_{i_\alpha}) < 0 \text{ for all } i_\alpha \in \bar{\mathscr{N}}_{\alpha\mathscr{B}}(\sigma_r,\sigma_s). \tag{8.6}$$

(ii) *The* s *-dimensional edge flow* $\mathbf{x}^{(\sigma_s;\mathscr{E})}(t)$ *in an edge* $\mathscr{E}_{\sigma_s}^{(s)}$ *to the edge* $\mathscr{E}_{\sigma_r}^{(r)}$ *is a leaving flow at point* $\mathbf{x}_m \in \mathscr{E}_{\sigma_r}^{(r)}$ *if and only if*

$$\hbar_{i_\alpha} G_{\partial\Omega_{\alpha i_\alpha}}^{(\sigma_s;\mathscr{E})}(\mathbf{x}_m, t_{m+}, \boldsymbol{\pi}_{\sigma_s}, \boldsymbol{\lambda}_{i_\alpha}) < 0 \text{ for all } i_\alpha \in \bar{\mathscr{N}}_{\alpha\mathscr{B}}(\sigma_r,\sigma_s). \tag{8.7}$$

(iii) *The* s *-dimensional edge flow* $\mathbf{x}^{(\sigma_s;\mathscr{E})}(t)$ *in an edge* $\mathscr{E}_{\sigma_s}^{(s)}$ *to the edge* $\mathscr{E}_{\sigma_r}^{(r)}$ *is a grazing edge flow at point* $\mathbf{x}_m \in \mathscr{E}_{\sigma_r}^{(r)}$ ($r \neq 0$) *if and only if*

$$\hbar_{i_\alpha} G^{(1,\sigma_s;\mathscr{E})}_{\partial\Omega_{\alpha i_\alpha}}(\mathbf{x}_m, t_{m\pm}, \boldsymbol{\pi}_{\sigma_s}, \boldsymbol{\lambda}_{i_\alpha}) > 0 \text{ for all } i_\alpha \in \bar{\mathscr{N}}_{\alpha\mathscr{B}}(\sigma_r, \sigma_s). \quad (8.8)$$

Proof. As in Theorem 3.1, this theorem can be proved similarly as the edge flow is treated as a special domain flow (also see Luo, 2008, 2009). ■

The foregoing discussion gives three basic coming, leaving and grazing edge flows. However, because the edge flows may possess higher order singularity to the specific edge, the coming, leaving and grazing edge flows with the higher-order singularity can be described as well herein.

Definition 8.2. For a discontinuous dynamical system in Eqs. (6.4)-(6.8), there is a point $\mathbf{x}^{(\sigma_r;\mathscr{E})}(t_m) \equiv \mathbf{x}_m \in \mathscr{E}^{(r)}_{\sigma_r}$ ($\sigma_r \in \mathscr{N}^r_{\mathscr{E}} = \{1, 2, \cdots, N_r\}$ and $r \in \{0, 1, 2, \cdots, n-2\}$) at time t_m with the related boundaries $\partial\Omega_{\alpha i_\alpha}(\sigma_r)$ and related domains $\Omega_\alpha(\sigma_r)$ ($\alpha \in \mathscr{N}_{\mathscr{D}}(\sigma_r) \subset \mathscr{N}_{\mathscr{D}} = \{1, 2, \cdots, N_d\}$ and $i_\alpha \in \mathscr{N}_{\alpha\mathscr{B}}(\sigma_r) \subset \mathscr{N}_{\mathscr{B}} = \{1, 2, \cdots, N_b\}$), where the boundary index subset of $\mathscr{N}_{\alpha\mathscr{B}}(\sigma_r)$ has n_{σ_r} -elements ($n_{\sigma_r} \geq (n-r)$). The edge of $\mathscr{E}^{(r)}_{\sigma_r}$ is the intersection of edges (including domains and boundaries) $\mathscr{E}^{(s)}_{\sigma_s}(\sigma_r)$ ($\sigma_s \in \mathscr{N}^s_{\mathscr{E}}(\sigma_r) \subset \mathscr{N}^s_{\mathscr{E}} = \{1, 2, \cdots, N_s\}$ and $s \in \{r+1, r+2, \cdots, n-1, n\}$). Suppose there is an s -dimensional edge $\mathscr{E}^{(s)}_{\sigma_s}(\sigma_r)$ with a boundary index set $\mathscr{N}_{\alpha\mathscr{B}}(\sigma_r, \sigma_s)$ relative to domain $\Omega_\alpha(\sigma_r)$. The complementary boundary index set is $\bar{\mathscr{N}}_{\alpha\mathscr{B}}(\sigma_r, \sigma_s) = \mathscr{N}_{\alpha\mathscr{B}}(\sigma_r) / \mathscr{N}_{\alpha\mathscr{B}}(\sigma_r, \sigma_s)$. For an arbitrarily small $\varepsilon > 0$, there are two time intervals $[t_{m-\varepsilon}, t_m)$ and $(t_m, t_{m+\varepsilon}]$. An edge flow $\mathbf{x}^{(\sigma_s;\mathscr{E})}(t)$ is $C^{r_{\sigma_s}}_{[t_{m-\varepsilon}, t_m)}$ and/or $C^{r_{\sigma_s}}_{(t_m, t_{m+\varepsilon}]}$-continuous for time t and $\| d^{r_{\sigma_s}+1}\mathbf{x}^{(\sigma_s;\mathscr{E})} / dt^{r_{\sigma_s}+1} \| < \infty$ ($r_{\sigma_s} \geq \max_{i_\alpha \in \mathscr{N}_{\alpha\mathscr{B}}}\{m_{i_\alpha}\}$).

(i) The s -dimensional edge flow $\mathbf{x}^{(\sigma_s;\mathscr{E})}(t)$ in an edge $\mathscr{E}^{(s)}_{\sigma_s}$ to the edge $\mathscr{E}^{(r)}_{\sigma_r}$ is a *coming flow* at point $\mathbf{x}_m \in \mathscr{E}^{(r)}_{\sigma_r}$ with the $(\mathbf{m}_{\sigma_s}; \mathscr{E})$ -singularity if

$$\left.\begin{aligned} &\hbar_{i_\alpha} G^{(s_{i_\alpha}, \sigma_s;\mathscr{E})}_{\partial\Omega_{\alpha i_\alpha}}(\mathbf{x}_m, t_{m-}, \boldsymbol{\pi}_{\sigma_s}, \boldsymbol{\lambda}_{i_\alpha}) = 0 \\ &\text{for } s_{i_\alpha} = 0, 1, 2, \cdots, m_{i_\alpha} - 1; \\ &\hbar_{i_\alpha} \mathbf{n}^{\mathrm{T}}_{\partial\Omega_{\alpha i_\alpha}}(\mathbf{x}^{(i_\alpha;\mathscr{B})}_{m-\varepsilon}) \cdot [\mathbf{x}^{(i_\alpha;\mathscr{B})}_{m-\varepsilon} - \mathbf{x}^{(\sigma_s;\mathscr{E})}_{m-\varepsilon}] < 0 \end{aligned}\right\} \text{for all } i_\alpha \in \bar{\mathscr{N}}_{\alpha\mathscr{B}}(\sigma_r, \sigma_s). \quad (8.9)$$

(ii) The s -dimensional edge flow $\mathbf{x}^{(\sigma_s;\mathscr{E})}(t)$ in an edge $\mathscr{E}^{(s)}_{\sigma_s}$ to the edge $\mathscr{E}^{(r)}_{\sigma_r}$ is a *leaving flow* at point $\mathbf{x}_m \in \mathscr{E}^{(r)}_{\sigma_r}$ with the $(\mathbf{m}_{\sigma_s}; \mathscr{E})$ -singularity if

$$\left.\begin{array}{l}\hbar_{i_\alpha} G_{\partial\Omega_{\alpha i_\alpha}}^{(s_{i_\alpha},\sigma_s;\mathscr{E})}(\mathbf{x}_m,t_{m+},\boldsymbol{\pi}_{\sigma_s},\boldsymbol{\lambda}_{i_\alpha})=0\\ \text{for } s_{i_\alpha}=0,1,2,\cdots,m_{i_\alpha}-1;\\ \hbar_{i_\alpha}\mathbf{n}_{\partial\Omega_{\alpha i_\alpha}}^{\mathrm{T}}(\mathbf{x}_{m+\varepsilon}^{(i_\alpha;\mathscr{B})})\cdot[\mathbf{x}_{m+\varepsilon}^{(\sigma_s;\mathscr{E})}-\mathbf{x}_{m+\varepsilon}^{(i_\alpha;\mathscr{B})}]>0\end{array}\right\}\text{for all } i_\alpha\in\bar{\mathscr{N}}_{\alpha\mathscr{B}}(\sigma_r,\sigma_s). \quad (8.10)$$

(iii) The s-dimensional edge flow $\mathbf{x}^{(\sigma_s;\mathscr{E})}(t)$ in an edge $\mathscr{E}_{\sigma_s}^{(s)}$ to the edge $\mathscr{E}_{\sigma_r}^{(r)}$ is a *grazing edge flow* at point $\mathbf{x}_m\in\mathscr{E}_{\sigma_r}^{(r)}$ ($r\neq 0$) with the $(2\mathbf{k}_{\sigma_s}+1;\mathscr{E})$-singularity if

$$\left.\begin{array}{l}\hbar_{i_\alpha} G_{\partial\Omega_{\alpha i_\alpha}}^{(s_{i_\alpha},\sigma_s;\mathscr{E})}(\mathbf{x}_m,t_{m\pm},\boldsymbol{\pi}_{\sigma_s},\boldsymbol{\lambda}_{i_\alpha})=0\\ \text{for } s_{i_\alpha}=0,1,2,\cdots,2k_{i_\alpha};\\ \hbar_{i_\alpha}\mathbf{n}_{\partial\Omega_{\alpha i_\alpha}}^{\mathrm{T}}(\mathbf{x}_{m-\varepsilon}^{(i_\alpha;\mathscr{B})})\cdot[\mathbf{x}_{m-\varepsilon}^{(i_\alpha;\mathscr{B})}-\mathbf{x}_{m-\varepsilon}^{(\sigma_s;\mathscr{E})}]<0\\ \hbar_{i_\alpha}\mathbf{n}_{\partial\Omega_{\alpha i_\alpha}}^{\mathrm{T}}(\mathbf{x}_{m+\varepsilon}^{(i_\alpha;\mathscr{B})})\cdot[\mathbf{x}_{m+\varepsilon}^{(\sigma_s;\mathscr{E})}-\mathbf{x}_{m+\varepsilon}^{(i_\alpha;\mathscr{B})}]>0\end{array}\right\}\text{for all } i_\alpha\in\bar{\mathscr{N}}_{\alpha\mathscr{B}}(\sigma_r,\sigma_s). \quad (8.11)$$

From the previous definition, the corresponding sufficient and necessary conditions for the coming, leaving and grazing edge flows to a specific edge will be developed as in Chapter 3 for each boundary of the specific edge.

Theorem 8.2. *For a discontinuous dynamical system in Eqs.* (6.4)-(6.8), *there is a point* $\mathbf{x}^{(\sigma_r;\mathscr{E})}(t_m)\equiv\mathbf{x}_m\in\mathscr{E}_{\sigma_r}^{(r)}$ ($\sigma_r\in\mathscr{N}_{\mathscr{E}}^{r}=\{1,2,\cdots,N_r\}$ *and* $r\in\{0,1,2,\cdots,n-2\}$) *at time* t_m *with the related boundaries* $\partial\Omega_{\alpha i_\alpha}(\sigma_r)$ *and related domains* $\Omega_\alpha(\sigma_r)$ ($\alpha\in\mathscr{N}_{\mathscr{D}}(\sigma_r)\subset\mathscr{N}_{\mathscr{D}}=\{1,2,\cdots,N_d\}$ *and* $i_\alpha\in\mathscr{N}_{\alpha\mathscr{B}}(\sigma_r)\subset\mathscr{N}_{\mathscr{B}}=\{1,2,\cdots,N_b\}$), *where the boundary index subset of* $\mathscr{N}_{\alpha\mathscr{B}}(\sigma_r)$ *has* n_{σ_r}*-elements* ($n_{\sigma_r}\geq(n-r)$). *The edge of* $\mathscr{E}_{\sigma_r}^{(r)}$ *is the intersection of edges (including domains and boundaries)* $\mathscr{E}_{\sigma_s}^{(s)}(\sigma_r)$ ($\sigma_s\in\mathscr{N}_{\mathscr{E}}^{s}(\sigma_r)\subset\mathscr{N}_{\mathscr{E}}^{s}=\{1,2,\cdots,N_s\}$ *and* $s\in\{r+1,r+2,\cdots,n-1,n\}$). *Suppose there is an* s*-dimensional edge* $\mathscr{E}_{\sigma_s}^{(s)}(\sigma_r)$ *with a boundary index set* $\mathscr{N}_{\alpha\mathscr{B}}(\sigma_r,\sigma_s)$ *relative to domain* $\Omega_\alpha(\sigma_r)$. *The complementary boundary index set is* $\bar{\mathscr{N}}_{\alpha\mathscr{B}}(\sigma_r,\sigma_s)=\mathscr{N}_{\alpha\mathscr{B}}(\sigma_r)/\mathscr{N}_{\alpha\mathscr{B}}(\sigma_r,\sigma_s)$. *For an arbitrarily small* $\varepsilon>0$, *there are two time intervals* $[t_{m-\varepsilon},t_m)$ *and* $(t_m,t_{m+\varepsilon}]$. *An edge flow* $\mathbf{x}^{(\sigma_s;\mathscr{E})}(t)$ *is* $C_{[t_{m-\varepsilon},t_m)}^{r_{\sigma_s}}$ *and/or* $C_{(t_m,t_{m+\varepsilon}]}^{r_{\sigma_s}}$*-continuous for time* t *and* $\|d^{r_{\sigma_s}+1}\mathbf{x}^{(\sigma_s;\mathscr{E})}/dt^{r_{\sigma_s}+1}\|<\infty$ (r_{σ_s} $\geq\max_{i_\alpha\in\bar{\mathscr{N}}_{\alpha\mathscr{B}}}\{m_{i_\alpha}+1\}$).

(i) *The* s*-dimensional edge flow* $\mathbf{x}^{(\sigma_s;\mathscr{E})}(t)$ *in an edge* $\mathscr{E}_{\sigma_s}^{(s)}$ *to the edge* $\mathscr{E}_{\sigma_r}^{(r)}$ *is a coming flow at point* $\mathbf{x}_m\in\mathscr{E}_{\sigma_r}^{(r)}$ *with the* $(\mathbf{m}_{\sigma_s};\mathscr{E})$*-singularity if*

$$\left.\begin{array}{l}\hbar_{i_\alpha} G^{(s_{i_\alpha},\sigma_s;\mathscr{E})}_{\partial\Omega_{\alpha i_\alpha}}(\mathbf{x}_m,t_{m-},\boldsymbol{\pi}_{\sigma_s},\boldsymbol{\lambda}_{i_\alpha})=0\\ \textit{for } s_{i_\alpha}=0,1,2,\cdots,m_{i_\alpha}-1;\\ (-1)^{m_{i_\alpha}}\hbar_{i_\alpha} G^{(m_{i_\alpha},\sigma_s;\mathscr{E})}_{\partial\Omega_{\alpha i_\alpha}}(\mathbf{x}_m,t_{m-},\boldsymbol{\pi}_{\sigma_s},\boldsymbol{\lambda}_{i_\alpha})<0\end{array}\right\}\textit{for all } i_\alpha\in\bar{\mathscr{N}}_{\alpha\mathscr{B}}(\sigma_r,\sigma_s). \quad (8.12)$$

(ii) *The* s *-dimensional edge flow* $\mathbf{x}^{(\sigma_s;\mathscr{E})}(t)$ *in an edge* $\mathscr{E}^{(s)}_{\sigma_s}$ *to the edge* $\mathscr{E}^{(r)}_{\sigma_r}$ *is a leaving flow at point* $\mathbf{x}_m\in\mathscr{E}^{(r)}_{\sigma_r}$ *with the* $(\mathbf{m}_{\sigma_s};\mathscr{E})$ *-singularity if*

$$\left.\begin{array}{l}\hbar_{i_\alpha} G^{(s_{i_\alpha},\sigma_s;\mathscr{E})}_{\partial\Omega_{\alpha i_\alpha}}(\mathbf{x}_m,t_{m+},\boldsymbol{\pi}_{\sigma_s},\boldsymbol{\lambda}_{i_\alpha})=0\\ \textit{for } s_{i_\alpha}=0,1,2,\cdots,m_{i_\alpha}-1;\\ \hbar_{i_\alpha} G^{(m_{i_\alpha},\sigma_s;\mathscr{E})}_{\partial\Omega_{\alpha i_\alpha}}(\mathbf{x}_m,t_{m+},\boldsymbol{\pi}_{\sigma_s},\boldsymbol{\lambda}_{i_\alpha})>0\end{array}\right\}\textit{for all } i_\alpha\in\bar{\mathscr{N}}_{\alpha\mathscr{B}}(\sigma_r,\sigma_s). \quad (8.13)$$

(iii) *The* s *-dimensional edge flow* $\mathbf{x}^{(\sigma_s;\mathscr{E})}(t)$ *in an edge* $\mathscr{E}^{(s)}_{\sigma_s}$ *to the edge* $\mathscr{E}^{(r)}_{\sigma_r}$ *is a grazing flow at point* $\mathbf{x}_m\in\mathscr{E}^{(r)}_{\sigma_r}$ ($r\neq 0$) *with the* $(2\mathbf{k}_{\sigma_s}+1;\mathscr{E})$ *-singularity if*

$$\left.\begin{array}{l}\hbar_{i_\alpha} G^{(s_{i_\alpha},\sigma_s;\mathscr{E})}_{\partial\Omega_{\alpha i_\alpha}}(\mathbf{x}_m,t_{m\pm},\boldsymbol{\pi}_{\sigma_s},\boldsymbol{\lambda}_{i_\alpha})=0\\ \textit{for } s_{i_\alpha}=0,1,2,\cdots,2k_{i_\alpha};\\ \hbar_{i_\alpha} G^{(2k_{i_\alpha}+1,\sigma_s;\mathscr{E})}_{\partial\Omega_{\alpha i_\alpha}}(\mathbf{x}_m,t_{m\pm},\boldsymbol{\pi}_{\sigma_s},\boldsymbol{\lambda}_{i_\alpha})>0\end{array}\right\}\textit{for all } i_\alpha\in\bar{\mathscr{N}}_{\alpha\mathscr{B}}(\sigma_r,\sigma_s). \quad (8.14)$$

Proof. As in Theorem 3.2, this theorem can be proved similarly as the edge flow is treated as a special domain flow (also see Luo, 2008, 2009). ■

To illustrate the above concepts of the coming, leaving and grazing edge flows, the s -dimensional edge flows to an $(s-3)$ -dimensional edge are used for illustration. The coming and leaving edge flows are presented through the s -dimensional edge flow to the $(s-3)$ -dimensional edge in Fig. 8.1(a) and (b), respectively. The black thick curve represents the $(s-2)$ -dimensional edge. The s -dimensional edge flow is a curve with arrow, and the circular symbol is the $(s-3)$ -dimensional edge. The coming and leaving edge flows with grazing singularity to the $(s-3)$ -dimensional edge are presented through the s -dimensional edge flow gazing to the $(s-2)$ -dimensional edge in Fig. 8.2(a) and (b), respectively. The grazing flow to the $(s-3)$ -dimensional edge is very difficult to be illustrated intuitively. However, such a grazing edge flow is similar to grazing to the $(s-2)$ -dimensional edge as shown in Fig. 8.3.

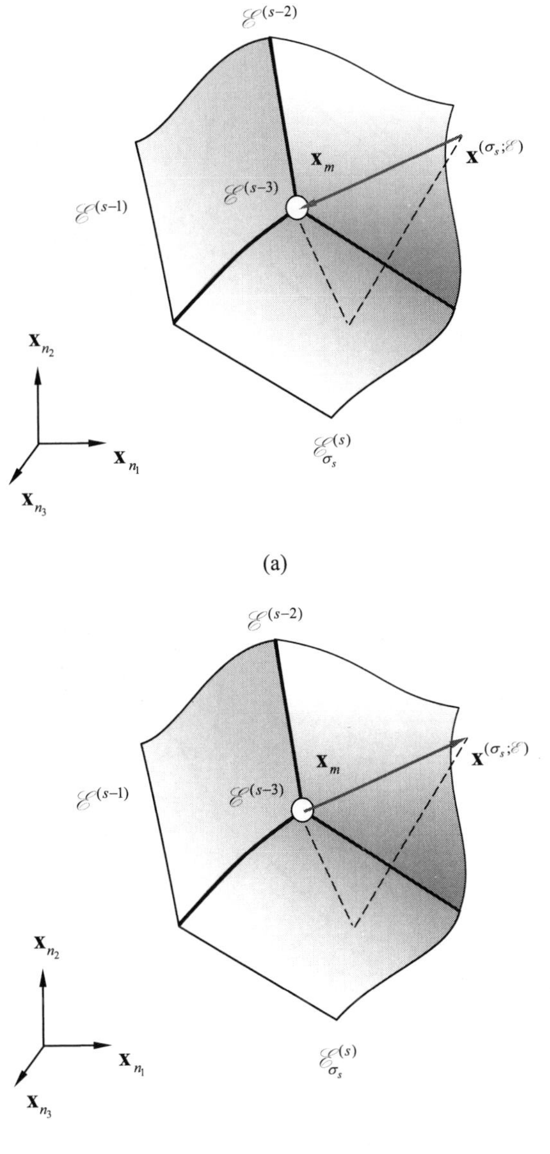

Fig. 8.1 An s-dimensional edge flow to an $(s-3)$-dimensional edge. **(a)** coming edge flow and **(b)** leaving edge flow. The black thick curve represents the $(s-2)$-dimensional edge. The s-dimensional edge flow is a curve with arrow, and the circular symbol is the $(s-3)$-dimensional edge. ($n_1+n_2+n_3=s$)

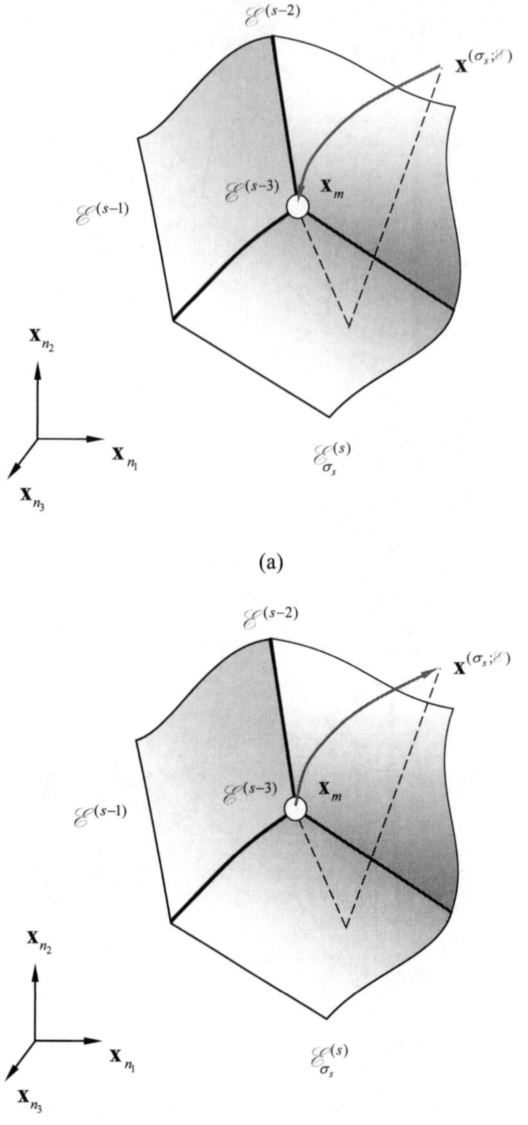

(b)

Fig. 8.2 An s-dimensional edge flow with grazing singularity to an $(s-3)$-dimensional edge. **(a)** coming edge flow and **(b)** leaving edge flow. The black thick curve represents the $(s-2)$-dimensional edge. The s-dimensional edge flow is a curve with arrow, and the circular symbol is the $(s-3)$-dimensional edge. ($n_1 + n_2 + n_3 = s$)

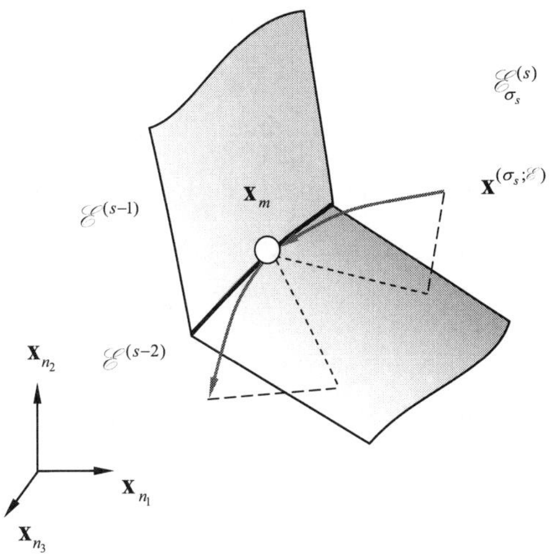

Fig. 8.3 An s-dimensional grazing edge flow to an $(s-2)$-dimensional edge. The black thick curve represents the $(s-2)$-dimensional edge, and the s-dimensional grazing edge flow is a curve with arrow. ($n_1+n_2+n_3=s$)

8.2. Edge flow switchability

As in Chapter 7, the switchability of an edge flow $\mathbf{x}^{(\sigma_s;\mathscr{E})}(t)$ to an r-dimensional edge $\mathscr{E}_{\sigma_r}^{(r)}$ can be discussed. The edge flow $\mathbf{x}^{(\sigma_s;\mathscr{E})}(t)$ *can be switched* to domain flow, boundary and any edge flow $\mathbf{x}^{(\sigma_p;\mathscr{E})}(t)$ ($p\in\{r,r+1,\cdots,n-1,n\}$). A similar presentation will be given for convenience.

Definition 8.3. For a discontinuous dynamical system in Eqs. (6.4)-(6.8), there is a point $\mathbf{x}^{(\sigma_r;\mathscr{E})}(t_m)\equiv\mathbf{x}_m\in\mathscr{E}_{\sigma_r}^{(r)}$ $\sigma_r\in\mathscr{N}_{\mathscr{E}}^{r}=\{1,2,\cdots,N_r\}$ and $r\in\{0,1,2,\cdots,n-2\}$) at time t_m with the related boundaries $\partial\Omega_{\alpha i_\alpha}(\sigma_r)$ and related domains $\Omega_\alpha(\sigma_r)$ ($\alpha\in\mathscr{N}_{\mathscr{D}}(\sigma_r)\subset\mathscr{N}_{\mathscr{D}}=\{1,2,\cdots,N_d\}$ and $i_\alpha\in\mathscr{N}_{\alpha\mathscr{B}}(\sigma_r)\subset\mathscr{N}_{\mathscr{B}}=\{1,2,\cdots,N_b\}$), where the boundary index subset of $\mathscr{N}_{\alpha\mathscr{B}}(\sigma_r)$ has n_{σ_r}-elements ($n_{\sigma_r}\geq(n-r)$). The edge of $\mathscr{E}_{\sigma_r}^{(r)}$ is the intersection of edges (including domains and boundaries) $\mathscr{E}_{\sigma_s}^{(s)}(\sigma_r)$ ($\sigma_s\in\mathscr{N}_{\mathscr{E}}^{s}(\sigma_r)\subset\mathscr{N}_{\mathscr{E}}^{s}=\{1,2,\cdots,N_s\}$ and $s\in\{r+1,r+2,\cdots,n-1,n\}$). Suppose there is an s-dimensional edge $\mathscr{E}_{\sigma_s}^{(s)}(\sigma_r)$ with a boundary index set

$\mathscr{N}_{\alpha\mathscr{B}}(\sigma_r,\sigma_s)$ relative to domain $\Omega_\alpha(\sigma_r)$. The complementary boundary index set is $\bar{\mathscr{N}}_{\alpha\mathscr{B}}(\sigma_r,\sigma_s)=\mathscr{N}_{\alpha\mathscr{B}}(\sigma_r)/\mathscr{N}_{\alpha\mathscr{B}}(\sigma_r,\sigma_s)$. For an arbitrarily small $\varepsilon>0$, there are two time intervals $[t_{m-\varepsilon},t_m)$ and $(t_m,t_{m+\varepsilon}]$. An edge flow $\mathbf{x}^{(\sigma_s;\mathscr{E})}(t)$ is $C^{r_{\sigma_s}}_{[t_{m-\varepsilon},t_m)}$ and/or $C^{r_{\sigma_s}}_{(t_m,t_{m+\varepsilon}]}$-continuous for time t and $\|d^{r_{\sigma_s}+1}\mathbf{x}^{(\sigma_s;\mathscr{E})}/dt^{r_{\sigma_s}+1}\|<\infty$ ($r_{\sigma_s}\geq 1$). An edge flow of $\mathbf{x}^{(\sigma_p;\mathscr{E})}(t)$ ($p\in\{r,r+1,\cdots,n-1,n\}$) is $C^{r_{\sigma_p}}_{[t_{m-\varepsilon},t_m)}$ and/or $C^{r_{\sigma_p}}_{(t_m,t_{m+\varepsilon}]}$-continuous for time t and $\|d^{r_{\sigma_p}+1}\mathbf{x}^{(\sigma_s;\mathscr{E})}/dt^{r_{\sigma_p}+1}\|<\infty$ ($r_{\sigma_p}\geq 1$). Suppose the switching rule $\mathbf{x}^{(\sigma_s;\mathscr{E})}(t_{m\pm})=\mathbf{x}_m=\mathbf{x}^{(\sigma_p;\mathscr{E})}(t_{m\pm})$ exist.

(i) The s-dimensional coming edge flow $\mathbf{x}^{(\sigma_s;\mathscr{E})}(t)$ in an edge $\mathscr{E}^{(s)}_{\sigma_s}$ and the p-dimensional leaving edge flow $\mathbf{x}^{(\sigma_p;\mathscr{E})}(t)$ in an edge $\mathscr{E}^{(p)}_{\sigma_p}$ at point $\mathbf{x}_m\in\mathscr{E}^{(r)}_{\sigma_r}$ are $(\mathscr{E}^{(s)}_{\sigma_s}:\mathscr{E}^{(p)}_{\sigma_p})$ *switchable* via the edge $\mathscr{E}^{(r)}_{\sigma_r}$ if

$$\begin{aligned}&\hbar_{i_\alpha}\mathbf{n}^{\mathrm{T}}_{\partial\Omega_{\alpha i_\alpha}}(\mathbf{x}^{(i_\alpha;\mathscr{B})}_{m-\varepsilon})\cdot[\mathbf{x}^{(i_\alpha;\mathscr{B})}_{m-\varepsilon}-\mathbf{x}^{(\sigma_s;\mathscr{E})}_{m-\varepsilon}]<0\ \text{ for all } i_\alpha\in\bar{\mathscr{N}}_{\alpha\mathscr{B}}(\sigma_r,\sigma_s);\\&\hbar_{j_\beta}\mathbf{n}^{\mathrm{T}}_{\partial\Omega_{\beta j_\beta}}(\mathbf{x}^{(j_\beta;\mathscr{B})}_{m+\varepsilon})\cdot[\mathbf{x}^{(\sigma_p;\mathscr{E})}_{m+\varepsilon}-\mathbf{x}^{(j_\beta;\mathscr{B})}_{m+\varepsilon}]>0\ \text{ for all } j_\beta\in\bar{\mathscr{N}}_{\beta\mathscr{B}}(\sigma_r,\sigma_p).\end{aligned}\tag{8.15}$$

(ii) The s-dimensional coming edge flow $\mathbf{x}^{(\sigma_s;\mathscr{E})}(t)$ in an edge $\mathscr{E}^{(s)}_{\sigma_s}$ and the p-dimensional coming edge flow $\mathbf{x}^{(\sigma_p;\mathscr{E})}(t)$ in an edge $\mathscr{E}^{(p)}_{\sigma_p}$ at point $\mathbf{x}_m\in\mathscr{E}^{(r)}_{\sigma_r}$ are $(\mathscr{E}^{(s)}_{\sigma_s}:\mathscr{E}^{(p)}_{\sigma_p})$ *non-switchable of the first kind* to the edge $\mathscr{E}^{(r)}_{\sigma_r}$ if

$$\begin{aligned}&\hbar_{i_\alpha}\mathbf{n}^{\mathrm{T}}_{\partial\Omega_{\alpha i_\alpha}}(\mathbf{x}^{(i_\alpha;\mathscr{B})}_{m-\varepsilon})\cdot[\mathbf{x}^{(i_\alpha;\mathscr{B})}_{m-\varepsilon}-\mathbf{x}^{(\sigma_s;\mathscr{E})}_{m-\varepsilon}]<0\ \text{ for all } i_\alpha\in\bar{\mathscr{N}}_{\alpha\mathscr{B}}(\sigma_r,\sigma_s);\\&\hbar_{j_\beta}\mathbf{n}^{\mathrm{T}}_{\partial\Omega_{\beta j_\beta}}(\mathbf{x}^{(j_\beta;\mathscr{B})}_{m-\varepsilon})\cdot[\mathbf{x}^{(j_\beta;\mathscr{B})}_{m-\varepsilon}-\mathbf{x}^{(\sigma_p;\mathscr{E})}_{m-\varepsilon}]<0\ \text{ for all } j_\beta\in\bar{\mathscr{N}}_{\beta\mathscr{B}}(\sigma_r,\sigma_p).\end{aligned}\tag{8.16}$$

(iii) The s-dimensional leaving edge flow $\mathbf{x}^{(\sigma_s;\mathscr{E})}(t)$ in an edge $\mathscr{E}^{(s)}_{\sigma_s}$ and the p-dimensional leaving edge flow $\mathbf{x}^{(\sigma_p;\mathscr{E})}(t)$ in an edge $\mathscr{E}^{(p)}_{\sigma_p}$ at point $\mathbf{x}_m\in\mathscr{E}^{(r)}_{\sigma_r}$ are $(\mathscr{E}^{(s)}_{\sigma_s}:\mathscr{E}^{(p)}_{\sigma_p})$ *non-switchable of the second kind* to the edge $\mathscr{E}^{(r)}_{\sigma_r}$ if

$$\begin{aligned}&\hbar_{i_\alpha}\mathbf{n}^{\mathrm{T}}_{\partial\Omega_{\alpha i_\alpha}}(\mathbf{x}^{(i_\alpha;\mathscr{B})}_{m+\varepsilon})\cdot[\mathbf{x}^{(\sigma_s;\mathscr{E})}_{m+\varepsilon}-\mathbf{x}^{(i_\alpha;\mathscr{B})}_{m+\varepsilon}]>0\ \text{ for all } i_\alpha\in\bar{\mathscr{N}}_{\alpha\mathscr{B}}(\sigma_r,\sigma_s);\\&\hbar_{j_\beta}\mathbf{n}^{\mathrm{T}}_{\partial\Omega_{\beta j_\beta}}(\mathbf{x}^{(j_\beta;\mathscr{B})}_{m+\varepsilon})\cdot[\mathbf{x}^{(\sigma_p;\mathscr{E})}_{m+\varepsilon}-\mathbf{x}^{(j_\beta;\mathscr{B})}_{m+\varepsilon}]>0\ \text{ for all } j_\beta\in\bar{\mathscr{N}}_{\beta\mathscr{B}}(\sigma_r,\sigma_p).\end{aligned}\tag{8.17}$$

From the foregoing definition, the four basic cases for two edge flows with different dimensions are presented. If $s=n$ and $p=n-1$, the above gives the switchability between domain and boundary flows via the a specific edge $\mathscr{E}^{(r)}_{\sigma_r}$ with $r\leq\min(s,p)$. If $s=p$, two edge flows possesses the same dimension. For $r=0$, the s-dimensional and p-dimensional edge flows will be switched via the vertex points. From the above definition, the corresponding conditions for such switchability are presented through the following theorem.

Theorem 8.3. *For a discontinuous dynamical system in Eqs.* (6.4)-(6.8), *there is a point* $\mathbf{x}^{(\sigma_r;\mathscr{E})}(t_m) \equiv \mathbf{x}_m \in \mathscr{E}_{\sigma_r}^{(r)}$ ($\sigma_r \in \mathscr{N}_{\mathscr{E}}^r = \{1,2,\cdots,N_r\}$ *and* $r \in \{0,1,2,\cdots,n-2\}$) *at time* t_m *with the related boundaries* $\partial\Omega_{\alpha i_\alpha}(\sigma_r)$ *and related domains* $\Omega_\alpha(\sigma_r)$ ($\alpha \in \mathscr{N}_{\mathscr{D}}(\sigma_r) \subset \mathscr{N}_{\mathscr{D}} = \{1,2,\cdots,N_d\}$ *and* $i_\alpha \in \mathscr{N}_{\alpha\mathscr{B}}(\sigma_r) \subset \mathscr{N}_{\mathscr{B}} = \{1,2,\cdots,N_b\}$), *where the boundary index subset of* $\mathscr{N}_{\alpha\mathscr{B}}(\sigma_r)$ *has* n_{σ_r}*-elements* ($n_{\sigma_r} \geq (n-r)$). *The edge of* $\mathscr{E}_{\sigma_r}^{(r)}$ *is the intersection of edges (including domains and boundaries)* $\mathscr{E}_{\sigma_s}^{(s)}(\sigma_r)$ ($\sigma_s \in \mathscr{N}_{\mathscr{E}}^s(\sigma_r) \subset \mathscr{N}_{\mathscr{E}}^s = \{1,2,\cdots,N_s\}$ *and* $s \in \{r+1,r+2,\cdots,n-1,n\}$). *Suppose there is an* s*-dimensional edge* $\mathscr{E}_{\sigma_s}^{(s)}(\sigma_r)$ *with a boundary index set* $\mathscr{N}_{\alpha\mathscr{B}}(\sigma_r,\sigma_s)$ *relative to domain* $\Omega_\alpha(\sigma_r)$. *The complementary boundary index set is* $\bar{\mathscr{N}}_{\alpha\mathscr{B}}(\sigma_r,\sigma_s) = \mathscr{N}_{\alpha\mathscr{B}}(\sigma_r) / \mathscr{N}_{\alpha\mathscr{B}}(\sigma_r,\sigma_s)$. *For an arbitrarily small* $\varepsilon > 0$, *there are two time intervals* $[t_{m-\varepsilon}, t_m)$ *and* $(t_m, t_{m+\varepsilon}]$. *An edge flow* $\mathbf{x}^{(\sigma_s;\mathscr{E})}(t)$ *is* $C_{[t_{m-\varepsilon},t_m)}^{r_{\sigma_s}}$ *and/or* $C_{(t_m,t_{m+\varepsilon}]}^{r_{\sigma_s}}$*-continuous for time* t *and* $\| d^{r_{\sigma_s}+1}\mathbf{x}^{(\sigma_s;\mathscr{E})} / dt^{r_{\sigma_s}+1} \| < \infty$ ($r_{\sigma_s} \geq 2$). *An edge flow* $\mathbf{x}^{(\sigma_p;\mathscr{E})}(t)$ ($p \in \{r,r+1,\cdots,n-1,n\}$) *is* $C_{[t_{m-\varepsilon},t_m)}^{r_{\sigma_p}}$ *and/or* $C_{(t_m,t_{m+\varepsilon}]}^{r_{\sigma_p}}$*-continuous for time* t *and* $\| d^{r_{\sigma_p}+1}\mathbf{x}^{(\sigma_s;\mathscr{E})} / dt^{r_{\sigma_p}+1} \| < \infty$ ($r_{\sigma_p} \geq 2$). *Suppose the switching rule* $\mathbf{x}^{(\sigma_s;\mathscr{E})}(t_{m\pm}) = \mathbf{x}_m = \mathbf{x}^{(\sigma_p;\mathscr{E})}(t_{m\pm})$ *exist.*

(i) *The* s*-dimensional coming edge flow* $\mathbf{x}^{(\sigma_s;\mathscr{E})}(t)$ *in an edge* $\mathscr{E}_{\sigma_s}^{(s)}$ *and the* p*-dimensional leaving edge flow* $\mathbf{x}^{(\sigma_p;\mathscr{E})}(t)$ *in an edge* $\mathscr{E}_{\sigma_p}^{(p)}$ *at point* $\mathbf{x}_m \in \mathscr{E}_{\sigma_r}^{(r)}$ *are* ($\mathscr{E}_{\sigma_s}^{(s)} : \mathscr{E}_{\sigma_p}^{(p)}$) *switchable from via edge* $\mathscr{E}_{\sigma_r}^{(r)}$ *if and only if*

$$
\begin{aligned}
&\hbar_{i_\alpha} G_{\partial\Omega_{\alpha i_\alpha}}^{(\sigma_s;\mathscr{E})}(\mathbf{x}_m, t_{m-}, \boldsymbol{\pi}_{\sigma_s}, \boldsymbol{\lambda}_{i_\alpha}) < 0 \;\; \text{for all } i_\alpha \in \bar{\mathscr{N}}_{\alpha\mathscr{B}}(\sigma_r,\sigma_s);\\
&\hbar_{j_\beta} G_{\partial\Omega_{\beta j_\beta}}^{(\sigma_p;\mathscr{E})}(\mathbf{x}_m, t_{m+}, \boldsymbol{\pi}_{\sigma_p}, \boldsymbol{\lambda}_{i_\beta}) > 0 \;\; \text{for all } j_\beta \in \bar{\mathscr{N}}_{\beta\mathscr{B}}(\sigma_r,\sigma_p).
\end{aligned}
\tag{8.18}
$$

(ii) *The* s*-dimensional coming edge flow* $\mathbf{x}^{(\sigma_s;\mathscr{E})}(t)$ *in an edge* $\mathscr{E}_{\sigma_s}^{(s)}$ *and the* p*-dimensional coming edge flow* $\mathbf{x}^{(\sigma_p;\mathscr{E})}(t)$ *in an edge* $\mathscr{E}_{\sigma_p}^{(p)}$ *at point* $\mathbf{x}_m \in \mathscr{E}_{\sigma_r}^{(r)}$ *are* ($\mathscr{E}_{\sigma_s}^{(s)} : \mathscr{E}_{\sigma_p}^{(p)}$) *non-switchable of the first kind to edge* $\mathscr{E}_{\sigma_r}^{(r)}$ *if and only if*

$$
\begin{aligned}
&\hbar_{i_\alpha} G_{\partial\Omega_{\alpha i_\alpha}}^{(\sigma_s;\mathscr{E})}(\mathbf{x}_m, t_{m-}, \boldsymbol{\pi}_{\sigma_s}, \boldsymbol{\lambda}_{i_\alpha}) < 0 \;\; \text{for all } i_\alpha \in \bar{\mathscr{N}}_{\alpha\mathscr{B}}(\sigma_r,\sigma_s);\\
&\hbar_{j_\beta} G_{\partial\Omega_{\beta j_\beta}}^{(\sigma_p;\mathscr{E})}(\mathbf{x}_m, t_{m-}, \boldsymbol{\pi}_{\sigma_p}, \boldsymbol{\lambda}_{j_\beta}) < 0 \;\; \text{for all } j_\beta \in \bar{\mathscr{N}}_{\beta\mathscr{B}}(\sigma_r,\sigma_p).
\end{aligned}
\tag{8.19}
$$

(iii) *The* s*-dimensional leaving edge flow* $\mathbf{x}^{(\sigma_s;\mathscr{E})}(t)$ *in an edge* $\mathscr{E}_{\sigma_s}^{(s)}$ *and the* p*-*

dimensional leaving edge flow $\mathbf{x}^{(\sigma_p;\mathscr{E})}(t)$ *in an edge* $\mathscr{E}_{\sigma_p}^{(p)}$ *at point* $\mathbf{x}_m \in \mathscr{E}_{\sigma_r}^{(r)}$ *are* $(\mathscr{E}_{\sigma_s}^{(s)} : \mathscr{E}_{\sigma_p}^{(p)})$ *non-switchable of the second kind to edge* $\mathscr{E}_{\sigma_r}^{(r)}$ *if and only if*

$$\begin{aligned} &\hbar_{i_\alpha} G_{\partial\Omega_{\alpha i_\alpha}}^{(\sigma_s;\mathscr{E})}(\mathbf{x}_m, t_{m+}, \boldsymbol{\pi}_{\sigma_s}, \boldsymbol{\lambda}_{i_\alpha}) > 0 \text{ for all } i_\alpha \in \bar{\mathscr{N}}_{\alpha\mathscr{B}}(\sigma_r, \sigma_s);\\ &\hbar_{j_\beta} G_{\partial\Omega_{\beta j_\beta}}^{(\sigma_p;\mathscr{E})}(\mathbf{x}_m, t_{m+}, \boldsymbol{\pi}_{\sigma_p}, \boldsymbol{\lambda}_{j_\beta}) > 0 \text{ for all } j_\beta \in \bar{\mathscr{N}}_{\beta\mathscr{B}}(\sigma_r, \sigma_p). \end{aligned} \tag{8.20}$$

Proof. As in Theorem 3.1, this theorem can be proved similarly as the edge flow is treated as a special domain flow (also see Luo, 2008, 2009). ■

With the higher order singularity, the switchability of the s-dimensional edge flow and the p-dimensional edge flow will be described.

Definition 8.4. For a discontinuous dynamical system in Eqs. (6.4)-(6.8), there is a point $\mathbf{x}^{(\sigma_r;\mathscr{E})}(t_m) \equiv \mathbf{x}_m \in \mathscr{E}_{\sigma_r}^{(r)}$ $\sigma_r \in \mathscr{N}_{\mathscr{E}}^{r} = \{1, 2, \cdots, N_r\}$ and $r \in \{0, 1, 2, \cdots, n-2\}$) at time t_m with the related boundaries $\partial\Omega_{\alpha i_\alpha}(\sigma_r)$ and related domains $\Omega_\alpha(\sigma_r)$ ($\alpha \in \mathscr{N}_{\mathscr{D}}(\sigma_r) \subset \mathscr{N}_{\mathscr{D}} = \{1, 2, \cdots, N_d\}$ and $i_\alpha \in \mathscr{N}_{\alpha\mathscr{B}}(\sigma_r) \subset \mathscr{N}_{\mathscr{B}} = \{1, 2, \cdots, N_b\}$), where the boundary index subset of $\mathscr{N}_{\alpha\mathscr{B}}(\sigma_r)$ has n_{σ_r}-elements ($n_{\sigma_r} \geq (n-r)$). The edge of $\mathscr{E}_{\sigma_r}^{(r)}$ is the intersection of edges (including domains and boundaries) $\mathscr{E}_{\sigma_s}^{(s)}(\sigma_r)$ ($\sigma_s \in \mathscr{N}_{\mathscr{E}}^{s}(\sigma_r) \subset \mathscr{N}_{\mathscr{E}}^{s} = \{1, 2, \cdots, N_s\}$ and $s \in \{r+1, r+2, \cdots, n-1, n\}$). Suppose there is an s-dimensional edge $\mathscr{E}_{\sigma_s}^{(s)}(\sigma_r)$ with a boundary index set $\mathscr{N}_{\alpha\mathscr{B}}(\sigma_r, \sigma_s)$ relative to domain $\Omega_\alpha(\sigma_r)$. The complementary boundary number set is $\bar{\mathscr{N}}_{\alpha\mathscr{B}}(\sigma_r, \sigma_s) = \mathscr{N}_{\alpha\mathscr{B}}(\sigma_r) / \mathscr{N}_{\alpha\mathscr{B}}(\sigma_r, \sigma_s)$. For an arbitrarily small $\varepsilon > 0$, there are two time intervals $[t_{m-\varepsilon}, t_m)$ and $(t_m, t_{m+\varepsilon}]$. An edge flow $\mathbf{x}^{(\sigma_s;\mathscr{E})}(t)$ is $C_{[t_{m-\varepsilon}, t_m)}^{r_{\sigma_s}}$ and/or $C_{(t_m, t_{m+\varepsilon}]}^{r_{\sigma_s}}$-continuous for time t *and* $\| d^{r_{\sigma_s}+1}\mathbf{x}^{(\sigma_s;\mathscr{E})} / dt^{r_{\sigma_s}+1} \| < \infty$ ($r_{\sigma_s} \geq \max_{i_\alpha \in \bar{\mathscr{N}}_{\alpha\mathscr{B}}}\{m_{i_\alpha}\}$). An edge flow $\mathbf{x}^{(\sigma_p;\mathscr{E})}(t)$ ($p \in \{r, r+1, \cdots, n-1, n\}$) is $C_{[t_{m-\varepsilon}, t_m)}^{r_{\sigma_p}}$ and/or $C_{(t_m, t_{m+\varepsilon}]}^{r_{\sigma_p}}$-continuous for time t and $\| d^{r_{\sigma_p}+1}\mathbf{x}^{(\sigma_p;\mathscr{E})} / dt^{r_{\sigma_p}+1} \| < \infty$ ($r_{\sigma_p} \geq \max_{i_\beta \in \bar{\mathscr{N}}_{\beta\mathscr{B}}}\{m_{i_\beta}\}$). Suppose $\mathbf{x}^{(\sigma_s;\mathscr{E})}(t_{m\pm}) = \mathbf{x}_m = \mathbf{x}^{(\sigma_p;\mathscr{E})}(t_{m\pm})$ exists.

(i) With the switching rule, the s-dimensional coming edge flow $\mathbf{x}^{(\sigma_s;\mathscr{E})}(t)$ in an edge $\mathscr{E}_{\sigma_s}^{(s)}$ and the p-dimensional leaving edge flow $\mathbf{x}^{(\sigma_p;\mathscr{E})}(t)$ in an edge $\mathscr{E}_{\sigma_p}^{(p)}$ at point $\mathbf{x}_m \in \mathscr{E}_{\sigma_r}^{(r)}$ are *switchable* with the $(\mathbf{m}_{\sigma_s};\mathscr{E} : \mathbf{m}_{\sigma_p};\mathscr{E})$-singularity from $\mathscr{E}_{\sigma_s}^{(s)}$ to $\mathscr{E}_{\sigma_p}^{(p)}$ via the edge $\mathscr{E}_{\sigma_r}^{(r)}$ if

$$\hbar_{i_\alpha} G_{\partial\Omega_{\alpha i_\alpha}}^{(s_{i_\alpha}, \sigma_s;\mathscr{E})}(\mathbf{x}_m, t_{m-}, \boldsymbol{\pi}_{\sigma_s}, \boldsymbol{\lambda}_{i_\alpha}) = 0 \text{ for } s_{i_\alpha} = 0, 1, 2, \cdots, m_{i_\alpha} - 1;$$

$$\hbar_{i_\alpha}\mathbf{n}^{\mathrm{T}}_{\partial\Omega_{\alpha i_\alpha}}(\mathbf{x}^{(i_\alpha;\mathscr{B})}_{m-\varepsilon})\cdot[\mathbf{x}^{(i_\alpha;\mathscr{B})}_{m-\varepsilon}-\mathbf{x}^{(\sigma_s;\mathscr{E})}_{m-\varepsilon}]<0 \text{ for all } i_\alpha\in\bar{\mathscr{N}}_{\alpha\mathscr{B}}(\sigma_r,\sigma_s); \tag{8.21}$$

$$\left.\begin{array}{l}\hbar_{j_\beta}G^{(s_{j_\beta},\sigma_s;\mathscr{E})}_{\partial\Omega_{\beta j_\beta}}(\mathbf{x}_m,t_{m+},\boldsymbol{\pi}_{\sigma_p},\boldsymbol{\lambda}_{j_\beta})=0\\ \text{for } s_{j_\beta}=0,1,2,\cdots,m_{j_\beta}-1,\\ \hbar_{j_\beta}\mathbf{n}^{\mathrm{T}}_{\partial\Omega_{\beta j_\beta}}(\mathbf{x}^{(j_\beta;\mathscr{B})}_{m+\varepsilon})\cdot[\mathbf{x}^{(\sigma_p;\mathscr{E})}_{m+\varepsilon}-\mathbf{x}^{(j_\beta;\mathscr{B})}_{m+\varepsilon}]>0\end{array}\right\}\text{for all } j_\beta\in\bar{\mathscr{N}}_{\beta\mathscr{B}}(\sigma_r,\sigma_p). \tag{8.22}$$

(ii) The s-dimensional coming edge flow $\mathbf{x}^{(\sigma_s;\mathscr{E})}(t)$ in an edge $\mathscr{E}^{(s)}_{\sigma_s}$ and the p-dimensional coming edge flow $\mathbf{x}^{(\sigma_p;\mathscr{E})}(t)$ in an edge $\mathscr{E}^{(p)}_{\sigma_p}$ at point $\mathbf{x}_m\in\mathscr{E}^{(r)}_{\sigma_r}$ are *non-switchable of the first kind* with the $(\mathbf{m}_{\sigma_s};\mathscr{E}:\mathbf{m}_{\sigma_p};\mathscr{E})$-singularity ($\mathbf{m}_{\sigma_s}\neq 2\mathbf{k}_{\sigma_s}+1$ and $\mathbf{m}_{\sigma_p}\neq 2\mathbf{k}_{\sigma_p}+1$) to the edge $\mathscr{E}^{(r)}_{\sigma_r}$ if

$$\left.\begin{array}{l}\hbar_{i_\alpha}G^{(s_{i_\alpha},\sigma_s;\mathscr{E})}_{\partial\Omega_{\alpha i_\alpha}}(\mathbf{x}_m,t_{m-},\boldsymbol{\pi}_{\sigma_s},\boldsymbol{\lambda}_{i_\alpha})=0\\ \text{for } s_{i_\alpha}=0,1,2,\cdots,m_{i_\alpha}-1,\\ \hbar_{i_\alpha}\mathbf{n}^{\mathrm{T}}_{\partial\Omega_{\alpha i_\alpha}}(\mathbf{x}^{(i_\alpha;\mathscr{B})}_{m-\varepsilon})\cdot[\mathbf{x}^{(i_\alpha;\mathscr{B})}_{m-\varepsilon}-\mathbf{x}^{(\sigma_s;\mathscr{E})}_{m-\varepsilon}]<0\end{array}\right\}\text{for all } i_\alpha\in\bar{\mathscr{N}}_{\alpha\mathscr{B}}(\sigma_r,\sigma_s); \tag{8.23}$$

$$\left.\begin{array}{l}\hbar_{j_\beta}G^{(s_{j_\beta},\sigma_s;\mathscr{E})}_{\partial\Omega_{\beta j_\beta}}(\mathbf{x}_m,t_{m-},\boldsymbol{\pi}_{\sigma_p},\boldsymbol{\lambda}_{j_\beta})=0\\ \text{for } s_{j_\beta}=0,1,2,\cdots,m_{j_\beta}-1,\\ \hbar_{j_\beta}\mathbf{n}^{\mathrm{T}}_{\partial\Omega_{\beta j_\beta}}(\mathbf{x}^{(j_\beta;\mathscr{B})}_{m-\varepsilon})\cdot[\mathbf{x}^{(j_\beta;\mathscr{B})}_{m-\varepsilon}-\mathbf{x}^{(\sigma_p;\mathscr{E})}_{m-\varepsilon}]<0\end{array}\right\}\text{for all } j_\beta\in\bar{\mathscr{N}}_{\beta\mathscr{B}}(\sigma_r,\sigma_p). \tag{8.24}$$

(iii) The s-dimensional leaving edge flow $\mathbf{x}^{(\sigma_s;\mathscr{E})}(t)$ in an edge $\mathscr{E}^{(s)}_{\sigma_s}$ and the p-dimensional leaving edge flow $\mathbf{x}^{(\sigma_p;\mathscr{E})}(t)$ in an edge $\mathscr{E}^{(p)}_{\sigma_p}$ at point $\mathbf{x}_m\in\mathscr{E}^{(r)}_{\sigma_r}$ are *non-switchable of the second kind* with the $(\mathbf{m}_{\sigma_s};\mathscr{E}:\mathbf{m}_{\sigma_p};\mathscr{E})$-singularity to the edge $\mathscr{E}^{(r)}_{\sigma_r}$ if

$$\left.\begin{array}{l}\hbar_{i_\alpha}G^{(s_{i_\alpha},\sigma_s;\mathscr{E})}_{\partial\Omega_{\alpha i_\alpha}}(\mathbf{x}_m,t_{m+},\boldsymbol{\pi}_{\sigma_s},\boldsymbol{\lambda}_{i_\alpha})=0\\ \text{for } s_{i_\alpha}=0,1,2,\cdots,m_{i_\alpha}-1,\\ \hbar_{i_\alpha}\mathbf{n}^{\mathrm{T}}_{\partial\Omega_{\alpha i_\alpha}}(\mathbf{x}^{(i_\alpha;\mathscr{B})}_{m+\varepsilon})\cdot[\mathbf{x}^{(\sigma_s;\mathscr{E})}_{m+\varepsilon}-\mathbf{x}^{(i_\alpha;\mathscr{B})}_{m+\varepsilon}]>0\end{array}\right\}\text{for all } i_\alpha\in\bar{\mathscr{N}}_{\alpha\mathscr{B}}(\sigma_r,\sigma_s); \tag{8.25}$$

$$\left.\begin{array}{l}\hbar_{j_\beta}G^{(s_{j_\beta},\sigma_s;\mathscr{E})}_{\partial\Omega_{\beta j_\beta}}(\mathbf{x}_m,t_{m+},\boldsymbol{\pi}_{\sigma_p},\boldsymbol{\lambda}_{j_\beta})=0\\ \text{for } s_{j_\beta}=0,1,2,\cdots,m_{j_\beta}-1,\\ \hbar_{j_\beta}\mathbf{n}^{\mathrm{T}}_{\partial\Omega_{\beta j_\beta}}(\mathbf{x}^{(j_\beta;\mathscr{B})}_{m+\varepsilon})\cdot[\mathbf{x}^{(\sigma_p;\mathscr{E})}_{m+\varepsilon}-\mathbf{x}^{(j_\beta;\mathscr{B})}_{m+\varepsilon}]>0\end{array}\right\}\text{for all } j_\beta\in\bar{\mathscr{N}}_{\beta\mathscr{B}}(\sigma_r,\sigma_p). \tag{8.26}$$

(iv) Without the switching rule, the s-dimensional coming edge flow $\mathbf{x}^{(\sigma_s;\mathscr{E})}(t)$ in an edge $\mathscr{E}_{\sigma_s}^{(s)}$ and the p-dimensional leaving edge flow $\mathbf{x}^{(\sigma_p;\mathscr{E})}(t)$ in an edge $\mathscr{E}_{\sigma_p}^{(p)}$ at point $\mathbf{x}_m \in \mathscr{E}_{\sigma_r}^{(r)}$ are *potentially switchable* with the $(\mathbf{m}_{\sigma_s};\mathscr{E}:\mathbf{m}_{\sigma_p};\mathscr{E})$-singularity from $\mathscr{E}_{\sigma_s}^{(s)}$ to $\mathscr{E}_{\sigma_p}^{(p)}$ via the edge $\mathscr{E}_{\sigma_r}^{(r)}$ if Eqs. (8.21) and (8.22) hold with $\mathbf{m}_{\sigma_s} \neq 2\mathbf{k}_{\sigma_s}+1$.

(v) Without the switching rule, the s-dimensional coming edge flow $\mathbf{x}^{(\sigma_s;\mathscr{E})}(t)$ in an edge $\mathscr{E}_{\sigma_s}^{(s)}$ and the p-dimensional leaving edge flow $\mathbf{x}^{(\sigma_p;\mathscr{E})}(t)$ in an edge $\mathscr{E}_{\sigma_p}^{(p)}$ at point $\mathbf{x}_m \in \mathscr{E}_{\sigma_r}^{(r)}$ are *non-switchable with the* $(\mathbf{m}_{\sigma_s};\mathscr{E}:\mathbf{m}_{\sigma_p};\mathscr{E})$-*singularity* from $\mathscr{E}_{\sigma_s}^{(s)}$ to $\mathscr{E}_{\sigma_p}^{(p)}$ via the edge $\mathscr{E}_{\sigma_r}^{(r)}$ if Eqs. (8.21) and (8.22) hold with $\mathbf{m}_{\sigma_s}=2\mathbf{k}_{\sigma_s}+1$ for at least one of coming flows.

(vi) With the switching rule, the s-dimensional grazing edge flow $\mathbf{x}^{(\sigma_s;\mathscr{E})}(t)$ in an edge $\mathscr{E}_{\sigma_s}^{(s)}$ and the p-dimensional grazing edge flow $\mathbf{x}^{(\sigma_p;\mathscr{E})}(t)$ in an edge $\mathscr{E}_{\sigma_p}^{(p)}$ at point $\mathbf{x}_m \in \mathscr{E}_{\sigma_r}^{(r)}$ are *switchable with the* $((2\mathbf{k}_{\sigma_s}+1);\mathscr{E}:(2\mathbf{k}_{\sigma_p}+1);\mathscr{E})$-*singularity* from $\mathscr{E}_{\sigma_s}^{(s)}$ to $\mathscr{E}_{\sigma_p}^{(p)}$ via the edge $\mathscr{E}_{\sigma_r}^{(r)}$ if

$$\left.\begin{array}{l}\hbar_{i_\alpha} G_{\partial\Omega_{\alpha i_\alpha}}^{(s_{i_\alpha},\sigma_s;\mathscr{E})}(\mathbf{x}_m,t_{m\pm},\boldsymbol{\pi}_{\sigma_s},\boldsymbol{\lambda}_{i_\alpha})=0\\ \text{for } s_{i_\alpha}=0,1,2,\cdots,2k_{i_\alpha},\\ \hbar_{i_\alpha}\mathbf{n}_{\partial\Omega_{\alpha i_\alpha}}^{\mathrm{T}}(\mathbf{x}_{m-\varepsilon}^{(i_\alpha;\mathscr{B})})\cdot[\mathbf{x}_{m-\varepsilon}^{(i_\alpha;\mathscr{B})}-\mathbf{x}_{m-\varepsilon}^{(\sigma_s;\mathscr{E})}]<0\\ \hbar_{i_\alpha}\mathbf{n}_{\partial\Omega_{\alpha i_\alpha}}^{\mathrm{T}}(\mathbf{x}_{m+\varepsilon}^{(i_\alpha;\mathscr{B})})\cdot[\mathbf{x}_{m+\varepsilon}^{(\sigma_s;\mathscr{E})}-\mathbf{x}_{m+\varepsilon}^{(i_\alpha;\mathscr{B})}]>0\end{array}\right\}\text{for all } i_\alpha\in\bar{\mathscr{N}}_{\alpha\mathscr{B}}(\sigma_r,\sigma_s); \tag{8.27}$$

$$\left.\begin{array}{l}\hbar_{j_\beta} G_{\partial\Omega_{\beta j_\beta}}^{(s_{j_\beta},\sigma_s;\mathscr{E})}(\mathbf{x}_m,t_{m\pm},\boldsymbol{\pi}_{\sigma_p},\boldsymbol{\lambda}_{j_\beta})=0\\ \text{for } s_{j_\beta}=0,1,2,\cdots,2k_{j_\beta},\\ \hbar_{j_\beta}\mathbf{n}_{\partial\Omega_{\beta j_\beta}}^{\mathrm{T}}(\mathbf{x}_{m-\varepsilon}^{(j_\beta;\mathscr{B})})\cdot[\mathbf{x}_{m-\varepsilon}^{(j_\beta;\mathscr{B})}-\mathbf{x}_{m-\varepsilon}^{(\sigma_p;\mathscr{E})}]<0\\ \hbar_{j_\beta}\mathbf{n}_{\partial\Omega_{\beta j_\beta}}^{\mathrm{T}}(\mathbf{x}_{m+\varepsilon}^{(j_\beta;\mathscr{B})})\cdot[\mathbf{x}_{m+\varepsilon}^{(\sigma_p;\mathscr{E})}-\mathbf{x}_{m+\varepsilon}^{(j_\beta;\mathscr{B})}]>0\end{array}\right\}\text{for all } j_\beta\in\bar{\mathscr{N}}_{\beta\mathscr{B}}(\sigma_r,\sigma_p). \tag{8.28}$$

(vii) Without the switching rule, the s-dimensional, grazing edge flow $\mathbf{x}^{(\sigma_s;\mathscr{E})}(t)$ in an edge $\mathscr{E}_{\sigma_s}^{(s)}$ and the p-dimensional, grazing edge flow $\mathbf{x}^{(\sigma_p;\mathscr{E})}(t)$ in an edge $\mathscr{E}_{\sigma_p}^{(p)}$ at point $\mathbf{x}_m \in \mathscr{E}_{\sigma_r}^{(r)}$ are a double tangential flow with the $((2\mathbf{k}_{\sigma_s}+1);\mathscr{E}:(2\mathbf{k}_{\sigma_p}+1);\mathscr{E})$-singularity to the edge $\mathscr{E}_{\sigma_r}^{(r)}$ if Eqs. (8.27) and (8.28) hold.

From the foregoing definition, the corresponding conditions are given through the following theorem.

Theorem 8.4. *For a discontinuous dynamical system in Eqs.* (6.4)-(6.8), *there is a point* $\mathbf{x}^{(\sigma_r;\mathscr{E})}(t_m)\equiv\mathbf{x}_m\in\mathscr{E}_{\sigma_r}^{(r)}$ ($\sigma_r\in\mathscr{N}_{\mathscr{E}}^{r}=\{1,2,\cdots,N_r\}$ *and* $r\in\{0,1,2,\cdots,n-2\}$) *at time* t_m *with the related boundaries* $\partial\Omega_{\alpha i_\alpha}(\sigma_r)$ *and related domains* $\Omega_\alpha(\sigma_r)$ ($\alpha\in\mathscr{N}_{\mathscr{D}}(\sigma_r)\subset\mathscr{N}_{\mathscr{D}}=\{1,2,\cdots,N_d\}$ *and* $i_\alpha\in\mathscr{N}_{\alpha\mathscr{B}}(\sigma_r)\subset\mathscr{N}_{\mathscr{B}}=\{1,2,\cdots,N_b\}$), *where the boundary index subset of* $\mathscr{N}_{\alpha\mathscr{B}}(\sigma_r)$ *has* n_{σ_r}*-elements* ($n_{\sigma_r}\geq(n-r)$). *The edge of* $\mathscr{E}_{\sigma_r}^{(r)}$ *is the intersection of edges (including domains and boundaries)* $\mathscr{E}_{\sigma_s}^{(s)}(\sigma_r)$ ($\sigma_s\in\mathscr{N}_{\mathscr{E}}^{s}(\sigma_r)\subset\mathscr{N}_{\mathscr{E}}^{s}=\{1,2,\cdots,N_s\}$ *and* $s\in\{r+1,r+2,\cdots,n-1,n\}$). *Suppose there is an* s*-dimensional edge* $\mathscr{E}_{\sigma_s}^{(s)}(\sigma_r)$ *with a boundary index set* $\mathscr{N}_{\alpha\mathscr{B}}(\sigma_r,\sigma_s)$ *relative to* $\Omega_\alpha(\sigma_r)$. *The complementary boundary index set is* $\bar{\mathscr{N}}_{\alpha\mathscr{B}}(\sigma_r,\sigma_s)=\mathscr{N}_{\alpha\mathscr{B}}(\sigma_r)/\mathscr{N}_{\alpha\mathscr{B}}(\sigma_r,\sigma_s)$. *For an arbitrarily small* $\varepsilon>0$, *there are two time intervals* $[t_{m-\varepsilon},t_m)$ *and* $(t_m,t_{m+\varepsilon}]$. *An edge flow* $\mathbf{x}^{(\sigma_s;\mathscr{E})}(t)$ *is* $C_{[t_{m-\varepsilon},t_m)}^{r_{\sigma_s}}$ *and/or* $C_{(t_m,t_{m+\varepsilon}]}^{r_{\sigma_s}}$*-continuous* ($r_{\sigma_s}\geq\max_{i_\alpha\in\mathscr{N}_{\alpha\mathscr{B}}^{C}}\{m_{i_\alpha}+1\}$) *for time* t *and* $\|d^{r_{\sigma_s}+1}\mathbf{x}^{(\sigma_s;\mathscr{E})}/dt^{r_{\sigma_s}+1}\|<\infty$. *An edge flow* $\mathbf{x}^{(\sigma_p;\mathscr{E})}(t)$ ($p\in\{r,r+1,\cdots,n-1,n\}$) *is* $C_{[t_{m-\varepsilon},t_m)}^{r_{\sigma_p}}$ *and/or* $C_{(t_m,t_{m+\varepsilon}]}^{r_{\sigma_p}}$ *-continuous* ($r_{\sigma_p}\geq\max_{i_\beta\in\mathscr{N}_{\beta\mathscr{B}}^{C}}\{m_{i_\beta}+1\}$) *for time* t *and* $\|d^{r_{\sigma_p}+1}\mathbf{x}^{(\sigma_p;\mathscr{E})}/dt^{r_{\sigma_p}+1}\|<\infty$. *Suppose* $\mathbf{x}^{(\sigma_s;\mathscr{E})}(t_{m\pm})=\mathbf{x}_m=\mathbf{x}^{(\sigma_p;\mathscr{E})}(t_{m\pm})$ *exist.*

(i) *With the switching rule, the* s*-dimensional coming edge flow* $\mathbf{x}^{(\sigma_s;\mathscr{E})}(t)$ *in an edge* $\mathscr{E}_{\sigma_s}^{(s)}$ *and the* p*-dimensional leaving edge flow* $\mathbf{x}^{(\sigma_p;\mathscr{E})}(t)$ *in an edge* $\mathscr{E}_{\sigma_p}^{(p)}$ *at point* $\mathbf{x}_m\in\mathscr{E}_{\sigma_r}^{(r)}$ *are switchable with the* $(\mathbf{m}_{\sigma_s};\mathscr{E}:\mathbf{m}_{\sigma_p};\mathscr{E})$*-singularity from* $\mathscr{E}_{\sigma_s}^{(s)}$ *to* $\mathscr{E}_{\sigma_p}^{(p)}$ *via the edge* $\mathscr{E}_{\sigma_r}^{(r)}$ *if and only if*

$$\left.\begin{array}{l}\hbar_{i_\alpha}G_{\partial\Omega_{\alpha i_\alpha}}^{(s_{i_\alpha},\sigma_s;\mathscr{E})}(\mathbf{x}_m,t_{m-},\boldsymbol{\pi}_{\sigma_s},\boldsymbol{\lambda}_{i_\alpha})=0\\ \text{for } s_{i_\alpha}=0,1,2,\cdots,m_{i_\alpha}-1,\\ (-1)^{m_{i_\alpha}}\hbar_{i_\alpha}G_{\partial\Omega_{\alpha i_\alpha}}^{(m_{i_\alpha},\sigma_s;\mathscr{E})}(\mathbf{x}_m,t_{m-},\boldsymbol{\pi}_{\sigma_s},\boldsymbol{\lambda}_{i_\alpha})<0\end{array}\right\}\text{for all } i_\alpha\in\bar{\mathscr{N}}_{\alpha\mathscr{B}}(\sigma_r,\sigma_s);\tag{8.29}$$

$$\left.\begin{array}{l}\hbar_{j_\beta}G_{\partial\Omega_{\beta j_\beta}}^{(s_{j_\beta},\sigma_s;\mathscr{E})}(\mathbf{x}_m,t_{m+},\boldsymbol{\pi}_{\sigma_p},\boldsymbol{\lambda}_{j_\beta})=0\\ \text{for } s_{j_\beta}=0,1,2,\cdots,m_{j_\beta}-1,\\ \hbar_{j_\beta}G_{\partial\Omega_{\beta j_\beta}}^{(s_{j_\beta},\sigma_s;\mathscr{E})}(\mathbf{x}_m,t_{m+},\boldsymbol{\pi}_{\sigma_p},\boldsymbol{\lambda}_{j_\beta})>0\end{array}\right\}\text{for all } j_\beta\in\bar{\mathscr{N}}_{\beta\mathscr{B}}(\sigma_r,\sigma_p).\tag{8.30}$$

(ii) *The* s*-dimensional coming edge flow* $\mathbf{x}^{(\sigma_s;\mathscr{E})}(t)$ *in an edge* $\mathscr{E}_{\sigma_s}^{(s)}$ *and the* p*-*

dimensional coming edge flow $\mathbf{x}^{(\sigma_p;\mathscr{E})}(t)$ *in an edge* $\mathscr{E}_{\sigma_p}^{(p)}$ *at point* $\mathbf{x}_m \in \mathscr{E}_{\sigma_r}^{(r)}$ *are non-switchable of the first kind with the* $(\mathbf{m}_{\sigma_s};\mathscr{E}:\mathbf{m}_{\sigma_p};\mathscr{E})$*-singularity* ($\mathbf{m}_{\sigma_s} \neq 2\mathbf{k}_{\sigma_s}+1$ *and* $\mathbf{m}_{\sigma_p} \neq 2\mathbf{k}_{\sigma_p}+1$) *to the edge* $\mathscr{E}_{\sigma_r}^{(r)}$ *if*

$$\left.\begin{array}{l}\hbar_{i_\alpha} G_{\partial\Omega_{\alpha i_\alpha}}^{(s_{i_\alpha},\sigma_s;\mathscr{E})}(\mathbf{x}_m,t_{m-},\boldsymbol{\pi}_{\sigma_s},\boldsymbol{\lambda}_{i_\alpha})=0\\ \text{for } s_{i_\alpha}=0,1,2,\cdots,m_{i_\alpha}-1,\\ (-1)^{m_{i_\alpha}}\hbar_{i_\alpha} G_{\partial\Omega_{\alpha i_\alpha}}^{(m_{i_\alpha},\sigma_s;\mathscr{E})}(\mathbf{x}_m,t_{m-},\boldsymbol{\pi}_{\sigma_s},\boldsymbol{\lambda}_{i_\alpha})<0\end{array}\right\}\text{for all } i_\alpha \in \bar{\mathscr{N}}_{\alpha\mathscr{B}}(\sigma_r,\sigma_s); \tag{8.31}$$

$$\left.\begin{array}{l}\hbar_{j_\beta} G_{\partial\Omega_{\beta j_\beta}}^{(s_{j_\beta},\sigma_s;\mathscr{E})}(\mathbf{x}_m,t_{m-},\boldsymbol{\pi}_{\sigma_p},\boldsymbol{\lambda}_{j_\beta})=0\\ \text{for } s_{j_\beta}=0,1,2,\cdots,m_{j_\beta}-1,\\ (-1)^{m_{j_\beta}}\hbar_{j_\beta} G_{\partial\Omega_{\beta j_\beta}}^{(m_{j_\beta},\sigma_s;\mathscr{E})}(\mathbf{x}_m,t_{m-},\boldsymbol{\pi}_{\sigma_p},\boldsymbol{\lambda}_{j_\beta})<0\end{array}\right\}\text{for all } j_\beta \in \bar{\mathscr{N}}_{\beta\mathscr{B}}(\sigma_r,\sigma_p). \tag{8.32}$$

(iii) *The* s*-dimensional leaving edge flow* $\mathbf{x}^{(\sigma_s;\mathscr{E})}(t)$ *in an edge* $\mathscr{E}_{\sigma_s}^{(s)}$ *and the* p*-dimensional leaving edge flow* $\mathbf{x}^{(\sigma_p;\mathscr{E})}(t)$ *in an edge* $\mathscr{E}_{\sigma_p}^{(p)}$ *at point* $\mathbf{x}_m \in \mathscr{E}_{\sigma_r}^{(r)}$ *are non-switchable of the second kind with the* $(\mathbf{m}_{\sigma_s};\mathscr{E}:\mathbf{m}_{\sigma_p};\mathscr{E})$*-singularity to the edge* $\mathscr{E}_{\sigma_r}^{(r)}$ *if and only if*

$$\left.\begin{array}{l}\hbar_{i_\alpha} G_{\partial\Omega_{\alpha i_\alpha}}^{(s_{i_\alpha},\sigma_s;\mathscr{E})}(\mathbf{x}_m,t_{m+},\boldsymbol{\pi}_{\sigma_s},\boldsymbol{\lambda}_{i_\alpha})=0\\ \text{for } s_{i_\alpha}=0,1,2,\cdots,m_{i_\alpha}-1,\\ \hbar_{i_\alpha} G_{\partial\Omega_{\alpha i_\alpha}}^{(m_{i_\alpha},\sigma_s;\mathscr{E})}(\mathbf{x}_m,t_{m+},\boldsymbol{\pi}_{\sigma_s},\boldsymbol{\lambda}_{i_\alpha})>0\end{array}\right\}\text{for all } i_\alpha \in \bar{\mathscr{N}}_{\alpha\mathscr{B}}(\sigma_r,\sigma_s); \tag{8.33}$$

$$\left.\begin{array}{l}\hbar_{j_\beta} G_{\partial\Omega_{\beta j_\beta}}^{(s_{j_\beta},\sigma_s;\mathscr{E})}(\mathbf{x}_m,t_{m+},\boldsymbol{\pi}_{\sigma_p},\boldsymbol{\lambda}_{j_\beta})=0\\ \text{for } s_{j_\beta}=0,1,2,\cdots,m_{j_\beta}-1,\\ \hbar_{j_\beta} G_{\partial\Omega_{\beta j_\beta}}^{(m_{j_\beta},\sigma_s;\mathscr{E})}(\mathbf{x}_m,t_{m+},\boldsymbol{\pi}_{\sigma_p},\boldsymbol{\lambda}_{j_\beta})>0\end{array}\right\}\text{for all } j_\beta \in \bar{\mathscr{N}}_{\beta\mathscr{B}}(\sigma_r,\sigma_p). \tag{8.34}$$

(iv) *Without the switching rule, the* s*-dimensional coming edge flow* $\mathbf{x}^{(\sigma_s;\mathscr{E})}(t)$ *in an edge* $\mathscr{E}_{\sigma_s}^{(s)}$ *and the* p*-dimensional leaving edge flow* $\mathbf{x}^{(\sigma_p;\mathscr{E})}(t)$ *in an edge* $\mathscr{E}_{\sigma_p}^{(p)}$ *at* $\mathbf{x}_m \in \mathscr{E}_{\sigma_r}^{(r)}$ *are potentially switchable with the* $(\mathbf{m}_{\sigma_s};\mathscr{E}:\mathbf{m}_{\sigma_p};\mathscr{E})$*-singularity from* $\mathscr{E}_{\sigma_s}^{(s)}$ *to* $\mathscr{E}_{\sigma_p}^{(p)}$ *via the edge* $\mathscr{E}_{\sigma_r}^{(r)}$ *if and only if Eqs.* (8.29) *and* (8.30) *hold with* $\mathbf{m}_{\sigma_s} \neq 2\mathbf{k}_{\sigma_s}+1$

(v) *Without the switching rule, the* s*-dimensional coming edge flow* $\mathbf{x}^{(\sigma_s;\mathscr{E})}(t)$ *in an edge* $\mathscr{E}_{\sigma_s}^{(s)}$ *and the* p*-dimensional leaving edge flow* $\mathbf{x}^{(\sigma_p;\mathscr{E})}(t)$ *in an edge*

$\mathscr{E}_{\sigma_p}^{(p)}$ at $\mathbf{x}_m \in \mathscr{E}_{\sigma_r}^{(r)}$ are potentially switchable with the $(\mathbf{m}_{\sigma_s};\mathscr{E}:\mathbf{m}_{\sigma_p};\mathscr{E})$*-singularity from $\mathscr{E}_{\sigma_s}^{(s)}$ to $\mathscr{E}_{\sigma_p}^{(p)}$ via the edge $\mathscr{E}_{\sigma_r}^{(r)}$ if and only if Eqs.* (8.33) *and* (8.34) *hold* $\mathbf{m}_{\sigma_s} = 2\mathbf{k}_{\sigma_s} + 1$ *for at least one of coming flows.*

(vi) *With the switching rule, the s-dimensional grazing edge flow $\mathbf{x}^{(\sigma_s;\mathscr{E})}(t)$ in an edge $\mathscr{E}_{\sigma_s}^{(s)}$ and the p-dimensional grazing edge flow $\mathbf{x}^{(\sigma_p;\mathscr{E})}(t)$ in an edge $\mathscr{E}_{\sigma_p}^{(p)}$ at point $\mathbf{x}_m \in \mathscr{E}_{\sigma_r}^{(r)}$ are switchable with the* $((2\mathbf{k}_{\sigma_s}+1);\mathscr{E}:(2\mathbf{k}_{\sigma_p}+1);\mathscr{E})$*-singularity from $\mathscr{E}_{\sigma_s}^{(s)}$ to $\mathscr{E}_{\sigma_p}^{(p)}$ via the edge $\mathscr{E}_{\sigma_r}^{(r)}$ if and only if*

$$\left.\begin{aligned}
&\hbar_{i_\alpha} G_{\partial\Omega_{\alpha i_\alpha}}^{(s_{i_\alpha},\sigma_s;\mathscr{E})}(\mathbf{x}_m, t_{m\pm}, \boldsymbol{\pi}_{\sigma_s}, \boldsymbol{\lambda}_{i_\alpha}) = 0\\
&\textit{for } s_{i_\alpha} = 0,1,2,\cdots,2k_{i_\alpha},\\
&\hbar_{i_\alpha} G_{\partial\Omega_{\alpha i_\alpha}}^{(2k_{i_\alpha}+1,\sigma_s;\mathscr{E})}(\mathbf{x}_m, t_{m\pm}, \boldsymbol{\pi}_{\sigma_s}, \boldsymbol{\lambda}_{i_\alpha}) > 0
\end{aligned}\right\}\textit{for all } i_\alpha \in \bar{\mathscr{N}}_{\alpha\mathscr{B}}(\sigma_r,\sigma_s); \tag{8.35}$$

$$\left.\begin{aligned}
&\hbar_{j_\beta} G_{\partial\Omega_{\beta j_\beta}}^{(s_{j_\beta},\sigma_s;\mathscr{E})}(\mathbf{x}_m, t_{m\pm}, \boldsymbol{\pi}_{\sigma_p}, \boldsymbol{\lambda}_{j_\beta}) = 0\\
&\textit{for } s_{j_\beta} = 0,1,2,\cdots,2k_{j_\beta},\\
&\hbar_{j_\beta} G_{\partial\Omega_{\beta j_\beta}}^{(2k_{j_\beta}+1,\sigma_s;\mathscr{E})}(\mathbf{x}_m, t_{m\pm}, \boldsymbol{\pi}_{\sigma_p}, \boldsymbol{\lambda}_{j_\beta}) > 0
\end{aligned}\right\}\textit{for all } j_\beta \in \bar{\mathscr{N}}_{\beta\mathscr{B}}(\sigma_r,\sigma_p). \tag{8.36}$$

(vii) *Without the switching rule, the s-dimensional, grazing edge flow $\mathbf{x}^{(\sigma_s;\mathscr{E})}(t)$ in an edge $\mathscr{E}_{\sigma_s}^{(s)}$ and the p-dimensional, grazing edge flow $\mathbf{x}^{(\sigma_p;\mathscr{E})}(t)$ in an edge $\mathscr{E}_{\sigma_p}^{(p)}$ at point $\mathbf{x}_m \in \mathscr{E}_{\sigma_r}^{(r)}$ are a double tangential flow with the* $((2\mathbf{k}_{\sigma_s}+1);\mathscr{E}:(2\mathbf{k}_{\sigma_p}+1);\mathscr{E})$*-singularity to the edge $\mathscr{E}_{\sigma_r}^{(r)}$ if and only if Eqs.* (8.35) *and* (8.36) *holds.*

Proof. As in Theorem 3.2, this theorem can be proved similarly as the edge flow is treated as a special domain flow (also see, Luo, 2008, 2009). ■

To help one understand the switchability between two of all the edge flows $\mathbf{x}^{(\sigma_s;\mathscr{E})}$ ($s = 0,1,2,\cdots,r-1$) to a specific edge $\mathscr{E}_{\sigma_r}^{(r)}$, consider $r = s-3$ to illustrate the switchability of the two edge flows in s-dimensional space. The non-switchable flows of the first and second kinds for the two flows of $\mathbf{x}^{(\sigma_s;\mathscr{E})}$ and $\mathbf{x}^{(\sigma_{s-1};\mathscr{E})}$ to the edge $\mathscr{E}^{(s-3)}$ are sketched in Figs. 8.4(a) and (b), respectively. The edge flows are presented by curves with arrows. In Fig. 8.5(a), the two edge flows of $\mathbf{x}^{(\sigma_s;\mathscr{E})}$ and $\mathbf{x}^{(\sigma_{s-1};\mathscr{E})}$ to the edge $\mathscr{E}^{(s-3)}$ are switchable. However, the two edge flows of $\mathbf{x}^{(\sigma_s;\mathscr{E})}$ and $\mathbf{x}^{(\sigma_{s-1};\mathscr{E})}$ grazing to the edge $\mathscr{E}^{(s-2)}$ are illustrated in Fig. 8.5(b). Because the edge $\mathscr{E}^{(s-3)}$ is a point, it is very difficult to illustrate the tangency of the edge flows to the specific edge. With a switching rule, the two grazing flows are switchable. Because a coming edge flow arrives to the specific edge, it has many possibilities to switch to the leaving flows. Thus, without the switching rule, a pair of coming and leaving edge flows is potentially switchable to the specific edge. Of course, only one leaving flow exists to the specific edge. Any coming edge flow to such a specific edge should be switched to the leaving flow if no any flow barriers exist on the specific edge.

In order to describe the global switchability of all edge flows relative to the specific edge $\mathscr{E}_{\sigma_r}^{(r)}$ at the same time, the accessible and inaccessible edge flows can be analogically described as the accessible and inaccessible domain flows in Chapter 2. The edge on which dynamical systems can be defined is called an accessible edge. Otherwise, the edge is called an inaccessible edge. Thus, the edge number set $\mathscr{N}_{\mathscr{E}}(\sigma_r)$ can be accessible and inaccessible edge index sets (i.e., $\mathscr{N}_{\mathscr{E}}^{A}(\sigma_r)$ and $\mathscr{N}_{\mathscr{E}}^{I}(\sigma_r)$). The accessible edge index set $\mathscr{N}_{\mathscr{E}}^{A}(\sigma_r)$ may possess three types of edge index sets for coming, leaving and grazing edge flows (i.e., $\mathscr{N}_{\mathscr{E}}^{C}(\sigma_r)$, $\mathscr{N}_{\mathscr{E}}^{L}(\sigma_r)$, and $\mathscr{N}_{\mathscr{E}}^{G}(\sigma_r)$). For instance, if $\mathscr{N}_{\mathscr{E}}^{A}(\sigma_r)=\mathscr{N}_{\mathscr{E}}^{C}(\sigma_r)$, all the edge flows are coming flows; and if $\mathscr{N}_{\mathscr{E}}^{A}(\sigma_r)=\mathscr{N}_{\mathscr{E}}^{C}(\sigma_r)\cup\mathscr{N}_{\mathscr{E}}^{L}(\sigma_r)$, the edge flows consists of the coming and leaving flows, and so on.

Definition 8.5. The total index set of edges relative to an edge $\mathscr{E}_{\sigma_r}^{(r)}$ ($\sigma_r\in\mathscr{N}_{\mathscr{E}}^{(r)}=\{1,2,\cdots,N_r\}$) is decomposed into four subsets of the coming, leaving, grazing and inaccessible edges, i.e.,

$$\begin{aligned}
&\mathscr{N}_{\mathscr{E}}(\sigma_r)=\mathscr{N}_{\mathscr{E}}^{A}(\sigma_r)\cup\mathscr{N}_{\mathscr{E}}^{I}(\sigma_r)=\cup_{j\in\mathscr{J}}\mathscr{N}_{\mathscr{E}}^{j}(\sigma_r),\\
&\mathscr{N}_{\mathscr{E}}^{A}(\sigma_r)=\mathscr{N}_{\mathscr{E}}^{C}(\sigma_r)\cup\mathscr{N}_{\mathscr{E}}^{L}(\sigma_r)\cup\mathscr{N}_{\mathscr{E}}^{G}(\sigma_r),\\
&\cap_{j\in\mathscr{J}}\mathscr{N}_{\mathscr{E}}^{j}(\sigma_r)=\varnothing,\\
&\mathscr{J}=\{C,L,G,I\};
\end{aligned}\tag{8.37}$$

where the *coming, leaving, grazing and inaccessible edges* are defined as

$$\begin{aligned}
&\mathscr{N}_{\mathscr{E}}^{C}(\sigma_r)=\cup_{s=r+1}^{n}\cup_{\sigma_s=1}^{N_s}\mathscr{N}_{\mathscr{E}}^{C}(\sigma_r,\sigma_s),\\
&\mathscr{N}_{\mathscr{E}}^{L}(\sigma_r)=\cup_{p=r+1}^{n}\cup_{\sigma_p=1}^{N_p}\mathscr{N}_{\mathscr{E}}^{L}(\sigma_r,\sigma_p),\\
&\mathscr{N}_{\mathscr{E}}^{G}(\sigma_r)=\cup_{q=r+1}^{n}\cup_{\sigma_q=1}^{N_q}\mathscr{N}_{\mathscr{E}}^{G}(\sigma_r,\sigma_q),\\
&\mathscr{N}_{\mathscr{E}}^{I}(\sigma_r)=\cup_{l=r+1}^{n}\cup_{\sigma_l=1}^{N_l}\mathscr{N}_{\mathscr{E}}^{I}(\sigma_r,\sigma_l).
\end{aligned}\tag{8.38}$$

In the foregoing definition, the total sets of $\mathscr{N}_{\mathscr{E}}(\sigma_r)$ includes all domain sets of $\mathscr{N}_{\mathscr{D}}(\sigma_r)$, boundary sets of $\mathscr{N}_{\mathscr{B}}(\sigma_r)$, and edges sets of $\mathscr{N}_{\mathscr{E}}^{(s)}$ ($s>r$). To the edge of $\mathscr{E}_{\sigma_r}^{(r)}$, the corresponding domains, boundaries and edges can be accessible and inaccessible. For the accessible domains, boundary and edges, the corresponding flows are classified into the coming, leaving and grazing flows.

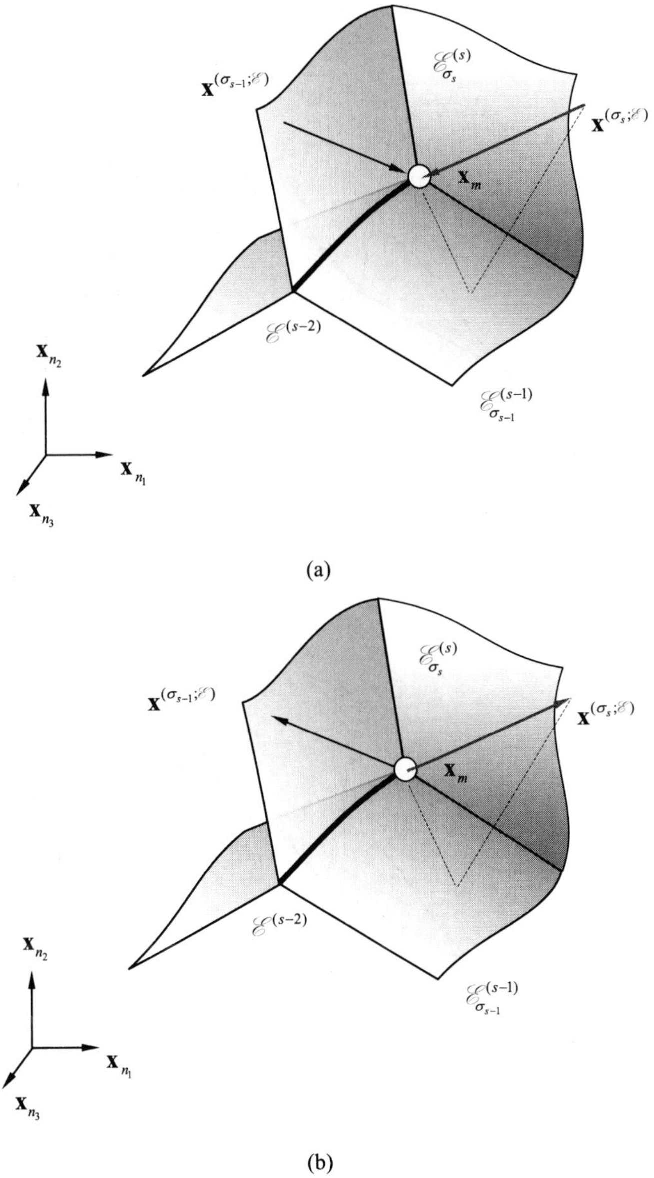

Fig. 8.4 The two flows of $\mathbf{x}^{(\sigma_s;\mathscr{E})}$ and $\mathbf{x}^{(\sigma_{s-1};\mathscr{E})}$. **(a)** the non-switchable flows of the first kind, and **(b)** the non-switchable flow of the second kind to the edge $\mathscr{E}^{(s-3)}$. The edge flows are presented by curves with arrows. ($n_1 + n_2 + n_3 = s$)

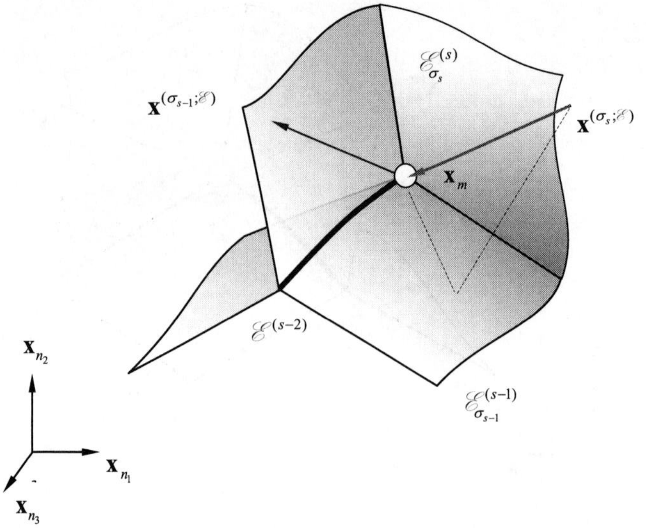

(a)

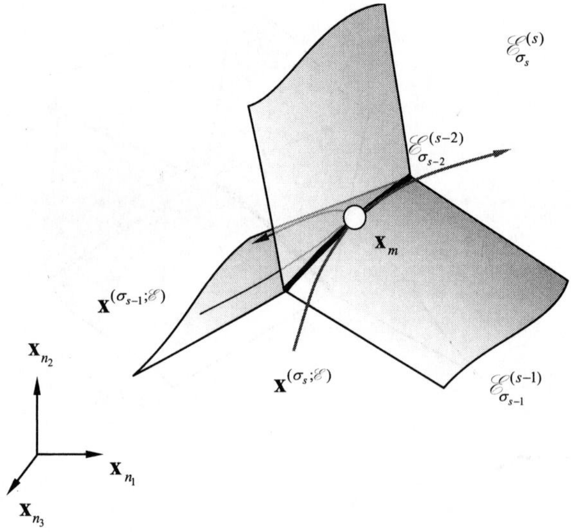

(b)

Fig. 8.5 The two flows of $\mathbf{x}^{(\sigma_s;\mathscr{E})}$ and $\mathbf{x}^{(\sigma_{s-1};\mathscr{E})}$. **(a)** the switchable edge flows to the edge $\mathscr{E}^{(s-3)}$, and **(b)** the grazing edge flows to the edge $\mathscr{E}^{(s-2)}$. The edge flows are presented by curves with arrows. ($n_1+n_2+n_3=s$)

Definition 8.6. For a discontinuous dynamical system in Eqs. (6.4)-(6.8), there is a point $\mathbf{x}^{(\sigma_r;\mathscr{E})}(t_m) \equiv \mathbf{x}_m \in \mathscr{E}_{\sigma_r}^{(r)}$ ($\sigma_r \in \mathscr{N}_{\mathscr{E}}^{(r)} = \{1,2,\cdots,N_r\}$ and $r \in \{0,1,2,\cdots,n-2\}$) at time t_m with the related boundaries $\partial\Omega_{\alpha i_\alpha}(\sigma_r)$ and related domains $\Omega_\alpha(\sigma_r)$ ($\alpha \in \mathscr{N}_{\mathscr{D}}(\sigma_r) \subset \mathscr{N}_{\mathscr{D}} = \{1,2,\cdots,N_d\}$ and $i_\alpha \in \mathscr{N}_{\alpha\mathscr{B}}(\sigma_r) \subset \mathscr{N}_{\mathscr{B}} = \{1,2,\cdots,N_b\}$), where the boundary index subset of $\mathscr{N}_{\alpha\mathscr{B}}(\sigma_r)$ has n_{σ_r}-elements ($n_{\sigma_r} \geq (n-r)$). The edge of $\mathscr{E}_{\sigma_r}^{(r)}$ is the intersection of edges (including domains and boundaries) $\mathscr{E}_{\sigma_s}^{(s)}(\sigma_r)$ ($\sigma_s \in \mathscr{N}_{\mathscr{E}}^{s}(\sigma_r) \subset \mathscr{N}_{\mathscr{E}}^{s} = \{1,2,\cdots,N_s\}$ and $s \in \{r+1,r+2,\cdots,n-1,n\}$). Suppose there is a s-dimensional edge $\mathscr{E}_{\sigma_s}^{(s)}(\sigma_r)$ with a boundary index set $\mathscr{N}_{\alpha\mathscr{B}}(\sigma_r,\sigma_s)$ relative to domain $\Omega_\alpha(\sigma_r)$. The complementary boundary index set is $\bar{\mathscr{N}}_{\alpha\mathscr{B}}(\sigma_r,\sigma_\rho) = \mathscr{N}_{\alpha\mathscr{B}}(\sigma_r)/\mathscr{N}_{\alpha\mathscr{B}}(\sigma_r,\sigma_\rho)$. Suppose the coming and leaving edge flows are $\mathbf{x}^{(\sigma_s;\mathscr{E})}(t)$ ($\sigma_s \in \mathscr{N}_{\mathscr{E}}^{C}(\sigma_r,\sigma_s)$) and $\mathbf{x}^{(\sigma_p;\mathscr{E})}(t)$ ($\sigma_p \in \mathscr{N}_{\mathscr{E}}^{L}(\sigma_r,\sigma_p)$) on the edge $\mathscr{E}_{\sigma_s}^{(s)}$ and $\mathscr{E}_{\sigma_p}^{(p)}$, respectively. For an arbitrarily small $\varepsilon > 0$, there are two time intervals $[t_{m-\varepsilon}, t_m)$ and $(t_m, t_{m+\varepsilon}]$. An edge flow $\mathbf{x}^{(\sigma_s;\mathscr{E})}(t)$ is $C_{[t_{m-\varepsilon},t_m)}^{r_{\sigma_s}}$ and/or $C_{(t_m,t_{m+\varepsilon}]}^{r_{\sigma_s}}$-continuous ($r_{\sigma_s} \geq 1$) for time t and $\| d^{r_{\sigma_s}+1}\mathbf{x}^{(\sigma_s;\mathscr{E})}/dt^{r_{\sigma_s}+1} \| < \infty$. An edge flow $\mathbf{x}^{(\sigma_p;\mathscr{E})}(t)$ ($p \in \{r,r+1,\cdots,n-1,n\}$) is $C_{[t_{m-\varepsilon},t_m)}^{r_{\sigma_p}}$ and/or $C_{(t_m,t_{m+\varepsilon}]}^{r_{\sigma_p}}$ continuous for time t and $\| d^{r_{\sigma_p}+1}\mathbf{x}^{(\sigma_p;\mathscr{E})}/dt^{r_{\sigma_p}+1} \| < \infty$ ($r_{\sigma_p} \geq 1$). Suppose the switching rule $\mathbf{x}^{(\sigma_s;\mathscr{E})}(t_{m\pm}) = \mathbf{x}_m = \mathbf{x}^{(\sigma_p;\mathscr{E})}(t_{m\pm})$ exists.

(i) The s-dimensional coming edge flow $\mathbf{x}^{(\sigma_s;\mathscr{E})}(t)$ in an edge $\mathscr{E}_{\sigma_s}^{(s)}$ ($\sigma_s \in \mathscr{N}_{\mathscr{E}}^{C}(\sigma_r)$) and the p-dimensional leaving edge flow $\mathbf{x}^{(\sigma_p;\mathscr{E})}(t)$ in an edge $\mathscr{E}_{\sigma_p}^{(p)}$ ($\sigma_s \in \mathscr{N}_{\mathscr{E}}^{L}(\sigma_r)$) at point $\mathbf{x}_m \in \mathscr{E}_{\sigma_r}^{(r)}$ are $(\mathscr{N}_{\mathscr{E}}^{C}(\sigma_r) : \mathscr{N}_{\mathscr{E}}^{L}(\sigma_r);\mathscr{E})$ *switchable* with from $\mathscr{E}_{\sigma_s}^{(s)}$ to $\mathscr{E}_{\sigma_p}^{(p)}$ via edge $\mathscr{E}_{\sigma_r}^{(r)}$ if

$$\left.\begin{array}{l}
\hbar_{i_\alpha}\mathbf{n}_{\partial\Omega_{\alpha i_\alpha}}^{\mathrm{T}}(\mathbf{x}_{m-\varepsilon}^{(i_\alpha;\mathscr{B})})\cdot[\mathbf{x}_{m-\varepsilon}^{(i_\alpha;\mathscr{B})} - \mathbf{x}_{m-\varepsilon}^{(\sigma_s;\mathscr{E})}] < 0 \\
\text{for all } i_\alpha \in \bigcup_{s=r+1}^{n}\bigcup_{\sigma_s \in \mathscr{N}_{\mathscr{E}}^{s}(\sigma_r)} \bar{\mathscr{N}}_{\alpha\mathscr{B}}(\sigma_r,\sigma_s); \\
\hbar_{j_\beta}\mathbf{n}_{\partial\Omega_{\beta j_\beta}}^{\mathrm{T}}(\mathbf{x}_{m+\varepsilon}^{(j_\beta;\mathscr{B})})\cdot[\mathbf{x}_{m+\varepsilon}^{(\sigma_p;\mathscr{E})} - \mathbf{x}_{m+\varepsilon}^{(j_\beta;\mathscr{B})}] > 0 \\
\text{for all } j_\beta \in \bigcup_{p=r+1}^{n}\bigcup_{\sigma_p \in \mathscr{N}_{\mathscr{E}}^{p}(\sigma_r)} \bar{\mathscr{N}}_{\beta\mathscr{B}}(\sigma_r,\sigma_p).
\end{array}\right\} \tag{8.39}$$

(ii) The edge $\mathscr{E}_{\sigma_r}^{(r)}$ for all the s-dimensional coming edge flows $\mathbf{x}^{(\sigma_s;\mathscr{E})}(t)$ in an edge $\mathscr{E}_{\sigma_s}^{(s)}$ ($\sigma_s \in \mathscr{N}_{\mathscr{E}}^{C}(\sigma_r) = \mathscr{N}_{\mathscr{E}}(\sigma_r)$) at point $\mathbf{x}_m \in \mathscr{E}_{\sigma_r}^{(r)}$ is $(\mathscr{N}_{\mathscr{E}}^{C}(\sigma_r);\mathscr{E})$ *non-switchable of the first kind* (sink) if

$$\hbar_{i_\alpha}\mathbf{n}_{\partial\Omega_{\alpha i_\alpha}}^{\mathrm{T}}(\mathbf{x}_{m-\varepsilon}^{(i_\alpha;\mathscr{B})})\cdot[\mathbf{x}_{m-\varepsilon}^{(i_\alpha;\mathscr{B})} - \mathbf{x}_{m-\varepsilon}^{(\sigma_s;\mathscr{E})}] < 0$$

$$\text{for all } i_\alpha \in \cup_{s=r+1}^{n} \cup_{\sigma_s \in \mathscr{N}_{\mathscr{E}}^{s}(\sigma_r)} \bar{\mathscr{N}}_{\alpha\mathscr{B}}(\sigma_r,\sigma_s). \tag{8.40}$$

(iii) The edge $\mathscr{E}_{\sigma_r}^{(r)}$ for the p-dimensional leaving edge flow $\mathbf{x}^{(\sigma_p;\mathscr{E})}(t)$ in an edge $\mathscr{E}_{\sigma_p}^{(p)}$ ($\sigma_p \in \mathscr{N}_{\mathscr{E}}^{L}(\sigma_r) = \mathscr{N}_{\mathscr{E}}(\sigma_r)$) at point $\mathbf{x}_m \in \mathscr{E}_{\sigma_r}^{(r)}$ is $(\mathscr{N}_{\mathscr{E}}^{L}(\sigma_r);\mathscr{E})$ *non-switchable of the second kind* (source) if

$$\begin{aligned} &\hbar_{j_\beta} \mathbf{n}_{\partial\Omega_{\beta j_\beta}}^{\mathrm{T}}(\mathbf{x}_{m+\varepsilon}^{(j_\beta;\mathscr{B})})\cdot[\mathbf{x}_{m+\varepsilon}^{(\sigma_p;\mathscr{E})} - \mathbf{x}_{m+\varepsilon}^{(j_\beta;\mathscr{B})}] > 0 \\ &\text{for all } j_\beta \in \cup_{p=r+1}^{n} \cup_{\sigma_p \in \mathscr{N}_{\mathscr{E}}^{p}(\sigma_r)} \bar{\mathscr{N}}_{\beta\mathscr{B}}(\sigma_r,\sigma_p). \end{aligned} \tag{8.41}$$

From the above description, the necessary and sufficient conditions for the switchability of all flows to the specific edge are given by the following theorem.

Theorem 8.5. *For a discontinuous dynamical system in Eqs.* (6.4)-(6.8), *there is a point* $\mathbf{x}^{(\sigma_r;\mathscr{E})}(t_m) \equiv \mathbf{x}_m \in \mathscr{E}_{\sigma_r}^{(r)}$ ($\sigma_r \in \mathscr{N}_{\mathscr{E}}^{(r)} = \{1,2,\cdots,N_r\}$ *and* $r \in \{0,1,2,\cdots,n-2\}$) *at time* t_m *with the related boundaries* $\partial\Omega_{\alpha i_\alpha}(\sigma_r)$ *and related domains* $\Omega_\alpha(\sigma_r)$ ($\alpha \in \mathscr{N}_{\mathscr{D}}(\sigma_r) \subset \mathscr{N}_{\mathscr{D}} = \{1,2,\cdots,N_d\}$ *and* $i_\alpha \in \mathscr{N}_{\alpha\mathscr{B}}(\sigma_r) \subset \mathscr{N}_{\mathscr{B}} = \{1,2,\cdots,N_b\}$), *where the boundary index subset of* $\mathscr{N}_{\alpha\mathscr{B}}(\sigma_r)$ *has* n_{σ_r}*-elements* ($n_{\sigma_r} \geq (n-r)$). *The edge of* $\mathscr{E}_{\sigma_r}^{(r)}$ *is the intersection of edges (including domains and boundaries)* $\mathscr{E}_{\sigma_s}^{(s)}(\sigma_r)$ ($\sigma_s \in \mathscr{N}_{\mathscr{E}}^{s}(\sigma_r) \subset \mathscr{N}_{\mathscr{E}}^{s} = \{1,2,\cdots,N_s\}$ *and* $s \in \{r+1,r+2,\cdots,n-1,n\}$). *Suppose there is an* s*-dimensional edge* $\mathscr{E}_{\sigma_s}^{(s)}(\sigma_r)$ *with a boundary index set* $\mathscr{N}_{\alpha\mathscr{B}}(\sigma_r,\sigma_s)$ *relative to domain* $\Omega_\alpha(\sigma_r)$. *The complementary boundary index set is* $\bar{\mathscr{N}}_{\alpha\mathscr{B}}(\sigma_r,\sigma_p) = \mathscr{N}_{\alpha\mathscr{B}}(\sigma_r)/\mathscr{N}_{\alpha\mathscr{B}}(\sigma_r,\sigma_p)$. *Suppose the coming and leaving edge flows are* $\mathbf{x}^{(\sigma_s;\mathscr{E})}(t)$ ($\sigma_s \in \mathscr{N}_{\mathscr{E}}^{C}(\sigma_r,\sigma_s)$) *and* $\mathbf{x}^{(\sigma_p;\mathscr{E})}(t)$ ($\sigma_p \in \mathscr{N}_{\mathscr{E}}^{L}(\sigma_r,\sigma_p)$) *on the edge* $\mathscr{E}_{\sigma_s}^{(s)}$ *and* $\mathscr{E}_{\sigma_p}^{(p)}$, *respectively. For an arbitrarily small* $\varepsilon > 0$, *there are two time intervals* $[t_{m-\varepsilon},t_m)$ *and* $(t_m,t_{m+\varepsilon}]$. *An edge flow* $\mathbf{x}^{(\sigma_s;\mathscr{E})}(t)$ *is* $C_{[t_{m-\varepsilon},t_m)}^{r_{\sigma_s}}$ *and/or* $C_{(t_m,t_{m+\varepsilon}]}^{r_{\sigma_s}}$*-continuous* ($r_{\sigma_s} \geq 2$) *for time* t *and* $\| d^{r_{\sigma_s}+1}\mathbf{x}^{(\sigma_s;\mathscr{E})}/dt^{r_{\sigma_s}+1} \| < \infty$. *An edge flow* $\mathbf{x}^{(\sigma_p;\mathscr{E})}(t)$ ($p \in \{r,r+1,\cdots,n-1,n\}$) *is* $C_{[t_{m-\varepsilon},t_m)}^{r_{\sigma_p}}$ *and/or* $C_{(t_m,t_{m+\varepsilon}]}^{r_{\sigma_p}}$ *continuous for time* t *and* $\| d^{r_{\sigma_p}+1}\mathbf{x}^{(\sigma_s;\mathscr{E})}/dt^{r_{\sigma_p}+1} \| < \infty$ ($r_{\sigma_p} \geq 2$). *Suppose the switching rule* $\mathbf{x}^{(\sigma_s;\mathscr{E})}(t_{m\pm}) = \mathbf{x}_m = \mathbf{x}^{(\sigma_p;\mathscr{E})}(t_{m\pm})$ *exists.*

(i) *The* s*-dimensional coming edge flow* $\mathbf{x}^{(\sigma_s;\mathscr{E})}(t)$ *in an edge* $\mathscr{E}_{\sigma_s}^{(s)}$ ($\sigma_s \in \mathscr{N}_{\mathscr{E}}^{C}(\sigma_r)$) *and the* p*-dimensional leaving edge flow* $\mathbf{x}^{(\sigma_p;\mathscr{E})}(t)$ *in an edge* $\mathscr{E}_{\sigma_p}^{(p)}$ ($\sigma_s \in \mathscr{N}_{\mathscr{E}}^{L}(\sigma_r)$) *at point* $\mathbf{x}_m \in \mathscr{E}_{\sigma_r}^{(r)}$ *are* $(\mathscr{N}_{\mathscr{E}}^{C}(\sigma_r):\mathscr{N}_{\mathscr{E}}^{L}(\sigma_r);\mathscr{E})$

switchable from $\mathscr{E}_{\sigma_s}^{(s)}$ *to* $\mathscr{E}_{\sigma_p}^{(p)}$ *via edge* $\mathscr{E}_{\sigma_r}^{(r)}$ *if and only if*

$$\left.\begin{aligned}&\hbar_{i_\alpha} G_{\partial\Omega_{\alpha i_\alpha}}^{(\sigma_s;\mathscr{E})}(\mathbf{x}_m, t_{m-}, \boldsymbol{\pi}_{\sigma_s}, \boldsymbol{\lambda}_{i_\alpha}) < 0\\ &\textit{for all } i_\alpha \in \textstyle\bigcup_{s=r+1}^{n}\bigcup_{\sigma_s\in\mathscr{N}_{\mathscr{E}}^{s}(\sigma_r)} \bar{\mathscr{N}}_{\alpha\mathscr{B}}(\sigma_r,\sigma_s);\\ &\hbar_{j_\beta} G_{\partial\Omega_{\beta j_\beta}}^{(\sigma_p;\mathscr{E})}(\mathbf{x}_m, t_{m+}, \boldsymbol{\pi}_{\sigma_p}, \boldsymbol{\lambda}_{i_\beta}) > 0\end{aligned}\right\}$$
$$\textit{for all } j_\beta \in \textstyle\bigcup_{p=r+1}^{n}\bigcup_{\sigma_p\in\mathscr{N}_{\mathscr{E}}^{p}(\sigma_r)} \bar{\mathscr{N}}_{\beta\mathscr{B}}(\sigma_r,\sigma_p). \tag{8.42}$$

(ii) *The edge* $\mathscr{E}_{\sigma_r}^{(r)}$ *for the* s*-dimensional coming edge flow* $\mathbf{x}^{(\sigma_s;\mathscr{E})}(t)$ *in an edge* $\mathscr{E}_{\sigma_s}^{(s)}$ ($\sigma_s \in \mathscr{N}_{\mathscr{E}}^{C}(\sigma_r) = \mathscr{N}_{\mathscr{E}}(\sigma_r)$) *at point* $\mathbf{x}_m \in \mathscr{E}_{\sigma_r}^{(r)}$ *is* $(\mathscr{N}_{\mathscr{E}}^{C}(\sigma_r);\mathscr{E})$ *non-switchable of the first kind (sink) if and only if*

$$\begin{aligned}&\hbar_{i_\alpha} G_{\partial\Omega_{\alpha i_\alpha}}^{(\sigma_s;\mathscr{E})}(\mathbf{x}_m, t_{m-}, \boldsymbol{\pi}_{\sigma_s}, \boldsymbol{\lambda}_{i_\alpha}) < 0\\ &\textit{for all } i_\alpha \in \textstyle\bigcup_{s=r+1}^{n}\bigcup_{\sigma_s\in\mathscr{N}_{\mathscr{E}}^{s}(\sigma_r)} \bar{\mathscr{N}}_{\alpha\mathscr{B}}(\sigma_r,\sigma_s).\end{aligned} \tag{8.43}$$

(iii) *The edge* $\mathscr{E}_{\sigma_r}^{(r)}$ *for all the* p*-dimensional leaving edge flows* $\mathbf{x}^{(\sigma_p;\mathscr{E})}(t)$ *in an edge* $\mathscr{E}_{\sigma_p}^{(p)}$ ($\sigma_p \in \mathscr{N}_{\mathscr{E}}^{L}(\sigma_r) = \mathscr{N}_{\mathscr{E}}(\sigma_r)$) *at point* $\mathbf{x}_m \in \mathscr{E}_{\sigma_r}^{(r)}$ *is* $(\mathscr{N}_{\mathscr{E}}^{L}(\sigma_r);\mathscr{E})$ *non-switchable of the second kind (source) if and only if*

$$\begin{aligned}&\hbar_{j_\beta} G_{\partial\Omega_{\beta j_\beta}}^{(\sigma_p;\mathscr{E})}(\mathbf{x}_m, t_{m+}, \boldsymbol{\pi}_{\sigma_p}, \boldsymbol{\lambda}_{i_\beta}) > 0\\ &\textit{for all } j_\beta \in \textstyle\bigcup_{p=r+1}^{n}\bigcup_{\sigma_p\in\mathscr{N}_{\mathscr{E}}^{p}(\sigma_r)} \bar{\mathscr{N}}_{\beta\mathscr{B}}(\sigma_r,\sigma_p).\end{aligned} \tag{8.44}$$

Proof. As similar to Theorem 3.1, this theorem can be proved similarly as the edge flow is treated as a special domain flow (also see, Luo, 2008, 2009). ■

Consider the higher-order singularity of flows to a specific edge, and the corresponding description of the switchability of all flows to the specific edge is given as follows.

Definition 8.7. For a discontinuous dynamical system in Eqs. (6.4)-(6.8), there is a point $\mathbf{x}^{(\sigma_r;\mathscr{E})}(t_m) \equiv \mathbf{x}_m \in \mathscr{E}_{\sigma_r}^{(r)}$ ($\sigma_r \in \mathscr{N}_{\mathscr{E}}^{(r)} = \{1,2,\cdots,N_r\}$ and $r \in \{0,1,2,\cdots,n-2\}$) at time t_m with the related boundaries $\partial\Omega_{\alpha i_\alpha}(\sigma_r)$ and related domains $\Omega_\alpha(\sigma_r)$ ($\alpha \in \mathscr{N}_{\mathscr{D}}(\sigma_r) \subset \mathscr{N}_{\mathscr{D}} = \{1,2,\cdots,N_d\}$ and $i_\alpha \in \mathscr{N}_{\alpha\mathscr{B}}(\sigma_r) \subset \mathscr{N}_{\mathscr{B}} = \{1,2,\cdots,N_b\}$), where the boundary index subset of $\mathscr{N}_{\alpha\mathscr{B}}(\sigma_r)$ has n_{σ_r}-elements ($n_{\sigma_r} \geq (n-r)$). The edge of $\mathscr{E}_{\sigma_r}^{(r)}$ is the intersection of edges (including domains and boundaries) $\mathscr{E}_{\sigma_s}^{(s)}(\sigma_r)$ ($\sigma_s \in \mathscr{N}_{\mathscr{E}}^{(s)}(\sigma_r) \subset \mathscr{N}_{\mathscr{E}}^{(s)} = \{1,2,\cdots,N_s\}$ and $s \in \{r+1, r+2,\cdots,n-1,n\}$). Suppose there is an s-dimensional edge $\mathscr{E}_{\sigma_s}^{(s)}(\sigma_r)$ with a boundary index set

$\mathscr{N}_{\alpha\mathscr{B}}(\sigma_r,\sigma_s)$ relative to domain $\Omega_\alpha(\sigma_r)$. The complementary boundary index set is $\bar{\mathscr{N}}_{\alpha\mathscr{B}}(\sigma_r,\sigma_\rho)=\mathscr{N}_{\alpha\mathscr{B}}(\sigma_r)/\mathscr{N}_{\alpha\mathscr{B}}(\sigma_r,\sigma_\rho)$. Suppose the coming and leaving edge flows are $\mathbf{x}^{(\sigma_s;\mathscr{E})}(t)$ ($\sigma_s\in\mathscr{N}_{\mathscr{E}}^{C}(\sigma_r,\sigma_s)$) and $\mathbf{x}^{(\sigma_p;\mathscr{E})}(t)$ ($\sigma_p\in\mathscr{N}_{\mathscr{E}}^{L}(\sigma_r,\sigma_p)$) on the edge $\mathscr{E}_{\sigma_s}^{(s)}$ and $\mathscr{E}_{\sigma_p}^{(p)}$, respectively. For an arbitrarily small $\varepsilon>0$, there are two time intervals $[t_{m-\varepsilon},t_m)$ and $(t_m,t_{m+\varepsilon}]$. An edge flow $\mathbf{x}^{(\sigma_s;\mathscr{E})}(t)$ is $C^{r_{\sigma_s}}_{[t_{m-\varepsilon},t_m)}$ and/or $C^{r_{\sigma_s}}_{(t_m,t_{m+\varepsilon}]}$-continuous for time t and $\|d^{r_{\sigma_s}+1}\mathbf{x}^{(\sigma_s;\mathscr{E})}/dt^{r_{\sigma_s}+1}\|<\infty$ ($r_{\sigma_s}\geq\max_{i_\alpha\in\mathscr{N}_{\alpha\mathscr{B}}^{C}}\{m_{i_\alpha}\}$). An edge flow $\mathbf{x}^{(\sigma_p;\mathscr{E})}(t)$ ($p\in\{r,r+1,\cdots,\ n-1,n\}$) is $C^{r_{\sigma_p}}_{[t_{m-\varepsilon},t_m)}$ and/or $C^{r_{\sigma_p}}_{(t_m,t_{m+\varepsilon}]}$-continuous for time t and $\|d^{r_{\sigma_p}+1}\mathbf{x}^{(\sigma_p;\mathscr{E})}/dt^{r_{\sigma_p}+1}\|<\infty$ ($r_{\sigma_p}\geq\max_{i_\alpha\in\mathscr{N}_{\alpha\mathscr{B}}^{L}}\{m_{j_\alpha}\}$). Suppose the switching rule $\mathbf{x}^{(\sigma_s;\mathscr{E})}(t_{m\pm})=\mathbf{x}_m=\mathbf{x}^{(\sigma_p;\mathscr{E})}(t_{m\pm})$ exists.

(i) With the switching rule, the s-dimensional edge flow $\mathbf{x}^{(\sigma_s;\mathscr{E})}(t)$ in an edge $\mathscr{E}_{\sigma_s}^{(s)}$ ($\sigma_s\in\mathscr{N}_{\mathscr{E}}^{C}(\sigma_r)$) and the p-dimensional edge flow $\mathbf{x}^{(\sigma_p;\mathscr{E})}(t)$ in an edge $\mathscr{E}_{\sigma_p}^{(p)}$ ($\sigma_s\in\mathscr{N}_{\mathscr{E}}^{L}(\sigma_r)$) at point $\mathbf{x}_m\in\mathscr{E}_{\sigma_r}^{(r)}$ are *switchable* with the singularity of $(\cup_{s=r+1}^{n}\cup_{\sigma_s}\mathbf{m}_{\sigma_s};\mathscr{E}:\cup_{p=r+1}^{n}\cup_{\sigma_p}\mathbf{m}_{\sigma_p};\mathscr{E})$ from $\mathscr{E}_{\sigma_s}^{(s)}$ to $\mathscr{E}_{\sigma_p}^{(p)}$ via edge $\mathscr{E}_{\sigma_r}^{(r)}$ if

$$\left.\begin{array}{l}\hbar_{i_\alpha}G_{\partial\Omega_{\alpha i_\alpha}}^{(s_{i_\alpha},\sigma_s;\mathscr{E})}(\mathbf{x}_m,t_{m-},\boldsymbol{\pi}_{\sigma_s},\boldsymbol{\lambda}_{i_\alpha})=0\\ \text{for } s_{i_\alpha}=0,1,2,\cdots,m_{i_\alpha}-1,\\ \hbar_{i_\alpha}\mathbf{n}_{\partial\Omega_{\alpha i_\alpha}}^{\mathrm{T}}(\mathbf{x}_{m-\varepsilon}^{(i_\alpha;\mathscr{B})})\cdot[\mathbf{x}_{m-\varepsilon}^{(i_\alpha;\mathscr{B})}-\mathbf{x}_{m-\varepsilon}^{(\sigma_s;\mathscr{E})}]<0\\ \text{for all } i_\alpha\in\bigcup_{s=r+1}^{n}\bigcup_{\sigma_s\in\mathscr{N}_{\mathscr{E}}^{s}(\sigma_r)}\bar{\mathscr{N}}_{\alpha\mathscr{B}}(\sigma_r,\sigma_s);\end{array}\right\}\tag{8.45}$$

$$\left.\begin{array}{l}\hbar_{j_\beta}G_{\partial\Omega_{\beta j_\beta}}^{(s_{j_\beta},\sigma_p;\mathscr{E})}(\mathbf{x}_m,t_{m-},\boldsymbol{\pi}_{\sigma_p},\boldsymbol{\lambda}_{j_\beta})=0\\ \text{for } s_{j_\beta}=0,1,2,\cdots,m_{j_\beta}-1,\\ \hbar_{j_\beta}\mathbf{n}_{\partial\Omega_{\beta j_\beta}}^{\mathrm{T}}(\mathbf{x}_{m+\varepsilon}^{(j_\beta;\mathscr{B})})\cdot[\mathbf{x}_{m+\varepsilon}^{(\sigma_p;\mathscr{E})}-\mathbf{x}_{m+\varepsilon}^{(j_\beta;\mathscr{B})}]>0\\ \text{for all } j_\beta\in\bigcup_{p=r+1}^{n}\bigcup_{\sigma_p\in\mathscr{N}_{\mathscr{E}}^{p}(\sigma_r)}\bar{\mathscr{N}}_{\beta\mathscr{B}}(\sigma_r,\sigma_p).\end{array}\right\}\tag{8.46}$$

(ii) The edge $\mathscr{E}_{\sigma_r}^{(r)}$ for the s-dimensional edge flow $\mathbf{x}^{(\sigma_s;\mathscr{E})}(t)$ in an edge $\mathscr{E}_{\sigma_s}^{(s)}$ ($\sigma_s\in\mathscr{N}_{\mathscr{E}}^{C}(\sigma_r)=\mathscr{N}_{\mathscr{E}}(\sigma_r)$) at point $\mathbf{x}_m\in\mathscr{E}_{\sigma_r}^{(r)}$ is *non-switchable of the first kind* (sink) with the $(\cup_{s=r+1}^{n}\cup_{\sigma_s}\mathbf{m}_{\sigma_s};\mathscr{E})$-singularity ($\mathbf{m}_{\sigma_s}\neq 2\mathbf{k}_{\sigma_s}+1$) if

$$\left.\begin{array}{l}\hbar_{i_\alpha} G_{\partial\Omega_{\alpha i_\alpha}}^{(s_{i_\alpha},\sigma_s;\mathscr{E})}(\mathbf{x}_m,t_{m-},\boldsymbol{\pi}_{\sigma_s},\boldsymbol{\lambda}_{i_\alpha})=0\\ \text{for } s_{i_\alpha}=0,1,2,\cdots,m_{i_\alpha}-1;\\ \hbar_{i_\alpha}\mathbf{n}_{\partial\Omega_{\alpha i_\alpha}}^{\mathrm{T}}(\mathbf{x}_{m-\varepsilon}^{(i_\alpha;\mathscr{B})})\cdot[\mathbf{x}_{m-\varepsilon}^{(i_\alpha;\mathscr{B})}-\mathbf{x}_{m-\varepsilon}^{(\sigma_s;\mathscr{E})}]<0\\ \text{for all } i_\alpha\in\bigcup_{s=r+1}^{n}\bigcup_{\sigma_s\in\mathscr{N}_{\mathscr{E}}^{s}(\sigma_r)}\bar{\mathscr{N}}_{\alpha\mathscr{B}}(\sigma_r,\sigma_s).\end{array}\right\}\tag{8.47}$$

(iii) The edge $\mathscr{E}_{\sigma_r}^{(r)}$ for the p-dimensional edge flow $\mathbf{x}^{(\sigma_p;\mathscr{E})}(t)$ in an edge $\mathscr{E}_{\sigma_p}^{(p)}$ ($\sigma_p\in\mathscr{N}_{\mathscr{E}}^{L}(\sigma_r)=\mathscr{N}_{\mathscr{E}}(\sigma_r)$) at point $\mathbf{x}_m\in\mathscr{E}_{\sigma_r}^{(r)}$ is *non-switchable of the second kind* (source) with the $(\cup_{p=r+1}^{n}\cup_{\sigma_p}\mathbf{m}_{\sigma_p};\mathscr{E})$ singularity if

$$\left.\begin{array}{l}\hbar_{j_\beta} G_{\partial\Omega_{\beta j_\beta}}^{(s_{j_\beta},\sigma_p;\mathscr{E})}(\mathbf{x}_m,t_{m-},\boldsymbol{\pi}_{\sigma_p},\boldsymbol{\lambda}_{j_\beta})=0\\ \text{for } s_{j_\alpha}=0,1,2,\cdots,m_{j_\beta}-1;\\ \hbar_{j_\beta}\mathbf{n}_{\partial\Omega_{\beta j_\beta}}^{\mathrm{T}}(\mathbf{x}_{m+\varepsilon}^{(j_\beta;\mathscr{B})})\cdot[\mathbf{x}_{m+\varepsilon}^{(\sigma_p;\mathscr{E})}-\mathbf{x}_{m+\varepsilon}^{(j_\beta;\mathscr{B})}]>0\\ \text{for all } j_\beta\in\bigcup_{p=r+1}^{n}\bigcup_{\sigma_p\in\mathscr{N}_{\mathscr{E}}^{p}(\sigma_r)}\bar{\mathscr{N}}_{\beta\mathscr{B}}(\sigma_r,\sigma_p).\end{array}\right\}\tag{8.48}$$

(iv) Without the switching rules, the s-dimensional edge flow $\mathbf{x}^{(\sigma_s;\mathscr{E})}(t)$ in an edge $\mathscr{E}_{\sigma_s}^{(s)}$ ($\sigma_s\in\mathscr{N}_{\mathscr{E}}^{C}(\sigma_r)$) and the p-dimensional edge flow $\mathbf{x}^{(\sigma_p;\mathscr{E})}(t)$ in an edge $\mathscr{E}_{\sigma_p}^{(p)}$ ($\sigma_s\in\mathscr{N}_{\mathscr{E}}^{L}(\sigma_r)$) at point $\mathbf{x}_m\in\mathscr{E}_{\sigma_r}^{(r)}$ are *potentially switchable* with the $(\cup_{s=r+1}^{n}\cup_{\sigma_s}\mathbf{m}_{\sigma_s};\mathscr{E}:\cup_{p=r+1}^{n}\cup_{\sigma_p}\mathbf{m}_{\sigma_p};\mathscr{E})$ singularity from $\mathscr{E}_{\sigma_s}^{(s)}$ to $\mathscr{E}_{\sigma_p}^{(p)}$ via edge $\mathscr{E}_{\sigma_r}^{(r)}$ if Eqs. (8.45) and (8.46) hold with $\mathbf{m}_{\sigma_s}\neq 2\mathbf{k}_{\sigma_s}+1$.

(v) Without the switching rules, the s-dimensional edge flow $\mathbf{x}^{(\sigma_s;\mathscr{E})}(t)$ in an edge $\mathscr{E}_{\sigma_s}^{(s)}$ ($\sigma_s\in\mathscr{N}_{\mathscr{E}}^{C}(\sigma_r)$) and the p-dimensional edge flow $\mathbf{x}^{(\sigma_p;\mathscr{E})}(t)$ in an edge $\mathscr{E}_{\sigma_p}^{(p)}$ ($\sigma_s\in\mathscr{N}_{\mathscr{E}}^{L}(\sigma_r)$) at point $\mathbf{x}_m\in\mathscr{E}_{\sigma_r}^{(r)}$ are *non-switchable* with the $(\cup_{s=r+1}^{n}\cup_{\sigma_s}\mathbf{m}_{\sigma_s};\mathscr{E}:\cup_{p=r+1}^{n}\cup_{\sigma_p}\mathbf{m}_{\sigma_p};\mathscr{E})$ singularity from $\mathscr{E}_{\sigma_s}^{(s)}$ to $\mathscr{E}_{\sigma_p}^{(p)}$ via edge $\mathscr{E}_{\sigma_r}^{(r)}$ if Eqs. (8.45) and (8.46) hold with $\mathbf{m}_{\sigma_s}=2\mathbf{k}_{\sigma_s}+1$ for at least one of coming flows.

(vi) With the switching rules, the s-dimensional grazing edge flow $\mathbf{x}^{(\sigma_s;\mathscr{E})}(t)$ in an edge $\mathscr{E}_{\sigma_s}^{(s)}$ ($\sigma_s\in\mathscr{N}_{\mathscr{E}}^{G}(\sigma_r)=\mathscr{N}_{\mathscr{E}}(\sigma_r)$) at point $\mathbf{x}_m\in\mathscr{E}_{\sigma_r}^{(r)}$ is *switchable* with the $(\cup_{s=r+1}^{n}\cup_{\sigma_s}(2\mathbf{k}_{\sigma_s}+1);\mathscr{E})$ singularity to edge $\mathscr{E}_{\sigma_r}^{(r)}$ if

$$\begin{array}{l}\hbar_{i_\alpha} G_{\partial\Omega_{\alpha i_\alpha}}^{(s_{i_\alpha},\sigma_s;\mathscr{E})}(\mathbf{x}_m,t_{m\pm},\boldsymbol{\pi}_{\sigma_s},\boldsymbol{\lambda}_{i_\alpha})=0\\ \text{for } s_{i_\alpha}=0,1,2,\cdots,2k_{i_\alpha};\end{array}$$

$$\begin{aligned}
&\hbar_{i_\alpha}\mathbf{n}^{\mathrm{T}}_{\partial\Omega_{\alpha i_\alpha}}(\mathbf{x}^{(i_\alpha;\mathscr{B})}_{m-\varepsilon})\cdot[\mathbf{x}^{(i_\alpha;\mathscr{B})}_{m-\varepsilon}-\mathbf{x}^{(\sigma_s;\mathscr{E})}_{m-\varepsilon}]<0,\\
&\hbar_{i_\alpha}\mathbf{n}^{\mathrm{T}}_{\partial\Omega_{\alpha i_\alpha}}(\mathbf{x}^{(i_\alpha;\mathscr{B})}_{m+\varepsilon})\cdot[\mathbf{x}^{(\sigma_s;\mathscr{E})}_{m+\varepsilon}-\mathbf{x}^{(i_\alpha;\mathscr{B})}_{m+\varepsilon}]>0\\
&\text{for all } i_\alpha\in\bigcup_{s=r+1}^{n}\bigcup_{\sigma_s\in\mathscr{N}^s_{\mathscr{E}}(\sigma_r)}\bar{\mathscr{N}}_{\alpha\mathscr{B}}(\sigma_r,\sigma_s).
\end{aligned} \tag{8.49}$$

(vii) Without the switching rules, the s-dimensional grazing edge flow $\mathbf{x}^{(\sigma_s;\mathscr{E})}(t)$ in an edge $\mathscr{E}^{(s)}_{\sigma_s}$ ($\sigma_s\in\mathscr{N}^G_{\mathscr{E}}(\sigma_r)=\mathscr{N}_{\mathscr{E}}(\sigma_r)$) at point $\mathbf{x}_m\in\mathscr{E}^{(r)}_{\sigma_r}$ is a *full* $\mathscr{N}^G_{\mathscr{E}}(\sigma_r)$*-grazing* edge flow with the $(\cup^n_{s=r+1}\cup_{\sigma_s}(2\mathbf{k}_{\sigma_s}+1);\mathscr{E})$-singularity to edge $\mathscr{E}^{(r)}_{\sigma_r}$ if Eq. (8.49) holds.

Similarly, from the foregoing definition, the corresponding theorem can be stated for the sufficient and necessary conditions for the switchability of all flows to the specific edge.

Theorem 8.6. *For a discontinuous dynamical system in Eqs.* (6.4)-(6.8)*, there is a point* $\mathbf{x}^{(\sigma_r;\mathscr{E})}(t_m)\equiv\mathbf{x}_m\in\mathscr{E}^{(r)}_{\sigma_r}$ ($\sigma_r\in\mathscr{N}^{(r)}_{\mathscr{E}}=\{1,2,\cdots,N_r\}$ *and* $r\in\{0,1,2,\cdots,n-2\}$) *at time* t_m *with the related boundaries* $\partial\Omega_{\alpha i_\alpha}(\sigma_r)$ *and related domains* $\Omega_\alpha(\sigma_r)$ ($\alpha\in\mathscr{N}_{\mathscr{D}}(\sigma_r)\subset\mathscr{N}_{\mathscr{D}}=\{1,2,\cdots,N_d\}$ *and* $i_\alpha\in\mathscr{N}_{\alpha\mathscr{B}}(\sigma_r)\subset\mathscr{N}_{\mathscr{B}}=\{1,2,\cdots,N_b\}$), *where the boundary index subset of* $\mathscr{N}_{\alpha\mathscr{B}}(\sigma_r)$ *has* n_{σ_r}*-elements* ($n_{\sigma_r}\geq(n-r)$). *The edge of* $\mathscr{E}^{(r)}_{\sigma_r}$ *is the intersection of edges (including domains and boundaries)* $\mathscr{E}^{(s)}_{\sigma_s}(\sigma_r)$ ($\sigma_s\in\mathscr{N}^{(s)}_{\mathscr{E}}(\sigma_r)\subset\mathscr{N}^{(s)}_{\mathscr{E}}=\{1,2,\cdots,N_s\}$ *and* $s\in\{r+1,r+2,\cdots,n-1,n\}$). *Suppose there is an* s*-dimensional edge* $\mathscr{E}^{(s)}_{\sigma_s}(\sigma_r)$ *with a boundary index set* $\mathscr{N}_{\alpha\mathscr{B}}(\sigma_r,\sigma_s)$ *relative to domain* $\Omega_\alpha(\sigma_r)$. *The complementary boundary index set is* $\bar{\mathscr{N}}_{\alpha\mathscr{B}}(\sigma_r,\sigma_p)=\mathscr{N}_{\alpha\mathscr{B}}(\sigma_r)/\mathscr{N}_{\alpha\mathscr{B}}(\sigma_r,\sigma_p)$. *Suppose the coming and leaving edge flows are* $\mathbf{x}^{(\sigma_s;\mathscr{E})}(t)$ ($\sigma_s\in\mathscr{N}^C_{\mathscr{E}}(\sigma_r,\sigma_s)$) *and* $\mathbf{x}^{(\sigma_p;\mathscr{E})}(t)$ ($\sigma_p\in\mathscr{N}^L_{\mathscr{E}}(\sigma_r,\sigma_p)$) *on the edge* $\mathscr{E}^{(s)}_{\sigma_s}$ *and* $\mathscr{E}^{(p)}_{\sigma_p}$, *respectively. For an arbitrarily small* $\varepsilon>0$, *there are two time intervals* $[t_{m-\varepsilon},t_m)$ *and* $(t_m,t_{m+\varepsilon}]$. *An edge flow* $\mathbf{x}^{(\sigma_s;\mathscr{E})}(t)$ *is* $C^{r_{\sigma_s}}_{[t_{m-\varepsilon},t_m)}$ *and/or* $C^{r_{\sigma_s}}_{(t_m,t_{m+\varepsilon}]}$*-continuous for time* t *and* $\|d^{r_{\sigma_s}+1}\mathbf{x}^{(\sigma_s;\mathscr{E})}/dt^{r_{\sigma_s}+1}\|<\infty$ ($r_{\sigma_s}\geq\max_{i_\alpha\in\mathscr{N}^C_{\alpha\mathscr{B}}}\{m_{i_\alpha}+1\}$). *An edge flow* $\mathbf{x}^{(\sigma_p;\mathscr{E})}(t)$ ($p\in\{r,r+1,\cdots,n-1,n\}$) *is* $C^{r_{\sigma_p}}_{[t_{m-\varepsilon},t_m)}$ *and/or* $C^{r_{\sigma_p}}_{(t_m,t_{m+\varepsilon}]}$*-continuous for time* t *and* $\|d^{r_{\sigma_p}+1}\mathbf{x}^{(\sigma_p;\mathscr{E})}/dt^{r_{\sigma_p}+1}\|<\infty$ ($r_{\sigma_p}\geq\max_{i_\alpha\in\mathscr{N}^L_{\alpha\mathscr{B}}}\{m_{j_\alpha}+1\}$). *Suppose the switching rule* $\mathbf{x}^{(\sigma_s;\mathscr{E})}(t_{m\pm})=\mathbf{x}_m=\mathbf{x}^{(\sigma_p;\mathscr{E})}(t_{m\pm})$ *exist.*

(i) *With the switching rule, the* s*-dimensional edge flow* $\mathbf{x}^{(\sigma_s;\mathscr{E})}(t)$ *in an edge*

$\mathscr{E}_{\sigma_s}^{(s)}$ ($\sigma_s \in \mathscr{N}_{\mathscr{E}}^{C}(\sigma_r)$) *and the* p*-dimensional edge flow* $\mathbf{x}^{(\sigma_p;\mathscr{E})}(t)$ *in an edge* $\mathscr{E}_{\sigma_p}^{(p)}$ ($\sigma_s \in \mathscr{N}_{\mathscr{E}}^{L}(\sigma_r)$) *at point* $\mathbf{x}_m \in \mathscr{E}_{\sigma_r}^{(r)}$ *are* ($\mathscr{N}_{\mathscr{E}}^{C}(\sigma_r):\mathscr{N}_{\mathscr{E}}^{L}(\sigma_r)$)*-switchable with the singularity of* $(\cup_{s=r+1}^{n}\cup_{\sigma_s}\mathbf{m}_{\sigma_s};\mathscr{E}:\cup_{p=r+1}^{n}\cup_{\sigma_p}\mathbf{m}_{\sigma_p};\mathscr{E})$ *from* $\mathscr{E}_{\sigma_s}^{(s)}$ *to* $\mathscr{E}_{\sigma_p}^{(p)}$ *via edge* $\mathscr{E}_{\sigma_r}^{(r)}$ *if and only if*

$$\left.\begin{array}{l}\hbar_{i_\alpha} G_{\partial\Omega_{\alpha i_\alpha}}^{(s_{i_\alpha},\sigma_s;\mathscr{E})}(\mathbf{x}_m,t_{m-},\boldsymbol{\pi}_{\sigma_s},\boldsymbol{\lambda}_{i_\alpha})=0\\ \text{for } s_{i_\alpha}=0,1,2,\cdots,m_{i_\alpha}-1,\\ (-1)^{m_{i_\alpha}}\hbar_{i_\alpha} G_{\partial\Omega_{\alpha i_\alpha}}^{(m_{i_\alpha},\sigma_s;\mathscr{E})}(\mathbf{x}_m,t_{m-},\boldsymbol{\pi}_{\sigma_s},\boldsymbol{\lambda}_{i_\alpha})<0\end{array}\right\}\tag{8.50}$$
$$\text{for all } i_\alpha \in \bigcup_{s=r+1}^{n}\bigcup_{\sigma_s\in\mathscr{N}_{\mathscr{E}}^{s}(\sigma_r)}\bar{\mathscr{N}}_{\alpha\mathscr{B}}(\sigma_r,\sigma_s);$$

$$\left.\begin{array}{l}\hbar_{j_\beta} G_{\partial\Omega_{\beta j_\beta}}^{(s_{j_\beta},\sigma_p;\mathscr{E})}(\mathbf{x}_m,t_{m+},\boldsymbol{\pi}_{\sigma_p},\boldsymbol{\lambda}_{j_\beta})=0\\ \text{for } s_{j_\alpha}=0,1,2,\cdots,m_{j_\beta}-1,\\ \hbar_{j_\beta} G_{\partial\Omega_{\beta j_\beta}}^{(m_{j_\beta},\sigma_p;\mathscr{E})}(\mathbf{x}_m,t_{m+},\boldsymbol{\pi}_{\sigma_p},\boldsymbol{\lambda}_{j_\beta})>0\end{array}\right\}\tag{8.51}$$
$$\text{for all } j_\beta \in \bigcup_{p=r+1}^{n}\bigcup_{\sigma_p\in\mathscr{N}_{\mathscr{E}}^{p}(\sigma_r)}\bar{\mathscr{N}}_{\beta\mathscr{B}}(\sigma_r,\sigma_p).$$

(ii) *The edge* $\mathscr{E}_{\sigma_r}^{(r)}$ *for the* s*-dimensional edge flow* $\mathbf{x}^{(\sigma_s;\mathscr{E})}(t)$ *in an edge* $\mathscr{E}_{\sigma_s}^{(s)}$ ($\sigma_s \in \mathscr{N}_{\mathscr{E}}^{C}(\sigma_r)=\mathscr{N}_{\mathscr{E}}(\sigma_r)$) *at point* $\mathbf{x}_m \in \mathscr{E}_{\sigma_r}^{(r)}$ *is an* $\mathscr{N}_{\mathscr{E}}^{C}(\sigma_r)$ *sink with the* $(\cup_{s=r+1}^{n}\cup_{\sigma_s}\mathbf{m}_{\sigma_s};\mathscr{E})$*-singularity* ($\mathbf{m}_{\sigma_s}\neq 2\mathbf{k}_{\sigma_s}+1$) *if and only if*

$$\left.\begin{array}{l}\hbar_{i_\alpha} G_{\partial\Omega_{\alpha i_\alpha}}^{(s_{i_\alpha},\sigma_s;\mathscr{E})}(\mathbf{x}_m,t_{m-},\boldsymbol{\pi}_{\sigma_s},\boldsymbol{\lambda}_{i_\alpha})=0\\ \text{for } s_{i_\alpha}=0,1,2,\cdots,m_{i_\alpha}-1,\\ (-1)^{m_{i_\alpha}}\hbar_{i_\alpha} G_{\partial\Omega_{\alpha i_\alpha}}^{(m_{i_\alpha},\sigma_s;\mathscr{E})}(\mathbf{x}_m,t_{m-},\boldsymbol{\pi}_{\sigma_s},\boldsymbol{\lambda}_{i_\alpha})<0\end{array}\right\}\tag{8.52}$$
$$\text{for all } i_\alpha \in \bigcup_{s=r+1}^{n}\bigcup_{\sigma_s\in\mathscr{N}_{\mathscr{E}}^{s}(\sigma_r)}\bar{\mathscr{N}}_{\alpha\mathscr{B}}(\sigma_r,\sigma_s).$$

(iii) *The edge* $\mathscr{E}_{\sigma_r}^{(r)}$ *for the* p*-dimensional edge flow* $\mathbf{x}^{(\sigma_p;\mathscr{E})}(t)$ *in an edge* $\mathscr{E}_{\sigma_p}^{(p)}$ ($\sigma_p \in \mathscr{N}_{\mathscr{E}}^{L}(\sigma_r)=\mathscr{N}_{\mathscr{E}}(\sigma_r)$) *at point* $\mathbf{x}_m \in \mathscr{E}_{\sigma_r}^{(r)}$ *is an* $\mathscr{N}_{\mathscr{E}}^{L}(\sigma_r)$ *source with the* $(\cup_{p=r+1}^{n}\cup_{\sigma_p}\mathbf{m}_{\sigma_p};\mathscr{E})$ *singularity if and only if*

$$\left.\begin{array}{l}\hbar_{j_\beta} G_{\partial\Omega_{\beta j_\beta}}^{(s_{j_\beta},\sigma_p;\mathscr{E})}(\mathbf{x}_m,t_{m+},\boldsymbol{\pi}_{\sigma_p},\boldsymbol{\lambda}_{j_\beta})=0\\ \text{for } s_{j_\beta}=0,1,2,\cdots,m_{j_\beta}-1,\\ \hbar_{j_\beta} G_{\partial\Omega_{\beta j_\beta}}^{(m_{j_\beta},\sigma_p;\mathscr{E})}(\mathbf{x}_m,t_{m+},\boldsymbol{\pi}_{\sigma_p},\boldsymbol{\lambda}_{j_\beta})>0\end{array}\right\}\tag{8.53}$$
$$\text{for all } j_\beta \in \bigcup_{p=r+1}^{n}\bigcup_{\sigma_p\in\mathscr{N}_{\mathscr{E}}^{p}(\sigma_r)}\bar{\mathscr{N}}_{\beta\mathscr{B}}(\sigma_r,\sigma_p).$$

(iv) *Without the switching rules, the* s *-dimensional edge flow* $\mathbf{x}^{(\sigma_s;\mathscr{E})}(t)$ *in an edge* $\mathscr{E}_{\sigma_s}^{(s)}$ ($\sigma_s \in \mathscr{N}_{\mathscr{E}}^{C}(\sigma_r)$) *and the* p *-dimensional edge flow* $\mathbf{x}^{(\sigma_p;\mathscr{E})}(t)$ *in an edge* $\mathscr{E}_{\sigma_p}^{(p)}$ ($\sigma_s \in \mathscr{N}_{\mathscr{E}}^{L}(\sigma_r)$) *at point* $\mathbf{x}_m \in \mathscr{E}_{\sigma_r}^{(r)}$ *are potentially switchable with the* $(\cup_{s=r+1}^{n} \cup_{\sigma_s} \mathbf{m}_{\sigma_s};\mathscr{E} : \cup_{p=r+1}^{n} \cup_{\sigma_p} \mathbf{m}_{\sigma_p};\mathscr{E})$ *singularity from* $\mathscr{E}_{\sigma_s}^{(s)}$ *to* $\mathscr{E}_{\sigma_p}^{(p)}$ *via edge* $\mathscr{E}_{\sigma_r}^{(r)}$ *if and only if Eqs.* (8.50) *and* (8.51) *hold with* $\mathbf{m}_{\sigma_s} \neq 2\mathbf{k}_{\sigma_s} + 1$.

(v) *Without the switching rules, the* s *-dimensional edge flow* $\mathbf{x}^{(\sigma_s;\mathscr{E})}(t)$ *in an edge* $\mathscr{E}_{\sigma_s}^{(s)}$ ($\sigma_s \in \mathscr{N}_{\mathscr{E}}^{C}(\sigma_r)$) *and the* p *-dimensional edge flow* $\mathbf{x}^{(\sigma_p;\mathscr{E})}(t)$ *in an edge* $\mathscr{E}_{\sigma_p}^{(p)}$ ($\sigma_s \in \mathscr{N}_{\mathscr{E}}^{L}(\sigma_r)$) *at point* $\mathbf{x}_m \in \mathscr{E}_{\sigma_r}^{(r)}$ *are potentially switchable with the* $(\cup_{s=r+1}^{n} \cup_{\sigma_s} \mathbf{m}_{\sigma_s};\mathscr{E} : \cup_{p=r+1}^{n} \cup_{\sigma_p} \mathbf{m}_{\sigma_p};\mathscr{E})$ *singularity from* $\mathscr{E}_{\sigma_s}^{(s)}$ *to* $\mathscr{E}_{\sigma_p}^{(p)}$ *via edge* $\mathscr{E}_{\sigma_r}^{(r)}$ *if and only if Eqs.* (8.50) *and* (8.51) *hold* $\mathbf{m}_{\sigma_s} = 2\mathbf{k}_{\sigma_s} + 1$ *for at least one of coming flows.*

(vi) *With the switching rules, the* s *-dimensional grazing edge flow* $\mathbf{x}^{(\sigma_s;\mathscr{E})}(t)$ *in an edge* $\mathscr{E}_{\sigma_s}^{(s)}$ ($\sigma_s \in \mathscr{N}_{\mathscr{E}}^{G}(\sigma_r) = \mathscr{N}_{\mathscr{E}}(\sigma_r)$) *at point* $\mathbf{x}_m \in \mathscr{E}_{\sigma_r}^{(r)}$ *is* $\mathscr{N}_{\mathscr{E}}^{G}(\sigma_r)$ *-switchable with the* $(\cup_{s=r+1}^{n} \cup_{\sigma_s} (2\mathbf{k}_{\sigma_s}+1);\mathscr{E})$ *-singularity to edge* $\mathscr{E}_{\sigma_r}^{(r)}$ *if and only if*

$$\left.\begin{aligned}
&\hbar_{i_\alpha} G_{\partial\Omega_{\alpha i_\alpha}}^{(s_{i_\alpha},\sigma_s;\mathscr{E})}(\mathbf{x}_m, t_{m\pm}, \boldsymbol{\pi}_{\sigma_s}, \boldsymbol{\lambda}_{i_\alpha}) = 0\\
&\text{for } s_{i_\alpha} = 0,1,2,\cdots,2k_{i_\alpha};\\
&\hbar_{i_\alpha} G_{\partial\Omega_{\alpha i_\alpha}}^{(2k_{i_\alpha}+1,\sigma_s;\mathscr{E})}(\mathbf{x}_m, t_{m\pm}, \boldsymbol{\pi}_{\sigma_s}, \boldsymbol{\lambda}_{i_\alpha}) > 0
\end{aligned}\right\} \tag{8.54}$$
$$\text{for all } i_\alpha \in \bigcup_{s=r+1}^{n} \bigcup_{\sigma_s \in \mathscr{N}_{\mathscr{E}}^{s}(\sigma_r)} \bar{\mathscr{N}}_{\alpha\mathscr{B}}(\sigma_r, \sigma_s).$$

(vii) *Without the switching rules, the* s *-dimensional grazing edge flow* $\mathbf{x}^{(\sigma_s;\mathscr{E})}(t)$ *in an edge* $\mathscr{E}_{\sigma_s}^{(s)}$ ($\sigma_s \in \mathscr{N}_{\mathscr{E}}^{G}(\sigma_r) = \mathscr{N}_{\mathscr{E}}(\sigma_r)$) *at point* $\mathbf{x}_m \in \mathscr{E}_{\sigma_r}^{(r)}$ *is a full* $\mathscr{N}_{\mathscr{E}}^{G}(\sigma_r)$ *-grazing edge flow with the* $(\cup_{s=r+1}^{n} \cup_{\sigma_s} (2\mathbf{k}_{\sigma_s}+1);\mathscr{E})$ *-singularity to edge* $\mathscr{E}_{\sigma_r}^{(r)}$ *if Eq.* (8.54) *holds.*

Proof. As in Theorem 3.2, this theorem can be proved similarly as the edge flow is treated as a special domain flow (also see, Luo, 2008, 2009). ■

As in Figs. 8.4 and 8.5, the switchability among all the edge flows of $\mathbf{x}^{(\sigma_s;\mathscr{E})}$ ($s = 0,1,2,\cdots,r-1$) to a specific edge $\mathscr{E}_{\sigma_r}^{(r)}$ can be illustrated to a specific edge $\mathscr{E}^{(s-3)}$ in s-dimensional space. If all the edge flows of $\mathbf{x}^{(\sigma_s;\mathscr{E})}$ ($s = 0,1,2,\cdots,r-1$)

to a specific edge $\mathscr{E}_{\sigma_r}^{(r)}$ are coming flows, then the specific edge $\mathscr{E}_{\sigma_r}^{(r)}$ is a sink edge. On the other hand, if all the edge flows of $\mathbf{x}^{(\sigma_s;\mathscr{E})}$ ($s=0,1,2,\cdots,r-1$) to a specific edge $\mathscr{E}_{\sigma_r}^{(r)}$ are leaving flows, then the specific edge $\mathscr{E}_{\sigma_r}^{(r)}$ is a source edge. All the edge flows of $\mathbf{x}^{(\sigma_s;\mathscr{E})}$, $\mathbf{x}^{(\sigma_{s-1};\mathscr{E})}$ and $\mathbf{x}^{(\sigma_{s-2};\mathscr{E})}$ to the edge $\mathscr{E}^{(s-3)}$ are used to illustrate the specific edge $\mathscr{E}^{(s-3)}$ in s-dimensional space as a sink and source edge, as shown in Fig. 8.6(a) and (b), respectively. If all the edge flows to the specific edge are divided into the coming and leaving flows, the edge flows to the special edge are switchable with the switching rule, but the edge flows to the special edge are potentially switchable without the switching rule, as shown in Fig. 8.7(a). In Fig. 8.7(b), all the edge flows of $\mathbf{x}^{(\sigma_s;\mathscr{E})}$ and $\mathbf{x}^{(\sigma_{s-1};\mathscr{E})}$ grazing to the edge $\mathscr{E}^{(s-2)}$ are illustrated in Fig. 8.7(b).

8.3. Edge flow attractivity

As for the domain flow attractivity to a specific edge in Chapter 6, the attractivity of an edge flow to a specific edge will be discussed herein.

Definition 8.8. For a discontinuous dynamical system in Eqs. (6.4)-(6.8), there is a point $\mathbf{x}^{(\sigma_r;\mathscr{E})}(t_m)\equiv\mathbf{x}_m\in\mathscr{E}_{\sigma_r}^{(r)}$ ($\sigma_r\in\mathscr{N}_{\mathscr{E}}^{(r)}=\{1,2,\cdots,N_r\}$ and $r\in\{0,1,2,\cdots,n-2\}$) at time t_m with the related boundaries $\partial\Omega_{\alpha i_\alpha}(\sigma_r)$ and related domains $\Omega_\alpha(\sigma_r)$ ($\alpha\in\mathscr{N}_{\mathscr{D}}(\sigma_r)\subset\mathscr{N}_{\mathscr{D}}=\{1,2,\cdots,N_d\}$ and $i_\alpha\in\mathscr{N}_{\alpha\mathscr{B}}(\sigma_r)\subset\mathscr{N}_{\mathscr{B}}=\{1,2,\cdots,N_b\}$), where the boundary index subset of $\mathscr{N}_{\alpha\mathscr{B}}(\sigma_r)$ has n_{σ_r}-elements ($n_{\sigma_r}\geq(n-r)$). The edge of $\mathscr{E}_{\sigma_r}^{(r)}$ is the intersection of edges (including domains and boundaries) $\mathscr{E}_{\sigma_s}^{(s)}(\sigma_r)$ ($\sigma_s\in\mathscr{N}_{\mathscr{E}}^{s}(\sigma_r)\subset\mathscr{N}_{\mathscr{E}}^{s}=\{1,2,\cdots,N_s\}$ and $s\in\{r+1,r+2,\cdots,n-1,n\}$). Suppose there is an s-dimensional edge $\mathscr{E}_{\sigma_s}^{(s)}(\sigma_r)$ with a boundary index set $\mathscr{N}_{\alpha\mathscr{B}}(\sigma_r,\sigma_s)$ relative to domain $\Omega_\alpha(\sigma_r)$. The complementary boundary index set is $\overline{\mathscr{N}}_{\alpha\mathscr{B}}(\sigma_r,\sigma_s)=\mathscr{N}_{\alpha\mathscr{B}}(\sigma_r)/\mathscr{N}_{\alpha\mathscr{B}}(\sigma_r,\sigma_s)$. If there is a flow $\mathbf{x}^{(\sigma_s;\mathscr{E})}(t)$ in edge $\mathscr{E}_{\sigma_s}^{(s)}(\sigma_r)$, at time t_m, $\mathbf{x}_m^{(\sigma_s;\mathscr{E})}=\mathbf{x}_\rho^{(\sigma_s;\mathscr{E})}(t_m)\in\mathscr{E}_{\sigma_r}^{(r,\sigma_s)}(\rho_m)$ ($\rho\in\mathbb{N}$) satisfying

$$
\begin{aligned}
&\dot{\mathbf{x}}_\rho^{(\sigma_s;\mathscr{E})}=\mathbf{F}^{(\sigma_s,\rho)}(\mathbf{x}_\rho^{(\sigma_s;\mathscr{E})},t,\boldsymbol{\pi}_{\sigma_s})\\
&\text{with } \varphi_{\alpha i_\alpha}(\mathbf{x}_\rho^{(\sigma_s;\mathscr{E})},t,\boldsymbol{\lambda}_{i_\alpha})=\varphi_{\alpha i_\alpha}(\mathbf{x}_{\rho m}^{(\sigma_s;\mathscr{E})},t_m,\boldsymbol{\lambda}_{i_\alpha})\equiv C_m^{(i_\alpha,\sigma_s)}\\
&\text{on } \mathscr{E}_{\sigma_s}^{(s)}(\sigma_r,\rho)\ (\alpha\in\mathscr{N}_{\mathscr{D}}(\sigma_s)\subset\mathscr{N}_{\mathscr{D}},i_\alpha\in\overline{\mathscr{N}}_{\alpha\mathscr{B}}(\sigma_r,\sigma_s)\subset\mathscr{N}_{\mathscr{B}};\\
&\sigma_s\in\mathscr{N}_{\mathscr{E}}^{s};s\in\{r+1,r+2,\cdots,n-2\}),
\end{aligned}
\tag{8.55}
$$

where $C_m^{(i_\alpha,\sigma_s)}$ is a non-zero constant.

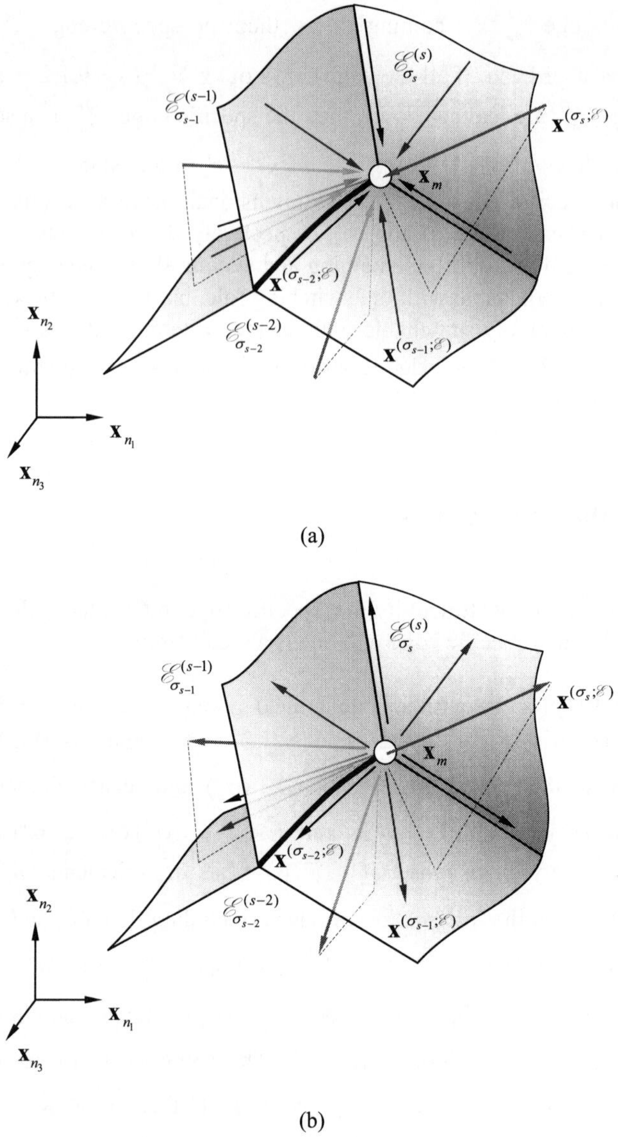

Fig. 8.6 The edge $\mathscr{E}^{(s-3)}$. **(a)** sink edge and **(b)** source edge for all domain, boundary and edge flows $\mathbf{x}^{(\sigma_p;\mathscr{E})}$ ($p = 0,1,2,\cdots,s-2$). All the edge flows are presented by curves with arrows. Only $\mathbf{x}^{(\sigma_s;\mathscr{E})}$, $\mathbf{x}^{(\sigma_{s-1};\mathscr{E})}$ and $\mathbf{x}^{(\sigma_{s-2};\mathscr{E})}$ can be illustrated. ($n_1 + n_2 + n_3 = s$) (color plot in the book end)

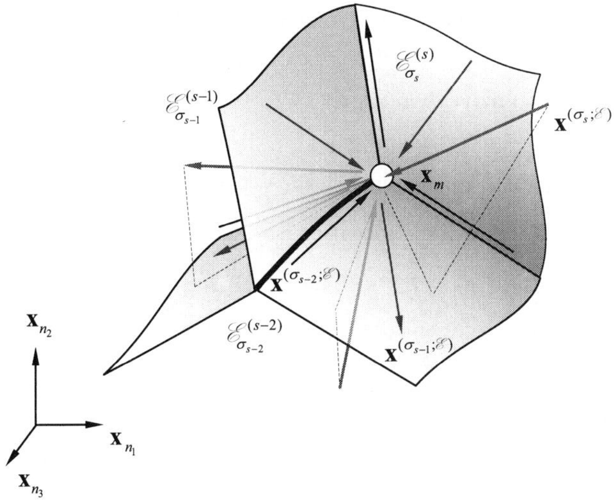

(a)

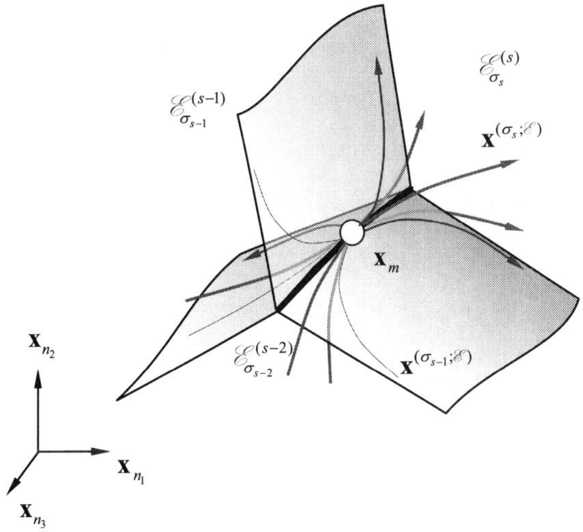

(b)

Fig. 8.7 **(a)** Potential switching flows to the edge $\mathscr{E}^{(s-3)}$ and **(b)** all tangential flows to the edge $\mathscr{E}^{(s-2)}$. $\mathbf{x}^{(\sigma_p;\mathscr{E})}$ ($p = 0,1,2,\cdots,s-2$). All the edge flows are presented by curves with arrows. Only $\mathbf{x}^{(\sigma_s;\mathscr{E})}$, $\mathbf{x}^{(\sigma_{s-1};\mathscr{E})}$ and $\mathbf{x}^{(\sigma_{s-2};\mathscr{E})}$ can be illustrated. ($n_1 + n_2 + n_3 = s$) (color plot in the book end)

Definition 8.9. For a discontinuous dynamical system in Eqs. (6.4)-(6.8), there is a point $\mathbf{x}^{(\sigma_r;\mathscr{E})}(t_m)\equiv\mathbf{x}_m\in\mathscr{E}_{\sigma_r}^{(r)}$ ($\sigma_r\in\mathscr{N}_{\mathscr{E}}^{(r)}=\{1,2,\cdots,N_r\}$ and $r\in\{0,1,2,\cdots,n-2\}$) at time t_m with the related boundaries $\partial\Omega_{\alpha i_\alpha}(\sigma_r)$ and related domains $\Omega_\alpha(\sigma_r)$ ($\alpha\in\mathscr{N}_{\mathscr{D}}(\sigma_r)\subset\mathscr{N}_{\mathscr{D}}=\{1,2,\cdots,N_d\}$ and $i_\alpha\in\mathscr{N}_{\alpha\mathscr{B}}(\sigma_r)\subset\mathscr{N}_{\mathscr{B}}=\{1,2,\cdots,N_b\}$), where the boundary index subset of $\mathscr{N}_{\alpha\mathscr{B}}(\sigma_r)$ has n_{σ_r}-elements ($n_{\sigma_r}\geq(n-r)$). The edge of $\mathscr{E}_{\sigma_r}^{(r)}$ is the intersection of edges (including domains and boundaries) $\mathscr{E}_{\sigma_s}^{(s)}(\sigma_r)$ ($\sigma_s\in\mathscr{N}_{\mathscr{E}}^{s}(\sigma_r)\subset\mathscr{N}_{\mathscr{E}}^{s}=\{1,2,\cdots,N_s\}$ and $s\in\{r+1,r+2,\cdots,n-1,n\}$). Suppose there is an s-dimensional edge $\mathscr{E}_{\sigma_s}^{(s)}(\sigma_r)$ with a boundary index set $\mathscr{N}_{\alpha\mathscr{B}}(\sigma_r,\sigma_s)$ relative to domain $\Omega_\alpha(\sigma_r)$. The complementary boundary index set is $\bar{\mathscr{N}}_{\alpha\mathscr{B}}(\sigma_r,\sigma_s)=\mathscr{N}_{\alpha\mathscr{B}}(\sigma_r)/\mathscr{N}_{\alpha\mathscr{B}}(\sigma_r,\sigma_s)$. For an arbitrarily small $\varepsilon>0$, there are two time intervals $[t_{m-\varepsilon},t_m)$ and $(t_m,t_{m+\varepsilon}]$. An edge flow $\mathbf{x}^{(\sigma_s;\mathscr{E})}(t)$ is $C^{r_{\sigma_s}}_{[t_{m-\varepsilon},t_m)}$ and/or $C^{r_{\sigma_s}}_{(t_m,t_{m+\varepsilon}]}$-continuous for time t and $\|d^{r_{\sigma_s}+1}\mathbf{x}^{(\sigma_s;\mathscr{E})}/dt^{r_{\sigma_s}+1}\|<\infty$ ($r_{\sigma_s}\geq1$). Suppose there is an $(n-s+r)$-dimensional measuring edge $\mathscr{E}_{\sigma_{n-s+r}}^{(\rho,n-s+r)}$ of a flow $\mathbf{x}^{(\sigma_s;\mathscr{E})}(t)$ in edge $\mathscr{E}_{\sigma_s}^{(s)}(\sigma_r,\alpha)$ with an equi-constant vector of the specific edge $\mathscr{E}_{\sigma_r}^{(r)}(\alpha)$, and the measuring edge $\mathscr{E}_{\sigma_{n-s+r}}^{(\rho,n-s+r)}(\sigma_r,\alpha)=\cap_{j_\alpha\in\bar{\mathscr{N}}_{\alpha\mathscr{B}}(\sigma_r,\sigma_s)}S_\rho^{(j_\alpha,\alpha)}$ is formed by an intersection of all measuring surfaces $S_{\sigma_m}^{(i_\alpha,\alpha)}$ for $i_\alpha\in\bar{\mathscr{N}}_{\alpha\mathscr{B}}(\sigma_r,\sigma_s)$. $\mathscr{E}_{\sigma_r}^{(r)}(\alpha)=\mathscr{E}_{\sigma_{n-s+r}}^{(n-s+r)}(\sigma_r,\alpha)\cap\mathscr{E}_{\sigma_s}^{(s)}(\sigma_r,\alpha)$ with $\mathscr{E}_{\sigma_{n-s+r}}^{(n-s+r)}(\sigma_r,\alpha)=\cap_{j_\alpha\in\bar{\mathscr{N}}_{\alpha\mathscr{B}}(\sigma_r,\sigma_s)}\partial\Omega_{\alpha j_\alpha}$ and $\mathscr{E}_{\sigma_s}^{(s)}(\sigma_r,\alpha)=\cap_{i_\alpha\in\mathscr{N}_{\alpha\mathscr{B}}(\sigma_r,\sigma_s)}\partial\Omega_{\alpha i_\alpha}\subset\mathscr{E}_{\sigma_s}^{(s)}(\sigma_r)$. At time t_m, $\mathbf{x}_m^{(\sigma_s;\mathscr{E})}=\mathbf{x}_\rho^{(\sigma_s;\mathscr{E})}(t_m)\in\mathscr{E}_{\sigma_r}^{(r,\sigma_s)}(\rho_m)$ ($\rho\in\mathbb{N}$) satisfies Eq. (8.55). Suppose $\mathbf{x}_{\rho(m\pm\varepsilon)}^{(j_\alpha,\alpha)}\in S_{\rho_m}^{(j_\alpha,\alpha)}$ and $\mathbf{x}_{m\pm\varepsilon}^{(\sigma_s;\mathscr{E})}=\mathbf{x}^{(\sigma_s;\mathscr{E})}(t_{m\pm\varepsilon})\in\mathscr{E}_{\sigma_{n-s+r}}^{(\rho_m,n-s+r)}$.

(i) The edge flow $\mathbf{x}^{(\sigma_s;\mathscr{E})}(t)$ at time t_m is *attractive* to the edge $\mathscr{E}_{\sigma_r}^{(r)}$ if

$$\left.\begin{aligned}&\hbar_{j_\alpha}\mathbf{n}_{\partial\Omega_{\alpha i_\alpha}}^{\mathrm{T}}(\mathbf{x}_{\rho(m-\varepsilon)}^{(j_\alpha,\alpha)})\cdot[\mathbf{x}_{\rho(m-\varepsilon)}^{(j_\alpha,\alpha)}-\mathbf{x}_{m-\varepsilon}^{(\sigma_s;\mathscr{E})}]<0,\\&\hbar_{j_\alpha}\mathbf{n}_{\partial\Omega_{\alpha i_\alpha}}^{\mathrm{T}}(\mathbf{x}_{\rho(m+\varepsilon)}^{(j_\alpha,\alpha)})\cdot[\mathbf{x}_{m+\varepsilon}^{(\sigma_s;\mathscr{E})}-\mathbf{x}_{\rho(m+\varepsilon)}^{(j_\alpha,\alpha)}]<0,\\&\hbar_{j_\alpha}C_{m-\varepsilon}^{(j_\alpha,\alpha)}>\hbar_{j_\alpha}C_{m}^{(j_\alpha,\alpha)}>\hbar_{j_\alpha}C_{m+\varepsilon}^{(j_\alpha,\alpha)}\\&\text{for all } j_\alpha\in\bar{\mathscr{N}}_{\alpha\mathscr{B}}(\sigma_r,\sigma_s).\end{aligned}\right\}\tag{8.56}$$

(ii) The edge flow $\mathbf{x}^{(\sigma_s;\mathscr{E})}(t)$ at time t_m is *repulsive to* the edge $\mathscr{E}_{\sigma_r}^{(r)}$ if

$$\left.\begin{aligned}&\hbar_{j_\alpha}\mathbf{n}_{\partial\Omega_{\alpha i_\alpha}}^{\mathrm{T}}(\mathbf{x}_{\rho(m-\varepsilon)}^{(j_\alpha,\alpha)})\cdot[\mathbf{x}_{\rho(m-\varepsilon)}^{(j_\alpha,\alpha)}-\mathbf{x}_{m-\varepsilon}^{(\sigma_s;\mathscr{E})}]>0,\\&\hbar_{j_\alpha}\mathbf{n}_{\partial\Omega_{\alpha i_\alpha}}^{\mathrm{T}}(\mathbf{x}_{\rho(m+\varepsilon)}^{(j_\alpha,\alpha)})\cdot[\mathbf{x}_{m+\varepsilon}^{(\sigma_s;\mathscr{E})}-\mathbf{x}_{\rho(m+\varepsilon)}^{(j_\alpha,\alpha)}]>0,\\&\hbar_{j_\alpha}C_{m-\varepsilon}^{(j_\alpha,\alpha)}<\hbar_{j_\alpha}C_{m}^{(j_\alpha,\alpha)}<\hbar_{j_\alpha}C_{m+\varepsilon}^{(j_\alpha,\alpha)}\end{aligned}\right\}$$

$$\text{for all } j_\alpha \in \bar{\mathscr{N}}_{\alpha\mathscr{B}}(\sigma_r,\sigma_s). \tag{8.57}$$

(iii) The edge flow $\mathbf{x}^{(\sigma_s;\mathscr{E})}(t)$ at time t_m to the edge $\mathscr{E}_{\sigma_r}^{(r)}$ is from the *attractive* to repulsive state if

$$\left.\begin{aligned}
&\hbar_{j_\alpha}\mathbf{n}^{\mathrm{T}}_{\partial\Omega_{\alpha i_\alpha}}(\mathbf{x}^{(j_\alpha,\alpha)}_{\rho(m-\varepsilon)})\cdot[\mathbf{x}^{(j_\alpha,\alpha)}_{\rho(m-\varepsilon)}-\mathbf{x}^{(\sigma_s;\mathscr{E})}_{m-\varepsilon}]<0,\\
&\hbar_{j_\alpha}\mathbf{n}^{\mathrm{T}}_{\partial\Omega_{\alpha i_\alpha}}(\mathbf{x}^{(j_\alpha,\alpha)}_{\rho(m+\varepsilon)})\cdot[\mathbf{x}^{(\sigma_s;\mathscr{E})}_{m+\varepsilon}-\mathbf{x}^{(j_\alpha,\alpha)}_{\rho(m+\varepsilon)}]>0,\\
&\hbar_{j_\alpha}C^{(j_\alpha,\alpha)}_{m}<\hbar_{j_\alpha}C^{(j_\alpha,\alpha)}_{m-\varepsilon}\text{ and }\hbar_{j_\alpha}C^{(j_\alpha,\alpha)}_{m}<\hbar_{j_\alpha}C^{(j_\alpha,\alpha)}_{m+\varepsilon}
\end{aligned}\right\} \tag{8.58}$$
$$\text{for all } j_\alpha \in \bar{\mathscr{N}}_{\alpha\mathscr{B}}(\sigma_r,\sigma_s).$$

(iv) The edge flow $\mathbf{x}^{(\sigma_s;\mathscr{E})}(t)$ at time t_m to the edge $\mathscr{E}_{\sigma_r}^{(r)}$ is from the *repulsive* to attractive state if

$$\left.\begin{aligned}
&\hbar_{j_\alpha}\mathbf{n}^{\mathrm{T}}_{\partial\Omega_{\alpha i_\alpha}}(\mathbf{x}^{(j_\alpha,\alpha)}_{\rho(m-\varepsilon)})\cdot[\mathbf{x}^{(j_\alpha,\alpha)}_{\rho(m-\varepsilon)}-\mathbf{x}^{(\sigma_s;\mathscr{E})}_{m-\varepsilon}]>0,\\
&\hbar_{j_\alpha}\mathbf{n}^{\mathrm{T}}_{\partial\Omega_{\alpha i_\alpha}}(\mathbf{x}^{(j_\alpha,\alpha)}_{\rho(m+\varepsilon)})\cdot[\mathbf{x}^{(\sigma_s;\mathscr{E})}_{m+\varepsilon}-\mathbf{x}^{(j_\alpha,\alpha)}_{\rho(m+\varepsilon)}]<0,\\
&\hbar_{j_\alpha}C^{(j_\alpha,\alpha)}_{m}>\hbar_{j_\alpha}C^{(j_\alpha,\alpha)}_{m-\varepsilon}\text{ and }\hbar_{j_\alpha}C^{(j_\alpha,\alpha)}_{m}>\hbar_{j_\alpha}C^{(j_\alpha,\alpha)}_{m+\varepsilon}
\end{aligned}\right\} \tag{8.59}$$
$$\text{for all } j_\alpha \in \bar{\mathscr{N}}_{\alpha\mathscr{B}}(\sigma_r,\sigma_s).$$

(v) The edge flow $\mathbf{x}^{(\sigma_s;\mathscr{E})}(t)$ at time t_m to the edge of $\mathscr{E}_{\sigma_r}^{(r)}$ is *invariant* with equi-measuring quantities of $C_m^{(i_\alpha,\alpha)}$ if for all $i_\alpha \in \mathscr{N}_{\alpha\mathscr{B}}(\sigma_r)$

$$\left.\begin{aligned}
&\mathbf{x}^{(\sigma_s;\mathscr{E})}_{m-\varepsilon}=\mathbf{x}^{(j_\alpha,\alpha)}_{\rho(m-\varepsilon)}\text{ and }\mathbf{x}^{(\sigma_s;\mathscr{E})}_{m+\varepsilon}=\mathbf{x}^{(j_\alpha,\alpha)}_{\rho(m+\varepsilon)}\\
&C^{(j_\alpha,\alpha)}_{m-\varepsilon}=C^{(j_\alpha,\alpha)}_{m}=C^{(j_\alpha,\alpha)}_{m+\varepsilon}\text{ for all } j_\alpha \in \bar{\mathscr{N}}_{\alpha\mathscr{B}}(\sigma_r,\sigma_s).
\end{aligned}\right. \tag{8.60}$$

From the definition, the corresponding conditions for such attractivity of the edge flow to the specific edge can be obtained through the following theorem.

Theorem 8.7. *For a discontinuous dynamical system in Eqs.* (6.4)-(6.8)*, there is a point* $\mathbf{x}^{(\sigma_r;\mathscr{E})}(t_m)\equiv\mathbf{x}_m\in\mathscr{E}_{\sigma_r}^{(r)}$ ($\sigma_r\in\mathscr{N}_{\mathscr{E}}^{(r)}=\{1,2,\cdots,N_r\}$ *and* $r\in\{0,1,2,\cdots,n-2\}$) *at time* t_m *with the related boundaries* $\partial\Omega_{\alpha i_\alpha}(\sigma_r)$ *and related domains* $\Omega_\alpha(\sigma_r)$ ($\alpha\in\mathscr{N}_{\mathscr{D}}(\sigma_r)\subset\mathscr{N}_{\mathscr{D}}=\{1,2,\cdots,N_d\}$ *and* $i_\alpha\in\mathscr{N}_{\alpha\mathscr{B}}(\sigma_r)\subset\mathscr{N}_{\mathscr{B}}=\{1,2,\cdots,N_b\}$), *where the boundary index subset of* $\mathscr{N}_{\alpha\mathscr{B}}(\sigma_r)$ *has* n_{σ_r}*-elements* ($n_{\sigma_r}\geq(n-r)$). *The edge of* $\mathscr{E}_{\sigma_r}^{(r)}$ *is the intersection of edges (including domains and boundaries)* $\mathscr{E}_{\sigma_s}^{(s)}(\sigma_r)$ ($\sigma_s\in\mathscr{N}_{\mathscr{E}}^{s}(\sigma_r)\subset\mathscr{N}_{\mathscr{E}}^{s}=\{1,2,\cdots,N_s\}$ *and* $s\in\{r+1,r+2,\cdots,n-1,n\}$). *Suppose there is an* s*-dimensional edge* $\mathscr{E}_{\sigma_s}^{(s)}(\sigma_r)$ *with a boundary index set* $\mathscr{N}_{\alpha\mathscr{B}}(\sigma_r,\sigma_s)$ *relative to domain* $\Omega_\alpha(\sigma_r)$. *The complementary boundary index* *set is* $\bar{\mathscr{N}}_{\alpha\mathscr{B}}(\sigma_r,\sigma_s)=\mathscr{N}_{\alpha\mathscr{B}}(\sigma_r)/\mathscr{N}_{\alpha\mathscr{B}}(\sigma_r,\sigma_s)$. *For an arbitrarily small* $\varepsilon>0$, *there are two time intervals* $[t_{m-\varepsilon},t_m)$ *and* $(t_m,t_{m+\varepsilon}]$. *An edge flow* $\mathbf{x}^{(\sigma_s;\mathscr{E})}(t)$ *is* $C^{r_{\sigma_s}}_{[t_{m-\varepsilon},t_m)}$ *and/or* $C^{r_{\sigma_s}}_{(t_m,t_{m+\varepsilon}]}$*-continuous for time* t *and* $\| d^{r_{\sigma_s}+1}\mathbf{x}^{(\sigma_s;\mathscr{E})}/dt^{r_{\sigma_s}+1}\|<\infty$ ($r_{\sigma_s}\geq 2$). *Suppose there is an* $(n-s+r)$*-dimensional measuring edge* $\mathscr{E}^{(\rho,n-s+r)}_{\sigma_{n-s+r}}$ *of a flow* $\mathbf{x}^{(\sigma_s;\mathscr{E})}(t)$ *in edge* $\mathscr{E}^{(s)}_{\sigma_s}(\sigma_r,\alpha)$ *with an equi-constant vector of the specific edge* $\mathscr{E}^{(r)}_{\sigma_r}(\alpha)$. *The measuring edge* $\mathscr{E}^{(\rho,n-s+r)}_{\sigma_{n-s+r}}(\sigma_r,\alpha)=\cap_{j_\alpha\in\bar{\mathscr{N}}_{\alpha\mathscr{B}}(\sigma_r,\sigma_s)}S^{(j_\alpha,\alpha)}_{\rho}$ *is formed by an intersection of all measuring surface* $S^{(i_\alpha,\alpha)}_{\sigma_m}$ *for* $i_\alpha\in\bar{\mathscr{N}}_{\alpha\mathscr{B}}(\sigma_r,\sigma_s)$. $\mathscr{E}^{(r)}_{\sigma_r}(\alpha)=\mathscr{E}^{(n-s+r)}_{\sigma_{n-s+r}}(\sigma_r,\alpha)\cap\mathscr{E}^{(s)}_{\sigma_s}(\sigma_r,\alpha)$ *with* $\mathscr{E}^{(n-s+r)}_{\sigma_{n-s+r}}(\sigma_r,\alpha)=\cap_{j_\alpha\in\bar{\mathscr{N}}_{\alpha\mathscr{B}}(\sigma_r,\sigma_s)}\partial\Omega_{\alpha j_\alpha}$ *and* $\mathscr{E}^{(s)}_{\sigma_s}(\sigma_r,\alpha)=\cap_{i_\alpha\in\mathscr{N}_{\alpha\mathscr{B}}(\sigma_r,\sigma_s)}\partial\Omega_{\alpha i_\alpha}\subset\mathscr{E}^{(s)}_{\sigma_s}(\sigma_r)$. *At time* t_m, $\mathbf{x}^{(\sigma_s;\mathscr{E})}_m=\mathbf{x}^{(\sigma_s;\mathscr{E})}_{\rho}(t_m)\in\mathscr{E}^{(r,\sigma_s)}_{\sigma_r}(\rho_m)$ ($\rho\in\mathbb{N}$) *satisfies Eq.* (8.55). *Suppose* $\mathbf{x}^{(j_\alpha,\alpha)}_{\rho(m\pm\varepsilon)}\in S^{(j_\alpha,\alpha)}_{\rho_m}$ *and* $\mathbf{x}^{(\sigma_s;\mathscr{E})}_{m\pm\varepsilon}=\mathbf{x}^{(\sigma_s;\mathscr{E})}(t_{m\pm\varepsilon})\in\mathscr{E}^{(\rho_m,n-s+r)}_{\sigma_{n-s+r}}$.

(i) *The edge flow* $\mathbf{x}^{(\sigma_s;\mathscr{E})}(t)$ *at time* t_m *is attractive to the edge* $\mathscr{E}^{(r)}_{\sigma_r}$ *if and only if*

$$\begin{gathered}\hbar_{j_\alpha}G^{(\sigma_s;\mathscr{E})}_{\partial\Omega_{\alpha j_\alpha}}(\mathbf{x}^{(\sigma_s;\mathscr{E})}_m,t_m,\boldsymbol{\pi}_{\sigma_s},\boldsymbol{\lambda}_{j_\alpha})<0\\ \textit{for all } j_\alpha\in\bar{\mathscr{N}}_{\alpha\mathscr{B}}(\sigma_r,\sigma_s).\end{gathered}\tag{8.61}$$

(ii) *The edge flow* $\mathbf{x}^{(\sigma_s;\mathscr{E})}(t)$ *at time* t_m *is repulsive to the edge* $\mathscr{E}^{(r)}_{\sigma_r}$ *if and only if*

$$\begin{gathered}\hbar_{j_\alpha}G^{(\sigma_s;\mathscr{E})}_{\partial\Omega_{\alpha j_\alpha}}(\mathbf{x}^{(\sigma_s;\mathscr{E})}_m,t_m,\boldsymbol{\pi}_{\sigma_s},\boldsymbol{\lambda}_{j_\alpha})>0\\ \textit{for all } j_\alpha\in\bar{\mathscr{N}}_{\alpha\mathscr{B}}(\sigma_r,\sigma_s).\end{gathered}\tag{8.62}$$

(iii) *The edge flow* $\mathbf{x}^{(\sigma_s;\mathscr{E})}(t)$ *at time* t_m *to the edge* $\mathscr{E}^{(r)}_{\sigma_r}$ *is from the attractive to repulsive state if and only if*

$$\begin{gathered}\hbar_{j_\alpha}G^{(\sigma_s;\mathscr{E})}_{\partial\Omega_{\alpha j_\alpha}}(\mathbf{x}^{(\sigma_s;\mathscr{E})}_m,t_m,\boldsymbol{\pi}_{\sigma_s},\boldsymbol{\lambda}_{j_\alpha})=0,\\ \hbar_{j_\alpha}G^{(1,\sigma_s;\mathscr{E})}_{\partial\Omega_{\alpha j_\alpha}}(\mathbf{x}^{(\sigma_s;\mathscr{E})}_m,t_m,\boldsymbol{\pi}_{\sigma_s},\boldsymbol{\lambda}_{j_\alpha})>0,\\ \textit{for all } j_\alpha\in\bar{\mathscr{N}}_{\alpha\mathscr{B}}(\sigma_r,\sigma_s).\end{gathered}\tag{8.63}$$

(iv) *The edge flow* $\mathbf{x}^{(\sigma_s;\mathscr{E})}(t)$ *at time* t_m *to the edge of* $\mathscr{E}^{(r)}_{\sigma_r}$ *is from the repulsive to attractive state if and only if*

$$\begin{gathered}\hbar_{j_\alpha}G^{(\sigma_s;\mathscr{E})}_{\partial\Omega_{\alpha j_\alpha}}(\mathbf{x}^{(\sigma_s;\mathscr{E})}_m,t_m,\boldsymbol{\pi}_{\sigma_s},\boldsymbol{\lambda}_{j_\alpha})=0,\\ \hbar_{j_\alpha}G^{(1,\sigma_s;\mathscr{E})}_{\partial\Omega_{\alpha j_\alpha}}(\mathbf{x}^{(\sigma_s;\mathscr{E})}_m,t_m,\boldsymbol{\pi}_{\sigma_s},\boldsymbol{\lambda}_{j_\alpha})<0,\\ \textit{for all } j_\alpha\in\bar{\mathscr{N}}_{\alpha\mathscr{B}}(\sigma_r,\sigma_s).\end{gathered}\tag{8.64}$$

(v) *The edge flow* $\mathbf{x}^{(\sigma_s;\mathscr{E})}(t)$ *at time* t_m *to the edge* $\mathscr{E}_{\sigma_r}^{(r)}$ *is invariant with equi-measuring quantities of* $C_m^{(i_\alpha,\alpha)}$ *if and only if*

$$\begin{aligned}&\hbar_{j_\alpha} G_{\partial\Omega_{\alpha j_\alpha}}^{(s_{j_\alpha},\sigma_s;\mathscr{E})}(\mathbf{x}_m^{(\sigma_s;\mathscr{E})},t_m,\boldsymbol{\pi}_{\sigma_s},\boldsymbol{\lambda}_{j_\alpha})=0 \;\; for\; s_{j_\alpha}=0,1,2,\cdots;\\ &for\; all\; j_\alpha\in\bar{\mathscr{N}}_{\alpha\mathscr{B}}(\sigma_r,\sigma_s).\end{aligned} \tag{8.65}$$

Proof. As in Theorem 3.1, this theorem can be proved similarly as the edge flow is treated as a special domain flow (also see, Luo, 2008, 2009). ■

Similarly, the edge flow attractivity to the specific edge with the higher-order singularity is described as follows.

Definition 8.10. For a discontinuous dynamical system in Eqs. (6.4)-(6.8), there is a point $\mathbf{x}^{(\sigma_r;\mathscr{E})}(t_m)\equiv\mathbf{x}_m\in\mathscr{E}_{\sigma_r}^{(r)}$ ($\sigma_r\in\mathscr{N}_{\mathscr{E}}^{(r)}=\{1,2,\cdots,N_r\}$ and $r\in\{0,1,2,\cdots,n-2\}$) at time t_m with the related boundaries $\partial\Omega_{\alpha i_\alpha}(\sigma_r)$ and related domains $\Omega_\alpha(\sigma_r)$ ($\alpha\in\mathscr{N}_{\mathscr{D}}(\sigma_r)\subset\mathscr{N}_{\mathscr{D}}=\{1,2,\cdots,N_d\}$ and $i_\alpha\in\mathscr{N}_{\alpha\mathscr{B}}(\sigma_r)\subset\mathscr{N}_{\mathscr{B}}=\{1,2,\;\cdots,N_b\}$), where the boundary index subset of $\mathscr{N}_{\alpha\mathscr{B}}(\sigma_r)$ has n_{σ_r}-elements ($n_{\sigma_r}\geq(n-r)$). The edge of $\mathscr{E}_{\sigma_r}^{(r)}$ is the intersection of edges (including domains and boundaries) $\mathscr{E}_{\sigma_s}^{(s)}(\sigma_r)$ ($\sigma_s\in\mathscr{N}_{\mathscr{E}}^{(s)}(\sigma_r)\subset\mathscr{N}_{\mathscr{E}}^{(s)}=\{1,2,\cdots,N_s\}$ and $s\in\{r+1,r+2,\cdots,n-1,n\}$). Suppose there is an s-dimensional edge $\mathscr{E}_{\sigma_s}^{(s)}(\sigma_r)$ with a boundary index set $\mathscr{N}_{\alpha\mathscr{B}}(\sigma_r,\sigma_s)$ relative to domain $\Omega_\alpha(\sigma_r)$. The complementary boundary index set is $\bar{\mathscr{N}}_{\alpha\mathscr{B}}(\sigma_r,\sigma_s)=\mathscr{N}_{\alpha\mathscr{B}}(\sigma_r)/\mathscr{N}_{\alpha\mathscr{B}}(\sigma_r,\sigma_s)$. For an arbitrarily small $\varepsilon>0$, there are two time intervals $[t_{m-\varepsilon},t_m)$ and $(t_m,t_{m+\varepsilon}]$. An edge flow $\mathbf{x}^{(\sigma_s;\mathscr{E})}(t)$ is $C_{[t_{m-\varepsilon},t_m)}^{r_{\sigma_s}}$ and/or $C_{(t_m,t_{m+\varepsilon}]}^{r_{\sigma_s}}$-continuous and $\| d^{r_{\sigma_s}+1}\mathbf{x}^{(\sigma_s;\mathscr{E})}/dt^{r_{\sigma_s}+1}\|<\infty$ for time t ($r_{\sigma_s}\geq\max_{j_\alpha\in\bar{\mathscr{N}}_{\alpha\mathscr{B}}}(2k_{j_\alpha}+1)$). Suppose there is an $(n-s+r)$-dimensional measuring edge $\mathscr{E}_{\sigma_{n-s+r}}^{(\rho,n-s+r)}$ of a flow $\mathbf{x}^{(\sigma_s;\mathscr{E})}(t)$ in edge $\mathscr{E}_{\sigma_s}^{(s)}(\sigma_r,\alpha)$ with an equi-constant vector of the specific edge $\mathscr{E}_{\sigma_r}^{(r)}(\alpha)$. The measuring edge $\mathscr{E}_{\sigma_{n-s+r}}^{(\rho,n-s+r)}(\sigma_r,\alpha)$ is formed by an intersection of all measuring surfaces $S_{\sigma_m}^{(j_\alpha,\alpha)}$ for $j_\alpha\in\bar{\mathscr{N}}_{\alpha\mathscr{B}}(\sigma_r,\sigma_s)$ (i.e., $\mathscr{E}_{\sigma_{n-s+r}}^{(\rho,n-s+r)}(\sigma_r,\alpha)=\bigcap_{j_\alpha\in\bar{\mathscr{N}}_{\alpha\mathscr{B}}(\sigma_r,\sigma_s)}S_\rho^{(j_\alpha,\alpha)}$). $\mathscr{E}_{\sigma_r}^{(r)}(\alpha)=\mathscr{E}_{\sigma_{n-s+r}}^{(n-s+r)}(\sigma_r,\alpha)\bigcap\mathscr{E}_{\sigma_s}^{(s)}(\sigma_r,\alpha)$ with $\mathscr{E}_{\sigma_{n-s+r}}^{(n-s+r)}(\sigma_r,\alpha)=\bigcap_{j_\alpha\in\bar{\mathscr{N}}_{\alpha\mathscr{B}}(\sigma_r,\sigma_s)}\partial\Omega_{\alpha j_\alpha}$ and $\mathscr{E}_{\sigma_s}^{(s)}(\sigma_r,\alpha)=\bigcap_{i_\alpha\in\mathscr{N}_{\alpha\mathscr{B}}(\sigma_r,\sigma_s)}\partial\Omega_{\alpha i_\alpha}$ $\subset\mathscr{E}_{\sigma_s}^{(s)}(\sigma_r)$. $\mathbf{x}_m^{(\sigma_s;\mathscr{E})}=\mathbf{x}_\rho^{(\sigma_s;\mathscr{E})}(t_m)\in\mathscr{E}_{\sigma_r}^{(r,\sigma_s)}(\rho_m)$ ($\rho\in\mathbb{N}$) satisfies Eq. (8.55) at time t_m, Suppose $\mathbf{x}_{m\pm\varepsilon}^{(\sigma_s;\mathscr{E})}=\mathbf{x}^{(\sigma_s;\mathscr{E})}(t_{m\pm\varepsilon})\in\mathscr{E}_{\sigma_{n-s+r}}^{(\rho_m,n-s+r)}$ and $\mathbf{x}_{\rho(m\pm\varepsilon)}^{(j_\alpha,\alpha)}\in S_{\rho_m}^{(j_\alpha,\alpha)}$.

(i) The edge flow $\mathbf{x}^{(\sigma_s;\mathscr{E})}(t)$ at time t_m is *attractive* with the $\mathbf{2k}_{(\sigma_r,\sigma_s)}$-order singularity to the edge $\mathscr{E}_{\sigma_r}^{(r)}$ if

$$\left.\begin{array}{l}\hbar_{j_\alpha} G_{\partial\Omega_{\alpha j_\alpha}}^{(s_{j_\alpha},\sigma_s;\mathscr{E})}(\mathbf{x}_m^{(\sigma_s;\mathscr{E})},t_m,\boldsymbol{\pi}_{\sigma_s},\boldsymbol{\lambda}_{j_\alpha})=0 \text{ for } s_{j_\alpha}=0,1,2,\cdots,2k_{j_\alpha}-1;\\ \hbar_{j_\alpha}\mathbf{n}_{\partial\Omega_{\alpha i_\alpha}}^{\mathrm{T}}(\mathbf{x}_{\rho(m-\varepsilon)}^{(j_\alpha,\alpha)})\cdot[\mathbf{x}_{\rho(m-\varepsilon)}^{(j_\alpha,\alpha)}-\mathbf{x}_{m-\varepsilon}^{(\sigma_s;\mathscr{E})}]<0,\\ \hbar_{j_\alpha}\mathbf{n}_{\partial\Omega_{\alpha i_\alpha}}^{\mathrm{T}}(\mathbf{x}_{\rho(m+\varepsilon)}^{(j_\alpha,\alpha)})\cdot[\mathbf{x}_{m+\varepsilon}^{(\sigma_s;\mathscr{E})}-\mathbf{x}_{\rho(m+\varepsilon)}^{(j_\alpha,\alpha)}]<0,\\ \hbar_{j_\alpha}C_{m-\varepsilon}^{(j_\alpha,\alpha)}>\hbar_{j_\alpha}C_{m}^{(j_\alpha,\alpha)}>\hbar_{j_\alpha}C_{m+\varepsilon}^{(j_\alpha,\alpha)}\\ \text{for all } j_\alpha\in\bar{\mathscr{N}}_{\alpha\mathscr{B}}(\sigma_r,\sigma_s).\end{array}\right\} \tag{8.66}$$

(ii) The edge flow $\mathbf{x}^{(\sigma_s;\mathscr{E})}(t)$ at time t_m is *repulsive* with the $2\mathbf{k}_{(\sigma_r,\sigma_s)}$-order singularity to the edge $\mathscr{E}_{\sigma_r}^{(r)}$ if

$$\left.\begin{array}{l}\hbar_{j_\alpha} G_{\partial\Omega_{\alpha j_\alpha}}^{(s_{j_\alpha},\sigma_s;\mathscr{E})}(\mathbf{x}_m^{(\sigma_s;\mathscr{E})},t_m,\boldsymbol{\pi}_{\sigma_s},\boldsymbol{\lambda}_{j_\alpha})=0 \text{ for } s_{j_\alpha}=0,1,2,\cdots,2k_{j_\alpha}-1;\\ \hbar_{j_\alpha}\mathbf{n}_{\partial\Omega_{\alpha i_\alpha}}^{\mathrm{T}}(\mathbf{x}_{\rho(m-\varepsilon)}^{(j_\alpha,\alpha)})\cdot[\mathbf{x}_{\rho(m-\varepsilon)}^{(j_\alpha,\alpha)}-\mathbf{x}_{m-\varepsilon}^{(\sigma_s;\mathscr{E})}]>0,\\ \hbar_{j_\alpha}\mathbf{n}_{\partial\Omega_{\alpha i_\alpha}}^{\mathrm{T}}(\mathbf{x}_{\rho(m+\varepsilon)}^{(j_\alpha,\alpha)})\cdot[\mathbf{x}_{m+\varepsilon}^{(\sigma_s;\mathscr{E})}-\mathbf{x}_{\rho(m+\varepsilon)}^{(j_\alpha,\alpha)}]>0,\\ \hbar_{j_\alpha}C_{m-\varepsilon}^{(j_\alpha,\alpha)}>\hbar_{j_\alpha}C_{m}^{(j_\alpha,\alpha)}>\hbar_{j_\alpha}C_{m+\varepsilon}^{(j_\alpha,\alpha)}\\ \text{for all } j_\alpha\in\bar{\mathscr{N}}_{\alpha\mathscr{B}}(\sigma_r,\sigma_s).\end{array}\right\} \tag{8.67}$$

(iii) The edge flow $\mathbf{x}^{(\sigma_s;\mathscr{E})}(t)$ at time t_m to the edge $\mathscr{E}_{\sigma_r}^{(r)}$ is *from the attractive to repulsive* state with the $(2\mathbf{k}_{(\sigma_r,\sigma_s)}+1)$-order singularity if

$$\left.\begin{array}{l}\hbar_{j_\alpha} G_{\partial\Omega_{\alpha j_\alpha}}^{(s_{j_\alpha},\sigma_s;\mathscr{E})}(\mathbf{x}_m^{(\sigma_s;\mathscr{E})},t_m,\boldsymbol{\pi}_{\sigma_s},\boldsymbol{\lambda}_{j_\alpha})=0 \text{ for } s_{j_\alpha}=0,1,2,\cdots,2k_{j_\alpha};\\ \hbar_{j_\alpha}\mathbf{n}_{\partial\Omega_{\alpha i_\alpha}}^{\mathrm{T}}(\mathbf{x}_{\rho(m-\varepsilon)}^{(j_\alpha,\alpha)})\cdot[\mathbf{x}_{\rho(m-\varepsilon)}^{(j_\alpha,\alpha)}-\mathbf{x}_{m-\varepsilon}^{(\sigma_s;\mathscr{E})}]<0,\\ \hbar_{j_\alpha}\mathbf{n}_{\partial\Omega_{\alpha i_\alpha}}^{\mathrm{T}}(\mathbf{x}_{\rho(m+\varepsilon)}^{(j_\alpha,\alpha)})\cdot[\mathbf{x}_{m+\varepsilon}^{(\sigma_s;\mathscr{E})}-\mathbf{x}_{\rho(m+\varepsilon)}^{(j_\alpha,\alpha)}]>0,\\ \hbar_{j_\alpha}C_{m}^{(j_\alpha,\alpha)}<\hbar_{j_\alpha}C_{m-\varepsilon}^{(j_\alpha,\alpha)} \text{ and } \hbar_{j_\alpha}C_{m}^{(j_\alpha,\alpha)}<\hbar_{j_\alpha}C_{m+\varepsilon}^{(j_\alpha,\alpha)}\\ \text{for all } j_\alpha\in\bar{\mathscr{N}}_{\alpha\mathscr{B}}(\sigma_r,\sigma_s).\end{array}\right\} \tag{8.68}$$

(iv) The flow $\mathbf{x}^{(\sigma_s;\mathscr{E})}(t)$ at time t_m to the edge $\mathscr{E}_{\sigma_r}^{(r)}$ is *from the repulsive to attractive* state with the $(2\mathbf{k}_{(\sigma_r,\sigma_s)}+1)$-order singularity if

$$\hbar_{j_\alpha} G_{\partial\Omega_{\alpha j_\alpha}}^{(s_{j_\alpha},\sigma_s;\mathscr{E})}(\mathbf{x}_m^{(\sigma_s;\mathscr{E})},t_m,\boldsymbol{\pi}_{\sigma_s},\boldsymbol{\lambda}_{j_\alpha})=0 \text{ for } s_{j_\alpha}=0,1,2,\cdots,2k_{j_\alpha};$$

$$\begin{aligned}
&\hbar_{j_\alpha} \mathbf{n}^{\mathrm{T}}_{\partial\Omega_{\alpha i_\alpha}} (\mathbf{x}^{(j_\alpha,\alpha)}_{\rho(m-\varepsilon)}) \cdot [\mathbf{x}^{(j_\alpha,\alpha)}_{\rho(m-\varepsilon)} - \mathbf{x}^{(\sigma_s;\mathscr{E})}_{m-\varepsilon}] > 0, \\
&\hbar_{j_\alpha} \mathbf{n}^{\mathrm{T}}_{\partial\Omega_{\alpha i_\alpha}} (\mathbf{x}^{(j_\alpha,\alpha)}_{\rho(m+\varepsilon)}) \cdot [\mathbf{x}^{(\sigma_s;\mathscr{E})}_{m+\varepsilon} - \mathbf{x}^{(j_\alpha,\alpha)}_{\rho(m+\varepsilon)}] < 0, \\
&\hbar_{j_\alpha} C^{(j_\alpha,\alpha)}_{m} > \hbar_{j_\alpha} C^{(j_\alpha,\alpha)}_{m-\varepsilon} \text{ and } \hbar_{j_\alpha} C^{(j_\alpha,\alpha)}_{m} > \hbar_{j_\alpha} C^{(j_\alpha,\alpha)}_{m+\varepsilon} \\
&\text{for all } j_\alpha \in \bar{\mathscr{N}}_{\alpha\mathscr{B}}(\sigma_r,\sigma_s).
\end{aligned} \tag{8.69}$$

To explain the above concept of the attractivity for an edge flow to a specific edge of $\mathscr{E}^{(r)}_{\sigma_r}$, consider a specific edge $\mathscr{E}^{(s-2)}_{\sigma_{s-2}}$ in the edge space of $\mathscr{E}^{(s)}_{\sigma_s}$, and the specific edge is an intersection of the two boundaries of $\partial\Omega_{\alpha i_\alpha}$ and $\partial\Omega_{\alpha j_\alpha}$. The measuring edge of $\mathscr{E}^{(\rho,s-2)}_{\sigma_{s-2}}$ can be formed by the intersection of two measuring boundaries of $S^{(i_\alpha;\alpha)}_{\rho}$ and $S^{(j_\alpha;\alpha)}_{\rho}$ in the s-dimensional edge space. This measuring edge possesses an equi-constant vector of the specific edge. If an edge flow $\mathbf{x}^{(\sigma_s;\mathscr{E})}$ arrives to the measuring edge, it corresponding attractivity to the specific edge can be determined, as shown in Fig. 8.8. The measuring edges and surfaces are given by the equi-constant edges and surfaces. The normal vectors of the two measuring surfaces are the same as the corresponding surfaces. The solid curve with arrow represents the edge flow of $\mathbf{x}^{(\sigma_s;\mathscr{E})}$, and the dashed curve with arrow gives a flow on the measuring surface. Through the measuring edge of $\mathscr{E}^{(\rho,s-2)}_{\sigma_{s-2}}$, the attractivity of an edge flow to the edge $\mathscr{E}^{(s-2)}_{\sigma_{s-2}}$ can be discussed.

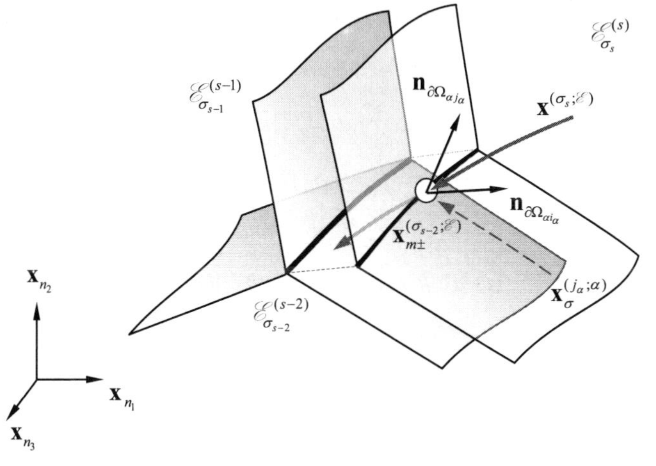

Fig. 8.8 A flow intersected with a measuring edge $\mathscr{E}^{(\rho,s-2)}_{\sigma_{s-2}}$ intersected between two equi-constant surfaces $S^{(i_\alpha;\alpha)}_{\rho}$ and $S^{(j_\alpha;\alpha)}_{\rho}$ in the s-dimensional edge space of $\mathscr{E}^{(s)}_{\sigma_s}$. The specific edge $\mathscr{E}^{(s-2)}_{\sigma_{s-2}}$ is the $(s-2)$-dimensional edge of two boundaries of $\partial\Omega_{\alpha i_\alpha}$ and $\partial\Omega_{\alpha j_\alpha}$. The corresponding normal vectors of the two boundaries are presented. $(n_1+n_2+n_3=s)$

The conditions for the attractivity of an edge flow to the specific edge are stated as follows.

Theorem 8.8. *For a discontinuous dynamical system in Eqs.* (6.4)-(6.8)*, there is a point* $\mathbf{x}^{(\sigma_r;\mathscr{E})}(t_m) \equiv \mathbf{x}_m \in \mathscr{E}_{\sigma_r}^{(r)}$ ($\sigma_r \in \mathscr{N}_{\mathscr{E}}^{(r)} = \{1,2,\cdots,N_r\}$ *and* $r \in \{0,1,2,\cdots,n-2\}$) *at time* t_m *with the related boundaries* $\partial\Omega_{\alpha i_\alpha}(\sigma_r)$ *and related domains* $\Omega_\alpha(\sigma_r)$ ($\alpha \in \mathscr{N}_{\mathscr{D}}(\sigma_r) \subset \mathscr{N}_{\mathscr{D}} = \{1,2,\cdots,N_d\}$ *and* $i_\alpha \in \mathscr{N}_{\alpha\mathscr{B}}(\sigma_r) \subset \mathscr{N}_{\mathscr{B}} = \{1,2,\cdots,N_b\}$)*, where the boundary index subset of* $\mathscr{N}_{\alpha\mathscr{B}}(\sigma_r)$ *has* n_{σ_r} *-elements* ($n_{\sigma_r} \geq (n-r)$)*. The edge of* $\mathscr{E}_{\sigma_r}^{(r)}$ *is the intersection of edges (including domains and boundaries)* $\mathscr{E}_{\sigma_s}^{(s)}(\sigma_r)$ ($\sigma_s \in \mathscr{N}_{\mathscr{E}}^{s}(\sigma_r) \subset \mathscr{N}_{\mathscr{E}}^{s} = \{1,2,\cdots,N_s\}$ *and* $s \in \{r+1,r+2,\cdots,n-1,n\}$)*. Suppose there is an* s *-dimensional edge* $\mathscr{E}_{\sigma_s}^{(s)}(\sigma_r)$ *with a boundary index set* $\mathscr{N}_{\alpha\mathscr{B}}(\sigma_r,\sigma_s)$ *relative to domain* $\Omega_\alpha(\sigma_r)$*. The complementary boundary index set is* $\bar{\mathscr{N}}_{\alpha\mathscr{B}}(\sigma_r,\sigma_s) = \mathscr{N}_{\alpha\mathscr{B}}(\sigma_r) / \mathscr{N}_{\alpha\mathscr{B}}(\sigma_r,\sigma_s)$*. For an arbitrarily small* $\varepsilon > 0$*, there are two time intervals* $[t_{m-\varepsilon},t_m)$ *and* $(t_m,t_{m+\varepsilon}]$*. An edge flow* $\mathbf{x}^{(\sigma_s;\mathscr{E})}(t)$ *is* $C_{[t_{m-\varepsilon},t_m)}^{r_{\sigma_s}}$ *and/or* $C_{(t_m,t_{m+\varepsilon}]}^{r_{\sigma_s}}$*-continuous and* $\| d^{r_{\sigma_s}+1}\mathbf{x}^{(\sigma_s;\mathscr{E})} / dt^{r_{\sigma_s}+1} \| < \infty$ *for time* t ($r_{\sigma_s} \geq \max_{j_\alpha \in \bar{\mathscr{N}}_{\alpha\mathscr{B}}}(2k_{j_\alpha}+1)$)*. Suppose there is an* $(n-s+r)$ *-dimensional measuring edge* $\mathscr{E}_{\sigma_{n-s+r}}^{(\rho,n-s+r)}$ *of a flow* $\mathbf{x}^{(\sigma_s;\mathscr{E})}(t)$ *in edge* $\mathscr{E}_{\sigma_s}^{(s)}(\sigma_r,\alpha)$ *with an equi-constant vector of the specific edge* $\mathscr{E}_{\sigma_r}^{(r)}(\alpha)$*. The measuring edge* $\mathscr{E}_{\sigma_{n-s+r}}^{(\rho,n-s+r)}(\sigma_r,\alpha)$ *is formed by an intersection of all measuring surfaces* $S_{\sigma_m}^{(i_\alpha,\alpha)}$ *for* $i_\alpha \in \bar{\mathscr{N}}_{\alpha\mathscr{B}}(\sigma_r,\sigma_s)$ *(i.e.,* $\mathscr{E}_{\sigma_{n-s+r}}^{(\rho,n-s+r)}(\sigma_r,\alpha) = \cap_{j_\alpha \in \bar{\mathscr{N}}_{\alpha\mathscr{B}}(\sigma_r,\sigma_s)} S_\rho^{(j_\alpha,\alpha)}$)*.* $\mathscr{E}_{\sigma_r}^{(r)}(\alpha) = \mathscr{E}_{\sigma_{n-s+r}}^{(n-s+r)}(\sigma_r,\alpha) \cap \mathscr{E}_{\sigma_s}^{(s)}(\sigma_r,\alpha)$ *with* $\mathscr{E}_{\sigma_{n-s+r}}^{(n-s+r)}(\sigma_r,\alpha) = \cap_{j_\alpha \in \bar{\mathscr{N}}_{\alpha\mathscr{B}}(\sigma_r,\sigma_s)} \partial\Omega_{\alpha j_\alpha}$ *and* $\mathscr{E}_{\sigma_s}^{(s)}(\sigma_r,\alpha) = \cap_{i_\alpha \in \mathscr{N}_{\alpha\mathscr{B}}(\sigma_r,\sigma_s)} \partial\Omega_{\alpha i_\alpha} \subset \mathscr{E}_{\sigma_s}^{(s)}(\sigma_r)$*.* $\mathbf{x}_m^{(\sigma_s;\mathscr{E})} = \mathbf{x}_\rho^{(\sigma_s;\mathscr{E})}(t_m) \in \mathscr{E}_{\sigma_r}^{(r,\sigma_s)}(\rho_m)$ ($\rho \in \mathbb{N}$) *satisfies Eq.* (8.55) *at time* t_m*. Suppose* $\mathbf{x}_{m\pm\varepsilon}^{(\sigma_s;\mathscr{E})} = \mathbf{x}^{(\sigma_s;\mathscr{E})}(t_{m\pm\varepsilon}) \in \mathscr{E}_{\sigma_{n-s+r}}^{(\rho_m,n-s+r)}$ *and* $\mathbf{x}_{\rho(m\pm\varepsilon)}^{(j_\alpha,\alpha)} \in S_{\rho_m}^{(j_\alpha,\alpha)}$*.*

(i) *The edge flow* $\mathbf{x}^{(\sigma_s;\mathscr{E})}(t)$ *at time* t_m *is attractive with the* $2\mathbf{k}_{(\sigma_r,\sigma_s)}$ *-order singularity to the edge* $\mathscr{E}_{\sigma_r}^{(r)}$ *if and only if*

$$\begin{aligned}
&\hbar_{j_\alpha} G_{\partial\Omega_{\alpha j_\alpha}}^{(s_{j_\alpha},\sigma_s;\mathscr{E})}(\mathbf{x}_m^{(\sigma_s;\mathscr{E})},t_m,\boldsymbol{\pi}_{\sigma_s},\boldsymbol{\lambda}_{j_\alpha}) = 0\\
&\text{for } s_{j_\alpha} = 0,1,2,\cdots,2k_{j_\alpha}-1;\\
&\hbar_{j_\alpha} G_{\partial\Omega_{\alpha j_\alpha}}^{(2k_{j_\alpha},\sigma_s;\mathscr{E})}(\mathbf{x}_m^{(\sigma_s;\mathscr{E})},t_m,\boldsymbol{\pi}_{\sigma_s},\boldsymbol{\lambda}_{j_\alpha}) < 0,\\
&\text{for all } j_\alpha \in \bar{\mathscr{N}}_{\alpha\mathscr{B}}(\sigma_r,\sigma_s).
\end{aligned} \tag{8.70}$$

(ii) *The edge flow* $\mathbf{x}^{(\sigma_s;\mathscr{E})}(t)$ *at time* t_m *is repulsive with the* $2\mathbf{k}_{(\sigma_r,\sigma_s)}$ *-order singu-*

larity to the edge $\mathscr{E}_{\sigma_r}^{(r)}$ *if and only if*

$$
\begin{aligned}
&\hbar_{j_\alpha} G_{\partial\Omega_{\alpha j_\alpha}}^{(s_{j_\alpha},\sigma_s;\mathscr{E})}(\mathbf{x}_m^{(\sigma_s;\mathscr{E})},t_m,\boldsymbol{\pi}_{\sigma_s},\boldsymbol{\lambda}_{j_\alpha})=0\\
&\text{for } s_{j_\alpha}=0,1,2,\cdots,2k_{j_\alpha}-1;\\
&\hbar_{j_\alpha} G_{\partial\Omega_{\alpha j_\alpha}}^{(2k_{j_\alpha},\sigma_s;\mathscr{E})}(\mathbf{x}_m^{(\sigma_s;\mathscr{E})},t_m,\boldsymbol{\pi}_{\sigma_s},\boldsymbol{\lambda}_{j_\alpha})0,\\
&\text{for all } j_\alpha \in \bar{\mathscr{N}}_{\alpha\mathscr{B}}(\sigma_r,\sigma_s).
\end{aligned}
\tag{8.71}
$$

(iii) *The edge flow* $\mathbf{x}^{(\sigma_s;\mathscr{E})}(t)$ *at time* t_m *to the edge* $\mathscr{E}_{\sigma_r}^{(r)}$ *is from the attractive to repulsive state with the* $(2\mathbf{k}_{(\sigma_r,\sigma_s)}+1)$*-order singularity if and only if*

$$
\begin{aligned}
&\hbar_{j_\alpha} G_{\partial\Omega_{\alpha j_\alpha}}^{(s_{j_\alpha},\sigma_s;\mathscr{E})}(\mathbf{x}_m^{(\sigma_s;\mathscr{E})},t_m,\boldsymbol{\pi}_{\sigma_s},\boldsymbol{\lambda}_{j_\alpha})=0\\
&\text{for } s_{j_\alpha}=0,1,2,\cdots,2k_{j_\alpha};\\
&\hbar_{j_\alpha} G_{\partial\Omega_{\alpha j_\alpha}}^{(2k_{j_\alpha}+1,\sigma_s;\mathscr{E})}(\mathbf{x}_m^{(\sigma_s;\mathscr{E})},t_m,\boldsymbol{\pi}_{\sigma_s},\boldsymbol{\lambda}_{j_\alpha})>0\\
&\text{for all } j_\alpha \in \bar{\mathscr{N}}_{\alpha\mathscr{B}}(\sigma_r,\sigma_s).
\end{aligned}
\tag{8.72}
$$

(iv) *The flow* $\mathbf{x}^{(\sigma_s;\mathscr{E})}(t)$ *at time* t_m *to the edge of* $\mathscr{E}_{\sigma_r}^{(r)}$ *is from the repulsive to attractive state with the* $(2\mathbf{k}_{(\sigma_r,\sigma_s)}+1)$*-order singularity if and only if*

$$
\begin{aligned}
&\hbar_{j_\alpha} G_{\partial\Omega_{\alpha j_\alpha}}^{(s_{j_\alpha},\sigma_s;\mathscr{E})}(\mathbf{x}_m^{(\sigma_s;\mathscr{E})},t_m,\boldsymbol{\pi}_{\sigma_s},\boldsymbol{\lambda}_{j_\alpha})=0\\
&\text{for } s_{j_\alpha}=0,1,2,\cdots,2k_{j_\alpha};\\
&\hbar_{j_\alpha} G_{\partial\Omega_{\alpha j_\alpha}}^{(2k_{j_\alpha}+1,\sigma_s;\mathscr{E})}(\mathbf{x}_m^{(\sigma_s;\mathscr{E})},t_m,\boldsymbol{\pi}_{\sigma_s},\boldsymbol{\lambda}_{j_\alpha})<0\\
&\text{for all } j_\alpha \in \bar{\mathscr{N}}_{\alpha\mathscr{B}}(\sigma_r,\sigma_s).
\end{aligned}
\tag{8.73}
$$

Proof. As in Theorem 3.2, this theorem can be proved similarly as the edge flow is treated as a special domain flow (also see, Luo, 2008, 2009). ■

It is very difficult to make an edge flow to all boundaries of the specific edge possess the same attractivity at time. Thus consider five types of the edge flow attractivity to a specific edge divided into five groups, and the general description for the edge flow attractivity to the specific edge is given as follows.

Definition 8.11. For a discontinuous dynamical system in Eqs. (6.4)-(6.8), there is a point $\mathbf{x}^{(\sigma_r;\mathscr{E})}(t_m)\equiv\mathbf{x}_m\in\mathscr{E}_{\sigma_r}^{(r)}$ ($\sigma_r\in\mathscr{N}_{\mathscr{E}}^{(r)}=\{1,2,\cdots,N_r\}$ and $r\in\{0,1,2,\cdots,n-2\}$) at time t_m with the related boundaries $\partial\Omega_{\alpha i_\alpha}(\sigma_r)$ and related domains $\Omega_\alpha(\sigma_r)$ ($\alpha\in\mathscr{N}_{\mathscr{D}}(\sigma_r)\subset\mathscr{N}_{\mathscr{D}}=\{1,2,\cdots,N_d\}$ and $i_\alpha\in\mathscr{N}_{\alpha\mathscr{B}}(\sigma_r)\subset\mathscr{N}_{\mathscr{B}}=\{1,2,\cdots,N_b\}$), where the boundary index subset of $\mathscr{N}_{\alpha\mathscr{B}}(\sigma_r)$ has n_{σ_r}-elements ($n_{\sigma_r}\geq(n-r)$).

The edge of $\mathscr{E}_{\sigma_r}^{(r)}$ is the intersection of edges (including domains and boundaries) $\mathscr{E}_{\sigma_s}^{(s)}(\sigma_r)$ ($\sigma_s \in \mathscr{N}_{\mathscr{E}}^{s}(\sigma_r) \subset \mathscr{N}_{\mathscr{E}}^{s} = \{1,2,\cdots,N_s\}$ and $s \in \{r+1, r+2, \cdots, n-1, n\}$). Suppose there is an s-dimensional edge $\mathscr{E}_{\sigma_s}^{(s)}(\sigma_r)$ with a boundary index set $\mathscr{N}_{\alpha\mathscr{B}}(\sigma_r,\sigma_s)$ relative to domain $\Omega_\alpha(\sigma_r)$. The complementary boundary index set is $\bar{\mathscr{N}}_{\alpha\mathscr{B}}(\sigma_r,\sigma_s) = \mathscr{N}_{\alpha\mathscr{B}}(\sigma_r) / \mathscr{N}_{\alpha\mathscr{B}}(\sigma_r,\sigma_s)$. For an arbitrarily small $\varepsilon > 0$, there are two time intervals $[t_{m-\varepsilon}, t_m)$ and $(t_m, t_{m+\varepsilon}]$. An edge flow $\mathbf{x}^{(\sigma_s;\mathscr{E})}(t)$ is $C_{[t_{m-\varepsilon},t_m)}^{r_{\sigma_s}}$ and/or $C_{(t_m,t_{m+\varepsilon}]}^{r_{\sigma_s}}$-continuous for time t and $\| d^{r_{\sigma_s}+1}\mathbf{x}^{(\sigma_s;\mathscr{E})} / dt^{r_{\sigma_s}+1} \| < \infty$ ($r_{\sigma_s} \ge 1$). Suppose there is an $(n-s+r)$-dimensional measuring edge $\mathscr{E}_{\sigma_{n-s+r}}^{(\rho,n-s+r)}$ of a flow $\mathbf{x}^{(\sigma_s;\mathscr{E})}(t)$ in edge $\mathscr{E}_{\sigma_s}^{(s)}(\sigma_r,\alpha)$ with an equi-constant vector of the specific edge $\mathscr{E}_{\sigma_r}^{(r)}(\alpha)$. The measuring edge $\mathscr{E}_{\sigma_{n-s+r}}^{(\rho,n-s+r)}(\sigma_r,\alpha)$ is formed by an intersection of all measuring surfaces $S_{\sigma_m}^{(i_\alpha,\alpha)}$ for $i_\alpha \in \bar{\mathscr{N}}_{\alpha\mathscr{B}}(\sigma_r,\sigma_s)$ (i.e., $\mathscr{E}_{\sigma_{n-s+r}}^{(\rho,n-s+r)}(\sigma_r,\alpha) = \bigcap_{j_\alpha \in \bar{\mathscr{N}}_{\alpha\mathscr{B}}(\sigma_r,\sigma_s)} S_\rho^{(j_\alpha,\alpha)}$). $\mathscr{E}_{\sigma_r}^{(r)}(\alpha) = \mathscr{E}_{\sigma_{n-s+r}}^{(n-s+r)}(\sigma_r,\alpha) \bigcap \mathscr{E}_{\sigma_s}^{(s)}(\sigma_r,\alpha)$ with $\mathscr{E}_{\sigma_{n-s+r}}^{(n-s+r)}(\sigma_r,\alpha)$ $= \bigcap_{j_\alpha \in \bar{\mathscr{N}}_{\alpha\mathscr{B}}(\sigma_r,\sigma_s)} \partial\Omega_{\alpha j_\alpha}$ and $\mathscr{E}_{\sigma_s}^{(s)}(\sigma_r,\alpha) = \bigcap_{i_\alpha \in \mathscr{N}_{\alpha\mathscr{B}}(\sigma_r,\sigma_s)} \partial\Omega_{\alpha i_\alpha} \subset \mathscr{E}_{\sigma_s}^{(s)}(\sigma_r)$. At time t_m, $\mathbf{x}_m^{(\sigma_s;\mathscr{E})} = \mathbf{x}_\rho^{(\sigma_s;\mathscr{E})}(t_m) \in \mathscr{E}_{\sigma_r}^{(r,\sigma_s)}(\rho_m)$ ($\rho \in \mathrm{N}$) satisfies Eq. (8.55). Suppose $\mathbf{x}_{m\pm\varepsilon}^{(\sigma_s;\mathscr{E})}$ $= \mathbf{x}^{(\sigma_s;\mathscr{E})}(t_{m\pm\varepsilon}) \in \mathscr{E}_{\sigma_{n-s+r}}^{(\rho_m,n-s+r)}$ and $\mathbf{x}_{\rho(m\pm\varepsilon)}^{(j_\alpha,\alpha)} \in S_{\rho_m}^{(j_\alpha,\alpha)}$. $\bar{\mathscr{N}}_{\alpha\mathscr{B}}(\sigma_r,\sigma_s) = \bigcup_{k=1}^{5} \bar{\mathscr{N}}_{\alpha\mathscr{B}}^{(k)}(\sigma_r,\sigma_s)$ and $\bigcap_{k=1}^{5} \bar{\mathscr{N}}_{\alpha\mathscr{B}}^{(k)}(\sigma_r,\sigma_s) = \varnothing$.

(i) The edge flow $\mathbf{x}^{(\sigma_s;\mathscr{E})}(t)$ at time t_m is *partially attractive* to the edge of $\mathscr{E}_{\sigma_r}^{(r)}$ for all the boundaries of $\partial\Omega_{\alpha j_\alpha}$ ($j_\alpha \in \bar{\mathscr{N}}_{\alpha\mathscr{B}}^{(1)}(\sigma_r,\sigma_s)$) if

$$\begin{aligned}
&\hbar_{j_\alpha} \mathbf{n}_{\partial\Omega_{\alpha i_\alpha}}^{\mathrm{T}}(\mathbf{x}_{\rho(m-\varepsilon)}^{(j_\alpha,\alpha)}) \cdot [\mathbf{x}_{\rho(m-\varepsilon)}^{(j_\alpha,\alpha)} - \mathbf{x}_{m-\varepsilon}^{(\sigma_s;\mathscr{E})}] < 0,\\
&\hbar_{j_\alpha} \mathbf{n}_{\partial\Omega_{\alpha i_\alpha}}^{\mathrm{T}}(\mathbf{x}_{\rho(m+\varepsilon)}^{(j_\alpha,\alpha)}) \cdot [\mathbf{x}_{m+\varepsilon}^{(\sigma_s;\mathscr{E})} - \mathbf{x}_{\rho(m+\varepsilon)}^{(j_\alpha,\alpha)}] < 0,\\
&\hbar_{j_\alpha} C_{m-\varepsilon}^{(j_\alpha,\alpha)} > \hbar_{j_\alpha} C_m^{(j_\alpha,\alpha)} > \hbar_{j_\alpha} C_{m+\varepsilon}^{(j_\alpha,\alpha)}.
\end{aligned} \tag{8.74}$$

(ii) The edge flow $\mathbf{x}^{(\sigma_s;\mathscr{E})}(t)$ at time t_m is *partially repulsive* to the edge of $\mathscr{E}_{\sigma_r}^{(r)}$ for all the boundaries of $\partial\Omega_{\alpha j_\alpha}$ ($j_\alpha \in \bar{\mathscr{N}}_{\alpha\mathscr{B}}^{(2)}(\sigma_r,\sigma_s)$) if

$$\begin{aligned}
&\hbar_{j_\alpha} \mathbf{n}_{\partial\Omega_{\alpha i_\alpha}}^{\mathrm{T}}(\mathbf{x}_{\rho(m-\varepsilon)}^{(j_\alpha,\alpha)}) \cdot [\mathbf{x}_{\rho(m-\varepsilon)}^{(j_\alpha,\alpha)} - \mathbf{x}_{m-\varepsilon}^{(\sigma_s;\mathscr{E})}] > 0,\\
&\hbar_{j_\alpha} \mathbf{n}_{\partial\Omega_{\alpha i_\alpha}}^{\mathrm{T}}(\mathbf{x}_{\rho(m+\varepsilon)}^{(j_\alpha,\alpha)}) \cdot [\mathbf{x}_{m+\varepsilon}^{(\sigma_s;\mathscr{E})} - \mathbf{x}_{\rho(m+\varepsilon)}^{(j_\alpha,\alpha)}] > 0,\\
&\hbar_{j_\alpha} C_{m-\varepsilon}^{(j_\alpha,\alpha)} > \hbar_{j_\alpha} C_m^{(j_\alpha,\alpha)} > \hbar_{j_\alpha} C_{m+\varepsilon}^{(j_\alpha,\alpha)}.
\end{aligned} \tag{8.75}$$

(iii) The edge flow $\mathbf{x}^{(\sigma_s;\mathscr{E})}(t)$ at time t_m to the edge of $\mathscr{E}_{\sigma_r}^{(r)}$ for all the boundaries of $\partial\Omega_{\alpha j_\alpha}$ ($j_\alpha \in \bar{\mathscr{N}}_{\alpha\mathscr{B}}^{(3)}(\sigma_r,\sigma_s)$) is *partially from the attractive to repulsive* state if

$$\hbar_{j_\alpha}\mathbf{n}^{\mathrm{T}}_{\partial\Omega_{\alpha i_\alpha}}(\mathbf{x}^{(j_\alpha,\alpha)}_{\rho(m-\varepsilon)})\cdot[\mathbf{x}^{(j_\alpha,\alpha)}_{\rho(m-\varepsilon)}-\mathbf{x}^{(\sigma_s;\mathscr{E})}_{m-\varepsilon}]<0,$$
$$\hbar_{j_\alpha}\mathbf{n}^{\mathrm{T}}_{\partial\Omega_{\alpha i_\alpha}}(\mathbf{x}^{(j_\alpha,\alpha)}_{\rho(m+\varepsilon)})\cdot[\mathbf{x}^{(\sigma_s;\mathscr{E})}_{m+\varepsilon}-\mathbf{x}^{(j_\alpha,\alpha)}_{\rho(m+\varepsilon)}]>0, \tag{8.76}$$
$$\hbar_{j_\alpha}C^{(j_\alpha,\alpha)}_{m}<\hbar_{j_\alpha}C^{(j_\alpha,\alpha)}_{m-\varepsilon}\text{ and }\hbar_{j_\alpha}C^{(j_\alpha,\alpha)}_{m}<\hbar_{j_\alpha}C^{(j_\alpha,\alpha)}_{m+\varepsilon}.$$

(iv) The flow $\mathbf{x}^{(\sigma_s;\mathscr{E})}(t)$ at time t_m to the edge of $\mathscr{E}^{(r)}_{\sigma_r}$ for all the boundaries of $\partial\Omega_{\alpha j_\alpha}$ ($j_\alpha\in\bar{\mathscr{N}}^{(4)}_{\alpha\mathscr{B}}(\sigma_r,\sigma_s)$) is *partially from the repulsive to attractive* state if

$$\hbar_{j_\alpha}\mathbf{n}^{\mathrm{T}}_{\partial\Omega_{\alpha i_\alpha}}(\mathbf{x}^{(j_\alpha,\alpha)}_{\rho(m-\varepsilon)})\cdot[\mathbf{x}^{(j_\alpha,\alpha)}_{\rho(m-\varepsilon)}-\mathbf{x}^{(\sigma_s;\mathscr{E})}_{m-\varepsilon}]>0,$$
$$\hbar_{j_\alpha}\mathbf{n}^{\mathrm{T}}_{\partial\Omega_{\alpha i_\alpha}}(\mathbf{x}^{(j_\alpha,\alpha)}_{\rho(m+\varepsilon)})\cdot[\mathbf{x}^{(\sigma_s;\mathscr{E})}_{m+\varepsilon}-\mathbf{x}^{(j_\alpha,\alpha)}_{\rho(m+\varepsilon)}]<0, \tag{8.77}$$
$$\hbar_{j_\alpha}C^{(j_\alpha,\alpha)}_{m}>\hbar_{j_\alpha}C^{(j_\alpha,\alpha)}_{m-\varepsilon}\text{ and }\hbar_{j_\alpha}C^{(j_\alpha,\alpha)}_{m}>\hbar_{j_\alpha}C^{(j_\alpha,\alpha)}_{m+\varepsilon}.$$

(v) The flow $\mathbf{x}^{(\sigma_s;\mathscr{E})}(t)$ at time t_m to the edge of $\mathscr{E}^{(r)}_{\sigma_r}$ for all the boundaries of $\partial\Omega_{\alpha j_\alpha}$ ($j_\alpha\in\bar{\mathscr{N}}^{(5)}_{\alpha\mathscr{B}}(\sigma_r,\sigma_s)$) is *partially invariant* with equi-measuring quantities of $C^{(i_\alpha,\alpha)}_{m}$ if

$$\mathbf{x}^{(\sigma_s;\mathscr{E})}_{m-\varepsilon}=\mathbf{x}^{(j_\alpha,\alpha)}_{\rho(m-\varepsilon)}\text{ and }\mathbf{x}^{(\sigma_s;\mathscr{E})}_{m+\varepsilon}=\mathbf{x}^{(j_\alpha,\alpha)}_{\rho(m+\varepsilon)}$$
$$C^{(j_\alpha,\alpha)}_{m-\varepsilon}=C^{(j_\alpha,\alpha)}_{m}=C^{(j_\alpha,\alpha)}_{m+\varepsilon}. \tag{8.78}$$

Similar to the Theorem 8.7, the conditions for the edge flow attractivity to the specific edge can be stated by the following theorem for convenience.

Theorem 8.9. *For a discontinuous dynamical system in Eqs.* (6.4)-(6.8)*, there is a point* $\mathbf{x}^{(\sigma_r;\mathscr{E})}(t_m)\equiv\mathbf{x}_m\in\mathscr{E}^{(r)}_{\sigma_r}$ ($\sigma_r\in\mathscr{N}^{(r)}_{\mathscr{E}}=\{1,2,\cdots,N_r\}$ *and* $r\in\{0,1,2,\cdots,n-2\}$) *at time* t_m *with the related boundaries* $\partial\Omega_{\alpha i_\alpha}(\sigma_r)$ *and related domains* $\Omega_\alpha(\sigma_r)$ ($\alpha\in\mathscr{N}_{\mathscr{D}}(\sigma_r)\subset\mathscr{N}_{\mathscr{D}}=\{1,2,\cdots,N_d\}$ *and* $i_\alpha\in\mathscr{N}_{\alpha\mathscr{B}}(\sigma_r)\subset\mathscr{N}_{\mathscr{B}}=\{1,2,\cdots,N_b\}$)*, where the boundary index subset of* $\mathscr{N}_{\alpha\mathscr{B}}(\sigma_r)$ *has* n_{σ_r} *-elements* ($n_{\sigma_r}\geq(n-r)$)*. The edge of* $\mathscr{E}^{(r)}_{\sigma_r}$ *is the intersection of edges (including domains and boundaries)* $\mathscr{E}^{(s)}_{\sigma_s}(\sigma_r)$ ($\sigma_s\in\mathscr{N}^{s}_{\mathscr{E}}(\sigma_r)\subset\mathscr{N}^{s}_{\mathscr{E}}=\{1,2,\cdots,N_s\}$ *and* $s\in\{r+1,r+2,\cdots,n-1,n\}$)*. Suppose there is an* s *-dimensional edge* $\mathscr{E}^{(s)}_{\sigma_s}(\sigma_r)$ *with a boundary index set* $\mathscr{N}_{\alpha\mathscr{B}}(\sigma_r,\sigma_s)$ *relative to domain* $\Omega_\alpha(\sigma_r)$*. The complementary boundary index set is* $\bar{\mathscr{N}}_{\alpha\mathscr{B}}(\sigma_r,\sigma_s)=\mathscr{N}_{\alpha\mathscr{B}}(\sigma_r)/\mathscr{N}_{\alpha\mathscr{B}}(\sigma_r,\sigma_s)$*. For an arbitrarily small* $\varepsilon>0$*, there are two time intervals* $[t_{m-\varepsilon},t_m)$ *and* $(t_m,t_{m+\varepsilon}]$*. An edge flow* $\mathbf{x}^{(\sigma_s;\mathscr{E})}(t)$ *is* $C^{r_{\sigma_s}}_{[t_{m-\varepsilon},t_m)}$ *and/or* $C^{r_{\sigma_s}}_{(t_m,t_{m+\varepsilon}]}$ *-continuous for time* t *and* $\|d^{r_{\sigma_s}+1}\mathbf{x}^{(\sigma_s;\mathscr{E})}/dt^{r_{\sigma_s}+1}\|<\infty$

($r_{\sigma_s} \geq 1$). *Suppose there is an* $(n-s+r)$*-dimensional measuring edge* $\mathscr{E}_{\sigma_{n-s+r}}^{(\rho,n-s+r)}$ *of a flow* $\mathbf{x}^{(\sigma_s;\mathscr{E})}(t)$ *in edge* $\mathscr{E}_{\sigma_s}^{(s)}(\sigma_r,\alpha)$ *with an eqi-constant vectors of the specific edge* $\mathscr{E}_{\sigma_r}^{(r)}(\alpha)$. *The measuring edge* $\mathscr{E}_{\sigma_{n-s+r}}^{(\rho,n-s+r)}(\sigma_r,\alpha)$ *is formed by an intersection of all measuring surfaces* $S_{\sigma_m}^{(i_\alpha,\alpha)}$ *for* $i_\alpha \in \bar{\mathscr{N}}_{\alpha\mathscr{B}}(\sigma_r,\sigma_s)$ (*i.e.,* $\mathscr{E}_{\sigma_{n-s+r}}^{(\rho,n-s+r)}(\sigma_r,\alpha) = \bigcap_{j_\alpha\in\bar{\mathscr{N}}_{\alpha\mathscr{B}}(\sigma_r,\sigma_s)} S_\rho^{(j_\alpha,\alpha)}$). $\mathscr{E}_{\sigma_r}^{(r)}(\alpha) = \mathscr{E}_{\sigma_{n-s+r}}^{(n-s+r)}(\sigma_r,\alpha) \bigcap \mathscr{E}_{\sigma_s}^{(s)}(\sigma_r,\alpha)$ *with* $\mathscr{E}_{\sigma_{n-s+r}}^{(n-s+r)}(\sigma_r,\alpha) = \bigcap_{j_\alpha\in\bar{\mathscr{N}}_{\alpha\mathscr{B}}(\sigma_r,\sigma_s)} \partial\Omega_{\alpha j_\alpha}$ *and* $\mathscr{E}_{\sigma_s}^{(s)}(\sigma_r,\alpha) = \bigcap_{i_\alpha\in\mathscr{N}_{\alpha\mathscr{B}}(\sigma_r,\sigma_s)} \partial\Omega_{\alpha i_\alpha} \subset \mathscr{E}_{\sigma_s}^{(s)}(\sigma_r)$. *At time* t_m, $\mathbf{x}_m^{(\sigma_s;\mathscr{E})} = \mathbf{x}_\rho^{(\sigma_s;\mathscr{E})}(t_m) \in \mathscr{E}_{\sigma_r}^{(r,\sigma_s)}(\rho_m)$ ($\rho\in\mathbb{N}$) *satisfies Eq.* (8.55). *Suppose* $\mathbf{x}_{m\pm\varepsilon}^{(\sigma_s;\mathscr{E})} = \mathbf{x}^{(\sigma_s;\mathscr{E})}(t_{m\pm\varepsilon}) \in \mathscr{E}_{\sigma_{n-s+r}}^{(\rho_m,n-s+r)}$ *and* $\mathbf{x}_{\rho(m\pm\varepsilon)}^{(j_\alpha,\alpha)} \in S_{\rho_m}^{(j_\alpha,\alpha)}$. $\bar{\mathscr{N}}_{\alpha\mathscr{B}}(\sigma_r,\sigma_s) = \bigcup_{k=1}^5 \bar{\mathscr{N}}_{\alpha\mathscr{B}}^{(k)}(\sigma_r,\sigma_s)$ *and* $\bigcap_{k=1}^5 \bar{\mathscr{N}}_{\alpha\mathscr{B}}^{(k)}(\sigma_r,\sigma_s) = \varnothing$.

(i) *The edge flow* $\mathbf{x}^{(\sigma_s;\mathscr{E})}(t)$ *at time* t_m *is partially attractive to the edge of* $\mathscr{E}_{\sigma_r}^{(r)}$ *for all the boundaries of* $\partial\Omega_{\alpha j_\alpha}$ ($j_\alpha \in \bar{\mathscr{N}}_{\alpha\mathscr{B}}^{(1)}(\sigma_r,\sigma_s)$) *if and only if*

$$\hbar_{j_\alpha} G_{\partial\Omega_{\alpha j_\alpha}}^{(\sigma_s;\mathscr{E})}(\mathbf{x}_m^{(\sigma_s;\mathscr{E})}, t_m, \boldsymbol{\pi}_{\sigma_s}, \boldsymbol{\lambda}_{j_\alpha}) < 0. \tag{8.79}$$

(ii) *The edge flow* $\mathbf{x}^{(\sigma_s;\mathscr{E})}(t)$ *at time* t_m *is partially repulsive to the edge of* $\mathscr{E}_{\sigma_r}^{(r)}$ *for all the boundaries of* $\partial\Omega_{\alpha j_\alpha}$ ($j_\alpha \in \bar{\mathscr{N}}_{\alpha\mathscr{B}}^{(2)}(\sigma_r,\sigma_s)$) *if and only if*

$$\hbar_{j_\alpha} G_{\partial\Omega_{\alpha j_\alpha}}^{(\sigma_s;\mathscr{E})}(\mathbf{x}_m^{(\sigma_s;\mathscr{E})}, t_m, \boldsymbol{\pi}_{\sigma_s}, \boldsymbol{\lambda}_{j_\alpha}) > 0. \tag{8.80}$$

(iii) *The edge flow* $\mathbf{x}^{(\sigma_s;\mathscr{E})}(t)$ *at time* t_m *to the edge of* $\mathscr{E}_{\sigma_r}^{(r)}$ *for all the boundaries of* $\partial\Omega_{\alpha j_\alpha}$ ($j_\alpha \in \bar{\mathscr{N}}_{\alpha\mathscr{B}}^{(3)}(\sigma_r,\sigma_s)$) *is partially from the attractive to repulsive state if and only if*

$$\begin{aligned} &\hbar_{j_\alpha} G_{\partial\Omega_{\alpha j_\alpha}}^{(\sigma_s;\mathscr{E})}(\mathbf{x}_m^{(\sigma_s;\mathscr{E})}, t_m, \boldsymbol{\pi}_{\sigma_s}, \boldsymbol{\lambda}_{j_\alpha}) = 0, \\ &\hbar_{j_\alpha} G_{\partial\Omega_{\alpha j_\alpha}}^{(1,\sigma_s;\mathscr{E})}(\mathbf{x}_m^{(\sigma_s;\mathscr{E})}, t_m, \boldsymbol{\pi}_{\sigma_s}, \boldsymbol{\lambda}_{j_\alpha}) > 0. \end{aligned} \tag{8.81}$$

(iv) *The flow* $\mathbf{x}^{(\sigma_s;\mathscr{E})}(t)$ *at time* t_m *to the edge of* $\mathscr{E}_{\sigma_r}^{(r)}$ *for all the boundaries of* $\partial\Omega_{\alpha j_\alpha}$ ($j_\alpha \in \bar{\mathscr{N}}_{\alpha\mathscr{B}}^{(4)}(\sigma_r,\sigma_s)$) *is partially from the repulsive to attractive state if and only if*

$$\begin{aligned} &\hbar_{j_\alpha} G_{\partial\Omega_{\alpha j_\alpha}}^{(\sigma_s;\mathscr{E})}(\mathbf{x}_m^{(\sigma_s;\mathscr{E})}, t_m, \boldsymbol{\pi}_{\sigma_s}, \boldsymbol{\lambda}_{j_\alpha}) = 0, \\ &\hbar_{j_\alpha} G_{\partial\Omega_{\alpha j_\alpha}}^{(1,\sigma_s;\mathscr{E})}(\mathbf{x}_m^{(\sigma_s;\mathscr{E})}, t_m, \boldsymbol{\pi}_{\sigma_s}, \boldsymbol{\lambda}_{j_\alpha}) < 0. \end{aligned} \tag{8.82}$$

(v) *The flow* $\mathbf{x}^{(\sigma_s;\mathscr{E})}(t)$ *at time* t_m *to the edge of* $\mathscr{E}_{\sigma_r}^{(r)}$ *for all the boundaries of* $\partial\Omega_{\alpha j_\alpha}$ ($j_\alpha \in \bar{\mathscr{N}}_{\alpha\mathscr{B}}^{(5)}(\sigma_r,\sigma_s)$) *is partially invariant with equi-measuring quan-*

tities of $C_m^{(i_\alpha,\alpha)}$ *if and only if*

$$\hbar_{j_\alpha} G_{\partial\Omega_{\alpha j_\alpha}}^{(s_{j_\alpha},\sigma_s;\mathscr{E})}(\mathbf{x}_m^{(\sigma_s;\mathscr{E})}, t_m, \boldsymbol{\pi}_{\sigma_s}, \boldsymbol{\lambda}_{j_\alpha}) = 0 \;\; for \; s_{j_\alpha} = 0,1,2\cdots. \tag{8.83}$$

Proof. The measuring surface of $\varphi_{\alpha j_\alpha}(\mathbf{x}_m^{(\sigma_s;\mathscr{E})}, t_m, \boldsymbol{\lambda}_{j_\alpha}) = C_m^{(j_\alpha,\alpha)}$ is as a boundary. As in Chapter 3, this theorem can be proved. ■

As in the previous subsection, the flow attractivity to the edge with the higher order singularity is also introduced herein.

Definition 8.12. For a discontinuous dynamical system in Eqs. (6.4)-(6.8), there is a point $\mathbf{x}^{(\sigma_r;\mathscr{E})}(t_m) \equiv \mathbf{x}_m \in \mathscr{E}_{\sigma_r}^{(r)}$ ($\sigma_r \in \mathscr{N}_{\mathscr{E}}^{(r)} = \{1,2,\cdots,N_r\}$ and $r \in \{0,1,2,\cdots,n-2\}$) at time t_m with the related boundaries $\partial\Omega_{\alpha i_\alpha}(\sigma_r)$ and related domains $\Omega_\alpha(\sigma_r)$ ($\alpha \in \mathscr{N}_{\mathscr{D}}(\sigma_r) \subset \mathscr{N}_{\mathscr{D}} = \{1,2,\cdots,N_d\}$ and $i_\alpha \in \mathscr{N}_{\alpha\mathscr{B}}(\sigma_r) \subset \mathscr{N}_{\mathscr{B}} = \{1,2, \;\cdots, N_b\}$), where the boundary index subset of $\mathscr{N}_{\alpha\mathscr{B}}(\sigma_r)$ has n_{σ_r} -elements ($n_{\sigma_r} \geq (n-r)$). The edge of $\mathscr{E}_{\sigma_r}^{(r)}$ is the intersection of edges (including domains and boundaries) $\mathscr{E}_{\sigma_s}^{(s)}(\sigma_r)$ ($\sigma_s \in \mathscr{N}_{\mathscr{E}}^{s}(\sigma_r) \subset \mathscr{N}_{\mathscr{E}}^{s} = \{1,2,\cdots,N_s\}$ and $s \in \{r+1, r+2,\cdots,n-1,n\}$). Suppose there is an s -dimensional edge $\mathscr{E}_{\sigma_s}^{(s)}(\sigma_r)$ with a boundary index set $\mathscr{N}_{\alpha\mathscr{B}}(\sigma_r,\sigma_s)$ relative to domain $\Omega_\alpha(\sigma_r)$. The complementary boundary index set is $\bar{\mathscr{N}}_{\alpha\mathscr{B}}(\sigma_r,\sigma_s) = \mathscr{N}_{\alpha\mathscr{B}}(\sigma_r) / \mathscr{N}_{\alpha\mathscr{B}}(\sigma_r,\sigma_s)$. For an arbitrarily small $\varepsilon > 0$, there are two time intervals $[t_{m-\varepsilon}, t_m)$ and $(t_m, t_{m+\varepsilon}]$. An edge flow $\mathbf{x}^{(\sigma_s;\mathscr{E})}(t)$ is $C_{[t_{m-\varepsilon},t_m)}^{r_{\sigma_s}}$ and/or $C_{(t_m,t_{m+\varepsilon}]}^{r_{\sigma_s}}$-continuous and $\| d^{r_{\sigma_s}+1}\mathbf{x}^{(\sigma_s;\mathscr{E})} / dt^{r_{\sigma_s}+1} \| < \infty$ for time t ($r_{\sigma_s} \geq \max_{j_\alpha \in \bar{\mathscr{N}}_{\alpha\mathscr{B}}}(2k_{j_\alpha}+1)$). Suppose there is an $(n-s+r)$ -dimensional measuring edge $\mathscr{E}_{\sigma_{n-s+r}}^{(\rho,n-s+r)}$ of a flow $\mathbf{x}^{(\sigma_s;\mathscr{E})}(t)$ in edge $\mathscr{E}_{\sigma_s}^{(s)}(\sigma_r,\alpha)$ with an equi-constant vector of the specific edge $\mathscr{E}_{\sigma_r}^{(r)}(\alpha)$. The measuring edge $\mathscr{E}_{\sigma_{n-s+r}}^{(\rho,n-s+r)}(\sigma_r,\alpha)$ is formed by an intersection of all measuring surfaces $S_{\sigma_m}^{(i_\alpha,\alpha)}$ for $i_\alpha \in \bar{\mathscr{N}}_{\alpha\mathscr{B}}(\sigma_r,\sigma_s)$ (i.e., $\mathscr{E}_{\sigma_{n-s+r}}^{(\rho,n-s+r)}(\sigma_r,\alpha) = \cap_{j_\alpha \in \bar{\mathscr{N}}_{\alpha\mathscr{B}}(\sigma_r,\sigma_s)} S_\rho^{(j_\alpha,\alpha)}$). $\mathscr{E}_{\sigma_r}^{(r)}(\alpha) = \mathscr{E}_{\sigma_{n-s+r}}^{(n-s+r)}(\sigma_r,\alpha) \cap \mathscr{E}_{\sigma_s}^{(s)}(\sigma_r,\alpha)$ with $\mathscr{E}_{\sigma_{n-s+r}}^{(n-s+r)}(\sigma_r,\alpha) = \cap_{j_\alpha \in \bar{\mathscr{N}}_{\alpha\mathscr{B}}(\sigma_r,\sigma_s)} \partial\Omega_{\alpha j_\alpha}$ and $\mathscr{E}_{\sigma_s}^{(s)}(\sigma_r,\alpha) = \cap_{i_\alpha \in \mathscr{N}_{\alpha\mathscr{B}}(\sigma_r,\sigma_s)} \partial\Omega_{\alpha i_\alpha}$ $\subset \mathscr{E}_{\sigma_s}^{(s)}(\sigma_r)$. $\mathbf{x}_m^{(\sigma_s;\mathscr{E})} = \mathbf{x}_\rho^{(\sigma_s;\mathscr{E})}(t_m) \in \mathscr{E}_{\sigma_r}^{(r,\sigma_s)}(\rho_m)$ ($\rho \in \mathbb{N}$) satisfies Eq. (8.55) at time t_m. Suppose $\mathbf{x}_{m\pm\varepsilon}^{(\sigma_s;\mathscr{E})} = \mathbf{x}^{(\sigma_s;\mathscr{E})}(t_{m\pm\varepsilon}) \in \mathscr{E}_{\sigma_{n-s+r}}^{(\rho_m,n-s+r)}$ and $\mathbf{x}_{\rho(m\pm\varepsilon)}^{(j_\alpha,\alpha)} \in S_{\rho_m}^{(j_\alpha,\alpha)}$. $\bar{\mathscr{N}}_{\alpha\mathscr{B}}(\sigma_r,\sigma_s) = \cup_{k=1}^{5} \bar{\mathscr{N}}_{\alpha\mathscr{B}}^{(k)}(\sigma_r,\sigma_s)$ and $\cap_{k=1}^{5} \bar{\mathscr{N}}_{\alpha\mathscr{B}}^{(k)}(\sigma_r,\sigma_s) = \varnothing$.

(i) The edge flow $\mathbf{x}^{(\sigma_s;\mathscr{E})}(t)$ at time t_m is *partially attractive* with the $2\mathbf{k}_{(\sigma_r,\sigma_s)}^{(1)}$-order singularity to the edge of $\mathscr{E}_{\sigma_r}^{(r)}$ for all the boundaries of $\partial\Omega_{\alpha j_\alpha}$ ($j_\alpha \in$

$\bar{\mathscr{N}}_{\alpha\mathscr{B}}^{(1)}(\sigma_r,\sigma_s)$) if

$$\begin{aligned}
&\hbar_{j_\alpha} G_{\partial\Omega_{\alpha j_\alpha}}^{(s_{j_\alpha},\sigma_s;\mathscr{E})}(\mathbf{x}_m^{(\sigma_s;\mathscr{E})},t_m,\boldsymbol{\pi}_{\sigma_s},\boldsymbol{\lambda}_{j_\alpha})=0 \text{ for } s_{j_\alpha}=0,1,2,\cdots,2k_{j_\alpha}-1;\\
&\hbar_{j_\alpha}\mathbf{n}_{\partial\Omega_{\alpha i_\alpha}}^{\mathrm{T}}(\mathbf{x}_{\rho(m-\varepsilon)}^{(j_\alpha,\alpha)})\cdot[\mathbf{x}_{\rho(m-\varepsilon)}^{(j_\alpha,\alpha)}-\mathbf{x}_{m-\varepsilon}^{(\sigma_s;\mathscr{E})}]<0,\\
&\hbar_{j_\alpha}\mathbf{n}_{\partial\Omega_{\alpha i_\alpha}}^{\mathrm{T}}(\mathbf{x}_{\rho(m+\varepsilon)}^{(j_\alpha,\alpha)})\cdot[\mathbf{x}_{m+\varepsilon}^{(\sigma_s;\mathscr{E})}-\mathbf{x}_{\rho(m+\varepsilon)}^{(j_\alpha,\alpha)}]<0,\\
&\hbar_{j_\alpha}C_{m-\varepsilon}^{(j_\alpha,\alpha)}>\hbar_{j_\alpha}C_{m}^{(j_\alpha,\alpha)}>\hbar_{j_\alpha}C_{m+\varepsilon}^{(j_\alpha,\alpha)}.
\end{aligned}\tag{8.84}$$

(ii) The edge flow $\mathbf{x}^{(\sigma_s;\mathscr{E})}(t)$ at time t_m is *partially repulsive* with the $2\mathbf{k}_{(\sigma_r,\sigma_s)}^{(2)}$-order singularity to the edge of $\mathscr{E}_{\sigma_r}^{(r)}$ for all the boundaries of $\partial\Omega_{\alpha j_\alpha}$ ($j_\alpha\in\bar{\mathscr{N}}_{\alpha\mathscr{B}}^{(2)}(\sigma_r,\sigma_s)$) if

$$\begin{aligned}
&\hbar_{j_\alpha} G_{\partial\Omega_{\alpha j_\alpha}}^{(s_{j_\alpha},\sigma_s;\mathscr{E})}(\mathbf{x}_m^{(\sigma_s;\mathscr{E})},t_m,\boldsymbol{\pi}_{\sigma_s},\boldsymbol{\lambda}_{j_\alpha})=0 \text{ for } s_{j_\alpha}=0,1,2,\cdots,2k_{j_\alpha}-1;\\
&\hbar_{j_\alpha}\mathbf{n}_{\partial\Omega_{\alpha i_\alpha}}^{\mathrm{T}}(\mathbf{x}_{\rho(m-\varepsilon)}^{(j_\alpha,\alpha)})\cdot[\mathbf{x}_{\rho(m-\varepsilon)}^{(j_\alpha,\alpha)}-\mathbf{x}_{m-\varepsilon}^{(\sigma_s;\mathscr{E})}]>0,\\
&\hbar_{j_\alpha}\mathbf{n}_{\partial\Omega_{\alpha i_\alpha}}^{\mathrm{T}}(\mathbf{x}_{\rho(m+\varepsilon)}^{(j_\alpha,\alpha)})\cdot[\mathbf{x}_{m+\varepsilon}^{(\sigma_s;\mathscr{E})}-\mathbf{x}_{\rho(m+\varepsilon)}^{(j_\alpha,\alpha)}]>0,\\
&\hbar_{j_\alpha}C_{m-\varepsilon}^{(j_\alpha,\alpha)}>\hbar_{j_\alpha}C_{m}^{(j_\alpha,\alpha)}>\hbar_{j_\alpha}C_{m+\varepsilon}^{(j_\alpha,\alpha)}.
\end{aligned}\tag{8.85}$$

(iii) The edge flow $\mathbf{x}^{(\sigma_s;\mathscr{E})}(t)$ at time t_m to the edge of $\mathscr{E}_{\sigma_r}^{(r)}$ for all the boundaries of $\partial\Omega_{\alpha j_\alpha}$ ($j_\alpha\in\bar{\mathscr{N}}_{\alpha\mathscr{B}}^{(3)}(\sigma_r,\sigma_s)$) is *partially from the attractive to repulsive* state with the $(2\mathbf{k}_{(\sigma_r,\sigma_s)}^{(3)}+1)$-order singularity if

$$\begin{aligned}
&\hbar_{j_\alpha} G_{\partial\Omega_{\alpha j_\alpha}}^{(s_{j_\alpha},\sigma_s;\mathscr{E})}(\mathbf{x}_m^{(\sigma_s;\mathscr{E})},t_m,\boldsymbol{\pi}_{\sigma_s},\boldsymbol{\lambda}_{j_\alpha})=0 \text{ for } s_{j_\alpha}=0,1,2,\cdots,2k_{j_\alpha};\\
&\hbar_{j_\alpha}\mathbf{n}_{\partial\Omega_{\alpha i_\alpha}}^{\mathrm{T}}(\mathbf{x}_{\rho(m-\varepsilon)}^{(j_\alpha,\alpha)})\cdot[\mathbf{x}_{\rho(m-\varepsilon)}^{(j_\alpha,\alpha)}-\mathbf{x}_{m-\varepsilon}^{(\sigma_s;\mathscr{E})}]<0,\\
&\hbar_{j_\alpha}\mathbf{n}_{\partial\Omega_{\alpha i_\alpha}}^{\mathrm{T}}(\mathbf{x}_{\rho(m+\varepsilon)}^{(j_\alpha,\alpha)})\cdot[\mathbf{x}_{m+\varepsilon}^{(\sigma_s;\mathscr{E})}-\mathbf{x}_{\rho(m+\varepsilon)}^{(j_\alpha,\alpha)}]>0,\\
&\hbar_{j_\alpha}C_{m}^{(j_\alpha,\alpha)}<\hbar_{j_\alpha}C_{m-\varepsilon}^{(j_\alpha,\alpha)} \text{ and } \hbar_{j_\alpha}C_{m}^{(j_\alpha,\alpha)}<\hbar_{j_\alpha}C_{m+\varepsilon}^{(j_\alpha,\alpha)}.
\end{aligned}\tag{8.86}$$

(iv) The flow $\mathbf{x}^{(\sigma_s;\mathscr{E})}(t)$ at time t_m to the edge of $\mathscr{E}_{\sigma_r}^{(r)}$ for all the boundaries of $\partial\Omega_{\alpha j_\alpha}$ ($j_\alpha\in\bar{\mathscr{N}}_{\alpha\mathscr{B}}^{(4)}(\sigma_r,\sigma_s)$) is *partially from the repulsive to attractive* state with the $(2\mathbf{k}_{(\sigma_r,\sigma_s)}^{(4)}+1)$-order singularity if

$$\begin{aligned}
&\hbar_{j_\alpha} G_{\partial\Omega_{\alpha j_\alpha}}^{(s_{j_\alpha},\sigma_s;\mathscr{E})}(\mathbf{x}_m^{(\sigma_s;\mathscr{E})},t_m,\boldsymbol{\pi}_{\sigma_s},\boldsymbol{\lambda}_{j_\alpha})=0 \text{ for } s_{j_\alpha}=0,1,2,\cdots,2k_{j_\alpha};\\
&\hbar_{j_\alpha}\mathbf{n}_{\partial\Omega_{\alpha i_\alpha}}^{\mathrm{T}}(\mathbf{x}_{\rho(m-\varepsilon)}^{(j_\alpha,\alpha)})\cdot[\mathbf{x}_{\rho(m-\varepsilon)}^{(j_\alpha,\alpha)}-\mathbf{x}_{m-\varepsilon}^{(\sigma_s;\mathscr{E})}]>0,\\
&\hbar_{j_\alpha}\mathbf{n}_{\partial\Omega_{\alpha i_\alpha}}^{\mathrm{T}}(\mathbf{x}_{\rho(m+\varepsilon)}^{(j_\alpha,\alpha)})\cdot[\mathbf{x}_{m+\varepsilon}^{(\sigma_s;\mathscr{E})}-\mathbf{x}_{\rho(m+\varepsilon)}^{(j_\alpha,\alpha)}]<0,\\
&\hbar_{j_\alpha}C_{m}^{(j_\alpha,\alpha)}>\hbar_{j_\alpha}C_{m-\varepsilon}^{(j_\alpha,\alpha)} \text{ and } \hbar_{j_\alpha}C_{m}^{(j_\alpha,\alpha)}>\hbar_{j_\alpha}C_{m+\varepsilon}^{(j_\alpha,\alpha)}.
\end{aligned}\tag{8.87}$$

From the foregoing definition, the theorem for the edge flow attractivity to the specific edge with the higher order singularity is presented as follows.

Theorem 8.10. *For a discontinuous dynamical system in Eqs.* (6.4)-(6.8)*, there is a point* $\mathbf{x}^{(\sigma_r;\mathscr{E})}(t_m) \equiv \mathbf{x}_m \in \mathscr{E}_{\sigma_r}^{(r)}$ ($\sigma_r \in \mathscr{N}_{\mathscr{E}}^{(r)} = \{1,2,\cdots,N_r\}$ *and* $r \in \{0,1,2,\cdots,n-2\}$) *at time* t_m *with the related boundaries* $\partial\Omega_{\alpha i_\alpha}(\sigma_r)$ *and related domains* $\Omega_\alpha(\sigma_r)$ ($\alpha \in \mathscr{N}_{\mathscr{D}}(\sigma_r) \subset \mathscr{N}_{\mathscr{D}} = \{1,2,\cdots,N_d\}$ *and* $i_\alpha \in \mathscr{N}_{\alpha\mathscr{B}}(\sigma_r) \subset \mathscr{N}_{\mathscr{B}} = \{1,2,\cdots,N_b\}$), *where the boundary index subset of* $\mathscr{N}_{\alpha\mathscr{B}}(\sigma_r)$ *has* n_{σ_r} *-elements* ($n_{\sigma_r} \geq (n-r)$). *The edge of* $\mathscr{E}_{\sigma_r}^{(r)}$ *is the intersection of edges (including domains and boundaries)* $\mathscr{E}_{\sigma_s}^{(s)}(\sigma_r)$ ($\sigma_s \in \mathscr{N}_{\mathscr{E}}^{s}(\sigma_r) \subset \mathscr{N}_{\mathscr{E}}^{s} = \{1,2,\cdots,N_s\}$ *and* $s \in \{r+1,r+2,\cdots,n-1,n\}$). *Suppose there is an* s *-dimensional edge* $\mathscr{E}_{\sigma_s}^{(s)}(\sigma_r)$ *with a boundary index set* $\mathscr{N}_{\alpha\mathscr{B}}(\sigma_r,\sigma_s)$ *relative to domain* $\Omega_\alpha(\sigma_r)$. *The complementary boundary index set is* $\bar{\mathscr{N}}_{\alpha\mathscr{B}}(\sigma_r,\sigma_s) = \mathscr{N}_{\alpha\mathscr{B}}(\sigma_r) / \mathscr{N}_{\alpha\mathscr{B}}(\sigma_r,\sigma_s)$. *For an arbitrarily small* $\varepsilon > 0$, *there are two time intervals* $[t_{m-\varepsilon}, t_m)$ *and* $(t_m, t_{m+\varepsilon}]$. *An edge flow* $\mathbf{x}^{(\sigma_s;\mathscr{E})}(t)$ *is* $C_{[t_{m-\varepsilon},t_m)}^{r_{\sigma_s}}$ *and/or* $C_{(t_m,t_{m+\varepsilon}]}^{r_{\sigma_s}}$*-continuous and* $\| d^{r_{\sigma_s}+1}\mathbf{x}^{(\sigma_s;\mathscr{E})} / dt^{r_{\sigma_s}+1} \| < \infty$ *for time* t ($r_{\sigma_s} \geq \max_{j_\alpha \in \bar{\mathscr{N}}_{\alpha\mathscr{B}}}(2k_{j_\alpha}+2)$). *Suppose there is an* $(n-s+r)$*-dimensional measuring edge* $\mathscr{E}_{\sigma_{n-s+r}}^{(\rho,n-s+r)}$ *of a flow* $\mathbf{x}^{(\sigma_s;\mathscr{E})}(t)$ *in edge* $\mathscr{E}_{\sigma_s}^{(s)}(\sigma_r,\alpha)$ *with an equi-constant vector of the specific edge* $\mathscr{E}_{\sigma_r}^{(r)}(\alpha)$. *The measuring edge* $\mathscr{E}_{\sigma_{n-s+r}}^{(\rho,n-s+r)}(\sigma_r,\alpha)$ *is formed by an intersection of all measuring surfaces* $S_{\sigma_m}^{(i_\alpha,\alpha)}$ *for* $i_\alpha \in \bar{\mathscr{N}}_{\alpha\mathscr{B}}(\sigma_r,\sigma_s)$ (*i.e.*, $\mathscr{E}_{\sigma_{n-s+r}}^{(\rho,n-s+r)}(\sigma_r,\alpha) = \cap_{j_\alpha \in \bar{\mathscr{N}}_{\alpha\mathscr{B}}(\sigma_r,\sigma_s)} S_\rho^{(j_\alpha,\alpha)}$). $\mathscr{E}_{\sigma_r}^{(r)}(\alpha) = \mathscr{E}_{\sigma_{n-s+r}}^{(n-s+r)}(\sigma_r,\alpha) \bigcap \mathscr{E}_{\sigma_s}^{(s)}(\sigma_r,\alpha)$ *with* $\mathscr{E}_{\sigma_{n-s+r}}^{(n-s+r)}(\sigma_r,\alpha) = \cap_{j_\alpha \in \bar{\mathscr{N}}_{\alpha\mathscr{B}}(\sigma_r,\sigma_s)} \partial\Omega_{\alpha j_\alpha}$ *and* $\mathscr{E}_{\sigma_s}^{(s)}(\sigma_r,\alpha) = \cap_{i_\alpha \in \mathscr{N}_{\alpha\mathscr{B}}(\sigma_r,\sigma_s)} \partial\Omega_{\alpha i_\alpha}$ $\subset \mathscr{E}_{\sigma_s}^{(s)}(\sigma_r)$. $\mathbf{x}_m^{(\sigma_s;\mathscr{E})} = \mathbf{x}_\rho^{(\sigma_s;\mathscr{E})}(t_m) \in \mathscr{E}_{\sigma_r}^{(r,\sigma_s)}(\rho_m)$ ($\rho \in \mathrm{N}$) *satisfies Eq.* (8.55) *at time* t_m. *Suppose* $\mathbf{x}_{m\pm\varepsilon}^{(\sigma_s;\mathscr{E})} = \mathbf{x}^{(\sigma_s;\mathscr{E})}(t_{m\pm\varepsilon}) \in \mathscr{E}_{\sigma_{n-s+r}}^{(\rho_m,n-s+r)}$ *and* $\mathbf{x}_{\rho(m\pm\varepsilon)}^{(j_\alpha,\alpha)} \in S_{\rho_m}^{(j_\alpha,\alpha)}$. $\bar{\mathscr{N}}_{\alpha\mathscr{B}}(\sigma_r,\sigma_s)$ $= \cup_{k=1}^{5} \bar{\mathscr{N}}_{\alpha\mathscr{B}}^{(k)}(\sigma_r,\sigma_s)$ *and* $\cap_{k=1}^{5} \bar{\mathscr{N}}_{\alpha\mathscr{B}}^{(k)}(\sigma_r,\sigma_s) = \varnothing$.

(i) *The edge flow* $\mathbf{x}^{(\sigma_s;\mathscr{E})}(t)$ *at time* t_m *is partially attractive with the* $2\mathbf{k}_{(\sigma_r,\sigma_s)}^{(1)}$ *-order singularity to the edge of* $\mathscr{E}_{\sigma_r}^{(r)}$ *for all the boundaries of* $\partial\Omega_{\alpha j_\alpha}$ ($j_\alpha \in \bar{\mathscr{N}}_{\alpha\mathscr{B}}^{(1)}(\sigma_r,\sigma_s)$) *if and only if*

$$
\begin{aligned}
&\hbar_{j_\alpha} G_{\partial\Omega_{\alpha j_\alpha}}^{(s_{j_\alpha},\sigma_s;\mathscr{E})}(\mathbf{x}_m^{(\sigma_s;\mathscr{E})}, t_m, \boldsymbol{\pi}_{\sigma_s}, \boldsymbol{\lambda}_{j_\alpha}) = 0 \text{ for } s_{j_\alpha} = 0,1,2,\cdots,2k_{j_\alpha}-1;\\
&\hbar_{j_\alpha} G_{\partial\Omega_{\alpha j_\alpha}}^{(2k_{j_\alpha},\sigma_s;\mathscr{E})}(\mathbf{x}_m^{(\sigma_s;\mathscr{E})}, t_m, \boldsymbol{\pi}_{\sigma_s}, \boldsymbol{\lambda}_{j_\alpha}) < 0.
\end{aligned}
\tag{8.88}
$$

(ii) *The edge flow* $\mathbf{x}^{(\sigma_s;\mathscr{E})}(t)$ *at time* t_m *is partially repulsive with the* $2\mathbf{k}_{(\sigma_r,\sigma_s)}^{(2)}$ -

order singularity to the edge of $\mathscr{E}_{\sigma_r}^{(r)}$ *for all the boundaries of* $\partial\Omega_{\alpha j_\alpha}$ ($j_\alpha \in \bar{\mathscr{N}}_{\alpha\mathscr{B}}^{(2)}(\sigma_r,\sigma_s)$) *if and only if*

$$\begin{aligned}&\hbar_{j_\alpha} G_{\partial\Omega_{\alpha j_\alpha}}^{(s_{j_\alpha},\sigma_s;\mathscr{E})}(\mathbf{x}_m^{(\sigma_s;\mathscr{E})},t_m,\boldsymbol{\pi}_{\sigma_s},\boldsymbol{\lambda}_{j_\alpha})=0 \text{ for } s_{j_\alpha}=0,1,2,\cdots,2k_{j_\alpha}-1;\\&\hbar_{j_\alpha} G_{\partial\Omega_{\alpha j_\alpha}}^{(2k_{j_\alpha},\sigma_s;\mathscr{E})}(\mathbf{x}_m^{(\sigma_s;\mathscr{E})},t_m,\boldsymbol{\pi}_{\sigma_s},\boldsymbol{\lambda}_{j_\alpha})>0.\end{aligned} \tag{8.89}$$

(iii) *The edge flow* $\mathbf{x}^{(\sigma_s;\mathscr{E})}(t)$ *at time* t_m *to the edge of* $\mathscr{E}_{\sigma_r}^{(r)}$ *for all the boundaries of* $\partial\Omega_{\alpha j_\alpha}$ ($j_\alpha \in \bar{\mathscr{N}}_{\alpha\mathscr{B}}^{(3)}(\sigma_r,\sigma_s)$) *is partially from the attractive to repulsive state with the* $(2\mathbf{k}_{(\sigma_r,\sigma_s)}^{(3)}+1)$*-order singularity if and only if*

$$\begin{aligned}&\hbar_{j_\alpha} G_{\partial\Omega_{\alpha j_\alpha}}^{(s_{j_\alpha},\sigma_s;\mathscr{E})}(\mathbf{x}_m^{(\sigma_s;\mathscr{E})},t_m,\boldsymbol{\pi}_{\sigma_s},\boldsymbol{\lambda}_{j_\alpha})=0 \text{ for } s_{j_\alpha}=0,1,2,\cdots,2k_{j_\alpha};\\&\hbar_{j_\alpha} G_{\partial\Omega_{\alpha j_\alpha}}^{(2k_{j_\alpha}+1,\sigma_s;\mathscr{E})}(\mathbf{x}_m^{(\sigma_s;\mathscr{E})},t_m,\boldsymbol{\pi}_{\sigma_s},\boldsymbol{\lambda}_{j_\alpha})>0.\end{aligned} \tag{8.90}$$

(iv) *The edge flow* $\mathbf{x}^{(\sigma_s;\mathscr{E})}(t)$ *at time* t_m *to the edge of* $\mathscr{E}_{\sigma_r}^{(r)}$ *for all the boundaries of* $\partial\Omega_{\alpha j_\alpha}$ ($j_\alpha \in \bar{\mathscr{N}}_{\alpha\mathscr{B}}^{(4)}(\sigma_r,\sigma_s)$) *is partially from the repulsive to attractive state with the* $(2\mathbf{k}_{(\sigma_r,\sigma_s)}^{(4)}+1)$*-order singularity if and only if*

$$\begin{aligned}&\hbar_{j_\alpha} G_{\partial\Omega_{\alpha j_\alpha}}^{(s_{j_\alpha},\sigma_s;\mathscr{E})}(\mathbf{x}_m^{(\sigma_s;\mathscr{E})},t_m,\boldsymbol{\pi}_{\sigma_s},\boldsymbol{\lambda}_{j_\alpha})=0 \text{ for } s_{j_\alpha}=0,1,2,\cdots,2k_{j_\alpha};\\&\hbar_{\mathbf{j}_\alpha} G_{\partial\Omega_{\alpha j_\alpha}}^{(2\mathbf{k}_{j_\alpha}+1,\sigma_s;\mathscr{E})}(\mathbf{x}_m^{(\sigma_s;\mathscr{E})},t_m,\boldsymbol{\pi}_{\sigma_s},\boldsymbol{\lambda}_{j_\alpha})<0.\end{aligned} \tag{8.91}$$

Proof. The measuring surface with $\varphi_{\alpha j_\alpha}(\mathbf{x}_m^{(\sigma_s;\mathscr{E})},t_m,\boldsymbol{\lambda}_{j_\alpha})=C_m^{(j_\alpha,\alpha)}$ is as a boundary. As in Chapter 3, this theorem can be proved. ■

As in Chapter 6, the measuring edge should not be intersected with the specific edge, as sketched in Fig. 8.9. The specific edge $\mathscr{E}_{\sigma_{s-2}}^{(s-2)}$ is the $(s-2)$-dimensional, intersection surface of three boundaries in the s-dimensional edge space of $\mathscr{E}_{\sigma_s}^{(s)}$. In edge $\mathscr{E}_{\sigma_s}^{(s)}$, the specific edge $\mathscr{E}_{\sigma_{s-2}}^{(s-2)}$ is an intersection of two edges $\mathscr{E}_{\sigma_{s-2}}^{(s-1)}$. The measuring edge of $\mathscr{E}_{\sigma_{s-2}}^{(\rho,s-2)}$ is also formed by the intersection of two $(s-1)$-dimensional measuring edges in the edge space of $\mathscr{E}_{\sigma_s}^{(s)}$. With different time, the edge flow of $\mathbf{x}^{(\sigma_s;\mathscr{E})}$ moves to different location, and the measuring surface in the edge space of $\mathscr{E}_{\sigma_s}^{(s)}$ is different. To describe the location of the measuring edge, the non-zero constant for each equi-constant measuring surface can be replaced by a variable. Thus, for a set of new constants, new measuring surfaces will give a new measuring edge. The measuring edge is expressed by a vector of non-zero constant variables. As in Chapter 6, the corresponding definition is given as follows.

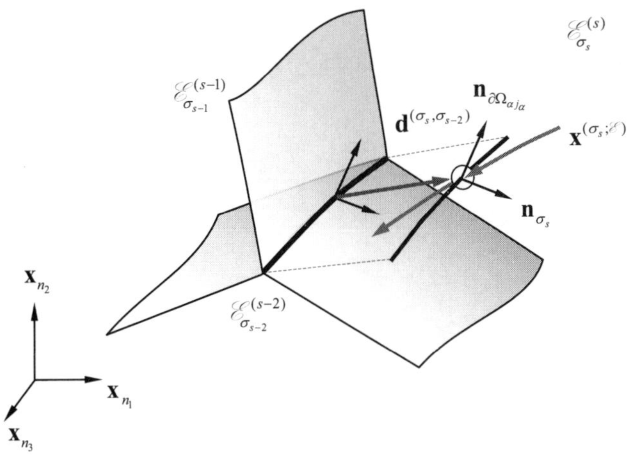

Fig. 8.9 The measuring edge $\mathscr{E}^{(\rho,s-2)}_{\sigma_{s-2}}$ paralleled with the specific edge $\mathscr{E}^{(s-2)}_{\sigma_{s-2}}$. The measuring edge $\mathscr{E}^{(\rho,s-2)}_{\sigma_{s-2}}$ is the intersection between the equi-constant edge $\mathscr{E}^{(\rho,s-1)}_{\sigma_{s-1}}$ and the equi-constant surface $S^{(j_\alpha;\alpha)}_{\rho}$ in edge $\mathscr{E}^{(s)}_{\sigma_s}$. The specific edge $\mathscr{E}^{(s-2)}_{\sigma_{s-2}}$ is the $(s-2)$-dimensional edge of two edges of $\mathscr{E}^{(s-1)}_{\sigma_{s-1}}$ and a boundary $\partial\Omega_{\alpha j_\alpha}$. The corresponding normal vectors of the boundary $\partial\Omega_{\alpha j_\alpha}$ and the edge $\mathscr{E}^{(s-1)}_{\sigma_{s-1}}$ are presented. $(n_1+n_2+n_3=s)$

Definition 8.13. For a discontinuous dynamical system in Eqs. (6.4)-(6.8), there is a point $\mathbf{x}^{(\sigma_r;\mathscr{E})}(t_m)\equiv\mathbf{x}_m\in\mathscr{E}^{(r)}_{\sigma_r}$ ($\sigma_r\in\mathscr{N}^{(r)}_{\mathscr{E}}=\{1,2,\cdots,N_r\}$ and $r\in\{0,1,2,\cdots,n-2\}$) at time t_m with the related boundaries $\partial\Omega_{\alpha i_\alpha}(\sigma_r)$ and related domains $\Omega_\alpha(\sigma_r)$ ($\alpha\in\mathscr{N}_{\mathscr{D}}(\sigma_r)\subset\mathscr{N}_{\mathscr{D}}=\{1,2,\cdots,N_d\}$ and $i_\alpha\in\mathscr{N}_{\alpha\mathscr{B}}(\sigma_r)\subset\mathscr{N}_{\mathscr{B}}=\{1,2,\cdots,N_b\}$), where the boundary index subset of $\mathscr{N}_{\alpha\mathscr{B}}(\sigma_r)$ has n_{σ_r}-elements ($n_{\sigma_r}\geq(n-r)$). The edge of $\mathscr{E}^{(r)}_{\sigma_r}$ is the intersection of edges (including domains and boundaries) $\mathscr{E}^{(s)}_{\sigma_s}(\sigma_r)$ ($\sigma_s\in\mathscr{N}^{s}_{\mathscr{E}}(\sigma_r)\subset\mathscr{N}^{s}_{\mathscr{E}}=\{1,2,\cdots,N_s\}$ and $s\in\{r+1,r+2,\cdots,n-1,n\}$). Suppose there is an s-dimensional edge $\mathscr{E}^{(s)}_{\sigma_s}(\sigma_r)$ with a boundary index set $\mathscr{N}_{\alpha\mathscr{B}}(\sigma_r,\sigma_s)$ relative to domain $\Omega_\alpha(\sigma_r)$. The complementary boundary index set is $\bar{\mathscr{N}}_{\alpha\mathscr{B}}(\sigma_r,\sigma_s)=\mathscr{N}_{\alpha\mathscr{B}}(\sigma_r)/\mathscr{N}_{\alpha\mathscr{B}}(\sigma_r,\sigma_s)$. Suppose there is an $(n-s+r)$-dimensional measuring edge $\mathscr{E}^{(\rho,n-s+r)}_{\sigma_{n-s+r}}$ of a flow $\mathbf{x}^{(\sigma_s;\mathscr{E})}(t)$ in edge of $\mathscr{E}^{(s)}_{\sigma_s}(\sigma_r,\alpha)$ with an equi-constant vector of the specific edge $\mathscr{E}^{(r)}_{\sigma_r}(\alpha)$, and the measuring edge $\mathscr{E}^{(\rho,n-s+r)}_{\sigma_{n-s+r}}(\sigma_r,\alpha)=\bigcap_{j_\alpha\in\bar{\mathscr{N}}_{\alpha\mathscr{B}}(\sigma_r,\sigma_s)}S^{(j_\alpha,\alpha)}_{\rho}$ is formed by an intersection of all measuring surfaces $S^{(j_\alpha,\alpha)}_{\sigma_m}$ for $j_\alpha\in\bar{\mathscr{N}}_{\alpha\mathscr{B}}(\sigma_r,\sigma_s)$. $\mathscr{E}^{(r)}_{\sigma_r}(\alpha)=\mathscr{E}^{(n-s+r)}_{\sigma_{n-s+r}}(\sigma_r,\alpha)\bigcap\mathscr{E}^{(s)}_{\sigma_s}(\sigma_r,\alpha)$ with $\mathscr{E}^{(n-s+r)}_{\sigma_{n-s+r}}(\sigma_r,\alpha)=\bigcap_{j_\alpha\in\bar{\mathscr{N}}_{\alpha\mathscr{B}}(\sigma_r,\sigma_s)}\partial\Omega_{\alpha j_\alpha}$ and $\bigcap_{i_\alpha\in\mathscr{N}_{\alpha\mathscr{B}}(\sigma_r,\sigma_s)}\partial\Omega_{\alpha i_\alpha}=$

$\mathscr{E}_{\sigma_s}^{(s)}(\sigma_r,\alpha)\subset\mathscr{E}_{\sigma_s}^{(s)}(\sigma_r)$.

(i) The *unit normal vector* of the boundary $\partial\Omega_{\alpha j_\alpha}(\sigma_r,\sigma_s)$ for $j_\alpha\in\bar{\mathscr{N}}_{\alpha\mathscr{B}}(\sigma_r,\sigma_s)$ is defined by

$$\mathbf{e}_{j_\alpha}=\frac{\mathbf{n}_{\partial\Omega_{\alpha j_\alpha}}}{|\mathbf{n}_{\partial\Omega_{\alpha j_\alpha}}|}=\frac{1}{\sqrt{g_{j_\alpha j_\alpha}}}\frac{\partial\varphi_{\alpha j_\alpha}}{\partial\mathbf{x}}=\frac{1}{\sqrt{g_{j_\alpha j_\alpha}}}\Big(\frac{\partial\varphi_{\alpha j_\alpha}}{\partial x_1},\frac{\partial\varphi_{\alpha j_\alpha}}{\partial x_2},\cdots,\frac{\partial\varphi_{\alpha j_\alpha}}{\partial x_n}\Big)^{\mathrm{T}} \tag{8.92}$$

where

$$g_{j_\alpha j_\alpha}=\Big(\frac{\partial\varphi_{\alpha j_\alpha}}{\partial x_1}\Big)^2+\Big(\frac{\partial\varphi_{\alpha j_\alpha}}{\partial x_2}\Big)^2+\cdots+\Big(\frac{\partial\varphi_{\alpha j_\alpha}}{\partial x_n}\Big)^2. \tag{8.93}$$

(ii) The *measuring vector* from the specific edge $\mathscr{E}_{\sigma_r}^{(r)}$ to the corresponding, measuring edge in the edge space of $\mathscr{E}_{\sigma_s}^{(s)}(\sigma_r)$ is defined as

$$\mathbf{d}^{(\sigma_s;\sigma_r)}=\sum_{j_\alpha=1}^{s-r}z^{j_\alpha}\mathbf{e}_{j_\alpha} \tag{8.94}$$

where the measuring variable z^{j_α} relative to $\partial\Omega_{\alpha j_\alpha}$ is

$$z^{j_\alpha}=\varphi_{\alpha j_\alpha}(\mathbf{x}^{(\sigma_s;\mathscr{E})},t,\boldsymbol{\lambda}_{j_\alpha}). \tag{8.95}$$

(iii) The *measuring function* is defined as

$$\mathscr{D}^{(\sigma_s;\sigma_r)}=\|\mathbf{d}^{(\sigma_s;\sigma_r)}\|^2=\sum_{j_\alpha=1}^{s-r}\sum_{k_\alpha=1}^{s-r}z^{k_\alpha}z^{j_\alpha}e_{k_\alpha j_\alpha} \tag{8.96}$$

where the metric tensor is

$$e_{k_\alpha j_\alpha}=\mathbf{e}_{k_\alpha}\cdot\mathbf{e}_{j_\alpha}=\frac{1}{\sqrt{g_{k_\alpha k_\alpha}g_{j_\alpha j_\alpha}}}\sum_{l=1}^{n}\frac{\partial\varphi_{\alpha k_\alpha}}{\partial x_l}\frac{\partial\varphi_{\alpha j_\alpha}}{\partial x_l}. \tag{8.97}$$

(iv) If $\mathscr{D}^{(\sigma_s;\sigma_r)}=\mathscr{C}^{(\sigma_s;\sigma_r)}$ is constant, the *measuring function surface* $\mathscr{M}^{(\sigma_s;\sigma_r)}$ is defined as

$$\sum_{j_\alpha=1}^{s-r}\sum_{k_\alpha=1}^{s-r}z^{j_\alpha}z^{k_\alpha}e_{j_\alpha k_\alpha}=\mathscr{C}^{(\sigma_s;\sigma_r)}. \tag{8.98}$$

(v) The *normal vector of the measuring surface* is defined as

$$\mathbf{n}_{\mathscr{M}^{(\sigma_s;\sigma_r)}}=\frac{\partial}{\partial\mathbf{x}}\sum_{j_\alpha=1}^{n-r}\sum_{k_\alpha=1}^{n-r}z^{j_\alpha}z^{k_\alpha}e_{j_\alpha k_\alpha}. \tag{8.99}$$

(vi) If $\mathscr{D}^{(\sigma_s;\sigma_r)}$ varies with time, the *derivative of the measuring function* is defined as

$$\begin{aligned}\frac{D}{Dt}\mathscr{D}^{(\sigma_s;\sigma_r)} &= \dot{\mathscr{D}}^{(\sigma_s;\sigma_r)} = \frac{\partial \mathscr{D}^{(\sigma_s;\sigma_r)}}{\partial \mathbf{x}}\dot{\mathbf{x}} + \frac{\partial \mathscr{D}^{(\sigma_s;\sigma_r)}}{\partial t} \\ &= \sum_{j_\alpha=1}^{s-r}\sum_{k_\alpha=1}^{s-r}(\dot{z}^{k_\alpha}z^{j_\alpha}e_{j_\alpha k_\alpha} + z^{k_\alpha}\dot{z}^{j_\alpha}e_{j_\alpha k_\alpha} + z^{j_\alpha}z^{k_\alpha}\dot{e}_{j_\alpha k_\alpha}).\end{aligned} \tag{8.100}$$

The G-functions for the measuring function surface can be defined similarly as in Eqs. (6.112) and (6.113), i.e., $G^{(\sigma_s;\mathscr{E})}_{\partial\Omega_{\alpha i_\alpha}} = G^{(\sigma_s;\mathscr{M})}_{\mathscr{M}^{(\sigma_s;\sigma_r)}}$ and $G^{(s_{i_\alpha},\sigma_s;\mathscr{E})}_{\partial\Omega_{\alpha i_\alpha}} = G^{(s_m,\sigma_s;\mathscr{M})}_{\mathscr{M}^{(\sigma_s;\sigma_r)}}$. The measuring vector can be decomposed by many measuring sub-vectors. The corresponding definition is given as follows.

Definition 8.14. For a discontinuous dynamical system in Eqs. (6.4)-(6.8), there is a point $\mathbf{x}^{(\sigma_r;\mathscr{E})}(t_m) \equiv \mathbf{x}_m \in \mathscr{E}^{(r)}_{\sigma_r}$ ($\sigma_r \in \mathscr{N}^{(r)}_{\mathscr{E}} = \{1,2,\cdots,N_r\}$ and $r \in \{0,1,2,\cdots,n-2\}$) at time t_m with the related boundaries $\partial\Omega_{\alpha i_\alpha}(\sigma_r)$ and related domains $\Omega_\alpha(\sigma_r)$ ($\alpha \in \mathscr{N}_{\mathscr{D}}(\sigma_r) \subset \mathscr{N}_{\mathscr{D}} = \{1,2,\cdots,N_d\}$ and $i_\alpha \in \mathscr{N}_{\alpha\mathscr{B}}(\sigma_r) \subset \mathscr{N}_{\mathscr{B}} = \{1,2,\cdots,N_b\}$), where the boundary index subset of $\mathscr{N}_{\alpha\mathscr{B}}(\sigma_r)$ has n_{σ_r}-elements ($n_{\sigma_r} \ge (n-r)$). The edge of $\mathscr{E}^{(r)}_{\sigma_r}$ is the intersection of edges (including domains and boundaries) $\mathscr{E}^{(s)}_{\sigma_s}(\sigma_r)$ ($\sigma_s \in \mathscr{N}^{s}_{\mathscr{E}}(\sigma_r) \subset \mathscr{N}^{s}_{\mathscr{E}} = \{1,2,\cdots,N_s\}$ and $s \in \{r+1,r+2,\cdots,n-1,n\}$). Suppose there is an s-dimensional edge $\mathscr{E}^{(s)}_{\sigma_s}(\sigma_r)$ with a boundary index set $\mathscr{N}_{\alpha\mathscr{B}}(\sigma_r,\sigma_s)$ relative to domain $\Omega_\alpha(\sigma_r)$. The complementary boundary index set is $\bar{\mathscr{N}}_{\alpha\mathscr{B}}(\sigma_r,\sigma_s) = \mathscr{N}_{\alpha\mathscr{B}}(\sigma_r)/\mathscr{N}_{\alpha\mathscr{B}}(\sigma_r,\sigma_s)$. Suppose there is an $(n-s+r)$-dimensional measuring edge $\mathscr{E}^{(\rho,n-s+r)}_{\sigma_{n-s+r}}$ of a flow $\mathbf{x}^{(\sigma_s;\mathscr{E})}(t)$ in the edge of $\mathscr{E}^{(s)}_{\sigma_s}(\sigma_r,\alpha)$ paralleled to the specific edge $\mathscr{E}^{(r)}_{\sigma_r}(\alpha)$, and the measuring edge $\mathscr{E}^{(\rho,n-s+r)}_{\sigma_{n-s+r}}(\sigma_r,\alpha) = \cap_{j_\alpha \in \bar{\mathscr{N}}_{\alpha\mathscr{B}}(\sigma_r,\sigma_s)} S^{(j_\alpha,\alpha)}_{\rho}$ is formed by an intersection of all measuring surfaces $S^{(i_\alpha,\alpha)}_{\sigma_m}$ for $j_\alpha \in \bar{\mathscr{N}}_{\alpha\mathscr{B}}(\sigma_r,\sigma_s)$. $\mathscr{E}^{(r)}_{\sigma_r}(\alpha) = \mathscr{E}^{(n-s+r)}_{\sigma_{n-s+r}}(\sigma_r,\alpha) \cap \mathscr{E}^{(s)}_{\sigma_s}(\sigma_r,\alpha)$ with $\mathscr{E}^{(n-s+r)}_{\sigma_{n-s+r}}(\sigma_r,\alpha) = \cap_{j_\alpha \in \bar{\mathscr{N}}_{\alpha\mathscr{B}}(\sigma_r,\sigma_s)} \partial\Omega_{\alpha j_\alpha}$ and $\mathscr{E}^{(s)}_{\sigma_s}(\sigma_r,\alpha) = \cap_{i_\alpha \in \mathscr{N}_{\alpha\mathscr{B}}(\sigma_r,\sigma_s)} \partial\Omega_{\alpha i_\alpha}$ $\subset \mathscr{E}^{(s)}_{\sigma_s}(\sigma_r)$. $\bar{\mathscr{N}}_{\alpha\mathscr{B}}(\sigma_r,\sigma_s) = \cup_{k=1}^{5} \bar{\mathscr{N}}^{(k)}_{\alpha\mathscr{B}}(\sigma_r,\sigma_s)$ and $\cap_{k=1}^{5} \bar{\mathscr{N}}^{(k)}_{\alpha\mathscr{B}}(\sigma_r,\sigma_s) = \varnothing$. The measuring vector from the common edge to the measuring edge in the edge space of $\mathscr{E}^{(s)}_{\sigma_s}(\sigma_r,\alpha)$ in Eq. (8.94) is composed of the measuring sub-vectors, i.e.,

$$\begin{aligned}\mathbf{d}^{(\sigma_s;\sigma_r)} &= \sum_{j_\alpha=1}^{s-r} z^{j_\alpha}\mathbf{e}_{j_\alpha} = \sum_{j=1}^{5}\mathbf{d}^{(\sigma_s;\sigma_r)}_j \\ \mathbf{d}^{(\sigma_s;\sigma_r)}_j &= \sum_{j_\alpha=1}^{s-r} z^{j_\alpha}_{(j)}\mathbf{e}^{(j)}_{j_\alpha} \text{ for } j_\alpha \in \bar{\mathscr{N}}^{(j)}_{\alpha\mathscr{B}}(\sigma_s,\sigma_r).\end{aligned} \tag{8.101}$$

Definition 8.15. For a discontinuous dynamical system in Eqs. (6.4)-(6.8), there is a point $\mathbf{x}^{(\sigma_r;\mathscr{E})}(t_m) \equiv \mathbf{x}_m \in \mathscr{E}^{(r)}_{\sigma_r}$ ($\sigma_r \in \mathscr{N}^{(r)}_{\mathscr{E}} = \{1,2,\cdots,N_r\}$ and $r \in \{0,1,2,\cdots,n-2\}$)

at time t_m with the related boundaries $\partial\Omega_{\alpha i_\alpha}(\sigma_r)$ and related domains $\Omega_\alpha(\sigma_r)$ ($\alpha \in \mathscr{N}_{\mathscr{D}}(\sigma_r) \subset \mathscr{N}_{\mathscr{D}} = \{1,2,\cdots,N_d\}$ and $i_\alpha \in \mathscr{N}_{\alpha\mathscr{B}}(\sigma_r) \subset \mathscr{N}_{\mathscr{B}} = \{1,2, \cdots,N_b\}$), where the boundary index subset of $\mathscr{N}_{\alpha\mathscr{B}}(\sigma_r)$ has n_{σ_r}-elements ($n_{\sigma_r} \geq (n-r)$). The edge of $\mathscr{E}_{\sigma_r}^{(r)}$ is the intersection of edges (including domains and boundaries) $\mathscr{E}_{\sigma_s}^{(s)}(\sigma_r)$ ($\sigma_s \in \mathscr{N}_{\mathscr{E}}^s(\sigma_r) \subset \mathscr{N}_{\mathscr{E}}^s = \{1,2,\cdots,N_s\}$ and $s \in \{r+1, r+2,\cdots,n-1,n\}$). Suppose there is an s-dimensional edge $\mathscr{E}_{\sigma_s}^{(s)}(\sigma_r)$ with a boundary index set $\mathscr{N}_{\alpha\mathscr{B}}(\sigma_r,\sigma_s)$ relative to domain $\Omega_\alpha(\sigma_r)$. The complementary boundary index set is $\bar{\mathscr{N}}_{\alpha\mathscr{B}}(\sigma_r,\sigma_s) = \mathscr{N}_{\alpha\mathscr{B}}(\sigma_r) / \mathscr{N}_{\alpha\mathscr{B}}(\sigma_r,\sigma_s)$. Suppose there is an $(n-s+r)$-dimensional measuring edge $\mathscr{E}_{\sigma_{n-s+r}}^{(\rho,n-s+r)}$ of a flow $\mathbf{x}^{(\sigma_s;\mathscr{E})}(t)$ in the edge of $\mathscr{E}_{\sigma_s}^{(s)}(\sigma_r,\alpha)$ with an equi-constant vector of the specific edge $\mathscr{E}_{\sigma_r}^{(r)}(\alpha)$, and the measuring edge $\mathscr{E}_{\sigma_{n-s+r}}^{(\rho,n-s+r)}(\sigma_r,\alpha)$ is formed by an intersection of all measuring surfaces $S_{\sigma_m}^{(i_\alpha,\alpha)}$ for $j_\alpha \in \bar{\mathscr{N}}_{\alpha\mathscr{B}}(\sigma_r,\sigma_s)$, i.e.

$$\mathscr{E}_{\sigma_{n-s+r}}^{(\rho,n-s+r)}(\sigma_r,\alpha) = \cap_{j_\alpha \in \bar{\mathscr{N}}_{\alpha\mathscr{B}}(\sigma_r,\sigma_s)} S_\rho^{(j_\alpha,\alpha)} \tag{8.102}$$

which lies in a measuring surface $\mathscr{M}^{(\sigma_s;\sigma_r)}$ given by

$$\mathscr{M}^{(\sigma_s;\sigma_r)} = \left\{ \mathbf{x}^{(\sigma_s;\mathscr{E})} \left| \begin{array}{l} \sum_{j_\alpha=1}^{s-r} \sum_{k_\alpha=1}^{s-r} z^{k_\alpha} z^{j_\alpha} e_{k_\alpha j_\alpha} = \mathscr{C}^{(\sigma_s,\sigma_r)} \\ \text{where } z^{j_\alpha} = \varphi_{\alpha j_\alpha}(\mathbf{x}^{(\sigma_s;\mathscr{E})}, t, \boldsymbol{\lambda}_{j_\alpha}) \\ j_\alpha \in \bar{\mathscr{N}}_{\alpha\mathscr{B}}(\sigma_r,\sigma_s) \end{array} \right. \right\}. \tag{8.103}$$

From the description of measuring vector and measuring surface, the flow attractivity to the edge can be described.

Definition 8.16. For a discontinuous dynamical system in Eqs. (6.4)-(6.8), there is a point $\mathbf{x}^{(\sigma_r;\mathscr{E})}(t_m) \equiv \mathbf{x}_m \in \mathscr{E}_{\sigma_r}^{(r)}$ ($\sigma_r \in \mathscr{N}_{\mathscr{E}}^r = \{1,2,\cdots,N_r\}$ and $r \in \{0,1,2,\cdots,n-2\}$) at time t_m with the related boundaries $\partial\Omega_{\alpha i_\alpha}(\sigma_r)$ and related domains $\Omega_\alpha(\sigma_r)$ ($\alpha \in \mathscr{N}_{\mathscr{D}}(\sigma_r) \subset \mathscr{N}_{\mathscr{D}} = \{1,2,\cdots,N_d\}$ and $i_\alpha \in \mathscr{N}_{\alpha\mathscr{B}}(\sigma_r) \subset \mathscr{N}_{\mathscr{B}} = \{1,2, \cdots,N_b\}$), where the boundary index subset of $\mathscr{N}_{\alpha\mathscr{B}}(\sigma_r)$ has n_{σ_r}-elements ($n_{\sigma_r} \geq (n-r)$). The edge of $\mathscr{E}_{\sigma_r}^{(r)}$ is the intersection of edges (including domains and boundaries) $\mathscr{E}_{\sigma_s}^{(s)}(\sigma_r)$ ($\sigma_s \in \mathscr{N}_{\mathscr{E}}^s(\sigma_r) \subset \mathscr{N}_{\mathscr{E}}^s = \{1,2,\cdots,N_s\}$ and $s \in \{r+1, r+2,\cdots,n-1,n\}$). Suppose there is an s-dimensional edge $\mathscr{E}_{\sigma_s}^{(s)}(\sigma_r)$ with a boundary index set $\mathscr{N}_{\alpha\mathscr{B}}(\sigma_r,\sigma_s)$ relative to domain $\Omega_\alpha(\sigma_r)$. The complementary boundary index set is $\bar{\mathscr{N}}_{\alpha\mathscr{B}}(\sigma_r,\sigma_s) = \mathscr{N}_{\alpha\mathscr{B}}(\sigma_r) / \mathscr{N}_{\alpha\mathscr{B}}(\sigma_r,\sigma_s)$. Suppose there is an $(n-s+r)$-

dimensional measuring edge $\mathscr{E}_{\sigma_{n-s+r}}^{(\rho,n-s+r)}$ of a flow $\mathbf{x}^{(\sigma_s;\mathscr{E})}(t)$ in the edge of $\mathscr{E}_{\sigma_s}^{(s)}(\sigma_r,\alpha)$ with an equi-constant vector the specific edge $\mathscr{E}_{\sigma_r}^{(r)}(\alpha)$, and the measuring edge and an equi-constant measuring surface $\mathscr{M}^{(\sigma_s;\sigma_r)}$ are determined by Eqs. (8.98) and (8.103), respectively. At time t_m, $\mathbf{x}_m^{(\sigma_s;\mathscr{M})}=\mathbf{x}_\sigma^{(\sigma_s;\mathscr{M})}(t_m)\in\mathscr{M}_m^{(\sigma_s;\sigma_r)}$ satisfies Eq. (8.55). An edge flow $\mathbf{x}^{(\sigma_s;\mathscr{E})}(t)$ is $C_{[t_{m-\varepsilon},t_m)}^{r_{\sigma_s}}$ and/or $C_{(t_m,t_{m+\varepsilon}]}^{r_{\sigma_s}}$-continuous for time t and $\| d^{r_{\sigma_s}+1}\mathbf{x}^{(\sigma_s;\mathscr{E})}/dt^{r_{\sigma_s}+1}\|<\infty(r_{\sigma_s}\geq 1)$. Suppose $\mathbf{x}_{m\pm\varepsilon}^{(\sigma_s;\mathscr{E})}=\mathbf{x}^{(\sigma_s;\mathscr{E})}(t_{m\pm\varepsilon})\in\mathscr{M}_{m\pm\varepsilon}^{(\alpha,\sigma_r)}$ and $\mathbf{x}_{\sigma(m\pm\varepsilon)}^{(\sigma_s;\mathscr{M})}=\mathbf{x}_\sigma^{(\sigma_s;\mathscr{M})}(t_{m\pm\varepsilon})\in\mathscr{M}_m^{(\sigma_s;\sigma_r)}$. $\mathscr{D}^{(\sigma_s;\sigma_r)}(t_{m\pm\varepsilon})\equiv\mathscr{C}_{m\pm\varepsilon}^{(\sigma_s;\sigma_r)}$ and $\mathscr{D}^{(\sigma_s;\sigma_r)}(t_m)\equiv\mathscr{C}_m^{(\sigma_s;\sigma_r)}$.

(i) The edge flow $\mathbf{x}^{(\sigma_s;\mathscr{E})}(t)$ at time t_m to the specific edge of $\mathscr{E}_{\sigma_r}^{(r)}$ is *attractive* in sense of the measuring surface of $\mathscr{M}^{(\sigma_s;\sigma_r)}$ if

$$\left.\begin{aligned}&\mathbf{n}^{\mathrm{T}}_{\mathscr{M}_m^{(\sigma_s,\sigma_r)}}(\mathbf{x}_{m-\varepsilon}^{(\sigma_s;\mathscr{M})})\cdot[\mathbf{x}_{m-\varepsilon}^{(\sigma_s;\mathscr{M})}-\mathbf{x}_{m-\varepsilon}^{(\sigma_s;\mathscr{E})}]<0,\\&\mathbf{n}^{\mathrm{T}}_{\mathscr{M}_m^{(\alpha,\sigma_r)}}(\mathbf{x}_{m+\varepsilon}^{(\sigma_s;\mathscr{M})})\cdot[\mathbf{x}_{m+\varepsilon}^{(\sigma_s;\mathscr{E})}-\mathbf{x}_{m+\varepsilon}^{(\sigma_s;\mathscr{M})}]<0,\\&\mathscr{C}_{m-\varepsilon}^{(\sigma_s,\sigma_r)}>\mathscr{C}_m^{(\sigma_s,\sigma_r)}>\mathscr{C}_{m+\varepsilon}^{(\sigma_s,\sigma_r)}>0.\end{aligned}\right\}\tag{8.104}$$

(ii) The edge flow $\mathbf{x}^{(\sigma_s;\mathscr{E})}(t)$ at time t_m to the specific edge of $\mathscr{E}_{\sigma_r}^{(r)}$ is *repulsive* in sense of the measuring surface of $\mathscr{M}^{(\sigma_s;\sigma_r)}$ if

$$\left.\begin{aligned}&\mathbf{n}^{\mathrm{T}}_{\mathscr{M}_m^{(\sigma_s,\sigma_r)}}(\mathbf{x}_{m-\varepsilon}^{(\sigma_s;\mathscr{M})})\cdot[\mathbf{x}_{m-\varepsilon}^{(\sigma_s;\mathscr{M})}-\mathbf{x}_{m-\varepsilon}^{(\sigma_s;\mathscr{E})}]>0,\\&\mathbf{n}^{\mathrm{T}}_{\mathscr{M}_m^{(\alpha,\sigma_r)}}(\mathbf{x}_{m+\varepsilon}^{(\sigma_s;\mathscr{M})})\cdot[\mathbf{x}_{m+\varepsilon}^{(\sigma_s;\mathscr{E})}-\mathbf{x}_{m+\varepsilon}^{(\sigma_s;\mathscr{M})}]>0,\\&0<\mathscr{C}_{m-\varepsilon}^{(\sigma_s,\sigma_r)}<\mathscr{C}_m^{(\sigma_s,\sigma_r)}<\mathscr{C}_{m+\varepsilon}^{(\sigma_s,\sigma_r)}.\end{aligned}\right\}\tag{8.105}$$

(iii) The edge flow $\mathbf{x}^{(\sigma_s;\mathscr{E})}(t)$ at time t_m to the specific edge of $\mathscr{E}_{\sigma_r}^{(r)}$ is *from the attractive to repulsive* state in sense of the measuring surface of $\mathscr{M}^{(\sigma_s;\sigma_r)}$ if

$$\left.\begin{aligned}&\mathbf{n}^{\mathrm{T}}_{\mathscr{M}_m^{(\sigma_s,\sigma_r)}}(\mathbf{x}_{m-\varepsilon}^{(\sigma_s;\mathscr{M})})\cdot[\mathbf{x}_{m-\varepsilon}^{(\sigma_s;\mathscr{M})}-\mathbf{x}_{m-\varepsilon}^{(\sigma_s;\mathscr{E})}]<0,\\&\mathbf{n}^{\mathrm{T}}_{\mathscr{M}_m^{(\alpha,\sigma_r)}}(\mathbf{x}_{m+\varepsilon}^{(\sigma_s;\mathscr{M})})\cdot[\mathbf{x}_{m+\varepsilon}^{(\sigma_s;\mathscr{E})}-\mathbf{x}_{m+\varepsilon}^{(\sigma_s;\mathscr{M})}]>0,\\&0<\mathscr{C}_m^{(\sigma_s,\sigma_r)}<\mathscr{C}_{m-\varepsilon}^{(\sigma_s,\sigma_r)}\text{ and }0<\mathscr{C}_m^{(\sigma_s,\sigma_r)}<\mathscr{C}_{m+\varepsilon}^{(\sigma_s,\sigma_r)}.\end{aligned}\right\}\tag{8.106}$$

(iv) The edge flow $\mathbf{x}^{(\sigma_s;\mathscr{E})}(t)$ at time t_m to the specific edge of $\mathscr{E}_{\sigma_r}^{(r)}$ is *from the repulsive to attractive* state in sense of the measuring surface of $\mathscr{M}^{(\sigma_s;\sigma_r)}$ if

$$\begin{aligned}&\mathbf{n}^{\mathrm{T}}_{\mathscr{M}_m^{(\sigma_s,\sigma_r)}}(\mathbf{x}_{m-\varepsilon}^{(\sigma_s;\mathscr{M})})\cdot[\mathbf{x}_{m-\varepsilon}^{(\sigma_s;\mathscr{M})}-\mathbf{x}_{m-\varepsilon}^{(\sigma_s;\mathscr{E})}]<0,\\&\mathbf{n}^{\mathrm{T}}_{\mathscr{M}_m^{(\alpha,\sigma_r)}}(\mathbf{x}_{m+\varepsilon}^{(\sigma_s;\mathscr{M})})\cdot[\mathbf{x}_{m+\varepsilon}^{(\sigma_s;\mathscr{E})}-\mathbf{x}_{m+\varepsilon}^{(\sigma_s;\mathscr{M})}]>0,\end{aligned}$$

$$0<\mathscr{C}_m^{(\sigma_s,\sigma_r)}<\mathscr{C}_{m-\varepsilon}^{(\sigma_s,\sigma_r)} \text{ and } 0<\mathscr{C}_m^{(\sigma_s,\sigma_r)}<\mathscr{C}_{m+\varepsilon}^{(\sigma_s,\sigma_r)}. \tag{8.107}$$

(v) The edge flow $\mathbf{x}^{(\sigma_s;\mathscr{E})}(t)$ at time t_m to the specific edge of $\mathscr{E}_{\sigma_r}^{(r)}$ is *invariant* in sense of the measuring surface of $\mathscr{M}^{(\sigma_s;\sigma_r)}$ if

$$\begin{gathered}\mathbf{x}_{m-\varepsilon}^{(\sigma_s;\mathscr{M})}=\mathbf{x}_{m-\varepsilon}^{(\sigma_s;\mathscr{E})} \text{ and } \mathbf{x}_{m+\varepsilon}^{(\sigma_s;\mathscr{E})}=\mathbf{x}_{m+\varepsilon}^{(\sigma_s;\mathscr{M})}\\ \mathscr{C}_m^{(\sigma_s,\sigma_r)}=\mathscr{C}_{m-\varepsilon}^{(\sigma_s,\sigma_r)}=\mathscr{C}_{m+\varepsilon}^{(\sigma_s,\sigma_r)}.\end{gathered} \tag{8.108}$$

Theorem 8.11. *For a discontinuous dynamical system in Eqs.* (6.4)-(6.8), *there is a point* $\mathbf{x}^{(\sigma_r;\mathscr{E})}(t_m)\equiv\mathbf{x}_m\in\mathscr{E}_{\sigma_r}^{(r)}$ ($\sigma_r\in\mathscr{N}_{\mathscr{E}}^{r}=\{1,2,\cdots,N_r\}$ *and* $r\in\{0,1,2,\cdots,n-2\}$) *at time* t_m *with the related boundaries* $\partial\Omega_{\alpha i_\alpha}(\sigma_r)$ *and related domains* $\Omega_\alpha(\sigma_r)$ ($\alpha\in\mathscr{N}_{\mathscr{D}}(\sigma_r)\subset\mathscr{N}_{\mathscr{D}}=\{1,2,\cdots,N_d\}$ *and* $i_\alpha\in\mathscr{N}_{\alpha\mathscr{B}}(\sigma_r)\subset\mathscr{N}_{\mathscr{B}}=\{1,2,\cdots,N_b\}$), *where the boundary index subset of* $\mathscr{N}_{\alpha\mathscr{B}}(\sigma_r)$ *has* n_{σ_r}*-elements* ($n_{\sigma_r}\geq(n-r)$). *The edge of* $\mathscr{E}_{\sigma_r}^{(r)}$ *is the intersection of edges (including domains and boundaries)* $\mathscr{E}_{\sigma_s}^{(s)}(\sigma_r)$ ($\sigma_s\in\mathscr{N}_{\mathscr{E}}^{s}(\sigma_r)\subset\mathscr{N}_{\mathscr{E}}^{s}=\{1,2,\cdots,N_s\}$ *and* $s\in\{r+1,r+2,\cdots,n-1,n\}$) *Suppose there is an* s*-dimensional edge* $\mathscr{E}_{\sigma_s}^{(s)}(\sigma_r)$ *with a boundary index set* $\mathscr{N}_{\alpha\mathscr{B}}(\sigma_r,\sigma_s)$ *relative to domain* $\Omega_\alpha(\sigma_r)$. *The complementary boundary number set is* $\overline{\mathscr{N}}_{\alpha\mathscr{B}}(\sigma_r,\sigma_s)=\mathscr{N}_{\alpha\mathscr{B}}(\sigma_r)/\mathscr{N}_{\alpha\mathscr{B}}(\sigma_r,\sigma_s)$. *Suppose there is an* $(n-s+r)$*-dimensional measuring edge* $\mathscr{E}_{\sigma_{n-s+r}}^{(\rho,n-s+r)}$ *of a flow* $\mathbf{x}^{(\sigma_s;\mathscr{E})}(t)$ *in the edge of* $\mathscr{E}_{\sigma_s}^{(s)}(\sigma_r,\alpha)$ *with an equi-constant vector of the specific edge* $\mathscr{E}_{\sigma_r}^{(r)}(\alpha)$. *The measuring edge and an equi-constant measuring surface* $\mathscr{M}^{(\sigma_s;\sigma_r)}$ *are determined by Eqs.* (8.98) *and* (8.103), *respectively. At time* t_m $\mathbf{x}_m^{(\sigma_s;\mathscr{M})}=\mathbf{x}_\sigma^{(\sigma_s;\mathscr{M})}(t_m)\in\mathscr{M}_m^{(\sigma_s;\sigma_r)}$ *satisfies Eq.* (8.55). *An edge flow* $\mathbf{x}^{(\sigma_s;\mathscr{E})}(t)$ *is* $C_{[t_{m-\varepsilon},t_m)}^{r_{\sigma_s}}$ *and/or* $C_{(t_m,t_{m+\varepsilon}]}^{r_{\sigma_s}}$*-continuous for time* t *and* $\|d^{r_{\sigma_s}+1}\mathbf{x}^{(\sigma_s;\mathscr{E})}/dt^{r_{\sigma_s}+1}\|<\infty$ ($r_{\sigma_s}\geq 2$). *Suppose* $\mathbf{x}_{m\pm\varepsilon}^{(\sigma_s;\mathscr{E})}=\mathbf{x}^{(\sigma_s;\mathscr{E})}(t_{m\pm\varepsilon})$ $\in\mathscr{M}_{m\pm\varepsilon}^{(\alpha,\sigma_r)}$ *and* $\mathbf{x}_{\sigma(m\pm\varepsilon)}^{(\sigma_s;\mathscr{M})}=\mathbf{x}_\sigma^{(\sigma_s;\mathscr{M})}(t_{m\pm\varepsilon})\in\mathscr{M}_m^{(\sigma_s;\sigma_r)}$. $\mathscr{D}^{(\sigma_s;\sigma_r)}(t_{m\pm\varepsilon})\equiv\mathscr{C}_{m\pm\varepsilon}^{(\sigma_s;\sigma_r)}$ *and* $\mathscr{D}^{(\sigma_s;\sigma_r)}(t_m)\equiv\mathscr{C}_m^{(\sigma_s;\sigma_r)}$.

(i) *The edge flow* $\mathbf{x}^{(\sigma_s;\mathscr{E})}(t)$ *at time* t_m *to the specific edge of* $\mathscr{E}_{\sigma_r}^{(r)}$ *is attractive in sense of the measuring surface of* $\mathscr{M}^{(\sigma_s;\sigma_r)}$ *if and only if*

$$G_{\mathscr{M}(\alpha;\sigma_r)}^{(\sigma_s;\mathscr{E})}(\mathbf{x}_m^{(\sigma_s;\mathscr{M})},t_m,\mathbf{p}_\alpha,\boldsymbol{\lambda})<0. \tag{8.109}$$

(ii) *The edge flow* $\mathbf{x}^{(\sigma_s;\mathscr{E})}(t)$ *at time* t_m *to the specific edge of* $\mathscr{E}_{\sigma_r}^{(r)}$ *is repulsive in sense of the measuring surface of* $\mathscr{M}^{(\sigma_s;\sigma_r)}$ *if and only if*

$$G^{(\sigma_s;\mathscr{E})}_{\mathscr{M}^{(\alpha;\sigma_r)}}(\mathbf{x}^{(\sigma_s;\mathscr{M})}_m, t_m, \mathbf{p}_\alpha, \boldsymbol{\lambda}) > 0. \tag{8.110}$$

(iii) *The edge flow* $\mathbf{x}^{(\sigma_s;\mathscr{E})}(t)$ *at time* t_m *to the specific edge of* $\mathscr{E}^{(r)}_{\sigma_r}$ *is from the attractive to repulsive state in sense of the measuring surface of* $\mathscr{M}^{(\sigma_s;\sigma_r)}$ *if and only if*

$$\begin{aligned} &G^{(\sigma_s;\mathscr{E})}_{\mathscr{M}^{(\alpha;\sigma_r)}}(\mathbf{x}^{(\sigma_s;\mathscr{M})}_m, t_m, \mathbf{p}_\alpha, \boldsymbol{\lambda}) = 0,\\ &G^{(1,\sigma_s;\mathscr{E})}_{\mathscr{M}^{(\alpha;\sigma_r)}}(\mathbf{x}^{(\sigma_s;\mathscr{M})}_m, t_m, \mathbf{p}_\alpha, \boldsymbol{\lambda}) > 0. \end{aligned} \tag{8.111}$$

(iv) *The edge flow* $\mathbf{x}^{(\sigma_s;\mathscr{E})}(t)$ *at time* t_m *to the specific edge of* $\mathscr{E}^{(r)}_{\sigma_r}$ *is from the repulsive to attractive state in sense of the measuring surface of* $\mathscr{M}^{(\sigma_s;\sigma_r)}$ *if and only if*

$$\begin{aligned} &G^{(\sigma_s;\mathscr{E})}_{\mathscr{M}^{(\alpha;\sigma_r)}}(\mathbf{x}^{(\sigma_s;\mathscr{M})}_m, t_m, \mathbf{p}_\alpha, \boldsymbol{\lambda}) = 0,\\ &G^{(1,\sigma_s;\mathscr{E})}_{\mathscr{M}^{(\alpha;\sigma_r)}}(\mathbf{x}^{(\sigma_s;\mathscr{M})}_m, t_m, \mathbf{p}_\alpha, \boldsymbol{\lambda}) < 0. \end{aligned} \tag{8.112}$$

(v) *The edge flow* $\mathbf{x}^{(\sigma_s;\mathscr{E})}(t)$ *at time* t_m *to the specific edge of* $\mathscr{E}^{(r)}_{\sigma_r}$ *is invariant in sense of the measuring surface of* $\mathscr{M}^{(\sigma_s;\sigma_r)}$ *if and only if*

$$G^{(s_{\sigma_s},\sigma_s;\mathscr{E})}_{\mathscr{M}^{(\alpha;\sigma_r)}}(\mathbf{x}^{(\sigma_s;\mathscr{M})}_m, t_m, \mathbf{p}_\alpha, \boldsymbol{\lambda}) = 0 \text{ for } s_{\sigma_s} = 0,1,2,\cdots. \tag{8.113}$$

Proof. The measuring surface of $\mathscr{M}^{(\sigma_s,\sigma_r)}$ is as a boundary. As in Chapter 3, this theorem can be proved. ■

Definition 8.17. For a discontinuous dynamical system in Eqs. (6.4)-(6.8), there is a point $\mathbf{x}^{(\sigma_r;\mathscr{E})}(t_m) \equiv \mathbf{x}_m \in \mathscr{E}^{(r)}_{\sigma_r}$ ($\sigma_r \in \mathscr{N}^r_{\mathscr{E}} = \{1,2,\cdots,N_r\}$ and $r \in \{0,1,2,\cdots,n-2\}$) at time t_m with the related boundaries $\partial\Omega_{\alpha i_\alpha}(\sigma_r)$ and related domains $\Omega_\alpha(\sigma_r)$ ($\alpha \in \mathscr{N}_{\mathscr{D}}(\sigma_r) \subset \mathscr{N}_{\mathscr{D}} = \{1,2,\cdots,N_d\}$ and $i_\alpha \in \mathscr{N}_{\alpha\mathscr{B}}(\sigma_r) \subset \mathscr{N}_{\mathscr{B}} = \{1,2,\cdots,N_b\}$), where the boundary index subset of $\mathscr{N}_{\alpha\mathscr{B}}(\sigma_r)$ has n_{σ_r}-elements ($n_{\sigma_r} \geq (n-r)$). The edge of $\mathscr{E}^{(r)}_{\sigma_r}$ is the intersection of edges (including domains and boundaries) $\mathscr{E}^{(s)}_{\sigma_s}(\sigma_r)$ ($\sigma_s \in \mathscr{N}^s_{\mathscr{E}}(\sigma_r) \subset \mathscr{N}^s_{\mathscr{E}} = \{1,2,\cdots,N_s\}$ and $s \in \{r+1, r+2,\cdots,n-1,n\}$). Suppose there is an s-dimensional edge $\mathscr{E}^{(s)}_{\sigma_s}(\sigma_r)$ with a boundary index set $\mathscr{N}_{\alpha\mathscr{B}}(\sigma_r,\sigma_s)$ relative to domain $\Omega_\alpha(\sigma_r)$. The complementary boundary index set is $\bar{\mathscr{N}}_{\alpha\mathscr{B}}(\sigma_r,\sigma_s) = \mathscr{N}_{\alpha\mathscr{B}}(\sigma_r) / \mathscr{N}_{\alpha\mathscr{B}}(\sigma_r,\sigma_s)$. Suppose there is an $(n-s+r)$-dimensional measuring edge $\mathscr{E}^{(\rho,n-s+r)}_{\sigma_{n-s+r}}$ of a flow $\mathbf{x}^{(\sigma_s;\mathscr{E})}(t)$ in the edge of $\mathscr{E}^{(s)}_{\sigma_s}(\sigma_r,\alpha)$ with an equi-constant vector of the specific edge $\mathscr{E}^{(r)}_{\sigma_r}(\alpha)$. The measuring edge and an equi-constant measuring surface $\mathscr{M}^{(\sigma_s;\sigma_r)}$ are determined by Eqs. (8.98) and (8.103), respectively. At time t_m, $\mathbf{x}_m^{(\sigma_s;\mathscr{M})} = \mathbf{x}_\sigma^{(\sigma_s;\mathscr{M})}(t_m) \in \mathscr{M}_m^{(\sigma_s;\sigma_r)}$ satisfies Eq. (8.55). An edge flow $\mathbf{x}^{(\sigma_s;\mathscr{E})}(t)$ is $C^{r_{\sigma_s}}_{[t_{m-\varepsilon},t_m)}$ and/or $C^{r_{\sigma_s}}_{(t_m,t_{m+\varepsilon}]}$-continuous for time t and $\| d^{r_{\sigma_s}+1}\mathbf{x}^{(\sigma_s;\mathscr{E})}/dt^{r_{\sigma_s}+1} \| < \infty$ ($r_{\sigma_s} \geq 2k_{\sigma_s} - 1$ or $2k_{\sigma_s}$). Suppose $\mathbf{x}_{m\pm\varepsilon}^{(\sigma_s;\mathscr{E})} = \mathbf{x}^{(\sigma_s;\mathscr{E})}(t_{m\pm\varepsilon}) \in \mathscr{M}_{m\pm\varepsilon}^{(\alpha,\sigma_r)}$ and $\mathbf{x}_{\sigma(m\pm\varepsilon)}^{(\sigma_s;\mathscr{M})} = \mathbf{x}_\sigma^{(\sigma_s;\mathscr{M})}(t_{m\pm\varepsilon}) \in \mathscr{M}_m^{(\sigma_s;\sigma_r)}$. $\mathscr{D}^{(\sigma_s;\sigma_r)}(t_m) \equiv \mathscr{C}_m^{(\sigma_s;\sigma_r)}$ and $\mathscr{D}^{(\sigma_s;\sigma_r)}(t_{m\pm\varepsilon}) \equiv \mathscr{C}_{m\pm\varepsilon}^{(\sigma_s;\sigma_r)}$.

(i) The flow $\mathbf{x}^{(\sigma_s;\mathscr{E})}(t)$ at time t_m to the specific edge of $\mathscr{E}_{\sigma_r}^{(r)}$ is *attractive* in sense of the measuring surface of $\mathscr{M}^{(\sigma_s;\sigma_r)}$ with the $2k_{\sigma_s}$-singularity if

$$\left.\begin{array}{l}
G_{\mathscr{M}^{(\alpha;\sigma_r)}}^{(s_{\sigma_s},\sigma_s;\mathscr{E})}(\mathbf{x}_m^{(\sigma_s;\mathscr{M})},t_m,\mathbf{p}_\alpha,\boldsymbol{\lambda}) = 0 \\
\text{for } s_{\sigma_s} = 0,1,2,\cdots,2k_{\sigma_s}-1; \\
\mathbf{n}^{\mathrm{T}}_{\mathscr{M}_m^{(\sigma_s,\sigma_r)}}(\mathbf{x}_{m-\varepsilon}^{(\sigma_s;\mathscr{M})})\cdot[\mathbf{x}_{m-\varepsilon}^{(\sigma_s;\mathscr{M})} - \mathbf{x}_{m-\varepsilon}^{(\sigma_s;\mathscr{E})}] < 0, \\
\mathbf{n}^{\mathrm{T}}_{\mathscr{M}_m^{(\alpha,\sigma_r)}}(\mathbf{x}_{m+\varepsilon}^{(\sigma_s;\mathscr{M})})\cdot[\mathbf{x}_{m+\varepsilon}^{(\sigma_s;\mathscr{E})} - \mathbf{x}_{m+\varepsilon}^{(\sigma_s;\mathscr{M})}] < 0, \\
\mathscr{C}_{m-\varepsilon}^{(\sigma_s,\sigma_r)} > \mathscr{C}_m^{(\sigma_s,\sigma_r)} > \mathscr{C}_{m+\varepsilon}^{(\sigma_s,\sigma_r)} > 0.
\end{array}\right\} \tag{8.114}$$

(ii) The flow $\mathbf{x}^{(\sigma_s;\mathscr{E})}(t)$ at time t_m to the specific edge of $\mathscr{E}_{\sigma_r}^{(r)}$ is *repulsive* in sense of the measuring surface of $\mathscr{M}^{(\sigma_s;\sigma_r)}$ if

$$\left.\begin{array}{l}
G_{\mathscr{M}^{(\alpha;\sigma_r)}}^{(s_{\sigma_s},\sigma_s;\mathscr{E})}(\mathbf{x}_m^{(\sigma_s;\mathscr{M})},t_m,\mathbf{p}_\alpha,\boldsymbol{\lambda}) = 0 \\
\text{for } s_{\sigma_s} = 0,1,2,\cdots,2k_{\sigma_s}-1; \\
\mathbf{n}^{\mathrm{T}}_{\mathscr{M}_m^{(\sigma_s,\sigma_r)}}(\mathbf{x}_{m-\varepsilon}^{(\sigma_s;\mathscr{M})})\cdot[\mathbf{x}_{m-\varepsilon}^{(\sigma_s;\mathscr{M})} - \mathbf{x}_{m-\varepsilon}^{(\sigma_s;\mathscr{E})}] > 0, \\
\mathbf{n}^{\mathrm{T}}_{\mathscr{M}_m^{(\alpha,\sigma_r)}}(\mathbf{x}_{m+\varepsilon}^{(\sigma_s;\mathscr{M})})\cdot[\mathbf{x}_{m+\varepsilon}^{(\sigma_s;\mathscr{E})} - \mathbf{x}_{m+\varepsilon}^{(\sigma_s;\mathscr{M})}] > 0, \\
0 < \mathscr{C}_{m-\varepsilon}^{(\sigma_s,\sigma_r)} < \mathscr{C}_m^{(\sigma_s,\sigma_r)} < \mathscr{C}_{m+\varepsilon}^{(\sigma_s,\sigma_r)}.
\end{array}\right\} \tag{8.115}$$

(iii) The flow $\mathbf{x}^{(\sigma_s;\mathscr{E})}(t)$ at time t_m to the specific edge of $\mathscr{E}_{\sigma_r}^{(r)}$ is *from the attractive to repulsive* state in sense of the measuring surface of $\mathscr{M}^{(\sigma_s;\sigma_r)}$ if

$$\left.\begin{array}{l}
G_{\mathscr{M}^{(\alpha;\sigma_r)}}^{(s_{\sigma_s},\sigma_s;\mathscr{E})}(\mathbf{x}_m^{(\sigma_s;\mathscr{M})},t_m,\mathbf{p}_\alpha,\boldsymbol{\lambda}) = 0 \text{ for } s_{\sigma_s} = 0,1,2,\cdots,2k_{\sigma_s}; \\
\mathbf{n}^{\mathrm{T}}_{\mathscr{M}_m^{(\sigma_s,\sigma_r)}}(\mathbf{x}_{m-\varepsilon}^{(\sigma_s;\mathscr{M})})\cdot[\mathbf{x}_{m-\varepsilon}^{(\sigma_s;\mathscr{M})} - \mathbf{x}_{m-\varepsilon}^{(\sigma_s;\mathscr{E})}] < 0, \\
\mathbf{n}^{\mathrm{T}}_{\mathscr{M}_m^{(\alpha,\sigma_r)}}(\mathbf{x}_{m+\varepsilon}^{(\sigma_s;\mathscr{M})})\cdot[\mathbf{x}_{m+\varepsilon}^{(\sigma_s;\mathscr{E})} - \mathbf{x}_{m+\varepsilon}^{(\sigma_s;\mathscr{M})}] > 0, \\
0 < \mathscr{C}_m^{(\sigma_s,\sigma_r)} < \mathscr{C}_{m-\varepsilon}^{(\sigma_s,\sigma_r)} \text{ and } 0 < \mathscr{C}_m^{(\sigma_s,\sigma_r)} < \mathscr{C}_{m+\varepsilon}^{(\sigma_s,\sigma_r)}.
\end{array}\right\} \tag{8.116}$$

(iv) The flow $\mathbf{x}^{(\sigma_s;\mathscr{E})}(t)$ at time t_m to the specific edge of $\mathscr{E}_{\sigma_r}^{(r)}$ is *from the repulsive to attractive* state in sense of the measuring surface of $\mathscr{M}^{(\sigma_s;\sigma_r)}$ if

$$
\begin{aligned}
& G^{(s_{\sigma_s},\sigma_s;\mathscr{E})}_{\mathscr{M}^{(\alpha;\sigma_r)}}(\mathbf{x}^{(\sigma_s;\mathscr{M})}_{m}, t_m, \mathbf{p}_\alpha, \boldsymbol{\lambda}) = 0 \\
& \text{for } s_{\sigma_s} = 0,1,2,\cdots,2k_{\sigma_s}; \\
& \mathbf{n}^{\mathrm{T}}_{\mathscr{M}_m^{(\sigma_s,\sigma_r)}}(\mathbf{x}^{(\sigma_s;\mathscr{M})}_{m-\varepsilon})\cdot[\mathbf{x}^{(\sigma_s;\mathscr{M})}_{m-\varepsilon} - \mathbf{x}^{(\sigma_s;\mathscr{E})}_{m-\varepsilon}] < 0, \\
& \mathbf{n}^{\mathrm{T}}_{\mathscr{M}_m^{(\alpha,\sigma_r)}}(\mathbf{x}^{(\sigma_s;\mathscr{M})}_{m+\varepsilon})\cdot[\mathbf{x}^{(\sigma_s;\mathscr{E})}_{m+\varepsilon} - \mathbf{x}^{(\sigma_s;\mathscr{M})}_{m+\varepsilon}] > 0, \\
& 0 < \mathscr{C}^{(\sigma_s,\sigma_r)}_{m} < \mathscr{C}^{(\sigma_s,\sigma_r)}_{m-\varepsilon} \text{ and } 0 < \mathscr{C}^{(\sigma_s,\sigma_r)}_{m} < \mathscr{C}^{(\sigma_s,\sigma_r)}_{m+\varepsilon}.
\end{aligned}
\tag{8.117}
$$

Theorem 8.12. *For a discontinuous dynamical system in Eqs.* (6.4)-(6.8)*, there is a point* $\mathbf{x}^{(\sigma_r;\mathscr{E})}(t_m) \equiv \mathbf{x}_m \in \mathscr{E}^{(r)}_{\sigma_r}$ ($\sigma_r \in \mathscr{N}^{(r)}_{\mathscr{E}} = \{1,2,\cdots,N_r\}$ *and* $r \in \{0,1,2,\cdots,n-2\}$) *at time* t_m *with the related boundaries* $\partial\Omega_{\alpha i_\alpha}(\sigma_r)$ *and related domains* $\Omega_\alpha(\sigma_r)$ ($\alpha \in \mathscr{N}_{\mathscr{D}}(\sigma_r) \subset \mathscr{N}_{\mathscr{D}} = \{1,2,\cdots,N_d\}$ *and* $i_\alpha \in \mathscr{N}_{\alpha\mathscr{B}}(\sigma_r) \subset \mathscr{N}_{\mathscr{B}} = \{1,2,\cdots,N_b\}$)*, where the boundary index subset of* $\mathscr{N}_{\alpha\mathscr{B}}(\sigma_r)$ *has* n_{σ_r} *-elements* ($n_{\sigma_r} \geq (n-r)$)*. The edge of* $\mathscr{E}^{(r)}_{\sigma_r}$ *is the intersection of edges (including domains and boundaries)* $\mathscr{E}^{(s)}_{\sigma_s}(\sigma_r)$ ($\sigma_s \in \mathscr{N}^{s}_{\mathscr{E}}(\sigma_r) \subset \mathscr{N}^{s}_{\mathscr{E}} = \{1,2,\cdots,N_s\}$ *and* $s \in \{r+1,r+2,\cdots,n-1,n\}$)*. Suppose there is an* s *-dimensional edge* $\mathscr{E}^{(s)}_{\sigma_s}(\sigma_r)$ *with a boundary index set* $\mathscr{N}_{\alpha\mathscr{B}}(\sigma_r,\sigma_s)$ *relative to domain* $\Omega_\alpha(\sigma_r)$*. The complementary boundary number set is* $\overline{\mathscr{N}}_{\alpha\mathscr{B}}(\sigma_r,\sigma_s) = \mathscr{N}_{\alpha\mathscr{B}}(\sigma_r) / \mathscr{N}_{\alpha\mathscr{B}}(\sigma_r,\sigma_s)$*. Suppose there is an* $(n-s+r)$*-dimensional measuring edge* $\mathscr{E}^{(\rho,n-s+r)}_{\sigma_{n-s+r}}$ *of a flow* $\mathbf{x}^{(\sigma_s;\mathscr{E})}(t)$ *in the edge of* $\mathscr{E}^{(s)}_{\sigma_s}(\sigma_r,\alpha)$ *with an equi-constant vector of the specific edge* $\mathscr{E}^{(r)}_{\sigma_r}(\alpha)$*. The measuring edge and an equi-constant measuring surface* $\mathscr{M}^{(\sigma_s;\sigma_r)}$ *is determined by Eqs.* (8.98) *and* (8.103)*, respectively. At time* t_m*,* $\mathbf{x}^{(\sigma_s;\mathscr{M})}_m = \mathbf{x}^{(\sigma_s;\mathscr{M})}_{\sigma}(t_m) \in \mathscr{M}^{(\sigma_s;\sigma_r)}_m$ *satisfies Eq.* (8.55)*. An edge flow* $\mathbf{x}^{(\sigma_s;\mathscr{E})}(t)$ *is* $C^{r_{\sigma_s}}_{[t_{m-\varepsilon},t_m)}$ *and/or* $C^{r_{\sigma_s}}_{(t_m,t_{m+\varepsilon}]}$*-continuous for time* t *and* $\| d^{r_{\sigma_s}+1}\mathbf{x}^{(\sigma_s;\mathscr{E})} / dt^{r_{\sigma_s}+1} \| < \infty$ ($r_{\sigma_s} \geq 2k_{\sigma_s}$ *or* $2k_{\sigma_s}+1$)*. Suppose* $\mathbf{x}^{(\sigma_s;\mathscr{E})}_{m\pm\varepsilon} = \mathbf{x}^{(\sigma_s;\mathscr{E})}(t_{m\pm\varepsilon}) \in \mathscr{M}^{(\alpha,\sigma_r)}_{m\pm\varepsilon}$ *and* $\mathbf{x}^{(\sigma_s;\mathscr{M})}_{\sigma(m\pm\varepsilon)} = \mathbf{x}^{(\sigma_s;\mathscr{M})}_{\sigma}(t_{m\pm\varepsilon}) \in \mathscr{M}^{(\sigma_s;\sigma_r)}_m$*.* $\mathscr{D}^{(\sigma_s;\sigma_r)}(t_m) \equiv \mathscr{C}^{(\sigma_s;\sigma_r)}_m$ *and* $\mathscr{D}^{(\sigma_s;\sigma_r)}(t_{m\pm\varepsilon}) \equiv \mathscr{C}^{(\sigma_s;\sigma_r)}_{m\pm\varepsilon}$*.*

(i) *The flow* $\mathbf{x}^{(\sigma_s;\mathscr{E})}(t)$ *at time* t_m *to the specific edge of* $\mathscr{E}^{(r)}_{\sigma_r}$ *is attractive in sense of the measuring surface of* $\mathscr{M}^{(\sigma_s;\sigma_r)}$ *with the* $2k_{\sigma_s}$*-singularity if and only if*

$$
\begin{aligned}
& G^{(s_{\sigma_s},\sigma_s;\mathscr{E})}_{\mathscr{M}^{(\alpha;\sigma_r)}}(\mathbf{x}^{(\sigma_s;\mathscr{M})}_{m}, t_m, \mathbf{p}_\alpha, \boldsymbol{\lambda}) = 0 \\
& \textit{for } s_{\sigma_s} = 0,1,2,\cdots,2k_{\sigma_s}-1; \\
& G^{(2k_{\sigma_s},\sigma_s;\mathscr{E})}_{\mathscr{M}^{(\alpha;\sigma_r)}}(\mathbf{x}^{(\sigma_s;\mathscr{M})}_{m}, t_m, \mathbf{p}_\alpha, \boldsymbol{\lambda}) < 0.
\end{aligned}
\tag{8.118}
$$

(ii) *The flow* $\mathbf{x}^{(\sigma_s;\mathscr{E})}(t)$ *at time* t_m *to the specific edge of* $\mathscr{E}_{\sigma_r}^{(r)}$ *is repulsive in sense of the measuring surface of* $\mathscr{M}^{(\sigma_s;\sigma_r)}$ *if and only if*

$$\begin{aligned} &G_{\mathscr{M}^{(\alpha;\sigma_r)}}^{(s_{\sigma_s},\sigma_s;\mathscr{E})}(\mathbf{x}_m^{(\sigma_s;\mathscr{M})},t_m,\mathbf{p}_\alpha,\boldsymbol{\lambda})=0\\ &\mathit{for}\ s_{\sigma_s}=0,1,2,\cdots,2k_{\sigma_s}-1;\\ &G_{\mathscr{M}^{(\alpha;\sigma_r)}}^{(2k_{\sigma_s},\sigma_s;\mathscr{E})}(\mathbf{x}_m^{(\sigma_s;\mathscr{M})},t_m,\mathbf{p}_\alpha,\boldsymbol{\lambda})>0. \end{aligned} \tag{8.119}$$

(iii) *The flow* $\mathbf{x}^{(\sigma_s;\mathscr{E})}(t)$ *at time* t_m *to the specific edge of* $\mathscr{E}_{\sigma_r}^{(r)}$ *is from the attractive to repulsive state in sense of the measuring surface of* $\mathscr{M}^{(\sigma_s;\sigma_r)}$ *if and only if*

$$\begin{aligned} &G_{\mathscr{M}^{(\alpha;\sigma_r)}}^{(s_{\sigma_s},\sigma_s;\mathscr{E})}(\mathbf{x}_m^{(\sigma_s;\mathscr{M})},t_m,\mathbf{p}_\alpha,\boldsymbol{\lambda})=0\\ &\mathit{for}\ s_{\sigma_s}=0,1,2,\cdots,2k_{\sigma_s};\\ &G_{\mathscr{M}^{(\alpha;\sigma_r)}}^{(2k_{\sigma_s}+1,\sigma_s;\mathscr{E})}(\mathbf{x}_m^{(\sigma_s;\mathscr{M})},t_m,\mathbf{p}_\alpha,\boldsymbol{\lambda})>0. \end{aligned} \tag{8.120}$$

(iv) *The flow* $\mathbf{x}^{(\sigma_s;\mathscr{E})}(t)$ *at time* t_m *to the specific edge of* $\mathscr{E}_{\sigma_r}^{(r)}$ *is from the repulsive to attractive state in sense of the measuring surface of* $\mathscr{M}^{(\sigma_s;\sigma_r)}$ *if and only if*

$$\begin{aligned} &G_{\mathscr{M}^{(\alpha;\sigma_r)}}^{(s_{\sigma_s},\sigma_s;\mathscr{E})}(\mathbf{x}_m^{(\sigma_s;\mathscr{M})},t_m,\mathbf{p}_\alpha,\boldsymbol{\lambda})=0\\ &\mathit{for}\ s_{\sigma_s}=0,1,2,\cdots,2k_{\sigma_s};\\ &G_{\mathscr{M}^{(\alpha;\sigma_r)}}^{(2k_{\sigma_s},\sigma_s;\mathscr{E})}(\mathbf{x}_m^{(\sigma_s;\mathscr{M})},t_m,\mathbf{p}_\alpha,\boldsymbol{\lambda})<0. \end{aligned} \tag{8.121}$$

Proof. The measuring surface of $\mathscr{M}^{(\sigma_s,\sigma_r)}$ is as a boundary. As in Chapter 3, this theorem can be proved. ∎

8.4. Edge dynamics with multi-valued vector fields

The multi-valued vector fields on edges to specific edges can be defined as in Section 6.7. If a vector field defined on an edge can be extended continuously to other edges (including boundaries and domains) via the specific edge, such an extended vector field generates an imaginary flow in the other edges (or domains). Equation (6.201) can be rewritten for real and imaginary edge flows via edges, i.e.,

$$\begin{aligned} &\dot{\mathbf{x}}_{\sigma_s}^{(\sigma_s;\mathscr{E})}=\mathbf{F}^{(\sigma_s;\mathscr{E})}(\mathbf{x}_{\sigma_s}^{(\sigma_s;\mathscr{E})},t,\boldsymbol{\pi}_{\sigma_s})\ \text{ on }\ \mathscr{E}_{\sigma_s}^{(s)}(\sigma_r),\\ &\dot{\mathbf{x}}_{\sigma_s}^{(\sigma_p;\mathscr{E})}=\mathbf{F}^{(\sigma_p;\mathscr{E})}(\mathbf{x}_{\sigma_s}^{(\sigma_p;\mathscr{E})},t,\boldsymbol{\pi}_{\sigma_p})\ \text{ on }\ \mathscr{E}_{\sigma_s}^{(s)}(\sigma_r),\\ &\dot{\mathbf{x}}_{\sigma_p}^{(\sigma_s;\mathscr{E})}=\mathbf{F}^{(\sigma_s;\mathscr{E})}(\mathbf{x}_{\sigma_p}^{(\sigma_s;\mathscr{E})},t,\boldsymbol{\pi}_{\sigma_s})\ \text{ on }\ \mathscr{E}_{\sigma_p}^{(p)}(\sigma_r),\\ &\text{from }\ \mathscr{E}_{\sigma_p}^{(p)}\ (p,s,r\in\{0,1,2,\cdots,n\}\text{ and }p,s>r), \end{aligned} \tag{8.122}$$

where $\mathbf{x}_{\sigma_s}^{(\sigma_s;\mathscr{E})}$ is a real boundary flow, but $\mathbf{x}_{\sigma_s}^{(\sigma_p;\mathscr{E})}$ and $\mathbf{x}_{\sigma_p}^{(\sigma_s;\mathscr{E})}$ are imaginary edge flows on edges $\mathscr{E}_{\sigma_s}^{(s)}(\sigma_r)$ and $\mathscr{E}_{\sigma_p}^{(p)}(\sigma_r)$, respectively. In Fig. 8.10, imaginary edge flows $\mathbf{x}_{\sigma_s}^{(\sigma_{s-1};\mathscr{E})}$ to an $(s-2)$-dimensional edge are an extensions of a real edge flow $\mathbf{x}_{\sigma_s}^{(\sigma_s;\mathscr{E})}$. The black curve represents the $(s-2)$ dimensional edge. The imaginary edge flows are represented by dashed curves with arrows. Similarly, the imaginary edge flow to the $(s-2)$ dimensional edge is an extension of the real edge flow $\mathbf{x}_{\sigma_s}^{(\sigma_s;\mathscr{E})}$ in Fig. 8.11. Thus, Axioms 7.1-7.4 can be extended to edges and vertexes.

Axiom 8.1 (Continuation axiom). For a discontinuous dynamical system, any *edge* flows on $\mathscr{E}_{\sigma_s}^{(s)}(\sigma_r)$ coming to (or leaving from) a specific edge of $\mathscr{E}_{\sigma_r}^{(r)}$ ($s,r \in \{0,1,2,\cdots,n\}$ and $s>r$) can be, forwardly (or backwardly) extended to the adjacent accessible edge $\mathscr{E}_{\sigma_p}^{(p)}(\sigma_r)$ ($p,r \in \{0,1,2,\cdots,n\}$ and $p>s>r$) without any switching rules, transport laws and any flow barriers at the edge.

Axiom 8.2 (Continuation axiom). For a discontinuous dynamical system, any imaginary *edge* flows on $\mathscr{E}_{\sigma_p}^{(p)}(\sigma_r)$ coming to (or leaving from) a specific edge of $\mathscr{E}_{\sigma_r}^{(r)}$ ($p,r \in \{0,1,2,\cdots,n\}$ and $p>r$) can be, forwardly (or backwardly) extended to the corresponding edge $\mathscr{E}_{\sigma_s}^{(s)}(\sigma_r)$ ($s,r \in \{0,1,2,\cdots,n\}$ and $p>s>r$) without any switching rules, transport laws and any flow barriers at the edge.

Axiom 8.3 (Non-extendibility axiom). For a discontinuous dynamical system, any *edge* flows on $\mathscr{E}_{\sigma_s}^{(s)}(\sigma_r)$ coming to (or leaving from) a specific edge of $\mathscr{E}_{\sigma_r}^{(r)}$ ($s,r \in \{0,1,2,\cdots,n\}$ and $s>r$) cannot be, forwardly (or backwardly) extended to the adjacent *inaccessible* edge $\mathscr{E}_{\sigma_p}^{(p)}(\sigma_r)$ ($p,r \in \{0,1,2,\cdots,n\}$ and $p>s>r$). Even if such edge flows are extendible to the inaccessible edges, then the corresponding extended flows are fictitious only.

Axiom 8.4 (Non-extendibility axiom). For a discontinuous dynamical system, any fictitious *edge* flows on an inaccessible edge $\mathscr{E}_{\sigma_p}^{(p)}(\sigma_r)$ coming to (or leaving from) a specific edge $\mathscr{E}_{\sigma_r}^{(r)}$ ($p,r \in \{0,1,2,\cdots,n\}$ and $p>r$) cannot be, forwardly (or backwardly) extended to the corresponding *edge* $\mathscr{E}_{\sigma_s}^{(s)}(\sigma_r)$ ($s,r \in \{0,1,2,\cdots,n\}$ and $p>s>r$) without any switching rules, transport laws and any flow barriers at the edge.

The multi-valued vector fields on an accessible edge in discontinuous dynamical systems can be defined as follows.

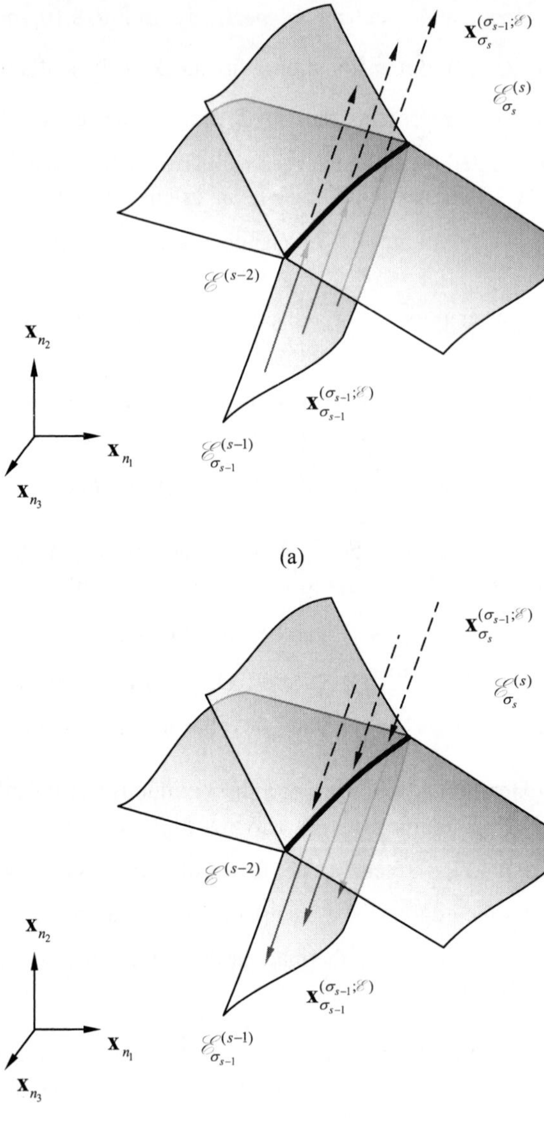

(a)

(b)

Fig. 8.10 Imaginary edge flows of $\mathbf{x}_{\sigma_s}^{(\sigma_{s-1};\mathscr{E})}$ from the edge $\mathscr{E}^{(s-1)}$. **(a)** coming edge flow extension of $\mathbf{x}_{\sigma_{s-1}}^{(\sigma_{s-1};\mathscr{E})}$ and **(b)** leaving edge flow extension of $\mathbf{x}_{\sigma_{s-1}}^{(\sigma_{s-1};\mathscr{E})}$. The black curve represents the $(s-2)$ -dimensional edge. The boundary flows are denoted by thin curves with arrows on the boundaries. The domain flows are represented by the thick curves with arrows in the domain. ($n_1 + n_2 + n_3 = s$)

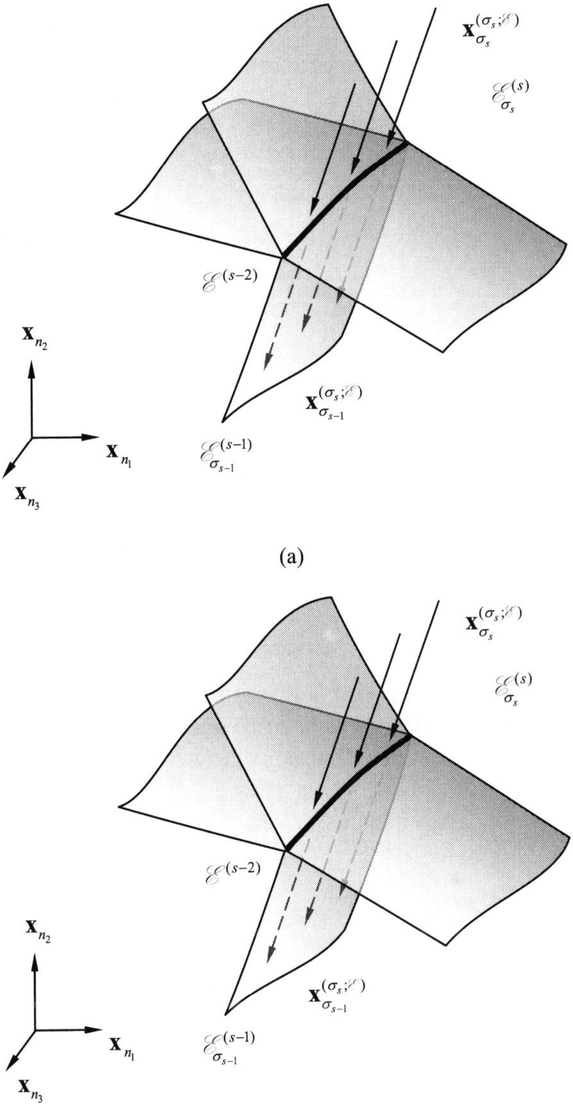

(b)

Fig. 8.11 Imaginary edge flows of $\mathbf{x}_{\sigma_{s-1}}^{(\sigma_s;\mathscr{E})}$ from the edge $\mathscr{E}_{\sigma_s}^{(s)}$ via the edge $\mathscr{E}^{(s-2)}$. **(a)** coming edge flow extension of $\mathbf{x}_{\sigma_s}^{(\sigma_s;\mathscr{E})}$ and **(b)** leaving edge flow extension of $\mathbf{x}_{\sigma_s}^{(\sigma_s;\mathscr{E})}$. The black curve represents the $(s-2)$-dimensional edge. The boundary flows are denoted by thin curves with arrows on the boundaries. The domain flows are represented by the thick curves with arrows in the domain. ($n_1+n_2+n_3=s$)

Definition 8.18. For a discontinuous system, there is an accessible edge $\mathscr{E}_{\sigma_s}^{(s)}(\sigma_r)$ ($\sigma_s \in \mathscr{N}_{\mathscr{E}}^{(s)} = \{1,2,\cdots,N_s\}$) where k_{σ_s}-vector fields of $\mathbf{F}^{(\sigma_{s(k)};\mathscr{E})}(\mathbf{x}^{(\sigma_{s(k)};\mathscr{E})},t,\boldsymbol{\pi}_{\sigma_{s(k)}})$ ($k=1,2,\cdots,k_{\sigma_s}$) are defined, and the corresponding dynamical system is

$$\dot{\mathbf{x}}^{(\sigma_{s(k)};\mathscr{E})} = \mathbf{F}^{(\sigma_{s(k)};\mathscr{E})}(\mathbf{x}^{(\sigma_{s(k)};\mathscr{E})},t,\boldsymbol{\pi}_{\sigma_{s(k)}}) \text{ on } \mathscr{E}_{\sigma_s}^{(s)}(\sigma_r). \tag{8.123}$$

The discontinuous dynamical system in edge $\mathscr{E}_{\sigma_s}^{(s)}(\sigma_r)$ is called to be *of the multi-valued vector fields*. A set of dynamical systems in edge $\mathscr{E}_{\sigma_s}^{(s)}(\sigma_r)$ is defined as

$$\begin{aligned} &\mathscr{E}_{\sigma_s}(\sigma_r) = \cup_{k=1}^{k_{\sigma_s}} \mathscr{E}_{\sigma_{s(k)}} \text{ on } \mathscr{E}_{\sigma_s}^{(s)}(\sigma_r), \\ &\mathscr{E}_{\sigma_{s(k)}}(\sigma_r) = \left\{ \dot{\mathbf{x}}^{(\sigma_{s(k)};\mathscr{E})} = \mathbf{F}^{(\sigma_{s(k)};\mathscr{E})}(\mathbf{x}^{(\sigma_{s(k)};\mathscr{E})},t,\boldsymbol{\pi}_{\sigma_{s(k)}}) \middle| k \in \{1,2,\cdots,k_{\sigma_s}\} \right\}. \end{aligned} \tag{8.124}$$

A total set of dynamical systems in the discontinuous dynamical system is

$$\mathscr{E} = \cup_{s=0}^{n} \cup_{\sigma_s \in N_{\mathscr{E}}^{s}} \mathscr{E}_{\sigma_s}(\sigma_r). \tag{8.125}$$

8.4.1. Bouncing edge flows

If multi-valued vector fields in each single edge exist, with a switching rule or transport law, the bouncing flow to a specific edge will exist as for the boundary. A definition for the bouncing edge flow is given herein.

Definition 8.19. For a discontinuous system with multi-valued vector fields in Eq. (8.125), there is a point $\mathbf{x}^{(\sigma_r;\mathscr{E})}(t_m) \equiv \mathbf{x}_m \in \mathscr{E}_{\sigma_r}^{(r)}$ ($\sigma_r \in \mathscr{N}_{\mathscr{E}}^{r} = \{1,2,\cdots,N_r\}$ and $r \in \{0,1,2,\cdots,n-2\}$) at time t_m with the related boundaries $\partial\Omega_{\alpha i_\alpha}(\sigma_r)$ and related domains $\Omega_\alpha(\sigma_r)$ ($\alpha \in \mathscr{N}_{\mathscr{D}}(\sigma_r) \subset \mathscr{N}_{\mathscr{D}} = \{1,2,\cdots,N_d\}$ and $i_\alpha \in \mathscr{N}_{\alpha\mathscr{B}}(\sigma_r) \subset \mathscr{N}_{\mathscr{B}} = \{1,2,\cdots,N_b\}$), where the boundary index subset of $\mathscr{N}_{\alpha\mathscr{B}}(\sigma_r)$ has n_{σ_r}-elements ($n_{\sigma_r} \geq (n-r)$). The edge of $\mathscr{E}_{\sigma_r}^{(r)}$ is the intersection of edges (including domains and boundaries) $\mathscr{E}_{\sigma_s}^{(s)}(\sigma_r)$ ($\sigma_s \in \mathscr{N}_{\mathscr{E}}^{s}(\sigma_r) \subset \mathscr{N}_{\mathscr{E}}^{s} = \{1,2,\cdots,N_s\}$ and $s \in \{r+1,r+2,\cdots,n-1,n\}$). Suppose there is an s-dimensional edge $\mathscr{E}_{\sigma_s}^{(s)}(\sigma_r)$ with a boundary index set $\mathscr{N}_{\alpha\mathscr{B}}(\sigma_r,\sigma_s)$ relative to domain $\Omega_\alpha(\sigma_r)$. The complementary boundary index set is $\overline{\mathscr{N}}_{\alpha\mathscr{B}}(\sigma_r,\sigma_s) = \mathscr{N}_{\alpha\mathscr{B}}(\sigma_r)/\mathscr{N}_{\alpha\mathscr{B}}(\sigma_r,\sigma_s)$. For an arbitrarily small $\varepsilon > 0$, there are two time intervals $[t_{m-\varepsilon},t_m)$ and/or $(t_m,t_{m+\varepsilon}]$. A coming edge flow $\mathbf{x}^{(\sigma_{s(k)};\mathscr{E})}$ is $C_{[t_{m-\varepsilon},t_m)}^{r_{\sigma_{s(k)}}}$-continuous ($r_{\sigma_{s(k)}} \geq 1$) for time t and $\| d^{r_{\sigma_{s(k)}}+1}\mathbf{x}^{(\sigma_{s(k)};\mathscr{E})} / dt^{r_{\sigma_{s(k)}}+1} \| < \infty$. A leaving edge flow $\mathbf{x}^{(\sigma_{s(l)};\mathscr{E})}$ is $C_{(t_m,t_{m+\varepsilon}]}^{r_{\sigma_{s(l)}}}$ continuous ($r_{\sigma_{s(l)}} \geq 1$) for time t and $\| d^{r_{\sigma_{s(l)}}+1}\mathbf{x}^{(\sigma_{s(k)};\mathscr{E})}/dt^{r_{\sigma_{s(l)}}+1} \| < \infty$. If a coming edge flow of $\mathbf{x}^{(\sigma_{s(k)};\mathscr{E})}$ arrives to edge of $\mathscr{E}_{\sigma_r}^{(r)}$, there is a switching rule between the two edge flows of $\mathbf{x}^{(\sigma_{s(k)};\mathscr{E})}$ and $\mathbf{x}^{(\sigma_{s(l)};\mathscr{E})}$ with $\mathbf{x}^{(\sigma_{s(k)};\mathscr{E})}(t_{m-}) = \mathbf{x}_m = \mathbf{x}^{(\sigma_{s(l)};\mathscr{E})}(t_{m+})$. The flows of $\mathbf{x}^{(\sigma_{s(k)};\mathscr{E})}$ and $\mathbf{x}^{(\sigma_{s(l)};\mathscr{E})}$ to the edge $\mathscr{E}_{\sigma_r}^{(r)}$ are called a *bouncing* flow in $\mathscr{E}_{\sigma_s}^{(s)}$ if

$$\left.\begin{aligned} &\hbar_{j_\alpha}\mathbf{n}^{\mathrm{T}}_{\partial\Omega_{\alpha j_\alpha}}(\mathbf{x}_{m-\varepsilon}^{(j_\alpha;\mathscr{B})})\cdot[\mathbf{x}_{m-\varepsilon}^{(j_\alpha;\mathscr{B})} - \mathbf{x}_{m-\varepsilon}^{(\sigma_{s(k)};\mathscr{E})}] < 0 \\ &\hbar_{j_\alpha}\mathbf{n}^{\mathrm{T}}_{\partial\Omega_{\alpha j_\alpha}}(\mathbf{x}_{m+\varepsilon}^{(j_\alpha;\mathscr{B})})\cdot[\mathbf{x}_{m+\varepsilon}^{(\sigma_{s(l)};\mathscr{E})} - \mathbf{x}_{m+\varepsilon}^{(j_\alpha;\mathscr{B})}] > 0 \end{aligned}\right\} \tag{8.126}$$

for all $j_\alpha \in \bar{\mathscr{N}}_{\alpha\mathscr{B}}(\sigma_r,\sigma_s)$.

Using the G-function to the boundary, the foregoing definition of the bouncing flow to the edge yields the corresponding theorem.

Theorem 8.13. *For a discontinuous system with multi-valued vector fields in Eq.* (8.125)*, there is a point* $\mathbf{x}^{(\sigma_r;\mathscr{E})}(t_m) \equiv \mathbf{x}_m \in \mathscr{E}_{\sigma_r}^{(r)}$ ($\sigma_r \in \mathscr{N}_{\mathscr{E}}^r = \{1,2,\cdots,N_r\}$ *and* $r \in \{0,1,2,\cdots,n-2\}$) *at time* t_m *with the related boundaries* $\partial\Omega_{\alpha i_\alpha}(\sigma_r)$ *and related domains* $\Omega_\alpha(\sigma_r)$ ($\alpha \in \mathscr{N}_{\mathscr{D}}(\sigma_r) \subset \mathscr{N}_{\mathscr{D}} = \{1,2,\cdots,N_d\}$ *and* $i_\alpha \in \mathscr{N}_{\alpha\mathscr{B}}(\sigma_r)$ $\subset \mathscr{N}_{\mathscr{B}} = \{1,2,\cdots,N_b\}$)*, where the boundary index subset of* $\mathscr{N}_{\alpha\mathscr{B}}(\sigma_r)$ *has* n_{σ_r}-*elements* ($n_{\sigma_r} \geq (n-r)$)*. The edge of* $\mathscr{E}_{\sigma_r}^{(r)}$ *is the intersection of edges (including domains and boundaries)* $\mathscr{E}_{\sigma_s}^{(s)}(\sigma_r)$ ($\sigma_s \in \mathscr{N}_{\mathscr{E}}^s(\sigma_r) \subset \mathscr{N}_{\mathscr{E}}^s = \{1,2,\cdots,N_s\}$ *and* $s \in \{r+1,r+2,\cdots,n-1,n\}$)*. Suppose there is an* s*-dimensional edge* $\mathscr{E}_{\sigma_s}^{(s)}(\sigma_r)$ *with a boundary index set* $\mathscr{N}_{\alpha\mathscr{B}}(\sigma_r,\sigma_s)$ *relative to domain* $\Omega_\alpha(\sigma_r)$*. The complementary boundary index set is* $\bar{\mathscr{N}}_{\alpha\mathscr{B}}(\sigma_r,\sigma_s) = \mathscr{N}_{\alpha\mathscr{B}}(\sigma_r)/\mathscr{N}_{\alpha\mathscr{B}}(\sigma_r,\sigma_s)$*. For an arbitrarily small* $\varepsilon > 0$*, there are two time intervals* $[t_{m-\varepsilon},t_m)$ *and/or* $(t_m,t_{m+\varepsilon}]$*. A coming edge flow* $\mathbf{x}^{(\sigma_{s(k)};\mathscr{E})}$ *is* $C_{[t_{m-\varepsilon},t_m)}^{r_{\sigma_{s(k)}}}$*-continuous* ($r_{\sigma_{s(k)}} \geq 2$) *for time* t *and* $\| d^{r_{\sigma_{s(k)}}+1}\mathbf{x}^{(\sigma_{s(k)};\mathscr{E})}/dt^{r_{\sigma_{s(k)}}+1} \| < \infty$*. A leaving edge flow* $\mathbf{x}^{(\sigma_{s(l)};\mathscr{E})}$ *is* $C_{(t_m,t_{m+\varepsilon}]}^{r_{\sigma_{s(l)}}}$ *continuous* ($r_{\sigma_{s(l)}} \geq 2$) *for time* t *and* $\| d^{r_{\sigma_{s(l)}}+1}\mathbf{x}^{(\sigma_{s(k)};\mathscr{E})}/dt^{r_{\sigma_{s(l)}}+1} \| < \infty$*. If a coming edge flow of* $\mathbf{x}^{(\sigma_{s(k)};\mathscr{E})}$ *arrives to edge* $\mathscr{E}_{\sigma_r}^{(r)}$*, there is a switching rule between two edge flows* $\mathbf{x}^{(\sigma_{s(k)};\mathscr{E})}$ *and* $\mathbf{x}^{(\sigma_{s(l)};\mathscr{E})}$ *with* $\mathbf{x}^{(\sigma_{s(k)};\mathscr{E})}(t_{m-}) = \mathbf{x}_m = \mathbf{x}^{(\sigma_{s(l)};\mathscr{E})}(t_{m+})$*. The flows* $\mathbf{x}^{(\sigma_{s(k)};\mathscr{E})}$ *and* $\mathbf{x}^{(\sigma_{s(l)};\mathscr{E})}$ *to the edge* $\mathscr{E}_{\sigma_r}^{(r)}$ *form a bouncing flow in* $\mathscr{E}_{\sigma_s}^{(s)}$ *if and only if*

$$\begin{aligned} &\hbar_{j_\alpha}G_{\partial\Omega_{\alpha j_\alpha}}^{(\sigma_{s(k)};\mathscr{E})}(\mathbf{x}_m,t_{m-},\boldsymbol{\pi}_{\sigma_{s(k)}},\boldsymbol{\lambda}_{j_\alpha}) < 0, \\ &\hbar_{j_\alpha}G_{\partial\Omega_{\alpha j_\alpha}}^{(\sigma_{s(l)};\mathscr{E})}(\mathbf{x}_m,t_{m+},\boldsymbol{\pi}_{\sigma_{s(l)}},\boldsymbol{\lambda}_{j_\alpha}) > 0 \end{aligned} \tag{8.127}$$

for all $j_\alpha \in \bar{\mathscr{N}}_{\alpha\mathscr{B}}(\sigma_r,\sigma_s)$*.*

Proof. Following the proof of Theorem 3.1, the theorem can be proved. ■

The bouncing flow with higher order singularity will be stated as follows.

Definition 8.20. For a discontinuous system with multi-valued vector fields in Eq. (8.125), there is a point $\mathbf{x}^{(\sigma_r;\mathscr{E})}(t_m) \equiv \mathbf{x}_m \in \mathscr{E}_{\sigma_r}^{(r)}$ ($\sigma_r \in \mathscr{N}_{\mathscr{E}}^r = \{1,2,\cdots,N_r\}$ and $r \in \{0,1,2,\cdots,n-2\}$) at time t_m with the related boundaries $\partial\Omega_{\alpha i_\alpha}(\sigma_r)$ and related domains $\Omega_\alpha(\sigma_r)$ ($\alpha \in \mathscr{N}_{\mathscr{D}}(\sigma_r) \subset \mathscr{N}_{\mathscr{D}} = \{1,2,\cdots,N_d\}$ and $i_\alpha \in \mathscr{N}_{\alpha\mathscr{B}}(\sigma_r) \subset \mathscr{N}_{\mathscr{B}} = \{1,2,\cdots,N_b\}$), where the boundary index subset of $\mathscr{N}_{\alpha\mathscr{B}}(\sigma_r)$ has n_{σ_r}-elements ($n_{\sigma_r} \ge (n-r)$). The edge of $\mathscr{E}_{\sigma_r}^{(r)}$ is the intersection of edges (including domains and boundaries) $\mathscr{E}_{\sigma_s}^{(s)}(\sigma_r)$ ($\sigma_s \in \mathscr{N}_{\mathscr{E}}^s(\sigma_r) \subset \mathscr{N}_{\mathscr{E}}^s = \{1,2,\cdots,N_s\}$ and $s \in \{r+1, r+2,\cdots,n-1,n\}$). Suppose there is an s-dimensional edge $\mathscr{E}_{\sigma_s}^{(s)}(\sigma_r)$ with a boundary index set $\mathscr{N}_{\alpha\mathscr{B}}(\sigma_r,\sigma_s)$ relative to domain $\Omega_\alpha(\sigma_r)$. The complementary boundary index set is $\bar{\mathscr{N}}_{\alpha\mathscr{B}}(\sigma_r,\sigma_s) = \mathscr{N}_{\alpha\mathscr{B}}(\sigma_r) / \mathscr{N}_{\alpha\mathscr{B}}(\sigma_r,\sigma_s)$. For an arbitrarily small $\varepsilon > 0$, there are two time intervals $[t_{m-\varepsilon}, t_m)$ and/or $(t_m, t_{m+\varepsilon}]$. A coming edge flow $\mathbf{x}^{(\sigma_{s(k)};\mathscr{E})}$ is $C_{[t_{m-\varepsilon},t_m)}^{r_{\sigma_{s(k)}}}$ continuous ($r_{\sigma_{s(k)}} \ge \max_{i_\alpha \in \bar{\mathscr{N}}_{\alpha\mathscr{B}}(\sigma_r,\sigma_s)}\{m_{i_\alpha}\}$) for time t and $\| d^{r_{\sigma_{s(k)}}+1}\mathbf{x}^{(\sigma_{s(k)};\mathscr{E})} / dt^{r_{\sigma_{s(k)}}+1} \| < \infty$. A leaving edge flow $\mathbf{x}^{(\sigma_{s(l)};\mathscr{E})}$ is $C_{(t_m,t_{m+\varepsilon}]}^{r_{\sigma_{s(l)}}}$-continuous for time t and $\| d^{r_{\sigma_{s(l)}}+1}\mathbf{x}^{(\sigma_{s(k)};\mathscr{E})} / dt^{r_{\sigma_{s(l)}}+1} \| < \infty$ ($r_{\sigma_{s(l)}} \ge \max_{j_\alpha \in \bar{\mathscr{N}}_{\alpha\mathscr{B}}(\sigma_r,\sigma_s)}\{m_{j_\alpha}\}$). If a coming edge flow of $\mathbf{x}^{(\sigma_{s(k)};\mathscr{E})}$ arrives to edge of $\mathscr{E}_{\sigma_r}^{(r)}$, there is a switching rule between the two edge flows of $\mathbf{x}^{(\sigma_{s(k)};\mathscr{E})}$ and $\mathbf{x}^{(\sigma_{s(l)};\mathscr{E})}$ with $\mathbf{x}^{(\sigma_{s(k)};\mathscr{E})}(t_{m-}) = \mathbf{x}_m = \mathbf{x}^{(\sigma_{s(l)};\mathscr{E})}(t_{m+})$. The flows of $\mathbf{x}^{(\sigma_{s(k)};\mathscr{E})}$ and $\mathbf{x}^{(\sigma_{s(l)};\mathscr{E})}$ to the edge $\mathscr{E}_{\sigma_r}^{(r)}$ are called an $(\mathbf{m}_{\sigma_{s(k)}} : \mathbf{m}_{\sigma_{s(l)}})$-*bouncing* flow in edge $\mathscr{E}_{\sigma_s}^{(s)}$ if

$$\left.\begin{array}{l}
\hbar_{j_\alpha} G_{\partial\Omega_{\alpha j_\alpha}}^{(s_{j_\alpha},\sigma_{s(k)};\mathscr{E})}(\mathbf{x}_m, t_{m-}, \boldsymbol{\pi}_{\sigma_{s(k)}}, \boldsymbol{\lambda}_{j_\alpha}) = 0 \\
\text{for } s_{j_\alpha} = 0,1,2,\cdots,m_{j_\alpha}, \\
\hbar_{j_\alpha} \mathbf{n}_{\partial\Omega_{\alpha i_\alpha}}^{\mathrm{T}}(\mathbf{x}_{m-\varepsilon}^{(j_\alpha;\mathscr{B})}) \cdot [\mathbf{x}_{m-\varepsilon}^{(i_\alpha;\mathscr{B})} - \mathbf{x}_{m-\varepsilon}^{(\sigma_{s(k)};\mathscr{E})}] < 0
\end{array}\right\} \tag{8.128}$$
$$\text{for all } j_\alpha \in \bar{\mathscr{N}}_{\alpha\mathscr{B}}(\sigma_r,\sigma_s);$$

$$\left.\begin{array}{l}
\hbar_{j_\alpha} G_{\partial\Omega_{\alpha j_\alpha}}^{(s_{j_\alpha},\sigma_{s(l)};\mathscr{E})}(\mathbf{x}_m, t_{m-}, \boldsymbol{\pi}_{\sigma_{s(l)}}, \boldsymbol{\lambda}_{j_\alpha}) = 0 \\
\text{for } s_{j_\alpha} = 0,1,2,\cdots,m_{j_\alpha}, \\
\hbar_{j_\alpha} \mathbf{n}_{\partial\Omega_{\alpha j_\alpha}}^{\mathrm{T}}(\mathbf{x}_{m+\varepsilon}^{(j_\alpha;\mathscr{B})}) \cdot [\mathbf{x}_{m+\varepsilon}^{(\sigma_{s(l)};\mathscr{E})} - \mathbf{x}_{m+\varepsilon}^{(j_\alpha;\mathscr{B})}] > 0 \\
\text{for all } j_\alpha \in \bar{\mathscr{N}}_{\alpha\mathscr{B}}(\sigma_r,\sigma_s).
\end{array}\right. \tag{8.129}$$

Theorem 8.14. *For a discontinuous system with multi-valued vector fields in Eq. (8.125), there is a point* $\mathbf{x}^{(\sigma_r;\mathscr{E})}(t_m) \equiv \mathbf{x}_m \in \mathscr{E}_{\sigma_r}^{(r)}$ ($\sigma_r \in \mathscr{N}_{\mathscr{E}}^r = \{1,2,\cdots,N_r\}$ *and* $r \in \{0,1,2,\cdots,n-2\}$) *at time* t_m *with the related boundaries* $\partial\Omega_{\alpha i_\alpha}(\sigma_r)$ *and related domains* $\Omega_\alpha(\sigma_r)$ ($\alpha \in \mathscr{N}_{\mathscr{D}}(\sigma_r) \subset \mathscr{N}_{\mathscr{D}} = \{1,2,\cdots,N_d\}$ *and* $i_\alpha \in \mathscr{N}_{\alpha\mathscr{B}}(\sigma_r) \subset \mathscr{N}_{\mathscr{B}} = \{1,2,\cdots,N_b\}$), *where the boundary index subset of* $\mathscr{N}_{\alpha\mathscr{B}}(\sigma_r)$ *has* n_{σ_r}*-elements* ($n_{\sigma_r} \geq (n-r)$). *The edge of* $\mathscr{E}_{\sigma_r}^{(r)}$ *is the intersection of edges (including domains and boundaries)* $\mathscr{E}_{\sigma_s}^{(s)}(\sigma_r)$ ($\sigma_s \in \mathscr{N}_{\mathscr{E}}^s(\sigma_r) \subset \mathscr{N}_{\mathscr{E}}^s = \{1,2,\cdots,N_s\}$ *and* $s \in \{r+1,r+2,\cdots,n-1,n\}$). *Suppose there is an* s*-dimensional edge* $\mathscr{E}_{\sigma_s}^{(s)}(\sigma_r)$ *with a boundary index set* $\mathscr{N}_{\alpha\mathscr{B}}(\sigma_r,\sigma_s)$ *relative to domain* $\Omega_\alpha(\sigma_r)$. *The complementary boundary index set is* $\bar{\mathscr{N}}_{\alpha\mathscr{B}}(\sigma_r,\sigma_s) = \mathscr{N}_{\alpha\mathscr{B}}(\sigma_r)/\mathscr{N}_{\alpha\mathscr{B}}(\sigma_r,\sigma_s)$. *For an arbitrarily small* $\varepsilon > 0$, *there are two time intervals* $[t_{m-\varepsilon},t_m)$ *and/or* $(t_m,t_{m+\varepsilon}]$. *A coming edge flow* $\mathbf{x}^{(\sigma_{s(k)};\mathscr{E})}$ *is* $C_{[t_{m-\varepsilon},t_m)}^{r_{\sigma_{s(k)}}}$ *continuous* ($r_{\sigma_{s(k)}} \geq \max_{i_\alpha \in \bar{\mathscr{N}}_{\alpha\mathscr{B}}(\sigma_r,\sigma_s)}\{m_{i_\alpha}\} + 1$) *for time* t *and* $\| d^{r_{\sigma_{s(k)}}+1}\mathbf{x}^{(\sigma_{s(k)};\mathscr{E})}/dt^{r_{\sigma_{s(k)}}+1} \| < \infty$. *A leaving edge flow* $\mathbf{x}^{(\sigma_{s(l)};\mathscr{E})}$ *is* $C_{(t_m,t_{m+\varepsilon}]}^{r_{\sigma_{s(l)}}}$*-continuous* ($r_{\sigma_{s(l)}} \geq \max_{j_\alpha \in \bar{\mathscr{N}}_{\alpha\mathscr{B}}(\sigma_r,\sigma_s)}\{m_{j_\alpha}\} + 1$) *for time* t *and* $\| d^{r_{\sigma_{s(l)}}+1}\mathbf{x}^{(\sigma_{s(k)};\mathscr{E})}/dt^{r_{\sigma_{s(l)}}+1} \| < \infty$. *If a coming edge flow of* $\mathbf{x}^{(\sigma_{s(k)};\mathscr{E})}$ *arrives to the edge of* $\mathscr{E}_{\sigma_r}^{(r)}$, *there is a switching rule between the two edge flows of* $\mathbf{x}^{(\sigma_{s(k)};\mathscr{E})}$ *and* $\mathbf{x}^{(\sigma_{s(l)};\mathscr{E})}$ *with* $\mathbf{x}^{(\sigma_{s(k)};\mathscr{E})}(t_{m-}) = \mathbf{x}_m = \mathbf{x}^{(\sigma_{s(l)};\mathscr{E})}(t_{m+})$. *The edge flows of* $\mathbf{x}^{(\sigma_{s(k)};\mathscr{E})}$ *and* $\mathbf{x}^{(\sigma_{s(l)};\mathscr{E})}$ *to the edge of* $\mathscr{E}_{\sigma_r}^{(r)}$ *form an* ($\mathbf{m}_{\sigma_{s(k)}} : \mathbf{m}_{\sigma_{s(l)}}$)*-bouncing flow in* $\mathscr{E}_{\sigma_s}^{(s)}$ *if and only if*

$$\left.\begin{aligned} &\hbar_{i_\alpha} G_{\partial\Omega_{\alpha i_\alpha}}^{(s_{i_\alpha},\sigma_{s(k)};\mathscr{E})}(\mathbf{x}_m,t_{m-},\boldsymbol{\pi}_{\sigma_{s(k)}},\boldsymbol{\lambda}_{i_\alpha}) = 0 \\ &\textit{for } s_{i_\alpha} = 0,1,2,\cdots,m_{i_\alpha}, \\ &\hbar_{i_\alpha} G_{\partial\Omega_{\alpha i_\alpha}}^{(m_{i_\alpha},\sigma_{s(k)};\mathscr{E})}(\mathbf{x}_m,t_{m-},\boldsymbol{\pi}_{\sigma_{s(k)}},\boldsymbol{\lambda}_{i_\alpha}) < 0 \end{aligned}\right\} \tag{8.130}$$
$$\textit{for all } i_\alpha \in \bar{\mathscr{N}}_{\alpha\mathscr{B}}(\sigma_r,\sigma_s);$$

$$\left.\begin{aligned} &\hbar_{j_\alpha} G_{\partial\Omega_{\alpha j_\alpha}}^{(s_{\sigma_{s(l)}},\sigma_{s(l)};\mathscr{E})}(\mathbf{x}_m,t_{m-},\boldsymbol{\pi}_{\sigma_{s(l)}},\boldsymbol{\lambda}_{j_\alpha}) = 0 \\ &\textit{for } s_{\sigma_{s(l)}} = 0,1,2,\cdots,m_{\sigma_{s(l)}}, \\ &\hbar_{j_\alpha} G_{\partial\Omega_{\alpha j_\alpha}}^{(m_{\sigma_{s(l)}},\sigma_{s(l)};\mathscr{E})}(\mathbf{x}_m,t_{m-},\boldsymbol{\pi}_{\sigma_{s(l)}},\boldsymbol{\lambda}_{j_\alpha}) > 0 \end{aligned}\right\} \tag{8.131}$$
$$\textit{for all } j_\alpha \in \bar{\mathscr{N}}_{\alpha\mathscr{B}}(\sigma_r,\sigma_s).$$

Proof. Following the proof of Theorem 3.2, the theorem can be proved. ■

The bouncing edge flow with two different vector fields in the edge of $\mathscr{E}_{\sigma_s}^{(s\text{-}2)}$ switches at the boundary of $\mathscr{E}^{(s\text{-}2)}$, which includes nine types bouncing flows. The cases are presented in Figs. 8.12-8.14. The solid curves give the real edge flows, and the dashed curves give the extension of the edge flows. In the same edge, such edge flow extensions are real flows.

8.4.2. Extended passable edge flows

The bouncing edge flow requires at least a coming flow and a leaving flow at the special edge. With the switching rule, the coming and leaving edge flows will form a bouncing edge flow. If two coming edge flows in edge $\mathscr{E}_{\sigma_s}^{(s)}$ at the same time arrive at the same location of the special edge of $\mathscr{E}_{\sigma_r}^{(r)}$, after switching, the edge flow in the edge of $\mathscr{E}_{\sigma_s}^{(s)}$ can extend to the edge of $\mathscr{E}_{\sigma_p}^{(p)}$ with the Continuation axiom. Such extended flows are discussed as follows.

Definition 8.21. For a discontinuous system with multi-valued vector fields in Eq. (8.125), there is a point $\mathbf{x}^{(\sigma_r;\mathscr{E})}(t_m) \equiv \mathbf{x}_m \in \mathscr{E}_{\sigma_r}^{(r)}$ ($\sigma_r \in \mathscr{N}_{\mathscr{E}}^{r} = \{1,2,\cdots,N_r\}$ and $r \in \{0,1,2,\cdots,n-2\}$) at time t_m with the related boundaries $\partial\Omega_{\alpha i_\alpha}(\sigma_r)$ and related domains $\Omega_\alpha(\sigma_r)$ ($\alpha \in \mathscr{N}_{\mathscr{D}}(\sigma_r) \subset \mathscr{N}_{\mathscr{D}} = \{1,2,\cdots,N_d\}$ and $i_\alpha \in \mathscr{N}_{\alpha\mathscr{B}}(\sigma_r) \subset \mathscr{N}_{\mathscr{B}} = \{1,2,\cdots,N_b\}$), where the boundary index subset of $\mathscr{N}_{\alpha\mathscr{B}}(\sigma_r)$ has n_{σ_r}-elements ($n_{\sigma_r} \geq (n-r)$). The edge of $\mathscr{E}_{\sigma_r}^{(r)}$ is the intersection of edges (including domains and boundaries) $\mathscr{E}_{\sigma_s}^{(s)}(\sigma_r)$ ($\sigma_s \in \mathscr{N}_{\mathscr{E}}^{s}(\sigma_r) \subset \mathscr{N}_{\mathscr{E}}^{s} = \{1,2,\cdots,N_s\}$ and $s \in \{r+1,r+2,\cdots,n-1,n\}$). Suppose there is an s-dimensional index edge $\mathscr{E}_{\sigma_s}^{(s)}(\sigma_r)$ with a boundary set $\mathscr{N}_{\alpha\mathscr{B}}(\sigma_r,\sigma_s)$ relative to domain $\Omega_\alpha(\sigma_r)$. The complementary boundary index set is $\bar{\mathscr{N}}_{\alpha\mathscr{B}}(\sigma_r,\sigma_s) = \mathscr{N}_{\alpha\mathscr{B}}(\sigma_r)/\mathscr{N}_{\alpha\mathscr{B}}(\sigma_r,\sigma_s)$. For an arbitrarily small $\varepsilon > 0$, there are two time intervals $[t_{m-\varepsilon}, t_m)$ and/or $(t_m, t_{m+\varepsilon}]$. A coming edge flow $\mathbf{x}^{(\sigma_{s(k)};\mathscr{E})}$ is $C_{[t_{m-\varepsilon},t_m)}^{r_{\sigma_{s(k)}}}$-continuous ($r_{\sigma_{s(k)}} \geq 1$) for time t and $\| d^{r_{\sigma_{s(k)}}+1}\mathbf{x}^{(\sigma_{s(k)};\mathscr{E})} / dt^{r_{\sigma_{s(k)}}+1} \| < \infty$. A leaving edge flow $\mathbf{x}^{(\sigma_{s(l)};\mathscr{E})}$ is $C_{(t_m,t_{m+\varepsilon}]}^{r_{\sigma_{s(l)}}}$ continuous ($r_{\sigma_{s(l)}} \geq 1$) for time t and $\| d^{r_{\sigma_{s(l)}}+1}\mathbf{x}^{(\sigma_{s(k)};\mathscr{E})} / dt^{r_{\sigma_{s(l)}}+1} \| < \infty$. If a coming edge flow of $\mathbf{x}^{(\sigma_{s(k)};\mathscr{E})}$ arrives to the edge of $\mathscr{E}_{\sigma_r}^{(r)}$, there is a switching rule between the two edge flows of $\mathbf{x}^{(\sigma_{s(k)};\mathscr{E})}$ and $\mathbf{x}^{(\sigma_{s(l)};\mathscr{E})}$ with $\mathbf{x}^{(\sigma_{s(k)};\mathscr{E})}(t_{m-}) = \mathbf{x}_m = \mathbf{x}^{(\sigma_{s(l)};\mathscr{E})}(t_{m+})$. The flows of $\mathbf{x}^{(\sigma_{s(k)};\mathscr{E})}$ and $\mathbf{x}^{(\sigma_{s(l)};\mathscr{E})}$ to the edge of $\mathscr{E}_{\sigma_r}^{(r)}$ are called an extended, passable edge flow in $\mathscr{E}_{\sigma_s}^{(s)}$ if

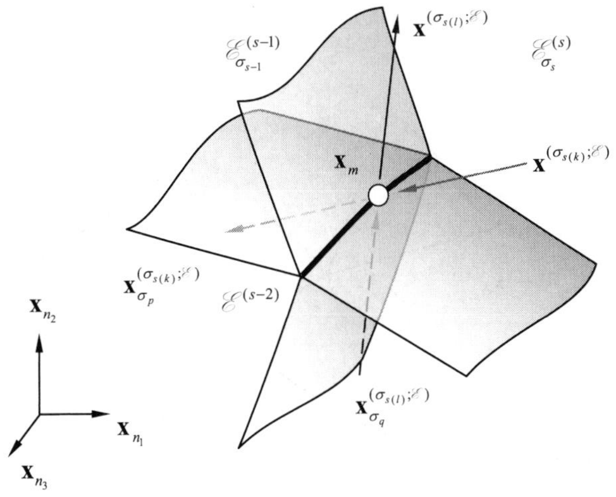

(a)

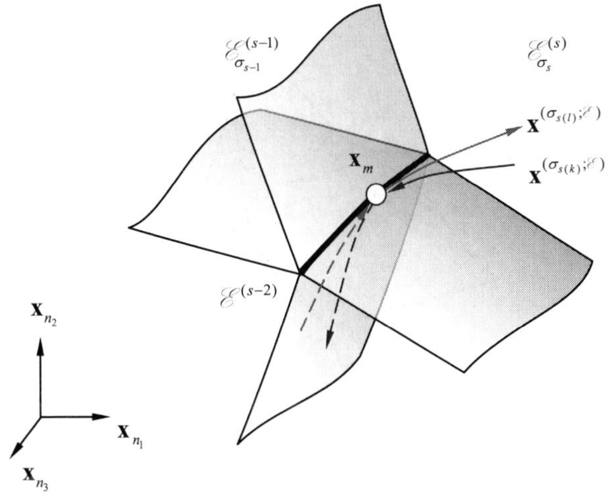

(b)

Fig. 8.12 Bouncing edge flows to the specific edge. **(a)** the $(2\mathbf{k}_{\sigma_{s(k)}} : 2\mathbf{k}_{\sigma_{s(l)}})$ singularity, **(b)** the $(2\mathbf{k}_{\sigma_{s(k)}} + 1 : 2\mathbf{k}_{\sigma_{s(l)}} + 1)$ singularity. The black curve represents the $(s-2)$-dimensional edge. The edge flows are represented by the thick curves with arrows in the edge of $\mathscr{E}^{(s)}_{\sigma_s}$. The dashed curves are the extensions of the real flows. ($n_1 + n_2 + n_3 = s$)

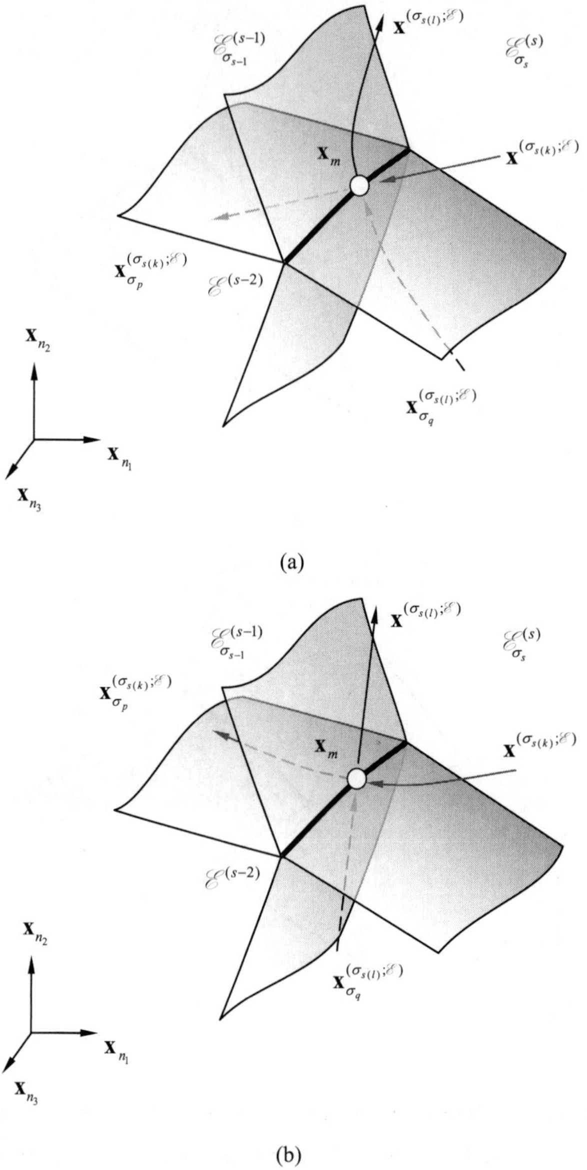

Fig. 8.13 Bouncing edge flows to the specific edge. **(a)** the $(2\mathbf{k}_{\sigma_{s(k)}} : \mathbf{m}_{\sigma_{s(l)}})$ singularity ($\mathbf{m}_{\sigma_{s(l)}} \neq 2\mathbf{k}_{\sigma_{s(l)}} + 1$), **(b)** the $(\mathbf{m}_{\sigma_{s(k)}} : 2\mathbf{k}_{\sigma_{s(l)}})$ singularity ($\mathbf{m}_{\sigma_{s(k)}} \neq 2\mathbf{k}_{\sigma_{s(k)}} + 1$). The black curve represents the $(s-2)$-dimensional edge. The edge flows are represented by the thick curves with arrows in the edge of $\mathscr{E}_{\sigma_s}^{(s)}$. The dashed curves are the extensions of the real flows. ($n_1 + n_2 + n_3 = s$)

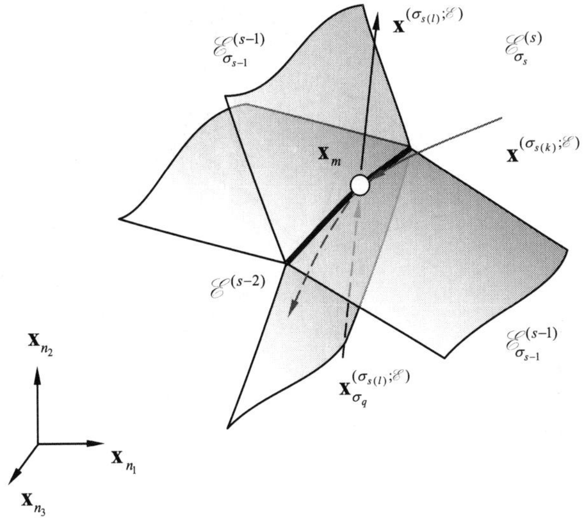

(a)

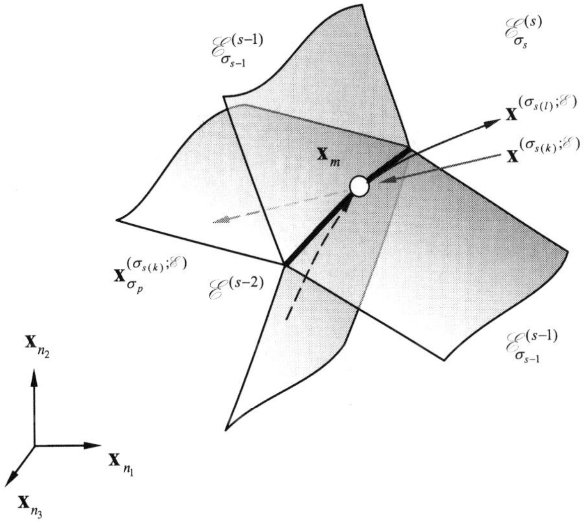

(b)

Fig. 8.14 Bouncing edge flows to the specific edge. **(a)** the $(2\mathbf{k}_{\sigma_{s(l)}}+1:2\mathbf{k}_{\sigma_{s(l)}})$ singularity, **(b)** the $(2\mathbf{k}_{\sigma_{s(k)}}:2\mathbf{k}_{\sigma_{s(k)}}+1)$ singularity. The black curve represents the $(s-2)$-dimensional edge. The edge flows are represented by the thick curves with arrows in the edge of $\mathscr{E}^{(s)}_{\sigma_s}$. The dashed curves are the extensions of the real flows. ($n_1+n_2+n_3=s$)

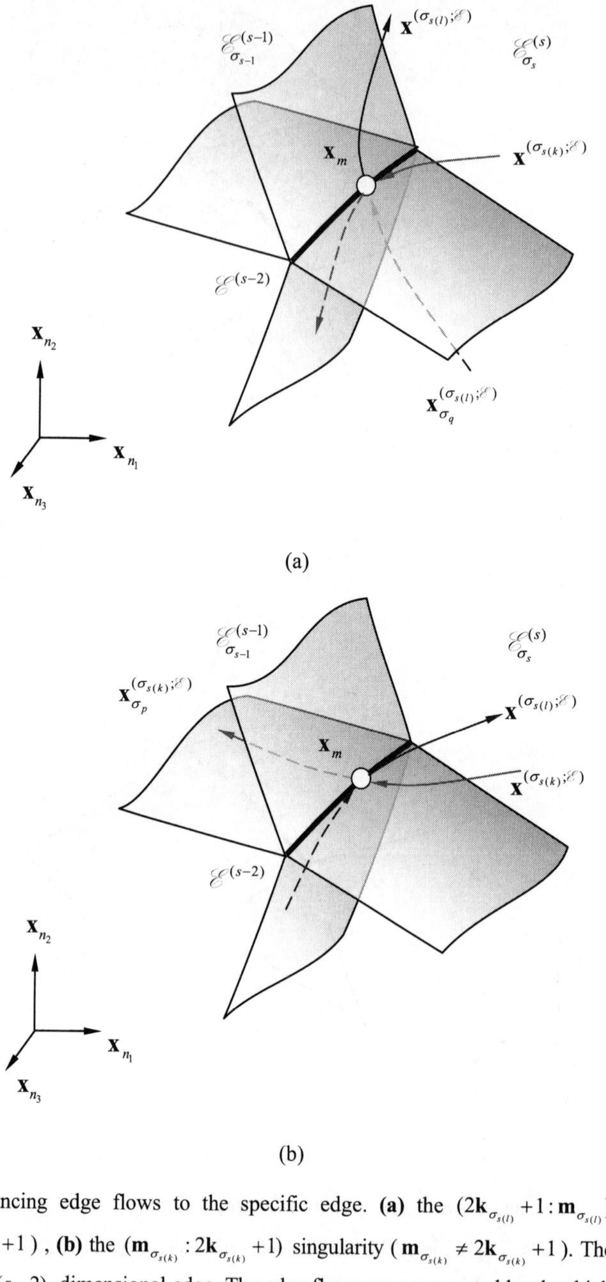

Fig. 8.15 Bouncing edge flows to the specific edge. **(a)** the $(2\mathbf{k}_{\sigma_{s(l)}}+1:\mathbf{m}_{\sigma_{s(l)}})$ singularity $(\mathbf{m}_{\sigma_{s(l)}}\neq 2\mathbf{k}_{\sigma_{s(l)}}+1)$, **(b)** the $(\mathbf{m}_{\sigma_{s(k)}}:2\mathbf{k}_{\sigma_{s(k)}}+1)$ singularity $(\mathbf{m}_{\sigma_{s(k)}}\neq 2\mathbf{k}_{\sigma_{s(k)}}+1)$. The black curve represents the $(s-2)$-dimensional edge. The edge flows are represented by the thick curves with arrows in the edge of $\mathscr{E}_{\sigma_s}^{(s)}$. The dashed curves are the extensions of the real flows. $(n_1+n_2+n_3=s)$

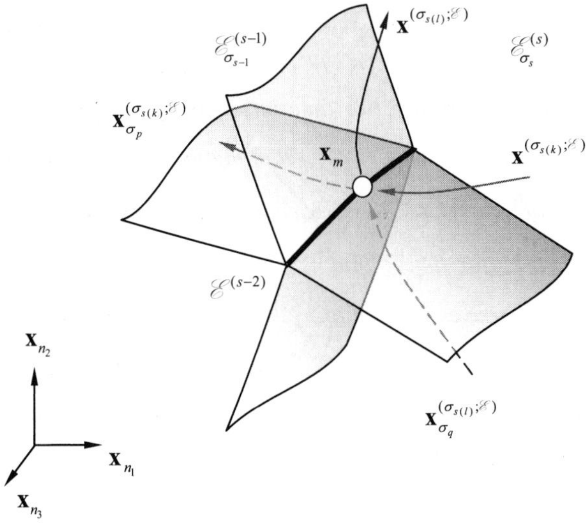

Fig. 8.16 Bouncing edge flows to the specific edge with the $(\mathbf{m}_{\sigma_{s(k)}} : \mathbf{m}_{\sigma_{s(l)}})$ singularity ($\mathbf{m}_{\sigma_{s(k)}} \neq 2\mathbf{k}_{\sigma_{s(k)}} + 1$ and $\mathbf{m}_{\sigma_{s(l)}} \neq 2\mathbf{k}_{\sigma_{s(l)}} + 1$). The black curve represents the $(s-2)$-dimensional edge. The edge flows are represented by the thick curves with arrows in the edge of $\mathscr{E}^{(s)}_{\sigma_s}$. The dashed curves are the extensions of the real flows. ($n_1 + n_2 + n_3 = s$)

$$\left.\begin{aligned}
&\hbar_{j_\alpha} \mathbf{n}^{\mathrm{T}}_{\partial\Omega_{\alpha i_\alpha}} (\mathbf{x}^{(j_\alpha;\mathscr{B})}_{m-\varepsilon}) \cdot [\mathbf{x}^{(j_\alpha;\mathscr{B})}_{m-\varepsilon} - \mathbf{x}^{(\sigma_{s(k)};\mathscr{E})}_{m-\varepsilon}] < 0, \\
&\hbar_{j_\alpha} \mathbf{n}^{\mathrm{T}}_{\partial\Omega_{\alpha j_\alpha}} (\mathbf{x}^{(j_\alpha;\mathscr{B})}_{m+\varepsilon}) \cdot [\mathbf{x}^{(\sigma_{s(l)};\mathscr{E})}_{m+\varepsilon} - \mathbf{x}^{(j_\alpha;\mathscr{B})}_{m+\varepsilon}] < 0
\end{aligned}\right\} \tag{8.132}$$
$$\text{for all } j_\alpha \in \bar{\mathscr{N}}_{\alpha\mathscr{B}}(\sigma_r, \sigma_s).$$

From the foregoing definition of an extended, passable flow from one edge to another edge, the corresponding conditions can be given as follows.

Theorem 8.15. *For a discontinuous system with multi-valued vector fields in Eq.* (8.125)*, there is a point* $\mathbf{x}^{(\sigma_r;\mathscr{E})}(t_m) \equiv \mathbf{x}_m \in \mathscr{E}^{(r)}_{\sigma_r}$ ($\sigma_r \in \mathscr{N}^{r}_{\mathscr{E}} = \{1, 2, \cdots, N_r\}$ *and* $r \in \{0, 1, 2, \cdots, n-2\}$) *at time* t_m *with the related boundaries* $\partial\Omega_{\alpha i_\alpha}(\sigma_r)$ *and related domains* $\Omega_\alpha(\sigma_r)$ ($\alpha \in \mathscr{N}_{\mathscr{D}}(\sigma_r) \subset \mathscr{N}_{\mathscr{D}} = \{1, 2, \cdots, N_d\}$ *and* $i_\alpha \in \mathscr{N}_{\alpha\mathscr{B}}(\sigma_r)$ $\subset \mathscr{N}_{\mathscr{B}} = \{1, 2, \cdots, N_b\}$)*, where the boundary index subset of* $\mathscr{N}_{\alpha\mathscr{B}}(\sigma_r)$ *has* n_{σ_r}*-elements* ($n_{\sigma_r} \geq (n-r)$)*. The edge of* $\mathscr{E}^{(r)}_{\sigma_r}$ *is the intersection of edges (including* *domains and boundaries)* $\mathscr{E}_{\sigma_s}^{(s)}(\sigma_r)$ ($\sigma_s \in \mathscr{N}_{\mathscr{E}}^{s}(\sigma_r) \subset \mathscr{N}_{\mathscr{E}}^{s} = \{1,2,\cdots,N_s\}$ *and* $s \in \{r+1, r+2, \cdots, n-1, n\}$). *Suppose there is an* s *-dimensional edge* $\mathscr{E}_{\sigma_s}^{(s)}(\sigma_r)$ *with a boundary set* $\mathscr{N}_{\alpha\mathscr{B}}(\sigma_r,\sigma_s)$ *relative to domain* $\Omega_\alpha(\sigma_r)$. *The complementary boundary number set is* $\bar{\mathscr{N}}_{\alpha\mathscr{B}}(\sigma_r,\sigma_s) = \mathscr{N}_{\alpha\mathscr{B}}(\sigma_r) / \mathscr{N}_{\alpha\mathscr{B}}(\sigma_r,\sigma_s)$. *For an arbitrarily small* $\varepsilon > 0$, *there are two time intervals* $[t_{m-\varepsilon}, t_m)$ *and/or* $(t_m, t_{m+\varepsilon}]$. *A coming edge flow* $\mathbf{x}^{(\sigma_{s(k)};\mathscr{E})}$ *is* $C_{[t_{m-\varepsilon},t_m)}^{r_{\sigma_{s(k)}}}$ *-continuous (* $r_{\sigma_{s(k)}} \geq 2$ *) for time* t *and* $\| d^{r_{\sigma_{s(k)}}+1} \mathbf{x}^{(\sigma_{s(k)};\mathscr{E})} / dt^{r_{\sigma_{s(k)}}+1} \| < \infty$. *A leaving edge flow* $\mathbf{x}^{(\sigma_{s(l)};\mathscr{E})}$ *is* $C_{(t_m,t_{m+\varepsilon}]}^{r_{\sigma_{s(l)}}}$ *-continuous (* $r_{\sigma_{s(l)}} \geq 2$*) for time* t *and* $\| d^{r_{\sigma_{s(l)}}+1} \mathbf{x}^{(\sigma_{s(k)};\mathscr{E})} / dt^{r_{\sigma_{s(l)}}+1} \| < \infty$. *If a coming edge flow of* $\mathbf{x}^{(\sigma_{s(k)};\mathscr{E})}$ *arrives to edge of* $\mathscr{E}_{\sigma_r}^{(r)}$, *there is a switching rule between the two edge flows of* $\mathbf{x}^{(\sigma_{s(k)};\mathscr{E})}$ *and* $\mathbf{x}^{(\sigma_{s(l)};\mathscr{E})}$ *with* $\mathbf{x}^{(\sigma_{s(k)};\mathscr{E})}(t_{m-}) = \mathbf{x}_m = \mathbf{x}^{(\sigma_{s(l)};\mathscr{E})}(t_{m+})$. *The flows of* $\mathbf{x}^{(\sigma_{s(k)};\mathscr{E})}$ *and* $\mathbf{x}^{(\sigma_{s(l)};\mathscr{E})}$ *to the edge of* $\mathscr{E}_{\sigma_r}^{(r)}$ *form an extended, passable edge flow in* $\mathscr{E}_{\sigma_s}^{(s)}$ *if and only if*

$$\left.\begin{aligned} &\hbar_{j_\alpha} G_{\partial\Omega_{\alpha j_\alpha}}^{(\sigma_{s(k)};\mathscr{E})}(\mathbf{x}_m, t_{m-}, \boldsymbol{\pi}_{\sigma_{s(k)}}, \boldsymbol{\lambda}_{j_\alpha}) < 0 \\ &\hbar_{j_\alpha} G_{\partial\Omega_{\alpha j_\alpha}}^{(\sigma_{s(l)};\mathscr{E})}(\mathbf{x}_m, t_{m+}, \boldsymbol{\pi}_{\sigma_{s(l)}}, \boldsymbol{\lambda}_{j_\alpha}) > 0 \end{aligned}\right\} \quad (8.133)$$
$$\text{for all } j_\alpha \in \bar{\mathscr{N}}_{\alpha\mathscr{B}}(\sigma_r,\sigma_s).$$

Proof. Consider a flow to all boundaries relative to the boundary. Following the proof of Theorem 3.1, the theorem can be proved. ■

Definition 8.22. For a discontinuous system with multi-valued vector fields in Eq. (8.125), there is a point $\mathbf{x}^{(\sigma_r;\mathscr{E})}(t_m) \equiv \mathbf{x}_m \in \mathscr{E}_{\sigma_r}^{(r)}$ ($\sigma_r \in \mathscr{N}_{\mathscr{E}}^{r} = \{1,2,\cdots,N_r\}$ and $r \in \{0,1,2,\cdots,n-2\}$) at time t_m with the related boundaries $\partial\Omega_{\alpha i_\alpha}(\sigma_r)$ and related domains $\Omega_\alpha(\sigma_r)$ ($\alpha \in \mathscr{N}_{\mathscr{D}}(\sigma_r) \subset \mathscr{N}_{\mathscr{D}} = \{1,2,\cdots,N_d\}$ and $i_\alpha \in \mathscr{N}_{\alpha\mathscr{B}}(\sigma_r) \subset \mathscr{N}_{\mathscr{B}} = \{1,2,\cdots,N_b\}$), where the boundary index subset of $\mathscr{N}_{\alpha\mathscr{B}}(\sigma_r)$ has n_{σ_r} -elements ($n_{\sigma_r} \geq (n-r)$). The edge of $\mathscr{E}_{\sigma_r}^{(r)}$ is the intersection of edges (including domains and boundaries) $\mathscr{E}_{\sigma_s}^{(s)}(\sigma_r)$ ($\sigma_s \in \mathscr{N}_{\mathscr{E}}^{s}(\sigma_r) \subset \mathscr{N}_{\mathscr{E}}^{s} = \{1,2,\cdots,N_s\}$ and $s \in \{r+1, r+2, \cdots, n-1, n\}$). Suppose there is an s -dimensional edge $\mathscr{E}_{\sigma_s}^{(s)}(\sigma_r)$ with a boundary index set $\mathscr{N}_{\alpha\mathscr{B}}(\sigma_r,\sigma_s)$ relative to domain $\Omega_\alpha(\sigma_r)$. The complementary boundary index set is $\bar{\mathscr{N}}_{\alpha\mathscr{B}}(\sigma_r,\sigma_s) = \mathscr{N}_{\alpha\mathscr{B}}(\sigma_r) / \mathscr{N}_{\alpha\mathscr{B}}(\sigma_r,\sigma_s)$. For an arbitrarily small $\varepsilon > 0$, there are two time intervals $[t_{m-\varepsilon}, t_m)$ and/or $(t_m, t_{m+\varepsilon}]$. A coming edge flow $\mathbf{x}^{(\sigma_{s(k)};\mathscr{E})}$ is $C_{[t_{m-\varepsilon},t_m)}^{r_{\sigma_{s(k)}}}$ continuous ($r_{\sigma_{s(k)}} \geq \max_{i_\alpha \in \bar{\mathscr{N}}_{\alpha\mathscr{B}}(\sigma_r,\sigma_s)} \{m_{i_\alpha}\}$

$+1$) for time t *and* $\| d^{r_{\sigma_{s(k)}}+1}\mathbf{x}^{(\sigma_{s(k)};\mathscr{E})} / dt^{r_{\sigma_{s(k)}}+1} \| < \infty$. A leaving edge flow $\mathbf{x}^{(\sigma_{s(l)};\mathscr{E})}$ is $C^{r_{\sigma_{s(l)}}}_{(t_m, t_{m+\varepsilon}]}$-continuous for time t *and* $\| d^{r_{\sigma_{s(l)}}+1}\mathbf{x}^{(\sigma_{s(k)};\mathscr{E})} / dt^{r_{\sigma_{s(l)}}+1} \| < \infty$ ($r_{\sigma_{s(l)}} \geq \max_{j_\alpha \in \bar{\mathscr{N}}_{\alpha\mathscr{B}}(\sigma_r, \sigma_s)} \{m_{j_\alpha}\} + 1$). If a coming edge flow of $\mathbf{x}^{(\sigma_{s(k)};\mathscr{E})}$ arrives to the edge of $\mathscr{E}^{(r)}_{\sigma_r}$, there is a switching rule between the two edge flows of $\mathbf{x}^{(\sigma_{s(k)};\mathscr{E})}$ and $\mathbf{x}^{(\sigma_{s(l)};\mathscr{E})}$ with $\mathbf{x}^{(\sigma_{s(k)};\mathscr{E})}(t_{m-}) = \mathbf{x}_m = \mathbf{x}^{(\sigma_{s(l)};\mathscr{E})}(t_{m+})$. The flows of $\mathbf{x}^{(\sigma_{s(k)};\mathscr{E})}$ and $\mathbf{x}^{(\sigma_{s(l)};\mathscr{E})}$ to the edge of $\mathscr{E}^{(r)}_{\sigma_r}$ are called $(\mathbf{m}_{\sigma_{s(k)}} : \mathbf{m}_{\sigma_{s(l)}})$-extended, passable flow in edge $\mathscr{E}^{(s)}_{\sigma_s}$ ($\mathbf{m}_{\sigma_{s(l)}} \neq 2\mathbf{k}_{\sigma_{s(l)}} + 1$) if

$$\left.\begin{aligned} &\hbar_{j_\alpha} G^{(s_{i_\alpha}, \sigma_{s(k)};\mathscr{E})}_{\partial\Omega_{\alpha j_\alpha}}(\mathbf{x}_m, t_{m-}, \boldsymbol{\pi}_{\sigma_{s(k)}}, \boldsymbol{\lambda}_{j_\alpha}) = 0 \\ &\text{for } s_{j_\alpha} = 0, 1, 2, \cdots, m_{i_\alpha}, \\ &\hbar_{j_\alpha} \mathbf{n}^{\mathrm{T}}_{\partial\Omega_{\alpha j_\alpha}}(\mathbf{x}^{(j_\alpha;\mathscr{B})}_{m-\varepsilon}) \cdot [\mathbf{x}^{(j_\alpha;\mathscr{B})}_{m-\varepsilon} - \mathbf{x}^{(\sigma_{s(k)};\mathscr{E})}_{m-\varepsilon}] < 0 \\ &\text{for all } j_\alpha \in \mathscr{N}_{\alpha\mathscr{B}}(\sigma_r, \sigma_s); \end{aligned}\right\} \tag{8.134}$$

$$\left.\begin{aligned} &\hbar_{j_\alpha} G^{(s_{j_\alpha}, \sigma_{s(l)};\mathscr{E})}_{\partial\Omega_{\alpha j_\alpha}}(\mathbf{x}_m, t_{m-}, \boldsymbol{\pi}_{\sigma_{s(l)}}, \boldsymbol{\lambda}_{j_\alpha}) = 0 \\ &\text{for } s_{j_\alpha} = 0, 1, 2, \cdots, m_{j_\alpha}, \\ &\hbar_{j_\alpha} \mathbf{n}^{\mathrm{T}}_{\partial\Omega_{\alpha j_\alpha}}(\mathbf{x}^{(j_\alpha;\mathscr{B})}_{m+\varepsilon}) \cdot [\mathbf{x}^{(\sigma_{s(l)};\mathscr{E})}_{m+\varepsilon} - \mathbf{x}^{(j_\alpha;\mathscr{B})}_{m+\varepsilon}] < 0 \\ &\text{for all } j_\alpha \in \bar{\mathscr{N}}_{\alpha\mathscr{B}}(\sigma_r, \sigma_s). \end{aligned}\right\} \tag{8.135}$$

From the foregoing description of the $(\mathbf{m}_{\sigma_{s(k)}} : \mathbf{m}_{\sigma_{s(l)}})$ extended, passable flow from one edge to another edge, the corresponding conditions can be given.

Theorem 8.16. *For a discontinuous system with multi-valued vector fields in Eq.* (8.125)*, there is a point* $\mathbf{x}^{(\sigma_r;\mathscr{E})}(t_m) \equiv \mathbf{x}_m \in \mathscr{E}^{(r)}_{\sigma_r}$ ($\sigma_r \in \mathscr{N}^r_{\mathscr{E}} = \{1, 2, \cdots, N_r\}$ *and* $r \in \{0, 1, 2, \cdots, n-2\}$) *at time* t_m *with the related boundaries* $\partial\Omega_{\alpha i_\alpha}(\sigma_r)$ *and related domains* $\Omega_\alpha(\sigma_r)$ ($\alpha \in \mathscr{N}_{\mathscr{D}}(\sigma_r) \subset \mathscr{N}_{\mathscr{D}} = \{1, 2, \cdots, N_d\}$ *and* $i_\alpha \in \mathscr{N}_{\alpha\mathscr{B}}(\sigma_r) \subset \mathscr{N}_{\mathscr{B}} = \{1, 2, \cdots, N_b\}$)*, where the boundary index subset of* $\mathscr{N}_{\alpha\mathscr{B}}(\sigma_r)$ *has* n_{σ_r}*-elements* ($n_{\sigma_r} \geq (n-r)$)*. The edge of* $\mathscr{E}^{(r)}_{\sigma_r}$ *is the intersection of edges (including domains and boundaries)* $\mathscr{E}^{(s)}_{\sigma_s}(\sigma_r)$ ($\sigma_s \in \mathscr{N}^s_{\mathscr{E}}(\sigma_r) \subset \mathscr{N}^s_{\mathscr{E}} = \{1, 2, \cdots, N_s\}$ *and* $s \in \{r+1, r+2, \cdots, n-1, n\}$)*. Suppose there is an* s*-dimensional edge* $\mathscr{E}^{(s)}_{\sigma_s}(\sigma_r)$ *with a boundary set* $\mathscr{N}_{\alpha\mathscr{B}}(\sigma_r, \sigma_s)$ *relative to domain* $\Omega_\alpha(\sigma_r)$*. The complementary boundary number set is* $\bar{\mathscr{N}}_{\alpha\mathscr{B}}(\sigma_r, \sigma_s) = \mathscr{N}_{\alpha\mathscr{B}}(\sigma_r) / \mathscr{N}_{\alpha\mathscr{B}}(\sigma_r, \sigma_s)$*. For an arbitrarily small* $\varepsilon > 0$*, there are two time intervals* $[t_{m-\varepsilon}, t_m)$ *and/or* $(t_m, t_{m+\varepsilon}]$*. A* *coming edge flow* $\mathbf{x}^{(\sigma_{s(k)};\mathscr{E})}$ *is* $C^{r_{\sigma_{s(k)}}}_{[t_{m-\varepsilon},t_m)}$ *-continuous* ($r_{\sigma_{s(k)}} \geq \max_{i_\alpha \in \bar{\mathscr{N}}_{\alpha\mathscr{B}}(\sigma_r,\sigma_s)}\{m_{i_\alpha}+1\}$) *for time* t *and* $\| d^{r_{\sigma_{s(k)}}+1}\mathbf{x}^{(\sigma_{s(k)};\mathscr{E})}/dt^{r_{\sigma_{s(k)}}+1}\| < \infty$. *A leaving edge flow* $\mathbf{x}^{(\sigma_{s(l)};\mathscr{E})}$ *is* $C^{r_{\sigma_{s(l)}}}_{(t_m,t_{m+\varepsilon}]}$ *-continuous for time* t *and* $\| d^{r_{\sigma_{s(l)}}+1}\mathbf{x}^{(\sigma_{s(k)};\mathscr{E})}/dt^{r_{\sigma_{s(l)}}+1}\| < \infty$ ($r_{\sigma_{s(l)}} \geq \max_{j_\alpha \in \bar{\mathscr{N}}_{\alpha\mathscr{B}}(\sigma_r,\sigma_s)}\{m_{j_\alpha}\}+1$). *If a coming edge flow of* $\mathbf{x}^{(\sigma_{s(k)};\mathscr{E})}$ *arrives to the edge of* $\mathscr{E}^{(r)}_{\sigma_r}$, *there is a switching rule between the two edge flows of* $\mathbf{x}^{(\sigma_{s(k)};\mathscr{E})}$ *and* $\mathbf{x}^{(\sigma_{s(l)};\mathscr{E})}$ *with* $\mathbf{x}^{(\sigma_{s(k)};\mathscr{E})}(t_{m-}) = \mathbf{x}_m = \mathbf{x}^{(\sigma_{s(l)};\mathscr{E})}(t_{m+})$. *The flows of* $\mathbf{x}^{(\sigma_{s(k)};\mathscr{E})}$ *and* $\mathbf{x}^{(\sigma_{s(l)};\mathscr{E})}$ *to the edge* $\mathscr{E}^{(r)}_{\sigma_r}$ *form an* $(\mathbf{m}_{\sigma_{s(k)}}:\mathbf{m}_{\sigma_{s(l)}})$ *-extended, passable flow in the edge* $\mathscr{E}^{(s)}_{\sigma_s}$ ($\mathbf{m}_{\sigma_{s(l)}} \neq 2\mathbf{k}_{\sigma_{s(l)}}+1$) *if and only if*

$$\left.\begin{array}{l}
\hbar_{i_\alpha} G^{(s_{i_\alpha},\sigma_{s(k)};\mathscr{E})}_{\partial\Omega_{\alpha i_\alpha}}(\mathbf{x}_m,t_{m-},\boldsymbol{\pi}_{\sigma_{s(k)}},\boldsymbol{\lambda}_{j_\alpha}) = 0 \\
\text{for } s_{j_\alpha} = 0,1,2,\cdots,m_{i_\alpha}, \\
\hbar_{i_\alpha} G^{(m_{i_\alpha},\sigma_{s(k)};\mathscr{E})}_{\partial\Omega_{\alpha i_\alpha}}(\mathbf{x}_m,t_{m-},\boldsymbol{\pi}_{\sigma_{s(k)}},\boldsymbol{\lambda}_{j_\alpha}) < 0
\end{array}\right\} \tag{8.136}$$
$$\text{for all } i_\alpha \in \bar{\mathscr{N}}_{\alpha\mathscr{B}}(\sigma_r,\sigma_s);$$

$$\left.\begin{array}{l}
\hbar_{j_\alpha} G^{(s_{j_\alpha},\sigma_{s(l)};\mathscr{E})}_{\partial\Omega_{\alpha j_\alpha}}(\mathbf{x}_m,t_{m-},\boldsymbol{\pi}_{\sigma_{s(l)}},\boldsymbol{\lambda}_{j_\alpha}) = 0 \\
\text{for } s_{j_\alpha} = 0,1,2,\cdots,m_{j_\alpha}, \\
\hbar_{j_\alpha} G^{(m_{j_\alpha},\sigma_{s(l)};\mathscr{E})}_{\partial\Omega_{\alpha j_\alpha}}(\mathbf{x}_m,t_{m-},\boldsymbol{\pi}_{\sigma_{s(l)}},\boldsymbol{\lambda}_{j_\alpha}) < 0
\end{array}\right\} \tag{8.137}$$
$$\text{for all } j_\alpha \in \bar{\mathscr{N}}_{\alpha\mathscr{B}}(\sigma_r,\sigma_s).$$

Proof. Consider a flow to all boundaries relative to the boundary. Following the proof of Theorem 3.2, the theorem can be proved. ■

To better understand the concept of extended flows in an edge with multi-vector fields, an extended edge flow $\mathbf{x}^{(\sigma_s,\mathscr{E})}$ in the edge $\mathscr{E}^{(s)}_{\sigma_s}$ to the specific edge $\mathscr{E}^{(s-2)}$ is sketched in Fig. 8.17. The black curve represents the $(s-2)$ -dimensional edge. The edge flows are represented by the thick curves with arrows in the edge of $\mathscr{E}^{(s)}_{\sigma_s}$. The dashed curves are the extensions of the real flows. In Fig. 8.17(a), an extended flow with the $(2\mathbf{k}_{\sigma_{s(k)}}:2\mathbf{k}_{\sigma_{s(l)}})$ -singularity is sketched, and an extended flow with the $(2\mathbf{k}_{\sigma_{s(k)}}:\mathbf{m}_{\sigma_{s(l)}})$ -singularity ($\mathbf{m}_{\sigma_{s(l)}} \neq 2\mathbf{k}_{\sigma_{s(l)}}+1$) is presented in Fig. 8.17(b). If $\mathbf{m}_{\sigma_{s(l)}} = 2\mathbf{k}_{\sigma_{s(l)}}+1$, an extended flow in the edge cannot be formed.

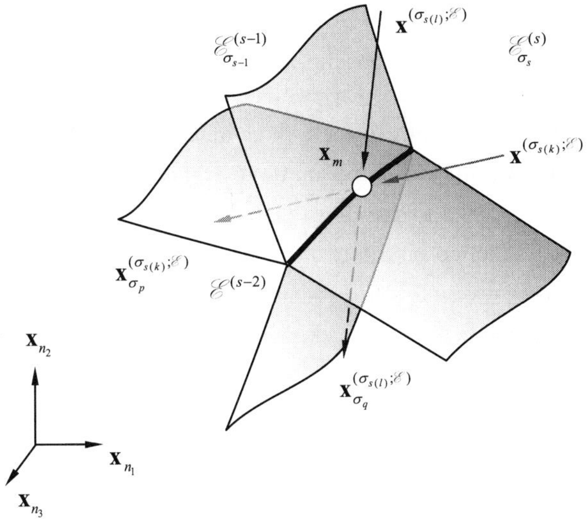

(a)

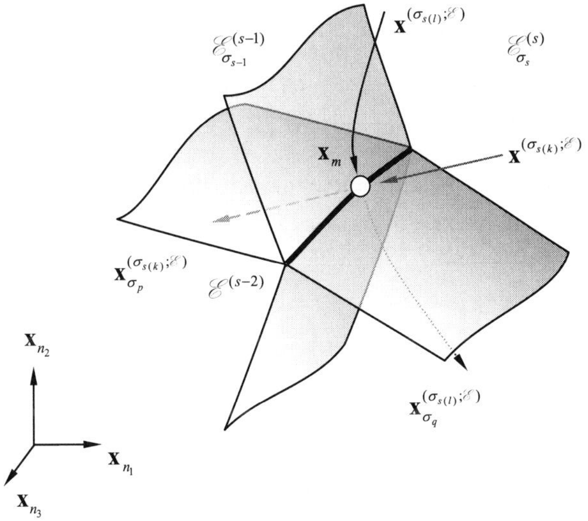

(b)

Fig. 8.17 Extended edge flows to the specific edge. **(a)** the $(2\mathbf{k}_{\sigma_{s(k)}} : 2\mathbf{k}_{\sigma_{s(l)}})$ -singularity, **(b)** the $(2\mathbf{k}_{\sigma_{s(k)}} : \mathbf{m}_{\sigma_{s(l)}})$ -singularity ($\mathbf{m}_{\sigma_{s(l)}} \neq 2\mathbf{k}_{\sigma_{s(l)}} + 1$), The black curve represents the $(s-2)$ -dimensional edge. The edge flows are represented by the thick curves with arrows in the edge of $\mathscr{E}^{(s)}_{\sigma_s}$. The dashed curves are the extensions of the real flows. ($n_1 + n_2 + n_3 = s$)

8.5. A frictional oscillator with two-degrees of freedom

As in Luo and Mao (2010), consider a two-degree of freedom, friction-induced oscillator as shown in Fig. 8.18. The system consists of two masses ($m_\alpha, \alpha = 1,2$), which are connected with two linear springs of k_α ($\alpha = 1,2$), and two dampers of d_α ($\alpha = 1,2$). Both of masses move on the two individual belts with constant velocity V_α ($\alpha = 1,2$). Two harmonic excitations with frequency Ω and amplitude Q_α ($\alpha = 1,2$) are exerted on the two masses, respectively. Since the masses are moving on the belt traveling with constant V_α ($\alpha = 1,2$), the kinetic friction exists between the mass and belt. Thus, the kinetic friction force in Fig. 8.19 is given by

$$F_f^{(\alpha)}(\dot{x}_\alpha)\begin{cases} = \mu_k F_N^{(\alpha)}, & \dot{x}_\alpha \in (V_\alpha, \infty) \\ \in [-\mu_k F_N^{(\alpha)}, \mu_k F_N^{(\alpha)}], & \dot{x}_\alpha = V_\alpha \\ = -\mu_k F_N^{(\alpha)}, & \dot{x}_\alpha \in (-\infty, V_\alpha) \end{cases} \tag{8.138}$$

where $\dot{x}_\alpha = dx_\alpha / dt$. μ_k and $F_N^{(\alpha)}$ ($\alpha = 1,2$) are a friction coefficient and a normal force to contact surface, respectively. For this case, $F_N^{(\alpha)} = m_\alpha g$ and g is the gravitational acceleration. Non-friction forces on two masses in the x-direction are

$$\begin{aligned} F_s^{(1)} &= Q_1 \cos\Omega t - k_1 x_1 - d_1 \dot{x}_1 - k_2(x_1 - x_2) - d_2(\dot{x}_1 - \dot{x}_2), \\ F_s^{(2)} &= Q_2 \cos\Omega t - k_2(x_1 - x_2) - d_2(\dot{x}_2 - \dot{x}_1). \end{aligned} \tag{8.139}$$

If the mass m_α sticks with the traveling belt, one obtains $\dot{x}_\alpha = V_\alpha$ ($\alpha \in \{1,2\}$). Before the non-friction force overcomes the friction force on the corresponding mass (i.e., $|F_s^{(\alpha)}| \le F_f^{(\alpha)}$ and $F_f^{(\alpha)} = \mu_k F_N^{(\alpha)}$), the two masses do not have any relative motion to the belt. Because of $V_\alpha = \text{const}$ for stick, the corresponding mass does not have any acceleration, i.e.,

$$\ddot{x}_\alpha = 0 \quad \text{for } \dot{x}_\alpha = V_\alpha \text{ and } \alpha \in \{1,2\}. \tag{8.140}$$

If $|F_s^{(\alpha)}| > F_f^{(\alpha)}$, the non-friction force overcomes the friction force and non-stick motion will appear. For the non-stick motion of mass m_α ($\alpha \in \{1,2\}$), the total forces acting on each mass are

$$\begin{aligned} F^{(1)} &= Q_1 \cos\Omega t - k_1 x_1 - d_1 \dot{x}_1 - k_2(x_1 - x_2) \\ &\quad - d_2(\dot{x}_1 - \dot{x}_2) - F_f^{(1)} \operatorname{sgn}(\dot{x}_1 - V_1) \quad \text{for } \dot{x}_1 \ne V_1, \\ F^{(2)} &= Q_2 \cos\Omega t - k_2(x_2 - x_1) \\ &\quad - d_2(\dot{x}_2 - \dot{x}_1) - F_f^{(2)} \operatorname{sgn}(\dot{x}_2 - V_2) \quad \text{for } \dot{x}_2 \ne V_2. \end{aligned} \tag{8.141}$$

From the above discussion, there are four cases of motions.

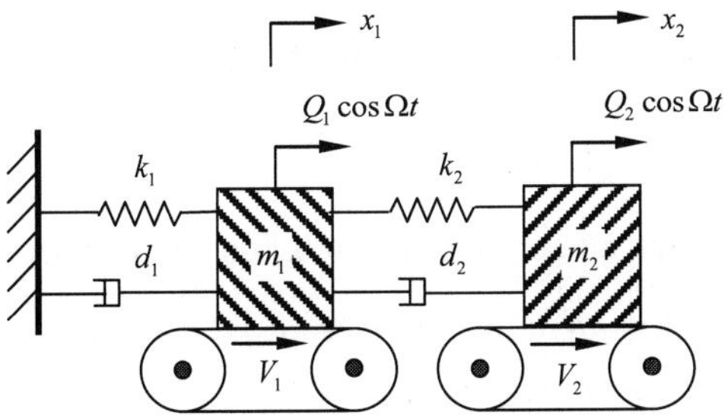

Fig. 8.18 Mechanical model.

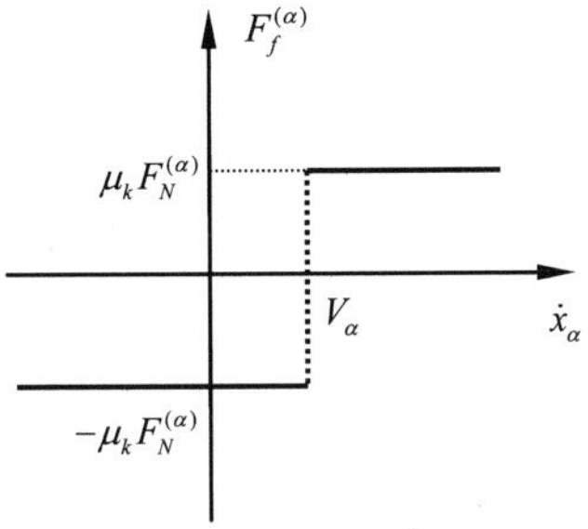

Fig. 8.19 Friction forces.

Case I. non-stick motion ($\dot{x}_\alpha \neq V_\alpha, \alpha = 1,2$)

The equations of nonstick motion for the two oscillators with friction are

$$
\begin{aligned}
& m_1\ddot{x}_1 + d_1\dot{x}_1 + d_2(\dot{x}_1 - \dot{x}_2) + k_1x_1 + k_2(x_1 - x_2) \\
& \quad = Q_1 \cos\Omega t - F_f^{(1)} \operatorname{sgn}(\dot{x}_1 - V_1), \\
& m_2\ddot{x}_2 + d_2(\dot{x}_2 - \dot{x}_1) + k_2(x_2 - x_1) \\
& \quad = Q_2 \cos\Omega t - F_f^{(2)} \operatorname{sgn}(\dot{x}_2 - V_2).
\end{aligned}
\tag{8.142}
$$

Case II. stick motion ($\dot{x}_1 = V_1$ and $\dot{x}_2 \neq V_2$)

The equations of nonstick motion for the two oscillators with friction are

$$
\begin{aligned}
& \dot{x}_1 = V_1, \\
& m_2\ddot{x}_2 + d_2(\dot{x}_2 - V_1) + k_2(x_2 - x_1) \\
& \quad = Q_2 \cos\Omega t - F_f^{(2)} \operatorname{sgn}(\dot{x}_2 - V_2),
\end{aligned}
\tag{8.143}
$$

with

$$\left|Q_1 \cos\Omega t - d_1 V_1 - d_2(V_1 - \dot{x}_2) - k_1 x_1 - k_2(x_1 - x_2)\right| \le F_f^{(1)}. \tag{8.144}$$

Case III. stick motion ($\dot{x}_1 \neq V_1$ and $\dot{x}_2 = V_2$)

The equations of nonstick motion for the two oscillators with friction are

$$\begin{aligned} &m_1\ddot{x}_1 + d_1\dot{x}_1 + d_2(\dot{x}_1 - V_2) + k_1 x_1 + k_2(x_1 - x_2) \\ &\quad = Q_1 \cos\Omega t - F_f^{(1)} \operatorname{sgn}(\dot{x}_1 - V_1), \\ &\dot{x}_2 = V_2; \end{aligned} \tag{8.145}$$

with

$$\left|Q_2 \cos\Omega t - d_2(V_2 - \dot{x}_1) - k_2(x_2 - x_1)\right| \le F_f^{(2)}. \tag{8.146}$$

Case IV. double-stick motion ($\dot{x}_1 = V_1$ and $\dot{x}_2 = V_2$)

The equations of nonstick motion for the two oscillators with friction are

$$\dot{x}_1 = V_1 \text{ and } \dot{x}_2 = V_2; \tag{8.147}$$

with

$$\begin{aligned} &\left|Q_1 \cos\Omega t - d_1 V_1 - d_2(V_1 - V_2) - k_1 x_1 - k_2(x_1 - x_2)\right| \le F_f^{(1)}, \\ &\left|Q_2 \cos\Omega t - d_2(V_2 - V_1) - k_2(x_2 - x_1)\right| \le F_f^{(2)}. \end{aligned} \tag{8.148}$$

8.5.1. Domains, edges and vector fields

Since the two friction forces exist on the two masses of the friction oscillator, the phase space of the system is divided into four 4-dimensional domains and four 3-dimensional boundaries with a 2-dimensional edge. The state variables and vector fields are introduced by

$$\begin{aligned} &\mathbf{x} = (x_1, \dot{x}_1, x_2, \dot{x}_2)^{\mathrm{T}} = (x_1, y_1, x_2, y_2)^{\mathrm{T}}, \\ &\mathbf{F} = (y_1, F_1, y_2, F_2)^{\mathrm{T}}. \end{aligned} \tag{8.149}$$

From the state variables, the four domains are defined as

$$\begin{aligned} \Omega_1 &= \left\{(x_1, y_1, x_2, y_2) \middle| y_1 \in (V_1, +\infty) \text{ and } y_2 \in (V_2, +\infty)\right\}, \\ \Omega_2 &= \left\{(x_1, y_1, x_2, y_2) \middle| y_1 \in (V_1, +\infty) \text{ and } y_2 \in (-\infty, V_2)\right\}, \\ \Omega_3 &= \left\{(x_1, y_1, x_2, y_2) \middle| y_1 \in (-\infty, V_1) \text{ and } y_2 \in (-\infty, V_2)\right\}, \\ \Omega_4 &= \left\{(x_1, y_1, x_2, y_2) \middle| y_1 \in (-\infty, V_1) \text{ and } y_2 \in (V_2, +\infty)\right\}, \end{aligned} \tag{8.150}$$

and the corresponding 3-dimensional boundaries are defined by $\partial\Omega_{\alpha_1\alpha_2} = \bar{\Omega}_{\alpha_1} \cap \bar{\Omega}_{\alpha_2}$ for ($\alpha_i \in \{1,2,3,4\}$, $i=1,2$; $\alpha_1 \neq \alpha_2$ without repeating), i.e.,

$$\left.\begin{aligned}
\partial\Omega_{12} &= \partial\Omega_{21} = \left\{(x_1,y_1,x_2,y_2)\middle|\varphi_{12} = y_2 - V_2 = 0, y_1 \geq V_1\right\},\\
\partial\Omega_{23} &= \partial\Omega_{32} = \left\{(x_1,y_1,x_2,y_2)\middle|\varphi_{24} = y_1 - V_1 = 0, y_2 \leq V_2\right\},\\
\partial\Omega_{34} &= \partial\Omega_{43} = \left\{(x_1,y_1,x_2,y_2)\middle|\varphi_{34} = y_2 - V_2 = 0, y_1 \leq V_1\right\},\\
\partial\Omega_{41} &= \partial\Omega_{14} = \left\{(x_1,y_1,x_2,y_2)\middle|\varphi_{41} = y_1 - V_1 = 0, y_2 \geq V_2\right\}.
\end{aligned}\right\} \tag{8.151}$$

The subscripts of the boundary $\partial\Omega_{\alpha_1\alpha_2}$ present the boundary between domains Ω_{α_1} and Ω_{α_2}. Finally, the 2-dimensional edge of the 3-diemsional boundaries is defined by

$$\angle\Omega_{\alpha_1\alpha_2\alpha_3} = \partial\Omega_{\alpha_1\alpha_2} \cap \partial\Omega_{\alpha_2\alpha_3} = \cap_{i=1}^{3}\Omega_{\alpha_i}. \tag{8.152}$$

for ($\alpha_i \in \{1,2,3,4\}$, $i=1,2,3$; $\alpha_1 \neq \alpha_2 \neq \alpha_3$ without repeating) and the union of four 2-dimeniaonal edges is

$$\begin{aligned}
\measuredangle\Omega_{1234} &= \cup\angle\Omega_{\alpha_1\alpha_2\alpha_3}\\
&= \left\{(x_1,y_1,x_2,y_2)\middle|\begin{aligned}\varphi_{12} = \varphi_{34} = y_2 - V_2 = 0\\ \varphi_{23} = \varphi_{41} = y_1 - V_1 = 0\end{aligned}\right\},
\end{aligned} \tag{8.153}$$

For illustration of the domains, the velocity plane is used for illustration and the four domains, boundary and vertex are illustrated in Fig. 8.20.

From the previous definitions, the equations of motion for the 2-DOF friction-induced oscillator are described by

$$\left.\begin{aligned}
&\dot{\mathbf{x}}^{(\alpha)} = \mathbf{F}^{(\alpha)}(\mathbf{x}^{(\alpha)},t,\mathbf{p}_\alpha) \text{ on } \Omega_\alpha,\\
&\dot{\mathbf{x}}^{(\alpha_1\alpha_2)} = \mathbf{F}^{(\alpha_1\alpha_2)}(\mathbf{x}^{(\alpha_1\alpha_2)},t,\mathbf{p}_{\alpha_1\alpha_2}) \text{ on } \partial\Omega_{\alpha_1\alpha_2},\\
&\dot{\mathbf{x}}^{(\alpha_1\alpha_2\alpha_3)} = \mathbf{F}^{(\alpha_1\alpha_2\alpha_3)}(\mathbf{x}^{(\alpha_1\alpha_2\alpha_3)},t,\mathbf{p}_{\alpha_1\alpha_2\alpha_3}) \text{ on } \angle\Omega_{\alpha_1\alpha_2\alpha_3};
\end{aligned}\right\} \tag{8.154}$$

and

$$\begin{aligned}
&\mathbf{x}^{(\alpha)} = \mathbf{x}^{(\alpha_1\alpha_2)} = \mathbf{x}^{(\alpha_1\alpha_2\alpha_3)} = (x_1,y_1,x_2,y_2)^{\mathrm{T}},\\
&\mathbf{F}^{(\alpha)} = (y_1,\frac{1}{m_1}F_1^{(\alpha)},y_2,\frac{1}{m_2}F_2^{(\alpha)})^{\mathrm{T}},\\
&\mathbf{F}^{(\alpha_1\alpha_2)} = (y_1,\frac{1}{m_1}F_1^{(\alpha_1\alpha_2)},y_2,\frac{1}{m_2}F_2^{(\alpha_1\alpha_2)})^{\mathrm{T}},\\
&\mathbf{F}^{(\alpha_1\alpha_2\alpha_3)} = (y_1,\frac{1}{m_1}F_1^{(\alpha_1\alpha_2\alpha_3)},y_2,\frac{1}{m_2}F_2^{(\alpha_1\alpha_2\alpha_3)})^{\mathrm{T}},
\end{aligned} \tag{8.155}$$

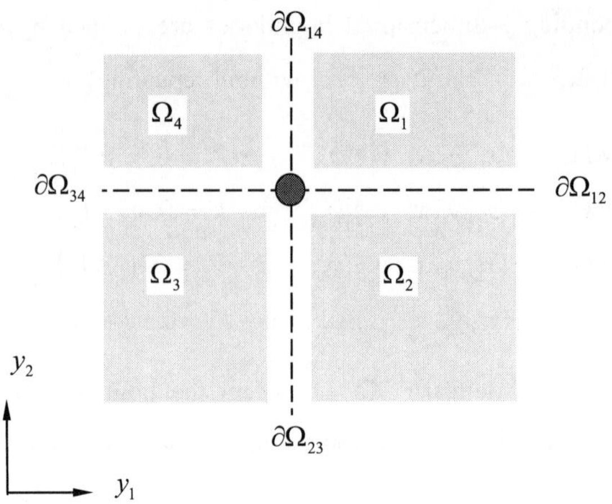

Fig. 8.20 Domain partition in velocity plane of 2-DOF oscillators with dry friction.

where the forces for the 2-DOF friction-induced oscillator in the domains Ω_α ($\alpha = 1,2,3,4$) are

$$\begin{aligned}
F_1^{(1)} &= F_1^{(2)} \\
&= Q_1 \cos\Omega t - F_f^{(1)} - d_1 y_1 - d_2(y_1 - y_2) - k_1 x_1 - k_2(x_1 - x_2), \\
F_1^{(3)} &= F_1^{(4)} \\
&= Q_1 \cos\Omega t + F_f^{(1)} - d_1 y_1 - d_2(y_1 - y_2) - k_1 x_1 - k_2(x_1 - x_2); \\
F_2^{(1)} &= F_2^{(4)} \\
&= Q_2 \cos\Omega t - F_f^{(2)} - d_2(y_2 - y_1) - k_2(x_2 - x_1), \\
F_2^{(2)} &= F_2^{(3)} \\
&= Q_2 \cos\Omega t + F_f^{(2)} - d_2(y_2 - y_1) - k_2(x_2 - x_1).
\end{aligned} \tag{8.156}$$

The forces for the 2-DOF friction-induced oscillator on the boundaries $\partial\Omega_{\alpha_1\alpha_2}$ for ($\alpha_i \in \{1,2,3,4\}$, $i = 1,2$; $\alpha_1 \neq \alpha_2$ without repeating) are

$$\left.\begin{aligned}
&F_1^{(12)} \equiv Q_1 \cos\Omega t - F_f^{(1)} - d_1 y_1 - d_2(y_1 - y_2) - k_1 x_1 - k_2(x_1 - x_2), \\
&F_2^{(12)} = 0 \ \text{ for stick on } \partial\Omega_{12}, \\
&F_2^{(12)} \in [F_2^{(1)}, F_2^{(2)}] \text{ for non-stick on } \partial\Omega_{12};
\end{aligned}\right\}$$

$$\left.\begin{aligned}
&F_1^{(23)} = 0 \ \text{ for stick on } \partial\Omega_{23}, \\
&F_1^{(23)} \in [F_1^{(2)}, F_1^{(3)}] \text{ for non-stick on } \partial\Omega_{23}, \\
&F_2^{(23)} = Q_2 \cos\Omega t + F_f^{(2)} - d_2(y_2 - y_1) - k_2(x_2 - x_1);
\end{aligned}\right\}$$

$$\left.\begin{aligned} &F_1^{(34)} = Q_1 \cos\Omega t + F_f^{(1)} - d_1 y_1 - d_2(y_1 - y_2) - k_1 x_1 - k_2(x_1 - x_2), \\ &F_2^{(34)} = 0 \text{ for stick on } \partial\Omega_{34}, \\ &F_2^{(34)} \in [F_2^{(4)}, F_2^{(3)}] \text{ for non-stick on } \partial\Omega_{34}; \end{aligned}\right\}$$
$$\left.\begin{aligned} &F_1^{(41)} = Q_2 \cos\Omega t - F_f^{(2)} - d_2(y_2 - y_1) - k_2(x_2 - x_1); \\ &F_2^{(41)} = 0 \text{ for stick on } \partial\Omega_{41}, \\ &F_2^{(41)} \in [F_2^{(1)}, F_2^{(4)}] \text{ for non-stick on } \partial\Omega_{41}. \end{aligned}\right\} \quad (8.157)$$

The forces for the 2-DOF friction-induced oscillator on the 2-D edges $\angle\Omega_{\alpha_1\alpha_2\alpha_3}$ for ($\alpha_i \in \{1,2,3,4\}$, $i = 1,2$; $\alpha_1 \neq \alpha_2 \neq \alpha_3$ without repeating) are

$$\begin{aligned} &(F_1^{(\alpha_1\alpha_2\alpha_3)}, F_2^{(\alpha_1\alpha_2\alpha_3)}) \in (F_1^{(\alpha_1\alpha_2)}, F_2^{(\alpha_2\alpha_3)}) \text{ on } \angle\Omega_{\alpha_1\alpha_2\alpha_3}, \\ &(F_1^{(\alpha_1\alpha_2\alpha_3)}, F_2^{(\alpha_1\alpha_2\alpha_3)}) = (0,0) \text{ for full stick on } \angle\Omega_{\alpha_1\alpha_2\alpha_3}. \end{aligned} \quad (8.158)$$

In other words,

$$\begin{aligned} &(F_1^{(123)}, F_2^{(123)}) \in (F_1^{(12)}, F_2^{(23)}) \text{ on } \angle\Omega_{123}, \\ &(F_1^{(123)}, F_2^{(123)}) = (0,0) \text{ for full stick on } \angle\Omega_{123}; \\ &(F_1^{(234)}, F_2^{(234)}) \in (F_1^{(23)}, F_2^{(34)}) \text{ on } \angle\Omega_{234}, \\ &(F_1^{(234)}, F_2^{(234)}) = (0,0) \text{ for full stick on } \angle\Omega_{234}; \\ &(F_1^{(341)}, F_2^{(341)}) \in (F_1^{(34)}, F_2^{(41)}) \text{ on } \angle\Omega_{341}, \\ &(F_1^{(341)}, F_2^{(341)}) = (0,0) \text{ for full stick on } \angle\Omega_{341}; \\ &(F_1^{(412)}, F_2^{(412)}) \in (F_1^{(41)}, F_2^{(12)}) \text{ on } \angle\Omega_{412}, \\ &(F_1^{(412)}, F_2^{(412)}) = (0,0) \text{ for full stick on } \angle\Omega_{412}. \end{aligned} \quad (8.159)$$

From the definition, at the 2-dimeniaonal edges, there are four possible states. (i) a passable motion to the edge, (ii) two passable-sliding motions to the edge, (iii) a full stick motion to the edge.

8.5.2. Analytical conditions

Before presenting the analytical conditions, the G-functions are introduced as

$$\begin{aligned} G^{(0,\alpha_1)}(t_{m\pm}) &= \mathbf{n}_{\partial\Omega_{\alpha_1\alpha_2}}^{\mathrm{T}} \cdot \mathbf{F}^{(\alpha_1)}(t_{m\pm}), \\ G^{(1,\alpha_1)}(t_{m\pm}) &= 2D\mathbf{n}_{\partial\Omega_{\alpha_1\alpha_2}}^{\mathrm{T}} \cdot [\mathbf{F}^{(\alpha_1)}(t_{m\pm}) - \mathbf{F}^{(\alpha_1\alpha_2)}(t_{m\pm})] \\ &\quad + \mathbf{n}_{\partial\Omega_{\alpha\beta}}^{\mathrm{T}} \cdot [D\mathbf{F}^{(\alpha)}(t_{m\pm}) - D\mathbf{F}^{(\alpha_1\alpha_2)}(t_{m\pm})] \end{aligned} \quad (8.160)$$

where $D(\cdot)=\Sigma_{i=1}^{2}\dot{x}_i\partial(\cdot)/\partial x_i+\dot{y}_i\partial(\cdot)/\partial y_i+\partial(\cdot)/\partial t$. If the boundary $\partial\Omega_{\alpha_1\alpha_2}$ is a n-dimensional plane independent of time t, one obtains $D\mathbf{n}_{\partial\Omega_{\alpha_1\alpha_2}}=0$. Form $\mathbf{n}^{\mathrm{T}}_{\partial\Omega_{\alpha_1\alpha_2}}\cdot\mathbf{F}^{(\alpha_1\alpha_2)}=0$, $D\mathbf{n}^{\mathrm{T}}_{\partial\Omega_{\alpha_1\alpha_2}}\cdot\mathbf{F}^{(\alpha_1\alpha_2)}+\mathbf{n}^{\mathrm{T}}_{\partial\Omega_{\alpha_1\alpha_2}}\cdot D\mathbf{F}^{(\alpha_1\alpha_2)}=0$ can be obtained. Thus, $\mathbf{n}^{\mathrm{T}}_{\partial\Omega_{\alpha_1\alpha_2}}\cdot D\mathbf{F}^{(\alpha_1\alpha_2)}=0$. Equation (8.160) reduces to

$$G^{(1,\alpha)}(t_{m\pm})=\mathbf{n}^{\mathrm{T}}_{\partial\Omega_{\alpha_1\alpha_2}}\cdot D\mathbf{F}^{(\alpha_1)}(t_{m\pm}). \tag{8.161}$$

For a general case, Equation (8.160) instead of Eq. (8.161) will be used. From the theory of discontinuous dynamical systems in Luo (2008, 2009), the non-stick motion (or called the passable motion) is guaranteed for a flow moving from domain Ω_{α_1} to domain Ω_{α_2} by

$$\left.\begin{aligned}G^{(0,\alpha_1)}(t_{m-})=\mathbf{n}^{\mathrm{T}}_{\partial\Omega_{\alpha_1\alpha_2}}\cdot\mathbf{F}^{(\alpha_1)}(t_{m-})<0\\ G^{(0,\alpha_2)}(t_{m+})=\mathbf{n}^{\mathrm{T}}_{\partial\Omega_{\alpha_1\alpha_2}}\cdot\mathbf{F}^{(\alpha_2)}(t_{m+})<0\end{aligned}\right\}\text{for } \mathbf{n}_{\partial\Omega_{\alpha_1\alpha_2}}\to\Omega_{\alpha_1};$$
$$\left.\begin{aligned}G^{(0,\alpha_1)}(t_{m-})=\mathbf{n}^{\mathrm{T}}_{\partial\Omega_{\alpha_1\alpha_2}}\cdot\mathbf{F}^{(\alpha_1)}(t_{m-})>0\\ G^{(0,\alpha_2)}(t_{m+})=\mathbf{n}^{\mathrm{T}}_{\partial\Omega_{\alpha_1\alpha_2}}\cdot\mathbf{F}^{(\alpha_2)}(t_{m+})>0\end{aligned}\right\}\text{for } \mathbf{n}_{\partial\Omega_{\alpha_1\alpha_2}}\to\Omega_{\alpha_2} \tag{8.162}$$

($\alpha_i\in\{1,2,3,4\}$, $i=1,2$; $\alpha_1\neq\alpha_2$ without repeating) with

$$\mathbf{n}_{\partial\Omega_{\alpha_1\alpha_2}}=\left.\left(\frac{\partial\varphi_{\alpha_1\alpha_2}}{\partial x_1},\frac{\partial\varphi_{\alpha_1\alpha_2}}{\partial y_1},\frac{\partial\varphi_{\alpha_1\alpha_2}}{\partial x_2},\frac{\partial\varphi_{\alpha_1\alpha_2}}{\partial y_2}\right)^{\mathrm{T}}\right|_{(x_{1m},y_{1m},x_{2m},y_{2m})}. \tag{8.163}$$

The time t_m represents the time for the motion on the velocity boundary, and $t_{m\pm}=t_m\pm 0$ reflects the responses on the regions rather than boundary.

In Luo (2008, 2009), the stick motion in physics (or called the sliding motion in mathematics) to the boundary $\partial\Omega_{\alpha_1\alpha_2}$ is guaranteed by

$$\left.\begin{aligned}G^{(0,\alpha_1)}(t_{m-})=\mathbf{n}^{\mathrm{T}}_{\partial\Omega_{\alpha_1\alpha_2}}\cdot\mathbf{F}^{(\alpha_1)}(t_{m-})<0\\ G^{(0,\alpha_2)}(t_{m-})=\mathbf{n}^{\mathrm{T}}_{\partial\Omega_{\alpha_1\alpha_2}}\cdot\mathbf{F}^{(\alpha_2)}(t_{m-})>0\end{aligned}\right\}\text{for } \mathbf{n}_{\partial\Omega_{\alpha_1\alpha_2}}\to\Omega_{\alpha_1};$$
$$\left.\begin{aligned}G^{(0,\alpha_1)}(t_{m-})=\mathbf{n}^{\mathrm{T}}_{\partial\Omega_{\alpha_1\alpha_2}}\cdot\mathbf{F}^{(\alpha_1)}(t_{m-})>0\\ G^{(0,\alpha_2)}(t_{m-})=\mathbf{n}^{\mathrm{T}}_{\partial\Omega_{\alpha_1\alpha_2}}\cdot\mathbf{F}^{(\alpha_2)}(t_{m-})<0\end{aligned}\right\}\text{for } \mathbf{n}_{\partial\Omega_{\alpha_1\alpha_2}}\to\Omega_{\alpha_2}. \tag{8.164}$$

The analytical conditions for the onset of the stick motion are

$$\left.\begin{aligned}G^{(0,\alpha_1)}(t_{m-})=\mathbf{n}^{\mathrm{T}}_{\partial\Omega_{\alpha_1\alpha_2}}\cdot\mathbf{F}^{(\alpha_1)}(t_{m-})<0\\ G^{(0,\alpha_2)}(t_{m+})=\mathbf{n}^{\mathrm{T}}_{\partial\Omega_{\alpha_1\alpha_2}}\cdot\mathbf{F}^{(\alpha_2)}(t_{m+})=0\\ G^{(1,\alpha_2)}(t_{m-})=\mathbf{n}^{\mathrm{T}}_{\partial\Omega_{\alpha_1\alpha_2}}\cdot D\mathbf{F}^{(\alpha_2)}(t_{m\pm})<0\end{aligned}\right\}\text{for } \mathbf{n}_{\partial\Omega_{\alpha_1\alpha_2}}\to\Omega_{\alpha_1};$$

$$\left.\begin{array}{l} G^{(0,\alpha_1)}(t_{m-}) = \mathbf{n}^{\mathrm{T}}_{\partial\Omega_{\alpha_1\alpha_2}} \cdot \mathbf{F}^{(\alpha_1)}(t_{m-}) > 0 \\ G^{(0,\alpha_2)}(t_{m-}) = \mathbf{n}^{\mathrm{T}}_{\partial\Omega_{\alpha_1\alpha_2}} \cdot \mathbf{F}^{(\alpha_2)}(t_{m+}) = 0 \\ G^{(1,\alpha_2)}(t_{m-}) = \mathbf{n}^{\mathrm{T}}_{\partial\Omega_{\alpha_1\alpha_2}} \cdot D\mathbf{F}^{(\alpha_2)}(t_{m\pm}) > 0 \end{array}\right\} \text{for } \mathbf{n}_{\partial\Omega_{\alpha_1\alpha_2}} \to \Omega_{\alpha_2}. \tag{8.165}$$

In Luo (2009), the analytical conditions for vanishing of the stick motion and entering domain Ω_{α_2} are

$$\begin{array}{l} \left.\begin{array}{l} G^{(0,\alpha_1)}(t_{m-}) = \mathbf{n}^{\mathrm{T}}_{\partial\Omega_{\alpha_1\alpha_2}} \cdot \mathbf{F}^{(\alpha_1)}(t_{m-}) < 0 \\ G^{(0,\alpha_2)}(t_{m\mp}) = \mathbf{n}^{\mathrm{T}}_{\partial\Omega_{\alpha_1\alpha_2}} \cdot \mathbf{F}^{(\alpha_2)}(t_{m\mp}) = 0 \\ G^{(1,\alpha_2)}(t_{m\mp}) = \mathbf{n}^{\mathrm{T}}_{\partial\Omega_{\alpha_1\alpha_2}} \cdot D\mathbf{F}^{(\alpha_2)}(t_{m\mp}) < 0 \end{array}\right\} \text{for } \mathbf{n}_{\partial\Omega_{\alpha_1\alpha_2}} \to \Omega_{\alpha_1}; \\ \left.\begin{array}{l} G^{(0,\alpha_1)}(t_{m-}) = \mathbf{n}^{\mathrm{T}}_{\partial\Omega_{\alpha_1\alpha_2}} \cdot \mathbf{F}^{(\alpha_1)}(t_{m-}) > 0 \\ G^{(0,\alpha_2)}(t_{m\mp}) = \mathbf{n}^{\mathrm{T}}_{\partial\Omega_{\alpha_1\alpha_2}} \cdot \mathbf{F}^{(\alpha_2)}(t_{m\mp}) = 0 \\ G^{(1,\alpha_2)}(t_{m\mp}) = \mathbf{n}^{\mathrm{T}}_{\partial\Omega_{\alpha_1\alpha_2}} \cdot D\mathbf{F}^{(\alpha_2)}(t_{m\mp}) > 0 \end{array}\right\} \text{for } \mathbf{n}_{\partial\Omega_{\alpha_1\alpha_2}} \to \Omega_{\alpha_2} \end{array} \tag{8.166}$$

and for vanishing of the stick motion and entering domain Ω_{α_1}

$$\begin{array}{l} \left.\begin{array}{l} G^{(0,\alpha_1)}(t_{m-}) = \mathbf{n}^{\mathrm{T}}_{\partial\Omega_{\alpha_1\alpha_2}} \cdot \mathbf{F}^{(\alpha_1)}(t_{m-}) = 0 \\ G^{(0,\alpha_2)}(t_{m\mp}) = \mathbf{n}^{\mathrm{T}}_{\partial\Omega_{\alpha_1\alpha_2}} \cdot \mathbf{F}^{(\alpha_2)}(t_{m\mp}) > 0 \\ G^{(1,\alpha_1)}(t_{m\mp}) = \mathbf{n}^{\mathrm{T}}_{\partial\Omega_{\alpha_1\alpha_2}} \cdot D\mathbf{F}^{(\alpha_1)}(t_{m\mp}) > 0 \end{array}\right\} \text{for } \mathbf{n}_{\partial\Omega_{\alpha_1\alpha_2}} \to \Omega_{\alpha_1}; \\ \left.\begin{array}{l} G^{(0,\alpha_1)}(t_{m-}) = \mathbf{n}^{\mathrm{T}}_{\partial\Omega_{\alpha_1\alpha_2}} \cdot \mathbf{F}^{(\alpha_1)}(t_{m-}) = 0 \\ G^{(0,\alpha_2)}(t_{m\mp}) = \mathbf{n}^{\mathrm{T}}_{\partial\Omega_{\alpha_1\alpha_2}} \cdot \mathbf{F}^{(\alpha_2)}(t_{m\mp}) < 0 \\ G^{(1,\alpha_1)}(t_{m\mp}) = \mathbf{n}^{\mathrm{T}}_{\partial\Omega_{\alpha_1\alpha_2}} \cdot D\mathbf{F}^{(\alpha_1)}(t_{m\mp}) < 0 \end{array}\right\} \text{for } \mathbf{n}_{\partial\Omega_{\alpha_1\alpha_2}} \to \Omega_{\alpha_2}. \end{array} \tag{8.167}$$

The conditions for grazing motion in domain Ω_{α_1} to the boundary $\partial\Omega_{\alpha_1\alpha_2}$ are

$$\begin{array}{l} \left.\begin{array}{l} G^{(0,\alpha_1)}(t_{m\pm}) = \mathbf{n}^{\mathrm{T}}_{\partial\Omega_{\alpha_1\alpha_2}} \cdot \mathbf{F}^{(\alpha_1)}(t_{m\pm}) = 0 \\ G^{(1,\alpha_1)}(t_{m\pm}) = \mathbf{n}^{\mathrm{T}}_{\partial\Omega_{\alpha_1\alpha_2}} \cdot D\mathbf{F}^{(\alpha_1)}(t_{m\pm}) > 0 \end{array}\right\} \text{for } \mathbf{n}_{\partial\Omega_{\alpha_1\alpha_2}} \to \Omega_{\alpha_1}; \\ \left.\begin{array}{l} G^{(0,\alpha_1)}(t_{m\pm}) = \mathbf{n}^{\mathrm{T}}_{\partial\Omega_{\alpha_1\alpha_2}} \cdot \mathbf{F}^{(\alpha_1)}(t_{m\pm}) = 0 \\ G^{(1,\alpha_1)}(t_{m\pm}) = \mathbf{n}^{\mathrm{T}}_{\partial\Omega_{\alpha_1\alpha_2}} \cdot D\mathbf{F}^{(\alpha_1)}(t_{m\pm}) < 0 \end{array}\right\} \text{for } \mathbf{n}_{\partial\Omega_{\alpha_1\alpha_2}} \to \Omega_{\alpha_2}. \end{array} \tag{8.168}$$

For the 2-dimensional edge $\angle\Omega_{\alpha_1\alpha_2\alpha_3}$ for ($\alpha_i \in \{1,2,3,4\}$, $i=1,2;$ $\alpha_1 \neq \alpha_2 \neq \alpha_3$ without repeating) are

$$\angle\Omega_{\alpha_1\alpha_2\alpha_3} = \partial\Omega_{\alpha_1\alpha_2} \cap \partial\Omega_{\alpha_2\alpha_3}. \tag{8.169}$$

The corresponding conditions for such 2-dimensional edge can be determined by the summation of the conditions of a flow to the two boundaries $\partial\Omega_{\alpha_1\alpha_2}$ and $\partial\Omega_{\alpha_2\alpha_3}$. In other words, there are four states. (i) non-stick motions for both of boundaries, (ii) stick on one-boundary and non-stick motion for another boundary (two cases), and (iii) stick motions on both of the boundaries. Three critical cases with grazing motion on the 2-dimensional edge include two single grazing motions to the boundary plus one double-grazing motion to the two boundaries. For a better understanding of the afore-presented conditions, the non-sliding flow at the boundary and the sliding flow on the boundary are sketched in Figs. 8.21 and 8.22, respectively. The *G*-functions in Eq. (8.162) and (8.164) are used to show the conditions. The vanishing conditions for the sliding flow on the boundary are also depicted in Fig. 8.21. Similarly, the onset conditions for the sliding motion on the boundary can be illustrated, and the conditions for the grazing flow to the boundary can be sketched as well.

Using Eq. (8.153), Eq. (8.163) gives

$$\mathbf{n}_{\partial\Omega_{23}} = \mathbf{n}_{\partial\Omega_{14}} = (0,1,0,0)^{\mathrm{T}} \text{ and } \mathbf{n}_{\partial\Omega_{12}} = \mathbf{n}_{\partial\Omega_{34}} = (0,0,0,1)^{\mathrm{T}}. \tag{8.170}$$

Thus,

$$\left.\begin{aligned}
&\mathbf{n}^{\mathrm{T}}_{\partial\Omega_{12}} \cdot \mathbf{F}^{(\alpha)}(t) = F_2^{(\alpha)} \text{ for } \alpha=1,2\\
&\mathbf{n}^{\mathrm{T}}_{\partial\Omega_{34}} \cdot \mathbf{F}^{(\alpha)}(t) = F_2^{(\alpha)} \text{ for } \alpha=3,4\\
&\mathbf{n}^{\mathrm{T}}_{\partial\Omega_{23}} \cdot \mathbf{F}^{(\alpha)}(t) = F_1^{(\alpha)} \text{ for } \alpha=2,3\\
&\mathbf{n}^{\mathrm{T}}_{\partial\Omega_{14}} \cdot \mathbf{F}^{(\alpha)}(t) = F_1^{(\alpha)} \text{ for } \alpha=1,4
\end{aligned}\right\} \tag{8.171}$$

and

$$\left.\begin{aligned}
&\mathbf{n}^{\mathrm{T}}_{\partial\Omega_{12}} \cdot D\mathbf{F}^{(\alpha)}(t) = DF_2^{(\alpha)} \text{ for } \alpha=1,2\\
&\mathbf{n}^{\mathrm{T}}_{\partial\Omega_{34}} \cdot D\mathbf{F}^{(\alpha)}(t) = DF_2^{(\alpha)} \text{ for } \alpha=3,4\\
&\mathbf{n}^{\mathrm{T}}_{\partial\Omega_{23}} \cdot D\mathbf{F}^{(\alpha)}(t) = DF_1^{(\alpha)} \text{ for } \alpha=2,3\\
&\mathbf{n}^{\mathrm{T}}_{\partial\Omega_{14}} \cdot D\mathbf{F}^{(\alpha)}(t) = DF_1^{(\alpha)} \text{ for } \alpha=1,4
\end{aligned}\right\} \tag{8.172}$$

where

$$\left.\begin{aligned}
DF_1^{(1)} &= DF_1^{(2)}\\
&= -Q_1\Omega\sin\Omega t - d_1\dot{y}_1 - d_2(\dot{y}_1 - \dot{y}_2) - k_1 y_1 - k_2(y_1 - y_2),\\
DF_1^{(3)} &= DF_1^{(4)}\\
&= -Q_1\Omega\sin\Omega t - d_1\dot{y}_1 - d_2(\dot{y}_1 - \dot{y}_2) - k_1 y_1 - k_2(y_1 - y_2);
\end{aligned}\right\}$$

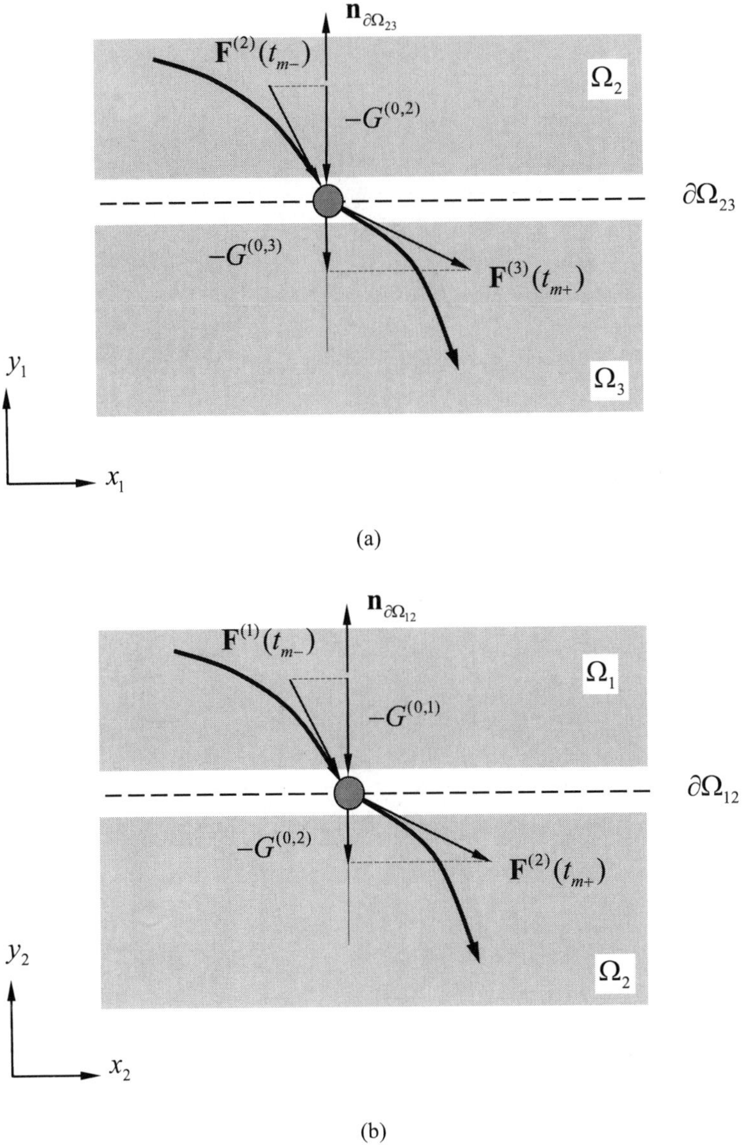

Fig. 8.21 Vector fields for non-stick motion. **(a)** first mass ($\partial\Omega_{23}$), **(b)** second mass ($\partial\Omega_{12}$), **(c)** first mass ($\partial\Omega_{14}$) and **(d)** second mass ($\partial\Omega_{34}$).

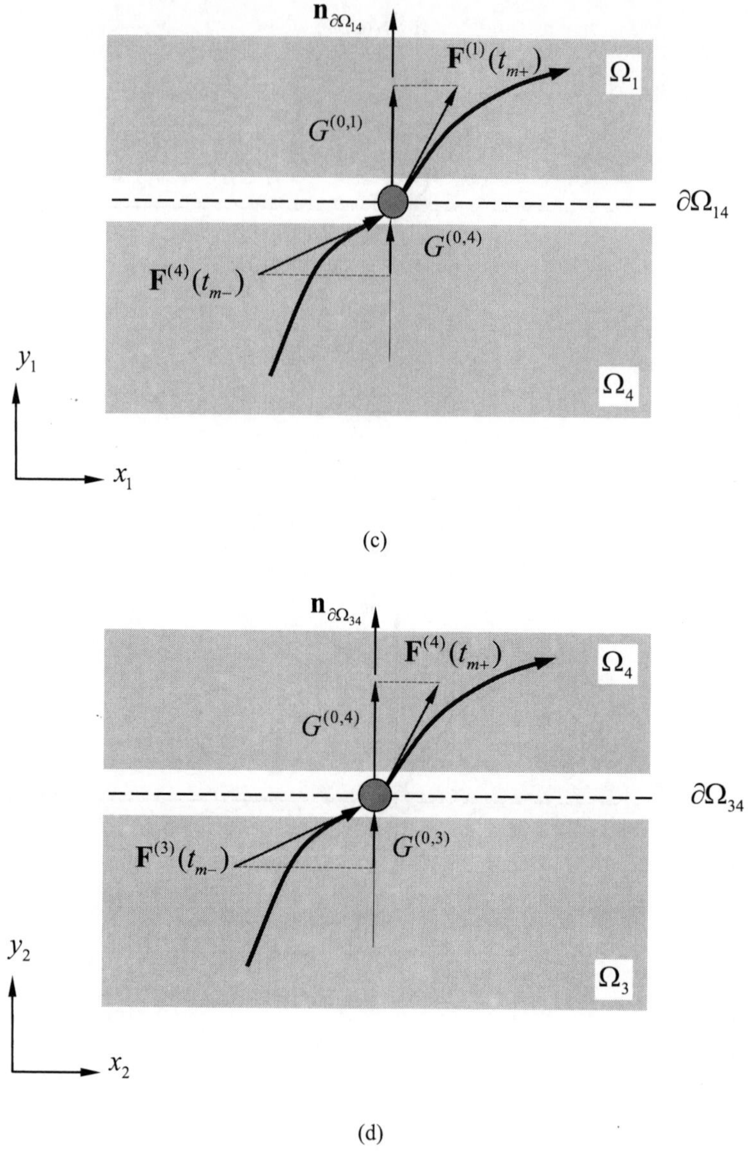

Fig. 8.21 Continued.

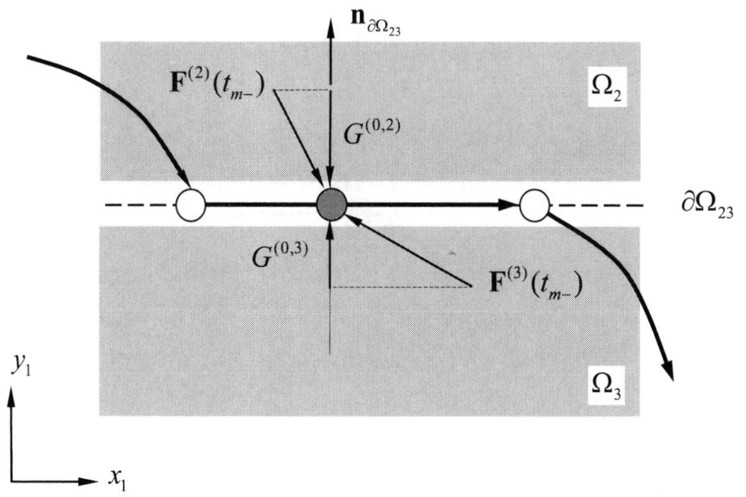

(a)

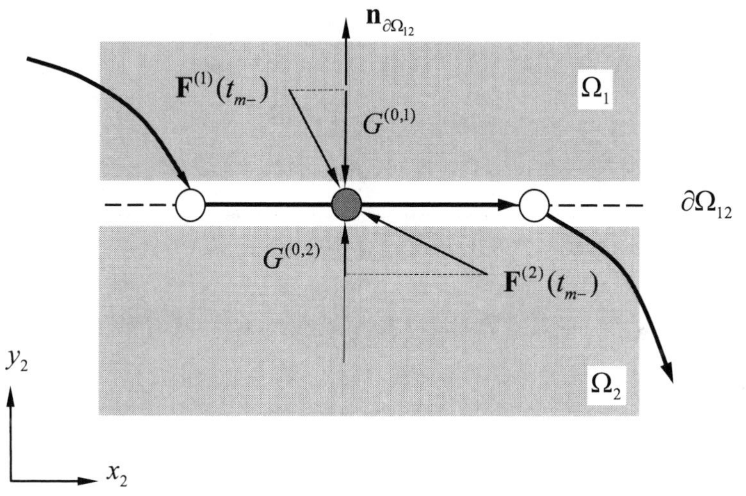

(b)

Fig. 8.22 Vector fields for stick motion. **(a)** first mass ($\partial\Omega_{23}$) , **(b)** second mass ($\partial\Omega_{12}$), **(c)** first mass ($\partial\Omega_{14}$) and **(d)** second mass ($\partial\Omega_{34}$).

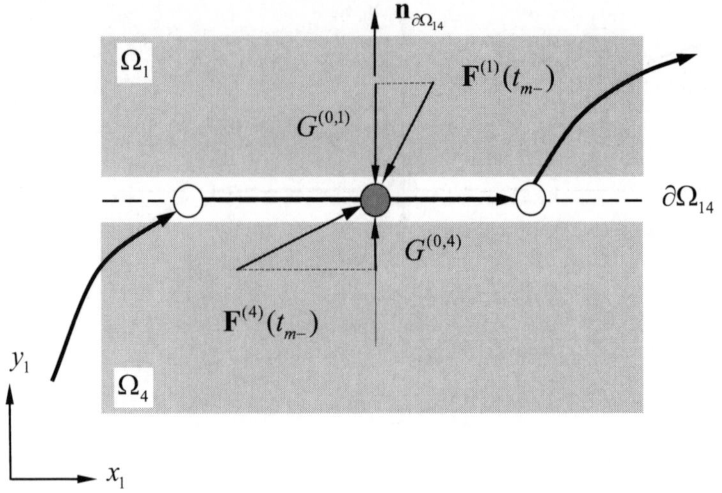

(c)

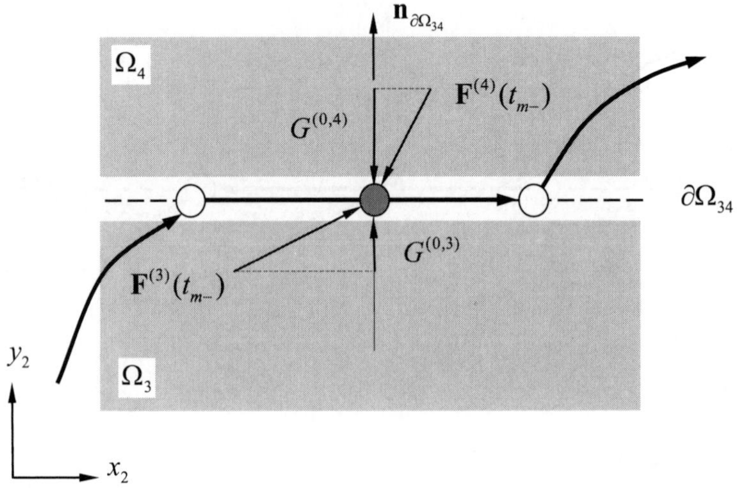

(d)

Fig. 8.22 Continued.

$$\left.\begin{aligned} DF_2^{(1)} &= DF_2^{(4)} = -Q_2\Omega\sin\Omega t - d_2(\dot{y}_2-\dot{y}_1) - k_2(y_2-y_1),\\ DF_2^{(2)} &= DF_2^{(3)} = -Q_2\Omega\sin\Omega t - d_2(\dot{y}_2-\dot{y}_1) - k_2(y_2-y_1).\end{aligned}\right\} \tag{8.173}$$

For stick motion with $y_\alpha = V_\alpha$ ($\alpha \in \{1,2\}$), one obtains $\dot{y}_\alpha = 0$.

From Eq. (8.171), the conditions for non-stick motions in Eq. (8.162) are

$$\begin{aligned} &F_2^{(2)}(t_{m-}) > 0 \text{ and } F_2^{(1)}(t_{m+}) > 0 \text{ from } \Omega_2 \to \Omega_1,\\ &F_2^{(1)}(t_{m-}) < 0 \text{ and } F_2^{(2)}(t_{m+}) < 0 \text{ from } \Omega_1 \to \Omega_2;\\ &F_2^{(3)}(t_{m-}) > 0 \text{ and } F_2^{(4)}(t_{m+}) > 0 \text{ from } \Omega_3 \to \Omega_4,\\ &F_2^{(4)}(t_{m-}) < 0 \text{ and } F_2^{(3)}(t_{m+}) < 0 \text{ from } \Omega_4 \to \Omega_3;\\ &F_1^{(4)}(t_{m-}) > 0 \text{ and } F_1^{(1)}(t_{m+}) > 0 \text{ from } \Omega_4 \to \Omega_1,\\ &F_1^{(1)}(t_{m-}) < 0 \text{ and } F_1^{(4)}(t_{m+}) < 0 \text{ from } \Omega_1 \to \Omega_4;\\ &F_1^{(2)}(t_{m-}) < 0 \text{ and } F_1^{(3)}(t_{m+}) < 0 \text{ from } \Omega_2 \to \Omega_3,\\ &F_1^{(3)}(t_{m-}) > 0 \text{ and } F_1^{(2)}(t_{m+}) > 0 \text{ from } \Omega_3 \to \Omega_2; \end{aligned} \tag{8.174}$$

and the conditions for stick motion in Eq. (8.164) are

$$\begin{aligned} &F_2^{(2)}(t_{m-}) > 0 \text{ and } F_2^{(1)}(t_{m-}) < 0 \text{ on } \partial\Omega_{12},\\ &F_2^{(3)}(t_{m-}) > 0 \text{ and } F_2^{(4)}(t_{m-}) < 0 \text{ on } \partial\Omega_{34},\\ &F_1^{(4)}(t_{m-}) > 0 \text{ and } F_1^{(1)}(t_{m-}) < 0 \text{ on } \partial\Omega_{14},\\ &F_1^{(2)}(t_{m-}) < 0 \text{ and } F_1^{(3)}(t_{m-}) > 0 \text{ on } \partial\Omega_{23}. \end{aligned} \tag{8.175}$$

The conditions for vanishing of stick motions in Eqs. (8.165) and (8.166) are

$$\left.\begin{aligned} &F_2^{(2)}(t_{m-}) > 0 \text{ and } F_2^{(1)}(t_{m-}) = 0,\\ &DF_2^{(1)}(t_{m-}) > 0 \end{aligned}\right\} \text{from } \partial\Omega_{12} \to \Omega_1,$$

$$\left.\begin{aligned} &F_2^{(2)}(t_{m-}) = 0 \text{ and } F_2^{(1)}(t_{m-}) < 0,\\ &DF_2^{(2)}(t_{m-}) < 0 \end{aligned}\right\} \text{from } \partial\Omega_{12} \to \Omega_2;$$

$$\left.\begin{aligned} &F_2^{(3)}(t_{m-}) > 0 \text{ and } F_2^{(4)}(t_{m-}) = 0,\\ &DF_2^{(4)}(t_{m-}) > 0 \end{aligned}\right\} \text{from } \partial\Omega_{34} \to \Omega_4,$$

$$\left.\begin{aligned} &F_2^{(3)}(t_{m-}) > 0 \text{ and } F_2^{(4)}(t_{m-}) < 0,\\ &DF_2^{(3)}(t_{m-}) < 0 \end{aligned}\right\} \text{from } \partial\Omega_{34} \to \Omega_3;$$

$$\left.\begin{aligned} &F_1^{(4)}(t_{m-}) > 0 \text{ and } F_1^{(1)}(t_{m-}) = 0,\\ &DF_1^{(1)}(t_{m-}) > 0 \end{aligned}\right\} \text{from } \partial\Omega_{14} \to \Omega_1,$$

$$\left.\begin{array}{l} F_1^{(4)}(t_{m-})=0 \text{ and } F_1^{(1)}(t_{m-})<0 \\ DF_1^{(4)}(t_{m-})<0 \end{array}\right\} \text{from } \partial\Omega_{14} \to \Omega_4;$$
$$\left.\begin{array}{l} F_1^{(3)}(t_{m-})>0 \text{ and } F_1^{(2)}(t_{m-})=0 \\ DF_1^{(3)}(t_{m-})>0 \end{array}\right\} \text{from } \partial\Omega_{23} \to \Omega_2, \tag{8.176}$$
$$\left.\begin{array}{l} F_1^{(3)}(t_{m-})=0 \text{ and } F_1^{(2)}(t_{m-})>0 \\ DF_1^{(3)}(t_{m-})<0 \end{array}\right\} \text{from } \partial\Omega_{23} \to \Omega_3.$$

The onset conditions for the stick motion can be presented as in Eq. (8.175). The conditions for the grazing motion are

$$\begin{array}{lll} F_2^{(1)}(t_{m\pm})=0 \text{ and } DF_2^{(1)}(t_{m\pm})>0 & \text{on } \partial\Omega_{12} & \text{in } \Omega_1\,, \\ F_2^{(2)}(t_{m\pm})=0 \text{ and } DF_2^{(2)}(t_{m\pm})<0 & \text{on } \partial\Omega_{12} & \text{in } \Omega_2; \\ F_2^{(4)}(t_{m\pm})=0 \text{ and } DF_2^{(4)}(t_{m\pm})>0 & \text{on } \partial\Omega_{34} & \text{in } \Omega_4, \\ F_2^{(3)}(t_{m\pm})=0 \text{ and } DF_2^{(3)}(t_{m\pm})<0 & \text{on } \partial\Omega_{34} & \text{in } \Omega_3; \\ F_1^{(1)}(t_{m\pm})=0 \text{ and } DF_1^{(1)}(t_{m\pm})>0 & \text{on } \partial\Omega_{14} & \text{in } \Omega_1, \\ F_1^{(4)}(t_{m\pm})=0 \text{ and } DF_1^{(4)}(t_{m\pm})<0 & \text{on } \partial\Omega_{14} & \text{in } \Omega_4; \\ F_1^{(2)}(t_{m\pm})=0 \text{ and } DF_1^{(2)}(t_{m\pm})>0 & \text{on } \partial\Omega_{23} & \text{in } \Omega_2, \\ F_1^{(3)}(t_{m\pm})=0 \text{ and } DF_1^{(3)}(t_{m\pm})<0 & \text{on } \partial\Omega_{34} & \text{in } \Omega_3. \end{array} \tag{8.177}$$

8.5.3. Mapping structures and numerical illustrations

To label the motion complexity, basic mappings, switching sets, and mapping structures will be introduced. From the boundary $\partial\Omega_{\alpha_1\alpha_2}$ in Eq. (8.151), the switching sets is defined by

$$\left.\begin{array}{l} \Sigma_1^{+}=\left\{(x_{1(k)},y_{1(k)},x_{2(k)},y_{2(k)},t_k)\,\middle|\,y_{2(k)}=V_2^{+},\ k\in\mathbb{N}\right\}, \\ \Sigma_1^{0}=\left\{(x_{1(k)},y_{1(k)},x_{2(k)},y_{2(k)},t_k)\,\middle|\,y_{2(k)}=V_2,\ k\in\mathbb{N}\right\}, \\ \Sigma_1^{-}=\left\{(x_{1(k)},y_{1(k)},x_{2(k)},y_{2(k)},t_k)\,\middle|\,y_{2(k)}=V_2^{-},\ k\in\mathbb{N}\right\}, \end{array}\right\} \tag{8.178}$$

and

$$\begin{array}{l} \Sigma_2^{+}=\left\{(x_{1(k)},y_{1(k)},x_{2(k)},y_{2(k)},t_k)\,\middle|\,y_{1(k)}=V_1^{+},k\in\mathbb{N}\right\}, \\ \Sigma_2^{0}=\left\{(x_{1(k)},y_{1(k)},x_{2(k)},y_{2(k)},t_k)\,\middle|\,y_{1(k)}=V_1,\ k\in\mathbb{N}\right\}, \\ \Sigma_2^{-}=\left\{(x_{1(k)},y_{1(k)},x_{2(k)},y_{2(k)},t_k)\,\middle|\,y_{1(k)}=V_1^{-},k\in\mathbb{N}\right\}, \end{array} \tag{8.179}$$

where

$$V_{\alpha}^{\pm}=\lim_{\varepsilon\to 0}(V_{\alpha}\pm\varepsilon),\tag{8.180}$$

and the switching set on the edge $\angle\Omega_{\alpha_1\alpha_2\alpha_3}$ is defined by

$$\Sigma_0^0=\left\{(x_{1(k)},y_{1(k)},x_{2(k)},y_{2(k)},t_k)\left|\begin{array}{l}y_{1(k)}=V_1 \text{ and}\\ y_{2(k)}=V_2,\ k\in\mathbb{N}\end{array}\right.\right\}.\tag{8.181}$$

From the switching sets, mappings are defined as

$$\begin{aligned}
&P_1:\Sigma_1^+\to\Sigma_1^+,\ P_2:\Sigma_1^-\to\Sigma_1^-,P_3:\Sigma_1^0\to\Sigma_1^0,\\
&P_6:\Sigma_2^+\to\Sigma_2^+,\ P_7:\Sigma_2^-\to\Sigma_2^-,P_8:\Sigma_2^0\to\Sigma_2^0;\\
&P_4:\Sigma_2^+\to\Sigma_1^+,\ P_5:\Sigma_1^-\to\Sigma_2^-;\\
&P_4:\Sigma_2^-\to\Sigma_1^-,\ P_5:\Sigma_1^+\to\Sigma_2^+;\\
&P_0:\Sigma_0\to\Sigma_0.
\end{aligned}\tag{8.182}$$

Since the switching set Σ_0 is the special case of the switching sets Σ_1^0 and Σ_2^0, the mappings P_3 and P_8 can apply to Σ_0, i.e.,

$$\begin{aligned}
&P_3:\Sigma_0\to\Sigma_1^0 \text{ and } P_3:\Sigma_1^0\to\Sigma_0;\\
&P_8:\Sigma_0\to\Sigma_2^0 \text{ and } P_8:\Sigma_2^0\to\Sigma_0.
\end{aligned}\tag{8.183}$$

In the total nine mappings, P_n are the local mappings for $n=0,1,2,3;6,7,8$ and the global mappings for $n=4,5$. The afore-defined, nine mappings are sketched in Fig. 8.23. The global and local mappings are clearly shown. Through all the mappings, the motions of the 2-DOF friction induced oscillator can be labeled if the motion is interacted at least with one of two velocity boundaries. From the defined mappings, the final switching points can be mapped from the initial switching points. Herein, the flow not intersected with the boundary will not be interested. For dynamical system in Eq. (8.154), if mapping P_n ($n=0,1,2,\cdots,8$) exist, then there are a set of nonlinear algebraic equations as

$$\mathbf{f}^{(n)}(\mathbf{x}_k,t_k,\mathbf{x}_{k+1},t_{k+1})=\mathbf{\Phi}(t_{k+1},\mathbf{x}_k,t_k)-\mathbf{x}_{k+1}=0\tag{8.184}$$

where

$$\begin{aligned}
&\mathbf{x}_k=(x_{1(k)},y_{1(k)},x_{2(k)},y_{2(k)})^{\mathrm{T}},\\
&\mathbf{f}^{(n)}=(f_1^{(n)},f_2^{(n)},f_3^{(n)},f_4^{(n)})^{\mathrm{T}},\\
&y_{2(k)}=y_{2(k+1)}=V_2 \qquad \text{for } n=1,2;\\
&y_{1(k)}=y_{1(k+1)}=V_1 \qquad \text{for } n=6,7;\\
&y_{2(k)}=V_2,\ y_{1(k+1)}=V_1 \quad \text{for } n=4;
\end{aligned}$$

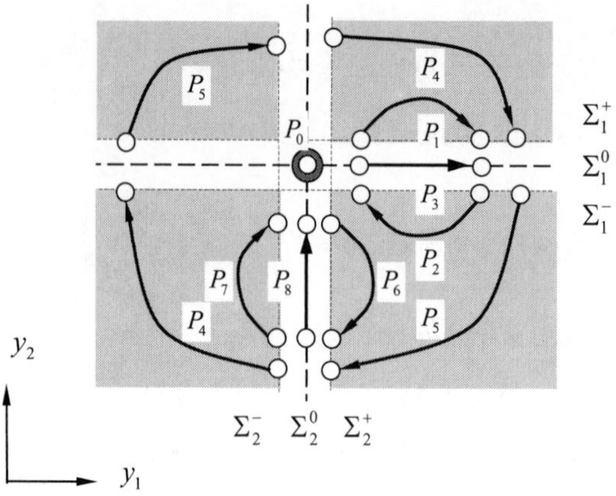

Fig. 8.23 Switching sets and mappings.

$$y_{1(k)} = V_1, \; y_{1(k+1)} = V_2 \quad \text{for } n = 5, \tag{8.185}$$

with the constraints for $\mathbf{x}_k$ and $\mathbf{x}_{k+1}$ from the boundaries. In other words, one has

$$\begin{aligned} &f_1^{(n)}(\mathbf{x}_k, t_k, \mathbf{x}_{k+1}, t_{k+1}) = 0, \\ &f_2^{(n)}(\mathbf{x}_k, t_k, \mathbf{x}_{k+1}, t_{k+1}) = 0, \\ &f_3^{(n)}(\mathbf{x}_k, t_k, \mathbf{x}_{k+1}, t_{k+1}) = 0, \\ &f_4^{(n)}(\mathbf{x}_k, t_k, \mathbf{x}_{k+1}, t_{k+1}) = 0. \end{aligned} \tag{8.186}$$

$$\begin{aligned} &\left.\begin{aligned} \mathbf{x}_k &= (x_{1(k)}, y_{1(k)}, x_{2(k)}, V_2)^{\mathrm{T}} \\ \mathbf{x}_{k+1} &= (x_{1(k+1)}, y_{1(k+1)}, x_{2(k+1)}, V_2)^{\mathrm{T}} \end{aligned}\right\} \text{for } P_n \; (n = 1,2,3); \\ &\left.\begin{aligned} \mathbf{x}_k &= (x_{1(k)}, V_1, x_{2(k)}, y_{2(k)})^{\mathrm{T}} \\ \mathbf{x}_{k+1} &= (x_{1(k+1)}, V_1, x_{2(k+1)}, y_{2(k+1)})^{\mathrm{T}} \end{aligned}\right\} \text{for } P_n \; (n = 6,7,8); \\ &\left.\begin{aligned} \mathbf{x}_k &= (x_{1(k)}, V_1, x_{2(k)}, y_{2(k)})^{\mathrm{T}} \\ \mathbf{x}_{k+1} &= (x_{1(k+1)}, y_{1(k+1)}, x_{2(k+1)}, V_2)^{\mathrm{T}} \end{aligned}\right\} \text{for } P_4 \text{ and } P_9; \\ &\left.\begin{aligned} \mathbf{x}_k &= (x_{1(k)}, y_{1(k)}, x_{2(k)}, V_2)^{\mathrm{T}} \\ \mathbf{x}_{k+1} &= (x_{1(k+1)}, V_1, x_{2(k+1)}, y_{2(k+1)})^{\mathrm{T}} \end{aligned}\right\} \text{for } P_5 \text{ and } P_6; \\ &\left.\begin{aligned} \mathbf{x}_k &= (x_{1(k)}, V_1, x_{2(k)}, V_2)^{\mathrm{T}} \\ \mathbf{x}_{k+1} &= (x_{1(k+1)}, V_1, x_{2(k+1)}, V_2)^{\mathrm{T}} \end{aligned}\right\} \text{for } P_0. \end{aligned} \tag{8.187}$$

To illustrate the switching mechanism of the frictional induced oscillator with two-degree of freedom, for simplicity, the following force variables are introduced

$$\left.\begin{aligned} F_{\alpha+} &= F_{\alpha} \ \text{ for } y_{\alpha} > V_{\alpha}, \\ F_{\alpha-} &= F_{\alpha} \ \text{ for } y_{\alpha} < V_{\alpha} \end{aligned}\right\} \ \alpha=1,2. \tag{8.188}$$

Consider a periodic motion with a mapping structure of P_{4651} with a set of system parameters

$$\begin{aligned} &m_1 = 4,\ m_2 = 1,\ d_1 = 0.05,\ d_2 = 0.5,\ k_1 = 4,\ k_2 = 1, \\ &\mu_k = 0.15,\ Q_1 = -15,\ Q_2 = 15,\ V_1 = V_2 = 2. \end{aligned} \tag{8.189}$$

The periodic motion of the frictional induced oscillator for $\Omega = 1.6$ is presented in Figs. 8.24-8.26 with the initial condition (i.e., $\Omega t_0 \approx 5.5840$, $x_{10} \approx 3.1002$, $y_{10} \approx -1.5826$, $x_{20} \approx -7.1437$ and $y_{20} = 2.0$). The solid and dashed curves give the real and imaginary responses. The velocity boundary is denoted by the straight line. The velocity and displacement planes for such a frictional oscillator are presented in Fig. 8.24 (a) and (b), respectively. The domains and boundaries for such a system are clearly shown through shaded regions and dashed lines, respectively. The mapping structure for the periodic motion of P_{4651} is labeled from basic mappings.

The responses of the second and first masses for the periodic motion of P_{4651} are shown in Figs. 8.25 and 8.26, respectively. The velocity-time history, force per unit mass-time history, phase plane, and force versus displacement are presented. Since an initial point is selected on the velocity boundary of the second mass, the motion of the second mass is discussed first. The initial condition is presented with a green circular symbol. Since $F_{2+} > 0$ and $F_{2-} > 0$ at the initial point are shown in Fig. 8.26(b) and (d). From the analytical condition in Eq. (8.174), the motion flow must move to the domain relative to $y_2 > V_2$, which means the system switching from the domain of $y_2 < V_2$ to the domain of $y_2 > V_2$ in Fig. 8.25(a) and (c). In such a domain of $y_2 > V_2$, the motion moves to the velocity boundary of the second mass, and the $F_{2+} < 0$ and $F_{2-} < 0$ at such a point. From the analytical condition in Eq. (8.174), the motion flow must move to the domain relative to $y_2 < V_2$. However, in such a domain, the motion moves to the velocity boundary of the first mass (i.e., $y_1 = V_1$), as shown in Fig. 8.26(a) and (c). Owing to the force conditions $F_{1+} > 0$ and $F_{1-} > 0$ in Fig. 8.27(b) and (d), the passable motion condition in Eq. (8.174) implies that the motion flow must enter the domain of $y_1 > V_1$ from the domain of $y_1 < V_1$. The motion flow returns back to the velocity boundary of $y_1 = V_1$. At this switching point, from the analytical condition in Eq. (8.174), the motion moves to the domain of $y_1 < V_1$ because of $F_{1+} < 0$ and $F_{1-} < 0$. Further, the motion will return back to the initial point to form a periodic motion.

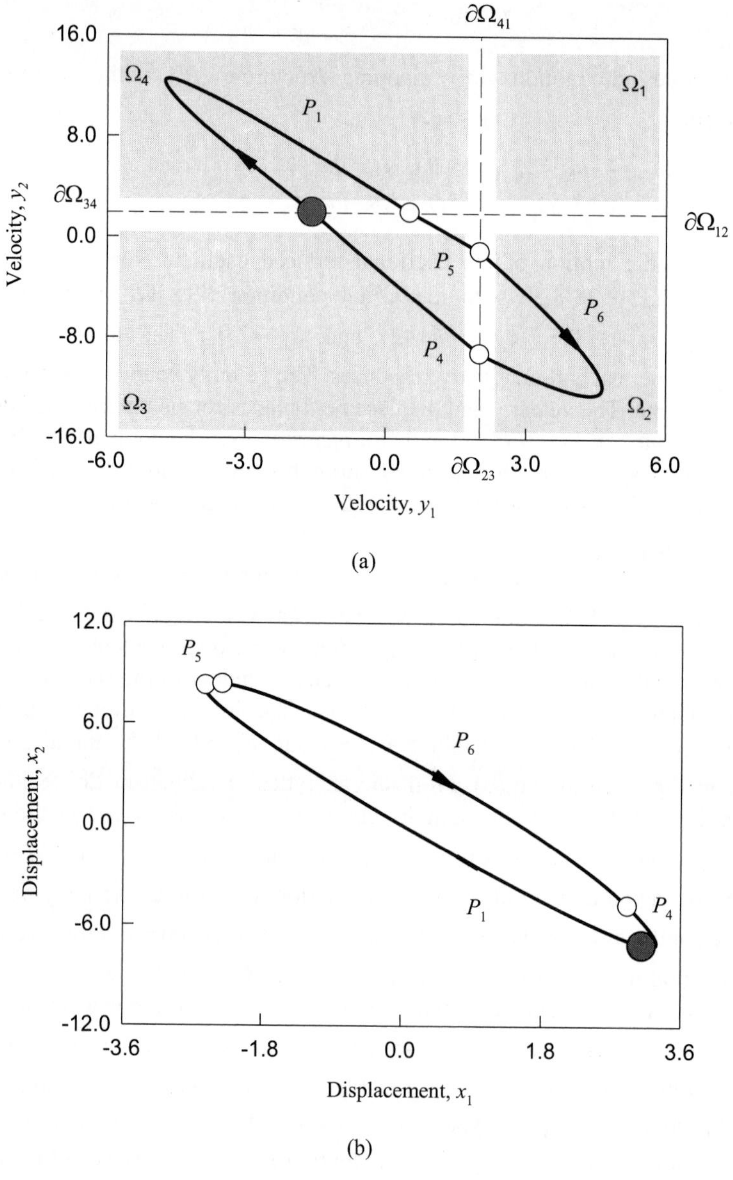

Fig. 8.24 Periodic motion of the frictional induced oscillator for P_{4651}. **(a)** velocity plane and **(b)** displacement plane. ($m_1 = 4,\ m_2 = 1,\ d_1 = 0.05,\ d_2 = 0.5,\ k_1 = 4,\ k_2 = 1,\ \mu_k = 0.15,\ Q_1 = -15,\ Q_2 = 15,\ V_1 = 2,\ V_2 = 2,\ \Omega = 1.6$). The initial condition is $\Omega t_0 \approx 5.5840$, $x_{10} \approx 3.1002,\ y_{10} \approx -1.5826,\ x_{20} \approx -7.1437,\ y_{20} = 2.0$.

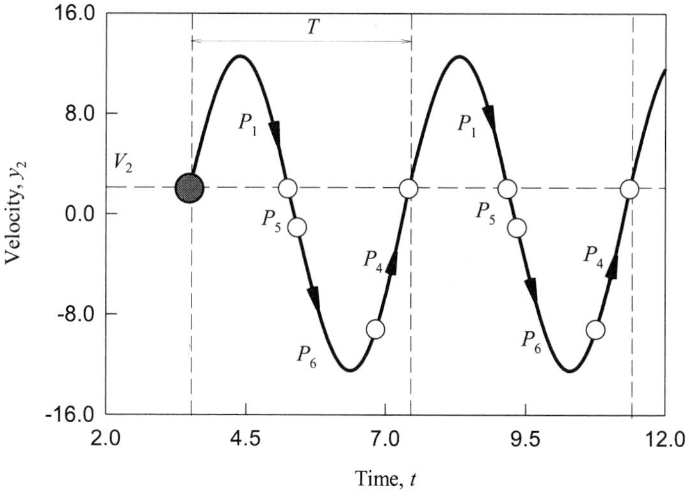

(a)

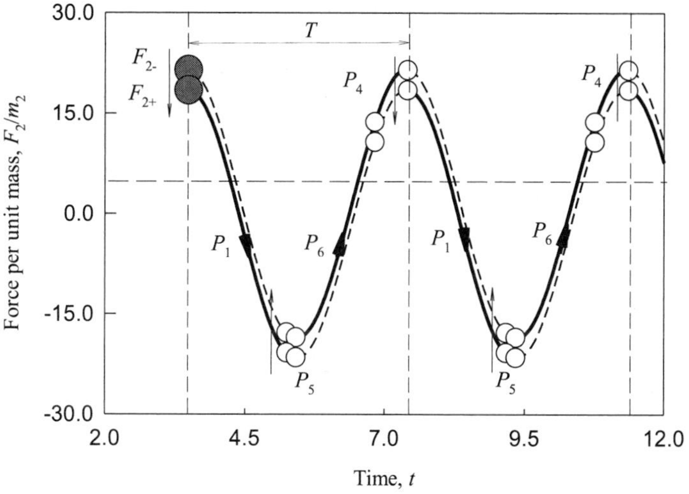

(b)

Fig. 8.25 Periodic responses of the second mass for P_{4651}. **(a)** velocity-time history, **(b)** force per unit mass-time history, **(c)** phase plane, and **(d)** force versus velocity. ($m_1 = 4,\ \ m_2 = 1,\ \ d_1 = 0.05,\ \ d_2 = 0.5,\ \ k_1 = 4,\ \ k_2 = 1,\ \ \mu_k = 0.15,\ \ Q_1 = -15,\ \ Q_2 = 15,\ \ V_1 = 2,\ \ V_2 = 2,\ \Omega = 1.6$). The initial condition is $\Omega t_0 \approx 5.5840,\ \ x_{10} \approx 3.1002,\ \ y_{10} \approx -1.5826,\ \ x_{20} \approx -7.1437,\ \ y_{20} = 2.0$.

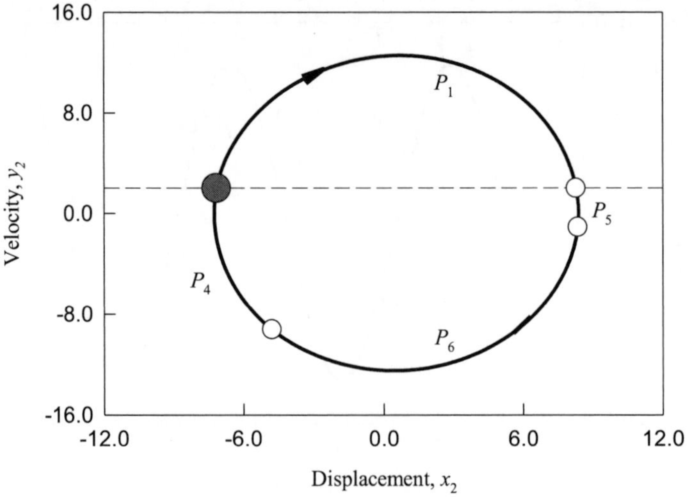

(c)

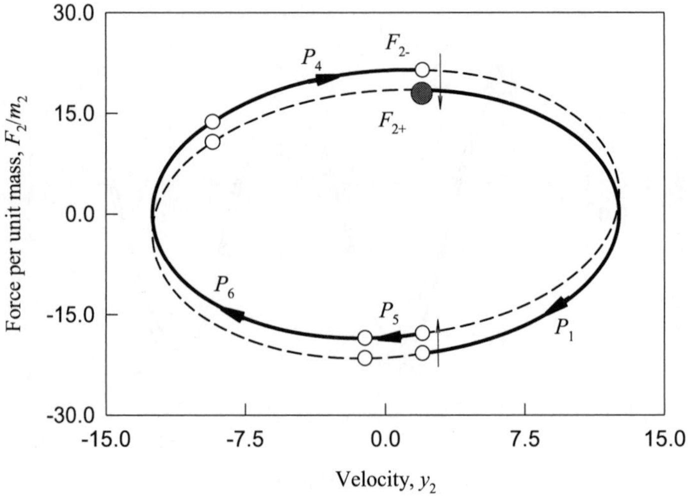

(d)

Fig. 8.25 Continued.

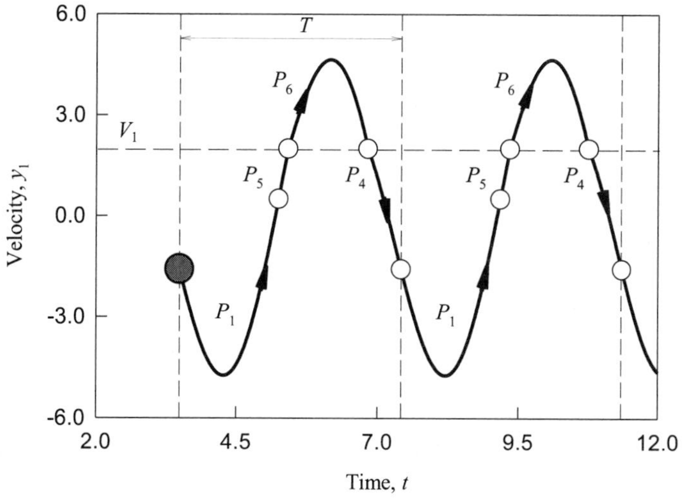

(a)

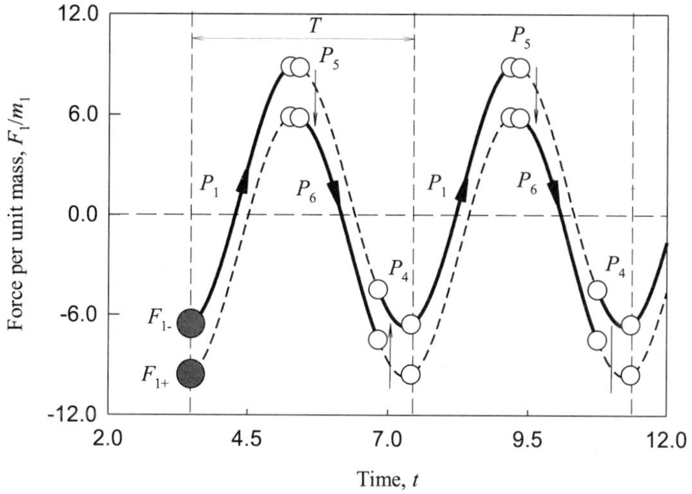

(b)

Fig. 8.26 Periodic responses of the first mass for P_{4651}. **(a)** velocity-time history, **(b)** force per unit mass-time history, **(c)** phase plane, and **(d)** force versus velocity. ($m_1 = 4$, $m_2 = 1$, $d_1 = 0.05$, $d_2 = 0.5$, $k_1 = 4$, $k_2 = 1$, $\mu_k = 0.15$, $Q_1 = -15$, $Q_2 = 15$, $V_1 = 2$, $V_2 = 2$, $\Omega = 1.6$). The initial condition is $\Omega t_0 \approx 5.5840$, $x_{10} \approx 3.1002$, $y_{10} \approx -1.5826$, $x_{20} \approx -7.1437$, $y_{20} = 2.0$.

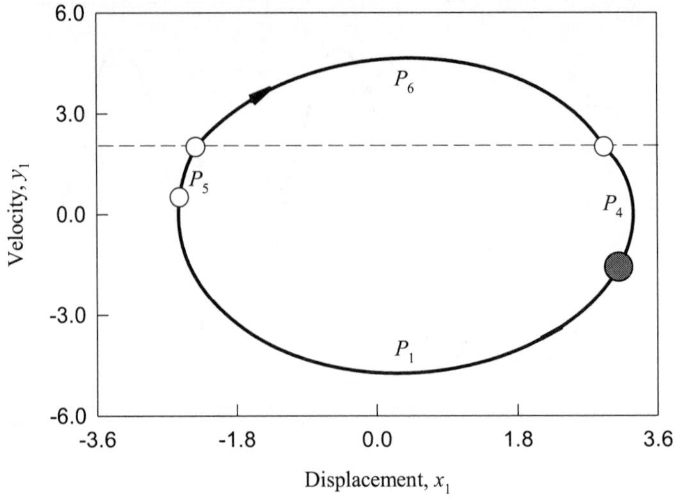

(c)

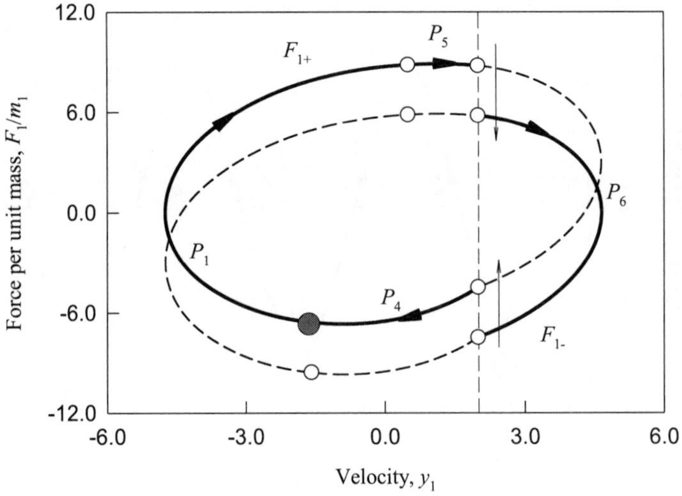

(d)

Fig. 8.26 Continued.

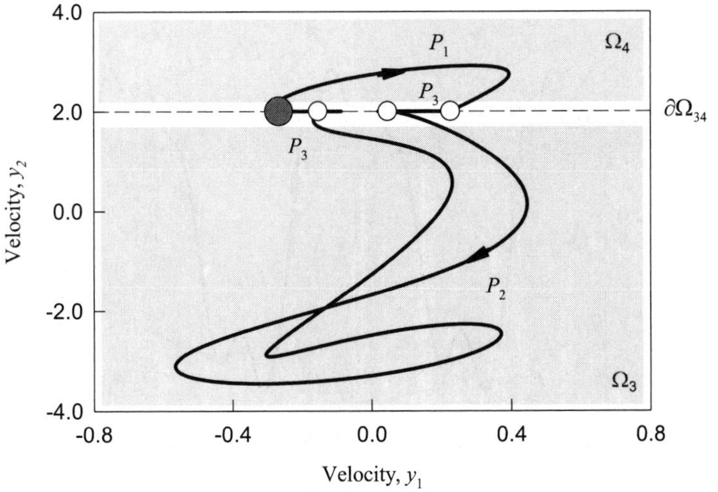

(a)

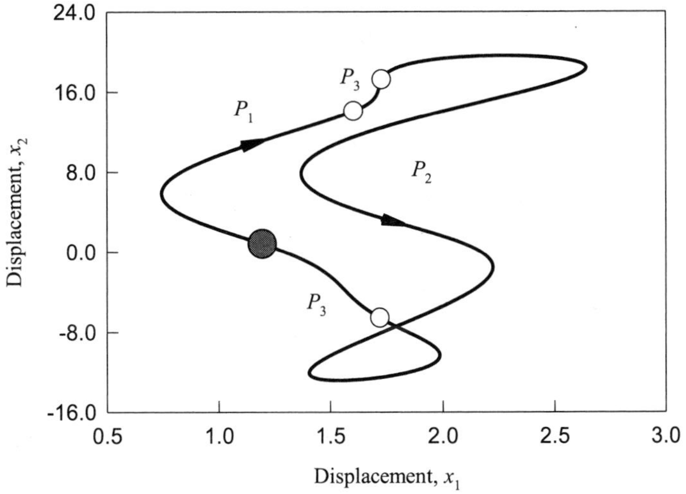

(b)

Fig. 8.27 Periodic motion of the frictional induced oscillator for P_{3231}. **(a)** velocity plane, **(b)** displacement plane. ($m_1 = 4,\ m_2 = 1,\ d_1 = 0.05,\ d_2 = 0.5,\ k_1 = 4,\ k_2 = 1,\ \mu_k = 0.15,\ Q_1 = -15,\ Q_2 = 15,\ V_1 = 2,\ V_2 = 2,\ \Omega = 0.2$). The initial condition is $\Omega t_0 \approx 4.8627$, $x_{10} \approx 1.1953$, $y_{10} \approx -0.2665$, $x_{20} \approx 0.8086$, $y_{20} = 2.0$.

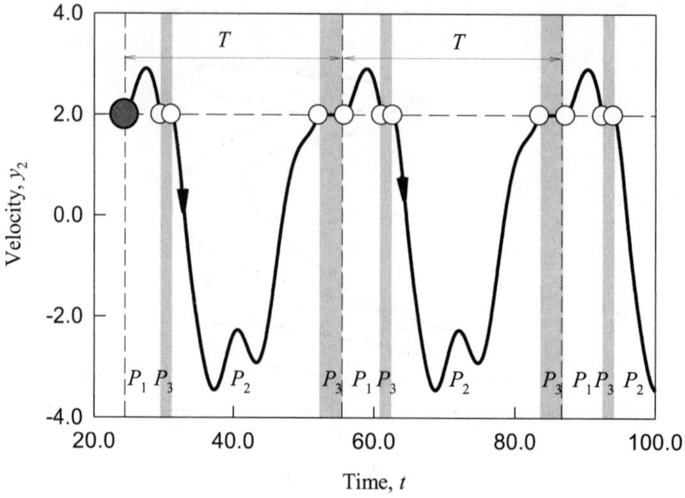

(a)

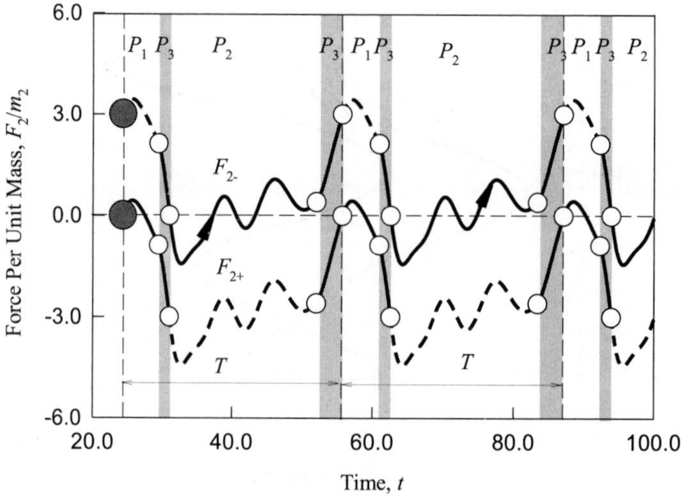

(b)

Fig. 8.28 The responses of the second mass for P_{3231}. **(a)** velocity-time history, **(b)** force per unit mass-time history, **(c)** phase plane, and **(d)** force versus velocity. ($m_1 = 4,\ m_2 = 1,\ d_1 = 0.05,\ d_2 = 0.5,\ k_1 = 4,\ k_2 = 1,\ \mu_k = 0.15,\ Q_1 = -15,\ Q_2 = 15,\ V_1 = 2,\ V_2 = 2,\ \Omega = 0.2$). The initial condition is $\Omega t_0 \approx 4.8627,\ x_{10} \approx 1.1953,\ y_{10} \approx -0.2665,\ x_{20} \approx 0.8086,\ y_{20} = 2.0$.

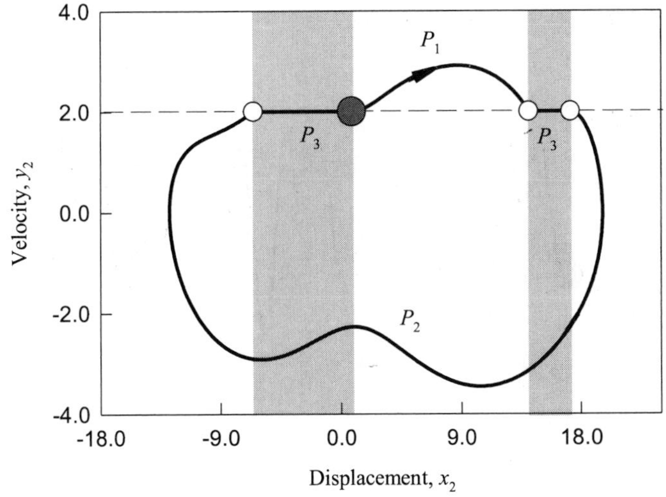

(c)

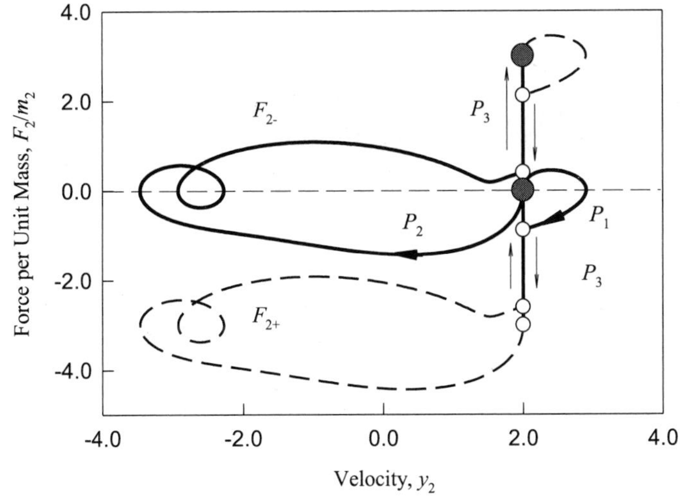

(d)

Fig. 8.28 Continued.

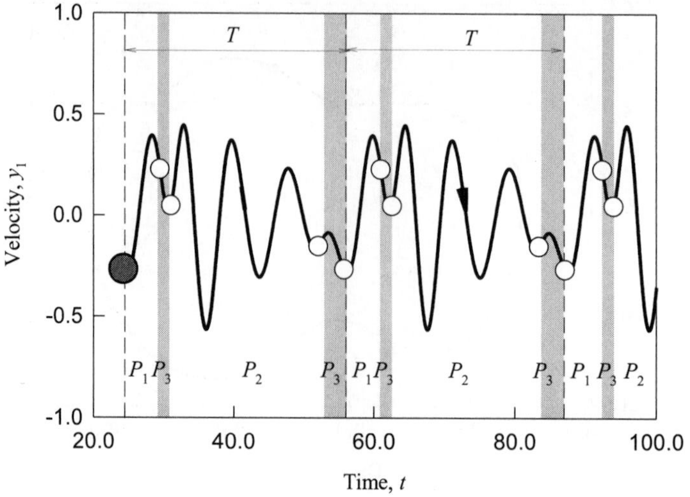

(a)

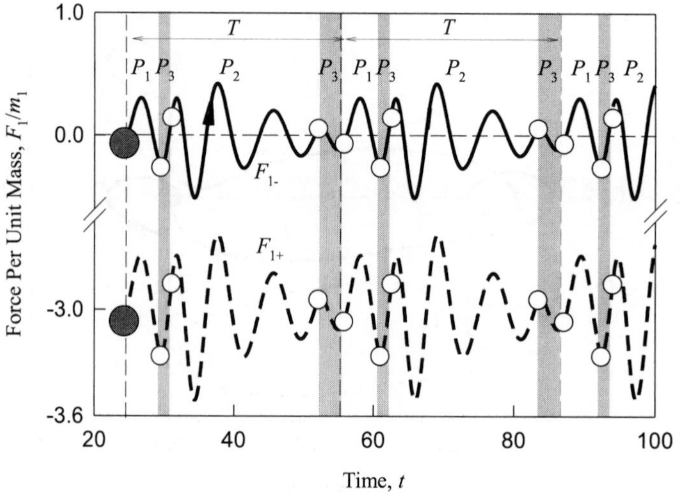

(b)

Fig. 8.29 The responses of the first mass for P_{3231}. **(a)** velocity-time history, **(b)** force per unit mass-time history, **(c)** phase plane, and **(d)** force versus velocity. ($m_1 = 4,\ m_2 = 1,\ d_1 = 0.05,\ d_2 = 0.5,\ k_1 = 4,\ k_2 = 1,\ \mu_k = 0.15,\ Q_1 = -15,\ Q_2 = 15,\ V_1 = 2,\ V_2 = 2,\ \Omega = 0.2$). The initial condition is $\Omega t_0 \approx 4.8627,\ x_{10} \approx 1.1953,\ y_{10} \approx -0.2665,\ x_{20} \approx 0.8086,\ y_{20} = 2.0.$

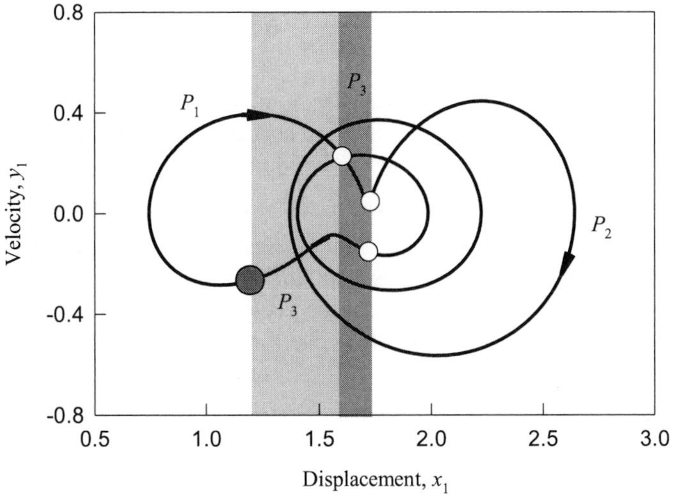

(c)

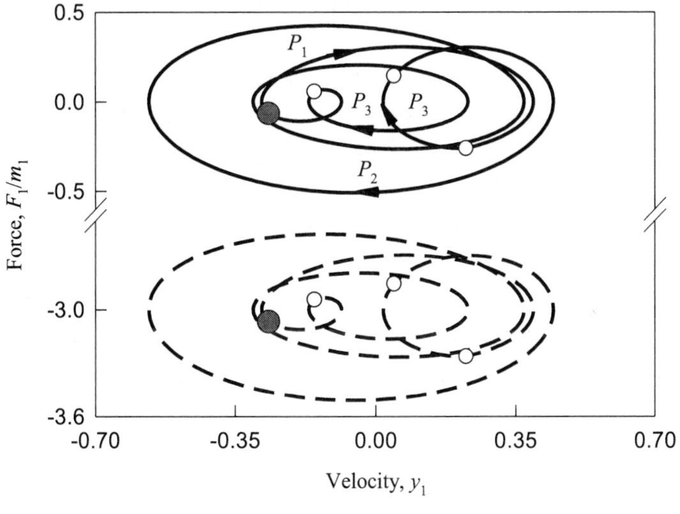

(d)

Fig. 8.29 Continued.

A periodic motion with two sliding portions for a mapping structure of P_{3231} is presented in Figs. 8.27-8.29. Choose the same system parameters and the initial conditions $\Omega t_0 \approx 4.8627,\ x_{10} \approx 1.1953,\ y_{10} \approx -0.2665,\ x_{20} \approx 0.8086,\ y_{20} = 2.0$ for $\Omega = 0.2$. The velocity plane is plotted in Fig. 8.27 to show domains, boundaries and mapping structures for such a periodic motion. The domains and boundaries for such a system are clearly shown through shaded regions and dashed lines, respectively. The mapping structure for the periodic motion of P_{4651} is labeled from basic mappings. The initial condition is at the velocity boundary of $y_2 = V_2$. The periodic responses for the first and second masses are presented in Figs. 8.28 and 8.29, respectively. For the second mass, the time-history of velocity and forces are presented in Fig. 8.28(a) and (b). The phase plane and the force varying with velocity is plotted in Figs. 8.29(c) and (d). The force conditions at the initial point are $F_{2+} = 0$ with $DF_{2+} > 0$ and $F_{2-} > 0$. From the analytical conditions in Eq. (8.176), the motion flow will enter the domain of $y_2 > V_2$. When the motion arrives to the boundary of $y_2 = V_2$, the force conditions are $F_{2+} < 0$ and $F_{2-} > 0$. From the analytical condition in Eq. (8.175), the motion flow will slide along this velocity boundary of $y_2 = V_2$. If the analytical conditions in Eq. (8.176) are satisfied, the motion flow will enter the corresponding domain from the boundary. From Fig. 8.28(a), the force and corresponding derivative with respect to time are $F_{2-} = 0$ and $DF_{2-} < 0$. The motion will enter the domain of $y_2 < V_2$ from the boundary. The motion relative to mapping P_2 returns back to the velocity boundary of $y_2 = V_2$, and the force conditions at this point are $F_{2+} < 0$ and $F_{2-} > 0$. From the analytical conditions in Eq. (8.175), the sliding flow along such velocity boundary will be formed with mapping P_3 again. Once one of two forces becomes zero with an appropriate condition in Eq. (8.176), the sliding flows on the boundary will disappear. The sliding flow on the boundary disappears and enters the domain of $y_2 > V_2$ and this point is the initial condition as well. For the first mass, the periodic motion will not interact with the velocity boundary of $y_1 = V_1$, and the periodic flow lies in the domain of $y_1 < V_1$, which can be found from Fig. 8.29.

References

Luo, A.C.J., 2008, *Global Transversality, Resonance and Chaotic Dynamics,* Singapore: World Scientific.

Luo, A.C.J., 2009, *Discontinuous Dynamical Systems on Time-Varying Domains*, Beijing: Higher Education Press and Springer.

Luo, A.C.J. and Mao, T.T., 2010, Analytical conditions for motion switchability in a 2-DOF friction-induced oscillator moving on two constant speed belts, *Canadian Applied Mathematics Quarterly*, **17**, 201-242.

Chapter 9
Dynamical System Interactions

In this chapter, discontinuous dynamical system theory will be applied to dynamical system interactions. The concept of interaction between two dynamical systems will be introduced. An interaction condition of two dynamical systems will be treated as a separation boundary, and such a boundary is time-varying. In other words, the boundary and domains for one of two dynamical systems are constrained by the other. The corresponding conditions for such an interaction will be presented via the theory for the switchability and attractivity of edge flows to the specific edges. The synchronization of two totally different dynamical systems will be presented as an application.

9.1. Introduction to system interactions

In this section, basic concepts of the dynamical system interactions will be presented. The discontinuous description of the interaction of two dynamical systems will be presented.

9.1.1. System interactions

Definition 9.1. Two dynamical systems are defined by

$$\dot{\mathbf{y}} = \mathbf{F}(\mathbf{y}, t, \mathbf{p}) \in \mathscr{R}^n \text{ and } \dot{\mathbf{x}} = \mathscr{F}(\mathbf{x}, t, \mathbf{q}) \in \mathscr{R}^m \tag{9.1}$$

If two flows $\mathbf{x}(t)$ and $\mathbf{y}(t)$ of the two systems in Eq.(9.1) satisfy

$$\varphi(\mathbf{x}(t), \mathbf{y}(t), t, \boldsymbol{\lambda}) = 0, \quad \boldsymbol{\lambda} \in \mathscr{R}^{n_0} \tag{9.2}$$

then the two systems are called to be interacted (or constrained) under such a condition at time t.

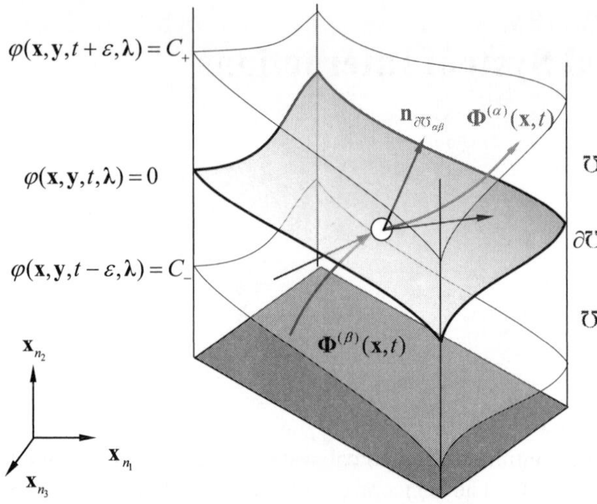

Fig.9.1 Interaction surface for the two dynamical systems in Eq.(9.1).

From the foregoing definition, the interaction (or constraint) of two dynamical systems in Eq.(9.1) occurs through $\varphi(\mathbf{x}(t),\mathbf{y}(t),t,\boldsymbol{\lambda})=0$ in Eq.(9.2). Such a condition may cause the discontinuity for two dynamical systems. If the interaction condition is the separation boundary, then the domain and boundary for the first dynamical system in Eq.(9.1) will be time-varying, which is controlled by a flow of the second dynamical system in Eq.(9.1) (i.e., $\mathbf{x}(t)$), vice versa. Suppose the interaction of two systems occurs at time t. For time $t\pm\varepsilon$ ($\varepsilon>0$), there are two constants with

$$\varphi(\mathbf{x},\mathbf{y},t\pm\varepsilon,\boldsymbol{\lambda})=C_\pm\neq 0. \tag{9.3}$$

If the flows of two systems in Eq.(9.1) satisfy Eq.(9.3), then the two systems will not be interacted, as shown in Fig.9.1. In fact, the interaction of two dynamical systems can occur under many constraints instead of Eq.(9.2), i.e.,

Definition 9.2. Consider l-non-identical functions of $\varphi_j(\mathbf{x}(t),\mathbf{y}(t),t,\boldsymbol{\lambda}_j)$ ($j\in\mathscr{L}$ and $\mathscr{L}=\{1,2,\cdots,l\}$). If two flows $\mathbf{x}(t)$ and $\mathbf{y}(t)$ of two systems in Eq.(9.1) satisfy for time t

$$\varphi_j(\mathbf{x}(t),\mathbf{y}(t),t,\boldsymbol{\lambda}_j)=0 \text{ for } \boldsymbol{\lambda}_j\in\mathscr{R}^{n_j} \text{ and } j\in\mathscr{L}, \tag{9.4}$$

then two systems in Eq.(9.1) are called to be constrained under the jth -condition at time t.

For the foregoing definition, two dynamical systems in Eqs.(9.1) possess l-conditions for interactions (or constraints). Thus, the l-separation boundaries relative to the interaction divide the corresponding phase space into many sub-domains for the two dynamical systems, and these sub-domains change with time.

For one of two dynamical systems, sub-domains and boundaries generated by the l-interaction conditions are sketched in Figs.9.2-9.5. Under the l-boundaries, because the interaction between two dynamical systems, there are $(N+1)$-pairs of sub-domains for two dynamical systems in Eqs.(9.1) in a pair of universal domains (i.e., $\mho \subset \mathscr{R}^n$ and $\underline{\mho} \subset \mathscr{R}^m$), and a pair of the universal domains in phase space for two dynamical systems is divided into *N-pairs of* accessible sub-domains $(\mho_\alpha, \underline{\mho}_\alpha)$ plus a pair of the inaccessible domain $(\mho_0, \underline{\mho}_0)$. The union of all the accessible sub-domains is $\cup_{\alpha=1}^{N} \mho_\alpha$ and the universal domain is $\mho = \cup_{\alpha=1}^{N} \mho_\alpha \cup \mho_0$ for the dynamical system in the first system of Eq.(9.1). However, for the second dynamical system in Eq.(9.1), the union of all the accessible sub-domains is $\cup_{\alpha=1}^{N} \underline{\mho}_\alpha$ and the universal domain is $\underline{\mho} = \cup_{\alpha=1}^{N} \underline{\mho}_\alpha \cup \underline{\mho}_0$. Both $\mho_0$ and $\underline{\mho}_0$ are the unions of the inaccessible domains for two systems. $\mho_0 = \mho \backslash \cup_{\alpha=1}^{N} \mho_\alpha$ and $\underline{\mho}_0 = \underline{\mho} \backslash \cup_{\alpha=1}^{N} \underline{\mho}_\alpha$ are the complements of the unions of the accessible sub-domain. From the definitions of accessible and inaccessible domains in Chapter 2 (also see, Luo (2006), a continuous dynamical system can be defined on an access-ible domain in phase space. On an inaccessible domain in phase space, no any dynamical systems can be defined. For the first dynamical system in Eq.(9.1), the boundary of two open domains $\mho_\alpha$ and $\mho_\beta$ are $\partial\mho_{\alpha\beta} = \bar{\mho}_\alpha \cap \bar{\mho}_\beta$. For the second dynamical system in Eq.(9.1), one has $\partial\underline{\mho}_{\alpha\beta} = \bar{\underline{\mho}}_\alpha \cap \bar{\underline{\mho}}_\beta$. Such boundaries are formed by the intersection of the closures of sub-domains. These interaction boundaries are time-varying and the corresponding domains will change with time. The sampled systems can be referred to Luo (2009b).

9.1.2. Discontinuous description

Without loss of generality, to avoid the complexity of domains and boundaries generated by the l-non-identical conditions in phase space, consider the jth -condition for two dynamical systems to interact. The boundary is determined by the jth -condition in Eq.(9.4), and the corresponding domain in phase space are divided into two domains $\mho_{(\alpha_j,j)}$ and $\underline{\mho}_{(\alpha_j,j)}$ ($\alpha_j = 1,2$) for two dynamical sys-tems in Eq.(9.1), respectively. Therefore, on the open sub-domain $\mho_{(\alpha_j,j)}$, there is a $C^{r_{\alpha_j}}$-continuous system ($r_{\alpha_j} \geq 1$) in a form of

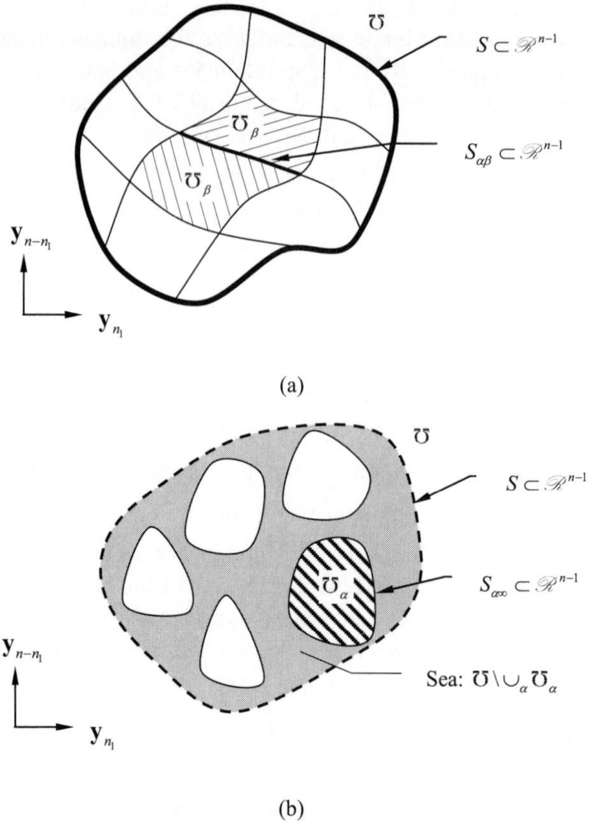

Fig.9.2. The first system in Eq.(9.1): **(a)** connectable and **(b)** separable domains.

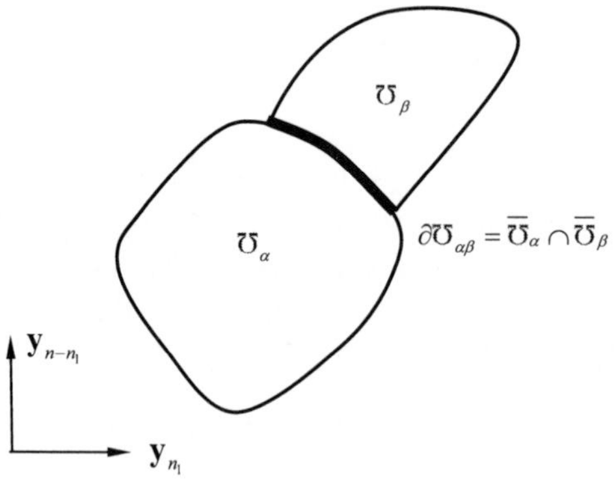

Fig.9.3 The boundary $\partial\mho_{\alpha\beta}$ between $\mho_\alpha$ and $\mho_\beta$ with for the first system in Eq.(9.1).

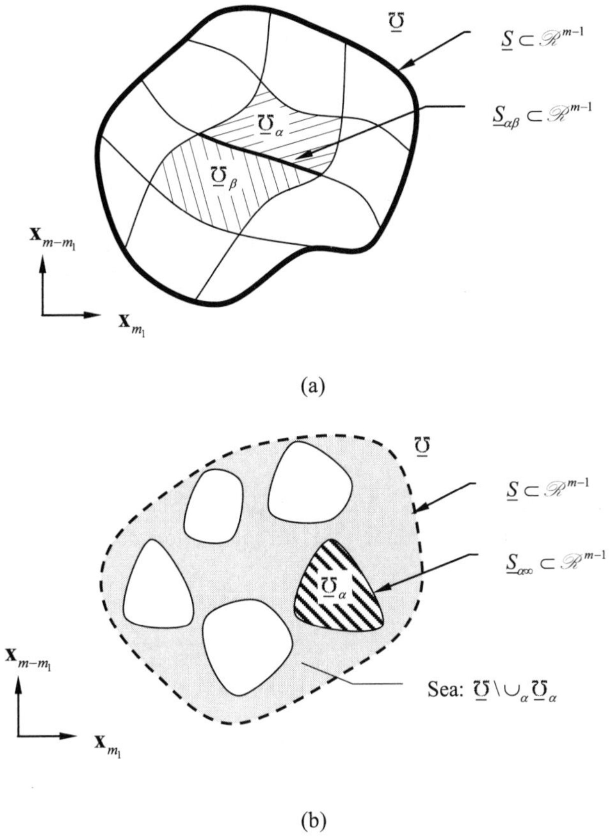

Fig.9.4 The second system in Eq.(9.1): **(a)** connectable and **(b)** separable domains.

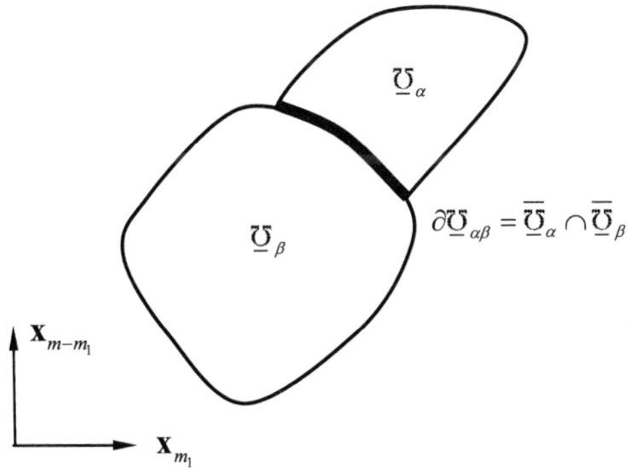

Fig.9.5 The boundary $\partial\underline{\mho}_{\alpha\beta}$ between $\underline{\mho}_\alpha$ and $\underline{\mho}_\beta$ with for the second system in Eq.(9.1).

$$\dot{\mathbf{y}}^{(\alpha_j,j)} = \mathbf{F}^{(\alpha_j,j)}(\mathbf{y}^{(\alpha_j,j)},t,\mathbf{p}^{(\alpha_j,j)}) \in \mathscr{R}^n,$$
$$\mathbf{y}^{(\alpha_j,j)} = (y_1^{(\alpha_j,j)}, y_2^{(\alpha_j,j)},\cdots, y_n^{(\alpha_j,j)})^{\mathrm{T}} \in \mho_{(\alpha_j,j)}. \tag{9.5}$$

In a sub-domain $\mho_{(\alpha_j,j)}$, the vector field of $\mathbf{F}^{(\alpha_j,j)}(\mathbf{y}^{(\alpha_j,j)},t,\mathbf{p}^{(\alpha_j,j)})$ with parameter $\mathbf{p}^{(\alpha_j,j)} = (p_1^{(\alpha_j,j)}, p_2^{(\alpha_j,j)},\cdots, p_k^{(\alpha_j,j)})^{\mathrm{T}} \in \mathscr{R}^{k_j}$ is $C^{r_{\alpha_j}}$-continuous ($r_{\alpha_j} \geq 1$) for a state vector $\mathbf{x}^{(\alpha_j,j)}$ and time t. The continuous flow of the first dynamical system in Eq.(9.1) (i.e., $\mathbf{y}^{(\alpha_j,j)}(t) = \mathbf{\Phi}^{(\alpha_j,j)}(\mathbf{y}^{(\alpha_j,j)}(t_0),t,\mathbf{p}^{(\alpha_j,j)})$) is $C^{r_{\alpha_j}+1}$-continuous for time t with initial condition $\mathbf{y}^{(\alpha_j,j)}(t_0) = \mathbf{\Phi}^{(\alpha_j,j)}(\mathbf{y}^{(\alpha_j,j)}(t_0),t_0,\mathbf{p}^{(\alpha_j,j)})$. The hypothesis (H2.1-H2.4) in Chapter 2 should hold.

The corresponding boundary relative to the jth non-connected interaction is defined as follows.

Definition 9.3. The interaction boundary in n-dimensional phase space for the first dynamical system in Eq.(9.1) under the jth interaction condition in Eq.(9.4) is defined as

$$\begin{aligned} S_{(\alpha_j\beta_j,j)} &\equiv \partial\mho_{(\alpha_j\beta_j,j)} = \overline{\mho}_{(\alpha_j,j)} \cap \overline{\mho}_{(\alpha_j,j)} \\ &= \left\{ \mathbf{y}^{(0,j)} \left| \begin{array}{l} \varphi_j(\mathbf{x}^{(0,j)},\mathbf{y}^{(0,j)},t,\boldsymbol{\lambda}_j) = 0, \\ \varphi_j \text{ is } C^{r_j}\text{-continuous } (r_j \geq 1) \end{array} \right. \right\} \subset \mathscr{R}^{n-1}. \end{aligned} \tag{9.6}$$

Similarly, the discontinuous system of the second system in Eq.(9.1), caused by the interaction at the jth-condition in Eq.(9.4), can be described. On the α_jth ($\alpha_j = 1,2$)-open sub-domain $\underline{\mho}_{(\alpha_j,j)}$, there is a $C^{s_{\alpha_j}}$-continuous system ($s_{\alpha_j} \geq 1$) in a form of

$$\dot{\mathbf{x}}^{(\alpha_j,j)} = \mathscr{F}^{(\alpha_j,j)}(\mathbf{x}^{(\alpha_j,j)},t,\mathbf{q}^{(\alpha_j,j)}) \in \mathscr{R}^m,$$
$$\mathbf{x}^{(\alpha_j,j)} = (x_1^{(\alpha_j,j)}, x_2^{(\alpha_j,j)},\cdots, x_m^{(\alpha_j,j)})^{\mathrm{T}} \in \underline{\mho}_{(\alpha_j,j)}. \tag{9.7}$$

In an sub-domain $\underline{\mho}_{(\alpha_j,j)}$, the vector field of $\mathscr{F}^{(\alpha_j,j)}(\mathbf{x}^{(\alpha_j,j)},t,\mathbf{q}^{(\alpha_j,j)})$ with parameter of $\mathbf{q}^{(\alpha_j,j)} = (q_1^{(\alpha_j,j)}, q_2^{(\alpha_j,j)},\cdots, q_m^{(\alpha_j,j)})^{\mathrm{T}} \in \mathscr{R}^{k_j}$ is $C^{s_{\alpha_j}}$-continuous ($s_{\alpha_j} \geq 1$) in a state vector $\mathbf{x}^{(\alpha_j,j)}$ and for time t. The continuous flow of the second dynamical systems in Eq.(9.1) (i.e., $\mathbf{x}^{(\alpha_j,j)}(t) = \underline{\mathbf{\Phi}}^{(\alpha_j,j)}(\mathbf{x}^{(\alpha_j,j)}(t_0),t,\mathbf{q}^{(\alpha_j,j)})$) is $C^{s_{\alpha_j}+1}$-continuous for time t with initial condition $\mathbf{x}^{(\alpha_j,j)}(t_0) = \underline{\mathbf{\Phi}}^{(\alpha_j,j)}(\mathbf{x}^{(\alpha_j,j)}(t_0),t_0,\mathbf{q}^{(\alpha_j,j)})$. The corresponding hypothesis (H2.1-H2.4) in Chapter 2 should also hold. The corresponding boundary relative to non-connected interaction is defined for the second dynamical system in Eq.(9.1).

Definition 9.4. The boundary in m-dimensional phase space for the second dynamical system in Eq.(9.1) is defined as

$$\begin{aligned} \underline{S}_{(\alpha_j\beta_j,j)} &\equiv \partial\underline{\mho}_{(\alpha_j\beta_j,j)} = \overline{\underline{\mho}}_{(\alpha_j,j)} \cap \overline{\underline{\mho}}_{(\beta_j,j)} \\ &= \left\{ \mathbf{x}^{(0,j)} \left| \begin{array}{l} \varphi_j(\mathbf{x}^{(0,j)},\mathbf{y}^{(0,j)},t,\boldsymbol{\lambda}_j)=0\,, \\ \varphi_j \text{ is } C^{r_j}\text{-continuous } (r_j \geq 1) \end{array} \right. \right\} \subset \mathscr{R}^{m-1}. \end{aligned} \tag{9.8}$$

On the boundaries $\partial\mho_{\alpha_j\beta_j}$ and $\partial\underline{\mho}_{\alpha_j\beta_j}$ with $\varphi_j(\mathbf{x}^{(0,j)},\mathbf{y}^{(0,j)},t,\boldsymbol{\lambda}_j)=0$, there is a dynamical system as

$$\begin{aligned} \dot{\mathbf{y}}^{(0,j)} &= \mathbf{F}^{(0,j)}(\mathbf{y}^{(0,j)},t,\boldsymbol{\lambda}_j), \\ \dot{\mathbf{x}}^{(0,j)} &= \mathscr{F}^{(0,j)}(\mathbf{x}^{(0,j)},t,\boldsymbol{\lambda}_j). \end{aligned} \tag{9.9}$$

where $\mathbf{y}^{(0,j)} = (y_1^{(0,j)}, y_2^{(0,j)}, \cdots, y_n^{(0,j)})^{\mathrm{T}}$ and $\mathbf{x}^{(0,j)} = (x_1^{(0,j)}, x_2^{(0,j)}, \cdots, x_m^{(0,j)})^{\mathrm{T}}$. The corresponding flow $\mathbf{y}^{(0,j)}$, (i.e., $\mathbf{y}^{(0,j)}(t) = \mathbf{\Phi}^{(0,j)}(\mathbf{y}^{(0,j)}(t_0),t,\boldsymbol{\lambda}_j)$) with an initial condition $\mathbf{y}^{(0,j)}(t_0) = \mathbf{\Phi}^{(0,j)}(\mathbf{y}^{(0,j)}(t_0),t_0,\boldsymbol{\lambda}_j)$ is C^{r_j+1}-continuous for time t. The corresponding flow $\mathbf{x}^{(0,j)}$ (i.e., $\mathbf{x}^{(0,j)}(t) = \underline{\mathbf{\Phi}}^{(0,j)}(\mathbf{x}^{(0,j)}(t_0),t,\boldsymbol{\lambda}_j)$) with an initial condition $\mathbf{x}^{(0,j)}(t_0) = \underline{\mathbf{\Phi}}^{(0,j)}(\mathbf{x}^{(0,j)}(t_0),t_0,\boldsymbol{\lambda}_j)$ is also $C^{\underline{r}_j+1}$-continuous.

9.1.3. Resultant dynamical systems

In the previous section, the discontinuity at the interaction boundary for two dynamical systems is described through two different systems. In this section, a resultant system will be introduced to describe such interaction between two dynamical systems. For doing so, a new vector of state variables of two dynamical systems in Eqs.(9.1) is introduced as

$$\boldsymbol{x} = (\mathbf{x};\mathbf{y})^{\mathrm{T}} = (x_1, x_2, \cdots, x_m; y_1, y_2, \cdots, y_n)^{\mathrm{T}} \in \mathscr{R}^{m+n}. \tag{9.10}$$

The notation $(\bullet;\bullet) \equiv (\bullet,\bullet)$ is for a combined vector of state vectors of two dynamical systems. From the interaction condition in Eq.(9.3) or (9.4), the interaction of two dynamical systems in Eq.(9.1) can be investigated through a discontinuous dynamical system, and the corresponding domain in phase pace is separated into two sub-domains by such an interaction boundary. The interaction boundary and domains are described as follows.

Definition 9.5. An interaction boundary in an $(n+m)$-dimensional phase space for the interaction of two dynamical systems in Eq.(9.1) to the interaction condition in Eq.(9.3) is defined as

$$
\begin{aligned}
\partial\Omega_{12} &= \bar{\Omega}_1 \cap \bar{\Omega}_2 \\
&= \left\{ \boldsymbol{x}^{(0)} \left| \begin{array}{l} \varphi(\boldsymbol{x}^{(0)},t,\boldsymbol{\lambda}) \equiv \varphi(\mathbf{x}^{(0)}(t),\mathbf{y}^{(0)}(t),t,\boldsymbol{\lambda}) = 0, \\ \varphi \text{ is } C^r\text{-continuous } (r \geq 1) \end{array} \right. \right\} \\
&\subset \mathscr{R}^{n+m-1};
\end{aligned}
\tag{9.11}
$$

and two corresponding domains for a resultant system of two dynamical systems in Eq.(9.1) are defined as

$$
\begin{aligned}
\Omega_1 &= \left\{ \boldsymbol{x}^{(1)} \left| \begin{array}{l} \varphi(\boldsymbol{x}^{(1)},t,\boldsymbol{\lambda}) \equiv \varphi(\mathbf{x}^{(1)}(t),\mathbf{y}^{(1)}(t),t,\boldsymbol{\lambda}) > 0, \\ \varphi \text{ is } C^r\text{-continuous } (r \geq 1) \end{array} \right. \right\} \\
&\subset \mathscr{R}^{m+n}; \\
\Omega_2 &= \left\{ \boldsymbol{x}^{(2)} \left| \begin{array}{l} \varphi(\boldsymbol{x}^{(2)},t,\boldsymbol{\lambda}) \equiv \varphi(\mathbf{x}^{(2)}(t),\mathbf{y}^{(2)}(t),t,\boldsymbol{\lambda}) < 0, \\ \varphi \text{ is } C^r\text{-continuous } (r \geq 1) \end{array} \right. \right\} \\
&\subset \mathscr{R}^{m+n}.
\end{aligned}
\tag{9.12}
$$

From the previous section, the boundary or domains can be expressed by the direct product of two boundaries or domains

$$
\left.
\begin{array}{l}
\Omega_\alpha = \mho_\alpha \otimes \underline{\mho}_\alpha \quad \text{for } \alpha \in \{1,2\} \\
\partial\Omega_{\alpha\beta} = \partial\mho_{\alpha\beta} \otimes \partial\underline{\mho}_{\alpha\beta} \quad \text{for } \alpha,\beta \in \{1,2\}
\end{array}
\right\}
\tag{9.13}
$$

On the two domains, a resultant system of two dynamical systems is discontinuous to the interaction boundary, defined by

$$
\dot{\boldsymbol{x}}^{(\alpha)} = \mathbb{F}^{(\alpha)}(\boldsymbol{x}^{(\alpha)},t,\boldsymbol{\pi}^{(\alpha)}) \text{ in } \Omega_\alpha \ (\alpha = 1,2)
\tag{9.14}
$$

where

$$
\begin{aligned}
\mathbb{F}^{(\alpha)} &= (\mathscr{F}^{(\alpha)};\mathbf{F}^{(\alpha)})^{\mathrm{T}} \\
&= (\mathscr{F}_1^{(\alpha)},\mathscr{F}_2^{(\alpha)},\cdots,\mathscr{F}_m^{(\alpha)};F_1^{(\alpha)},F_2^{(\alpha)}\cdots,F_n^{(\alpha)})^{\mathrm{T}} \\
\boldsymbol{\pi}^{(\alpha)} &= (\mathbf{q}_\alpha,\mathbf{p}_\alpha)^{\mathrm{T}}.
\end{aligned}
\tag{9.15}
$$

Suppose there is a vector field $\mathbb{F}^{(0)}(\boldsymbol{x}^{(0)},t,\boldsymbol{\lambda})$ on the interaction boundary with $\varphi(\boldsymbol{x}^{(0)},t,\boldsymbol{\lambda}) = 0$, and the corresponding dynamical system on such a boundary is expressed by

$$
\dot{\boldsymbol{x}}^{(0)} = \mathbb{F}^{(0)}(\boldsymbol{x}^{(0)},t,\boldsymbol{\lambda}) \text{ on } \partial\Omega_{12}.
\tag{9.16}
$$

The domains Ω_α ($\alpha = 1,2$) are separated by the constraint boundary of $\partial\Omega_{12}$, as shown in Fig.9.6. For a point $(\mathbf{x}^{(1)};\mathbf{y}^{(1)}) \in \Omega_1$ at time t, $\varphi(\mathbf{x}^{(1)},\mathbf{y}^{(1)},t,\boldsymbol{\lambda}) > 0$. For a point $(\mathbf{x}^{(2)},\mathbf{y}^{(2)})\in\Omega_2$ at time t, $\varphi(\mathbf{x}^{(2)},\mathbf{y}^{(2)},t,\boldsymbol{\lambda})<0$. However, on the boundary $(\mathbf{x}^{(0)},\mathbf{y}^{(0)})\in\partial\Omega_{12}$ at time t, the condition for interaction should be satisfied (i.e., $\varphi(\mathbf{x}^{(0)},\mathbf{y}^{(0)},t,\boldsymbol{\lambda})=0$). If the interaction condition is time-independent, the interaction boundary determined by the interaction condition is invariant. If the interaction condition is time-dependent, the interaction boundary determined by the interaction condition is time-varying, and the corresponding domain for the resultant system is time-varying. As in Eq.(9.4), there are many conditions for interactions of two dynamical systems, Suppose only the jth interaction boundary occurs for two system for time t. The above definitions can be extended accordingly.

Definition 9.6. The jth interaction boundary in an $(n+m)$-dimensional phase space for the interaction of two dynamical systems in Eq.(9.1), relative to the jth constraint of the interaction conditions in Eq.(9.4), is defined as

$$\begin{aligned}\partial\Omega_{(\alpha_j\beta_j,j)}&=\bar{\Omega}_{(\alpha_j,j)}\cap\bar{\Omega}_{(\beta_j,j)}\\&=\left\{\boldsymbol{x}^{(0,j)}\left|\begin{array}{l}\varphi_j(\boldsymbol{x}^{(0,j)},t,\boldsymbol{\lambda}_j)\\ \equiv\varphi_j(\mathbf{x}^{(0,j)}(t),\mathbf{y}^{(0,j)}(t),t,\boldsymbol{\lambda}_j)=0,\\ \varphi_j \text{ is } C^{r_j}\text{-continuous } (r_j\geq 1)\end{array}\right.\right\}\\&\subset\mathscr{R}^{m+n-1};\end{aligned}\tag{9.17}$$

and two domains pertaining to the jth boundary for a resultant system of two dynamical systems in Eq.(9.1) are defined as

$$\begin{aligned}\Omega_{(1,j)}&=\left\{\boldsymbol{x}^{(1,j)}\left|\begin{array}{l}\varphi_j(\boldsymbol{x}^{(1,j)},t,\boldsymbol{\lambda}_j)\\ \equiv\varphi_j(\mathbf{x}^{(1,j)}(t),\mathbf{y}^{(1,j)}(t),t,\boldsymbol{\lambda}_j)>0\,,\\ \varphi_j \text{ is } C^{r_j}\text{-continuous } (r_j\geq 1)\end{array}\right.\right\}\subset\mathscr{R}^{m+n};\\ \Omega_{(2,j)}&=\left\{\boldsymbol{x}^{(2,j)}\left|\begin{array}{l}\varphi_j(\boldsymbol{x}^{(2,j)},t,\boldsymbol{\lambda}_j)\\ \equiv\varphi_j(\mathbf{x}^{(2,j)}(t),\mathbf{y}^{(2,j)}(t),t,\boldsymbol{\lambda}_j)<0\,,\\ \varphi_j \text{ is } C^{r_j}\text{-continuous } (r_j\geq 1)\end{array}\right.\right\}\subset\mathscr{R}^{m+n}.\end{aligned}\tag{9.18}$$

Such a boundary and domains for the jth interaction condition are sketched in Fig.9.7, which can be expressed by the direct product of two corresponding individual boundaries or domains

$$\left.\begin{aligned}\Omega_{(\alpha_j,j)}&=\mho_{(\alpha_j,j)}\otimes\underline{\mho}_{(\alpha_j,j)} \text{ for } \alpha_j\in\{1,2\},\\ \partial\Omega_{(\alpha_j\beta_j,j)}&=\partial\mho_{(\alpha_j\beta_j,j)}\otimes\partial\underline{\mho}_{(\alpha_j\beta_j,j)} \text{ for } \alpha_j,\beta_j\in\{1,2\}\end{aligned}\right\}\tag{9.19}$$

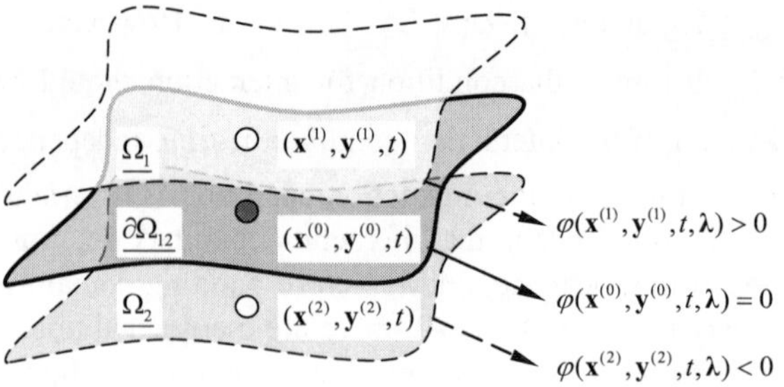

Fig. 9.6 Interaction boundary and domains in $(n+m)$ -dimensional state space.

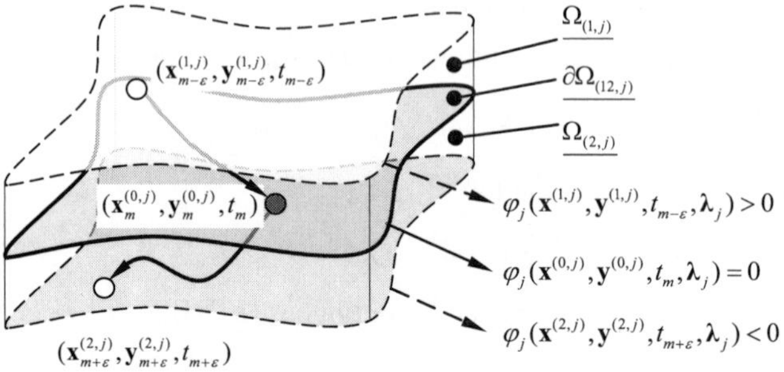

Fig.9.7 The jth -boundary and domains of a resultant flow to the interaction boundary in $(n+m)$ -dimensional state space.

On the two domains relative to the jth interaction boundary, a discontinuous resultant system of two dynamical systems in Eqs.(9.1) with the jth interaction in Eq.(9.4) is defined by

$$\dot{\boldsymbol{x}}^{(\alpha_j,j)} = \mathbb{F}^{(\alpha_j,j)}(\boldsymbol{x}^{(\alpha_j,j)},t,\boldsymbol{\pi}_j^{(\alpha_j)}) \text{ in } \Omega_{(\alpha_j,j)} \tag{9.20}$$

where $\mathbb{F}^{(\alpha_j,j)} = (\mathscr{F}^{(\alpha_j,j)};\mathbf{F}^{(\alpha_j,j)})^{\mathrm{T}}$ and $\boldsymbol{\pi}_j^{(\alpha_j)} = (\mathbf{q}_j^{(\alpha_j)},\mathbf{p}_j^{(\alpha_j)})^{\mathrm{T}}$. Suppose there is a vector field $\mathbb{F}^{(0,j)}(\boldsymbol{x}^{(0,j)},t,\boldsymbol{\lambda}_j)$ on the jth interaction boundary with $\varphi_j(\boldsymbol{x}^{(0,j)},t,\boldsymbol{\lambda}_j)$ $=0$, and the corresponding dynamical system on the jth interaction boundary is expressed by

$$\dot{\boldsymbol{x}}^{(0,j)} = \mathbb{F}^{(0,j)}(\boldsymbol{x}^{(0,j)},t,\boldsymbol{\lambda}_j) \text{ on } \partial\Omega_{(12,j)}. \tag{9.21}$$

Based on the discontinuous description, the dynamical behaviors between two discontinuous dynamical systems with interaction boundary can be investigated through the theory in Chapter 3. In next section, a generalized relative coordinate system to each interaction will be introduced, and the fundamental interaction of two dynamical systems will be discussed.

9.2. Fundamental interactions

In this section, the interaction behaviors between two dynamical systems will be discussed in the vicinity of interaction boundary. Herein, new variables are introduced in domain $\Omega_{(\alpha_j,j)}$ as

$$z^{(\alpha_j,j)} = \varphi_j(\mathbf{x}^{(\alpha_j,j)}(t),\mathbf{y}^{(\alpha_j,j)}(t),t,\boldsymbol{\lambda}_j) \text{ for } j \in \mathscr{L}. \tag{9.22}$$

On the boundary $\partial\Omega_{(\alpha_j\beta_j,j)}$,

$$z^{(0,j)} = \varphi_j(\mathbf{x}^{(0,j)}(t),\mathbf{y}^{(0,j)}(t),t,\boldsymbol{\lambda}_j) = 0 \text{ for } j \in \mathscr{L}. \tag{9.23}$$

If the two systems do not interact each other, the new variables ($z_j \neq 0$, $j \in \mathscr{L}$) will change with time t. The corresponding time-change rate is given by

$$\begin{aligned}
\dot{z}^{(\alpha_j,j)} &= D\varphi_j(\mathbf{x}^{(\alpha_j,j)},\mathbf{y}^{(\alpha_j,j)},t,\boldsymbol{\lambda}_j) \\
&= \frac{\partial\varphi_j}{\partial\mathbf{x}^{(\alpha_j,j)}}\dot{\mathbf{x}}^{(\alpha_j,j)} + \frac{\partial\varphi_j}{\partial\mathbf{y}^{(\alpha_j,j)}}\dot{\mathbf{y}}^{(\alpha_j,j)} + \frac{\partial\varphi_j}{\partial t} \\
&= \sum_{p=1}^{m}\frac{\partial\varphi_j}{\partial x_p^{(\alpha_j,j)}}\dot{x}_p^{(\alpha_j,j)} + \sum_{q=1}^{n}\frac{\partial\varphi_j}{\partial y_q^{(\alpha_j,j)}}\dot{y}_q^{(\alpha_j,j)} + \frac{\partial\varphi_j}{\partial t}
\end{aligned} \tag{9.24}$$

Substitution of Eqs. (9.5) and (9.7) into Eq.(9.24) yields

$$\begin{aligned}
\dot{z}^{(\alpha_j,j)} &= \sum_{p=1}^{m}\frac{\partial\varphi_j}{\partial x_p^{(\alpha_j,j)}}\mathscr{F}_p^{(\alpha_j,j)}(\mathbf{x}^{(\alpha_j,j)},t,\mathbf{p}^{(\alpha_j,j)}) \\
&\quad + \sum_{q=1}^{n}\frac{\partial\varphi_j}{\partial y_q^{(\alpha_j,j)}}F_q^{(\alpha_j,j)}(\mathbf{y}^{(\alpha_j,j)}t,\mathbf{q}^{(\alpha_j,j)}) + \frac{\partial\varphi_j}{\partial t}
\end{aligned} \tag{9.25}$$

Two new normal vectors are defined as

$$\begin{aligned}
\underline{\mathbf{n}}_{\varphi_j} &= \frac{\partial\varphi_j}{\partial\mathbf{x}^{(\alpha_j,j)}} = \Big(\frac{\partial\varphi_j}{\partial x_1^{(\alpha_j,j)}},\frac{\partial\varphi_j}{\partial x_2^{(\alpha_j,j)}},\cdots,\frac{\partial\varphi_j}{\partial x_m^{(\alpha_j,j)}}\Big)^{\mathrm{T}}, \\
\mathbf{n}_{\varphi_j} &= \frac{\partial\varphi_j}{\partial\mathbf{y}^{(\alpha_j,j)}} = \Big(\frac{\partial\varphi_j}{\partial y},\frac{\partial\varphi_j}{\partial y_2^{(\alpha_j,j)}},\cdots,\frac{\partial\varphi_j}{\partial y_n^{(\alpha_j,j)}}\Big)^{\mathrm{T}}.
\end{aligned} \tag{9.26}$$

Using Eq.(9.26), equation (9.25) becomes

$$\begin{aligned}
\dot{z}^{(\alpha_j,j)} &= \underline{\mathbf{n}}_{\varphi_j}\cdot\mathbf{F}^{(\alpha_j,j)}(\mathbf{x}^{(\alpha_j,j)},t,\mathbf{p}^{(\alpha_j,j)}) \\
&\quad + \mathbf{n}_{\varphi_j}\cdot\mathscr{F}^{(\alpha_j,j)}(\mathbf{y}^{(\alpha_j,j)},t,\mathbf{q}^{(\alpha_j,j)}) + \frac{\partial\varphi_j}{\partial t}.
\end{aligned} \tag{9.27}$$

If the vector fields in different domains $\Omega_{(\alpha_j,j)}$ ($\alpha_j = 1,2$) are distinguishing,

$\dot{z}^{(\alpha_j,j)}$ is discontinuous. Similarly, for each domain $\Omega_{(\alpha_j,j)}$, we have

$$\ddot{z}^{(\alpha_j,j)} = \frac{D}{Dt}\Big[\underline{\mathbf{n}}_{\varphi_j} \cdot \mathscr{F}^{(\alpha_j,j)}(\mathbf{x}^{(\alpha_j,j)}, t, \mathbf{q}^{(\alpha_j,j)}) + \mathbf{n}_{\varphi_j} \cdot \mathbf{F}^{(\alpha_j,j)}(\mathbf{y}^{(\alpha_j,j)}, t, \mathbf{p}^{(\alpha_j,j)}) + \frac{\partial \varphi_j}{\partial t}\Big]. \tag{9.28}$$

The combination of Eqs.(9.24) and (9.28) gives a dynamical system in phase space of $(z, \dot{z})$, i.e., for $j \in \mathscr{L}$

$$\begin{aligned}
\dot{z}^{(\alpha_j,j)} &= g_1^{(\alpha_j,j)}(\mathbf{z}^{(\alpha_j,j)}, t) \equiv \underline{\mathbf{n}}_{\varphi_j} \cdot \mathscr{F}^{(\alpha_j,j)}(\mathbf{x}^{(\alpha_j,j)}, t, \mathbf{p}^{(\alpha_j,j)}) \\
&\quad + \mathbf{n}_{\varphi_j} \cdot \mathbf{F}^{(\alpha_j,j)}(\mathbf{y}^{(\alpha_j,j)}, t, \mathbf{q}^{(\alpha_j,j)}) + \frac{\partial \varphi_j}{\partial t}, \\
\ddot{z}^{(\alpha_j,j)} &= g_2^{(\alpha_j,j)}(\mathbf{z}^{(\alpha_j,j)}, t) \equiv \frac{D}{Dt} g_1^{(\alpha_j,j)}(\mathbf{z}^{(\alpha_j,j)}, t) \\
&= \frac{D}{Dt}\Big[\underline{\mathbf{n}}_{\varphi_j} \cdot \mathscr{F}^{(\alpha_j,j)}(\mathbf{x}^{(\alpha_j,j)}, t, \mathbf{p}^{(\alpha_j,j)}) \\
&\quad + \mathbf{n}_{\varphi_j} \cdot \mathbf{F}^{(\alpha_j,j)}(\mathbf{y}^{(\alpha_j,j)}, t, \mathbf{q}^{(\alpha_j,j)}) + \frac{\partial \varphi_j}{\partial t}\Big].
\end{aligned} \tag{9.29}$$

where $\mathbf{z}^{(\alpha_j,j)} = (z^{(\alpha_j,j)}, \dot{z}^{(\alpha_j,j)})^{\mathrm{T}}$. Letting $\mathbf{g}^{(\alpha_j,j)} = (g_1^{(\alpha_j,j)}, g_2^{(\alpha_j,j)})^{\mathrm{T}}$, one obtains

$$\left.\begin{aligned}
\dot{\mathbf{z}}^{(\alpha_j,j)} &= \mathbf{g}^{(\alpha_j,j)}(\mathbf{z}^{(\alpha_j,j)}, t) \ \text{for}\ j \in \mathscr{L}; \\
\dot{\mathbf{x}}^{(\alpha_j,j)} &= \mathscr{F}^{(\alpha_j,j)}(\mathbf{x}^{(\alpha_j,j)}, t, \mathbf{q}^{(\alpha_j,j)}) \in \mathscr{R}^m, \\
\dot{\mathbf{y}}^{(\alpha_j,j)} &= \mathbf{F}^{(\alpha_j,j)}(\mathbf{y}^{(\alpha_j,j)}, t, \mathbf{p}^{(\alpha_j,j)}) \in \mathscr{R}^n.
\end{aligned}\right\} \tag{9.30}$$

For a better understanding of such a discontinuous dynamical system, the boundary and domains in phase space are defined as

$$\begin{aligned}
\partial\Xi_{(\alpha_j\beta_j,j)} &= \bar{\Xi}_{(\alpha_j,j)} \cap \bar{\Xi}_{(\beta_j,j)} \\
&= \left\{(z^{(0,j)}, \dot{z}^{(0,j)}) \middle| \psi_j(z^{(0,j)}, \dot{z}^{(0,j)}) = z^{(0,j)} = 0\right\} \\
&\subset \mathscr{R};
\end{aligned} \tag{9.31}$$

and

$$\begin{aligned}
\Xi_{(1,j)} &= \left\{(z^{(1,j)}, \dot{z}^{(1,j)}) \middle| z^{(1,j)} > 0\right\} \subset \mathscr{R}^2; \\
\Xi_{(2,j)} &= \left\{(z^{(1,j)}, \dot{z}^{(1,j)}) \middle| z^{(2,j)} < 0\right\} \subset \mathscr{R}^2.
\end{aligned} \tag{9.32}$$

$\varphi_j(\mathbf{x}^{(0,j)}(t), \mathbf{y}^{(0,j)}(t), t, \boldsymbol{\lambda}_j) = 0$ on the boundary yields

$$\frac{d^s z^{(0,j)}}{dt^s} = D^s \varphi_j(\mathbf{x}^{(0,j)}(t), \mathbf{y}^{(0,j)}(t), t, \boldsymbol{\lambda}_j) = 0 \text{ for } s = 1, 2, \cdots \tag{9.33}$$

Thus, the interaction boundary is governed by

$$\left.\begin{aligned} z^{(0,j)} &= \varphi_j(\mathbf{x}^{(0,j)}, \mathbf{y}^{(0,j)}, t, \boldsymbol{\lambda}_j) = 0, \\ \dot{z}^{(0,j)} &= D\varphi_j(\mathbf{x}^{(0,j)}, \mathbf{y}^{(0,j)}, t, \boldsymbol{\lambda}_j) = 0 \end{aligned}\right\} \text{ for } j \in \mathcal{L};$$
$$\begin{aligned} \dot{\mathbf{x}}^{(0,j)} &= \mathscr{F}^{(0,j)}(\mathbf{x}^{(0,j)}, t, \boldsymbol{\lambda}_j) \in \mathscr{R}^n, \\ \dot{\mathbf{y}}^{(0,j)} &= \mathbf{F}^{(0,j)}(\mathbf{y}^{(0,j)}, t, \boldsymbol{\lambda}_j) \in \mathscr{R}^m. \end{aligned} \tag{9.34}$$

The domains and boundary in phase space of $(z^{(j)}, \dot{z}^{(j)})$ are sketched in Fig.9.8 and the location for switching may not be continuous (i.e., $\mathbf{z}^{(\alpha_j,j)} \neq \mathbf{z}^{(\beta_j,j)} \neq \mathbf{z}^{(0,j)} = 0$) because the vector fields of the resultant system are discontinuous (or $\dot{z}^{(\alpha_j,j)} \neq \dot{z}^{(\beta_j,j)} \neq \dot{z}^{(0,j)} = 0$), but the boundary in such phase space is independent of time. However, the boundaries and domains in phase space of dynamical systems in Eqs.(9.5) and (9.7) are shown in Fig.9.9. The interaction boundary varying with time is presented, but switching points for a flow are continuous (i.e., $\boldsymbol{x}^{(\alpha_j,j)}(t_m) = \boldsymbol{x}^{(\beta_j,j)}(t_m) = \boldsymbol{x}^{(0,j)}(t_m)$). So the dynamical response will be completely determined by Eq.(9.14). However, such flows will be controlled by the vector fields $\mathbf{g}^{(1,j)}(\mathbf{z}^{(1,j)}, t)$ and $\mathbf{g}^{(2,j)}(\mathbf{z}^{(2,j)}, t)$. The dynamical systems in phase space $(z, \dot{z})$ are summarized as follows.

$$\begin{aligned} \dot{\mathbf{z}}^{(\Lambda_j,j)} &= \mathbf{g}^{(\Lambda_j,j)}(\mathbf{z}^{(\Lambda_j,j)}, t) \text{ for } j \in \mathcal{L}, \Lambda_j = 0, \alpha_j \\ \dot{\mathbf{x}}^{(\Lambda_j,j)} &= \mathscr{F}^{(\Lambda_j,j)}(\mathbf{x}^{(\Lambda_j,j)}, t, \boldsymbol{\lambda}_j) \in \mathscr{R}^n, \\ \dot{\mathbf{y}}^{(\Lambda_j,j)} &= \mathbf{F}^{(\Lambda_j,j)}(\mathbf{y}^{(\Lambda_j,j)}, t, \boldsymbol{\lambda}_j) \in \mathscr{R}^m. \end{aligned} \tag{9.35}$$

where

$$\left.\begin{aligned} &\mathbf{g}^{(\alpha_j,j)}(\mathbf{z}^{(\alpha_j,j)}, t) = (g_1^{(\alpha_j,j)}(\mathbf{z}^{(\alpha_j,j)}, t), g_2^{(\alpha_j,j)}(\mathbf{z}^{(\alpha_j,j)}, t))^{\mathrm{T}} \\ &\qquad \text{in } \Xi_{\alpha_j} \ (\alpha_j \in \{1,2\}); \\ &\mathbf{g}^{(0,j)}(\mathbf{z}^{(\alpha_j,j)}, t) \in [\mathbf{g}^{(\alpha_j,j)}(\mathbf{z}^{(\alpha_j,j)}, t),\ \mathbf{g}^{(\beta_j,j)}(\mathbf{z}^{(\beta_j,j)}, t)] \\ &\qquad \text{on } \partial\Xi_{(\alpha_j\beta_j,j)} \text{ for non-stick,} \\ &\mathbf{g}^{(0,j)}(\mathbf{z}^{(\alpha_j,j)}, t) = (0,0)^T \text{ on } \partial\Xi_{(\alpha_j\beta_j,j)} \text{ for stick.} \end{aligned}\right\} \tag{9.36}$$

The normal vector of $\partial\Xi_{(\alpha_j\beta_j,j)}$ is computed from Eq.(9.31), i.e.,

$$\mathbf{n}_{\partial\Xi_{(\alpha_j\beta_j,j)}} = (1,0)^{\mathrm{T}} \text{ and } D\mathbf{n}_{\partial\Xi_{(\alpha_j\beta_j,j)}} = (0,0)^{\mathrm{T}}, \tag{9.37}$$

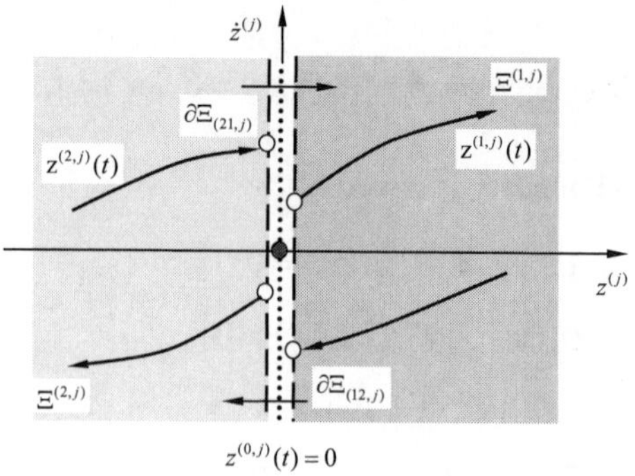

Fig.9.8 A partition of phase space in $(z,\dot{z})$ for the jth interaction boundary. Two dashed lines are infinitesimally close to the boundary with the dotted line.

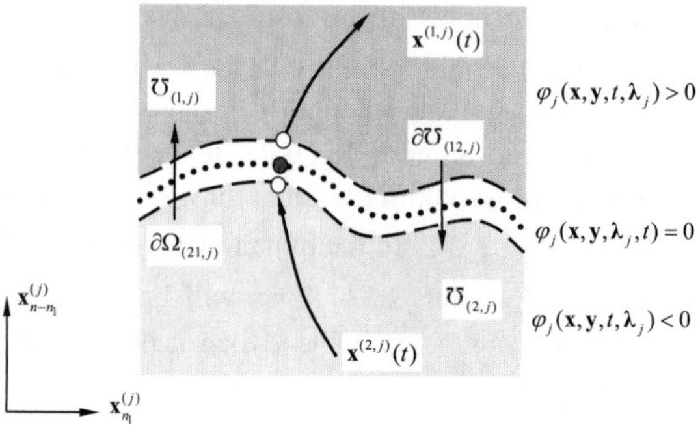

(a)

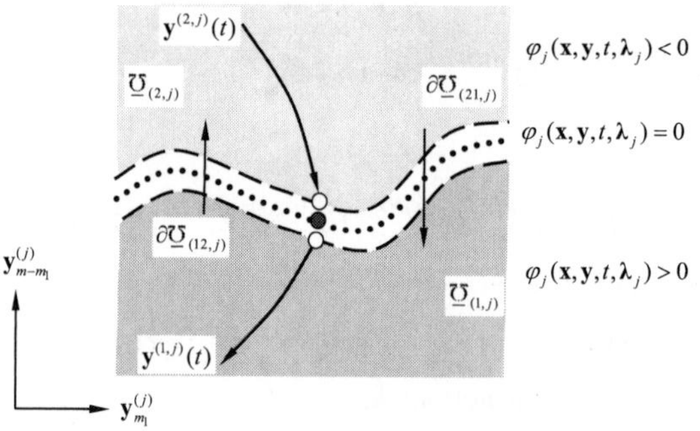

(b)

Fig. 9.9 Phase space partitions: **(a)** the first system and **(b)** the second system. Two dashed curves are infinitesimally close to the boundary with the dotted curves.

where $D(\cdot) = D(\cdot)/Dt$. From Luo(2008a,b), the corresponding two G-functions are computed by

$$\begin{aligned} G^{(0,\alpha_j)}_{\partial\Xi_{(12,j)}}(\mathbf{z}^{(\alpha_j,j)},t) &= \mathbf{n}_{\partial\Xi_{(12,j)}} \cdot \mathbf{g}^{(\alpha_j,j)}(\mathbf{z}^{(\alpha_j,j)},t) \\ &= g_1^{(\alpha_j,j)}(\mathbf{z}^{(\alpha_j,j)},t), \\ G^{(1,\alpha_j)}_{\partial\Xi_{(12,j)}}(\mathbf{z}^{(\alpha_j,j)},t) &= \mathbf{n}_{\partial\Xi_{(12,j)}} \cdot D\mathbf{g}^{(\alpha_j,j)}(\mathbf{z}^{(\alpha_j,j)},t) \\ &= g_2^{(\alpha_j,j)}(\mathbf{z}^{(\alpha_j,j)},t). \end{aligned} \tag{9.38}$$

With G-functions, the sufficient and necessary conditions for a passable flow at $(\mathbf{z}_m^{(0,j)},t_m)$ and $(\mathbf{z}_m^{(\alpha_j,j)},t_m)$ for the boundary $\partial\Xi_{(12,j)}$ are given from Chapter 3 (also see, Luo 2005, 2006), i.e.,

$$\left.\begin{aligned} G^{(0,1)}_{\partial\Xi_{(12,j)}}(\mathbf{z}_m^{(1,j)},t_{m-}) = g_1^{(1,j)}(\mathbf{z}_m^{(1,j)},t_{m-}) < 0, \\ G^{(0,2)}_{\partial\Xi_{(12,j)}}(\mathbf{z}_m^{(2,j)},t_{m+}) = g_1^{(2,j)}(\mathbf{z}_m^{(2,j)},t_{m+}) < 0 \end{aligned}\right\} \text{from } \Xi_{(1,j)} \to \Xi_{(2,j)}$$
$$\left.\begin{aligned} G^{(0,1)}_{\partial\Xi_{(12,j)}}(\mathbf{z}_m^{(1,j)},t_{m+}) = g_1^{(1,j)}(\mathbf{z}_m^{(1,j)},t_{m+}) > 0, \\ G^{(0,2)}_{\partial\Xi_{(12,j)}}(\mathbf{z}_m^{(2,j)},t_{m-}) = g_1^{(2,j)}(\mathbf{z}_m^{(2,j)},t_{m-}) > 0 \end{aligned}\right\} \text{from } \Xi_{(2,j)} \to \Xi_{(1,j)} \tag{9.39}$$

where

$$\begin{aligned} g_1^{(\alpha_j,j)}(\mathbf{z}_m^{(\alpha_j,j)},t_{m\pm}) &= \underline{\mathbf{n}}_{\varphi_j} \cdot \mathscr{F}^{(\alpha_j,j)}(\mathbf{x}_m^{(\alpha_j,j)},t_{m\pm},\mathbf{q}^{(\alpha_j,j)}) \\ &\quad + \mathbf{n}_{\varphi_j} \cdot \mathbf{F}^{(\alpha_j,j)}(\mathbf{y}_m^{(\alpha_j,j)},t_{m\pm},\mathbf{p}^{(\alpha_j,j)}) + \frac{\partial\varphi_j}{\partial t}. \end{aligned} \tag{9.40}$$

The foregoing condition gives the sufficient and necessary conditions for two systems to interact under the jth interaction condition and to switch the current states through such an interaction condition. Such a flow to the boundary is called an *instantaneous interaction* between two systems.

The sufficient and necessary conditions for a stick flow (or sink flow) on the boundary $\partial\Xi_{(\alpha_j\beta_j,j)}$ are obtained from Chapter 3 (also see, Luo, 2005, 2006), i.e.,

$$\left.\begin{aligned} G^{(0,1)}_{\partial\Xi_{(12,j)}}(\mathbf{z}_m^{(1,j)},t_{m-}) = g_1^{(1,j)}(\mathbf{z}_m^{(1,j)},t_{m-}) < 0, \\ G^{(0,2)}_{\partial\Xi_{(12,j)}}(\mathbf{z}_m^{(2,j)},t_{m-}) = g_1^{(2,j)}(\mathbf{z}_m^{(2,j)},t_{m-}) > 0 \end{aligned}\right\} \text{on } \partial\Xi_{(12,j)}. \tag{9.41}$$

From the foregoing condition, the two systems will stick together under the jth interaction condition, which is called a *stick interaction*.

Similarly, the sufficient and necessary conditions for a source flow on the boundary $\partial\Xi_{(12,j)}$ are given in Chapter 3 (also see, Luo, 2005, 2006), i.e.,

$$\left.\begin{aligned} G^{(0,1)}_{\partial\Xi_{(12,j)}}(\mathbf{z}_m^{(1,j)},t_{m+}) &= g_1^{(1,j)}(\mathbf{z}_m^{(1,j)},t_{m+})>0,\\ G^{(0,2)}_{\partial\Xi_{(12,j)}}(\mathbf{z}_m^{(2,j)},t_{m+}) &= g_1^{(2,j)}(\mathbf{z}_m^{(2,j)},t_{m+})<0 \end{aligned}\right\}\text{ on }\partial\Xi_{(12,j)}. \tag{9.42}$$

For this case, the two dynamical systems will not interact at $(\mathbf{z}_m^{(0,j)},t_m)$ for the boundary $\partial\Xi_{(\alpha_j\beta_j,j)}$ relative to the jth interaction condition. The phenomenon is called the *source* interaction (or de-interaction).

As in Chapter 3, an L-function is introduced to measure the above three interaction states

$$\begin{aligned} L_{12}^{(j)}(t_{m\pm}) &= G^{(0,\alpha_j)}_{\partial\Xi_{(\alpha_j\beta_j,j)}}(\mathbf{z}_m^{(\alpha_j,j)},t_{m-})\times G^{(0,\beta_j)}_{\partial\Xi_{(\alpha_j\beta_j,j)}}(\mathbf{z}_m^{(\beta_j,j)},t_{m+})\\ &= g_1^{(\alpha_j,j)}(\mathbf{z}_m^{(\alpha_j,j)},t_{m-})\times g_1^{(\beta_j,j)}(\mathbf{z}_m^{(\beta_j,j)},t_{m+});\\ L_{12}^{(j)}(t_{m-}) &= G^{(0,1)}_{\partial\Xi_{(12,j)}}(\mathbf{z}_m^{(1,j)},t_{m-})\times G^{(0,2)}_{\partial\Xi_{(12,j)}}(\mathbf{z}_m^{(2,j)},t_{m-})\\ &= g_1^{(1,j)}(\mathbf{z}_m^{(1,j)},t_{m-})\times g_1^{(2,j)}(\mathbf{z}_m^{(2,j)},t_{m-}),\\ L_{12}^{(j)}(t_{m+}) &= G^{(0,1)}_{\partial\Xi_{(12,j)}}(\mathbf{z}_m^{(1,j)},t_{m+})\times G^{(0,2)}_{\partial\Xi_{(12,j)}}(\mathbf{z}_m^{(2,j)},t_{m+})\\ &= g_1^{(1,j)}(\mathbf{z}_m^{(1,j)},t_{m+})\times g_1^{(2,j)}(\mathbf{z}_m^{(2,j)},t_{m+}). \end{aligned} \tag{9.43}$$

No matter what interaction exists, the same quantity can be used for measuring the three interaction states. Such a quantity can be easily embedded in computer program (e.g., Luo and Gegg, 2006a,b; Luo, 2006). With Eq.(9.43), equations (9.39), (9.41) and (9.42) yield three new forms of the necessary and sufficient conditions for three interaction states.

$$\begin{aligned} L_{12}^{(j)}(t_{m\pm}) &= g_1^{(\alpha_j,j)}(\mathbf{z}_m^{(\alpha_j,j)},t_{m\mp})\times g_1^{(\beta_j,j)}(\mathbf{z}_m^{(\beta_j,j)},t_{m\pm})>0,\\ L_{12}^{(j)}(t_{m-}) &= g_1^{(1,j)}(\mathbf{z}_m^{(1,j)},t_{m-})\times g_1^{(2,j)}(\mathbf{z}_m^{(2,j)},t_{m-})<0,\\ L_{12}^{(j)}(t_{m+}) &= g_1^{(1,j)}(\mathbf{z}_m^{(1,j)},t_{m+})\times g_1^{(2,j)}(\mathbf{z}_m^{(2,j)},t_{m+})<0. \end{aligned} \tag{9.44}$$

The appearance and disappearance of three interaction states of the two dynamical systems to the jth interaction condition in Eq.(9.4) can be determined from Luo (2008a,b), which are for the switching bifurcation of three states of interactions between the two dynamical systems.

(i) For appearance of the stick interaction from the instantaneous interaction, the sufficient and necessary conditions are given by

$$\begin{aligned} &(-1)^{\alpha_j}G^{(0,\alpha_j)}_{\partial\Xi_{(\alpha_i\beta_j,j)}}(\mathbf{z}_m^{(\alpha_j,j)},t_{m-}) = (-1)^{\alpha_j}g_1^{(\alpha_j,j)}(\mathbf{z}_m^{(\alpha_j,j)},t_{m-})>0,\\ &G^{(0,\beta_j)}_{\partial\Xi_{(\alpha_i\beta_j,j)}}(\mathbf{z}_m^{(\beta_j,j)},t_{m\pm}) = g_1^{(\beta_j,j)}(\mathbf{z}_m^{(\beta_j,j)},t_{m\pm})=0,\\ &(-1)^{\beta_j}G^{(1,\beta_j)}_{\partial\Xi_{(\alpha_i\beta_j,j)}}(\mathbf{z}_m^{(\beta_j,j)},t_{m\pm}) = (-1)^{\beta_j}g_2^{(\beta_j,j)}(\mathbf{z}_m^{(\beta_j,j)},t_{m\pm})<0. \end{aligned} \tag{9.45}$$

For the vanishing of the stick interaction to form the instantaneous interaction on the *j*th interaction boundary, the sufficient and necessary conditions are

$$\left.\begin{aligned}&(-1)^{\alpha_j} G^{(0,\alpha_j)}_{\partial\Xi_{(\alpha_i\beta_j,j)}}(\mathbf{z}^{(\alpha_j,j)}_m,t_{m-})=(-1)^{\alpha_j} g^{(\alpha_j,j)}_1(\mathbf{z}^{(\alpha_j,j)}_m,t_{m-})>0,\\&G^{(0,\beta_j)}_{\partial\Xi_{(\alpha_i\beta_j,j)}}(\mathbf{z}^{(\beta_j,j)}_m,t_{m\mp})=g^{(\beta_j,j)}_1(\mathbf{z}^{(\beta_j,j)}_m,t_{m\mp})=0,\\&(-1)^{\beta_j} G^{(1,\beta_j)}_{\partial\Xi_{(\alpha_i\beta_j,j)}}(\mathbf{z}^{(\beta_j,j)}_m,t_{m\mp})=(-1)^{\beta_j} g^{(\beta_j,j)}_2(\mathbf{z}^{(\beta_j,j)}_m,t_{m\mp})<0.\end{aligned}\right\} \quad (9.46)$$

The foregoing appearance and vanishing conditions for the stick interaction relative to the instantaneous interaction in Eq.(9.45) are also the vanishing and appearance conditions for the instantaneous interaction relative to the stick interaction, respectively. As in Chapter 3 (also see Luo, 2006, 2008a,b), such appearance and vanishing conditions are the switching bifurcations between the stick and instantaneous interactions of the two dynamical system under the *j*th interaction condition. Once the *L*-function in Eq.(9.43) is used, equations (9.45) and (9.46) become

$$\left.\begin{aligned}&L^{(j)}_{12}(t_{m\pm})=g^{(\alpha_j,j)}_1(\mathbf{z}^{(\alpha_j,j)}_m,t_{m\mp})\times g^{(\beta_j,j)}_1(\mathbf{z}^{(\beta_j,j)}_m,t_{m\pm})=0,\\&(-1)^{\beta_j} G^{(1,\beta_j)}_{\partial\Xi_{(\alpha_i\beta_j,j)}}(\mathbf{z}^{(\beta_j,j)}_m,t_{m\pm})=(-1)^{\beta_j} g^{(\beta_j,j)}_2(\mathbf{z}^{(\beta_j,j)}_m,t_{m\pm})<0,\\&(-1)^{\alpha_j} G^{(0,\alpha_j)}_{\partial\Xi_{(\alpha_i\beta_j,j)}}(\mathbf{z}^{(\alpha_j,j)}_m,t_{m-})=(-1)^{\alpha_j} g^{(\alpha_j,j)}_1(\mathbf{z}^{(\alpha_j,j)}_m,t_{m-})>0.\end{aligned}\right\} \quad (9.47)$$

From the foregoing equation, the *L*-function for such switching bifurcation is zero.

(ii) From Chapter 3 (also see, Luo, 2008a,b), the sufficient and necessary conditions for appearance and vanishing of the *source* interaction, pertaining to the instantaneous interaction are obtained. For the appearance of source interaction, one gets

$$\left.\begin{aligned}&(-1)^{\alpha_j} G^{(0,\alpha_j)}_{\partial\Xi_{(\alpha_i\beta_j,j)}}(\mathbf{z}^{(\alpha_j,j)}_m,t_{m+})=(-1)^{\alpha_j} g^{(\alpha_j,j)}_1(\mathbf{z}^{(\alpha_j,j)}_m,t_{m+})<0,\\&G^{(0,\beta_j)}_{\partial\Xi_{(\alpha_i\beta_j,j)}}(\mathbf{z}^{(\beta_j,j)}_m,t_{m\mp})=g^{(\beta_j,j)}_1(\mathbf{z}^{(\beta_j,j)}_m,t_{m\mp})=0,\\&(-1)^{\beta_j} G^{(1,\beta_j)}_{\partial\Xi_{(\alpha_i\beta_j,j)}}(\mathbf{z}^{(\beta_j,j)}_m,t_{m\mp})=(-1)^{\beta_j} g^{(\beta_j,j)}_2(\mathbf{z}^{(\beta_j,j)}_m,t_{m\mp})<0.\end{aligned}\right\} \quad (9.48)$$

and for the vanishing of the source interaction,

$$\left.\begin{aligned}&(-1)^{\alpha_j} G^{(0,\alpha_j)}_{\partial\Xi_{(\alpha_i\beta_j,j)}}(\mathbf{z}^{(\alpha_j,j)}_m,t_{m+})=(-1)^{\alpha_j} g^{(\alpha_j,j)}_1(\mathbf{z}^{(\alpha_j,j)}_m,t_{m+})<0,\\&G^{(0,\beta_j)}_{\partial\Xi_{(\alpha_i\beta_j,j)}}(\mathbf{z}^{(\beta_j,j)}_m,t_{m\pm})=g^{(\beta_j,j)}_1(\mathbf{z}^{(\beta_j,j)}_m,t_{m\pm})=0,\\&(-1)^{\beta_j} G^{(1,\beta_j)}_{\partial\Xi_{(\alpha_i\beta_j,j)}}(\mathbf{z}^{(\beta_j,j)}_m,t_{m\pm})=(-1)^{\beta_j} g^{(\beta_j,j)}_2(\mathbf{z}^{(\beta_j,j)}_m,t_{m\pm})<0.\end{aligned}\right\} \quad (9.49)$$

Such a switching bifurcation between the source interaction and the instantaneous interaction can be expressed as similar to Eq.(9.47).

(iii) From Chapter 3 (also see, Luo, 2008a,b), the sufficient and necessary conditions for the switching between the stick and source interactions on the jth interaction boundary are

$$\left.\begin{aligned}&G^{(0,\alpha_j)}_{\partial\Xi_{(\alpha_i\beta_j,j)}}(\mathbf{z}^{(\alpha_j,j)}_m,t_{m\mp})=g^{(\alpha_j,j)}_1(\mathbf{z}^{(\alpha_j,j)}_m,t_{m\mp})=0\\&(-1)^{\alpha_j}G^{(1,\alpha_j)}_{\partial\Xi_{(\alpha_i\beta_j,j)}}(\mathbf{z}^{(\alpha_j,j)}_m,t_{m\mp})=(-1)^{\alpha_j}g^{(\alpha_j,j)}_2(\mathbf{z}^{(\alpha_j,j)}_m,t_{m\mp})<0;\\&G^{(0,\beta_j)}_{\partial\Xi_{(\alpha_i\beta_j,j)}}(\mathbf{z}^{(\beta_j,j)}_m,t_{m\mp})=g^{(\beta_j,j)}_1(\mathbf{z}^{(\beta_j,j)}_m,t_{m\mp})=0,\\&(-1)^{\beta_j}G^{(1,\beta_j)}_{\partial\Xi_{(\alpha_i\beta_j,j)}}(\mathbf{z}^{(\beta_j,j)}_m,t_{m\mp})=(-1)^{\beta_j}g^{(\beta_j,j)}_2(\mathbf{z}^{(\beta_j,j)}_m,t_{m\mp})<0.\end{aligned}\right\}\tag{9.50}$$

Similarly, the sufficient and necessary conditions for the switching between two instantaneous interaction states on the jth interaction boundary for $\alpha_j\neq\beta_j$

$$\left.\begin{aligned}&G^{(0,\alpha_j)}_{\partial\Xi_{(\alpha_i\beta_j,j)}}(\mathbf{z}^{(\alpha_j,j)}_m,t_{m\mp})=g^{(\alpha_j,j)}_1(\mathbf{z}^{(\alpha_j,j)}_m,t_{m\mp})=0\text{ for }\alpha_j\in\{1,2\};\\&(-1)^{\alpha_j}G^{(1,\alpha_j)}_{\partial\Xi_{(\alpha_i\beta_j,j)}}(\mathbf{z}^{(\alpha_j,j)}_m,t_{m\mp})=(-1)^{\alpha_j}g^{(\alpha_j,j)}_2(\mathbf{z}^{(\alpha_j,j)}_m,t_{m\mp})<0;\\&G^{(0,\beta_j)}_{\partial\Xi_{(\alpha_i\beta_j,j)}}(\mathbf{z}^{(\beta_j,j)}_m,t_{m\pm})=g^{(\beta_j,j)}_1(\mathbf{z}^{(\beta_j,j)}_m,t_{m\pm})=0\text{ for }\beta_j\in\{1,2\};\\&(-1)^{\beta_j}G^{(1,\beta_j)}_{\partial\Xi_{(\alpha_i\beta_j,j)}}(\mathbf{z}^{(\beta_j,j)}_m,t_{m\pm})=(-1)^{\beta_j}g^{(\beta_j,j)}_2(\mathbf{z}^{(\beta_j,j)}_m,t_{m\pm})<0.\end{aligned}\right\}\tag{9.51}$$

In the foregoing equation, both of the zero-order G-functions should be zero. For such two interactions, equation (9.47) becomes

$$\left.\begin{aligned}&L^{(j)}_{12}(t_{m\pm})=g^{(\alpha_j,j)}_1(\mathbf{z}^{(\alpha_j,j)}_m,t_{m\mp})\times g^{(\beta_j,j)}_1(\mathbf{z}^{(\beta_j,j)}_m,t_{m\pm})=0,\\&(-1)^{\alpha_j}G^{(1,\alpha_j)}_{\partial\Xi_{(\alpha_i\beta_j,j)}}(\mathbf{z}^{(\alpha_j,j)}_m,t_{m\mp})=(-1)^{\alpha_j}g^{(\alpha_j,j)}_2(\mathbf{z}^{(\alpha_j,j)}_m,t_{m\mp})<0;\\&(-1)^{\beta_j}G^{(1,\beta_j)}_{\partial\Xi_{(\alpha_i\beta_j,j)}}(\mathbf{z}^{(\beta_j,j)}_m,t_{m\pm})=(-1)^{\beta_j}g^{(\beta_j,j)}_2(\mathbf{z}^{(\beta_j,j)}_m,t_{m\pm})<0\\&\text{for }\alpha_j,\beta_j=1,2\text{ and }\alpha_j\neq\beta_j.\end{aligned}\right\}\tag{9.52}$$

In addition to the instantaneous, stick and source interactions with the corresponding singularity, a flow tangential to the boundary $\partial\Xi_{(\alpha_j\beta_j,j)}$ is another instantaneous interaction (or tangential interaction), and the corresponding sufficient and necessary conditions are

$$\left.\begin{aligned}&G^{(0,\alpha_j)}_{\partial\Xi_{(\alpha_i\beta_j,j)}}(\mathbf{z}^{(\alpha_j,j)}_m,t_{m\pm})=g^{(\alpha_j,j)}_1(\mathbf{z}^{(\alpha_j,j)}_m,t_{m\pm})=0\text{ for }\alpha_j\in\{1,2\};\\&(-1)^{\alpha_j}G^{(1,\alpha_j)}_{\partial\Xi_{(\alpha_i\beta_j,j)}}(\mathbf{z}^{(\alpha_j,j)}_m,t_{m\pm})=(-1)^{\alpha_j}g^{(\alpha_j,j)}_2(\mathbf{z}^{(\alpha_j,j)}_m,t_{m\pm})<0.\end{aligned}\right\}\tag{9.53}$$

9.3. Interactions with singularity

With the higher-order singularity to the interaction boundary $\partial\Xi_{(\alpha_j\beta_j,j)}$, the higher-order G-function should be defined. That is,

$$
\begin{aligned}
&G^{(k_{\alpha_j},\alpha_j)}_{\partial\Xi_{(\alpha_i\beta_j,j)}}(\mathbf{z}^{(\alpha_j,j)},t)\\
&=\mathbf{n}_{\partial\Xi_{(\alpha_i\beta_j,j)}}\cdot D^{k_{\alpha_j}}\mathbf{g}^{(\alpha_j,j)}(\mathbf{z}^{(\alpha_j,j)},t)=g^{(\alpha_j,j)}_{k_{\alpha_j}+1}(\mathbf{z}^{(\alpha_j,j)},t)
\end{aligned}
\tag{9.54}
$$

where

$$
g^{(\alpha_j,j)}_{k_{\alpha_j}+1}(\mathbf{z}^{(\alpha_j,j)},t)\equiv D^{k_{\alpha_j}}\varphi^{(\alpha_j,j)}_j(\mathbf{z}^{(\alpha_j,j)},t). \tag{9.55}
$$

With the higher order G-functions, the sufficient and necessary conditions for a $(2k_{\alpha_j}:2k_{\beta_j})$-instantaneous interaction at $(\mathbf{z}^{(0,j)}_m,t_m)$ for the boundary $\partial\Xi_{(\alpha_j\beta_j,j)}$ are obtained from Chapter 3 (also see, Luo, 2006, 2008a, b),

$$
\begin{aligned}
&G^{(s_{\alpha_j},\alpha_j)}_{\partial\Xi_{(\alpha_i\beta_j,j)}}(\mathbf{z}^{(\alpha_j,j)}_m,t_{m-})=g^{(\alpha_j,j)}_{s_{\alpha_j}+1}(\mathbf{z}^{(\alpha_j,j)}_m,t_{m-})=0\\
&\quad\text{for } s_{\alpha_j}=0,1,\cdots,2k_{\alpha_j}-1,\\
&G^{(s_{\beta_j},\beta_j)}_{\partial\Xi_{(\alpha_i\beta_j,j)}}(\mathbf{z}^{(\beta_j,j)}_m,t_{m+})=g^{(\beta_j,j)}_{s_{\beta_j}+1}(\mathbf{z}^{(\beta_j,j)}_m,t_{m+})=0\\
&\quad\text{for } s_{\beta_j}=0,1,\cdots,2k_{\beta_j}-1,\\
&(-1)^{\alpha_j}G^{(2k_{\alpha_j},\alpha_j)}_{\partial\Xi_{(\alpha_i\beta_j,j)}}(\mathbf{z}^{(\alpha_j,j)}_m,t_{m-})=(-1)^{\alpha_j}g^{(\alpha_j,j)}_{2k_{\alpha_j}+1}(\mathbf{z}^{(\alpha_j,j)}_m,t_{m-})>0,\\
&(-1)^{\beta_j}G^{(2k_{\beta_j},\beta_j)}_{\partial\Xi_{(\alpha_i\beta_j,j)}}(\mathbf{z}^{(\beta_j,j)}_m,t_{m+})=(-1)^{\beta_j}g^{(\beta_j,j)}_{2k_{\beta_j}+1}(\mathbf{z}^{(\beta_j,j)}_m,t_{m+})<0\\
&\text{from } \Xi_{(1,j)}\to\Xi_{(2,j)},\alpha_j,\beta_j\in\{1,2\},\alpha_j\neq\beta_j.
\end{aligned}
\tag{9.56}
$$

From Chapter 3 (or Luo (2006, 2008a,b), the sufficient and necessary conditions for a $(2k_{\alpha_j}:2k_{\beta_j})$-stick interaction at $(\mathbf{z}^{(0,j)}_m,t_m)$ on the boundary $\partial\Xi_{(\alpha_j\beta_j,j)}$ are

$$
\begin{aligned}
&G^{(s_{\alpha_j},\alpha_j)}_{\partial\Xi_{(\alpha_i\beta_j,j)}}(\mathbf{z}^{(\alpha_j,j)}_m,t_{m-})=g^{(\alpha_j,j)}_{s_{\alpha_j}+1}(\mathbf{z}^{(\alpha_j,j)}_m,t_{m-})=0\\
&\quad\text{for } s_{\alpha_j}=0,1,\cdots,2k_{\alpha_j}-1,\\
&G^{(s_{\beta_j},\beta_j)}_{\partial\Xi_{(\alpha_i\beta_j,j)}}(\mathbf{z}^{(\beta_j,j)}_m,t_{m-})=g^{(\beta_j,j)}_{s_{\beta_j}+1}(\mathbf{z}^{(\beta_j,j)}_m,t_{m-})=0\\
&\quad\text{for } s_{\beta_j}=0,1,\cdots,2k_{\beta_j}-1,\\
&(-1)^{\alpha_j}G^{(2k_{\alpha_j},\alpha_j)}_{\partial\Xi_{(\alpha_i\beta_j,j)}}(\mathbf{z}^{(\alpha_j,j)}_m,t_{m-})=(-1)^{\alpha_j}g^{(\alpha_j,j)}_{2k_{\alpha_j}+1}(\mathbf{z}^{(\alpha_j,j)}_m,t_{m-})>0,\\
&(-1)^{\beta_j}G^{(2k_{\beta_j},\beta_j)}_{\partial\Xi_{(\alpha_i\beta_j,j)}}(\mathbf{z}^{(\beta_j,j)}_m,t_{m-})=(-1)^{\beta_j}g^{(\beta_j,j)}_{2k_{\beta_j}+1}(\mathbf{z}^{(\beta_j,j)}_m,t_{m-})>0\\
&\text{for } \alpha_j,\beta_j\in\{1,2\},\alpha_j\neq\beta_j.
\end{aligned}
\tag{9.57}
$$

Similarly, the sufficient and necessary conditions for a $(2k_{\alpha_j}:2k_{\beta_j})$-source interaction (or de-interaction) at $(\mathbf{z}_m^{(0,j)},t_m)$ on the interaction boundary $\partial\Xi_{(\alpha_j\beta_j,j)}$ are

$$\begin{aligned}
&G_{\partial\Xi_{(\alpha_j\beta_j,j)}}^{(s_{\alpha_j},\alpha_j)}(\mathbf{z}_m^{(\alpha_j,j)},t_{m+})=g_{s_{\alpha_j}+1}^{(\alpha_j,j)}(\mathbf{z}_m^{(\alpha_j,j)},t_{m+})=0\\
&\quad\text{for } s_{\alpha_j}=0,1,\cdots,2k_{\alpha_j}-1,\\
&G_{\partial\Xi_{(\alpha_j\beta_j,j)}}^{(s_{\beta_j},\beta_j)}(\mathbf{z}_m^{(\beta_j,j)},t_{m+})=g_{s_{\beta_j}+1}^{(\beta_j,j)}(\mathbf{z}_m^{(\beta_j,j)},t_{m+})=0\\
&\quad\text{for } s_{\beta_j}=0,1,\cdots,2k_{\beta_j}-1,\\
&(-1)^{\alpha_j}G_{\partial\Xi_{(\alpha_j\beta_j,j)}}^{(2k_{\alpha_j},\alpha_j)}(\mathbf{z}_m^{(\alpha_j,j)},t_{m+})=(-1)^{\alpha_j}g_{2k_{\alpha_j}+1}^{(\alpha_j,j)}(\mathbf{z}_m^{(\alpha_j,j)},t_{m+})<0,\\
&(-1)^{\beta_j}G_{\partial\Xi_{(\alpha_j\beta_j,j)}}^{(2k_{\beta_j},\beta_j)}(\mathbf{z}_m^{(\beta_j,j)},t_{m+})=(-1)^{\beta_j}g_{2k_{\beta_j}+1}^{(\beta_j,j)}(\mathbf{z}_m^{(\beta_j,j)},t_{m+})<0\\
&\text{for } \alpha_j,\beta_j\in\{1,2\},\alpha_j\neq\beta_j.
\end{aligned}\tag{9.58}$$

As in Chapter 3, similar to Eq.(9.43), the $(2k_{\alpha_j}:2k_{\beta_j})$-order L-function is defined as follows.

$$\left.\begin{aligned}
L_{12}^{((2k_{\alpha_j}:2k_{\beta_j}),j)}(t_{m\pm})&=G_{\partial\Xi_{(\alpha_j\beta_j,j)}}^{(2k_{\alpha_j},\alpha_j)}(\mathbf{z}_m^{(\alpha_j,j)},t_{m-})\times G_{\partial\Xi_{(\alpha_j\beta_j,j)}}^{(2k_{\beta_j},\beta_j)}(\mathbf{z}_m^{(\beta_j,j)},t_{m+}),\\
L_{12}^{((2k_{\alpha_j}:2k_{\beta_j}),j)}(t_{m-})&=G_{\partial\Xi_{(\alpha_j\beta_j,j)}}^{(2k_{\alpha_j},\alpha_j)}(\mathbf{z}_m^{(\alpha_j,j)},t_{m-})\times G_{\partial\Xi_{(\alpha_j\beta_j,j)}}^{(2k_{\beta_j},\beta_j)}(\mathbf{z}_m^{(\beta_j,j)},t_{m-}),\\
L_{12}^{((2k_{\alpha_j}:2k_{\beta_j}),j)}(t_{m+})&=G_{\partial\Xi_{(\alpha_j\beta_j,j)}}^{(2k_{\alpha_j},\alpha_j)}(\mathbf{z}_m^{(\alpha_j,j)},t_{m+})\times G_{\partial\Xi_{(\alpha_j\beta_j,j)}}^{(2k_{\beta_j},\beta_j)}(\mathbf{z}_m^{(\beta_j,j)},t_{m+}).
\end{aligned}\right\}\tag{9.59}$$

From the $(2k_{\alpha_j}:2k_{\beta_j})$-order L-function, the sufficient and necessary conditions for three interactions with the $(2k_{\alpha_j}:2k_{\beta_j})$-order singularity are

$$\left.\begin{aligned}
L_{12}^{((2k_{\alpha_j}:2k_{\beta_j}),j)}(t_{m\pm})&=g_{2k_{\alpha_j}+1}^{(\alpha_j,j)}(\mathbf{z}_m^{(\alpha_j,j)},t_{m-})\times g_{2k_{\beta_j}+1}^{(\beta_j,j)}(\mathbf{z}_m^{(\beta_j,j)},t_{m+})>0,\\
L_{12}^{((2k_{\alpha_j}:2k_{\beta_j}),j)}(t_{m-})&=g_{2k_{\alpha_j}+1}^{(\alpha_j,j)}(\mathbf{z}_m^{(\alpha_j,j)},t_{m-})\times g_{2k_{\beta_j}+1}^{(\beta_j,j)}(\mathbf{z}_m^{(\beta_j,j)},t_{m-})<0,\\
L_{12}^{((2k_{\alpha_j}:2k_{\beta_j}),j)}(t_{m+})&=g_{2k_{\alpha_j}+1}^{(\alpha_j,j)}(\mathbf{z}_m^{(\alpha_j,j)},t_{m+})\times g_{2k_{\beta_j}+1}^{(\beta_j,j)}(\mathbf{z}_m^{(\beta_j,j)},t_{m+})<0.
\end{aligned}\right\}\tag{9.60}$$

The conditions for the appearance and vanishing of the $(2k_{\alpha_j}:2k_{\beta_j})$-stick interaction relative to the $(2k_{\alpha_j}:2k_{\beta_j})$-instantaneous interaction are

$$\begin{aligned}
&G_{\partial\Xi_{(\alpha_j\beta_j,j)}}^{(s_{\alpha_j},\alpha_j)}(\mathbf{z}_m^{(\alpha_j,j)},t_{m-})=g_{s_{\alpha_j}+1}^{(\alpha_j,j)}(\mathbf{z}_m^{(\alpha_j,j)},t_{m-})=0\\
&\quad\text{for } s_{\alpha_j}=0,1,\cdots,2k_{\alpha_j}-1,\\
&G_{\partial\Xi_{(\alpha_j\beta_j,j)}}^{(s_{\beta_j},\beta_j)}(\mathbf{z}_m^{(\beta_j,j)},t_{m\pm})=g_{s_{\beta_j}+1}^{(\beta_j,j)}(\mathbf{z}_m^{(\beta_j,j)},t_{m\pm})=0\\
&\quad\text{for } s_{\beta_j}=0,1,\cdots,2k_{\beta_j},
\end{aligned}$$

$$
\begin{aligned}
&(-1)^{\alpha_j} G^{(2k_{\alpha_j},\alpha_j)}_{\partial\Xi_{(\alpha_i\beta_j,j)}}(\mathbf{z}_m^{(\alpha_j,j)},t_{m-}) = (-1)^{\alpha_j} g^{(\alpha_j,j)}_{2k_{\alpha_j}+1}(\mathbf{z}_m^{(\alpha_j,j)},t_{m-}) > 0,\\
&(-1)^{\beta_j} G^{(2k_{\beta_j}+1,\beta_j)}_{\partial\Xi_{(\alpha_i\beta_j,j)}}(\mathbf{z}_m^{(\beta_j,j)},t_{m\pm}) = (-1)^{\beta_j} g^{(\beta_j,j)}_{2k_{\beta_j}+2}(\mathbf{z}_m^{(\beta_j,j)},t_{m\pm}) < 0\\
&\text{for } \alpha_j,\beta_j \in \{1,2\}, \alpha_j \neq \beta_j.
\end{aligned} \tag{9.61}
$$

and

$$
\begin{aligned}
&G^{(s_{\alpha_j},\alpha_j)}_{\partial\Xi_{(\alpha_i\beta_j,j)}}(\mathbf{z}_m^{(\alpha_j,j)},t_{m-}) = g^{(\alpha_j,j)}_{s_{\alpha_j}+1}(\mathbf{z}_m^{(\alpha_j,j)},t_{m-}) = 0\\
&\quad \text{for } s_{\alpha_j} = 0,1,\cdots,2k_{\alpha_j}-1,\\
&G^{(s_{\beta_j},\beta_j)}_{\partial\Xi_{(\alpha_i\beta_j,j)}}(\mathbf{z}_m^{(\beta_j,j)},t_{m\mp}) = g^{(\beta_j,j)}_{s_{\beta_j}+1}(\mathbf{z}_m^{(\beta_j,j)},t_{m\mp}) = 0\\
&\quad \text{for } s_{\beta_j} = 0,1,\cdots,2k_{\beta_j},\\
&(-1)^{\alpha_j} G^{(2k_{\alpha_j},\alpha_j)}_{\partial\Xi_{(\alpha_i\beta_j,j)}}(\mathbf{z}_m^{(\alpha_j,j)},t_{m-}) = (-1)^{\alpha_j} g^{(\alpha_j,j)}_{2k_{\alpha_j}+1}(\mathbf{z}_m^{(\alpha_j,j)},t_{m-}) > 0,\\
&(-1)^{\beta_j} G^{(2k_{\beta_j}+1,\beta_j)}_{\partial\Xi_{(\alpha_i\beta_j,j)}}(\mathbf{z}_m^{(\beta_j,j)},t_{m\mp}) = (-1)^{\beta_j} g^{(\beta_j,j)}_{2k_{\beta_j}+2}(\mathbf{z}_m^{(\beta_j,j)},t_{m\mp}) < 0\\
&\text{for } \alpha_j,\beta_j \in \{1,2\}, \alpha_j \neq \beta_j.
\end{aligned} \tag{9.62}
$$

The conditions in both of two foregoing equations can be expressed by

$$
\begin{aligned}
&L_{12}^{((2k_{\alpha_j}:2k_{\beta_j}),j)}(t_{m\pm}) = g^{(\alpha_j,j)}_{2k_{\alpha_j}+1}(\mathbf{z}_m^{(\alpha_j,j)},t_{m-}) \times g^{(\beta_j,j)}_{2k_{\beta_j}+1}(\mathbf{z}_m^{(\beta_j,j)},t_{m+}) = 0,\\
&(-1)^{\alpha_j} G^{(2k_{\alpha_j},\alpha_j)}_{\partial\Xi_{(\alpha_i\beta_j,j)}}(\mathbf{z}_m^{(\alpha_j,j)},t_{m-}) = (-1)^{\alpha_j} g^{(\alpha_j,j)}_{2k_{\alpha_j}+1}(\mathbf{z}_m^{(\alpha_j,j)},t_{m-}) > 0,\\
&(-1)^{\beta_j} G^{(2k_{\beta_j}+1,\beta_j)}_{\partial\Xi_{(\alpha_i\beta_j,j)}}(\mathbf{z}_m^{(\beta_j,j)},t_{m\pm}) = (-1)^{\beta_j} g^{(\beta_j,j)}_{2k_{\beta_j}+2}(\mathbf{z}_m^{(\beta_j,j)},t_{m\pm}) < 0.
\end{aligned} \tag{9.63}
$$

The sufficient and necessary conditions for appearance and vanishing of the $(2k_{\alpha_j}:2k_{\beta_j})$-*source* interaction pertaining to the $(2k_{\alpha_j}:2k_{\beta_j})$- instantaneous interaction are

$$
\begin{aligned}
&G^{(s_{\alpha_j},\alpha_j)}_{\partial\Xi_{(\alpha_i\beta_j,j)}}(\mathbf{z}_m^{(\alpha_j,j)},t_{m+}) = g^{(\alpha_j,j)}_{s_{\alpha_j}+1}(\mathbf{z}_m^{(\alpha_j,j)},t_{m+}) = 0\\
&\quad \text{for } s_{\alpha_j} = 0,1,\cdots,2k_{\alpha_j}-1,\\
&G^{(s_{\beta_j},\beta_j)}_{\partial\Xi_{(\alpha_i\beta_j,j)}}(\mathbf{z}_m^{(\beta_j,j)},t_{m\mp}) = g^{(\beta_j,j)}_{s_{\beta_j}+1}(\mathbf{z}_m^{(\beta_j,j)},t_{m\mp}) = 0\\
&\quad \text{for } s_{\beta_j} = 0,1,\cdots,2k_{\beta_j},\\
&(-1)^{\alpha_j} G^{(2k_{\alpha_j},\alpha_j)}_{\partial\Xi_{(\alpha_i\beta_j,j)}}(\mathbf{z}_m^{(\alpha_j,j)},t_{m+}) = (-1)^{\alpha_j} g^{(\alpha_j,j)}_{2k_{\alpha_j}+1}(\mathbf{z}_m^{(\alpha_j,j)},t_{m+}) < 0,\\
&(-1)^{\beta_j} G^{(2k_{\beta_j}+1,\beta_j)}_{\partial\Xi_{(\alpha_i\beta_j,j)}}(\mathbf{z}_m^{(\beta_j,j)},t_{m\mp}) = (-1)^{\beta_j} g^{(\beta_j,j)}_{2k_{\beta_j}+2}(\mathbf{z}_m^{(\beta_j,j)},t_{m\mp}) < 0\\
&\text{for } \alpha_j,\beta_j \in \{1,2\}, \alpha_j \neq \beta_j.
\end{aligned} \tag{9.64}
$$

$$
\begin{aligned}
&G_{\partial\Xi_{(\alpha_i\beta_j,j)}}^{(s_{\alpha_j},\alpha_j)}(\mathbf{z}_m^{(\alpha_j,j)},t_{m+}) = g_{s_{\alpha_j}+1}^{(\alpha_j,j)}(\mathbf{z}_m^{(\alpha_j,j)},t_{m+}) = 0\\
&\quad \text{for } s_{\alpha_j} = 0,1,\cdots,2k_{\alpha_j}-1,\\
&G_{\partial\Xi_{(\alpha_i\beta_j,j)}}^{(s_{\beta_j},\beta_j)}(\mathbf{z}_m^{(\beta_j,j)},t_{m\pm}) = g_{s_{\beta_j}+1}^{(\beta_j,j)}(\mathbf{z}_m^{(\beta_j,j)},t_{m\pm}) = 0\\
&\quad \text{for } s_{\beta_j} = 0,1,\cdots,2k_{\beta_j},\\
&(-1)^{\alpha_j} G_{\partial\Xi_{(\alpha_i\beta_j,j)}}^{(2k_{\alpha_j},\alpha_j)}(\mathbf{z}_m^{(\alpha_j,j)},t_{m+}) = (-1)^{\alpha_j} g_{2k_{\alpha_j}+1}^{(\alpha_j,j)}(\mathbf{z}_m^{(\alpha_j,j)},t_{m+}) < 0,\\
&(-1)^{\beta_j} G_{\partial\Xi_{(\alpha_i\beta_j,j)}}^{(2k_{\beta_j}+1,\beta_j)}(\mathbf{z}_m^{(\beta_j,j)},t_{m\pm}) = (-1)^{\beta_j} g_{2k_{\beta_j}+2}^{(\beta_j,j)}(\mathbf{z}_m^{(\beta_j,j)},t_{m\pm}) < 0\\
&\text{for } \alpha_j,\beta_j \in \{1,2\}, \beta_j \neq \alpha_j.
\end{aligned}
\tag{9.65}
$$

The sufficient and necessary for the switching between the $(2k_{\alpha_j}:2k_{\beta_j})$-stick and source interactions on the jth interaction boundary

$$
\begin{aligned}
&G_{\partial\Xi_{(\alpha_i\beta_j,j)}}^{(s_{\alpha_j},\alpha_j)}(\mathbf{z}_m^{(\alpha_j,j)},t_{m\pm}) = g_{s_{\alpha_j}+1}^{(\alpha_j,j)}(\mathbf{z}_m^{(\alpha_j,j)},t_{m\pm}) = 0\\
&\quad \text{for } s_{\alpha_j} = 0,1,\cdots,2k_{\alpha_j} \text{ and } \alpha_j = 1,2,\\
&(-1)^{\alpha_j} G_{\partial\Xi_{(\alpha_i\beta_j,j)}}^{(2k_{\alpha_j}+1,\alpha_j)}(\mathbf{z}_m^{(\alpha_j,j)},t_{m\pm}) = (-1)^{\alpha_j} g_{2k_{\alpha_j}+2}^{(\alpha_j,j)}(\mathbf{z}_m^{(\alpha_j,j)},t_{m\pm}) < 0.
\end{aligned}
\tag{9.66}
$$

Similarly, the sufficient and necessary for the switching between two $(2k_{\alpha_j}:2k_{\beta_j})$-instantaneous states on the jth interaction boundary

$$
\begin{aligned}
&G_{\partial\Xi_{(\alpha_i\beta_j,j)}}^{(s_{\alpha_j},\alpha_j)}(\mathbf{z}_m^{(\alpha_j,j)},t_{m\mp}) = g_{s_{\alpha_j}+1}^{(\alpha_j,j)}(\mathbf{z}_m^{(\alpha_j,j)},t_{m\mp}) = 0\\
&\text{for } s_{\alpha_j} = 0,1,\cdots,2k_{\alpha_j} \text{ and } \alpha_j \in \{1,2\},\\
&(-1)^{\alpha_j} G_{\partial\Xi_{(\alpha_i\beta_j,j)}}^{(2k_{\alpha_j}+1,\alpha_j)}(\mathbf{z}_m^{(\alpha_j,j)},t_{m\mp}) = (-1)^{\alpha_j} g_{2k_{\alpha_j}+2}^{(\alpha_j,j)}(\mathbf{z}_m^{(\alpha_j,j)},t_{m\mp}) < 0;\\
&G_{\partial\Xi_{(\alpha_i\beta_j,j)}}^{(s_{\beta_j},\beta_j)}(\mathbf{z}_m^{(\beta_j,j)},t_{m\pm}) = g_{s_{\beta_j}+1}^{(\beta_j,j)}(\mathbf{z}_m^{(\beta_j,j)},t_{m\pm}) = 0\\
&\text{for } s_{\beta_j} = 0,1,\cdots,2k_{\beta_j} \text{ and } \beta_j \in \{1,2\} \text{ and } \alpha_j \neq \beta_j,\\
&(-1)^{\beta_j} G_{\partial\Xi_{(\alpha_i\beta_j,j)}}^{(2k_{\beta_j}+1,\beta_j)}(\mathbf{z}_m^{(\beta_j,j)},t_{m\pm}) = (-1)^{\beta_j} g_{2k_{\beta_j}+2}^{(\beta_j,j)}(\mathbf{z}_m^{(\beta_j,j)},t_{m\pm}) < 0.
\end{aligned}
\tag{9.67}
$$

Both of the G-functions with the $(2k_{\alpha_j}:2k_{\beta_j})$-order singularity are zero, thus, one obtains for $\alpha_j \neq \beta_j$

$$
\begin{aligned}
&L_{12}^{((2k_{\alpha_j}:2k_{\beta_j}),j)}(t_{m\pm}) = g_{2k_{\alpha_j}+1}^{(\alpha_j,j)}(\mathbf{z}_m^{(\alpha_j,j)},t_{m-}) \times g_{2k_{\beta_j}+1}^{(\beta_j,j)}(\mathbf{z}_m^{(\beta_j,j)},t_{m+}) = 0,\\
&(-1)^{\alpha_j} G_{\partial\Xi_{(\alpha_i\beta_j,j)}}^{(2k_{\alpha_j}+1,\alpha_j)}(\mathbf{z}_m^{(\alpha_j,j)},t_{m\mp}) = (-1)^{\alpha_j} g_{2k_{\alpha_j}+2}^{(\alpha_j,j)}(\mathbf{z}_m^{(\alpha_j,j)},t_{m\mp}) < 0,\\
&(-1)^{\beta_j} G_{\partial\Xi_{(\alpha_i\beta_j,j)}}^{(2k_{\beta_j}+1,\beta_j)}(\mathbf{z}_m^{(\beta_j,j)},t_{m\pm}) = (-1)^{\beta_j} g_{2k_{\beta_j}+2}^{(\beta_j,j)}(\mathbf{z}_m^{(\beta_j,j)},t_{m\pm}) < 0.
\end{aligned}
\tag{9.68}
$$

For the $(2k_{\alpha_j}+1)$th-order tangential interaction to the boundary $\partial\Xi_{(\alpha_j\beta_j,j)}$, the corresponding sufficient and necessary conditions are

$$\left.\begin{aligned} &G^{(s_{\alpha_j},\alpha_j)}_{\partial\Xi_{(\alpha_j\beta_j,j)}}(\mathbf{z}^{(\alpha_j,j)}_m,t_{m\pm})=g^{(\alpha_j,j)}_{s_{\alpha_j}+1}(\mathbf{z}^{(\alpha_j,j)}_m,t_{m\pm})=0\\ &\quad\text{for } s_{\alpha_j}=0,1,\cdots,2k_{\alpha_j} \text{ and } \alpha_j\in\{1,2\}\\ &(-1)^{\alpha_j}G^{(2k_{\alpha_j}+1,\alpha_j)}_{\partial\Xi_{(\alpha_j\beta_j,j)}}(\mathbf{z}^{(\alpha_j,j)}_m,t_{m\pm})=(-1)^{\alpha_j}g^{(\alpha_j,j)}_{2k_{\alpha_j}+2}(\mathbf{z}^{(\alpha_j,j)}_m,t_{m\pm})<0\,.\end{aligned}\right\}\tag{9.69}$$

9.4. Interactions at edges

Suppose there are s-linear-independent interaction conditions among l-interaction conditions in Eq.(9.4), such s-interaction conditions form an edge (or a singular edge), which is defined in Chapter 6. On the edge, the conditions presented in this section should be checked for all s-interaction boundaries via

$$z^{(\alpha_j,j)}=\varphi_j(\mathbf{x}^{(\alpha_j,j)},\mathbf{y}^{(\alpha_j,j)},t,\boldsymbol{\lambda}_j)\text{ for all } j\in\mathcal{L}.\tag{9.70}$$

On the boundary $\partial\Omega_{(\alpha_j\beta_j,j)}$, one obtains

$$z^{(0,j)}=\varphi_j(\mathbf{x}^{(0,j)},\mathbf{y}^{(0,j)},t,\boldsymbol{\lambda}_j)=0\text{ for all } j\in\mathcal{L}.\tag{9.71}$$

Without loss of generality, suppose the l linearly independent interaction constraints form an edge and there are three subsets $\mathcal{L}_i\subseteq\mathcal{L}$ ($i=1,2,3$) with l_i-interaction conditions among l-conditions and $l_1+l_2+l_3=l$. For $j\in\mathcal{L}_i$, the conditions for l_1-stick interaction, l_2-non-interaction and l_3-instantaneous interaction are given.

Theorem 9.1. *Consider two dynamical systems in Eqs.(9.1) with constraints in Eq.(9.4). Let* $\mathcal{L}_i=\varnothing\cup\{k_1^{(i)},k_2^{(i)},\cdots,k_{l_i}^{(i)}\}$ *(*$i=1,2,3$*) with* $k_\kappa^{(i)}\in\mathcal{L}$ $(\kappa=1,2,\cdots l_i)$ *and* $l_1+l_2+l_3=l$. *For* $j\in\mathcal{L}_i\subseteq\mathcal{L}$ *and* $\mathcal{L}=\cup_{i=1}^3\mathcal{L}_i$, $\boldsymbol{x}_m^{(\alpha_j,j)}=(\mathbf{x}^{(\alpha_j,j)},\mathbf{y}^{(\alpha_j,j)})^{\mathrm{T}}$ $\in\Omega_{(\alpha_j,j)}$ *(*$\alpha_j\in\mathcal{I}$ *with* $\mathcal{I}=\{1,2\}$*) and* $\boldsymbol{x}_m^{(0,j)}=(\mathbf{x}_m^{(0,j)},\mathbf{y}_m^{(0,j)})^{\mathrm{T}}$ $\in\partial\Omega_{(12,j)}$ *at time* t_m, $\boldsymbol{x}_m^{(\alpha_j,j)}=\boldsymbol{x}_m^{(0,j)}$. *For any small* $\varepsilon>0$, *there is a time interval* $[t_{m-\varepsilon},t_m)$ *or* $(t_m,t_{m+\varepsilon}]$. *At* $\boldsymbol{x}^{(\alpha_j,j)}\in\Omega^{\pm\varepsilon}_{(\alpha_j,j)}$ *for time* $t\in[t_{m-\varepsilon},t_m)$ *or* $(t_m,t_{m+\varepsilon}]$, $z^{(\alpha_j,j)}(t)=$ $\varphi^{(\alpha_j,j)}(\boldsymbol{x}^{(\alpha_j,j)}(t),t,\boldsymbol{\lambda}_j)$ *is* $C^{r_{\alpha_j}}$ *continuous and* $|D^{(r_{\alpha_j}+1)}z^{(\alpha_j,j)}(t)|<\infty$ $(r_{\alpha_j}\geq 3)$. *For* $\boldsymbol{x}^{(\alpha_j,j)}\in\Omega_{(\alpha_j,j)}$ *and* $\boldsymbol{x}^{(0,j)}\in\partial\Omega_{(12,j)}$. $\mathbb{F}^{(\alpha_j,j)}(\boldsymbol{x}^{(\alpha_j,j)},t,\boldsymbol{\pi}^{(\alpha_j,j)})\neq\mathbb{F}^{(0,j)}(\boldsymbol{x}^{(0,j)},t,\boldsymbol{\lambda}_j)$ *at* $\boldsymbol{x}^{(\alpha_j,j)}=\boldsymbol{x}^{(0,j)}$. *The two dynamical systems in Eq.(9.1) to the corner points of* l *-*

interaction conditions in Eq.(9.4) are of the (l_1, l_2, l_3) *-stick interaction, non-interaction and instantaneous interaction for time* t_m *if and only if*
(i) for all $j \in \mathcal{L}_1$ *with* $\alpha_j = 1,2$ *at time* t_m

$$\left.\begin{array}{l} \boldsymbol{x}_{m-}^{(\alpha_j,j)} = \boldsymbol{x}_m^{(0,j)} \ (\text{or } z_{m-}^{(\alpha_j,j)} = z_m^{(0,j)}), \\ (-1)^{\alpha_j} G_{\partial\Xi_{(\alpha_j\beta_j,j)}}^{(0,j)}(\mathbf{z}_m^{(\alpha_j,j)}, t_{m-}) = (-1)^{\alpha_j} g_1^{(\alpha_j,j)}(\mathbf{z}_m^{(\alpha_j,j)}, t_{m-}) > 0. \end{array}\right\} \tag{9.72}$$

(ii) for all $j \in \mathcal{L}_2$ *with* $\alpha_j = 1,2$ *at time* t_m

$$\left.\begin{array}{l} \boldsymbol{x}_{m+}^{(\alpha_j,j)} = \boldsymbol{x}_m^{(0,j)} \ (\text{or } z_{m+}^{(\alpha_j,j)} = z_m^{(0,j)}), \\ (-1)^{\alpha_j} G_{\partial\Xi_{(\alpha_j\beta_j,j)}}^{(0,j)}(\mathbf{z}_m^{(\alpha_j,j)}, t_{m+}) = (-1)^{\alpha_j} g_1^{(\alpha_j,j)}(\mathbf{z}_m^{(\alpha_j,j)}, t_{m+}) < 0. \end{array}\right\} \tag{9.73}$$

(iii) for all $j \in \mathcal{L}_3$ *with* $\alpha_j, \beta_j = 1,2$ *and* $\alpha_j \neq \beta_j$ *at time* t_m

$$\left.\begin{array}{l} \boldsymbol{x}_{m-}^{(\alpha_j,j)} = \boldsymbol{x}_m^{(0,j)} \ (\text{or } z_{m-}^{(\alpha_j,j)} = z_m^{(0,j)}), \\ (-1)^{\alpha_j} G_{\partial\Xi_{(\alpha_j\beta_j,j)}}^{(0,j)}(\mathbf{z}_m^{(\alpha_j,j)}, t_{m-}) = (-1)^{\alpha_j} g_1^{(\alpha_j,j)}(\mathbf{z}_m^{(\alpha_j,j)}, t_{m-}) > 0; \\ \boldsymbol{x}_{m+}^{(\beta_j,j)} = \boldsymbol{x}_m^{(0,j)} \ (\text{or } z_{m+}^{(\beta_j,j)} = z_m^{(0,j)}), \\ (-1)^{\beta_j} G_{\partial\Xi_{(\alpha_j\beta_j,j)}}^{(0,j)}(\mathbf{z}_m^{(\beta_j,j)}, t_{m+}) = (-1)^{\beta_j} g_1^{(\beta_j,j)}(\mathbf{z}_m^{(\beta_j,j)}, t_{m+}) < 0. \end{array}\right\} \tag{9.74}$$

(iv) The switching bifurcation conditions for one of three cases (i)-(iii) for $j \in \mathcal{L}_i$ *(*$i = 1,2,3$*) with time* $t = t_m$ *with* $\alpha_j \in \{1,2\}$,

$$\left.\begin{array}{l} \boldsymbol{x}_{m\pm}^{(\alpha_j,j)} = \boldsymbol{x}_m^{(0,j)} \ (\text{or } z_{m\pm}^{(\alpha_j,j)} = z_m^{(0,j)}), \\ G_{\partial\Xi_{(\alpha_j\beta_j,j)}}^{(0,j)}(\mathbf{z}_m^{(\alpha_j,j)}, t_{m\pm}) = g_1^{(\alpha_j,j)}(\mathbf{z}_m^{(\alpha_j,j)}, t_{m\pm}) = 0; \\ (-1)^{\alpha_j} G_{\partial\Xi_{(\alpha_j\beta_j,j)}}^{(1,j)}(\mathbf{z}_m^{(\alpha_j,j)}, t_{m\pm}) = (-1)^{\alpha_j} g_2^{(\alpha_j,j)}(\mathbf{z}_m^{(\alpha_j,j)}, t_{m\pm}) < 0. \end{array}\right\} \tag{9.75}$$

and for $\beta_j = \{1,2\}$ *but* $\beta_j \neq \alpha_j$

$$\begin{array}{l} \boldsymbol{x}_{m\pm}^{(\beta_j,j)} = \boldsymbol{x}_m^{(0,j)} \ (\text{or } z_{m\pm}^{(\beta_j,j)} = z_m^{(0,j)}), \\ G_{\partial\Xi_{(\alpha_j\beta_j,j)}}^{(0,j)}(\mathbf{z}_m^{(\beta_j,j)}, t_{m\pm}) = g_1^{(\beta_j,j)}(\mathbf{z}_m^{(\beta_j,j)}, t_{m\pm}) \neq 0, \end{array} \tag{9.76}$$

or

$$\left.\begin{array}{l} \boldsymbol{x}_{m\pm}^{(\beta_j,j)} = \boldsymbol{x}_m^{(0,j)} \ (\text{or } z_{m\pm}^{(\beta_j,j)} = z_m^{(0,j)}), \\ G_{\partial\Xi_{(\alpha_j\beta_j,j)}}^{(0,j_1)}(\mathbf{z}_m^{(\beta_j,j)}, t_{m\pm}) = g_1^{(\beta_j,j)}(\mathbf{z}_m^{(\beta_j,j)}, t_{m\pm}) = 0; \\ (-1)^{\beta_j} G_{\partial\Xi_{(\alpha_j\beta_j,j)}}^{(1,j)}(\mathbf{z}_m^{(\beta_j,j)}, t_{m\pm}) = (-1)^{\beta_j} g_2^{(\beta_j,j)}(\mathbf{z}_m^{(\beta_j,j)}, t_{m\pm}) < 0. \end{array}\right\} \tag{9.77}$$

Proof. For each interaction, the proof is similar to Luo (2008a,b). ∎

Theorem 9.2. *Consider two dynamical systems in Eqs.(9.1) with constraints in Eq.(9.4). Let* $\mathcal{L}_i = \varnothing \cup \{k_1^{(i)}, k_2^{(i)}, \cdots, k_{l_i}^{(i)}\}$ $(i=1,2,3)$ *with* $k_\kappa^{(i)} \in \mathcal{L}$ $(\kappa = 1,2,\cdots l_i)$ *and* $l_1 + l_2 + l_3 = l$. *For* $j \in \mathcal{L}_i \subseteq \mathcal{L}$ *and* $\mathcal{L} = \cup_{i=1}^{3} \mathcal{L}_i$, $\boldsymbol{x}_m^{(\alpha_j,j)} = (\mathbf{x}^{(\alpha_j,j)}, \mathbf{y}^{(\alpha_j,j)})^{\mathrm{T}} \in \Omega_{(\alpha_j,j)}$ $(\alpha_j \in \mathcal{I}$ *with* $\mathcal{I} = \{1,2\})$ *and* $\boldsymbol{x}_m^{(0,j)} = (\mathbf{x}_m^{(0,j)}, \mathbf{y}_m^{(0,j)})^{\mathrm{T}} \in \partial\Omega_{(12,j)}$ *at time* t_m, $\boldsymbol{x}_m^{(\alpha_j,j)} = \boldsymbol{x}_m^{(0,j)}$. *For any small* $\varepsilon > 0$, *there is a time interval* $[t_{m-\varepsilon}, t_m)$ *or* $(t_m, t_{m+\varepsilon}]$. *At* $\boldsymbol{x}^{(\alpha_j,j)} \in \Omega_{(\alpha_j,j)}^{\pm\varepsilon}$ *for time* $t \in [t_{m-\varepsilon}, t_m)$ *or* $(t_m, t_{m+\varepsilon}]$, $z^{(\alpha_j,j)}(t) = \varphi^{(\alpha_j,j)}(\boldsymbol{x}^{(\alpha_j,j)}(t), t, \boldsymbol{\lambda}_j)$ *is* $C^{r_{\alpha_{j1}}}$*-continuous and* $|D^{(r_{\alpha_j}+1)} z^{(\alpha_j,j)}(t)| < \infty$ $(r_{\alpha_j} \ge 2k_{\alpha_j} + 2)$. $\mathbb{F}^{(\alpha_j,j)}(\boldsymbol{x}^{(\alpha_j,j)}, t, \boldsymbol{\pi}^{(\alpha_j,j)}) \ne \mathbb{F}^{(0,j)}(\boldsymbol{x}^{(0,j)}, t, \boldsymbol{\lambda}_j)$ *at* $\boldsymbol{x}^{(\alpha_j,j)} = \boldsymbol{x}^{(0,j)}$. *The two dynamical systems in Eq.(9.1) to the corner points of* l*-interaction conditions in Eq.(9.4) are of the* (l_1, l_2, l_3)*-stick interaction, non-interaction and instantaneous interaction for time* t_m *if and only if*

(i) for time t_m *for all* $j \in \mathcal{L}_1$ *with* $\alpha_j = 1,2$

$$\left.\begin{aligned}
&\boldsymbol{x}_{m-}^{(\alpha_j,j)} = \boldsymbol{x}_m^{(0,j)} \;(\text{or } z_{m-}^{(\alpha_j,j)} = z_m^{(0,j)}),\\
&G_{\partial\Xi_{(\alpha_j\beta_j,j)}}^{(s_{\alpha_j},j)}(\mathbf{z}_m^{(\alpha_j,j)}, t_{m-}) = g_{s_{\alpha_j}+1}^{(\alpha_j,j)}(\mathbf{z}_m^{(\alpha_j,j)}, t_{m-}) = 0\\
&\text{for } s_{\alpha_j} = 0,1,\cdots,2k_{\alpha_j}-1;\\
&(-1)^{\alpha_j} G_{\partial\Xi_{(\alpha_j\beta_j,j)}}^{(2k_{\alpha_j},j)}(\mathbf{z}_m^{(\alpha_j,j)}, t_{m-}) = (-1)^{\alpha_j} g_{2k_{\alpha_j}+1}^{(\alpha_j,j)}(\mathbf{z}_m^{(\alpha_j,j)}, t_{m-}) > 0.
\end{aligned}\right\} \tag{9.78}$$

(ii) for time t_m for all $j \in \mathcal{L}_2$ with $\alpha_j = 1,2$

$$\left.\begin{aligned}
&\boldsymbol{x}_{m+}^{(\alpha_j,j)} = \boldsymbol{x}_m^{(0,j)} \;(\text{or } z_{m+}^{(\alpha_j,j)} = z_m^{(0,j)}),\\
&G_{\partial\Xi_{(\alpha_j\beta_j,j)}}^{(s_{\alpha_j},j)}(\mathbf{z}_m^{(\alpha_j,j)}, t_{m+}) = g_{s_{\alpha_j}+1}^{(\alpha_j,j)}(\mathbf{z}_m^{(\alpha_j,j)}, t_{m+}) = 0\\
&\text{for } s_{\alpha_j} = 0,1,2,\cdots,2k_{\alpha_j}-1;\\
&(-1)^{\alpha_j} G_{\partial\Xi_{(\alpha_j\beta_j,j)}}^{(2k_{\alpha_j},j)}(\mathbf{z}_m^{(\alpha_j,j)}, t_{m+}) = (-1)^{\alpha_j} g_{2k_{\alpha_j}+1}^{(\alpha_j,j)}(\mathbf{z}_m^{(\alpha_j,j)}, t_{m+}) < 0.
\end{aligned}\right\} \tag{9.79}$$

(iii) for time t_m for all $j \in \mathcal{L}_3$ with $\alpha_j, \beta_j \in \{1,2\}$ and $\alpha_j \ne \beta_j$,

$$\begin{aligned}
&\boldsymbol{x}_{m-}^{(\alpha_j,j)} = \boldsymbol{x}_m^{(0,j)} \;(\text{or } z_{m-}^{(\alpha_j,j)} = z_m^{(0,j)}),\\
&G_{\partial\Xi_{(\alpha_j\beta_j,j)}}^{(s_{\alpha_j},j)}(\mathbf{z}_m^{(\alpha_j,j)}, t_{m-}) = g_{s_{\alpha_j}+1}^{(\alpha_j,j)}(\mathbf{z}_m^{(\alpha_j,j)}, t_{m-}) = 0\\
&\text{for } s_{\alpha_j} = 0,1,2,\cdots,2k_{\alpha_j}-1;\\
&(-1)^{\alpha_j} G_{\partial\Xi_{(\alpha_j\beta_j,j)}}^{(2k_{\alpha_j},j)}(\mathbf{z}_m^{(\alpha_j,j)}, t_{m-}) = (-1)^{\alpha_j} g_{2k_{\alpha_j}+1}^{(\alpha_j,j)}(\mathbf{z}_m^{(\alpha_j,j)}, t_{m-}) > 0;
\end{aligned}$$

$$\left.\begin{aligned}&\mathbf{x}_{m+}^{(\beta_j,j)} = \mathbf{x}_m^{(0,j)} \text{ (or } z_{m+}^{(\beta_j,j)} = z_m^{(0,j)} \text{)},\\&G_{\partial\Xi_{(\alpha_j\beta_j,j)}}^{(s_{\beta_j},j)}(\mathbf{z}_m^{(\beta_j,j)}, t_{m+}) = g_{s_{\beta_j}+1}^{(\beta_j,j)}(\mathbf{z}_m^{(\beta_j,j)}, t_{m+}) = 0\\&s_{\beta_j} = 0,1,2,\cdots,2k_{\beta_j}-1\\&(-1)^{\beta_j} G_{\partial\Xi_{(\alpha_j\beta_j,j)}}^{(2k_{\beta_j},j_{13})}(\mathbf{z}_m^{(\beta_j,j)}, t_{m+}) = (-1)^{\beta_j} g_{2k_{\beta_j}+1}^{(\beta_j,j)}(\mathbf{z}_m^{(\beta_j,j)}, t_{m+}) < 0.\end{aligned}\right\} \tag{9.80}$$

(iv) The switching bifurcation conditions for one of three cases (i)-(iii) for $j \in \mathcal{L}_i$ ($i = 1,2,3$) with time $t = t_m$ with $\alpha_j \in \{1,2\}$,

$$\left.\begin{aligned}&\mathbf{x}_{m\pm}^{(\alpha_j,j)} = \mathbf{x}_m^{(0,j)} \text{ (or } z_{m\pm}^{(\alpha_j,j)} = z_m^{(0,j)} \text{)},\\&G_{\partial\Xi_{(\alpha_j\beta_j,j)}}^{(s_{\alpha_j},j)}(\mathbf{z}_m^{(\alpha_j,j)}, t_{m\pm}) = g_{s_{\alpha_j}+1}^{(\alpha_j,j)}(\mathbf{z}_m^{(\alpha_j,j)}, t_{m\pm}) = 0\\&\text{for } s_{\alpha_j} = 0,1,2,\cdots,2k_{\alpha_j};\\&(-1)^{\alpha_j} G_{\partial\Xi_{(\alpha_j\beta_j,j)}}^{(2k_{\alpha_j}+1,j)}(\mathbf{z}_m^{(\alpha_j,j)}, t_{m\pm}) = (-1)^{\alpha_j} g_{2k_{\alpha_j}+2}^{(\alpha_j,j)}(\mathbf{z}_m^{(\alpha_j,j)}, t_{m\pm}) < 0.\end{aligned}\right\} \tag{9.81}$$

and for $\beta_j \in \{1,2\}$ and $\beta_j \neq \alpha_j$

$$\left.\begin{aligned}&\mathbf{x}_{m\pm}^{(\beta_j,j)} = \mathbf{x}_m^{(0,j)} \text{ (or } z_{m\pm}^{(\beta_j,j)} = z_m^{(0,j)} \text{)},\\&G_{\partial\Xi_{(\alpha_j\beta_j,j)}}^{(s_{\beta_j},j)}(\mathbf{z}_m^{(\beta_j,j)}, t_{m\pm}) = g_{s_{\beta_j}+1}^{(\beta_j,j)}(\mathbf{z}_m^{(\beta_j,j)}, t_{m\pm}) = 0\\&\text{for } s_{\beta_j} = 0,1,2,\cdots 2k_{\beta_j}-1;\\&G_{\partial\Xi_{(\alpha_j\beta_j,j)}}^{(2k_{\beta_j},j)}(\mathbf{z}_m^{(\beta_j,j)}, t_{m\pm}) = g_{2k_{\beta_j}+1}^{(\beta_j,j)}(\mathbf{z}_m^{(\beta_j,j)}, t_{m\pm}) \neq 0.\end{aligned}\right\} \tag{9.82}$$

or

$$\left.\begin{aligned}&\mathbf{x}_{m\pm}^{(\beta_j,j)} = \mathbf{x}_m^{(0,j)} \text{ (or } z_{m\pm}^{(\beta_j,j)} = z_m^{(0,j)} \text{)},\\&G_{\partial\Xi_{(\alpha_j\beta_j,j)}}^{(s_{\beta_j},j)}(\mathbf{z}_m^{(\beta_j,j)}, t_{m\pm}) = g_{s_{\beta_j}+1}^{(\beta_j,j)}(\mathbf{z}_m^{(\beta_j,j)}, t_{m\pm}) = 0\\&\text{for } s_{\beta_j} = 0,1,2,\cdots 2k_{\beta_j};\\&(-1)^{\beta_j} G_{\partial\Xi_{(\alpha_j\beta_j,j)}}^{(2k_{\beta_j}+1,j)}(\mathbf{z}_m^{(\beta_j,j)}, t_{m\pm}) = (-1)^{\beta_j} g_{2k_{\beta_j}+2}^{(\beta_j,j)}(\mathbf{z}_m^{(\beta_j,j)}, t_{m\pm}) < 0.\end{aligned}\right\} \tag{9.83}$$

Proof. For each interaction, the proof is similar to Luo (2008a,b). ■

If the two system interaction with flow barriers is for the flows not to be passed, the transport laws should be used as in Chapter 5.

9.5. Application to system synchronization

The synchronization of two dynamical systems is a special case of uni-interaction instead of bi-interaction. To illustrate the theory of dynamical system interactions, the synchronization of two totally distinct dynamical systems is addressed herein. As in Luo and Min (2010a,b), consider a periodically forced, damped Duffing oscillator with a twin well potential as a master system, i.e.,

$$\ddot{x} + d\dot{x} - a_1 x + a_2 x^3 = A_0 \cos \omega t. \tag{9.84}$$

A periodically driven pendulum is considered as a slave system, i.e.,

$$\ddot{y} + a_0 \sin y = Q_0 \cos \Omega t. \tag{9.85}$$

For the Duffing oscillator, periodic and chaotic motions exist. To enforce the chaotic pendulum to synchronize with periodical and chaotic motions in the Duffing oscillator, the control laws should be exerted to the slave system. Before the control laws are exerted, the state variables are introduced for the Duffing oscillator and pendulum as

$$\mathbf{x} = (x_1, x_2)^{\mathrm{T}} \text{ and } \mathbf{y} = (y_1, y_2)^{\mathrm{T}} \tag{9.86}$$

and the vector fields are defined as

$$\mathscr{F}(\mathbf{x},t) = (x_2, \mathscr{F}(\mathbf{x},t))^{\mathrm{T}} \text{ and } \mathbf{F}(\mathbf{y},t) = (y_2, F(\mathbf{y},t))^{\mathrm{T}}. \tag{9.87}$$

So the Duffing oscillator is described by

$$\dot{\mathbf{x}} = \mathscr{F}(\mathbf{x},t) \tag{9.88}$$

where

$$x_2 \equiv \dot{x}_1 \text{ and } \mathscr{F}(\mathbf{x},t) = -d_1 x_2 + a_1 x_1 - a_2 x_1^3 + A_0 \cos \omega t. \tag{9.89}$$

The equation of motion for the controlled pendulum is

$$\dot{\mathbf{y}} = \mathbf{F}(\mathbf{y},t) \tag{9.90}$$

where

$$y_2 \equiv \dot{y}_1 \text{ and } F(\mathbf{y},t) = -a_0 \sin y_1 + Q_0 \cos \Omega t. \tag{9.91}$$

With a control law, equations of motion for the controlled pendulum become

$$\dot{\mathbf{y}} = \mathbf{F}(\mathbf{y},t) - \mathbf{u}(\mathbf{x},\mathbf{y},t) \tag{9.92}$$

where

$$\begin{aligned} &\mathbf{u}(\mathbf{x},\mathbf{y},t) = (u_1, u_2)^{\mathrm{T}} \\ &u_1 = k_1 \operatorname{sgn}(y_1 - x_1) \text{ and } u_2 = k_2 \operatorname{sgn}(y_2 - x_2). \end{aligned} \tag{9.93}$$

Thus, the interaction condition are $\varphi_1 = y_1 - x_1 = 0$ and $\varphi_2 = y_2 - x_2 = 0$.

9.5.1. Discontinuous description

In despite of periodic or chaotic motions in the Duffing oscillator, the controlled pendulum under a control will follow the master system (i.e., the Duffing oscillator) to be synchronized. Under the control, the slave system will become discontinuous. The two control laws make the controlled pendulum in Eq.(9.92) possess four regions, and the corresponding vector field components of $\mathbf{F} = (F_1, F_2)^{\mathrm{T}}$ are given as follows.

(i) For $y_1 > x_1$ and $y_2 > x_2$,

$$\begin{aligned} F_1(\mathbf{y}, t) &= y_2 - k_1, \\ F_2(\mathbf{y}, t) &= -a_0 \sin y_1 + Q_0 \cos \Omega t - k_2. \end{aligned} \tag{9.94}$$

(ii) For $y_1 > x_1$ and $y_2 < x_2$,

$$\begin{aligned} F_1(\mathbf{y}, t) &= y_2 - k_1, \\ F_2(\mathbf{y}, t) &= -a_0 \sin y_1 + Q_0 \cos \Omega t + k_2. \end{aligned} \tag{9.95}$$

(iii) For $y_1 < x_1$ and $y_2 < x_2$,

$$\begin{aligned} F_1(\mathbf{y}, t) &= y_2 + k_1, \\ F_2(\mathbf{y}, t) &= -a_0 \sin y_1 + Q_0 \cos \Omega t + k_2. \end{aligned} \tag{9.96}$$

(iv) For $y_1 < x_1$ and $y_2 > x_2$,

$$\begin{aligned} F_1(\mathbf{y}, t) &= y_2 + k_1, \\ F_2(\mathbf{y}, t) &= -a_0 \sin y_1 + Q_0 \cos \Omega t - k_2. \end{aligned} \tag{9.97}$$

From the above four regions, four domains in phase space of the controlled pendulum are defined as

$$\begin{aligned} \Omega_1 &= \{(y_1, y_2) | y_1 - x_1(t) > 0, y_2 - x_2(t) > 0\}, \\ \Omega_2 &= \{(y_1, y_2) | y_1 - x_1(t) > 0, y_2 - x_2(t) < 0\}, \\ \Omega_3 &= \{(y_1, y_2) | y_1 - x_1(t) < 0, y_2 - x_2(t) < 0\}, \\ \Omega_4 &= \{(y_1, y_2) | y_1 - x_1(t) < 0, y_2 - x_2(t) > 0\}. \end{aligned} \tag{9.98}$$

The corresponding boundaries of the four domains are

$$\begin{aligned}
\partial\Omega_{12} &= \{(y_1,y_2)\,|\,y_2 - x_2(t) = 0, y_1 - x_1(t) > 0\},\\
\partial\Omega_{23} &= \{(y_1,y_2)\,|\,y_1 - x_1(t) = 0, y_2 - x_2(t) < 0\},\\
\partial\Omega_{34} &= \{(y_1,y_2)\,|\,y_2 - x_2(t) = 0, y_1 - x_1(t) < 0\},\\
\partial\Omega_{14} &= \{(y_1,y_2)\,|\,y_1 - x_1(t) = 0, y_2 - x_2(t) > 0\}.
\end{aligned} \tag{9.99}$$

From the above definition, the velocity and displacement boundaries are shown in Fig.9.10(a) and (b), respectively. The dashed curves are the boundaries. The intersection point of the two boundaries is labeled by a filled circular symbol, which will change with time. From the afore-defined domains, the equation of motion for the controlled pendulum in domain Ω_α ($\alpha = 1,2,3,4$) is

$$\dot{\mathbf{y}}^{(\alpha)} = \mathbf{F}^{(\alpha)}(\mathbf{y}^{(\alpha)},t) \tag{9.100}$$

where

$$\begin{aligned}
&F_1^{(\alpha)}(\mathbf{y}^{(\alpha)},t) = y_2^{(\alpha)} - k_1 \ \text{ for } \alpha\text{=1,2;}\\
&F_1^{(\alpha)}(\mathbf{y}^{(\alpha)},t) = y_2^{(\alpha)} + k_1 \ \text{ for } \alpha\text{=3,4;}\\
&F_2^{(\alpha)}(\mathbf{y}^{(\alpha)},t) = -a_0 \sin y_1^{(\alpha)} + Q_0 \cos\Omega t - k_2 \ \text{ for } \alpha\text{=1,4;}\\
&F_2^{(\alpha)}(\mathbf{y}^{(\alpha)},t) = -a_0 \sin y_1^{(\alpha)} + Q_0 \cos\Omega t + k_2 \ \text{ for } \alpha\text{=2,3.}
\end{aligned} \tag{9.101}$$

From Luo (2008a,b), the dynamical systems on the boundaries are

$$\dot{\mathbf{y}}^{(\alpha\beta)} = \mathbf{F}^{(\alpha\beta)}(\mathbf{y}^{(\alpha\beta)},\mathbf{x}(t),t) \text{ with } \dot{\mathbf{x}} = \mathscr{F}(\mathbf{x},t) \tag{9.102}$$

where

$$F_1^{(\alpha\beta)}(\mathbf{y}^{(\alpha\beta)},t) = y_2(t) = x_2(t) \text{ and } F_2^{(\alpha\beta)}(\mathbf{y}^{(\alpha\beta)},t) = \dot{x}_2(t) \tag{9.103}$$

with

$$\left.\begin{aligned}
&y_1^{(\alpha\beta)} = x_1(t) \ \text{ and } \ y_2^{(\alpha\beta)} = x_2(t) \ \text{ on } \partial\Omega_{\alpha\beta} \text{ for } (\alpha,\beta)\text{=(2,3), (1,4);}\\
&y_1^{(\alpha\beta)} = x_1(t) + C \ \text{ and } \ y_2^{(\alpha\beta)} = x_2(t) \text{ on } \partial\Omega_{\alpha\beta} \text{ for } (\alpha,\beta)\text{=(1,2), (3,4).}
\end{aligned}\right\} \tag{9.104}$$

The boundary flow of $\mathbf{x}(t)$ is controlled by the Duffing system, which implies that the boundaries change with time. From such boundaries, it is very difficult to develop the analytical conditions for the synchronization of the controlled pendulum with the Duffing oscillator. Without loss of generality, the relative coordinates are introduced by

$$z_1 = y_1 - x_1 \ \text{ and } \dot{z}_1 \equiv z_2 = y_2 - x_2. \tag{9.105}$$

The domains and boundaries in the relative coordinates become

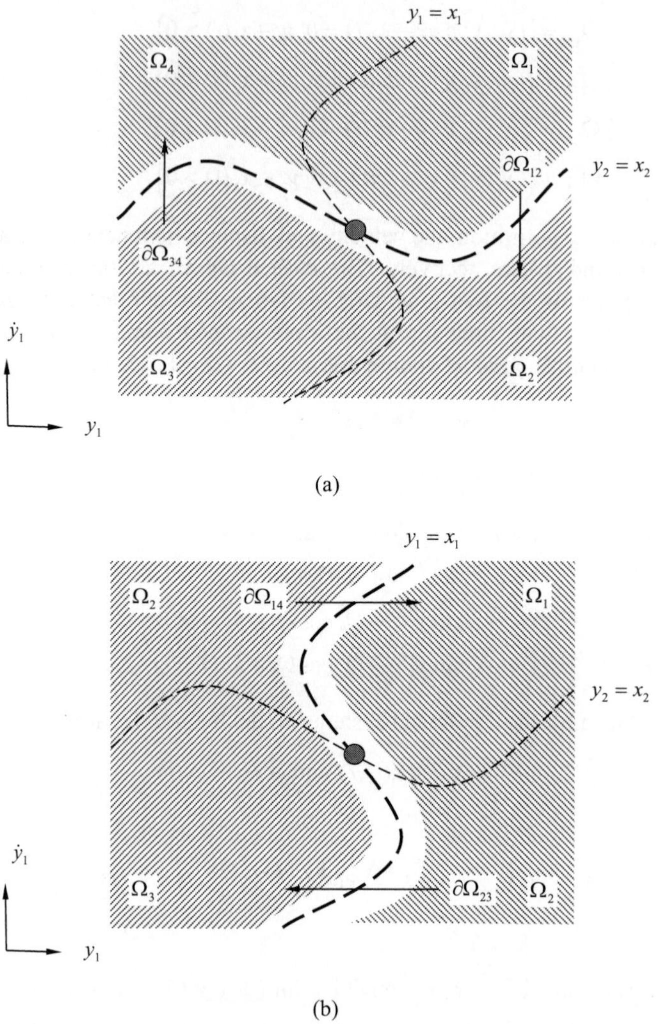

Fig.9.10 Two boundaries in the absolute coordinates: **(a)** velocity, and **(b)** displacement.

$$\begin{aligned}
\Omega_1 &= \{(z_1, z_2) | z_1 > 0, z_2 > 0\},\\
\Omega_2 &= \{(z_1, z_2) | z_1 > 0, z_2 < 0\},\\
\Omega_3 &= \{(z_1, z_2) | z_1 < 0, z_2 < 0\},\\
\Omega_4 &= \{(z_1, z_2) | z_1 < 0, z_2 > 0\}.
\end{aligned} \tag{9.106}$$

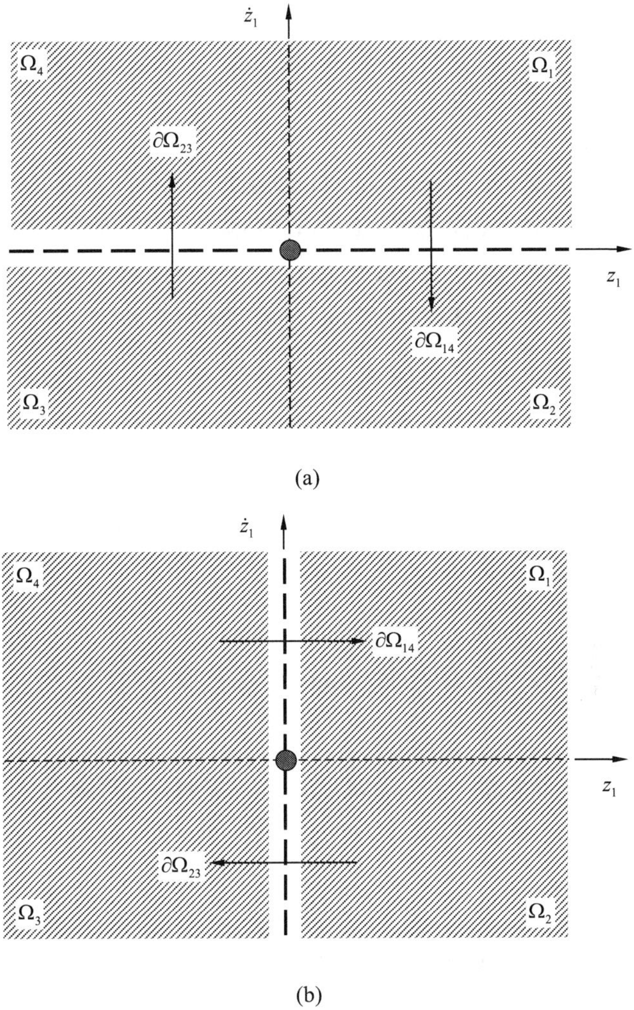

Fig.9.11 Two boundaries in the relative coordinates: **(a)** velocity, and **(b)** displacement.

$$\begin{aligned}
\partial\Omega_{12} &= \{(z_1, z_2) | z_2 = 0, z_1 > 0\}, \\
\partial\Omega_{23} &= \{(z_1, z_2) | z_1 = 0, z_2 < 0,\}, \\
\partial\Omega_{34} &= \{(z_1, z_2) | z_2 = 0, z_1 < 0\}, \\
\partial\Omega_{14} &= \{(z_1, z_2) | z_1 = 0, z_2 > 0,\}.
\end{aligned} \tag{9.107}$$

The velocity and displacement boundaries in the relative coordinates are constant, as shown in Fig.9.11. The controlled pendulum system in the relative coordinates in Ω_α ($\alpha = 1,2,3,4$) is expressed by

$$\dot{\mathbf{z}}^{(\alpha)} = \mathbf{g}^{(\alpha)}(\mathbf{z}^{(\alpha)}, \mathbf{x}, t) \text{ with } \dot{\mathbf{x}} = \mathscr{F}(\mathbf{x}, t) \tag{9.108}$$

where

$$\begin{aligned}
&\mathbf{g}^{(\alpha)}(\mathbf{z}^{(\alpha)}, \mathbf{x}, t) = (g_1^{(\alpha)}, g_2^{(\alpha)})^{\mathrm{T}};\\
&g_1^{(\alpha)}(\mathbf{z}^{(\alpha)}, \mathbf{x}, t) = z_2^{(\alpha)} - k_1 \quad \text{for } \alpha\text{=1,2};\\
&g_1^{(\alpha)}(\mathbf{z}^{(\alpha)}, \mathbf{x}, t) = z_2^{(\alpha)} + k_1 \quad \text{for } \alpha\text{=3,4};\\
&g_2^{(\alpha)}(\mathbf{z}^{(\alpha)}, \mathbf{x}, t) = \mathscr{G}(\mathbf{z}^{(\alpha)}, \mathbf{x}, t) - k_2 \quad \text{for } \alpha\text{=1,4};\\
&g_2^{(\alpha)}(\mathbf{z}^{(\alpha)}, \mathbf{x}, t) = \mathscr{G}(\mathbf{z}^{(\alpha)}, \mathbf{x}, t) + k_2 \quad \text{for } \alpha\text{=2,3}
\end{aligned} \tag{9.109}$$

with

$$\begin{aligned}
\mathscr{G}(\mathbf{z}^{(\alpha)}, \mathbf{x}, t) = &-a_0 \sin(z_1^{(\alpha)} + x_1) + Q_0 \cos \Omega t\\
&+ d_1 x_2 - a_1 x_1 + a_2 x_1^3 - A_0 \cos \omega t.
\end{aligned} \tag{9.110}$$

The equation of motion on the boundary in the relative coordinates becomes

$$\dot{\mathbf{z}}^{(\alpha\beta)} = \mathbf{g}^{(\alpha\beta)}(\mathbf{z}^{(\alpha\beta)}, \mathbf{x}, t) \text{ with } \dot{\mathbf{x}} = \mathscr{F}(\mathbf{x}, t) \tag{9.111}$$

where

$$g_1^{(\alpha\beta)}(\mathbf{z}^{(\alpha\beta)}, \mathbf{x}, t) = z_2 = 0 \text{ and } g_2^{(\alpha\beta)}(\mathbf{z}^{(\alpha\beta)}, \mathbf{x}, t) = 0 \tag{9.112}$$

with

$$\begin{aligned}
&z_1^{(\alpha\beta)} = 0 \text{ and } z_2^{(\alpha\beta)} = 0 \text{ on } \partial\Omega_{\alpha\beta} \text{ for } (\alpha, \beta)\text{=(2,3), (1,4)};\\
&z_1^{(\alpha\beta)} = C \text{ and } z_2^{(\alpha\beta)} = 0 \text{ on } \partial\Omega_{\alpha\beta} \text{ for } (\alpha, \beta)\text{=(1,2), (3,4).}
\end{aligned} \tag{9.113}$$

9.5.2. Flow switchability on boundaries

To make the controlled pendulum synchronize with the Duffing oscillator, as in Luo (2009a), the synchronization state of the controlled pendulum with the Duffing oscillator requires a sliding flow on a boundary (or stick motion in physics). The non-synchronization state requires either a passable flow to a boundary (penetration state) or a flow in a domain. The de-synchronization requires a source flow to a boundary. From the theory of the discontinuous dynamical system in Luo (2008b, 2009b) and the synchronization theory of two dynamical systems in Luo (2009a), the necessary and sufficient conditions for synchronization, penetration and de-synchronization of the controlled pendulum with the Duffing oscillator will be presented herein. Before the analytical conditions are developed, the G-functions will be introduced from Luo (2008b, 2009b) in the relative coordinates for $\alpha = i, j$ and $(i, j) \in \{(1,2), (2,3), (3,4), (1,4)\}$, i.e., for $\mathbf{z}_m \in \partial\Omega_{ij}$ at $t = t_m$,

$$G_{\partial\Omega_{ij}}^{(\alpha)}(\mathbf{z}_m,\mathbf{x},t_{m\pm})=\mathbf{n}_{\partial\Omega_{ij}}^{\mathrm{T}}\cdot[\mathbf{g}^{(\alpha)}(\mathbf{z}_m,\mathbf{x},t_{m\pm})-\mathbf{g}^{(ij)}(\mathbf{z}_m,\mathbf{x},t_{m\pm})],\tag{9.114}$$

$$G_{\partial\Omega_{ij}}^{(1,\alpha)}(\mathbf{z}_m,\mathbf{x},t_{m\pm})=\mathbf{n}_{\partial\Omega_{ij}}^{\mathrm{T}}\cdot[D\mathbf{g}^{(\alpha)}(\mathbf{z}_m,\mathbf{x},t_{m\pm})-D\mathbf{g}^{(ij)}(\mathbf{z}_m,\mathbf{x},t_{m\pm})].\tag{9.115}$$

From Eq.(9.107), the normal vectors of the relative boundaries are

$$\mathbf{n}_{\partial\Omega_{12}}=\mathbf{n}_{\partial\Omega_{34}}=(0,1)^{\mathrm{T}}\text{ and }\mathbf{n}_{\partial\Omega_{23}}=\mathbf{n}_{\partial\Omega_{14}}=(1,0)^{\mathrm{T}}.\tag{9.116}$$

From Eqs.(9.108)-(9.113), the corresponding G-functions for the boundary are

$$\begin{aligned}&G_{\partial\Omega_{12}}^{(\alpha)}(\mathbf{z}_m,\mathbf{x},t_{m\pm})=G_{\partial\Omega_{34}}^{(\alpha)}(\mathbf{z}_m,\mathbf{x},t_{m\pm})=g_2^{(\alpha)}(\mathbf{z}_m,\mathbf{x},t_{m\pm}),\\&G_{\partial\Omega_{23}}^{(\alpha)}(\mathbf{z}_m,\mathbf{x},t_{m\pm})=G_{\partial\Omega_{14}}^{(\alpha)}(\mathbf{z}_m,\mathbf{x},t_{m\pm})=g_1^{(\alpha)}(\mathbf{z}_m,\mathbf{x},t_{m\pm});\end{aligned}\tag{9.117}$$

$$\begin{aligned}&G_{\partial\Omega_{12}}^{(1,\alpha)}(\mathbf{z}_m,\mathbf{x},t_{m\pm})=G_{\partial\Omega_{34}}^{(1,\alpha)}(\mathbf{z}_m,\mathbf{x},t_{m\pm})=Dg_2^{(\alpha)}(\mathbf{z}_m,\mathbf{x},t_{m\pm}),\\&G_{\partial\Omega_{23}}^{(1,\alpha)}(\mathbf{z}_m,\mathbf{x},t_{m\pm})=G_{\partial\Omega_{14}}^{(1,\alpha)}(\mathbf{z}_m,\mathbf{x},t_{m\pm})=Dg_1^{(\alpha)}(\mathbf{z}_m,\mathbf{x},t_{m\pm});\end{aligned}\tag{9.118}$$

where

$$\begin{aligned}&Dg_1^{(\alpha)}(\mathbf{z}^{(\alpha)},\mathbf{x},t)=g_2^{(\alpha)}(\mathbf{z}^{(\alpha)},\mathbf{x},t)\text{ for }\alpha\text{=1,2,3,4;}\\&Dg_2^{(\alpha)}(\mathbf{z}^{(\alpha)},\mathbf{x},t)=D\mathscr{G}(\mathbf{z}^{(\alpha)},\mathbf{x},t)\\&\qquad=-a_0(z_2^{(\alpha)}+x_2)\cos(z_1^{(\alpha)}+x_1)-Q_0\Omega\sin\Omega t\\&\qquad+d_1F_2(\mathbf{x},t)-a_1x_2+3a_2x_1^2x_2+\omega A_0\sin\omega t,\\&\qquad\text{for }\alpha\text{=1,2,3,4.}\end{aligned}\tag{9.119}$$

The G-functions in Ω_α ($\alpha=1,2,3,4$) to the boundary are defined as

$$\begin{aligned}&G_{\partial\Omega_{12}}^{(\alpha)}(\mathbf{z}^{(\alpha)},\mathbf{x},t)=G_{\partial\Omega_{34}}^{(\alpha)}(\mathbf{z}^{(\alpha)},\mathbf{x},t)=g_2^{(\alpha)}(\mathbf{z}^{(\alpha)},\mathbf{x},t),\\&G_{\partial\Omega_{23}}^{(\alpha)}(\mathbf{z}^{(\alpha)},\mathbf{x},t)=G_{\partial\Omega_{14}}^{(\alpha)}(\mathbf{z}^{(\alpha)},\mathbf{x},t)=g_1^{(\alpha)}(\mathbf{z}^{(\alpha)},\mathbf{x},t).\end{aligned}\tag{9.120}$$

From Luo (2008b, 2009a, b), the analytical conditions of a flow sliding on the boundaries of $\partial\Omega_{12}$, $\partial\Omega_{34}$, $\partial\Omega_{23}$ and $\partial\Omega_{14}$ for the controlled pendulum are

$$\begin{aligned}&\left.\begin{aligned}&G_{\partial\Omega_{12}}^{(1)}(\mathbf{z}_m,\mathbf{x},t_{m-})=g_2^{(1)}(\mathbf{z}_m,\mathbf{x},t_{m-})<0,\\&G_{\partial\Omega_{12}}^{(2)}(\mathbf{z}_m,\mathbf{x},t_{m-})=g_2^{(2)}(\mathbf{z}_m,\mathbf{x},t_{m-})>0\end{aligned}\right\}\text{ for }\mathbf{z}_m\in\partial\Omega_{12};\\&\left.\begin{aligned}&G_{\partial\Omega_{34}}^{(3)}(\mathbf{z}_m,\mathbf{x},t_{m-})=g_2^{(3)}(\mathbf{z}_m,\mathbf{x},t_{m-})>0,\\&G_{\partial\Omega_{34}}^{(4)}(\mathbf{z}_m,\mathbf{x},t_{m-})=g_2^{(4)}(\mathbf{z}_m,\mathbf{x},t_{m-})<0\end{aligned}\right\}\text{ for }\mathbf{z}_m\in\partial\Omega_{34}.\end{aligned}\tag{9.121}$$

$$\left.\begin{aligned}&G_{\partial\Omega_{23}}^{(2)}(\mathbf{z}_m,\mathbf{x},t_{m-})=g_1^{(2)}(\mathbf{z}_m,\mathbf{x},t_{m-})<0,\\&G_{\partial\Omega_{12}}^{(3)}(\mathbf{z}_m,\mathbf{x},t_{m-})=g_1^{(3)}(\mathbf{z}_m,\mathbf{x},t_{m-})>0\end{aligned}\right\}\text{ for }\mathbf{z}_m\in\partial\Omega_{23};$$

$$\left.\begin{aligned} G_{\partial\Omega_{14}}^{(1)}(\mathbf{z}_m, \mathbf{x}, t_{m-}) = g_1^{(1)}(\mathbf{z}_m, \mathbf{x}, t_{m-}) < 0, \\ G_{\partial\Omega_{14}}^{(4)}(\mathbf{z}_m, \mathbf{x}, t_{m-}) = g_1^{(4)}(\mathbf{z}_m, \mathbf{x}, t_{m-}) > 0 \end{aligned}\right\} \text{for } \mathbf{z}_m \in \partial\Omega_{14}. \tag{9.122}$$

From Luo (2008b, 2009a,b), analytical conditions of a flow passing through the boundaries of $\partial\Omega_{12}$, $\partial\Omega_{34}$, $\partial\Omega_{23}$ and $\partial\Omega_{14}$ for the controlled pendulum are

$$\begin{aligned} &\left.\begin{aligned} G_{\partial\Omega_{12}}^{(1)}(\mathbf{z}_m, \mathbf{x}, t_{m-}) = g_2^{(1)}(\mathbf{z}_m, \mathbf{x}, t_{m-}) < 0, \\ G_{\partial\Omega_{12}}^{(2)}(\mathbf{z}_m, \mathbf{x}, t_{m+}) = g_2^{(2)}(\mathbf{z}_m, \mathbf{x}, t_{m+}) < 0 \end{aligned}\right\} \text{for } \mathbf{z}_m \in \partial\Omega_{12}; \\ &\left.\begin{aligned} G_{\partial\Omega_{34}}^{(3)}(\mathbf{z}_m, \mathbf{x}, t_{m-}) = g_2^{(3)}(\mathbf{z}_m, \mathbf{x}, t_{m-}) > 0, \\ G_{\partial\Omega_{34}}^{(4)}(\mathbf{z}_m, \mathbf{x}, t_{m+}) = g_2^{(4)}(\mathbf{z}_m, \mathbf{x}, t_{m+}) > 0 \end{aligned}\right\} \text{for } \mathbf{z}_m \in \partial\Omega_{34}. \end{aligned} \tag{9.123}$$

$$\begin{aligned} &\left.\begin{aligned} G_{\partial\Omega_{23}}^{(2)}(\mathbf{z}_m, \mathbf{x}, t_{m-}) = g_1^{(2)}(\mathbf{z}_m, \mathbf{x}, t_{m-}) < 0, \\ G_{\partial\Omega_{23}}^{(3)}(\mathbf{z}_m, \mathbf{x}, t_{m+}) = g_1^{(3)}(\mathbf{z}_m, \mathbf{x}, t_{m+}) < 0 \end{aligned}\right\} \text{for } \mathbf{z}_m \in \partial\Omega_{23}; \\ &\left.\begin{aligned} G_{\partial\Omega_{14}}^{(1)}(\mathbf{z}_m, \mathbf{x}, t_{m-}) = g_1^{(1)}(\mathbf{z}_m, \mathbf{x}, t_{m-}) > 0, \\ G_{\partial\Omega_{14}}^{(4)}(\mathbf{z}_m, \mathbf{x}, t_{m+}) = g_1^{(4)}(\mathbf{z}_m, \mathbf{x}, t_{m+}) > 0 \end{aligned}\right\} \text{for } \mathbf{z}_m \in \partial\Omega_{14}. \end{aligned} \tag{9.124}$$

From Luo (2008b, 2009a,b), the analytical conditions of a flow grazing to the boundaries of $\partial\Omega_{12}$, $\partial\Omega_{34}$, $\partial\Omega_{23}$ and $\partial\Omega_{14}$ for the controlled pendulum are

$$\begin{aligned} &\left.\begin{aligned} &G_{\partial\Omega_{12}}^{(0,\alpha)}(\mathbf{z}_m, \mathbf{x}, t_{m\pm}) = g_2^{(\alpha)}(\mathbf{z}_m, \mathbf{x}, t_{m\pm}) = 0, \\ &(-1)^{\alpha} G_{\partial\Omega_{12}}^{(1,\alpha)}(\mathbf{z}_m, \mathbf{x}, t_{m\pm}) = (-1)^{\alpha} Dg_2^{(\alpha)}(\mathbf{z}_m, \mathbf{x}, t_{m\pm}) < 0 \end{aligned}\right\} \\ &\text{for } \mathbf{z}_m \in \partial\Omega_{12} \text{ in } \Omega_{\alpha} (\alpha \in \{1,2\}); \\ &\left.\begin{aligned} &G_{\partial\Omega_{34}}^{(0,\alpha)}(\mathbf{z}_m, \mathbf{x}, t_{m\pm}) = g_2^{(\alpha)}(\mathbf{z}_m, \mathbf{x}, t_{m\pm}) = 0, \\ &(-1)^{\alpha} G_{\partial\Omega_{34}}^{(1,\alpha)}(\mathbf{z}_m, \mathbf{x}, t_{m\pm}) = (-1)^{\alpha} Dg_2^{(\alpha)}(\mathbf{z}_m, \mathbf{x}, t_{m\pm}) > 0 \end{aligned}\right\} \\ &\text{for } \mathbf{z}_m \in \partial\Omega_{34} \text{ in } \Omega_{\alpha} \ (\alpha \in \{3,4\}); \end{aligned} \tag{9.125}$$

$$\begin{aligned} &\left.\begin{aligned} &G_{\partial\Omega_{23}}^{(0,\alpha)}(\mathbf{z}_m, \mathbf{x}, t_{m\pm}) = g_1^{(\alpha)}(\mathbf{z}_m, \mathbf{x}, t_{m\pm}) = 0 \\ &(-1)^{\alpha} G_{\partial\Omega_{23}}^{(1,\alpha)}(\mathbf{z}_m, \mathbf{x}, t_{m\pm}) = (-1)^{\alpha} Dg_1^{(\alpha)}(\mathbf{z}_m, \mathbf{x}, t_{m\pm}) > 0 \end{aligned}\right\} \\ &\text{for } \mathbf{z}_m \in \partial\Omega_{23} \text{ in } \Omega_{\alpha} (\alpha \in \{2,3\}); \\ &\left.\begin{aligned} &G_{\partial\Omega_{14}}^{(0,\alpha)}(\mathbf{z}_m, \mathbf{x}, t_{m\pm}) = g_1^{(\alpha)}(\mathbf{z}_m, \mathbf{x}, t_{m\pm}) = 0 \\ &(-1)^{\alpha} G_{\partial\Omega_{14}}^{(1,\alpha)}(\mathbf{z}_m, \mathbf{x}, t_{m\pm}) = (-1)^{\alpha} Dg_1^{(\alpha)}(\mathbf{z}_m, \mathbf{x}, t_{m\pm}) < 0 \end{aligned}\right\} \\ &\text{for } \mathbf{z}_m \in \partial\Omega_{14} \text{ in } \Omega_{\alpha} (\alpha \in \{1,4\}). \end{aligned} \tag{9.126}$$

From Luo (2008b, 2009a,b), the analytical conditions for onset of a sliding flow on the boundaries of $\partial\Omega_{12}$, $\partial\Omega_{34}$, $\partial\Omega_{23}$ and $\partial\Omega_{14}$ in the controlled pendulum are

$$\left.\begin{aligned}&G_{\partial\Omega_{12}}^{(0,1)}(\mathbf{z}_m,\mathbf{x},t_{m-})=g_2^{(1)}(\mathbf{z}_m,\mathbf{x},t_{m-})<0,\\&G_{\partial\Omega_{12}}^{(0,2)}(\mathbf{z}_m,\mathbf{x},t_{m\pm})=g_2^{(2)}(\mathbf{z}_m,\mathbf{x},t_{m\pm})=0,\\&G_{\partial\Omega_{12}}^{(1,2)}(\mathbf{z}_m,\mathbf{x},t_{m\pm})=Dg_2^{(2)}(\mathbf{z}_m,\mathbf{x},t_{m\pm})<0\end{aligned}\right\}\text{ from }\Omega_1\to\partial\Omega_{12};$$
$$\left.\begin{aligned}&G_{\partial\Omega_{34}}^{(0,3)}(\mathbf{z}_m,\mathbf{x},t_{m-})=g_2^{(3)}(\mathbf{z}_m,\mathbf{x},t_{m-})>0,\\&G_{\partial\Omega_{34}}^{(0,4)}(\mathbf{z}_m,\mathbf{x},t_{m\pm})=g_2^{(4)}(\mathbf{z}_m,\mathbf{x},t_{m\pm})=0,\\&G_{\partial\Omega_{34}}^{(1,4)}(\mathbf{z}_m,\mathbf{x},t_{m\pm})=Dg_2^{(4)}(\mathbf{z}_m,\mathbf{x},t_{m\pm})>0\end{aligned}\right\}\text{ from }\Omega_3\to\partial\Omega_{34}. \tag{9.127}$$

$$\left.\begin{aligned}&G_{\partial\Omega_{23}}^{(0,2)}(\mathbf{z}_m,\mathbf{x},t_{m-})=g_1^{(2)}(\mathbf{z}_m,\mathbf{x},t_{m-})<0,\\&G_{\partial\Omega_{23}}^{(0,3)}(\mathbf{z}_m,\mathbf{x},t_{m\pm})=g_1^{(3)}(\mathbf{z}_m,\mathbf{x},t_{m\pm})=0,\\&G_{\partial\Omega_{23}}^{(1,3)}(\mathbf{z}_m,\mathbf{x},t_{m\pm})=Dg_1^{(3)}(\mathbf{z}_m,\mathbf{x},t_{m\pm})<0\end{aligned}\right\}\text{ from }\Omega_2\to\partial\Omega_{23};$$
$$\left.\begin{aligned}&G_{\partial\Omega_{14}}^{(0,4)}(\mathbf{z}_m,\mathbf{x},t_{m-})=g_1^{(4)}(\mathbf{z}_m,\mathbf{x},t_{m-})>0,\\&G_{\partial\Omega_{14}}^{(0,1)}(\mathbf{z}_m,\mathbf{x},t_{m\pm})=g_1^{(1)}(\mathbf{z}_m,\mathbf{x},t_{m\pm})=0,\\&G_{\partial\Omega_{14}}^{(1,1)}(\mathbf{z}_m,\mathbf{x},t_{m\pm})=Dg_1^{(1)}(\mathbf{z}_m,\mathbf{x},t_{m\pm})>0\end{aligned}\right\}\text{ from }\Omega_4\to\partial\Omega_{14}. \tag{9.128}$$

From Luo (2008b, 2009a,b), the analytical conditions for vanishing of a sliding flow from the boundaries of $\partial\Omega_{12}$, $\partial\Omega_{34}$, $\partial\Omega_{23}$ and $\partial\Omega_{14}$ to a domain Ω_α ($\alpha = 1,2,3,4$) for the controlled pendulum are

$$\left.\begin{aligned}&(-1)^{\beta}G_{\partial\Omega_{12}}^{(0,\beta)}(\mathbf{z}_m,\mathbf{x},t_{m-})=(-1)^{\beta}g_2^{(\beta)}(\mathbf{z}_m,\mathbf{x},t_{m-})>0\\&G_{\partial\Omega_{12}}^{(0,\alpha)}(\mathbf{z}_m,\mathbf{x},t_{m\mp})=g_2^{(\alpha)}(\mathbf{z}_m,\mathbf{x},t_{m\mp})=0\\&(-1)^{\alpha}G_{\partial\Omega_{12}}^{(1,\alpha)}(\mathbf{z}_m,\mathbf{x},t_{m\mp})=(-1)^{\alpha}Dg_2^{(\alpha)}(\mathbf{z}_m,\mathbf{x},t_{m\mp})<0,\\&\text{for }\mathbf{z}_m\in\partial\Omega_{12};\alpha,\beta\in\{1,2\}\text{ and }\beta\neq\alpha\end{aligned}\right\}$$
$$\text{from }\partial\Omega_{12}\to\Omega_\alpha;$$
$$\left.\begin{aligned}&(-1)^{\beta}G_{\partial\Omega_{34}}^{(0,\beta)}(\mathbf{z}_m,\mathbf{x},t_{m-})=(-1)^{\beta}g_2^{(\beta)}(\mathbf{z}_m,\mathbf{x},t_{m-})<0\\&G_{\partial\Omega_{34}}^{(0,\alpha)}(\mathbf{z}_m,\mathbf{x},t_{m\mp})=g_2^{(\alpha)}(\mathbf{z}_m,\mathbf{x},t_{m\mp})=0\\&(-1)^{\alpha}G_{\partial\Omega_{34}}^{(1,\alpha)}(\mathbf{z}_m,\mathbf{x},t_{m\mp})=(-1)^{\alpha}Dg_2^{(\alpha)}(\mathbf{z}_m,\mathbf{x},t_{m\mp})>0,\\&\text{for }\mathbf{z}_m\in\partial\Omega_{34};\alpha,\beta\in\{3,4\}\text{ and }\beta\neq\alpha\end{aligned}\right\} \tag{9.129}$$
$$\text{from }\partial\Omega_{34}\to\Omega_\alpha;$$

$$\left.\begin{aligned}&(-1)^{\beta}G_{\partial\Omega_{23}}^{(0,\beta)}(\mathbf{z}_m,\mathbf{x},t_{m-})=(-1)^{\beta}g_1^{(\beta)}(\mathbf{z}_m,\mathbf{x},t_{m-})<0,\\&G_{\partial\Omega_{23}}^{(0,\alpha)}(\mathbf{z}_m,\mathbf{x},t_{m\mp})=g_1^{(\alpha)}(\mathbf{z}_m,\mathbf{x},t_{m\mp})=0\\&(-1)^{\alpha}G_{\partial\Omega_{23}}^{(1,\alpha)}(\mathbf{z}_m,\mathbf{x},t_{m\mp})=(-1)^{\alpha}Dg_1^{(\alpha)}(\mathbf{z}_m,\mathbf{x},t_{m\mp})>0\\&\text{for }\mathbf{z}_m\in\partial\Omega_{23};\alpha,\beta\in\{2,3\}\text{ and }\beta\neq\alpha\end{aligned}\right\}$$

$$\left.\begin{array}{l}\text{from } \partial\Omega_{23} \to \Omega_{\alpha};\\ (-1)^{\beta} G^{(0,\beta)}_{\partial\Omega_{14}}(\mathbf{z}_m,\mathbf{x},t_{m-}) = (-1)^{\beta} g_1^{(\beta)}(\mathbf{z}_m,\mathbf{x},t_{m-}) > 0,\\ G^{(0,\alpha)}_{\partial\Omega_{14}}(\mathbf{z}_m,\mathbf{x},t_{m\mp}) = g_1^{(\alpha)}(\mathbf{z}_m,\mathbf{x},t_{m\mp}) = 0\\ (-1)^{\alpha} G^{(1,\alpha)}_{\partial\Omega_{14}}(\mathbf{z}_m,\mathbf{x},t_{m\mp}) = (-1)^{\alpha} Dg_1^{(\alpha)}(\mathbf{z}_m,\mathbf{x},t_{m\mp}) < 0\\ \text{for } \mathbf{z}_m \in \partial\Omega_{14}; \alpha,\beta \in \{1,4\} \text{ and } \beta \neq \alpha\\ \text{from } \partial\Omega_{14} \to \Omega_{\alpha}.\end{array}\right\} \tag{9.130}$$

9.5.3. Synchronization invariant sets and mechanism

As in Luo and Min (2010a), the synchronization of the controlled slave system with the master system occurs at the intersection of the two separation boundaries ($\mathbf{z}_m = \mathbf{0}$). The corresponding conditions are

$$\begin{array}{ll}\left.\begin{array}{l} G^{(1)}_{\partial\Omega_{14}}(\mathbf{z}_m,\mathbf{x},t_{m-}) = g_1^{(1)}(\mathbf{z}_m,\mathbf{x},t_{m-}) < 0,\\ G^{(1)}_{\partial\Omega_{12}}(\mathbf{z}_m,\mathbf{x},t_{m-}) = g_2^{(1)}(\mathbf{z}_m,\mathbf{x},t_{m-}) < 0\end{array}\right\} & \text{for } \mathbf{z}_m \in \partial\Omega_{12}\cap\partial\Omega_{14} \text{ on } \Omega_1;\\ \left.\begin{array}{l} G^{(2)}_{\partial\Omega_{12}}(\mathbf{z}_m,\mathbf{x},t_{m-}) = g_2^{(2)}(\mathbf{z}_m,\mathbf{x},t_{m-}) > 0,\\ G^{(2)}_{\partial\Omega_{23}}(\mathbf{z}_m,\mathbf{x},t_{m-}) = g_1^{(2)}(\mathbf{z}_m,\mathbf{x},t_{m-}) < 0\end{array}\right\} & \text{for } \mathbf{z}_m \in \partial\Omega_{12}\cap\partial\Omega_{23} \text{ on } \Omega_2;\\ \left.\begin{array}{l} G^{(3)}_{\partial\Omega_{23}}(\mathbf{z}_m,\mathbf{x},t_{m-}) = g_1^{(3)}(\mathbf{z}_m,\mathbf{x},t_{m-}) > 0,\\ G^{(3)}_{\partial\Omega_{34}}(\mathbf{z}_m,\mathbf{x},t_{m-}) = g_2^{(3)}(\mathbf{z}_m,\mathbf{x},t_{m-}) > 0\end{array}\right\} & \text{for } \mathbf{z}_m \in \partial\Omega_{23}\cap\partial\Omega_{34} \text{ on } \Omega_3;\\ \left.\begin{array}{l} G^{(4)}_{\partial\Omega_{34}}(\mathbf{z}_m,\mathbf{x},t_{m-}) = g_2^{(4)}(\mathbf{z}_m,\mathbf{x},t_{m-}) < 0,\\ G^{(4)}_{\partial\Omega_{14}}(\mathbf{z}_m,\mathbf{x},t_{m-}) = g_1^{(4)}(\mathbf{z}_m,\mathbf{x},t_{m-}) > 0\end{array}\right\} & \text{for } \mathbf{z}_m \in \partial\Omega_{34}\cap\partial\Omega_{14} \text{ on } \Omega_4.\end{array} \tag{9.131}$$

From Eq.(9.109), four basic functions are introduced as

$$\begin{array}{l} g_1(\mathbf{z}^{(\alpha)},\mathbf{x},t) \equiv g_1^{(\alpha)}(\mathbf{z}^{(\alpha)},\mathbf{x},t) = z_2^{(\alpha)} - k_1 \text{ in } \Omega_\alpha \text{ for } \alpha\text{=1,2};\\ g_2(\mathbf{z}^{(\alpha)},\mathbf{x},t) \equiv g_1^{(\alpha)}(\mathbf{z}^{(\alpha)},\mathbf{x},t) = z_2^{(\alpha)} + k_1 \text{ in } \Omega_\alpha \text{ for } \alpha\text{=3,4};\\ g_3(\mathbf{z}^{(\alpha)},\mathbf{x},t) \equiv g_2^{(\alpha)}(\mathbf{z}^{(\alpha)},\mathbf{x},t) = \mathscr{G}(\mathbf{z}^{(\alpha)},\mathbf{x},t) - k_2 \text{ in } \Omega_\alpha \text{ for } \alpha\text{=1,4};\\ g_4(\mathbf{z}^{(\alpha)},\mathbf{x},t) \equiv g_2^{(\alpha)}(\mathbf{z}^{(\alpha)},\mathbf{x},t) = \mathscr{G}(\mathbf{z}^{(\alpha)},\mathbf{x},t) + k_2 \text{ in } \Omega_\alpha \text{ for } \alpha\text{=2,3}.\end{array} \tag{9.132}$$

where

$$\begin{aligned}\mathscr{G}(\mathbf{z}^{(\alpha)},\mathbf{x},t) = -a_0\sin(z_1^{(\alpha)} + x_1) + Q_0\cos\Omega t\\ +d_1x_2 - a_1x_1 + a_2x_1^3 - A_0\cos\omega t.\end{aligned} \tag{9.133}$$

The synchronization conditions in Eq.(9.131) become

$$\begin{aligned} g_1(\mathbf{z}_m, \mathbf{x}, t_{m-}) &= z_{2m} - k_1 < 0, \\ g_2(\mathbf{z}_m, \mathbf{x}, t_{m-}) &= z_{2m} + k_1 > 0, \\ g_3(\mathbf{z}_m, \mathbf{x}, t_{m-}) &= \mathcal{G}(\mathbf{z}_m, \mathbf{x}, t_{m-}) - k_2 < 0, \\ g_4(\mathbf{z}_m, \mathbf{x}, t_{m-}) &= \mathcal{G}(\mathbf{z}_m, \mathbf{x}, t_{m-}) + k_2 > 0. \end{aligned} \tag{9.134}$$

Letting $\mathbf{z}_m = \mathbf{0}$, the synchronization conditions of the controlled pendulum with the Duffing oscillator are

$$\begin{aligned} g_1(\mathbf{z}_m, \mathbf{x}, t_{m-}) &= -k_1 < 0, \\ g_2(\mathbf{z}_m, \mathbf{x}, t_{m-}) &= +k_1 > 0, \\ g_3(\mathbf{z}_m, \mathbf{x}, t_{m-}) &= \mathcal{G}(\mathbf{x}, t_{m-}) - k_2 < 0, \\ g_4(\mathbf{z}_m, \mathbf{x}, t_{m-}) &= \mathcal{G}(\mathbf{x}, t_{m-}) + k_2 > 0. \end{aligned} \tag{9.135}$$

where

$$\mathcal{G}(\mathbf{x}, t) = -a_0 \sin x_1 + Q_0 \cos \Omega t + d_1 x_2 - a_1 x_1 + a_2 x_1^3 - A_0 \cos \omega t. \tag{9.136}$$

If $k_1 > 0$ and $k_2 > 0$, the first two equations of Eq.(9.135) can be automatically satisfied, and the third and fourth equations give *the synchronization invariant set* as

$$-k_2 < \mathcal{G}(\mathbf{x}, t_{m-}) < k_2. \tag{9.137}$$

In the small neighborhood of the synchronization of $\mathbf{z}_m = \mathbf{0}$, the attractivity conditions can be given for $|\mathbf{z} - \mathbf{z}_m| < \varepsilon$, i.e.,

$$\begin{aligned} 0 \le z_2 < k_1 \text{ and } \mathcal{G}(\mathbf{z}, \mathbf{x}, t) < k_2 \;\; &\text{for } z_1 \in [0, \infty) \text{ in } \Omega_1, \\ 0 \le z_2 < k_1 \text{ and } -k_2 < \mathcal{G}(\mathbf{z}, \mathbf{x}, t) \;\; &\text{for } z_1 \in [0, \infty) \text{ in } \Omega_2, \\ -k_1 < z_2 \le 0 \text{ and } -k_2 < \mathcal{G}(\mathbf{z}, \mathbf{x}, t) \;\; &\text{for } z_1 \in (-\infty, 0] \text{ in } \Omega_3, \\ -k_1 < z_2 \le 0 \text{ and } \mathcal{G}(\mathbf{z}, \mathbf{x}, t) < k_2 \;\; &\text{for } z_1 \in (-\infty, 0] \text{ in } \Omega_4 \end{aligned} \tag{9.138}$$

from which z_1^* and z_2^* are obtained. The initial conditions for the controlled pendulum synchronization can be determined by

$$y_1 = z_1^* + x_1 \text{ and } y_2 = z_2^* + x_2. \tag{9.139}$$

The conditions for vanishing of synchronization are for $\mathbf{z}^{(\alpha)}(t_{m\mp}) = \mathbf{z}_m^{(\alpha)} = \mathbf{z}_m$

$$\left.\begin{aligned} g_1(\mathbf{z}_m^{(\alpha)}, \mathbf{x}, t_{m\mp}) &= z_{2m}^{(\alpha)} - k_1 = 0, \\ Dg_1(\mathbf{z}_m^{(\alpha)}, \mathbf{x}, t_{m\mp}) &= \mathcal{G}(\mathbf{z}_m^{(\alpha)}, \mathbf{x}, t_{m\mp}) > 0, \\ g_2(\mathbf{z}_m^{(\beta)}, \mathbf{x}, t_{m-}) &= z_{2m}^{(\beta)} + k_1 > 0 \end{aligned}\right\} \tag{9.140}$$

for $(\alpha, \beta) = \{(1,4), (2,3)\}$

from $z_{m+\varepsilon} = y_1 - x_1 > 0$, and

$$\left.\begin{aligned} g_1(\mathbf{z}_m^{(\alpha)}, \mathbf{x}, t_{m-}) &= z_{2m}^{(\alpha)} - k_1 < 0; \\ g_2(\mathbf{z}_m^{(\beta)}, \mathbf{x}, t_{m\mp}) &= z_{2m}^{(\beta)} + k_1 = 0, \\ Dg_2(\mathbf{z}_m^{(\beta)}, \mathbf{x}, t_{m\mp}) &= \mathscr{G}(\mathbf{z}_m^{(\beta)}, \mathbf{x}, t_{m\mp}) < 0 \end{aligned}\right\} \tag{9.141}$$
$$\text{for } (\alpha, \beta) = \{(1,4),(2,3)\}$$

from $z_{m+\varepsilon} = y_1 - x_1 < 0$.

The conditions for vanishing of synchronization are for $\mathbf{z}^{(\alpha)}(t_{m\mp}) = \mathbf{z}_m^{(\alpha)} = \mathbf{z}_m$

$$\left.\begin{aligned} g_3(\mathbf{z}_m^{(\alpha)}, \mathbf{x}, t_{m\mp}) &= \mathscr{G}(\mathbf{z}_m^{(\alpha)}, \mathbf{x}, t_{m\mp}) - k_2 = 0, \\ Dg_3(\mathbf{z}_m^{(\alpha)}, \mathbf{x}, t_{m\mp}) &= D\mathscr{G}(\mathbf{z}_m^{(\alpha)}, \mathbf{x}, t_{m\mp}) > 0; \\ g_4(\mathbf{z}_m^{(\beta)}, \mathbf{x}, t_{m-}) &= \mathscr{G}(\mathbf{z}_m^{(\beta)}, \mathbf{x}, t_{m-}) + k_2 > 0 \end{aligned}\right\} \tag{9.142}$$
$$\text{for } (\alpha, \beta) = \{(1,2),(4,3)\}$$

from $\dot{z}_{m+\varepsilon} = y_2 - x_2 > 0$, and

$$\left.\begin{aligned} g_3(\mathbf{z}_m^{(\alpha)}, \mathbf{x}, t_{m-}) &= \mathscr{G}(\mathbf{z}_m^{(\alpha)}, \mathbf{x}, t_{m-}) - k_2 < 0; \\ g_4(\mathbf{z}_m^{(\beta)}, \mathbf{x}, t_{m\mp}) &= \mathscr{G}(\mathbf{z}_m^{(\beta)}, \mathbf{x}, t_{m\mp}) + k_2 = 0, \\ g_4(\mathbf{z}_m^{(\beta)}, \mathbf{x}, t_{m\mp}) &= D\mathscr{G}(\mathbf{z}_m^{(\beta)}, \mathbf{x}, t_{m\mp}) < 0 \end{aligned}\right\} \tag{9.143}$$
$$\text{for } (\alpha, \beta) = \{(1,2),(4,3)\}$$

from $\dot{z}_{m+\varepsilon} = y_2 - x_2 < 0$.

The conditions for onset of synchronization are for $\mathbf{z}^{(\alpha)}(t_{m\mp}) = \mathbf{z}_m^{(\alpha)} = \mathbf{z}_m$

$$\left.\begin{aligned} g_1(\mathbf{z}_m^{(\alpha)}, \mathbf{x}, t_{m\pm}) &= z_{2m}^{(\alpha)} - k_1 = 0, \\ Dg_1(\mathbf{z}_m^{(\alpha)}, \mathbf{x}, t_{m\pm}) &= \mathscr{G}(\mathbf{z}_m^{(\alpha)}, \mathbf{x}, t_{m\pm}) > 0, \\ g_2(\mathbf{z}_m^{(\beta)}, \mathbf{x}, t_{m-}) &= z_{2m}^{(\beta)} + k_1 > 0; \end{aligned}\right\} \tag{9.144}$$
$$\text{for } (\alpha, \beta) = \{(1,4),(2,3)\}$$

from $z_{m-\varepsilon} = y_1 - x_1 > 0$, and

$$\left.\begin{aligned} g_1(\mathbf{z}_m^{(\alpha)}, \mathbf{x}, t_{m-}) &= z_{2m}^{(\alpha)} - k_1 < 0; \\ g_2(\mathbf{z}_m^{(\beta)}, \mathbf{x}, t_{m\pm}) &= z_{2m}^{(\beta)} + k_1 = 0, \\ Dg_2(\mathbf{z}_m^{(\beta)}, \mathbf{x}, t_{m\pm}) &= \mathscr{G}(\mathbf{z}_m^{(\beta)}, \mathbf{x}, t_{m\pm}) < 0 \end{aligned}\right\} \tag{9.145}$$
$$\text{for } (\alpha, \beta) = \{(1,4),(2,3)\}$$

from $z_{m-\varepsilon} = y_1 - x_1 < 0$.

The conditions for onset of synchronization are for $\mathbf{z}^{(\alpha)}(t_{m\pm}) = \mathbf{z}_m^{(\alpha)} = \mathbf{z}_m$

$$\left.\begin{aligned} g_3(\mathbf{z}_m^{(\alpha)},\mathbf{x},t_{m\pm}) &= \mathscr{G}(\mathbf{z}_m^{(\alpha)},\mathbf{x},t_{m\pm}) - k_2 = 0, \\ Dg_3(\mathbf{z}_m^{(\alpha)},\mathbf{x},t_{m\pm}) &= D\mathscr{G}(\mathbf{z}_m^{(\alpha)},\mathbf{x},t_{m\pm}) > 0; \\ g_4(\mathbf{z}_m^{(\beta)},\mathbf{x},t_{m-}) &= \mathscr{G}(\mathbf{z}_m^{(\beta)},\mathbf{x},t_{m-}) + k_2 > 0 \end{aligned}\right\} \tag{9.146}$$
$$\text{for } (\alpha,\beta) = \{(1,2),(4,3)\}$$

from $\dot{z}_{m-\varepsilon} = y_2 - x_2 > 0$, and

$$\left.\begin{aligned} g_3(\mathbf{z}_m^{(\alpha)},\mathbf{x},t_{m-}) &= \mathscr{G}(\mathbf{z}_m^{(\alpha)},\mathbf{x},t_{m-}) - k_2 < 0; \\ g_4(\mathbf{z}_m^{(\beta)},\mathbf{x},t_{m\pm}) &= \mathscr{G}(\mathbf{z}_m^{(\beta)},\mathbf{x},t_{m\pm}) + k_2 = 0, \\ g_4(\mathbf{z}_m^{(\beta)},\mathbf{x},t_{m\pm}) &= D\mathscr{G}(\mathbf{z}_m^{(\beta)},\mathbf{x},t_{m\pm}) < 0 \end{aligned}\right\} \tag{9.147}$$
$$\text{for } (\alpha,\beta) = \{(1,2),(4,3)\}$$

from $\dot{z}_{m-\varepsilon} = y_2 - x_2 < 0$.

9.5.4. Illustrations for synchronization

Consider a periodic motion of the Duffing oscillator with a set of parameters and initial condition, i.e.,

$$\begin{aligned} &\text{Duffing}: \quad a_1 = a_2 = 1.0, d_1 = 0.25, \omega = 1.0; \\ &\text{Pendulum: } a_0 = 1.0, Q_0 = 0.275, \Omega = 2.18517. \end{aligned} \tag{9.148}$$

To understand the synchronization of the controlled pendulum with periodic and chaotic motions in the Duffing oscillator, the synchronization switching scenario of the controlled slave system (i.e., controlled pendulum) varying with control parameter k_2 is presented in Fig.9.12 first with ($k_1 = 1, A_0 = 0.454$) plus the other parameters in Eq.(9.148) and initial conditions ($x_1 = y_1 \approx 0.3646916$ and $x_2 = y_2$ ≈ 1.2598329 for $t_0 = 0$). Acronyms "FS", "PS" and "NS" represent *f*ull, *p*artial and *n*on synchronizations, respectively. A and V denote synchronization appearance and vanishing, respectively. The switching displacement, velocity and phase plane for the controlled pendulum to synchronize with a periodic motion of the Duffing oscillator are illustrated in Fig.9.12(a)-(c), respectively. The distribution of synchronization appearance and vanishing on the trajectory of periodic motion in phase plane is presented in Fig.9.12(d). From such a synchronization scenario ($k_1 = 1$), the partial synchronization of the controlled pendulum with the periodic motion of the Duffing oscillator lies in $k_2 \in (0.045, 2.775)$. If $k_2 \in (0, 0.045)$, no synchronization can be observed. If $k_2 \in (2.275, \infty)$, the controlled pendulum will fully synchronize with the periodic motion in the Duffing oscillator.

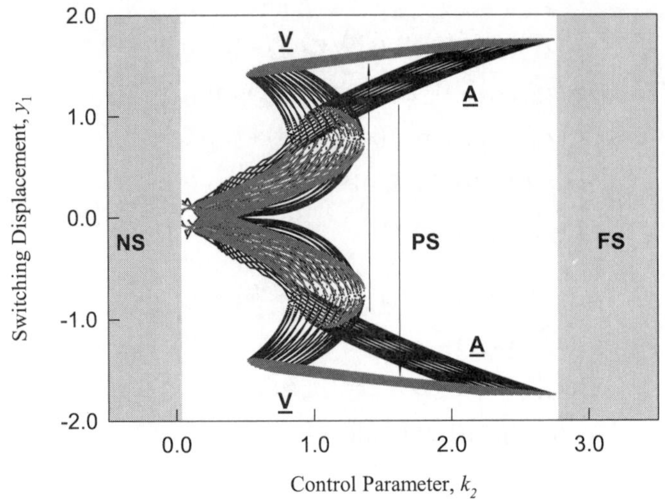

(b)

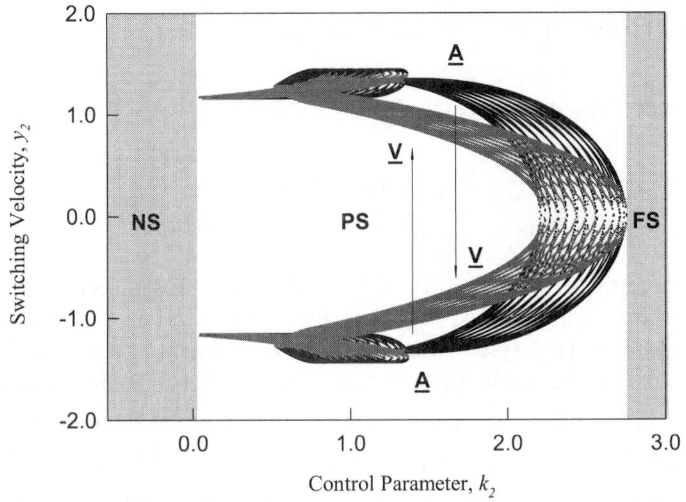

(b)

Fig.9.12 Synchronization scenario of switching points versus control parameter k_2 : **(a)** switching displacement, **(b)** switching velocity, **(c)** switching phase, and **(d)** switching points on the synchronized periodic orbit. (Control parameter. $k_1 = 1$. Duffing: $a_1 = a_2 = 1.0$, $d_1 = 0.25$, $A_0 =$ 0.454, $\omega = 1.0$. Pendulum: $a_0 = 1.0$, $Q_0 = 0.275$, $\Omega = 2.18519$). (Initial condition. $x_1 = y_1 \approx$ 0.3646916 and $x_2 = y_2 \approx 1.2598329$). (FS. Full synchronization, PS. Partial synchronization, NS. Non-synchronization). (A. Synchronization appearance; V. Synchronization vanishing).

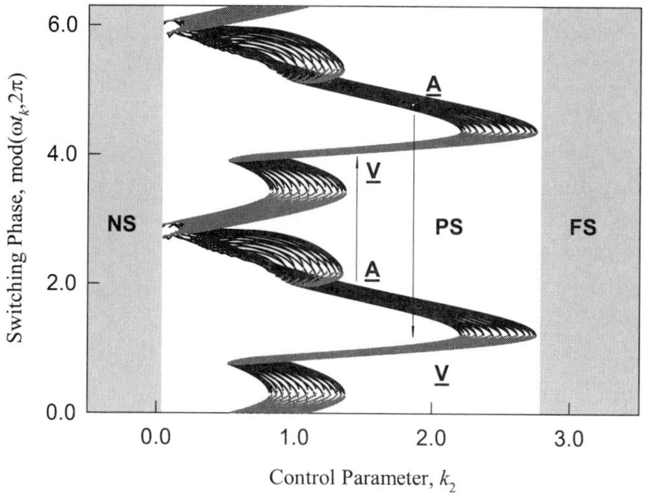

(c)

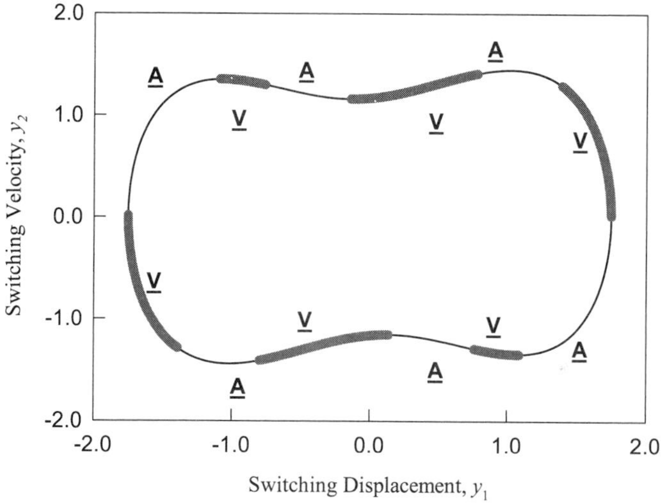

(d)

Fig.9.12 Continued.

For such a chaotic motion, the synchronization switching scenario of the controlled slave system (i.e., controlled pendulum system) varying with control parameter k_2 is presented in Fig.9.13(a)-(f) with ($k_1 = 1$, and $A_0 = 0.265$) plus the other parameters in Eq.(9.148) and initial conditions ($x_1 = y_1 \approx -0.4180597$ and $x_2 = y_2 \approx 0.2394332$ for $t_0 = 0$). Similarly, acronyms "FS", "PS" and "NS" represent *f*ull, *p*artial and *n*on synchronizations, respectively. In Figs.9.13(a)-(c), switching displacement, velocity and switching phase versus the control parameter k_2 are presented for synchronization appearance points. In Figs.9.13(d)-(f), switching displacement, velocity and switching phase versus control parameter k_2 are presented for synchronization vanishing points. Compared to synchronization with periodic motions, the switching scenario of synchronization is chaotic. From such a synchronization scenario ($k_1 = 1$), the partial synchronization of the controlled slave system with the chaotic motion lies in $k_2 \in (0.126, 1.408)$. If $k_2 \in (0, 0.126)$, no synchronization can be observed. If $k_2 \in (1.408, \infty)$, the controlled pendulum fully synchronize with the chaotic motion in the Duffing oscillator.

With varying control parameters k_1 and k_2, the parameter map (k_1, k_2) for the synchronization of the controlled pendulum with the periodic and chaotic motions in the Duffing oscillator are presented in Figs.9.14, respectively. The shade area is a partial synchronization area. The boundary of the partial synchronization with chaotic motions is rougher than the boundary for periodic motions. The parameter maps are based on the control parameters to specific periodic and chaotic motions. To consider the synchronization to all possible motions in master system, the parameters of the master system should be varied. As in Luo (2008), the excitation amplitude and damping coefficient in the Duffing oscillator will be varied, and all possible periodic and chaotic motions in the Duffing oscillator will be captured. For clearer illustrations, the control parameter $k_1 = 1$ is fixed. The control parameter k_2 will be varied to control the slave system synchronizing with motions in the master system. Thus, the parameter maps of (A_0, k_2) and (d_1, k_2) are presented in Fig.9.15. The smooth boundary of synchronization in parameter maps is for periodic motions, but the rough boundary of synchronization is for chaotic motions. The shade area is for partial synchronization. In Fig.9.15(a), the ranges of synchronization are different with increasing and decreasing excitation amplitude of the Duffing oscillator. The different area is hatched, which is from the nonlinearity of the Duffing oscillator. As in Luo (2008a,b), with increasing excitation amplitude, the Duffing oscillator experiences periodic motions inside potential wells, periodic and chaotic motions near separatrix, and periodic motions outside potential wells. For small k_2, the synchronization onset is not smooth because an excited pendulum system possesses Hamiltonian chaos. The Duffing oscillator possesses Hamiltonian chaos at $d_1 = 0$. Thus, the control parameter k_2 for $d_1 = 0$ is different from $d_1 \neq 0$. The control parameter k_2 for other periodic and chaotic motions with damping coefficient d_1 are presented in Fig.9.15(b).

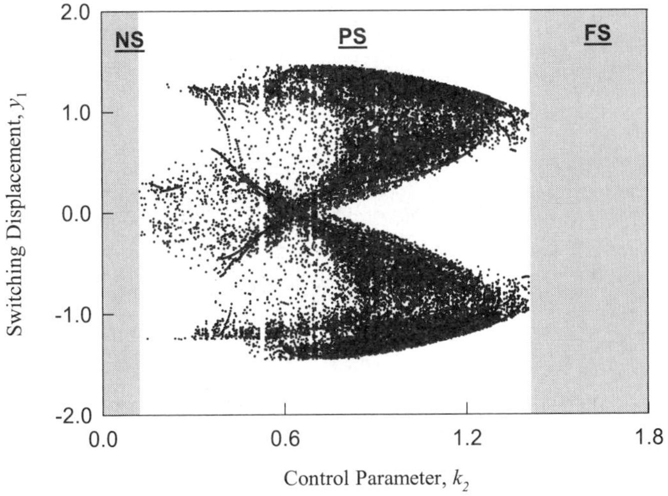

(a)

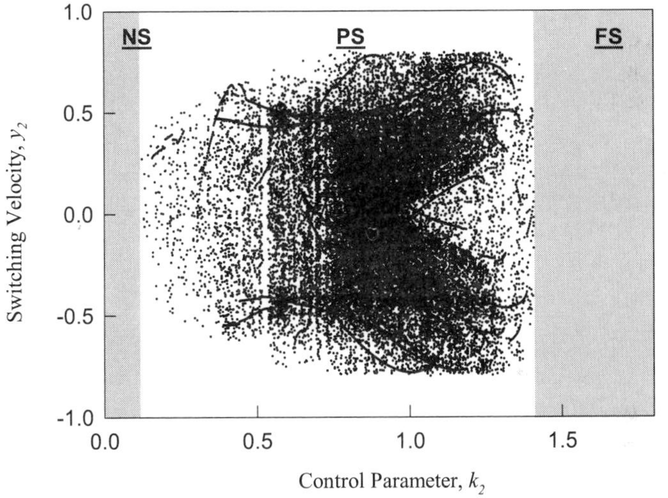

(b)

Fig.9.13 Synchronization scenario of switching points versus control parameter k_2 . Appearance: **(a)** switching displacement, **(b)** switching velocity, **(c)** switching phase. Disappearance: **(d)** switching displacement, **(e)** switching velocity, and **(f)** switching phase. (Control parameter. $k_1 = 1$. Duffing: $a_1 = a_2 = 1.0$, $d_1 = 0.25$, $A_0 = 0.265$, $\omega = 1.0$. Pendulum: $a_0 = 1.0$, $Q_0 =$ 0.275, $\Omega = 2.18519$). (Initial condition. $x_1 = y_1 \approx -0.4180597$ and $x_2 = y_2 \approx 0.2394332$). (FS. full synchronization, PS. partial synchronization, NS. Non-synchronization).

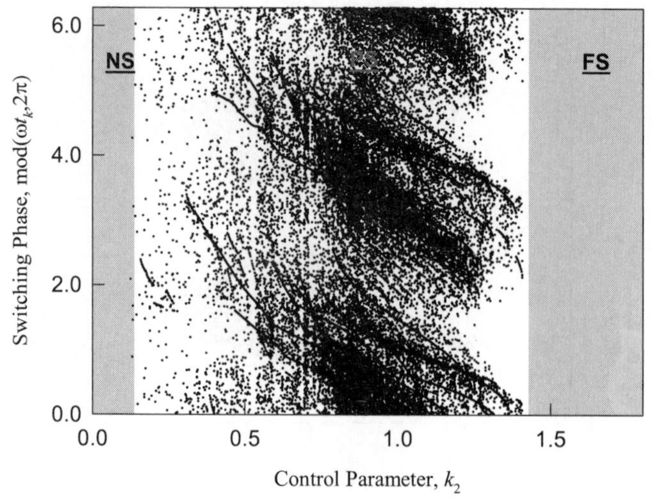

(c)

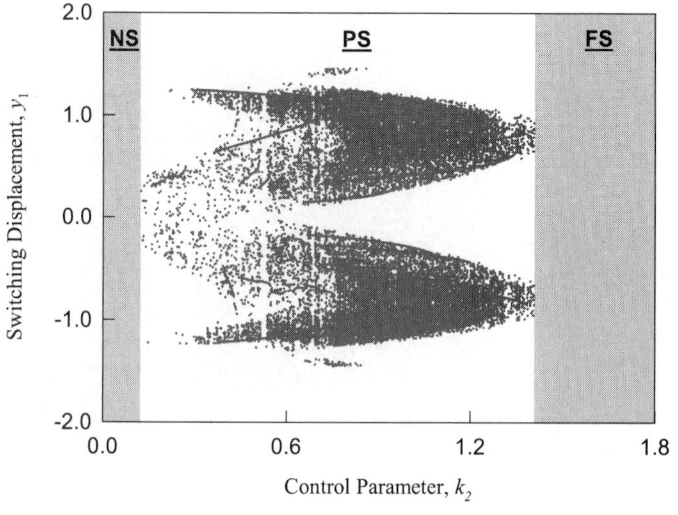

(d)

Fig.9.13 Continued.

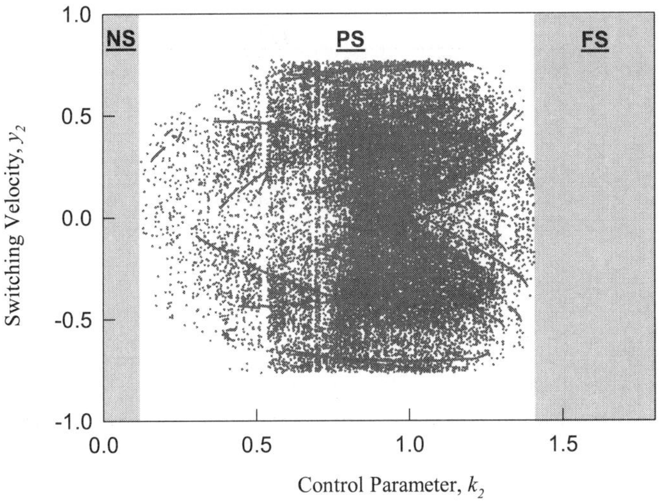

(e)

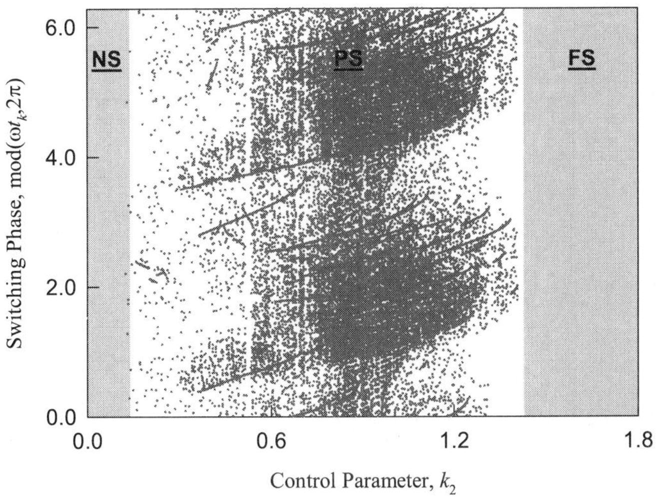

(f)

Fig.9.13 Continued.

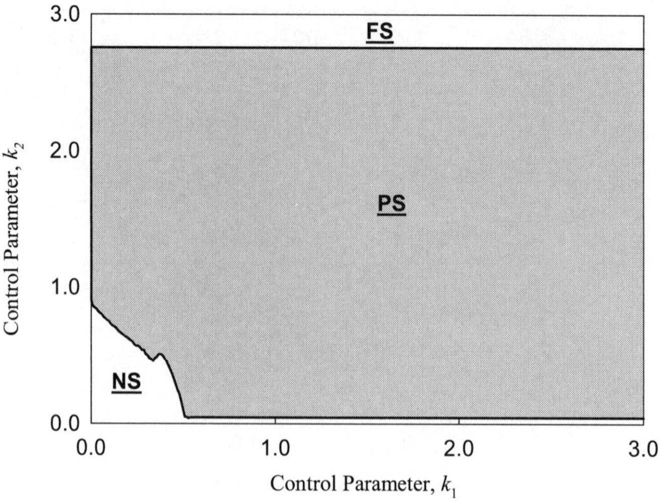

(a)

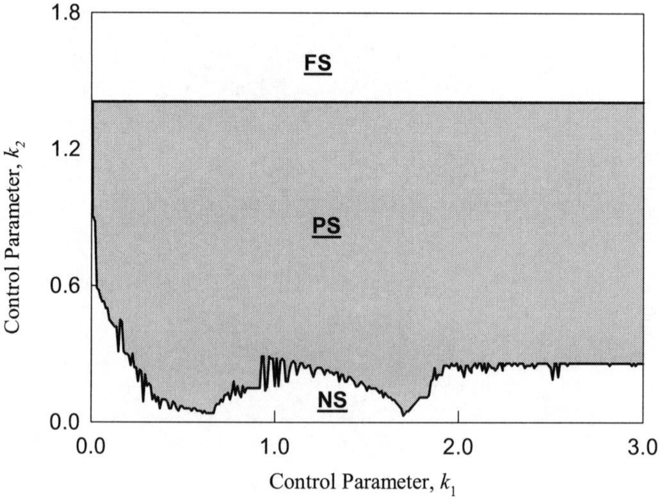

(b)

Fig.9.14 Control parameter maps of (k_1, k_2) for motion synchronization: **(a)** periodic motion ($A_0 = 0.454$) with initial condition ($x_1 = y_1 \approx 0.3646916$ and $x_2 = y_2 \approx 1.2598329$), **(b)** chaotic motion ($A_0 = 0.265$) with initial condition ($x_1 = y_1 \approx -0.4180597$ and $x_2 = y_2 \approx 0.2394332$). (Control parameters for Duffing: $a_1 = a_2 = 1.0$, $d_1 = 0.25$, $\omega = 1.0$; for pendulum: $a_0 = 1.0$, $Q_0 = 0.275$, $\Omega = 2.18519$). (FS. full synchronization, PS. partial synchronization, NS. Non-synchronization).

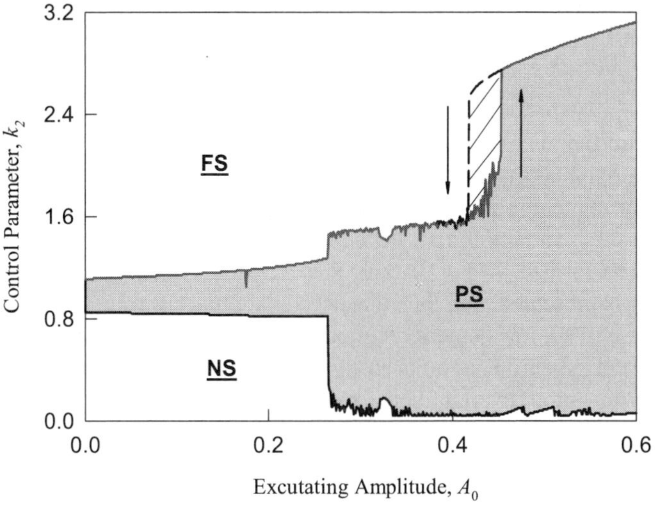

(a)

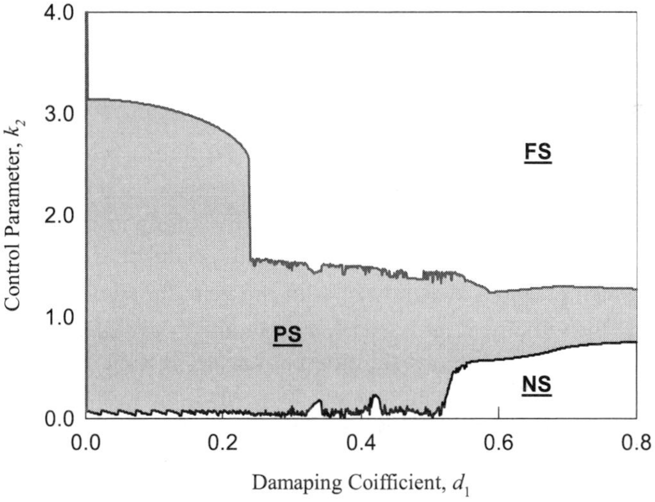

(b)

Fig.9.15 Control parameter maps for motion synchronization with $k_1 = 1$: **(a)** (A_0, k_2) with $d_1 = 0.25$, **(b)** (d_1, k_2) with $A_0 = 0.4$. (Control parameters for Duffing: $a_1 = a_2 = 1.0$, $\omega = 1.0$; for pendulum: $a_0 = 1.0$, $Q_0 = 0.275$, $\Omega = 2.18519$). (FS. full synchronization, PS. partial synchronization, NS. Non-synchronization).

To understand the synchronization of two dynamical systems, the synchronization invariant set of the master system is produced from Eq.(9.137) for the control parameters of $k_1 = 1$ and $k_2 = 1.5$, as shown in Fig.9.16. The synchronization invariant domain is shaded, and the other regions are for non-synchronization. The boundaries are the maximum and minimum values for the onset and vanishing of synchronization. For the controlled pendulum to synchronize with the Duffing oscillator, the synchronized portion of the trajectory of the Duffing oscillator should be in the invariant domain. Otherwise, the synchronization of the two systems cannot be formed. In Fig.9.16(a), the overview of the synchronization invariant domain is presented, and the velocity can approach infinity. Once the control parameters are fixed, for the periodic motion with large orbits in the Duffing oscillator, the partial synchronization is easily observed. In Fig.9.16(b), a zoomed view of the synchronization invariant domain is presented.

For a better understanding of synchronization between two different dynamical systems, consider the partial synchronization of the controlled pendulum with a periodic motion in the Duffing oscillator in Figs.9.17 and 9.18. In Fig.9.17(a), the time-history of velocity for the periodic motion of the Duffing oscillator is depicted by the solid curve, but the velocity response for the controlled pendulum is given by the dashed curve. The corresponding G-functions for controlled pendulum are plotted in Fig.9.17(b). The shaded portions are for synchronization. The non-shaded regions are for non-synchronization, and the G-functions for non-synchronization are presented by dashed curves. If the G-function of g_4 is dashed curve, the controlled pendulum is in domain Ω_α ($\alpha = 1,4$), which does not synchronize with the periodic motion in the Duffing oscillator, and the velocity of the controlled pendulum is *greater* than that of the Duffing oscillator. If the G-function of g_3 is dashed curve, the controlled pendulum is in domain Ω_α ($\alpha = 2,3$), which does not synchronize with the periodic motion in the Duffing oscillator, and the velocity of the controlled pendulum is *less* than that of the Duffing oscillator. If initial conditions are different for the Duffing oscillator and controlled pendulum, the non-synchronization for displacement can be observed. To check whether the synchronized portion of the trajectory is in the synchronization invariant domain or not, the invariant domain is superimposed on phase plane, as shown in Fig.9.17(c). It is clearly observed that the synchronized portions of the trajectory lie in the invariant domain indeed. Even if the non-synchronized portion is in the invariant domain, the controlled pendulum should satisfy the corresponding conditions for synchronization appearance and vanishing in the previous section (i.e., Eqs.(9.139)-(9.147)). To observe the existence of the synchronization pattern for long time, the switching points for synchronization and non-synchronization are presented in Fig.9.17(d) for 10,000 periods of the Duffing oscillator. To observe synchronization and non-synchronization properties, phase planes for the Duffing oscillator and controlled pendulum, and G-function distributions along the displacement are also plotted in Fig.9.18(a)-(d), respectively. The synchronicity of two systems to a periodic motion is clearly presented.

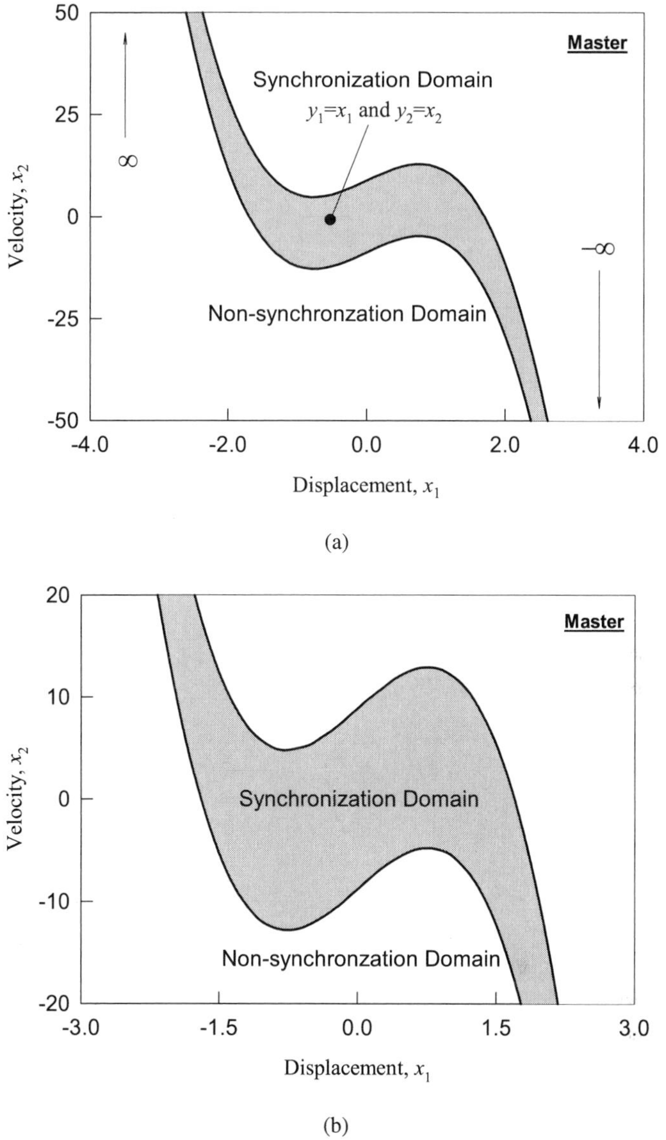

Fig.9.16 Synchronization invariant sets of the Duffing oscillator for the controlled pendulum.: **(a)** wide range view, and **(b)** zoomed view. (Control parameters: $k_1 = 1$ and $k_2 = 1.5$. Duffing: $a_1 = a_2 = 1.0$, $d_1 = 0.25$, $A_0 = 0.454$, $\omega = 1.0$. Pendulum: $a_0 = 1.0$, $Q_0 = 0.275$, $\Omega = 2.18519$).

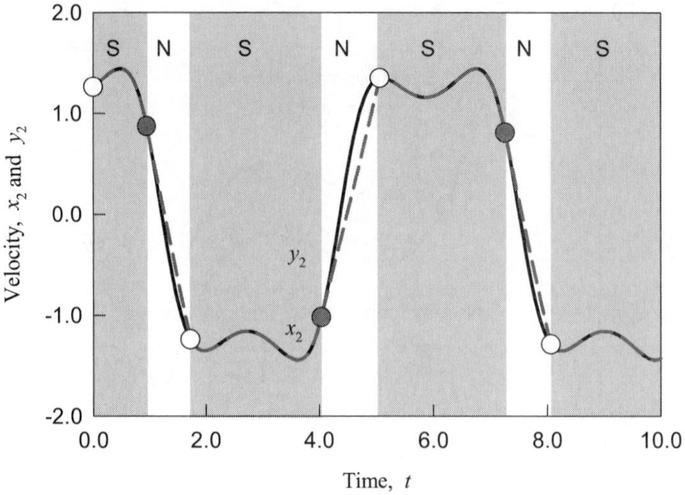

(a)

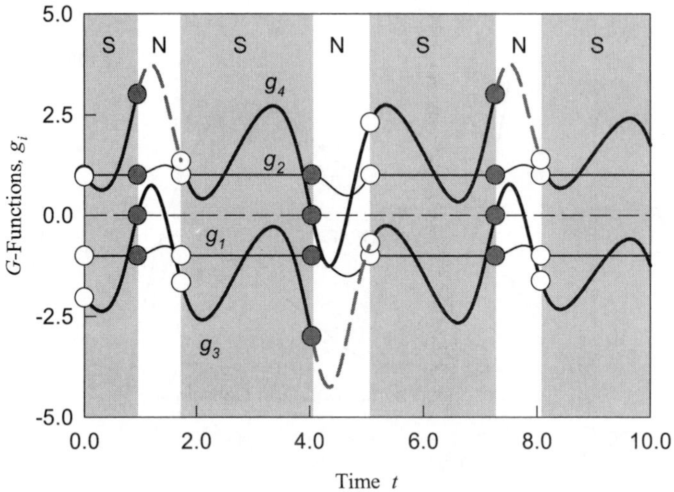

(b)

Fig.9.17 Partial synchronization of the Duffing oscillator and controlled pendulum: **(a)** velocity responses, **(b)** G-function responses, **(c)** phase plane with the invariant domain, and **(d)** switching points for synchronization appearance and vanishing. (Control parameters. $k_1 = 1$ and $k_2 = 1.5$. Duffing: $a_1 = a_2 = 1.0$, $d_1 = 0.25$, $A_0 = 0.454$, $\omega = 1.0$. Pendulum: $a_0 = 1.0$, $Q_0 = 0.275$, $\Omega = 2.18517$). (Initial condition. $x_1 = y_1 \approx 0.3646916$ and $x_2 = y_2 \approx 1.2598329$). (S. Synchronization; N. Non-synchronization). Hollow and filled circular symbols are synchronization appearance (A) and vanishing (V), respectively.

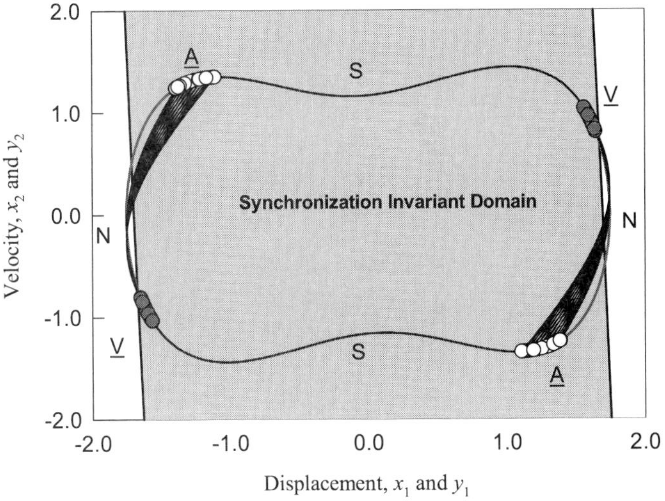

(c)

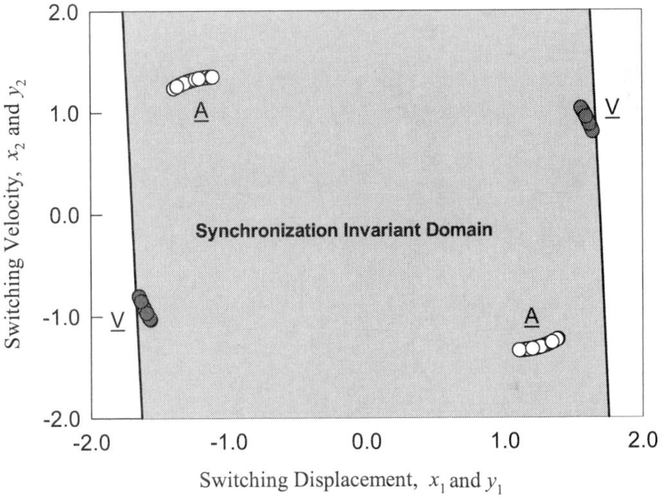

(d)

Fig.9.17 Continued.

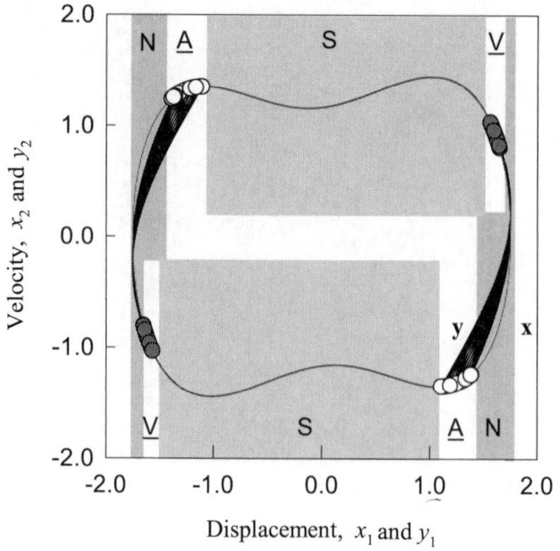

(a)

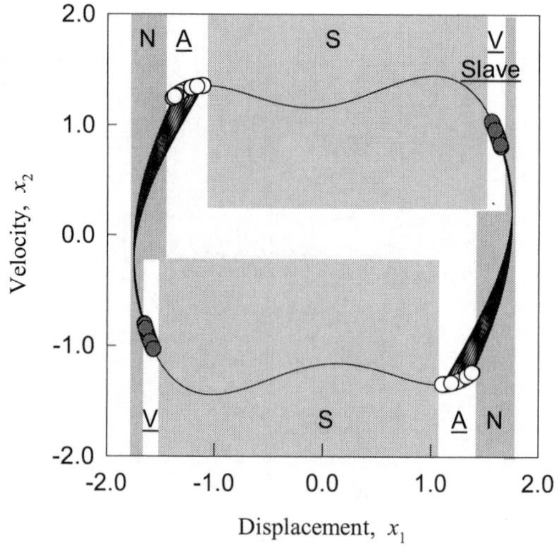

(b)

Fig.9.18 Partial synchronization of the Duffing oscillator and controlled pendulum. Trajectories in phase plane: **(a)** slave and master systems and **(b)** controlled pendulum. G-function distribution along displacement: **(c)** G-function ($g_{1,2}$) and **(d)** G-function ($g_{3,4}$). (Control parameters: $k_1 = 1$ and $k_2 = 3$ Duffing: $a_1 = a_2 = 1.0$, $d_1 = 0.25$, $A_0 = 0.454$, $\omega = 1.0$. Pendulum: $a_0 = 1.0$, $Q_0 =$ 0.275, $\Omega = 2.18519$). (Initial condition. $x_1 = y_1 \approx 0.3646916$ and $x_2 = y_2 \approx 1.2598329$). (S. Synchronization; N. Non-synchronization). Hollow and filled circular symbols are synchronization appearance (A) and vanishing (V), respectively.

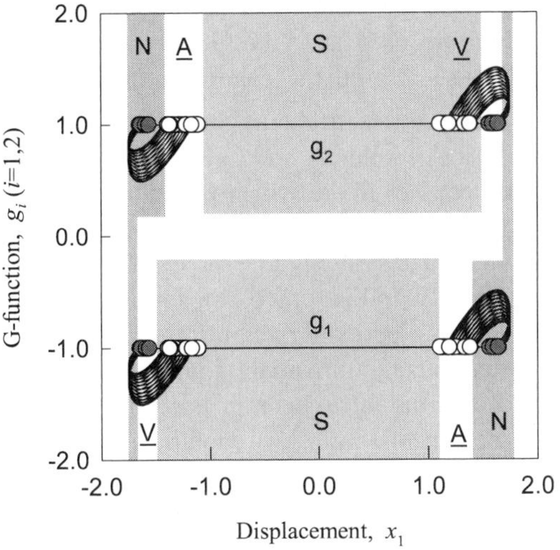

(c)

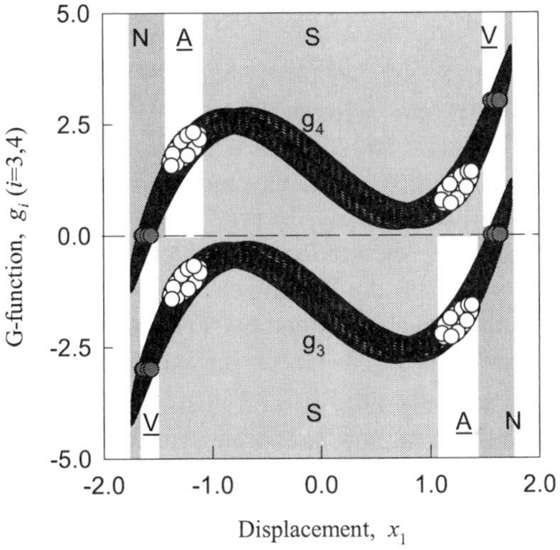

(d)

Fig.9.18 Continued.

For full-synchronizations, the same initial conditions for the Duffing oscillator and controlled pendulum are used as in Eq.(9.148), and the control parameters ($k_1 = 1$ and $k_2 = 3$) are used. From the parameter map (k_1, k_2), the controlled pendulum will fully synchronize with the periodic motion of the Duffing oscillator. Thus, the time-histories of velocity and G-functions, and the synchronization invariant domain and trajectories in the synchronization invariant domain in phase plane are presented in Fig.9.19(a)-(d), respectively. In Fig.9.19(a) and (d), the solid curves and hollow-circular symbols are for the Duffing oscillator and controlled pendulum, respectively. In Fig.9.19(b), the G-function tells the two systems should be synchronized with a periodic motion from Eq.(9.135). Under such control parameters, the synchronization invariant domain for the Duffing oscillator is presented in Fig.9.19(c). For the full synchronization of the Duffing oscillator and controlled pendulum, the periodic trajectory should be in the corresponding invariant domain. Thus, the synchronization invariant domain is superimposed on phase plane in Fig.9.19(d). Indeed, the trajectory for the fully synchronized periodic motion is in the invariant domain.

To illustrate the partial synchronization of the controlled pendulum with a chaotic motion in the Duffing oscillator, as in Luo and Min (2010b), the parameters ($k_1 = 1$, $k_2 = 0.9$, and $A_0 = 0.265$) plus the other parameters in Eq.(9.148) and initial conditions ($x_1 = y_1 \approx -0.4180597$ and $x_2 = y_2 \approx 0.2394332$ for $t_0 = 0$) are used for illustrations. The velocity time history, G-function time history, the trajectories in phase plane and switching sections for synchronization appearance and disappearance of the controlled pendulum and Duffing oscillator are also presented in Fig.9.20(a)-(d), respectively. In Fig.9.20(a), the velocity response does not show periodicity because the motion of the Duffing oscillator is chaotic. The solid and dashed curves are for the Duffing oscillator and the controlled systems, respectively. The synchronization portion of chaos in the Duffing oscillator always requires ($g_1 < 0$ and $g_2 > 0$) for displacement and ($g_3 < 0$ and $g_4 > 0$) for velocity. For the non-synchronization, one of such conditions at least will not be satisfied even if the initial conditions are in the synchronization invariant domain. In a similar fashion, the corresponding time histories of G-functions are presented in Fig.9.20(b). The trajectories and switching section of synchronization appearance and vanishing are placed in the synchronization invariant domain, and such a domain is shaded. In Fig.9.20(c), the solid curve gives the trajectory of chaos for the Duffing oscillator, and the dashed curve shows the trajectory of the controlled pendulum. To observe the switching of the synchronization and non-synchronization between the Duffing oscillator and controlled pendulum, the switching points of the partial synchronization are presented for 10,000 periods of the Duffing oscillator. The switching points for synchronization appearance and vanishing in phase plane to the chaotic motion in the Duffing oscillator are presented for synchronization appearance in Fig.9.20(d).

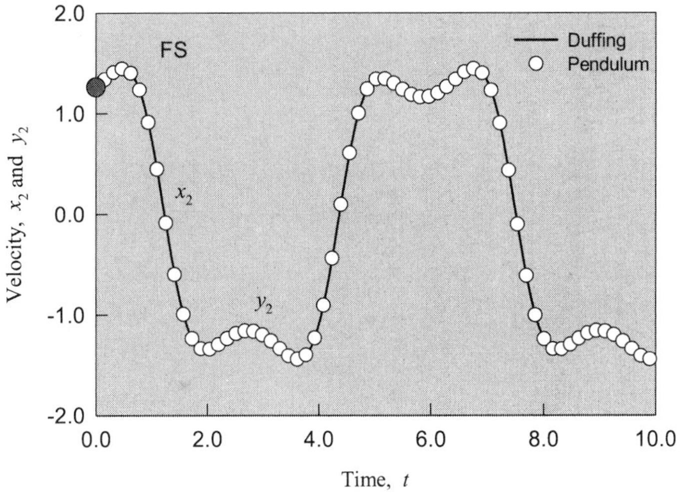

(a)

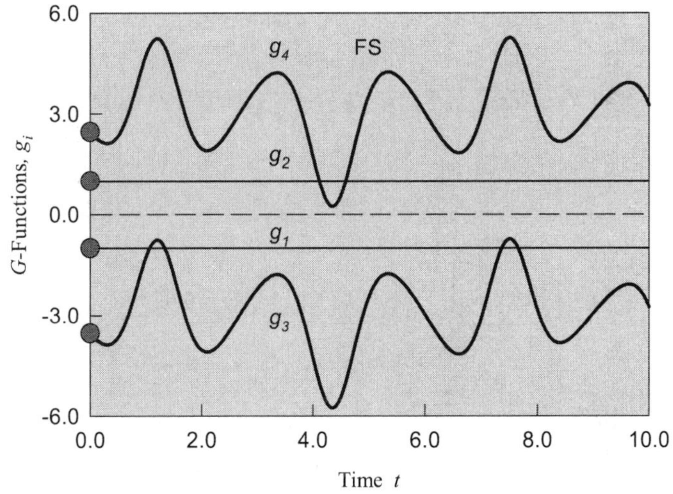

(b)

Fig.9.19 Full synchronization of the Duffing oscillator and controlled pendulum: **(a)** time-history of velocity, **(b)** time-history of G-function, **(c)** synchronization invariant domain, and **(d)** trajectories in phase plane. (Control parameters: $k_1 = 1$ and $k_2 = 3$. Duffing: $a_1 = a_2 = 1.0$, $d_1 = 0.25$, $A_0 = 0.454$, $\omega = 1.0$. Pendulum: $a_0 = 1.0$, $Q_0 = 0.275$, $\Omega = 2.18519$). (Initial condition. $x_1 = y_1 \approx 0.3646916$ and $x_2 = y_2 \approx 1.2598329$). (FS. Full synchronization).

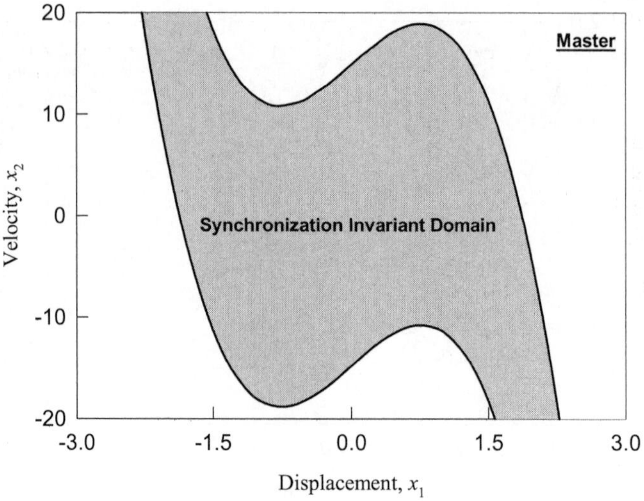

(c)

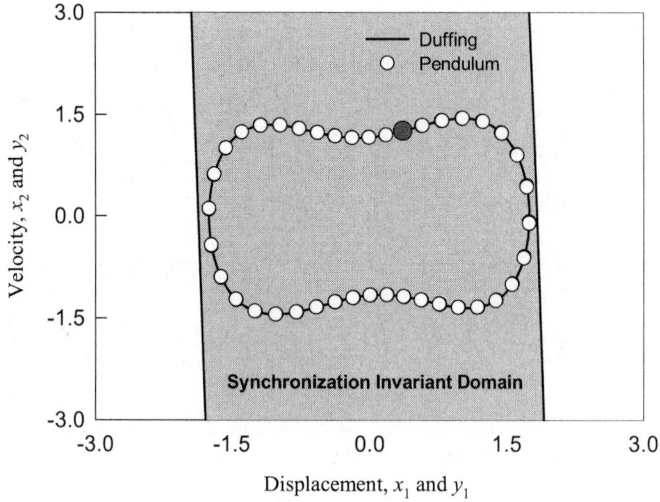

(d)

Fig.9.19 Continued.

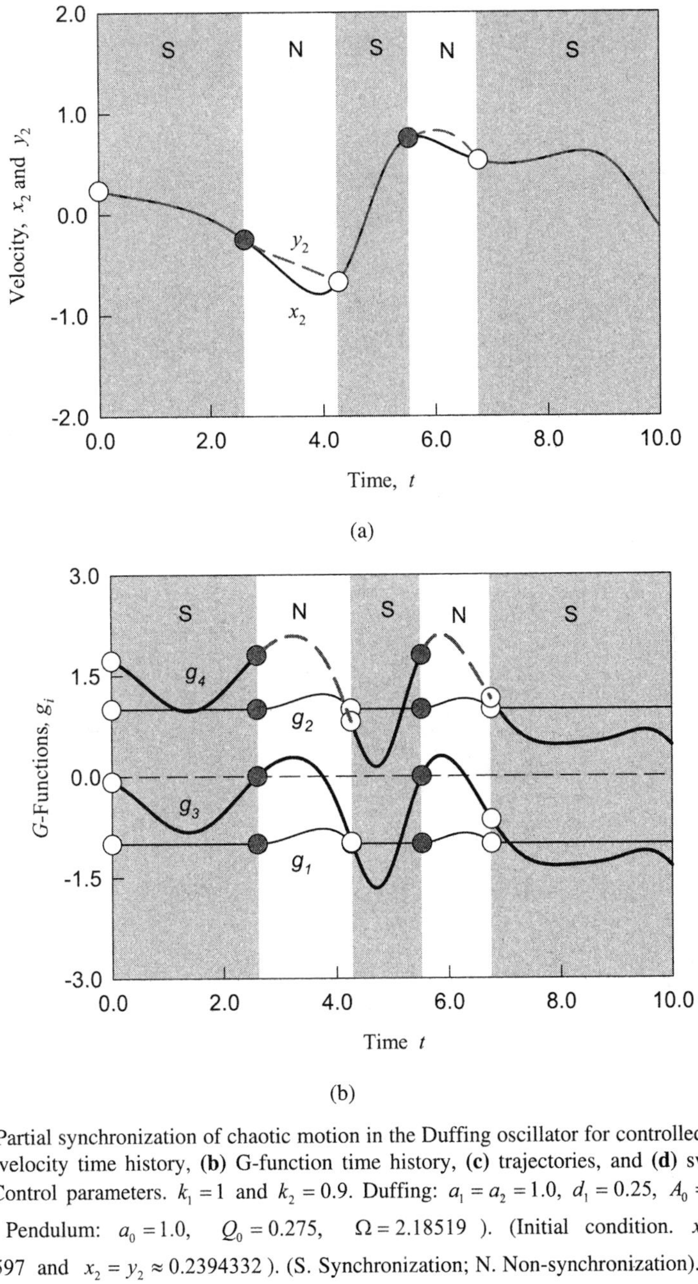

Fig.9.20 Partial synchronization of chaotic motion in the Duffing oscillator for controlled pendulum: **(a)** velocity time history, **(b)** G-function time history, **(c)** trajectories, and **(d)** switching points. (Control parameters. $k_1 = 1$ and $k_2 = 0.9$. Duffing: $a_1 = a_2 = 1.0$, $d_1 = 0.25$, $A_0 = 0.265$, $\omega = 1.0$. Pendulum: $a_0 = 1.0$, $Q_0 = 0.275$, $\Omega = 2.18519$). (Initial condition. $x_1 = y_1 \approx -0.4180597$ and $x_2 = y_2 \approx 0.2394332$). (S. Synchronization; N. Non-synchronization). Hollow and filled circular symbols are synchronization appearance (A) and vanishing (V), respectively.

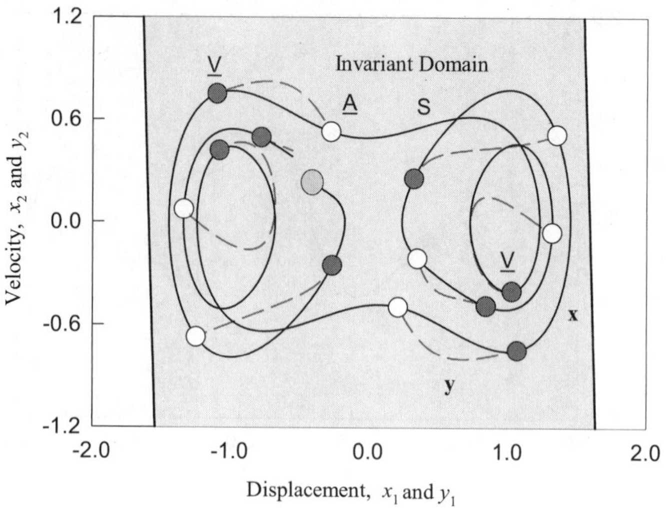

(c)

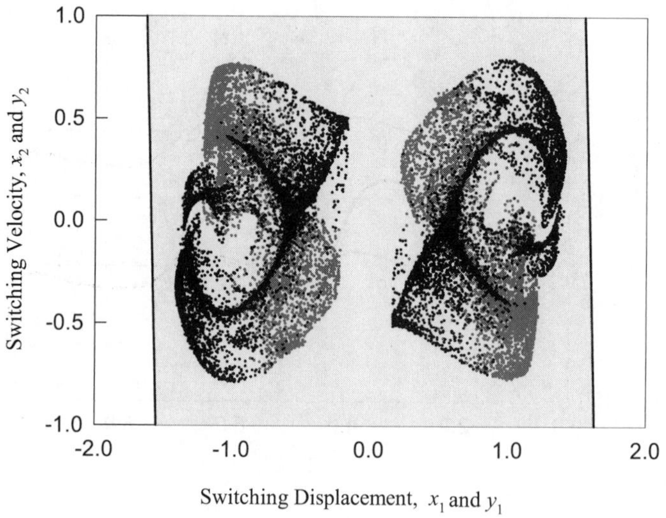

(d)

Fig.9.20 Continued.

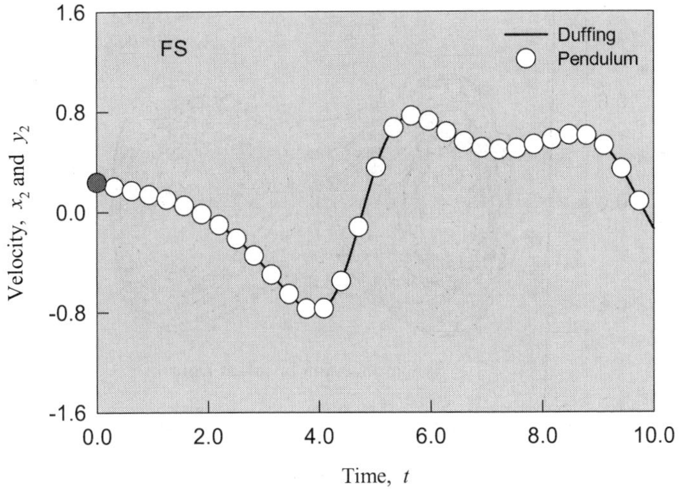

(a)

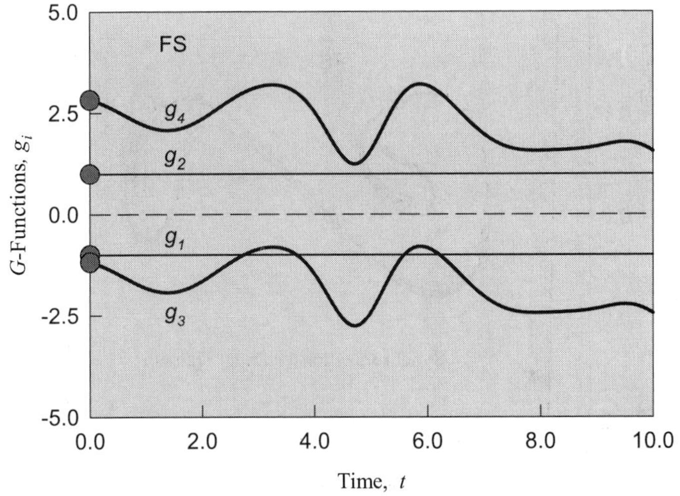

(b)

Fig.9.21 Full synchronization with chaos in the Duffing oscillator for controlled pendulum: **(a)** velocity-time history, **(b)** G-function-time history, **(c)** trajectories in phase plane, and **(d)** Poincare mapping section in the invariant domain. (Control parameters: $k_1 = 1$ and $k_2 = 2$, Duffing: $a_1 = a_2 = 1.0$, $d_1 = 0.25$, $A_0 = 0.265$, $\omega = 1.0$. Pendulum: $a_0 = 1.0$, $Q_0 = 0.275$, $\Omega = 2.18519$). (Initial condition. $x_1 = y_1 \approx -0.4180597$ and $x_2 = y_2 \approx 0.2394332$). (FS. Full synchronization).

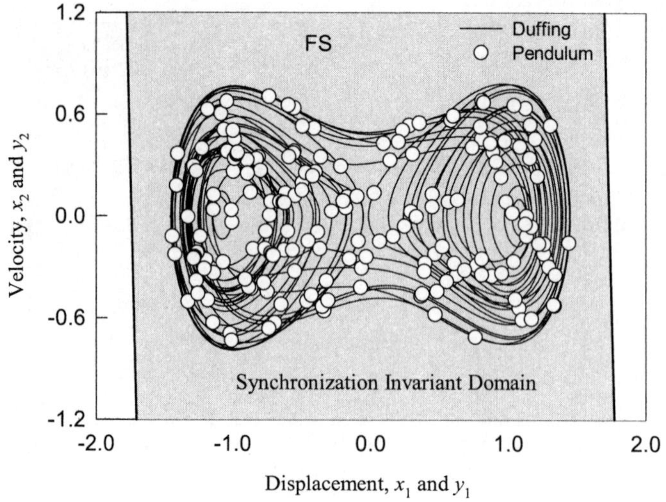

(c)

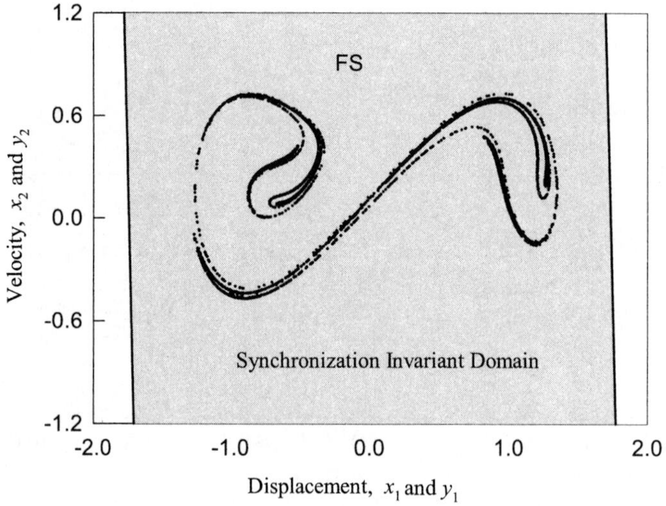

(d)

Fig.9.21 Continued.

For full synchronizations with chaotic motions in the Duffing oscillator, the same initial conditions for the slave and master systems are used, but the control parameter ($k_1 = 1$ and $k_2 = 2$) are adopted. From such parameters, the full synchronization of the controlled pendulum with chaos in the Duffing oscillator will be observed in Fig. 9.21. The velocity responses for such a full synchronization are presented in Fig. 9.21(a). Because both of velocity responses are identical, the Duffing oscillator and the controlled pendulum are presented by solid curve and circular symbols, respectively. The full synchronization of chaos in the Duffing oscillator always requires ($g_1 < 0$ and $g_2 > 0$) for displacement and ($g_3 < 0$ and $g_4 > 0$) for velocity. Thus, the time histories of the *G*-functions are presented in Fig. 9.21(b). The full synchronization of chaos for the Duffing oscillator and the controlled pendulum should be in the synchronization invariant domain. The trajectories of the controlled pendulum and Duffing oscillator are identical, as shown in Fig. 9.21(c). The Duffing oscillator and the controlled pendulum are also represented by the solid curve and circular symbols, respectively. In Fig. 9.21(d), the Poincare mapping section for chaotic motions for 10,000 periods of the Duffing oscillator and the controlled pendulum is presented. The Poincare mapping section is fully synchronized in the synchronization invariant domain.

References

Luo, A.C.J., 2005, A theory for non-smooth dynamical systems on connectable domains, *Communication in Nonlinear Science and Numerical Simulation,* **10**, pp.1-55.

Luo, A.C.J., 2006, *Singularity and Dynamics on Discontinuous Vector Fields*, Amsterdam: Elsevier.

Luo, A.C.J., 2008a, A theory for flow switchability in discontinuous dynamical systems, *Nonlinear Analysis. Hybrid Systems*, **2**(4), pp.1030-1061.

Luo, A.C.J., 2008b, *Global Transversality, Resonance and Chaotic Dynamics*, Singapore: World Scientific.

Luo, A.C.J., 2009a, A theory for dynamical system synchronization, *Communications in Nonlinear Science and Numerical Simulation*, **14**, pp.1901-1951.

Luo, A.C.J., 2009b, *Discontinuous Dynamical Systems on Time-Varying Domains*, Higher Education Press and Springer. Beijing.

Luo, A.C.J. and Min, F.H., 2011a, The mechanism of a controlled pendulum synchronizing with periodic motions in a periodically forced, damped Duffing oscillator, *International Journal of Bifurcation and Chaos*, in press.

Luo, A.C.J. and Min, F.H., 2011b, Synchronization dynamics of two different dynamical systems, *Chaos, Solitons and Fractals*, **44**, pp.362-380.

Subject Index

E

F

G

H

I

L

O

P

R

S

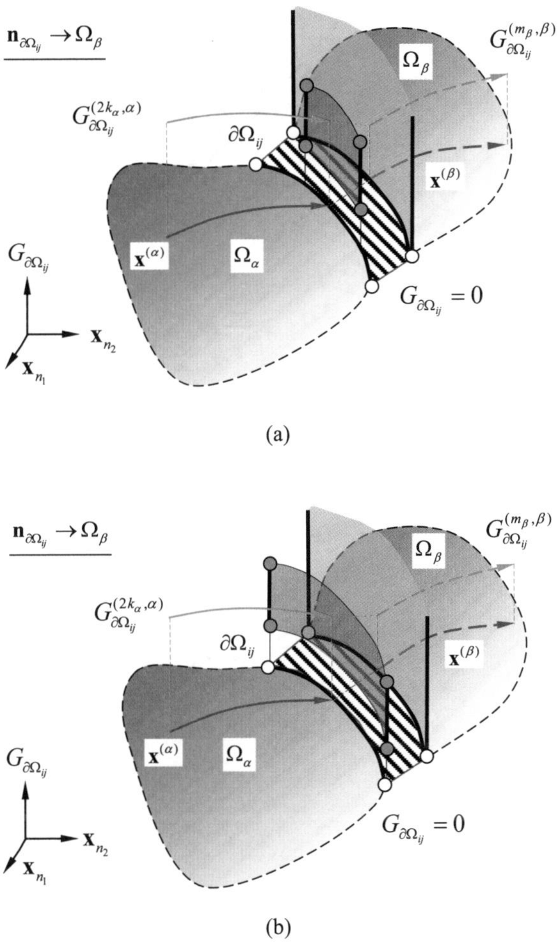

Fig. 4.2 A coming flow barrier on the α-side in the $(2k_\alpha : m_\beta)$-semi-passable flow: **(a)** partial flow barrier and **(b)** full flow barrier. The dark gray surface is for the flow barrier on $\partial\Omega_{ij}$. The upper solid and dashed curves with arrows are *G*-functions of flows on α and β-domains. The lower curves are semi-passable flows. The vertical surface behind the dark gray surface represents "no flow barrier". ($k_\alpha, m_\beta \in \{0,1,2,\cdots\}$).

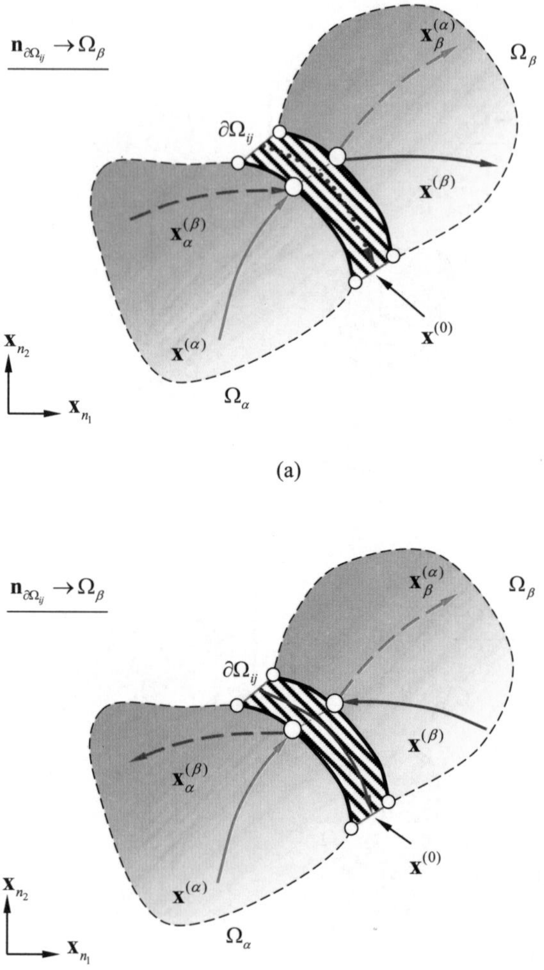

(a)

(b)

Fig. 5.43 Real and imaginary flows: **(a)** semi-passable flows and **(b)** sink flows. The hatched area with two thick solid curves is the boundary. The boundary flows for the sink and passable flows are presented by the solid and dotted curves with arrows, respectively. The solid and dashed curves in the domains are the real and imaginary flows.

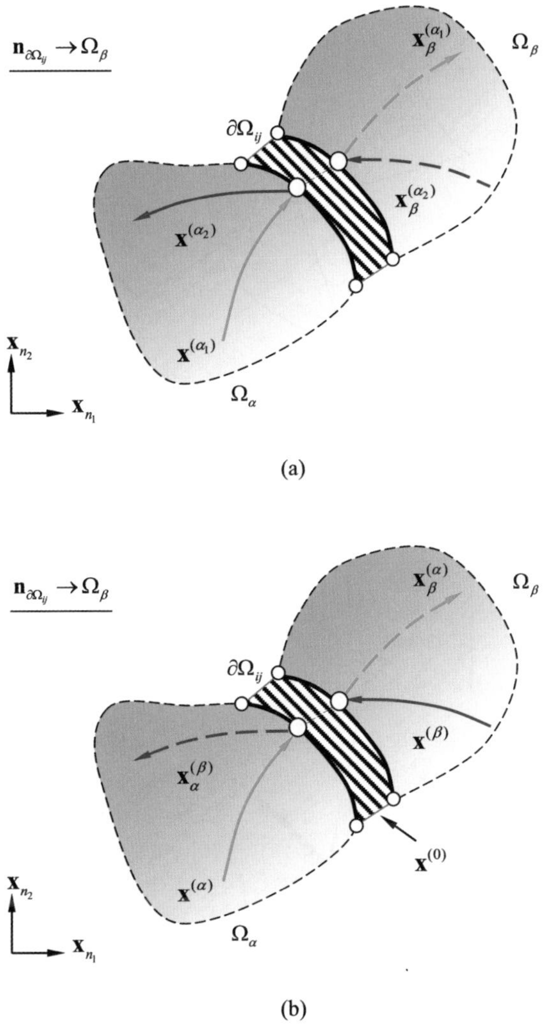

Fig. 5.44 Bouncing flows: **(a)** $(2k_{\alpha_1} : 2k_{\alpha_2})$-bouncing flows and **(b)** $(2k_\alpha : 2k_\beta)$-sink flows. The hatched area with two thick solid curves is the boundary. The solid and dashed curves are the real and imaginary flows in the domains. The dotted curves are the forward and backward extensions of the real flows without switching.

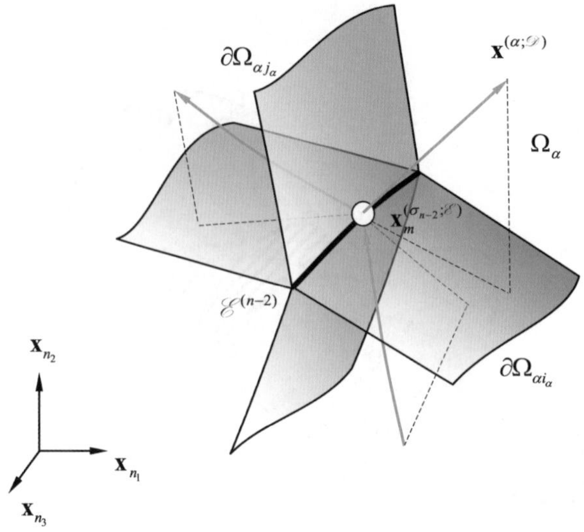

(a)

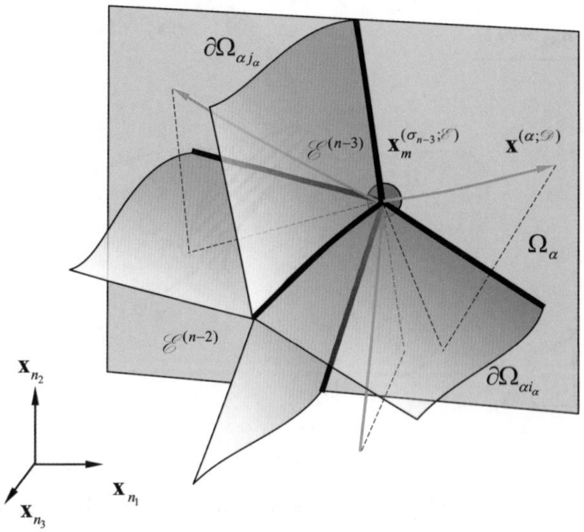

(b)

Fig. 6.3 The domain flows $\mathbf{x}^{(\alpha;\mathscr{D})}$ in accessible domains to edges: **(a)** $\mathscr{E}^{(n-2)}$ and **(b)** $\mathscr{E}^{(n-3)}$. The solid curves with arrows represent domain flows in accessible domains. On the vertical wall, the $(n-3)$-dimensional edge is specified. ($n_1+n_2+n_3=n$)

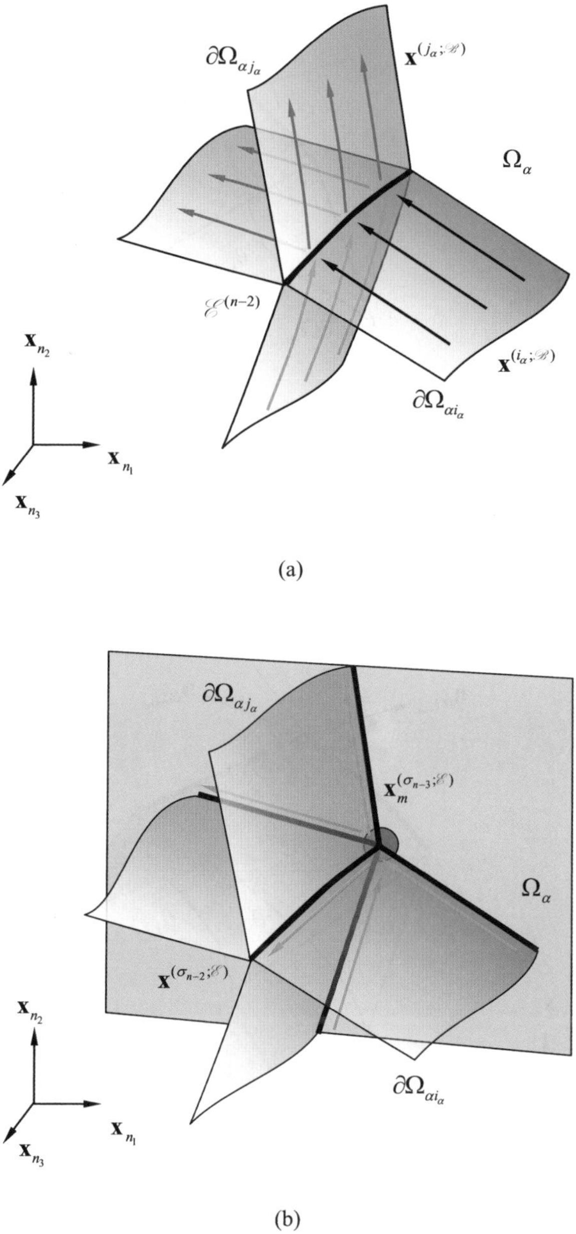

Fig. 6.4 (a) The boundary flows $\mathbf{x}^{(i_\alpha;\mathscr{B})}$ on the boundaries of $\partial\Omega_{\alpha i_\alpha}$ to the edges of $\mathscr{E}^{(n-2)}$, and **(b)** the edge flows $\mathbf{x}^{(\sigma_{n-2};\mathscr{E})}$ on the edges of $\mathscr{E}^{(n-2)}$ to an edge of $\mathscr{E}^{(n-3)}$. The curves with arrows represent the boundary and edge flows. On the vertical wall, the $(n-3)$-dimensional edge is specified. ($n_1+n_2+n_3=n$)

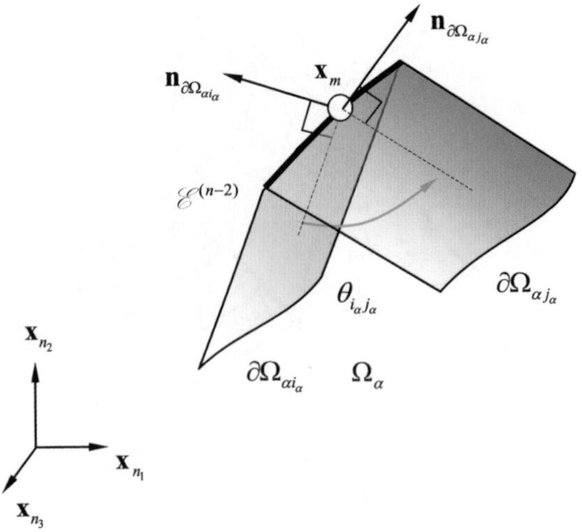

(a)

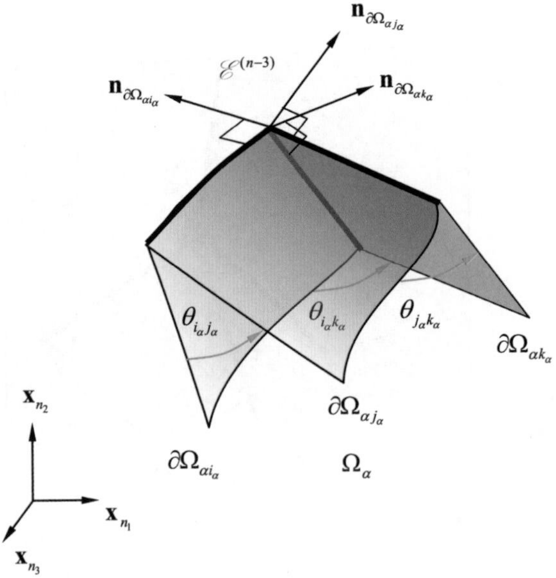

(b)

Fig. 6.5 Edge angles and convex edges in domain Ω_α : **(a)** an $(n-2)$ -dimensional edge, and **(b)** an $(n-3)$ -dimensional edge. The angle $\theta_{i_\alpha j_\alpha}$ of the two boundaries $\partial\Omega_{\alpha i_\alpha}$ and $\partial\Omega_{\alpha j_\alpha}$ is depicted. The $(n-1)$ -dimensional boundaries are shaded. The $(n-2)$ -dimensional edges are represented by thick curves. The $(n-3)$ -edge is shown by a corner.

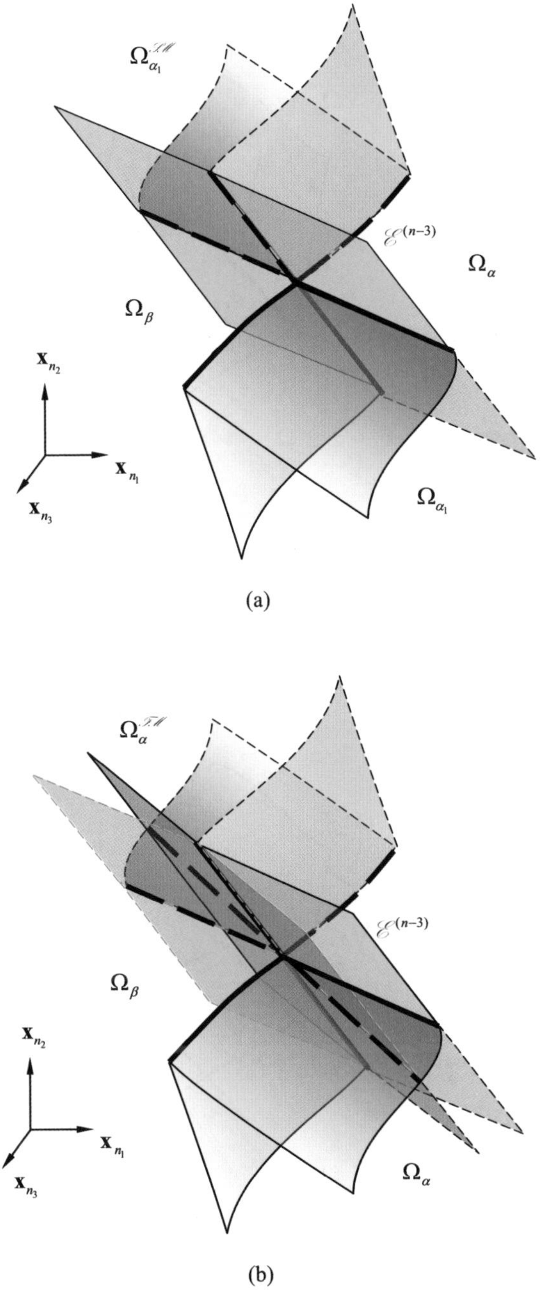

Fig. 6.8 An $(n-3)$-dimensional convex-concave edge: **(a)** self-mirror imaginary domain in its own domain, and **(b)** scaring mirror imaginary domains. The boundaries are shaded, and the $(n-2)$-dimensional edges are depicted by thick curves. The $(n-3)$-dimensional edge is presented by a corner point. Dashed curves are for minor imaginary domains.

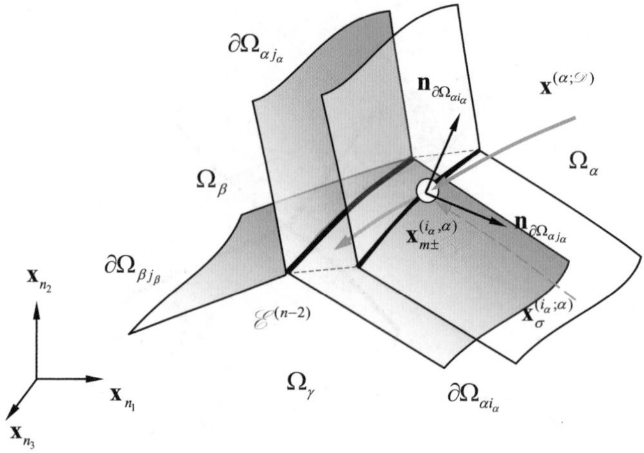

Fig. 6.18 A flow intersected with an edge between two separatrix measuring surfaces in domain Ω_α. The two measuring surfaces are two equi-constant surfaces relative to the two separation boundaries of $\partial\Omega_{\alpha i_\alpha}$ and $\partial\Omega_{\alpha j_\alpha}$. The common edge $\mathscr{E}^{(n\text{-}2)}$ is the $(n-2)$ -dimensional, intersection surface of three separation boundaries. The corresponding normal vectors are presented.

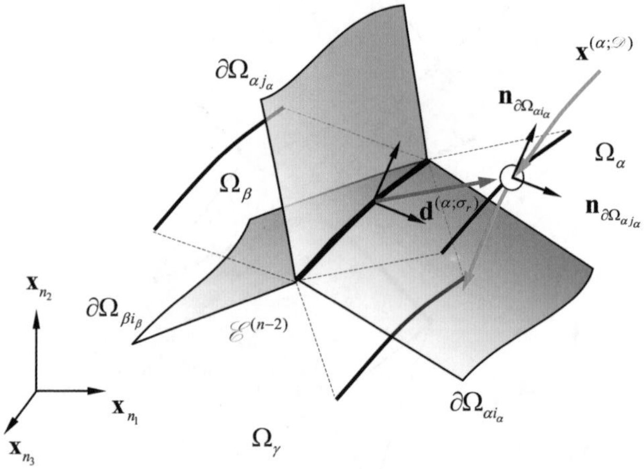

Fig. 6.19 Three measuring edges paralleled with the common edge $\mathscr{E}^{(n\text{-}2)}$ in three different domains. Each of the three measuring edges is the intersection of two separatrix measuring surfaces in each domain. The two measuring surfaces are two equi-constant surfaces relative to the two corresponding separation boundaries. The common edge $\mathscr{E}^{(n\text{-}2)}$ is the $(n-2)$ -dimensional, intersection surface of three separation boundaries. The corresponding normal vectors are presented. ($n_1+n_2+n_3=n$)

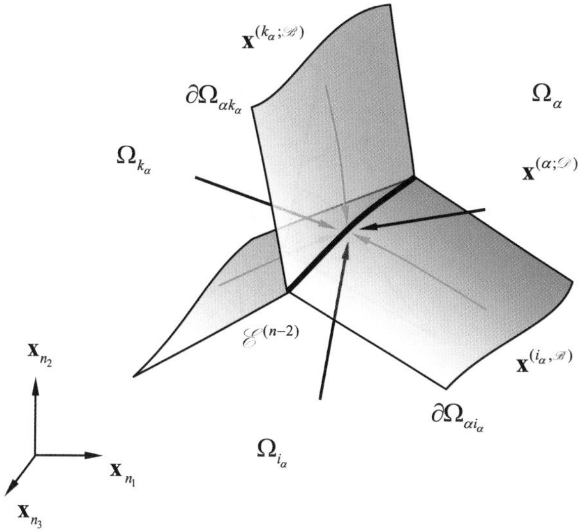

(a)

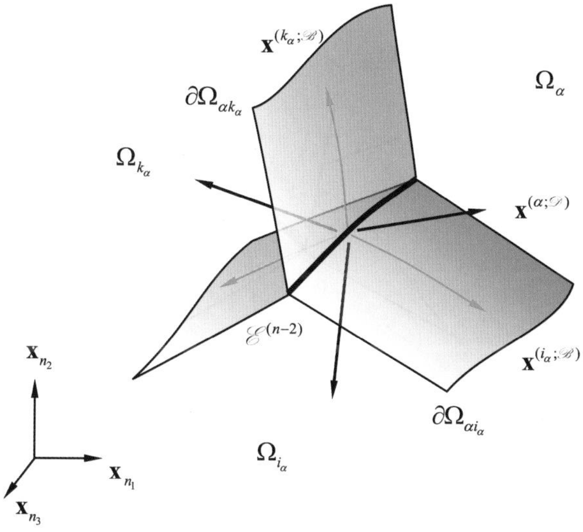

(b)

Fig. 7.18 (a) The complete sink edge and **(b)** complete source edge for domain and boundary flows. The black curve represents the $(n-2)$ -dimensional edge. The boundary flows are denoted by thin curves with arrows on the boundaries. The domain flows are represented by the thick curves with arrows in the domain. ($n_1 + n_2 + n_3 = n$)

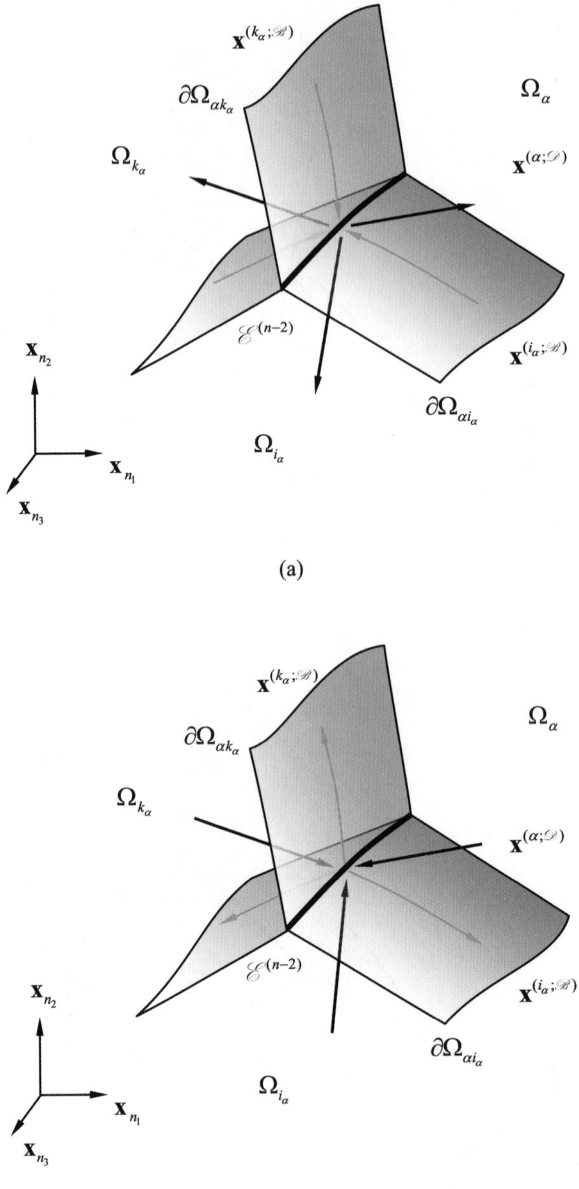

Fig. 7.19 (a) The sink boundary flows to source domain flows via the edge, and **(b)** the sink domain flows to source boundary flows via the edge. The black curve represents the $(n-2)$-dimensional edge. The two blue points are two $(n-3)$-dimensional edges. The boundary flows are denoted by thin curves with arrows on the boundaries. The domain flows are represented by the thick curves with arrows in the domain. ($n_1+n_2+n_3=n$)

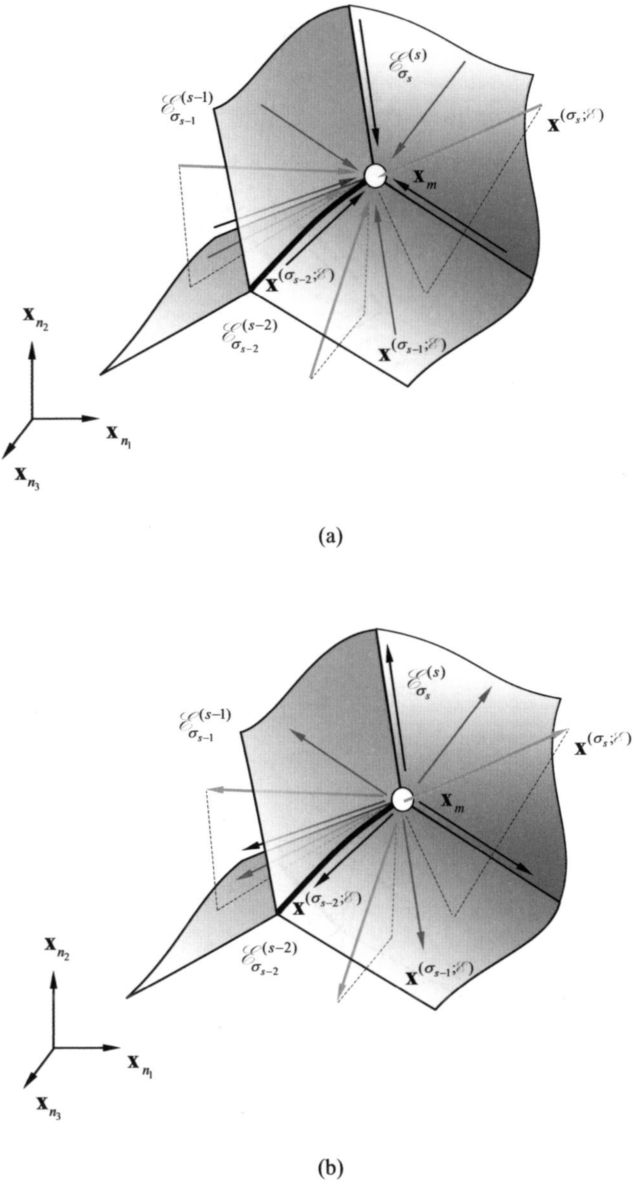

Fig. 8.6 The edge $\mathscr{E}^{(s-3)}$. **(a)** sink edge and **(b)** source edge for all domain, boundary and edge flows $\mathbf{x}^{(\sigma_p;\mathscr{E})}$ ($p=0,1,2,\ \ ,s-2$). All the edge flows are presented by curves with arrows. Only $\mathbf{x}^{(\sigma_s;\mathscr{E})}$, $\mathbf{x}^{(\sigma_{s-1};\mathscr{E})}$ and $\mathbf{x}^{(\sigma_{s-2};\mathscr{E})}$ can be illustrated. ($n_1+n_2+n_3=s$)

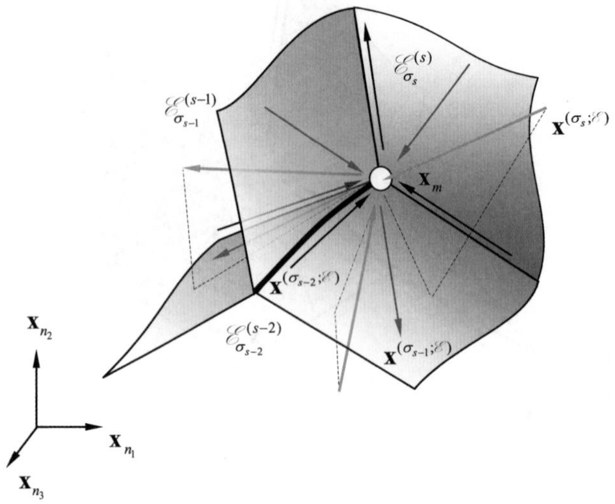

(a)

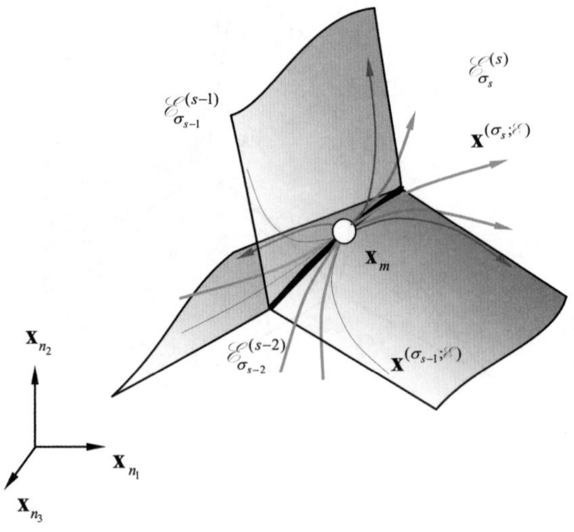

(b)

Fig. 8.7 (a) Potential switching flows to the edge $\mathscr{E}^{(s-3)}$ and **(b)** all tangential flows to the edge $\mathscr{E}^{(s-2)}$. $\mathbf{x}^{(\sigma_p;\mathscr{E})}$ ($p = 0,1,2,\ \ ,s-2$). All the edge flows are presented by curves with arrows. Only $\mathbf{x}^{(\sigma_s;\mathscr{E})}$, $\mathbf{x}^{(\sigma_{s-1};\mathscr{E})}$ and $\mathbf{x}^{(\sigma_{s-2};\mathscr{E})}$ can be illustrated. ($n_1 + n_2 + n_3 = s$)